WALTER GILBERT

SELECTED WORKS

WALTER GILBERT
SELECTED WORKS

Walter Gilbert
Harvard University, USA

Edited by
Manyuan Long
The University of Chicago, USA

World Scientific

NEW JERSEY • LONDON • SINGAPORE • BEIJING • SHANGHAI • HONG KONG • TAIPEI • CHENNAI • TOKYO

Published by

World Scientific Publishing Co. Pte. Ltd.
5 Toh Tuck Link, Singapore 596224
USA office: 27 Warren Street, Suite 401-402, Hackensack, NJ 07601
UK office: 57 Shelton Street, Covent Garden, London WC2H 9HE

British Library Cataloguing-in-Publication Data
A catalogue record for this book is available from the British Library.

WALTER GILBERT
Selected Works

ISBN 978-981-120-329-9

For any available supplementary material, please visit
https://www.worldscientific.com/worldscibooks/10.1142/11355#t=suppl

Typeset by Stallion Press
Email: enquiries@stallionpress.com

Printed in Singapore

Contents

Introduction **xi**

Part One Selected Papers **1**

Section I: How DNA Works, mRNA and Protein Synthesis **2**

1. Unstable ribonucleic acid revealed by pulse labelling of *Escherichia coli*, F. Gros, H. Hiatt, W. Gilbert, C.G. Kurland, R.W. Risebrough and J.D. Watson. *Nature 190*, 581–585 (1961). 3
2. Polypeptide synthesis in *Escherichia coli*. I. Ribosomes and the active complex, W. Gilbert. *J. Mol. Biol. 6*, 374–388 (1963). 8
3. Polypeptide synthesis in *Escherichia coli*. II. The polypeptide chain and S-RNA, W. Gilbert. *J. Mol. Biol. 6*, 389–403 (1963). 23
4. Streptomycin, suppression and the code, J. Davies, W. Gilbert and L. Gorini. *Proc. Natl. Acad. Sci. 51*, 883–890 (1964). 38

Section II: How DNA Is Controlled 46

5. Isolation of the *lac* repressor, W. Gilbert and B. Müller-Hill. *Proc. Natl. Acad. Sci. 56*, 1891–1898 (1966). 47
6. The *lac* operator is DNA, W. Gilbert and B. Müller-Hill. *Proc. Natl. Acad. Sci. 58*, 2415–2421 (1967). 55
7. Mutants that make more *lac* repressor, B. Müller-Hill, L. Crapo and W. Gilbert. *Proc. Natl. Acad. Sci. 59*, 1259–1264 (1968). 62
8. The nucleotide sequence of the *lac* operator, W. Gilbert and A. Maxam. *Proc. Natl. Acad. Sci. USA 70*, 3581–3584 (1973). 68
9. Sequences of controlling regions of the lactose operon, W. Gilbert, N. Maizels and A. Maxam. *Cold Spring Harb. Symp. Quant. Biol.* 38, 845–855 (1974). 72
10. Contacts between the *lac* repressor and DNA revealed by methylation, W. Gilbert, A. Maxam and A. Mirzabekov. In *Control of Ribosome Synthesis* (N. Kjeldgaard and O. Maaloe, eds.), Munksgaard, Copenhagen, 139–148 (1976). 83

Section III: The Rolling Circle Model 94

11. DNA replication: The rolling circle model, W. Gilbert and D. Dressler. *Cold Spring Harb. Symp. Quant. Biol. 33*, 473–484 (1968). 95

Section IV: DNA Sequencing **107**

12. A new method for sequencing DNA, A.M. Maxam and W. Gilbert. *Proc. Natl. Acad. Sci. USA 74*, 560–564 (1977). 108
13. Sequence of a mouse germ-line gene for a variable region of an immunoglobulin light chain, S. Tonegawa, A.M. Maxam, R. Tizard, O. Bernard and W. Gilbert. *Proc. Natl. Acad. Sci. USA 75*, 1485–1489 (1978). 113
14. Why genes in pieces? W. Gilbert. *Nature 271*, 501 (1978). 118
15. DNA sequencing and gene structure: Nobel lecture, 8 December 1980, W. Gilbert. *Biosci. Rep. 1*, 353–375 (1981). 119
16. Genomic sequencing, G.M. Church and W. Gilbert. *Proc. Natl. Acad. Sci. USA 81*, 1991–1995 (1984). 142
17. Direct genomic sequencing of bacterial DNA: The pyruvate kinase I gene of *Escherichia coli*, O. Ohara, R.L. Dorit and W. Gilbert. *Proc. Natl. Acad. Sci. USA 86*, 6883–6887 (1989). 147
18. Genome sequencing: Creating a new biology for the twenty-first century, W. Gilbert. *Issues in Science & Technology III(3)*, 26–35 (1987). 152

Section V: Recombinant DNA **162**

19. Construction of plasmids carrying the *c*I gene of bacteriophage λ, K. Backman, M. Ptashne and W. Gilbert. *Proc. Natl. Acad. Sci. USA 73*, 4174–4178 (1976). 163
20. Immunological screening method to detect specific translation products, S. Broome and W. Gilbert. *Proc. Natl. Acad. Sci. USA 75*, 2746–2749 (1978). 168
21. A bacterial clone synthesizing proinsulin, L. Villa-Komaroff, A. Efstratiadis, S. Broome, P. Lomedico, R. Tizard, S.P. Naber, W.L. Chick and W. Gilbert. *Proc. Natl. Acad. Sci. USA 75*, 3727–3731 (1978). 172

Section VI: A Variety of Topics **177**

22. Molecular basis of base substitution hotspots in *Escherichia coli,* C. Coulondre, J.H. Miller, P.J. Farabaugh and W. Gilbert. *Nature 274*, 775–780 (1978). 178
23. Parabiosis as a model system for network interactions, K. Fischer Lindahl, W. Gilbert and K. Rajewsky. In *The Immune System,* Vol. 2, 24–32, Basel (1981). 184
24. Chicken triosephosphate isomerase complements an *Escherichia coli* deficiency, D. Straus and W. Gilbert. *Proc. Natl. Acad. Sci. USA 82*, 2014–2018 (1985). 193
25. The monoclonal antibody B30 recognizes a specific neuronal cell surface antigen in the developing mesencephalic trigeminal nucleus of the mouse, D.Y. Stainier and W. Gilbert. *J. Neurosci. 9(7)*, 2468–2485 (1989). 198

26. One-sided polymerase chain reaction: The amplification of cDNA, O. Ohara, R.L. Dorit and W. Gilbert. *Proc. Natl. Acad. Sci. USA 86*, 5673–5677 (1989). 216
27. Pioneer neurons in the mouse trigeminal sensory system, D.Y.R. Stainier and W. Gilbert. *Proc. Natl. Acad. Sci. USA 87*, 923–927 (1990). 221
28. RNA editing as a source of genetic variation, L.F. Landweber and W. Gilbert. *Nature 363*, 179–182 (1993). 226

Section VII: G4 DNA **230**

29. Formation of parallel four-stranded complexes by guanine-rich motifs in DNA and its implications for meiosis, D. Sen and W. Gilbert. *Nature 334*, 364–366 (1988). 231
30. Identification and characterization of a nuclease activity specific for G4 tetrastranded DNA, Z. Liu, J.D. Frantz, W. Gilbert and B.K. Tye. *Proc. Natl. Acad. Sci. USA 90*, 3157–3161 (1993). 234
31. The yeast *KEM1* gene encodes a nuclease specific for G4 tetraplex DNA: Implication of *in vivo* functions for this novel DNA structure, Z. Liu and W. Gilbert. *Cell 77*, 1083–1092 (1994). 239

Section VIII: The RNA World **249**

32. Origin of life: The RNA world, W. Gilbert. *Nature 319*, 618 (1986). 250
33. Basic protein enhances the incorporation of DNA into lipid vesicles: Model for the formation of primordial cells, D.G. Jay and W. Gilbert. *Proc. Natl. Acad. Sci. USA 84*, 1978–1980 (1987). 251

Section IX: Introns, Exons, and Gene Evolution **254**

34. Introns and exons: Playgrounds of evolution, W. Gilbert. In *Eucaryotic Gene Regulation* (R. Axel, T. Maniatis and C.F. Fox, eds.), Academic Press, 1–12 (1979). 255
35. The structure and evolution of the two nonallelic rat preproinsulin genes, P. Lomedico, N. Rosenthal, A. Efstratiadis, W. Gilbert, R. Kolodner and R. Tizard. *Cell 18*, 545–558 (1979). 267
36. The evolution of genes: The chicken preproinsulin gene, F. Perler, A. Efstratiadis, P. Lomedico, W. Gilbert, R. Kolodner and J. Dodgson. *Cell 20*, 555–566 (1980). 281
37. Intron/exon structure of the chicken pyruvate kinase gene, N. Lonberg and W. Gilbert. *Cell 40*, 81–90 (1985). 293
38. Genes-in-pieces revisited, W. Gilbert. *Science 228*, 823–824 (1985). 303
39. The triosephosphate isomerase gene from maize: Introns antedate the plant–animal divergence, M. Marchionni and W. Gilbert. *Cell 46*, 133–141 (1986). 305
40. On the antiquity of introns, W. Gilbert, M. Marchionni and G. McKnight. *Cell 46*, 151–154 (1986). 314
41. The exon theory of genes, W. Gilbert. *Cold Spring Harb. Symp. Quant. Biol. 52*, 901–905 (1987). 317

42. How big is the universe of exons? R.L. Dorit, L. Schoenbach and W. Gilbert. *Science 250*, 1377–1382 (1990). 322
43. On the ancient nature of introns, W. Gilbert and M. Glynias. *Gene 135*, 137–144 (1993). 328
44. Tests of the exon theory of genes, W. Gilbert, M. Long, C. Rosenberg and M. Glynias. In *Tracing Biological Evolution in Protein and Gene Structures*, Proceedings of the 20th Taniguchi International Symp., Div. of Biophysics (M. Go and P. Schimmel, eds.), Elsevier Science B.V., 237–247 (1995). 336
45. Intron phase correlations and the evolution of the intron/exon structure of genes, M. Long, C. Rosenberg and W. Gilbert. *Proc. Natl. Acad. Sci. USA 92,* 12495–12499 (1995). 347
46. Exon shuffling and the origin of the mitochondrial targeting function in plant cytochrome c1 precursor, M. Long, S.J. de Souza, C. Rosenberg and W. Gilbert. *Proc. Natl. Acad. Sci. USA 93,* 7727–7731 (1996). 352
47. Intron positions correlate with module boundaries in ancient proteins, S.J. de Souza, M. Long, L. Schoenbach, S.W. Roy and W. Gilbert. *Proc. Natl. Acad. Sci. USA 93*, 14632–14636 (1996). 357
48. Origin of genes, W. Gilbert, S.J. de Souza and M. Long. *Proc. Natl. Acad. Sci. USA 94,* 7698–7703 (1997). 362
49. Toward a resolution of the introns early/late debate: Only phase zero introns are correlated with the structure of ancient proteins, S.J. de Souza, M. Long, R.J. Klein, S. Roy, S. Lin, and W. Gilbert. *Proc. Natl. Acad. Sci. USA 95,* 5094–5099 (1998). 368
50. Centripetal modules and ancient introns, S.W. Roy, M. Nosaka, S.J. de Souza and W. Gilbert. *Gene 238*, 85–91 (1999). 374
51. Large-scale comparison of intron positions in mammalian genes shows intron loss but no gain, S.W. Roy, A. Fedorov and W. Gilbert. *Proc. Natl. Acad. Sci. USA 100*, 7158–7162 (2003). 381
52. The universe of exons revisited, S. Saxonov and W. Gilbert. *Genetica 118*, 267–278 (2003). 386
53. The pattern of intron loss, S. W. Roy and W. Gilbert. *Proc. Natl. Acad. Sci. USA 102*, 713–718 (2005). 398

Section X: Paradigm Shift and Computing 404

54. Towards a paradigm shift in biology, W. Gilbert. *Nature 349*, 99 (1991). 405
55. Large scale bacterial gene discovery by similarity search, K. Robison, W. Gilbert and G.M. Church. *Nature Genetics 7*, 205–214 (1994). 406

Section XI: Physics **416**

56. On generalized dispersion relations II, A. Salam and W. Gilbert. *Nuovo Cimento 3*, 607–611 (1956). 417
57. New dispersion relations for pion-nucleon scattering, W. Gilbert. *Phys. Rev. 108*, 1078–1083 (1957). 422
58. Structure of the vertex function, S. Deser, W. Gilbert and E.C.G. Sudarshan. *Phys. Rev. 115*, 731–735 (1959). 428

Introduction

This volume, entitled *Walter Gilbert*: *Selected Works*, collected documents including 58 research articles selected by Walter Gilbert from his previous publications, 11 memorial articles composed by Walter Gilbert for 11 fields he has worked in, 7 memorial chapters by the members of the Walter Gilbert laboratory that recalled their life, academic and secular, in the lab, numerous images that recorded the experiences, scientific research and academic activities, and multiple pieces of art works created by Walter Gilbert.

In 2005, a symposium was dedicated to the celebration of the scientific career of Walter Gilbert in the Cold Spring Harbor Laboratory. In total 140 scientists from the former Gilbert laboratory or the Gilbert–Watson laboratory gathered in the celebration. Soon after this symposium, I came up with a desire to edit a volume to record the academic accomplishment with the most diverse and colorful intellectual life of Walter Gilbert. With the documents collected and presented in this volume, I do not think it necessary to tell any reasons to edit the volume. However, I did want to tell a desire that these documents should be seen by the future generations and help them to shape a sense of history with a concrete idea of what a scientist and his laboratory responsible for something important in science in the second half of the 20th century look like.

It was lucky that, through the enthusiastic support of a friend, Bailin Hao in Shanghai (a theoretical physicist and computational biologist in his later life impacted by Walter Gilbert's works), I received a generous letter from Kok Khoo Phua (潘国驹, Chairman and Editor-in-Chief of World Scientific Publishing), telling me that he would like to publish this volume. I am grateful to Bailin Hao and Phua Kok Khoo for their strong support that eventually led to this publication project.

Editing this volume received supports from many who have worked with Walter Gilbert, including Sandro DeSouza, Juergen Brosius, Karen Talmadge, Marty Kreitman, Lydia Villa-Komaroff, Benno Muller-Hill and James Watson. I am thankful to Jianhai Chen and UnJin Lee in Manyuan Long lab for their great efforts to dig up the papers before the digital era and made them into PDF files. I would like to thank Emily Mortola for editing, proofreading, image preparation and helping to format all chapters. I enjoyed very much when Jürgen Brosius joined me in reading the manuscript of selected paper 1 and deciphering the handwritings of Walter Gilbert and James Watson. I am grateful for the support of Bairong Shen for his persistent help during the publication process. I am especially graceful to Jan A. Witkowski of the Cold Spring Harbor Laboratory for generously sharing his precious experiences in editing and publishing similar types of books. I would like to show my highest respect for the editors at World Scientific Publishing for their professionalism in the entire process: Yubing Zhai and Ling Xiao. I am grateful to Jingqiu Cheng at the West China School of Medicine for his enthusiastic support to my efforts during the pursuit of editing and publishing of this book. Finally, I want to express my gratitude to my wife, Liming Li, for her persistent support while I was working on the project for six years.

Manyuan Long
Chicago, USA
June 2019

Part One
Selected Papers

Section I:
How DNA Works, mRNA and Protein Synthesis

UNSTABLE RIBONUCLEIC ACID REVEALED BY PULSE LABELLING OF *ESCHERICHIA COLI*

By Drs. FRANCOIS GROS and H. HIATT
The Institut Pasteur, Paris
Dr. WALTER GILBERT
Departments of Physics, Harvard University
AND
Dr. C. G. KURLAND, R. W. RISEBROUGH and Dr. J. D. WATSON
The Biological Laboratories, Harvard University

WHEN *Escherichia coli* cells are infected with *T* even bacteriophage particles, synthesis of host proteins stops[1], and much if not all new protein synthesis is phage specific[2]. This system thus provides an ideal model for observing the synthesis of new proteins following the introduction of specific DNA. In particular, we should expect the appearance of phage-specific RNA, since it is generally assumed that DNA is not a direct template for protein synthesis but that its genetic information is transmitted to a specific sequence of bases in RNA. It was thus considered paradoxical when it was first noticed[3] that, following infection by the *T* even phages, net RNA synthesis stops even though protein synthesis continues at the rate of the uninfected bacterium. This could mean that DNA sometimes serves as a direct template for protein synthesis. Alternatively, net RNA synthesis may not be necessary so long as there exists the synthesis of a genetically specific RNA that turns over rapidly. This possibility was first suggested by experiments of Hershey[4], who, in 1953, reported that *T*2 infected cells contain a metabolically active RNA fraction comprising about 1 per cent of the total RNA. Several years later, Volkin and Astrachan[5] reported that this metabolic RNA possessed base ratios similar, if not identical (considering uracil formally equivalent to thymine), to those of the infecting *T*2 DNA. By 1958 they[6] extended their observation to *T*7 infected cells, where again the RNA synthesized after phage infection had base ratios similar to those of the phage DNA.

During these years, evidence[7] accumulated that the sites of much, if not all, protein synthesis are the ribosomal particles, and it was thought most likely that ribosomal RNA was genetically specific, with each ribosome possessing a base sequence which coded for a specific amino-acid sequence (one ribosome–one protein hypothesis). Direct verification of this hypothesis was lacking, and its proponents[8] were troubled by the fact that, except for phage-specific RNA, it was impossible to find any correlation within a given organism between the base ratios of DNA and RNA. Moreover, there was no evidence that phage-specific RNA was ribosomal RNA.

Nomura, Hall and Spiegelman[9] have recently discovered that following *T*2 infection there is no synthesis of typical (see below) ribosomal RNA and that the phage-specific RNA sediments at a slower rate (8*s*) than ribosomal RNA (16*s* and 23*s*). The genetic information for the synthesis of phage-specific proteins does not reside in the usual ribosomal RNA. Instead, if we assume that the synthesis of phage-specific proteins also occurs on ribosomes, then the phage-specific RNA might be viewed as a 'messenger' (to use the terminology of Monod and Jacob[10]) which carries the genetic information to the ribosomes. Furthermore, unless we postulate that there exist two different mechanisms for protein synthesis, there should also exist within uninfected normal cells RNA molecules physically similar to the phage-specific RNA and having base ratios similar to its specific DNA.

582 NATURE May 13, 1961 Vol. 190

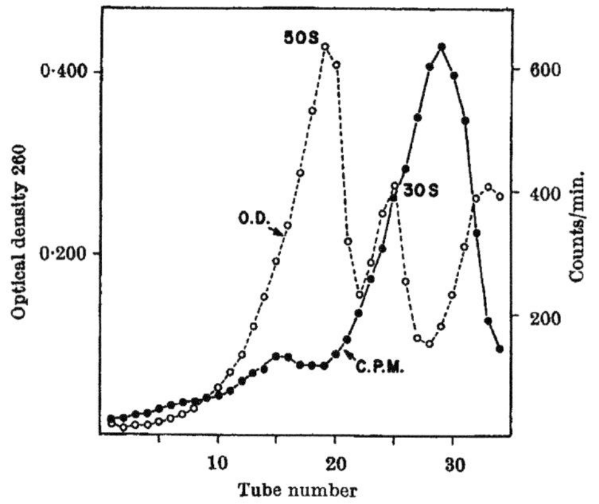

Fig. 1. Sedimentation of pulse-labelled 10^{-4} Mg^{2+} extract. 1 ml. of a crude extract labelled by a 20-sec. pulse of ^{14}C-uracil and made in 10^{-4} *M* Mg^{2+} and 5×10^{-3} *tris* buffer (*p*H 7·4) was carefully layered on top of a 24-ml. sucrose gradient (5–20 per cent) in a Spinco *SW*25 swinging-bucket tube. It was spun 2 hr. 45 min. at 25,000 r.p.m. at 4° C. The tube bottom was then punctured and 10 drop samples taken to measure 2600 Å. absorption (2 drops) and the radioactivity in 5 per cent trichloracetic acid precipitates (8 drops). All points were counted to at least 1,000 counts

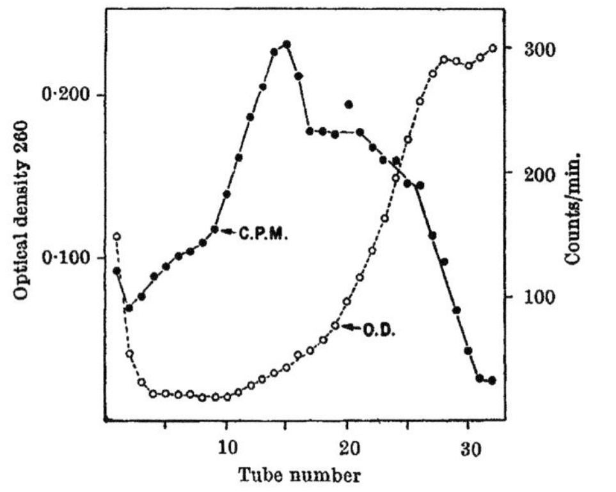

Fig. 2. Sedimentation of pulse-labelled 10^{-4} Mg^{2+} extract, long run. An extract as in Fig. 1 run for 11 hr. at 25,000 r.p.m., 15° C., on a sucrose gradient. Drop collection and radioactivity measurements were made as described in Fig. 1

Here we present evidence that RNA molecules physically similar to phage-specific RNA exist in normal *E. coli* cells and that, under suitable ionic conditions, they are associated with ribosomal particles.

Experimental plan. Most (80–85 per cent) RNA in actively growing *Escherichia coli* cells is found in ribosomal particles composed of 64 per cent RNA and 36 per cent protein[11]. There are two sizes[12] of ribosomal RNA, 16*s* (molecular weight = $5 \cdot 5 \times 10^5$) and 23*s* (molecular weight = $1 \cdot 1 \times 10^6$). The 16*s* RNA is derived from both 30*s* and 50*s* ribosomes, while 23*s* RNA is only found in 50*s* ribosomes. The other principal (10–15 per cent) form of RNA in *E. coli* is soluble RNA (now more appropriately called transfer RNA), which functions in the movement of activated amino-acids to the ribosomes. At least 20 (one for each amino-acid) different transfer RNA molecules exist[13], all of which have molecular weights about 25,000 and sedimentation constants of 4*s*.

Collectively, ribosomal and transfer RNA comprise at least 95 per cent of *E. coli* RNA. Thus messenger RNA, if present, can amount to at most only several per cent of the total RNA. Now if the messenger were stable, only a corresponding fraction of newly synthesized RNA could be messenger RNA; the great majority of new RNA being the metabolically stable ribosomal and transfer RNA's. If, however, messenger RNA is turning over (as is suggested by the original Hershey experiments) then a much larger fraction of newly made RNA must be messenger. For example, if the messenger functions only once for the synthesis of a single protein molecule, its lifetime might be only several seconds. In this event, if we look at the RNA synthesis occurring during a very short interval, then most of the newly synthesized RNA would be messenger even though this fraction may comprise only 1 per cent of the total RNA.

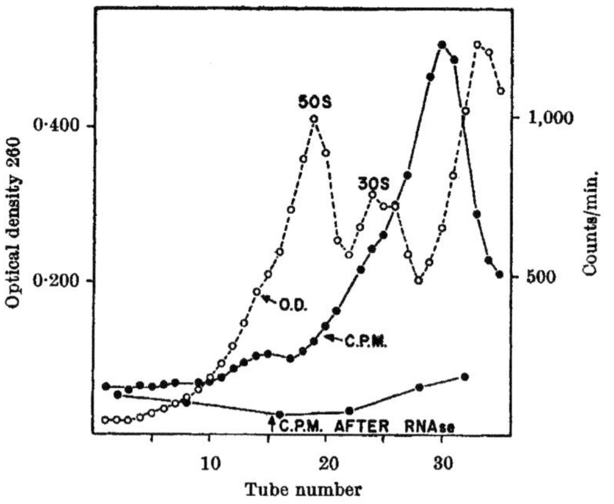

Fig. 3. Sedimentation of *T*2-infected 10^{-4} Mg^{2+} extract. An extract in 10^{-4} Mg^{2+} of cells infected with *T*2 at a multiplicity of 20 and labelled with phosphorus-32 between the second and fifth minutes after infection. Run as in Fig. 1. Several samples were treated with ribonuclease (10 γ/c.c.) to determine the background of label not in RNA

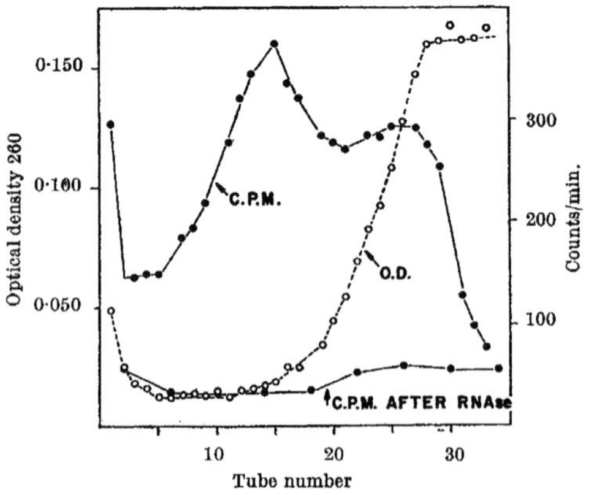

Fig. 4. Sedimentation of *T*2-infected 10^{-4} Mg^{2+} extract, long run. An extract as in Fig. 3 run for 11 hr. at 25,000 r.p.m. 20° C. Measured as in Fig. 1

No. 4776 May 13, 1961 NATURE 583

We have therefore exposed *E. coli* cells to short pulses (10–20 sec. at 25° C. where the time of generation is about 90 min.) of radioactive RNA precursors (^{32}P or ^{14}C-uracil), rapidly chilled the cells with crushed ice and *M*/100 azide, prepared cell-free extracts by alumina grinding, added deoxyribonuclease at 5γ/ml., and examined the newly made RNA using the sucrose-gradient centrifugation technique[14]. In some experiments ^{14}C-uracil was given to cells of a pyrimidine-requiring mutant (*B*148) which was briefly starved (30 min. at 25° C.) for uracil. No difference has been seen between the properties of RNA labelled in these two ways. Starved cells incorporate more radioactivity, and they were used in most of the experiments reported below.

Results. Radioactive uracil (or phosphorus-32) is incorporated in RNA within several seconds after addition of the isotope. The RNA labelled by 10–20 sec. pulses is stable at 4° C. in cell-free extracts where ribonuclease and polynucleotide phosphorylase are

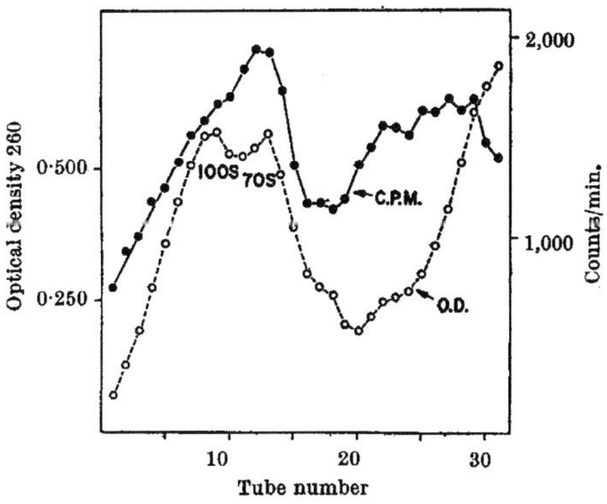

Fig. 7. Sedimentation of *T*2-infected 10^{-2} Mg^{2+} extract. An extract in 10^{-2} *M* Mg^{2+} of cells infected with *T*2 and labelled with phosphorus-32 from the second to the fifth minutes after infection. Run on a sucrose gradient in 10^{-2} *M* Mg^{2+} for 2 hr. 45 min. at 25,000 r.p.m. Drop collection and radioactivity measurements as for Fig. 1

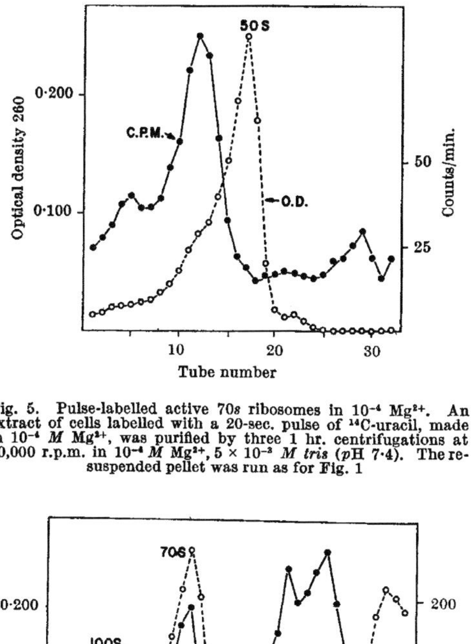

Fig. 5. Pulse-labelled active 70*s* ribosomes in 10^{-4} Mg^{2+}. An extract of cells labelled with a 20-sec. pulse of ^{14}C-uracil, made in 10^{-4} *M* Mg^{2+}, was purified by three 1 hr. centrifugations at 40,000 r.p.m. in 10^{-4} *M* Mg^{2+}, 5 × 10^{-3} *M tris* (*p*H 7·4). The re-suspended pellet was run as for Fig. 1

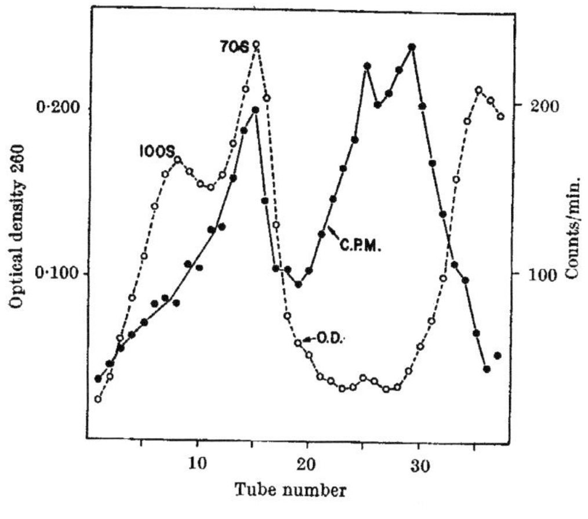

Fig. 6. Sedimentation of pulse-labelled 10^{-4} Mg^{2+} extract in 10^{-2} *M* Mg^{2+}. Magnesium was added to the 10^{-4} Mg^{2+} extract of Fig. 1 to bring it to 10^{-2} *M* Mg^{2+}. The sample was then run on a sucrose gradient in 10^{-2} *M* Mg^{2+}, 5 × 10^{-3} *M tris*, for 2 hr. 45 min. The measurements were made as described for Fig. 1

not active. Traces (1γ/ml.) of ribonuclease degrade it under conditions where RNA in ribosomes is untouched. This suggests that it exists less protected than bound ribosomal RNA. Similarly, addition of phosphate permits the polynucleotide phosphorylase[15] in the cell extract to degrade preferentially pulse-labelled RNA. Our experiments thus as a matter of routine avoid the use of phosphate buffers.

Figs. 1 and 2 show how RNA labelled by a 20-sec. exposure of uninfected cells to ^{14}C-uracil sediments in a sucrose gradient. The cell-free extract contains 10^{-4} *M* magnesium ions in which 30*s* and 50*s* ribosomes predominate. The majority of the radioactivity is not associated with ribosomes but moves with a 14–16*s* peak. There is also RNA which moves more slowly, in addition to a faster forward fraction moving at 70*s*. The slower sedimenting fraction shows up more clearly in the longer centrifugation shown in Fig. 2. Here a sharp 14–16*s* component is seen together with material sedimenting at 4–8*s*. Figs. 3 and 4 illustrate experiments with extracts from *T*2 infected cells exposed to phosphorus-32 from 2 to 5 min. after infection. They are similar, if not identical, to Figs. 1 and 2. Both have a major 14–16*s* component, a slower trailing fraction, and about 10 per cent of the material moving at 70*s*. The 70*s* component can be purified by three 1-hr. centrifugations at 40,000 r.p.m. with 10^{-4} *M* magnesium ions to concentrate the faster-moving ribosomes of Fig. 1. Fig. 5 shows the purified ribosomes consisting largely of 50*s* particles together with a now visible 70*s* component. The radioactivity, however, sediments as 70*s* ribosomes.

Still more label is attached to 70*s* particles when extracts are made in 10^{-2} *M* magnesium ions or when additional magnesium ions are added to 10^{-4} *M* extracts. Figs. 6 and 7 again illustrate the parallel appearance of normal and phage-infected extracts. About 30 per cent of label sediments with the 70*s* and 100*s* ribosomes, with the specific activity of 70*s* ribosomes generally greater than that of 100*s* ribosomes. When slightly lower concentrations of magnesium ions are used to give similar amounts of 30*s*, 50*s* and 70*s* ribosomes, label is specifically associated only with 70*s* particles and no label is in 50*s* ribosomes. The radioactivity sedimenting about

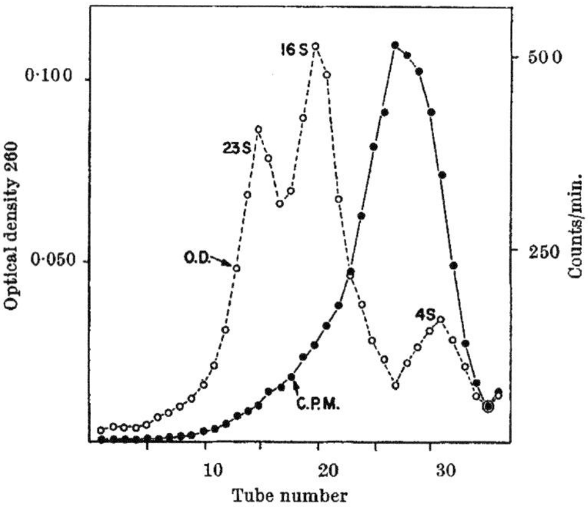

Fig. 8. Sedimentation of pulse-labelled RNA. An extract of cells labelled with a 20-sec. pulse of ^{14}C-uracil was treated with 0·5 per cent sodium lauryl sulphate ('Duponol') and then extracted three times with phenol. The RNA was then precipitated three times with alcohol and finally resuspended in 10^{-2} acetate buffer (*p*H 5·1). The RNA was run on a sucrose gradient, containing 10^{-2} acetate (*p*H 5·1) and *M*/10 sodium chloride, for 10 hr. at 25,000 r.p.m., 4° C. The drop collection and radioactivity measurements were made as described for Fig. 1

30*s* shows a very broad peak (20–40*s*) and it is probable that high concentrations of magnesium ions cause the 14–16*s* material either to aggregate or to assume a more compact shape.

Purified RNA, prepared by sodium lauryl sulphate ('Duponol') and phenol treatment of 20-sec. pulse extracts, sediments as shown in Fig. 8. Again the pattern is identical to that of labelled RNA from *T*2 infected cells (Fig. 9). Most newly synthesized RNA has a sedimentation constant about 8*s*; there is no appreciable synthesis of typical 16*s* or 23*s* ribosomal RNA. Since much new RNA in crude extracts moves at 14–16*s*, we considered the possibility that pulse-RNA is more sensitive than ribosomal RNA to degradation during the phenol extractions. Sodium lauryl sulphate (0·5 per cent) alone was therefore added to labelled crude extracts. This detergent treatment[12] both separates RNA from protein and completely inhibits action of the latent ribonuclease[16] release upon ribosome breakdown. When such

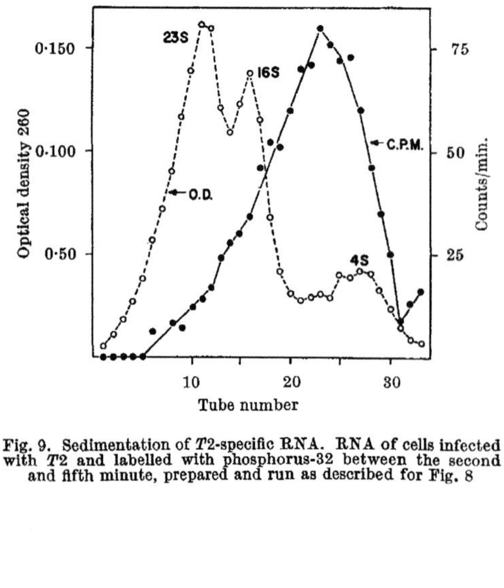

Fig. 9. Sedimentation of *T*2-specific RNA. RNA of cells infected with *T*2 and labelled with phosphorus-32 between the second and fifth minute, prepared and run as described for Fig. 8

treated extracts were run in sucrose gradients, the results were identical to those with phenol-prepared RNA.

The turnover of pulse-RNA is seen in chase experiments in which an excess of cold uracil is added after a 10-sec. ^{14}C-uracil pulse. After 15 min. at 37° C., all the label in crude extracts leaves the 14–16*s* component and is incorporated in the metabolically stable ribosomal and soluble RNA. In our kinetics experiments, the 16*s* RNA molecules always become labelled before appreciable labelling of 23*s* RNA molecules; likewise the 30*s* ribosomes become labelled before the 50*s* ribosomes. Since the base ratios of 16*s* and 23*s* chains are identical[17], it is likely that the 23*s* molecules form from two 16*s* chains.

Discussion. Our pulse experiments show that uninfected cells contain unstable RNA with sedimentation constants and attachment properties similar to those of *T*2-specific RNA. It is tempting to believe that these unstable molecules convey genetic information and are 'messenger' RNA. Complete homology with phage-specific RNA, however, will be demonstrated only if the base ratios of pulse-RNA are those of DNA. Unfortunately, in *E. coli*, the DNA base ratios do not differ greatly from ribosomal RNA, and as yet the base ratios of *E. coli* pulse-RNA are not precisely known. But we do have preliminary results (Hayes, D., and Gros, F., unpublished results) from *Staphylococcus aureus* which indicate an RNA component rapidly turning over and possessing DNA-like base ratios. Moreover, in yeast, there is also reported an unstable RNA resembling DNA in base composition[18]. We thus believe that our current measurements with *E. coli* will extend this fact and rule out the possibility that pulse-RNA is a precursor sub-unit from which ribosomal RNA is built up.

A messenger role for pulse-RNA fits nicely with its specific attachment to 70*s* ribosomes, the sites of protein synthesis[19]. Our experiments reveal two types of attachment. In one, the labelled RNA moves reversibly on or off 70*s* ribosomes depending on the concentration of magnesium ions. The second type of attachment binds pulse-RNA irreversibly to 'active' 70*s* ribosomes. These are 70*s* ribosomes which do not break apart with 10^{-4} *M* magnesium ions and which Tissières *et al.*[19] have shown to be the principal, if not sole, site of *in vitro* protein synthesis. They can be washed several times with 10^{-4} *M* magnesium ions without losing their ability to act in protein synthesis, and so they must have their genetic RNA firmly attached. It is reassuring that they can contain pulse-RNA. At least two stages, one reversible, the other not, thus exist in active 70*s* formation.

So far as we can tell, pulse-RNA, when associated with ribosomes, sediments at exactly 70*s* or 100*s*. Many experiments show that the label sediments neither faster nor slower than ordinary ribosomes. We find it hard to believe that the 14–16*s* RNA component which exists in 10^{-4} *M* magnesium ion extracts can combine with 70*s* or 100*s* ribosomes without altering their rate of sedimentation. Another possibility is that with 10^{-2} *M* magnesium ions, the 14–16*s* component assumes a more compact configuration similar to that of the RNA in a 30*s* particle. It could then associate with a free 50*s* ribosome to form a 70*s* ribosome which is transformed to an active 70*s* ribosome.

In conclusion, we state our findings: bacteria contain an RNA component turning over rapidly which

No. 4776 May 13, 1961 NATURE 585

is physically distinct from ribosomal or soluble (transfer) RNA. This fraction behaves, in its range of sedimentation constants and its attachment to ribosomes in high magnesium ion concentrations, exactly as does the phage-specific RNA made after *T*2 infection. Furthermore, it is associated with the active 70*s* ribosomes, the site of protein synthesis.

Our working hypothesis is that no fundamental difference exists between protein synthesis in phage-infected and uninfected bacteria. In both cases typical ribosomal RNA does not carry genetic information, but has another function, perhaps to provide a stable surface on which transfer RNA's can bring their specific amino-acids to the messenger RNA template.

These experiments were initiated when F. Gros was visiting the Biological Laboratories (May–August 1960). The pyrimidine-requiring strain *B*148 was kindly provided by Dr. Martin Lubin. The financial support of the National Science Foundation and the National Institutes of Health is gratefully appreciated.

[1] Cohen, S. S., *Bact. Rev.*, **13**, 1 (1949).

[2] Koch, G., and Hershey, A. D., *J. Mol. Biol.*, **1**, 260 (1959). Kornberg, A., Zimmerman, S. B., Kornberg, S. R., and Josse, J., *Proc. U.S. Nat. Acad. Sci.*, **45**, 772 (1959).

[3] Cohen, S. S., *J. Biol. Chem.*, **174**, 271 (1948).

[4] Hershey, A. D., Dixon, J., and Chase, M., *J. Gen. Physiol.*, **36**, 777 (1953).

[5] Volkin, E., and Astrachan, L., *Virology*, **2**, 149 (1956).

[6] Volkin, E., Astrachan, L., and Countryman, J. L., *Virology*, **6**, 545 (1958).

[7] Zamecnik, P. Z., *The Harvey Lectures* (1958–59), 256 (Academic Press, New York, 1960).

[8] Crick, F. H. C., *Brookhaven Symposia in Biology*, **12**, Structure and Function of Genetic Elements, 35 (1959).

[9] Nomura, M., Hall, B. D., and Spiegelman, S., *J. Mol. Biol.*, **2**, 306 (1960).

[10] Jacob, F., and Monod, J., *J. Mol. Biol.* (in the press).

[11] Tissières, A., Watson, J. D., Schlessinger, D., and Hollingworth, B. R., *J. Mol. Biol.*, **1**, 221 (1959).

[12] Kurland, C. G., *J. Mol. Biol.*, **2**, 83 (1960).

[13] Berg, P., and Ofengand, E. J., *Proc. U.S. Nat. Acad. Sci.*, **44**, 78 (1958). Tissières, A., *J. Mol. Biol.*, **1**, 365 (1959).

[14] McQuillen, K., Roberts, R. B., and Britten, R. J., *Proc. U.S. Nat. Acad. Sci.*, **45**, 1437 (1959).

[15] Grunberg-Manogo, M., Ortiz, P. J., and Ochoa, S., *Science*, **122**, 907 (1955).

[16] Elson, D., *Biochim. Biophys. Acta*, **27**, 217 (1958). Spahr, P. F., and Hollingworth, B., *J. Biol. Chem.* (in the press).

[17] Spahr, P. F., and Tissières, A., *J. Mol. Biol.*, **1**, 237 (1959).

[18] Yčas, M., and Vincent, W. S., *Proc. U.S. Nat. Acad. Sci.*, **46**, 804 (1960).

[19] Tissières, A., Schlessinger, D., and Gros, Françoise, *Proc. U.S. Nat. Acad. Sci.*, **46**, 1450 (1960).

J. Mol. Biol. (1963) **6**, 374–388

Polypeptide Synthesis in *Escherichia coli*.
I. Ribosomes and the Active Complex

Walter Gilbert

Jefferson Laboratory of Physics, Harvard University, Cambridge, Mass., U.S.A.

(Received 4 January 1963)

The poly U-directed† synthesis of polyphenylalanine in the cell-free system from *Escherichia coli* is used as a model system in which to investigate the interaction of messenger RNA with ribosomes. It is shown that both the 50 s and the 30 s ribosome are necessary for polypeptide synthesis. The ribosomes accept the messenger RNA under conditions in which the dominant form is the 70 s particles, but the particles involved will dimerize more readily than the rest of the 70 s particles.

All of the synthetic activity of the poly U–ribosome mixture appears as a rapidly sedimenting complex, 140 to 200 s. This active complex depends upon RNA for its integrity and contains an amount of poly U consistent with one molecule for several ribosomes. Furthermore, all of the synthetic capacity of the usual crude extract from *E. coli* is in the form of rapidly sedimenting complexes in the 100 to 200 s range.

1. Introduction

There is abundant evidence that protein synthesis occurs on ribosomes (Hoagland, 1960). At first it was thought that the template for the amino acid sequence was the ribosomal RNA. Now it is known that an additional RNA molecule, messenger RNA, attaches to the ribosomes and directs the ordered assembly of amino acids into proteins (Jacob & Monod, 1961; Nomura, Hall & Spiegelman, 1960; Brenner, Jacob & Meselson, 1961; Gros *et al.*, 1961). A ribosome is a non-specific work-bench which holds the messenger, the growing peptide chain, the amino acyl S-RNAs and the necessary enzymes in the correct position for peptide bond formation.

How the ribosomes work is unclear. Ribosomes, from all sources, exist in several different states of aggregation, but the significance of their subunits is unknown. The ribosomes of *E. coli* (Tissières, Watson, Schlessinger & Hollingworth, 1959) are constructed from two fundamental subunits characterized by their sedimentation coefficients as 30 s and 50 s. These subunits are easily observed in low magnesium ion concentrations (10^{-4} M-Mg^{2+}). At the levels of magnesium necessary for protein synthesis (on the order of 10^{-2} M-Mg^{2+}) one 30 s and one 50 s ribosome associate reversibly to form a 70 s ribosome, which in turn may dimerize to a 100 s particle. These same high concentrations of magnesium ions are needed for the attachment of messenger RNA to the ribosomes. In 10^{-4} M-Mg^{2+} most of the *E. coli* messenger molecules sediment free of the ribosomes with an average sedimentation coefficient of 14 s (Gros *et al.*, 1961). Early experiments suggested that ribosomes carrying messenger

† Abbreviations used: poly U = polyuridylic acid; poly UG = polyuridylic-guanidylic acid; PEP = phosphoenolpyruvate; PK = pyruvate kinase; TCA = trichloroacetic acid.

molecules sedimented in coincidence with the bulk of the 70 s and 100 s ribosomes. Better technique, keeping the samples always near 0°C, revealed that the ribosomes which carried the messenger sedimented more rapidly than the bulk of the ribosomes and permitted the delineation of a class of "heavy ribosomes" (Risebrough, Tissières & Watson, 1962).

Here we shall report experiments that ask which ribosomes function and what is the nature of the functioning complex. They utilize the discovery by Nirenberg & Matthaei (1961) that poly U is a synthetic messenger capable of directing the polymerization of phenylalanine into polyphenylalanine.

2. Materials and Methods

Cells. E. coli strain B was grown at 37°C in a medium composed of 0·005 M-potassium phosphate, pH 6·8, 2 g dextrose, 2 g vitamin-free Casamino acids (Difco), 2 g $(NH_4)_2SO_4$, 160 mg $MgCl_2$, 3 mg $CaCl_2$, 1 mg $FeSO_4$, and 1 g NaCl/l. The cells were harvested during the exponential phase at an O.D. (550 mμ) of 0·4, washed twice in 0·01 M-magnesium acetate and 0·005 M-tris buffer, pH 7·3, and frozen until use. *Crude extracts* were made by grinding frozen cells with alumina (2·5/1 wt./wet wt.) and extracting with buffer (3/1 wt./wet wt. cells) containing 10 μg/ml. DNase. All manipulations were carried out at 0°C. The crude extract was clarified by two low-speed centrifugations (10 min at 5000 **g**). *The buffers* used were all 0·005 M-tris, pH 7·3, containing Mg^{2+} and KCl as specified. *The supernatant solution* was made by taking the upper two-thirds of the supernatant fraction from a 5 hr 100,000 **g** centrifugation of a crude extract in 0·01 M-Mg^{2+}. *Dialysed supernatant* solution was dialysed overnight against 0·01 M-Mg^{2+} and 0·006 M-mercaptoethanol.

The incorporation system used was basically that of Tissières, Schlessinger & Gros (1960) modified to contain more KCl and Mg^{2+}. *For phenylalanine incorporation* the reaction mixture contained 0·001 M-ATP, 0·005 M-PEP, 40 μg/ml. PK, 0·086 M-KCl, 0·0175 M-magnesium acetate, 0·005 M-tris buffer, pH 7·3, 0·1 to 0·5 ml./ml. crude extract, poly U, and [^{14}C]phenylalanine. (If the supernatant solution and ribosomes were added separately, the supernatant solution was always 0·1 ml./ml.) The total vol. was generally 0·1 or 0·2 ml. The mixtures were usually incubated at 30°C for 30 min. The reactions were stopped with 5% TCA and carrier albumin was added, if necessary. The precipitates were washed twice by centrifugation with 5% TCA, resuspended and held at 90°C for 15 min in 5% TCA, then filtered onto Millipore filters (Millipore Filter Co., Bedford, Mass.), glued to aluminum planchets and counted in gas-flow counters at efficiencies of the order of 30%. *The alanine assay* for protein synthesis was identical except that the reaction mixture contained only 0·01 M-Mg^{2+}. [^{14}C]Alanine was used as a label and from 5 to 20 μg/ml. of each of the other 18 common amino acids were added. *Pre-incubated crude extracts*: The crude extracts used for phenylalanine incorporation were pre-incubated 10 to 15 min at 30°C with ATP, PEP and PK in order to exhaust the synthetic activity of the natural messenger and to permit this natural messenger, present in these extracts, to be degraded. If the pre-incubation was to be followed directly by a poly U assay, it was carried out at 0·0175 M-Mg^{2+} and 0·086 M-KCl. The extract was chilled after pre-incubation, the poly U and other ingredients added, and then the incubation continued. *Gradients* were exponential gradients from 5 to 20% (w/w) sucrose (Merck reagent), equilibrated for 6 to 12 hr at 4°C, loaded with a 1 ml. sample on a 24 ml. gradient and run at 25,000 rev./min at the 10°F setting of the Spinco model L. Ten-drop samples were collected; 2 drops diluted with 3 ml. of water were used to read the optical density at 260 mμ and 8 drops to provide a sample for further assay. The sucrose did not interfere with the incorporation assays. *The sedimentation coefficients* reported on the figures are nominal values given to identify the particle types. The sedimentation values for unknown material were inferred by assuming that the behavior along the gradient was linear. Both [^{12}C] and [^{14}C]poly U were provided by Robert Thach of the Chemistry Department, Harvard University. The [^{12}C]poly U had a bulk sedimentation coefficient of 5·3 and would saturate the synthetic capacity of the incorporation system in amounts in excess of 2% (wt. poly U/wt. ribosomes).

3. Results

(a) *Which ribosomes can accept the messenger?*

If we sediment a crude extract through a sucrose gradient to display the various ribosomes and assay samples along the gradient with excess poly U, supernatant fraction, [^{14}C]phenylalanine and an ATP generating system, then the amount of synthesis is a measure of the kinds of ribosomes that can accept the poly U and function. We can control the type of ribosomes that will appear by varying the concentration of magnesium ions and of salt. Raising the magnesium ion concentration moves the equilibria toward the formation of 70 s ribosomes and their dimerization into 100 s particles. Raising the salt concentration, at fixed magnesium concentration, moves the equilibria in the opposite direction and prevents the formation of

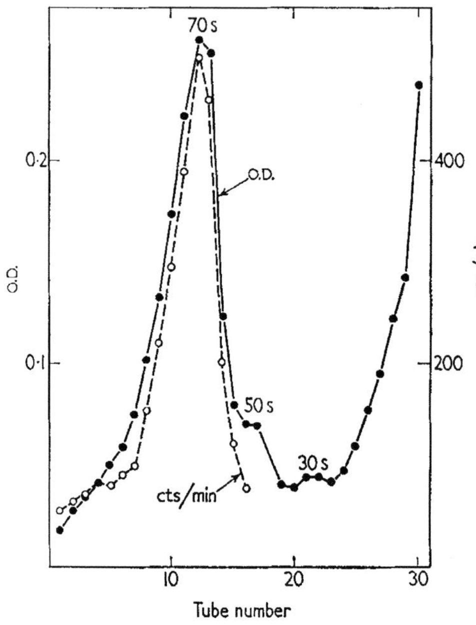

FIG. 1. *A sucrose gradient of a crude extract in 0·0175 M-Mg^{2+} and 0·086 M-KCl assayed with poly U and [^{14}C]phenylalanine after the gradient.* 0·25 ml. of crude extract made in 0·0175 M-Mg^{2+}, 0·086 M-KCl and 0·005 M-tris, pH 7·3, was run for 4 hr at 10°F on a sucrose gradient in the same buffer. 0·1 ml. samples from the tubes were first pre-incubated with the generating system and supernatant solution for 15 min at 30°C, then 5 μg of poly U and 0·3 μc of [^{14}C]phenylalanine (3·8 μg/μc) were added and the tubes were incubated for a further 30 min at 30°C and assayed as described in Materials and Methods.

100 s particles. Figure 1 shows the results of such an experiment. Here the magnesium and salt concentrations have been chosen to be near the optimum for polyphenylalanine synthesis (0·0175 M-Mg^{2+} and 0·086 M-KCl) and the ribosomes are predominantly in the 70 s form. The response to the added poly U follows the optical density of the 70 s ribosomes. The extract used in this experiment had not been pre-incubated to destroy any natural messenger before being run on the gradient. As we know from the work of Risebrough *et al.* (1962) (and as we shall show later, Figs. 11 and 12), the fraction of the ribosomes that carry messenger RNA would run ahead of the 70 s ribosomes. Since we observe no preferential activity in this region, we conclude that

the ribosomes that have carried the messenger are *not* especially competent in later polyphenylalanine synthesis.

Figure 2 shows the analysis of a pre-incubated crude extract run on a gradient in 10^{-2} M-Mg^{2+}. The 100 s, 70 s, 50 s and 30 s ribosomes are resolved on the gradient, but predominantly the 100 s ribosomes have accepted the poly U and functioned. The assay is conducted in the presence of salt; we cannot infer that the ribosomes are still in the 100 s form when they function. Together these two experiments show that only a fraction of the 70 s ribosomes function in accepting poly U and making polyphenylalanine and that this fraction is more "sticky", dimerizes more readily than the average 70 s ribosomes. We offer no real explanation for this behavior. It may be due to some of the particles being blocked by inactive messenger or incomplete protein

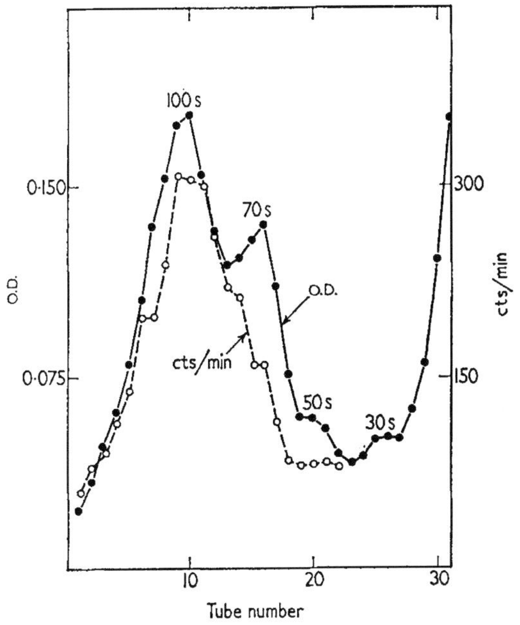

Fig. 2. *A sucrose gradient of a pre-incubated crude extract in* 0·01 *M-Mg^{2+} assayed with poly U and [^{14}C]phenylalanine after the gradient.* 0·3 ml. of crude extract was pre-incubated in 0·014 M-KCl and 0·01 M-Mg^{2+} for 15 min at 30°C and run for 4 hr at 20°F on a gradient in 0·01 M-Mg^{2+}. A 0·1 ml. sample from each tube was assayed with 5 μg poly U, 0·3 μc [^{14}C]phenylalanine (3·8 μg/μc), and supernatant solution as described in Materials and Methods.

chains, or it may be that the site for acceptance of the messenger is related to the site at which the 70 s particles bind together to form 100 s particles and that this site is in better condition on those particles that can function.

Figure 3 shows the analysis of a pre-incubated crude extract in 10^{-3} M-Mg^{2+}. In the incorporation assay with added poly U the 70 s particles function, but there is also a secondary maximum between the 50 s and 30 s peaks. That this is due to the re-formation of 70 s particles from 50 s and 30 s material is confirmed by the experiment shown in Fig. 4. A gradient in 10^{-4} M-Mg^{2+} was run to display the 50 s and 30 s ribosomes. An excess of 30 s ribosomes taken from the 30 s peak was added to each tube of a series of samples taken across both peaks, and these tubes were incubated with poly U, [^{14}C]phenylalanine, supernatant solution and the generating system. The amount of

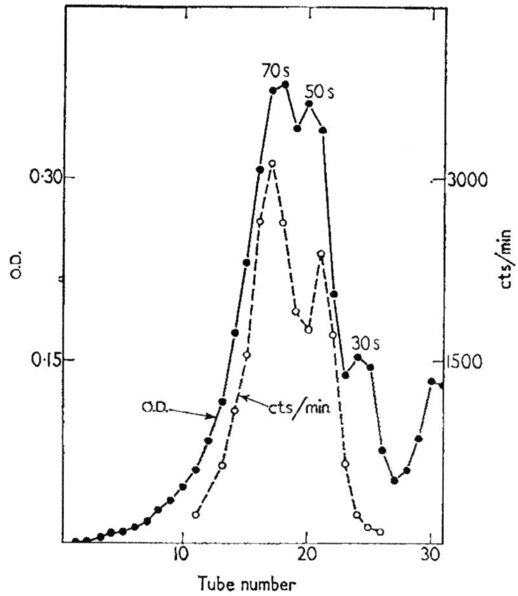

FIG. 3. *A sucrose gradient of a crude extract in 0·001 M-Mg²⁺ assayed with poly U and [¹⁴C]phenylalanine.* 0·8 ml. of crude extract was pre-incubated for 15 min at 30°C, dialysed overnight against 0·001 M-Mg^{2+} and 0·005 M-tris, pH 7·3, and run for 4 hr on a gradient in this buffer. 0·1 ml. samples from each tube were brought to 0·0125 M-Mg^{2+} and assayed at 0·0125 M-Mg^{2+}, 0·086 M-KCl with 5 μg of poly U, 0·2 μc of [^{14}C]phenylalanine (0·78 μg/μc) and supernatant solution.

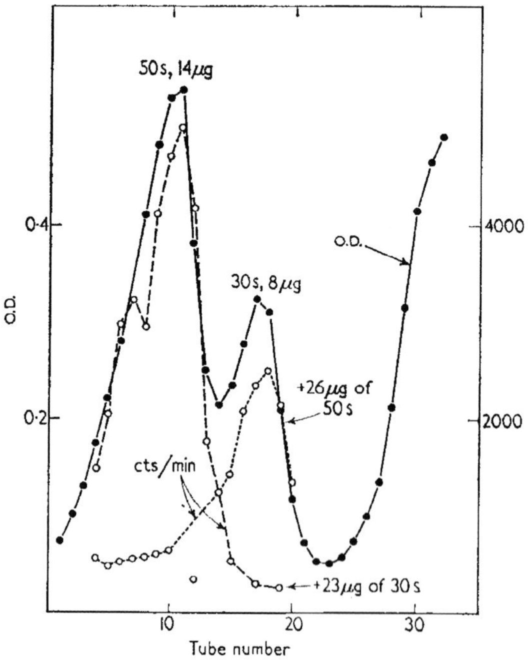

FIG. 4. *Reconstruction of acceptors for poly U from 30 S and 50 S ribosomes.* One ml. of crude extract made in 10^{-4} M-Mg^{2+} and 0·005 M-tris, pH 7·3, was run on a gradient in this buffer for 7 hr at 15°F. A series of 0·02 ml. samples taken across both peaks (containing 14 μg of 50 S ribosomes and 8 μg of 30 S ribosomes at the respective peaks) was assayed with 23 μg of 30 S ribosomes from the 30 S, 5 μg of poly U, 0·2 μc of [^{14}C]phenylalanine (0·78 μg/μc) and supernatant solution. A second series was similarly assayed with 26 μg of 50 S ribosomes.

incorporation followed the optical density of the 50 s peak, showing that there was activity only when the particles could recombine to form 70 s ribosomes and that the 30 s alone cannot function. The experiment was repeated using an excess of 50 s ribosomes. The curve showing the incorporation into polyphenylalanine follows the distribution of 30 s material, showing that the 50 s ribosome alone will not function but must be combined with a 30 s particle.

(b) *The active complex*

We have used three types of experiments to study the behavior of the poly U–ribosome complexes. The first and most direct way to see the active complexes is to analyse for phenylalanine incorporation after running a gradient of a poly U–ribosome

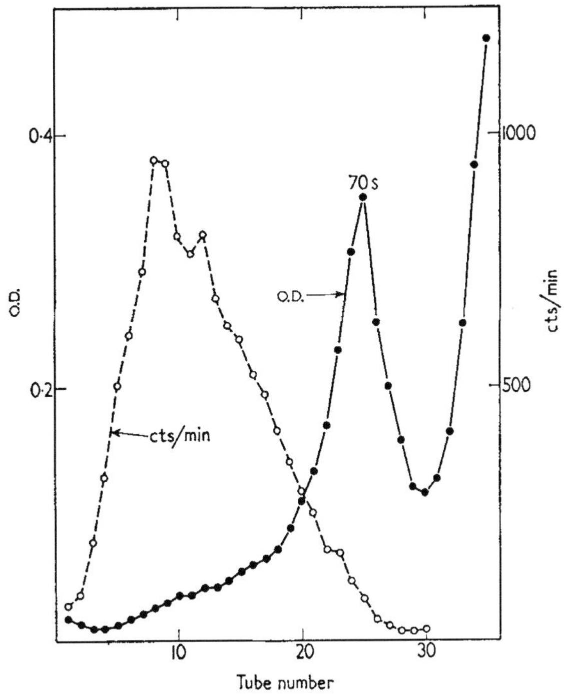

FIG. 5. *A sucrose gradient of a mixture of poly U and crude extract assayed with* [^{14}C]*phenylalanine after the gradient.* 0·5 ml. of crude extract was pre-incubated for 15 min at 30°C in 0·0125 M-Mg^{2+} and 0·086 M-KCl, and then chilled. 150 μg of poly U was added. The mixture was kept at 0°C for 30 min, then run on a gradient in 0·0125 M-Mg^{2+} and 0·086 M-KCl for 110 min. 0·1 ml. samples were assayed with 0·2 μc of [^{14}C]phenylalanine (0.78 μg/μc) and supernatant solution.

mixture. The gradient separates the poly U–ribosome complexes from the rest of the ribosomes, the unattached poly U and the supernatant factors. If we add back the supernatant factors along with [^{14}C]phenylalanine we will observe only those complexes that are able to function in the synthesis of polyphenylalanine. The second approach is to sediment a reaction mixture after phenylalanine incorporation, observing those structures that have functioned. A third measurement can be made by following the attachment of labeled poly U; observing a class of structures that includes those that can function.

The results of an experiment that involves the incorporation assay after the gradient is shown in Fig. 5. For this experiment we have mixed poly U and a pre-incubated crude extract, sedimented the mixture on a sucrose gradient and analysed a sample from each tube with [^{14}C]phenylalanine and supernatant solution. The structures that polymerize the phenylalanine sediment extremely rapidly in a broad peak with no corresponding peak in the optical density. The peak of synthesizing activity has moved almost three times as far down the gradient as have the 70 s particles; thus it represents material sedimenting in the neighborhood of 200 s. This rapidly sedimenting complex of poly U and ribosomes was formed in the cold during the 30 minutes before the sample was put on the gradient. Let us examine what has happened half-way through the synthetic reaction. A similar mixture of poly U and a pre-incubated crude extract was incubated for 15 minutes with cold phenylalanine.

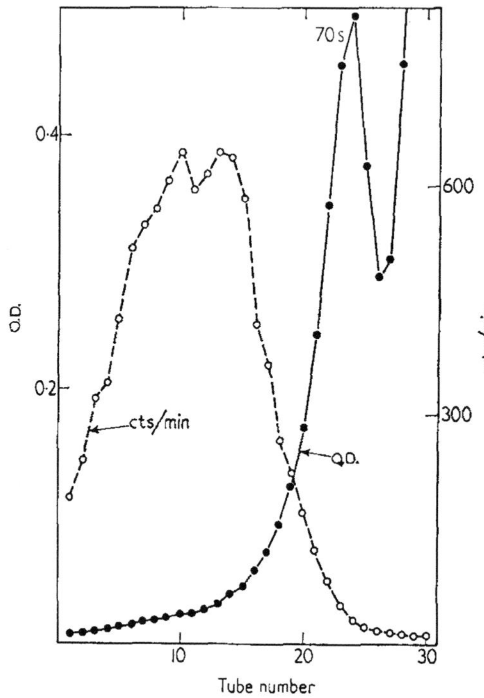

FIG. 6. *A sucrose gradient of an incubated mixture of poly U and crude extract assayed with [^{14}C]phenylalanine after the gradient.* 0·45 ml. of crude extract was pre-incubated for 15 min at 30°C in 0·0175 M-Mg^{2+} and 0·086 M-KCl; then 150 μg of poly U and 50 μg of [^{12}C]phenylalanine were added and the incubation was continued for 15 min at 30°C. The reaction mixture was run on a gradient in 0·0175 M-Mg^{2+} and 0·086 M-KCl for 110 min. 0·1 ml. samples were assayed with 0·2 μC [^{14}C]phenylalanine (0·78 μg/μC) and supernatant solution.

(The reaction takes about 30 minutes at 30°C in the crude extract.) The mixture was then chilled and run on a gradient at the same salt and magnesium concentration as the incubation. Samples from each tube were re-incubated with [^{14}C]phenylalanine and supernatant solution. The results are shown in Fig. 6. The structures that will synthesize sediment in a broad peak from 140 to 200 s. An identical reaction mixture, incubated for 15 minutes with [^{14}C]phenylalanine, was run in parallel and is shown in Fig. 7. The curve of the radioactivity in this figure shows the sedimentation of the complexes that have synthesized polyphenylalanine. The curve in Fig. 6 shows where the complexes that will continue to synthesize sediment. The curves agree closely except for a shoulder at the 70 s ribosomes in Fig. 7. These ribosomes carry new chains of polyphenylalanine but are no longer synthetically active.

We performed several controls. To see whether or not the poly U itself was aggregated in this high salt buffer, poly U alone was run on a gradient and occasional tubes were analysed with ribosomes, supernatant solution and [^{14}C]phenylalanine. There was no indication of any rapidly sedimenting activity. To see if the insolubility of the polyphenylalanine was responsible, a reaction mixture of crude extract, poly U and [^{14}C]phenylalanine was incubated for 15 minutes, treated with 5 μg/ml. of RNase,

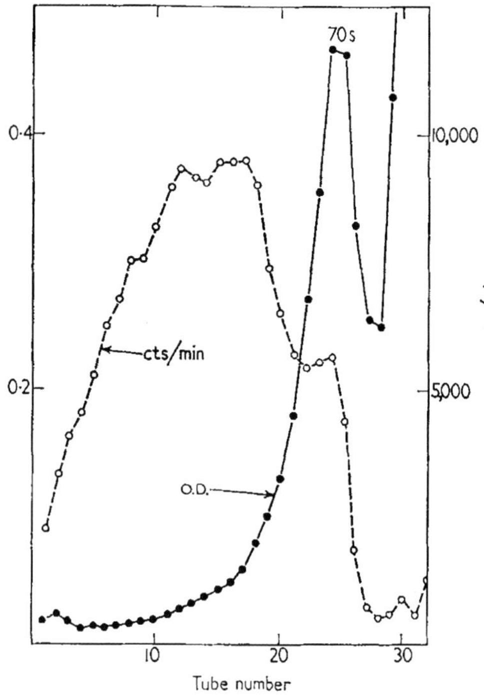

FIG. 7. *A sucrose gradient analysis of a [^{14}C]phenylalanine incorporation.* 0·45 ml. of crude extract was pre-incubated for 15 min at 30°C. 150 μg of poly U and 1 μc of [^{14}C]phenylalanine (0·78 μg/μc) were added and the mixture was incubated for 15 min at 30°C and run on a gradient in parallel with that of Fig. 6. The 8 drop samples were precipitated with 5% TCA, heated to 90°C for 15 min and plated.

then run on a gradient. As we see in Fig. 8, the rapidly sedimenting complexes have been broken up, the polyphenylalanine now sediments with the 70 s ribosomes while the sedimentation behavior of the ribosomes themselves has not been affected by this treatment. The optical density of the 70 s ribosomes rides up on the peak of soluble u.v.-absorbing material and is displaced from the true position of the particles. A longer centrifuge run reveals that the two peaks, 70 s ribosomes and ribosomes carrying labeled polyphenylalanine, are coincident.

We have seen that the structures that either will synthesize or are in process of synthesizing polyphenylalanine are complexes of poly U and ribosomes which have sedimentation coefficients in a broad range from 140 to 200 s. These structures depend on RNA for their integrity. The most likely explanation of this behavior is that these are complexes of from four to eight 70 s ribosomes with one strand of poly U. The

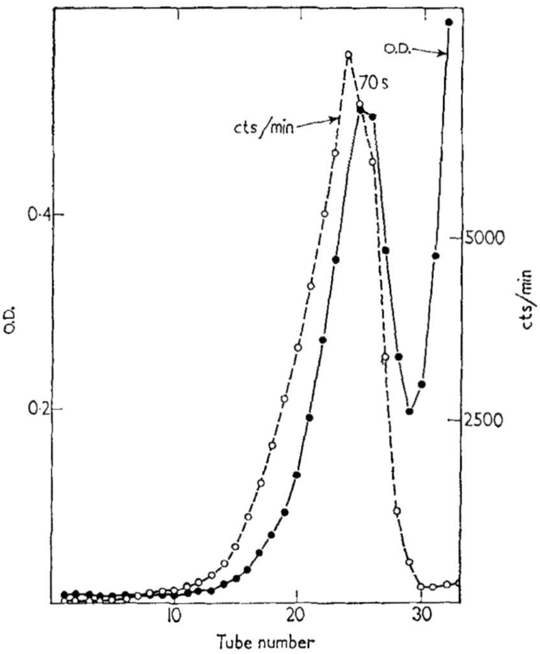

FIG. 8. *A sucrose gradient analysis of [^{14}C]phenylalanine incorporation after RNase treatment.* An experiment similar to that of Fig. 7 except that after the second incubation the sample was treated in the cold with 5 μg of RNase for 5 min.

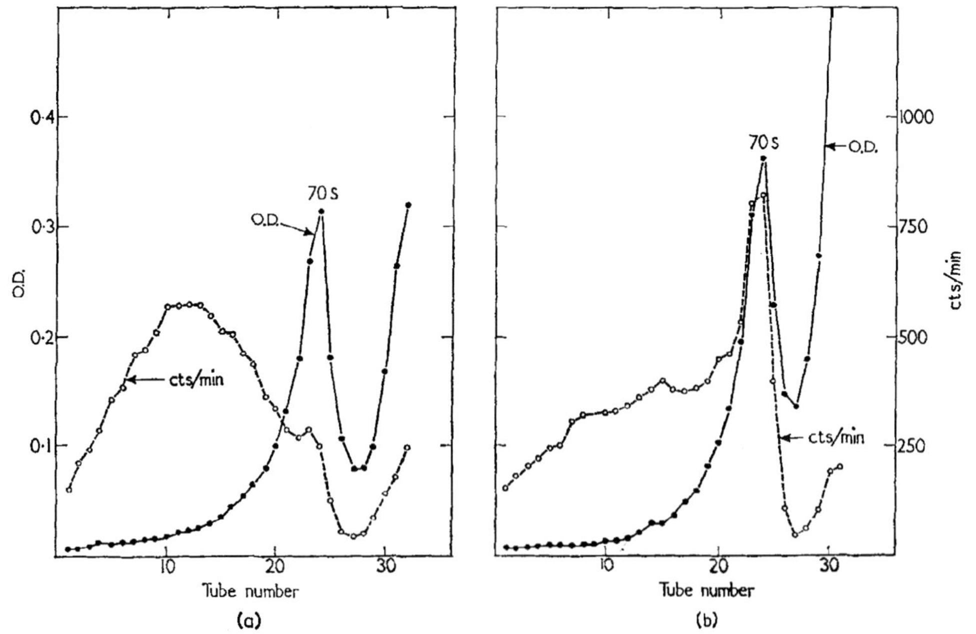

FIG. 9. *Sucrose gradient analysis of a chase of [^{14}C]phenylalanine with [^{12}C]phenylalanine.* Two samples containing 0·2 ml. of crude extract were pre-incubated for 15 min at 30°C, then 50 μg of poly U and 1 μc of [^{14}C]phenylalanine (3·6 μg/μc) were added and incubated for a further 10 min, then chilled. One sample was then run for 120 min on a gradient in 0·0175 M-Mg^{2+} and 0·086 M-KCl, assayed as described for Fig. 7 and shown in Fig. 9(a). To the other sample was added 1 mg [^{12}C]phenylalanine and 600 μg of charged S-RNA, and the reaction was continued for 20 min at 30°C. This material was then run in parallel and is shown in Fig. 9(b).

difference between the experiments shown in Figs. 6 and 7 suggests that, after synthesis, 70 s ribosomes carrying the new polyphenylalanine chain are released from the active complex. This is consistent with a chase experiment shown in Fig. 9. A reaction mixture containing poly U and [^{14}C]phenylalanine was incubated for 10 minutes at 30°C. At the end of this time half of the mixture was chilled while the other half was incubated for a further 20 minutes with a great excess of unlabeled phenylalanine and S-RNA charged with the unlabeled amino acid. Figure 9(a) shows the phenylalanine label still predominantly on the active complexes, while Fig. 9(b) shows that after further synthesis the originally labeled material has been chased out of the complexes and appears on the 70 s particles.

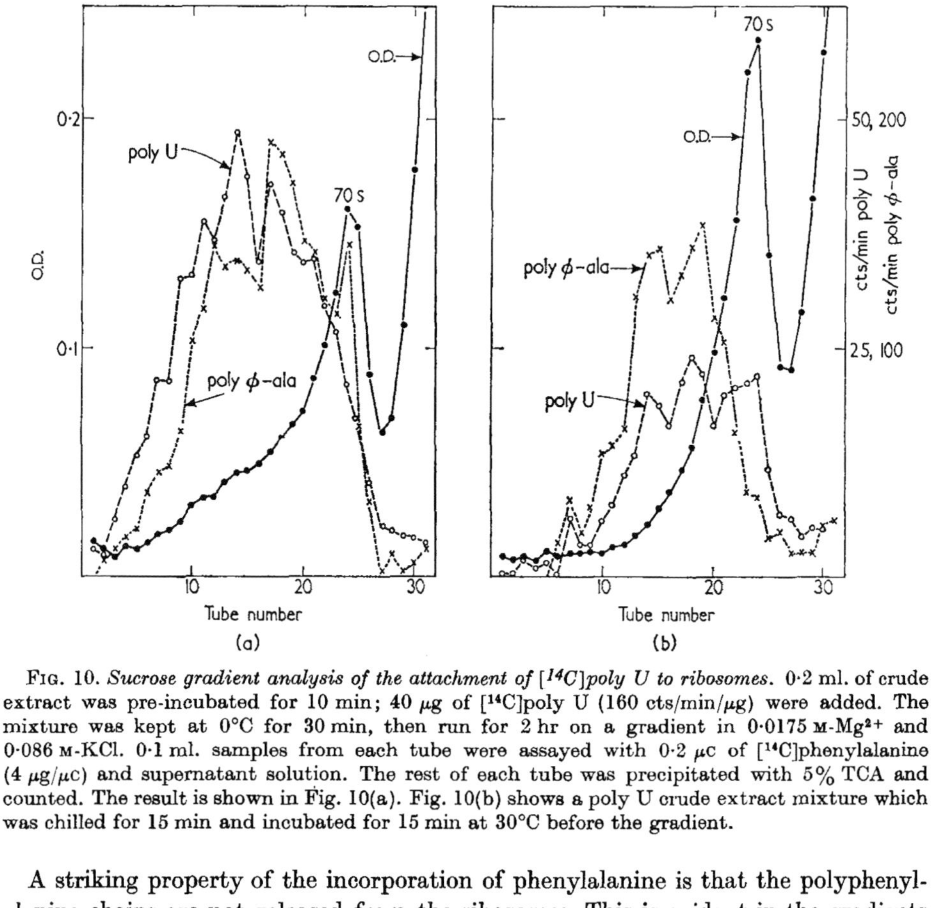

FIG. 10. *Sucrose gradient analysis of the attachment of* [^{14}C]*poly U to ribosomes*. 0·2 ml. of crude extract was pre-incubated for 10 min; 40 μg of [^{14}C]poly U (160 cts/min/μg) were added. The mixture was kept at 0°C for 30 min, then run for 2 hr on a gradient in 0·0175 M-Mg^{2+} and 0·086 M-KCl. 0·1 ml. samples from each tube were assayed with 0·2 μc of [^{14}C]phenylalanine (4 μg/μc) and supernatant solution. The rest of each tube was precipitated with 5% TCA and counted. The result is shown in Fig. 10(a). Fig. 10(b) shows a poly U crude extract mixture which was chilled for 15 min and incubated for 15 min at 30°C before the gradient.

A striking property of the incorporation of phenylalanine is that the polyphenylalanine chains are not released from the ribosomes. This is evident in the gradients shown in Figs. 7, 8 and 9. The simplest explanation for this is that the signal to release the chain is given by a code word, since one would expect that any non-specific mechanism, such as the end of the messenger being the signal to release the chain, would also work for poly U.

We have examined these complexes using labeled poly U. Figure 10 illustrates a pair of experiments in which labeled poly U, 160 cts/min/μg, was mixed with a pre-incubated crude extract and run on a gradient. In this experiment samples from each tube were treated with [^{14}C]phenylalanine and supernatant solution and the remainder was precipitated with 5% TCA. For Fig. 10(a) the poly U reaction mixture was kept chilled for 35 minutes while for Fig. 10(b) the mixture was chilled for 20 minutes and incubated at 30°C for 15 minutes. Before the incubation both the synthetic activity and the TCA-precipitable poly U run ahead of the 70 s particles. After the incubation a fraction of the poly U appears on the 70 s ribosomes, while the synthetic activity still runs ahead. The incubation has decreased the amount of precipitable poly U, the amount of polyphenylalanine synthesis, and has shifted, in this case, about 10% of the u.v.-absorbing material from the region of 100 to 200 s to the 70 s peak. If the inactive poly U associated with the 70 s peak is not counted, after incubation only 30% of the poly U, originally attached, accounts for 65% of the original synthesis. However, only 20% of the original poly U was recovered in this experiment. TCA is not a good precipitant for poly U; the poly U strands precipitated only where they were tightly associated with other material. If the poly U is precipitated with 70% ethanol and 0·2 M-NaCl, only half of the precipitable material is found associated with the ribosomes.

The ratio by weight of poly U to ribosomes in the most rapidly sedimenting peak in this experiment is 1:100 before incubation, 1:140 after. This corresponds to 27,000 daltons of poly U per 70 s ribosome on the first gradient and 19,000 per 70 s on the second. We do not know the molecular weight of the poly U, either before it is added or after it has been partially destroyed by the extract, but we expect it to be greater than 50,000. Thus this experiment rules out the possibility that the active complexes contain many poly U molecules bound to a single ribosome. These numbers show that there is at least one 70 s ribosome for each poly U molecule (more probably, several) in this region of the gradient. This estimate suffers from the fact that we do not know that all the ribosomes in this region are complexed with the poly U. In addition, the active complexes seen in this experiment do not sediment as rapidly as those observed with the ^{12}C preparation of poly U.

(c) *The behavior of natural messenger*

We have analysed gradients of crude extracts from *E. coli* by adding [^{14}C]alanine, cold amino acids, supernatant solution and the ATP generating system to a sample from each tube to study the sedimentation of the naturally occurring active complex of messenger RNA and ribosomes. Figure 11 shows the results of such an experiment in 10^{-2} M-Mg^{2+}. Ninety per cent of the original incorporating activity of the crude extract was recovered on this gradient, sedimenting ahead of the 100 s and 70 s ribosomes. This behavior agrees exactly with that found by Risebrough *et al.* (1962) for the complex formed by pulse-labeled RNA or T2 messenger RNA and ribosomes, and confirms the hypothesis that only those ribosomes carrying a piece of rapidly labeled RNA function. In order to compare the behavior of the natural messenger more directly with that of poly U, a similar assay was performed on a gradient of crude extract in 10^{-2} M-Mg^{2+} and 0·086 M-KCl. The results are shown in Fig. 12. The salt prevents the dimerization of the 70 to 100 s but enhances the active complexes which appear as a broad peak sedimenting from 100 to 200 s.

Let us emphasize that this salt effect shows that the involvement of 100 s ribosomes as such in protein synthesis is probably an artefact. Although nascent protein has been found on 100 s ribosomes (Tissières *et al.*, 1960), those experiments examined the products of the reaction by running gradients in 10^{-2} M-Mg^{2+} with no salt. Under these conditions a large fraction of the particles appear in the 100 s form, whether or

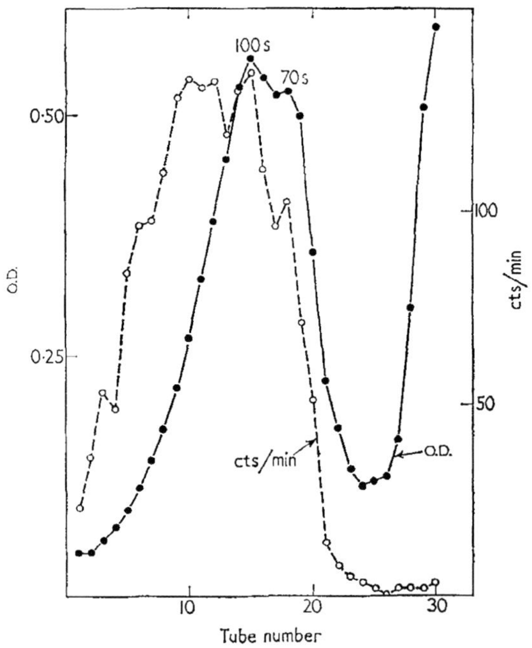

FIG. 11. *Sucrose gradient analysis of natural active complexes in 0·01 M-Mg²⁺*. One ml. of crude extract was run on a gradient in 0·01 M-Mg^{2+} for 3·5 hr. Each tube was assayed by adding supernatant solution, 1 μc of [^{14}C]alanine (1·25 μg/μc), 5 μg each of the other 18 amino acids, the generating system, and then incubating for 12 min at 34°C. A sample of the original crude extract was tested under the same conditions to determine the original level of activity; 90% of this activity was recovered on the gradient.

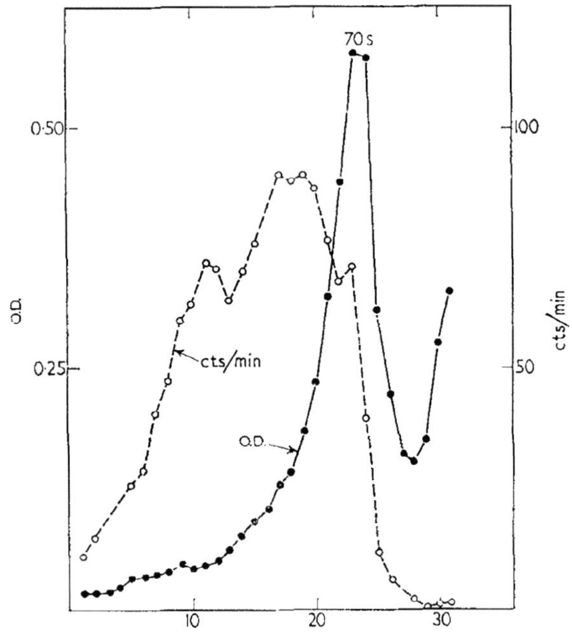

FIG. 12. *Sucrose gradient analysis of natural active complexes in 0·01 M-Mg^{2+} and 0·086 M-KCl*. One ml. of crude extract was run for 2 hr on a gradient in 0·01 M-Mg^{2+} and 0·086 M-KCl. 0·1 ml. samples were assayed with dialysed supernatant solution, 20 μg S-RNA, 0·3 μc of [^{14}C]alanine (1·25 μg/μc), and 4 μg of each of the other amino acids, in a vol. of 0·2 ml. as described in Materials and Methods.

not any messenger is still present. However, at the salt concentrations that are used during incubation in the *in vitro* system, which approach the physiological concentrations of the cell, the dominant particle form is the 70 s ribosome, with only 5% of the material occurring as 100 s. There is no evidence that the actual synthesis is a co-operative phenomenon involving more than one 70 s ribosome.

4. Discussion

We have shown that the complex of messenger RNA and ribosomes that is active in protein synthesis is a very rapidly sedimenting structure. In order to produce a sedimentation coefficient in the range from 140 to 200 s, from four to eight 70 s ribosomes would be required if the packing is similar to that involved in the 70 to 100 s dimerization. The experiment with RNase shows that this structure is held together by RNA. There are two possibilities left open by our experiment with labeled poly U. Either the active complex is a series of ribosomes attached to a single messenger molecule or it is a non-essential aggregate of functional units. The experiment with labeled poly U shows that the amount of poly U per ribosome is small and suggests that the first hypothesis is correct. In addition, the fact that high salt concentrations which prevent the aggregation of 70 to 100 s particles do not affect the active complexes argues that these structures are not just aggregates and furthermore suggests that the ribosomal species involved is the 70 s ribosome.

We thus take the view that the active complex for polypeptide synthesis is most probably a series of 70 s ribosomes functioning along a single molecule of messenger RNA. We expect that each 70 s ribosome synthesizes a single polypeptide chain as the messenger moves over or through the ribosome. The reason for this is the following. In order to read out the information that is arranged in sequence along the messenger, the growing point of the polypeptide chain must move along the messenger molecule. We shall present elsewhere evidence that there is only one active center for peptide bond formation on the 70 s ribosome (Cannon, Krug & Gilbert, 1963; Gilbert, 1963). Hence, the ribosomes must move along the messenger as the polypeptide chain is synthesized. This is supported also by the chase experiment (Fig. 9) that suggests that, after synthesis, 70 s ribosomes bearing nascent chains are released from the active complex.

This model provides an answer for an old question: why is there so little messenger? If the average messenger has a molecular weight of half a million, then one messenger for each 70 s ribosome would require that 25% of the high molecular weight RNA in the cell be messenger RNA. But if one messenger can occupy simultaneously from four to eight ribosomes, then only 6 to 3% of the RNA need be messenger RNA. This is within the range of estimates for the amount of messenger (Levinthal, Keynan & Higa, 1962).

However, this model raises a new problem. How can the string of ribosomes form in the cold, under conditions where we do not observe any synthesis? We expect that the string is formed by the messenger threading its way through a series of ribosomes. This process could proceed in the cold by diffusion if the mechanism that moves the ribosomes along the messenger does not require irreversible steps such as the formation of peptide bonds. Another possibility is that the ribosomes attach to the messenger at many points along its length. In both cases the mechanism that starts the synthesis of the polypeptide chain, about which we know nothing, must be required to start chains only on those ribosomes that are at the beginning of the messenger in order to prevent the production of partial chains.

Palade (1955), in his original observations of the ribosomes using the electron microscope, noted that the particles attached to the endoplasmic reticulum are "frequently disposed in linear series" and that these series then form patterns "among which parallel double rows, loops, spirals, circles and rosettes appear to be predominant". This is consistent with the hypothesis that the ribosomes are arranged along a messenger molecule, since, if the hypothesis is correct, one should see such forms.

Other recent work supports this picture of large active complexes. Barondes & Nirenberg (1962) have observed active complexes in the 100 to 130 s range using labeled poly U and much more rapidly sedimenting complexes with poly UG. Spyrides & Lipmann (1962) have observed the poly U complexes. In reticulocytes this phenomenon has been observed by three groups using *in vivo* protein labeling, *in vitro* labeling and incorporation assays after gradients of reticulocyte ribosomes. Marks, Burka & Schlessinger (1963) and Gierer (1963) have shown that all the synthetic ability sediments more rapidly than 140 s. Warner, Knopf & Rich (1963) have shown that the *in vivo* label can be found on complexes containing up to five ribosomes.

In conclusion, our evidence that the active complex for polypeptide synthesis is a series of 70 s ribosomes using the same messenger molecule is that:

(1) both the 50 s and the 30 s sub-units of the 70 s are necessary for synthesis;

(2) the ribosomes accept poly U under conditions in which the 70 s ribosomes predominate, although the ribosomes that accept the messenger will dimerize to 100 s more readily than the rest of the 70 s ribosomes;

(3) after the ribosomes accept the poly U they form a rapidly sedimenting complex (140 to 200 s) which contains all of the synthesizing activity; this is also true for the natural messenger;

(4) this complex is destroyed by RNase under conditions that leave the ribosomes undamaged; and

(5) this complex contains an amount of poly U consistent with one molecule for several ribosomes.

I should like to thank Dr. J. D. Watson for having introduced me to molecular biology, Mrs. Anne Nevins for invaluable technical assistance, Robert Thach for the poly U, and the National Institutes of Health and the National Science Foundation for their support of this work. These experiments were performed in Dr. J. D. Watson's laboratories in the Biology Department of Harvard University.

REFERENCES

Barondes, S. H. & Nirenberg, M. W. (1962). *Science*, **138**, 813.

Brenner, S., Jacob, F. & Meselson, M. (1961). *Nature*, **190**, 576.

Cannon, M., Krug, R. & Gilbert, W. (1963). *J. Mol. Biol.* To be published.

Gierer, A. (1963). *J. Mol. Biol.* **6**, 148.

Gilbert, W. (1963). *J. Mol. Biol.* **6**, 389.

Gros, F., Hiatt, H., Gilbert, W., Kurland, C. G., Risebrough, R. W. & Watson, J. D. (1961). *Nature*, **190**, 581.

Hoagland, M. (1960). In *The Nucleic Acids*, ed. by E. Chargaff & J. N. Davidson, vol. 3, p. 349. New York: Academic Press.

Jacob, F. & Monod, J. (1961). *J. Mol. Biol.* **3**, 318.

Levinthal, C., Keynan, A. & Higa, A. (1962). *Proc Nat. Acad. Sci., Wash.* **48**, 1631.

Marks, P. A., Burka, E. K. & Schlessinger, D. (1963). *Proc. Nat. Acad. Sci., Wash.* In the press.
Nirenberg, M. W. & Matthaei, J. H. (1961). *Proc. Nat. Acad. Sci., Wash.* **47**, 1588.
Nomura, M., Hall, B. D. & Spiegelman, S. (1960). *J. Mol. Biol.* **2**, 306.
Palade, G. E. (1955). *J. Biophys. Biochem. Cytol.* **1**, 59.
Risebrough, R. W., Tissières, A. & Watson, J. D. (1962). *Proc. Nat. Acad. Sci., Wash.* **48**, 430.
Spyrides, G. J. & Lipmann, F. (1962). *Proc. Nat. Acad. Sci., Wash.* **48**, 1977.
Tissières, A., Schlessinger, D. & Gros, Fe. (1960). *Proc. Nat. Acad. Sci., Wash.* **46**, 1450.
Tissières, A., Watson, J. D., Schlessinger, D. & Hollingworth, B. R. (1959). *J. Mol. Biol.* **1**, 221.
Warner, J. R., Knopf, P. M. & Rich, A. (1963). *Proc. Nat. Acad. Sci., Wash.* In the press.

J. Mol. Biol. (1963) 6, 389–403

Polypeptide Synthesis in *Escherichia coli*.
II. The Polypeptide Chain and S-RNA

Walter Gilbert

Jefferson Laboratory of Physics, Harvard University, Cambridge, Mass., U.S.A.

(*Received 4 January 1963*)

The polyphenylalanine chain, made by poly U-directed† synthesis in the cell-free system from *Escherichia coli*, is studied. This nascent polypeptide chain is bound to the 50 s subunit of the 70 s ribosome. Furthermore, the growing chain is covalently linked to an S-RNA molecule. The bond between the polypetide chain and the S-RNA is similar to that in amino acyl S-RNA but is an order of magnitude more stable against alkaline hydrolysis or hydroxylamine treatment. When puromycin releases the chain from the ribosome, it breaks the bond to the S-RNA. This terminal S-RNA is used as an end group method to measure the molecular weight of the nascent chain. Chains of the order of 40 amino acids are made, and new chains are initiated throughout the reaction.

The behavior of the nascent proteins found after cell-free synthesis in *E. coli* differs from that of the polyphenylalanine only in that the bound nascent proteins stabilize the 70 s ribosome to a higher degree and bind less tightly to the 50 s subunit at low magnesium ion concentrations.

1. Introduction

The S-RNA of *Escherichia coli* consists of molecules of molecular weights around 25,000 (Tissières, 1959). They sediment at 4 s and constitute 15% of the total RNA of the cell. The role of these molecules in the synthesis of proteins is that of the translator from the nucleotide code to the amino acid, the adaptor suggested by Crick (1958). An amino acid is first activated (Hoagland, Zamecnik & Stephenson, 1957) and then transferred to a specific S-RNA by a specific enzyme (Hoagland, Stephenson, Scott, Hecht & Zamecnik, 1958; Berg & Ofengand, 1958), which leaves the amino acid bound by an ester linkage between its carboxyl group and the 2 or 3 position on the terminal adenosine of the S-RNA (Zachau, Acs & Lipmann, 1958). There are several species of S-RNA for some amino acids (Doctor, Apgar & Holley, 1961; Sueoka & Yamane, 1962), and Weisblum, Benzer & Holley (1962) have shown that the two different species for leucine read different code words. There is probably one species for each word in a highly degenerate triplet code (Crick, Barnett, Brenner & Watts-Tobin, 1961).

There are two lines of argument that lead to the suggestion that the growing polypeptide chain is terminated by an S-RNA molecule. One is based on the

† Abbreviations used: CCA = cytidylic cytidylic adenylic acid; poly U = polyuridylic acid; PEP = phosphoenylpyruvate; PK = pyruvate kinase; SDS = sodium dodecyl sulphate; TCA = trichloroacetic acid.

observation by Bishop, Leahy & Schweet (1960) and Dintzis (1961) that the protein chain grows from the amino end. This means that the formation of each peptide bond involves an attack by an incoming amino group upon the carboxyl end of the chain. The S-RNA attached to the carboxyl group of the incoming amino acid need not enter into the reaction and could be left on the end of the chain. (The S-RNA previously on the chain end would be split off by this attack on the ester linkage (Nathans & Lipmann, 1961).) The other line of argument is based on the observation that puromycin releases nascent peptide chains from the ribosomes (Morris & Schweet, 1961; Allen & Zamecnik, 1962). Puromycin resembles an amino acid bound to adenosine (Yarmolinsky & de la Haba, 1959). Allen & Zamecnik (1962) have shown that a radioactive label originally in the amino acid-like part of the puromycin molecule appears bound to the released peptide chain. This has led them to the suggestion that the nascent chain is bound to the ribosome by a terminal S-RNA and that the mechanism of puromycin action is to mimic the next amino acyl S-RNA and form a peptide bond to the carboxyl end of the chain. The chain would then terminate on a puromycin molecule which could not fit into all of the binding site available to the S-RNA, and hence the chain would be freed from the ribosome. We shall show, explicitly for polyphenylalanine synthesis, that these suggestions are correct.

2. Materials and Methods

Most of the methods have been described previously (Gilbert, 1963). *All buffers* contain 0·005 M-tris, pH 7·3, and Mg^{2+} and KCl as indicated. *SDS gradients* were exponential gradients from 5 to 20% sucrose containing 0·1 M-NaCl, 0·005 M-tris, pH 7·3, and 0·5% SDS (Matheson, Coleman & Bell, Norwood, Ohio, U.S.A.). These gradients were run at 25,000 rev./min at 15°C (37°F setting on the Spinco model L) in order to keep the SDS in solution. Twice the number of drops were collected in each sample (four drops for the optical density measurement and sixteen for plating) because the SDS lowers the surface tension and makes the drops half the usual size. Albumin was used as a carrier and the samples were precipitated with 15% TCA. The samples were heated to 90°C for 15 min in order to destroy any label in amino acyl S-RNA.

^{32}P-labeled S-RNA was made by growing *E. coli* B in a low phosphate medium containing 10^{-4} M-Na_2HPO_4, 10^{-4} M-$CaCl_2$, 10^{-5} M-$FeCl_3$, 10^{-3} M-$MgSO_4$, 0·02 M-tris, pH 7·3, and 4 ml. glycerol, 1 g NH_4Cl, 5 g NaCl, and 1 g Casamino acids (vitamin-free)/l. In this medium the bacteria run out of inorganic phosphate at an optical density (550 mμ) near 0·4 and slow their growth. 100 ml. of medium containing 10 mc ^{32}P were inoculated, and the cells were grown at 37°C with shaking until the growth of a parallel culture indicated that the labeled bacteria had taken up all the phosphate. The cells were harvested and washed once with 0·01 M-Mg^{2+}, 0·005 M-tris, pH 7·3. The cells were suspended in 1 ml. and then lysed by 1% SDS at 30°C for 30 min. The lysate was extracted three times with an equal vol. of phenol, the phenol layers were washed with 0·5 ml. of buffer, and the combined aqueous layers were layered on a sucrose gradient containing 0·1 M-NaCl and 0·005 M-tris, pH 7·3. The gradient was run at 25,000 rev./min for 16 hr at 10°F. Ten-drop samples were collected from the top half of the tube by piercing the side of the tube with a hypodermic needle. The position of the 4 S peak was determined by counting 0·005 ml. samples. The tubes from the peak were pooled, dialysed against 0·086 M-KCl, 0·0175 M-Mg^{2+} and 0·005 M-tris, pH 7·3, and frozen until use.

G-200 Sephadex (140 to 400 mesh) (Pharmacia, Uppsala, Sweden) was made into a column using 0·1 M-NaCl, 0·25% SDS and 0·01 M-tris buffer, pH 7·3. The column used for the final analysis had a volume of 12 ml. and a length of 13 cm. It was loaded with volumes of the order of 0·2 to 0·4 ml. Samples of 0·5 ml. were collected by a drop counter directly into scintillation vials. Then 15 ml. of a dioxane based scintillation fluid were added, and the samples counted.

3. Results

(a) *The polypeptide strand after synthesis*

We attempted to observe the stuck 70 s particles of Tissières, Schlessinger & Gros (1960) (70 s ribosomes stabilized in low magnesium by the presence of a nascent protein chain) with polyphenylalanine. We dialysed an incorporation mixture overnight against 10^{-4} M-Mg^{2+}, 0·005 M-tris buffer, and examined the particles on a gradient in this buffer. As is shown in Fig. 1, the particles appear in the 50 s and 30 s form, and

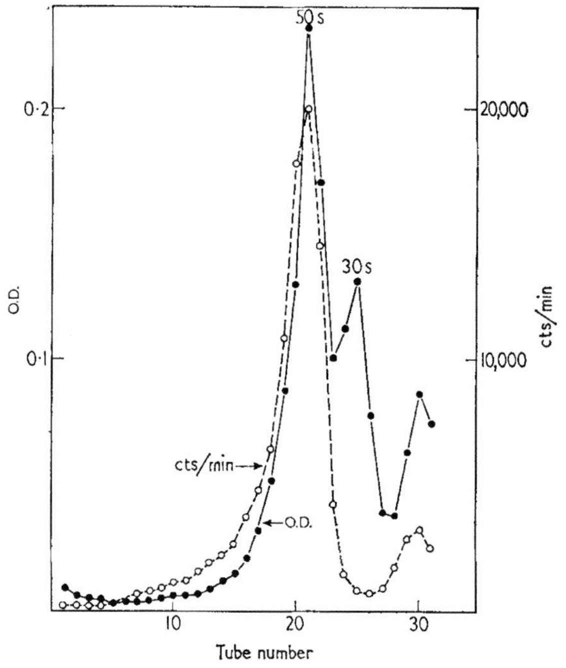

FIG. 1. *Sucrose gradient analysis of a polyphenylalanine incorporation after dialysis against* 10^{-4} *M-Mg^{2+}*. 0·25 ml. of crude extract was pre-incubated for 15 min at 30°C with the ATP generating system in 0·086 M-KCl and 0·0175 M-Mg^{2+}. Then 25 μg of poly U and 1 μc of [^{14}C]phenylalanine (0·78 μg/μc) were added and the incubation continued for 30 min. The reaction mixture was dialysed against 10^{-4} M-Mg^{2+} and 0·005 M-tris, pH 7·3, overnight and then run on a gradient in this buffer for 4 hr at 25,000 rev./min. The samples were precipitated with 5% TCA, heated to 90°C for 15 min and plated.

the polyphenylalanine occurs attached solely to the 50 s ribosomes. We do not observe any stuck 70 s ribosomes. Very little of the label has been released to the supernatant fluid. In 10^{-5} M-Mg^{2+} the only difference is that a greater fraction of the counts are released, about 30% appear at the top of the gradient tube. If the magnesium concentration is lowered still further, the particles begin to break up, but the 50 s ribosomes carrying the polyphenylalanine chain are more stable than the bulk of the 50 s ribosomes and break up last.

In order to see if the factor that binds the chain to the ribosome could be identified, the ribosomes in a similar preparation of polyphenylalanine in low magnesium were broken up with 0·5% SDS and run on a gradient in 0·1 M-NaCl, 0·005 M-tris and

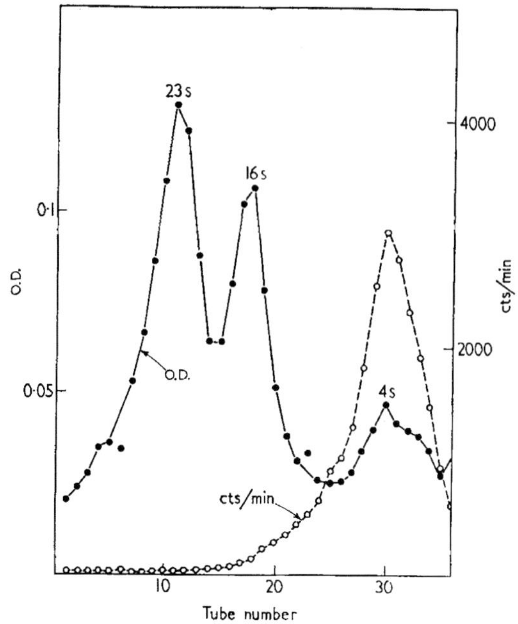

Fig. 2. *Sucrose gradient analysis of the polypeptide chain released with SDS.* This was an incorporation like that described for Fig. 1, except that 1 μc of 3·6 μg/μc [^{14}C]phenylalanine was used, and the incubation was 60 min. After the incorporation the reaction mixture was dialysed against 0·001 M-Mg^{2+} to remove the KCl and Mg^{2+} which could cause the SDS to precipitate. The sample was then brought to 0·5% SDS and run on a gradient containing 0·1 M-NaCl, 0·005 M-tris, pH 7·3, and 0·5% SDS for 11 hr at 25,000 rev./min at 15°C (37°F setting). The tubes were processed as described in Materials and Methods.

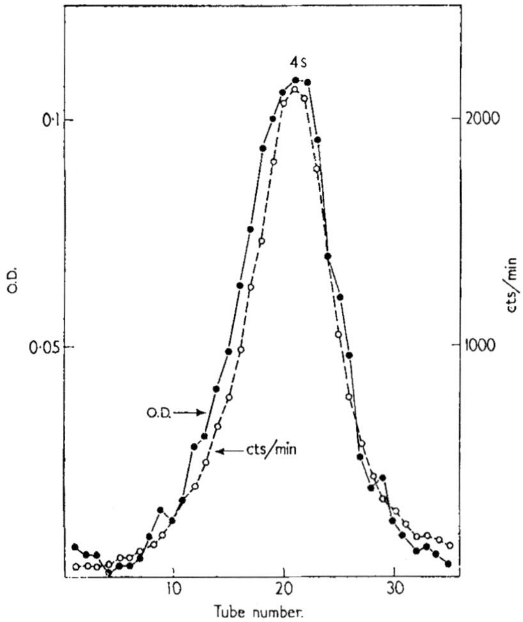

Fig. 3. *Long sucrose gradient analysis of the polypeptide chain released with SDS.* A reaction mixture, like that used for Fig. 2, was dialysed, brought to 0·5% SDS, and mixed with 1·2 mg of S-RNA in order to provide an optical density marker. The material was run on a gradient for 36 hr under the same conditions as were used in the experiment of Fig. 2. The total recovery of radioactive material from the gradient was compared with a sample of the original reaction mixture; the recovery was 87%.

0·5% SDS. The result of this experiment is shown in Fig. 2. The 16 and 23 s peaks of ribosomal RNA are visible, and the polyphenylalanine sediments with the 4 s S-RNA. This phenomenon was examined more closely on the 36 hr gradient shown in Fig. 3. Here we have run the polyphenylalanine in SDS with 1 mg of purified S-RNA as a marker. We recovered 87% of the hot TCA-insoluble counts that were in the original reaction mixture, and the radioactive label coincides with the u.v. absorption of the marker. This coincidence suggests that the polyphenylalanine chain is attached to an S-RNA molecule. This is partially confirmed by the experiments shown in Fig. 4. For the experiment shown in Fig. 4(a), the ribosomes were broken up

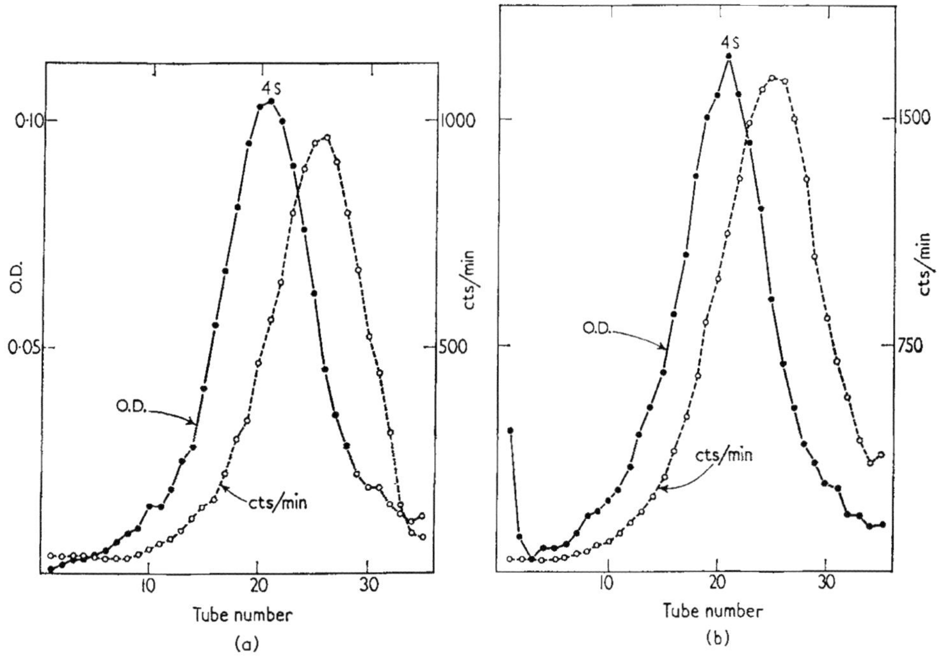

FIG. 4. *The sedimentation of the polypeptide chain after RNase treatment or after puromycin stripping.* Incorporation mixtures like that used for Fig. 3 were made. For Fig. 4(a) the sample was prepared by adding 10 μg/ml. of RNase at the end of the incorporation and dialysing for 12 hr against 10^{-6} M-Mg^{2+} and 0·01 M-NaCl at room temperature. This treatment destroys the ribosomes and leaves a precipitate of ribosomal protein. The material was dissolved in 0·5% SDS (which inhibits the RNase). 1·2 mg of S-RNA was added as a marker, and the sample was run on a gradient for 34 hr. The recovery was 88%. For Fig. 4(b), after a 60 min incorporation, 35 μg/ml. of puromycin was added to a reaction mixture and the incubation was continued for 10 min more at 30°C. The mixture was dialysed against 10^{-4} M-Mg^{2+}, then brought to 0·5% SDS. 1·2 mg of S-RNA were added, and the sample was run on a gradient for 36 hr. 96% of the radioactive material was recovered on the gradient.

in the presence of RNase by a 6 hour dialysis against 10^{-6} M-Mg^{2+} and 0·01 M-NaCl at room temperature. After the digestion, which left a precipitate of ribosomal protein in the dialysis tube, the precipitate and the polyphenylalanine were solubilized with 0·5% SDS, 1 mg of S-RNA was added as a marker, and the material run on a gradient for 36 hours. The sedimentation of the polyphenylalanine has been changed by this treatment, the chain now moving near 2·6 s. This experiment supports the view that the original sedimentation at 4 s is due to an attached piece of RNA. Allen & Zamecnik (1962) have shown that puromycin will remove the nascent chain from the ribosome and that part of the puromycin molecule appears attached to the chain. Figure 4(b) shows the sedimentation of polyphenylalanine chains removed from the ribosomes by treatment with puromycin. Again the label is displaced from the marker; the puromycin stripping and the RNase treatment yield identical pictures. The recovery of the original ^{14}C material was about 90% on these gradients. We do not put any weight on the value for the sedimentation coefficient of the polypeptide chain, since the chain is probably complexed with SDS. If the material released by puromycin is run on a long gradient in the absence of SDS, there is aggregation and most of the polyphenylalanine goes to the bottom of the tube.

(b) *Studies with labeled S-RNA*

If the factor binding the phenylalanine chain to the 50 s ribosome is S-RNA, then in a double label experiment, with [^{32}P]S-RNA and [^{14}C]phenylalanine, we should see both labels associated with the 50 s ribosomes. We made a small amount

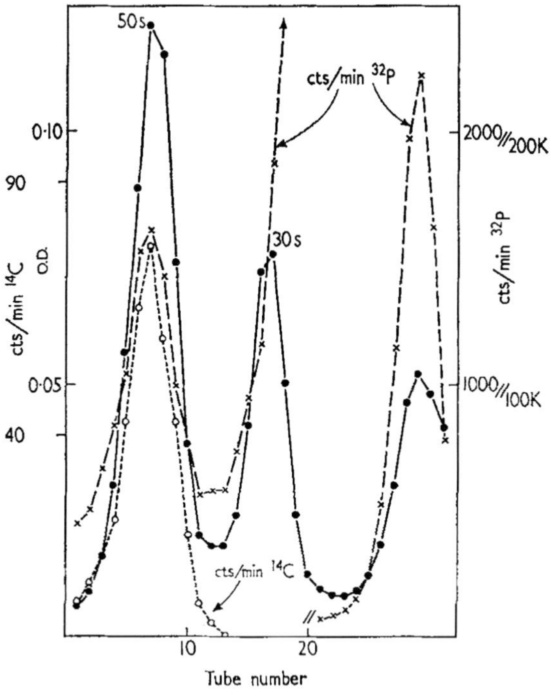

FIG. 5. *A gradient of [^{14}C]polyphenylalanine and [^{32}P]S-RNA in low magnesium.* 0·5 ml. of crude extract was pre-incubated with the ATP generating system for 10 min at 30°C in 0·0175 M-Mg^{2+} and 0·086 M-KCl. Then 70 μg of poly U, 4 μc of [^{14}C]phenylalanine (7·1 μg/μc), and [^{32}P]S-RNA (4×10^7 cts/min) were added. The incubation was continued and samples were stopped by chilling at 5, 10 and 20 min. Each sample was dialysed against 3×10^{-5} M-Mg^{2+} for 12 hr and then run on a gradient in this magnesium for 10·5 hr at 23,000 rev./min at the 10°F setting. 0·050 ml. samples from each tube were put on disks cut from Whatman 3MM paper, the disks were dried, washed twice with 5% TCA, twice with absolute alcohol, and then placed in scintillation vials and counted with a toluene-based scintillator. This method was suggested by M. Nomura. The final specific activity of the S-RNA was measured by taking the tubes from the 4 s peak of these gradients and re-isolating the S-RNA by extracting three times with phenol and then precipitating the RNA 3 times with 0·1 M-NaCl and 75% EtOH. Figure 5 shows the 20 min point of this experiment. The [^{32}P]S-RNA is shown on two scales.

of very highly labeled S-RNA (about 4 μc $^{32}P/\mu g$), incubated it for 20 minutes with a pre-incubated crude extract, poly U and [^{14}C]phenylalanine, dialysed the mixture for 12 hours against 3×10^{-5} M-Mg^{2+}, and ran this material on a gradient in this buffer. The results are shown in Fig. 5. There is a peak of S-RNA on the 50 S ribosomes, coincident with the polyphenylalanine, containing 1% of the S-RNA that was in the extract. This represents rather efficient use of the S RNA. There may be as many as 50 to 60 S-RNA species, and thus the total amount of phenylalanine S-RNA may be only 2%.

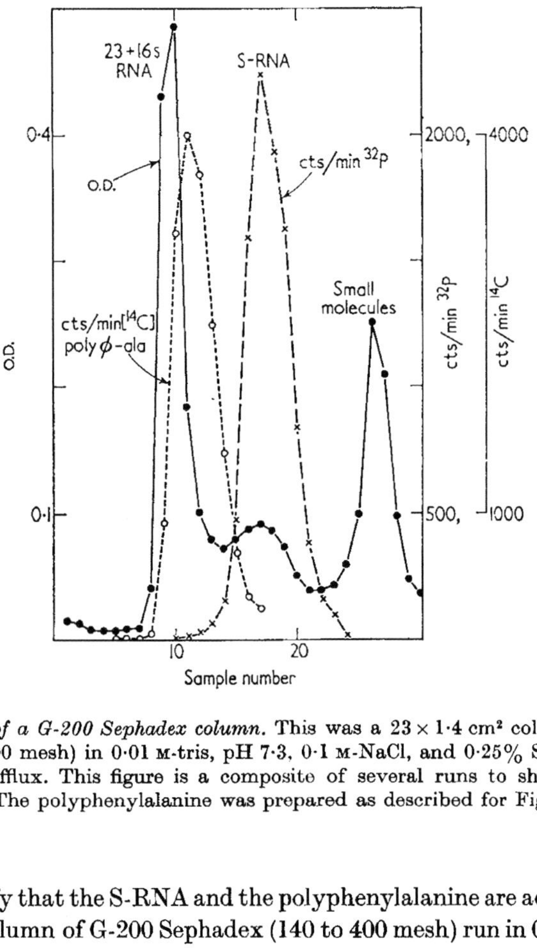

FIG. 6. *Behavior of a G-200 Sephadex column.* This was a 23 × 1·4 cm² column made of G-200 Sephadex (140 to 400 mesh) in 0·01 M-tris, pH 7·3, 0·1 M-NaCl, and 0·25% SDS. 1·3 ml. samples were taken of the efflux. This figure is a composite of several runs to show the behavior of several substances. The polyphenylalanine was prepared as described for Fig. 2 or for Fig. 4(b).

In order to verify that the S-RNA and the polyphenylalanine are actually associated, we have used a column of G-200 Sephadex (140 to 400 mesh) run in 0·1 M-NaCl, 0·25% SDS and 0·01 M-tris buffer, pH 7·3. The pore size of this gel is large enough to retard the S-RNA. The behavior of such a column is shown in Fig. 6. The ribosomal RNA is not retained at all by the gel and leaves the column with the last of the void volume. Polyphenylalanine is partially retained, S-RNA still more, and the small molecules leave the column last. A sample from the peak of the 50 s material on the gradient shown in Fig. 5 was broken up with SDS and run on a Sephadex column. The result is shown in Fig. 7. Only the two radioactive labels were measured and they moved together. The complex of S-RNA and polyphenylalanine (and possibly SDS) is larger than the S-RNA alone.

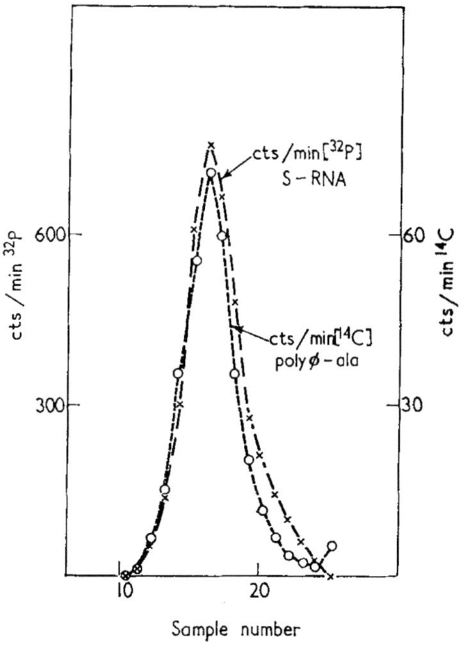

FIG. 7. *Sephadex column analysis of polyphenylalanine S-RNA from the 50 S peak of the gradient shown in Fig. 5.* 0·2 ml. from tube number 7 of the gradient shown in Fig. 5 was brought to 0·5% SDS and then run on a G-200 Sephadex column as described in Materials and Methods. Only the two radioactive labels were followed: [^{14}C]polyphenylalanine and [^{32}P]S-RNA.

(c) *The stability of polyphenylalanine S-RNA*

This column provides a test for the association of S-RNA and polyphenylalanine. We have used this test to study the stability of the S-RNA–polyphenylalanine bond under several conditions. This experiment is illustrated in Fig. 8. Figure 8(a) is a control, a sample from the 50 S region of a gradient similar to that of Fig. 5. Figure 8(b) shows that treating this material at pH 10 (0·5 M-$NaHCO_3$:Na_2CO_3) in the presence of 0·5% SDS for 1 hour at 30°C breaks the bond and permits the polyphenylalanine and the S-RNA to run separately on the column. This also verifies that the association of the counts in a single peak is not some artefact. Figure 8(c) shows that treatment at pH 8·8 (0·25 M-tris) in 0·5% SDS releases only 20% of the S-RNA after 1 hour at 30°C. For comparison we have charged S-RNA with phenylalanine, stopped the reaction with SDS, and followed the hydrolysis at different temperatures and pH values. As is shown in Fig. 9, at 30°C the amino acyl S-RNA hydrolyses with a half-life of 55 minutes at pH 7·3 and 15 minutes at pH 8·8. These results are essentially the same as those of Berg & Ofengand (1958). The 20% hydrolysis of the polypeptide S-RNA at pH 8·8 and 30°C corresponds to a half-life of 100 minutes. Thus the polypeptide–S-RNA bond is 6 or 7 times more stable than the single amino acid–S-RNA bond. Hydroxylamine also breaks the polypeptide–S-RNA bond, 50% of the S-RNA was released after 1 hour at 30°C in 0·5 M-NH^2OH and 0·5% SDS at pH 7·0. This is also an order of magnitude slower than the breakdown of amino acyl S-RNA which has a half-life of the order of 5 minutes under these conditions (Berg & Ofengand, 1958).

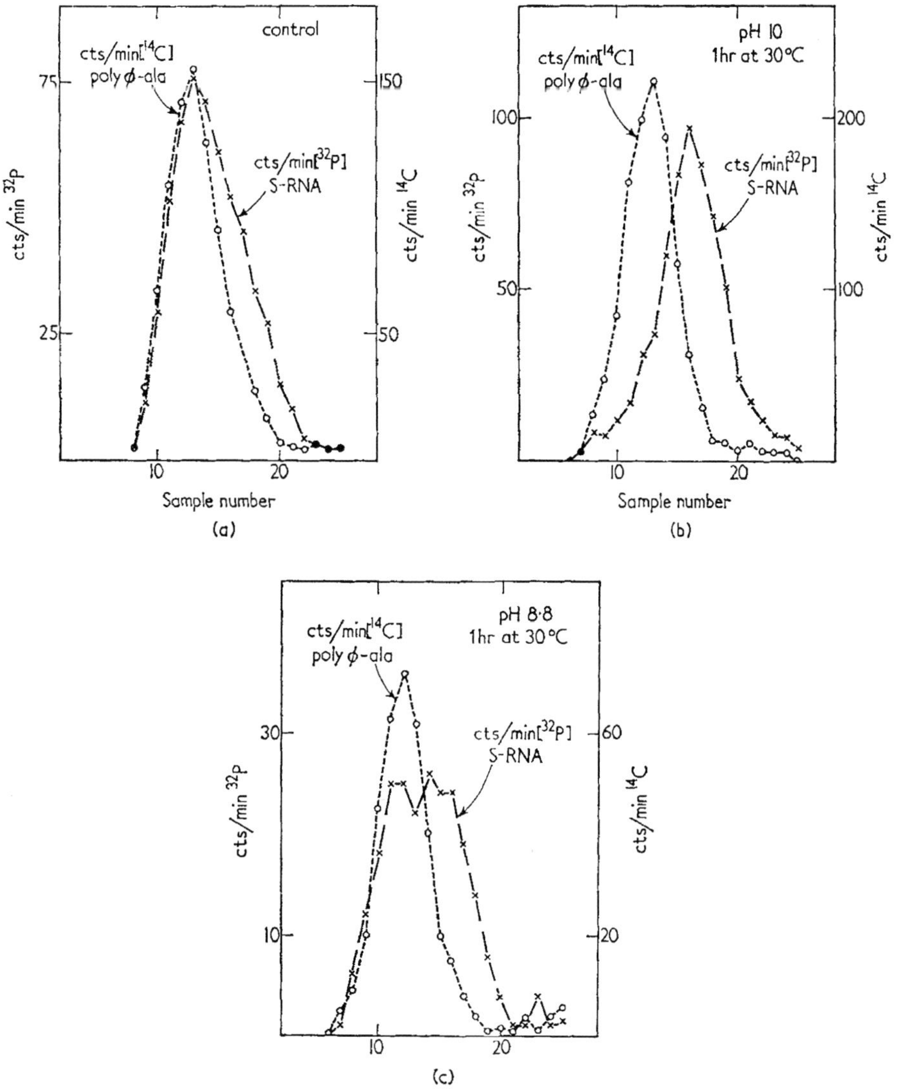

FIG. 8. *Sephadex column analysis of the stability of polyphenylalanine S-RNA.* The samples of [^{14}C]polyphenylalanine [^{32}P]S-RNA were taken from the 50 s peak of a gradient like that shown in Fig. 5. Figure 8(a) is a control, 0·2 ml. from the 50 s peak tube run on a G-200 Sephadex column as described in Materials and Methods. For Fig. 8(b) 0·2 ml. from the gradient was brought to 0·5% SDS and mixed with 0·2 ml. of 1 M-Na_2CO_3—$NaHCO_3$ at pH 10. The mixture was held at 30°C for 1 hr and then run on the Sephadex column. For Fig. 8(c) an 0·1 ml. sample was brought to 0·5% SDS, mixed with 0·1 ml. of 0·5 M-tris, pH 8·8, and held at 30°C for 1 hr before being run on the column.

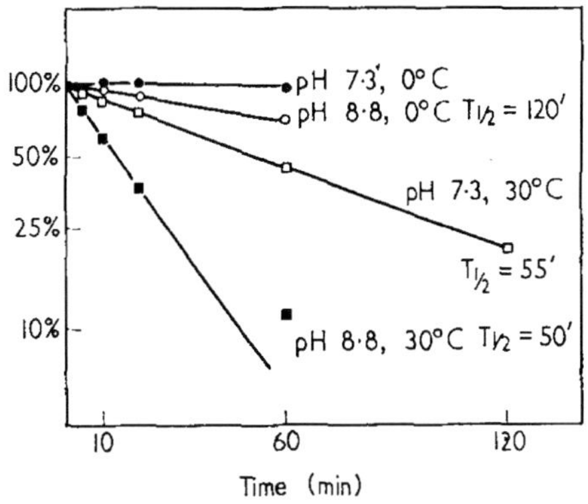

FIG. 9. *The hydrolysis of phenylalanine amino acyl S-RNA*. Several series of tubes, each containing 50 μg S-RNA, 0·01 ml. dialysed supernatant fluid, 0·05 μc [^{14}C]phenylalanine (3·6 μg/μc), 0·06 μM-ATP, 0·5 μM-PEP, 4 μg PK, 1·4 μM-KCl, 1 μM-Mg^{2+}, and 0·25 μM-tris, pH 7·3, in a vol. of 0·1 ml., were incubated for 20 min at 30°C to charge the S-RNA. This reaction was stopped by bringing the tubes to 0·5% SDS. Then one series of tubes was incubated at 30°C, another at 0°C, and two further series were brought to pH 8·8 (0·5 M-tris) and incubated at these two temperatures. The hydrolysis was stopped with 15% TCA, carrier RNA was added, and the precipitates were filtered onto Millipore filters, washed and counted. 2·2% of the S-RNA was charged with phenylalanine.

(d) *Chain length and chain growth*

It is evident that these techniques enable us, by knowing the specific activity of the S-RNA and of the phenylalanine and by assuming that there is one molecule of S-RNA attached to the end of each polypeptide chain, to estimate the length of the polypeptide chain formed. Furthermore, we can estimate the number of chains bound to the 50 S ribosomes. If we assume that there is one binding site for this polypeptide–S-RNA molecule on a 50 S ribosome, this number of chains is directly the number of 50 S ribosomes that have engaged in the synthetic process. By making these measurements at several times during the period of incorporation, we can distinguish between two possible processes: (1) the slow growth of the polypeptide chain: the increase in the total amount synthesized would be due to an increase in chain length associated with a constant number of ribosomes; and (2) the rapid growth of the chain. The increase in incorporation would be the result of more and more ribosomes becoming associated with the poly U and carrying chains. Figure 5 is the 20 minute point of such an experiment: other portions of this reaction mixture were stopped by chilling at 5 and 10 minutes. The amount of [^{14}C]polyphenylalanine and [^{32}P]S-RNA on the 50 S ribosomes was measured by integrating over the 50 S peak. The specific activity of the S-RNA was measured after a phenol purification of the material in 4 S S-RNA peak of the gradients. The results are shown in Fig. 10 where we have plotted the total polypeptide synthesis as moles of phenylalanine per mole of 50 S ribosome, the molar fraction of ribosomes carrying polypeptide chains as moles of bound S-RNA per mole of ribosome assuming a molecular weight of 25,000 for the S-RNA (Tissières, 1959), and $1 \cdot 8 \times 10^6$ for the 50 S particle (Tissières, Watson, Schlessinger & Hollingworth, 1959) and the ratio of these numbers: an estimate of the length of the polypeptide chain. Our greatest source of error is contamination with unbound S-RNA. The column analysis of the 20 minute point, Fig. 7, shows that at this time there is very little contamination. The earlier points are more in doubt, but this error would cause us to underestimate the chain length, especially at early times. Our results are that the chain length saturates very rapidly, and that the increase in the amount of synthesis during this period is due largely to the increase in the number of ribosomes bearing chains. During this period, from 5 to 20 minutes, the total synthesis rose by a factor of 5·5, and the number of ribosomes involved rose by a factor of 3·6, from 3% to 11%. The average chain length, however, rose only 60%, from 24 to 38.

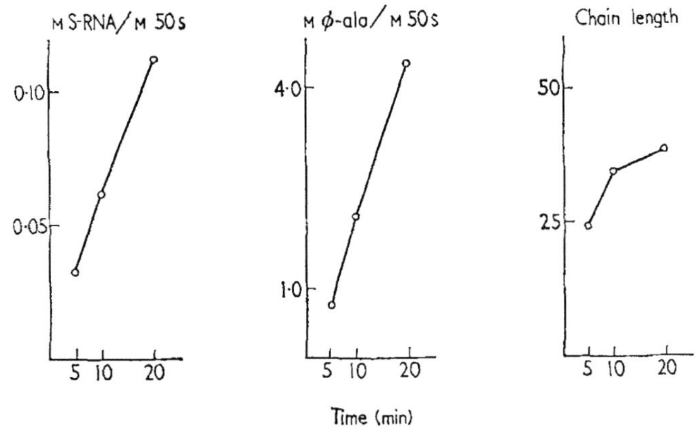

FIG. 10. *The number of active centers and the length of the polypeptide chains as a function of time.* The data were collected in the experiment described in the caption of Fig. 5. Each of the reaction mixtures was run on a gradient in 3 × 10^{-5} M-Mg^{2+} to separate the 50 S ribosomes bearing the polypeptide S-RNA from the bulk of the S-RNA. The optical density and the counts were integrated over the 50 S peak in each case. The specific activity of the phenylalanine was known from the input. The specific activity of the S-RNA was measured by re-isolating the S-RNA from the gradient. The chain length was calculated assuming that there is one S-RNA molecule on the end of each polypeptide chain.

(e) *The situation in normal synthesis*

The polypeptide chain occurring bound to the 50 S ribosome is not unique to polyphenylalanine synthesis. Although during the synthesis of normal proteins by these extracts the 70 S ribosome is stabilized at 10^{-4} M-Mg^{2+} by the presence of a nascent chain, Schlessinger & Gros (manuscript in preparation, 1963) have shown that this "stuck" 70 S is only a relative phenomenon and that, if the magnesium concentration is lowered, these particles break up to their 50 S and 30 S subunits and release some of the nascent protein to the supernatant fluid. We have repeated one of their experiments to see if the dominant behavior was that the nascent material would appear bound to the 50 S ribosome. This experiment is shown in Fig. 11. A crude extract was labeled with [^{14}C]alanine. The particles were washed 3 times in 5 × 10^{-4} M-Mg^{2+} to concentrate the stuck 70 S ribosomes, and then one sample was run directly on a gradient in 5 × 10^{-4} M-Mg^{2+} while another was dialysed against 10^{-5} M-Mg^{2+} and run on a gradient at that magnesium concentration. Figure 11(a) shows the material before dialysis; the counts in the nascent chains are on the 70 S ribosomes, but a good fraction has gone to the bottom of the tube. After dialysis, Fig. 11(b), the optical density of the 70 S material has disappeared, most of the counts now appear in the supernatant solution, but about one-third of the radioactivity remains bound to the 50 S ribosomes, while there is no comparable binding to the 30 S material. Thus normal synthesis differs from polyphenylalanine synthesis only in that the 70 S structure is more stable and that the chains are less firmly bound to the 50 S subunit.

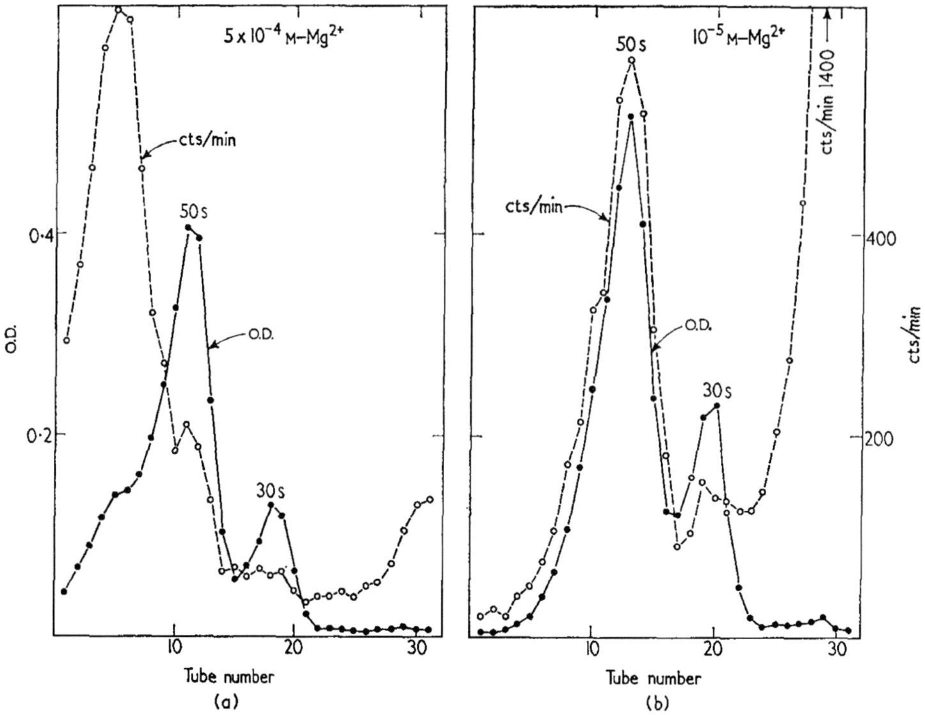

FIG. 11. *Sucrose gradient analysis of the breakdown of stuck 70 S ribosomes in low magnesium.* 5 ml. of a crude extract, made without DNase, was incubated with the ATP generating system and 15 μc of [^{14}C]alanine in 0·086 M-KCl and 0·01 M-Mg^{2+}. The ribosomes were pelleted and then twice resuspended in 5×10^{-4} M-Mg^{2+} and centrifuged for 90 min in the SW39 head. The final pellet was resuspended in 5×10^{-4} M-Mg^{2+}. One half of this material was dialysed for 6 hr against 5×10^{-4} M-Mg^{2+} and then run on a gradient in this buffer. The samples for the radioactivity measurement were treated for 15 min at 90°C in 5% TCA and then plated. The result is shown in Fig. 11(a). The other half of the material was dialysed for 6 hr against 10^{-5} M-Mg^{2+}. The parallel gradient of this material, run in 10^{-5} M-Mg^{2+}, is shown in Fig. 11(b).

4. Discussion

We have shown that the growing polyphenylalanine chain terminates in an S-RNA molecule. The bond between the chain, which contains about forty residues, and the S-RNA has properties similar to those of the amino acyl S-RNA. Thus we believe that this bond is also an ester linkage between the carboxyl end of the chain and the 2′ or 3′ position on the terminal adenosine of the S-RNA. This linkage is an order of magnitude more stable than that between a single amino acid and S-RNA. The experiment shown in Fig. 4(b) demonstrates that the result of treatment with puromycin is to free the polypeptide chain from this terminal S-RNA. Takanami (1962) has shown that the nascent chains are resistant to carboxypeptidase but become sensitive after treatment at pH 10 for 1 hour at 30°C. He has suggested that this sensitization reflects the cleaving off of a terminal S-RNA molecule. This interpretation is in complete agreement with our observations.

The polypeptide S-RNA occurs bound to the 50 S subunit of the 70 S ribosome. Elsewhere it will be shown (Cannon, Krug & Gilbert, 1963) that there is exactly one site on the 70 S ribosome, on the 50 S subunit, which will bind S-RNA in high magnesium concentrations in the absence of protein synthesis. This site behaves as an exchange site:

S-RNA on the site will not wash off, but can be displaced by S-RNA in the medium. S-RNA on this site is freed in low magnesium concentrations. We presume that it is at this same site that the S-RNA of the polypeptide S-RNA is bound. This S-RNA cannot be washed off in low magnesium; the polypeptide stabilizes the S-RNA binding, and the S-RNA helps bind the polypeptide chain. The existence of a unique site for S-RNA binding and the fact that the growing chain ends in S-RNA suggest that there is only one active site for peptide bond synthesis on a 70 s ribosome.

We have followed the number of polypeptide–S-RNA molecules that are bound to the ribosomes during the time course of the reaction. This number increases steadily during the period of synthesis and is a direct measurement of the number of active sites in use. With the additional assumption, justified above, that there is one active site on a 70 s ribosome, this measurement reveals that more and more ribosomes become involved with the messenger as the synthesis continues.

Let us put all these facts together into a hypothetical but consistent picture of the process of polypeptide synthesis. We begin with the picture of peptide bond formation that we described in the introduction, an attack by an amino acyl S-RNA on the ester linkage joining the polypeptide chain to its terminal S-RNA. This attack splits off the S-RNA that was previously on the end of the chain and leaves the chain bound, through the new amino acid, to the new S-RNA. The specificity of this process is dictated by messenger RNA, and the process takes place on the ribosome. Thus the ribosome must have a site capable of holding the S-RNA polypeptide. It must also have a structure that holds the messenger so that a new amino acyl S-RNA is forced to base pair correctly in order to bring its amino acid to the active site. When the peptide bond is formed, the S-RNA that was on the end of the chain is ejected and the chain will end on the new S-RNA. During this ejection, the new S-RNA would move to the site that the old S-RNA occupied. That is, the site is an exchange site; a site that will hold one molecule firmly, but in such a way that it can be displaced by a similar molecule. For want of a better mechanism, we may picture this motion of the S-RNA as it slips into the site and displaces the previous tenant as forcing the messenger to move simultaneously over the surface of the ribosome. At the end of this process we have the S-RNA polypeptide bound to the original site on the ribosome with the S-RNA still hydrogen bonded to the messenger (keeping its finger on the place). The messenger has been moved one reading unit over the surface of the ribosome. The next section along the messenger is now in such a position that any incoming S-RNA must hydrogen bond correctly in order to place its amino acid correctly and the process continues. As this process repeats, the polypeptide chain grows, and the ribosome moves along the messenger. Eventually the leading end of the messenger gets far enough away from the ribosome so that a second ribosome can attach. Then a second polypeptide chain is begun and the second ribosome is moved along the messenger by this reading device. In this way the idea that the ribosome has one active site leads in a natural fashion to the picture of a single messenger servicing many ribosomes.

Two subsidiary points are as follows. (1) What happens if an uncharged S-RNA molecule reads the messenger and approaches the active site? If there is no peptide chain the displacement mechanism works and the S-RNA previously on the site is ejected. But if there is a chain on the site, the structure is stabilized and the terminal S-RNA is held in place—the uncharged S-RNA will diffuse away. (2) We are postulating that the step that requires energy and enzymes is the formation of the peptide bond. If no bonds are being formed, and no polypeptide chain exists, the exchange of S-RNA molecules and the motion of the messenger over the ribosome can proceed by diffusion—but in this case there could not be any specificity in the direction of motion.

We view the binding of the S-RNA to the ribosome as being non-specific. These bonds are probably magnesium bridges to the ribosomal RNA and possibly hydrogen bonds to the common terminal CCA group. The specificity is imposed by the messenger; the hydrogen bonded fit between complementary bases on the messenger and the incoming S-RNA acts as the wards of a lock to permit only the correct S-RNA to slip into place. After the motion accompanying the peptide bond formation, the S-RNA now bound covalently to the chain and by magnesium bridges to the ribosome pins the messenger against the ribosome by these same hydrogen bonds. The messenger is also attached by magnesium bridges to the ribosomal structure—the hydrogen bonding making this over-all fit possible. This fit preserves the register between the polypeptide chain and the messenger.

It is clear that the initiation of chains may be a separate problem since we have not provided any specific mechanism that makes the beginning of the messenger pair with an S-RNA in the correct register. If we read the first triplet one or two bases off from the correct register, then our model will read all of the triplets incorrectly. Our model is adapted to read the code in the fashion suggested by Crick *et al.* (1961), the steric relations at the reading site determining the next sequence to be read.

The ending and releasing of chains is a separate mechanism. The fact that the polyphenylalanine chain is not released indicates that the mechanism for release is not a non-specific one, such as the end of the messenger being the signal to end the chain, but involves some coded information.

The S-RNA molecule has been shown to be a double helix, a twisted hairpin, 100 Å long (Spencer, Fuller, Wilkins & Brown, 1962). The peptide chain is probably attached at the CCA end. The correct shape at the CCA end is necessary for the binding of the S-RNA to the ribosome (Cannon *et al.*, 1963) (specifically to the 50 s subunit). This does not mean that the messenger need be bound anywhere near this point. The sequence that reads the messenger could be as far as 100 Å away from the point of growth of the peptide chain. The 50 s particles are about 140 × 170 Å, the 30 s are 90 × 170 Å, and the 70 s ribosomes are 200 × 170 Å (Hall & Slayter, 1959; Huxley & Zubay, 1960). Thus the S-RNA is long enough to reach from the surface to the center of the ribosome or from the join between the 50 and the 30 to the other side of the 30 s particle. We emphasize the 30 s particle: although the polypeptide chain is bound to the 50 s ribosome, both the 50 s and the 30 s are necessary for synthesis (Gilbert, 1963).

Our hypothesis provides a role for the complicated structure of the ribosome. One way of looking at our suggestion is to think of the ribosome as passing through a series of shape changes and moving like an inchworm over the messenger.

Note added in proof.

Later experiments using a fractionated system have produced polyphenylanine chains of 100 to 120 residues attached to 50 to 60% of the ribosomes.

I should like to express my gratitude to Dr. J. D. Watson for many conversations, to Mrs. Anne Nevins for invaluable technical assistance, and to the National Institutes of Health and the National Science Foundation for their support of this work. These experiments were performed in Dr. J. D. Watson's laboratories in the Biology Department of Harvard University.

REFERENCES

Allen, D. W. & Zamecnik, P. C. (1962). *Biochim. biophys. Acta*, **55**, 865.
Berg, P. & Ofengand, E. J. (1958). *Proc. Nat. Acad. Sci., Wash.* **44**, 78.
Bishop, J., Leahy, J. & Schweet, R. (1960). *Proc. Nat. Acad. Sci., Wash.* **46**, 1030.
Cannon, M., Krug, R. & Gilbert, W. (1963). *J. Mol. Biol.* To be published.
Crick, F. H. C. (1958). *Symp. Soc. Exp. Biol.* **12**, 138.
Crick, F. H. C., Barnett, L., Brenner, S. & Watts-Tobin, R. J. (1961). *Nature*, **192**, 1227.
Dintzis, H. (1961). *Proc. Nat. Acad. Sci., Wash.* **47**, 247.
Doctor, B. P., Apgar, J. & Holley, R. M. (1961). *J. Biol. Chem.* **236**, 1117.
Gilbert, W. (1963). *J. Mol. Biol.* **6**, 374.
Hall, C. E. & Slayter, H. S. (1959). *J. Mol. Biol.* **1**, 329.
Hoagland, M. B., Stephenson, M. L., Scott, J. F., Hecht, L. I. & Zamecnik, P. C. (1958). *J. Biol. Chem.* **231**, 241.
Hoagland, M. B., Zamecnik, P. C. & Stephenson, M. L. (1957). *Biochim. biophys. Acta*, **24**, 215.
Huxley, H. E. & Zubay, G. (1960). *J. Mol. Biol.* **2**, 10.
Morris, A. J. & Schweet, R. S. (1961). *Biochim. biophys. Acta*, **47**, 415.
Nathans, D. & Lipmann, F. (1961). *Proc. Nat. Acad. Sci., Wash.* **47**, 497.
Spencer, M., Fuller, W., Wilkins, M. H. F. & Brown, G. L. (1962). *Nature*, **194**, 1014.
Sueoka, N. & Yamane, I. (1962). *Proc. Nat. Acad. Sci., Wash.* **48**, 1454.
Takanami, M. (1962). *Biochim. biophys. Acta*, **61**, 432.
Tissières, A. (1959). *J. Mol. Biol.* **1**, 365.
Tissières, A., Schlessinger, D. & Gros, Fe. (1960). *Proc. Nat. Acad. Sci., Wash.* **46**, 1450.
Tissières, A., Watson, J. D., Schlessinger, D. & Hollingworth, B. R. (1959). *J. Mol. Biol.* **1**, 221.
Weisblum, B., Benzer, S. & Holley, R. W. (1962). *Proc. Nat. Acad. Sci., Wash.* **48**, 1449.
Yarmolinsky, M. B. & de la Haba, G. L. (1959). *Proc. Nat. Acad. Sci., Wash.* **45**, 1721.
Zachau, H. G., Acs, G. & Lipmann, F. (1958). *Proc. Nat. Acad. Sci., Wash.* **44**, 885.

STREPTOMYCIN, SUPPRESSION, AND THE CODE*

By Julian Davies, Walter Gilbert, and Luigi Gorini

DEPARTMENT OF BACTERIOLOGY AND IMMUNOLOGY, HARVARD MEDICAL SCHOOL, AND JEFFERSON LABORATORY OF PHYSICS, HARVARD UNIVERSITY

Communicated by James D. Watson, March 23, 1964

We have found that an external agent, streptomycin, can upset the genetic code, producing specific misreadings during *in vitro* polypeptide synthesis. This interference is at the level of the ribosome-messenger RNA-sRNA complex, for a modification in the ribosome makes the *in vitro* system insensitive to this effect.

Streptomycin[1] is a bacteriocidal agent. It is a basic molecule that can bind strongly to nucleic acids.[2] It interferes with and finally blocks protein synthesis while permitting continued RNA and DNA synthesis.[3] The mechanism of its killing is not known, but the existence of single mutations to high-level resistance suggests a unitary cause, a single, vital point of attack. Spotts and Stanier[4] hypothesized that the ribosomes were the sensitive elements, and work with the *in vitro* system, in which the poly U-directed incorporation of phenylalanine was shown to be inhibited by streptomycin,[5, 6] further implicated the ribosomes as the site of the shift from sensitivity to resistance. The sensitivity to streptomycin has just been shown by Davies[7] and by Cox, White, and Flaks[8] to reside on the 30s subunit of the 70s ribosome. If these 30s subunits are taken from a sensitive strain, the reconstructed *in vitro* system is sensitive, while if they are derived from a resistant strain, the system is resistant.

The Effects of Streptomycin on the Specificity of Amino Acid Incorporation.—Although the incorporation of phenylalanine into hot TCA-insoluble material is blocked (often by 50–75%) by streptomycin, the incorporation of other amino acids, not normally coded for by poly U, is stimulated, Table 1 shows that, using a purified system, the incorporation of isoleucine (UUA, UAA, CAU),[9] and to a much lesser extent serine (UUC, UCC, AGC) and leucine (UUA, UUC, UUG, UCC), is stimulated in the presence of streptomycin. The same is true for crude extracts (Table 2). Tyrosine (UUA) is not noticeably stimulated, nor are the other amino acids. Asparagine (CAA, CUA, UAA) was examined by damping the incorpo-

TABLE 1

C^{14} amino acid	− Poly U	+ Poly U	+ Sm, poly U
Ala	6	8	8
Arg	8	5	7
Asp	9	8	5
Cys	693	380	396
Glu	24	11	19
Gly	9	8	6
His	30	17	13
Ileu	6	17	340
Leu	17	96	130
Lys	28	21	33
Met	228	102	185
Phe	47	949	452
Pro	10	16	21
Ser	18	11	41
Thr	8	14	7
Try	480	418	482
Tyr	511	396	438
Val	9	13	17

Each amino acid was tested in a 0.050-ml reaction mixture like that for Fig. 2*a*, containing 0.4 μg of each of the 20 amino acids and 0.1 μC of the labeled amino acid, at 0.19 *M* NH_4^+ and 0.0175 *M* Mg^{++}. 0.3 μg poly U and 1 μg Sm were added as indicated. Results are expressed as cpm/reaction mixture. Significant changes produced by the antibiotic are underlined.

TABLE 2

C^{14} amino acid	− Poly U	− Poly U, Sm	+ Poly U	+ Poly U, Sm	+ Poly U, Km	+ Poly U, NeoB	+ Poly U, NeoC	+ Poly U, DHSm	+ Poly U, Om
Phe	80	99	1035	1270	1050	735	690	1210	730
Leu	25	24	400	775	763	545	590	740	275
Ileu	13	13	88	925	985	805	895	860	50
Ser	20	30	36	140	545	515	565	104	23
Tyr	255	240	207	198	286	341	335	212	225

The reaction mixtures contained in 0.5 ml, 0.1 *M* Tris, pH 7.8; 0.01 *M* MgAc; 0.075 *M* NH_4Ac; 6×10^{-3} *M* mercaptoethanol; 1×10^{-3} *M* ATP; 5×10^{-3} *M* PEP; 20 μg PK; 3×10^{-5} *M* GTP; 2×10^{-4} *M* of each of 19 L-amino acids minus the C^{14} amino acid; 10^{-5} *M* C^{14} amino acid (approx. 50,000 cpm per reaction mixture); 20 μg of poly U, and 0.025 ml of dialyzed, preincubated *E. coli* crude extract[7] containing *ca.* 150–200 μg ribosomes. Antibiotics (5 μg/0.5 ml) were added before the poly U. Mixtures were incubated at 34°C for 30 min. The cpm per reaction mixture are given.

ration of a C^{14}-chlorella hydrolysate with the cold amino acid, with negative results. The experiments with purified ribosomes were done at the optimum ion concentration for phenylalanine incorporation, 17.5 mM Mg^{++} and 0.19 *M* NH_4^+.

The incorporation of isoleucine and phenylalanine as a function of the magnesium concentration is shown in Figure 1. The ammonium concentration was 0.086 *M* and all twenty amino acids were present. In Figure 1*a* we see that the incorporation of isoleucine is stimulated by streptomycin by more than a factor of 10 at all magnesium levels, and reaches the same level as the phenylalanine incorporation. The optimum magnesium concentration for this incorporation is very high (about 30 mM at this particular ammonium ion concentration, higher than that for phenylalanine), and at the optimum more isoleucine than phenylalanine is inserted. There is a significant background incorporation of isoleucine in the absence of streptomycin at very high magnesium levels. Figure 1*b* compares the streptomycin-stimulated incorporation of sensitive and resistant ribosomes, using a supernatant from a sensitive strain. The streptomycin-resistant ribosomes are essentially resistant to this miscoding.

The stimulation of isoleucine incorporation by magnesium ions alone is akin to the effects observed by Szer and Ochoa,[10] who have shown that the magnesium optimum for the poly U-directed incorporation of leucine was higher than that for

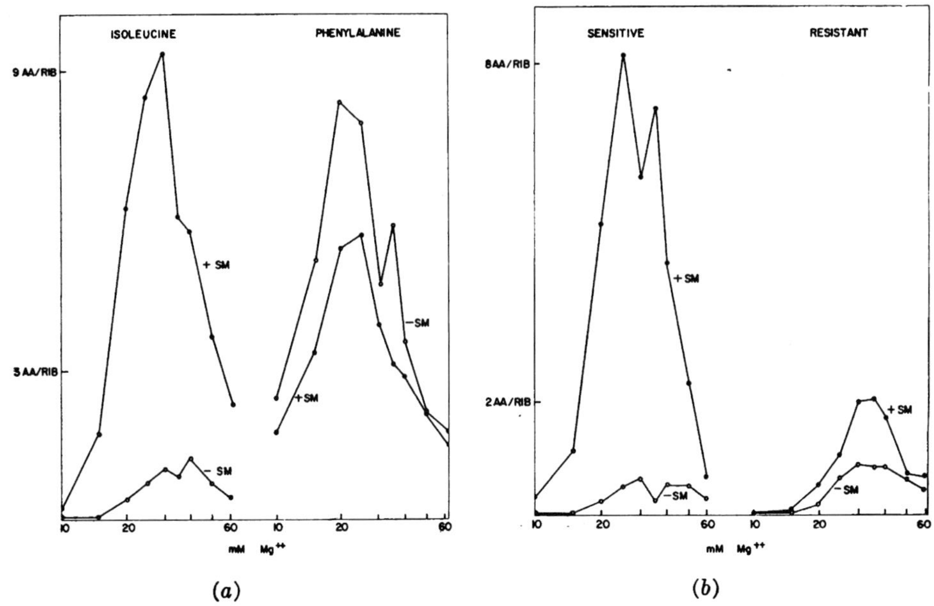

FIG. 1.—Streptomycin effect on the incorporation of isoleucine and phenylalanine at 0.086 M NH_4^+. 0.050-ml reaction mixtures contained 0.010 ml supernatant fraction from *E. coli* B (Sm^s) (the upper two thirds of a 5-hr, 100,000 g centrifugation of a crude extract), 10 μg ribosomes from *E. coli* C600 Sm^s or from a resistant mutant of this strain (the ribosomes were 70s ribosomes isolated on a gradient from an extract preincubated with puromycin (50 μg/ml) and the ATP generating system), 0.1 μg of each of the 20 amino acids and 0.25 μC of the labeled amino acid (0.14 μg of ileu or 0.12 μg of phe), 0.3 μg of poly U, 20 μg ATP, 12 μg GTP, 0.25 μM of PEP, 2 μg PK, 0.10 M Tris pH 7.5, 0.086 M NH_4Ac, and MgAc as specified. The mixtures were incubated for 90 min at 30°C. The results are given as moles of amino acid/mole of 70s ribosome; the ribosomes are limiting. Streptomycin (20 μg/ml) was added after messenger. (*a*) The magnesium dependence of the streptomycin stimulation of isoleucine incorporation and inhibition of phenylalanine incorporation on sensitive ribosomes. (*b*) Comparison of the stimulation of isoleucine incorporated on streptomycin-sensitive and -resistant ribosomes.

phenylalanine and that there were slight incorporations of isoleucine, serine, and tyrosine in high magnesium, when the amino acids were supplied singly. Our findings differ in that we are following incorporation in the presence of all the amino acids, and we are exploring a still higher (20–60 mM) range of magnesium concentrations.

There is also a monovalent ion effect: if we raise the ammonium ion concentration from 0.086 M to 0.19 M, much more isoleucine is incorporated in high magnesium. Figure 2*a* shows the streptomycin and magnesium effects in 0.19 M NH_4^+ ion. At 30–40 mM Mg^{++}, the code is perturbed sufficiently to put in equal parts of isoleucine and phenylalanine. The additional effect of streptomycin is now only a 2.5-fold stimulation of the isoleucine incorporation near the optimum, but at lower Mg^{++} levels the stimulation is much greater. Again, in Figure 2*b*, we see that resistant ribosomes do not miscode under the influence of streptomycin.

The Effect of Other Aminoglycoside Antibiotics.—Kanamycin and neomycins B and C are aminoglycoside antibiotics structurally related to streptomycin and appear to have a similar mechanism of action.[1, 11] The streptomycin-resistant strain used in this investigation was highly sensitive, like the streptomycin-sensitive parent, to the effect of these antibiotics *in vivo*. In the poly U system, kanamycin

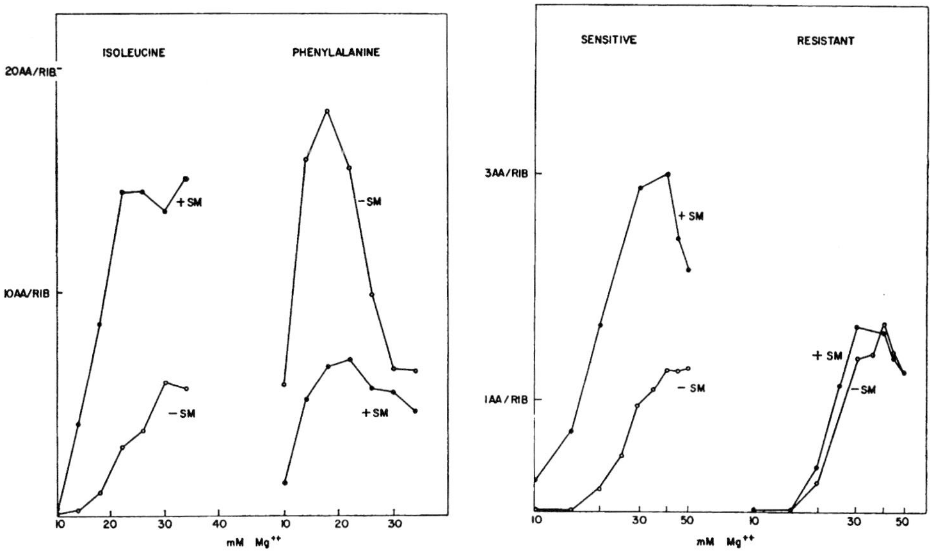

(a) (b)

FIG. 2.—Streptomycin effect on the incorporation of isoleucine and phenylalanine at 0.19 M NH_4^+. The reaction mixtures were as for Fig. 1, but those for (a) contained 0.4 μg of each amino acid plus 0.1 μC of the labeled amino acid, and 10 μg of ribosomes from *E. coli* B. Those for (b) contained 0.1 μg of each amino acid plus 0.1 μC of the labeled amino acid, and 10 μg of ribosomes from *E. coli* C600 Sm^s or Sm^r. Both were at 0.19 M NH_4^+ and the specified Mg^{++} concentration. (a) The stimulation of isoleucine incorporation and the inhibition of phenylalanine incorporation by streptomycin acting on sensitive ribosomes. (b) Comparison of the stimulation of isoleucine incorporation on streptomycin-sensitive and -resistant ribosomes.

and neomycins B and C all stimulate the incorporation of isoleucine. In addition, they stimulate tyrosine, leucine, and serine as is shown in Table 2. They are indifferent to whether or not the ribosomes are streptomycin-sensitive or -resistant. One such experiment is shown in Figure 3, where the stimulation of serine incorporation by either streptomycin or neomycin C is compared on streptomycin-sensitive and -resistant ribosomes. We observe a small stimulation with streptomycin on sensitive ribosomes, no effect on resistant ribosomes, while with neomycin C there is a marked, identical stimulation on both. This experiment underscores the specificity of the change from sensitive to resistant ribosomes. We expect that the ribosomes from high-level kanamycin- and neomycin-resistant strains may be resistant to the code shifts induced by their respective antibiotics.

These coding alterations have also been observed with a polymer that does not contain U. We tested the effect of the antibiotics on incorporation directed by poly CA (2:1). This polymer would be expected to code well for proline (CAC, CCC, CUC), to a lesser extent for threonine (ACA, CCA, UCA) and histidine (ACC, AUC), and for aspartic acid (GCA, GUA) and glutamic acid (AAC, UAC).[12] Table 3 shows that neomycin and kanamycin stimulated the incorporation of all three amino acids tested—threonine, proline, and histidine. Streptomycin produced a smaller stimulation of histidine and threonine, while causing an inhibition of proline incorporation. It is not possible to identify specific code changes in this system, but it is obvious that such changes do occur, particularly in the presence of neomycin and kanamycin.

Discussion.—The fact that an external agent, acting on the ribosomes, can perturb the code leads to a change in our view of the sources of specificity in protein synthesis. The translation mechanism involves not only the specific hydrogen bonds formed between the sRNA adaptor and the messenger, but also the conformation of the site on the ribosome that holds the sRNA to the messenger: this site must be such as to permit or require only the correct pairing to take place. A modification in this site, for example by the binding of streptomycin, permits a "wrong" sRNA to fit so well against the messenger that a "wrong" amino acid is entered into the polypeptide chain. A further modification, by mutation, changes the structure of the ribosomal site so that the correct sRNA is paired, whether or not streptomycin is present. Still a further modification might make the site require the presence of a streptomycin molecule in order to function correctly; this would constitute a mechanism for the classical streptomycin dependence.

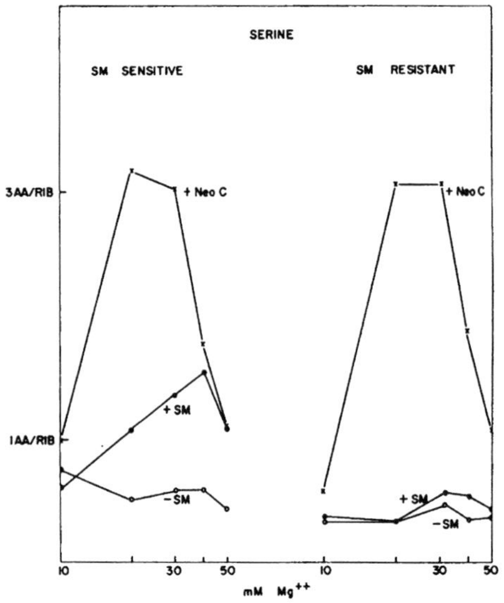

FIG. 3.—Effects of streptomycin and neomycin C on serine incorporation with streptomycin-sensitive or -resistant ribosomes. The reaction mixtures were as for Fig. 1, but contained 0.1 μC of C^{14} serine (0.0876 μg) in addition to 0.1 μg of all 20 cold and either 20 μg/ml streptomycin or neomycin C.

The streptomycin effect gives us confidence that the point of contact between the sRNA and the messenger is on the 30s subunit, as is suggested by messenger binding experiments,[13] although the growing point of the polypeptide chain is probably on the 50s subunit, along with the site that binds the sRNA.[14]

The large perturbations induced by ion shifts, and the results of Szer and Ochoa,[10] raise questions about the validity of the code determined by the *in vitro* system. One would expect that a valid *in vitro* system would display a code insensitive to small changes in the conditions of assay. The ion effects on the code may be due to a direct influence on the secondary structure of the messenger and the sRNA, as is suggested by Szer and Ochoa.[10] Another possibility is that these higher ionic strengths produce a relaxation in the structure of the ribosomes, making them impose less stringent conditions on the pairing between the messenger and sRNA, in analogy to the streptomycin effect.

The phenomenon of intergenic suppression is thought to involve modifications in the translation process. The suppressor genes produce a restoration of enzy-

TABLE 3

C^{14} amino acid	−Poly CA		+Poly CA			
	—	Sm	—	Sm	Km	NeoB
Thr	351	340	427	450	710	822
Pro	14	18	111	91	215	196
Hist	44	—	71	105	201	177

Reaction conditions as in Table 2, except that 0.05 ml of dialyzed crude extract was used in each incubation. 20 μg of poly CA was used.

matic activity through inducing either an amino acid replacement[15] or a transition from "nonsense" to sense or from the command to "end the chain" to an amino acid.[16] The models that have been proposed for such suppressors involve altered sRNA's or activating enzymes.

Recently, however, Gorini and Kataja[17] have demonstrated the existence of a streptomycin-activated suppressor phenotype, which suggests a new mechanism for suppression. This phenotype describes streptomycin-resistant strains that display suppression only when grown in streptomycin. Only in the presence of streptomycin do these cells make a small amount of functional protein. It is evident that the miscoding property of streptomycin offers an explanation for this suppression. We need only imagine that in some cases the mutation, in the structure of the ribosomes, to streptomycin resistance is not complete, that a residual error frequency, a few parts in a thousand in the presence of streptomycin, remains. This error frequency, involving the specific misreading of a subset of code words, would produce the shift of a number of amino acids. The argument is this: we can interpret our finding of a specific shift, phenylalanine to isoleucine, as a forced misreading of U as A in a specific position in a triplet. We might then expect that we could force U to read as A in that position of any triplet (unless there are sequence effects). This converts 16 triplets into other triplets and produces changes in up to 16 amino acids ("up to" because of degeneracy). Since streptomycin upsets the reading of other bases, still a larger number of replacements is possible. Thus, we might expect the streptomycin-activated suppressor to cause a number of different misreadings of the code, resulting in a variety of amino acid replacements or changes in other coded functions.

The involvement of the ribosomes in the accuracy of the reading, and the interpretation we have put upon the streptomycin-activated suppressor suggests that many suppressors may be modifications in the structure of the ribosomes; that is, the product of the suppressor gene is an altered component of the ribosome, whose incorporation in a complete ribosome makes that ribosome alter the code. Such a model would also explain one characteristic of the suppressor genes: since each suppressor would affect a number of different amino acids, different suppressors would have wide and overlapping spectra of repair. If we wished to provide an explanation of very efficient suppression along these lines, we would require the assumption that the triplets that are being misread by this mechanism are uncommon, as has been suggested for other high suppression mechanisms.[16] The assumption that these triplets are associated with rare sRNA's would then permit the suppressed reading to occur faster than the normal reading and would yield large amounts of the suppressed protein.[18]

We conclude by observing that the gross misreading that we find *in vitro* could be the basic mechanism of streptomycin killing: flooding the cell with nonfunctional proteins would be lethal and would perturb, in an unpredictable way, all other cellular functions. To explain the requirement for growth in the killing phenomenon, we assume that the presence of the messenger prevents the attachment of streptomycin to the ribosomes.[5] During the recycling of ribosomes as the messenger moves through the polyribosome, the ribosomes are free to bind streptomycin. If streptomycin binds irreversibly to the ribosomes when they are exposed, then its action will be irreversible. In order to explain the dominance of sensitivity over resistance[19] we need only observe that an equal mixture of good and bad ribosomes would produce mostly bad proteins because of the multimeric form of many proteins. Such bacteria should not grow in streptomycin but might survive a pulsed exposure, a few good ribosomes enabling the cell to throw off viable daughter cells on subsequent incubation in the absence of the drug. This phenomenon has been shown to occur in Pneumococcus.[20, 21]

Summary.—Streptomycin and related antibiotics cause extensive and specific alterations in the coding properties of synthetic polynucleotides *in vitro*. Streptomycin-resistant ribosomes are resistant only to the shift in the code induced by streptomycin. These findings provide evidence that the ribosomes control the accuracy of the reading and may have a role in suppression.

Experiments by W. Gilbert were carried out in the Biological Laboratories, Harvard University. We wish to thank Mrs. Anne Nevins and Mrs. Susanne Armour for their technical assistance. We thank Dr. J. D. Watson and Dr. B. D. Davis for their interest and support. We are grateful to Dr. M. Lubin for a gift of poly CA, and Dr. W. T. Sokolski for neomycins B and C.

The following abbreviations are used: Sm = streptomycin; Km = kanamycin; Neo = neomycin; CM = chloramphenicol; DHSm = dihydrostreptomycin; poly U = polyuridylic acid; poly CA = cytidylic acid-adenylic acid copolymer; PEP = phosphoenolpyruvate; PK = phosphoenolpyruvate kinase.

* This investigation was supported by grants from the American Cancer Society (E-226-B), National Science Foundation (GB-1307), and U.S. Public Health Service (AI-02011-06; GM-09541-03).

[1] For a general review of streptomycin effects see Davis, B. D., and D. S. Feingold, *The Bacteria*, ed. I. C. Gunsalus and R. Stanier (New York: Academic Press, 1962), vol. 4, p. 343.

[2] Cohen, S. S., *J. Biol. Chem.*, **166**, 393 (1946).

[3] Anand, N., and B. D. Davis, *Nature*, **185**, 22 (1960).

[4] Spotts, C. R., and R. Y. Stanier, *Nature*, **192**, 633 (1961).

[5] Flaks, J. G., E. C. Cox, M. L. Witting, and J. R. White, *Biochem. Biophys. Res. Commun.*, **7**, 385 and 390 (1962).

[6] Speyer, J. F., P. Lengyel, and C. Basilio, these Proceedings, **48**, 684 (1962).

[7] Davies, J., these Proceedings, **51**, 659 (1964).

[8] Cox, E. C., J. R. White, and J. G. Flaks, these Proceedings, **51**, 703 (1964).

[9] We give currently suggested codons for reference. See Nirenberg, M. W., O. W. Jones, P. Leder, B. F. C. Clark, W. S. Sly, and S. Pestka, in *Synthesis and Structure of Macromolecules*, Cold Spring Harbor Symposia on Quantitative Biology, vol. 28 (1963), p. 549; and Speyer, J. F., P. Lengyel, C. Basilio, A. J. Wahba, R. S. Gardner, and S. Ochoa, in *Synthesis and Structure of Macromolecules*, Cold Spring Harbor Symposia on Quantitative Biology, vol. 28 (1963), p. 559.

[10] Szer, W., and S. Ochoa, *J. Mol. Biol.*, in press.

[11] Feingold, D. S., and B. D. Davis, *Biochim. Biophys. Acta*, **55**, 787 (1962).

[12] Jones, O. W., and M. W. Nirenberg, these Proceedings, **48**, 2115 (1962).

[13] Okamoto, T., and M. Takanami, *Biochim. Biophys. Acta*, **68**, 325 (1963).

[14] Gilbert, W., *J. Mol. Biol.*, **6**, 389 (1963); Cannon, M., R. Krug, and W. Gilbert, *J. Mol. Biol.*, **7**, 360 (1963).

[15] Brody, S., and C. Yanofsky, these Proceedings, **50**, 9 (1963).

[16] Garen, A., and O. Siddiqi, these Proceedings, **48**, 1121 (1962); Benzer, S., and S. Champe, these Proceedings, **47**, 1025 (1961). These suppressible mutations release fragments in the unsuppressed host (Sarabhai, A. S., A. O. W. Stretton, S. Brenner, and A. Bolle, *Nature*, **201**, 13 (1964)).

[17] Gorini, L., and E. Kataja, these Proceedings, **51**, 487 (1964); see also Lederberg, E. M., L. Cavalli-Sforza, and J. Lederberg, these Proceedings, **51**, 678 (1964).

[18] This is the modulation hypothesis of Ames, B. N., and P. E. Hartman, in *Synthesis and Structure of Macromolecules*, Cold Spring Harbor Symposia on Quantitative Biology, vol. 28 (1963), p. 349.

[19] Lederberg, J., *J. Bact.*, **61**, 549 (1951).
[20] Hotchkiss, R. D., in *Enzymes: Units of Biological Structure and Function* (New York: Academic Press, 1956), p. 119.
[21] Ephrussi-Taylor, H., *Nature*, **196**, 748 (1962).

Section II:
How DNA Is Controlled

ISOLATION OF THE LAC REPRESSOR

By Walter Gilbert and Benno Müller-Hill

DEPARTMENTS OF PHYSICS AND BIOLOGY, HARVARD UNIVERSITY

Communicated by J. D. Watson, October 24, 1966

The realization that the synthesis of proteins is often under the control of repressors[1, 2] has posed a central question in molecular biology: What is the nature of the controlling substances? The scheme of negative control proposed by Jacob and Monod envisages that certain genes, regulatory genes, make products that can act through the cytoplasm to prevent the functioning of other genes. These other genes are organized into operons with cis-dominant operators, such operators behaving as acceptors for the repressor. Appropriate small molecules act either as inducers, by preventing the repression, or as corepressors, leading to the presence of active repressor. The simplest explicit hypothesis for inducible systems is that the direct product of the control gene is itself the repressor and that this repressor binds to the operator site on a DNA molecule to prevent the transcription of the operon. The inducer would combine with the repressor to produce a molecule which can no longer bind to the operator, and the synthesis of the enzymes made by the operon would begin. However, other models will also fit the data. Repressors could have almost any target that would serve as a block to any of the initiation processes required to make a protein. A molecular understanding of the control process has waited on the isolation of one or more repressors.

We have developed an assay for the lactose repressor, the product of the control gene (*i* gene) of the lactose operon. The assay detects and quantitates this repressor by measuring its binding to an inducer, as seen in this case by equilibrium dialysis against radioactive IPTG (isopropyl-thio-galactoside).

In order to seek the lactose repressor, we desired some means that would not depend on the specific models that might be imagined for the actual mechanism of repression. The minimal assumption on which the assay is based is that there should be some interaction between the repressor and the inducer. However, inducing substances added to the cell are often modified before they can trigger induction. Such is the case with lactose which must be split by β-galactosidase in order to induce,[3] but the thiogalactosides appear to be true gratuitous inducers; no chemical modification has yet been detected associated with their action as inducers. The ability of IPTG to stabilize certain temperature-sensitive *i*-gene mutants[4] argues strongly that the *i*-gene product interacts directly with this inducer. Furthermore, the existence of a competitive inhibitor of induction, ONPF (o-nitrophenylfucoside), which can also stabilize certain leaky *i* mutants and which behaves as though it drives the repressor into the form that shuts off the operator, also supports this thesis.[5]

Design of the Experiment.—In order to detect the binding of IPTG to the repressor by equilibrium dialysis, one must achieve concentrations of repressor that are comparable to the dissociation constant of the complex. What affinity does the repressor have for the inducer? Half-maximal induction occurs at $2 \cdot 10^{-4}$ *M* IPTG in a permeaseless strain. This fact alone does not lead directly to an estimate of the equilibrium constant because the enzyme level changes by a factor of 1,000 on induction. Since the rate of enzyme synthesis varies inversely with the first power of the repressor concentration,[4] the concentration of free repressor must have dropped by a factor of 1,000. Because the interaction between the repressor and inducer is quadratic in the inducer concentration, as is shown by the shape of the induction curve,[6] the inducer concentration must be a factor of the square root of 1,000 above the equilibrium constant at half-maximal induction. Thus, the wild-type repressor should have an affinity of the order of $6 \cdot 10^{-6}$ M for IPTG, and if any binding is to be detectable, the repressor concentration must exceed a few times 10^{-6} M. Since the cell pellet is about 10^{-9} M in cells, and since one does not expect many copies of the repressor per operator site, it seems likely that one would have to fractionate and concentrate the repressor by at least a factor of 100, working blindly, before detecting any effect. To improve the chances for success, we decided to isolate a mutant bacterium in which the i gene produces a product that binds more tightly to the inducer. Fortunately, such a mutant, an i^t (tight-binding) mutant, was found.

The i^t Mutant.—We enriched for an i^t mutant by using a technique which will select preinduced cells out of an uninduced population. A challenge with a very low level of inducer will trigger only those cells that are unusually sensitive. Such cells will be selected for along with constitutive mutants, but the constitutives can then be selected against by growth in the presence of TONPG, a compound that inhibits cells with an expressed permease (see *Experimental Details*). A super-inducible mutant was found. The induction curve of a permeaseless derivative of this strain is shown in Figure 1. The basal level is raised, in comparison to the wild type, but the midpoint of the induction curve is pulled lower than would be expected on the basis of the changed basal level alone. Furthermore, in contrast to the wild type, the induction kinetics are *linear* at the lowest levels of inducer (shown in the insert), the basal level being doubled at $7 \cdot 10^{-7}$ M IPTG, and the behavior only becoming quadratic as the level of inducer rises toward that necessary for full induction. The linear behavior suggests that this range shows single-site binding of IPTG to the repressor and that the point at which the basal level is doubled corresponds to the condition that half the repressor is bound to IPTG and half is free to repress. This yields a naïve estimate for the K_m of $7 \cdot 10^{-7}$ M. The rest of the curve can be interpreted as showing that there are two (or more) sites for the binding of IPTG on the molecule and that in order to scavenge the last repressor

TABLE 1

THE REVERSIBLE BINDING OF C^{14} IPTG

Sample	Cpm/0.100 ml	Corrected value	Excess bound as % of outside concentration
Outside concentration of C^{14} IPTG	475	—	—
1—Inside after 30 min dialysis	841	1025	116
2—Inside after 1 hr dialysis	950	1125	137
3—Inside a sac after 1 hr further of dialysis against buffer without IPTG	78	—	—
4—Inside after another 30 min dialysis against C^{14} IPTG	795	1035	118

Four identical 0.1-ml samples of protein were dialyzed against C^{14} IPTG in TMS buffer at 4°. At the end of 30 min, the IPTG concentration inside sac #1 was measured. At 1 hr the concentration inside sac #2 was measured, and the others were transferred to unlabeled buffer. At the end of the next hour, sac #3 was read, and sac #4 was transferred back to the original flask for another 30 min. The raw data, cpm/0.100 ml, are given in the first column. The volume of each sample was measured, to correct for the water uptake during the dialysis and that carried on the walls of the dialysis sac, and the corrected numbers, expressed as cpm/0.100 ml, are given in the second column.

off the operator, both IPTG sites must be fully loaded. (If the two-site nature of the curve is taken into account, the estimate for the binding constant would rise to about $1.2 \cdot 10^{-6}$ *M*.)

Detection of an Effect.—Since, even for the mutant strain, the repressor must be more concentrated than it is in the cell in order to display any binding of IPTG, we made a diploid derivative of the i^t mutant strain and proceeded to fractionate cell extracts with spermine and ammonium sulfate and to examine the concentrated protein fractions. The first detection of any effect was marginal; the spermine precipitate from the mutant strain yielded a 4 per cent binding (1000 cpm/0.1 ml outside a dialysis sac, 1040 cpm within). Further purification, however, immediately produced greater effects. The procedure described in the *Experimental Details* yields material that will draw the inducer into a dialysis sac to a concentration 1.5–2 times that outside the sac (50–100% excess binding) at a protein concentration of 10 mg/ml.

The labeled IPTG is bound reversibly; it can be freely dialyzed into, out of, and even into the sac again. This is illustrated in Table 1. At 4°C, the dialysis goes essentially to completion in about 30 minutes; assays were run for periods ranging up to eight hours. That the observed binding is not due to a contaminant in the radioactive IPTG was confirmed by using two different preparations of labeled IPTG (a C^{14}-labeled commercial preparation and a H^3-labeled Wilzbach preparation) and by examining the competition with unlabeled IPTG.

Negative Controls.—Is the material that binds IPTG really the *i*-gene product? The most critical controls are to examine a diploid amber-suppressor-sensitive i^- strain, which should have only a fragment of the *i*-gene product, and to examine a diploid i^s strain, which should have an *i*-gene product that is unable to recognize the inducer.[7] The i^s strain that was used, like many such strains, is slightly inducible at 0.1 *M* IPTG. This affinity of the i^s repressor for IPTG should be completely undetectable by the assay. The necessary diploid strains, isogenic to the i^t mutant strain, were constructed by F-duction. An identical parallel purification was run on 50-gm lots of control cells and i^t mutant cells. At the end of the purification, samples of each protein solution were dialyzed against several different IPTG concentrations. Figure 2 shows the result for the i^s control; no binding was observed. This finding implies that the substance to which the IPTG binds is either the *i*-gene product or else some other material of the *lac* operon whose synthesis would be blocked in this strain. This second possibility is ruled out by the control shown in Figure 3. No binding was found with the fraction isolated from the i^- strain. Since this strain is wild type with respect to the lactose enzymes, no binding to any of them was being observed.

Other less specific negative controls have been done. No binding was found in an isogenic diploid deletion strain carrying the Beckwith deletion M116, which has cut out the *i* gene as well as the beginning of the β-galactosidase gene. Mixing experiments were done with the extracts from this delection and the *sus* i^- strains mixed with the i^t extract to show that no inhibitors were present in the negative controls. Furthermore, we examined, in haploid strains, a deletion of the entire *lac* region carried in the Lederberg strain W-4032 and the i^- strains ML 308 and 2.340; no binding was found. The haploid amount of wild-type *i*-gene product is detectable. Our yields, however, are not sufficiently consistent to guarantee a factor of two

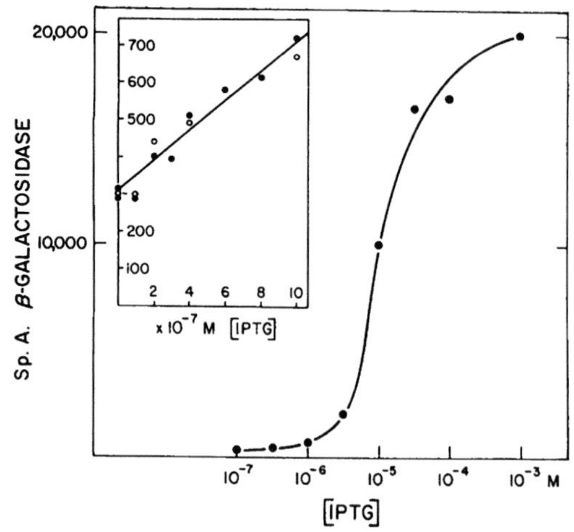

FIG. 1.—The induction of β-galactosidase in the i^t mutant. A permease-negative derivative of the i^t mutant, carrying an F' gal^+ factor, was grown at 35°C in M56-glycerol medium for ten generations in the presence of various concentrations of IPTG. The β-galactosidase activity was measured on toluenized samples. The insert shows that the induction kinetics are linear at the lowest concentrations of inducer. The open and closed circles represent independent experiments done with different permeaseless derivatives.

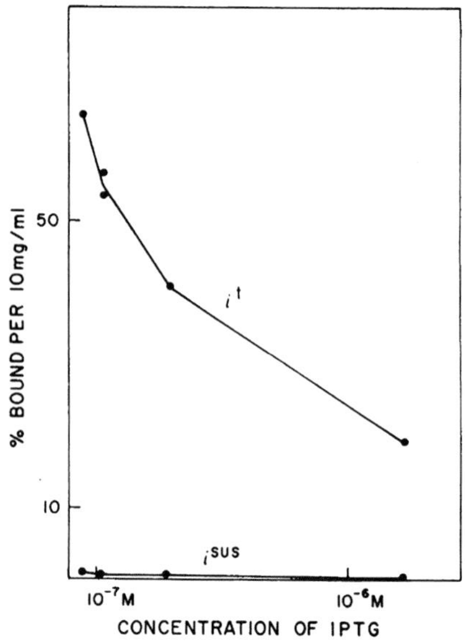

FIG. 3.—The binding ability of a suppressible i^- strain. The repressor fraction was isolated in parallel from isogenic $F'i^{sus}/i^{sus}$ and $F'i^t/i^t$ cells and assayed as described for Fig. 2.

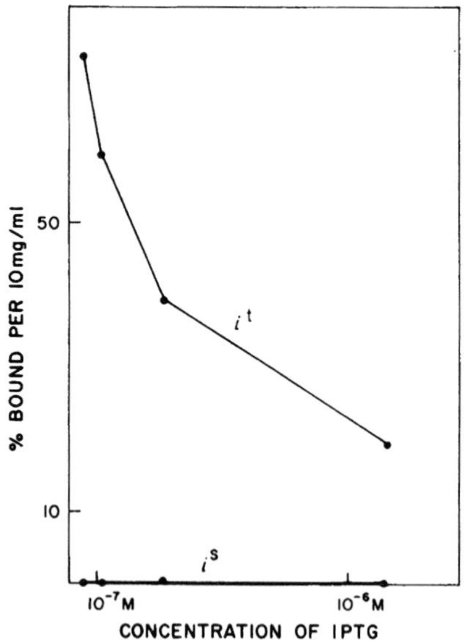

FIG. 2.—The binding ability of an i^s strain. The repressor fraction was isolated in parallel from 50-gm lots of isogenic F' i^s/i^s and F' i^t/i^t cells. Samples of each extract were dialyzed against different concentrations of C^{14} IPTG for 8 hr at 4°C in TMS buffer as described in the *Experimental Details*. The excess label bound inside each dialysis sac, normalized to a protein concentration of 10 mg/ml, is plotted against the IPTG concentration outside the sac. The corrections for protein concentration are of the order of 20%.

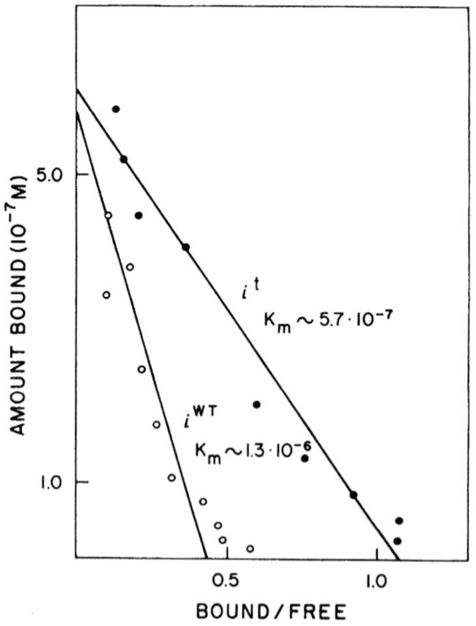

FIG. 4.—The binding constants of the wild-type and i^t mutant strains. Samples isolated from isogenic wild-type and i^t diploid strains were dialyzed against many different concentrations of C^{14} IPTG as described under Fig. 2. The amount of IPTG bound at each external IPTG concentration, normalized to a protein concentration of 10 mg/ml, is plotted against the ratio of bound to free IPTG. Single-side binding should obey the relation: (Amount bound) = (number of sites) − K_m (bound/free).

difference that might be due to gene dosage, so we cannot usefully compare haploid and diploid strains.

The amount of repressor is not increased if the cells are fully induced. Neither is the purification affected if done in the presence of a high concentration of IPTG.

Positive Controls.—Since the binding of IPTG to the wild-type repressor is detectable, the binding constants of the i^t mutant and wild-type strains can be compared *in vitro*. By measuring the amount of IPTG bound at various IPTG concentrations and plotting the amount bound against the ratio of bound to free IPTG, one would get a straight line if there is a single type of noninteracting binding site. The intercept on the y axis will be the molarity of binding sites and the slope will be the negative of the binding constant. Such plots are shown in Figure 4 for the mutant and wild strains. The binding is linear, and the measured binding constants (K_m) *in vitro* at 4°C are $6 \cdot 10^{-7}$ M for the i^t mutant and $1.3 \cdot 10^{-6}$ M for the wild type. The *in vivo* estimates led us to expect that these binding constants would differ by a factor of 4–10. We find only a factor of two difference and can only suggest that temperature effects, buffer effects, or the indirect nature of the *in vivo* estimates might account for this discrepancy. The important finding is that the mutation in the i gene that produces the superinducible phenotype, and presumably a repressor more sensitive to inducer, alters similarly the substance observed *in vitro*.

What affinities does the binding site have for other compounds? One can measure K_i's by dialyzing the repressor against a mixture of radioactive IPTG and varying amounts of unlabeled competitors. Such estimates are compiled in Table 2. Competition with cold IPTG yields essentially the same binding constant as was obtained with the labeled IPTG. This confirms that it is truly the IPTG that is binding. TMG (thio-methyl-galactoside), which is a weaker inducer than IPTG, has a weaker binding constant. ONPF, which is a competitive inhibitor of induction (and behaves as though it drives an allosteric equilibrium toward the form that binds to the operator), shows a 40-fold weaker binding than IPTG, again about the value that would be anticipated from its behavior *in vivo*.[8]

The interaction with galactose is of interest because the i^t represssor *in vivo* shows a sensitivity to galactose. The enhanced basal level of β-galactosidase in the i^t strain is doubled if the i^t gene is in a strain that lacks galactokinase. Such strains have internal galactose pools of the order of $2.5 \cdot 10^{-4}$ M produced from UDPGal.[9] Thus one expects an interaction between the mutant repressor and galactose in this concentration range.

Glucose shows very little affinity for this site. There is a 100-fold discrimination between glucose and galactose. The magnitudes of all these binding constants further support the identification of this material as the repressor.

Properties of the Repressor.—The ability to bind IPTG is not attacked by RNase or DNase. It is destroyed by pronase. The binding site can be inactivated by temperatures above 50°C. These are properties of the part of the molecule that interacts with the inducer. As yet there is no way of seeing the rest of the molecule.

TABLE 2

COMPETITION FOR THE INDUCER BINDING SITE

Competitor	K_i for the i^t gene product
IPTG	$5 \cdot 10^{-7}$ M
TMG	$2 \cdot 10^{-6}$ M
ONPF	$2 \cdot 10^{-5}$ M
Galactose	$5 \cdot 10^{-4}$ M
Glucose	$\geqslant 3 \cdot 10^{-2}$ M

Binding constants inferred by competition of the unlabeled material with labeled IPTG. Protein samples were dialyzed against $1.2 \cdot 10^{-7}$ M C^{14} IPTG and three different concentrations of each of the competitors for 2 hr at 4°C. The amount of IPTG bound was measured, corrected for any changes in the protein concentration, and plotted to permit an estimate of the K..

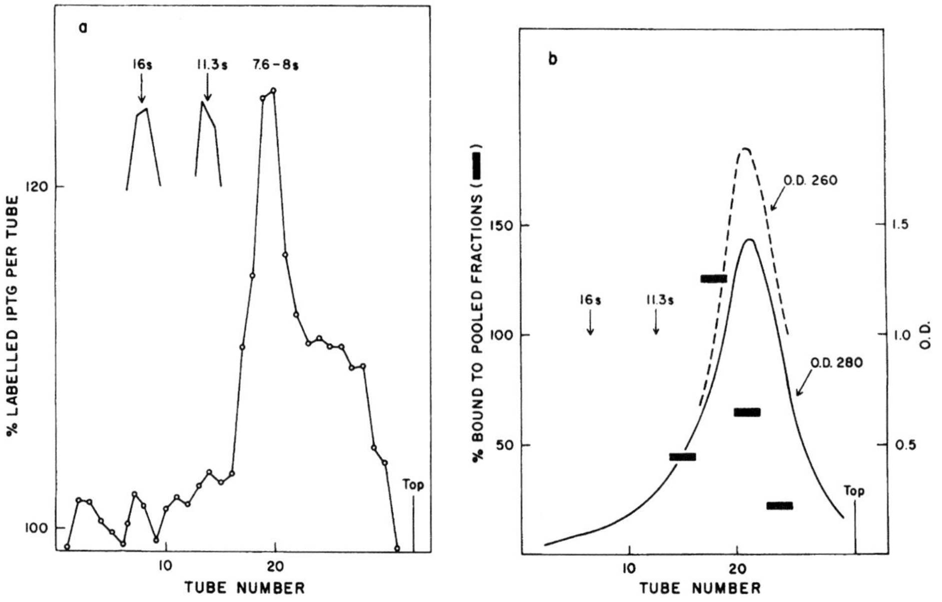

FIG. 5.—Sedimentation of the repressor. Samples of 0.2 ml of a concentrated repressor fraction were layered on 5-ml gradients, from 5 to 30% glycerol, in TMS buffer. The gradients were spun for 8.5 hr at 50,000 rpm at 4°C. In one tube, catalase and β-galactosidase were run as sedimentation markers. For the experiment shown in (*a*) the repressor fraction was brought to $2.5 \cdot 10^{-7}$ *M* C^{14} IPTG and layered on a gradient containing $1.2 \cdot 10^{-7}$ *M* C^{14} IPTG. After the centrifugation, two-drop samples were collected directly into scintillation vials and counted with Bray's solution. The background level in the gradient is about 1000 cpm; the counts are plotted as a per cent of the background level (*open circles*). For the experiment shown in (*b*), a parallel gradient was run without IPTG. The optical density was measured on each two-drop sample at 260 mμ (- - - - -) and 280 mμ (———). Three-tube samples were pooled and placed in dialysis sacs. Their volumes were reduced by burying the sacs under dry G 200 Sephadex for 1 hr. Then the samples were dialyzed against C^{14} IPTG. The excess bound, corrected to 0.100 ml final volume inside the dialysis sac, is plotted as a bar. Superimposed on both experimental curves is the position of the markers taken from a parallel gradient.

The sedimentation of the repressor was followed on glycerol gradients. By sedimenting the material through a gradient in the presence of a uniform level of labeled IPTG, one can observe a peak of label bound in equilibrium. This is shown in Figure 5*a*. In a parallel tube, catalase and β-galactosidase were used as markers. The repressor sediments at 7–8*S*; we estimate its molecular weight to be 150,000–200,000. Figure 5*b* shows the repressor seen by assaying after a gradient without IPTG. The tubes were pooled; the samples were concentrated and dialyzed against radioactive IPTG. The 280- and 260-mμ absorption is indicated, to underline the impurity of our preparation. The material that binds IPTG sediments again at 7–8*S*.

We estimate the amount of repressor in the cell by the yield that is obtained after the first steps of purification. The most recent procedure, breaking the cells with glass beads in a Waring Blendor, and fractionating directly with ammonium sulfate, yields all of the binding ability in the 0–35 per cent cut. The highest recoveries are about 100 sites per cell for the binding of IPTG. These are diploid cells in a tryptone-yeast extract medium harvested late in the log phase. These cells may have four to six gene copies, so we interpret the figure as 20 sites per gene copy and probably ten molecules per gene if there are two sites for the inducer on each molecule. With this last assumption, the repressor corresponds to about one part in 10^4 of the cell's proteins.

Conclusions and Outlook.—Our findings that the *i*-gene product is a protein, that it is uninducible, and that it occurs in a small number of copies serve to confirm many of the expectations that have grown up over the years. The discovery of temperature-sensitive mutants in the *i* gene implied that the *i* gene coded for a protein.[10, 11] The isolation of amber-suppressor-sensitive i^- mutants further proved the point.[12, 13] The estimate of a small number of copies of the repressor has been the traditional explanation of the phenomenon of escape synthesis.[14] The positioning of the *i* gene outside the operon[15] and an *in vivo* experiment on *i*-gene induction[16] both argue that the level of the *i* product would not rise and fall with the state of induction of the lactose enzymes.

An explicit assay, however, unambiguously demonstrates these points and opens the way to a full physical and chemical characterization of the *i*-gene product. Furthermore, experiments designed to ask which steps are blocked by the repressor are now possible *in vitro*.

Summary.—The *lac* repressor binds radioactive IPTG strongly enough to be visible by equilibrium dialysis. This property serves as an assay to detect the repressor, to quantitate it, and to guide a purification. It is a protein molecule, about 150,000–200,000 in molecular weight, occurring in about ten copies per gene. That the assay detects the product of the regulatory *i* gene is confirmed by the unusually high affinity shown for IPTG, by the difference in affinity of the substances isolated from the wild-type and a superinducible *i*-gene mutant, and by the absence of binding in fractions from i^-, *i*-deletion, and i^s strains.

Experimental Details.—Buffers: Two buffers were used. TMEM buffer contained 0.01 *M* tris HCl, pH 7.4, 0.01 *M* Mg^{++}, 0.006 *M* β-mercaptoethanol, and 10^{-4} *M* EDTA; TMS buffer was 0.2 *M* KCl in TMEM buffer. Double-distilled water was used throughout.

The assay: A 0.1-ml protein sample was dialyzed at 4°C with shaking against 10 ml of TMS buffer containing $1.2 \cdot 10^{-7}$ *M* C^{14} IPTG (about 500 cpm/0.1 ml). Number 20 visking tubing was prepared by boiling three times in 10^{-3} *M* EDTA and stored in 10^{-4} *M* EDTA. A sac about 3–5 cm long was used for the dialysis. After the dialysis had gone to completion, the contents of the sac were squeezed out into a tube. A 0.100-ml portion was put directly into Bray's[17] solution and counted in a scintillation counter. A standard was made by taking a 0.100-ml sample of the external fluid. An additional 0.020 ml of the dialyzed protein sample was used for Biuret determination of the protein concentration, to correct for variations in liquid uptake from sample to sample. C^{14} IPTG, 25 mC/mM, was obtained from Calbiochem.

The Purification: Cells were grown in 8 gm Tryptone, 1 gm yeast extract, and 5 gm NaCl per liter of medium. They were harvested in late log phase at $1.5 \cdot 10^9$ cells/ml. The cultures were poured on ice; the cells were washed in TMEM and stored frozen in 50–60-gm lots. For most of the experiments described in this paper, the following purification was used. A 50-gm lot of cells was broken in the Hughes Press, and taken up in 200 ml of TMEM. After a DNase treatment (Worthington, electrophoretically pure) at 2 μg/ml and a low-speed spin, the extract was brought to 2 mg/ml with spermine. The spermine precipitate was dissolved in 50 ml of TMS buffer and the ribosomes were removed by a 90-min, 40,000-rpm spin. The high-speed supernatant was brought to 35% saturation with solid ammonium sulfate, at pH 7.0, and the precipitate collected, dissolved in, and dialyzed against TMS buffer. All these operations were carried out at 4°C.

Our latest procedure consists in breaking the cells by blending with glass beads in a Waring Blendor. Cells (150 gm) are blended with 150 ml of TMS buffer and 450 gm of glass beads for 15 min. The extract is made up to 5 vol with TMS buffer. After a DNase treatment and a ow-speed spin, two ammonium sulfate precipitations are done: the 0–23% cut is discarded; the

23–33% cut is saved. The repressor is then applied to a DEAE Sephadex column in 0.075 *M* KCl in TMEM and eluted during a gradient at 0.15 *M* KCl.

Selection of the i^t mutant: A Sm^R derivative of W3102 (F^- *gal* k^-), obtained from M. Meselson, was used. After mutagenesis in N-methyl-N′-nitro-N-nitrosoguanidine (100 μg/ml in mineral medium M56[18] for 1 hr at 37°) and segregation, the cells were given a maintenance challenge. They were grown in M56 with 10^{-2} *M* glycerol, $8 \cdot 10^{-4}$ *M* leucine, and 10 μg/ml B_1 at 42° for six generations, challenged with 10^{-6} *M* IPTG during the last two generations of growth, then chilled at $5 \cdot 10^8$ cells/ml, washed twice with preconditioned medium (made by growing W3102 to glycerol exhaustion in M56), and inoculated at $5 \cdot 10^6$ cells/ml into 100 ml M56, preconditioned, containing 10^{-3} *M* melibiose and $1.5 \cdot 10^{-3}$ *M* ONPF.[8, 13] After 36 hr the bacteria came up, about 50% constitutives. For the back selection, the bacteria were deadapted in M56-glycerol, then inoculated at 10^6/ml into M56 containing 10^{-2} *M* glycerol and $3\text{–}6 \cdot 10^{-3}$ *M* TONPG which, as we found, inhibits the growth of bacteria whose *lac* permease is expressed. After three cycles of forward and backward selection, the bacterial colonies were screened with ONPG on plates containing 10^{-6} *M* IPTG.

Enzyme assays: The lactose enzymes were assayed as described previously.[8]

We wish to thank Christina Weiss and Susan Michener for their technical assistance; Drs. Jonathan Beckwith, Salvador Luria, and Matthew Meselson for providing bacterial strains; and the National Institutes of Health (GM 09541-05) and the NSF (GB 4369) for their support of this work.

Abbreviations used: IPTG, isopropyl-1-thio-β-D-galactopyranoside; ONPF, o-nitrophenyl-β-D-fucopyranoside; ONPG, o-nitrophenyl-β-D-galactopyranoside; TMG, methyl-1-thio-β-D-galactopyranoside; TONPG, o-nitrophenyl-1-thio-β-D-galactopyranoside.

1 Pardee, A. B., F. Jacob, and J. Monod, *J. Mol. Biol.*, **1,** 165 (1959).

2 Jacob, F., and J. Monod, *J. Mol. Biol.*, **3,** 318 (1961).

3 Burstein, C., M. Cohn, A. Kepes, and J. Monod, *Biochim. Biophys. Acta,* **95,** 634 (1965).

4 Sadler, J. R., and A. Novick, *J. Mol. Biol.*, **12,** 305 (1965).

5 Kunthala Jayaraman, B. Müller-Hill, and H. V. Rickenberg, *J. Mol. Biol.*, **18,** 339 (1966).

6 Boezi, J. A., and D. B. Cowie, *Biophys. J.*, **1,** 639 (1961).

7 Willson, C., D. Perrin, M. Cohn, F. Jacob, and J. Monod, *J. Mol. Biol.*, **8,** 582 (1964).

8 Müller-Hill, B., H. V. Rickenberg, and K. Wallenfels, *J. Mol. Biol.*, **10,** 303 (1964).

9 Wu, H. C. P., and H. M. Kalckar, manuscript in preparation.

10 Horiuchi, T., and A. Novick, in *Cold Spring Harbor Symposia on Quantitative Biology,* vol. 26 (1961), p. 247.

11 Novick, A., E. S. Lennox, and F. Jacob, in *Cold Spring Harbor Symposia on Quantitative Biology,* vol. 28 (1963), p. 397.

12 Bourgeois, S., M. Cohn, and L. E. Orgel, *J. Mol. Biol.*, **14,** 300 (1965).

13 Müller-Hill, B., *J. Mol. Biol.*, **15,** 374 (1966).

14 Revel, H. R., and S. E. Luria, in *Cold Spring Harbor Symposia on Quantitative Biology,* vol. 28 (1963), p. 403.

15 Jacob, F., and J. Monod, *Biochem. Biophys. Res. Commun.*, **18,** 693 (1965).

16 Novick, A., J. M. McCoy, and J. R. Sadler, *J. Mol. Biol.* **12,** 328 (1965).

17 Bray, G. A., *Anal. Biochem.*, **1,** 279 (1960).

18 Monod, J., G. Cohen-Bazire, and M. Cohn, *Biochim. Biophys. Acta,* **7,** 585 (1951).

THE LAC OPERATOR IS DNA

By Walter Gilbert and Benno Müller-Hill

DEPARTMENTS OF PHYSICS AND BIOLOGY, HARVARD UNIVERSITY

Communicated by J. D. Watson, October 25, 1967

How repressors act at the molecular level to turn off genes is only now beginning to be worked out. Most vital to this understanding is whether the operator, defined genetically as the site for the action of a repressor, would turn out to be part of a DNA molecule, a region of a messenger RNA molecule, or even a protein. Now that two specific repressors (lactose and λ) are available,[1, 2] it is possible to attack this problem directly. This was first done by Ptashne,[3] who showed that the λ phage repressor, a 30,000-mol-wt protein, binds specifically only to that region of a λ-DNA molecule where the genetic receptors (operators) lie. Here we report experiments, with the lactose repressor, that further show that the operator is DNA. This repressor binds specifically to DNA molecules that carry the lactose operon, attaching only to that unique region of the DNA molecule where the mutations that characterize the operator lie. Furthermore, this repressor is *released* from the operator by inducers, such as IPTG (isopropyl-1-thio-β-D-galactoside).

The Principle of the Experiment.—The assay for the *lac* repressor used the fact that this repressor could bind radioactive IPTG tightly enough to be detected by equilibrium dialysis. Since the relevant affinity is on the order of 10^{-6} *M*, only repressor concentrations in this range are detectable. This assay cannot be used immediately to study the interaction of the repressor with the operator because attainable gene concentrations are so small. Even if one uses *lac* genes carried on the DNA isolated from a defective phage, one set of genes for each 3×10^7 mol wt, a 3 mg/ml solution of DNA is only 10^{-7} *M*. The binding of repressor to such DNA would only be barely visible by the IPTG binding assay. An alternative approach is to prepare radioactive repressor, to follow the molecule directly. The IPTG binding assay has been used to guide a several thousandfold purification of unlabeled repressor. With this knowledge one could try to mimic this purification on a small scale with very highly labeled proteins—a blind purification, since the specific labeling is so high and the physical scale of the radioactive preparation so small that one cannot follow the purification by the IPTG binding assay. A complete purification is unnecessary; all that is required is a sufficient enrichment of the *lac* repressor so that it represents a reasonable fraction of the labeled material, while other proteins that bind to DNA are removed so that specific effects can be observed. By including a sizing step, isolating only 7–8*S* material that includes the *lac* repressor, one can easily distinguish later a small fraction of the label binding to and sedimenting with 35*S dlac* phage DNA.

The details of the purification are given in the experimental methods. Sulfur-labeled proteins from a triploid strain, carrying three copies of the *lac* genes, are fractionated with ammonium sulfate and then run on a DEAE Sephadex column using a step elution. The material is then concentrated and run upon a glycerol gradient. Since the repressor, as determined by its binding to IPTG, sediments near 7.6*S*, samples are taken from this region of the gradient, determined by an aldolase marker. When this radioactive material is mixed with phage DNA carry-

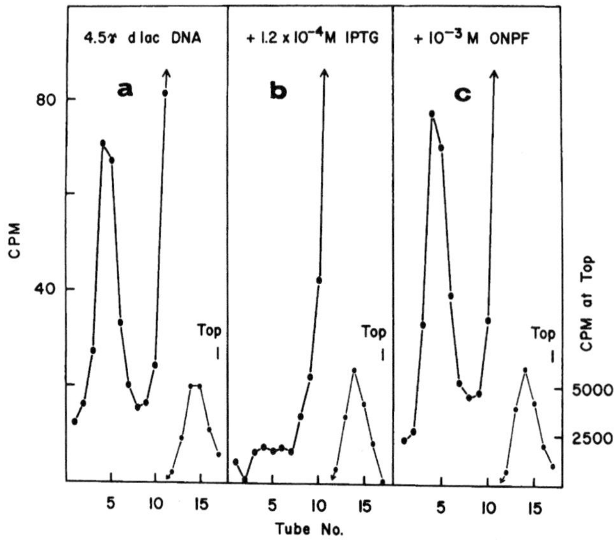

FIG. 1.—The binding of the *lac* repressor to *dlac* phage DNA and its release by inducer. Three identical mixtures of 4.5 γ of *dlac* phage DNA and 8*S* radioactive protein were sedimented on glycerol gradients containing TMEM buffer and run for 2 hr and 20 min at 65,000 rpm as described in the *Experimental Details*. The DNA by itself would form a sharp peak at tubes 4 and 5. The left panel (*a*) shows that a distinct peak of label sticks to the DNA and sediments down the gradient. The DNA is in at least tenfold excess; all the radioactivity that can bind to DNA at this concentration has bound to it. The center panel (*b*) shows that if the gradient solution contains 1.2 × 10^{-4} *M* IPTG, this binding is abolished. The right-hand panel (*c*) shows that if the gradient contains 10^{-3} *M* ONPF, there is no effect on the binding.

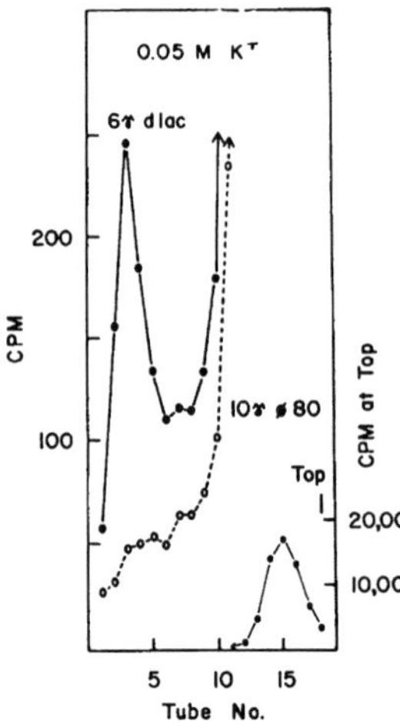

FIG. 2.—The specificity of the binding. In parallel gradients the same repressor preparation was run with two different DNA's. One reaction mixture contained 6 γ of pure *dlac* phage DNA; the other contained 10 γ of the parental φ80–λ hybrid. The gradients contained 0.05 *M* KCl in TMEM.

ing the *lac* region, the mixture incubated and then sedimented on a glycerol gradient, a small peak of radioactivity moves out of the 8*S* region and sediments with the DNA around 35 to 40*S*. Figure 1*a* shows such a gradient pattern in 0.01 *M* Mg^{++}. Only 1 per cent of the radioactivity moves with the DNA, even though the DNA is in excess. That the label binding to DNA represents the *lac* repressor, and not just sticky proteins, is shown by the material being released by IPTG. Figure 1*b* shows that if 10^{-4} *M* IPTG is put throughout the gradient, no binding is observed. This effect of IPTG is specific: ONPF (*o*-nitrophenyl-β-D-fucoside), a substance which binds to the repressor but does not induce, has no effect. Figure 1*c* shows that even 10^{-3} *M* ONPF does not interfere with the binding.

RNase has no effect on this binding. When 75 γ/ml of RNase was added to the binding mixture, for a 20-minute incubation at 30°C, no effect on the binding to DNA was observed. Unlabeled, purified *lac* repressor competes for this binding.

The Specificity of the Binding to DNA.—If the repressor interacts with the operator region, the repressor should bind only to DNA carrying the lactose operon itself and, specifically, only to that region at the beginning of the lactose operon which is characterized through mutations as the genetic operator. In fact, no binding is found with phage DNA not carrying the *lac* genes. Figure 2 shows such an experiment in 0.05 *M* KCl. Furthermore, one can ask the more specific question, Does the repressor bind to the operator region by using operator-constitutive (o^c) mutants carried on the phage DNA? We have examined two such mutants: one has a level of enzyme activity in the absence of inducer 20 per cent of that attainable in the presence of inducer; the other, an extremely low-level constitutive, has an enzyme level in the absence of inducer only 1 per cent of the full level. If the active operator region itself is a region of the DNA molecule, the affinity of the repressor for DNA would be changed in both of these o^c mutants. Since the basal level of enzyme is only 0.1 per cent of the fully induced level, the affinity should be at least a factor of 10 weaker for the 1 per cent o^c and a factor of 200 weaker for the 20 per cent o^c.

When DNA isolated from purified defective phages carrying these o^c mutations is used in the experiment, one observes the patterns shown in Figure 3. Figure 3*a* shows the control binding of the radioactive repressor to wild-type DNA, while Figure 3*b* shows the binding to the 20 per cent o^c. No peak is visible, but some radioactive material has been pulled down from the top of the gradient. The residual affinity of the repressor for the DNA is still detectable. Figure 3*c* shows the affinity of the repressor for the 1 per cent o^c. In this case, still more label moves down the gradient, but the affinity of the repressor for this mutant DNA is less than the affinity for the wild-type DNA.

These experiments demonstrate that the repressor binds to a unique sequence on this DNA molecule, the operator region. Furthermore, all attempts to demonstrate binding to denatured DNA have failed. One infers that the binding is to double-stranded DNA.

The Magnitude of the Binding Constants.—One can obtain rough estimates for the affinity of the repressor for the operator region by observing the shape of the peaks riding on the DNA. These experiments have all been done with an excess of DNA, but the DNA concentration falls as the band moves down the gradient, dropping by a factor of 6 from its initial value in the reaction mixture to its final value when the gradient is collected. If the DNA is run separately in a parallel gradient, the recovery is about 70 per cent and the peak concentration, with 4.5 μg of DNA as an input, is only 2.5 μg/ml (only 8×10^{-11} *M*). From the sharpness of the peaks shown in Figure 1, one would infer that the DNA concentration must be at least a factor of 10 higher than the binding constant. Since a peak of similar sharpness is obtained when only 0.9 μg of DNA is used, one would estimate that the affinity is on the order of 2×10^{-12} *M* in 0.01 *M* Mg^{++}.

The tightness of the binding is influenced by the salt concentration. As Ptashne has observed for the λ repressor, when the salt concentration rises, the affinity for the DNA weakens. Since the DNA concentrations that are used are close to the affinity constants, dissociation plays a role, and a slight change in the salt alters the experimental picture. The profile shown in Figure 2, taken in 0.05 *M* KCl, can be interpreted as showing that 20 per cent of the bound material trails immediately behind the DNA peak due to a weakened affinity. At 0.15 *M* KCl, the binding to our standard amount of DNA has been essentially abolished. Figure 4 shows, however, that the binding can be easily observed again by raising the DNA concentration a factor of 4. Table 1 collects estimates for the binding constants of the wild type and of operator-constitutive mutants.

What affinities does one expect *in vivo*? The affinity of the repressor for the operator can be estimated from the magnitude of the basal level of enzyme synthe-

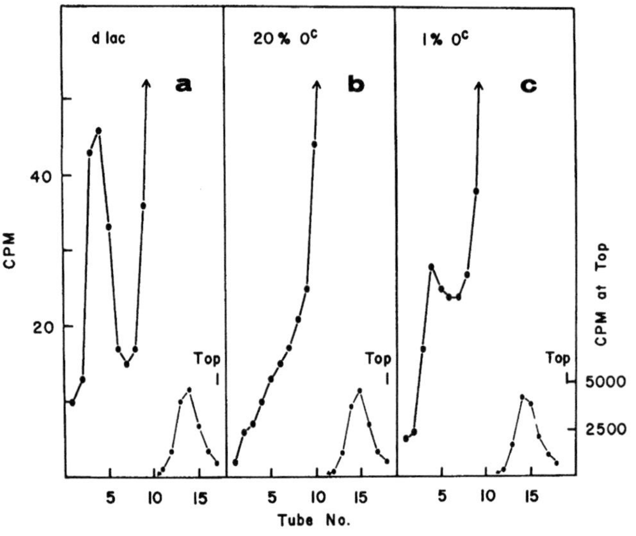

FIG. 3.—The *lac* repressor binds specifically to the *lac* operator. In parallel gradients the same repressor preparation was run with three different purified *dlac* phage DNA's in TMEM. The left-hand panel (*a*) shows the profile obtained with 4.5 γ of wild-type *dlac* DNA. The center panel (*b*) shows that the same amount of DNA carrying an o^c mutation that produces 20% of the full level of enzyme does not bind repressor. The third panel (*c*) shows the binding to an o^c that produces 1% of the full amount (10 times the basal level).

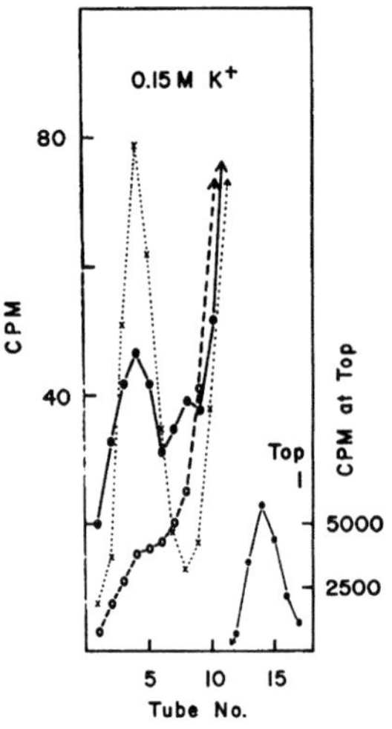

FIG. 4.—The effect of salt. The three superimposed profiles correspond to three parallel gradients: (1) repressor bound to 4.5 γ of *dlac* DNA run in 0.15 *M* KCl and TMEM, ○---○--○; (2) repressor bound to 18 γ of *dlac* DNA run in 0.15 *M* KCl and TMEM, ●—●—●; and (3) a control of 4.5 γ of *dlac* DNA run in TMEM, x···x···x.

sis. To the extent that the rate of enzyme synthesis is simply proportional to the amount of DNA free of repressor and obeys mass action kinetics,[4] the basal rate's being only one thousandth of the full rate means that the repressor concentration is about one thousand times the dissociation constant for the operator. On the basis of the isolation procedures, our current estimates are of the order of 10–20 repressor molecules per haploid cell. Thus the repressor concentration in the cell is of the order of $1\text{–}2 \times 10^{-8}$ *M*, and the affinity of the repressor for the operator should be $1\text{–}2 \times 10^{-11}$ *M*. One does not know how to duplicate the ionic conditions within the cell, nor does one know that the interpretation of the basal level as simply determined by the repressor concentration is true. Furthermore, the *in vitro* estimates are made at high pressure on a gradient. Nonetheless, the estimates *in vitro* for the affinity, ranging from 2×10^{-12} *M* in low salt to several times 10^{-10} *M* in higher salt, are in reasonable agreement with the *in vivo* estimate. The weakening of the binding with the o^c's is in the right direction and consistent, roughly, with the difference of the basal rates of synthesis of these two mutations.

TABLE 1

ESTIMATED REPRESSOR-OPERATOR DISSOCIATION CONSTANTS *in vitro*

DNA	0.01 *M* Mg^{++}	0.05 *M* K^+ + 0.01 *M* Mg^{++}	0.15 *M* K^+ + 0.01 *M* Mg^{++}
dlac o^+	$2\text{–}4 \times 10^{-12}$ *M*	2×10^{-11} *M*	3×10^{-10} *M*
1% o^c	10^{-10} *M*	$>10^{-9}$ *M*	
20% o^c	4×10^{-10} *M*	$>10^{-9}$ *M*	

Is the magnitude of this binding physically reasonable? A binding of the order of 10^{-11} *M* requires some 15 or 16 kcal of binding energy. This energy could arise through the formation of many weak or four to five moderately strong bonds. The repressor must be able to recognize a stretch of at least 11 to 12 bases to select a unique site on the *E. coli* chromosome. (A 12-base sequence selects one out of 1.6×10^7 possible locations, while there are about 3×10^6 base pairs in the chromosome.) To recognize this number of bases individually would require at least 11 or 12 bonds, and thus a free energy change easily in the 15-kcal range. The recognition region will span a considerable distance along the DNA molecule, at least one turn of the helix, a 35-Å stretch; however, the *lac* repressor, 150,000 mol wt, is large enough. If the recognition were not as efficient as possible, the region would be larger.

These dissociation constants imply that the repressor takes hours to fall off the operator. Because the forward rate of formation of the operator-repressor complex will be limited by diffusion and steric factors to be only 10^8 M^{-1} sec^{-1} at the most, a dissociation constant in the 10^{-11} to 10^{-12} *M* range requires that the rate of decay of the complex be 10^{-3} to 10^{-4} sec^{-1}. How then is it possible that enzyme synthesis begins only minutes after the addition of inducer? Clearly the inducer will bind to the repressor-DNA complex in times that are short compared to 10^4 seconds and trigger the release of the repressor from the operator. Since the rate of release will depend on the amount of inducer, one expects a lag in enzyme induction at low levels of inducer. This lag can be estimated on very general principles and shown to depend only on the slow decay rate of the repressor-operator complex (k_s) and on the ratio of the final rate of enzyme synthesis to the basal rate (r_E/r_B). The argument given in the appendix shows that the time delay (T_c) in the induction curve can be written

$$T_c = (1/k_s)(r_B/r_E) + \text{a constant}.$$

Such time delays have been observed for the induction of the lactose enzymes by Boezi and Cowie.[5] Their data fit this formula with a k_s of 2.2×10^{-4} sec^{-1}, an entirely independent estimate for the decay time *in vivo.*

Summary.—The experiments reported here demonstrate that the *lac* repressor binds specifically to the operator region, that its binding to the operator is weakened by mutations in that region which produce o^c's, and that it is released from the operator by the inducer. These experiments completely support the model of repression which proposes that the repressor, on binding to the operator, hinders the transcription of the adjacent genes into RNA and thus prevents their functioning.

Experimental Details.—Buffers and general methods: TMEM buffer is 0.01 *M* tris, pH 7.4, 0.01 *M* magnesium acetate, 10^{-4} *M* ethylenediaminetetraacetate (EDTA), and 0.007 *M* β-mercaptoethanol. TMS buffer is TMEM with 0.2 *M* KCl. All tubes and centrifuge tubes were boiled in EDTA. Other methods were described previously.[1]

Labeled repressor: The bacterial strain carried three sets of *lac* genes, one on φ80 *dlac* carried as a single defective lysogen at the φ80 attachment site, one at the normal *lac* site, and one on an F *lac* episome. For the sulfur labeling the cells were grown in minimal medium: 0.1 *M* potassium phosphate buffer, pH 7.4, 2 gm/liter NH_4Cl, 3 gm/liter NaCl, 2×10^{-4} *M* Mg^{++}, and 10^{-4} *M* sulfate, to glycerol starvation at a few times 10^8/ml. They were diluted back and grown to sulfur starvation in 6×10^{-5} *M* radioactive sulfate. Ten ml of cells were labeled with 20 mc of S^{35}, harvested, and diluted with 1 gm of unlabeled cells. The cells were ground with alumina, the extract suspended in TMS buffer to a final volume of 5 ml, and the debris spun out. The extract was brought to 35% saturation with solid ammonium sulfate, and the precipitate collected and back-extracted with 2 × 1-ml portions of TMS buffer at 28% and 23% of saturation with ammonium sulfate. The two 23% extractions were pooled, and the ammonium sulfate dialyzed out against TMEM containing 0.1 *M* KCl. The sample was applied to a 2-ml DEAE Sephadex column in the same buffer, and a cut was taken between 0.12 *M* and 0.17 *M* KCl in TMEM. Column fractions (0.5 ml) were collected into tubes containing 0.1 mg of aldolase to provide protective protein and a marker during the centrifugation. The material was concentrated by drying down a dialysis sac with G200 Sephadex, layered on a 5–30% glycerol gradient in TMS containing 0.1 mg/ml BSA, and centrifuged for 16 hr at 45,000 rpm, 4°C. The aldolase marker was located by optical density, and samples from the tubes at and immediately following the aldolase peak were used for the binding experiment.

Phage DNA: The *lac* DNA that was used was isolated from a defective *lac* phage made by V. Rybtchine and Ethan Signer. The phage is derived from a ϕ80–λ hybrid,[6] $h_{80}i^{\lambda}$, contains the c_I857 temperature-inducible mutation, and carries the *lac* genes as a replacement of late phage functions. The o^c mutants used were also made in this phage by Signer. They all have a functioning *i* gene. The defective phages are 0.005 gm/cc denser than the parental hybrid. A double lysogen was grown at 34°C in a glucose casein–amino acid medium buffered with 0.1 *M* tris, pH 7.5, and containing more magnesium ion than phosphate. At 5–8 × 10^8 cells/ml the culture was heat-shocked to 42°C for 15 min, chilled to 37°C, and shaken at 37°C until lysis. The pH was maintained at 7.5. Titers were 1–2 × 10^{11}/ml, after chloroforming. The phage were harvested in the Spinco 30 head, purified on a block CsCl gradient, and then banded in an equilibrium CsCl gradient overnight in the 40 head. DNA was prepared from the purified defective phage by rolling the phage stocks with phenol. The DNA was dialyzed overnight against 0.01 *M* tris, pH 7.4, 0.05 *M* KCl, and 10^{-4} *M* EDTA, and stored at about 100 γ/ml at 4°C.

DNA binding assay: The DNA, handled in 0.1-ml pipettes, was heated to 70°C for 5 min and chilled quickly, to break aggregates. The reaction mixture contained, in a final volume of 0.25–0.3 ml, TMEM buffer with 0.05 *M* KCl, 150 γ of BSA, generally about 5 γ of DNA, and radioactive repressor. After a 20-min incubation at 30°C, the mixture was layered on a 5-ml, 5–30% glycerol gradient containing 500 γ/ml BSA and the specified buffer, either TMEM or TMEM with KCl. After a 2-hr 20-min spin at 65,000 rpm, 8°C, 4-drop samples were collected into scintillation vials and counted with Bray's solution.

Mathematical Appendix.—One can calculate how rapidly the repressor will be driven off the operator by the inducer in a variety of specific models for the induction process. The results are identical so we shall give a general, rough argument. The rate of change of the amount of the repressor-operator complex ($[D\text{-}R]$) after the addition of inducer will be given by

$$\frac{d}{dt}[D\text{-}R] = k_s [D\text{-}R] + k_f [D\text{-}R\text{-}I],$$

where $[D\text{-}R\text{-}I]$ is the amount of the repressor-inducer-operator complex and k_s and k_f are the slow and fast decay rates of the two complexes, respectively. One expects k_f to be at least 10^4 times larger than k_s. Furthermore, the equilibration of the inducer, a small molecule, with the various complexes should be very rapid. As we shall see, $[D\text{-}R\text{-}I]$ itself is in general negligible compared to $[D\text{-}R]$. One does not know directly what the affinity of the inducer for the DNA-repressor complex would be, but each of these complexes has a dissociation constant

$$\frac{[D][R]}{[D\text{-}R]} = K_0 \quad \text{and} \quad \frac{[D][R\text{-}I]}{[D\text{-}R\text{-}I]} = \tilde{K}_0.$$

Each of these dissociation constants is the ratio of a decay rate to an association rate. To the extent that the shape of the repressor is not greatly changed by complexing with the inducer, the diffusion constant and steric factors will not be changed and the two association rates, k_1 and k_2, will be equal. Thus

$$\frac{[D\text{-}R\text{-}I]}{[D\text{-}R]} = \frac{K_0[R\text{-}I]}{\tilde{K}_0\,[R]} = \frac{k_s/k_1}{k_f/k_2}\,\frac{[R\text{-}I]}{[R]} \approx \frac{k_s}{k_f}\,\frac{[R\text{-}I]}{[R]}.$$

This last is the crucial statement: the amount of DNA-repressor-inducer complex is small just in proportion to its instability. The decay of the repressor-operator complex can be rewritten as

$$\frac{d}{dt}[D\text{-}R] = [D\text{-}R]\, k_s \left(1 + \frac{[R\text{-}I]}{[R]}\right).$$

Thus the complex decays exponentially with a time constant

$$T = \frac{1}{k_s}\left(\frac{[R]}{[R] + [R\text{-}I]}\right).$$

Since there is a reasonable excess of repressor over operators, $[R] + [R\text{-}I]$ is approximately the total amount of repressor in the cell, and the quantity in parentheses is just $(r_B/r_E) - r_B$ because the rate of enzyme synthesis is given by

$$r_E = 1/(1 + [R]/K_0),$$

if one assumes that this rate is proportional to the amount of DNA free of repressor.

We wish to thank Christine Weiss for her excellent technical assistance, Drs. Ethan Signer and Jonathan Beckwith for providing bacteria and phage strains, and the National Institutes of Health (GM 09541-06) for their support of this work.

[1] Gilbert, W., and B. Müller-Hill, these PROCEEDINGS, **56**, 1891 (1966).

[2] Ptashne, M., these PROCEEDINGS, **57**, 306 (1967).

[3] Ptashne, M., *Nature*, **214**, 232 (1967).

[4] Sadler, J. R., and A. Novick, *J. Mol. Biol.*, **12**, 305 (1965).

[5] Boezi, J. A., and D. B. Cowie, *Biophys. J.*, **1**, 639 (1961).

[6] Signer, E., *Virology*, **22**, 650 (1964).

MUTANTS THAT MAKE MORE LAC REPRESSOR

By Benno Müller-Hill, Lawrence Crapo, and Walter Gilbert

DEPARTMENTS OF PHYSICS AND BIOLOGY, HARVARD UNIVERSITY, CAMBRIDGE, MASSACHUSETTS

Communicated by J. D. Watson, January 24, 1968

The gene that makes the *lac* repressor functions at an extremely low rate. It synthesizes only a few thousandths of a per cent of the total protein of an *E. coli* cell: about 5–10 molecules each generation.[1] In order to study this molecule more easily, we have sought to enhance the amount present in a cell. Two approaches have been successful. The repressor gene, the *i* gene, itself can be mutated to a form that produces more repressor, or the number of copies of the gene can be drastically increased by incorporating the gene into a phage chromosome which will multiply in the cell. The combination of these two approaches yields a cell strain which can make 0.5 per cent of its protein *lac* repressor.

The i^q Mutant.—Several mutants that make about tenfold more repressor than the wild type have been found. These i^q mutants (*q* for quantity) were selected by forcing temperature-sensitive repressor mutants, strains constitutive at 43° and inducible at 30°, to revert to an inducible phenotype at 43°C. At the high temperature, the temperature-sensitive mutants make too little repressor, either because the protein itself is unstable (i^{TL}) or because the final assembly of active repressor is blocked (i^{TSS}).[2] Although the reversion could simply correct the original defect, the temperature effects can also be overcome by producing *more* of the protein (or a more active protein: so that the small amount left at the high temperature will suffice). However, the ability of the *lac* repressor to bind IPTG (isopropyl-1-thio-β-D-galactopyranoside) gives one a direct, quantitative measure of the number of molecules present.[1] By screening extracts from "revertant" strains, we could identify those mutants which, in contrast to the wild type, gave an easily detectable binding in the crude extract.

The revertants were isolated using TONPG (o-nitrophenyl-1-thio-β-D-galactopyranoside), which will inhibit the growth of *lac* constitutive cells that have a functioning *lac* permease. y^- (permease$^-$) mutants will arise and can be counterselected against by growth on lactose in the presence of IPTG. Both i^{TSS} and i^{TL} strains have yielded overproducing derivatives. Table 1 shows the IPTG-binding data for some of these strains; the data suggest that there is about ten times more repressor in the i^q strains. One of these mutants, a derivative of an i^{TSS} strain, has been studied in detail, and was used to produce pure repressor for physical characterization (studies that will be reported elsewhere). The affinity of this i^q repressor for IPTG has been checked and is the same as the wild type; therefore, the increase in the binding truly means an increase in amount of repressor.

Dominance and Complementation.—To study further the properties of this i^q mutant, we constructed heterozygous diploids with a variety of *i* and *o* (operator)

TABLE 1. *Amount of repressor present in various strains.*

Strain	Excess IPTG bound (%)	Mg protein/ml	Specific activity of repressor	Number of repressor molecules/cell
i^+	≦2	46	≦0.05	≦20
$i^+/F'i^+$	≦2	40	≦0.05	≦20
i^{qTSS}	12	49	0.25	100
i^{qTL}	8	40	0.20	80
$i^{qTSS}/F'i^{qTSS}$	22	42	0.50	200

Cells were grown in yeast tryptone medium. Repressor was determined as described in.[1] Specific activity of the *lac* repressor is defined as excess IPTG bound in %/mg protein/ml. The number of repressor molecules per cell is based on the specific activity of pure repressor:[5] 2000.

mutants. Although an i gene with an i^s mutation, producing a repressor that does not easily recognize the inducer, is dominant over the wild-type i^+, and the diploid $F'i^s/i^+$ is *lac*$^-$, an $F'i^s/i^q$ diploid is *lac*$^+$! Table 2 shows this dominance quantitatively. The *lac* genes are inducible, even in the presence of the i^s repressor. The dominance is not total: the induction is only to 25 per cent of the full diploid level.

The simplest explanation is intragenic complementation between subunits of an oligomeric repressor, as suggested by Sadler and Novick.[2] Purified *lac* repressor can indeed be shown to be a tetramer, of molecular weight 150,000 daltons.[5] In an i^q/i^s heteromerozygote most of the i^s subunits would occur in the mixed tetramer $i^q{}_3i^s$. These mixed oligomers must be sensitive to the inducer and driven off the DNA.

One would expect that in a diploid containing both the i^q and an o^c mutation, the o^c would synthesize β-galactosidase at a lower rate than in the presence of a wild-type i^+ gene, being sensitive to the greater amount of repressor. An $F'i^+/i^+o^c$ diploid generally functions at about half the level of the parental i^+o^c. However, when such diploids are constructed with the i^q, the i^q and i^+ genes behave comparably, as is shown in Table 3. The i^q, while making more repressor, must make a repressor with lessened affinity for the operator. This interpretation is supported by the i^q's mutation's not fully depressing the basal level to the normal value, even at low temperatures (also shown in Table 3). We infer that the affinity for the operator has decreased by a factor of 5–10. This can be confirmed *in vitro:* when radioactive repressor is prepared from the i^q, it binds to the operator region on a DNA molecule with an affinity that is about fivefold weaker than that of the wild type.[5] The i^q gene here is derived from an i^{TSS}, and the repressor involved contains two changes with respect to the wild type. Presumably, the defect that results in lessened affinity is in the original TSS repressor.

TABLE 2. *Dominance of i^q over i^s.*

Strain	Specific activity of β-galactosidase
i^sz^+	20
i^qz^+	10,000
$i^qz^+/F'i^sz^+$	5,500

Cells were grown in minimal medium M56[3] in the presence of $10^{-3}M$ IPTG with glycerol as carbon source at 37°. The β-galactosidase was determined as described previously.[4] The i^s strain used is AB 785 derivative Al-42 isolated by us.

TABLE 3. *The i^q repressor does not repress as well as wild-type i^+ repressor.*

Strain	Specific activity of β-galactosidase	Strain	Specific activity of β-galactosidase
$i^+o^+z^+$	15	$i^+o^cz^+/F'i^+o^+z^+$	750
$i^qo^+z^+$	140	$i^+o^cz^+/F'i^qo^+z^+$	1050
$i^+o^cz^+$	1300		

Cells were grown in the absence of inducer and assayed for β-galactosidase as described under Table 2. All strains were grown at 37° with exception of the $i^qo^+z^+$, which was grown at 20°. The o^c strain used is strain 2000 o^c obtained from C. Willson.

There is an anomalous class of *i*-gene mutants called originally "i^-o^c" and thought at one time to be deletions cutting into both the *i* gene and the operator. Davies[6] has mapped these and discovered that some of them are point mutants which map in the middle of the *i*-gene ("i^-o^c" nos. 24, 198, 522). When diploids are constructed with these mutants and the rates of β-galactosidase synthesis compared in these "i^-o^c" strains carrying either an $F'i^+$ or an $F'i^q$ episome, the galactosidase level is lowered from the wild-type level by a factor of nine in the presence of the i^q. These mutants clearly do not behave as operator constitutives. They are sensitive to the amount rather than to the quality of the repressor. The dominance of the constitutive character in diploids is thus a complementation phenomenon. These are trans-dominant i^-'s which we shall call i^{-d} (for dominant). The i^{-d} repressor is not only incapable itself of binding to the operator, but even the presence of one or more i^{-d} subunits in a mixed tetramer renders the whole structure incapable of functioning as a repressor. The i^q making an excess of subunits then dominates over the i^{-d}. That the i^{-d} is in fact trans-dominant has been confirmed by inserting an $i^{-d}z^-$ episome in an i^+z^+ *rec*$^-$ cytoplasm. The chromosomal z^+ (β-galactosidase) gene then functions constitutively, as expected.

The Mechanism of the i^q Mutation.—Although the very low rate of synthesis of *lac* repressor is presumably only one extreme of a spectrum of rates that span four orders of magnitude, how in fact is this low rate achieved in the i^+ gene, and how was it altered in the i^q mutant? Some rate-limiting step in the over-all synthesis has been changed. One possibility is that the *i* gene has an inefficient promoter:[13, 14] an attachment and initiation site for the RNA polymerase with low affinity for the enzyme. The alteration of a base may have increased the affinity for the polymerase by 1 or 2 kc, and thus increased the rate by a factor of 10. Alternatively, some limiting step in the translation process might be altered. The rate at which ribosomes attach and initiate may be influenced by sequences near the chain-initiating codon; the messenger may contain sequences of codons that are inefficiently read, either because the sRNA's are in too low an amount or because they cannot sit down next to each other with any ease; or possibly, the final amino acid sequence itself might be unusually sensitive to proteolytic attack before the protein folds into its final form.

There is no definitive understanding of this problem at present, but we believe that the i^q mutation is most likely an enhanced promoter, a mutation leading to an increased rate of messenger RNA synthesis. The very low rate of normal synthesis would be due to an extremely inefficient promoter: the probability of initiating the messenger RNA for the i gene being so low that only one or two mRNA molecules are synthesized per gene for each generation. Each messenger would then make five or ten molecules of repressor.

This assumption allows a simple explanation for the classical observation that if an i^+z^+ piece of DNA is introduced into an i^-z^- cytoplasm, the β-galactosidase gene functions at a high rate initially and appears to be only slowly turned off.[7] One can interpret this slow turn-off as being an exponential decrease in the number of β-galactosidase-producing cells: the cell population shuts off heterogeneously as each cell succeeds in making a messenger for the repressor and then a burst of repressor molecules. The same explanation applies to the behavior of the i^{TL} mutation.[2] The published curves for the i^{TL} can be plotted as exponential decays, yielding a time constant corresponding to 2.4 i-gene messengers per generation in haploid cells, i.e., about one from each nucleus for each generation.

If this interpretation is correct, a comparison of the i^+ and i^q genes in a mating experiment should show whether or not more i-gene messenger is made by the i^q. Only in that case would the period of constitutive synthesis be less extensive, for if the mutation increased the yield per messenger, i.e., if it had an effect on translation, the cells must still wait the same time before the synthesis of the first messenger. Figure 1 shows such a mating experiment: the i^q shuts off β-galactosidase synthesis about seven times more effectively than does the wild type.

Still More Repressor.—The production of *lac* repressor is subject to gene dosage, the rate increasing appropriately in diploid and triploid cells. Thus one expects the rate of synthesis to increase markedly if several hundred gene copies could be put in the same cell. This can be achieved by incorporating the i^q mutation into the genome of an inducible prophage. We have used a derivative of the ϕ80-λ hybrid, carrying the *lac* genes as a replacement of late functions. The phage used as a single defective lysogen is heat-inducible and carries also a specific defective mutation t_{68}. This mutation prevents lysis (although not the production of phage lysozyme) and furthermore prevents the shutoff of all functions that normally occurs about an hour after induction, whether or not the cell lyses.

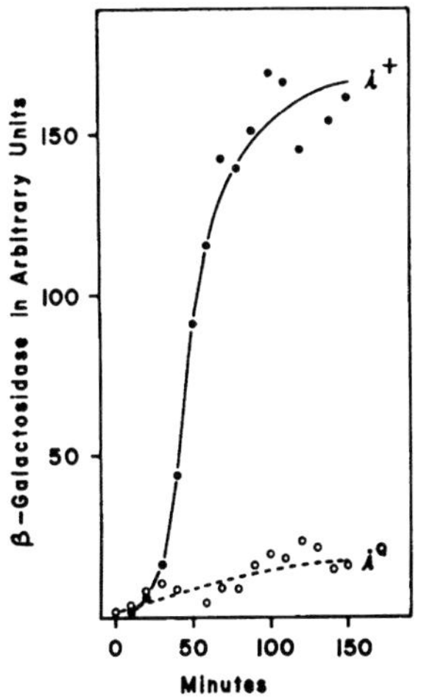

FIG. 1.—Onset of repression in a mating experiment. At time t = 0, Sm^s male cells growing exponentially at 37° and carrying either the $F'i^+z^+$ or the $F'i^qz^+$ episome were mixed with a 2 to 1 excess of Sm^r female cells carrying the *RV lac* deletion. After 10 min of further growth at 37°, in the casamino-acid-M56[3]-glycerol medium, mating was interrupted by vortexing and adding 30γ/ml of sodium dodecyl sulfate. Streptomycin (250γ/ml) was added to stop further growth of the male cells. At 10-min intervals, samples were withdrawn to measure β-galactosidase.[4] The efficiency of transfer of the lac^+ character was determined by plating. The β-galactosidase levels are given in arbitrary units normalized to equal transfer after subtraction of a male cell blank. ● ●● $F'i^+z^+$ males; ○○○ $F'i^qz^+$ males.

When cells carrying this *dlac* prophage are heat-induced, returned to low temperature, and harvested six hours after the induction, there is a 10- to 20-fold increase in the amount of repressor (Table 4). If the i^q mutation has been crossed into the phage genome, this increase, starting from the enhanced level of the i^q, now yields extracts in which 0.5 per cent of the protein is *lac* repressor.

TABLE 4. *Amount of repressor after phage induction.*

Strain	Specific activity of repressor	Strain	Specific activity of repressor
i^+	≦0.05	i^q	0.25
ϕi^+	0.5	ϕi^q	10

Cells were grown in yeast tryptone medium at room temperature. The phage (λ *h80* C_{I857} *lac* t68) was induced by heating for 15 min to 43°. The cells were then cooled to 32° and grown for 6–7 hr. Repressor was determined as described under Table 1.

Summary—An *i*-gene mutant isolated from *E. coli* makes ten times the normal amount of *lac* repressor. After heat induction of a prophage carrying this mutated *i* gene, 0.5 per cent of the soluble protein of the cell is repressor.

Experimental Details.—Isolation of the repressor overproducing mutants: Bacteria, either a strain carrying an i^{TSS} mutation (isolated by us) or a strain carrying an i^{TL} mutation (strain E103 from Sadler and Novick[2]), were treated with N-methyl-N′-nitro-N-nitrosoguanidine as described previously.[8] After segregation in rich medium and adaptation to mineral medium at 44°, 5 × 10^8 bacteria were inoculated at 44° into 100 ml of mineral medium M56[3] containing 2 × 10^{-2} *M* glycerin and 3 × 10^{-3} *M* TONPG (o-nitrophenyl-1-thio-β-D-galactopyranoside). The TONPG inhibits lac^+ constitutive bacteria from growing but allows growth of lac^+ wild type. The selection was repeated with 10^7 bacteria in the same volume of fresh M56, glycerin, and TONPG. The bacteria were then plated on minimal plates containing 10^{-2} *M* lactose and 10^{-4} *M* IPTG at 44°. Large colonies were picked and their β-galactosidase level was determined after growth at 44°. These strains were tested *in vitro* for the presence of excess repressor. Three out of six in the case of the i^{TSS} gave better binding to IPTG than the wild type. One of them was examined in greater detail and was called i^q.

Construction of the strain carrying the phage λ h80 c_{I857} dlac i^q t_{68}: A phage stock of λt_{68}[9] was obtained from E. Signer. This mutation prevents lysis (although not the production of phage lysozyme) and furthermore prevents the shut-off of all functions that normally occurs about an hour after induction, whether or not the cell lyses. A temperature-sensitive allele of this mutation, ts_{96}, leads to phage overproduction: one infers that the totally defective form does so too. We constructed the single lysogen of the defective phage carrying both the *dlac* and the t_{68} character in the following way: the t_{68} mutation was introduced into a *dlac* prophage by the recombination act associated with the formation of a double lysogen upon infection of λt_{68} into the single lysogen, *K* 12 *RV* (λ *h*80 c_{I857} *dlac*). The double lysogen was selected for by growth at 41°C. The single lysogen (λ *h*80 cI_{857} t_{68} *dlac*) was isolated by transduction from an Hft obtained from this double lysogen. The λ *h*80 c_{I857} *dlac* single lysogen was constructed from a double lysogen[10–12] obtained from E. Signer. The i^q marker was introduced by inserting an *F′lac* $i^q z^+ y^-$ into the single lysogen. Lac^- segregants were isolated and scored for the i^q/i^q property.

Growth and heat induction of phage carrying strains: If not indicated otherwise, bacteria were grown in rich medium containing 16 gm bactotryptone, 10 gm yeast extract, and 5 gm NaCl/l to saturation at room temperature. For phage induction, an exponential culture at an OD_{550} of 1 was heated to 43° for 15 min, then cooled to 32°. The bacteria were harvested after 6–7 hr further growth.

Assay of repressor and β-galactosidase: As described previously.[1, 4]

We wish to thank June Andersen, Katherine Crolius Rhee, and Christine Weiss for their technical assistance; Drs. Julian Davies, Aaron Novick, Ethan Signer, and Clyde Willson for providing strains; and the National Institutes of Health (GM 09541-06) and the National Science Foundation (GB-4369) for their support of this work. L. Crapo was supported by a National Science Foundation postdoctoral fellowship.

[1] Gilbert, W., and B. Müller-Hill, these Proceedings, **56**, 1891 (1966).
[2] Sadler, J. R., and A. Novick, *J. Mol. Biol.*, **12**, 305 (1965).
[3] Monod, J., G. Cohen-Bazire, and M. Cohn, *Biochim. Biophys. Acta*, **7**, 585 (1961).
[4] Müller-Hill, B., H. V. Rickenberg, and K. Wallenfels, *J. Mol. Biol.*, **10**, 303 (1964).
[5] Gilbert, W., unpublished.
[6] Davies, J., personal communication.
[7] Pardee, A. B., F. Jacob, and J. Monod, *J. Mol. Biol.*, **1**, 165 (1959).
[8] Müller-Hill, B., *J. Mol. Biol.*, **15**, 374 (1966).
[9] Harris, A. W., D. W. A. Mount, C. R. Fuerst, and L. Siminovitch, *Virology*, **32**, 553 (1967).
[10] Signer, E., *Virology*, **22**, 650 (1964).
[11] Signer, E., and J. Beckwith, *J. Mol. Biol.*, **22**, 33 (1966).
[12] Signer, E., personal communication.
[13] Jacob, F., A. Ullman, and J. Monod, *Compt. Rend.*, **258**, 3125 (1964).
[14] Ippen, K., J. Miller, J. Scaife, and J. Beckwith, *Nature*, **217**, 825 (1968).

Proc. Nat. Acad. Sci. USA
Vol. 70, No. 12, Part I, pp. 3581–3584, December 1973

The Nucleotide Sequence of the *lac* Operator

(regulation/protein–nucleic acid interaction/DNA–RNA sequencing/oligonucleotide priming)

WALTER GILBERT AND ALLAN MAXAM

Department of Biochemistry and Molecular Biology, Harvard University, Cambridge, Massachusetts 02138

Communicated by J. D. Watson, August 9, 1973

ABSTRACT **The *lac* repressor protects the *lac* operator against digestion with deoxyribonuclease. The protected fragment is double-stranded and about 27 base-pairs long. We determined the sequence of RNA transcription copies of this fragment and present a sequence for 24 base pairs. It is:**

5′--T G G A A T T G T G A G C G G A T A A C A A T T 3′
3′--A C C T T A A C A C T C G C C T A T T G T T A A 5′

The sequence has 2-fold symmetry regions; the two longest are separated by one turn of the DNA double helix.

The lactose repressor selects one out of six million nucleotide sequences in the *Escherichia coli* genome and binds to it to prevent the expression of the genes for lactose metabolism. How does this protein, a 150,000-dalton tetramer of identical subunits, recognize its target? To answer this question we have determined the sequence of the repressor-binding site: the operator.

Genetically the operator is the locus of operator constitutive (o^c) mutations, *cis*-dominant changes that render the adjacent structural genes less sensitive to repressor control (1). The isolation of repressors (2, 3) and the demonstration that o^c mutations interfere with the binding of these proteins to DNA (4, 5) showed that the operator is the DNA binding site for the repressor protein.

We have isolated the lactose operator as a double-stranded fragment of DNA that has been protected by the lactose repressor against digestion by pancreatic DNase (6). The first step in this purification isolated DNA fragments, about 1000 base-pairs long, that carry the *lac* promoter–operator region. The *lac* repressor, when mixed with sonicated fragments of DNA from a phage that carries the *lac* genes, binds to a cellulose nitrate filter only those fragments containing the operator. IPTG (isopropyl-β,D-thiogalactoside, a synthetic inducer of the *lac* operon) specifically elutes from the filter those fragments that are found to the repressor (7). The *lac* repressor binds to these DNA fragments, as it does to whole phage DNA, with a dissociation constant of 10^{-13} M in 0.01 M salt and 0.01 M Mg^{++} (8). [The repressor binds IPTG with a constant of 1.3×10^{-6} M (2).] Since the repressor dissociates from the operator with a half-life (under our conditions) of 15 min, we treat a mixture of repressor and sonic fragments with DNase for a very short time, 1 min, and isolate any protected region before the repressor can release the DNA. The protected fragment we obtain is double-stranded, can bind again to the repressor, and is about 27 base-pairs long. Here we shall describe its sequence.

Abbreviations: TES, *N*-tris(hydroxymethyl)methyl-2-aminoethanesulfonic acid; CMCT, 1-cyclohexyl-3-(2-morpholinoethyl)-carbodiimide metho-*p*-toluenesulfonate; IPTG, isopropyl-β, D-thiogalactoside.

METHODS

Sonicated DNA Fragments. Sonicated [^{32}P]DNA fragments were made by growing a temperature-inducible lysogen of λcI857p*lac*5S7 at 34° in a glucose–50 mM Tris·HCl or TES (pH 7.4) medium in 3 mM phosphate, heating at 42° for 15 min at a cell density of 4×10^8/ml, then washing and resuspending the cells at a density of 8×10^8/ml in the same medium with 0.1 mM phosphate. 100 mCi of neutralized $H_3{}^{32}PO_4$ was added to 10 ml of cells, and the incorporation was continued for 2 hr at 34°. The cells were washed, suspended in 2 ml of TE buffer [10 mM Tris·HCl (pH 7.5)–1 mM EDTA], sonicated with six 15-sec bursts, and extracted with phenol. The aqueous phase was extracted with ether, and the residual ether was removed with a stream of N_2. The mixture of radioactive DNA and RNA was then made up to about 6 ml with binding buffer [containing 10 mM $MgCl_2$, 10 mM KCl, 10 mM Tris·HCl (pH 7.5), 0.1 mM EDTA, 0.1 mM dithiothreitol, 5% dimethylsulfoxide, and 50 μg/ml of bovine-serum albumin]. 20 μg of *lac* repressor was added and 1-ml aliquots of this mixture were filtered on six separate 25-mm Schleicher and Schuell B-6 filters. The filters were presoaked in binding buffer without bovine-serum albumin; the filtration was fast. Each filter was rinsed twice with 0.3 ml of binding buffer, then transferred to a second filtration apparatus and eluted twice with 0.25 ml of 0.1 mM IPTG in binding buffer. The eluate was extracted with phenol and dialyzed against TE buffer. This one-step process yields sonic fragments containing the *lac* operator, which repressor will bind again to filters with about 50–70% efficiency. The overall yield is about 0.2% of the input label.

Unlabeled sonicated fragments were made by purifying 100–300-mg amounts of phage. A 150-liter fermentor culture of a λcI857S7p*lac*5 lysogen was heat-induced at $A_{550} = 0.6$ in M63 minimal medium supplemented with glucose and casamino acids. After 3 hr the bacteria were harvested and frozen. 300–700 g of bacteria were thawed, lysed by suspension in phage buffer [0.1 M NaCl–10 mM Tris·HCl (pH 7.4)–10 mM $MgSO_4$] containing 0.1 μg/ml of DNase, and centrifuged at 17,000 × *g* for 30 min. The phages were precipitated with 2.1% NaCl and 6.5% polyethylene glycol. The precipitated phages were dissolved in phage buffer, run on a block CsCl gradient, and then on an equilibrium CsCl gradient. After brief dialysis to remove CsCl, purified phages were made up to 20 mM EDTA and sonicated with eight 15-sec bursts at full power, and extracted with phenol. The DNA was dialyzed against TE buffer.

The DNA solution was made up to binding buffer, and 5-ml

TABLE 1. *Pyrimidine tracts from the lac operator*

Tract	Moles	Yields
pCp	4–5	(4.6)
pTp	7	(7.0)
pTpTp	4	(3.8)
pTpCpCp	1	(2.0)
pCpTpCp	1	
pTpTpCpCp	1	(0.7)

Pyrimidine tracts were isolated and fingerprinted. The sequences were determined by partial digestion of phosphatase-treated material by spleen and by venom phosphodiesterase. The relative molar yields are the averages of three experiments, taking the TCC and CTC isostichs together as 2 mol/mol of operator.

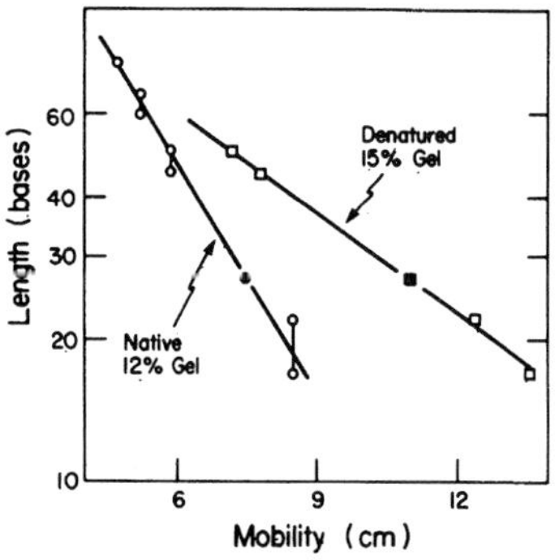

FIG. 1. Gel electrophoresis of operator fragments. ^{32}P-Labeled operator fragments were subjected to electrophoresis: native fragments (●) on a 12% polyacrylamide gel in 90 mM Tris·borate (pH 8.3)–5 mM $MgCl_2$ with double-stranded synthetic DNA markers of length (○) 77/77, 66/60, 50/45, and 22/17, or after denaturation (■) on a 15% polyacrylamide gel in 90 M Tris-borate (pH 8.3)–2.5 mM EDTA–7 M urea with DNA strands of length (□) 50, 45, 22, and 17. This experiment was done by Dr. T. Maniatis (14); the synthetic DNAs were the generous gift of Drs. H. Van de Sande and H. G. Khorana.

aliquots of DNA with repressor were filtered on a 47-mm B-6 filter. The filter capacity is 25 mg of DNA with 0.25 mg of active repressor (a 2-fold excess). The filter was washed with 3 ml of binding buffer and then eluted with 1 mM IPTG.

Operator Fragments. ^{32}P-Labeled operator was made by mixing sonic ^{32}P-labeled fragments with 2 μg of pure unlabeled *lac* fragments and 1 μg of repressor in 0.5 ml of binding buffer, digesting with 0.05–0.1 mg of DNase for 1 min, filtering rapidly on a B-6 filter, washing with 0.5 ml of 0° binding buffer without Mg^{++} and containing 20 mM EDTA, and shaking the filter on a Vortex mixer in 0.2 ml of 50 mM NH_4HCO_3 with 0.1% Na dodecyl sulfate.

Unlabeled operator was prepared by mixing 300 μg of sonic *lac* fragments, 10^6 cpm of ^{32}P-labeled sonic *lac* fragments (as tracer), and 150 μg of active repressor (300–600 μg of repressor) in 3 ml of binding buffer, and treating with 1 mg of DNase for 1 min at room temperature. The mixture was filtered on a 47-mm Schleicher and Schuell B-6 filter, washed with 3 ml of 20 mM EDTA, and shaken on a Vortex mixer in 1 ml of 50 mM NH_4HCO_3 with 1% Na dodecyl sulfate.

In both cases, the Na dodecyl sulfate rinses were run on a Sephadex G-100 column in 50 mM NH_4HCO_3 and 0.1% Na dodecyl sulfate. The operator eluted as a peak at 1.4 times the void volume. The peak fractions were pooled, the Na dodecyl sulfate was extracted with phenol, the phenol was extracted with ether, and the material was concentrated by lyophilization, which removed the NH_4HCO_3. There was about a 0.2–0.5% yield of material from the sonic *lac* fragment stage. We finally recover 100,000–200,000 Cerenkov cpm from 100-mCi incorporations.

Pyrimidine Tracts. DNA was depurinated according to Burton (9). Operator [^{32}P]DNA (10,000–100,000 cpm) and 50 μg of calf-thymus DNA were incubated with 66% formic acid and 2% diphenylamine in 200 μl for 18 hr at 37° in the dark. 200 μl of H_2O was added, and the mixture was extracted four times with 2 ml of ice-cold ether to remove diphenylamine, lyophilized repeatedly to remove formic acid, and dried to a compact residue on a polyethylene sheet.

RNA Synthesis. Reaction mixtures contained, in a volume of 20–50 μl: 10 mM $MgCl_2$, 20 mM Tris·HCl (pH 7.8), 70 mM KCl, 0.1 mM K_2HPO_4, 0.1 mM dithiothreitol, 250 μg/ml of bovine-serum albumin, generally 3–5 pmol of denatured operator strands, and 10–30 pmol of RNA polymerase (holoenzyme or core). For labeling with a single NTP, the mixtures also contained one [α-^{32}P]NTP (5–50 μM) at specific activities of 20–150 mCi/μmol (New England Nuclear Corp.), and three other cold NTPs at 300 μM. Reactions using these NTP concentrations were quite efficient, incorporating 20–70% of the label, about a 30-fold net synthesis. For primed synthesis, the mixtures contained either all four labeled NTPs at 5 μM or only GTP at 5 μM, the others at 20 μM, and dinucleotide primers at 100–500 μM or oligonucleotide primers at 50–100 μM. Reactions using these lower NTP concentrations with primers were less efficient, incorporating about 10% of the label, but synthesized better-defined species of RNA.

The reaction mixtures were incubated at 35° for 8–15 hr. 5 μl of 0.1 M EDTA and 20 μg of tRNA (Schwarz) were added and the reaction mixture was extracted with phenol. The RNA was isolated either by passage through a Sephadex G-25 or G-50 column in 10 mM NH_4HCO_3, followed by lyophilization, or by addition of 0.1 volume of 20% NaOAc and precipitation with 2 volumes of ethanol in a silicated tube, chilling in dry ice–acetone. The precipitate was washed with ice-cold 95% ethanol and dried.

Primers. Dinulceoside monophosphates were obtained from Sigma. The pentanucleotide ApUpCpCpG was synthesized enzymatically with primer-dependent polynucleotide phosphorylase by the method of Thach (10). GpGpApApU was recovered from a two-dimensional RNase A fingerprint of $tRNA_{IA}{}^{Gly}$ isolated from *Staphylococcus epidermis* by Dr. Richard Roberts.

Polyacrylamide Gels. RNA molecules were resolved on 12% or 18% (w/v) polyacrylamide gels made in TBE buffer [90 mM Tris-borate (pH 8.3)–2.5 mM EDTA] and 7 M urea. Samples were heated at 100° for 3 min in $^1/_{10}$ TBE buffer with 7 M urea. Slabs, 20 × 20 × 0.2 cm, were run at 30 mA (300 V) for 3–4 hr at room temperature. After autoradiography, the films were aligned with the gels; sections corresponding to bands were cut out, crushed, suspended in 1 ml of

TABLE 2. *Sequence analysis of some larger fragments*

Fragment	Source	Digestion products		
		RNase A	RNase U2	RNase T1
AUAACAAUU$_{OH}$	RNase T1	AU(A) AAC(A) AAU(U)	A CAA UAA	
CUCACAAUUCCA$_{OH}$	RNase T1	2C(A) 2U(C) C(U) AC(A) AAU(U) C(C)	CAA (C,C,U)A (C,C,U,U)N$_{OH}$	
UUAUCCG(C)	RNase T1	U(U) U(A) C(C) C(G) G(C) AU(C)		
AAUUG(U)	RNase T1	AAU(U) U(G) G(U)	AA UUG	
GGAAU(U)	RNase A			G(G) G(A) AAU(U)
UGAGCG(G)	partial RNase T1	U(G) GAGC(G) G(G)		UG AG CG(G)
AAUUGUG	partial RNase T1	U G GU AAU		UG AAUUG
GGAUAAC	CMCT-blocked RNase A	GGAU AAC		
GUGAGC	CMCT-blocked RNase A	GU GAGC		

The long fragments with 3′-hydroxyl termini were isolated from RNase T1 digests of gel bands from primed syntheses. The other oligonucleotides were isolated from unprimed total syntheses. The fingerprinting and sequencing techniques are as described by Barrell (11) and in the *Methods*, except that the CMCT blocking was 25 mg/ml of CMCT in 10 mM borate (pH 8.5)–3.5 M urea overnight at 32°. Nucleotides determined through nearest-neighbor analysis are shown in *parentheses*.

0.5 M NH_4OAc, 10 mM $Mg(OAc)_2$ 0.1% Na dodecyl sulfate, and 20 μg/ml of tRNA, and left overnight at 25°. The polyacrylamide fragments were filtered off by passing the solution, by gentle centrifugation, twice through a plug of glass-wool packed into the bottom of a small, punctured, plastic tube fitted into the top of a silicated centrifuge tube. The RNA was then precipitated with 2 volumes of ethanol.

Fingerprinting. The fingerprinting and sequencing techniques are those described in Barrall (11), except that the first dimension on cellulose acetate was in 5% acetic acid–0.1% pyridine–7 M urea–1–5 mM EDTA; the EDTA eliminated streaking, especially of pyrimidine tracts. Transfer to DEAE–paper for the second dimension was facilitated by placing a pad of dry Whatman 3MM paper strips beneath the usual layers of DEAE–paper, cellulose acetate, and wet strips, and washing through extensively with H_2O. All the radioactive label is found on the DEAE–paper, and all the urea is absorbed by the bottom pad and need not be washed away afterwards. For the first dimension, the yellow dye is run to 38 cm for a pancreatic fingerprint or to 42 cm for a T1 fingerprint. The second dimension is run in 7% formic acid until the blue dye has migrated 21 cm (pancreatic) or 27 cm (T1).

RESULTS

The operator fragment, protected by the repressor from digestion with pancreatic DNase, is a double-stranded DNA molecule that has a T_m of 67° in 0.15 M NaCl–0.015 M Na-citrate. (This is easily measured because *lac* repressor does not bind the denatured operator to a cellulose nitrate filter, and the denatured fragments themselves do not stick to the filter.) It is 60% A + T. Fig. 1 shows that the operator fragments move in gel electrophoresis as though they are 27 base-pairs long, in both the native and denatured states.

We began to determine the sequence of the operator by isolating pyrimidine tracts from ^{32}P-labeled material. Table 1 lists these pyrimidine tracts and their relative amounts. Unfortunately the longest pyrimidine run is pTpTpCpCp—not long enough to establish much sequence. The number of tracts is roughly consistent with our estimated size of th operator.

The final yields of ^{32}P-labeled operator were too limited for us to make progress with a direct attack on the sequence of the DNA; instead, we transcribed the operator fragment with RNA polymerase and determined the sequence of the RNA product. To do this, we scaled up the purification, beginning with 300 mg of purified phage to obtain, ultimately, micrograms of operator. A typical RNA reaction mixture contained 3 pmol of denatured operator DNA and 20 pmol of RNA polymerase, either holoenzyme or core. With the nearest-neighbor information from α-^{32}P transfer experiments, we were able to work out the structure of the oligonucleotides from complete digests, as well as to obtain some overlapping sequences from partial RNase T1 and 1-cyclohexyl-3-(2-morpholinoethyl)-carbodiimide metho-*p*-toluenesulfonate (CMCT)-blocked RNase A digests. Table 2 lists some of the longer fragments. However, because the synthesis produces a mixture of products initiating and terminating at many different points on both strands, we were not able to establish a sequence.

Partial Synthesis. We resolved the RNA product into single molecular species by using oligonucleotides to direct the synthesis of specific regions. Downey and So (12) have shown that when the NTP concentration is lowered to 5 μM, the RNA polymerase can no longer initiate normally with a triphosphate (because the K_m for initiation is much greater than the K_m for elongation). Under these conditions, on denatured DNA, the polymerase will initiate if it is presented with a dinucleotide or oligonucleotide complementary to the template which it can incorporate into the beginning of an RNA chain. (The oligonucleotide does not have to form a stable association with its complementary DNA sequence.) We exploited this fact to force the polymerase to synthesize from a single strand of our denatured operator DNA and to begin synthesis at a specific point. Although the oligonucleotide priming selects one strand rather than the other, the product still has ragged 3′ ends. These products resolve into a

TABLE 3. *Products of oligonucleotide-primed synthesis*

One strand
*GGAAU*UGUGAGCGGAUAACAAUU$_{OH}$
The other strand
*UA*UCCGCUCACAAUUCCA$_{OH}$
*AUCCG*CUCACAAUUCCA$_{OH}$
*GU*UAUCCGCUCACAAUUCCA$_{OH}$

Single bands on polyacrylamide gels were isolated from primed syntheses. The primers are *italicized.* The GpU-primed molecule was isolated from a two-dimensional gel as described by de Wachter and Fiers (20), except that the first dimension was a 4% gel in 10 mM citric acid and the second dimension a 12% TBE-7 M urea gel.

series of bands during electrophoresis on 12% or 18% polyacrylamide gels in 7 M urea. The oligonucleotide GpGpApApU isolated from *Staphylococcus epidermis* $tRNA_{IA}{}^{Gly}$ (a gift and a suggestion from Dr. Richard Roberts), primed the synthesis of the RNA molecule from one strand. The dinucleotide UpA and the pentanucleotide ApUpCpCpG primed synthesis from the other strand. GpU primed on both strands. Table 3 shows these results. The two sequences overlap and are complementary. Matching them up, written as DNA sequences, and filling out the gaps by complementary base pairing, produces a 24-base long sequence:

5′ TGGAATTGTGAGCGGATAACAATT 3′
3′ ACCTTAACACTCGCCTATTGTTAA 5′

The squared ends of the sequence are hypothetical; we do not know exactly where the DNase cuts. There are a few more bases on the DNA fragment to the left of this sequence. The pyrimidine tracts can be accommodated in the sequence (with the exception that the yields of pCp and pTp are too high); although they fail to provide any overlaps, they serve to confirm the correctness of some of the oligonucleotide sequences. We believe the orientation of the fragment to be correct as written, with the *i* gene to the left and the *z* gene to the right, in order to be consistent with the *lac* messenger sequence described by Maizels (13).

DISCUSSION

The sequence for the operator fragment shows symmetries. A total of 16 bases lie in a 2-fold symmetrical pattern, which is not centered on the fragment. Six base pairs to the left are symmetrically oriented with respect to six base pairs to the extreme right of the fragment. There are additional symmetrically arranged base pairs in a nine-base stretch between the two major regions. These symmetry regions are boxed below:

```
TGG[AATTGT]G[A]G[C]G[G]A[T]A[AGAATT]
ACC[TTAACA]C[T]C[G]C[C]T[A]T[TGTTAA]
```

(However, two to three of these matched base pairs might be symmetrical by chance.) This symmetry could permit the repressor protein to interact with DNA on a 2-fold symmetry axis: either two subunits could interact, each with the appropriate half of the symmetry region; or (as suggested by Dr. Thomas Steitz) all four subunits could interact, two with the left-hand region, the other two, related by a 2-fold axis, with the right-hand region. A 2-fold symmetry was suggested on genetic grounds by Sadler and Smith (14). The two longer regions are about a turn apart and can be approached from one side of the DNA helix. If these regions do reflect the interaction of the protein, then the DNase has shown some sequence specificity in cutting asymmetrically.

The structure of the *lac* operator is different from the λ operators isolated by Pirrotta (15) and analyzed by Maniatis and Ptashne (16). Maniatis, and Ptashne find a series of concatenated binding sites, 35–100 base pairs in length, while we isolate, under all conditions, only a single small fragment protected by the *lac* repressor.

The operator sequence does not immediately show how the repressor interacts with DNA. Mechanisms by which the repressor might detect the operator include: hydrogen bonding to the outside of the bases in the large or small groove of the DNA double helix, feeling variations in the positions of the phosphates that might be dictated by the A + T content affecting the structure (17), or opening up the DNA to see the bases directly. We discuss these possibilities more fully elsewhere (18).

The knowledge of both the nucleotide sequence of the operator and the complete amino-acid sequence of the repressor (19) should lead toward an understanding of this DNA-protein interaction. We anticipate that the sequences of o^c mutant operators will ultimately reveal which bases determine the structure that the repressor recognizes and will cast light on the actual interaction.

We thank Mary Archer and Joanna Knobler for technical assistance, Terry Platt and Ron Ogata for *lac* repressor, Charlotte Hering for RNA polymerase, and Richard Roberts for conversations about sequencing. This work was supported by USPHS Grant GM 09541 from the National Institute of General Medical Sciences. W.G. is an American Cancer Society Professor of Molecular Biology.

1. Jacob, F. & Monod, J. (1961) *J. Mol. Biol.* **3**, 318–356.
2. Gilbert, W. & Müller-Hill, B. (1966) *Proc. Nat. Acad. Sci. USA* **56**, 1891–1898.
3. Ptashne, M. (1967) *Proc. Nat. Acad. Sci. USA* **57**, 306–313.
4. Ptashne, M. (1967) *Nature* **214**, 232–234.
5. Gilbert, W. & Müller-Hill, B. (1967) *Proc. Nat. Acad. Sci. USA* **58**, 2415–2421.
6. Gilbert, W. (1972) in *Polymerization in Biological Systems,* Ciba Foundation Symposium (ASP, Amsterdam), Vol. 7 (new series), pp. 245–256.
7. Bourgeois, S. & Riggs, A. D. (1970) *Biochem. Biophys. Res. Commun.* **38**, 348–354.
8. Riggs, A. D., Suzuki, H. & Bourgeois, S. (1970) *J. Mol. Biol.* **48**, 67–83.
9. Burton, K. (1967) in *Methods in Enzymology,* eds. Grossman, L. & Moldave, K. (Academic Press, New York), Vol. XIIA, pp. 222–224.
10. Thach, R. (1966) in *Procedures in Nucleic Acid Research,* eds. Cantoni, G. L. & Davies, D. R. (Harper and Row, New York), Vol. 1, pp. 520–534.
11. Barrell, B. G. (1971) in *Procedures in Nucleic Acid Research,* eds. Cantoni, G. L. & Davies, D. R. (Harper and Row, New York), Vol. 2, pp. 751–779.
12. Downey, K. M. & So, A. G. (1970) *Biochemistry* **9**, 2520–2525.
13. Maizels, N. (1973) *Proc. Nat. Acad. Sci. USA* **70**, 3585–3589.
14. Sadler, J. R. & Smith, T. F. (1971) *J. Mol. Biol.* **62**, 139–169.
15. Pirrotta, V. (1973) *Nature New Biol.* **244**, 13–16.
16. Maniatis, T. & Ptashne, M. (1973) *Proc. Nat. Acad. Sci. USA* **70**, 1531–1535.
17. Bram, S. (1971) *Nature New Biol.* **232**, 174–176.
18. Gilbert, W., Maizels, N. & Maxam, A. (1973) *Cold Spring Harbor Symp. Quant. Biol.* **38**, in press.
19. Beyreuther, K., Adler, K., Geisler, N. & Klemm, A. (1973) *Proc. Nat. Acad. Sci. USA* **70**, 3576–3580.
20. de Wachter, R. & Fiers, W. (1972) *Anal. Biochem.* **49**, 184–197.

Sequences of Controlling Regions of the Lactose Operon

WALTER GILBERT, NANCY MAIZELS, AND ALLAN MAXAM
Department of Biochemistry and Molecular Biology and Committee on Biophysics, Harvard University, Cambridge, Massachusetts 02138

The control region of the lactose operon of *E. coli* provides an unusually appropriate DNA segment for sequence analysis. Here lie the sites at which the RNA polymerase and the catabolite activator protein act to potentiate synthesis of *lac* mRNA and the site at which the *lac* repressor acts to control mRNA synthesis. The repressor protein, a 150,000-dalton tetramer of identical subunits, discerns a unique, short region of DNA among the six million nucleotide sequences on the *E. coli* genome. The site at which the repressor acts, named the *operator* by Jacob and Monod (1961), is defined genetically as the locus of mutations that render the adjacent structural genes less sensitive to repressor control: these are cis-dominant constitutive mutations (o^c for "operator constitutive"). After the isolation of repressors and the demonstration that these proteins bind to specific sites on DNA (Gilbert and Müller-Hill, 1966; Ptashne, 1967a) and that operator constitutive mutations interfere with their binding (Ptashne, 1967b; Gilbert and Müller-Hill, 1967), one could alternatively define the operator as the binding site for the repressor protein on DNA.

The *lac* repressor prevents transcription of the *lac* operon in the absence of a specific inducer. The natural inducer is allolactose, an isomer of lactose made by β-galactosidase (Jobe and Bourgeois, 1972). In addition to this operon-specific negative control, the *lac* operon responds to the cellular, pleiotropic positive control mediated by the catabolite activator protein (CAP), a 45,000-dalton protein that senses the level of cyclic AMP within the cell (Zubay et al., 1970; Emmer et al., 1970). When the cell is growing on glucose, the level of cyclic AMP is low (Makman and Sutherland, 1965), and this prevents the synthesis of a variety of enzymes involved in the metabolism of other sources of energy, such as lactose, galactose, arabinose, or maltose. When the cell runs out of glucose, the level of cyclic AMP rises and activates the CAP protein, which then potentiates RNA synthesis from any catabolite-controlled operon (Ullmann and Monod, 1968; Perlman and Pastan, 1968). Silverstone et al. (1970) have shown that the *lac* promoter, originally pictured as a binding and initiation site for the RNA polymerase, is also the site for interaction with CAP protein. (See also Beckwith et al., 1972.)

The control region lies on the DNA between the repressor gene (the *i* gene) and the first structural gene of the operon (the *z* gene for β-galactosidase). The genetic loci are, in order, the end of the *i* gene, the CAP interaction site, the RNA polymerase site, the operator, and the beginning of the *z* gene. Our goal is to know the nucleotide sequence for this entire region in order to understand the nature of the interactions involved in these controls. Here we shall describe our present knowledge.

Lac Promoter-operator Fragments

Our experiments use fragments of DNA, approximately one thousand base pairs long, purified to contain the *lac* promoter-operator region. To isolate these fragments, we sonicate the DNA of a phage that carries the *lac* regions (*λplac5* or *λh80dlac*), add *lac* repressor to this mixture of fragments, and pass the mixture through a cellulose nitrate filter. The *lac* repressor binds to the filter only those fragments that contain the *lac* operator. Rinsing the filter with IPTG (isopropyl-β, D-thio-galactoside, a synthetic inducer of the *lac* operon) specifically elutes those fragments that are bound to the repressor. [This procedure was devised by Bourgeois and Riggs (1970).] We obtain these sonic fragments in about 1% yield starting from whole phage DNA or in about 0.2% yield starting from induced cells in which the RNA and DNA has been labeled with ^{32}P during the period of DNA replication after prophage induction. These fragments are small compared to the size of the *lac* operon. On the average, they will contain a C-terminal region from the *i* gene and an N-terminal region of the *z* gene. The *i* gene itself is 1000 base pairs long, the *z* gene 4000 base pairs long. The size of the region between the end of the *i* gene and the beginning of the *z* gene is not yet known, but might be estimated to be about 100 base pairs (Miller et al., 1968).

The *Lac* Operator

Starting with the thousand-base-pair DNA fragments containing the *lac* operator and promoter regions, we can isolate the lactose operator as a double-stranded fragment of DNA that *lac* repressor protects against digestion by pancreatic DNase (Gilbert, 1972).

Cold Spring Harb Symp Quant Biol 1974 38: 845–855

The *lac* repressor binds to the thousand-base-pair fragments with a dissociation constant (in 0.01 M KCl, 0.01 M Mg^{++}) of the order of 10^{-13} M (Riggs et al., 1970a). This binding is noncovalent; the repressor will eventually leave its binding site on the DNA. However, the repressor dissociates from the operator rather slowly; the half-time for dissociation from the fragments, under our conditions, is 15 min. We exploit this fact by digesting with DNase for a short time and stopping the treatment before the repressor dissociates from the operator: 1–2 min of digestion at room temperature at DNase concentrations ranging from 0.1–1 mg/ml. During filtration, the repressor retains the protected fragment on the filter, whereas digested material passes through. An EDTA wash inhibits the further action of DNase, and the fragments are finally eluted from the filter with 0.1–1% sodium dodecyl sulfate (SDS). Upon Sephadex G-100 gel filtration in 0.05 M ammonium bicarbonate, 0.1% SDS, the protected material elutes as a peak at about 1.4 times the void volume of the column, and the small oligonucleotides left after the DNase digestion are retarded and appear as a trail. We pool the peak fractions, remove the SDS by shaking with water-saturated phenol, remove the phenol with ether, and concentrate by lyophilization. *Lac* repressor protects about 1% of the input label in purified *lac* sonic fragments against the action of DNase, as measured by filter binding. We recover about 20–50% of the protected material after handling on the column and lyophilization. Our final yield from a 100-mCi preparation is 100,000 to 200,000 cpm in operator fragments.

Operator fragments can again be bound to a filter by repressor (with about 50% efficiency), and they are released by IPTG. These DNA molecules are double-stranded with a T_m of 67° in SSC. (The T_m is measured by repressor binding: *lac* repressor does not bind denatured operator fragments to a cellulose nitrate filter, and denatured fragments do not stick to the filter themselves.) Electrophoresis on polyacrylamide gels yields estimates of the length of the molecule. Our earliest estimates on SDS gels indicated that the strands were 20 bases long; our current estimate, based on comparison with synthetic DNA fragments of known size, is that the molecule is 27 base pairs long (Gilbert and Maxam, 1973).

We began sequencing this fragment by studying ^{32}P-labeled material. Its composition was 60% A·T. Pyrimidine tracts in the operator fragment are 5 C, 7 T, 4 TT, 1 TCC, 1 CTC, and 1 TTCC. We did not learn much of the sequence from these tracts because the longest is only a tetranucleotide. The number of pyrimidine tracts is roughly consistent with the estimated size of the operator. However,

Table 1. Products of Oligonucleotide-primed Transcription from Denatured *Lac* Operator

One strand	$GGAAUUGUGAGCGGAUAACAAUU_{OH}$
The other strand	$UAUCCGCUCACAAUUCCA_{OH}$
	$AUCCGCUCACAAUUCCA_{OH}$
	$GUUAUCCGCUCACAAUUCCA_{OH}$

The priming oligonucleotides are underlined. Primed RNA was synthesized in 20 μl containing 0.01 M $MgCl_2$, 0.02 M Tris pH 7.8, 0.07 M KCl, 10^{-4} M K_2HPO_4, 10^{-4} M DTT, 250 μg/ml BSA, 3 pmoles denatured operator strands, 10 pmoles RNA polymerase (holoenzyme or core), α-^{32}P-NTP's (GTP at 5 μM, the others at 5 or 20 μM, at specific activities near 100 mCi/μmole), and dinucleotide primers at 100–500 μM or oligonucleotide primers at 50–100 μM. After 8–15 hr at 35°C, 5 μl 0.01 M EDTA and 20 μg tRNA were added, the reaction mixture was phenol-extracted and made 2% in NaAc, and RNA precipitated with 2 vol ethanol. The product was run on a 12% polyacrylamide gel, and, after radioautography, portions of the gel containing bands were cut out, crushed, and eluted with 1 ml of 0.5 M NH_4Ac, 0.01 M $Mg(Ac)_2$, 0.1% SDS, and 20 μg/ml tRNA at 25°C. Gel fragments were removed by filtration and the RNA precipitated with 2 vol ethanol. The RNase digestion, fingerprinting, and sequencing techniques were as described by Barrell (1971). Further details are in Gilbert and Maxam (1973). The GpU-primed molecule was isolated from a two-dimensional gel, because GpU primes on both strands.

the amounts of ^{32}P-labeled operator were too limited for us to pursue direct sequencing of this DNA molecule. We turned to transcribing the operator fragment with RNA polymerase and sequencing the RNA product. To do this, we scaled up the purification, beginning with about 300 mg of purified phage to prepare the sonic fragments, digesting milligram amounts of purified sonic fragments, and obtaining ultimately microgram amounts of operator. Typically, a transcription reaction mixture contains on the order of 3 pmoles of denatured operator strands and 10–20 pmoles of RNA polymerase, either holoenzyme or core, as described in the legend of Table 1. With one triphosphate labeled and present at low concentrations, whereas the other three are present at high concentrations, we observe repeated initiations and 30-fold net synthesis incorporating 20–70% of the label over 8 to 12 hr. With nearest-neighbor information from α-^{32}P transfer experiments, we were able to determine the structure of the oligonucleotides from complete digests as well as to obtain some overlapping sequences from partial RNase T1 and CMCT-blocked RNase A digests. However, because the synthesis produces a mixture of products initiating and terminating at many different points on both strands, we were not able to establish a sequence.

Partial synthesis. We resolved the RNA product into single molecular species by a technique of partial synthesis, using oligonucleotides to direct the synthesis of specific regions. Downey and So (1970) have shown that when the nucleoside triphosphate concentration is lowered to 1–5 μM, the RNA polymerase can no longer initiate normally with a triphosphate (because the K_m for initiation is much greater than the K_m for elongation). Under these conditions, on denatured DNA, the polymerase will initiate if it is presented with a dinucleotide or oligonucleotide that it can incorporate at the beginning of an RNA chain. (The oligonucleotide does not have to form a stable association with its complementary DNA sequence.) We exploited this fact to force the polymerase to synthesize from only one strand or the other of our denatured operator material and to begin synthesis at a specific point. We used our knowledge of the oligonucleotide fragments and a trial sequence in order to suggest which oligonucleotides and which dinucleotides might be most useful (selecting dinucleotides that occurred rarely in the sequence). Although the oligonucleotide priming selects one strand rather than the other, the product still has ragged 3′ ends. These products resolve into a series of bands during electrophoresis on 12 or 18% polyacrylamide gels in 7 M urea. The oligonucleotide GGAAU, which was isolated from a *Staphylococcus epidermis* gly I_A tRNA, a suggestion and a gift of Dr. Richard Roberts, primed the synthesis of the RNA molecule whose structure is given in Table 1. We have used the dinucleotide UpA and the pentanucleotide AUCCG to prime synthesis from the other strand. The dinucleotide GpU primes synthesis from both strands. Table 1 shows the results. The two sequences overlap and are complementary. Matching them up, written as DNA sequences, yields the structure:

G G A A T T G T G A G C G G A T A A C A A T T
A C C T T A A C A C T C G C C T A T T G

Filling out the gaps by complementary base pairing produces a 24-base-long sequence:

T G G A A T T G T G A G C G G A T A A C A A T T
A C C T T A A C A C T C G C C T A T T G T T A A

The squared ends of the sequence are hypothetical, because we do not know exactly where the DNase cuts. The pyrimidine tracts can be accommodated in the sequence (with the exception that the yields of pCp and pTp are too high), and although they fail to provide any overlaps, they serve to confirm the correctness of some of the oligonucleotide sequences. The messenger transcription results described later will show that the orientation of the fragment is correct as written, with the *i* gene to the left and the *z* gene to the right.

There are symmetries in this sequence. A total of 16 bases fit in a twofold symmetrical pattern, which is not centered on the fragment. Six base pairs to the left are symmetrically oriented with respect to six base pairs to the extreme right of the fragment. There are additional symmetrically arranged A·T base pairs and G·C base pairs in a nine-base stretch between these two major symmetry regions around a central G·C base pair. These symmetry regions are boxed below:

T G G |A A T T G T| G |A| G |C| G |G| A |T| A |A C A A T T|
A C C |T T A A C A| C |T| C |G| C |C| T |A| T |T G T T A A|

These symmetries would permit the repressor protein to interact on a twofold symmetry axis: either two subunits could interact, each with the appropriate half of the symmetry region; or (as suggested by Dr. Thomas Steitz) all four subunits could interact, two with the left-hand region, the other two, related by a twofold axis, with the right-hand region. We will discuss some of the possible interactions later.

Lac mRNA Transcribed from the UV5 Promoter

We have sequenced the 5′ end of *lac* messenger RNA transcribed in vitro. In order to simplify our work, we have used a mutant promoter which bypasses the requirement for the CAP function. Arditti et al. (1968) isolated this mutant as a phenotypic revertant of a point mutation in the promoter, L8 (presumably a mutation in the CAP binding site), which expresses the *lac* genes at a high level. This double mutant, called UV5, expresses the *lac* genes independently of the CAP function both in vivo (Silverstone et al., 1970) and in vitro (Eron and Block, 1971) and is controlled solely by the *lac* repressor. In vitro it can be transcribed directly with RNA polymerase, independent of CAP and cyclic AMP. Thus, this promoter provides an ideal simple model system to study the control of transcription in vitro. As template, we have used 1000-base-pair long sonic fragments, purified by repressor binding, which contain the *lac* promoter-operator region from the $\lambda h80cI857S7dlac$UV5 phage. These fragments provide, in principle, a way of synthesizing a single messenger species, since they should contain only a single promoter. We can use these fragments at high concentrations to incorporate large amounts of label at the commercially available specific activities. As a technical device for most of these experiments in which transcription initiates at a natural promoter, we

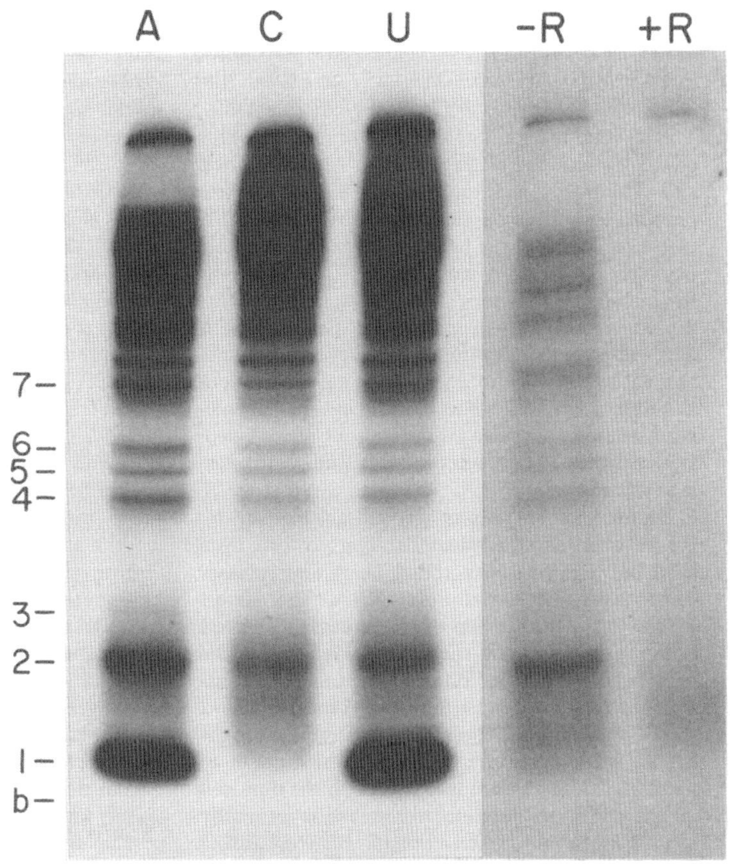

Figure 1. The picture on the left displays the products of A-, C-, and U-labeled syntheses electrophoresed on a 12% polyacrylamide gel. The gel band numbering used in the text is indicated in the column at the left; "b" indicates a bromphenol blue dye marker. The panel on the right shows the products of A-labeled reactions run with (+R) and without (−R) *lac* repressor. Discrete bands are absent in the repressed reaction. A T1 fingerprint of repressed material revealed no characteristic oligonucleotides, showing that the diffuse background is the product of random synthesis. In a typical reaction, 5 μg *lac* operator-promoter fragments and 40 μg RNA polymerase [purified as described by Berg et al. (1971) through the glycerol gradient] are preincubated for 6 min at 25°C in a 100-μl reaction mixture containing 0.03 M Tris pH 8.0, 0.1 M KCl, 5 mM $MgCl_2$, 4 mM EDTA, 0.1 mM K_2HPO_4, 0.1 mM DTT, 10% glycerol, 0.05% BSA, 0.2 mM GpA (Sigma), and 5 μM triphosphates, all four or only one triphosphate labeled in the alpha position with ^{32}P (New England Nuclear, SA > 100 mCi/μmole). Absence of free divalent cations inhibits initiation of RNA synthesis until addition of $\frac{1}{10}$ vol 0.15 M $MgCl_2$ + 95 μg/ml rifamycin; reactions done in the presence of repressor or CAP factor are preincubated with 15 mM $MgCl_2$ in the absence of triphosphates, and synthesis initiated by adding triphosphates and rifamycin. The rifamycin reduces background synthesis very slightly, but since the polymerase/promoter ratio is low (0.4/1, assuming 10% active polymerase) and the reaction very quick, the background without rifamycin is not significant. After synthesis for 3–9 min at 25°C, the reaction is quenched by adding $\frac{1}{10}$ vol 0.8 M EDTA. After adding 20–40 μg carrier tRNA (Schwarz), ^{32}P-labeled RNA is separated from unincorporated label either by: (1) chromatography on a 2-ml Sephadex G-100 column in 0.05 M NH_4HCO_3 (incorporated label, about 3–8% of the total, elutes in the void volume and unincorporated label is retarded), followed by phenol extraction, ether extraction to remove residual phenol, and concentration by lyophilization; or (2) increasing the reaction volume to 0.4 ml, extracting with phenol, transferring the aqueous phase to a silicated disposable tube containing $\frac{1}{10}$ vol 20% sodium acetate, precipitating with 2 vol 95% EtOH, pelleting the RNA for 30 min at 20,000 *g*, and finally washing with 1 ml 95% EtOH. (These two ethanol precipitations remove over 99% of the unincorporated label and are carried out rapidly by chilling the suspension at −70°C for 15 min rather than for several hours at −20°C.) The RNA is boiled for 3 min in 7 M urea, quick chilled, and electrophoresed on a 20 × 20 × 0.2 cm, 7 M urea, 12% acrylamide slab gel in 90 mM Tris borate pH 8.3, 2.5 mM EDTA for 2.5 hr at 25 mA.

used the fact that, at very low nucleotide triphosphate concentrations (less than 5 μM), the polymerase cannot initiate with a triphosphate. Under these conditions, on native DNA, specific dinucleotides or trinucleotides serve to stimulate synthesis of RNA at specific promoters (Downey et al., 1971; Minkley and Pribnow, 1973). One infers that the polymerase recognizes correct promoter regions, but can initiate with the dinucleotide only at that promoter at which the dinucleotide can be incorporated into the beginning of the messenger. Presumably, the polymerase opens up the DNA so that a dinucleotide that can match the sequence revealed will serve as primer. The dinucleotides that function in this way normally contain A or G and compete appropriately with γ-^{32}P-labeled ATP or GTP for initiation. For the UV5 promoter, we found that only the dinucleotide GpA stimulates synthesis appreciably (5- to 15-fold over background). There are several benefits from this technique: we can use extremely low triphosphate levels and thus incorporate the radioactive triphosphate efficiently; the dinucleotides increase the specificity of initiation and help eliminate non-specific background; and the low triphosphate levels slow down the reaction, so that short RNA molecules are synthesized in reasonable times. We hoped to isolate short regions from the beginning of the mRNA by stopping synchronized syntheses at various times and recovering progressively longer mRNA fragments. In fact, we found that the polymerase tarries at specific places along the DNA. The reaction mixture builds up a variety of lengths of RNA molecules, all beginning at the same point and extending to different pausing places. Electrophoresis on polyacrylamide gels reveals these RNA molecules as a series of bands. Figure 1 shows these patterns, which are characteristic of, and different for, different promoters. These patterns are quite reproducible. The pausing of the polymerase is only temporary; for example, a higher triphosphate concentration will arouse the polymerase to continue along the DNA. This pausing aids in sequencing because it provides an ordering of oligonucleotides produced by RNase digestion and some overlap information.

Figure 1 shows that in the presence of *lac* repressor there is no distinct banding pattern from UV5, only a diffuse background, therefore repressor controls the synthesis of the RNA. The fastest moving band (band 1) on the gel shown in Figure 1 is the seven-base-long sequence GAAUUGU. The next band is 18 bases long and contains the same starting sequence, with 11 more bases. Table 2 shows sequences for these and longer RNAs made from the UV5 promoter. Although the amount of material in larger bands increases with the duration of the synthesis reaction, some bands generally contain much more material than others, indicating that certain pausing places delay the polymerase much longer than others. Band 2 is usually the strongest stop. The bands often contain several closely related species, differing in length by one or two bases at the 3′ end. There are no obvious sequence similarities at the 3′ ends of pausing RNA molecules. For sequence analysis, each band was cut from the gel and fingerprinted by standard Sanger procedures described in detail by Barrell (1971). Phosphate transfer from RNA synthesized using a single α-^{32}P-labeled triphosphate gave nearest-neighbor information for each nucleotide in the sequence. (Further information on the sequencing can be found in Maizels, 1973.)

Triphosphate initiation. ^{32}P-labeled phosphate is never transferred to the first G in the sequence, showing that the cold GpA dinucleotide is incorporated at the 5′ ends of the RNA products. One might worry that the dinucleotide initiation had in some way modified the natural start at the UV5 promoter. High ATP or GTP concentrations (200 μM) permit the polymerase to initiate normally with a triphosphate. The gel electrophoresis pattern of RNA synthesized with high (200 μM) ATP and GTP and low (5 μM) CTP and UTP is the same as that of RNA initiated with GpA. Since high triphosphate initiation does not perceptibly alter the lengths of the RNA molecules as seen on gels, the GpA must initiate at or very near the natural triphosphate initiation point. Because triphosphate ends streak in our two-dimensional fingerprinting system, we identified the positions of the 5′ triphosphate by an indirect experiment. We compared the fingerprints of molecules that had or had not been treated with alkaline phosphatase before the gel electrophoresis and fingerprinting. Since the oligonucleotide bearing the terminal triphosphate is lost, we expected to find a spot on the fingerprint of the alkaline phosphatase-treated sample that would be missing on the fingerprint of the sample carrying the triphosphate end. We examined corresponding bands from gels of a reaction labeled with [α-^{32}P]UTP and primed by a high level of ATP and GTP. On the T1 fingerprint of band 4, an AAUUG

Table 2. Sequences of Paused Transcripts from UV5 *Lac* DNA

Gel band	
1	$GAAUUGU_{OH}$
2	$GAAUUGUGAGCGGAUA^{A_{OH}}_{AC_{OH}}$
3	GAAUUGUGAGCGGAUAACAAU . . .
4	$GAAUUGUGAGCGGAUAACAAUUUCACACAGGAAACAGCUA_{OH}$
5	$GAAUUGUGAGCGGAUAACAAUUUCACACAGGAAACAGCUAUGACC^{A_{OH}}_{AU_{OH}}$
6	GAAUUGUGAGCGGAUAACAAUUUCACACAGGAAACAGCUAUGACCAUGAUU . . .
7	GAAUUGUGAGCGGAUAACAAUUUCACACAGGAAACAGCUAUGACCAUG(AUUACGG,AUUCACUGG) . . .

Individual bands were eluted from a 7 M urea, 12% acrylamide slab gel as described in the legend to Table 1 and sequenced by standard fingerprinting techniques described by Barrell (1971) with modifications (Gilbert and Maxam, 1973). For all two-dimensional fingerprints, the first dimension of electrophoresis is on cellulose acetate strips in 7 M urea, 0.2% pyridine, 5% acetic acid; 1 mM EDTA pH 3.5, and the second dimension is on DEAE paper in 7% formic acid.

The ordering of the last two long T1 fragments in band 6 [AUUACG(G) and AUUCACUG(G)] could not be clearly determined by sequencing a minor intermediate gel band. They are written in the order implied by that sequence, which agrees with the known amino acid sequence of β-galactosidase (Zabin and Fowler, 1972).

```
       1         10        20        30        40        50        60
5'- pppAAUUGUGAGCGGAUAACAAUUUCACACAGGAAACAGCUAUGACCAUGAUUACGGAUUCACUGG...-3'
                                                  Thr Met Ile Thr Asp Ser Leu Ala
     lac operator                                 β-galactosidase
```

Figure 2. The first 64 bases of the messenger RNA transcribed from the UV5 *lac* promoter. The *lac* operator sequence is transcribed into bases 1–21. About 15% of the messenger molecules initiate at the G before the first A as pppGAAUUG. β-galactosidase synthesis initiates at the AUG codon at position 39, and the amino-terminal sequence of β-galactosidase (Zabin and Fowler, 1972) is matched to the triplets on the messenger that code for those amino acids.

spot appears in the phosphatase-treated sample and is absent in the control with an intact triphosphate end. In addition, a new AAU sequence appears in the pancreatic RNase digest of band 2 after phosphase treatment and is missing from the control. Thus, we infer that in the presence of high ATP and GTP, transcription begins at the first A in our sequence: pppAAUUGU . . . . A small amount of GAAU also appears on the pancreatic RNase fingerprint of the alkaline phosphatase-treated RNA; it is absent if the triphosphate end is still intact. Thus, the messenger can begin either on the G or on the A, at the position at which the GpA dinucleotide primes, with a 5- or 6-fold preference for the ATP start. (The reaction initiates efficiently with either high ATP or high GTP alone.)

The operator is transcribed. The initial sequence of the mRNA matches the sequence that we found for the operator fragment. Messenger RNA begins at the first GA of the operator sequence and runs across the rest of the protected fragment. A few bases on the left-hand side of the operator are not transcribed. The operator fragment was isolated from DNA molecules that carried the wild-type *lac* promoter, and thus the UV5 mutation does not lie inside the initial region transcribed by the RNA polymerase.

The initial sequence of the z gene. The first AUG occurs on this messenger at position 39, counting from the triphosphate end. If this AUG codes for formylmethionine, the translation initiation signal, then reading out the ensuing series of bases in triplets determines the amino-terminal sequence of β-galactosidase. Figure 2 shows this matching. The first nine amino acids were determined in this laboratory by Charlotte Hering using the micro Dansyl-Edman technique of Weiner et al. (1972) and agree with those determined by Zabin and Fowler (1972). Seventeen bases separate the operator from the initiation triplet for the *z* gene. Part of this region is presumably the ribosome binding site for the initiation of β-galactosidase synthesis. Robertson et al., (1973) have tallied similar sequences in untranslated regions of different ribosome binding sites: these include PuPuUUUPuPu, PyAGGA, and a nonsense codon just before the AUG initiator triplet. The UV5 *lac* mRNA contains the sequences AAUUUCA and CAGGA, but the only nonsense codons in the untranslated region are close to the 5′ end—at positions 6–8 and 14–16. Two nonsense codons occur out of phase in the translated region, at positions 40–42 and 46–48. There is little secondary structure in this messenger RNA. The operator sequence at the beginning of the messenger can be folded into a loop, with six base pairs in the stem and nine bases in the loop. This loop would be expected to be only marginally stable.

That messenger transcription begins on the protected region of DNA, and thus presumably at a site hidden by repressor as it binds to the operator, provides us with a simple picture of how the repressor prevents the RNA polymerase from initiating on the UV5 promoter.

Operator Constitutive Mutations

How do repressor and operator interact? By identifying the molecular changes within the protected fragment that weaken the binding of repressor to DNA, we should, in principle, be able to delineate fully those regions of the DNA that maintain the structure of the recognition site. Mutations in the operator that decrease the affinity of operator for repressor, called operator constitutive, o^c, are available. Smith and Sadler (1971) isolated a large number of these mutants and examined the in vivo constitutive level of β-galactosidase synthesis. The constitutive level of synthesis by an o^c mutant reflects the degree of damage to the repressor-operator interaction. They found that these mutants fell into a small number of classes, and, from fine structure mapping data, argued that these classes are symmetrically arranged in the operator region (Sadler and Smith, 1971). The symmetry that they predicted is the dyad symmetry that we find. We hope eventually to sequence their genetically mapped o^c mutants, but those mutants are in a strain with a wild-type promoter, and we have not yet sequenced mRNA from the wild-type promoter. However, we have isolated a new set of operator constitutive mutations on the phage carrying the UV5 promoter mutation. We selected for growth on phenylgalactoside, a substrate of β-galactosidase that does not induce the *lac* operon.

We then sequenced mRNA from five such independently isolated operator constitutive mutants. Unfortunately, four of these mutations turned out to be identical changes, so to date we know only two different o^c changes.

Since the o^c mutation weakens the ability of the repressor to bind to DNA, one would expect it to be very difficult, if not impossible, to isolate from such mutants any fragment protected by repressor against DNase digestion. Therefore, we isolated the 1000-base-pair sonic fragments from the mutant DNA and transcribed them into messenger RNA;

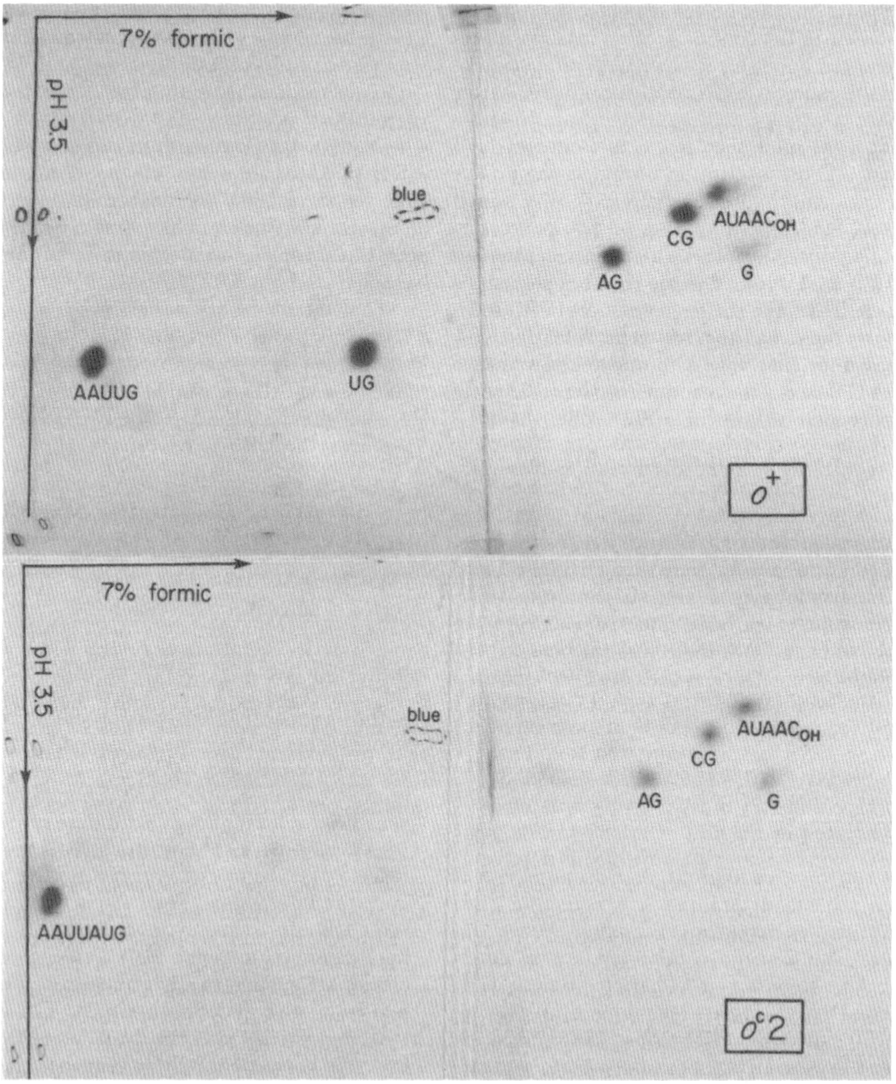

Figure 3. This figure compares T1 fingerprints of RNA synthesized from UV5 template DNA with either a wild-type operator (o^+) or an operator containing a spontaneous mutation to the operator constitutive phenotype (o^c2). RNA was synthesized as described in the legend to Figure 1, except that all four triphosphates were ^{32}P-labeled. Synthesis was for 4 min at 25°C. After gel electrophoresis, RNA was eluted from gel band 2, digested with RNase T1, and fingerprinted as described in the legend to Table 2. The right halves of both the o^+ and o^c2 fingerprints are identical, whereas the two spots in the left half of the fingerprint of o^+ mRNA (AAUUG and UG) are replaced in o^c2 by a single spot which moves slower than AAUUG in the second dimension of electrophoresis. Loss of a T1 oligonucleotide implies change of a G to some other base. Analysis of RNase A digests of oligonucleotides on the T1 fingerprint, as well as analysis of two-dimensional RNase A fingerprints of gel band 2, indicated the change is from G to A at position 5 of the mRNA.

since we knew that the operator sequence was located at the beginning of the messenger, we hoped that the sequence of this RNA would reflect the o^c base change. We purified sonic fragments containing the *lac* operator region from o^c mutants by lowering the magnesium concentration to 4 mM in the filter binding step and filtering rapidly. The fragments we obtained, although not as pure as those of the o^+ phage, are still adequate to synthesize specific messengers using the GpA dinucleotide initiation.

An o^c change in the symmetry region. The first o^c mutation that we examined (UV5o^c2, a 15% constitutive) produces a transcription pausing pattern similar to the wild type. Figure 3 compares the T1 fingerprint of the second gel band (18 bases long) from o^c2 DNA with that from the wild type. The mutant fingerprint has one slow-moving spot, replacing two spots from the fingerprint of the wild type. The two wild-type spots that disappeared have the sequences AAUUG and UG; the new spot has the sequence AAUUAUG. Thus we deduce that this o^c mutation is a transition from G·C to A·T at position 5 in the left-hand symmetry region. Figure 4 shows this change. Clearly, if the symmetric sequences are involved in contacts with the protein, this mutation damages a contact.

An o^c change not in a symmetry region. The second o^c (UV5o^c5, a 3–5% constitutive) eliminates gel band number 2 by abolishing the second halting place. The pancreatic fingerprint of the second strong gel band from this mutant (equivalent to gel band 4) shows that a characteristic fragment GAGC has been replaced by GAU. A new T1 fragment, AUCG, also appears. Thus, the G at position 9 in o^+ messenger RNA has been mutated to U in o^c5, which is a transversion, a G·C to T·A change. This change occurred in four independently isolated o^c's and is illustrated in Figure 4. Curiously, although this transversion is in a base pair which is not symmetrically placed with respect to another base pair in the sequence, this change increases the symmetry of the operator: the mutation generates a T·A base pair that is symmetric to the A·T base pair at position 13. Thus, both mutations that decrease *and* that increase the symmetry can lead to defective operators. That this base change at position 9 eliminates the pausing place at position 17 suggests that the places at which the polymerase stops and waits reflect a subtle secondary structure in the messenger being synthesized, or some interaction between the enzyme and the DNA sequence that it has copied. We have noticed that there is often a short run of G·C base pairs, about 8 to 11 positions, before the end of the pausing products. We speculate that these G·C regions make it difficult for the polymerase to peel nascent RNA off the DNA so that further synthesis can proceed.

Do We Know All of the Operator?

Does our protected fragment include the entire region of the DNA that is involved in repressor binding? The isolation of the protected fragment by DNase digestion raises the possibility that the DNase cuts off some DNA that is part of the biological operator, but is more susceptible to digestion than the rest of the region. The fragment does bind repressor less strongly than do longer pieces of DNA—the half-time for dissociation of repressor from ^{32}P-labeled protected fragments is 2 min compared to 15 min for whole phage DNA (Gilbert, 1972), so the affinity is probably one to two orders of magnitude weaker. One of many possible explanations for this weaker binding is that a few bases have been cut off the true operator.

One could question whether the transcripts that we have sequenced include the very ends of the template DNA strands. We think that the transcription products do include the entire region at the right-hand end of the protected fragment: from the messenger sequence we know that the ATT sequence at the right end of the protected fragment is part of an ATTTCA sequence on the DNA, but we do not observe any pyrimidine tracts that would arise from this longer sequence if it were part of the ^{32}P-labeled protected DNA fragment. In early fingerprints of the transcribed RNA, we did see the GAAAUUG(U) sequence that would come from the complementary strand in this region, but we did not recover this sequence from later preparations of template in which the DNase digestion may have gone further. Thus we think it likely that we have, on occasion, rescued larger regions from the DNA shown in Figure 5, but that the minimal protected region is that given before.

How Significant Are the Symmetries?

The minimal operator fragment has sixteen bases arranged symmetrically. The longer region, shown in Figure 5, has four more bases that are arranged with twofold symmetry (and a further purine-purine symmetry between a C·G and a T·A base pair). Thus 20 out of the 27 bases are arranged symmetrically on the DNA. Around any given point there is a one in four chance that any base will have a symmetric partner, and, therefore, we would expect 2 or 3 of the 10 base pairs in each half of the symmetry region to occur by chance. Thus, not every symmetrically placed base need be a contact for a protein. Figure 5 also shows another set of symmetries, for which we know no role, centered between positions 26 and 27 and

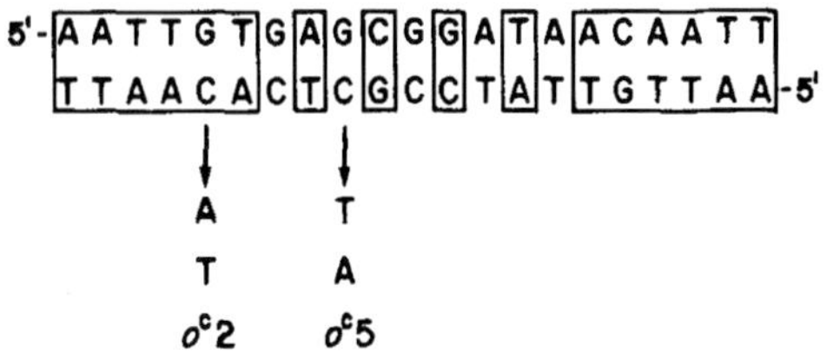

Figure 4. o^c mutant changes in the *lac* operator sequence. Symmetric regions in the operator are boxed. The mutation in o^c2, a 15% constitutive, is a transition from G·C to A·T in a major symmetry region of the operator. Four other independently isolated mutants, including o^c5, are 3–5% constitutives caused by a transversion from G·C to T·A in a pair of bases not in a symmetry region. This latter mutation increases the symmetry in the operator sequence.

containing 16 symmetric bases out of 28 (over twice the random frequency). Ultimately, only the o^c mutation sites will correctly identify the bounds of the operator region.

How Does Repressor Interact with the Operator?

The sequence that we have found is amply long enough to explain the specificity of the lactose repressor-operator interaction. Even if we were to assume that the primary contacts were only with the symmetrically arranged base pairs, the repressor would recognize 16 base pairs: since 12 base pairs is the minimum number necessary to comprise a unique sequence in the *Escherichia coli* genome, any longer sequence would certainly do. The two major contact regions are likely to be the six-base-long sequences from 1–6 and 16–21 in Figure 5. If one examines the large groove on a molecular model of the operator, one is struck by the prominent methyl groups of the thymines; especially the pairs at positions 3–4 and at 18–19 and the two methyl groups that lie across from each other in the large groove at positions 6 and 8 and at positions 14 and 16.

Repressor would interact with these regions on a twofold axis, either two or four subunits touching the DNA. The amino-terminal region of these subunits may make some of the contacts. Adler et al. (1972) argued that the amino-terminal region of the repressor binds to DNA because a very large number of mutations that damage DNA binding map at this end of the molecule. Furthermore, trypsin can cut fragments off the amino and carboxy termini; this abolishes DNA binding but leaves a tetrameric protein core that still binds IPTG (Platt et al., 1973). Adler et al. (1972) proposed a specific interaction between the known amino-terminal sequence of the repressor and a predicted DNA sequence. They folded the amino-terminal region of the protein into an α-helix and fitted that helix into the large groove of DNA, as had been done for histones (Sung and Dixon, 1970; Zubay, 1964). They then argued that specific amino acids could hydrogen-bond with certain base pairs; these contacts predicted (with ambiguities) a four-base-long sequence:

(C,A)	T	A	(G,T)
(G,T)	A	T	(C,A)
I	II	III	IV

A tyrosine at position 17 in the amino acid chain was to make contact with either a C·G or an A·T base pair at position I. Glutamine 18 and serine 21 were to contact the T·A pair at position II. Aspartic acid 25 was to contact the third position, the A·T pair; and glutamine 26 was to bond either to a G·C or T·A pair at the fourth position (or if it was not able to do that, it was to bond to a fifth position instead).

The similarity between their operator sequence and ours is no better than chance. However, two further arguments are relevant. On one hand, the

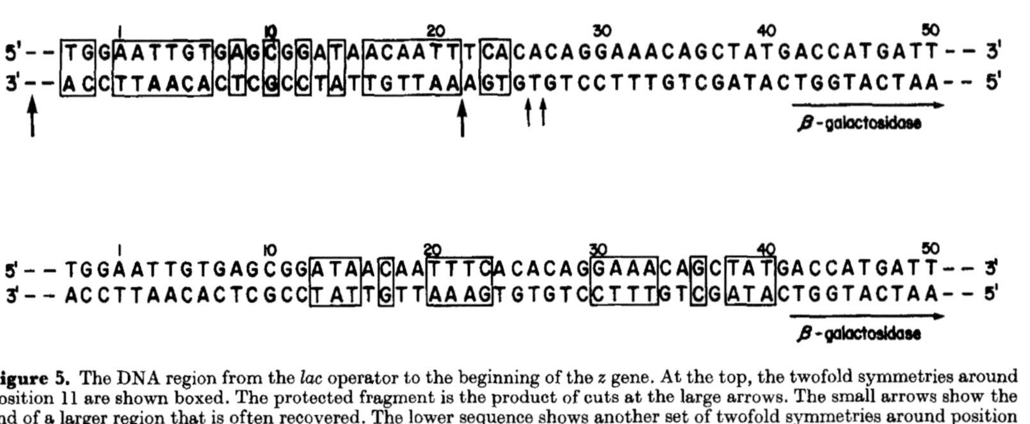

Figure 5. The DNA region from the *lac* operator to the beginning of the *z* gene. At the top, the twofold symmetries around position 11 are shown boxed. The protected fragment is the product of cuts at the large arrows. The small arrows show the end of a larger region that is often recovered. The lower sequence shows another set of twofold symmetries around position 26.5, which have no known function.

prediction can be made to look a little more striking: there was one error in their amino acid sequence—position 25 should be asparagine, not aspartic acid. An asparagine can donate a hydrogen bond to the T of a T·A base pair, and this would permit the protein to be matched to the sequence ATTG which occurs in our six-base-long symmetry regions. (However, there is an 80% chance that a sequence of the form (C,A)TT(G,T) would occur once on a *lac*-operator-sized fragment.) On the other hand, the known i^{-d} mutations do not support the detailed models and, by the use of suppressors, glutamine 26 can be replaced by a variety of amino acids, including the bulky hydrophobic leucine, and still result in an active repressor (Weber et al., 1972). Thus, although Adler et al. (1972) do specify possible amino acid–nucleotide contacts, there is no proof that *those* contacts, in fact, pick out the operator sequence.

More generally, there are three models for the interaction of proteins with DNA sequences. The first hypothesizes that the protein interacts with DNA in a rigid double-helical form—groups on the protein forming hydrogen-bond contacts either in the large or small groove, hydrophobic contacts with methyl groups of thymines, stacked complexes with bases, and electrostatic contacts to the phosphates. In this way, a protein can read a DNA sequence from the outside directly. A second class of models suggests that a protein uses contacts with the phosphates of the DNA molecule in order to detect a structural pattern in the DNA, sensing some subtle shift, perhaps in the DNA structure in A·T-rich regions compared to G·C-rich regions (Bram, 1971), or some breathing property of the DNA dictated ultimately by sequence. The third class of models argues that proteins can only recognize DNA molecules if the DNA contains extensive secondary structure or unpaired bases. Although open regions and loops are always unstable with respect to double-stranded DNA, a regulatory protein could stabilize these configurations by binding to the exposed bases. One of these models proposes that a protein recognizes a symmetric sequence, such as that of the operator, by pulling out the sequence into an alternative pairing arrangement of the bases. In the *lac* operator, the sequence from 1–6 could base-pair to the sequence from 16–21 on the same strand, leaving nine bases in the middle unpaired. Similar re-pairing of the other strand leads to "ears" protruding from the DNA molecule. Such models were first proposed by Gierer (1966) and more recently discussed by Sobell (1972). The overall structure of the operator fragments does not support this model: the formation of ears requires disruption of the G·C-rich region, and the ear stems contain mainly weak A·T base pairs.

All three models easily explain the mutation at position 5. This change introduces a methyl group into the large groove, where it could prevent a close contact by the protein. (Alternatively, the hydrogen-bonding opportunities in the small groove are changed.) The creation of a new ApT sequence might change electrostatic interactions, by permitting the binding of a small charged ion in the small groove, as recently suggested by Rosenberg et al. (1973). Or, this replacement of a G·C by an A·T base pair might be seen as increasing the run of A·T residues and thus changing the structure of the helix or the flexibility of this region. In the Gierer model, this change would interfere with the base pairing in the stem of the pulled out lobe, and thus interrupt the structure.

The mutation at position 9 is, of course, the type most desired by the Gierer model, because it is a mutation in the unpaired loops which would be in contact with the protein. This mutation increases the symmetry but weakens the interaction. One might infer that the A·T pair contact at position 13 is already a bad contact, and creating another such contact at position 9 further damages the binding. However, it is often the case that in protein–protein interactions with twofold symmetry, the symmetry is not perfectly obeyed close to the dyad axis. Thus, it is possible that near the axis, groups on the protein must make nonsymmetric contacts with the DNA and cannot make good contacts with the symmetric sequence of the o^c-mutant DNA.

The single base changes in these two o^c's reveal different aspects of the protein-nucleic acid interaction, but the sequences of many more o^c's will be needed to define the specific features which beckon the repressor to the DNA.

Acknowledgments

We thank Mary Archer and Joanna Knobler for technical assistance, Jim Files, Terry Platt, and Ron Ogata for *lac* repressor, and Charlotte Hering for some of the RNA polymerase used in these experiments. This work was supported by USPHS grant GM-09541 from the National Institute of General Medical Sciences. W. G. is an American Cancer Society Professor of Molecular Biology. N. M. has been supported on NIGMS Biophysics training grant TOI GM-782-15.

References

Adler, K., K. Beyreuther, E. Fanning, N. Geisler, B. Gronenborn, A. Klemm, B. Müller-Hill, M. Pfahl, and A. Schmitz. 1972. How *lac* repressor binds to DNA. *Nature* **237:** 322.

Arditti, R. R., J. G. Scaife, and J. R. Beckwith. 1968. The nature of mutants in the *lac* promoter region. *J. Mol. Biol.* **38:** 421.

Barrell, B. G. 1971. Fractionation and sequence analysis

of radioactive nucleotides, p. 751. In *Procedures in nucleic acid research*, vol. 2, ed. G. L. Cantoni and D. R. Davies. Harper & Row, New York.

Beckwith, J., T. Grodzicker, and R. R. Arditti. 1972. Evidence for two sites in the *lac* promoter region. *J. Mol. Biol.* **69:** 155.

Berg, D., K. Barrett, and M. Chamberlin. 1971. Purification of two forms of *Escherichia coli* RNA polymerase and of sigma component, p. 506. In *Methods in enzymology*, vol. 21, part 2, ed. L. Grossman and K. Moldave. Academic Press, New York.

Bourgeois, S. and A. D. Riggs. 1970. The *lac* repressor-operator interaction. IV. Assay and purification of operator DNA. *Biochem. Biophys. Res. Commun.* **38:** 348.

Bram, S. 1971. Secondary structure of DNA depends on base composition. *Nature New Biol.* **232:** 174.

Downey, K. M. and A. G. So. 1970. Studies on the kinetics of ribonucleic acid chain initiation and elongation. *Biochemistry* **9:** 2520.

Downey, K. M., B. S. Jurmark, and A. G. So. 1971. Determination of nucleotide sequences at promoter regions by the use of dinucleotides. *Biochemistry* **10:** 4970.

Emmer, M., B. DeCrombrugghe, I. Pastan, and R. L. Perlman. 1970. Cyclic AMP receptor protein of *Escherichia coli*: Its role in the synthesis of inducible enzymes. *Proc. Nat. Acad. Sci.* **66:** 480.

Eron, L. and R. Block. 1971. Mechanisms of initiation and repression of *in vitro* transcription of the *lac* operon of *Escherichia coli*. *Proc. Nat. Acad. Sci.* **68:** 1828.

Gierer, A. 1966. Model for DNA and protein interactions and the function of the operator. *Nature* **212:** 1480.

Gilbert, W. 1972. The *lac* repressor and the *lac* operator. *Ciba Found. Symp.* **7:** 245.

Gilbert, W. and A. Maxam. 1973. The sequence of the *lac* operator. *Proc. Nat. Acad. Sci.* In press.

Gilbert, W. and B. Müller-Hill. 1966. Isolation of the *lac* repressor. *Proc. Nat. Acad. Sci.* **56:** 1891.

———. 1967. The *lac* operator is DNA. *Proc. Nat. Acad. Sci.* **58:** 2415.

Jacob, F. and J. Monod. 1961. Genetic regularity mechanisms in the synthesis of proteins. *J. Mol. Biol.* **3:** 318.

Jobe, A. and S. Bourgeois. 1972. The natural inducer of the *lac* operon. *J. Mol. Biol.* **69:** 397.

Maizels, N. 1973. The sequence of the lactose mRNA transcribed from the UV5 promoter mutant of *E. coli*. *Proc. Nat. Acad. Sci.* In press.

Makman, R. S. and F. W. Sutherland. 1965. Adenosine 3′,5′-phosphate in *Escherichia coli*. *J. Biol. Chem.* **240:** 1309.

Miller, J. H., K. Ippen, J. G. Scaife, and J. R. Beckwith. 1968. The promoter-operator region of the *lac* operon of *Escherichia coli*. *J. Mol. Biol.* **38:** 413.

Minkley, E. G. and D. Pribnow. 1973. Transcription of the early region of bacteriophage T7: Selective initiation with dinucleotides. *J. Mol. Biol.* **77:** 255.

Perlman, R. L. and I. Pastan. 1968. Regulation of β-galactosidase synthesis in *Escherichia coli* by cAMP. *J. Biol. Chem.* **243:** 5420.

Platt, T., J. G. Files, and K. Weber. 1973. *Lac* repressor: Specific proteolytic destruction of the NH_2-terminal region and loss of the deoxyribonucleic acid binding. *J. Biol. Chem.* **248:** 110.

Ptashne, M. 1967a. Isolation of the λ phage repressor. *Proc. Nat. Acad. Sci.* **57:** 306.

———. 1967b. Specific binding of the λ phage repressor to λ DNA. *Nature* **214:** 232.

Riggs, A. D., H. Suzuki, and S. Bourgeois. 1970. *Lac* repressor-operator interaction. I. Equilibrium studies. *J. Mol. Biol.* **48:** 67.

Robertson, H. D., B. G. Barrell, H. L. Weith, and J. E. Donelson. 1973. Isolation and sequence analysis of a ribosome-protected fragment from bacteriophage ϕX174 DNA. *Nature New Biol.* **241:** 38.

Rosenberg, J. M., N. C. Seeman, J. J. P. King, F. L. Suddath, H. B. Nicholas, and A. Rich. 1973. Double helix at atomic resolution. *Nature* **243:** 150.

Sadler, J. R. and T. F. Smith. 1971. Mapping of the lactose operator. *J. Mol. Biol.* **62:** 139.

Silverstone, A. E., R. R. Arditti, and B. Magasanik. 1970. Catabolite-insensitive revertants of *lac* promoter mutants. *Proc. Nat. Acad. Sci.* **66:** 773.

Smith, T. F. and J. R. Sadler. 1971. The nature of lactose operator constitutive mutations. *J. Mol. Biol.* **59:** 273.

Sobell, H. M. 1972. Molecular mechanism for genetic recombination. *Proc. Nat. Acad. Sci.* **69:** 2483.

Sung, M. T. and G. H. Dixon. 1970. Modification of histones during spermiogenesis in trout: A molecular mechanism for altering histone binding to DNA. *Proc. Nat. Acad. Sci.* **67:** 1616.

Ullmann, A. and J. Monod. 1968. Cyclic AMP as an antagonist of catabolite repression in *E. coli*. *FEBS Letters* **2:** 57.

Weber, K., T. Platt, D. Ganem, and J. H. Miller. 1972. Altered sequences changing the operator-binding properties of the *lac* repressor: Colinearity of the repressor protein with the *i*-gene map. *Proc. Nat. Acad. Sci.* **69:** 3624.

Weiner, A. M., T. Platt, and K. Weber. 1972. Amino-terminal sequence analysis of proteins purified on a nanomole scale by gel electrophoresis. *J. Biol. Chem.* **247:** 3242.

Zabin, I. and A. V. Fowler. 1972. The amino acid sequence of β-galactosidase. III. The sequences of NH_2- and COOH-terminal tryptic peptides. *J. Biol. Chem.* **247:** 5432.

Zubay, G. 1964. Nucleohistone structure and function, p. 95. In *The nucleohistones*, ed. J. Bonner and P. Ts'o. Holden-Day, San Francisco.

Zubay, G., D. Schwartz, and J. Beckwith. 1970. Mechanism of activation of catabolite-sensitive genes: A positive control system. *Proc. Nat. Acad. Sci.* **66:** 104.

Contacts Between the *LAC* Repressor and DNA Revealed by Methylation

Walter Gilbert, Allan Maxam* & Andrei Mirzabekov[+]*

How does the lactose repressor recognize the operator sequence? Does this protein lie in the major or the minor groove of DNA? We have begun to answer these questions by methylating the DNA with dimethylsulfate and blocking that methylation specifically with the lactose repressor.

Dimethylsulfate methylates the purines in DNA. The most reactive site is the N–7 position on guanine, but there is also attack at the N–3 position of adenine in double-stranded DNA (The N–1 of adenine is more sensitive to attack but is involved in base-pairing and thus not exposed in native DNA) (Lawley & Brooks 1963, Mirzabekov & Kolschinsky 1974, Singer 1975, is a recent review). Since the N–7 position is exposed in the major groove of DNA, and the N–3 position is exposed only in the minor groove, the methylation of guanine or adenine will probe the major and minor grooves respectively. This approach was used by Mirzabekov & Melinkova (1974) to study the disposition of histones in chromatin. Can we recognize which bases are methylated in the operator sequence?

METHODS

A typical reaction contained about 0.1 microgram of P^{32}-labelled DNA fragment that had been kinased during the isolation procedure using gamma-labelled ATP at a specific activity greater than 1000 C/mmol. These fragments were mixed with 5 micrograms of unlabeled p*lac* DNA and treated with demithylsulfate (25 m*M*) in 0.400 ml of 50 *mM* sodium cacodylate buffer, pH 8.0, 10 m*M* Mg^{++}, and 0.25 mg/ml BSA for 12 hours at 0°C. The reaction was quenched with 0.200 ml of 1 *M* Tris. MCl, pH 7.4, 4 *M* NH_4Ac, 0.7 *M* mer-

* Department of Biochemistry and Molecular Biology, Harvard University, Cambridge, Massachusetts 02138 U. S. A. [+] Institute of Molecular Biology, Academy of Sciences of the U. S. S. R., Moscow B-312, U. S. S. R.

"Control of Ribosome Synthesis", Alfred Benzon Symposium IX, Munksgaard 1976 captoethanol and 0.15 mg/ml tRNA. After precipitation and washing with ethyl alcohol, the sample was dissolved in 20 microliters of 0.02 *M* potassium phosphate buffer, pH 7.0 and heated to 90°C for 10 minutes. Then 2 microliters of 1 *N* NaOH was added and the sample was heated in a sealed capillary tube for 5 minutes at 90°C. The sample was added to 10 mg of urea; then electrophoresed for 12 hours at 1000 volts on a 40 cm gel (20 % polyacrylamide in 7 *M* urea, 0.05 *M* tris-borate, pH 8.3, and 0.5 m*M* EDTA) using xylene cyanol and brom-phenol blue as markers. The protection experiments with *lac* repressor used a 4 to 20-fold excess of X86 (a tight-binding mutant) repressor.

Table I shows the sequence of a 55/53 base long restriction enzyme fragment of DNA that carries the lactose operator. This fragment lies between cuts made by two different restriction enzymes: the right end is produced by the Alu 1 enzyme from *Arthrobacter luteus,* cutting at the middle of an AGCT sequence; the left end of this fragment is made by the Hap enzyme from *Haemophilus aphrophilus,* making a two base staggered cut in a CCGG sequence. We isolate this fragment in 5 to 10 microgram amounts from *lac* phage DNA by a process involving filter-binding to the lactose repressor, cutting with restriction enzymes and sizing on polyacrylamide gels (Gilbert, Gralla, Maxam to be published). Since this fragment is terminated by different restriction cuts, we can place a P^{32} label at either the right or the left end of this sequence by using the T4 polynucleotide kinase to transfer P^{32}

Table I

The sequence of the 55/53 long Hap-Alu fragment. The boxes show the symmetry regions around l a c operator.

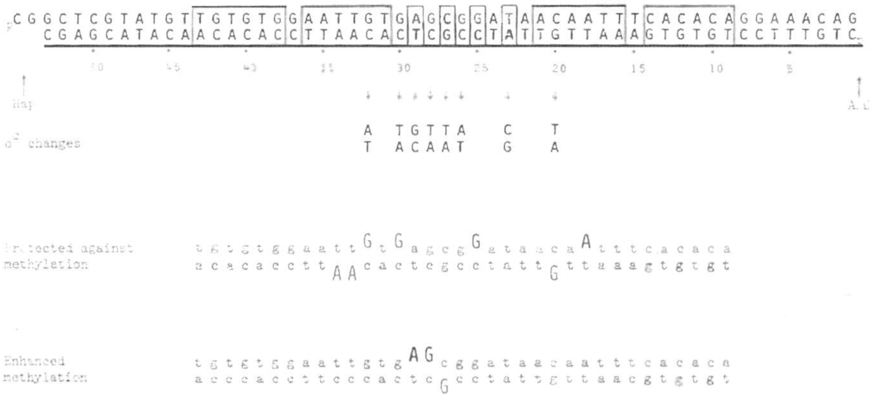

to the 5′ ends of fragments at the penultimate stage of the isolation, before the final restriction cut. We treat a DNA fragment, which has a P^{32} label at the end of one strand, with dimethylsulfate at a level and for times such that about 1 base per strand methylates. The glycosidic bond of the methylated purine is unstable, and the purine will come off the sugar during a heat treatment at neutral pH. Then alkali will break the chain by hydrolyzing the sugar-phosphate linkage, releasing single-stranded, phosphate-ended fragments. The P^{32}-labelled fragment's length is precisely the distance between the restriction cut and the position of the methylated purine. We resolve these fragments by electrophoresis on a 20 % polyacrylamide gel in 7 *M* urea. Fig. 1* shows an autoradiograph of such a gel. Column B displays the pattern obtained by methylating the fragment labelled at the right end with P^{32}. The full-length chain, 53 bases long, is at the top; the shorter fragments are at the bottom, and the very shortest were allowed to run off the gel. In this pattern of light and dark bands, the dark bands correspond to fragments methylated originally at a guanine while the lighter bands correspond to fragments methylated at adenines, since guanine N–7 reacts about 5 times more readily than adenine N–3. (There is a still lighter pattern of broken fragments of unknown origin providing a background). This identification of light and dark bands can be confirmed by observing the time course of the release of the methylated purine bases; guanines release during the heat step at about one-fourth the rate of the adenines. The pattern of bands is distinctive enough to be easily correlated with the sequence. For example, in Column B, reading from the bottom, the first dark band corresponds to the fragment released following the methylation of the guanine at position 14 in Table I. Then, reading up the gel, (right to left on the bottom strand of Table I), there are three adenines which react less strongly, a space, a strong band corresponding to guanine 20, a space, and a faint adenine 23. The next very dark band, which runs just below the xylene cyanol dye, is from guanine 27. Above this there are three bands, adenines 31, 33, and 34. Thus the sequence read up this column might be thought of as:

– – – GAAA . . G . . A . . . G . . . A . AA . . . . A – – – 3′

The repressor blocks the methylation of three purines: guanine 20, adenine

---

* See figure insert opposite p. 144.

33, and adenine 34 (Column A). Furthermore, the band corresponding to guanine 27 is visibly darker; its methylation is enhanced. (If these bands are cut out and counted, the label has increased from 50 to 100 %.) The repressor survives methylation, presumably because the reaction is quite limited, attacking about one of every 10 guanines, and so methylating only a comparable fraction of the reactive groups on the protein.

Fig. 1 also shows (D and C) the methylation pattern on the fragment labelled at the left end. Again, the repressor protects a number of bases: it blocks guanines at positions 32, 30, and 25, and the adenine at position 18. Furthermore, methylation of adenine 29 and guanine 28 is enhanced. The lower section of Table I shows these results.

We conclude that repressor does not wholly block either the major or the minor groove of DNA but makes specific contacts in both grooves. Although most contacts are in the larger groove, near the center of the operator, the repressor touches DNA at points in both grooves, almost opposite one another, out toward the ends of the original DNase-protected fragment (Gilbert & Maxam 1973). We see no contacts in the outer symmetry regions; thus we have no reason to regard them as important. The bases affected in the methylation pattern, whether inhibited or enhanced, correspond to many sites at which we have identified operator constitutive mutations, again suggesting that the major points of contact with the repressor are in the middle of the operator sequence. The methylation pattern is not symmetrical in detail; this reinforces the conclusion that we have drawn earlier from the operator constitutive mutations (Gilbert *et al.* 1975; Gilbert & Maxam, to be published) that the interaction does not fully exploit the symmetry.

How might methylation be enhanced? There are sequence-dependent effects on the methylation. The very darkest bands arise from sequences of the form TGT, in which the guanines are most reactive. In the sequence GGA the reactivity of the middle guanine is suppressed, while in a sequence of the form AAA, the reactivity of the middle adenine is enhanced. These effects of neighboring bases are either steric or represent electronic effects that enhance the reactivity of the ring nitrogens. Thus the enhanced methylation could reflect a slight change in electronic structure of the DNA, a slight opening of the structure to remove a quenching of the reaction, or a hydrophobic pocket created between the protein and DNA that would have some affinity for dimethylsulfate.

Overall, the methylation pattern suggests that the repressor interacts with

DNA in two separate regions, the region in the center being unusually exposed or accessible. This finding of contacts between the DNA and the repressor in the major groove is in distinction to the recent observation by Richmond & Steitz (to be published) that in the binding of *lac* repressor to poly dAT the repressor does not detect blocking groups in the major groove. We think it likely that the binding to dAT represents a reaction more similar to the non-specific binding of the repressor to DNA, which presumably observes only the phosphate backbone and the general shape of the DNA molecule, and conclude that many of the contacts for specific DNA recognition lie in the major groove.

This methylation technique provides a probe for the detailed study of binding sites of proteins on nucleic acid molecules and can be applied to any nucleic fragment whose sequence is known.

ACKNOWLEDGMENTS

We thank Edith Butler for excellent technical assistance. This work was supported by the NIHGMS, grant #GM 09541. W.G. is an American Cancer Society Professor of Molecular Biology.

REFERENCES

Gilbert, W. & Maxam, A. (1973) The nucleotide sequence of the *lac* operator. *Proc. nat. Acad. Sci. (Wash.) 70,* 3581–3584.

Gilbert, W., Gralla, J., Majors, J. & Maxam, A. (1975) Lactose operator sequences and the action of lac repressor. In *Protein-Ligand Interactions,* ed. Sund, H. & G. Blauer, pp. 193–210. Walter de Gruyter & Co, Berlin.

Lawley, P. D. & Brookes, P. (1963) Further studies on the alkylation of nucleic acids and their constitutent nucleotides. *Biochem. J. 89,* 127–138.

Mirzabekov, A. D. & Kolchinsky, A. M. (1974) Localization of some molecules within the grooves of DNA by modification of their complexes with dimethyl sulphate. *Molec. Biol. Rep. 1,* 385–390.

Mirzabekov, A. D. & Melnikova, A. F. (1974) Localization of chromatin proteins within DNA grooves by methylation of chromatin with dimethyl sulphate. *Molec. Biol. Rep. 1,* 379–384.

Singer, B. (1975) The chemical effects of nucleic acid alkylation and their relation to mutagenesis and carcinogenesis. In *Progress in Nucleic Acid Research & Molecular Biology,* vol. 15, ed. Cohn, W., pp. 219–284. Academic Press, New York.

DISCUSSION

MAALØE: Would you say that the 'pinching' by the repressor argues for interaction largely between repressor and a stretch of closed double helix or would you not go that far?

GILBERT: I can not rigorously argue that, although I believe the contacts are to a closed double helix. If the double helix were open, it would not change the methylation pattern, even though the N1 position of the A would also be sensitive. That modified purine is not unstable on DNA, so we could not see that reaction point. It is known from the work of Suzanne Bourgeois and Jim Wang that the DNA is not unwound to a major degree: the greatest unwinding is only 90° across the entire region. That is the amount of unwinding you might get if you intercalated 1 or 2 tyrosine or if the angle of each base changed by a few degrees.

There is a secondary binding site for the *lac* repressor, a pseudooperator with one-tenth the affinity of the operator, located about 400 bases into the *z-gene*. Its sequence is:

C A A C A T T A A A T G T G A G C G A G T A A C A A C C C G T C G G A
G T T G T A A T T T A C A C T C G C T C A T T G T T G G G C A G C C T

with homology: A A T G T G A G C G T A A C A A

compared with
the operator:

T G T G T G G A A T T G T G A G C G G A T A A C A A T T T C A C A C A
A C A C A C C T T A A C A C T C G C C T A T T G T T A A A G T G T G T

The outside symmetry wings do not appear in the second operator, and also are not implicated by the methylation experiment as points of contact.

BAUTZ: Does the structure of the pseudo-operator not argue for closed structure of DNA because that sequence has very little homology that could go into a Gierer-type structure?

GILBERT: That is absolutely right. There is almost no symmetry in the pseudo-operator.

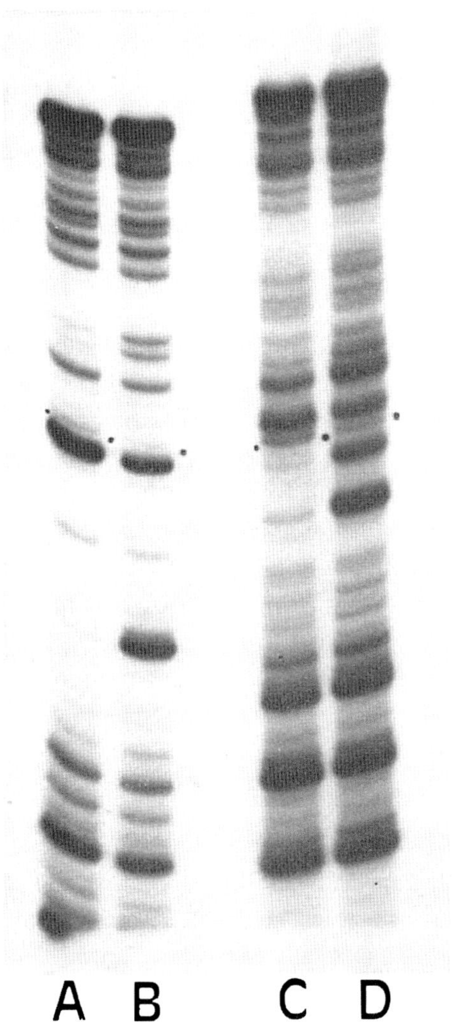

Fig. 1. An autoradiograph of the pattern of DNA fragments after methylation, cleavage, and electrophoresis on a 20 % polyacrylamide 7 *M* urea gel. The xylene cyanol dye (at the dots) marks the position of material 27 bases long. A and B are right-end labelled fragments of the 53 long strand, treated with and without repressor. C and D are left-end labelled fragments from the 55 long strand, with and without repressor.

YANOFSKY: Since the repressor is a multimer and since you have symmetrical sites in the operator, would you not expect the protected sites to also show more symmetry than is apparent?

GILBERT: That is a peculiarity we do not have an answer to. Tom Steitz argues that the repressor multimer has 2–2–2 symmetry. If it has that symmetry then one would expect it to try to interact with DNA on a 2-fold symmetry axis. There is not any complete 2-fold axis in the DNA. The methylation pattern of the first operator and the positions and strengths of 0^c mutants show that the interaction is not symmetric in detail. All we can conclude, however, is that the actual contacts of the protein and DNA are not symmetric. The protein may still be a perfectly symmetrical structure that tries to get up to a non-symmetric object and naturally fails to make symmetrical contacts. The alternative possibility is that the symmetry is a total illusion (the symmetry that we notice in the DNA structure not being structurally significant), and the protein does not in fact have 2–2–2 symmetry but has some other structure and binds to the DNA through contacts between different subunits and different parts of the sequence.

CHAMBERLIN: Does the methylated fragment bind repressor?

GILBERT. Yes. We can methylate the DNA first and then bind and release repressor and thus ask which methylated bases do interfere with the binding. We can also methylate the DNA first, warm it up to knock the bases out, and then take that DNA, which I call a "missing-tooth molecule", and ask where the repressor binds. The repressor *will* bind to the methylated DNA; however, it is released more rapidly from molecules methylated at any of the bases we identify as close contact points between the protein and DNA. The interaction between the protein and the DNA is not so critical that the slightest change in DNA structure totally abolished the binding; the extra methyl group weakens the binding by a factor from 10 to 100. In the missing tooth experiment, with greater defects in the DNA, almost all the bases across the central 20-base region are required.

COLLINS: I was wondering about the interpretation when you talk about enhancement of methylation. I do not see quite what you are normalizing it to. If you have reduced cutting in some of these sites then of course the bands that are left will be there in larger amounts.

GILBERT: We are putting one hit into each fragment, as you can tell by the fact that there is unmodified material left. If we methylate to a greater extent, if we put in several hits, we eliminate the full length material. When I speak of enhancement, we are comparing that to a normalized intensity across all the other bands. Doing the experiment at a one-hit level, the label that is lost at a protected site appears only in the uncut material; the increase of label in a specific band reflects an absolute increase in rate.

COLLINS: If all the bands that you were normalizing were of the same intensities, and if they were protected to the same extent, you could still say that the enhanced band was attacked in the normal way.

GILBERT: We had about the same amount of label in both tubes, therefore, the intensities of the bands represent very closely the absolute amount of methylated material. The relative intensities of most of the bands are unaffected by the presence of repressor.

BAUTZ: Would the methylation enhancement argue that DNA is not an entirely tight double stranded region at that point? Is the reaction rate for DNA much lower than for the free bases?

GILBERT: In the literature, the reaction rates are not very different between native DNA and the free bases. The methylation of guanine with DMS does not change greatly whether free or in DNA. However, at some guanine positions we see very specific sequence effects, e.g. methylation is tremedously enhanced by the presense of T's on both sides, while in the sequence GGA the middle guanine is suppressed. The central A in an AAA sequence is enhanced. If you like complicated models, you could say that we can not even prove that the reason we fail to methylate at the N7 position is that the repressor blocks there, as opposed to the repressor simply pushing electrons around the DNA enough to block methylation another way.

SCHLEIF: The electronmicrographs of Jay Hirsch show that with both RNA polymerase and *lac* repressor about 25 % of the 100 base pair *lac* molecules are slightly hooked under the binding protein. This suggested that there is a weakening there.

GILBERT: In fact the photographs you showed yesterday had a very distinct bend at the position where the repressor sits on the DNA.

MAALØE: The next problem in that region, I guess, concerns the CAP protein. Where would you think it sits?

GILBERT: We now know a little bit about where the CAP factor sits, and we are trying to use these methylation methods for the RNA polymerase, for the CAP protein and everything else we can get our hands on. The DMS method has now been applied by Dennis Kleid to the lambda repressor, which binds in the major groove, blocking only G's. A DMS experiment for the RNA polymerase has not worked yet.

Let me discuss the general properties of promoter regions. We know the sequence across the entire *lac* promoter-operator region from the work of Barnes, Reznikoff, Abelson and Dickson. In terms of distances before the beginning of the messenger RNA, we now identify three regions of interest: a CAP binding site around –60 base pairs, a "contact region" at –35 base pairs, and a RNA polymerase-protected fragment which extends from about 25 bases before to 20 bases after the beginning of the messenger. In this protected region Pribnow identified a region of homology, the "Pribnow box", a seven-base region homologous in the sequence TATAATG centered at –10 base pairs, one turn, before the start of the message.

The –60 cap region is identified by two mutant changes, L8 and L29, both GC to AT changes, which, as John Majors has shown, abolish the binding of the CAP factor. However, the CAP protein, on binding here, might be thought of as covering a region from –70 to –50, a long way from the RNA polymerase-protected region.

The –35 base region lies outside the RNA polymerase protected region. This region contains down promoter mutations, two in *lac* and one in lambda. Furthermore, this region contains Hin enzyme cuts in lambda and in SV40, and these cuts destroy the promoter. For these reasons I like to think of the sequence here as being necessary for the ultimate formation of the initiation complex and refer to this "–35" region as a "contact region".

The up promoter mutants that have been sequenced, mutations that increase the promotion and render the CAP factor unnecessary, lie in the RNAP polymerase protected region. The UV-5 change, the stronegst up promoter mutation, changes a . . GT . . sequence in the wild-type *lac* promoter to a . . AA . . sequence in the mutant, producing the "best" Pribnow box sequence. All the up promoter mutants involve G · C to A · T changes; this supports the notion that the polymerase melts out the DNA in this region, immediately before the start of the mRNA, in forming the initiation complex. The other mutations, those in the CAP site and those in the "contact" site, are not melting mutations, but are sequence changes whose effects can be most easily interpreted as recognition defects, preventing recognition of the outside of the DNA.

Although the mechanism for the CAP factor action is unknown, its requirement can be bypassed by mutation or, *in vitro,* by the presence of dimethylsulfoxide or glycerol (as was first shown for the *gal* operon by Max Gottesman and his coworkers). A restriction fragment that is cut in the CAP site, but contains all the wild-type sequence between –60 base pairs and +35 base pairs will not be transcribed. Mutant fragments will transcribe, as will the wild-type fragment if the polymerase is tricked with dimethylsulfoxide or glycerol.

There are three models for these interactions: one is a gross melting picture in which one assumes that the CAP factor melts the DNA all across the region from −60 base pairs up to the RNA start and the polymerase touches down onto this altered DNA and binds. A second picture is the wandering polymerase: the enzyme molecule touches CAP, makes contact with the −35 region and then slides along the DNA to the initiation point. A third model would suggest that the proteins are all large enough to touch each other as well as the DNA. The CAP protein, bound to DNA, makes contact with part of the polymerase, the sigma subunit say, which in turn touches the −35 base "contact" region while the body of the polymerase covers and melts the DNA at the initiation site. This third model invokes weak protein-protein contacts to explain the CAP stimulation of transcription. My prejudice is for this simpler model, but there is not yet any firm evidence. We have not been able to identify any larger protected fragment – or any protected fragment corresponding to a weak binding of the polymerase. So far the only DNase resistant fragments that have been seen correspond to the final initiation complex, even though others have been looked for. For example, CAP and polymerase on the wild-type promoter still produce the same protected fragment as the polymerase alone does on the UV5 promoter.

Section III:
The Rolling Circle Model

DNA Replication: The Rolling Circle Model

Walter Gilbert[1] and David Dressler[2]
Departments of Physics,[1] Biochemistry and Molecular Biology,[1] and Biology,[1,2]
Harvard University, Cambridge, Massachusetts

Why do DNA circles occur? The rolling circle model for DNA replication suggests that DNA must be circular in order to be copied completely; the basic mode of reproducing an entire genome is to copy it from a circular template, using the circularity in an intrinsic way to guarantee that all of the genetic information is preserved. The guarantee is enforced by copying always more than one full genome's worth: copying the circle plus a bit. The model uses an asymmetric mode of replication and thus can employ the *E. coli* DNA polymerase or analogous enzymes. The synthesis begins by opening one strand of the original circle at a specific point. We imagine that the positive strand is opened, that the newly exposed 5′ end is attached to the 'membrane,' and that a new copy of this strand is synthesized by chain elongation of the 3′ end of the old positive strand, using the negative strand, which remains always closed, as a template. The old positive strand is peeled off as a single strand, but a new negative strand, as is required by the 5′ to 3′ growth of DNA, is synthesized upon it in short pieces that are ultimately tied together by the ligase. This mode of synthesis is shown in Fig. 1. The synthesis is continuous, one daughter molecule being peeled off endlessly as the growing point continues around the circle. The long strand, or tail, which could in principle contain several genome's worth of information, can be used to construct daughter circular molecules by any recombination process between like sequences a genome's length apart. This could be either general recombination or some site specific recombination: specific endonucleases introduce staggered nicks producing overlaps like the sticky ends of phage λ, cutting out genomes that can circularize. This replication leads from circles to circles, but the asymmetry resolves the swivel problem (Cairns, 1963), because one parental strand is always open.

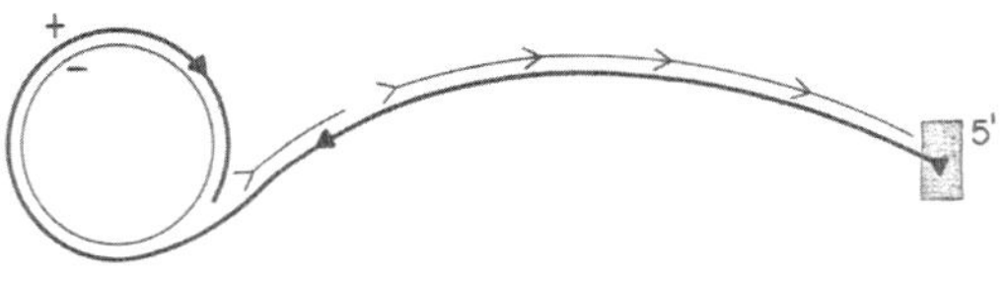

Figure 1. *The Rolling Circle.* The closed inner circle represents the negative strand serving as a template for the elongation of the positive strand. The arrow tails represent the 5′ ends of the strands; the direction of the arrows indicates the direction of chain growth. The positive strand (outer heavy line) is growing by the addition of nucleotides to its 3′ hydroxyl end. The 5′ end of the positive strand is attached to a site (stippled rectangle), and just over a genome's length has been displaced from the circle. On the positive strand, synthesized in short pieces, a new copy of the negative strand grows. The closed triangles represent the nickase sensitive sequences on the positive strand.

The specific transfer of the 5′ end of the positive strand to some site, presumably a membrane site, is required to prevent the ligase from repairing the original nick. At any interruption in one strand of a double-stranded region, the free 3′ end would be subject either to attack by exo III (exonuclease III, [DNA phosphatase-exonuclease], Richardson, Schildkraut, and Kornberg, 1963) or to chain addition by the *E. coli* DNA polymerase. These two enzymes should nibble away and regrow the 3′ end until the ligase acts, repairing the continuity of the strand. To permit net *new* synthesis, by end addition, we propose that there will be a specific transferase that recognizes the specific 5′ end released by the positive strand nickase, and ties the 5′ phosphate to some object in the cell, for instance to a site on the cell membrane. It is this act that distinguishes the main process of replication from the repair process, for once this has happened the positive strand can grow past the nick by chain elongation, the DNA polymerase displacing the old positive strand. (One might assume that the displaced single-strand material sticks to hydrophobic regions of the membrane, thus not interfering with the action of the polymerase at the growing point.)

The synthesis could be imagined as driven by a torque exerted around the circumference of the circle. This would make the closure of the circle a prerequisite for DNA synthesis, and any break would interrupt that synthesis (Cairns and Davern, 1966; Sly, Eisen, and Siminovitch, 1968). We prefer to imagine the synthesis being driven by the growth of the positive strand, displacing the old strand. This mode requires the integrity of the strands ahead of, but not behind, the growing point. If the enzyme runs into a nick in the positive strand, the tail is displaced entirely and the strands seal off, leaving a supercoil. If the enzyme hits a nick in the

Cold Spring Harb Symp Quant Biol 1968 33: 473–484

negative strand, it falls off, leaving an extra long rod (which can recircularize). In both cases, the main synthesis can only re-initiate at the origin. Thus for the main synthesis to continue, the repair processes must be very active, with the equilibria far enough over so that all breaks are repaired before the growing point arrives. Any sequence of events that interferes with the repair mechanisms, by upsetting the balance between the ligase action and the nicking and excision processes, will stop the main synthesis, while allowing repair synthesis to continue. This is an alternative explanation of the Cairns and Davern (1966) result. The observation of Howard-Flanders, Rupp, Wilkins, and Cole (this volume) that the replication machinery can jump over thymine dimers and continue synthesis to produce a discontinuous strand suggests that nicks behind the growing point do not matter.

SPECIFIC PHAGES: ΦX-174

This model can be most explicitly presented as a description of ΦX-174 replication. After infection with the single-stranded circle, the cell goes through a period of double-stranded DNA synthesis, followed by the synthesis of single-stranded circles. The rolling circle model, with one further *ad hoc* assumption, yields a detailed description of this process.

The infecting single-stranded circle is seized by the DNA polymerase which initiates the synthesis of the negative strand. The polymerase synthesizes a complete complementary copy of the positive strand, and the ligase promptly closes the circle, resulting in a covalently sealed supercoil (Fig. 2). This process has been demonstrated in vitro (Goulian and Kornberg, 1967), and is known to involve only pre-existing host enzymes (Tessman, 1966).

We do not wish to conjecture too far about the initiation of the synthesis of the negative strand. The enzyme probably uses an oligonucleotide fragment as a primer (see Goulian, this volume). All we shall assume is that there is at least *one* sequence on the positive strand of ΦX at which negative strand synthesis can be initiated. For later convenience, we assume that this initiation sequence is about 10 or 20 bases away from the point at which the positive strand nickase will function, in the 3′ direction from that point.

The phage functions which are necessary for replication are then

(a) a positive strand nickase, that acts at a specific, unique sequence on the molecule, nicking the positive strand *only* if the region is double-stranded; and

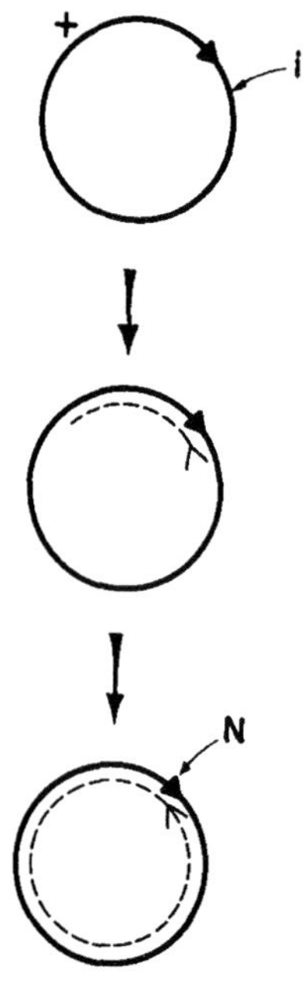

FIGURE 2. *The Conversion of the Infecting ΦX Ring to a Double-stranded Form:* The point on the positive strand at which the synthesis of the negative strand will start is shown as "i". The nickase sensitive sequence (once in a double-stranded region) is shown as —▲—. A polymerase begins the negative strand, probably using an oligonucleotide primer complementary to the sequence at "i", and grows the strand around in the 5′ to 3′ direction. The negative circle is ultimately closed by the ligase. The double-stranded circle is opened by the nickase (N).

(b) a specific transferase that transfers the 5′ end of the positive strand to a site on the membrane. In starved cells, one would presume there are very few such sites available (Yarus and Sinsheimer, 1967) (and if these sites are also used for *coli* synthesis, their occupation by the phage will block the next round of host synthesis).

The DNA polymerase now elongates the parental positive strand. After the growing point has proceeded once around the circle, it next displaces a new copy of the negative strand initiation point: the displaced positive strand now reveals a starting region, identical to that on the original circle, and the DNA polymerase can initiate a new negative strand.

After the negative strand synthesis has proceeded far enough, the region in which the nickase will act becomes double-stranded, the nickase acts, and the overlapping strands are not sufficient to hold the circle to the parental strand (Fig. 3). The freed parental form will circularize, because the overlapping

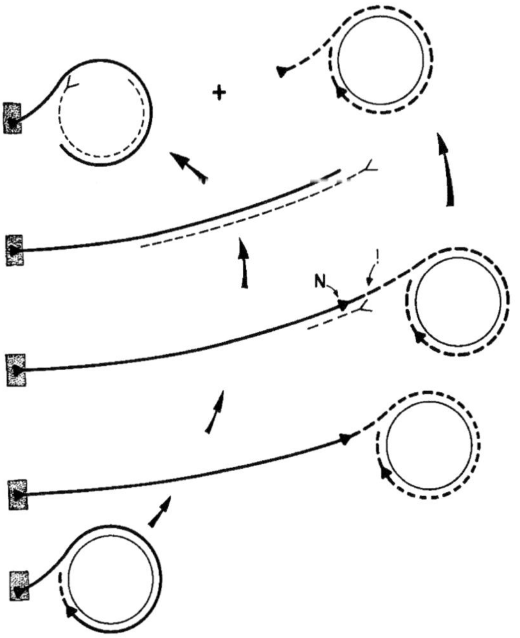

FIGURE 3. *The ΦX Replicating Intermediates.* After the double-stranded circle is opened, the 5′ end of the positive strand is transferred to a site on the membrane, and new material (– – –) is added to the 3′ end of the old positive strand. After the parental strand has been completely peeled off, the new copy of the site "i" is displaced, and a new negative strand is initiated. The nickase acts again, and cuts the replicating structure free from the parental strand left attached to the membrane site. The parental form can now circularize using the complementary sequences on the 5′ end of the negative strand and the 5′ end of the positive strand. The negative strand will continue to grow, finally to be closed.

5′ end of the negative strand provides the redundant information to recircularize. As shown in Fig. 3, this process results in a progeny replicating structure, which is still synthesizing, and a parental replicating structure, attached to the membrane site, always capable of continued synthesis. After the next round of replication (Fig. 4), the progeny structure throws off a molecule that can circularize into a supercoil.

Whether single- or multiple-length circles are formed depends on the ratio of the rate at which the nickase acts to the rate at which the polymerase runs around the circle. A few per cent of multiple circles occur after ΦX infection (Rush and Warner, 1967, and also this volume). This model would ascribe these to replication, not to recombination.

The action of the nickase and the ligase must be balanced in rates so that the DNA is almost always closed at the origin. Otherwise, when the polymerase returns it will displace a unit length that cannot circularize. On the tail, the shortness of the overlap region will ensure that the cut ends will drift away from each other once the nickase acts, before the ligase repairs.

The order of events in the circularization is not critical. Once a structure of the form shown in Fig. 3 has been made, it will eventually circularize. The polymerase can fill out both ends, to make a double stranded rod, but exo III will chew back the 3′ ends leaving the 5′ ends to provide overlaps. We assume that the 5′ ends are sacrosanct; the 3′ ends readily subject to digestion and regrowth. The distance between the 5′ ends is more than one genome (just as in λ): there is always enough information to circularize. The reversible growth and nibbling of the 3′ ends continues until the irreversible act, the action of the ligase, intervenes.

This mode of synthesis throws off progeny replicating structures which in turn throw off supercoils. To a first approximation the supercoils are dead: they are not capable of replication unless acted upon by the nickase and the transferase and unless there are enough sites on the membrane. The progeny replicating structures can die as synthetic units by losing their open regions and becoming supercoiled. Only those structures that are attached to the membrane sites are guaranteed to keep functioning.

In the case of ΦX, if the deposition of a positive strand on the membrane is irreversible, and if there are very few sites on the membrane, then one expects the parental strand not to appear in the burst (Sinsheimer, 1961; Kozinski, 1961); the parental RF to remain attached to the membrane, although progeny RF are free (Knippers and Sinsheimer, 1968); and that ^{32}P decays in the parental strand would be uniquely efficacious in

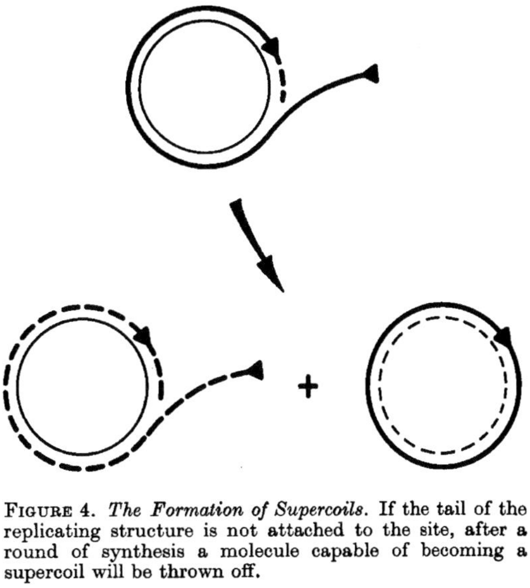

FIGURE 4. *The Formation of Supercoils.* If the tail of the replicating structure is not attached to the site, after a round of synthesis a molecule capable of becoming a supercoil will be thrown off.

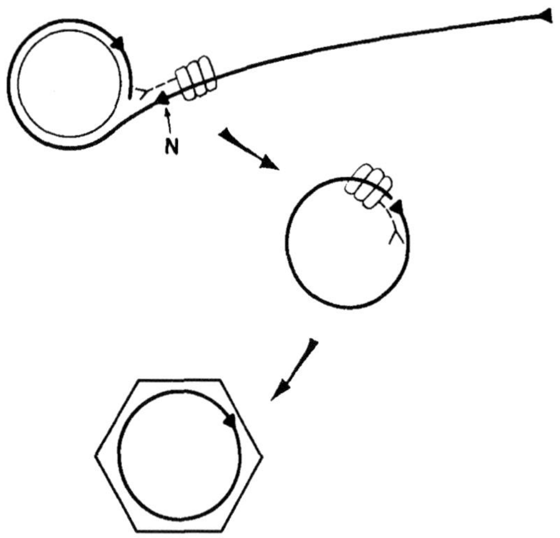

Figure 5. *The Formation of Single-stranded Circles.* The coat protein is imagined to block the growth but not initiation of the negative strand. The nickase (N) acts and cuts off a unit length strand with the 'sticky end' provided by the fragment. The ligase seals the circle, as the circle wraps up in a phage coat.

stopping phage production (Denhardt and Sinsheimer, 1965). To the extent that any functioning intermediate will ultimately throw off many single-stranded circles, the appearance of mutant phage should obey a stamping-machine distribution of clone sizes (Denhardt and Silver, 1966).

For other phages, the thought either that there are many sites available for the phage-specific transferase, or that the enzyme may exchange molecules on the site suggests the more usually accepted kinetics, i.e., no loss of a unique parental strand, and a large enough set of multiplying centers to mimic the exponential clonal distribution of progeny.

Single-strand synthesis. These same replicating structures will turn out single-stranded positive DNA, if the synthesis of the negative strand is blocked. We assume that the coat proteins, building up in amount later in the infection, finally begin to bind to the single-stranded material as it peels off the circle. The central assumption is that this complexing prevents the growth, but not the initiation, of negative strands. If a small piece of negative strand DNA is synthesized, the positive strand nickase and ligase then yield a single-stranded circle, the information for circularizing coming from the negative strand fragment. This is pictured in Fig. 5. Should the short negative strand remain in the phage particle it could, during the next cycle of phage growth, prime the formation of the first double-stranded circle. The invoking of coat protein for the purpose of blocking negative strand synthesis is done to make the model fit the observation that coat mutants do not make single-stranded DNA (Dowell and Sinsheimer, 1966) and, moreover, to place single-stranded and double-stranded DNA synthesis under one unified mechanism.

The unique predictions of this model are that sufficiently short pulse label going into positive strands should always go into open strands whose length is *greater* than the original input strand.

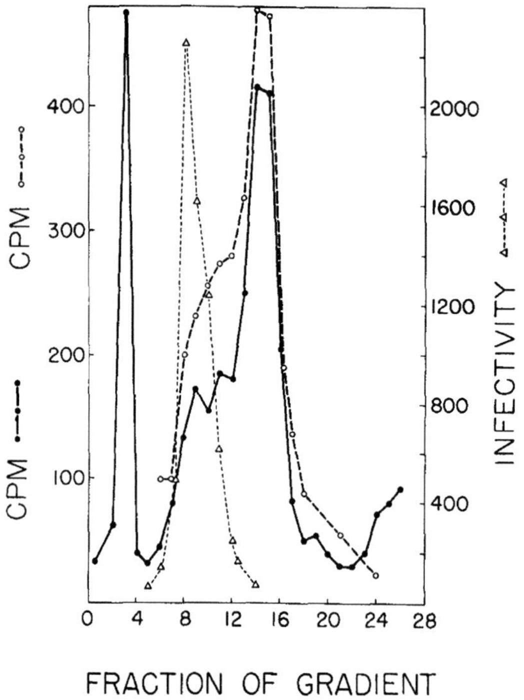

Figure 6. ΦX *Replicating Intermediates, Pulse Labeled for 3 minutes Late in Infection.* Cells were treated with mitomycin to block host DNA synthesis, and then infected with a lysis-defective mutant of ΦX (*am*-3). Tritiated thymidine was present between the 45th and 48th minutes at 29°C, during the period when single-stranded DNA rings are being synthesized. The cells were lysed by an EDTA-lysozyme, sarkosyl, and pronase procedure, and the unfractionated cell lysate was sedimented, preparatively, on a D_2O-H_2O gradient in 1 M NaCl and 0.05 M Tris, pH 8.0 for 4 hr at 8°C and 50,000 rpm. Part of each fraction was counted (solid line) showing that the radioactivity moves *from* 16 S (the velocity of double-stranded circles with at least one nick) to about 27 S (the velocity of compact single-stranded rings, the position of which is marked [– – – △ – – –] by the infectivity to spheroplasts of cosedimented $\Phi X\rho^-$ DNA circles). The recovery of label is high, 89%. Shorter pulses give the same gradient profile, except that in pulses nominally shorter than 1 min, there is no label in mature phage particles which, here, have sedimented onto a CsCl cushion.

That all of the labeled DNA present in the 16–27 S range is in ΦX positive strands (and not in *E. coli* DNA fragments) is shown by hybridization. A constant percentage, 90%, of the label hybridizes to filters containing the separated strands from purified ΦX supercoils (the counts hybridized are shown - - ○ - - ○ - -). No annealing was observed to filters containing only positive phage strands.

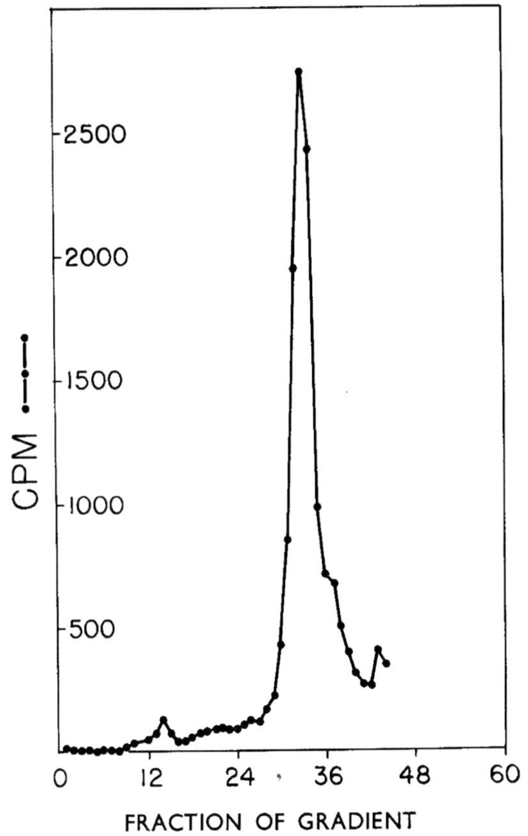

FIGURE 7. *The 3 minute Pulse in Alkali.* A part of the lysate was denatured and run on an alkaline sucrose gradient in 0.2 M NaOH, 0.8 M NaCl for 4 hr at 36,000 rpm, 8°C, to display supercoils which would sediment at 54 S (around tube 14). The recovery of label is close to 100%.

Pulse label going into negative strands should go into strands whose length is *less* than the mature strand.

A critical test of this model has been performed. During the period of synthesis of single-stranded circles, labeled thymidine enters positive strand material in ΦX replicative forms (Lindqvist and Sinsheimer, 1968; Dressler and Denhardt, 1968; Knippers, Komano, and Sinsheimer, 1968; Komano, Knippers, and Sinsheimer, 1968) and ΦX replicating intermediates (Dressler and Denhardt, 1968). The label is displaced from these structures to give, ultimately, single-stranded circles. The total pulse label of the lysate can be displayed as sedimenting, in high salt, in a distribution between the 16 S position of nicked double-stranded circles and the 27 S position of compact single-stranded circles. Such a pattern is shown in Fig. 6: a preparative gradient of ΦX intracellular DNA forms labeled for three minutes. The lysis procedure does not disrupt any phage particles, labeled during this longish time, which sediment to a cushion at the bottom of the gradient. All of the label is in ΦX-specific DNA, shown by hybridization. Although the forward shoulder of label is running in the 21 S position of supercoiled DNA, when the entire pulse is denatured in alkali, very little supercoil appears (Fig. 7). Individual tubes of the preparative gradient of Fig. 6 were taken, the DNA denatured in alkali, reneutralized, and sedimented on neutral gradients in high salt. The denatured strands were sedimented in these neutral conditions so that circles (marked by the infectivity of ΦX rings to spheroplasts) and rods of unit length would sediment together. The sedimentation patterns of denatured label from tubes 9, 11, and 15 of the preparative gradient (Fig. 6) are shown in panels A, B, and C of Fig. 8. The denatured labeled strands isolated from the gradient at tube 15 sediment as unit-length strands, but the denatured label from the heavier material, tubes 9 and 11 of Fig. 6, sediments ahead of the single-stranded circles. The front of the labeled peak in Fig. 8A runs as would strands of twice the unit length. That this label after denaturation is in extra-long open strands has been further confirmed by examining the material on alkaline gradients on which label again moves as would open strands longer than unit length.

Thus the label has been added to a strand that is greater in contour length than that found in the phage. We conclude that during the period of single-strand synthesis the DNA strands are not synthesized as short pieces, growing toward the full length, but by end addition as suggested by the rolling circle model.

When purified intermediate is denatured and re-run in a high salt, neutral gradient *without* the addition of marker positive strand circles, a peak of infectivity nonetheless appears at the position of unit length genomes. We believe this infectivity is due to a circular negative strand template in the intermediate, released during denaturation.

PHAGE λ

A scenario for λ would envision the infecting phage first as circularizing, using the host ligase, and then synthesizing a specific nickase and a transferase to initiate phage DNA synthesis. Genes *O* and *P* are candidates for these functions. The transfer of λ to a membrane site cannot be an irreversible event, because a unique strand is not lost from the burst (Fox and Meselson, 1963). The membrane site must be used many times, and many replicative structures can be maintained.

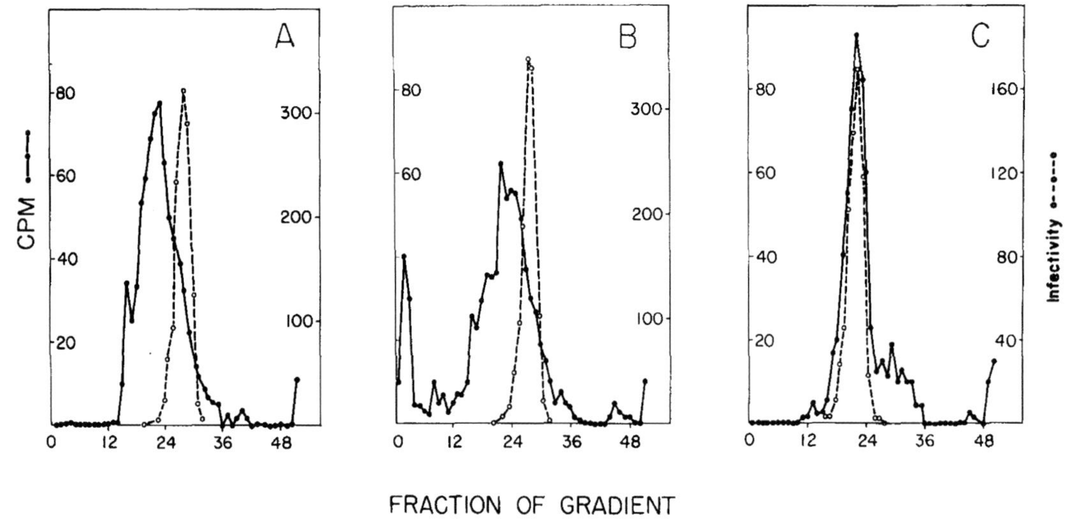

FIGURE 8. *An Analysis of the ΦX replicating intermediates.* Frames A, B, and C show the profiles obtained from tubes 9, 11, and 15 of the gradient shown in Fig. 6. To samples from each tube, marker circles of $\Phi X\rho$-DNA were added. Each sample was made 0.5% in SDS, and the labeled DNA denatured by 0.2 M NaOH, reneutralized quickly, and sedimented on a 1 M NaCl, 1 mg/ml Sarkosyl, 0.05 M Tris, pH 8.0, 10–30% sucrose gradient for 4 hr (5 hr for C) at 50,000 rpm. Sedimentation is from right to left. The radioactivity is shown by the solid line, the infectivity of the marker circles is shown dashed. The recovery was 60%. (For further details see Dressler and Gilbert, 1969.)

One can guess that the strand that is nicked, for λ, would be the 'Crick' strand (the poly U,G-binding strand, Hradecna and Szybalski, 1967), because the induction of the prophage produces some escape of the *gal* genes from repression (Buttin, 1963; Echols et al., 1967). We attribute this escape to a multiplication of the *gal* genes. The initiation of DNA synthesis can occur before prophage excision, and would then lead to the multiple copying of genes to one side of the prophage attachment site. If this is to be the *gal* side, then the 'Crick' strand is the positive strand (Fig. 9) and the replication would be from right to left on the usual phage map. (This last is the case; see LePecq and Baldwin, this volume.)

If the initiation sequence for the synthesis of the negative strand occurs once in ΦX (once in 5500 nucleotides), on a random basis it would occur ten times along λ (or a thousand times along *E. coli*). Thus one expects to see the negative strand synthesis going through 'small' pieces, of an average length of about a million. These may be among the fragments observed by Okazaki and his co-workers (Sakabe and Okazaki, 1966; Okazaki et al., 1968, and this volume). However, for the simple enzymology we are considering, the only fragments we expect are those released in alkali, corresponding initially to exactly *half* the pulse label and complementary to a unique strand. We interpret the claim of Okazaki et al. that most of the pulse label can be recovered as fragments in neutral conditions as revealing either a unique frangibility of the replication point (and thus an artifact of isolation) or, possibly, a different enzymology for the actual growth of the chain.

Where might the positive strand nickase of λ act? Although, if it were to act at a sticky end, one enzyme could be spared in the final formation of the ends, it is more likely to cut somewhere in the $x - y - C_{II} - O - P$ operon. Replication is under the direct control of the phage repressor (Thomas and Bertani, 1964); thus the origin should be either in the immunity region or in one of the two operons extending from that region (see Ptashne and Hopkins, 1968). Replication inhibition by the

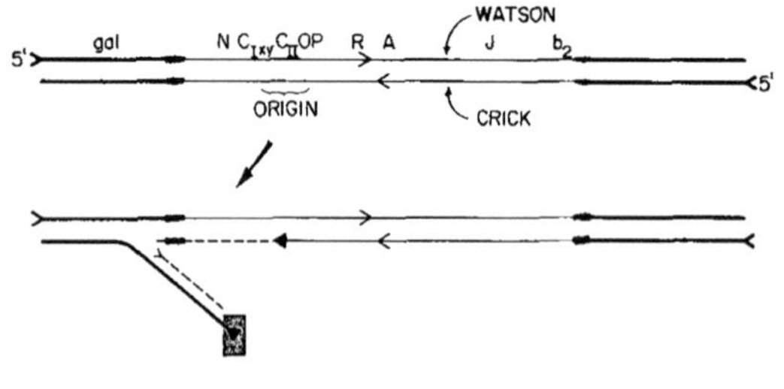

FIGURE 9. *A Prophage Map of λ.* The orientation of the strands is shown (Wu and Kaiser, 1967). A possible position of the origin for DNA synthesis is marked. The 5′ ends of the bacterial chain as well as the 5′ terminus of the sticky ends (joined between *R* and *A*) are shown as arrow tails. DNA replication beginning before the prophage is excised is shown proceeding toward the *gal* region.

repressor is relieved in cis by changes at the operator in x (Ptashne and Hopkins, 1968), the ri^c (replication constitutive) site, located between C_{II} and O (Dove, unpubl.), or by changes in y (in c_{17}, a mutation in y that is also constitutive for C_{II}, O, and P; Packman and Sly, 1968; Pereira da Silva and Jacob, 1968). Either these last changes introduce new origins, and the normal initiation takes place at the repressor binding site in x (Ptashne and Hopkins, 1968), or the origin is downstream from the operator, and one might argue that when the operon is being read into messenger, the DNA is open enough for replication to initiate. The origin of DNA replication for λ has been observed by physical experiments to be on the right arm of the phage map (Makover, 1968, and this volume; LePecq and Baldwin, this volume).

The replication can throw off a long concatomer. Recombination would make circles, as could the timing described earlier for ΦX. Whether circles or concatomers build up depends on the rates at which the recombination processes work relative to the rate of synthesis around the circle. The model views the long strands observed by Smith and Skalka (1966) and Salzman and Weissbach (1967) as the *primary* product of replication.

The maturation process involves, then, putting in the staggered single-stranded nicks to generate the sticky ends and to cut out phage length pieces either from a concatomer or from a supercoil. Since a unit circle can be matured, we see no reason to expect the unique loss of the parental negative strand circle. Normally the ligase would act to seal immediately the new nicks. The ligase activity must be turned off, at maturation, or the DNA must be sequestered from the ligase, presumably by being encapsulated.

We picture T3 and T7, which have unique but repetitious double-stranded ends (Ritchie, Thomas, MacHattie, and Wensink, 1967), as behaving just like λ, but circularizing through the action of exo III. The rolling circle then throws off concatomers. The maturation would again involve specific nicks—to generate the analog of the sticky ends—and also a hyperactive polymerase to fill out the ends and generate the double-stranded termini (Thomas, 1967). Again, the rolling circle *itself*, containing the parental negative strand template, can be matured.

In our view, the model is supported by the existence of concatomers (Smith and Skalka, 1966; Salzman and Weissbach, 1967; Thomas, Kelly and Rhoades, this volume; Weissbach et al., this volume). The experiments of LePecq and Baldwin (this volume) are strikingly in agreement with the rolling circle model. However, the pictures obtained by Tomizawa and Ogawa (this volume) are not what one might have hoped for. The forms that they observe may be the result of a single-strand exchange between the 5′ end of the tail and the positive strand at the origin. (The λ exonuclease [Little, 1967] can chew up the positive strand, once the origin has been nicked, in the direction opposite to that of exo III, to reveal a sequence complementary to the 5′ end of the tail.) Such a partial recombination will produce the Cairns structures (Cairns, 1963; Bode and Morowitz, 1967). Alternatively, some link, unresolved in the Kleinschmidt procedure, might hold the tail near the origin.

Jacob has also elaborated a rolling circle model. The evidence, based on λ, is discussed elsewhere in this volume (Eisen, Pereira da Silva, and Jacob).

PHAGE T4

Although at first sight it is very appealing to think of T4 as circularizing and then peeling off a long strand, to be cut up into headfuls of DNA (Streisinger et al., 1964), circularization, using exo III, is unlikely to be the first act. The Doermann and Boehner (1963) experiment showed that in the progeny from a single infection with a multiply marked heterozygote there is a gradient in the recovery of the terminal heterozygous markers: those markers closest to the ends are preferentially lost. If the entire phage DNA is turned into a circle before any DNA replication, then even if the heterozygous region is maintained, the physical distinction between markers close to or far from the ends cannot be made. Furthermore, the overlapping region is long (ca. 5000 bases; MacHattie, Ritchie, Thomas, and Richardson, 1967) and would not be expected to be easily peeled back. One might invoke a reciprocal recombination to create a circle and to leave a fragment. Then the fragment must persist in the cell and recombine actively with the progeny molecules to create the gradient of markers.

The model, however, demands only that a redundant template, such as a closed circle, be ultimately read. If DNA synthesis begins before the circularization, then the model would predict that synthesis is initiated at a specific sequence that would occur internally on the permuted rods in the phage population. The first mode of replication would build up a series of copies of one end of the molecule. Eventually, recombination will create a circle, and then the circle will throw off daughters.

Since the clonal distributions suggest many replicating centers, we expect a rapid re-use of membrane sites. If the nickase and transferase act more rapidly than the growing center moves around the molecule, there will be many initiations

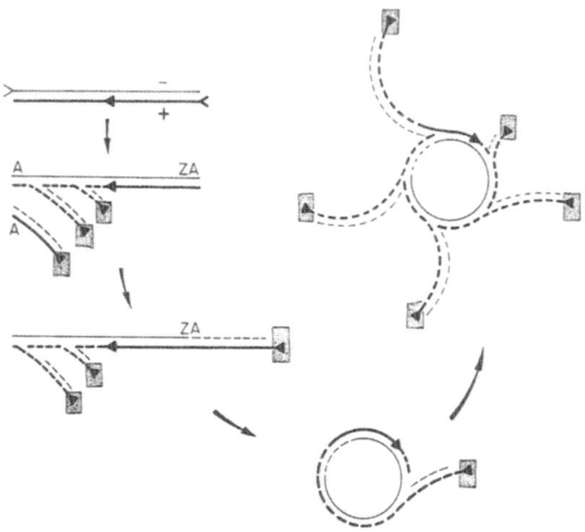

FIGURE 10. *A Life Cycle for T4.* The infecting phage rod is nicked at the nickase sensitive sequence (—▶—), the newly opened 5′ end transferred to the membrane (stippled rectangle), and synthesis begun. New DNA is indicated by a dashed line to distinguish it from the parental solid line. After a sufficient number of copies of the end build up, a recombination will finally occur between the repetitious ends, (between ZA and A) to make a longer molecule. Now there is more homology, and recombination will quickly make a circle. Further replication, which earlier ran off the end, will now make long strands; repeated initiations will lead to the pinwheel (and still more complicated forms).

on a single circle and the growing structures will be the 'pinwheel' forms shown in Fig. 10. Werner's (1968, and this volume) experiments correspond, then, to sixty growing points distributed on ten circles. As the arms grow ever longer, late in infection they will be encapsulated and trimmed to headfuls.

This description corresponds to the replication involving the dispersion of parental strands (Roller, 1961, 1964; Kozinski, 1961); producing shorter than full-length pieces very early in infection (the model predicts a distribution of single-stranded lengths over the infecting phage population, with average length one-half) (short, single strands have been seen after denaturation by Frankel, 1968); followed by a shift to a population of larger material, neither wholly rods nor circles, with a variety of single-strand lengths (Frankel, 1966, 1968, and this volume).

The negative strand of the parental DNA cannot mature directly, because the *unit* length circle cannot mature by the headful mechanism. Recombination, however, will break up all the parental material and permit material from both strands to appear in the progeny.

Recombination between the pinwheels will create a tangle (see Huberman's pictures, this volume). Such tangles will not interfere either with replication or with maturation. The replication only requires that growing points are running along *redundant* templates (a circle is only a trivial way of forming such a redundant template). The maturation process, in general, simply involves a condensation: as those DNA molecules that can fit (roughly) into heads condense, they will be drawn out of the tangled complex. The same condensation principle may be illustrated by the problem: how do the ends of a single λ molecule find each other (rather than tie onto the ends of other molecules in the cell)? The DNA must condense (charge neutalization by polyamines, for example) so that the ends of the same molecule are kept within several hundred angstroms of each other, rather than being totally free to wander throughout the cell volume.

PHAGE P22

P22 combines the properties of λ with those of the T-even phages in a life cycle easily described by the rolling circle hypothesis. This phage has a permuted sequence and double-stranded redundant ends (Rhoades, MacHattie, and Thomas, 1968). However, it has a lysogenic mode, which implies that it should form supercoiled circles in order to integrate into the host chromosome. Here an initial formation of circles (after infection or excision) is followed by the formation of concatomers through the rolling circle process. Maturation cuts up the long DNA strands into headfuls, generating the permuted sequences. (P22 is also a generalized transducing phage: the maturation process cuts out and wraps up headfuls of host DNA rather than viral DNA [see Ikeda and Tomizawa, 1965].)

Botstein's experiments (Botstein, 1968; Botstein and Levine, this volume) provide a clear suggestion that the replicating phage DNA first undergoes a membrane attachment and then throws off long strands which are subsequently cut up to phage length.

E. COLI

Mating. The rolling circle model makes an immediate prediction about bacterial mating: the strand that is peeled off the circle should be the one transferred to the female. This strand would enter with its 5′ end first; and DNA synthesis upon it, in the female, would be in the fragment-ligase mode. (One would have expected the opposite result: the 3′ end to enter first, so that a continuous polymerization on it would drive the transfer.) This prediction has been verified. Rupp and Ihler, and Ohki and Tomizawa (this volume) have shown, using zygotic induction, that a unique old strand of DNA enters, during mating, and that this strand enters 5′ end first.

The transfer process could be driven by diffusion. Once one end of the DNA strand has been tied to some structure in the female, the condensation of DNA will serve to move the strand across. Possibly the single-stranded region, in the male and between the cells, sticks hydrophobically to the membrane, and it is the synthesis of double-stranded material, hydrophilic material, that makes the female a 'sink' for DNA. Some synthesis might even occur on both sides of the bridge. Because it is not the direction of growth of the new strand itself that is used to drive the transfer, the process would not care if a small amount of double-stranded material is synthesized in the male, before moving across and being completed in the female.

Cyclic replication and segregation. The general model is a prescription for continuous replication. In order to have a resting phase, some provision must be made to stop the synthesis after having copied more than a genome. Some suitably timed event could be invoked. The rolling circle principle, however, can be specialized to produce a very simple model for cyclic replication and segregation, shown in Fig. 11. This special model economically uses only a single control. We assume that the resting state consists of a rolling circle: open, tied to the membrane, but with further synthesis blocked by a 'repressor' of DNA synthesis. This repressor is some element which, on binding to a single-stranded region on the negative strand, prevents the progress of the polymerase. In this resting state, when a new membrane site appears, the nickase and the transferase (which are always present) put a strand end upon it. However, only abortive chain elongation occurs, from the origin to the repressor, displacing a short piece of DNA attached to the new site. Only when the repressor has been removed, does the major synthesis begin. The repressor becomes active again but has no single-stranded target available. The next membrane site to appear receives a newly cut 5′ end of the elongating positive strand. Since the peeled-back positive strand exposes the site for repressor, the new synthesis from the origin continues only up to the repressor site. Meanwhile, the main growing point runs all the way around the circle to the repressor, displacing a more than complete copy of the chromosome. At the end there is a circular molecule, in a repressed state, on the new membrane site, and a linear molecule on the old membrane site. The old molecule can recircularize and, with the aid of the repressor, return to the resting state.

This model has only one control—the level of 'inducer' for the repressor. The nickase and transferase can always be present, and membrane sites can be made at a uniform rate. If, for instance, the inducer were the level of one of the DNA precursors, then there would be a cyclic triggering of DNA synthesis by the rise in that level after the end of the last round of synthesis. The triggering of multiple re-initiations at the origin is easily understood. (All the secondary starting points involve the reduplication of one, but not both, of the daughter chromosomes.) The segregation is automatic because the chromosome is linked to the membrane (Jacob, Brenner, and Cuzin, 1963).

Although such a model is extremely simple, it does not encompass F-factor integration to form Hfr strains, nor does it automatically explain the amino acid starvation experiments, which suggest that new protein synthesis is needed for the initiation of DNA synthesis. Furthermore, one might ask the experimental question: is the *resting* state of bacterial or F-factor DNA a supercoil? (See Freifelder, this volume.) If it is, this model must be made more complicated by specific enzymatic steps that cut the DNA off the membrane and close the circle. The induction of membrane sites, or the specific induction of the

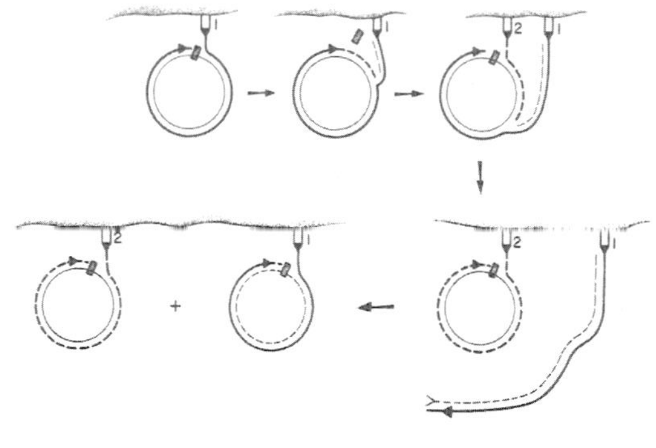

FIGURE 11. *An Hypothesis for Replication and Segregation.* The replication of the bacterial chromosome is shown as starting from a resting state. The negative strand is closed, the positive strand opened, with its 5′ end tied to a site. Synthesis has proceeded from the origin (—▶—) down to a 'repressor', indicated by the rectangle, where continued movement of the polymerase is blocked. When the repression is lifted, synthesis continues, displacing the old positive strand. The next site to appear receives a 5′ end cut from the origin. Again the 3′ end of the positive strand starts to elongate, but is stopped by the repressor after the strand has grown a short distance past the nickase sensitive region. The growing center, displacing the strand attached to site (1), moves around the circle until it comes to the repressor, at which point the entire strand has been displaced, leaving a resting circle on site (2).

In order to recircularize, the 3′ ends of the displaced tail will be digested. The distance between 5′ ends (the one on the site and the one synthesized on the interval between the nickase sensitive origin and the repressor) is more than a genome. After the circle is reformed, the 3′ end of the positive strand grows up to the repressor and stops, leaving a resting circle on site (1).

nickase or transferase, then could be used to initiate DNA replication.

Pro and con. The rolling circle model, as applied to a structure as large as the bacterial chromosome, probably can only be tested in terms of its local topological statements. The most central statement here is that the replication treats the origin asymmetrically. The origin is recreated on the circle, as soon as replication begins, but the terminus on the membrane is not an origin. Thus if there is premature initiation of a new round of replication before the old round is finished, only one side of the 'Y' can fork again. This is the case for premature re-initiation upon thymine starvation (Lark, 1966). However, rapid growth involves initiation before the termination of the previous cycle of replication. We expect this re-initiation to be asymmetric, and consequently the distribution of forked daughter chromosomes to daughter cells to be asymmetric (leading to daughter cells that are not equivalent during the first generation). To distinguish this from symmetric segregation demands extremely accurate quantitative experiments (see Bird and Lark; Helmstetter, Cooper, Pierucci, and Revelas; and Yoshikawa and Haas; all in this volume).

The model asserts that, at the origin, *one* old strand is elongated with new material, while the other is not. Yoshikawa (1967) has observed new material covalently linked to old in *B. subtilis;* but he interprets his experiments as showing that *both* new strands are extensions of old material.

HIGHER CELLS

Because the size of daughter circles is determined by the relation between the rate of synthesis around the template circle and the rate of the recombination act that makes a daughter circle, we can expect circles that are multiples of the template, for instance, the multiple length circles of polyoma (Dulbecco, this volume) or the multiple length and interlocking circles of mitochondrial DNA (Hudson and Vinograd, 1967; also Hudson, Clayton, and Vinograd, this volume). We view these interlocking circles as a result of rapid synthesis leading first to a large circle and then to smaller circles. Each recombination act involved in cutting the large circle up into two smaller circles has a 50-50 chance of leaving the circles intertwined.

The most striking example of the change from a small circle to a large one may involve the replication of a single gene; this appears to be the case when the ribosomal DNA cistrons undergo massive multiplication in amphibian oocytes (Peacock, 1965; Miller, 1966; Brown and Dawid, 1968). Here the ribosomal DNA multiplies a thousandfold, and the iterated genes appear as DNA circles of a variety of sizes from a few up to several hundred microns in circumference.

HISTORICAL

Many versions of the idea of continuous copying off circles have been suggested. Chandler Fulton (1965) suggested that the transferred strand in mating was continuously copied from a circular template chromosome, in order to explain an observation of linkage between the initial and terminal markers. Brown and Martin (1965) suggested that the RNA phages might replicate as a long strand copied from a circular negative-strand template, then chopped to unit length. (The current understanding of the enzymology of the RNA viral synthetase would rule out such models, because the detection of a 5′ triphosphate terminus for both positive and negative strands argues that neither of these has been cut from a longer piece.)

Lark (1966) suggested a rolling circle variant for the bacterial chromosome, to explain segregation and to resolve the swivel problem. He coupled the model, however, to a rule that the strand to be opened should alternate for each round of replication in order to fit a model of segregation that has now been ruled out (Ryter, 1968, and Ryter, Hirota, and Jacob, this volume).

The thought has certainly occurred to many others that concatomers or double circles might have arisen by copying from circles among other possibilities (such as recombination) (see, for instance, Frankel 1968). However, the questions raised by Cairns' pictures (1963) suggesting that the copying was symmetric were not resolved.

The important new realization is that copying circles *resolves* the bewildering multiplicity of appearances of DNA molecules to show that they are all simple variants on an underlying structural expression. Furthermore, one must accommodate the asymmetry involved in this copying of circles. Jacob has elaborated the consequences of such a model for λ (Eisen, Pereira da Silva, and Jacob, this volume), as we have done for ΦX.

Acknowledgments

We wish to thank the National Institutes of Health and the John Simon Guggenheim Foundation for their support.

REFERENCES

Bode, H. R., and H. J. Morowitz. 1967. Size and structure of the *Mycoplasma hominis* H39 chromosome. J. Mol. Biol. *23:* 191.

BOTSTEIN, D. 1968. Synthesis and maturation of phage P22 DNA. I. Identification of intermediates. J. Mol. Biol. *34:* 621.

BROWN, D., and B. DAWID. 1968. Specific gene amplification in oocytes. Science *160:* 272.

BROWN, F., and S. J. MARTIN. 1965. A new model for virus ribonucleic acid replication. Nature *208:* 861.

BUTTIN, G. 1963. Mechanismes regulateurs dans la biosynthese des enzymes du metabolisme du galactose chez *Escherichia coli* K12. III. L'"Effet de derepression" provoque par le developpement du phage λ. J. Mol. Biol. *7:* 610.

CAIRNS, J. 1963. The chromosome of *Escherichia coli*. Cold Spring Harbor Symp. Quant. Biol. *28:* 43.

CAIRNS, J., and C. I. DAVERN. 1966. Effect of P^{32} decay upon DNA synthesis by a radiation-sensitive strain of *Escherichia coli*. J. Mol. Biol. *17:* 418.

DENHARDT, D. T., and R. B. SILVER. 1966. An analysis of the clone size distribution of ΦX-174 mutants and recombinants. Virology *30:* 10.

DENHARDT, D. T., and R. L. SINSHEIMER. 1965. The process of infection with bacteriophage ΦX-174. V. Inactivation of the phage-bacterium complex by decay of ^{32}P incorporated in the infecting particle. J. Mol. Biol. *12:* 663.

DOERMANN, A. H., and L. BOEHNER. 1963. An experimental analysis of bacteriophage T4 heterozygotes. I. Mottled phages from crosses involving six *rII* loci. Virology *21:* 551.

DOWELL, C. E., and R. L. SINSHEIMER. 1966. The process of infection with bacteriophage ΦX-174. IX. Studies on the physiology of three ΦX-174 temperature-sensitive mutants. J. Mol. Biol. *16:* 374.

DRESSLER, D., and D. DENHARDT. 1968. Mechanism of replication of ΦX-174 single-stranded DNA. Nature *219:* 346.

DRESSLER, D., and W. GILBERT. 1969. A rolling circle model for ΦX-174 DNA replication. Nature, in press.

ECHOLS, H., B. BUTLER, A. JOYNER, M. WILLARD, and L. Pilarski. 1967. The regulation of viral genes, and the uncontrolled expression of the galactose genes during development. *In* J. S. Colter and W. Paranchych [ed.] The molecular biology of viruses. Academic Press, New York.

FOX, E., and M. MESELSON. 1963. Unequal photosensitivity of the two strands of DNA in bacteriophage λ. J. Mol. Biol. *7:* 583.

FRANKEL, F. R. 1966. Studies on the nature of replicating DNA in T4-infected *Escherichia coli*. J. Mol. Biol. *18:* 127.

FRANKEL, F. R. 1968. Evidence for long DNA strands in the replicating pool after T4 infection. Proc. Nat. Acad. Sci. *59:* 131.

FULTON, C. 1965. Continuous chromosome transfer in *Escherichia coli*. Genetics *52:* 55.

GOULIAN, M., and A. KORNBERG. 1967. Enzymatic synthesis of DNA. XXIII. Synthesis of circular replicative form of phage ΦX-174 DNA. Proc. Nat. Acad. Sci. *58:* 1723.

HRADECNA, Z., and W. SZYBALSKI. 1967. Fractionation of the complementary strands of coli phage λ DNA based on the asymmetric distribution of the poly I, G binding sites. Virology *32:* 633.

HUDSON, B., and J. VINOGRAD. 1967. Catenated circular DNA molecules in HeLa cell mitochondria. Nature *216:* 647.

IKEDA, H., and J. TOMIZAWA. 1965. Transducing fragments in generalized transduction by phage P1. III. Studies with small phage particles. J. Mol. Biol. *14:* 120.

JACOB, F., S. BRENNER, and F. CUZIN. 1963. On the regulation of DNA replication in bacteria. Cold Spring Harbor Symp. Quant. Biol. *28:* 329.

KNIPPERS, R., T. KOMANO, and R. L. SINSHEIMER. 1968. The process of infection with bacteriophage ΦX174. XXI. Replication and fate of the replicative form. Proc. Nat. Acad. Sci. *59:* 577.

KNIPPERS, R., and R. L. SINSHEIMER. 1968. Process of infection with bacteriophage ΦX-174. XX. Attachment of the parental DNA of bacteriophage ΦX-174 to a fast-sedimenting cell component. J. Mol. Biol. *34:* 17.

KOMANO, T., R. KNIPPERS, and R. L. SINSHEIMER. 1968. The process of infection with bacteriophage ΦX174. XXII. Synthesis of progeny single-stranded DNA. Proc. Nat. Acad. Sci. *59:* 911.

KOZINSKI, A. W. 1961. Fragmentary transfer of P^{32}-labeled parental DNA to progeny phage. Virology *13:* 124.

KOZINSKI, A. W. 1961. Uniform sensitivity to P^{32} decay among progeny of P^{32}-free phage ΦX-174 grown on P^{32}-labeled bacteria. Virology *13:* 377.

LARK, G. 1966. Regulation of chromosome replication and segregation in bacteria. Bact. Rev. *30:* 3.

LINDQVIST, B. H., and R. L. SINSHEIMER. 1967. The process of infection with bacteriophage ΦX174. XVI. Synthesis of the replicative form and its relationship to viral single-stranded DNA synthesis. J. Mol. Biol. *32:* 285.

LITTLE, J. W. 1967. An exonuclease induced by bacteriophage λ. II. Nature of the enzymatic reaction. J. Biol. Chem. *242:* 679.

MACHATTIE, L. A., D. A. RITCHIE, C. A. THOMAS, and C. C. RICHARDSON. 1967. Terminal repetition in permuted T2 bacteriophage DNA molecules. J. Mol. Biol. *23:* 355.

MAKOVER, S. 1968. A preferred origin for the replication of lambda DNA. Proc. Nat. Acad. Sci. *59:* 1345.

MILLER, O. L. 1966. Structure and composition of peripheral nucleoli and salamander oocytes. Nat. Cancer. Inst. Monograph *23:* 53.

OKAZAKI, R., T. OKAZAKI, K. SAKABE, K. SUGIMOTO, and A. SUGINO. 1968. Mechanism of DNA chain growth. I. Possible discontinuity and unusual secondary structure of newly synthesized chains. Proc. Nat. Acad. Sci. *59:* 598.

PACKMAN, S., and W. S. SLY. 1968. Constitutive DNA replication by λc_{17}, a regulatory mutant related to virulence. Virology *34:* 778.

PEACOCK, W. J. 1965. Chromosome replication. Nat. Cancer Inst. Monograph *18:* 101.

PEREIRA DA SILVA, L. H., and F. JACOB. 1968. Etude génétique d'une mutation modifiant la sensibilité a l'immunité chez le bactériophage lambda. Annales de l'Institut Pasteur *115:* 148.

PTASHNE, M., and N. HOPKINS. 1968. The operators controlled by the λ phage repressor. Proc. Nat. Acad. Sci. *60:* 1282.

RHOADES, M., L. A. MACHATTIE, and C. A. THOMAS. 1968. The P22 bacteriophage DNA molecule. I. The mature form. J. Mol. Biol., *37:* 21.

RICHARDSON, C. C., C. L. SCHILDKRAUT, and A. KORNBERG. 1963. Studies on the replication of DNA by DNA polymerases. Cold Spring Harbor Symp. Quant. Biol. *28:* 9.

RITCHIE, D. A., C. A. THOMAS, L. A. MACHATTIE, and P. C. WENSINK. 1967. Terminal repetition in non-permuted T3 and T7 bacteriophage DNA molecules. J. Mol. Biol. *23:* 365.

ROLLER, A. 1961. Studies on the replication and transfer to progeny of the DNA of bacteriophage T4. Thesis, Cal. Inst. Tech.

ROLLER, A. 1964. Replication and transfer of the DNA of phage T4. J. Mol. Biol. *9:* 260.
RUSH, M. G., and R. C. WARNER. 1967. Multiple-length rings of ΦX-174 replicative form. II. Infectivity. Proc. Nat. Acad. Sci. *58:* 2372.
RYTER, A. 1968. Association of the nucleus and the membrane of bacteria: a morphological study. Bact. Rev. *32:* 39.
SAKABE, K., and R. OKAZAKI. 1966. A unique property of the replicating region of chromosomal DNA. Biochim. Biophys. Acta *129:* 651.
SALZMAN, L. A., and A. WEISSBACH. 1967. Formation of intermediates in the replication of phage lambda DNA. J. Mol. Biol. *28:* 53.
SINSHEIMER, R. L. 1961. Replication of bacteriophage ΦX-174. Proc. R. A. Welch Found. Conf. on Chem. Res. *V.* 227.
SLY, W. S., H. A. EISEN, and L. SIMINOVITCH. 1968. Host survival following infection with or induction of bacteriophage lambda mutants. Virology *34:* 112.
SMITH, M. G., and A. SKALKA. 1966. Some properties of DNA from phage-infected bacteria. J. Gen. Phys. *49:* 127.
STREISINGER, G., R. S. EDGAR, and G. H. DENHARDT. 1964. Chromosome structure in phage T4. I. Circularity of the linkage map. Proc. Nat. Acad. Sci. *51:* 775.
TESSMAN, E. S. 1966. Mutants of bacteriophage S 13 blocked in infectious DNA synthesis. J. Mol. Biol. *17:* 218.
THOMAS, C. A. 1967. The rule of the ring. J. Cell Phys. *70:* Supp. 1, 13.
THOMAS, R., and L. E. BERTANI. 1964. On the control of the replication of temperate bacteriophages superinfecting immune hosts. Virology *24:* 241.
WERNER, R. 1968. Distribution of growing points in DNA of bacteriophage T4. J. Mol. Biol. *33:* 679.
WU, R., and A. D. KAISER. 1967. Mapping the 5′ terminal nucleotides of the DNA of bacteriophage λ and related phages. Proc. Nat. Acad. Sci. *57:* 170.
YARUS, M. J., and R. L. SINSHEIMER. 1967. The process of infection with bacteriophage ΦX-174. XIII. Evidence for an essential bacterial "Site". J. Virol. *1:* 135.
YOSHIKAWA, H. 1967. The initiation of DNA replication in *Bacillus subtilis*. Proc. Nat. Acad. Sci. *58:* 312.

DISCUSSION

O. MAALØE: If the rolling circle model is applied to the phage replication, certain predictions concerning the material transfer from parental to progeny phage can be made. *If* the 'nickase' recognizes a particular base sequence on *one* of the DNA strands, for instance, of T4, this is the strand that will be displaced from the parental duplex during the first turn of the wheel. Consequently, one could argue that this strand only would contribute material to the progeny, and the observed transfer of about 50% of the parental label would be accounted for. On this model, 100% of the label might be transferred from the first to the second progeny, because F_1 particles would be labeled exclusively in the strand recognized by the 'nickase'. This expectation is *not* fulfilled; the transfer from F_1 to F_2 particles is again about 50% (Maaløe and Watson, Proc. Nat. Acad. Sci. 1951, *37:* 507).

Two opposing effects would tend to obscure the simple prediction: First, by recombination, material from the 'resident' strand of the wheel could enter the DNA pool from which mature particles are drawn; this would permit the transfer to exceed 50%. Second, if the DNA in the pool is not quantitatively incorporated into progeny particles, the full potential transfer will not be realized.

The fact that the second generation transfer falls much below 100% could reflect incomplete maturation; however, in that case one would have to expect the transfer to keep increasing during lysis inhibition.

A. J. CLARK: The replication which leads to conjugational transfer is thought to be characteristic of the replication of F and not of the vegetative replication of the bacterial chromosome (Jacob, Brenner and Cuzin, 1963, Cold Spring Harbor Symp. Quant. Biol. *28:* 329). Hence, strictly speaking, a prediction of the nature of conjugationally transferred DNA, if verified, can be interpreted to indicate the nature of F specific replication but not the nature of normal vegetative chromosomal replication. Are you therefore in agreement that the discovery that conjugational transfer has some properties predicted from the rolling circle model does not indicate the validity of this model to explain vegetative chromosomal replication?

W. GILBERT: Maybe.

Section IV:
DNA Sequencing

Proc. Natl. Acad. Sci. USA
Vol. 74, No. 2, pp. 560–564, February 1977
Biochemistry

A new method for sequencing DNA

(DNA chemistry/dimethyl sulfate cleavage/hydrazine/piperidine)

ALLAN M. MAXAM AND WALTER GILBERT

Department of Biochemistry and Molecular Biology, Harvard University, Cambridge, Massachusetts 02138

Contributed by Walter Gilbert, December 9, 1976

ABSTRACT DNA can be sequenced by a chemical procedure that breaks a terminally labeled DNA molecule partially at each repetition of a base. The lengths of the labeled fragments then identify the positions of that base. We describe reactions that cleave DNA preferentially at guanines, at adenines, at cytosines and thymines equally, and at cytosines alone. When the products of these four reactions are resolved by size, by electrophoresis on a polyacrylamide gel, the DNA sequence can be read from the pattern of radioactive bands. The technique will permit sequencing of at least 100 bases from the point of labeling.

We have developed a new technique for sequencing DNA molecules. The procedure determines the nucleotide sequence of a terminally labeled DNA molecule by breaking it at adenine, guanine, cytosine, or thymine with chemical agents. Partial cleavage at each base produces a nested set of radioactive fragments extending from the labeled end to each of the positions of that base. Polyacrylamide gel electrophoresis resolves these single-stranded fragments; their sizes reveal *in order* the points of breakage. The autoradiograph of a gel produced from four different chemical cleavages, each specific for a base in a sense we will describe, then shows a pattern of bands from which the sequence can be read directly. The method is limited only by the resolving power of the polyacrylamide gel; in the current state of development we can sequence inward about 100 bases from the end of any terminally labeled DNA fragment.

We attack DNA with reagents that first damage and then remove a base from its sugar. The exposed sugar is then a weak point in the backbone and easily breaks; an alkali- or amine-catalyzed series of β-elimination reactions will cleave the sugar completely from its 3′ and 5′ phosphates. The reaction with the bases is a limited one, damaging only 1 residue for every 50 to 100 bases along the DNA. The second reaction to cleave the DNA strand must go to completion, so that the molecules finally analyzed do not have hidden damages. The purine-specific reagent is dimethyl sulfate; the pyrimidine-specific reagent is hydrazine.

The sequencing requires DNA molecules, either double-stranded or single-stranded, that are labeled at one end of one strand with ^{32}P. This can be a 5′ or a 3′ label. A restriction fragment of any length is labeled at both ends—for example, by being first treated with alkaline phosphatase to remove terminal phosphates and then labeled with ^{32}P by transfer from γ-labeled ATP with polynucleotide kinase. There are then two strategies: either (*i*) the double-stranded molecule is cut by a second restriction enzyme and the two ends are resolved on a polyacrylamide gel and isolated for sequencing or (*ii*) the doubly labeled molecule is denatured and the strands are separated on a gel (1), extracted, and sequenced.

THE SPECIFIC CHEMISTRY

A Guanine/Adenine Cleavage (2). Dimethyl sulfate methylates the guanines in DNA at the N7 position and the adenines at the N3 (3). The glycosidic bond of a methylated purine is unstable (3, 4) and breaks easily on heating at neutral pH, leaving the sugar free. Treatment with 0.1 M alkali at 90° then will cleave the sugar from the neighboring phosphate groups. When the resulting end-labeled fragments are resolved on a polyacrylamide gel, the autoradiograph contains a pattern of dark and light bands. The dark bands arise from breakage at guanines, which methylate 5-fold faster than adenines (3).

This strong guanine/weak adenine pattern contains almost half the information necessary for sequencing; however, ambiguities can arise in the interpretation of this pattern because the intensity of isolated bands is not easy to assess. To determine the bases we compare the information contained in this column of the gel with that in a parallel column in which the breakage at the guanines is suppressed, leaving the adenines apparently enhanced.

An Adenine-Enhanced Cleavage. The glycosidic bond of methylated adenosine is less stable than that of methylated guanosine (4); thus, gentle treatment with dilute acid releases adenines preferentially. Subsequent cleavage with alkali then produces a pattern of dark bands corresponding to adenines with light bands at guanines.

Cleavage at Cytosines and Thymines. Hydrazine reacts with thymine and cytosine, cleaving the base and leaving ribosylurea (5–7). Hydrazine then may react further to produce a hydrazone (5). After a partial hydrazinolysis in 15–18 M aqueous hydrazine at 20°, the DNA is cleaved with 0.5 M piperidine. This cyclic secondary amine, as the free base, displaces all the products of the hydrazine reaction from the sugars and catalyzes the β-elimination of the phosphates. The final pattern contains bands of similar intensity from the cleavages at cytosines and thymines.

Cleavage at Cytosine. The presence of 2 M NaCl preferentially supresses the reaction of thymines with hydrazine. Then, the piperidine breakage produces bands only from cytosine.

AN EXAMPLE

Consider a 64-base-pair DNA fragment, cut from *lac* operon DNA by the *Alu I* enzyme from *Arthrobacter luteus*, which cleaves flush at an AGCT sequence between the G and the C (8). After dephosphorylation, the two 5′ ends of this fragment were labeled with ^{32}P. The autoradiograph in Fig. 1 shows that the two strands separate during electrophoresis, after denaturation, on a neutral polyacrylamide gel (1); they can be easily excised and extracted. For each strand, aliquots of the four cleavage reactions (strong G/weak A, strong A, strong C, and C + T) were electrophoresed at 600–1000 V on a 40-cm 20% polyacrylamide/7 M urea gel. Twelve hours later, a second portion of each sample was loaded on the gel and electrophoresis was continued. Fig. 2 displays autoradiographs showing two regions of the sequence of each strand derived from this single gel: one close to the labeled end of the molecule in those samples that had been electrophoresed in a short time, and a region further into the molecule expanded by electrophoresis for a longer time. The sequence can easily be read from the pattern of bands. The spacing between fragments decreases (roughly as an inverse square) from the bottom toward the top of the gel. The slight variations in the spacing are sequence-specific and reflect the last nucleotide added, a T or G decreasing the mobility more than an A or C. The fragments on the gel end with the base just before the one destroyed by the chemical attack; the labels on the bands in the figure represent the attacked bases. In Fig. 2, 62 bases can be read for both strands, the last 2 bases at the two 5′ ends not being determined by this gel. The sequence of each strand is consistent with and confirms that of the other:

pXXGGCACGACAGGTTTCCCGACTGGAAAGCGGGCAGTGAGCGCAACGCAATTAATGTGAGTTAG
GACCGTGCTGTCCAAAGGGCTGACCTTTCGCCCGTCACTCGCGTTGCGTTAATTACACTCAAXX$_p$

Fig. 3 is an expansion of one region of the sequencing gel to show the base specificity of the cleavage reactions.

DISCUSSION

The chemical sequencing method has certain specific advantages. First, the chemical treatment is easy to control; the ideal chemical attack, one base hit per strand, produces a rather even distribution of labeled material across the sequence. Second, each base is attacked, so that in a run of any single base all those are displayed. The chemical distinction between the different bases is clear, and, as in our example, the sequence of both strands provides a more-than-adequate check.

We have chosen this specific set of chemical reactions to provide more than enough information for the sequencing. The *Techniques* section describes another reaction that displays the Gs alone as well as an alternative reaction for breaking at As and Cs. However, it is more useful to have a strong G/weak A display, in which there is generally enough information to distinguish both the Gs and the As, than just a pure G pattern alone, because redundant information serves as a check on the identifications. In principle, one could sequence DNA with three chemical reactions, each of single-base specificity, using the absence of a band to identify the fourth position. This would be a nonredundant method in which every bit of information was required. Such an approach is subject to considerable error, and any hesitation in the chemistry would be misinterpreted as a different base. For that reason we have chosen redundant displays, which increase one's confidence in the sequencing.

5-Methylcytosine and N^6-methyladenine are occasionally found in DNA. 5-Methylcytosine can be recognized by our method because the methyl group interferes with the action of hydrazine [thymine reacts far more slowly than does uracil (5)]; thus, a 5-methylcytosine cleavage does not appear in the pattern, producing a gap in the sequence opposite a guanine (observed in this laboratory by J. Tomizawa and H. Ohmori). However, we do not expect to recognize an N^6-methyladenine; the glycosidic bond should not be unstable, and an earlier methylation of adenine at the N^6 position should not prevent the later methylation at N3.

These methods work equally well on double- or single-stranded DNA. There are sequence-specific effects on the methylation reaction with double-stranded DNA that do not appear with single-stranded DNA: in the sequence GGA the reactivity of the middle G is suppressed; in the sequence AAA the reactivity of the central A is enhanced. Since these effects are absent with single-stranded DNA, they must arise through steric hindrance or stacking interactions; however, they do not interfere with sequencing because they appear equivalently in both displays of the base. Although in single-stranded DNA the N1 of adenine is exposed to methylation and should methylate as readily as the N7 of guanine, methylation at this position does not destabilize adenine on the sugar. Under our conditions the methyl group will migrate to the N^6 position and the extra charge will disappear (3).

The sequencing method is limited only by the resolution attainable in the gel electrophoresis. On 40-cm gels we can, without ambiguities, sequence out to 100 bases from the point of labeling. If there is other information available to support the sequencing, such as an amino acid sequence, one can often read further. The availability of restriction endonucleases is now such that any DNA molecule, obtainable from a phage, virus, or plasmid, can be sequenced.

TECHNIQUES

[γ-^{32}P]ATP Exchange Synthesis (9). The specific activity routinely attains 1200 Ci/mmol. Dialyze glyceraldehyde-3-phosphate dehydrogenase against 3.2 M ammonium sulfate, pH 8/50 mM Tris·HCl, pH 8/10 mM mercaptoethanol/1 mM EDTA/0.1 mM NAD^+; and dialyze 3-phosphoglycerate kinase (ATP:3-phospho-D-glycerate 1-phosphotransferase, EC 2.7.2.3) against the same solution minus NAD^+ (enzymes from Calbiochem). Combine 50 μl of the dialyzed dehydrogenase and 25 μl of the dialyzed kinase, sediment at 12,000 × g, and redissolve the pellet in 75 μl of twice-distilled water to remove ammonium sulfate. Dissolve 25 mCi (2.7 nmol) of HCl-free, carrier-free $^{32}P_i$ in 50 μl of 50 mM Tris·HCl, pH 8.0/7 mM $MgCl_2$/ 0.1 mM EDTA/2 mM reduced glutathione/1 mM sodium 3-phosphoglycerate/0.2 mM ATP (10 nmol); add 2 μl of the dialyzed, desalted enzyme mixture, and allow to react at 25°. Follow the reaction by thin-layer chromatography on PEI cellulose in 0.75 M sodium phosphate, pH 3.5, by autoradiography of the plate. At the plateau, usually 30 min, add 250 μl of twice-distilled water and 5 μl of 0.1 M EDTA, mix, and heat at 90° for 5 min to inactivate the enzymes. Then chill, add 700 μl of 95% ethanol, mix well, and store at −20°. The theoretical limit of conversion is 79%, and if this is achieved the [γ-^{32}P]ATP would have a specific activity near 2000 Ci/mmol.

Labeling 5′ Ends. 5′-Phosphorylation (10, 11) includes a heat-denaturation in spermidine which increases the yield 15-fold with flush-ended restriction fragments. Dissolve dephosphorylated DNA in 75 μl of 10 mM glycine·NaOH, pH 9.5/1 mM spermidine/0.1 mM EDTA; heat at 100° for 3 min and chill in ice water. Then add 10 μl of 500 mM glycine·NaOH, pH 9.5/100 mM $MgCl_2$/50 mM dithiothreitol, 10 μl of [γ-^{32}P]ATP (100 pmol or molar equivalent of DNA 5′ ends, 1000 Ci/mmol), and several units of polynucleotide kinase to a final volume of 100 μl. Heat at 37° for 30 min; add 100 μl of 4 M ammonium acetate, 20 μg of tRNA, and 600 μl of ethanol, mix well, chill at −70°, centrifuge at 12,000 × g, remove the supernatant phase, rinse the pellet with ethanol, and dry under vacuum.

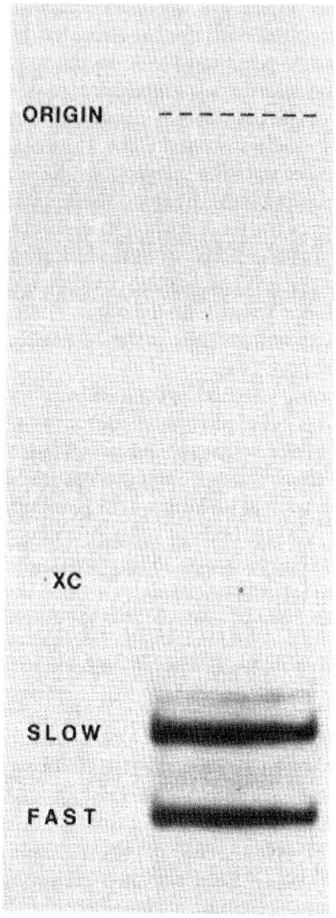

FIG. 1. Strand separation of a restriction fragment: 1.5 μg of a 64-base-pair DNA fragment (75 pmol of 5′ ends) was phosphorylated with [γ-^{32}P]ATP (800 Ci/mmol) and polynucleotide kinase, denatured in alkali, layered onto a 0.3 cm × 3 cm surface of an 8% polyacrylamide slab gel (see under *Techniques*), and electrophoresed at 200 V (regulated) and 20 mA (average), until the xylene cyanol (XC) dye moved 9 cm. The gel on one glass plate was then tightly covered with Saran Wrap and exposed to Kodak XR-5 x-ray film for 10 min.

Labeling 3′ Ends. To adenylate with [α-^{32}P]ATP and terminal transferase (12), dissolve DNA in 70 μl of 10 mM Tris·HCl, pH 7.5/0.1 mM EDTA, heat at 100° for 3 min, and chill at 0°. Then add, in order, 10 μl of 1.0 M sodium cacodylate (pH 6.9), 2 μl of 50 mM $CoCl_2$, mix, 2 μl of 5 mM dithiothreitol, 10 μl of [α-^{32}P]ATP (500 pmol, 100 Ci/mmol), and several units of terminal transferase to a final volume of 100 μl. Heat at 37° for several hours, add 100 μl of 4 M ammonium acetate, 20 μg of tRNA, and 600 μl of ethanol, and precipitate, centrifuge, rinse, and dry the DNA as described above. Dissolve the pellet in 40 μl of 0.3 M NaOH/1 mM EDTA, heat at 37° for 16 hr, and either add glycerol and dyes for strand separation or neutralize, ethanol precipitate, and renature the DNA for secondary restriction cleavage.

Strand Separation. Dissolve the DNA in 50 μl of 0.3 M NaOH/10% glycerol/1 mM EDTA/0.05% xylene cyanol/0.05% bromphenol blue. Load on a 5–10% acrylamide/0.16–0.33% bisacrylamide/50 mM Tris·borate, pH 8.3/1 mM EDTA gel and electrophorese. The concentration of DNA entering the gel is critical and must be minimized to prevent renaturation. Use thick gels with wide slots (0.3- to 1-cm-thick slabs with 3-cm to full width slots), and run cool (at 25°).

Gel Elution. Insert an excised segment of the gel into a 1000 μl (blue) Eppendorf pipette tip, plugged tightly with siliconized glass wool and heat-sealed at the point. Grind the gel to a paste with a siliconized 5-mm glass rod, add 0.6 ml of 0.5 M ammonium acetate/0.01 M magnesium acetate/0.1% sodium dodecyl sulfate/0.1 mM EDTA (and 50 μg of tRNA carrier if the DNA has already been labeled); seal with Parafilm and hold at 37° for 10 hr. Cut off the sealed point, put the tip in a siliconized 10 × 75 mm tube, centrifuge for a few minutes, rinse with 0.2 ml of fresh gel elution solution, and alcohol precipitate twice.

Partial Methylation of Purines. Combine 1 μl of sonicated carrier DNA, 10 mg/ml, with 5 μl of ^{32}P-end-labeled DNA in 200 μl of 50 mM sodium cacodylate, pH 8.0/10 mM $MgCl_2$/0.1 mM EDTA. Mix and chill in ice. Add 1 μl of 99% (10.7 M) dimethyl sulfate, mix, cap, and heat at 20° for 15 min. To stop the reaction, add 50 μl of a stop solution (1.0 M mercaptoethanol/1.0 M Tris·acetate, pH 7.5/1.5 M sodium acetate/0.05 M magnesium acetate/0.001 M EDTA), 1 mg/ml of tRNA, and mix. Add 750 μl (3 volumes) of ethanol, chill, and spin. Reprecipitate from 250 μl of 0.3 M sodium acetate, rinse with alcohol, and dry.

Strong Guanine/Weak Adenine Cleavage. Dissolve methylated DNA in 20 μl of 10 mM sodium phosphate, pH 7.0/1 mM EDTA, and collect the liquid on the bottom of the tube with a quick low-speed spin. Close the tube and heat in a water bath at 90° for 15 min. Chill in ice and collect the condensate with a quick low-speed spin. Add 2 μl of 1.0 M NaOH, mix, and draw the liquid up into the middle of a pointed glass capillary tube, seal with a flame, and hold at 90° for 30 min. Open the capillary and empty into 20 μl of urea-dye mixture, heat, and layer on the gel.

Strong Adenine/Weak Guanine Cleavage. Dissolve methylated DNA in 20 μl of distilled water. Chill to 0°, add 5 μl of 0.5 M HCl, mix, and keep the sample at 0° in ice, mixing occasionally. After 2 hr, add 200 μl of 0.3 M sodium acetate and 750 μl of ethanol, chill, spin, rinse, and dry. Then dissolve in 10 μl of 0.1 M NaOH/1 mM EDTA and heat at 90° for 30 min in a sealed capillary. Add contents to urea-dye mixture, heat, and layer.

An Alternative Guanine Cleavage. Dissolve methylated DNA in 20 μl of freshly diluted 1.0 M piperidine. Heat at 90° for 30 min in a sealed capillary. [This reaction opens 7-MeG adjacent to the glycosidic bond (13), displaces the ring-opened product from the sugar, and eliminates both phosphates to cleave the DNA wherever G was methylated.] Return the contents of the capillary to the reaction tube, lyophilize, wet the residue, and lyophilize again. Finally, dissolve the last residue in 10 μl of 0.1 M NaOH/1 mM EDTA and prepare for the gel.

An Alternative Strong Adenine/Weak Cytosine Cleavage. Combine 20 μl of 1.5 M NaOH/1 mM EDTA with 1 μl of sonicated carrier DNA (10 mg/ml) and 5 μl of ^{32}P end-labeled DNA, and heat at 90° for 30 min in a sealed capillary. [The strong alkali opens the adenine and cytosine rings (13); then, the ring-opened products can be displaced and phosphates eliminated with piperidine.] Rinse the capillary into 100 μl of 1.0 M sodium acetate, add 5 μl of tRNA (10 mg/ml), add 750 μl of ethanol, chill, spin, rinse, and dry. Dissolve the pellet in 20 μl of freshly diluted 1.0 M piperidine, and heat at 90° for 30 min in a sealed capillary. Lyophilize twice, dissolve the last residue in 10 μl of 0.1 M NaOH/1 mM EDTA, and add urea-dye mixture.

Cleavage at Thymine and Cytosine. Combine 20 μl of distilled water, 1 μl of sonicated carrier DNA (10 mg/ml), and 5 μl of ^{32}P end-labeled DNA. Mix and chill at 0°. Add 30 μl of

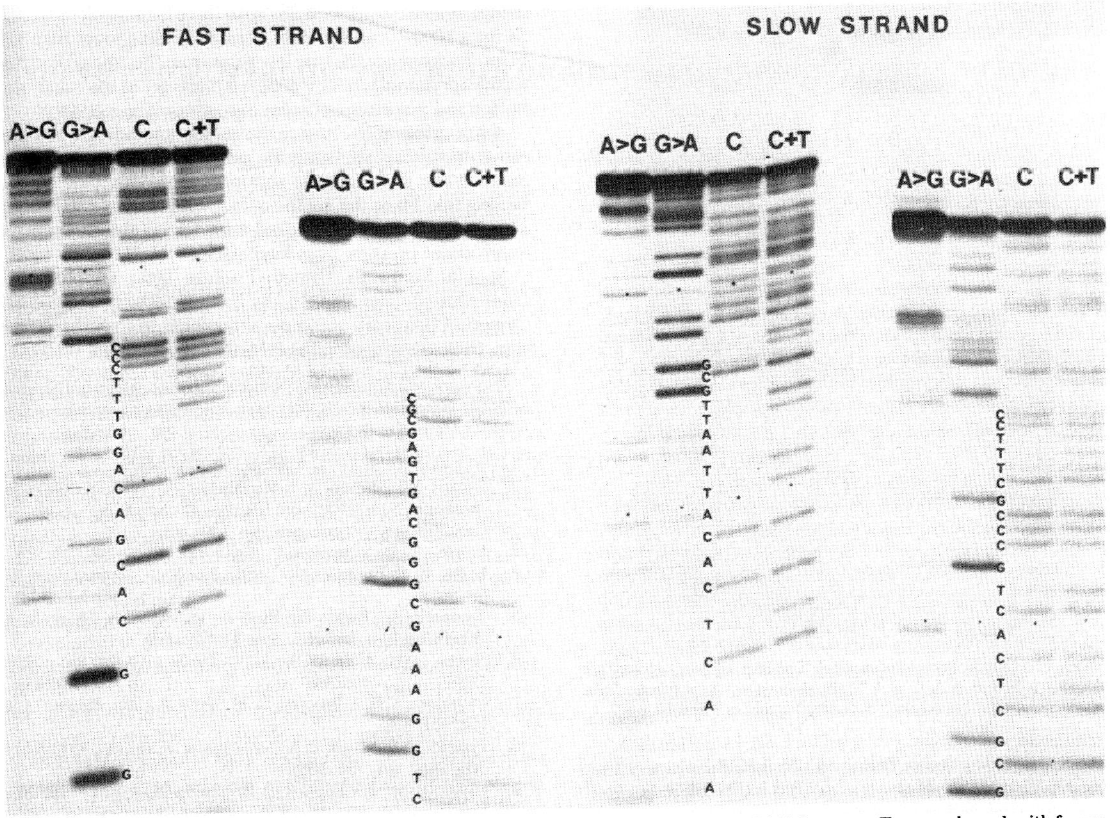

FIG. 2. Autoradiograph of a sequencing gel of the complementary strands of a 64-base-pair DNA fragment. Two panels, each with four reactions, are shown for each strand; cleavages proximal to the 5′ end are at the bottom on the left. A strong band in the first column with a weaker band in the second arises from an A; a strong band in the second column with a weaker band in the first is a G; a band appearing in both the third and fourth columns is a C; and a band only in the fourth column is a T. To derive the sequence of each strand, begin at the bottom of the left panel and read upward until the bands are not resolved; then, pick up the pattern at the bottom of the right panel and continue upward. One-tenth of each strand, isolated from the gel of Fig. 1, was used for each of the base-modification reactions. The dimethyl sulfate treatment was 50 mM for 30 min to react with A and G; hydrazine treatment was 18 M for 30 min to react with C and T and 18 M with 2 M NaCl for 40 min to cleave C. After strand breakage, half of the products from the four reactions were layered on a 1.5 × 330 × 400 mm denaturing 20% polyacrylamide slab gel, pre-electrophoresed at 1000 V for 2 hr. Electrophoresis at 20 W (constant power), 800 V (average), and 25 mA (average) proceeded until the xylene cyanol dye had migrated halfway down the gel. Then the rest of the samples were layered and electrophoresis was continued until the new bromphenol blue dye moved halfway down. Autoradiography of the gel for 8 hr produced the pattern shown.

95% (30 M) hydrazine*, mix well, and keep at 0° for several minutes. Close the tube and heat at 20° for 15 min. Add 200 μl of cold 0.3 M sodium acetate/0.01 M magnesium acetate/0.1 mM EDTA/0.25 mg/ml tRNA, vortex mix, add 750 μl of ethanol, chill, spin, dissolve the pellet in 250 μl of 0.3 M sodium acetate, add 750 μl of ethanol, chill, spin, rinse with ethanol, and dry. Dissolve the pellet and rinse the walls with 20 μl of freshly diluted 0.5 M piperidine. Heat for 30 min at 90° in a sealed capillary. Lyophilize twice, dissolve in 10 μl of 0.1 M NaOH/1 mM EDTA, add urea-dye mixture, heat, and layer on gel.

Cleavage at Cytosine. Replace the water in the hydrazinolysis reaction mixture with 20 μl of 5 M NaCl, and increase the reaction time to 20 min. The freshness and the concentration of the hydrazine are critical for base-specificity.

Reaction Times. The reaction conditions provide a uniformly labeled set of partial products of chain length 1 to 100. To distribute the label over a shorter region, increase the reaction time, and vice versa.

* CAUTION: Hydrazine is a volatile neurotoxin. Dispense with care in a fume hood, and inactivate it with concentrated ferric chloride.

Reaction Vessels. We use 1.5-ml Eppendorf conical polypropylene tubes with snap caps, treated with 5% (vol/vol) dimethyldichlorosilane in CCl_4 and rinsed with distilled water.

Alcohol Precipitation, Wash, and Rinse. Unless otherwise specified, the initial ethanol precipitation is from 0.3 M sodium acetate/0.01 M magnesium acetate/0.1 mM EDTA, with 50 μg of tRNA as carrier. Add 3 volumes of ethanol, cap and invert to mix, chill at −70° in a Dry Ice-ethanol bath for 5 min, and spin in the Eppendorf 3200/30 microcentrifuge at 15,000 rpm (12,000 × *g*) for 5 min. Reprecipitate with 0.3 M sodium acetate and 3 volumes of ethanol, chill, and spin. Rinse the final pellet with 1 ml of cold ethanol, spin, and dry in a vacuum for several minutes.

Gel Samples. All samples for sequencing gels are in 10 or 20 μl of 0.1 M NaOH/1 mM EDTA to which is added an equal volume of 10 M urea/0.05% xylene cyanol/0.05% bromphenol blue. Heat the sample at 90° for 15 sec, then layer on the gel.

Sequencing Gels. These are commonly slabs 1.5 mm × 330 mm × 400 mm with 18 sample wells 10 mm deep and 13 mm

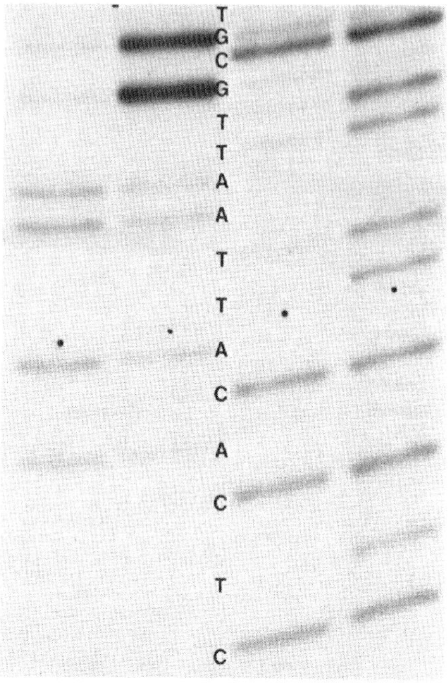

FIG. 3. Detail of the sequence gel. The four lanes are (from left to right) A > G, G > A, C, C + T; the dots show the position of the bromphenol blue dye marker, between fragments 9 and 10 long.

wide separated by 3 mm (fitting on a 35.5 × 43 cm x-ray film). They are 20% (wt/vol) acrylamide (Bio-Rad)/0.67% (wt/vol) methylene bisacrylamide/7 M urea/50 mM Tris-borate, pH 8.3/1 mM EDTA/3 mM ammonium persulfate; 300 ml of gel solution is polymerized with TEMED within 30 min (generally 50 μl of TEMED). Age the gel at least 10 hr before using it. Electrophorese with some heating (30–40°), to help keep the DNA denatured, between 800 and 1200 V. Load successively whenever the previous xylene cyanol has moved halfway down the gel. Bromphenol blue runs with 10-nucleotide-long fragments, xylene cyanol with 28. With three loadings at 0, 12, and 24 hr, a 1000-V run for 36 hr permits reading more than 100 bases. To sequence the first few bases from the labeled end, use a 25% acrylamide/0.83% bisacrylamide gel in the usual urea buffer and pre-electrophorese this gel for 2 hr at 1000 V.

Autoradiography. Freeze the gel for autoradiography. Remove one glass plate, wrap the gel and supporting plate with Saran Wrap, and mark the positions of the dyes with ^{14}C-containing ink. Place the gel in contact with film in a light-tight x-ray exposure holder (backed with lead and aluminum) at −20° under pressure from lead bricks.

Special Materials. Dimethyl sulfate (99%) was purchased from Aldrich Chemical Co., hydrazine (95%) from Eastman Organic Chemicals, and piperidine (99%) from Fisher Scientific; these were used without further purification.

This work was supported by National Institute of General Medical Sciences, Grant GM 09541. W.G. is an American Cancer Society Professor of Molecular Biology.

1. Hayward, G. S. (1972) *Virology* **49,** 342–344.
2. Gilbert, W., Maxam, A. & Mirzabekov, A. (1976) in *Control of Ribosome Synthesis,* eds. Kjeldgaard, N. O. & Maaløe, O. (Munksgaard, Copenhagen), pp. 139–148.
3. Lawley, P. D. & Brookes, P. (1963) *Biochem. J.* **89,** 127–128
4. Kriek, E. & Emmelot, P. (1964) *Biochim. Biophys. Acta* **91,** 59–66.
5. Temperli, A., Turler, H., Rust, P., Danon, A. & Chargaff, E. (1964) *Biochim. Biophys. Acta* **91,** 462–476.
6. Hayes, D. H. & Hayes-Baron, F. (1967) *J. Chem. Soc.,* 1528–1533.
7. Cashmore, A. R. & Petersen, G. B. (1969) *Biochim. Biophys. Acta* **174,** 591–603.
8. Roberts, R. J., Myers, P. A., Morrison, A. & Murray, K. (1976) *J. Mol. Biol.* **102,** 157–165.
9. Glynn, I. M. & Chappell, J. B. (1964) *Biochem. J.* **90,** 147-149.
10. van de Sande, J. H., Kleppe, K. & Khorana, H. G. (1973) *Biochemistry* **12,** 5050–5055.
11. Lillehaug, J. R. & Kleppe, K. (1975) *Biochemistry* **14,** 1225–1229.
12. Roychoudhury, R., Jay, E. & Wu, R. (1976) *Nucleic Acids Res.* **3,** 863–878.
13. Kochetkov, N. K. & Budovskii, E. I. (1972) *Organic Chemistry of Nucleic Acids* (Plenum, New York), Part B, pp. 381–397.

Proc. Natl. Acad. Sci. USA
Vol. 75, No. 3, pp. 1485–1489, March 1978
Genetics

Sequence of a mouse germ-line gene for a variable region of an immunoglobulin light chain

(λ light chain/hypervariable region/DNA sequencing/interspersed noncoding sequences)

SUSUMU TONEGAWA*, ALLAN M. MAXAM†, RICHARD TIZARD†, ORA BERNARD*, AND WALTER GILBERT†

* Basel Institute for Immunology, 487 Grenzacherstrasse, CH-4005 Basel, Switzerland; and † Department of Biochemistry and Molecular Biology, Harvard University, Cambridge, Massachusetts 02138

Contributed by Walter Gilbert, January 9, 1978

ABSTRACT We have determined the sequence of the DNA of a germ-line gene for the variable region of a mouse immunoglobulin light chain, the $V\lambda_{II}$ gene. The sequence confirms that the variable region gene lies on the DNA separated from the constant region. Hypervariable region codons appear in the germ-line sequence. A sequence for the hydrophobic leader, 19 amino acids that are cleaved from the amino terminus of the protein, appears near, but not continuous with, the light chain structural sequence: most of the leader sequence is separated from the rest of the gene by 93 bases of untranslated DNA.

Immunoglobulin molecules are synthesized by special differentiated cells which produce and export a protein specialized to bind an antigen. That binding site is a pocket between two subunits, a light chain and a heavy chain. The part of the light chain in contact with the antigen, the amino-terminal region, termed the variable region, differs extensively from one immunoglobulin to another; the carboxy-terminal half of the light chain, the constant region, is invariant and, in mice, falls into three classes, one κ and two λ. The variable region itself, though different in different immunoglobulins, shows certain consistencies in structure: three hypervariable regions embedded in a framework (1). The hypervariable regions span five to ten amino acids each around positions 30, 55, and 90 in the immunoglobulin chain and contain most of the variation. In the three-dimensional structure of an immunoglobulin, they are the segments in contact with antigen. Differences in sequence in framework regions provide a classification of light chains into different subgroups (2). Some of these differences are a change in only a single amino acid; others are more extensive. Estimates of the largest number of different variable region subgroups that can appear with single constant regions range from 10 to 100. (A similar structure holds for the heavy chains; the first 110 amino acids constitute a variable domain and the next 330, a series of constant domains.)

Hozumi and Tonegawa (3) showed, by restriction enzyme analysis and hybridization mapping techniques, that in the germ line, in embryonic tissue, the sequences coding for the variable region and those coding for the constant region of an immunoglobulin light chain lie separate from one another. These two genetic regions move during differentiation of lymphocyte precursors and come closer together to form a single transcription unit that will produce the final immunoglobulin molecule. Experiments with cloned immunoglobulin genes have since demonstrated this movement more directly and revealed a surprising structure for the active gene. Brack, Lenhard-Schuller, and Tonegawa (ref. 4; unpublished data) analyzed a gene isolated from a myeloma, by R-loop mapping and by heteroduplex comparisons with embryonic genes. They showed that the variable and constant region genes move from distant positions in the embryonic DNA to less distant positions in the producing cell but do not become contiguous. The variable and constant region coding sequences in the myeloma are separated still by 1250 bases. Presumably the DNA rearranges to bring these two regions close enough to lie on a single messenger precursor. One hypothesizes that this precursor then loses the additional sequences as the messenger matures in the nucleus.

In this paper we report the sequence of a germ-line gene for a mouse λ light chain variable region. We determined the sequence of this DNA for two reasons: to establish that the DNA coded for a variable region truly in isolation from any constant portion, and to determine whether a germ-line gene contained hypervariable as well as framework sequences.

A leading model for the behavior of the hypervariable regions, suggested by Kabat and his coworkers (1, 5), is that hypervariable region DNA is extraneous to the germ-line gene and inserts, like a small version of phage lambda, into receptor sites in the gene. This model would explain why different immunoglobulins in different subgroups occasionally have identical hypervariable regions. If this model were true, we would expect a germ-line gene to be a framework surrounding attachment sites rather than hypervariable sequences. Alternative models for the production of different immunoglobulins within one subgroup envisage the germ-line variable gene as a complete, inherited sequence that could be expressed with a unique antigenic specificity, a germ-line idiotype. This specificity will then change by mutation in somatic cells. Either these changes are localized in hypervariable regions by some process that makes the DNA in these areas unusually labile or, alternatively, mutations might arise anywhere in the variable gene but appear only in the hypervariable areas because the selection pressure matching immunoglobulin to antigen will select primarily for mutations in the antigen combining site. Determination of the sequence of a germ-line immunoglobulin light chain gene answers some of these questions.

MATERIALS AND METHODS

The recombinant phage λgt-WES-Ig 13, containing a 4800-base-pair insert of mouse DNA carrying a Vλ gene (6, 7), was grown under P3-EK2 conditions in Basel. A total of 5 mg of purified phage DNA was used for the sequence determination in Cambridge. Milligram amounts of the phage DNA were cleaved with a mixture of *Eco*RI and *Hae* III restriction endonucleases and the three cleavage products of the mouse insert, 750, 1500, and 2500 base pairs long, were separated on a 6 × 200 × 400 mm slab gel polymerized from 5% (wt/vol) acrylamide/0.17% (wt/vol) bisacrylamide/50 mM Tris-borate (pH

The costs of publication of this article were defrayed in part by the payment of page charges. This article must therefore be hereby marked "*advertisement*" in accordance with 18 U. S. C. §1734 solely to indicate this fact.

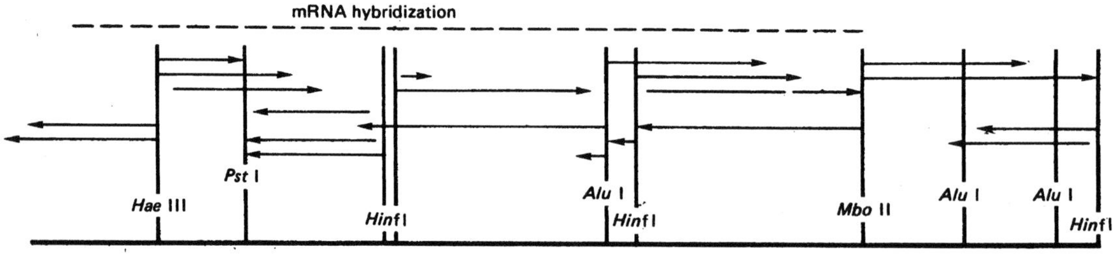

FIG. 1. Restriction endonuclease (7), mRNA hybridization (7), and DNA sequence map of the mouse embryonic $V\lambda_{II}$ region. The direction and the extent of the sequence determinations are shown by the arrows.

8.3)/1 mM EDTA, and extracted as described (8), except that the eluate was passed through a Millipore filter before ethanol precipitation.

The two larger *Hae* III fragments were then cleaved with *Hinf*, *Mbo* II, *Alu* I, or *Pst* I (9), end-labeled, and cleaved again, and their sequences were determined precisely as described (8) except for the following modifications. Tris·HCl buffer (50 mM, pH 9.5) was used instead of glycine/NaOH in the kinase reaction. One-tenth the carrier DNA (1μg) and tRNA (5μg) in all reactions for sequence determination produced sharper bands on autoradiograms. Addition of only sodium acetate, EDTA, and tRNA (no magnesium acetate) after the hydrazinolysis reactions and careful rinsing of the final ethanol precipitate to remove any residual sodium acetate eliminated occasional aberrations in the pyrimidine cleavage patterns. The "alternative guanine cleavage" (8) was used for guanine-only base-specificity.

For determination of long sequences, 200 nucleotides or so from the labeled end, products of sequencing reactions were loaded once or twice on the 20% gel described (8) and once on a 1.5 × 200 × 400 mm 10% gel polymerized from 9.5% (wt/vol) acrylamide/0.5% (wt/vol) bisacrylamide/7 M urea/100 mM Tris-borate (pH 8.3)/2 mM EDTA and electrophoresed at a high enough voltage (800–1100 V) and current to keep the surface of the gel at or above 50° during the run.

RESULTS AND DISCUSSION

Tonegawa *et al.* (6) isolated a derivative of phage lambda containing a 4800-base-pair insert of mouse DNA that hybridizes to the variable region half of a λ_I mRNA. A partial restriction map was constructed for the right third of this fragment, from a *Hae* III cut to the *Eco*RI end: the region in which hybridization to a messenger shows the putative variable region gene to lie (7). We chemically determined the sequence (8) of the relevant restriction fragments. In so doing, we could immediately establish the position of the variable region gene by comparison of the DNA sequence to the known protein sequences. Fig. 1 shows the restriction map of this area and indicates the experimental strategy that developed and checked the sequence. Fig. 3 shows the sequence of some 750 base pairs around the variable region gene.

Identification of Variable Region. Although the original clone was identified by hybridization to a message for a λ_I immunoglobulin, the sequence on the DNA corresponds more closely to a λ_{II} variable region. λ_I is a minor group in the mouse; about 5% of the light chains have this variable region attached to a λ_I constant region. The sequences of 18 λ_I chains from mouse myelomas have been determined (2, 10–12). Twelve of these are identical and the six others differ by one, two, or three amino acids in the hypervariable regions. λ_{II} accounts for only 1–2% of mouse immunoglobulin (H. N. Eisen, personal communication); the sequence of only one example, MOPC 315, has been determined (13). The variable regions of λ_I and λ_{II} chains are very similar; their frameworks differ in seven places. [The two λ constant regions differ at about 29 places (13).] Fig. 2 schematically compares the protein sequence determined by the DNA and those of the λ_I and λ_{II} variable regions, while Fig. 3 shows the entire sequence. Specifically, at framework residues 16 (Glu or Gly), 19 (Thr or Ile), 62 (Ala or Val), 71 (Asn or Asp), 85 (Glu or Asp), and 87 (Ile or Met), the DNA sequence corresponds to a λ_{II} chain (underlined). Thus we believe that we have determined the sequence of the λ_{II} germ-line gene.

There is one exception to this framework agreement: at position 38, λ_I has Val and λ_{II} has Ile, while the DNA sequence at this point predicts Val. We regard the DNA sequence as unambiguous. Since the protein sequence suggests that the expressed gene in MOPC 315 has Ile at this point, we have an excellent candidate for a mutational change in a *framework* region. This interpretation would suggest that single amino acid changes in the framework regions of immunoglobulins are mutations and do not represent different germ-line genes, reducing sharply the number of different subgroups of κ chains that one might estimate to exist in the germ line.

The λ subgroups in the mouse may represent a degenerate example: there appear to be only single variable-region genes for λ_I and λ_{II} (restriction enzyme analysis by M. Hirama and S. Tonegawa, unpublished). Presumably the other possible λ frameworks have been lost in the mouse.

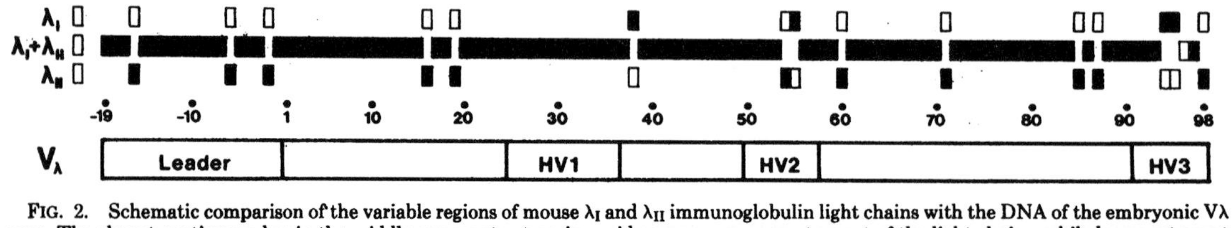

FIG. 2. Schematic comparison of the variable regions of mouse λ_I and λ_{II} immunoglobulin light chains with the DNA of the embryonic Vλ gene. The almost continuous bar in the middle represents an amino acid sequence common to most of the light chains, while boxes set apart from it mark positions at which they differ. Wherever the bar or a box is shaded, the sequence of that protein corresponds with the sequence of the DNA. Of 15 positions that distinguish the two variable regions, the DNA matches λ_{II} (MOPC 315) at 11 and λ_I (MOPC 104E) at 4. The actual sequences are given in Fig. 3.

CCCATACTAAGAGTTATATTATGTCTGTCTCACTGCCTGCTGCTGACCAATATTGAAAATAATAGACTTGGTTTGTGA
GGGTATGATTCTCAATATAATACAGACAGAGTGACGGACGACGACTGGTTATAACTTTTATTATCTGAACCAAACACT

λ_I *MetAlaTrpIleSerLeuIleLeuSerLeuLeuAlaLeuSerSer*
λ_{II} *MetAlaTrpThrSerLeuIleLeuSerLeuLeuAlaLeuCysSer*

MetAlaTrpThrSerLeuIleLeuSerLeuLeuAlaLeuCysSerGly UGAPhe
ATTATGGCCTGGACTTCACTTATACTCTCTCTCCTGGCTCTCTGCTCAGGTCAGCAGCCTTTCTACACTGCAGTGGGTATGCAACAATACACATCTTGTCTCTCATTT
TAATACCGGACCTGAAGTGAATATGAGAGAGAGGACCGAGAGACGAGTCCAGTCGTCGGAAAGATGTGACGTCACCCATACGTTGTTATGTGTAGAACAGAGAGTAAA

λ_I *GlyAlaIleSerGlpAlaValValThrGlnGluSerAlaLeuThrThrSerProGlyGluThrValThrLeuThr*
λ_{II} *GlyAlaSerSerGlpAlaValValThrGlnGluSerAlaLeuThrThrSerProGlyGlyThrValIleLeuThr*

1 10 20
AlaThrAspAspTrpIleSerTyrLeuPheAlaGlyAlaSerSerGlnAlaValValThrGlnGluSerAlaLeuThrThrSerProGlyGlyThrValIleLeuThr
GCTACTGATGACTGGATTTCTTACCTGTTTGCAGGAGCCAGTTCCCAGGCTGTTGTGACTCAGGAATCTGCACTCACCACATCACCTGGTGGAACAGTCATACTCACT
CGATGACTACTGACCTAAAGAATGGACAAACGTCCTCGGTCAAGGGTCCGACAACACTGAGTCCTTAGACGTGAGTGGTGTAGTGGACCACCTTGTCAGTATGAGTGA

AsnThr *Gly* *Leu* *Asn* *Val*
λ_I *CysArgSerSerThrGlyAlaValThrThrSerAsnTyrAlaAsnTrpValGlnGlnLysProAspHisLeuPheThrGlyLeuIleGlyGlyThrAsnAsnArgAla*
λ_{II} *CysArgSerSerThrGlyAlaValThrThrSerAsnTyrAlaAsnTrpIleGlnGluLysProAspHisLeuPheThrGlyLeuIleGlyGlyThrSerAspArgAla*

30 40 50
CysArgSerSerThrGlyAlaValThrThrSerAsnTyrAlaAsnTrpValGlnGluLysProAspHisLeuPheThrGlyLeuIleGlyGlyThrSerAsnArgAla
TGTCGCTCAAGTACTGGGGCTGTTACAACTAGTAACTATGCCAACTGGGTTCAAGAAAAACCAGATCATTTATTCACTGGTCTAATAGGTGGTACCAGCAACCGAGCT
ACAGCGAGTTCATGACCCCGACAATGTTGATCATTGATACGGTTGACCCAAGTTCTTTTTGGTCTAGTAAATAAGTGACCAGATTATCCACCATGGTCGTTGGCTCGA
HV1 HV2

λ_I *ProGlyValProAlaArgPheSerGlySerLeuIleGlyAsnLysAlaAlaLeuThrIleThrGlyAlaGlnThrGluAspGluAlaIleTyrPheCysAlaLeuTrp*
λ_{II} *ProGlyValProValArgPheSerGlySerLeuIleGlyAspLysAlaAlaLeuThrIleThrGlyAlaGlnThrGluAspAspAlaMetTyrPheCysAlaLeuTrp*

60 70 80 90
ProGlyValProValArgPheSerGlySerLeuIleGlyAspLysAlaAlaLeuThrIleThrGlyAlaGlnThrGluAspAspAlaMetTyrPheCysAlaLeuTrp
CCAGGTGTTCCTGTCAGATTCTCAGGCTCCCTGATTGGAGACAAGGCTGCCCTCACCATCACAGGGGCACAGACTGAGGATGATGCAATGTATTTCTGTGCTCTATGG
GGTCCACAAGGACAGTCTAAGAGTCCGAGGGACTAACCTCTGTTCCGACGGGAGTGGTAGTGTCCCCGTGTCTGACTCCTACTACGTTACATAAAGACACGAGATACC
HV3

Cys *Arg*
λ_I *TyrSerAsnHisTrp* | *ValPheGlyGlyGlyThrLysLeuThrValLeuGlyGlnPro* ¦ *LysSerSerProSerValThrLeuPheProProSerSerGluGluLeuThr*
100 110 120
λ_{II} *PheArgAsnHisPhe* | *ValPheGlyGlyGlyThrLysValThrValLeuGlyGlnPro* ¦ *LysSerThrProThrLeuThrValPheProProSerSerGluGluLeuLys*
gtxttyggxggxggxacxaargtxacxgtxttrggxcarccx ¦ aartcxacxccxacxttracxgtxttyccxccxtcxtcxgargarttraar
ctx ctx agyagy ctx

TyrSerThrHisPhe | HisAsnAspMetCysArgTrpGlySerArgThrArgThrLeu ¦ TrpTyrSerLeuThrThrIlePheLeuThrGlyGlyTyrMetSerLeuVal
TACAGCACCCATTTC | CACAATGACATGTGTAGATGGGGAAGTAGAACAAGAACACTC ¦ TGGTACAGTCTCACTACCATCTTCTTAACAGGTGGCTACATGTCCCTAGTC
ATGTCGTGGGTAAAG | GTGTTACTGTACACATCTACCCCTTCATCTTGTTCTTGTGAG ¦ ACCATGTCAGAGTGATGGTAGAAGAATTGTCCACCGATGTACAGGGATCAG
HV3 V | C V ¦ C

λ_I *GluAsnLysAlaThrLeuValCysThrIleThrAspPheTyrProGlyVal....*
130 140
λ_{II} *GluAsnLysAlaThrLeuValCysLeuIleSerAsnPheSerProGlySer....*
garaayaargcxacxttrgtxtgyttratytcxaayttytcxccxggxtcx
ctx ctxataagy agy agy

CysSerLeuLeuLeuUAG
TGTTCTCTTTTACTATAGAGAAATTTATAAAAGCTGTTGTCTCGAGCAACAAAAAGTTTTATTCAACAAATTGTATAATAATTATGCCTTGATGACAAGCTTTGTTTA
ACAAGAGAAAATGATATCTCTTTAAATATTTTCGACAACAGAGCTCGTTGTTTTTCAAAATAAGTTGTTTAACATATTATTAATACGGAACTACTGTTCGAAACAAAT

FIG. 3. DNA sequence of the mouse embryonic $V\lambda_{II}$ immunoglobulin light chain gene in direct comparison with all known mouse λ light chains. The nucleotide sequence of the gene coincides best with the amino acid sequence of the MOPC 315 λ_{II} light chain. Coding begins with the first amino acid of the hydrophobic leader, the Met at position 19, is interrupted at Ser 5 by a 93-base-pair intron (see text), resumes with Gly 4, and (with five single-base-change exceptions) continues from this point to Phe 98, where it ends abruptly. Positions which distinguish λ_I (MOPC 104E) and λ_{II} (MOPC 315) light chains are indicated (*), as are known substitutions (↑) in λ_I hypervariable regions (HV1, HV2, and HV3). Amino acid numbering begins with the cyclized glutamine (Glp) found at the amino terminus of mature light chains. [The Gln-Glu at 6–7 is predicted by the DNA, confirming a correction (10) for both λ_I and λ_{II}.] Underlined amino acids are encoded by the DNA below, while underlined DNA matches codons for at least one of the proteins above. V/C indicates variable-constant junctions based on the DNA (solid line) and on all light chain protein sequences (broken line).

Variable-Constant Junction. We hoped to find an isolated variable region on the DNA. The conventional assignment of the junction between variable and constant regions is at amino acid 112; however, the DNA sequence deviates from the protein sequence sharply after amino acid 98. There is no agreement between λ light-chain amino acid codons and the DNA sequence over the next 240 bases past this point. We conclude that the embryonic DNA contains this variable region gene in iso-

1488 Genetics: Tonegawa *et al.* *Proc. Natl. Acad. Sci. USA 75 (1978)*

HV1	T	A	C	**T**	A	G	T	T	G	T	A	A	C	A	G	C	C	C	C	A	G	T	A	**C**	T
HV2	A	**A**	T	A	G	G	T	**G**	**G**	T	A	C	C	A	G	C	**A**	A	C	C	G	A	G	**C**	T
HV3	G	C	T	C	T	A	T	G	G	T	**A**	-	C	**A**	G	**C**	A	**C**	C	C	**A**	T	T	T	C

FIG. 4. Double and triple DNA sequence homologies in the $V\lambda_{II}$ gene corresponding to immunoglobulin light chain hypervariable regions. The bottom strand of HV1 and top strands of HV2 and HV3 (Fig. 3) have been aligned 5′ to 3′, with the bases that could change to produce known λ_I and inferred λ_{II} amino acid substitutions shown in boldface.

lation from the constant region. No obvious element identifies the boundary as unusual; we cannot interpret a feature of the DNA sequence that shows how the structures are brought together. The identification of amino acid 112 as the variable-constant junction is based on analogy with human sequences. We do not know, for the mouse λ_{II} chains, that the junction should be there: we interpret our sequence as showing that the variable-constant junction of λ_{II} gene follows amino acid 98.

Hypervariable Regions. Sequences corresponding to hypervariable regions appear in this germ-line gene. The first hypervariable region in the DNA sequence fits both the λ_I and λ_{II} proteins (Fig. 3). The second hypervariable region corresponds to λ_{II} at position 54 (Ser rather than Asn) to λ_I at 55 (Asn rather than Asp) and to both elsewhere. However, at the third hypervariable region, the MOPC 315 protein deviates from our DNA sequence at three amino acids, two of which correspond to λ_I chains. Our finding that the hypervariable regions are in the germ-line DNA rules out models in which DNA is inserted into a pre-existing germ-line sequence. Differences among mouse λ light chains in the hypervariable regions are thus most likely due to the gene for the final protein having accrued one or several mutations before or during the antigen-driven expansion of the pool of precursor cells specialized to make that particular immunoglobulin.

There are no features in these hypervariable regions that are so dramatic as to define a mechanism by which these regions might be more labile than the rest of the DNA. We do not find extensive palindromes surrounding these hypervariable regions (14, 15). (Of the four palindromes suggested in ref. 14, we find only one in the second hypervariable region.) There is, however, a similarity in the DNA structure of the three hypervariable regions. Fig. 4 matches the bottom strand of the first hypervariable region with the top strand of the second and third. These similarities may be simply evolutionary relics, but it is not impossible that they could serve as a recognition site for an enzyme that would cleave or modify the DNA in order to make the sequence labile.

At three positions in the hypervariable regions and one in the framework, this $V\lambda_{II}$ gene deviates from λ_{II} (MOPC 315) expectations and agrees with the most common λ_I residues. This may reflect a past duplication that gave rise to both the λ_I and λ_{II} germ-line genes.

$V_{\lambda_{II}}$ Gene Contains a Genetic Discontinuity—93 Nontranslated Bases in Leader Region. The light chain is synthesized as a longer precursor (16, 17), containing at its amino terminus a hydrophobic peptide that, presumably, is involved in moving this protein through the cell membrane (18, 19). As the protein matures, this precursor is cleaved at a glutamine (20), which cyclizes into a pyrrolidone carboxylic acid (10). Sequences of the 19-amino-acid long signal peptides for mouse λ light chains have been worked out by Schecter and his coworkers (20, 21) by translating purified immunoglobulin messenger *in vitro*. The DNA, however, codes for only four of the expected amino acids preceding the initial glutamine (Fig. 3) and then deviates entirely from the precursor protein. Ninety-three bases earlier in the DNA, we find a sequence for the first 15 amino acids of the light chain precursor, including an ATG for the initial methionine. When this match was first made, only the λ_I precursor sequence was available, and that deviates from the DNA in three places. However, after the DNA sequence was established, Burstein and Schecter (ref. 20; personal communication) worked out the hydrophobic leader sequence for the λ_{II} light chain; that protein sequence agrees with the DNA at all points. Therefore the DNA sequence was a predictor of the protein sequence. The extra region of 93 bases cannot be translated into protein; it contains a stop signal in phase. This 93-base region is not an artifact of the cloning. It contains a unique *Pst* I cut that can be exhibited in uncloned embryonic DNA (M. Hirama and S. Tonegawa, unpublished).

We believe the functioning gene in the myeloma will consist of the precursor region followed first by the 93-base-long interspersed DNA, followed by the variable region gene, then by a 1250-base-long piece of noncoding DNA (4), and last, by the constant region gene. We call such an additional piece of DNA that arises within a gene an ***intron*** (for intragenic region or intracistron) and thus look upon the structure of this gene as leader(45)-intron(93)-variable (306)-intron(1250)-constant(348). The current level of analysis cannot exclude still more small introns within the constant region nor specify the exact location of the second intron.

Maturation. Does the sequence for the 93-base intron show us how this area can be skipped or excised from the message? There is only a short repetition at its ends, a CAGG sequence repeated in the correct phase. A four-base repeat, however, is not specific enough to be used by the RNA polymerase as a signal to stop and start up again on an RNA-priming model. The signals to eliminate the unwanted region may be contained in the base sequence and structure of the messenger. A hairpin structure could hold the point of splicing in its stem, but that would necessitate ligation from one chain across to the opposite side of the helix. Fig. 5 shows the one well-placed hairpin we find in the sequence; it is neither convincing nor impossible. An alternative structure would be to use a sequence elsewhere in the messenger as a template to bring into juxtaposition the bonds that are to be split and rejoined.

Genes in Pieces. We hypothesize that most higher cell genes will consist of informational DNA interspersed with silent sequences. The eukaryotic cistron is a transcription unit containing alternate regions to be excised from the messenger, the introns, and regions left to be expressed, exons. [There are many recent examples of this structure (4, 22–26); others are reviewed in ref. 27.] Thus the gene is a mosaic: sequences corresponding to expressed functions held in a matrix, the latter possibly 10 times larger than the coding components. This model immediately resolves two puzzles. Heterogeneous nuclear RNA would be the long transcription products from which the much smaller ultimate messengers are spliced. The extra DNA in higher cells would arise because the genes, the transcription units, are much larger than those for a single polypeptide product.

Assume that the splicing mechanism is general, independent of the gene structure, and recognizes simply a unique secondary structure in the RNA. Of what benefit then are the infilling sequences? They speed evolution. Single base changes not only

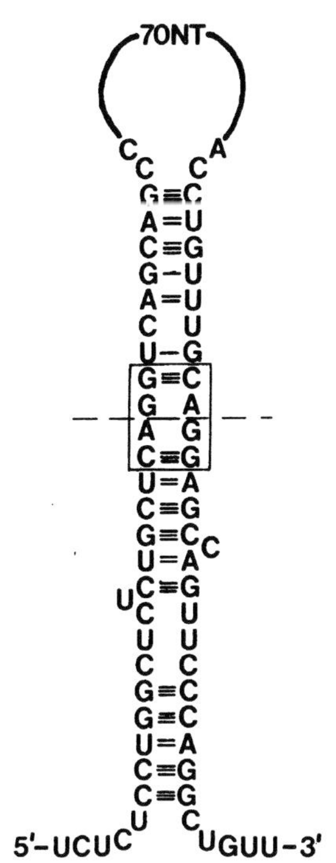

FIG. 5. Hypothetical hairpin structure in λ_{II} light chain precursor mRNA with the nontranslated sequences to be excised (intron) above the dashed line and coding sequences to be joined (exon) below. The base-paired sequences are from the ends of the 93-base noncoding segment in Fig. 3. Proper splicing of this precursor RNA would require a symmetric pair of cuts in the antiparallel sequences boxed in the stem, followed by ligation of the strands below to preserve one copy of the CAGG sequence. The strands to be joined, however, are not favorably oriented for ligation.

can alter single amino acids, but now, if they occur at the boundaries of introns, can change the splicing to add or delete a string of amino acids. This generates a more rapid search through the space of protein molecules. Second, the splicing process need not be 100% efficient. For example, changes in silent positions, third base positions in codons, can alter both the pattern and the efficiency of splicing so that the product of a single transcription unit can be two polypeptide chains, one being the original gene product and the second, also synthesized at high frequency, the new product. Evolution can seek new solutions without destroying old. This resolves a classic problem: one thought that the organism had to create a second copy of an essential gene in order to mutate it to a new function. The intronic model eliminates any special duplication. The extra material, scattered widespread across the genome, can be called into action at any time. After a new gene function appears, there can be selective pressure for duplication. One consequence of the intron model is that the dogma of one gene, one polypeptide chain disappears. A gene, a contiguous region on DNA, now corresponds to one transcription unit, but that transcription unit can correspond to many polypeptide chains, of related or differing functions.

Since the gene is now spread out over a larger region of DNA, recombination between exons will be enhanced. Furthermore, if the exons correspond to identifiable functions put together by splicing to form a special combination in a finished protein, then recombination between these regions can sort their functions independently. In the λ light chain a hydrophobic precursor sequence is separated by an intron from its follower region; recombination could combine this leader sequence with some other protein. One might anticipate that middle repetitive sequences within introns will provide hot spots for recombination to reassort the exonic sequences.

Still another picture emerges if we hypothesize that specific new splicing patterns can be turned on by special gene products, providing developmental control. A differentiation pathway can be defined by the appearance of a new splicing enzyme.

One striking interpretation, with this new picture, would be that the simultaneous expression of IgM and IgD with the same idiotype by a single cell (28) will turn out to be due to a V_H gene translocated near the C_H genes and transcribed together with C_μ and C_δ into a single V_H-μ-δ precursor, alternate splicings then providing the two products. The switch from IgM to IgG may be a new translocation of a V_H gene, or it may be a new processing of a V_H-μ-δ-γ precursor to produce a V_H-γ product.

We are grateful for the expert assistance of Rita Lenhard-Schuller. We thank I. Schecter for the communication of results before publication and P. Slonimsky, B. Blomberg, D. Wiley, and S. Harrison for discussions. W.G. is an American Cancer Society Research Professor. Part of this work was supported by National Institutes of Health Grant GM 09541-16.

1. Wu, T. T. & Kabat, E. A. (1970) *J. Exp. Med.* **132,** 211–250.
2. Cohn, M., Blomberg, B., Geckeler, W., Raschke, W., Riblet, R. & Weigert, M. (1974) in *The Immune System; Genes, Receptors, Signals*, eds. Sercarz, E. E., Williams, A. R. & Fox, C. F. (Academic, New York), pp. 89–117.
3. Hozumi, N. & Tonegawa, S. (1976) *Proc. Natl. Acad. Sci. USA* **73,** 3628–3632.
4. Brack, C. & Tonegawa, S. (1977) *Proc. Natl. Acad. Sci. USA* **74,** 5652–5656.
5. Wu, T. T., Kabat, E. A. & Bilofsky, H. (1975) *Proc. Natl. Acad. Sci. USA* **72,** 5107–5110.
6. Tonegawa, S., Brack, C., Hozumi, N. & Schuller, R. (1977) *Proc. Natl. Acad. Sci. USA* **74,** 3518–3522.
7. Tonegawa, S., Brack, C., Hozumi, N. & Pirrotta, V. (1977) in *Cold Spring Harbor Symp. Quant. Biol.*, in press.
8. Maxam, A. M. & Gilbert, W. (1977) *Proc. Natl. Acad. Sci. USA* **74,** 560–564.
9. Roberts, R. J. (1977) in *Critical Reviews in Biochemistry*, ed. Fasman, G. D. (CRC Press, Cleveland, OH), pp. 123–164.
10. Weigert, M. G., Cesali, I. M., Yonkovich, S. J. & Cohn, M. (1970) *Nature* **228,** 1045–1047.
11. Apella, E. (1971) *Proc. Natl. Acad. Sci. USA* **68,** 590–594.
12. Cesari, I. M. & Weigert, M. (1973) *Proc. Natl. Acad. Sci. USA* **70,** 2112–2116.
13. Dugan, E. S., Bradshaw, R. A., Simms, E. S. & Eisen, H. N. (1973) *Biochemistry* **12,** 5400–5416.
14. Leder, P., Honjo, T., Seidman, J. & Swan, D. (1976) *Cold Spring Harbor Symp. Quant. Biol.* **41,** 855–862.
15. Wuilmart, C., Urbain, J. & Givol, D. (1977) *Proc. Natl. Acad. Sci. USA* **74,** 2526–2530.
16. Milstein, C., Brownlee, G., Harrison, T. M. & Mathews, M. B. (1972) *Nature New Biol.* **239,** 117–120.
17. Swan, D., Aviv, H. & Leder, P. (1972) *Proc. Natl. Acad. Sci. USA* **69,** 1967–1972.
18. Blobel, G. & Dobberstein, B. (1975) *J. Cell Biol.* **67,** 835–851.
19. Blobel, G. & Dobberstein, B. (1975) *J. Cell Biol.* **67,** 852–862.
20. Burstein, Y. & Schecter, I. (1977) *Biochem. J.* **165,** 347–354.
21. Burstein, Y. & Schecter, I. (1977) *Proc. Natl. Acad. Sci. USA* **74,** 716–760.
22. Berget, S. M., Moore, C. & Sharp, P. A. (1977) *Proc. Natl. Acad. Sci. USA* **74,** 3171–3175.
23. Chow, L. T., Gelinas, R. E., Broker, T. R. & Roberts, R. J. (1977) *Cell* **12,** 1–8.
24. Klessig, D. F. (1977) *Cell* **12,** 9–21.
25. Breathmark, R., Mandel, J. L. & Chambon, P. (1977) *Nature* **270,** 314–319.
26. Doel, M. T., houghton, M., Cook, E. A. & Carey, N. H. (1977) *Nucleic Acids Res.* **4,** 3701–3713.
27. Williamson, B. (1977) *Nature* **270,** 295–297..w
28. Rowe, D. S., Hug, K., Forni, L. & Permis, B. (1973) *J. Exp. Med.* **138,** 965–972.

Nature Vol. 271 9 February 1978 **501**

news and views

Why genes in pieces?

from Walter Gilbert

Our picture of the organisation of genes in higher organisms has recently undergone a revolution. Analyses of eukaryotic genes in many laboratories[1–10], studies of globin, ovalbumin, immunoglobulin, SV40 and polyoma, suggest that in general the coding sequences on DNA, the regions that will ultimately be translated into amino acid sequence, are not continuous but are interrupted by 'silent' DNA. Even for genes with no protein product such as the tRNA genes of yeast and the rRNA genes in *Drosophila*, and also for viral messages from adenovirus, Rous sarcoma virus and murine leukaemia virus, the primary RNA transcript contains internal regions that are excised during maturation, the final tRNA or messenger being a spliced product.

The notion of the cistron, the genetic unit of function that one thought corresponded to a polypeptide chain, now must be replaced by that of a transcription unit containing regions which will be lost from the mature messenger—which I suggest we call introns (for intragenic regions)—alternating with regions which will be expressed—exons. The gene is a mosaic: expressed sequences held in a matrix of silent DNA, an intronic matrix. The introns seen so far range from 10 to 10,000 bases in length; I expect the amount of DNA in introns will turn out to be five to ten times the amount in exons.

This model immediately accommodates two aspects of the genetic structure of higher cells. Heterogeneous nuclear RNA clearly is the long transcription products out of which the much smaller ultimate messengers for expressed polypeptide sequences are spliced. The unexpected extra DNA in higher cells, the excess of DNA over that needed to code for the number of products defined genetically, now is ascribed to the introns.

What are the benefits of this intronic/exonic structure for genes? For the sake of argument let us assume that the splicing mechanism is general and independent of the specific gene or the state of the cell, reflecting simply some secondary structure in the RNA. For example, base-pairing in the messenger could generate sites which would serve as signals for enzymes, such as those that excise tRNAs from their precursors, to cut out a section. The cut would be resealed by an RNA ligase. Even if RNA processing is general, the presence of infilling sequences can speed evolutiony

Single base changes, the elementar. mutational events, not only can change protein sequences by the alteration of single amino acids but now, if they occur at the boundaries of the regions to be spliced out, can change the splicing pattern, resulting in the deletion or addition of whole sequences of amino acids. During the course of evolution relatively rare single mutations can generate novel proteins much more rapidly than would be possible if no splicing occurred.

Furthermore, the splicing need not be a hundred per cent efficient; changes in sequence can alter the process so that base pairing and splicing occurs only some of the time. Even mutations in silent third base positions, could modify the joining so that the products of a single transcription unit can be both the original gene product and a new product, also synthesised at a high rate. Evolution can seek new solutions without destroying the old. A classic problem is resolved: the genetic material does not have to duplicate to provide a second copy of an essential gene in order to mutate to a new function. Rather than a special duplication, the extra material is scattered in the genome, to be called into action at any time. After a new gene function appears, if a higher level of product is needed, there will be selective pressure for gene duplication (as well as pressure for the loss of the introns in highly repeated genes). One consequence of the intronic model is that the dogma of one gene, one polypeptide chain disappears. A gene, a contiguous region of DNA, now corresponds to one transcription unit, but that transcription unit can correspond to many polypeptide chains, of related or differing functions.

Recombination now becomes more rapid. Since the gene is spread out over a larger region of DNA, recombination, which should be hampered in higher cells by the inability of DNA molecules to get together, will be enhanced. Furthermore, if exonic regions correspond to functions put together by splicing to form special combinations in the finished protein, then recombination within introns will assort these functions independently. Middle repetitious sequences within introns may create hot spots for recombination to rearrange the exonic sequences.

Recombination within introns will generate curious genetic structures for eukaryotic genes. Structural mutations should be clustered, separated by long distances from mutations in other exons. Mutations in different functions may be interspersed, when one product's intron becomes another's exon.

According to this view, introns are both frozen remnants of history and as the sites of future evolution. Nevertheless, they could also have other roles. Specific recombinations between introns can bring together exons into a transcription unit to make special differentiation products. Specific new splicing patterns could be turned on by special gene products. A differentiation pathway may be determined by the appearance of a new splicing enzyme, calling forth new proteins out of the heterogeneous nuclear RNA.

On this can be based a striking hypothesis to explain the behaviour of immunoglobulin heavy chains. At an early stage of the immune response a single lymphocyte can synthesise two different immunoglobulins, IgM and IgD, with the same idiotype; two different constant portions attached to the same V_H region. This may be the result of a V_H region translocating by recombination within an intron near the constant genes so that a trranscription unit is formed for a V_H-C_μ-C_δ message. Splicing can then create contiguous messenger sequences for V_HC_μ and V_HC_δ chains. The switch from IgM to IgG might be a new translocation of the V_H gene, but, alternatively, it may be a new enzyme that changes the processing of a V_H-$C_{\mu\gamma}$ C_δ-C_γ message to produce a V_HC-product. □

Walter Gilbert is American Cancer Society Professor of Molecular Biology at Harvard University.

1. *News and Views, Nature* **268**, 101 (1977).
2. Berget *et al. Proc. natn. Acad. Sci. U.S.A.* **74**, 3171 (1977).
3. Papers in *Cell* **12** (1) (1977).
4. Aloni *et al. Proc. natn. Acad. Sci. U.S.A.* **74**, 3686 (1977).
5. Leder *et al. Cold Spring Harbor Symp. quant. Biol.* (in the press).
6. *News and Views, Nature* **270**, 295 (1977).
7. Breathmark *et al. Nature* **270**, 314 (1977).
8. Deol *et al. Nucleic Acids Res.* **4**, 3701 (1977).
9. Jeffreys & Flavell *Cell* **12**, 1097 (1977).
10. Brack & Tonegawa *Proc. natn. Acad. Sci. U.S.A.* **74**, 5652 (1977).

0028-0836/78/0271—0501$01.00

Bioscience Reports 1, 353-375 (1981)
Printed in Great Britain

353

DNA sequencing and gene structure

Nobel lecture, 8 December 1980

Walter GILBERT

Department of Biochemistry and Molecular Biology,
Harvard University, The Biological Laboratories,
Cambridge, Massachusetts 02138, U.S.A.

When we work out the structure of DNA molecules, we examine the fundamental level that underlies all processes in living cells. DNA is the information store that ultimately dictates the structure of every gene product, delineates every part of the organism. The order of the bases along DNA contains the complete set of instructions that make up the genetic inheritance. We do not know how to interpret those instructions; like a child, we can spell out the alphabet without understanding more than a few words on a page.

I came to the chemical DNA sequencing by accident. Since the middle sixties my work had focussed on the control of genes in bacteria, studying a specific gene product, a protein repressor made by the control gene for the *lac* operon (the cluster of genes that metabolize the sugar lactose). Benno Müller-Hill and I had isolated and characterized this molecule during the late sixties and demonstrated that this protein bound to bacterial DNA immediately at the beginning of the first gene of the three-gene cluster that this repressor controlled (1,2). In the years since then, my laboratory had shown that this protein acted by preventing the RNA polymerase from copying the *lac* operon genes into RNA. I had used the fact that the *lac* repressor bound to DNA at a specific region, the operator, to isolate the DNA of this region by digesting all of the rest of the DNA with DNase to leave only a small fragment bound to the repressor, protected from the action of the enzyme. This isolated a twenty-five-base-pair fragment of DNA out of the 3 million base pairs in the bacterial chromosome. In the early seventies, Allan Maxam and I worked out the sequence of this small fragment (3) by copying this DNA into short fragments of RNA and using on these RNA copies the sequencing methods that had been developed by Sanger and his colleagues in the late sixties. This was a laborious process that took several years. When a student, Nancy Maizels, then determined the sequence of the first 63 bases of the messenger RNA for the *lac* operon genes, we discovered that the *lac* repressor bound to DNA immediately after the start of the messenger RNA (4), in a region that lies under the RNA polymerase when it binds to DNA to initiate RNA synthesis. We continued to characterize the *lac* operator by sequencing a number of mutations (operator constitutive mutations) that damaged the ability of the repressor to bind to DNA. We wanted to determine more DNA sequence in the region to define the polymerase binding site and other elements involved in *lac* gene control; however, that sequence was worked out in another laboratory by Dickson, Abelson, Barnes, and Reznikoff (5). Thus by the middle seventies I knew all the sequences that I had been curious about, and my students (David Pribnow and John Majors) and I were trying to answer questions about the interaction of the RNA polymerase and other control factors with DNA.

At this point, another line of experiments was opened up by a new suggestion. Andrei Mirzabekov came to visit me in early 1975. The purpose of his visit was twofold: to describe experiments that he had been doing using dimethyl sulfate to methylate the guanines and the adenines in DNA and to urge me to do a similar experiment with the *lac* repressor. Dimethyl sulfate methylates the guanines uniquely at the *N7* position, which is exposed in the major groove of the DNA double helix, while it methylates the adenines at the *N3* position which is exposed in the minor groove (Fig. 1). Mirzabekov had used this property to attempt to determine the disposition of histones and of certain antibiotics on the DNA molecule by observing the blocking of the incorporation of radioative methyl groups onto the guanines and adenines of bulk DNA. He urged me to use this groove specificity to learn something about the interaction of the *lac* repressor with the *lac* operator. However, the amounts of *lac* operator available were extremely small, and there was no obvious way of examining the protein sitting on DNA to ask which bases in the sequence the protein would protect against attack by the dimethyl sulfate reagent.

It was not until after a second visit by Mirzabekov that an idea finally emerged. He and I, and Allan Maxam and Jay Gralla, had lunch together. During our conversation I had an idea for an experiment, which ultimately underlies our sequencing method. We knew we could obtain a defined DNA fragment, 55 base-pairs long, which carried near its center the region to which the *lac* repressor bound. This fragment was made by cutting the DNA sequentially with two different restriction enzymes, each defining one end of the fragment (see Fig. 2). Secondly, I knew that at every base along the DNA at which methylation occurred, that base could be removed by heat. Furthermore, once that had happened, only a sugar would be left holding the DNA chain together, and that sugar could be hydrolysed, in principle, in alkali to break the DNA chain. I put these ideas together by conjecturing that if we labelled one end of one strand of the DNA fragment with radioactive phosphate, we might determine the point of methylation by measuring the distance between the labelled end and the point of breakage. We could get such labelled DNA by isolating a DNA fragment (by length by electrophoresis through polyacrylamide gels) made by cutting with one restriction enzyme, labelling both ends of that fragment, and then cutting it again with a second restriction enzyme to release two separable double-stranded fragments, each having a label at one end but not the other. Using polynucleotide kinase this procedure would introduce a radioactive label into the 5' end of one of the DNA strands of the fragment bearing the operator while leaving the other unlabelled (Fig. 2). If we then modified that DNA with dimethyl sulfate so that only an occasional adenine or guanine would be methylated, heated, and cleaved the DNA with alkali at the point of depurination, we would release among other fragments a labelled fragment extending from the unique point of labelling to the first point of breakage. Fig. 3 shows this idea. Any fragments from the

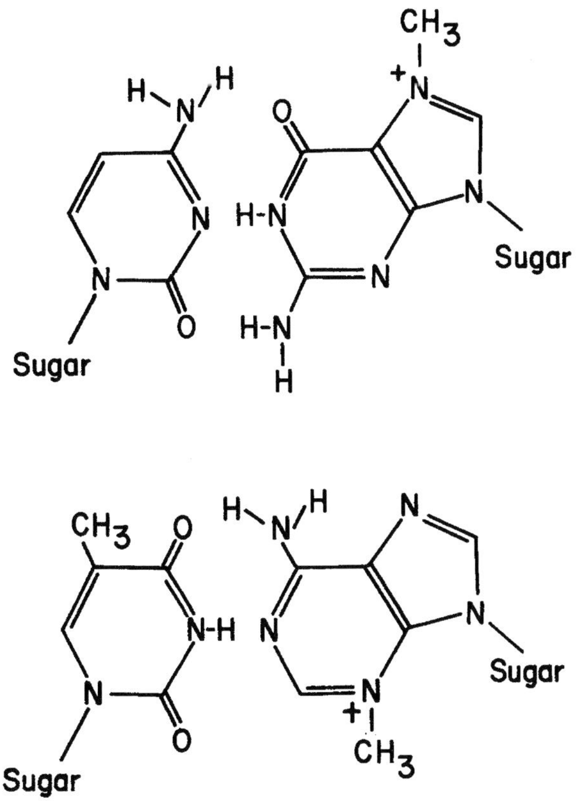

Fig. 1. Methylated cytosine-guanine and thymine-adenine base pairs. The top of the figure shows a cytosine-guanine base pair methylated at the *N7* position of guanine. The bottom of the figure shows a thymine-adenine base pair methylated at the *N3* position of adenine. The region above each of the base pairs is exposed in the major groove of DNA. The region below each of the base pairs lies between the sugar phosphate backbones in the minor groove of the DNA double helix.

other strand would be unlabelled, as would any fragments arising beyond the first point of breakage. If we could separate these fragments by size, as we could in principle by electrophoresis on a polyacrylamide gel, we might be able to associate the labelled fragments back to the known sequence and thus identify each guanine and adenine in the operator that had been modified by dimethyl sulfate. If we could do the modification in the presence of the *lac* repressor protein bound to the DNA fragment, then if the repressor lay close to the *N7* of a guanine, we would not modify the DNA at that base, and the corresponding fragment would not appear in the analytical pattern.

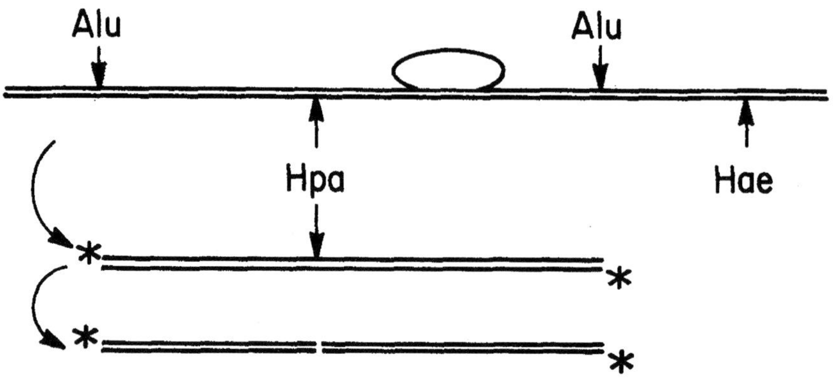

Fig. 2. Procedure for obtaining a double-stranded DNA fragment uniquely labelled at one end of one strand. The figure shows the restriction cuts for the enzymes *Alu*I (target AG/CT) and *Hpa* (target C/CGG producing an uneven end) in the neighborhood of the *lac* operator. The *lac* repressor is shown bound to the DNA. By cutting the DNA from this region first with the enzyme *Alu*, then labelling with radioactive P^{32} the 5' termini of both strands of the DNA with polynucleotide kinase, and then cutting in turn with the enzyme *Hpa*, we can isolate a DNA fragment that carries the binding site for the repressor uniquely labelled at one end of one strand.

I set out to do this experiment. Allan Maxam made the labelled DNA fragments, and I began to learn how to modify and to break the DNA. This involved analysing the release of the bases from DNA and the breakage steps separately. Finally the experiment was put together. Fig. 4 shows the results: an autoradiogram of the electrophoretic pattern displays a series of bands extending downward in size from the full-length fragment, each caused by the cleavage of the DNA at an adenine or a guanine. The same treatment of the DNA fragment with dimethyl sulfate, now carried out in the presence of the repressor, produced a similar pattern, except that some of the bands were missing (lane one versus lane two in Fig. 4). The experiment was clearly a success in that the presence of the repressor blocked the attack by dimethyl sulfate on some of the guanines and some of the adenines in the operator (6). I hoped that the size discrimination would be accurate enough to permit the assignment of each band in the pattern to a specific base in the sequence. This proved true because the spacing in the pattern, and the presence of light and dark bands, the dark bands corresponding to guanines and the light ones to adenines, were sufficiently characteristic to correlate the two. The guanines react about five times more rapidly with dimethyl sulfate, while the methylated adenines are released from DNA more rapidly than the guanines during heating; the shift in intensities as a function

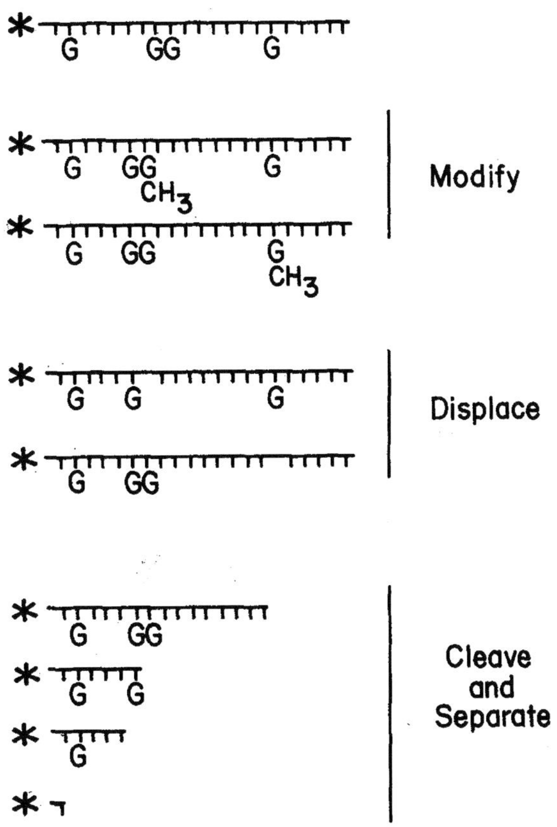

Fig. 3. Outline of procedure to produce fragments of DNA by breaking the DNA at guanines. Consider an end-labelled strand of DNA. We modify an occasional guanine by methylation with dimethyl sulfate. Heating the DNA will then displace that guanine from the DNA strand, leaving behind the bare sugar; cleaving the DNA with alkali will break the DNA at the missing guanine; the fragments are then separated by size, the actual size of the fragment (followed because it carries the radioactive label) being determined by the position of the modified guanine.

of the time of heat treatment could be used to establish unambiguously which base was which originally. Furthermore, the gel pattern is so clear, bands corresponding to fragments differing by one base were resolved. At this point it was clear that this technique could determine the adenines and guanines along DNA for distances of the order of 40 nucleotides. By determining the purines on one strand and the purines on the complementary strand (as Fig. 4 does), one has in principle a complete sequencing method.

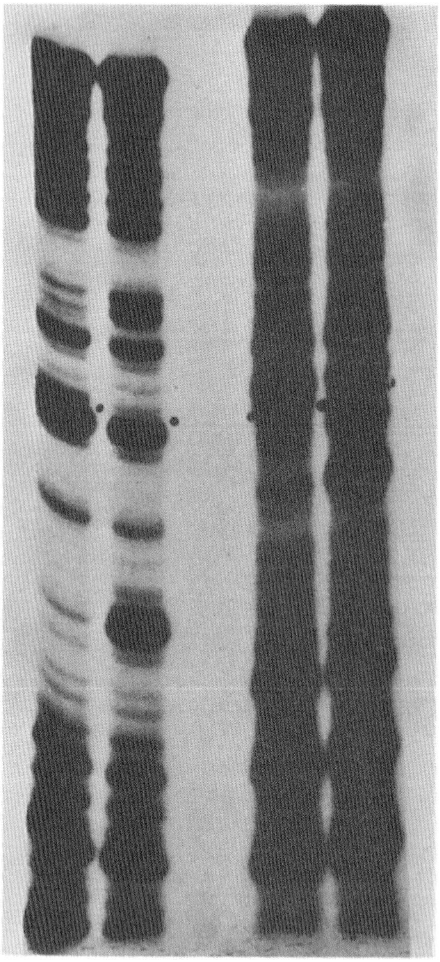

Having in hand a reaction that will determine and distinguish adenines and guanines, could we find reactions that would distinguish cytosines and thymines? Allan Maxam and I turned our attention to this end. (First we examined a second binding site for the *lac* repressor that lies a few hundred bases further along the DNA, under the first gene of the operon. This binding site has no physiological function. We could locate this binding site on a restriction fragment by repeating the methylation-protection experiment and identifying bases protected by the *lac* repressor. I used the methylation pattern to attempt to predict the positions of the adenines and the guanines in the unknown DNA sequence; Allan Maxam then used the wandering spot sequencing method of Sanger and his coworkers to determine the DNA sequence of this region to verify that we had made a successful prediction.) Allan Maxam then went on to do the next part of the development. We knew that hydrazine would attack the cytosines

Fig. 4. Methylation protection experiment with the *lac* repressor. The columns show the pattern of cleavage along each strand of the 53- to 55-base-long fragment bearing the *lac* operator. The second column from the left represents the DNA (labelled at the 5' end of the 53-base-long strand) treated by dimethyl sulfate, cleaved by heat and alkali. The dark bands correspond to breaks at guanines; the light bands are breaks at adenines. The second column of the figure reads from the bottom:

-G-GAAA--G--A---G---A-AA----A-A-AA-A-A-....

The first column shows this same double-stranded piece of DNA treated with dimethyl sulfate in the presence of the *lac* repressor. The repressor prevents the interaction of dimethyl sulfate with the guanine a third of the way up the pattern and blocks the reaction with two adenines in the upper third of the pattern. The bands corresponding to these two adenines represent fragments that differ in length by one base, 30 and 31 long. The right-hand side of this pattern shows the same experiment done with the label at the end of the other DNA strand. The sequence at the far right would be:

-G-G-GGAA--G-GAG-GGA-AA-AA---....

and thymines in DNA and damage them sufficiently, or eliminate them to form a hydrazone, so that a further treatment of the DNA with benzaldehyde followed by alkali (or a treatment with an amine), would cleave at the damaged base. This soon gave us a similar pattern, but broke the DNA at the cytosines and thymines without discrimination. Allan Maxam then discovered that salt, one-molar salt in the 15-molar hydrazine, altered the reaction to suppress the reactivity of the thymines. The two reactions together then positioned and distinguished the thymines and the cytosines in a DNA sequence. This last discrimination conceptually completed the method. To improve the discrimination between the purines, and to provide redundant information which would serve to make the sequencing more secure against errors, we used the fact that the methylated adenosines depurinate more rapidly, especially in acid, to release the adenosines preferentially and thus to obtain four reactions: one for A's, one preferential for G's, one for C's and T's, and one for C's determining the T's by difference. This stage of the work was completed within a few months. As the range of resolution on the gels was extended toward 100 bases, the cleavage at the pyrimidines was not satisfactory; the result of the incomplete cleavage was that the longer fragments contained a variety of internal damages and the pattern blurred out. After many months of searching, an answer was found. A primary amine, aniline, will displace the hydrazine products and produce a beta elimination that releases the phosphate from the 3' position on the sugar, but it will not release the other phosphate, and the mobility of a DNA fragment with a blocked 3' phosphate bearing a sugar-aniline residue is different from the free-phosphate-ended chains from

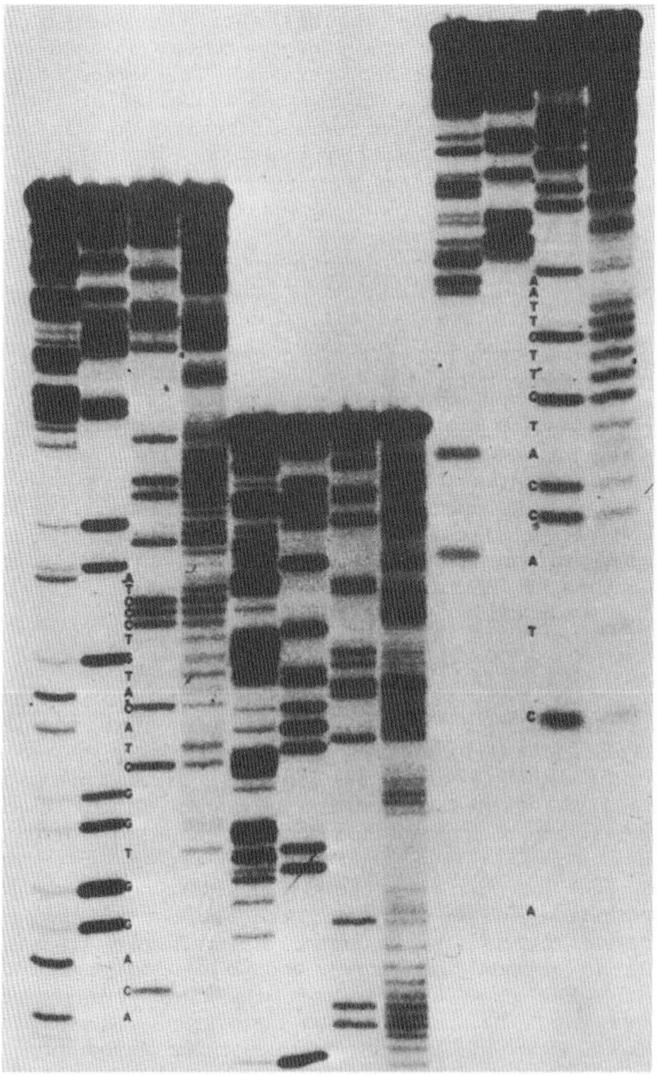

the other reactions. A secondary amine, piperidine, is far more effective and triggers both beta eliminations as well as eliminating all the breakdown products of the hydrazine reaction from the sugar. This reagent completed the DNA sequencing techniques (7). Although the development of the techniques continued for another nine months, they were distributed freely to other groups that wanted to use them. Fig. 5 shows an actual sequencing pattern from the 1978 period, used in the work described in (8). Fig. 6 shows two examples of the chemistry (9).

The logic behind the chemical method is to divide the attack into two steps. In the first we use a reagent that carries the specificity,

Fig. 5. Actual sequencing pattern from the 1978 period. Products of four different chemical reactions, applied to a DNA fragment about 150 bases long, are electrophoresed on a polyacrylamide gel; three loadings produce sets of patterns that have moved different distances down the gel. The four columns correspond to reactions that break the DNA: 1) primarily at the adenines, 2) only at the guanines, 3) at the cytosines but not the thymines, and 4) at both the cytosines and thymines. The very shortest fragments are at the bottom right-hand side of the picture and the sequence is read up the gel recognizing first the band in the left-hand column corresponding to A, a band in the two right-hand columns corresponding to C, a band in the far right-hand column corresponding to T, a band in the left-hand column corresponding to A, and so forth. After reading up as far as possible, the sequence continues in the sets of bands at the left-hand side of the gel and then still further in the pattern in the center of the gel. From the original photograph the sequence of the entire fragment can be read. The fragment is from the genomic DNA corresponding to the variable region of the lambda light chain of mouse immunoglobulin (8).

but we limit the extent of that reaction - to only one base out of several hundred possible targets in each DNA fragment. This permits the reaction to be used in the domain of greatest specificity: only the very initial stages of a chemical reaction are involved. The second step, the cleavage of the DNA strand, must be complete. Since the target has already been distinguished from the other bases along the DNA chain by the preliminary damage, we can use vigorous, quantitative reaction conditions. The result is a clean break, releasing a fragment without hidden damages, which is required if the mobilities of the fragments are to be very closely correlated so that the bands will not blur. (The specificity need be only about a factor of ten for the sequence to be read unambiguously.)

Today, later developments of the technique (9) have modified the guanine reaction and replaced the dimethyl sulfate adenosine reaction with a direct depurination reaction that releases both the adenines and guanines equally. These changes, and the introduction of the very thin gels by Sanger's group (10), now make it possible to read sequences out between 200-400 bases from the point of labelling. The actual chemical workup, the analysis on gels, and the autoradiography are the short part of the process. The major time spent in DNA sequencing is spent in the preparation of the DNA fragments and on the elements of strategy. The speed of the sequencing comes only in part from the ability to read off quickly several hundred bases of DNA - at a glance. The more important element is the linear presentation of the problem. Rather than sequence randomly, one can begin at one end of a restriction map and move rationally through a gene - or construct the restriction map as one goes.

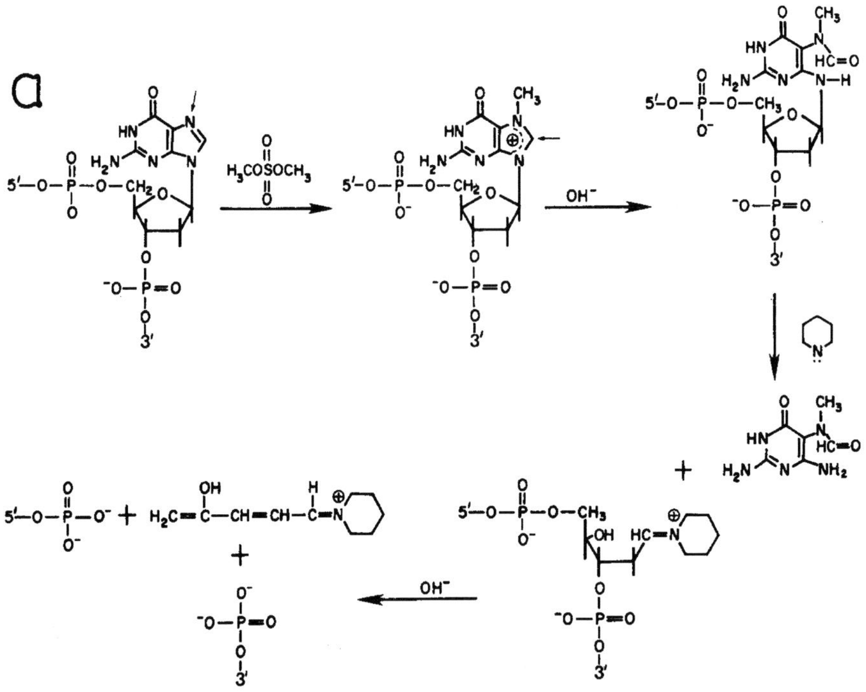
a
$H_3COSOCH_3$
OH⁻

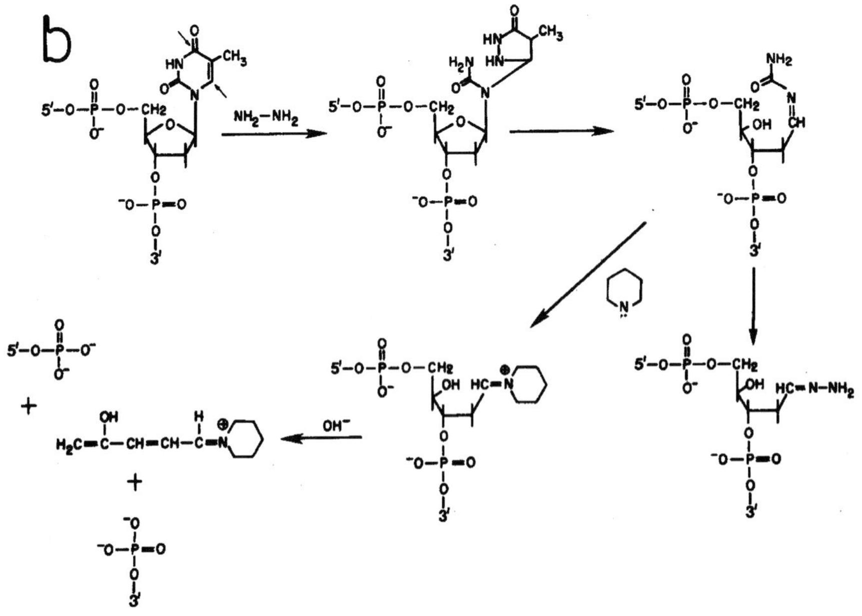
b
NH_2-NH_2
OH⁻

Fig. 6. Examples of the detailed chemistry involved in breaking the DNA. (a) The guanine breakage. The guanines are first methylated with dimethyl sulfate. The imidazole ring is opened by treatment with alkali (during the piperidine treatment). Piperidine displaces the base and then triggers two beta eliminations that release both phosphates from the sugar and cleave the DNA strand leaving a 3' and a 5' phosphate. (b) The hydrazine attack on a thymine that breaks the DNA at the pyrimidines.

The first long sequence was done by a graduate student, Phillip Farabaugh, who used the new techniques to sequence the gene for the *lac* repressor (11). The protein sequence of this gene product had been worked out in the early seventies by Beyreuther and his co-workers (12). Since the amino-acid sequence was known, he could quickly (a few months) establish the DNA sequence. However, the DNA sequence showed that there were errors in the protein sequence, two amino acids dropped at one place and eleven at another. Since the protein sequence contains 360 amino acids, he had to work out a gene of 1080 bases. DNA sequencing is faster and more accurate than protein sequencing. The reason for this is that DNA is a linear information store. Because the chemistry of each restriction fragment is like any other - they differ only in length - there is no particular reason for losing track of them, except for the very smallest. By sequencing across the joins between the fragments, one establishes an unambiguous order. Proteins, on the other hand, are strings of amino acids used by Nature to create a wide variety of chemistries. When a protein is fragmented, the fragments can exhibit quite different properties, some of which may be unusually unfortunate in terms of solubility or loss. There is no simple way of keeping account of the total content of amino acids, or of the order of fragments, as there is for DNA, where the length of the restriction fragments can easily be measured.

Jeffrey Miller and his coworkers had done an extensive analysis of the appearance of mutations in the *lac* repressor gene. Three sites in the gene are hotspots, at which the mutation rate is some 10 times higher than at other sites. DNA sequencing showed that at each of these sites there was a modified base, a 5-methyl cytosine, in the sequence (13). (The chemical sequencing detects the presence of the 5-methyl cytosine directly, because the methyl group suppresses completely the reactivity of this base in the hydrazine reaction. A blank space appears in the sequence, but on the other strand is a guanine.) The high mutation rate is a transition to a thymine. 5-Methyl cytosine occurs at a low frequency in DNA: this observation shows that it is a mutagen. What is the explanation? Deamination of cytosine to uracil occurs naturally. If this occurred in DNA it could lead to a transition; however, it usually does not, since there is an enzyme that scans DNA examining it for deoxyuridine (14). When it finds this base in DNA, mismatched or not, it breaks the glycosidic bond and removes the uracil. This is then recognized as a defect in

DNA, and another group of enzymes then repair the depyrimidinated spot. However, 5-methyl cytosine deaminates to thymine - a natural component of DNA. On repair or resynthesis a transition will ensue. This whole argument explains why thymine is used in DNA - the extra methyl group serves to suppress the effects of the natural rate of deamination.

To find out how easy and how accurate DNA sequencing was, I asked a student, Gregor Sutcliffe, to sequence the ampicillin resistance gene, the beta-lactamase gene, of *Escherichia coli*. This gene is carried on a variety of plasmids, including a small constructed plasmid, pBR322, in *E. coli*. All that he knew about the protein was an approximate molecular weight, and that a certain restriction cut on the plasmid inactivated that gene. He had no previous experience with DNA sequencing when he set out to work out the structure of DNA for this gene. After seven months he had worked out about 1000 bases of double-stranded DNA, sequencing one strand and then sequencing the other for confirmation. The unique long reading frame determined the sequence of the protein product of this gene, a protein of 286 residues (15). We thought that the DNA sequence was unambiguous. Luckily there was available, from Ambler's laboratory, partial sequence information about the protein which had been obtained as a result of several years' work attempting to develop a sequence for the beta-lactamase (16). This information, while not sufficient to determine the protein sequence directly, was adequate to confirm that the prediction of the DNA sequencing was correct. Sutcliffe then became very enthusiastic and sequenced the rest of the plasmid pBR322 during the next six months, to finish his thesis. He sequenced both strands of this 4362-base-pair-long plasmid in order to confirm the sequence (17). The chemical sequencing is unambiguous, except for an occasional characteristic feature in the DNA fragment itself that causes it to move anomalously during the gel electrophoresis. As longer and longer strands are being analysed on the gel, a hairpin loop can form at one end of the fragment if the sequence is sufficiently self-complementary. As the fragmentation passes through this portion of the molecule, the mobilities on the gel do not decrease uniformly as a function of length, but some of the molecules move aberrantly, a feature called compression, because the bands on an autoradiograph become close together, or can overlap to conceal one or more bases. This rare feature occurs about once every thousand bases. It is resolved by sequencing the opposite strand in the other direction along the double-stranded molecule (or the same strand in the opposite chemical direction) because the hairpin will form when a different region of the sequence is exposed and the compression feature will occur in a different place in the sequence. If both strands of the DNA helix are sequenced, the sequence can be unambiguous.

The Structure of Genes

The first genes to be sequenced, those in bacteria, yielded an expected structure: a contiguous series of codons lying upon the DNA between an initiation signal and one of the terminator signals. Before the position at which the RNA copy will start, there lies a site for the RNA polymerase, interacting with the Pribnow box, a region of sequence homology lying one turn of the helix before the initial base of the messenger RNA, and also with another region of homology, thirty-five bases before the start. Thus one could understand the bacterial gene in terms of a binding site for the RNA polymerase, and further binding sites for repressors and activator proteins around and under the polymerase. Alternatively, the control on transcription could be exercised by a control of the termination function: new proteins or an elegant translation control (18) could determine whether or not the polymerase would read past a stop signal into a new gene.

When the first genes from vertebrates were transferred into bacteria by the recombinant DNA techniques and sequenced, an entirely different structure emerged. The coding sequences for globin (19,20), for immunoglobulin (21), and for ovalbumin (22) did not lie on the DNA as a continuous series of codons but rather were interrupted by long stretches of non-coding DNA. The discovery of RNA splicing in adenovirus by Sharp and his coworkers (23) and Broker and Roberts and their coworkers (24) paved the way for this new structure. They had shown that after the original transcription of DNA into a long RNA, regions of this RNA are spliced out: some stretches are excised and the remaining portions are fused together by an as yet undefined enzymatic process. The exons (25), regions of the DNA that will be expressed in mature message, are separated from each other by introns, regions of DNA that lie within the genetic element but whose transcripts will be spliced out of the message. Fig. 7 shows this process: the original transcript of a gene (now thought of as a transcription unit) will undergo a series of splices before being able to function as a mature message in the cytoplasm. Fig. 8 shows a few examples. Vertebrate genes can have many - 8, 15, even 50 - exons (29,30), and the exons are for the most part short coding stretches separated by hundreds to several thousands of base pairs of intron DNA. The rapid sequencing has meant that we can work out the DNA sequence of any of these complex gene structures. But can we understand them?

The emerging generalization is that prokaryotic genes have contiguous coding sequences while the genes for the highest eukaryotes are characterized by a complex exon/intron structure. As we move up from prokaryotes, the simplest eukaryotes, such as yeasts, have few introns; further up the evolutionary ladder the genes are more broken up. (Yeast mitochondria have introns; are they an exception to this pattern?) Are we seeing the emergence of the intron-exon structure rising to ever greater degrees of complexity as we move up to the vertebrates, or the loss of preexisting intron-exon structures as we move down to the simplest invertebrates and the prokaryotes? One view considers the splicing as an adaptation that becomes ever more necessary in more highly structured organisms. The other view considers the splicing as lost if the organism makes a choice to simplify and to replicate its DNA more rapidly, to go through more generations in a short time, and thus to be under a significant pressure to restrict its DNA content (31).

What role can this general intron-exon structure play in the genes of the higher organisms? Although most genes that have been studied have this structure, there are two notable exceptions: the genes for the histones and those for the interferons. This last demonstrates that

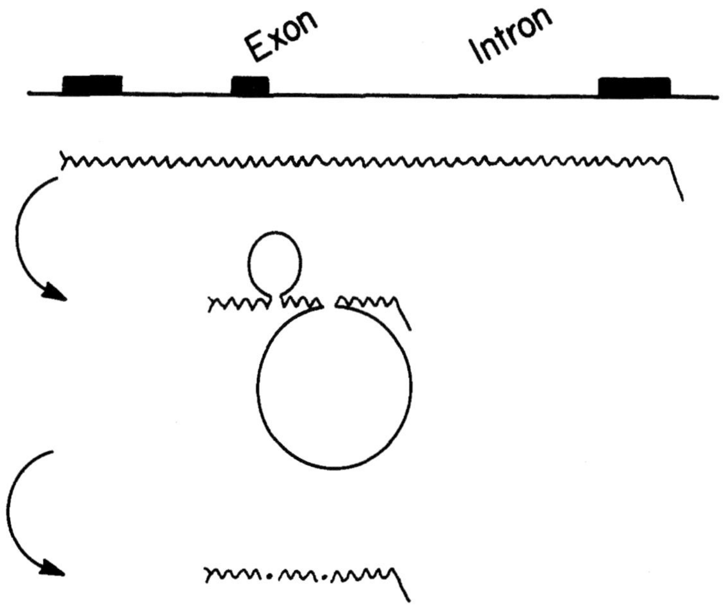

Fig. 7. A transcription unit corresponding to alternating exons and introns. The whole gene, a transcription unit, is copied into RNA terminating in a poly(A) tail. The regions corresponding to introns are spliced out leaving a messenger RNA made up of the three exons, the regions that are expressed in the mature message.

there can be no absolutely essential role that the introns must play, there can be no absolute need for splicing in order to express a protein in mammalian cells. Although there is a line of experiments that shows that some messengers must have at least a single splice made before they can be expressed, there is no evidence that the great multiplicity of splices are needed. There is a pair of genes for insulin in the rat, that differ in the number of introns; both are expressed - which demonstrates that the intron that splits the coding region of one of them has no essential role, in cis, in the expression of that gene (32). Although a common conjecture is that the splicing might have a regulatory role, so far there is no tissue-dependent splicing pattern that could be interpreted as showing the existence of a gene (or tissue) specific splicing enzyme.

The introns are much longer than the exons. Their DNA sequence drifts rapidly by mutation and small additions and deletions (accumulating changes as rapidly as possible at the same rate as the silent changes in codons). This suggests that it is not their sequence that is relevant, but their length. Their function is to move the exons apart along the chromosome.

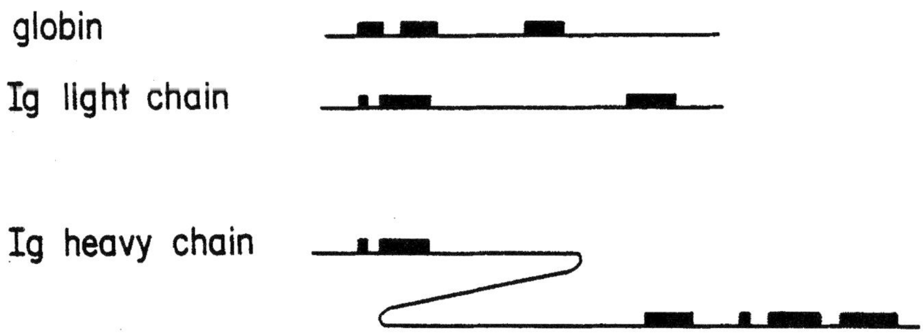

Fig. 8. Examples of the intron-exon structure of a few genes. (1) The gene for globin is broken up by two introns into three exons (20). (2) The functional gene in a myeloma cell for the immunoglobulin lambda light chain is broken up into a short exon corresponding to the hydrophobic leader sequence, an exon corresponding to the V region, and then, after an intron of some thousand bases, an exon corresponding to the 112 amino acids of the constant region (26). (3) A typical gene for a gamma heavy chain of immunoglobulin (27,28). The mature gene corresponds to a hydrophobic leader sequence, an exon corresponding to the variable region, and then, after a long intron, a series of exons: the first corresponding to the first domain of the constant region, the second corresponding to a 15-amino-acid hinge region, the third corresponding to the second domain of the constant region, and the fourth exon corresponding to the third domain of the constant region.

A consequence of the separation of exons by long introns is that the recombination frequency, both illegitimate and legitimate, between exons will be higher (25). This will increase the rate, over evolutionary time, at which the exons, representing parts of the protein structures, will be shuffled and reassorted to make new combinations. Consider the process by which a structural domain is duplicated to make the two-domain structure of the light chain of the immunoglobulins (or duplicated again to make the four-domain structure of the heavy chain, or combined to make the triple structure represented by ovomucoid (29)). Classically, this involved a precise unequal crossing-over that fused the two copies of the original gene, in phase, to make a double-length gene. As Fig. 9 shows, this process involves an extremely rare, precise illegitimate event (a recombination event that leads to the fusing of two DNA sequences at a point where there is no matching of sequence) that has as its consequence the synthesis at a high level of the new, presumably more useful, double-length gene product. Consider the same process against the background of a general splicing mechanism. Again, the process of forming the double gene must involve an illegitimate recombination event, but now that

a

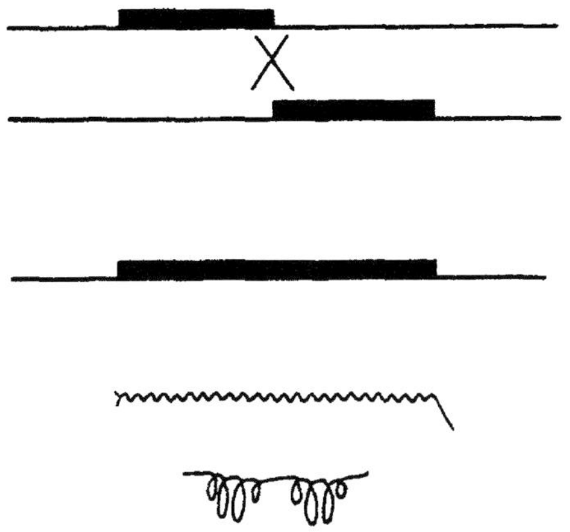

b

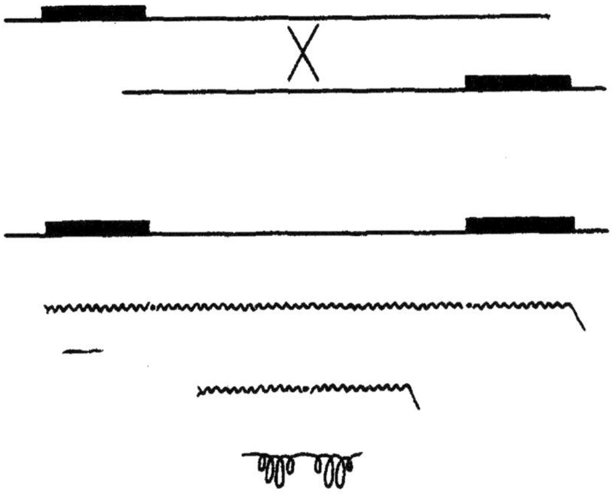

event can occur anywhere within a stretch of 1000 to 10 000 bases flanking the 3' side of one copy and the 5' side of the other to form an intron separating the genes for the two domains. From a long transcript across this region, even inefficient splicing may produce the new double-length gene product. This will happen some 10^6 to 10^8 times more rapidly than the classical process because of all of the different combinations of sites at which the recombination can occur. If the long transcript can be spliced, even at a low frequency, some of the double-length product can be made. This is a faster way for evolution to form the final gene: proceeding through a rapid step to a structure that can produce a small amount of the useful gene product. Small mutational steps can be selected to produce better splicing signals and thus more of the gene product. If the splicing

Fig. 9. A double-length gene product arises through unequal crossing over. (a) The classical process by which a gene corresponding to a single polypeptide chain might have its length doubled by a crossing over. The top two lines indicate two coding regions, brought into accidental apposition by some act of illegitimate recombination which fuses the carboxy terminal region (the 3' end) of one copy of the gene to the amino terminal region (the 5' end) of the other. This rare illegitimate event (involving no sequence matching) would, if it occurs in phase, produce a double-length gene which could code for a double-length RNA which in turn translates into a double-length protein containing the reiteration of a basic domain. (b) The same process occurring in the presence of the splicing function. Now the unequal crossing over can occur anywhere to the 3' side of one copy of the gene and anywhere in front of the 5' end of the other copy of the gene to produce a gene containing two exons separated by a long intron. I conjecture that the long transcript of this region now will be spliced at some low frequency to produce a mature message encoding the reiterated protein.

signals already exist, recombination within introns provides an immediate way to build polymeric structures out of simpler units. One would predict that polymeric structures, made up of simpler units, will be found to have genes in which the intron-exon structure of the primitive unit is repeated, separated again by introns. That is the case.

The rate of legitimate recombination between the exons of a gene will be increased by the introns. Consider two mutations to better functioning, arising in different parts of a gene and spreading, by selection, through the population. Classically, both mutations could end up in a single polypeptide chain, after both genes find their way into a single diploid individual, by homologous recombination within the gene. Fig. 10 shows that this process also should be speeded some 10- to 100-fold by spreading the exons apart. This effect will be strongest if the exons can evolve separately - if they represent structures that can accumulate successful changes independently.

Furthermore, one can change the pattern of exons by changing the initiation or termination of the RNA transcript, to add extra exons or to tie together exons from one region of the DNA to exons from another. This has been observed in adenovirus, and is found in notable examples in the immunoglobulins in which exons can be added to or subtracted from the carboxy terminus of the heavy chain to modify the protein. Hood's laboratory has shown that this process is used to switch between two different forms of an IgM heavy chain (33). A membrane-bound form is synthesized by a longer transcript, which splices on two additional exons and splices out part of the last exon of the shorter transcript. The shorter transcript synthesizes a secreted form of the protein. In a similar way the switch of the V

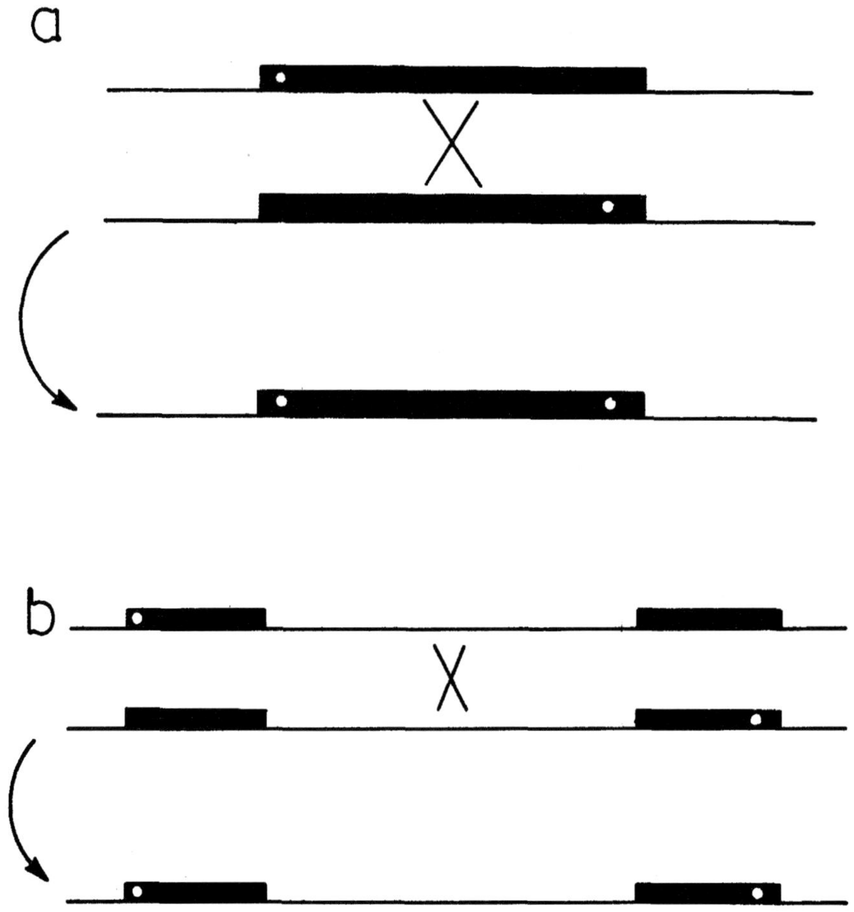

Fig. 10. Introns speed legitimate recombination. (a) The classical pattern by which two mutations, one occurring in one copy of the gene, at the left end, and the other occurring in the other copy of the gene, at the right-hand end, might get together by recombination happening in the homologous stretch of DNA that separates the two mutations. This recombination can create a single gene carrying both mutations. (b) The same process happening in a gene in which the mutations occur in separate exons separated by an intron. Now the recombination can occur anywhere, either in the exon or within the intron, to produce a new gene carrying both mutations. Since the rate of recombination will be directly proportional to the distance along the DNA between the mutations, it will be faster.

region from IgM to an IgD constant region is probably the result of a different, still longer, transcript which splices across to attach the V-region exons to the new constant-region exons of the delta class. These combinations of genes have certainly been created by recombination events within the DNA that ultimately becomes the intron of the longer transcription unit.

The most striking prediction of this evolutionary view is that separate elements defined by the exons have some functional significance, that these elements have been assorted and put together in new combinations to make up the proteins that we know. Gene products are assembled out of previously achieved solutions of the structure-function problem. Clear examples of this are still meager. The hydrophobic leader sequence which is involved in the transfer of proteins through membranes, and which is trimmed off after the secretion, is often on separate exons - most notably in the immunoglobulins (see Fig. 8), but also in ovomucoid (29). In the pair of genes for insulin in the rat, a product of recent duplication (32), the two chains of insulin lie on separate exons in one gene, on a single exon in the other. The ancestral gene (the common structure in other species (34)) has the additional intron - suggesting that the gene was put together originally from separate pieces. The gene for lysozyme is broken up into four exons; the second one carries the critical amino acids of the active site and most of the substrate contacts (35). In the gene for globin the central exon encodes almost all of the heme contacts. Fig. 11 shows a schematic dissection of the molecule. A recent experiment (36) has shown that the polypeptide that corresponds to the central exon in itself is a heme-binding 'miniglobin'; the side exons have provided polypeptide material to stabilize the protein.

At the same moment that the rapid sequencing methods and the molecular cloning gave us the promise of being able to work out the structure of any gene, the ability to achieve a complete understanding of the genetic material, Nature revealed herself to be more complex than we had imagined. We could not read the gene product directly from the chromosome by DNA sequencing alone. We must appeal to the sequence of the actual protein, or at least the sequence of the mature messenger RNA, to learn the intron-exon structure of the gene. Nonetheless the hope exists, that as we look down on the sequence of DNA in the chromosome, we will not learn simply the primary structure of the gene products, but we will learn aspects of the functional structure of the proteins - put together over evolutionary time as exons linked through introns.

My interest in biology has always centered on two problems: how is the genetic information made manifest? and how is it controlled? We have learned much about the way in which a gene is translated into protein. The control of genes in prokaryotes is well understood, but for eukaryotes the critical mechanisms of control are still not known. The purpose of research is to explore the unknown. The desire for new knowledge calls forth the answers to new questions.

I owe a great debt to my students and collaborators over the years; the greatest to Jim Watson, who stimulated my interest in molecular biology, to Benno Muller-Hill, with whom I worked on the *lac* repressor, and to Allan Maxam, with whom I developed the DNA sequencing.

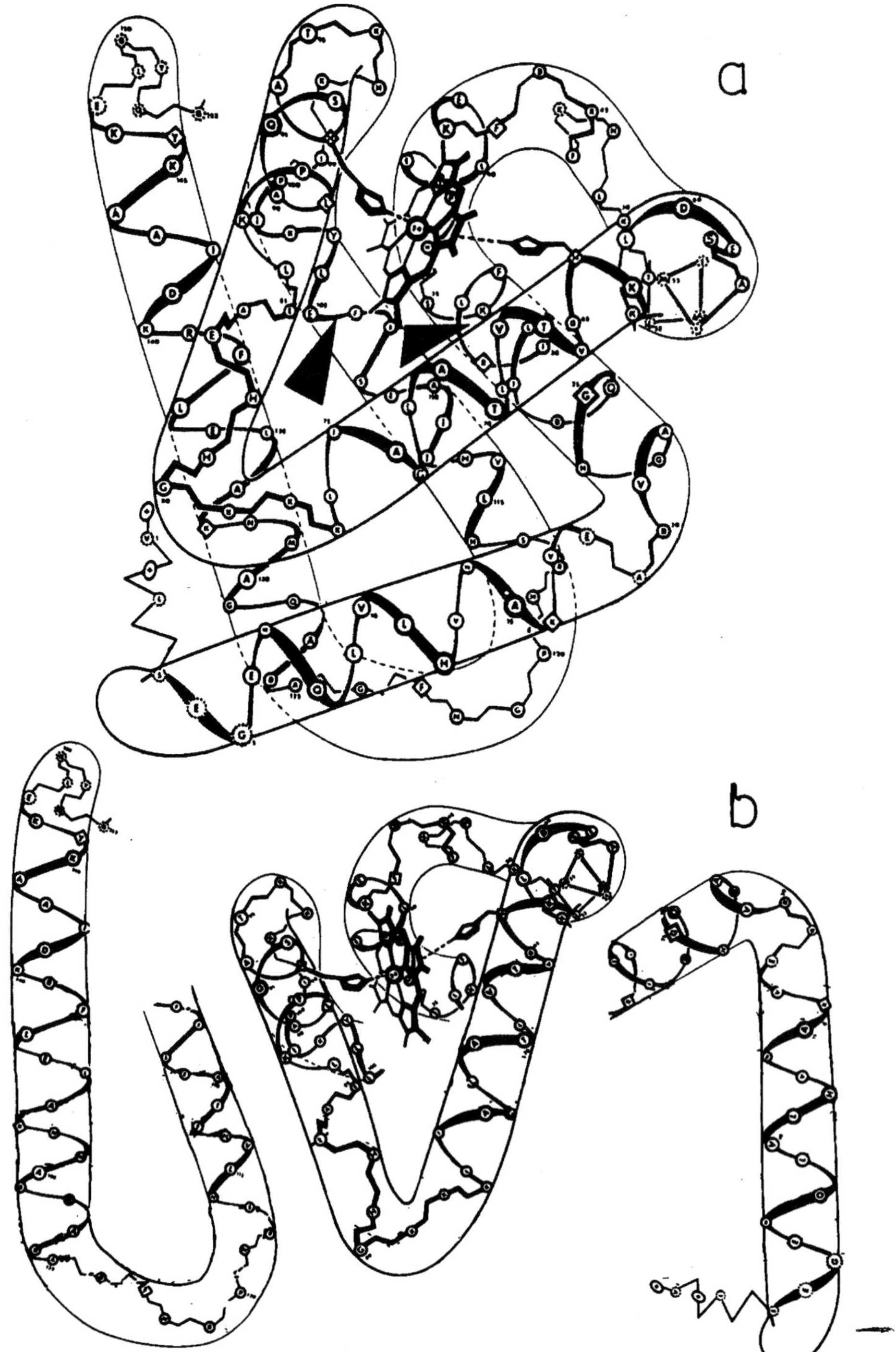
a
b

Fig. 11. A schematic dissection of globin into the product of the separate exons. (a) The black arrows show the points at which the structure of a chain of globin is interrupted by the introns. (The structures of the chains of globin and of myoglobin are very similar; the schematic structure shown is myoglobin.) The introns interrupt the protein in the alpha-helical regions to break the protein into three portions, shown in (b). The product of the central exon surrounds the heme; the products of the other two exons, I conjecture, wrap around and stabilize the protein.

References

1. Gilbert W & Müller-Hill B (1966) Isolation of the *lac* repressor. Proc. Natl. Acad. Sci. U.S.A. **56**, 1891-1898.
2. Gilbert W & Müller-Hill B (1967) The *lac* operator is DNA. Proc. Natl. Acad. Sci. U.S.A. **58**, 2415-2421.
3. Gilbert W & Maxam A (1973) The nucleotide sequence of the *lac* operator. Proc. Natl. Acad. Sci. U.S.A. **70**, 3581-3584.
4. Gilbert W, Maizels N & Maxam A (1975) Sequences of controlling regions of the *E. coli* lactose operator. Genetics **79**, 227.
5. Dickson RC, Abelson J, Barnes WM & Reznikoff WS (1975) Genetic regulation: the lac control region. Science **187**, 27-35.
6. Gilbert W, Maxam A & Mirzabekov A (1976) Contacts between the *lac* repressor and DNA revealed by methylation. In *Control of Ribosome Synthesis*, pp 139-148, Alfred Benzon Symposium 9, Munksgaard.
7. Maxam AM & Gilbert W (1977) A new method for sequencing DNA. Proc. Natl. Acad. Sci. U.S.A. **74**, 560-564.
8. Tonegawa S, Maxam AM, Tizard R, Bernard O & Gilbert W (1978) Sequence of a mouse germ-line gene for a variable region of an immunoglobulin light chain. Proc. Natl. Acad. Sci U.S.A. **75**, 1485-1489.
9. Maxam AM & Gilbert W (1980) Sequencing end-labeled DNA with base-specific chemical cleavages. Methods in Enzymology **65**, 499-560.
10. Sanger F & Coulson AR (1978) The use of thin acrylamide gels for DNA sequencing. FEBS Lett. **87**, 107-110.
11. Farabaugh PJ (1978) Sequence of the *lacI* gene. Nature **274**, 765-769.
12. Beyreuther K, Adler K, Fanning E, Murray C, Klemm A & Geisler N (1975) Amino-acid sequence of *lac* repressor from *Escherichia coli*. Eur. J. Biochem. **59**, 491-509.
13. Coulondre C, Miller JH, Farabaugh PJ & Gilbert W (1978) Molecular basis of base substitution hotspots in *Escherichia coli*. Nature **274**, 775-780.

14. Lindahl T, Ljungquist S, Siegert W, Nyberg B & Sperens B (1977) DNA *N*-glycosidases. J. Biol. Chem. **252**, 3286-3294.
15. Sutcliffe JG (1978) Nucleotide sequence of the ampicillin resistance gene of *Escherichia coli* plasmid pBR322. Proc. Natl. Acad. Sci. U.S.A. **75**, 3737-3741.
16. Ambler RP & Scott GK (1978) Partial amino acid sequence of penicillinase coded by *Escherichia coli* plasmid R6K. Proc. Natl. Acad. Sci. U.S.A. **75**, 3732-3736.
17. Sutcliffe JG (1978) Complete nucleotide sequence of the *Escherichia coli* plasmid pBR322. Cold Spring Harbor Symposia on Quantitative Biology **43**, 77-90.
18. For a review, see: Yanofsky C (1981) Attenuation in the control of expression of bacterial operons. Nature **289**, 751-758.
19. Tilghman SM, Tiemeier DC, Seidman JG, Peterlin BM, Sullivan M, Maizel JV & Leder P (1978) Intervening sequence of DNA identified in the structural portion of a mouse β-globin gene. Proc. Natl. Acad. Sci. U.S.A. **75**, 725-729.
20. Konkel DA, Tilghman SM & Leder P (1978) The sequence of the chromosomal mouse β-globin major gene: homologies in capping, splicing and poly(A) sites. Cell **15**, 1125-1132.
21. Brack C & Tonegawa S (1977) Variable and constant parts of the immunoglobulin light chain gene of a mouse myeloma cell are 1250 nontranslated bases apart. Proc. Natl. Acad. Sci. U.S.A. **74**, 5652-5656.
22. Breathnach R, Mandel JL & Chambon P (1977) Ovalbumin gene is split in chicken DNA. Nature **270**, 314-319.
23. Berget SM, Moore C & Sharp PA (1977) Spliced segments at the 5' terminus of adenovirus 2 late mRNA. Proc. Natl. Acad. Sci. U.S.A. **74**, 3171-3175.
24. Chow LT, Gelinas RE, Broker TR & Roberts RJ (1977) An amazing sequence arrangement at the 5' ends of adenovirus 2 messenger RNA. Cell **12**, 1-8.
25. Gilbert W (1978) Why genes in pieces? Nature **271**, 501.
26. Bernard O, Hozumi N & Tonegawa S (1978) Sequences of mouse immunoglobulin light chain genes before and after somatic changes. Cell **15**, 1133-1144.
27. Sakano H, Rogers JH, Hüppi K, Brack C, Traunecker A, Maki R, Wall R & Tonegawa S (1979) Domains and the hinge region of an immunoglobulin heavy chain are encoded in separate DNA segments. Nature **277**, 627-633.
28. Honjo T, Obata M, Yamawaki-Kataoka Y, Kataoka T, Kawakami T, Takahashi N & Mano Y (1979) Cloning and complete nucleotide sequence of mouse immunoglobin γ1 chain gene. Cell **18**, 559-568.
29. Stein JP, Catterall JF, Kristo P, Means AR & O'Malley BW (1980) Ovomucoid intervening sequences specify functional domains and generate protein polymorphism. Cell **21**, 681-687.
30. Yamado Y, Avvedimento VE, Mudryj M, Ohkubo H, Vogeli G, Irani M, Pastan I & de Crombrugghe B (1980) The collagen gene: evidence for its evolutionary assembly by amplification of a DNA segment containing an exon of 54 bp. Cell **22**, 887-892.

31. Doolittle WF (1978) Genes in pieces: were they ever together? Nature **272**, 581.
32. Lomedico P, Rosenthal N, Efstratiadis A, Gilbert W, Kolodner R & Tizard R (1979) The structure and evolution of the two nonallelic rat preproinsulin genes. Cell **18**, 545-558.
33. Early P, Rogers J, Davis M, Calame K, Bond M, Wall R & Hood L (1980) Two mRNAs can be produced from a single immunoglobin μ gene by alternative RNA processing pathways. Cell **20**, 313-319.
34. Perler F, Efstratiadis A, Lomedico P, Gilbert W, Kolodner R & Dodgson J (1980) The evolution of genes: the chicken preproinsulin gene. Cell **20**, 555-566.
35. Jung A, Sippel AE, Grez M & Schütz G (1980) Exons encode functional and structural units of chicken lysozyme. Proc. Natl. Acad. Sci. U.S.A. **77**, 5759-5763.
36. Craik CS, Buchman SR & Beychok S (1980) Characterization of globin domains: heme binding to the central exon product. Proc. Natl. Acad. Sci. U.S.A. **77**, 1384-1388.

Proc. Natl. Acad. Sci. USA
Vol. 81, pp. 1991–1995, April 1984
Biochemistry

Genomic sequencing

(DNA methylation/UV crosslinking/filter hybridization/immunoglobulin genes)

GEORGE M. CHURCH* AND WALTER GILBERT*†

*Biological Laboratories, Harvard University, Cambridge, MA 02138; and †Biogen, Inc., 14 Cambridge Center, Cambridge, MA 02142

Contributed by Walter Gilbert, December 19, 1983

ABSTRACT **Unique DNA sequences can be determined directly from mouse genomic DNA. A denaturing gel separates by size mixtures of unlabeled DNA fragments from complete restriction and partial chemical cleavages of the entire genome. These lanes of DNA are transferred and UV-crosslinked to nylon membranes. Hybridization with a short ^{32}P-labeled single-stranded probe produces the image of a DNA sequence "ladder" extending from the 3′ or 5′ end of one restriction site in the genome. Numerous different sequences can be obtained from a single membrane by reprobing. Each band in these sequences represents 3 fg of DNA complementary to the probe. Sequence data from mouse immunoglobulin heavy chain genes from several cell types are presented. The genomic sequencing procedures are applicable to the analysis of genetic polymorphisms, DNA methylation at deoxycytidines, and nucleic acid-protein interactions at single nucleotide resolution.**

How can we visualize the state of individual nucleotides within large chromosomes? During recombinant DNA cloning, information about DNA methylation and chromatin structure is lost. Direct chemical modification of the genome combined with complete restriction enzyme digestion and separation by size on a denaturing gel preserves some of this information in the form of numerous comigrating sets of DNA sequence "ladders." To access one sequence at a time, the lanes of DNA are transferred and crosslinked to a nylon membrane and hybridized to a short single-stranded ^{32}P-labeled probe specific for one end of one restriction fragment within the genome. Fig. 1 illustrates why this works. The probe called "3′ lower" will hybridize to only three classes of DNA reaction products: the appropriate fragments extending from the 3′ end of the lower strand of the restriction fragment, the longest fragments from the 5′ end of the upper strand (which will only affect the top of the sequence ladder), and the middle fragments with both ends produced by chemical cleavage ("zigzags"). The abundance of the appropriate fragments is proportional to the probability (P) that the chemical reaction cleaves at any given target. Because the abundance of the middle fragments is proportional to P^2, interference can be diminished by decreasing the extent of the reaction. About one cleavage every 500 nucleotides is optimal. Decreasing the length of the probe also alleviates the problem of hybridization to fragments from the wrong end and the middle; however, very short probes (less than 20 nucleotides long) and low stringency washes can produce a crossreacting background. Probes 100 to 200 nucleotides long work well.

DNA Methylation. Up to 12% of all cytosines in vertebrate genomes are methylated mainly at C-G sequences. In plants, up to 50% of all cytosines are methylated mainly at C-G and C-N-G sequences (reviewed in ref. 1). Only a small subset of these methylation sites can be assayed by restriction analyses (2), and flanking sequences can severely affect the relative efficiencies of restriction enzyme recognition (3, 4). Transfection of *in vitro* methylated DNA into mammalian cells (5, 6) can demonstrate sites of cytosine methylation that affect RNA transcription, as long as the maintenance of the methylation at such sites can be followed. The ability to analyze every cytosine should aid the analyses of correlations between gene expression and DNA methylation (1).

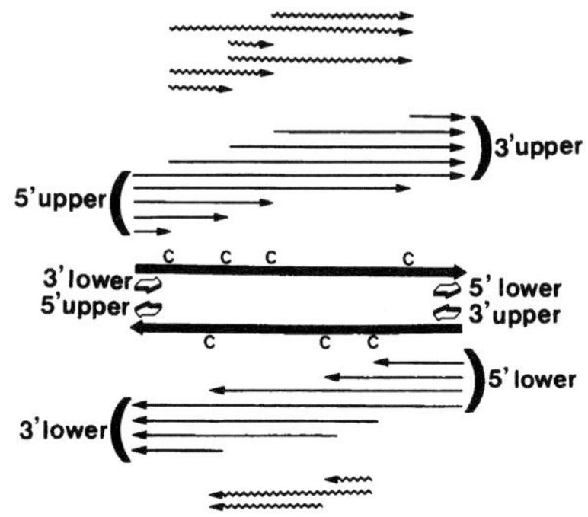

FIG. 1. All the single-stranded DNA reaction products expected in the complete restriction cleavage and partial cytosine-specific cleavage reactions for a restriction fragment. The cytosine residues are indicated by Cs on each of the central genomic DNA strands. Between these strands, four short white arrows indicate the four possible probes for this restriction fragment. The probes are referred to by the type of sequence they will produce. For example, hybridization with the 3′ lower probe produces the image of the 3′ end-labeled sequence of the lower strand. The straight arrows represent reaction products that will contribute to readable sequences only when the probe indicated by adjacent brackets is used. Zigzag lines represent internal reaction products that can only deteriorate the sequence when the probes are made long enough to overlap many of these.

Hydrazine reacts poorly with 5-methylcytosine relative to cytosine and thymine residues (7, 8). Thus, methylation levels of individual cytosine bases in DNA from various tissues can be quantitated from genomic sequence determination autoradiographs. The same DNA replicated in *Escherichia coli* acts as an unmethylated control.

MATERIALS AND METHODS

DNA Samples. To determine the sequence of a specific region of DNA, we use a restriction enzyme that cuts about 100–200 base pairs (bp) away on one side of the region and a greater distance away on the other side. Genomic DNA samples are cut to completion, precipitated by ethanol, and resuspended at 10 μg/μl in 10 mM Tris·HCl, pH 7.5/1 mM EDTA. Five 5-μl aliquots of the DNA are treated with standard G, A+G, T+C, T, and C chemistries (9, 10). After the final lyophilizations, the samples are resuspended at 10

The publication costs of this article were defrayed in part by page charge payment. This article must therefore be hereby marked "*advertisement*" in accordance with 18 U.S.C. §1734 solely to indicate this fact.

Abbreviations: C_μ, IgM heavy chain constant region gene; bp, base pair(s).

μg/μl in 94% formamide/0.05% xylene cyanol/0.05% bromophenol blue/10 mM Na_2EDTA, pH 7.2. With a Hamilton syringe (1701SNWG, 31-gauge, 5-cm, point 3 needle), 2.5 μl of each reaction is loaded on a 50 × 50 × 0.076 cm 6% acrylamide/0.15% bisacrylamide/7 M urea/50 mM Tris borate/EDTA, pH 8.3/0.1% ammonium persulfate/0.1% *N,N,N′,N′*-tetramethylethylenediamine gel in 5-mm-wide slots formed by a series of polyacrylamide or acetal plastic dividers 0.5–1 mm wide.

Nylon Membranes. Many filters retain small DNA fragments and resist damage better than does nitrocellulose. DBM (11), DPT (12), Millipore poly(vinylidene fluoride), Bio-Rad Zetaprobe, New England Nuclear GeneScreen plus, AMF Cuno Zetabind, Pall RU, and Pall Biodyne A have been used to produce sequence patterns with loadings that range from nanogram to 20-fg amounts of hybridizable DNA per band. Certain nylon membranes allow us to detect 3 fg of hybridizable DNA per band: among two lots of Pall NR (32 × 50 cm from Chisholm (Cranston, RI) and three lots of GeneScreen (New England Nuclear), we have noticed no significant variation in the signal-to-noise ratio for single-copy sequences.

Electrophoretic Transfer. The following protocol deviates from previous protocols (13–21). We built a 38 × 46 × 20 cm transfer device from Plexiglas egg-crate louver panels (from AIN Plastics (Mt. Vernon, NY), Scotch-Brite pads (96 type-industrial; 3M, Inc., Minneapolis, MN), and 32-gauge platinum wire. Replicas of this device can be obtained from Charles Barbagallo (Harvard Biological Laboratories, Cambridge, MA).

The gel and nylon membrane are kept thoroughly wet with 50 mM Tris borate/EDTA, pH 8.3. The gel can be transferred immediately after completion of the electrophoretic separation of the DNA fragments (removal of urea from the gel, DNA depurination, and cleavage of bisacrylamide cross-linkers are not necessary). The gel, if 0.76 mm thick, is lowered directly from the glass plate onto one Scotch-Brite surface of the transfer device. Thinner gels are lifted with dry Whatman 540 paper and placed immediately on the Scotch-Brite surface. The nylon membrane is placed on the gel without trapping pockets of air or buffer at the interface. Keeping the gel plane horizontal during transfer prevents sagging of the large thin gels and membranes and allows application of pressure from a 2-kg mass to the central region of the transfer assembly to aid tight contact between the gel and nylon surfaces. The power for electrophoretic transfer is supplied by line current with a bridge rectifier, which delivers 108 V at 120 pulses per sec direct current output up to 4 A (available from J. Skare, 665 North St., Tewksbury, MA 01876). The rate of transfer depends on DNA size, but in 30 min with a 108-V potential over the 10 cm between electrodes in 50 mM Tris borate/EDTA, greater than 90% of the DNA fragments, 26 through 516 nucleotides in length, transfer from a 1.5-mm thick 6% polyacrylamide gel.

UV Irradiation. The lower left corner of the wet membrane is clipped to designate that the DNA-coated side is facing up. This side is placed on taut Saran Wrap and irradiated through the Saran Wrap at a distance of 35 cm from one or more germicidal UV bulbs. The UV flux can be measured with a Blak-Ray 260-nm UV meter (from American Scientific Products) or can be determined empirically by using genomic sequence analysis transfers. The optimal UV dose for cross-linking DNA to the filter is 1.6 kJ/m^2, which on our device using six bulbs is 1200 $\mu W/cm^2$ for 2 min. DNA images on completely dry membranes have a 90% lower optimal UV dose.

We dried [α-^{32}P]-labeled NTPs onto nylon membranes and UV-irradiated these at 0.16 kJ/m^2. Nucleotide binding was stabilized 130-fold for TTP and 30-, 20-, and 10-fold for dGTP, dCTP, and dATP, respectively. These bonds were stable for over 14 hr at 65°C at pH 2 through 11. Primary amino groups (which are present on nylon) are highly reactive with 254-nm light-activated thymines (22).

DNA Probe Synthesis. A gel-purified DNA fragment, homologous to sequences to one side of the restriction site selected for the total genomic digests, is subcloned into *Sma* I-cut mp8 vector (23). Large quantities (>500 μg) of single-stranded phage DNA are prepared from individual white plaques, avoiding residual traces of bacterial DNA and polyethylene glycol. The single-stranded DNA is resuspended at 2 mg/ml in 10 mM Tris·HCl/1 mM Na_2EDTA, pH 7.5, and 30 μl of this DNA, 7 μl of synthetic 17-nucleotide-long sequence assay primer from Collaborative Research (Waltham, MA) at 10 ng/μl, and 2 μl of 100 mM $MgCl_2$ are incubated at 50°C for 40 min. Then, 1 μl of crystalline grade bovine serum albumin at 10 mg/ml, 1 μl of 3 mM dCTP/3 mM dGTP/3 mM TTP, 0.1 μl of 200 mM dithiothreitol, 50 μl of [α-^{32}P]dATP at 10 mCi/ml (>5000 Ci/mmol; 1 Ci = 37 GBq), and 3 μl of DNA polymerase I large fragment at 5 units/μl from New England BioLabs or Bethesda Research Laboratories are mixed and incubated at 25°C for 40 min. After addition of 200 μl of 94% formamide/0.05% xylene cyanol/0.05% bromophenol blue/10 mM Na_2EDTA, pH 7.2, the reaction is heated in a boiling water bath, 10 μl of 1 M NaOH is added, and the entire sample is loaded directly (without cooling) onto a 15 × 0.6 × 0.15 cm slot on a 20 × 10 × 0.15 cm 6% acrylamide/0.15% bisacrylamide/7 M urea/50 mM Tris borate/EDTA, pH 8.3/0.1% ammonium persulfate/0.1% *N,N,N′,N′*-tetramethylethylenediamine gel that has been preheated to a 50°C surface temperature by preelectrophoresis at 250 V/10 cm and 90 mA. When the bromophenol blue dye reaches 1.4 cm from the origin (about 10 min), the xylene cyanol dye-stained region of the gel is excised, ground with a thick glass rod, and eluted in 10 ml of hybridization buffer (see below) for 45 min at 50°C. Small scale reactions should show greater than 90% of the label incorporated into fragments 90–130 nucleotides in length. If most of the extension reaction products are longer than 150 bp, the primer and template DNA concentrations should be increased and the reaction retested.

RNA Probe Synthesis. The SP6 RNA polymerase transcription is done as described by Zinn *et al.* (24) and modified by K. Zinn. The SP63 vector used contains an SP6 promoter very near a *Hin*dIII cloning site (P. Kreig and D. Melton, personal communication). The following are mixed in order at 24°C: 8 μl of mix [90 mM Tris·HCl, pH 7.5/14 mM $MgCl_2$/5 mM spermidine·HCl/1.3 mM ATP/1.3 mM CTP/1.3 mM GTP/22 mM dithiothreitol/0.6 units of Promega Biotec (Madison, WI) RNasin per ml], 10 μl of [α-^{32}P]rUTP at 50 mCi/ml (3000 Ci/mmol; partially lyophilized from 50 μl), 2 μl of restricted DNA template at 1 mg/ml in H_2O, and 0.7 μl of SP6 RNA polymerase at 7 units/μl. The reaction is incubated at 37°C for 1 hr, then 8 μl of mix, 10 μl of H_2O, 2 μl of DNA template, and 0.5 μl of enzyme are added, and the reaction mixture is incubated an additional hour. Fifty-percent incorporation of the label should be achieved. The reaction mixture is extracted with phenol, and the extracted DNA is precipitated and rinsed with ethanol, dried, and resuspended in 50 μl of 10 mM Tris·HCl/1 mM EDTA before dilution in 5 ml of hybridization buffer. The enzyme is available from New England Nuclear and Promega Biotec.

Hybridization. Gel elution and hybridization buffer is 1% crystalline grade bovine serum albumin/1 mM EDTA/0.5 M $NaHPO_4$, pH 7.2/7% $NaDodSO_4$ [1 M $NaHPO_4$ (pH 7.2) stock is composed of 134 g of $Na_2HPO_4 \cdot 7H_2O$ and 4 ml of 85% H_3PO_4 per liter]. Either Pierce 28365 or Bio-Rad 161-0302 $NaDodSO_4$ is adequate. Carrier DNA, RNA, pyrophosphate, Dextran sulfate, Ficoll, and polyvinylpyrrolidone are not required in these hybridizations. Bovine serum albumin has only a 2-fold effect on background. The UV-irradiated nylon membrane is placed in a polyethylene/polyester-laminated bag (18 × 15 inch Scotchpak 229 from Spec-Fab (Riverton, NJ) and rinsed with water. Excess water is removed before addition of 30 ml of hybridization buffer. The bag is heat-sealed (model 254-B 1/4 × 24 inch Teflon jaws from Clamco, Cleveland, OH) for 4 sec at 250°F and placed at 65°C for 5 min. The gel particles are removed from the eluted probe by centrifugation at 3000 × *g* for 3 min and by rapid filtration of the supernatant through a 0.22-μm Millex-GV filter. The 30 ml of hybridization buffer in the bag is replaced by the probe. The bag is resealed, incubated at 65°C for 8–24 hr in a water bath, reopened, and submerged in 1 liter of 0.5% fraction V-grade bovine serum albumin/1 mM Na_2EDTA/40 mM $NaHPO_4$, pH 7.2/5% $NaDodSO_4$. After 5 min with agitation, the membrane is transferred to another such wash, followed by eight washes (1 liter each) in 1 mM Na_2EDTA/40 mM $NaHPO_4$, pH 7.2/1% $NaDodSO_4$ for 5 min each. For the RNA probes, three additional washes in 100 mM $NaHPO_4$ (pH 7.2) to remove $NaDodSO_4$ and treatment with 30 ml of RNase A (10 μg/ml) in 0.3 M NaCl/10 mM Tris·HCl/1 mM Na_2EDTA, pH 7.5, at 37°C for 15 min were done.

The stated volumes are intended for one nylon sheet measuring 30 × 40 cm. When the total membrane surface area is different, the probe synthesis, hybridization, and wash volumes are adjusted proportionally. The wash solutions are kept at 65°C prior to use, but the wash agitations are done at room temperature for convenience. The last wash is brought to 65°C by sealing the membrane and wash solution in a bag and submerging this in a water bath for 20 min. The total Na^+ concentration in this wash is 76 mM. The dry membrane is autoradiographed on preflashed XAR-5 film at −80°C with an intensifying screen (25) for 2 or more days. The probe can be eluted by washing in 500 ml of 2 mM Tris/EDTA, pH 8.2/0.1% $NaDodSO_4$ for 15 min at 65°C, and then the membrane can be reprobed.

RESULTS AND DISCUSSION

An Example. We have studied the methylation of cytosines in a region from the mouse IgM heavy chain constant region gene C_μ in five cell types. Fig. 2 shows the probes for the ends of a 511-bp *Mbo* I fragment that covers the third intron of C_μ. Fig. 3 shows the genomic sequencing patterns. DNA in lanes of Fig. 3 *Left* are hybridized to the 110-nucleotide long 5′ lower probe (Fig. 2); then the membrane was stripped and rehybridized (Fig. 3 *Right*) with the 256-nucleotide-long RNA transcript corresponding to the 3′ lower probe (Fig. 2). The lanes in Fig. 3 marked l (liver) display the base-specific (guanine, purine, pyrimidine, thymine, and cytosine) reactions on 25 μg of liver DNA. The subsequent lanes displayed are cytosine-specific reactions on DNA from thymus (t), spleen cells sorted for surface IgM, positive and negative fractions (m and n), and RAW 8.1 lymphosarcoma cells (r).

The genomic clone (pIgMC1) used as the unmethylated control (Fig. 3, lanes p) was derived from the BALB/c cell line 18-48 (29). We analyzed tissues from 3-month-old female mice of the same inbred BALB/c strain. Three groups independently have reported sequence data for the C_μ region of BALB/c mice (30–32). Although these three sequences differ at some positions, they agree at the positions of the C-G dinucleotides that we have analyzed. Interlane variations in cytosine band intensities are apparent at the C-G dinucleotides at positions 192, 325, and 401 (see arrowheads in Figs. 2 and 3). The cytosine at 192 is normally reactive (unmethylated) in liver and in the IgM-positive spleen cells and is hyporeactive (methylated) in the thymus, a lymphosarcoma, and the surface IgM-negative spleen cells. The cytosine at 401, part of a *Hpa* II site, appears methylated in all cell types, but

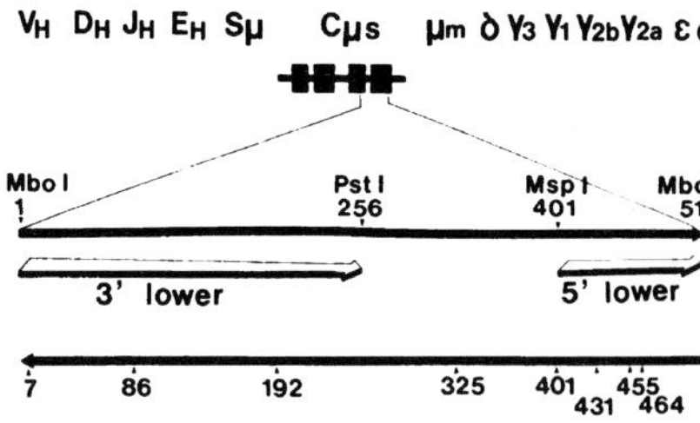

FIG. 2. Mouse C_μ-specific probes. The four exons of the C_μ DNA (solid black blocks) are placed in the context of the other gene segments required for heavy chain gene expression and recognition diversity [not to scale (V, variable; D, diversity; J, joining; E, enhancer; S, switch; $C_{\mu s}$–α, constant domains); see ref. 26 for a review]. The 511-bp *Mbo* I fragment at the 3′ end of C_μ is shown enlarged as two long black arrows. Between these strands the extents and polarities of the two probes are indicated by white arrows. The numerals beneath the lower strand indicate the positions of C-G dinucleotides relative to the 3′ end of this strand. Three of these are present in the sequences in Fig. 3: 192, 325, and 401. The plasmid pIgMC1 was constructed from pBR322 and a 5.0-kbp *Bam*HI–*Eco*RI fragment from genomic DNA of Abelson virus-transformed 18-48 BALB/c cell line (provided by A. Perlmutter). The 5′ lower probe was synthesized as described in *Materials and Methods* on a single stranded template from an mp8 recombinant phage, pIgM511Mbo-5′lower, containing a 110-bp *Msp* I–*Sau*3A fragment from pIgMC1 cloned into mp8 with the *Hpa* II site 33 nucleotides from the synthetic primer binding site. The 3′ lower RNA probe was synthesized from the plasmid, pIgM511Mbo3′lower, which was constructed from *Hin*dIII-cut pSP63 (provided by P. Krieg and D. Melton) and the 511-bp *Sau*3AI fragment of pIgMC1. All 3′ ends were filled in before ligation. The recombinant plasmid cut with *Pst* I acted as template for transcription.

in the IgM-positive cells (where this site is less clear), a small degree of unmethylation is not ruled out. The cytosine at 325 produces no bands (less that 10% reactivity by densitometry compared to normal intensities in the control plasmid cytosine-specific reaction—lane p) in all mouse cell types analyzed, consistent with high levels of methylation. Thus, within a space of 133 nucleotides, we find two C-G dinucleotides (positions 192 and 325) with different tissue-specific DNA methylation. Previous studies on C_μ gene DNA methylation (33–35) were limited to *Hpa* II cleavage sites.

Unambiguous sequence can be read for the liver DNA at positions 368–399 when hybridized with the 110 nucleotide long pIgM511Mbo5′ lower probe (Fig. 3 *Left*). When the DNA was rehybridized with the 256-nucleotide-long RNA probe (Fig. 3 *Right*), the lane labeled G became less readable, even though the C lanes were clear in both hybridizations. This probably represents hybridization to internal fragments with both ends produced by chemical cleavages (zigzag lines in Fig. 1), which affected one G lane because of the combined effects of overreaction (about one hit per 80 nucleotides rather than the recommended one per 500) and a long probe.

Other Applications. The genomic sequencing techniques will allow analysis of enzymatic and chemical inhibition enhancement patterns ("footprinting") of the chromatin over single genes in whole cells or nuclei at single nucleotide resolution. By determining the sequence of genomic DNA with the single-base-specific chemical reactions, genetic polymorphisms and somatic mutations should be detectable even in heterozygotes. The UV irradiation of nucleic acids on pure nylon membranes and the use of high $NaDodSO_4$ washes should be helpful in a variety of filter hybridization studies that require high sensitivity and reprobing. Gel fractionation of genomic restriction digests will permit sequence determination of repetitive elements or DNA from organisms

1994 Biochemistry: Church and Gilbert Proc. Natl. Acad. Sci. USA 81 (1984)

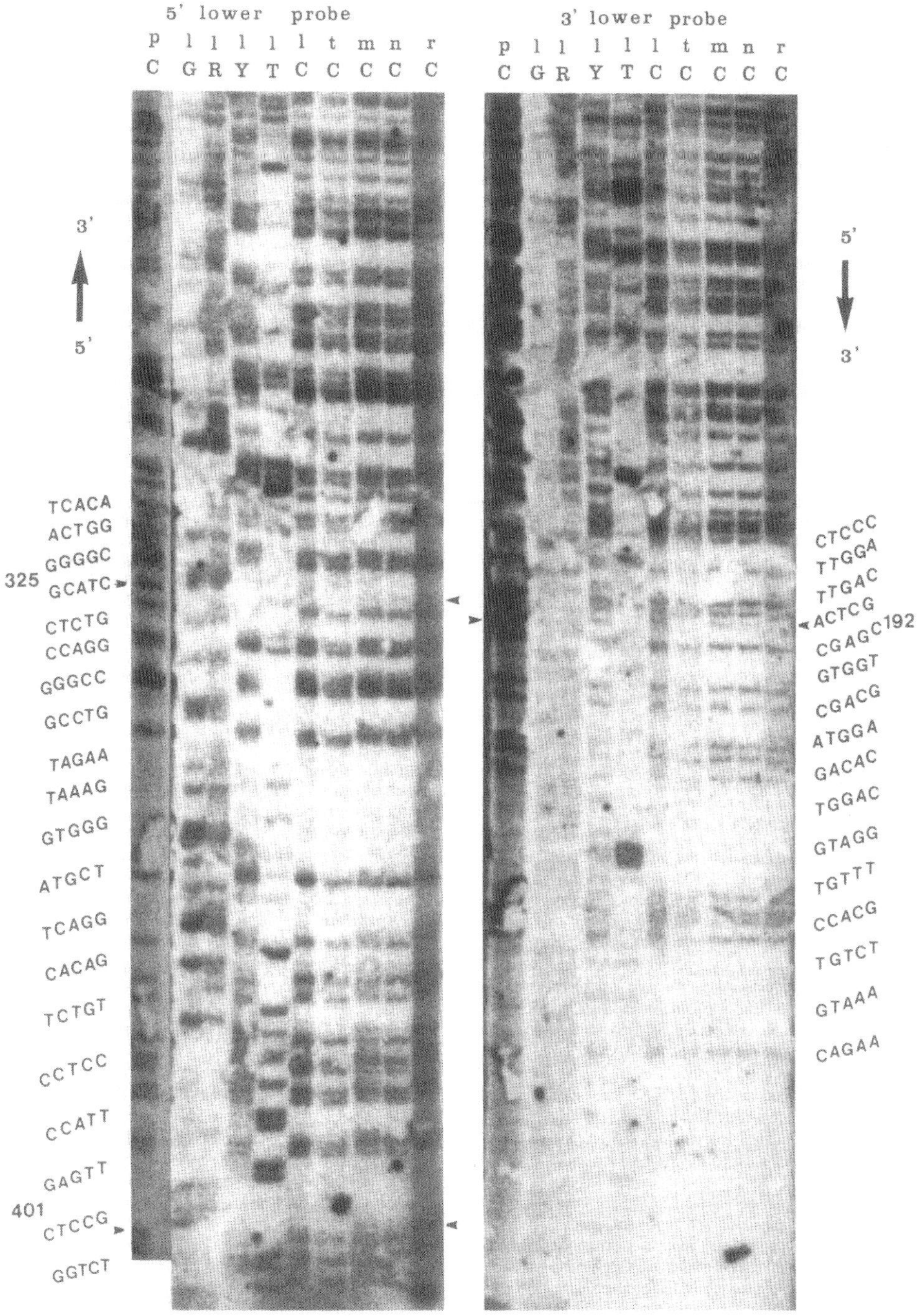

FIG. 3. Genomic sequences of mouse C_μ genes. The symbols p, l, t, m, n, and r represent DNA samples from plasmid pIgMC1, liver cells, thymus cells, spleen cells sorted for presence of surface IgM (27), spleen cells sorted for lack of surface IgM, and BALB/c lymphosarcoma RAW 8.1 TIB 50 cells (28), respectively, which were cut to completion with *Mbo* I or *Sau*3A I. The symbols C, G, R (A+G), Y (C+T), and T ($KMnO_4$) indicate the standard chemical sequence determination reactions used (9, 10). Twenty-five micrograms of genomic DNA or 200 pg of plasmid DNA were loaded per lane. Arrows indicate the positions of C-G dinucleotides; the sequences (from 27–29) are aligned along the extreme right and left. (*Left*) DNA in these 10 lanes was probed with the 5′ lower DNA probe (Fig. 2). The probe concentration during hybridization was 6 μCi/ml = 10 ng/ml = 1.3×10^7 dpm/ml. Exposure time was 21 days. (*Right*) The membrane was stripped of the first probe and rehybridized with the 3′ lower RNA probe. The probe concentration during hybridization was 24 μCi/ml. The exposure time was 10 days.

with more than 3×10^9 bp per haploid genome equivalent. In conjunction with enzyme-linked probes (36), it may be possible to determine the sequence of DNA without using radioactive compounds.

Genomic DNA sequence determination has been successfully applied to other systems: rat insulin II gene DNA methylation (H. Nick and R. Cate, personal communication); DNA methylation maintenance of *in vitro* methylated human fetal globin gene transfected into mouse L cells (ref. 6; unpublished data); *lac* operator DNA accessibility to dimethyl sulfate in intact *E. coli* cells (H. Nick, personal communication); and accessibility of a human β-interferon gene–bovine papillomavirus construct to dimethyl sulfate in mouse C127 cells (K. Zinn and T. Maniatis, personal communication).

We thank Richard Cate, Wei Chung Goh, Winship Herr, Marty Kreitman, Harry Nick, Larry Peck, Aaron Perlmutter, Gary Ruvkun, Dennis Schwartz, Richard Tizard, Carl Wu, Chao-Ting Wu, and Kai Zinn for sharing their ideas and their experiences with early versions of these procedures. We thank Paul Krieg, Doug Melton, Michael Rebagliati, and Kai Zinn for help with the SP6 system and James Skare for help with his power supplies. This work was supported by National Institutes of Health Grant GM09541-22 and by Biogen N.V.

1. Doefler, W. (1983) *Annu. Rev. Biochem.* **52,** 93–124.
2. Bird, A. P. & Southern, E. W. (1978) *J. Mol. Biol.* **118,** 27–47.
3. Busslinger, M., deBoer, E., Wright, S., Grosveld, F. G. & Flavell, R. A. (1983) *Nucleic Acids Res.* **11,** 3559–3569.
4. Keshet, E. & Cedar, H. (1983) *Nucleic Acids Res.* **11,** 3571–3580.
5. Simon, D., Stuhlmann, H., Jahner, D., Wagner, H., Werner, H. & Jaenisch, R. (1983) *Nature (London)* **304,** 275–277.
6. Busslinger, M., Hurst, J. & Flavell, R. A. (1983) *Cell* **34,** 197–206.
7. Ohmori, H., Tomizawa, J. & Maxam, A. M. (1978) *Nucleic Acids Res.* **8,** 1479–1486.
8. Miller, J. R., Cartwright, E. M., Brownlee, G. G., Federoff, N. V. & Brown, D. D. (1978) *Cell* **13,** 717–725.
9. Maxam, A. M. & Gilbert, W. (1980) *Methods Enzymol.* **65,** 497–560.
10. Rubin, C. M. & Schmid, C. W. (1980) *Nucleic Acids Res.* **8,** 4613–4619.
11. Alwine, J. C., Kemp, D. J. & Stark, G. R. (1977) *Proc. Natl. Acad. Sci. USA* **74,** 5350–5354.
12. Seed, B. (1982) *Nucleic Acids Res.* **10,** 1799–1810.
13. Southern, E. M. (1975) *J. Mol. Biol.* **98,** 503–517.
14. Arnheim, N. & Southern, E. M. (1977) *Cell* **11,** 363–370.
15. Reiser, J., Renart, J. & Stark, G. (1978) *Biochem. Biophys. Res. Commun.* **85,** 1104–1112.
16. Kutateladze, T. V., Axelrod, V. D., Gorbulev, V. G., Belzhelarskaya, S. N. & Vartikyan, R. M. (1979) *Anal. Biochem.* **100,** 129–135.
17. Bittner, M., Kupferer, P. & Morris, C. F. (1980) *Anal. Biochem.* **102,** 459–471.
18. Stellwag, E. J. & Dahlberg, A. E. (1980) *Nucleic Acids Res.* **8,** 299–317.
19. Smith, G. E. & Summers, M. D. (1980) *Anal. Biochem.* **109,** 123–129.
20. Levy, A., Frei, E. & Noll, M. (1980) *Gene* **11,** 283–290.
21. Thomas, P. S. (1980) *Proc. Natl. Acad. Sci. USA* **77,** 5201–5205.
22. Saito, I., Sugiyama, H., Furukawa, N. & Matsuura, T. (1981) *Tetrahedron Lett.* **22,** 3265–3268.
23. Messing, J. (1983) *Methods Enzymol.* **101,** 20–79.
24. Zinn, K., DiMiao, D. & Maniatis, T. (1983) *Cell* **34,** 865–879.
25. Swanstrom, R. & Shank, P. R. (1978) *Anal. Biochem.* **86,** 184–192.
26. Tonegawa, S. (1983) *Nature (London)* **302,** 575–581.
27. Wysocki, L. & Sato, V. L. (1978) *Proc. Natl. Acad. Sci. USA* **75,** 2844–2848.
28. Ralph, P. & Nakoinz, I. (1974) *Nature (London)* **249,** 49–51.
29. Siden, E. J., Baltimore, D., Clark, D. & Rosenberg, N. E. (1979) *Cell* **16,** 389–396.
30. Kawakami, T., Takahashi, N. & Honjo, T. (1980) *Nucleic Acids Res.* **8,** 3933–3945.
31. Auffray, C. & Rougeon, F. (1980) *Gene* **12,** 77–86.
32. Goldberg, G. I., Vanin, E. F., Zrolka, A. M. & Blattner, F. R. (1981) *Gene* **15,** 33–42.
33. Yagi, M. & Koshland, M. E. (1981) *Proc. Natl. Acad. Sci. USA* **78,** 4907–4911.
34. Rogers, J. & Wall, R. (1981) *Proc. Natl. Acad. Sci. USA* **78,** 7497–7501.
35. Storb, U. & Arp, B. (1983) *Proc. Natl. Acad. Sci. USA* **80,** 6642–6646.
36. Leary, J. J., Brigati, D. J. & Ward, D. C. (1983) *Proc. Natl. Acad. Sci. USA* **80,** 4045–4049.

Proc. Natl. Acad. Sci. USA
Vol. 86, pp. 6883–6887, September 1989
Biochemistry

Direct genomic sequencing of bacterial DNA: The pyruvate kinase I gene of *Escherichia coli*

(polymerase chain reaction/expression/multiplex oligomer walking)

Osamu Ohara*, Robert L. Dorit†, and Walter Gilbert

Department of Cellular and Developmental Biology, Harvard University, 16 Divinity Avenue, Cambridge, MA 02138

Contributed by Walter Gilbert, June 14, 1989

ABSTRACT The genomic sequencing procedure is applied to the direct sequencing of uncharacterized regions of bacterial DNA by a "multiplex walking" approach. Samples of bulk *Escherichia coli* DNA are cut with various restriction enzymes, subjected to chemical sequencing degradations, run in a sequencing gel, and transferred to nylon membranes. When a labeled oligomer is hybridized to a membrane, a sequence ladder appears wherever the probe lies near a restriction cut. New probes, based on sequence that lies beyond other restriction sites, are then synthesized, and the membranes are reprobed to reveal new sequence. Repeated cycles of oligomer probe synthesis and subsequent reprobing permit rapid sequence walking along the genome. This oligomer walking technique was used to sequence the pyruvate kinase (EC 2.7.1.40) gene in *E. coli* without resorting to cloning or to library construction. The sequenced region was amplified by the polymerase chain reaction and subsequently transcribed and translated using both *in vivo* and *in vitro* systems, and the resultant gene product characterized to show that the gene encodes the type I isoform of pyruvate kinase.

Genomic sequencing detects the DNA sequence of a particular molecule present in a heterogeneous mixture by exhibiting the sequence on a Southern blot, using a hybridizing labeled probe as an end label (1). The method examines the original genomic DNA molecules directly, without an intervening stage of amplification or cloning. It can detect sequence features in DNA from organisms with large, complex genomes and has been used to study the methylation patterns of animal (2–8) and plant (9) DNA. Footprinting and DNA interaction experiments in both prokaryotic and eukaryotic cells, *in vivo* (10–12) and *in vitro* (3, 13–15), have also taken advantage of this method. After a single set of chemical degradations and a single electrophoretic run, the DNA sequence lanes are transferred onto a membrane that can then be reprobed many times to display particular sequence ladders, making this a very fast sequencing method. Such an adaptation was devised to sequence inserts in both bacteriophage and plasmids by Richard Tizard and Harry Nick (personal communication, used in ref. 16). George Church has developed a very high-throughput, shotgun pattern of "multiplex" sequencing (17), which derives ≈50 times the usual sequence information from each gel and each set of chemistries. Here we wish to explore techniques to sequence bacterial DNA directly, moving step-by-step along the bacterial chromosome.

We hoped to extend the utility of genomic sequencing into previously unsequenced regions by using a "multiplex walking" strategy along the *Escherichia coli* chromosome. Aliquots of bulk genomic DNA from *E. coli* are digested with a variety of restriction enzymes, subjected to Maxam–Gilbert chemical sequencing reactions, run out in polyacrylamide sequencing gels, and finally transferred and crosslinked onto nylon filters. The filters thus contain adjacent tracks of restricted, chemically cleaved genomic DNA. When these filters are hybridized to an extremely radioactive probe, made by adding a tail of [^{32}P]AMP to a synthetic oligomer with terminal deoxynucleotidyltransferase, those lanes containing DNA that has been restricted at or near the hybridization site of the probe yield legible sequence.

This sequence information is then used to design new probes, derived from the 3′-most or 5′-most regions of the newly obtained sequence. The same nylon filters can then be reprobed with the new oligomers, providing additional new sequence. Repeated cycles of probe synthesis and reprobing of the nylon membranes enable one to walk briskly along the DNA in both 5′ and 3′ directions (Fig. 1).

To exemplify this oligomer walking method, we sequenced a pyruvate kinase (PK; EC 2.7.1.40) gene in *E. coli*. The comparison of all available PK sequences reveals regions of high amino acid sequence conservation (18–21). By synthesizing a long oligomer corresponding to one such region, we gained entry into the *E. coli* PK gene by carrying out genomic sequencing on an enriched genomic DNA mixture, using cross-hybridization with this heterologous probe. Repeated cycles of genomic sequencing and probe synthesis then provided the entire sequence of the gene.

Previous workers had suggested, on biochemical (22, 23) and on electrophoretic and genetic (24) grounds, that two noninterconvertible forms of PK are present in *E. coli*: type I (fructose 1,6-bisphosphate-activated) and type II (AMP-activated). Using the direct method, we could not *a priori* determine which of the two genes we had sequenced. Examination of the sequence could not resolve this question, because of an ambiguity created by the presence of two potential initiation sites. Thus, we amplified, transcribed, and translated the sequenced region to examine the gene product directly. On the basis of amino acid composition, molecular weight, and protein activity, we conclude that the gene we have sequenced encodes type I PK (PK-I).

MATERIALS AND METHODS

Materials. *E. coli* strains HB101, DH5α, and XL1-Blue were obtained from BRL and Stratagene. Terminal deoxynucleotidyltransferase was obtained from Boehringer Mannheim; *Thermus aquaticus* (*Taq*) polymerase was from Perkin–Elmer/Cetus; mung bean nuclease and exonuclease III were from BRL; all other DNA-modifying enzymes came from New England Biolabs. Buffers and conditions followed suppliers' specifications. Oligodeoxynucleotide primers and

The publication costs of this article were defrayed in part by page charge payment. This article must therefore be hereby marked "*advertisement*" in accordance with 18 U.S.C. §1734 solely to indicate this fact.

Abbreviations: PK, pyruvate kinase; PCR, polymerase chain reaction.

*Present address: Shionogi Research Laboratories, Fukushima-ku, Osaka 553, Japan.
†To whom reprint requests should be addressed.

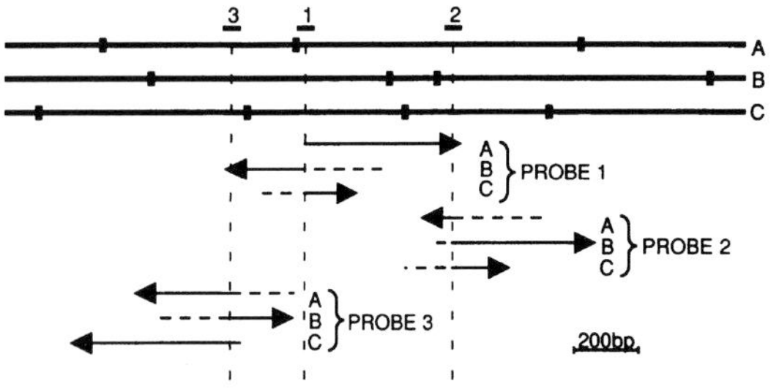

FIG. 1. The multiplex walking strategy. Total *E. coli* DNA is digested with an arsenal of 23 restriction enzymes (*Acc* I, *Alu* I, *Ava* II, *Apa*LI, *Ban* I, *Ban* II, *Bst*NI, *Cal* I, *Hae* II, *Hae* III, *Hin*cII, *Hin*dIII, *Kpn* I, *Mbo* II, *Pst* I, *Pvu* II, *Sau*3AI, *Nci* I, *Rsa* I, *Sty* I, *Taq* I, *Xho* I, *Xmn* I). Each restriction digest is subjected to a full set of Maxam–Gilbert sequencing chemistries, electrophoresed in denaturing polyacrylamide gels, electrotransferred to nylon filters, and immobilized by crosslinking to the filters. Each filter holds the chemical sequencing products of 8–10 different restriction digests. Filters are then probed with short, complementary, radioactively tailed oligonucleotides. Depending on the site of hybridization of the probe relative to a restriction cut, up to 300 base pairs (bp) of sequence can be directly visualized. In this diagram, the heavy lines represent three molecules of *E. coli* genomic DNA restricted with enzymes A, B, and C, respectively. The first cluster of three arrows shows the overall extent and orientation of sequence information that can be visualized from a single filter hybridized with probe 1. Each arrow shows the information obtained from each of the sequenced restriction digests (A, B, and C), given the indicated restriction sites. The solid portion of the arrow indicates legible sequence; the dashed portion originates at the relevant restriction cut. Sequence will be read in either the 5′ or 3′ direction, depending on the position of the probe relative to the restriction cuts. Sequence can be read along the length of a restriction fragment, beginning at the hybridization site and extending downstream to the position at which extraneous sequences, originating from the downstream restriction site, obscure the sequence ladder. The 3′-most and 5′-most sequence thus obtained is used to design probes 2 and 3, respectively. Sequence filters are subsequently stripped and reprobed with the newly designed oligomer probes, yielding additional sequence information as shown. This technique of filter probing, new probe design, and subsequent reprobing of the same nylon filters permits one to move rapidly in both directions along the length of a sequence of interest.

probes were synthesized on a Milligen 6500 synthesizer and were purified by electrophoresis in 20% polyacrylamide denaturing gels. Radiolabeled nucleotides, GeneScreen membranes, and the *in vitro* coupled transcription/translation system were purchased from NEN/DuPont and used according to their recommendations. Vector pBluescript KS(−) was obtained from Stratagene.

All computer analyses of DNA and amino acid sequences were carried out on a MicroVAX II, using the 1986 University of Wisconsin Genetics Computer Group (UWGCG) software package (41).

DNA Blot Analysis of Restriction Digests. Total *E. coli* genomic DNA was isolated from the HB101 strain (25) and subsequently cut with a variety of restriction enzymes (4- and 6-base-specific). Restriction digests of *E. coli* genomic DNA were run out in a 1% agarose gel containing 89 mM Tris/89 mM boric acid/2 mM EDTA (TBE buffer) and were subsequently transferred onto GeneScreen nylon membrane by standard capillary methods (26). After overnight transfer, DNA was crosslinked to the membrane by UV irradiation. Heterologous probes (e.g., C-1, Table 1) were radioactively labeled with [γ-^{32}P]dATP (5000–6000 Ci/mmol; 1 Ci = 37 GBq) by using T4 polynucleotide kinase. The membrane was probed at 45°C for 4 hr with 10 pmol of labeled oligonucleotide in 10 ml of genomic sequencing hybridization buffer [1% bovine serum albumin/0.5 M sodium phosphate (as Na^+ concentration), pH 7.2/7% $NaDodSO_4$/1 mM EDTA]. The

Table 1. Genomic sequencing probes and polymerase chain reaction (PCR) primers

Name	Sequence (5′ to 3′)	Position*
	Oligonucleotides for genomic sequencing	
C-1	GCAGGGATCTCAATACCCAGGTCACCAC-GGGCCACCATAA	−1112
E-1	CGAGGATTTCGTCGAAGTTGTTG	−1058
E-2	TTTTCGATTTTGGAGATGTGGAT	−1023
E-3	CGCCTTGTTCGCAACCAAAGATC	−889
E-4	CTTCTGCGCGAGTCGGGCGTGGG	−1238
E-5	CAGTTTACGGTTGTCATTGTTG	−1411
E-6	GCCAACAGACAGGTCAGTAGTGA	−715
E-7	GCTCGCGCAGTACGTAAATAC	+1505
E-8	TAACATCTCTTCAGATTCGGT	−415
E-9	GATCTCTTTAACAAGCTGCGGCAC	−1624
E-10	CTCGCTCTAAGGATAGGTGAC	−260
E-11	ACGTCACCTTTGTGTGCCAGACCG	−1698
	Oligonucleotides for PCR	
E-12	GAGCTCTTCGATATACAAATTAATTC	−1802
E-13	AAGCTTGCGTAACCTTTTCCC	+5
C-2	CGTGGTGACCTGGGTATTGAGATCCC	+1085
C-3	GCGGTCTCCCCAGACAGCAT	−1302
C-4	ACCAAGGGACCTGAAATCCG	+554

*Relative to final PK sequence.

membrane was washed three times (20 min per wash) at 45°C with 0.4 M NaCl/40 mM sodium phosphate, pH 7.2/1% $NaDodSO_4$/1 mM EDTA. Two final 10-min washes were carried out at 50°C using a solution containing 3.2 M tetramethylammonium chloride and 1% $NaDodSO_4$ (27). The wet nylon membrane was then wrapped in plastic wrap and exposed for 4–24 hr to Kodak X-Omat AR film.

Fractionation and Enrichment of *Pst* I-Digested Target Fragment. To obtain legible sequence with cross-hybridizing probes, we enriched the DNA mixture for the PK target (see *Results*). DNA blot analysis showed that the 4.5- to 4.8-kilobase (kb) size range of a *Pst* I digest contained the entire PK sequence. *E. coli* DNA digested to completion with *Pst* I was run out in a 1% agarose gel, and fragments in the 4.5-kb range were isolated from the gel by transfer onto DEAE membrane (NA-45; Schleicher & Schuell) (28). The DNA in this size range (including the target) was eluted from the membrane by standard protocols, ethanol-precipitated, desalted, dried under vacuum, resuspended in water to a concentration of ≈20 ng/ml, and ligated into the dephosphorylated *Pst* I site of the pBluescript vector. The resultant plasmid mixture was amplified in *E. coli* strain DH5α and subsequently purified according to standard methods (29).

Genomic Sequencing. We sequenced DNA of interest by chemical degradation methods (30–32) using six reactions (G, A + G, A > C, C + T, C, and T). Approximately 5 μg of the appropriate DNA (plasmid mixture or total *E. coli* DNA) was used in each chemical sequencing reaction. Samples were then electrophoresed in either isocratic or gradient acrylamide denaturing sequencing gels (0.4 mm thick) (33), electrotransferred onto nylon membranes, and immobilized by UV crosslinking (1). Filters were probed with oligonucleotides that had been tailed with [α-^{32}P]dATP (6000 Ci/mmol) and subsequently purified on oligo(dT)-cellulose (34). Hybridizations were carried out as described above for restriction digests, except for a lower final probe concentration (0.2 nM) and the presence in the hybridization mixture of polyadenylic acid (20 μg/ml), added to prevent spurious probe binding due to the oligo(dA) tail. Sequencing filters were washed seven times (10 min per wash) at room temperature with a solution (preheated to 45°C) containing 0.2 M NaCl, 40 mM sodium phosphate (pH 7.2), 1% $NaDodSO_4$, and 1 mM EDTA. When imperfect probes were used, filters were also subjected to two final 3.2 M tetramethylammonium chloride washes as previously described. Wet filters were wrapped in plastic wrap and exposed for 1–6 days to Kodak XAR-5 film without intensifying screens. Probes were stripped from the membranes by washing with 50 mM NaOH at room temperature for ≈20 min. Membranes were then rinsed with 50 mM Tris/EDTA buffer (pH 8.3), wrapped in plastic wrap, and stored at 4°C until the next reprobing.

Expression and Characterization of *E. coli* PK. After determining the DNA sequence of the PK gene, we designed and synthesized two flanking primers (E-12 and E-13, Table 1) to amplify a DNA fragment containing the entire PK gene by the PCR (35, 36). For ease in cloning, the primers were designed explicitly to contain restriction enzyme recognition sites at their 5′ ends (*Sac* I site for E-12; *Hin*dIII site for E-13). The PCR was carried out in a 100-μl volume containing 1 μg of *E. coli* total DNA, 100 pmol of each primer, 200 μM each dNTP, 2.5 units of *Taq* polymerase, 50 mM KCl, 10 mM Tris·HCl (pH 8.8), 10 μg of bovine serum albumin, 3 mM dithiothreitol, and 1.5 mM $MgCl_2$. Thirty cycles of amplification were carried out as follows: annealing, 50°C, 1 min; extension, 72°C, 2 min; denaturation, 94°C, 1 min; final extension, 72°C, 5 min. The amplified product of expected size (1.8 kb) was recovered from a 1% agarose gel by adsorption onto a DEAE membrane (28), subsequently eluted, and cloned into the *Pst* I site of the pBluescript vector by the "G-C tailing" method (37). The resultant construct is referred to as pEPK. Subsequently, we modified pEPK by deleting from the 5′ end of the insert with exonuclease III and mung bean nuclease (38). A new plasmid construct, determined by genomic sequencing to involve the deletion of 212 bp from the 5′ end of the PK gene, is referred to as pdEPK. Both of the plasmids were prepared using *E. coli* strain XL1-Blue, under standard procedures (29). *E. coli* cells harboring pBluescript, pEPK, or pdEPK were grown for 8 hr at 37°C in M9 medium supplemented with Casamino acids (0.2%) and ampicillin (100 μg/ml). The cells were ruptured by sonication and soluble protein fractions were prepared. We assayed PK activity by a modified lactate dehydrogenase coupled assay. The different regulatory properties of the two isozymes allowed us to measure the activity of type I and type II isozymes (PK-I and PK-II) separately (39). The soluble protein fractions were also analyzed by $NaDodSO_4$/PAGE (40). Protein concentrations were determined by Coomassie blue G250 binding (Bio-Rad protein assay kit).

The protein products encoded by the various plasmids were generated in the presence of [^{35}S]methionine by using an *in vitro* coupled transcription/translation system (NEN/DuPont) derived from *E. coli*. The products synthesized *in vitro* were analyzed by $NaDodSO_4$/PAGE and detected by autoradiography. Apparent molecular weights of the PK gene products were determined by comparison with protein molecular weight standards from Sigma and BRL.

RESULTS

Our intention was to sequence a gene directly on bacterial DNA by the genomic sequencing method. We initiated the sequencing of a PK gene in *E. coli* by using a heterologous probe derived from the chicken PK sequence. We first synthesized a 40-base-long oligomer (C-1) from a highly conserved region of amino acid sequence and tested whether this oligomer would hybridize specifically to a single region of the *E. coli* genome. A single band appeared in certain lanes when probe C-1 was hybridized to a DNA blot containing various restriction digests of total *E. coli* DNA (Fig. 2). This suggests that probe C-1 most likely cross-hybridizes to a bacterial PK sequence. However, when C-1 was used to probe membranes containing total *E. coli* DNA subjected to chemical sequencing reactions, using 5 μg of total DNA per

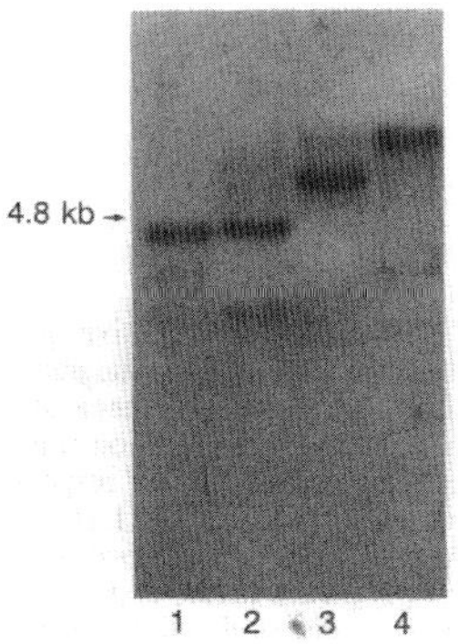

FIG. 2. Blot of digested total *E. coli* genomic DNA probed with heterologous oligomer C-1. Lanes: 1, *Pvu* II; 2, *Pst* I; 3, *Hin*dIII; 4, *Eco*RI. Autoradiogram was obtained by a 90-hr exposure with an intensifying screen, following a tetramethylammonium chloride wash.

sequencing reaction, a clear sequence ladder could not be discerned due to a poor signal-to-noise ratio.

We sought to improve this ratio by enriching for the signal. The DNA size range (4.5–4.8 kb) containing the target sequence was isolated from a *Pst* I digest of total *E. coli* DNA, inserted into a conventional vector, and then amplified in mass culture. The resultant amplified mixture of plasmids was digested with seven different restriction enzymes (*Apa*LI, *Hae* II, *Hae* III, *Hin*cII, *Pvu* II, *Stu* I, *Sty* I), since we did not know *a priori* which restriction sites would lie near the hybridization site of probe C-1. Using 5 μg of DNA (derived from the plasmid mixture) per sequencing reaction, we found that *Apa*LI-digested DNA produced a legible sequence ladder. This DNA sequence translated to a protein fragment similar to previously published eukaryotic PK sequences. Based on this initial fragment of *E. coli* PK sequence, we designed two perfect oligonucleotide probes (E-1 and E-2), which we then used for a new round of genomic sequencing (Fig. 3). Thus, by designing new probes as new sequence became available, and by reprobing previously used filters, we were able to move both upstream and downstream of our original entry point. The signal-to-noise problem encountered with the heterologous C-1 probe disappeared when we used perfect probes. In this manner, we were able to read sequences of 200–400 bp with overnight to 5-day exposures. The synthesis of new probes and the labeling and reprobing took as little as 3 days per cycle. Fig. 4 shows a representative fragment of the PK sequence, visualized directly either from total *E. coli* DNA or from an enriched mixture by using perfect probes.

Using this walking strategy for genomic sequencing, we determined a 1.8-kb nucleotide sequence that included a long open reading frame. The predicted amino acid sequence shows extensive similarity to all previously sequenced PKs. However, two important issues could not be resolved on the

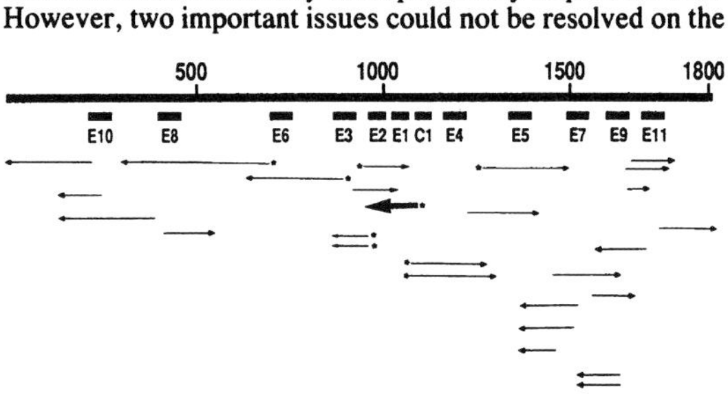

FIG. 3. Sequencing strategy, showing the extent of sequence information obtained using each of the listed oligomer probes. Positions of the oligomers are shown relative to the final numbering of the PK sequence (oligomers not to scale). Asterisks indicate sequences obtained from enriched plasmid mixture; others were obtained from total genomic DNA. Heavy arrow marks point of entry into sequence with heterologous probe.

6886 Biochemistry: Ohara *et al.* Proc. Natl. Acad. Sci. USA 86 (1989)

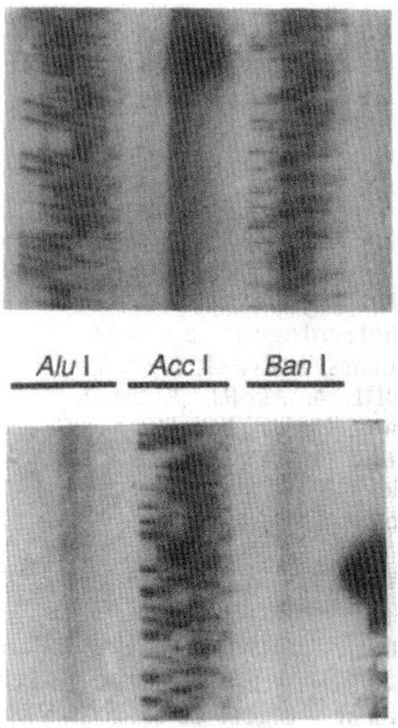

FIG. 4. Representative genomic sequence autoradiogram. (*Upper*) Detail of sequence obtained using probe E-3 on total genomic DNA restricted respectively with *Alu* I and *Ban* I. (*Lower*) The same filter, probed with E-8; sequence ladder appears in an *Acc* I digest of enriched plasmid mixture. Lanes show products of Maxam–Gilbert reactions, in following order: G, A + G, A > C, T + C, C, and T. Six-day exposures were made without intensifying screens.

basis of the DNA sequence alone: (*i*) the actual initiation codon, since two possible candidates for the initiating methionine are present (bases 140 and 355), and (*ii*) the identity of the PK isotype (I or II).

The two isotypes of PK in *E. coli* have been reported to have molecular weights of 56,000 and 51,000, respectively. Their overall amino acid compositions have been published and their sizes estimated at 522 residues (type I) and 488 residues (type II) (23). The *E. coli* sequence we obtained predicts possible proteins of 534 and 462 amino acids, while the amino acid composition suggests that the sequenced gene corresponds to type I.

To resolve the identity of this gene, we examined the gene product. Fig. 5 shows two expression plasmids constructed for the purpose: pEPK contains the full (1.8-kb) PCR-amplified PK gene, inserted in pBluescript in the opposite direction to the *lac* promoter; pdEPK is a truncated derivative. To determine the site of translation initiation, we attempted to delete between 140 and 355 bp from pEPK. After analyzing the deletion products, we selected one, pdEPK, from which nucleotides 1–212 had been deleted, thus removing the first methionine codon at base 140.

When plasmids pEPK and pdEPK were translated in a cell-free transcription/translation system derived from *E. coli*, two polypeptides (M_r 56,000 and 54,000) appeared on NaDodSO$_4$/PAGE, along with the ampicillin-resistance gene product (Fig. 6*A*), although pdEPK produced smaller amounts of both polypeptides. Further, when extracts of *E. coli* cells transformed by these two plasmids were run out on NaDodSO$_4$/PAGE, they consistently showed a strong doublet that comigrated with the M_r 56,000/54,000 *in vitro* products of pEPK (Fig. 6*B*). Since the deletion plasmid and the intact construct produce polypeptides of the same length,

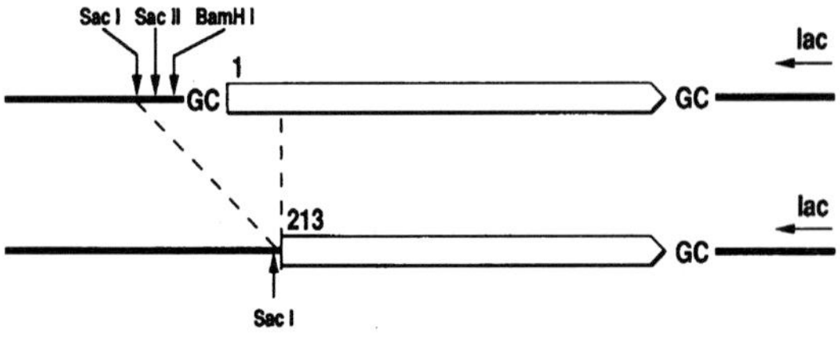

FIG. 5. pEPK and pdEPK constructs, derived from pBluescript vector. Open bars indicate the PCR-amplified PK gene either ligated directly into pBluescript by homopolymer tailing or after partial deletion at the 5′ end. Orientation of *lac* promoter is indicated.

the methionine codon at position 140 can be ruled out as the initiation codon.

Table 2 shows the PK enzymatic activity in crude extracts of *E. coli* cells harboring pEPK, pdEPK, or pBluescript. Since PK-I has only about 1% of its activity in the absence of fructose 1,6-bisphosphate, we could separately estimate PK-I and PK-II activity in the crude cell extracts. Both plasmids significantly stimulated PK-I activity (200-fold and 20-fold, respectively). We attribute the apparent increase in PK-II activities to the residual PK-I activity present at the 1% level in the PK-II fraction. We conclude that pEPK contains both the promoter and coding sequence of the *E. coli* PK-I gene, while the 212-bp deletion at the 5′ end of the insert in pdEPK eliminates much of the PK-I promoter activity but does not cut into the coding sequence.

We were surprised not to see any evidence by cross-hybridization for a PK-II gene. We then sought to find such a gene by PCR amplification of *E. coli* DNA, using two primers constructed from conserved regions of chicken PK. However, that amplification yielded only one major band, which, on sequencing, turned out to be the PK-I gene. We cut that band with two restriction enzymes (*Apa* I, *Bcl* I) to see whether any PK-II DNA fragments might be comigrating with the PK-I fragment; only PK-I sequences were detected.

DISCUSSION

These results extend the utility of genomic sequencing by demonstrating that it can be used, with a multiplex walking strategy, to decipher bacterial DNA sequences. Repeated cycles of sequencing, probe synthesis, and hybridization allowed us to advance rapidly both upstream and downstream from our original point of entry onto the *E. coli* genome and yielded the sequence of the entire PK-I gene.

However, we were not able to obtain readable sequence patterns from bacterial DNA by using imperfect, cross-hybridizing primers. We solved this problem by size-selecting particular fragment lengths (in this case, 4.5-kb *Pst* I fragments containing the entire PK gene) and thus reducing the background. Alternatively, we could have used imperfect primers in conjunction with the PCR. In fact, two PCR primers derived from conserved regions of the chicken PK sequence were successfully used to amplify a region of the *E.*

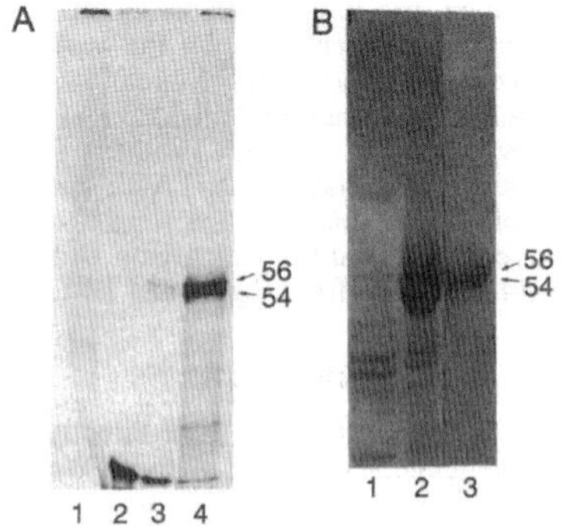

FIG. 6. (*A*) Autoradiogram of *in vitro* translation products labeled with [^{35}S]methionine. Lanes: 1, control, no DNA added; 2, control, intact pBluescript vector; 3, deletion plasmid pdEPK products; 4, plasmid pEPK products. (*B*) Autoradiogram of *in vivo* translation products obtained by using the soluble fraction of *E. coli* cell extracts. Total protein loaded, determined by Coomassie blue G250 binding assay, is indicated in parentheses. Lanes: 1, control, intact pBluescript vector (100 μg); 2, pEPK (150 μg); 3, pEPK (40 μg). Products were run in 8% NaDodSO$_4$/polyacrylamide gels and stained with Coomassie brilliant blue R250. The M_r 56,000 and 54,000 bands are indicated.

Table 2. PK-I and PK-II activities of plasmid constructs

	Activity, units/mg of protein		
Plasmid	PK-I	PK-II	Adjusted PK-II
pBluescript	0.24 (1.0)	0.04 (1.0)	—
pEPK	>50 (>208)	0.74 (18.5)	0.04
pdEPK	5.0 (20.8)	0.12 (3.0)	0.05

Numbers in parentheses indicate activities relative to pBluescript control. Adjusted PK-II activity: observed PK-II activity − residual PK-I activity (assumed to be 1.4% of total activity). Unit activity is proportional to amount of enzyme present.

coli PK gene for sequencing. Heterologous PCR, when successful, may represent the fastest route of entry into new sequences. Once a novel sequence is entered, either by using imperfect probes on enriched target DNA or by PCR, genomic sequencing provides a rapid, general method for brisk walking along sequences of interest. A single set of nylon membranes, containing genomic DNA digested with a battery of restriction enzymes and subjected to Maxam–Gilbert chemistries, can be probed, stripped, and reprobed up to 40 times with newly synthesized oligomers, yielding sequence in both directions from the point of entry. By obviating the need for library construction and screening and by reusing the products—already fixed onto a nylon filter—of a set of sequencing chemistries, our methods facilitate and speed the sequencing process and eliminate the risk of artifacts generated during library construction or subcloning.

Although this multiplex oligomer walking technique could be applied to genomes of much higher complexity, the practical limitation is the time involved in the exposures needed to visualize the sequence. Methods of labeling probes to higher specific activities or more sensitive detection techniques would eliminate the bottleneck.

The results of sequencing can be coupled, via PCR, to the analysis of gene products. We inserted PCR-generated sequences of the complete PK gene into plasmids that could be expressed both *in vivo* and *in vitro*. The PCR product also can function directly in an *in vitro* coupled transcription/translation system to yield the same protein products as the plasmid construct.

Translation, both *in vivo* and *in vitro*, of the pEPK plasmid consistently produced two bands on $NaDodSO_4$/PAGE; *in vitro*, the PK-I sequence directs the synthesis of two polypeptides of M_r 56,000 and 54,000. Posttranslational modification of the protein is not likely in the cell-free system, so these two distinct products probably arise by an inappropriate initiation of translation from an internal methionine codon. The smaller band appears too large to be the PK-II product and does not show any PK-II activity. Thus PK-I is 462 amino acids long yet has an apparent molecular weight of 56,000, in contradiction to the length (522 amino acids) predicted by previous authors. We attribute this discrepancy to the anomalous migration of PK-I in $NaDodSO_4$/PAGE. The amino acid composition deduced from our PK-I sequence is in close accord with that reported by other workers, although we do find a single tryptophan residue not identified previously. It is unclear whether the two gene products appearing in the doublet are made by *E. coli* under normal circumstances. The second gene product has not been observed before, although if it does not exhibit PK-I activity, it is unlikely to have been noticed previously.

Table 3. PK amino acid sequence comparisons

	E. coli	Yeast	Chicken
Yeast (*Saccharomyces cerevisiae*)	43%		
Chicken (*Gallus gallus*)	47%	47%	
Rat (*Rattus norvegicus*)	44%	48%	68%

Values indicate the percentage of identical amino acids shared in a given pairwise comparison. Sequences were aligned using the algorithm of Needleman and Wunsch as implemented in the UWGCG software package (41).

PK-I is homologous to the PKs of chicken, rat, cat, and yeast. Table 3 shows the extent of amino acid sequence conservation. However, we could not detect the PK-II gene by cross-hybridization or PCR, suggesting that it is highly divergent from, or perhaps unrelated to, the PK-I gene.

Our results show that the power and relative ease of genomic sequencing can be used to decipher unknown sequences directly from bacterial DNA, providing a rapid, general method for the characterization of genes.

We thank Milligen for the use of their 6500 synthesizer, Tom Maniatis for critical reading of the manuscript, members of the Gilbert lab for helpful discussions, and L. Schoenbach for the illustrations. This work was supported by the National Institutes of Health (Grant GM37997-02). O.O. was supported by Shionogi Research Laboratories.

1. Church, G. M. & Gilbert, W. (1984) *Proc. Natl. Acad. Sci. USA* **81**, 1991–1995.
2. Ephrussi, A., Church, G. M., Tonegawa, S. & Gilbert, W. (1985) *Science* **227**, 134–140.
3. Church, G. M., Ephrussi, A., Gilbert, W. & Tonegawa, S. (1985) *Nature (London)* **313**, 798–801.
4. Nick, H. & Gilbert, W. (1985) *Nature (London)* **313**, 795–798.
5. Saluz, H. P., Jiricny, J. & Jost, J. P. (1986) *Proc. Natl. Acad. Sci. USA* **83**, 7167–7171.
6. Saluz, H. & Jost, J. P. (1986) *Gene* **42**, 151–157.
7. Becker, P. B., Ruppert, S. & Schutz, G. (1987) *Cell* **51**, 435–443.
8. Saluz, H. P., Feavers, I. M., Jiricny, J. & Jost, J. P. (1988) *Proc. Natl. Acad. Sci. USA* **85**, 6697–6700.
9. Nick, H., Bowen, B., Perl, R. J. & Gilbert, W. (1986) *Nature (London)* **319**, 243–246.
10. Selleck, S. B. & Majors, J. (1987) *Nature (London)* **325**, 173–177.
11. Pauli, V., Chrysogelos, S., Stein, G., Stein, J. & Nick, H. (1987) *Science* **236**, 1308–1311.
12. Bushman, F. P. & Ptashne, M. (1988) *Cell* **54**, 191–197.
13. Richet, E., Abcarian, P. & Nash, H. (1986) *Cell* **46**, 1011–1021.
14. Fujiwara, T. & Mizuuchi, K. (1988) *Cell* **54**, 497–504.
15. Hromas, R., Pauli, U., Marcuzzi, A., Lafrenz, D., Nick, H., Stein, J., Stein, G. & Van-Ness, B. (1988) *Nucleic Acids Res.* **16**, 953–967.
16. Marchionni, M. & Gilbert, W. (1986) *Cell* **46**, 133–141.
17. Church, G. M. & Kieffer-Higgins, S. (1988) *Science* **240**, 185–188.
18. Lonberg, N. & Gilbert, W. (1983) *Proc. Natl. Acad. Sci. USA* **80**, 3661–3665.
19. Burke, R. L., Tekamp-Olson, P. & Najarian, R. (1983) *J. Biol. Chem.* **258**, 2193–2201.
20. Lone, Y. C., Simon, M. P., Kahn, A. & Marig, J. (1986) *FEBS Lett.* **195**, 97–100.
21. Muirhead, H., Clayden, D. A., Barford, D., Lorimer, C. G., Fothergill-Gilmore, L. A., Schiltz, E. & Schmitt, W. (1986) *EMBO J.* **5**, 475–481.
22. Somani, B. L., Valentini, G. & Malcovati, M. (1977) *Biochim. Biophys. Acta* **482**, 52–63.
23. Valentini, G., Iadarola, P., Somani, B. L. & Malcovati, M. (1979) *Biochim. Biophys. Acta* **570**, 248–258.
24. Garrido, P. A. & Cooper, R. A. (1983) *FEBS Lett.* **162**, 420–422.
25. Sato, H. & Miure, K. I. (1963) *Biochim. Biophys. Acta* **72**, 619–629.
26. Southern, E. (1975) *J. Mol. Biol.* **98**, 503–517.
27. Wood, W. I., Gitschier, J., Lasky, L. A. & Lawn, R. M. (1985) *Proc. Natl. Acad. Sci. USA* **82**, 1585–1588.
28. Dretzen, G., Bellard, M., Sassone-Corsi, P. & Chambon, P. (1981) *Anal. Biochem.* **112**, 295–298.
29. Maniatis, T., Fritsch, E. F. & Sambrook, J. (1982) *Molecular Cloning: A Laboratory Manual* (Cold Spring Harbor Lab., Cold Spring Harbor, NY).
30. Maxam, A. M. & Gilbert, W. (1977) *Proc. Natl. Acad. Sci. USA* **74**, 560–564.
31. Rubin, C. M. & Schmid, C. W. (1980) *Nucleic Acids Res.* **8**, 4613–4619.
32. Maxam, A. M. & Gilbert, W. (1980) *Methods Enzymol.* **65**, 499–560.
33. Biggin, M. D., Gibson, T. J. & Hong, G. F. (1983) *Proc. Natl. Acad. Sci. USA* **80**, 3963–3965.
34. Collins, M. L. & Hunsker, W. R. (1985) *Anal. Biochem.* **151**, 211–224.
35. Scharf, S. J., Horn, G. T. & Erlich, H. A. (1986) *Science* **233**, 1076–1078.
36. Saiki, R. K., Gelfand, D. H., Stoffel, S., Scharf, S. J., Higuchi, R., Horn, G. T., Mullis, K. B. & Erlich, H. A. (1988) *Science* **239**, 487–491.
37. Eschenfeldt, W. H., Puskas, R. S. & Berger, S. L. (1987) *Methods Enzymol.* **152**, 337–342.
38. Henikoff, S. (1984) *Gene* **28**, 351–359.
39. Malcovati, M. & Valentini, G. (1982) *Methods Enzymol.* **90**, 170–179.
40. Laemmli, U. K. (1970) *Nature (London)* **227**, 680–685.
41. Devereux, J., Haeberli, P. & Smithies, O. (1984) *Nucleic Acids Res.* **12**, 387–395.

GENOME SEQUENCING: Creating a New Biology for the Twenty-First Century

Walter Gilbert

PROLOGUE: *Walter Gilbert, a Nobel laureate and one of the chief proponents of the human genome project, believes that the sequence of the human genome would create a Rosetta stone, a complete library of information that biologists could search as they strive to understand human genes. In the clinical area, the genome sequence would enable researchers to pinpoint the mechanisms of genetic diseases and, eventually, develop therapies to treat them.*

Here Gilbert describes the enormous benefits he believes the project would yield for both science and medicine and urges the immediate establishment of a center devoted exclusively to mapping and sequencing the human genome. Only as a centralized effort will this massive project become practically and economically feasible, he says. He predicts that, with anticipated improvements in technology and expected economies of scale, the entire sequence could be obtained for $300 million (representing a 10-year effort by some 300 scientists). While expensive compared to the rest of biological research, he says, the human genome project would provide the basis for a new biology for the twenty-first century, revealing, at last, how genes determine form and function.

Walter Gilbert is H. H. Timken Professor of Science in the department of cellular and developmental biology at Harvard University. He received his A.B. in chemistry and physics and his A.M. in physics from Harvard in 1953 and 1954. He did his doctoral work at Cambridge University, where he received a Ph.D. in mathematics in 1957. He has taught at Harvard since 1958 in the fields of physics, biophysics, biochemistry, and biology. Gilbert was awarded the Nobel Prize for Chemistry in 1980.

SEQUENCING: A NEW BIOLOGY

Over the past year biologists have been debating whether or not to engage in a massive effort to work out the entire sequence of the DNA, or genetic material, of a human being: the human genome project. Such a project would create a Rosetta stone, a complete library of information that biologists could search as they move forward to understand the totality of genes that make up a human.

Several hundred thousand genes specify all aspects of the structure of our bodies, determining the patterns of growth that turn a single cell into a functioning adult. DNA is the ultimate repository of all that information; one can ask no more fundamental question about a gene than its DNA structure. The DNA sequence, or the order of the nucleotide bases that make up the coiled strands of DNA, not only specifies which protein a gene produces but also contains all of the elements that control when and where that gene functions. The genes determine the potentiality of each cell, and of the entire organism, in the sense that they specify all the rules by which the cells multiply, move, and specialize as the individual develops. They determine the ground plan of the organism, even though the individual may then be formed by the interaction of that pattern of development with the world around it. Today, we understand only one-tenth of one percent of human genes. The human genome project would dramatically speed the development of our knowledge.

What do we want to know about genes? In general, for animals and plants we want to know how genes control form and function. The specification of the form of an animal is a mystery arising from the process of development, in which the turning on and off of specific genes in specific cells takes an organism from a single cell to its specialized final form. How do the genes function to set up structures in the organism: patterns of limbs and fingers, muscles and bones, nerves and arteries? To understand that, we ask questions about the control of genes: How does one gene act to turn on or off another or, more broadly, how does the structure of the chromosomes on which genes are located affect the way in which genes are expressed?

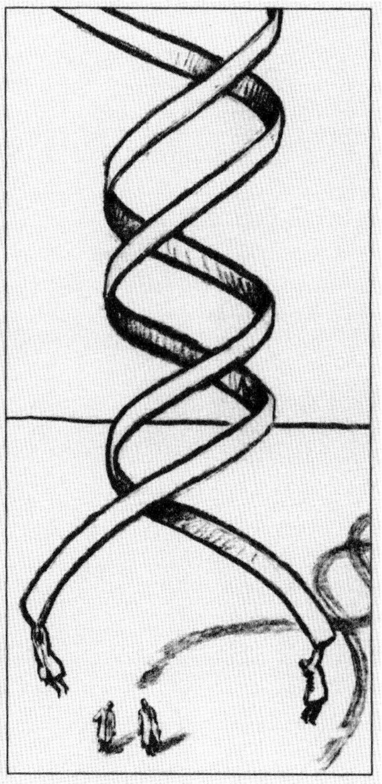

Robert Alan Soulé

For humans we want to know far more detail. The medical impulse—the study of causes of human disease and functioning with the goal of ameliorating the human condition—stimulates us to know all human genes, not just in principle but in practice, to enumerate them and understand their functions. Identifying specific genes with specific malfunctions permits us to understand the mechanisms behind that malfunction and ultimately to try to develop rational treatments. The first step to this path is to list the myriad genes that make up an individual. This will serve as a reference sequence. No two people have identical sequences, and there is no perfect set of genes for the human. Each of us is made up of interacting genes whose compensating effects endow our bodies with shape and function.

Over the past decade we have made significant progress toward these goals; we have identified some genes of interest and have determined their exact structure, or sequence, to study their function. But the human genome project would transform the nature of the endeavor—indeed, the nature of biological science. Advances in technology have given us the ability to gain access to all the information contained within the human genome—to determine the entire genetic sequence (the order of the 3 billion nucleotide base pairs that constitute our genetic material)—at unprecedented speed and with incalculable benefits to both science and medicine.

But the proposal to sequence the human genome is controversial. It would be an effort of unprecedented size in biology. Some biologists are concerned that the project will distort research priorities and bring biology, like physics, into the realm of "big science." Some are questioning whether this is the best use of limited research funds, and some even question whether knowing the sequence of human DNA will be useful.

I have no doubt that it will be immensely useful. For the first time, the goal of understanding the human genome is within reach. We should make a concerted effort to attain it.

One can ask no more fundamental question about a gene than its DNA structure.

The human genome, the entire complement of genetic material inherited from one's parents, is carried in a sequence of chemical groups, called nucleotides or bases, arranged linearly along a DNA strand. Four different nucleotides make up the DNA chain: adenosine, guanosine, thymidine, and cytosine, often represented by the letters A, G, T, and C. The two strands of DNA are held together by bonds between these four bases: A binds with T, and G binds with C. The information in DNA can be thought of as a long sequence of these four letters.

A gene is a sequence of several thousand nucleotides that function as a coherent unit. Each gene carries the instructions for a specific protein or an RNA molecule. This genetic information is expressed in a two-step process: a region of the DNA chain is copied into an RNA molecule, which is then translated by special machinery into a protein molecule. The translation of DNA and RNA into protein involves the decoding of the information written out in a four-letter code of the nucleotide bases, the genetic code. A series of three nucleotides codes for a specific amino acid, the building blocks of proteins; proteins are long sequences made of a combination of the 20 different amino acids. The nucleotide sequence ATG, for example, codes for the amino acid methionine. The protein molecules, whose structures are ultimately dictated by the DNA, then create all of the other molecules and functions of the cell. Proteins provide a cell with some enzymatic activity, control the functioning of some other gene, or provide a structural component of a cell or tissue.

The genetic material of a human consists of 3 billion nucleotide pairs of DNA arranged in 23 chromosome pairs: written out, that would be 3 billion letters from the four-letter genetic alphabet (the four initials repeated over and over again). One copy of a newspaper like the *New York Times* contains 3 million letters. So, the entire human sequence could be written out in three years worth of daily newspapers or a thousand large telephone books.

A typical gene product is a protein molecule some 300 amino acids long. It would take a sequence of 1,000 bases to specify such a protein. In mammals, such a DNA sequence that encodes a protein is not contiguous along the DNA but is stretched out over some 10,000 bases of DNA. That DNA sequence contains alternating regions—short ones, called exons, encoding part of the protein, and long ones, called introns, whose information is skipped over in producing the protein. The 10,000 bases of DNA are copied out into a 10,000-base-long RNA strand, and then that strand is cut and spliced together to remove the introns. Reduced in size to 1,000 bases, the RNA then moves from the cell's nucleus to the cytoplasm to be translated to make many copies of the corresponding protein. Although a typical gene is 10,000 to 20,000 bases long, the shortest genes are less than 1,000 bases and the longest range from 200,000 to 1,000,000 bases. It is not known exactly how many genes make up a human being, but the number is likely to be between 100,000 and 300,000.

Today, by genetic techniques involving studies of inheritance across populations, we can identify some 3,000 genes that cause clearly defined human diseases. For such a gene to be recognized, however, there must be a very obvious pattern of inheritance, and the gene must have a simple relationship to the disease, as is the case in sickle-cell anemia. Diseases that might be caused by the interaction of several genes, or that may have many alternative genetic causes, cannot be recognized. Only a few hundred genes have been isolated and fully characterized—that is, their entire nucleotide sequence worked out—by molecular biological techniques.

Once we have the sequence, human biology can shift from the question of how to find the genes to the question of what genes do.

The molecular approach to determining what genes do is to identify a specific gene, isolate it, work out its sequence, understand what protein is involved, and then synthesize that protein and study its function in the test tube or interrupt its functioning in the cell. This process goes on today over and over again as each new gene is studied.

When the protein product of a gene of biological or medical interest is identified, years are spent first to find and clone the corresponding DNA fragments and then to work out their structure by sequencing before one can turn to the interesting biology of what that gene does. If a new gene is identified solely by genetics—by recognizing from an inheritance pattern that some gene controls a human disease or human function—the process is even longer. First one must struggle to locate that gene on the DNA by finding ever closer genetic markers, and then one must go through DNA isolation and sequencing before being able to turn again to the biological problem. All of these intermediate steps would be simplified or eliminated by having available the sequence of the genome. Knowing the full sequence would mean that at the moment one isolates a new protein and works out a bit of its amino acid sequence, one could identify in a data base a portion of its gene and the region of the genome from which it came. This leads directly to the isolation and full characterization of the gene for the protein, shortcutting much of today's major efforts.

The human genome project is a proposal to isolate and sequence human genes once and for all so that the study of human biology can shift from the question of how to find the genes to the question of what the genes do.

How is it possible to contemplate working out the sequence of the 3 billion bases that make up the human genome? In the early 1970s it was not possible to sequence an entire gene; one could work out only about 100 bases a year. But, with the introduction a decade ago of rapid sequencing methods, the sequencing rate increased 100-fold, so that a single person could work out 10,000 to 20,000 bases of finished DNA sequence in a year. Worldwide over the last 10 years, close to 10 million bases of DNA have been sequenced, about 2 million of which are human DNA. Today such information is being collected at an even faster rate, now approaching about 3 million bases a year. One might expect the normal course of technological advance to increase that rate by another factor of 10 over the next decade, to a level of about 30 million bases a year worldwide. However, it will take the creation of focused centers of sequencing to bring the rate of sequence

The development of these maps will totally change the way biologists identify human genes of medical interest.

acquisition up to 300 million bases per year, a rate that would be needed to do the human genome in a reasonable time, say 10 years.

The human genome project would be a concerted effort to develop first a physical map of the genome and then to determine the entire DNA sequence. The first phase, the physical map, entails breaking the DNA of a chosen human genome—which will serve as a reference—into a series of fragments, each about 40,000 bases (40 kilobases) long. These fragments are called cosmids. By use of recombinant DNA methods, each cosmid can be grown in a bacterial strain to produce endless copies of itself, called clones. By ordering the cosmids with respect to each other along each chromosome, a physical map is created, known as a cosmid map, that specifies the location of each of the 100,000 cosmids that is needed to cover the entire genome. Each of these cosmids would then be sequenced, beginning with regions that correspond to the most interesting genes, then moving on to sequence one of the smaller chromosomes, and moving up through the 23 chromosomes. (The smallest chromosome is 50 million bases long; the largest is 250 million.) Sequencing is done by breaking the cosmids into even smaller pieces. Then molecular biological techniques are used to work out the nucleotide sequence of each of them (see "DNA Sequencing," p. 31).

Several types of maps are needed to understand the human genome: the cosmid map, described above; a restriction enzyme map, which is necessary to complete the cosmid map; and a genetic map, or polymorphism map, which shows the location of specific genes along the chromosome. Each is in varying stages of development (see "Mapping," p. 32).

The development of these maps will totally change the way biologists identify human genes of medical interest. Currently, we are able to identify some of the genes that cause clearly defined genetic diseases, such as Huntington's chorea or Duchenne muscular dystrophy. But many of the more common ailments, specifically heart disease and cancer, most likely have multifactorial genetic backgrounds in which many genes cooperate or many alternative genes are involved in creating a predisposition to disease. The genetic polymorphism map will make it possible to analyze such diseases, and the cosmid map will make it possible to isolate and examine the relevant genes directly.

No technological advances are needed to create the cosmid map or other maps of the human genome, although an organized effort is needed to isolate and order all the cosmids. Such a cosmid map is already being completed for the nematode, a small eukaryote that contains 1/40th the DNA complement of the human being.

However, technological developments are needed to contemplate sequencing the entire genome. Some of these developments are already under way, but some will not occur without the stimulus of a specific project. Advances in sequencing technology promise to increase the rate of sequence acquisition dramatically. An automatic DNA sequencer will be introduced in 1987. This machine, in principle, can work at the rate of about 20,000 bases a day for total rough sequence. (The ratio of rough to final sequence depends on the sequencing strategy used but ranges from 10:1 to 2:1). Recently, multiplexing techniques for sequencing by hand have been developed. These can also produce rough sequence at the rate of 20,000 to 30,000 bases a day.

A great deal of the time and effort in DNA sequencing, however, is not in the actual sequencing but in handling the DNA pieces and preparing and selecting materials for later workup. These are the steps that would benefit most from a large-scale, centrally organized project.

DNA Sequencing

The sequence of the human genome will be developed by first breaking the genome, 3 billion bases long, into an ordered set of some 100,000 fragments (known as cosmids), each about 40,000 bases long. The next step is to break those cosmids into smaller pieces, each about 400 bases long, and then work out the order of bases along each of the pieces.

The order of the bases is determined by breaking up a 400- to 1,000-base-long DNA strand into a set of nested fragments, each fragment running from a reference position out to one occurrence of one of the four bases. By measuring the sizes of one set of each of the fragments (for example, the fragments that correspond to G, or guanosine), one establishes the positions of each of the guanosines in the sequence. The sizes are determined by forcing the fragments to move through a gel-like substance under the influence of an electric field. The shorter fragments move faster; the longer, slower; and the gel will distinguish a fragment 300 bases long from a fragment 301 bases long.

There are two techniques for generating these nested fragments. One is an enzymatic procedure that synthesizes DNA chains of varying length. The second is a chemical procedure that breaks the DNA chain at each of the bases. The electrophoretic analysis by size permits the reading of some 300 to 600 bases of DNA in a single pass.

There are two basic strategies for putting together these short sequences to make a longer sequence. In "shotgun" sequencing, random 400-base-long fragments of a longer DNA strand are isolated and each of these is sequenced. A computer is then used to seek overlaps between these sequences and thus to assemble these fragments together to generate a sequence of a longer region. Because of the random element, some regions are sequenced over and over again; about 10 times more DNA must be sequenced than would be present in the final assembled sequence. However, this redundancy provides an error correction: the individual error rate in the short sequences that are used in the shotgun method is not terribly important.

An alternative approach is to order the short pieces of DNA chosen for sequencing along the longer DNA strand by other techniques, and then to sequence the molecule deliberately from one end to the other. Both strands of the DNA double helix are sequenced independently to provide for error correction.

Shotgun sequencing is attractive because it takes little preparation, and the initial information is obtained very readily. However, the final information needed to complete the sequence is extremely difficult to obtain. The ordered approach requires a greater investment in time before the sequencing, but it completes the sequence in a straightforward fashion and involves about one-fifth of the sequencing work. The longest single DNA that has been sequenced to date is the genome of Epstein-Barr virus—172,282 base pairs. ■

Robert Alan Soulé

At this level the project becomes practically and economically feasible; the total cost would be on the order of $300 million.

There are other benefits to establishing a center or centers devoted entirely to sequencing the human genome, as opposed to a piecemeal effort in a number of laboratories around the country. A centralized project would be important in ensuring the accuracy of the sequence. Today, the error rate for sequencing is roughly 1 base in 10,000 to 1 base in 100,000. The difficulties lie not in statistical errors but in weakness of technique: rare, short regions of DNA are extremely hard to sequence and require unusual time and effort. I expect the sequencing error rate could be reduced to between 1 part in 100,000 to 1 part in 1,000,000, but that will require organized quality control. The development of the technology, including the strategies for faster sequencing, is also best accomplished in a single location.

The challenge to that centralized group is to demonstrate rapid and accurate sequencing at the rate of about one megabase (1 million bases) per year per person at a cost on the order of 10 cents per base. At this level the project becomes practically and economically feasible; the total cost would be on the order of $300 million. This would represent an effort carried out by some 300 people over a 10-year period in a 300-megabase-per-year institute.

Mapping

Three kinds of maps are involved in the analysis of the human genome. As described, the physical map is a set of ordered cosmids that overlap and span the human chromosomes. Where the cosmids overlap, the cosmid map provides a detailed ordering along the chromosome. But inescapably, some cosmids will be missing, and thus the map will not be complete. That is where the restriction enzyme map comes in. It can identify the distances between the various cosmids and thus be used to complete the map.

The restriction enzyme map is an abstract map of the distances along each chromosome. This map identifies the distances between specific sites—the places that restriction enzymes cut DNA molecules—along each of the chromosomes. Such a map will identify sites 10,000 to 1,000,000 bases apart along each chromosome and determine the sizes of the chromosomes. The largest restriction map to date has been constructed for a bacterium that contains 1/1,000th the DNA complement of a human.

The third type of map is a genetic map that shows the location of known human genes on the chromosomes. We now have a rudimentary genetic map with about 2,000 genes on it. The most useful genetic map would be a fine-structure map called a DNA polymorphism map. The polymorphism map is made by hybridizing DNA probes that can detect polymorphisms (genetic variation within the population) to samples of human DNA. It depends on the fact that the DNA sequence between individuals differs enough that the pattern of inheritance of parts of chromosomes can be traced by molecular biological techniques. Work is already under way on this map, which today has some 150 informative markers. When it is completed, the map will have about 3,000 markers, which roughly translates into a genetic marker every million base pairs apart along each chromosome. Each marker will allow the tracing of that region of the chromosome as it is passed to subsequent generations to see if it correlates with a disease or other property of interest. ■

SEQUENCING: A NEW BIOLOGY

The sequence would be a monumental step in the study of human biology. It is one ultimate answer to the commandment "know thyself." However, the DNA sequence itself will not identify all of the genes or tell us directly, with today's knowledge, what any of the genes do. The sequence will need interpretation, an understanding of the several hundred thousand genes it contains. This understanding will be the fruit of human biology in the twenty-first century. The development of the human sequence is an investment to speed that understanding.

We are not engineers proceeding by design, but rather untrained mechanics, hammering on a shaft to make it fit.

On the industrial side the sequence would speed up the isolation of human gene products for use in the pharmaceutical and biotechnology industries. And the development of the cosmid map will produce many probes for the diagnosis of genetic diseases.

A major challenge in the next decade will be the solution of the structure-function problem—that is, how a gene determines the structure and function of its protein product. Today, we can isolate enzymes (specialized proteins) and their genes from microbes, plants, animals, and humans. We can also make these proteins to order in bacteria and other cells, and we can modify them (by changing the DNA) to try to make new products that may be more useful. However, we cannot predict what we are doing in today's protein engineering. We are not engineers proceeding by design, but rather untrained mechanics, hammering on a shaft to make it fit. The underlying theoretical problem is how to predict from a sequence of amino acids how that sequence will fold up in space to create the three-dimensional structure of a protein. The next question is how that total structure will then carry out a chemical reaction or serve another function. We know that there is an answer; nature folds these structures all the time. A theoretical solution should begin to emerge over the next few years. With the solution of the structure-function problem, the human genome sequence becomes amenable to theoretical work. One would then try to predict from the DNA sequence the function of the gene products.

Other biological questions, however, cannot be approached through the human sequence alone. The key questions in evolutionary or population biology lie in how the diversity of a species is encoded in all its genomes. In these two spheres of biology, the questions are not related to the structure of a typical individual of a species, but rather relate to the variation among individuals that make up a viable species that can survive and evolve. These questions cannot be attacked directly by developing a single reference sequence but would need several comparison sequences.

Our understanding of the human body does not develop in isolation. Our experimental biology is done on animals other than human. Critical developments in molecular or developmental genetics are made most easily by working with simpler animals such as the fruit fly. The details of mammalian development will be most fully worked out in the mouse. However, the human sequence will not exist in isolation. Development of techniques to sequence DNA rapidly will lead over the next decades to working out the sequence of the fruit fly and much of the sequence of the mouse or other mammals, allowing biologists to analyze the parts most relevant for comparison with the human and for understanding the process of development.

We have the choice of continuing biology as we do it today, putting no special effort into developing a human sequence and counting instead on the gradual development of the science. It is true that biologists will continue to

If we do the human map and genome early, we speed all the rest of biological research and make each dollar spent on research more effective.

discover and isolate, one at a time, genes involved in development and in human diseases. As technology gradually improves, these gains will come more quickly. However, today's scattered efforts are wasteful and do not yield the benefits, such as efficiencies of scale, that will accrue with a focused approach. If we take advantage of those efficiencies to do the human map and genome early, we speed all the rest of biological research and make each dollar spent on research more effective by avoiding reiterated steps. But even more important, the existence of the human sequence would itself provide an object of study, leading to a new science that would try to understand directly the information contained in the genome and to predict how the genetic structure works. This new science will not exist without the development of the full sequence of the organism.

Some biologists have questioned the usefulness of sequencing the entire genome, asserting that some 90 percent of the effort will be wasted. Not all of the DNA codes for proteins—only about 5 to 10 percent does. One might sequence only the coding regions because most of the noncoding regions serve no genetic function and are essentially "junk."

I believe that it will be faster to sequence the entire genome and examine it later than it would be to first isolate the "relevant" material. The effort to filter out the relevant material beforehand is far larger than the total effort involved in the sequencing. On a deeper level, we have no way of knowing whether portions of DNA we think are irrelevant today may not turn out to be deeply relevant tomorrow.

Controversy over the project, however, runs far deeper. Some biologists feel that the project is incommensurate with the traditional way biology is conducted. Despite changes in other fields of science, biology has remained essentially a cottage industry: the problems now being studied can be tackled by small groups, essentially by single, independent investigators. As has often happened in the history of science, the human genome project seems wrong to many biologists solely because of its scale, with its high costs and demand for a large number of investigators operating within a centralized organization. But the human genome project is actually consistent with other biological efforts, for example, with the effort required to develop a new pharmaceutical drug, which involves a program of directed medical research that ultimately costs $50 million to $100 million.

The direct analogy to biology's human genome project is the development of the large synchrotrons in the field of physics. These large accelerators required resources far greater than a single laboratory or a single university could mount; they could be built only as a group effort. But they represent research tools, built by specialists, by which experimental physics can be done. The human genome project will create such research tools: the physical map and the sequence. Understanding the sequence, and understanding the genes it contains, is a goal of human biology. The sequence itself is only an aid to that end.

The best way to attack the human genome project would be to develop major centers immediately, each one containing about 300 people and capable of sequencing some 300 megabases a year. Three such centers—one in the United States, one in Europe, and one in Japan—would move the world level of DNA sequencing to 1 billion bases a year by the late 1990s. Thus, we would be able to know not just a human sequence, but other sequences for comparison: primate, mouse, and some of the simpler eukaryotes. By comparing the human sequence with its close relatives among the primates, we would be able to determine all of the genes, because the sequence of intergenic regions changes and drifts more rapidly between neighboring species than does the sequence corresponding to functioning genes. By looking across mammals and birds, we will learn which genes are old and which new, what makes the human similar to, yet different from, the other mammals.

By the year 2000 we can know our entire genetic structure and have a grasp on the question of how our genes make us unique. ■

Section V:
Recombinant DNA

Proc. Natl. Acad. Sci. USA
Vol. 73, No. 11, pp. 4174–4178, November 1976
Genetics

Construction of plasmids carrying the *c*I gene of bacteriophage λ

(repressor/DNA cloning/operon construction)

KEITH BACKMAN*, MARK PTASHNE†, AND WALTER GILBERT†

* Committee on Higher Degrees in Biophysics and † Department of Biochemistry and Molecular Biology, The Biological Laboratories, Harvard University, Cambridge, Massachusetts 02138

Contributed by Walter Gilbert, September 9, 1976

ABSTRACT **By techniques of recombination *in vitro*, we have constructed a plasmid bearing the repressor gene (*c*I) of bacteriophage λ fused to the promoter of the *lac* operon. Strains carrying this plasmid overproduce λ repressor. This functional *c*I gene was reconstituted by joining DNA fragments bearing different parts of that gene. Flush end fusion techniques, involving no sequence overlap, were necessary for the construction; in certain cases, the abutting of the DNA molecules bearing ends generated by different restriction endonucleases creates a sequence at the junction which is recognized by one of the restriction endonucleases.**

We have constructed a plasmid in which the promoter of the *lac* operon has been placed adjacent to the repressor gene (*c*I) of bacteriophage λ, by techniques of recombination *in vitro*. *Escherichia coli* strains carrying this plasmid overproduce repressor because transcription of *c*I originates mainly at the *lac* promoters and because each cell contains multiple copies of the plasmid. Certain novel aspects of our construction may be applied to the construction *in vitro* of other hybrid operons. Other λ repressor overproducing strains have been created by Gronenborn and Müller-Hill (1), who have isolated *lac* promoter-*c*I fusions *in vivo*.

MATERIALS AND METHODS

E. coli K12 strain 294 (endo I$^-$, B_1^-, r_K^-, m_K^+; obtained from M. Meselson) was the host used in most of these experiments. A *lac* repressor overproducing strain (2), V2000 [F′ *pro lac* $i^{Q1}Z^-{}_{U118}$/Δ(*lac-pro*) Sm^R(λ*imm*^{21} *plac*5)], was constructed and used in some experiments. The plasmid vectors pCR11 (3) and pMB9 (4) were used. Plasmid DNA was prepared by the method of Clewell and Helinski (5). DNA fragments generated by restriction endonucleases were prepared by electrophoresis on polyacrylamide gels as described by Maniatis *et al.* (6). *Eco*RI, *Hin*dIII, and *Hae*III digestions were performed in *Hin* buffer (7) and *Hpa*II digestions were performed in *Hpa* buffer (7). Reactions using *E. coli* DNA polymerase I (DNA nucleotidyltransferase, deoxynucleosidetriphosphate:DNA deoxynucleotidyltransferase, EC 2.7.7.7) (a gift of W. McClure) were performed in 50 mM Tris·HCl at pH 7.8, 5 mM $MgCl_2$, 1 mM 2-mercaptoethanol, 50 μg/ml of bovine serum albumin plus 2 μM each of dATP, dCTP, dGTP, and dTTP for 1 hr at 15° (8). T4 polynucleotide ligase [polynucleotide synthetase, poly(deoxyribonucleotide):poly(deoxyribonucleotide) ligase(AMP-forming), EC 6.5.1.1] reactions were performed in 6.6 mM Tris·HCl at pH 7.4, 6.6 mM $MgCl_2$, 6.6 mM 2-mercaptoethanol, 100 μM ATP at 4°; DNA fragments were present at 100–300 μg/ml. For joining DNA fragments bearing flush ends (9), 5-fold higher concentrations of ligase were used than were required to join an equivalent number of staggered ends, and the reactions were incubated at room temperature for 4 hr. Cells were transformed with DNA as described by Cohen *et al.* (10). Drug resistant transformants were selected on drug supplemented agar plates, and λ immune transformants were selected on agar plates seeded with 10^9 λKH54 (11) and 10^9 λKH54h$_{80}$; in some cases, both selections were performed simultaneously. The immunity of plasmid containing strains was further tested by streaking single colonies across a streak of phage on an agar plate. Colonies of transformants which constitutively synthesized β-galactosidase appeared blue on agar plates containing a chromogenic, noninducing substrate, 5-chloro-4-bromo-3-indolyl-β-D-galactoside (40 μg/ml). Assays of λ repressor (12), β-galactosidase (β-D-galactoside galactohydrolase, EC 3.2.1.23) (13), and the isolation of operator containing DNA fragments on nitrocellulose filters (14) have been described previously. Purified λ repressor was a gift of R. Sauer and purified *lac* repressor was a gift of A. Maxam. Experiments were carried out in a P1 (EK1) facility.

RESULTS

Preliminary considerations

We wished to proceed in two steps; first, to clone *c*I flanked by as little extraneous phage DNA as possible on a plasmid and then to insert a DNA fragment bearing the *lac* promoter near the beginning of *c*I. Two problems arose. No single restriction endonuclease cleaves λ DNA just outside the ends of *c*I without also cleaving within it, and no restriction endonuclease site suitable for inserting the *lac* promoter exists in λ DNA near the beginning of *c*I. We adopted a strategy based on the following considerations: gene *c*I can be neatly isolated on two DNA fragments, one of which bears two *Hin*dIII ends and the other of which bears one *Hin*dIII end and one *Hae*III end (see Fig. 1). Proper joining of these fragments reconstitutes *c*I on a larger fragment bearing one *Hin*dIII end and one *Hae*III end. *Hin*dIII ends are staggered and readily anneal to each other, whereas *Hae*III ends are flush. Staggered ends which anneal to each other can be joined by T4 polynucleotide ligase (15), and flush ends can be joined to other flush ends by that enzyme (9). To clone our *c*I fragment, we sought a plasmid which could be opened so as to produce one *Hin*dIII end and one flush end. The plasmids we used, pCR11 and pMB9, each have a single *Hin*dIII site and a single *Eco*RI site. Although *Eco*RI produces staggered ends, we anticipated that these ends could be converted to flush ends by treatment with DNA polymerase I and the four deoxyribonucleotide triphosphates, because the recessed 3′ end can be extended by copying the protruding 5′ end of the complementary strand (a process we hereafter refer to as filling-in). Precise joining of a filled-in *Eco*RI end to a *Hae*III end should produce a molecule recognized by *Eco*RI at the junction (see Fig. 2). This regenerated *Eco*RI site near the beginning of *c*I could then be used as a site to insert the *lac* promoter.

The plasmids pCR11 and pMB9 are derived from Col E1, which is normally present in about twenty copies per cell (16). pCR11 and pMB9 carry drug resistance determinants (kanamycin and tetracycline, respectively).

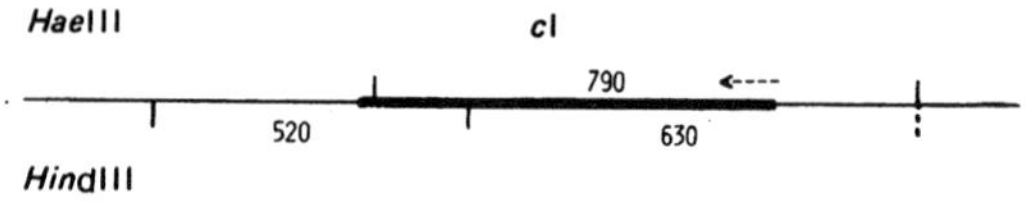

FIG. 1. Restriction endonuclease sites in the vicinity of the *c*I gene of λ. *Hae*III cleavage sites are indicated above, and *Hin*dIII sites below, the map of the region including and flanking *c*I. (The extreme right *Hae*III site is also indicated below the line.) Distances in base pairs between various cleavage sites are given. The extent of *c*I is indicated by the heavy line. The extreme right *Hae*III site is in the *cro* gene, and the extreme left *Hin*dIII site is in the gene which lies to the left of *c*I, *rex*. The arrow indicates the direction of transcription of *c*I.

Cloning *c*I

With these considerations in mind, we proceeded as follows (see Fig. 2): plasmid pCR11 DNA was digested with *Eco*RI and the protruding single-stranded ends converted to double-stranded flush ends by filling-in with DNA polymerase I. The product was then digested with *Hin*dIII to produce a vector DNA molecule bearing one *Hin*dIII end and one flush end. This procedure removes at least part of the kanamycin resistance gene(s). Two DNA fragments bearing portions of *c*I were produced as follows (see Fig. 1): λ DNA was digested with *Hae*III, and a 790 base pair fragment was isolated that carries the right portion of *c*I plus about 150 base pairs to the right of *c*I. This fragment was further digested with *Hin*dIII to produce a fragment 630 base pairs long. Separately, λ DNA was digested with *Hin*dIII and a 520 base pair fragment bearing the left portion of *c*I was isolated. Together these two fragments span the entire *c*I gene. Gene *c*I was then simultaneously reconstituted and inserted in the plasmid by treating a mixture of the specially prepared plasmid and the fragments which span *c*I with T4 polynucleotide ligase (see Fig. 2). Five λ immune clones were isolated from bacteria transformed with the products of this reaction, and their plasmids were analyzed by restriction endonuclease digestion. One isolate, pKB155, produced three fragments (5000, 630, and 520 base pairs long) when cleaved with *Eco*RI and *Hin*dIII, as predicted by the construction shown in Fig. 2 (E and F). In particular, the fused *Hae*III–*Eco*RI junction to the right of *c*I was cleaved again by *Eco*RI.

The *c*I gene was transferred from pKB155, which bears no drug resistance marker, to the plasmid pMB9, which carries a tetracycline resistance gene, as follows: an 1150 base pair piece of DNA bearing the entire *c*I gene was excised from pKB155 by complete digestion with *Eco*RI and partial digestion with *Hin*dIII. Plasmid pMB9 was digested with *Eco*RI and *Hin*dIII, which removes a nonessential 350 base pair fragment; the 1150 base pair *c*I gene fragment was added, and the mixture was treated with ligase. Bacteria were transformed with the products of ligation, and tetracycline resistant lambda immune clones were readily isolated. Restriction endonuclease analysis of several isolates (not shown), typical of which is pKB158, confirmed the structure shown in Fig. 3.

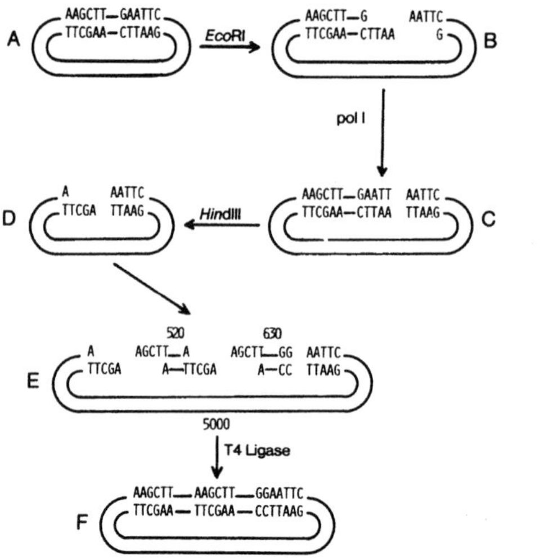

FIG. 2. Cloning the *c*I gene. Gene *c*I, carried on two DNA fragments, was inserted into the plasmid pCR11 as diagrammed (see *text*). The plasmid pCR11 (A) was digested with *Eco*RI (B), and the protruding 5′ ends were filled in (see *text*) with DNA polymerase I (pol I) (C); the product was digested further with *Hin*dIII (D). The *Hae*III 790 base pair fragment digested with *Hin*dIII and the *Hin*dIII 520 base pair fragment (see Fig. 1) were added to the prepared plasmid (E). No attempt was made to remove the extra pieces of DNA generated by the *Hin*dIII digestions of the plasmid and of the *Hae*III 790 fragment. The mixture was treated with polynucleotide ligase, and the desired recombinant (F) was selected as described in the *text*.

Construction of a plasmid fusing the *lac* promoter to the *c*I gene

A plasmid bearing a (nominally) 205 base pair fragment of DNA which carries the *lac* UV5 promoter-operator region was constructed by F. Fuller. He used a 203 base pair fragment of known sequence generated by *Hae*III. The *Hae*III ends of this

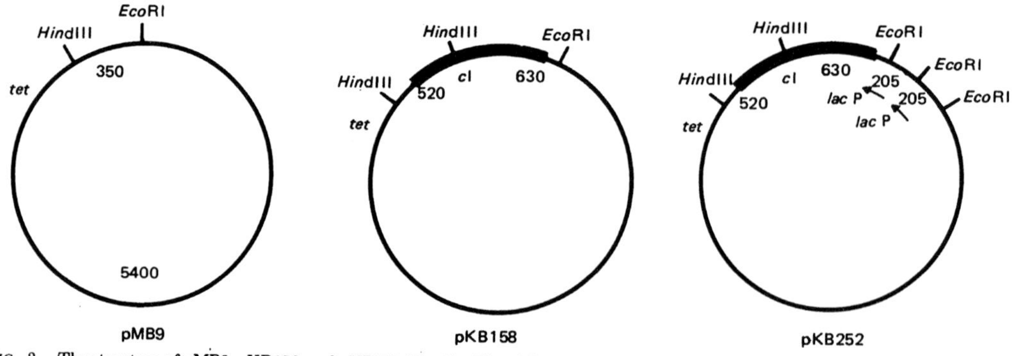

FIG. 3. The structure of pMB9, pKB158, and pKB252. The *Eco*RI and *Hin*dIII sites, the sizes of the cleavage products, and the location of *c*I in the various plasmids are shown. The location and orientation of the *lac* promoters in pKB252 are indicated.

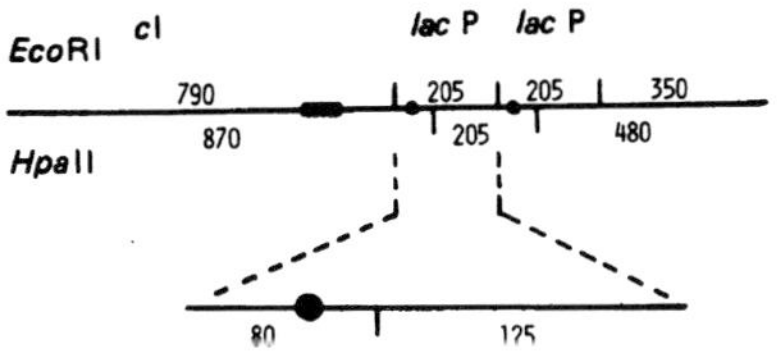

FIG. 4. Structure of pKB252 in the *lac* promoter-cI gene region. The location of *Eco*RI and *Hpa*II cleavage sites are shown, respectively, above and below the map of the 1550 base pair fragment produced by *Hae*III digestion of pKB252. The locations of the *lac* and lambda operators on this fragment are indicated by the solid circles and solid box, respectively. Distances between various endonuclease sites and the position of the *Hpa*II site in the *lac* promoter are given in base pairs.

fragment were converted to *Eco*RI ends by joining the fragment to filled-in *Eco*RI ends on a plasmid by using the methods described above. Colonies harboring plasmids which carry the *lac* promoter-operator were identified by their constitutive synthesis of β-galactosidase, which renders them blue on agar plates containing 5-chloro-4-bromo-3-indolyl-β-D-galactoside. The *lac* promoter fragment serves as a "portable promoter" with an *Eco*RI site 65 base pairs from the startpoint of transcription (corresponding to the ninth amino-acid residue of the *lac z* gene). This fragment was excised from its carrier with *Eco*RI and inserted near the beginning of *c*I at the single *Eco*RI site in pKB158. About one half of the tetracycline resistant clones which were isolated carried the *lac* promoter fragment on the plasmid, as manifested by their constitutive synthesis of β-galactosidase. All of these clones were immune to λ*vir* (unlike the strain carrying pKB158) and sensitive to $\lambda imm^{434}cI^-$ in cross-streak tests, although many clones exhibited some cell death at the junction of the cross-streak with λ*vir*. Three clones which exhibited no apparent sensitivity to λ*vir* were selected for further study. Typical of the plasmids in these clones is pKB252.

Structure of pKB252

An *Eco*RI plus *Hin*dIII digest of pKB252 yielded the 205 base pair promoter fragment as well as the fragments produced by a similar digestion of pKB158. When pKB252 was digested with *Hae*III, a 1550 base pair fragment was produced. This fragment, which extends from the *Hae*III site in *c*I to a *Hae*III site which is 350 base pairs beyond the *Eco*RI site in the plasmid DNA (17), was expected to contain the *lac* promoter. Analysis of this 1550 base pair fragment confirmed the structure of pKB252 as shown in Fig. 3. The products of *Hpa*II digestion of this fragment were 205, 480, and 870 base pairs long (see the diagram of Fig. 4). Because the *c*I region of λ does not have any *Hpa*II sites (11), and the *lac* promoter fragment does have a *Hpa*II site about 85 bases from one end, this pattern of fragments suggested the structure shown in Fig. 4. In particular, the 205 base pair product suggested the presence of multiple *lac* promoters. If only one promoter were present, then no such piece would have been produced. Moreover, all such promoters must have the same orientation; if any two promoter regions were oppositely oriented, then *Hpa*II digestion would have produced a 160 or a 250 base pair fragment.

The exact number of *lac* promoters in pKB252 was determined as follows: uniformly ^{32}P labeled plasmid DNA was digested with *Hae*III and the 1550 base pair fragment was isolated. This fragment was digested with *Eco*RI and the products separated on a polyacrylamide gel. The resulting fragments (205, 360, and 790 base pairs) were cut from the gel and the

Table 1. The number of *lac* promoter fragments in pKB252

Fragment size in base pairs	Radioactivity ^{32}P counts per minute	Stoichiometry
790	3110	1
360	1345	1
205	1350	2

Stoichiometry is calculated by comparing the ratios of radioactivity/fragment size.

radioactivity in each fragment was determined (Table 1). The 205 base pair fragment was present in twice molar quantities compared to the 360 and 790 base pair fragments which indicated that there are two *lac* promoters in pKB252.

To prove that the *lac* promoters transcribe toward *c*I, we took advantage of the fact that *Hpa*II cuts near the *lac* operator in the *lac* promoter. The *Hae*III 1550 fragment was digested with *Hpa*II and the fragments produced were mixed with either *lac* or lambda repressor and passed through nitrocellulose filters. The fragments retained on the filters were eluted and analyzed on a polyacrylamide gel. It was found that the 870 base pair fragment was retained by both *lac* and λ repressors, the 205 base pair fragment was retained only by *lac* repressor, and the 480 base pair fragment was not retained by either repressor. These results are consistent only with the orientation shown in Fig. 4 in which the *lac* promoters transcribe toward *c*I.

Control of repressor synthesis *in vivo*

Cell extracts of strains containing various plasmids were assayed for λ repressor and, in some cases, an aliquot of the culture was assayed for β-galactosidase. The results are listed in Table 2. Strains carrying plasmids bearing *c*I but not the *lac* promoter (pKB155 and pKB158) contained about five times more repressor than did single lysogens. Strains carrying pKB252, which has two *lac* promoters, produced 35 times more repressor than did single lysogens.

Synthesis of lambda repressor in pKB252 strains is regulated by *lac* repressor (Table 2). A strain bearing a wild-type *lac* operon does not make enough *lac* repressor to repress significantly the synthesis of λ repressor from the *lac* promoters on pKB252 (Table 2). When this plasmid was transferred to a strain which

Table 2. Repressor synthesis in plasmid containing strains

Strain	Repressor level × 10^{-3}	β-Galactosidase level
28(λ)	0.3	N.D.
294/pKB155	1.0	N.D.
294/pKB158	1.5	3
294/pKB252	9.7	2.8 × 10^3
294/pKB252 (+IPTG)	10.7	2.8 × 10^3
V2000/pKB252	1.3	10
V2000/pKB252 (+IPTG)	2.9	1.9 × 10^3

Various strains were grown overnight at 37° in M9 salts and glucose medium, and were assayed for lambda repressor and β-galactosidase. The results are given in units of repressor per milligram of protein. One unit is the amount of repressor which gives half-maximal binding in the nitrocellulose filter binding assay and corresponds to approximately 0.5 ng of repressor. β-Galactosidase results are given in the units of Miller (13). A single lysogen, 28 (λ), is included for comparison purposes. Strain 294 makes wild-type *lac* repressor levels, whereas strain V2000 over-produces *lac* repressor about 100-fold (2). ITPG, isopropyl-thiogalactoside. N.D., not determined.

Table 3. Use of the filling-in method with various restriction endonucleases

Restriction endonuclease	Recognition sequence	Requirement for site regeneration	Subsite	Subsite Recognition sequence
*Eco*RI	↓ GAATTC CTTAAG ↑	5′ C	*Eco*RI*	↓ AATT TTAA ↑
*Eco*RII	↓ CCAGG GGTCC ↑	None	*Eco*RII	↓ CCAGG GGTCC ↑
*Hind*III	↓ AAGCTT TTCGAA ↑	5′ T	*Alu*I	↓ AGCT TCGA ↑
*Bam*I	↓ GGATCC CCTAGG ↑	5′ C	*Mbo*I	↓ GATC CTAG ↑
*Bgl*II	↓ AGATCT TCTAGA ↑	5′ T	*Mbo*I	↓ GATC CTAG ↑

Several endonucleases which produce 5′ protruding ends are listed with the DNA sequence which they recognize (20–24, R. J. Roberts, G. Wilson, and F. E. Young, submitted for publication). After filling in these ends with DNA polymerase I, the original recognition sequence can be regenerated by joining the filled-in end to a flush ended DNA molecule which meets the requirement listed in the third column. When joined to a flush ended fragment not meeting that requirement, the filled-in end generates a sub-specificity, which is described in the fourth and fifth columns (20, 25, R. E. Gelinas, P. A. Myers, K. Murray, and R. J. Roberts, unpublished). The top line of each recognition sequence reads from the 5′ to the 3′ end; the bottom line reads in the opposite direction.

overproduces *lac* repressor, the synthesis of λ repressor was reduced to the basal (i.e., pKB158) level. IPTG (isopropyl-thiogalactoside), an inducer of the *lac* operon, partially induces β-galactosidase synthesis in cells which overproduce *lac* repressor. Table 2 shows that isopropyl-thiogalactoside also partially induces λ repressor synthesis when such cells carry pKB252.

DISCUSSION

We have described the use of techniques of DNA recombination *in vitro* to construct a plasmid on which the λ repressor gene (*c*I) is transcribed largely from the promoter of the *lac* operon. This plasmid has two copies of the *lac* promoter which transcribe toward *c*I. As predicted by this arrangement, the synthesis of λ repressor in strains bearing this plasmid is regulated by the *lac* repressor. The *lac* promoter fragments in pKB252 contain the startpoint of translation of the *lac z* gene; however, several translational stop signals immediately precede *c*I (18), and repressor synthesized in strains carrying pKB252 is not a fusion product initiated at the *lac z* translational startpoint. Strains carrying this plasmid produce about 35 times more repressor than do single lysogens, as measured by DNA binding activity; however, these levels are variable and are often as much as 3-fold higher. The source of this variation is not understood. These repressor levels are 10- to 25-fold lower than would be expected on the basis of the known strength of the *lac* promoter (19). This inefficient expression might occur because transcription beginning at the *lac* promoter attenuates frequently before transcribing *c*I, because the progress of RNA polymerase is blocked by λ repressor bound to the operator just to the right of *c*I, or because the mRNA produced is inefficiently translated. The level of repressor was not higher when plasmid pKB252 was carried by a strain bearing an *SuA* mutation. Nevertheless, the high levels of repressor made by strains carrying pKB252 have greatly facilitated structural studies of repressor.

Strains carrying plasmids in which *c*I is transcribed from its own promoter produce about five times more repressor than does a single lysogen, even though these plasmids should be present in about 20 copies per cell (16). Gene *c*I is known to regulate its own synthesis (18), and these strains are exemplary of such autogenous regulation.

The expression of tetracycline resistance in strains carrying pKB158 and pKB252 may be under the control of the promoter for *c*I (P_{RM}) and the *lac* promoter. H. W. Boyer (personal communication) has found evidence suggesting that the *Hin*dIII site in pMB9 is within the promoter of the tetracycline resistance gene. In our experiments, tetracycline resistant λ immune clones were readily isolated which were missing the pMB9 specific material between the *Eco*RI and *Hin*dIII sites; presumably the expression of tetracycline resistance in these cases results from an extension of the *c*I mRNA.

Our construction required the use of new techniques which are generally applicable. We have assembled a hybrid operon from separate pieces; some junctions involved sequence overlap, some did not. Furthermore, the DNA polymerase I filling-in technique allowed us to join precisely DNA fragments which could not be joined directly by treatment with ligase. These techniques greatly expand the spectrum of restriction endonucleases which are useful in cloning experiments. An especially important feature of these techniques is the ability to regenerate endonuclease recognition sites at junctions involving filled-in ends. We demonstrate this explicitly in one case (the joining of a filled-in *Eco*RI end to a *Hae*III end), and available sequence data suggest that it will be a general feature of the method (see Table 3).

W. G. is American Cancer Society Professor of Molecular Biology. We wish to thank F. Fuller and L. Johnsrud, who constructed and characterized the *lac* promoter containing plasmid, and R. Brent, who prepared the *lac* promoter fragment. We also wish to thank R. T. Sauer for λ repressor and for performing some of the repressor assays, A. Jeffrey for restriction endonucleases, A. Maxam for *lac* repressor, and W. McClure for DNA polymerase I. This work was supported by grants from the National Science Foundation and the National Institutes of Health.

1. Gronenborn, B. & Müller-Hill, B. (1976) *Mol. Gen. Genet.*, in press.
2. Müller-Hill, B., Fanning, T., Geisler, N., Gho, D., Kania, J., Kathman, P., Meissner, H., Schlotmann, M., Schmitz, A., Treisch, I. & Beyreuther, K. (1975) in *Protein Ligand Interactions*, eds. Sund, H. & Blauer, G. (Walter de Gruyter, Berlin), pp. 211–224.
3. Covey, C., Richardson, D. & Carbon, J. (1976) *Mol. Gen. Genet.* **145**, 155–158.
4. Rodriguez, R. L., Bolivar, F., Goodman, H. M., Boyer, H. W. & Betlach, M. (1976) "Molecular mechanisms in control of gene expression," eds. Nierlich, D. P., Rutter, W. S. & Fox, C. F., in *Proceedings of the ICN-UCLA Symposia on Molecular and Cellular Biology* (Academic Press, San Francisco, California), Vol. 5, in press.
5. Clewell, D. B. & Helinski, D. R. (1970) *Biochemistry* **9**, 4428–4440.
6. Maniatis, T., Jeffrey, A. & van deSande, H. (1975) *Biochemistry* **14**, 3787–3794.

7. Maniatis, T., Ptashne, M. & Maurer, R. (1973) *Cold Spring Harbor Symp. Quant. Biol.* **38,** 857–868.
8. Maniatis, T., Jeffrey, A. & Kleid, D. (1975) *Proc. Natl. Acad. Sci. USA* **72,** 1184–1188.
9. Sgaramella, V., van deSande, J. & Khorana, H. G. (1970) *Proc. Natl. Acad. Sci. USA* **67,** 1468–1475.
10. Cohen, S. N., Chang, A. C. Y. & Hsu, L. (1972) *Proc. Natl. Acad. Sci. USA* **09,** 2110–2114.
11. Blattner, F. R., Fiandt, M., Hass, K. K., Twose, P. A. & Szybalski, W. (1974) *Virology* **62,** 458–471.
12. Chadwick, P., Pirrotta, V., Steinberg, R., Hopkins, N. & Ptashne, M. (1970) *Cold Spring Harbor Symp. Quant. Biol.* **35,** 283–294.
13. Miller, J. H. (1972) *Experiments in Molecular Genetics* (Cold Spring Harbor Laboratory, Cold Spring Harbor, N.Y.)
14. Maniatis, T. & Ptashne, M. (1973) *Nature* **246,** 133–136.
15. Sgaramella, V. & Khorana, H. G. (1972) *J. Mol. Biol.* **72,** 427–444.
16. Clewell, D. B. & Helinski, D. R. (1972) *J. Bacteriol.* **110,** 1135–1146.
17. Maniatis, T., Kee, S. G., Efstratiadis, A. & Kafatos, F. C. (1976) *Cell* **8,** 163–182.
18. Ptashne, M., Backman, K., Humayun, M. Z., Jeffrey, A., Maurer, R., Meyer, B. & Sauer, R. T. (1976) *Science,* **194,** 156–161.
19. Cohn, M. (1957) *Bacteriol. Rev.* **21,** 140–168.
20. Polisky, B., Greene, P., Garfin, D., McCarthy, B., Goodman, H. & Boyer, H. W. (1975) *Proc. Natl. Acad. Sci. USA* **72,** 3310–3314.
21. Bigger, C. H., Murray, K. & Murray, N. E. (1973) *Nature New Biol.* **244,** 7–10.
22. Boyer, H. W., Chow, L. T., Dugaiczyk, A., Hedgpeth, J. & Goodman, H. M. (1973) *Nature New Biol.* **244,** 40–43.
23. Old, R., Murray, K. & Roizes, G. (1975) *J. Mol. Biol.* **92,** 331–339.
24. Pirrotta, V. (1976) *Nucleic Acids Res.* **3,** 1747–1760.
25. Roberts, R. J., Myers, P. A., Morrison, A. & Murray, K. (1976) *J. Mol. Biol.* **102,** 157–165.

Proc. Natl. Acad. Sci. USA
Vol. 75, No. 6, pp. 2746–2749, June 1978
Biochemistry

Immunological screening method to detect specific translation products

(solid-phase radioimmunodetection/bacterial colonies/phage λ/autoradiography/β-galactosidase)

STEPHANIE BROOME AND WALTER GILBERT

Department of Biochemistry and Molecular Biology, Harvard University, Cambridge, Massachusetts 02138

Contributed by Walter Gilbert, April 10, 1978

ABSTRACT We describe a very sensitive method to detect as antigens the presence of specific proteins within phage plaques or bacterial colonies. We coat plastic sheets with antibody molecules, expose the sheet to lysed bacteria so that a released antigen can bind, and then label the immobilized antigen with radioiodinated antibodies. Thus, the antigen is sandwiched between the antibodies attached to the plastic sheet and those carrying the radioactive label. Autoradiography then shows the positions of antigen-containing colonies or phage plaques. A few molecules of antigen released from each bacterial cell generate an adequate signal.

How can we detect small amounts of proteins made in a bacterial cell without an enzymatic assay? Immunological methods offer an approach: one should be able to detect not only a complete protein, but even a fragment of a protein sequence by virtue of its antigenic determinants. This problem is posed in recombinant DNA experiments in which one might desire to identify a bacterial cell containing a fragment of a gene from a higher cell. If that gene fragment were inserted within a bacterial protein, in phase, antigenic determinants on the higher cell protein could be synthesized and detected. Two immunological screening techniques have been reported (1, 2): they depended on precipitation of the antigen by antibodies included in the agar of the plate or in an agarose overlay. We have devised an extremely sensitive method for screening plaques or colonies that detects antigen-containing areas by a solid-phase "sandwich" assay.

Antibody molecules adsorb strongly to plastics such as polystyrene or polyvinyl and are not significantly dislodged by washing. This is the basis of very sensitive and simple two-site radioimmune assays (3, 4). Thus, we coat a flat disk of flexible polyvinyl with the IgG fraction from an immune serum and press this disk onto an agar plate so that antigen released from bacterial cells during the formation of a phage plaque or through *in situ* lysis of a colony can bind to the fixed antibody. We then incubate the plastic disk with the same total IgG fraction labeled with radioactive iodine so that other determinants on the bound antigen can in turn bind the iodinated antibody. The radioactive areas on the disk expose x-ray film during autoradiography. Since the polyvinyl disk can be thoroughly washed and treated with carrier serum, the background labeling is low; we have detected as little as 5 pg of antigen distributed over an area somewhat larger than a bacterial colony. Microgram amounts of antibody saturate the plastic disk, so an Ig fraction prepared from 5 ml of immune serum will serve to screen 1000 plates of colonies or plaques.

We describe the application of this technique to the detection of an independently assayable protein whose level of production is well characterized and subject to manipulation, *Escherichia coli* β-galactosidase (β-galactoside galactohydrolase; EC 3.2.1.23).

MATERIALS AND METHODS

Solid-phase radioimmunodetection

Preparation of Solid-Phase IgG. Press 8.25-cm diameter disks cut from clear, flexible polyvinyl, 8 mil thickness (Dora May Co., New York), between sheets of smooth paper to flatten them. Place 10 ml of 0.2 M $NaHCO_3$ (pH 9.2) containing 60 μg of IgG per ml in a glass petri dish and set a flat polyvinyl disk upon the liquid surface. After 2 min at room temperature, remove the disk and wash it twice with 10 ml of cold wash buffer: phosphate-buffered saline (PBS), 0.5% normal rabbit serum (vol/vol), and 0.1% bovine serum albumin (wt/vol). Wash by gently swirling the buffer over the disk, then pouring and aspirating off the wash solution. We coat 10 polyvinyl disks successively with the same IgG solution and use each disk immediately after it is coated.

Release of Antigen from Cells in Bacterial Colonies. Heat induction of λcI857 prophages conveniently lyses cells in colonies (2). After growth of colonies of lysogens for 24 hr at 32°, incubate the plates for 2 hr at 42°.

Alternatively, apply 2 μl of 10 mM $MgSO_4$ containing 10^5 λvir to each colony and leave the plates for 3 hr at 37°, to effect a direct phage-mediated lysis.

Finally, chloroform vapor will lyse the bacteria *in situ*. Place a small volume of chloroform in the bottom of a tightly coverable container and set open petri dishes on a glass support above the liquid. After 10 min, transfer the plates to a desiccator and apply a vacuum to remove any residual chloroform.

Immunoadsorption of Antigen onto Solid Phase. Gently place the IgG-coated surface of a polyvinyl disk in contact with the agar and lysed colonies or the top agar and phage plaques within a petri dish. Smooth any air bubbles that form between disk and agar to the side, since the plastic is flexible. Leave the plates for 3 hr at 4°; then remove the disks and wash them three times with 10 ml of cold wash buffer, aspirating off any adhering cellular material during the washing.

Reaction of ^{125}I-Labeled Antibodies with Solid-Phase Antigen and Autoradiography. Pipet 1.5 ml of wash buffer containing 5×10^6 cpm (γ emission) of ^{125}I-labeled IgG (^{125}I-IgG) onto the center of an 8.25-cm diameter flat disk of nylon mesh (carried by most fabric stores) placed in the bottom of a petri dish. The mesh serves a necessary spacer function. Set a polyvinyl disk on the mesh and the solution so that the entire lower polyvinyl surface is accessible to the radioactive antibody. By repeating the layering process, generating a stack of alternating nylon mesh and polyvinyl disks, 15–20 polyvinyl disks

The costs of publication of this article were defrayed in part by the payment of page charges. This article must therefore be hereby marked "*advertisement*" in accordance with 18 U. S. C. §1734 solely to indicate this fact.

Abbreviations: PBS, phosphate-buffered saline; wash buffer, PBS containing 0.5% normal rabbit serum and 0.1% bovine serum albumin; ^{125}I-IgG, ^{125}I-labeled IgG.

can be incubated in a single petri dish with 5 × 10^6 cpm of ^{125}I-IgG each. Incubate overnight at 4°; then wash each disk twice with 10 ml of cold wash buffer and twice with water. Lightly blot the disks to remove water droplets and let them dry at room temperature. Finally, autoradiograph the disks with either Kodak No Screen film or Kodak X-OMAT R film and a Du Pont Cronex Lighting Plus intensifying screen (5).

Procedures specific for β-galactosidase detection

Plating of Bacteria and Phage. Colonies of FMA-10 (W3102 r^- *thy*$^-$; from F. Ausubel) (λcI857) or RV(Δ*lac*) (λcI857) were grown for 24 hr at 32° on YT plates or on Minimal A plates containing 0.2% glycerol and 40 μg of 5-bromo-4-chloro-3-indolyl-β-D-galactoside per ml (6). Colonies of nonlysogenic FMA-10 were grown overnight at 37° on YT plates.

Phage strains were λvir (kindly provided by J. G. Sutcliffe) and λ*plac5* cI857 *Sam7*. Host cells and phage were plated in 2.5 ml of H soft agar (0.8% agar) over H bottom agar (1% agar) (6), and phage plaques were allowed to form overnight at 37° on lawns of QD5003(*supIII*$^+$) or at 32° on FMA-10(λcI857). Because FMA-10 is *thy*$^-$, all plates and media contained 10 μg of thymidine per ml.

Antiserum. New Zealand White rabbits were immunized with 1 mg of electrophoretically pure β-galactosidase (7, 8) in complete Freund's adjuvant (Difco). Booster injections were administered in incomplete Freund's adjuvant (Difco) 2 and 3 weeks after the initial injection, and the rabbits were bled 1 week later. Ten microliters of immune serum precipitated 10 μg of pure β-galactosidase.

The IgG fractions of rabbit pre-immune and rabbit anti-β-galactosidase immune sera were prepared by ammonium sulfate precipitation followed by DEAE-cellulose (Whatman, DE-52) chromatography (9) in 25 mM potassium phosphate, pH 7.3/1% glycerol. Fractions containing the bulk of the flow-through material were pooled, and protein was precipitated by adding ammonium sulfate to 40% saturation. The resulting pellet was resuspended in one-third the original serum volume of 25 mM potassium phosphate, pH 7.3/0.1 M NaCl/1% glycerol, and dialyzed against the same buffer. After dialysis, any residual precipitate was removed by centrifugation. IgG fractions were stored in aliquots at −70°.

Iodination of IgG. IgG fractions were radioiodinated by the method of Hunter and Greenwood (10). The 25-μl reaction mixture contained 0.5 M potassium phosphate (pH 7.5), 2 mCi of carrier-free Na^{125}I, 150 μg of IgG, and 2 μg of chloramine T. After 3 min at room temperature, 8 μg of sodium metabisulfite in 25 μl of PBS was added, followed by 200 μl of PBS containing 2% normal rabbit serum. The ^{125}I-labeled IgG was purified by chromatography on a Sephadex G-50 column equilibrated with PBS containing 2% normal rabbit serum. The ^{125}I-IgG elution fraction was diluted to 5 ml with PBS containing 10% normal rabbit serum, filtered through a sterile Millipore VC filter (0.1 μm pore size), divided into aliquots, and stored at −70°. The specific activities were 1.5 × 10^7 cpm per μg.

RESULTS

Specificity of solid-phase ^{125}I-antibody binding

Lac$^+$ colonies growing on a minimal plate containing glycerol as the carbon source and the β-galactosidase indicator substrate 5-bromo-4-chloro-3-indolyl-β-D-galactoside develop a blue coloration while *lac*$^-$ colonies remain white. Fig. 1*A* shows such a plate bearing colonies of either FMA-10(λcI857) (*lac*$^+$) or RV(λcI857) (*lac* deletion) cells, after 24 hr of colony growth at 32°. Since no inducer of *lac* transcription was present in the plate, each cell in the five dark (*lac*$^+$) colonies contained approximately 10–20 molecules of β-galactosidase (7).

We lysed the cells within the *lac*$^+$ and *lac*$^-$ colonies by heat induction of the λcI857 prophage and adsorbed any β-galactosidase released onto a polyvinyl disk coated with anti-β-galactosidase antibodies. Fig. 1*B* shows an autoradiograph of this plastic disk after labeling with I^{125}-anti-β-galactosidase. The regions of ^{125}I-antibody binding clearly show the positions of *lac*$^+$ colonies, while the *lac*$^-$ colonies do not label.

This detection of the basal level of β-galactosidase represents essentially a full signal for this assay. If the cells had been fully induced, the labeled spots would have been larger but not more intense. Since only background labeling is observed when uninduced *lac*$^+$ colonies are screened with uncoated polyvinyl or coated with pre-immune serum IgG, the fixed, specific antibody is required. Faintly detectable amounts of antigen can be adsorbed, however, by uncoated polyvinyl exposed to lysed, fully induced *lac*$^+$ colonies in which approximately 2% of total protein is β-galactosidase.

Fig. 2*A* shows that ^{125}I-antibody binding is dependent upon *lac*$^+$ cell lysis and demonstrates an application of this solid-phase radioimmunodetection to phage plaques. This disk was applied to a plate bearing plaques of λvir phage on a lawn of

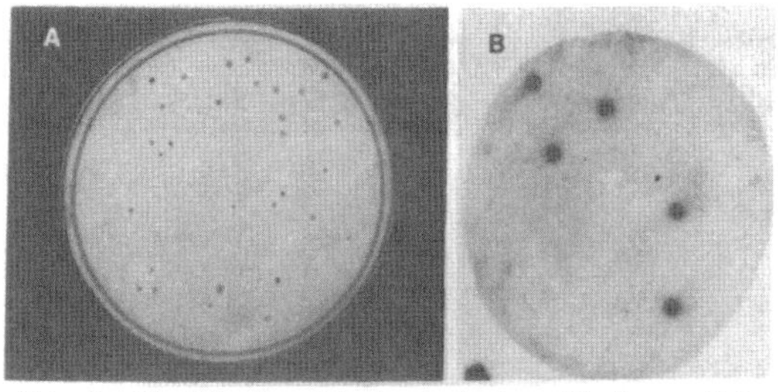

FIG. 1. Identification of *lac*$^+$ colonies by solid-phase radioimmunodetection. A mixture of RV (Δ*lac*) (λcI857) and FMA-10 (λcI857) cells was spread on a glycerol Minimal A plate containing 5-bromo-4-chloro-3-indolyl-β-D-galactoside. (*A*) Colonies formed after 24 hr of growth at 32°. Due to the presence of hydrolyzed indicator substrate, the five *lac*$^+$ strain colonies appear darker than the *lac*$^-$ strain colonies. Cells within each colony on this plate were lysed by prophage induction; released antigen was adsorbed to a polyvinyl disk that had been coated with anti-β-galactosidase IgG. Immobilized antigen was labeled by incubating the polyvinyl disk with radioiodinated anti-β-galactosidase IgG. (*B*) Autoradiograph of this polyvinyl disk exposed on No Screen film for 48 hr.

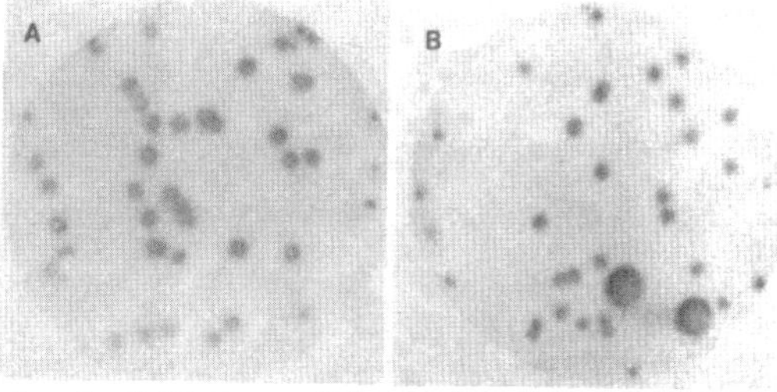

FIG. 2. Solid-phase radioimmunodetection of β-galactosidase present in phage plaques. Antigen released from host cells lysed during phage plaque formation was immobilized on a polyvinyl disk and labeled as described in the legend to Fig. 1. The autoradiographs are of disks imprinted on plates bearing (*A*) λvir plaques on a lawn of FMA-10(λcI857) cells or (*B*) λvir and λ*plac5 Sam7* plaques on QD5003 cells, exposed on No Screen film for 48 hr.

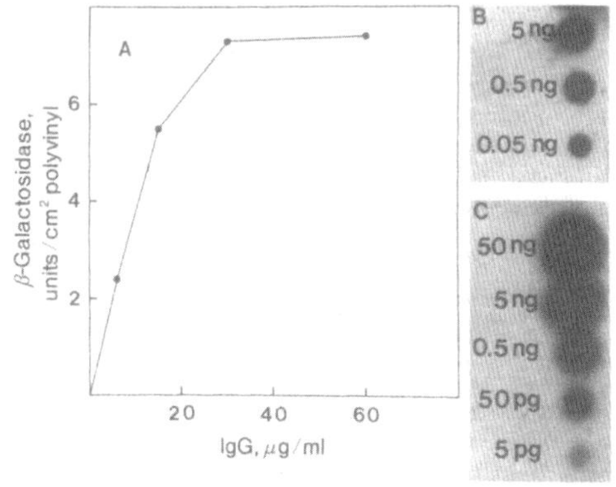

FIG. 3. Assays of IgG adsorption to polyvinyl and solid-phase radioimmunodetection of two different antigens. (*A*) Polyvinyl disks (8.25-cm) were coated with various dilutions of anti-β-galactosidase IgG in 20 mM $NaHCO_3$ (pH 9.2) for 1 min at room temperature. The coated disks were washed twice with 10 ml of cold wash buffer, and 1-cm^2 pieces cut from these disks were each incubated with 1 ml of wash buffer containing 1 mM $MgSO_4$, 500 mM 2-mercaptoethanol, and 1 μg of β-galactosidase for 4 hr at 4°. Unbound antigen was then removed by washing each 1-cm^2 piece of polyvinyl twice with 5 ml of cold wash buffer. Solid-phase-bound β-galactosidase was measured by *o*-nitrophenyl-β-D-galactoside hydrolysis (6). One unit of enzyme hydrolyzes 1 nmol of *o*-nitrophenyl-β-D-galactoside per min at 28°. (*B*) Aliquots (1-μl) of wash buffer containing the indicated amounts of β-galactosidase were applied to the surface of an agar plate. After the liquid had entered the agar, the antigen was adsorbed in the usual way to a polyvinyl disk that had been coated with anti-β-galactosidase IgG. Immobilized antigen was labeled and detected as described in the legend to Fig. 1, except autoradioagraphy was for 18 hr with an intensifying screen at −70°. (*C*) Aliquots (1-μl) of wash buffer containing the indicated amounts of *E. coli* penicillinase (EC 3.5.2.6) were used in a similar experiment. The anti-penicillinase IgG was prepared from an immune serum of a titer comparable to that of our rabbit anti-β-galactosidase immune serum.

uninduced FMA-10(λcI857). The IgG-coated polyvinyl disk detects the 10 β-glactosidase molecules per cell released as the cells lyse during the formation of the plaque but does not pick up any antigen from the unlysed lawn. The control labeling of IgG-coated polyvinyl disks placed upon λvir plaques on the *lac*⁻ strain RV(λcI857) is completely uniform and of background intensity (data not shown).

The autoradiographic image is larger than the original colony or phage plaque because the antigen diffuses in the plate. The spread from phage plaques is especially apparent since infection and lysis of host cells (embedded in 0.8% top agar) occurs over an extensive period of time. Fig. 2*B* shows the distribution of antigen released during λ*plac5 Sam7* and λvir growth on a *lac*⁺ *suIII*⁺ host, QD5003. The large exposed areas on the autoradiograph correspond to λ*plac5 Sam7* plaques and reflect the high level of β-galactosidase produced during the I^-, Z^+ *lac* phage infection. Each λ*plac5 Sam7* plaque contained on the order of 10–20 ng of β-galactosidase (assuming 10^6 cells lysed per plaque and 10^4 molecules of enzyme per cell). The smaller exposed areas on the film correspond in both size and position to λvir plaques, each of which contained approximately 0.1 ng of β-galactosidase (10^7 cells lysed per plaque, 10 molecules of enzyme per cell). The identity of the phage within each plaque was confirmed since only phage isolated from ostensible λvir plaques produced "macroplaques" on the lysogenic, nonsuppressing host RV(λcI857). Fig. 2*B* demonstrates that the enlargement of the autoradiographic image is proportional to the amount of β-galactosidase released from lysed cells, and further shows, therefore, that the antigen recognized by ^{125}I-antibody is β-galactosidase.

Selection of solid-phase assay conditions

We determined the appropriate polyvinyl coating conditions by measuring the amounts of β-galactosidase immobilized on 1-cm^2 pieces of polyvinyl cut from disks that had been incubated with various dilutions of anti-β-galactosidase IgG. The bound β-galactosidase retained enzymatic activity and was assayed by *o*-nitrophenyl-β-D-galactoside hydrolysis (7). Fig. 3*A* shows that a 1-min incubation with 30 μg of IgG per ml saturates the antibody-adsorbing capacity of an 8.25-cm diameter polyvinyl disk. In addition, the same amount of β-galactosidase is bound by polyvinyl coated for 1 min or for 5 hr with 30 μg of IgG per ml. By the same procedure, we found that 10 disks coated successively with one 10-ml solution of 60 μg of IgG per ml possess identical capacities to bind β-galactosidase. A decrease in binding was observed, however, as disks were coated successively in 10 ml of 30 μg of IgG per ml, from which we estimate that each disk binds about 20 μg of antibody.

In order to estimate the minimum amount of protein detectable using 5×10^6 cpm of ^{125}I-IgG, we applied a series of dilutions of antigen directly to the surface of a typical agar plate. The antigen then was adsorbed to an IgG-coated polyvinyl disk and labeled with ^{125}I-antibodies. Fig. 3 *B* and *C* shows that 50 pg of β-galactosidase in one experiment and 5 pg of penicillinase (EC 3.5.2.6) in another were detected, spread over an area somewhat larger than a phage plaque or a bacterial colony. Because the antigen had diffused into the agar, these

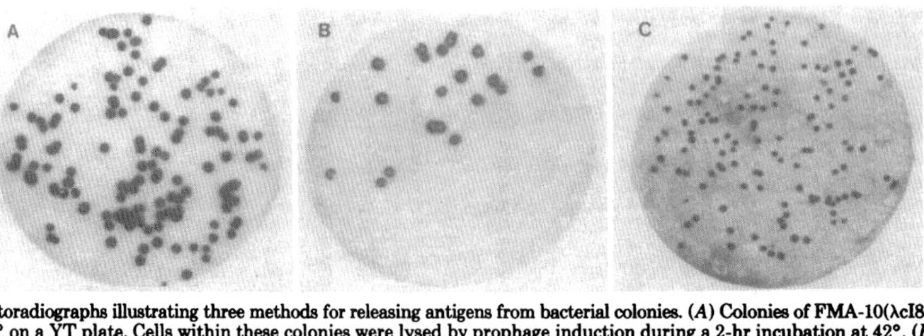

FIG. 4. Autoradiographs illustrating three methods for releasing antigens from bacterial colonies. (*A*) Colonies of FMA-10(λcI857) were grown for 24 hr at 32° on a YT plate. Cells within these colonies were lysed by prophage induction during a 2-hr incubation at 42°. (*B*) Cells within FMA-10 colonies, formed overnight at 37° on YT plates, were lysed *in situ* by applying 2 μl of 10 mM $MgSO_4$ containing 10^5 λvir to each colony on one-half of a plate and then incubating this plate for 3 hr at 37°. The other half of this plate serves as a control. (*C*) Similar colonies were lysed by a 10-min exposure to chloroform vapor in a tightly sealed glass container, as described in *Materials and Methods*.

experiments detected only a fraction of the initial sample.

The specific labeling of the immobilized antigen is maximal after an overnight incubation with ^{125}I-IgG. One-half maximal labeling is reached after 5 hr.

Alternative methods of lysing colonies

As Fig. 4 shows, comparable amounts of β-galactosidase are released from colonies by prophage induction or by direct application of λvir. Chloroform vapor also will lyse cells sufficiently to permit detection of β-galactosidase present in colonies of cells uninduced for *lac* expression (Fig. 4*C*). Colonies of RV(λcI857) (*lac* deletion) cells lysed by any of these methods did not release any material that reacted with ^{125}I-anti-β-galactosidase IgG.

Viable bacteria exist within a colony of λcI857 lysogens after a 2-hr incubation at 42°. Many of the survivors are lysogens and can be recovered, to confirm a positive response, by picking from the site of a colony. Replica plates must be used if colonies are to be treated with λvir or with chloroform vapor.

DISCUSSION

This solid-phase screening method is simple and sensitive. It detects a few picograms of protein antigen, a few molecules from each bacterial cell, using the IgG fraction from an immune serum of moderate titer. The only clear requirement for this approach is that the antigen bind at least two antibody molecules simultaneously.

Enzymatic assays suggest that about 2×10^8 molecules of β-galactosidase can bind per mm^2 of coated plastic. This is consistent with our estimate of the IgG-adsorbing capacity of the polyvinyl, 2×10^{10} molecules per mm^2, since the specific antibodies constitute only a few percent of the immune IgG fraction. Direct counting of samples of labeled plastic showed that only about 5×10^7 labeled antibodies were bound to each mm^2 of fixed antigen under the conditions described; at least 10-fold more label could bind, but at a price in terms of a higher background. The lower limit of detection presumably could be extended by using affinity-purified antibodies.

Uses for this immunological screening procedure include direct identification of clones containing specific foreign DNA segments, if they express a translation product either fortuitously or after *in vitro* genetic manipulations to that end. Furthermore, this technique provides a simple way to follow the movement of antigen on columns or on slab gels.

This two-site detection is particularly suited for the recognition of certain novel genetic constructions which are much less easily assayed by *in situ* immunoprecipitation approaches. For instance, by coating polyvinyl disks with an IgG fraction prepared from an immune serum directed against one protein and labeling the immobilized antigen with ^{125}I-antibodies directed against another protein, only hybrid polypeptide molecules, synthesized as the result of *in vitro* or *in vivo* DNA sequence rearrangement, would produce an autoradiographic response.

We thank Dr. Abe Fuks for advice about radioiodination, Drs. Jeremy Knowles and Alan Hall for samples of *E. coli* penicillinase and rabbit anti-penicillinase serum, and Roger Brent for suggesting the chloroform release. This work was supported by the National Institutes of Health, Grant GM09541-17. W.G. is an American Cancer Society Professor. S.B. was supported by a National Institutes of Health training grant.

1. Sanzey, B., Mercereau, O., Ternynck, T. & Kourilsky, P. (1976) *Proc. Natl. Acad. Sci. USA* **73**, 3394–3397.
2. Skalka, A. & Shapiro, L. (1976) *Gene* **1**, 65–79.
3. Catt, K. & Tregear, G. W. (1967) *Science* **158**, 1570–1571.
4. Miles L. E. M. (1977) in *Handbook of Radioimmunoassay*, ed. Abraham, G. E. (Dekker, New York), pp. 131–177.
5. Laskey, R. A. & Mills, A. D. (1977) *FEBS Lett.* **82**, 314–316.
6. Miller, J. M. (1972) in *Experiments in Molecular Genetics*, ed. Miller, J. M. (Cold Spring Harbor Laboratory, Cold Spring Harbor, New York), pp. 432–434.
7. Platt, T. (1972) in *Experiments in Molecular Genetics*, ed. Miller, J. M. (Cold Spring Harbor Laboratory, Cold Spring Harbor, New York), pp. 398–404.
8. Fowler, A. V. (1972) *J. Bacteriol.* **112**, 856–860.
9. Livingston, D. M. (1974) in *Methods in Enzymology*, ed. Jakoby, W. B. & Wilchek, M. (Academic, New York), Vol. 34, 723–731.
10. Hunter, W. M. Greenwood, F. C. (1964) *Biochem. J.* **91**, 43–46.

Proc. Natl. Acad. Sci. USA
Vol. 75, No. 8, pp. 3727–3731, August 1978
Biochemistry

A bacterial clone synthesizing proinsulin

(rat preproinsulin/cDNA cloning/solid-phase radioimmunoassay/DNA sequence/fused proteins)

LYDIA VILLA-KOMAROFF*, ARGIRIS EFSTRATIADIS*, STEPHANIE BROOME*, PETER LOMEDICO*, RICHARD TIZARD*, STEPHEN P. NABER†, WILLIAM L. CHICK†, AND WALTER GILBERT*

* Biological Laboratories, Harvard University, Cambridge, Massachusetts 02138; and † Elliot P. Joslin Research Laboratory, Harvard Medical School, and the Peter Bent Brigham Hospital, Boston, Massachusetts 02215

Contributed by Walter Gilbert, June 9, 1978

ABSTRACT We have cloned double-stranded cDNA copies of a rat preproinsulin messenger RNA in *Escherichia coli* χ1776, using the unique *Pst* endonuclease site of plasmid pBR322 that lies in the region encoding amino acids 181–182 of penicillinase. This site was reconstructed by inserting the cDNA with an oligo(dG)-oligo(dC) joining procedure. One of the clones expresses a fused protein bearing both insulin and penicillinase antigenic determinants. The DNA sequence of this plasmid shows that the insulin region is read in phase; a stretch of six glycine residues connects the alanine at position 182 of penicillinase to the fourth amino acid, glutamine, of rat proinsulin.

Can the structural information for the production of a higher cell protein be inserted into a plasmid in such a way as to be expressed in a transformed bacterium? To attack this problem, we used as a model rat insulin, an interesting protein that can be identified by immunological and biological means.

Although mature insulin contains two chains, A and B, it is the product of a single longer polypeptide chain. The hormone is initially synthesized as a preproinsulin structure (1, 2). A hydrophobic leader sequence of 23 amino acids at the amino terminus of the nascent chain is cleaved off, presumably as the polypeptide chain moves through the endoplasmic reticulum (2–4), producing a proinsulin molecule. The proinsulin chain folds up and then the C peptide is cleaved from its middle (5). Thus each of the two (nonallelic) insulin genes in the rat (6–8) encodes a polypeptide 109 amino acids long, whose initial structure is NH_2—leader sequence—B chain—C peptide—A chain.

Ullrich *et al.* (9) have cloned double-stranded cDNA copies of rat preproinsulin mRNA isolated from pancreatic islets and determined sequences covering much of those two genes. We have made double-stranded cDNA copies of mRNA from a rat insulinoma (10) and cloned these in the *Pst* (*Providencia stuartii* endonuclease) site of pBR322 (11), which lies within the penicillinase gene.

The *Escherichia coli* penicillinase is a periplasmic protein, the gene for which was recently sequenced (12). Penicillinase is synthesized as a preprotein with a 23 amino acid leader sequence (12, 13), which presumably serves as a signal to direct the secretion of the protein to the periplasmic space, and is removed as the protein traverses the membrane. Insertion of the structural information for insulin into the penicillinase gene should cause expression of the insulin sequence as a fusion product transported outside the cell.

MATERIALS AND METHODS

Bacterial Strains. *E. coli* K-12, strain HB101 [*hsm*$^-$, *hrs*$^-$, *recA*$^-$, *gal*$^-$, *pro*$^-$, *str*^{r} (14)] was initially obtained from H. Boyer. *E. coli* K-12 strain χ1776 (15) (F$^-$, *tonA53, dapD8, minA1, supE42,* Δ40[*gal-uvrB*], λ$^-$, *minB2, rfb-2, nalA25, oms-2, thyA57, metC65, oms-1,* Δ29[*bioH-asd*], *cycB2, cycA1, hsdR2*) was provided by R. Curtiss.

DNA and Enzymes. pBR322 DNA, a gift from A. Poteete, was used to transform *E. coli* HB101. Plasmid DNA was purified according to the procedure of Clewell (16). Avian myeloblastosis virus reverse transcriptase (RNA-dependent DNA polymerase), *E. coli* DNA polymerase I, and terminal transferase were gifts from T. Papas, M. Goldberg, and J. Wilson, respectively. Restriction enzymes were purchased from Bethesda Research Labs and New England BioLabs.

RNA Purification. An x-ray-induced, transplantable rat beta cell tumor (10) was used as source of preproinsulin mRNA. Tumor slices (20 g per preparation) were homogenized, and a cytoplasmic RNA (about 2 mg/g of tissue) was purified from a postnuclear supernatant by Mg^{2+} precipitation (17), followed by extraction with phenol and chloroform, and enriched for poly(A)-containing RNA by oligo(dT)-cellulose chromatography (18). About 4% of the material binds to the column (data from eight preparations). Further purification of the oligo(dT)-cellulose-bound material by sucrose gradient centrifugation and/or polyacrylamide gel electrophoresis showed that the preproinsulin mRNA was a minor component of the preparation.

Double-Stranded cDNA Synthesis. Oligo(dT)-cellulose-bound RNA was used directly as template for double-stranded cDNA synthesis (19), except that a specific $p(dT)_8dG$-dC primer (Collaborative Research) was utilized for reverse transcription. The concentrations of RNA and primer were 7 mg/ml and 1 mg/ml, respectively. All four [α-^{32}P]dNTPs were at 1.25 mM (final specific activity 0.85 Ci/mmol). The reverse transcript was 2% of the input RNA, and 25% of it was finally recovered in the double-stranded DNA product.

Construction of Hybrid DNA Molecules. pBR322 DNA (5.0 μg) was linearized with *Pst*, and approximately 15 dG residues were added per 3′ end by terminal transferase at 15° in the presence of 1 mM Co^{2+} (20) and autoclaved gelatin at 100 μg/ml. Similarly, dC residues were added to 2.0 μg of double-stranded cDNA, which was then electrophoresed in a 6% polyacrylamide gel. Following autoradiography, molecules in the size range of 300 to 600 base pairs (0.5 μg) were eluted from the gel (21). Size selection was done after tailing rather than before because previous experience had indicated that occasionally impurities contaminating DNA extracted from gels inhibits terminal transferase. The eluted double-stranded cDNA was concentrated by ethanol precipitation, redissolved in 10 mM Tris·HCl at pH 8, mixed with 4 μg of dG-tailed pBR322, and dialyzed versus 0.1 M NaCl/10 mM EDTA/10 mM Tris, pH 8. The mixture (4 ml) was then heated at 56° for 2 min, and annealing was performed at 42° for 2 hr. The hybrid DNA was used to transform *E. coli* χ1776.

Transformation and Identification of Clones. Transformation of *E. coli* χ1776 (an EK2 host) with pBR322 (an EK2

The costs of publication of this article were defrayed in part by the payment of page charges. This article must therefore be hereby marked "*advertisement*" in accordance with 18 U. S. C. §1734 solely to indicate this fact.

vector) was performed in a biological safety cabinet in a P3 physical containment facility in compliance with NIH guidelines for recombinant DNA research published in the *Federal Register* [(1976) **41**, 27902–27943].

χ1776 was transformed by a transfection procedure (22) adapted to χ1776 by A. Bothwell (personal communication) and slightly modified as follows: χ1776 was grown in L broth (23) supplemented with diaminopimelic acid at 10 μg/ml and thymidine (Sigma) at 40 μg/ml to OD_{590} of 0.5. Cells (200 ml) were sedimented at 500 × *g* and resuspended by swirling in 1/10th vol of cold buffer containing 70 mM $MnCl_2$, 40 mM Na acetate at pH 5.6, 30 mM $CaCl_2$, and kept on ice for 20 min. The cells were repelleted and resuspended in 1/30th of the original volume in the same buffer. Two milliliters of the annealed DNA preparation was added to the cells. Aliquots of this mixture (0.3 ml) were placed in sterile tubes and incubated on ice for 60 min. The cells were then placed at 37° for 2 min. Broth was added to each tube (0.7 ml) and the tubes were incubated at 37° for 15 min; 200 μl of the cells was spread on sterile nitrocellulose filters (Millipore) overlaying agar plates containing tetracycline at 15 μg/ml. (The filters were boiled to remove detergents before use.) The plates were incubated at 37° for 48 hr. Replicas of the filters were made by a procedure developed by D. Hanahan (personal communication): The nitrocellulose filters containing the transformants were removed from the agar and placed on a layer of sterile Whatman filter paper. A new sterile filter was placed on top of the filter containing the colonies and pressure was applied with a sterile velvet cloth and a replica block. A sterile needle was used to key the filters. The second filter was placed on a new agar plate and incubated at 37° for 48 hr. The colonies on the first filter were screened by the Grunstein–Hogness technique (24), using as probe an 80-nucleotide-long fragment produced by *Hae* III digestion of high specific activity cDNA (9). Positive colonies were rescreened by hybrid-arrested translation (25) as described in the legend of Table 1.

Radioimmunoassays. Two-site solid-phase radioimmunoassays were performed (28). Cells from colonies to be tested were transferred with an applicator stick onto 1.5% agarose containing 30 mM Tris-HCl, pH 8, lysozyme at 0.5 mg/ml, and 10 mM EDTA; released antigen was adsorbed to an IgG-coated polyvinyl disk during a 1-hr incubation at 4°. The wash buffer contained streptomycin sulfate at 300 μg/ml and normal guinea pig serum (Grand Island Biological Co.) instead of normal rabbit serum. Guinea pig antiserum to bovine insulin was purchased from Miles Laboratories.

Standard (liquid) radioimmunoassays were performed using the back titration procedure employing alcohol precipitation of insulin–antibody complexes (29).

DNA Sequencing. DNA sequencing was performed as described by Maxam and Gilbert (30).

RESULTS

Construction and Identification of cDNA Clones. We isolated poly(A)-containing RNA from a transplantable rat insulinoma. This preparation contained preproinsulin mRNA, because it directed the synthesis in a cell-free system of a product precipitable with anti-insulin antibody (data not shown). However, the mRNA yield after further purification was not sufficient for cloning, and therefore we decided to clone cDNA synthesized from the total preparation. In an attempt to enrich the reverse transcript for insulin sequences, we utilized the DNA sequence reported by Ullrich *et al.* (9) to choose a specific primer, $(dT)_8$dG-dC. The product of double-stranded cDNA synthesis (19) was extended by a short oligo(dC) tail about 15 nucleotides in length, and sized on a polyacrylamide gel. A broad size cut averaging 500 base pairs was selected in order to enrich for full-length sequences. We inserted these molecules into the *Pst* site of pBR322 after elongating the 3′-terminal extension of the cleavage site with oligo(dG). We used this oligo(dG)·oligo(dC) joining procedure in order to reconstruct the *Pst* recognition sequence (ref. 31; W. Rowenkamp and R. Firtel, personal communication); approximately 40% of the inserts were excisable with *Pst* after cloning. From about 0.25 μg of tailed cDNA we obtained 2355 transformants in *E. coli* strain χ1776. To identify clones containing insulin sequences, we first screened one-third of the transformants, using as a probe an 80-nucleotide-long *Hae* III fragment of cDNA synthesized from oligo(dT)-bound RNA because the results of Ullrich *et al.* (9) suggested that such a fragment should be insulin specific. About 20% of the clones were positive, but restriction analysis of plasmid DNA from a few candidates showed that the inserts were not insulin sequences. We concluded that our probe was not pure and rescreened some of the positive clones, using hybrid-arrested translation (25). This method is based on the principle that mRNA in the form of an RNA·DNA hybrid does not direct cell-free protein synthesis. We incubated aliquots of oligo(dT)-bound RNA with linearized

Table 1. Hybrid-arrested translation and immunoprecipitation of the cell-free products

	Radioactivity, cpm/20 μl			
		Immunoprecipitable		
Source of arresting DNA	Acid insoluble	− Insulin	+ Insulin	% Immunoprecipitable*
Control I				
(−DNA, −RNA)†	2,570			
Control II				
(−DNA, +RNA)‡	35,700	12,300	310	36.2
pBR322	28,800	7,850	245	29.0
Clone 3	15,100	3,630	264	26.9
Clone 13	19,600	5,190	350	28.4
Clone 15	18,600	4,850	252	28.7
Clone 16	29,200	8,830	247	32.2
Clone 17	24,000	6,700	316	30.0
Clone 18	15,900	3,690	251	25.8
Clone 19	8,650	587	277	5.0
Clone 20	15,100	4,070	231	30.6
Clone 21	21,100	5,170	223	26.7

Plasmid DNA (about 3 μg) was digested with *Pst*, precipitated with ethanol, and dissolved directly in 20 μl of deionized formamide. After heating for one minute at 95° each sample was placed on ice. Following the addition of 1.5 μg of oligo(dT)-cellulose-bound RNA, piperazine-*N,N′*-bis(2-ethanesulfonic acid) (Pipes) at pH 6.4 to 10 mM, and NaCl to 0.4 M, the mixtures were incubated for 2 hr at 50°. They were then diluted by the addition of 75 μl of H_2O and ethanol precipitated in the presence of 10 μg of wheat germ tRNA, washed with 70% (vol/vol) ethanol, dissolved in H_2O, and added to a wheat germ cell-free translation mixture (26) containing 10 μCi of [^{3}H]leucine (60 Ci/mmol). Fifty-microliter reaction mixtures were incubated at 23° for 3 hr and then duplicate 2-μl aliquots were removed for trichloroacetic acid precipitation. From the remainder two 20-μl aliquots were treated with ribonuclease, diluted with immunoassay buffer, and analyzed for the synthesis of immunoreactive preproinsulin by means of a double antibody immunoprecipitation (27) in the absence or presence of 10 μg of bovine insulin. The washed immunoprecipitates were dissolved in 1 ml of NCS (Amersham) and assayed in 10 μl of Omnifluor (New England Nuclear) by liquid scintillation counting.

* Calculated using the formula [(immunoprecipitable radioactivity in the absence of insulin) − (immunoprecipitable radioactivity in the presence of insulin)]/[(acid-insoluble radioactivity) − (acid-insoluble radioactivity of control I)].

† Reaction mixture incubated in the absence of added RNA.

‡ Cell-free translation by the direct addition of oligo(dT)-cellulose-bound RNA into the reaction mixture.

Biochemistry: Villa-Komaroff *et al.* *Proc. Natl. Acad. Sci. USA 75 (1978)* 3729

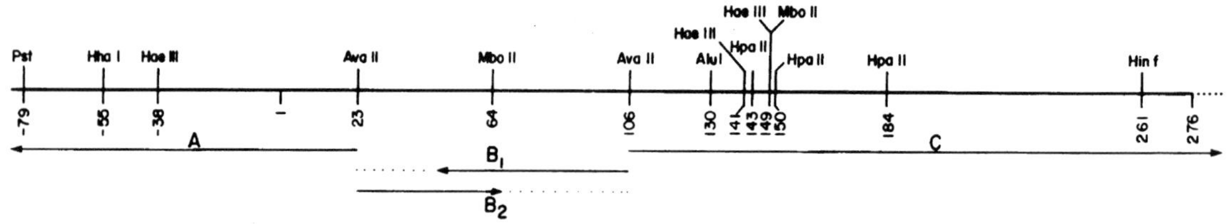

FIG. 1. Restriction map of the insertion in clone pI19. Each restriction site is identified by a number indicating the 5′-terminal nucleotide generated by cleavage at the message strand. Nucleotides are numbered beginning with the first base of the sequence encoding proinsulin. Nucleotides in the 5′ direction from position 1 in the message strand are identified by negative numbers, beginning with −1. Arrows indicate the sequenced fragments; those pointing to the left indicate sequences derived from the antimessage strand, and those pointing to the right indicate sequences derived from the message strand. The uniquely labeled restriction fragments were generated as follows: Following excision with *Pst*, DNA of the insertion was digested with *Ava* II and end labeled. Fragments A and C purified from a polyacrylamide gel were sequenced directly because the *Pst* ends do not label significantly. Fragment B was strand separated on a polyacrylamide gel and sequenced in both directions. The exact number of C·G pairs in the right-hand tail before the *Pst* site could not be counted.

DNA from nine clones under conditions favoring DNA·RNA hybridization (32), added them to cell-free translation systems, and assayed for a specific inhibition of insulin synthesis. Table 1 shows that one of the plasmids, pI19, inhibited the synthesis of immunoprecipitable material. Restriction endonuclease digestions of the *Pst*-excised insert of pI19 with several enzymes generated fragments whose sizes were consistent with the sequence of Ullrich *et al.* (9). We confirmed the presence of insulin DNA in pI19 by direct DNA sequence analysis and screened the rest of the clones with purified pI19 insert labeled by nick translation. About 2.5% (48/1745) of the clones hybridized strongly to this probe. There must have been enrichment for insulin sequence at some step of our procedure, because hybridization analysis using cloned insulin DNA as probe showed the presence of only 0.3% insulin mRNA in the original oligo(dT)-bound RNA.

Sequence Information. Fig. 1 shows the restriction map of the insertion in clone pI19 and Fig. 2 shows the sequence of the insert. It corresponds to rat insulin I (5, 33) and encodes the entire preproinsulin chain with the exception of the first two amino acid residues of the reported preregion (1). It therefore extends the sequence determined by Ullrich *et al.* (9) by twenty-five 5′-terminal nucleotides. It also verifies the reported amino acid residues for positions −14, −17, −18, and −20; it identifies the previously uncertain residue −15; and it identifies the unknown residue −19. However, the residues at positions −16 and −21 differ from those reported (1).

The sequence deviates from that determined by Ullrich *et al.* (9) at the region immediately after the UGA terminator, where a GAGTC sequence occurs, predicting a *Hinf* cleavage site that we have experimentally verified. Furthermore, only moderate agreement exists between the two sequences for the next 15 nucleotides of the 3′ untranslated region.

Expression. Almost two-thirds of the clones carrying inserts were ampicillin resistant; thus the active site of penicillinase must lie between amino acid residues 23 and 182 (12). The degree of resistance was variable, suggesting the expression of different sequences from the inserts in the form of fused translation products, probably differing in length and stability.

We therefore screened colonies of the 48 clones containing insulin sequence for the presence of insulin antigenic determinants, using a solid-phase radioimmunoassay (28). Polyvinyl sheets coated with antibody molecules will bind specific antigens released from bacteria. The immobilized antigen can then be detected by autoradiography following exposure of the sheets to ^{125}I-labeled antibody. This method permits detection of as little as 10 pg of insulin in a colony. We coated plastic disks with anti-insulin antibody and used ^{125}I-labeled anti-insulin to detect solely insulin antigenic determinants. Disks coated with anti-penicillinase antibody and exposed to ^{125}I-anti-insulin detect the presence of a fused protein, as do disks coated with anti-insulin and exposed to radioiodinated anti-penicillinase.

One clone, pI47, gave positive responses with all of the combinations described above; this indicates the presence of a penicillinase–insulin hybrid polypeptide. Fig. 3 shows some of the results. To determine whether this fused protein is secreted, we grew clone pI47 in liquid culture and extracted the proteins in the periplasmic space by osmotic shock, a method that does not lyse bacteria (34). Fig. 4 shows that the insulin

```
                          -21                                                    -1↓
                          TrpMetArgPheLeuProLeuLeuAlaLeuLeuValLeuTrpGluProLysProAlaGlnAla
             (C)TGCAGGGGGGGGGGGGGGGGTGGATGCGCTTCCTGCCCCTGCTGGCCCTGCTCGTCCTCTGGGAGCCCAAGCCTGCCCAGGCT
             (G)ACGTCCCCCCCCCCCCCCCCACCTACGCGAAGGACGGGGACGACCGGGACGAGCAGGAGACCCTCGGGTTCGGACGGGTCCGA   -1
                 Pst                   HhaI           HaeIII

   1                                                                                ↓        33
   PheValLysGlnHisLeuCysGlyProHisLeuValGluAlaLeuTyrLeuValCysGlyGluArgGlyPhePheTyrThrProLysSerArgArgGlu
   TTTGTCAAACAGCACCTTTGTGGTCCTCACCTGGTGGAGGCTCTGTACCTGGTGTGTGGGGAACGTGGTTTCTTCTACACACCCAAGTCCCGTCGTGAA
 1 AAACAGTTTGTCGTGGAAACACCAGGAGTGGACCACCTCCGAGACATGGACCACACACCCCTTGCACCAAAGAAGATGTGTGGGTTCAGGGCAGCACTT   99
                     AvaII                                            MboII

   34                                                                                        ↓66
   ValGluAspProGlnValProGlnLeuGluLeuGlyGlyGlyProGluAlaGlyAspLeuGlnThrLeuAlaLeuGluValAlaArgGlnLysArgGly
   GTGGAGGACCCGCAAGTGCCACAACTGGAGCTGGGTGGAGGCCCGGAGGCCGGGGATCTTCAGACCTTGGCACTGGAGGTTGCCCGGCAGAAGCGTGGC
100 CACCTCCTGGGCGTTCACGGTGTTGACCTCGACCCACCTCCGGGCCTCCGGCCCCTAGAAGTCTGGAACCGTGACCTCCAACGGGCCGTCTTCGCACCG  198
       AvaII                AluI       HaeIII  HaeIII  MboII                      HpaII
                                         HpaII   HpaII

   67                                                 86↓
   IleValAspGlnCysCysThrSerIleCysSerLeuTyrGlnLeuGluAsnTyrCysAsn
   ATTGTGGATCAGTGCTGCACCAGCATCTGCTCCCTCTACCAACTGGAGAACTACTGCAACTGAGTCCACCACTCCCCGCCCCC·····
199 TAACACCTAGTCACGACGTGGTCGTAGACGAGGGAGATGGTTGACCTCTTGATGACGTTGACTCAGGTGGTGAGGGGCGGGGG·····
                                                              Hinf
```

FIG. 2. DNA sequence of the insertion in clone pI19. Nucleotides are numbered using the convention described in Fig. 1. Accordingly, amino acids are numbered beginning with the first amino acid of proinsulin, while the last amino acid of the leader sequence (pre region) is numbered as −1. Restriction endonuclease cleavage sites experimentally verified are underlined and identified. The arrows indicate, in order, the ends of the leader sequence and the peptides B, C, and A. Two nucleotides indicated by double underlining are uncertain.

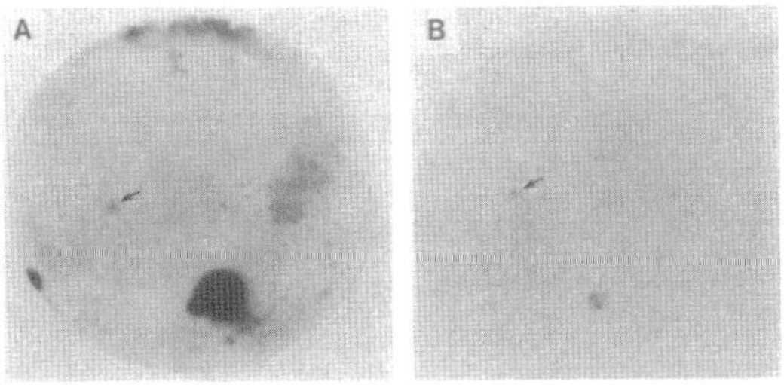

FIG. 3. Initial detection of penicillinase-insulin hybrid polypeptides in an insulin cDNA clone. Cells from colonies of the 48 insulin cDNA clones and from control colonies, χ1776 and χ1776-pBR322, were applied to an agarose/lysozyme/EDTA plate. Positive controls, 5 ng of insulin and 5 ng of penicillinase, each in 1 μl of wash buffer, also were spotted on plate. Antigen was adsorbed to an IgG-coated polyvinyl disk during a 1-hr incubation at 4°. Immobilized antigen was labeled by setting the plastic disk on a solution containing radioiodinated anti-insulin IgG. The autoradiographs are of disks precoated with anti-insulin IgG (*A*) or anti-pencillinase IgG (*B*), exposed on Kodak X-Omat R film using a Du Pont Cronex Lightning Plus intensifying screen for 12 hr at −70°. The arrows indicate the signal generated by clone pI47. The large exposed area in the lower right of (*A*) is the positive control for insulin detection.

antigen was recovered in the distilled water wash of the shock procedure. Table 2 shows that the insulin antigen in the wash is also detectable and quantifiable by a standard radioimmunoassay. The yield of antigen depended on the growth medium; antigen was released by cells grown in M9/glucose/amino acids medium but not by cells grown in brain/heart infusion. We estimate a recovery of about 100 molecules per cell.

Structure of the Fused Protein. We sequenced pI47 to determine the sequence around the junctions. Fig. 5 shows that a proinsulin I cDNA lies in the *Pst* site in the correct orientation and in phase, so that a fused protein can be synthesized. In pI19, the insert is in the correct orientation, but not in phase. In pI47 the oligo(dG)-oligo(dC) region encodes six glycines that connect the penicillinase sequence, ending at amino acid 182 (alanine), to the fourth amino acid (glutamine), of the proinsulin sequence. The cDNA sequence in pI47 extends 26 base pairs past the UGA terminator. Thus, we infer the structure of the fused protein to be penicillinase(24–182)-$(Gly)_6$-proinsulin(4–86).

DISCUSSION

The coding regions of eukaryotic structural genes are often interrupted by introns (35–38), whose transcripts are spliced out of the mature mRNA. Because prokaryotes do not appear to process their messengers, double-stranded cDNA made from a mature messenger is the material of choice to carry eukaryotic structural information into bacteria.

By using cDNA cloning technology and an extremely sensitive method to assay expression, we were able to construct a derivative of *E. coli* strain χ1776 carrying an insulin gene sequence and to detect the synthesis and secretion into the periplasmic space of a fused protein carrying antigenic determinants of both insulin and penicillinase. This was accomplished simply by inserting double-stranded cDNA carrying the structural information for insulin into a restriction site within the structural gene for penicillinase. Not only is the fused DNA sequence expressed as a chain of amino acids, but also the polypeptide folds so as to reveal insulin antigenic shapes. Thus we expect soon to be able to demonstrate biological function for this, or for a similar, fused protein.

We anticipate that the joining of cDNA sequences to nucleotides that lie ahead of the *Pst* site in the penicillinase gene will also produce fused and secreted molecules. Moreover, if the fusion replaces the preproinsulin leader with that of penicillinase it is likely that the new protein will also be secreted by the *E. coli* cell and may even be correctly matured by cleavage of the leader sequence.

Clearly, we have exploited a general method that should lead to the expression and secretion of any eukaryotic protein provided another protein, such as penicillinase, will serve as a carrier, by virtue of its leader sequence. Moreover, the secretion of the eukaryotic protein sequence to the periplasm or extracellular space will both permit its harvest in a purified form and probably eliminate intracellular sources of instability.

Often just an expression of antigens is the goal. In a "shotgun" screening, the existence of a fused protein antigen could be used to identify transformants carrying desired eukaryotic gene fragments. On the other hand, the insertion of a DNA fragment coding for surface antigenic determinants of a virus into a carrier protein should lead to the secretion of a fused protein that could serve as a vaccine, even though no entirely correct virus product is ever produced.

Table 2. Immunoreactive insulin concentration in distilled water wash of osmotic shock procedure

Exp.	Insulin, μunits/ml	Cells/ml
1	318	1.5×10^{10}
2	166	6.0×10^{9}
3	386	4.2×10^{10}

Duplicate 0.1-ml aliquots of each sample prepared as described in the legend to Fig. 4 were assayed (29) in a final volume of 0.4 ml using rat insulin standard, a gift from J. Schlichtkrull. One unit = 48 μg. The NaCl/Tris wash, the 20% sucrose wash, and the media of χ1776-pI47 as well as the water wash from osmotic shock of χ1776-pBR322 gave values below the sensitivity of the assay (25 μunits/ml).

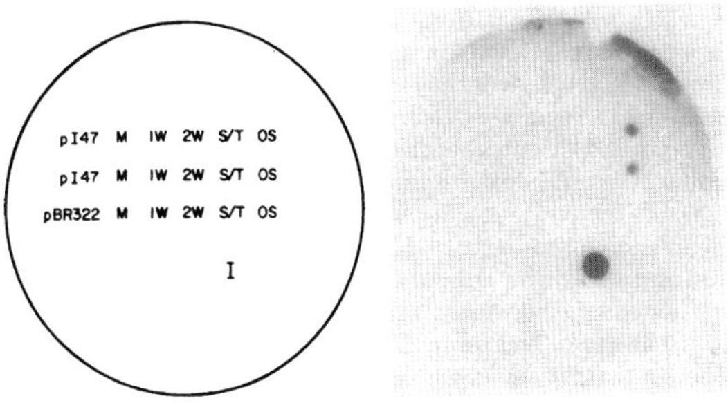

FIG. 4. Release of insulin antigen from χ1776-pI47 cells by osmotic shock. One liter of χ1776-pI47 cells growing at 37° in M9 medium supplemented with 1 g of tryptone, 0.5 g of yeast extract, and 0.5% glucose was harvested at a density of 5×10^7 cells per ml and washed two times in 10 ml of cold 10 mM Tris·HCl, pH 8/30 mM NaCl. The cells were then osmotically shocked (34) in the following manner: The final wash pellet was resuspended in 10 ml of 20% sucrose per 30 mM Tris·HCl, pH 8, at room temperature, made 1 mM in EDTA, shaken at room temperature for 10 min, centrifuged out, resuspended in 10 ml of cold distilled water, shaken in an ice bath for 10 min, and again pelleted. The resulting supernatant was termed the "water wash." As a control, 1 liter of χ1776-pBR322 was grown and treated in a similar manner. Aliquots (1 μl) of each fraction to be assayed for the presence of insulin antigen were applied to the surface of a 1.5% agar plate. (*A*) Positions of each fraction on the plate. M, medium; 1W, first wash supernatant; 2W, second wash supernatant; S/T, sucrose/Tris supernatant; OS, distilled water wash; I, insulin. (*B*) Autoradiograph showing results of a two-site radioimmunoassay of these fractions. Antigen was adsorbed to a polyvinyl disk and labeled by using anti-insulin IgG. The labeled areas correspond to the water washes and the positive control (1 ng insulin). A spectrophotometric assay for β-galactosidase (23) indicated that no more than 4% of cells lyse during this procedure.

```
                       182                  4
.....AspThrThrMetProAlaGlyGlyGlyGlyGlyGlyGlnHisLeuCysGlyProHisLeuValGluAlaLeuTyr.....
.....GACACCACGATGCCTGCAGGGGGGGGGGGGGGGGGGACAGCACCTTTGTGGTCCTCACCTGGTGGAGGCTCTGTAC.....
.....CTGTGGTGCTACGGACGTCCCCCCCCCCCCCCCCCTGTCGTGGAAACACCAGGAGTGGACCACCTCCGAGACATG.....
                  Pst                                AvaII
                  <-------------------A--------------

                   86                                          183
.....LeuGluAsnTyrCysAsn                                        AlaMetAlaThrThrLeu.....
.....CTGGAGAACTACTGCAA[CTGA]GTCCACCACTCCCCGCCCACCCCTCTCCCCCTGCAGCAATGGCAACAACGTTG.....
.....GACCTCTTGATGACGTTGACTCAGGTGGTGAGGGGCGGGTGGGGAGAGGGGGACGTCGTTACCGTTGTTGCAAC.....
                       Hinf                             Pst
                          ---------------------B-------------------------------->
```

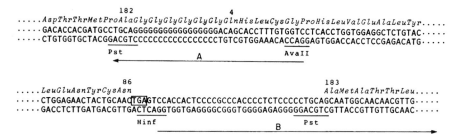

FIG. 5. Partial DNA sequence of the insertion in clone pI47. Clone pI47 DNA was digested with *Hinf* and two fragments, H1 and H2 (ca. 1700 and 280 base pairs long, respectively) were isolated. H1 contains the amino-terminal portion of the penicillinase gene and the bulk of the cDNA insert. H1 was digested with *Ava* II, end labeled, and digested again with *Pst*. A fragment 39 nucleotides long (fragment A, arrow) was isolated and sequenced. Fragment H2 was end labeled and digested with *Alu* I (which cuts at the region corresponding to amino acid 200 of penicillinase). A fragment 88 base pairs long (fragment B, arrow) was isolated and sequenced. The termination sequence TGA is boxed.

We thank David Baltimore, Philip Sharp, and Salvador Luria for the use of the Massachusetts Institute of Technology P3 laboratory. We thank Macy Koehler for help with the figures; Fotis Kafatos for use of facilities; Philip Sharp, Al Bothwell, Shirley Tilghman, Doug Hanahan, and Richard Firtel for discussions. W.G. is an American Cancer Society Professor of Molecular Biology. W.L.C. is an Established Investigator of the American Diabetes Association. This work was supported by National Institutes of Health Grants AM 21240 and GM 09541-17 to W.G. and AM 15398 to W.L.C.

1. Chan, S. J., Keim, P. & Steiner, D. F. (1976) *Proc. Natl. Acad. Sci. USA* **73**, 1964–1968.
2. Chan, S. J. & Steiner, D. F. (1977) *Trends Biochem. Sci.* **2**, 254–256.
3. Blobel, G. & Dobberstein, B. (1975) *J. Cell Biol.* **67**, 835–851.
4. Blobel, G. & Dobberstein, B. (1975) *J. Cell Biol.* **67**, 852–862.
5. Steiner, D. F., Kemmler, W., Tager, H. S. & Peterson, J. D. (1974) *Fed. Proc. Fed. Am. Soc. Exp. Biol.* **33**, 2105–2115.
6. Smith, L. F. (1966) *Am. J. Med.* **40**, 662–666.
7. Clark, J. L. & Steiner, D. F. (1969) *Proc. Natl. Acad. Sci. USA* **62**, 278–285.
8. Markussen, J. & Sundby, F. (1972) *Eur. J. Biochem.* **25**, 153–162.
9. Ullrich, A., Shine, J., Chirgwin, J., Pictet, R., Tischer, E., Rutter, W. J. & Goodman, H. M. (1977) *Science* **196**, 1313–1319.
10. Chick, W. L., Warren, S., Chute, R. N., Like, A. A., Lauris, V. & Kitchen, K. C. (1977) *Proc. Natl. Acad. Sci. USA* **74**, 628–632.
11. Bolivar, F., Rodriguez, R. L., Greene, P. J., Betlach, M. C., Heyneker, H. L., Boyer, H. W., Crossa, J. H. & Falkow, S. (1977) *Gene* **2**, 95–113.
12. Sutcliffe, J. G. (1978) *Proc. Natl. Acad. Sci. USA* **75**, 3737–3741.
13. Ambler, R. P. & Scott, G. K. (1978) *Proc. Natl. Acad. Sci. USA* **75**, 3732–3736.
14. Boyer, H. W. & Rouland-Dussoix, D. (1969) *J. Mol. Biol.* **41**, 459–472.
15. Curtiss, R., III, Pereira, D. A., Hsu, J. C., Hull, S. C., Clarke, J. E., Maturin, L. J., Sr., Goldschmidt, R., Moody, R., Inoue, M. & Alexander, L. (1977) in *Recombinant Molecules: Impact on Science and Society. Proceedings of the 10th Miles International Symposium*, eds. Beers, R. F., Jr., & Bassett, E. G. (Raven, New York), pp. 45–56.
16. Clewell, D. B. (1972) *J. Bacteriol.* **110**, 667–676.
17. Palmiter, R. (1974) *Biochemistry* **13**, 3603–3615.
18. Aviv, H. & Leder, P. (1972) *Proc. Natl. Acad. Sci. USA* **69**, 1408–1412.
19. Efstratiadis, A., Kafatos, F. C., Maxam, A. M. & Maniatis, T. (1976) *Cell* **7**, 279–288.
20. Roychoudhury, R., Jay, E. & Wu, R. (1976) *Nucleic Acid Res.* **3**, 101–116.
21. Gilbert, W. & Maxam, A. M. (1973) *Proc. Natl. Acad. Sci. USA* **70**, 3581–3584.
22. Enea, V., Vovis, G. F. & Zinder, N. D. (1975) *J. Mol. Biol.* **96**, 495–509.
23. Miller, J. M. (1972) *Experiments in Molecular Genetics* (Cold Spring Harbor Laboratory, Cold Spring Harbor, New York), pp. 431–435.
24. Grunstein, M. & Hogness, D. S. (1975) *Proc. Natl. Acad. Sci. USA* **72**, 3961–3965.
25. Paterson, B. M., Roberts, B. E. & Kuff, E. L. (1977) *Proc. Natl. Acad. Sci. USA* **74**, 4370–4374.
26. Roberts, B. E. & Paterson, B. M. (1973) *Proc. Natl. Acad. Sci. USA* **70**, 2330–2334.
27. Lomedico, P. T. & Saunders, G. F. (1976) *Nucleic Acids Res.* **3**, 381–391.
28. Broome, S. & Gilbert, W. (1978) *Proc. Natl. Acad. Sci. USA* **75**, 2746–2749.
29. Makula, D. R., Vichnuk, D., Wright, P. H., Sussman, K. E. & Yu, P. L. (1969) *Diabetes* **18**, 660–689.
30. Maxam, A. M. & Gilbert, W. (1977) *Proc. Natl. Acad. Sci. USA* **74**, 560–564.
31. Boyer, H. W., Betlach, M., Bolivar, F., Rodriguez, R. L., Heyneker, H. L., Shine, J. & Goodman, H. M. (1977) in *Recombinant Molecules: Impact on Science and Society. Proceedings of the 10th Miles International Symposium*, eds. Beers, R. F., Jr. & Bassett, E. G. (Raven, New York), pp. 9–20.
32. Casey, J. & Davidson, N. (1977) *Nucleic Acids Res.* **4**, 1539–1552.
33. Humbel, R. E., Bosshard, H. R. & Zahn, H. (1972) in *Handbook of Physiology, Section 7 (Endocrinology)*, eds. Steiner, D. F. & Freinkel, N., (American Physiological Society, Washington, DC), Vol. 1, pp. 111–132.
34. Neu, H. C. & Heppel, L. A. (1965) *J. Biol. Chem.* **240**, 3685–3692.
35. Tilghman, S. M., Tiemeier, D. C., Seidman, J. G., Peterlin, B. M., Sullivan, M., Maizel, J. V. & Leder, P. (1978) *Proc. Natl. Acad. Sci. USA* **75**, 725–729.
36. Jeffreys, A. J. & Flavell, R. A. (1977) *Cell* **12**, 1097–1108.
37. Breathnach, R., Mandel, J. L. & Chambon, P. (1977) *Nature* **270**, 314–319.
38. Gilbert, W. (1978) *Nature* **271**, 501.

Section VI:
A Variety of Topics

Molecular basis of base substitution hotspots in *Escherichia coli*

Christine Coulondre & Jeffrey H. Miller
Département de Biologie Moléculaire, Université de Genève, Geneva, Switzerland

Philip J. Farabaugh & Walter Gilbert
Biological Laboratories, Harvard University, Cambridge, Massachusetts 02138

In the lacI *gene of* Escherichia coli *spontaneous base substitution hotspots occur at 5-methylcytosine residues. The hotspots disappear when the respective cytosines are not methylated. We suggest that the hotspots may result from the spontaneous deamination of 5-methylcytosine to thymine, which is not excised by the enzyme DNA-uracil glycosidase.*

IN 1961 Benzer demonstrated that all sites on DNA are not equally mutable, and termed highly mutable sites 'hotspots'[1]. We have studied base substitution mutations in the *lacI* gene of *E. coli*. Because deletions, insertions, and small frameshifts constitute ~90% of the mutations arising spontaneously in the *I* gene[2], we selected directly for base substitutions by focusing on nonsense mutations[3]. Over 80 of the characterised nonsense sites are correlated with specific codons[4,5]; thus in each of these cases the conversion to UAG or UAA occurs by a known base substitution. This system has been used to analyse the specificities of base substitutions detected spontaneously and after induction by various mutagens[3].

Several striking hotspots appear in this collection of mutations, particularly among spontaneous base substitutions, and among those generated by 2-aminopurine (2-AP) and ultraviolet light. The elucidation of the full nucleotide sequence of the *lacI* gene[6,7] now allows the examination of the DNA sequence surrounding each site of mutation. It is clear that 5-methylcytosine residues are associated with high mutability for the G:C→A:T change, and we describe here several experiments which demonstrate the direct involvement of 5-methylcytosine in causing base substitution hotspots.

Spontaneous hotspots

Figures 1 and 2 depict the nonsense mutations which arise through a single base change from wild-type, listing the sites according to the type of base substitution involved. In order to

Table 1 Spontaneous amber and ochre mutations

Site	No. of occurrences	Nucleotide sequence 5′ 3′
*A*5	0	A T *C* A G A
*A*6	51	A C *C** A G G
*A*9	8	A A *C* A G T
*A*15	37	A C *C** A G G
*A*16	9	A C *C* A G A
*A*19	11	A C *C* A G C
*A*21	8	G C T *G* G C
*A*23	8	T T *C* A G C
*A*24	10	A C T *G* G A
*A*26	6	A T *C* A G A
*A*31	1	A A *C* A G G
*A*33	5	C T *C* A G G
*A*34	12	G C *C** A G G
*A*35	1	A T *C* A G C
*O*9	2	A A *C* A A C
*O*10	7	C A *C* A A C
*O*11	2	C G *C* A A A
*O*13	1	A T *C* A A C
*O*17	3	C G *C* A A C
*O*21	4	A G *C* A A A
*O*24	0	C G *C* A A T
*O*27	0	T T *C* A A C
*O*28	2	A A *C* A A A
*O*29	5	T G *C* A A A
*O*34	2	G G *C* A A A
*O*35	4	T G *C* A A C

The nucleotide sequence[5] surrounding each base (italic print) that generates a nonsense codon by a transition is shown. The anti-sense strand is depicted. Asterisks (*) indicate 5-methylcytosine. The number of occurrences refer to the amber and ochre mutations found spontaneously in a collection of 222 amber I^- and 76 ochre I^- mutations. The characterisation of these mutations has been described previously[2,3,12]. Only the transition sites are shown here. Table 2 lists the transversions found among the amber and ochre mutations from this collection.

0028–0836/78/0274—0775$01.00

776 Nature Vol. 274 24 August 1978

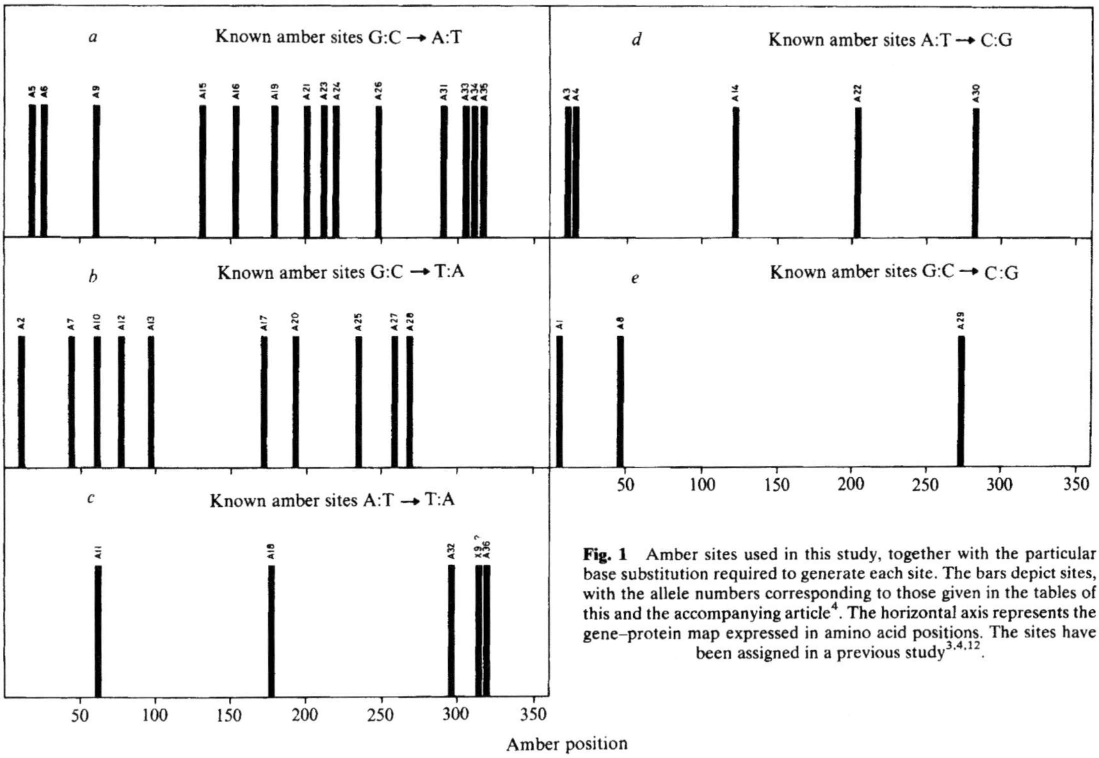

Fig. 1 Amber sites used in this study, together with the particular base substitution required to generate each site. The bars depict sites, with the allele numbers corresponding to those given in the tables of this and the accompanying article[4]. The horizontal axis represents the gene–protein map expressed in amino acid positions. The sites have been assigned in a previous study[3,4,12].

determine the frequency of mutations at each site, we picked independent occurrences of *I* mutations, without any selection bias, and assigned them to one of the known sites. Figure 3 shows the relative frequencies of spontaneous amber and ochre mutations. Although amber and ochre mutations arise at similar frequencies, more amber mutations (222) were analysed than ochre sites (76), so we have magnified the ochre scale by a factor of three to allow a direct comparison with the amber bar heights. The amber spectrum is dominated by two hotspots, both involving a G:C→A:T transition. Out of approximately 70 single base substitutions that generate nonsense codons, 26 of which involve the G:C→A:T transition, mutations at these two sites are favoured by nearly a 10:1 ratio (over the average frequencies at the remaining transition sites).

Table 1 lists the sequences surrounding each G:C→A:T transition site. Both hotspots occur at the second C (cytosine) of a CCAGG sequence. One other CCAGG sequence appears

Fig. 2 Ochre sites resulting from different base substitutions. (See legend to Fig. 1).

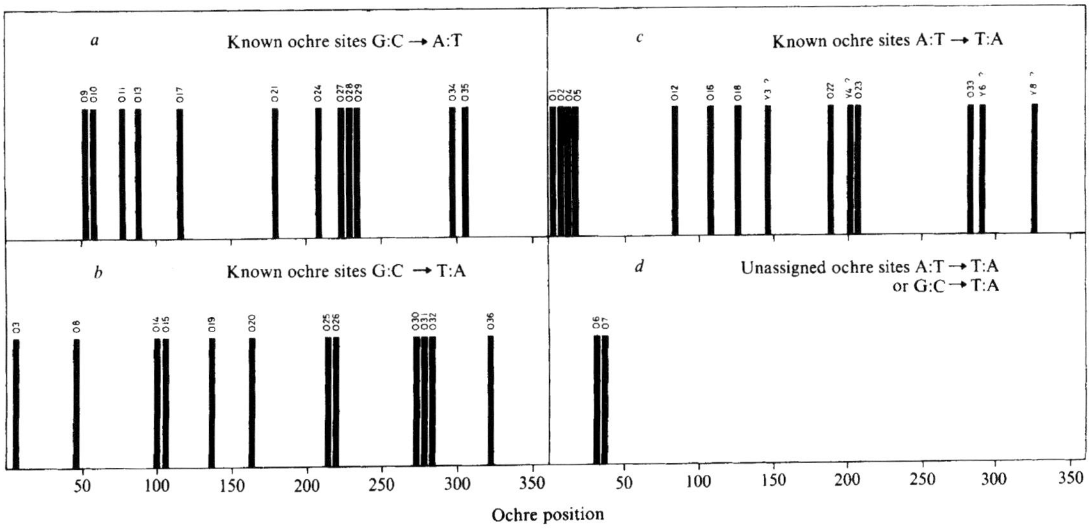

Nature Vol. 274 24 August 1978 **777**

Fig. 3 The distribution of 222 spontaneous amber mutations (*a*, *b*, *c*) and 76 spontaneous ochre mutations (*d*, *e*, *f*). The mutations are of independent origin and have been characterised previously[2,3,12]. Although the ochre scale has been magnified to enable a direct comparison of amber and ochre mutations, hotspots can be detected with a greater degree of assurance by considering the amber and ochre distributions as separate groups. The arrows in *a* and *d* indicate transition sites not found in this collection. (Only half of the spontaneous collection was assayed for the missing site, *A*5.) Transition sites not represented in this collection can be recognised by consulting Figs 1 and 2.

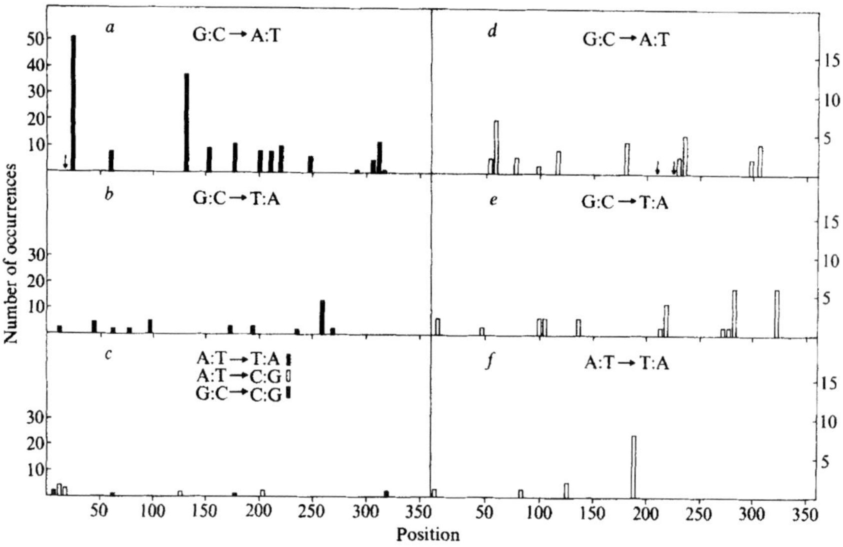

among the remaining 24 transition sites, but mutations here are not as frequent. Because *E. coli* K12 carries the chromosomal *mec* gene[8], this sequence, the normal cutting site for the RII restriction endonuclease[9], should be methylated at the second C from the 5′-end of each strand, generating 5-methylcytosine (C-5meC-A- G -G-) / (G- G -T-5meC-C-). This suggests that 5-methylcytosine may be the cause of the spontaneous hotspots. The presence of 5-methylcytosine at these positions in the *I* gene DNA was verified directly by chemical DNA sequencing[7], as this substituted residue leaves a characteristic gap in the sequence, with a G (guanine) residue appearing at the corresponding position on the complimentary strand.

Creation of a new CCAGG sequence

To test whether the appearance of hotspots at two of the methylated bases was fortuitous, we introduced a new CCAGG sequence into the gene. The amber mutation *A*28 is derived from a codon normally specifying serine 269 (refs 4, 5) which lies in the sequence CTCGG, changed to CTAGG in *A*28. Reversion to a sense codon by a transition would generate either CCAGG or CTGGG, specifying glutamine or tryptophan, respectively. Tryptophan at this site should not produce an active repressor, as leucine and tyrosine, introduced in place of serine by nonsense suppressors produce i^- proteins[4]; while the replacement of serine by glutamine results in a temperature sensitive repressor[4]. We used 2-AP to induce i^+ revertants. As all these revertants displayed temperature sensitive repression, they most probably represented the conversion of the amber UAG triplet to a CAG codon. We proved the construction by crossing the mutation onto a plasmid and sequencing the relevant region. We found a new 5-methylcytosine in the CCAGG sequence that replaced the original CTCGG. Figure 4 summarises the steps involved in this construction.

Amber mutations arising from the new CCAGG sequence

We collected spontaneous I^- amber mutations using an episome carrying the new CCAGG sequence at position 269, in our standard strain background[3]. Figure 5 presents the new distribution of amber mutations. In addition to the two original hotspots, a third hotspot appears at the new CCAGG sequence.

Table 2 Amber mutations detected in *E. coli* B and K12

Base substitution	Site	Number of occurrences: *E. coli* B	*E. coli* K12: *a*	*b*	*c*
G:C→A:T	*A*5	3	0	0	0
	*A*6	4	51	13	64
	*A*9	3	8	2	10
	*A*15	3	37	15	52
	*A*16	1	9	0	9
	*A*19	5	11	0	11
	*A*21	6	8	0	8
	*A*23	0	8	0	8
	*A*24	4	10	3	13
	*A*26	0	6	0	6
	*A*31	1	1	1	2
	*A*33	2	5	0	5
	*A*34	3	12	5	17
	*A*35	3	1	0	1
	*A*28*	—	—	11	—
G:C→T:A	*A*2	0	2	0	2
	*A*7	2	4	0	4
	*A*10	0	2	0	2
	*A*12	0	2	0	2
	*A*13	2	4	0	4
	*A*17	2	3	0	3
	*A*20	1	3	2	5
	*A*25	0	2	0	2
	*A*27	5	13	0	13
	*A*28	2	2	—	—
A:T→T:A	*A*11	0	1	0	1
	*A*18	4	1	0	1
	*A*32	0	0	0	0
	*X*9	0	0	0	0
	*A*36	3	2	0	2
A:T→C:G	*A*3	0	4	0	4
	*A*4	0	3	0	3
	*A*14	1	2	0	2
	*A*22	0	2	0	2
	*A*30	0	0	1	1
G:C→C:G	*A*1	0	2	0	2
	*A*8	1	0	0	0
	*A*29	2	0	0	0

The amber mutations found spontaneously in *E. coli* B are compared with those found in *E. coli* K12. *a*, The collection in K12 using the wild-type episome from strain GM1 (ref. 12); *b*, that for the episome carrying the additional CCAGG sequence; *c*, combined distribution showing both of these collections together (except for site *A*28, which is derived by a transition in *b* but by a transversion in *a*).

Wild-type	5' CTCGG 3'
	↓
A28	CTAGG
	2AP ↓
Revertant	CC*AGG
	↓
A28	CTAGG

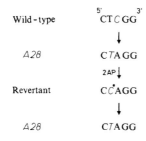

Fig. 4 The nucleotide sequence surrounding the codon at position 269. The amber mutation *A*28 is generated by a C→A transversion from this sequence[3,4]. The resulting CTAGG sequence can be converted by 2-AP to a CCAGG sequence, which is methylated at the second cytosine (marked by asterisk; see text). This restores wild-type activity at 37 °C and allows the generation of *A*28 by a C→T transition at the 5-methylcytosine.

Strains lacking methylation activity

E. coli B strains do not methylate the CCAGG sequence[10]. We crossed the *lacproB* deletion, XIII, into an *E. coli* B derivative in such a manner as to avoid transferring the region of the chromosome carrying the *mec* locus. We checked this by transforming this strain with a plasmid carrying the wild-type *I* gene[11] and sequencing one of the CCAGG sites in the *I* gene. It had unmethylated cytosines. After introducing the *lacpro* episome from strain GM1 (ref. 12) into this *E. coli* B strain, we derived and analysed amber mutations on the episome. Table 2 compares these results with those from the K12 strain. The two hotspots disappear when the *E. coli* B derivative is used. (This is particularly evident when the K12 results obtained with the altered sequence at position 269 are considered, see column *b*, since a similar sample size was analysed). Thus, the spontaneous hotspots are the result of methylation of cytosine at the 5 position.

2-Aminopurine induced mutations

The base analogue 2-AP specifically induces transitions at frequencies more than 100-fold above the spontaneous background[12–14]. Figure 6 shows the distribution of over 1,000 ochre and amber mutations induced by 2-AP (Fig. 6*a* and *b*, respectively), all arising by the G:C→A:T transition. There are significant hotspots at both amber and ochre sites. The three major amber hotspots are at the CCAGG sequences discussed above, and when we rederived the 2-AP spectrum

Fig. 5 The distribution of 50 spontaneous amber mutations derived from a G:C→A:T transition. This collection was obtained using a derivative carrying an *F'lacpro* episome with an additional CCAGG sequence in the *lacI* gene (see text). The asterisk marks the position of amber mutations derived from this sequence. In addition to the mutations shown above, three amber mutations generated by transversions were also found (see Table 2).

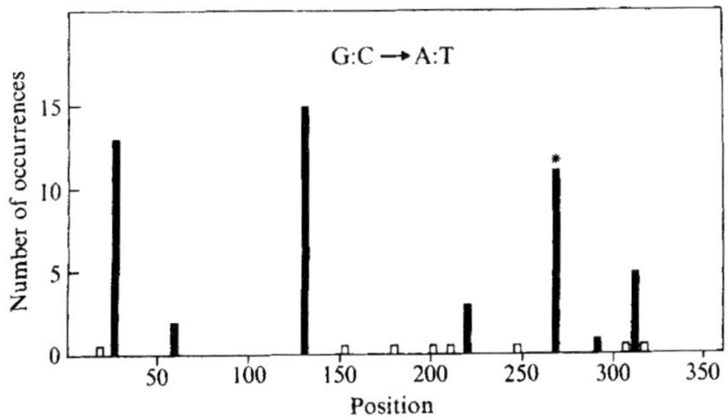

using the episome carrying the additional CCAGG sequence at 269, a fourth hotspot appeared at the new position (Fig. 6*c*). Therefore, 2-AP mutagenesis also favours the methylation CCAGG sequence. However, in the *E. coli* B derivative which lacks the *mec* system that methylates the CCAGG stretches, the 2-AP amber hotspots are reduced but not eliminated. In *E. coli.* K12 the two major hotspots are about 27 times more frequent than the average of the other sites; in *E. coli* B this ratio falls by a factor of 4 to 6.7:1. Both the methylation and the sequence seem to affect 2-AP mutability.

None of the ochre sites involves wild-type codons which are part of a methylation sequence, and yet there is still a wide

Table 3 Nucleotide sequence surrounding each mutational site

Site	Sequence (5'–3')	No. of occurrences 2-AP	EMS	NG	UV	NQO
*A*6	G A A C C* A G G C C T T G G T C* C G	270	14	20	15	61
*A*15	T G A C C* A G G A A C T G G T C* C T	207	7	4	7	54
*A*34	G G G C C* A G G C C C C G G T C* C G	66	15	12	1	55
*A*9	C A A A C A G T C G T T T G T C A G	8	30	34	1	61
*A*31	C A A A C A G G A G T T T G T C C T	9	19	18	2	20
*A*16	T G A C C A G A C A C T G G T C T G	11	14	21	20	29
*A*19	T C A C C A G C A A G T G G T C G T	23	30	37	16	13
*A*21	A T G C C A G C C T A C G G T C G G	15	12	22	5	16
*A*24	A C T C C A G T C T G A G G T C A G	13	39	34	9	51
*A*5	T T A T C A G A C A A T A G T C T G	3	17	13	19	23
*A*23	A A T T C A G C C T T A A G T C G G	2	30	32	80	24
*A*26	C G C T C A G A T G C G A G T C T A	5	20	12	2	36
*A*33	C T C T C A G G G G A G A G T C C C	7	38	24	36	33
*A*35	C A A T C A G C T G T T A G T C G A	1	27	9	6	25
*O*11	G T C G C A A A T C A G C G T T T A	41	56	28	2	11
*O*17	C G C G C A A C G G C G C G T T G C	44	27	26	3	28
*O*21	C C A G C A A A T G G T C G T T T'A	26	57	15	2	43
*O*29	C A T G C A A A T G T A C G T T T A	77	36	14	4	83
*O*34	G G G G C A A A C C C C C G T T T G	112	46	19	2	47
*O*35	G C T G C A A C T C G A C G T T G A	81	49	22	3	48
*O*9	G G C A C A A C A C C G T G T T G T	3	23	18	3	15
*O*10	A C A A C A A C T T G T T G T T G A	11	17	10	4	31
*O*28	T C A A C A A A C A G T T G T T T G	11	13	12	4	24
*O*13	C A A T C A A A T G C T A G T T G A	5	13	9	6	37
*O*24	C A A T C A A A T G T T A G T T T A	2	16	16	61	22
*O*27	T T T T C A A C A A A A A G T T G T	5	15	20	86	10

Nature Vol. 274 24 August 1978 779

variation in the frequency of the G:C→A:T transition at these points. Table 3 shows the DNA sequence surrounding each mutational site. As described below, the C in a GCA sequence is more highly mutable with 2-AP.

Effect of neighbouring nucleotide sequence on mutagen-induced mutation rates

In Table 3 we present an analysis of the nucleotide sequence surrounding each mutational site which generates an amber (UAG) or ochre (UAA) codon by a G:C→A:T transition. The base involved in the transition is italicised. The columns give the relative frequencies of occurrence of each amber or ochre site in a collection of mutagen induced I^- mutations. One amber and one ochre mutation for each mutagenised culture was analysed[1,2], and therefore each mutation is of independent origin. A correction factor is necessary if amber and ochre mutations are to be considered together, as these two types of mutation did not always arise at identical frequencies, and different sample sizes were analysed. To compare 2-AP-induced ochres with the 2-AP-induced amber mutations, the ochre values should be multiplied by 0.9. For 4-nitroquinoline-1-oxide (NQO), ultraviolet light, ethyl methanesulphonate (EMS) and N′-methyl-*N*′-nitro-*N*-nitroso-guanidine (NG) ochres the respective factors are 0.6, 0.7, 2.3 and 1.9.

Even without placing too much emphasis on normalising amber and ochre frequencies, some patterns seem to emerge from these data. 2-aminopurine clearly prefers methylated cytosines for the G:C→A:T transition (see ref. 4). Otherwise, the different frequencies of occurrence do follow the nature of the base on the 5′ side of the cytosine residue, the tendency being G>C>A>T.

Fig. 6 Distributions of 418 ochre (*a*) and 640 amber mutations (*b*) induced by 2-AP. These mutations were characterised in a previous study[12]. Fifty-two amber mutations were derived by 2-AP using the episome carrying an additional CCAGG sequence surrounding position 269 (see text). *c*, The distribution of these mutations. The asterisk (*) marks the position of site *A*28 which is derived by a C→T transition at the new CCAGG sequence. 2-AP-induced amber mutations derived in *E. coli* B are represented in *d*. The open bars in *c* and *d* refer to amber sites derived by C→T transitions which are not represented in these particular collections.

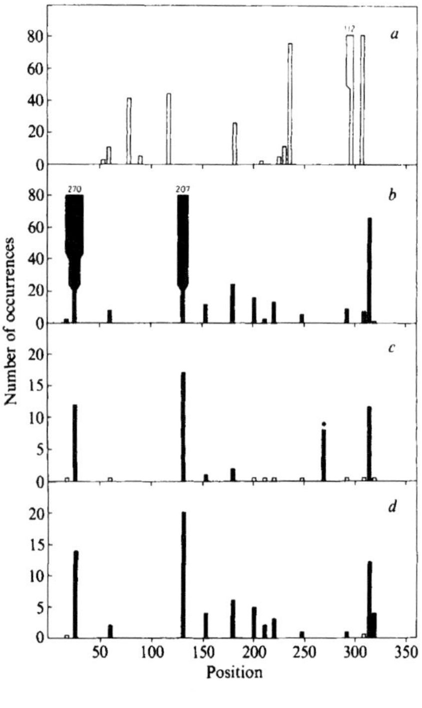

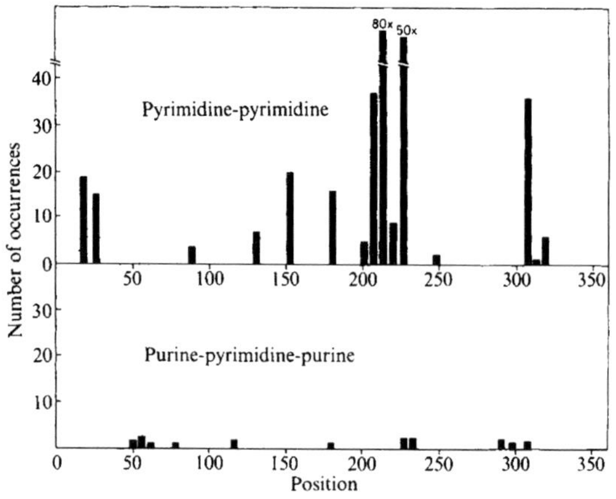

Fig. 7 Correlation of UV-induced mutations with the surrounding nucleotide sequence (see text).

Figure 7 shows a strong correlation between being well induced by ultraviolet and being part of a pyrimidine–pyrimidine sequence (on one strand or the other), and a strong correlation for being poorly induced and not being part of such a sequence. All of the amber sites are in the pyrimidine–pyrimidine category except for *A*9 and *A*31, whereas only the ochre sites *O*13, *O*24, and *O*27 are part of a pyrimidine–pyrimidine sequence.

The EMS, NG and NQO data do not seem to allow any general rules to be deduced at this stage.

Hotspots involve methylated cytosines

By using a set of characterised nonsense mutations and the full nucleotide sequence of the *lacI* gene, we have demonstrated that base substitution hotspots occur at 5-methylcytosines. Two of the three methylated cytosines, which occur among 26 cytosines involved in the creation of amber and ochre codons, generate transitions at a 10-fold higher rate than the average. A new methylated cytosine introduced by creating an additional CCAGG sequence also results in a hotspot for the G:C→AT transition. The elimination of methylation at these sites, demonstrated in an *E. coli* B derivative lacking the methylating enzyme, results in the disappearance of the transition hotspots.

What mechanism could account for these results? 5-methylcytosine does not cause any direct mispairing[15], nor do *in vitro* measurements indicate a significant difference between cytosine and 5-methylcytosine in the tautomeric equilibria[16]. An attractive explanation is the deamination of cytosine to uracil, which occurs spontaneously at a high rate *in vitro* and presumably *in vivo*[17]. Uracil residues in DNA are rapidly excised by the enzyme uracil-DNA glycosidase[18]. Any uracil residues which are not removed would eventually result in G:C→A:T transitions, unless otherwise repaired. (Drake and coworkers have shown that spontaneous transition mutations occur in T4 DNA on extended storage in the cold, as a result of deamination of cytosine[17]. T4 DNA contains glucosylated 5-hydroxymethylcytosine instead of unsubstituted cytosine; after deamination, the resulting 5-hydroxymethyluracil residues are not subject to the action of uracil-DNA glycosidase, probably accounting for the high rate of transition mutations.) Although most spontaneous deaminations of cytosine to uracil would be efficiently repaired by uracil-DNA glycosidase, those occurring at 5-methylcytosine residues would yield thymine (5-methyluracil), which will not be released from DNA by this enzyme.

780 *Nature Vol. 274 24 August 1978*

The G:T base pair would still be subject to various types of mismatch repair, as would any G:U base pairs which escape the action of the glycosidase. The greatly increased number of unexcised G:T base pairs which originated from G:5meC would then appear as a hotspot relative to other transition sites. This argues that deamination is a major cause of spontaneous transitions in *E. coli*, at least in the direction G:C→A:T. This model predicts that these hotspots should disappear in mutant strains lacking uracil-DNA glycosidase, since deaminations at all cytosines should then be immune to this excision repair. Such mutants, termed Ung^-, have been described[20]. Hotspots among nitrous acid-induced amber mutations should also occur at 5-methylcytosines, since this mutagen preferentially stimulates deamination of cytosine[15].

Thus T is used in place of U in DNA to lower the mutation rate, the role of the methyl group being to distinguish one natural base from the deamination product of the other[18]. Possibly the structures of A and G are also selected to allow enzymatic repair of deamination.

The base analogue 2-AP shows a remarkable selectivity for certain sites. All three CCAGG sequences are hotspots for amber mutations, and a fourth CCAGG sequence introduced into the gene results in the appearance of a new hotspot. However, the elimination of methylation at these sites does not entirely remove the hotspots, although it does reduce their relative intensity by a factor of 4. One possibility is that a residual methylation exists in these strains, too low to detect by our methods. Alternatively, a different enzyme in the cell could be operating on these sites, or some subtle aspect of the base sequence may be responsible for the high mutability in the presence of 2-AP. The same situation is seen for the 2-AP-induced ochre mutations, where certain unmethylated sequences clearly result in significant hotspots. The elucidation of the mechanism that produces this phenomenon requires additional experiments. Other mutagens, such as ultraviolet light, NQO, NG, and EMS also result in considerable differences in mutation rate from site to site. The hotspots favoured by these mutagens, particularly ultraviolet light, are different from those described above. At least for 2-AP some rules are evident. Higher mutation rates occur when G is the adjacent base, in accordance with both predictions and *in vitro* findings of M. Bessman and coworkers (personal communication).

We thank S. Brenner, S. Hattman, W. Arber, J. D. Watson, M. Radman, T. Lindahl and A. Maxam for helpful discussions, and M. Hofer for technical assistance. This work was supported by a grant from the Swiss National Fund (3.179.77) to J.H.M. and an NIGMS grant to W.G. W.G. is an American Cancer Society Professor of Molecular Biology.

Note added in proof: Experiments with this system (B. K. Duncan and J. H. Miller, unpublished) show that G:C→A:T transitions occur at a high rate in an Ung^- strain, and that the 5-methylcytosine residues are no longer hotspots relative to the other cytosines.

Received 27 February; accepted 26 June, 1978.

1. Benzer, S. *Proc. natn. Acad. Sci. U.S.A.* **47,** 403–416 (1961).
2. Farabaugh, P. J., Schmeissner, U., Hofer, M. & Miller, J. H. *J. molec. Biol.* (in the press).
3. Coulondre, C. & Miller, J. H. *J. molec. Biol.* **117,** 577–606 (1977).
4. Coulondre, C. & Miller, J. H. *J. molec. Biol.* **117,** 525–575 (1977).
5. Miller, J. H., Coulondre, C. & Farabaugh, P. J. *Nature* **274,** 770–775 (1978).
6. Steege, D. A. *Proc. natn. Acad. Sci. U.S.A.* **74,** 4163–4167 (1977).
7. Farabaugh, P. J. *Nature* **274,** 765–769 (1978).
8. Gold, M., Gefter, R., Hausmann, R. & Hurwitz, J. in *Macromolecular Metabolism* 5–28 (Little Brown & Co., Boston, 1966).
9. Boyer, H. W., Chow, L. T., Dugaiczyk, A., Hedgpeth, J. & Goodman, H. M. *Nature new biol.* **244,** 40–48 (1973).
10. Lederberg, S. *J. molec. Biol.* **17,** 293–297 (1966).
11. Calos, M. *Nature* **274,** 762–765 (1978).
12. Miller, J. H., Ganem, D., Lu, P. & Schmitz, A. *J. molec. Biol.* **109,** 245–302 (1977).
13. Yanofsky, C., Ito, J. & Horn, V. *Cold Spring Harb. Symp. quant. Biol.* **31,** 151–162 (1966).
14. Osborn, M., Person, S., Phillips, S. & Funk, F. *J. molec. Biol.* **26,** 437–447 (1967).
15. Drake, J. W. & Baltz, R. H. *A. Rev. Biochem.* **45,** 11–37 (1976).
16. Jencks, W. P. & Regenstein, J. *CRC Handbook of Biochemistry* 5187–5226 (1976).
17. Shapiro, R. & Klein, R. S. *Biochemistry* **5,** 2358–2362 (1966).
18. Lindahl, T., Ljungquist, S., Siegert, W., Nyberg, B. & Sperens, B. *J. biol. Chem.* **252,** 3286–3294 (1977).
19. Baltz, R. H., Bingham, P. M. & Drake, J. W. *Proc. natn. Acad. Sci. U.S.A.* **73,** 1269–1273 (1976).
20. Duncan, B. K., Rockstroh, P. A. & Warner, H. R. *Fedn Proc.* **35,** 1493 (1977).

The Immune System, vol. 2, pp. 24–32 (Karger, Basel 1981)

Parabiosis as a Model System for Network Interactions

K. Fischer Lindahl[1], *W. Gilbert*, *K. Rajewsky*[1]

Institute for Genetics, University of Cologne, Cologne, FRG

Jerne [6] sees the immune system as a network of idiotopes and anti-idiotopes, reflecting the immunological history of an individual. Networks may thus vary among genetically identical members of an inbred strain. Since such networks represent a steady state of suppressive and stimulatory interactions, any two might be incompatible, and their fusion might perturb the system and render parts non-functional.

Parabiosis might offer a working model for network interactions. By coelomic parabiosis, the peritoneal cavities and the skin of a pair of mice are joined surgically. Serum antibody levels equilibrate within days, and by the end of a week one full blood volume is exchanged every hour between the parabionts [9]. The system thus allows fusion of two complete immune networks.

Materials and Methods

Mice. Female A/J and male BALB/c mice were immunized against ectromelia upon arrival and rested for 2 weeks.

Parabiosis. Mice matched for sex, age and, as far as possible, for weight were anesthetized with tribromoethanol. They were parabiosed as described [11], with the modification that the shoulder blades were tied together with a suture [*J.G. Howard*, personal communication], but the mice were not restricted after the operation. The pairs were kept in

[1] *Kirsten Fischer Lindahl* has been a member of the Basel Institute for Immunology since 1978, and *Klaus Rajewsky* started to collaborate with *Niels Jerne* when *Jerne* was in Frankfurt from 1966 to 1969.

individual cages with filter bonnets or in a laminar flow hood and given antibiotics in the drinking water for at least a week. They were bled from the retroorbital plexus or the tail vein.

Strep A Responses. Mice received i.p. injections of group A streptococcal vaccine thrice a week for several weeks [3]. Their sera were analyzed by isoelectric focusing on 5% polyacrylamide gels, containing 2 *M* urea and LKB Ampholines, pH range 5–9. The gels were stained with ^{131}I-labeled tyraminated group A carbohydrate, fixed and autoradiographed [4]. Sera were also titrated in a plastic plate assay [10].

Anti-NIP Responses. Mice were primed i.p. with an alum-precipitated NIP-carrier conjugate mixed with *Bordetella pertussis* as adjuvant. Anti-NIP sera were titrated in a modified Farr assay at 10^{-8} *M* NIP-caproic acid, and titers are expressed as molar serum binding capacities × 10^8, calculated from the dilution(s) closest to binding 20% of the antigen [8].

Results

Anti-Strep A Responses in BALB/c Mice. On the network hypothesis a response of restricted heterogeneity represents a steady state with some clones expanded and the rest suppressed. The hyperimmune response of BALB/c mice to Streptococcus A vaccine is of restricted heterogeneity, as seen by isoelectric focusing (IEF). Many mice display unique IEF patterns, constant over weeks after immunization and recalled by boosting [3]. If different IEF patterns reflect different idiotypic connectivities, their stability in single mice should allow us to detect changes due to parabiosis.

As a control, we established that parabiosed mice can mount an immune response to Strep A. Normal mice were parabiosed, and hyperimmunization was started 3–5 days later. At day 40 two pairs were separated; the union was complete and vascularized at this time. Figure 1a shows the IEF patterns of the parabiosed mice and of a sham-operated, immunized control. The patterns and titers were comparable to each other and to those of hyperimmunized single mice (fig. 1c, d).

A second experiment tested whether mice parabiosed for 6 weeks and then separated would respond more uniformly than randomly chosen mice. Three pairs of parabiosed mice were separated and hyperimmunization was started 4 days later. IEF spectra were compared on days 25 and 32 (fig. 1b). Though not identical, the IEF patterns of the paired mice were very similar, while the two pairs differed from each other. Comparing these spectra to those in figure 1a, it is clear that spectra may overlap between randomly chosen pairs of mice (Nos. 1 and 2, Nos. 3 and 4), yet each is distinct.

Figure 1d shows the spectra of hyperimmunized mice which were bled, rested for 4 weeks, boosted and bled again. The 10 mice represent at least seven different spectrotypes. The IEF patterns are stable – only No. 3 seems to lose the most alkaline set of bands, and in No. 2 some bands fade. In none of the mice do completely new bands appear.

The main experiment tested how parabiosis would affect an established antibody response. Mice were hyperimmunized with Strep A, bled and their IEF spectra compared (fig. 1c shows an example). Pairs with distinct (different or similar) patterns were selected and parabiosed, while other mice were sham-operated or parabiosed with normal mice. The mice were bled weekly and boosted with Strep A after the 3rd, 4th and 5th bleeding. The IEF patterns before and after parabiosis were then compared.

The spectra of four hyperimmune-normal pairs are shown in figure 2a. In each case, the pattern established before parabiosis persisted or faded slightly. The pairs were always bled from the non-immunized parabiont; its serum contained the full spectrum of antibodies in concentrations comparable to the hyperimmune partner 7–9 days after the parabiosis. In sham-operated controls (fig. 2b) the bands remained in the same position for 6 weeks after the operation, fading at first and then intensifying with boosting.

A total of 20 pairs of hyperimmune mice was followed for 1–6 weeks after parabiosis. Figure 2c shows spectra from eight of these. The spectrum of antibodies in the parabionts closely matched the spectrum of a mixture of sera taken from each mouse before the operation. Only one of the pairs (No. 22 + 27) showed an obvious change: a completely new, strong set of bands appeared after boosting. Other changes were all minor – fading or disappearance of bands which could usually be recalled by boosting.

Fig. 1. IEF spectra of anti-Strep A antibodies from BALB/c mice. The mice are numbered within each experiment in larger characters below the gel, and the individual bleedings are numbered on the gel in smaller characters. *a* Mice immunized during parabiosis. The pairs were bled 25 (1), 32 (2) and 40 (3) days after the operation. Number 4 is a sham-operated control. Sheep hemoglobin (SH) was used as a marker (pI = ~6.7 and ~7.0). *b* Mice immunized after parabiosis. The pairs 1–2, 3–4, 5–6 were parabionts. Each mouse was bled 25 and 32 days after the separation. Mouse hemoglobin (MH, pI = ~7.0 and ~7.2) shows up in some sera (e.g. 3/32). *c* 15 normal mice primed with Strep A vaccine, boosted a week later with three injections weekly for 2 weeks, and bled 3 days later. *d* 10 normal mice immunized with Strep A as above. After the first bleeding (1), the mice were rested for 4 weeks, boosted for 1 week and bled again 5 days later (2).

Parabiosis of Immune Networks

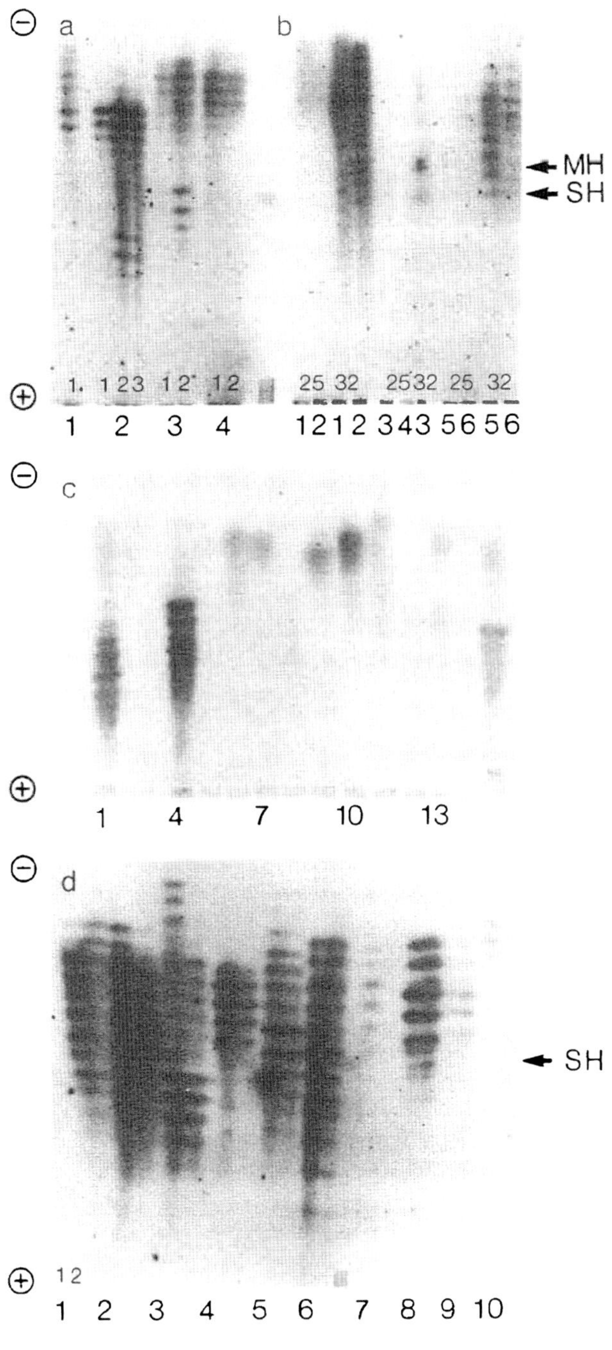

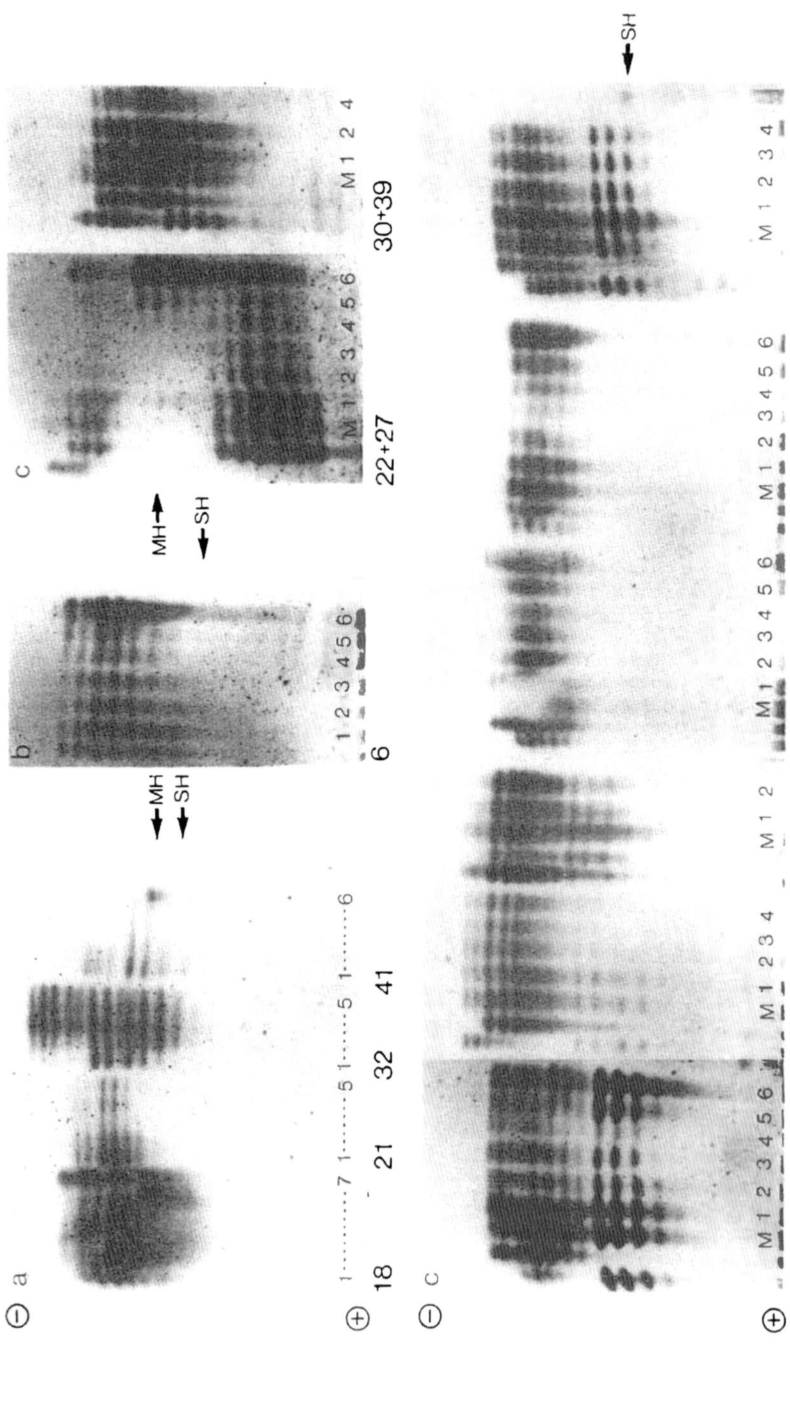
a
b
c
MH
SH
18
21
32
41
6
22+27
30+39
28+43
8+11
4+12
3+9
14+15
5+10

We also studied five pairs of mice immune to LDH_B. Within 2 weeks of parabiosis they maintained a mixture of both spectra: a hint of minor shifts in isoelectric point is of questionable significance.

Adoptive Transfer. Immune lymphocytes transferred into a syngeneic host and challenged with antigen respond far better if the recipient has been irradiated (table I) [1] or is devoid of T lymphocytes [7]. This 'isogeneic barrier' may represent suppression by the immune network of the recipient. The suppression could rest either on general incompatibility of the networks, or it could affect specifically the antigen-primed memory cells.

Since parabiosis could conceivably lower the 'isogeneic barrier', we parabiosed A/J mice for 3–4 weeks, a period considered long enough to establish a common immune network. The pairs were then separated, and one of the partners was immunized with NIP-OA within the next month. 4 weeks later a cell transfer, followed by challenge with NIP-OA, was made from the immune into the non-immune partner, into a recipient from another parabiosed pair, and into normal and irradiated syngeneic controls. Comparable antibody titers were produced by the transferred cells in the matched partner and in the 'third-party' parabiont (table II): these titers were somewhat lower than those of the normal and irradiated recipients, which were almost equal.

Discussion

Parabiosed mice are immunologically competent (fig. 1a), and there is a rapid exchange of antibodies between immune parabionts (fig. 2). Yet, parabiosis had no effect on established immune responses of restricted heterogeneity. Fading of parts of a spectrum was observed also in controls, and

Fig. 2. IEF spectra of anti-Strep A antibodies from BALB/c mice hyperimmunized before parabiosis. Mice are numbered below the gel, bleedings on the gel. *a* Parabiosis of a hyperimmune and a normal mouse. The first two samples were taken (a month apart) from the immune mouse before parabiosis, the rest from the non-immunized partner weekly after parabiosis. Serum 41/6 has a strong pair of mouse hemoglobin bands. *b* A sham-operated control. *c* Eight pairs of hyperimmune mice. The two thin strips on the left are the sera taken before parabiosis (the mouse with the higher number was always put on the right side). The third strip is a 1:1 mixture (M) of the same sera, followed by weekly bleedings (1–6) from alternate sides of the pair. Serum 22+27/1 has a strong set of mouse hemoglobin bands.

Table I. Adoptive immunity in irradiated recipients

Dose of cells transferred (× 10^6)	Normal recipients	Irradiated recipients	Ratio irrad.:normal
1	1 (0.5–1.5)	7 (4–18)	7
3	4 (3–5)	47 (28–60)	11.8
10	29 (13–54)	113 (23–341)	3.9

3 A/J mice were immunized i.p. with 100 μg $NIP_{10}CG$. 4 weeks later their pooled spleen cells were transferred into normal or irradiated (500 rad) syngeneic recipients which were immediately challenged with 0.2 μg $NIP_{10}CG$. The figures give the geometric mean titer and range for groups of 4–6 recipients on day 11. Comparable results but lower titers were obtained with sera taken on day 7.

missing bands could in most cases be recalled by boosting with antigen. Only in a single instance did a new set of bands appear after 5 weeks of parabiosis. While such an event was not observed among our controls, it is certainly not a general effect of parabiosis, nor is it clear that it results from parabiosis at all.

There are two obvious reasons why we may have failed to detect network interactions. Of necessity, we had to choose a stable antibody response of restricted heterogeneity for our analysis. Such responses may well represent dominant clones particularly insensitive to anti-idiotypic regulation. It is equally possible that, even though the antibodies made by different mice had different isoelectric points, they shared the idiotope that was the target for regulation. Precedents are known in the dextran [5], T15 [2] and NP^b systems [*T. Imanishi-Kari*, personal communication].

Although the sample was small, our preliminary experiment suggested that mice previously parabiosed made very similar antibodies (fig. 1b). While this may reflect regulatory interactions, it is more likely due to an exchange of precursor cells during parabiosis. In fact, since a low level of immunization with streptococcal carbohydrate probably occurs all the time, the mice may have been primed already before and during their parabiosis. Use of an antigen not present in the environment might have been more informative.

Table II. Adoptive immunity after parabiosis

Donor	Interval weeks	Ex-parabionts		Controls		Ratio irrad.:normal
		partner	3rd party	normal	irradiated	
1	4	188	64	298 (222–434)	385 (326–426)	1.3
2	4	42	41	68 (47–112)	234 (194–263)	3.4
3	3	101	340	651 (628–675)	493 (384–612)	0.8
4	1	123	–	75 (37–374)	278 (222–348)	3.7
5	1	233	–	315 (259–414)	684 –	2.2
6	1	262	–	359 (179–794)	687 (550–897)	1.9
7	1	10	–	43 (27–66)	93 (68–110)	2.2
8	1	704	362	–	1,339	–

A/J mice were parabiosed for 3–4 weeks. After separation (interval in column 2) 1 mouse from each pair was immunized with 100 μg $NIP_{3,4}OA$. 4 weeks later 1/8 of their spleen (7–11 × 10^6 cells) was transferred to each of 8 recipients, and these were challenged with 1 μg NIP_7OA (donor 8 supplied 34 × 10^6 cells per recipient). The figures give the geometric mean titer and range for 2–4 normal or irradiated (350 rad) recipients on day 12. Comparable results but lower titers were obtained with sera taken on day 7.
– = Not done.

Parabiosis did not improve adoptive transfer of immune lymphocytes between erstwhile partners (table II), the poor acceptance due perhaps to their indifferent health and loss of weight. Unexpectedly, adoptive transfer into normal recipients was rather successful in this experiment, compared with the example in table I. While this might be due to minor differences in the technique, it is equally possible that parabiosis had lowered the sensitivity of the donor lymphocytes to the 'isogeneic barrier'. The problem deserves further investigation.

In sum, the advantage of being able to fuse two complete immune systems, lock, stock and barrel, hardly makes up for the technical difficulties of keeping parabiosed mice alive and kicking. The more so as interest in network regulation today is focused on the identification and characterization of its individual active components.

Acknowledgements. This work was supported by the Deutsche Forschungsgemeinschaft through SFB 74. *K.F.L.* was a recipient of an EMBO long-term fellowship, and *W.G.* was on sabbatical leave from Harvard University. The assistance of *G. von Hesberg* in bleeding mice was essential, as was the help, reagents and advice of *M. Cramer, W. Fastenrath, T. Imanishi-Kari, I. Melchers, M. Reth* and *I. Wilke.*

References

1 Celada, F.: J. exp. Med. *124:* 1–14 (1966).
2 Claflin, J.L.; Cubberley, M.: J. Immun. *121:* 1410–1415 (1978).
3 Cramer, M.; Braun, D.G.: Scand. J. Immunol. *4:* 63–70 (1974).
4 Cramer, M.; Braun, D.G.: J. exp. Med. *139:* 1513–1528 (1974).
5 Hansburg, D.; Perlmutter, R.M.; Briles, D.E.; Davie, J.M.: Eur. J. Immunol. *8:* 352–359 (1978).
6 Jerne, N.K.: Annls Immunol. *125C:* 373–389 (1974).
7 Kobow, U.; Weiler, E.: Eur. J. Immunol. *5:* 628–632 (1975).
8 Mitchison, N.A.: Eur. J. Immunol. *1:* 10–17 (1971).
9 Nisbet, N.W.: Transplant. Rev. *15:* 123–161 (1973).
10 Rajewsky, K.; von Hesberg, G.; Lemke, H.; Hämmerling, G.J.: Annls Immunol. *129C:* 389–400 (1978).
11 Simonsen, M.; Christensen, R.: Acta path. microbiol. scand. *27:* 325–337 (1950).

Kirsten Fischer Lindahl, Basel Institute for Immunology, Postfach, CH–4005 Basel (Switzerland)

Proc. Natl. Acad. Sci. USA
Vol. 82, pp. 2014–2018, April 1985
Biochemistry

Chicken triosephosphate isomerase complements an *Escherichia coli* deficiency

(glycolytic enzymes/immunoselection of polysomes/cDNA cloning/expression/genomic cloning)

DONALD STRAUS* AND WALTER GILBERT*†

*The Biological Laboratories, Harvard University, Cambridge, MA 02138; and †Biogen, Inc., 14 Cambridge Center, Cambridge, MA 02142

Contributed by Walter Gilbert, December 4, 1984

ABSTRACT **We present the sequence of full-length chicken triosephosphate isomerase (D-glyceraldehyde 3-phosphate ketol-isomerase, EC 5.3.1.1) mRNA based on the analysis of cDNA and genomic clones. To isolate cDNA clones encoding the enzyme, we screened a muscle cDNA library with radioactively labeled cDNA made from RNA that had been enriched by immunoselection of polysomes. We blocked the signal caused by contaminating species in the probe with cloned DNA corresponding to the contaminants. Screening a chicken genomic library with cDNA coding for triosephosphate isomerase led to the isolation of phage containing the entire gene, which we used to map the transcriptional start. When placed downstream from a hybrid *trp–lac* promoter, the cDNA encoding the chicken enzyme programs the synthesis of functional protein, as judged by enzymatic criteria and by complementation of an *Escherichia coli* mutant that is deficient in bacterial triosephosphate isomerase.**

Triosephosphate isomerase (TIM; D-glyceraldehyde-3-phosphate ketol-isomerase, EC 5.3.1.1) is a glycolytic enzyme that catalyzes the interconversion of dihydroxyacetone phosphate and glyceraldehyde 3-phosphate (1). TIM, a dimer of 53,000 Da, is a particularly well characterized enzyme. High resolution crystal structures for the chicken and yeast proteins have been solved (2, 3) and the Gibbs free energy profile has been determined for catalysis by chicken TIM (4). The enzyme is so highly conserved in structure, sequence, and mechanism across both prokaryotes and eukaryotes, that it is certain that all *TIM* genes are descendants of a single primordial gene. Our goals are to use the *TIM* gene as a paradigm for the evolution of gene structure and to characterize further the molecular basis of the isomerization reaction by site-directed mutagenesis. To initiate these studies, we needed genomic and cDNA clones, the isolation of which we report in this paper. We chose to clone the chicken gene, as opposed to one from a different species, because the chicken protein is well characterized both enzymatically and structurally.

MATERIALS AND METHODS

Strains and Media. DF502, Δ(*rha-pfkA-tpi*), *pfkB1*, *his*$^-$, *pyrd*$^-$, *edd*$^-$, *galB*, *strA*, *sup*, was constructed and provided by D. Fraenkel. DH1, provided by D. Hanahan, is *endA1*, *hsdR17*, (r_k^-, m_k^+) *supE44*, *thi-1*, *recA1*, *gyrkA96*, *relA1*?. M63 medium (5) was supplemented with 0.2% glycerol or lactate.

DNA. Plasmids were prepared by either the alkaline or boiling method (6–8). We prepared phage by following procedures described in ref. 9. Plasmid pKK233-2 was constructed and provided by J. Brosius.

Enzymes. DNA polymerase I and all restriction enzymes, except *Pst* I, were from New England Biolabs. *Pst* I, terminal transferase, and S1 nuclease were from Boehringer-Mannheim. J. Beard supplied the avian myeloblastosis virus reverse transcriptase (Life Sciences, St. Petersburg, FL).

RNA Isolation. Total nucleic acid was isolated from fresh adult chicken breast muscle by phenol extraction and ethanol precipitation (10). After oligo(dT)-cellulose purification (11), poly(A)$^+$ RNA was precipitated from 300 mM NaOAc with 3 vol of ethanol, washed with 100% ethanol, dried, and resuspended in water. Whenever possible we used disposable plastic tubes and pipettes; otherwise, they were either heat-baked for 2 hr at 190°C, or washed first with 50% NaOH, and then with diethylpyrocarbonate-treated H_2O. Solutions (except for extraction buffer) were treated with diethylpyrocarbonate (12). Typically, we recovered ≈2 mg of poly(A)$^+$ RNA from 80 g of muscle tissue.

Polysome Isolation. All steps were carried out at 4°C. Approximately 80 g of breast tissue was dissected from four 5-week-old chickens and minced with scissors in buffer A: 10 mM Hepes, pH 7.5/25 mM NaCl/15 mM $MgCl_2$/140 mM sucrose/250 mM KCl/10 mM vanadyl ribonucleotide complex (13)/2 μg of cyclohexamide per ml (50 ml of buffer per 5 g of tissue). We homogenized the tissue in a motor driven Potter–Elvehjem grinder in buffer A (25 ml of solution per 5 g of tissue) containing 2% Triton X-100. Pooled homogenates were spun for 10 min at 12,000 rpm in a Sorvall GSA rotor. The supernatant was brought to 0.5% Na·deoxycholate and 2 mg of heparin per ml. Sixty milliliters of supernatant was layered onto 20 ml of 60% sucrose/10 mM Hepes, pH 7.5/25 mM NaCl/15 mM $MgCl_2$/250 mM KCl/2 mg of heparin per ml. After spinning for 5.5 hr at 37,000 rpm in an IEC A-170 rotor, supernatants were aspirated, leaving pellets containing polysomes. One-half of the pellets were frozen, and one-half were used for immunoselection. We resuspended the latter in 10 mM Hepes, pH 7.5/150 mM NaCl/5 mM $MgCl_2$/0.1% Nonidet P-40/2 mg of heparin per ml/2 μg of cyclohexamide per ml.

Antibody. We immunized rabbits with chicken TIM (gift of J. Belasco), that had been crosslinked by the procedure of Reichlin (14). Anti-TIM antibody titers were 0.38 mg of antibody equivalents per ml. Twenty milliliters of antiserum was loaded onto a 5.25-ml protein A-Sepharose column (Pharmacia), that had been washed in diethylpyrocarbonate-treated 20 mM sodium phosphate, pH 7.5/150 mM NaCl. The column was washed with the same solution, and IgG was eluted with 20 ml of 1 M HOAc. Fractions containing protein were pooled and dialyzed against 4 liters of phosphate-buffered saline. IgG (7 mg/ml) was stored at −20°C.

Immunoselection of TIM Polysomes. The polysomes from 40 g of tissue were incubated with 7 ml of anti-TIM IgG fraction (7 mg/ml). We then immunoselected the polysomes according to the procedure of Shapiro and Young (15).

cDNA Library. Double-stranded cDNA (30 μg) was synthesized from poly(A)$^+$ muscle RNA (25 μg) (16), tailed with

The publication costs of this article were defrayed in part by page charge payment. This article must therefore be hereby marked "*advertisement*" in accordance with 18 U.S.C. §1734 solely to indicate this fact.

Abbreviations: TIM, triosephosphate isomerase; bp, base pair(s).

dCTP (17), and fractionated on a 2% low-melting agarose gel (SeaKem Laboratories, Rockland, ME). The region of the gel containing DNA >650 base pairs (bp) long was melted and then extracted with phenol. Transformation of strain DH1 with dG-tailed *Pst* I-cut pBR322 (180 ng) that had been annealed to dC-tailed sized cDNA (30 ng) yielded ≈6000 clones (18).

Translation and Immunoprecipitation. mRNA was translated in micrococcal nuclease-treated rabbit reticulocyte lysate (19). ^{35}S-labeled proteins were electrophoresed on 10% polyacrylamide gels according to the method of Laemmli and Favre (20) and were detected by fluorography (EN^3-HANCE, New England Nuclear). Immunoprecipitation was carried out by the method of Kessler (21). *Staphylococcus aureus* was purchased from Enzo Biochemicals (New York).

Probing the cDNA Library. We prepared labeled cDNA from the immunoselected polysomal poly(A)$^+$ RNA by using reverse transcriptase and a mixture of random primers (22, 23); 1.4×10^7 cpm, or 10 ng, of probe was recovered. This result indicates a yield of ≥130 ng of selected polysomal poly(A)$^+$ RNA from 40 g of chicken muscle.

Filters were probed with either cDNA synthesized to selected polysomal RNA, or with nick-translated inserts from clones corresponding to non-TIM species present in the cDNA probe (24–26). When probing with cDNA, both prehybridization and hybridization solutions contained 0.5 μg of poly(A) per ml, 0.5 μg of poly(G) per ml, and ≈500 ng of unlabeled *Pst* I-linearized plasmids per ml. The latter were a mixture of DNA from 15 independent clones of three classes, each class capable of selecting RNA from a different contaminant of the selected polysome probe. The radioactive cDNA (7×10^4 cpm/ml) or the nick-translated probe (1×10^6 cpm/ml) was hybridized with filters (27, 28).

Hybridization Selection. DNA, prepared from 10-ml overnight cultures, was isolated by the boiling minipreparation method (8), treated with RNase A, and then spermine-precipitated (29). Plasmids were bound to nitrocellulose (30) using an inverted Pasteur pipette. After baking, filters were moistened and cut with a heat-baked hole punch. Twenty filter circles were hybridized in a tube with chicken muscle poly(A)$^+$ RNA (200 μg per 400 μl) for 2 hr at 42°C (31). We used glycogen (Calbiochem or Sigma) as a carrier during the ethanol precipitation of the eluted RNA. One-half of the selected RNA was translated.

Genomic Cloning. The nick-translated (26) 1-kbp *Pst* I fragment of cDNA clone pcT7 hybridized to six Charon 4A plaques in a screening of a chicken genomic library (32, 33). The *TIM* gene resides on a 9.3-kbp fragment, which we isolated from phage φgt3. We subcloned this fragment into dG-tailed *Pst* I-cut pBR322 by blunt-ending with a brief BAL-31 digestion and dC-tailing using terminal transferase. Plasmid pgT9 contains the *TIM* gene in one orientation and pgT7 contains the gene in the other orientation.

DNA Sequencing. Sequence was obtained using the Maxam–Gilbert method (34).

TIM Assay. We used D-glyceraldehyde 3-phosphate as a substrate for TIM in a glycerol-phosphate dehydrogenase/NADH coupled assay (35). We confirmed that the coupling enzyme was in excess by doubling the amount of TIM in the most active sample and determining that the rate of NADH oxidation also doubled. NADH oxidase activity that was not dependent on the triosephosphate substrate was subtracted from the total activity.

RESULTS AND DISCUSSION

cDNA Cloning. TIM is abundant in muscle, a tissue that is rich in the enzymes needed for anaerobic glycolysis (36); it constitutes ≈2% of muscle protein (35, 37) and ≈2% of the total [^{35}S]methionine-labeled *in vitro* translation products of muscle poly(A)$^+$ RNA. In the final analysis, however, we found TIM clones account for only ≈0.5% of our muscle cDNA clones. Our strategy was to make a cDNA library from total muscle mRNA and then to probe it with cDNA made to RNA that we had enriched for TIM message. We synthesized double-stranded cDNA from muscle poly(A)$^+$ RNA, tailed it with dCTP, and annealed the products of 650 bp or longer to dG-tailed *Pst* I-cut pBR322. Transformation of DH1 by the method of Hanahan (18) yielded tetracycline-resistant transformants at an efficiency of ≈240,000 per μg of starting mRNA.

To enrich for chicken TIM RNA, we passed muscle polysomes over a column of anti-TIM antibody bound to protein A-Sepharose. Fig. 1 (lane 1) shows the *in vitro* translation products of the immunoselected polysomal poly(A)$^+$ RNA. TIM is a major band, but there are three contaminants, each of which comigrates with an abundant muscle contractile protein (actin, 43 kDa; tropomyosin, 35 kDa; troponin C and myosin light chain 2, 18 kDa). The upper band in lane 1 is an artifact of the *in vitro* translation reaction. Assuming equal translation efficiencies for all the RNAs and using the methionine contents of the tentatively assigned contaminating proteins, we calculate that TIM accounts for ≈40% of the *in vitro* translation products—a 20-fold enrichment. We made ^{32}P-labeled reverse transcripts of the selected RNA to probe our chicken cDNA library. In a previous experiment, we had isolated several clones corresponding to each of the three contaminants in the selected RNA. By including unlabeled linearized plasmids from these clones when we screened with the polysomal probe, we effectively prevented signal caused by the contaminants without affecting the signal from TIM clones. Screening 6000 cDNA clones yielded 28 positives, none of which corresponded to any of the three non-TIM contaminants in the probe. Of the 21 positive clones tested in a hybridization-selection experiment (Fig. 2), 19 clones selected TIM mRNA and 2 clones did not select any mRNA.

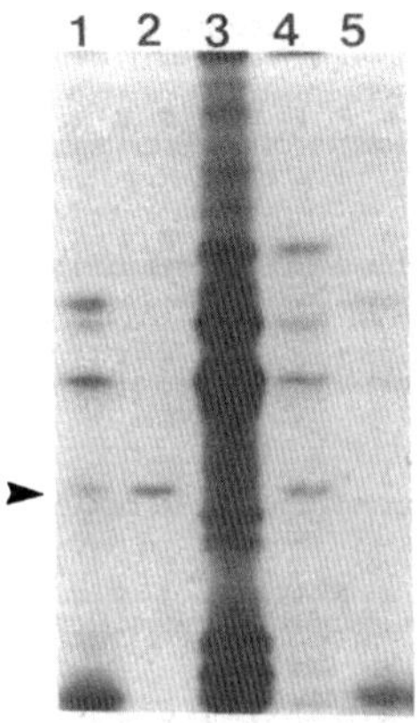

FIG. 1. Translation of immunoselected polysomal poly(A)$^+$ RNA. ^{35}S-labeled products of translation, in rabbit reticulocyte lysate, were separated by $NaDodSO_4$/polyacrylamide electrophoresis (10% gel). Lanes 1, 3, and 5 show total products from 1.5 μl of 25-μl reaction mixtures. Lanes 2 and 4 show the protein immunoprecipitated by anti-TIM serum from 10 μl of the reaction mixtures corresponding to lanes 1 and 3, respectively. Reaction depicted in lanes 1 and 2 contained immunoselected mRNA from ≈130 μg of polysomal mRNA; reaction shown in lanes 3 and 4 was programmed with 0.4 μg of total muscle poly(A)$^+$ RNA. Lanes: 1, translation of immunoselected mRNA; 2, immunoprecipitation of products in lane 1; 3, protein translated from total muscle poly(A)$^+$ RNA; 4, immunoprecipitation of products in lane 3; 5, no RNA added to the reaction mixture.

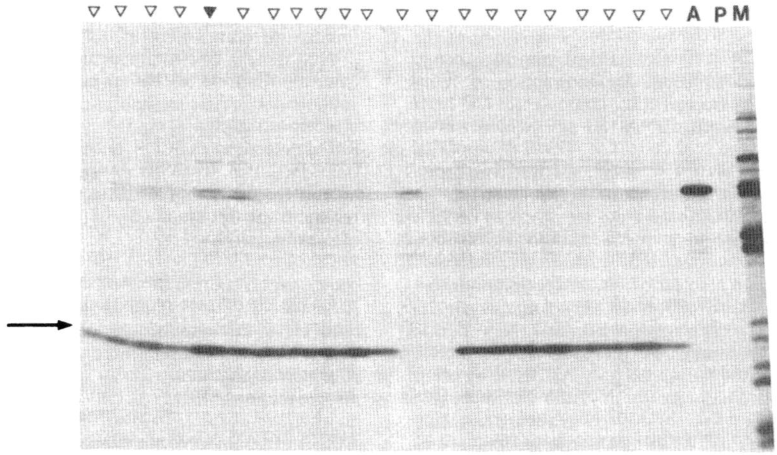

FIG. 2. Hybridization selection of cDNA clones. Linearized plasmids from boiling minipreparations were bound to small nitrocellulose filters and hybridized in a tube containing total chicken muscle poly(A)$^+$ RNA. Selected RNA was released by boiling individual filters. After translation in reticulocyte lysate, ^{35}S-labeled samples were separated on a 10% polyacrylamide gel in the presence of NaDodSO$_4$. Solid triangle indicates lane containing protein synthesized from RNA selected by clone pcT7. Control lanes: A, pα-actin 2 (38) bound to filter; P, pBR322 bound to filter; M, translated total chicken muscle poly(A)$^+$ RNA. Arrow, TIM.

Sequence and Genomic Cloning. The sequence of one of the clones, pcT7, showed that it contains a cDNA insert coding for all but 95 nucleotides at the 5′ end of TIM mRNA. Sequence analysis of the chromosomal *TIM* gene provided the remaining 5′ information. We obtained genomic clones by probing a Charon 4A chicken genomic library with the nick-translated insert of pcT7. Three phage contained the *TIM* gene (as defined by hybridization with the ends of cDNA clone pcT7) on a 9.3-kbp *Eco*RI fragment. We subcloned the fragment in both orientations into the *Pst* I site of pBR322 (pgT9 and pgT7) and used a nested deletion strategy (39) to sequence the gene and its flanking regions. (The structure of the chicken *TIM* gene will be published elsewhere.) We mapped the transcriptional start site by hybridizing muscle poly(A)$^+$ RNA to a 5′-end-labeled, single stranded, *Nco* I/*Bst*EII fragment, which extends 340 bp in the 5′ direction from the AUG initiation codon. The RNA protected a fragment that extends 51 bp upstream from the translational start (50; data not shown). Thus, the coding sequence is flanked by 51 bp of untranslated RNA at the 5′ end and 425

```
A.A.                                                                                                              BASE
                                                                     -->
     {GGGGCTACGGGGGGTGGGGCTTCGCGCCCCGCCGGCCTATAAAAGCGGCCGCCGCGGCTCCGTGCC} GTTGCCGACCTTCGCCTGCGCCGCTGCTGCTTCGCGCCCGTCGCCTCCGCC   -1
                                                     *K**R*                          *D*
  1   M  A  P  R  K  F  F  V  G  G  N  W  K  M  N  G  D  K  K  S  L  G  E  L  I  H  T  L  N  G  A  K  L  S  A  D  T  E  V  V
     ATGGCTCCCAGGAAGTTCTTCGTGGGTGGCAACTGGAAGATGAACGGCGACAAGAAGAGCTTGGGCGAGCTCATCCACACGCTGAATGGCGCCAAGCTCTCGGCCGACACCGAGGTGGTT  120

 41   C  G  A  P  S  I  Y  L  D  F  A  R  Q  K  L  D  A  K  I  G  V  A  A  Q  N  C  Y  K  V  P  K  G  A  F  T  G  E  I  S  P
     TGCGGAGCCCCTTCAATCTACCTTGATTTTGCCCGCCAGAAGCTTGATGCAAAGATTGGAGTTGCAGCACAAAACTGTTACAAGGTACCGAAGGGTGCTTTCACAGGAGAGATCAGCCCA  240

 81   A  M  I  K  D  I  G  A  A  W  V  I  L  G  H  S  E  R  R  H  V  F  G  E  S  D  E  L  I  G  Q  K  V  A  H  A  L  A  E  G
     GCAATGATCAAAGATATTGGAGCTGCATGGGTGATCCTGGGCCACTCAGAGCGGAGGCATGTTTTTGGAGAGTCTGATGAGTTGATTGGGCAGAAGGTGGCTCATGCTCTTGCTGAAGGC  360
                                                                            *Q**E*
121   L  G  V  I  A  C  I  G  E  K  L  D  E  R  E  A  G  I  T  E  K  V  V  F  E  Q  T  K  A  I  A  D  N  V  K  D  W  S  K  V
     CTCGGTGTCATCGCCTGCATTGGGGAGAAGCTGGATGAGAGAGAAGCTGGCATAACGGAGAAGGTGGTCTTTGAACAGACCAAAGCTATTGCTGATAACGTGAAGGACTGGAGTAAGGTG  480
                                                                                                    *T*
161   V  L  A  Y  E  P  V  W  A  I  G  T  G  K  T  A  T  P  Q  Q  A  Q  E  V  H  E  K  L  R  G  W  L  K  S  H  V  S  D  A  V
     GTTCTTGCCTATGAGCCAGTTTGGGCTATCGGAACTGGTAAAACTGCTACTCCCCAACAGGCTCAGGAGGTTCATGAGAAGCTGAGAGGCTGGCTCAAAAGCCACGTGTCTGATGCTGTT  600
        *V**Q**S*
201   A  Q  S  T  R  I  I  Y  G  G  S  V  T  G  G  N  C  K  E  L  A  S  Q  H  D  V  D  G  F  L  V  G  G  A  S  L  K  P  E  F
     GCTCAGTCAACTAGGATCATCTATGGAGGTTCAGTCACTGGTGGCAACTGTAAGGAACTGGCCTCCCAGCATGATGTGGATGGCTTCCTTGTTGGTGGGGCTTCTCTCAAGCCAGAGTTT  720

241   V  D  I  I  N  A  K  H ---
     GTGGATATTATCAATGCAAAACATTAAAGCAGCCTGTGAGGAGCAGTCCCTTACGGTTAAGAGCAAGAAACTGAAGCAAGAAGGGACCTTGTGTTGCACGTCTCTCGGTACAGAGGCTTC  840

     TTCTGAGGCTTTCCCCCACCACCACAATTATTGTTCTAGCTGTGCTGCTAACCCCCACCACCTTGTTGGAGTCCCATTAGTGTGAGCCCATCTCAGCAGAGTCTCCTTTCTGAACTGGCA  960

     AAATCCTTGGTTATCTGTTGAGCTGTTAGAGCCCACAGTCAACCTGGCCATTGCCTCTCTCTCTTTGCAGCCCTGCAGGGAGGGAGGGCCACTAGTACTGGGGGAAGAAAAAGGAACCA  1080

     CCATCTTCTGCATCTTTCAGCTCCATCCGCAAGGAGCCTGGCAGCTTAGACCCTTGTGAGACACCTTACCTCACCAATGTCCTGTATTGAACAATAAATGAGAAGGAAAAAAAAAAAAAA  1200
```

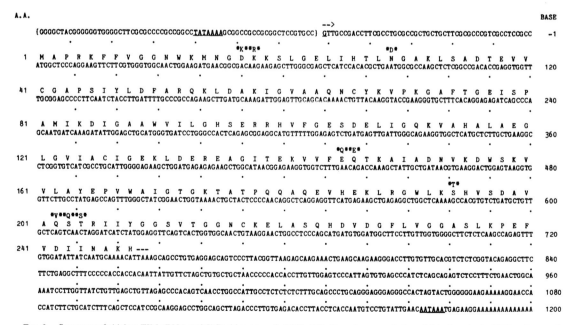

FIG. 3. Sequence of chicken TIM cDNA and 5′ flanking genomic DNA. This figure is a compilation of data from both cDNA and genomic clones. G is underlined at position −51 and an arrow lies over bases corresponding to strong bands in the S1 nuclease protection experiment (see text). The "TATA" box (40) and poly(A)$^+$ addition signal (A-A-U-A-A-A) are also underlined. Predicted amino acid sequence (indicated by standard one-letter abbreviations) is shown above the nucleotide sequence. At sites where our predicted sequence is in conflict with a published amino acid sequence, we have noted the published residue above the predicted residue.

bp at the 3′ end, the latter being defined by the poly(A)$^+$ addition signal A-A-U-A-A-A.

Fig. 3 depicts the sequence of TIM cDNA, the deduced protein sequence, and the 5′ flanking DNA sequence. Where our protein sequence conflicts with the published sequence, the published residue is noted above the predicted sequence. In their tentative amino acid sequence, Furth *et al.* (41) include Lys-Arg at position 17, where our sequence predicts an aspartic acid. Banner *et al.* (2) correctly deduced from the electron density map that there were too many residues in the proposed primary sequence in this region. To reconcile the sequence and diffraction data, Banner *et al.* eliminated a residue, but they incorrectly chose lysine-19, suggesting that the Lys-Lys dipeptide of Furth *et al.* was an artifact of trypsin digestion. The next four discrepant residues in our sequence are homologues of the amino acids previously reported. At positions 202 and 204 we predict, relative to the earlier reports, deletion of a valine and insertion of a threonine, respectively. With the exception of one, our new assignments change residues in the chicken sequence to amino acids that match the corresponding rabbit TIM residues (42). The exception, aspartic acid-17, matches the aspartic acid-17 reported for coelocanth TIM (43). Reexamination of the electron density map of the protein indicates that our sequence is consistent with the map (P. Artymiuk, personal communication).

Expression. To facilitate *in vitro* mutagenesis studies on chicken TIM, we wanted to express the enzyme in *E. coli*. Fig. 4 diagrams our approach. There are 44 nucleotides of coding sequence missing from the 5′ end of cDNA clone pcT7. We fused a fragment that contains this region, derived from genomic clone pgT9, to pcT7. The resultant plasmid, pTF2, has DNA encoding the entire TIM mRNA in addition to 5′ flanking genomic DNA. The AUG initiator codon fortuitously falls within an *Nco* I site. We had available to us cloning vector pKK233-2, with an *Nco* I cloning site just downstream from a hybrid *trp–lac* promoter and a ribosome binding site. The promoter, called *trc*, is identical to the *tac* promoter (44), except that the spacing between the −10 and −35 consensus sequences is 17 bp for *trc*, as opposed to 16 bp for *tac* (J. Brosius, personal communication). Our goal, in the following steps, was to position the 1-kbp *Nco* I/*Pst* I fragment of pTF2, which carries the complete coding sequence of chicken TIM, in the *Nco* I site of vector pKK233-2. The vector has a *Pst* I site, which is, unfortunately, contiguous with the *Nco* I site, making double cutting impossible. We inserted a *Pst* I fragment, containing the desired region plus 5′ flanking genomic DNA, into the *Pst* I site of pKK233-2, creating plasmid pX0. Ligation of the 1-kbp *Nco* I fragment of pX0 into the *Nco* I site of pKK233-2 positioned the gene correctly for expression from the *trc* promoter in plasmid pX1. The TIM coding sequence, starting with the initiation AUG, is thus conveniently located on an *Nco* I fragment that can be easily moved to other vectors. pX1 directs expression of chicken TIM, as demonstrated by the complementation studies and enzyme assays outlined below.

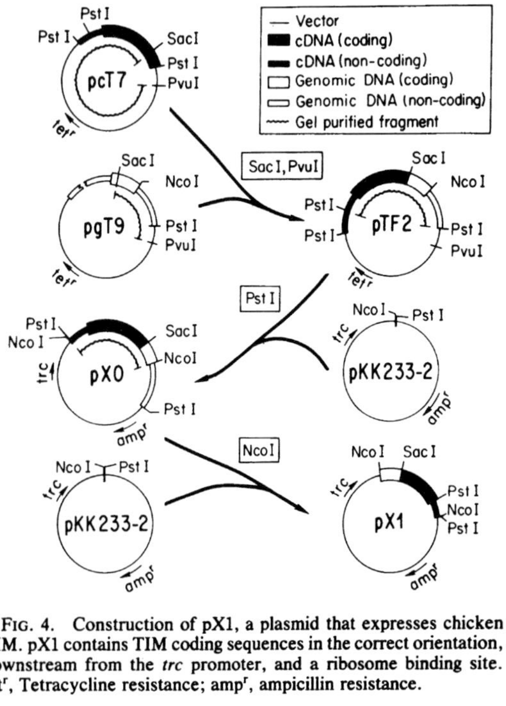

FIG. 4. Construction of pX1, a plasmid that expresses chicken TIM. pX1 contains TIM coding sequences in the correct orientation, downstream from the *trc* promoter, and a ribosome binding site. tet^r, Tetracycline resistance; amp^r, ampicillin resistance.

Complementation. *E. coli* strain DF502 is deficient in TIM because of a deletion mutation (45–47). DF502 cannot grow on plates with either lactate or glycerol as sole carbon sources, whereas it does grow on plates containing both. Presumably, this is because lactate can be metabolized to glyceraldehyde 3-phosphate and glycerol to dihydroxyacetone phosphate. When transformed with pX1, however, the strain grows on lactate or glycerol alone. Thus, the chicken gene complements the *E. coli* deficiency. Enzymatic assays, on French press lysates of these transformants, routinely indicate ≈3 units of TIM activity $OD_{550}^{-1}\cdot ml^{-1}$. (Wild-type *E. coli* produce ≈15 units TIM activity $OD_{550}^{-1}\cdot ml^{-1}$.) Since the specific activity of pX1 encoded TIM is 9000 units/mg (48), the recombinant bacteria produce ≈0.3 μg of TIM at $OD_{550}^{-1}\cdot ml^{-1}$, corresponding to 9000 monomers per cell or 0.3% of soluble bacterial protein. DF502 also grows in selective medium when it harbors either plasmid pLC16-4, which contains the *E. coli TIM* gene (49) or pYTPIC10up, which expresses the gene for yeast TIM (45).

We thank the members of the Gilbert laboratory, especially N. Lonberg, G. Church, and R. Cate for their advice and support. We also acknowledge A. Korman, R. Freund, J. Kadonaga, M. Kreitman, V. Tate, and D. Hanahan for their assistance and/or advice. Thanks are due to J. B. Dodgson for providing the genomic chicken library, C. P. Ordahl for making pα-actin 2 available to us, D. Fraenkel for supplying strain DF502, J. Belasco for his gift of purified chicken TIM, J. Brosius for plasmid pKK233-2, and J. Pustell and R. Staden for the use of their respective computer programs. This work was supported by National Institutes of Health Grant GM09541-22 and by Biogen N.V.

1. Meyerhof, O. & Lohmann, K. (1934) *Biochem. Z.* **273**, 413–418.
2. Banner, D. W., Bloomer, A. C., Petsko, G. A., Phillips, D. C., Pogson, C. I., Wilson, I. A., Corran, P. H., Furth, A. J., Milman, J. D., Offord, R. E., Priddle, J. D. & Waley, S. G. (1975) *Nature (London)* **255**, 609–614.
3. Alber, T., Banner, D. W., Bloomer, A. C., Petsko, G. A., Phillips, D. C., Rivers, P. S. & Wilson, I. A. (1981) *Philos. Trans. R. Soc. London Ser. B* **293**, 159–171.
4. Albery, W. J. & Knowles, J. R. (1976) *Biochemistry* **15**, 5627–5631.
5. Miller, J. H. (1972) *Experiments in Molecular Genetics* (Cold Spring Harbor Laboratory, Cold Spring Harbor, NY).
6. Birnboim, H. C. & Doly, J. (1979) *Nucleic Acids Res.* **7**, 1513–1523.
7. Ish-Horowicz, D. & Burke, J. F. (1981) *Nucleic Acids Res.* **9**, 2989–2998.
8. Holmes, D. S. & Quigly, M. (1981) *Anal. Biochem.* **114**, 193–197.
9. Maniatis, T., Fritsch, E. F. & Sambrook, J. (1982) *Molecular Cloning: A Laboratory Manual* (Cold Spring Harbor Laboratory, Cold Spring Harbor, NY).
10. Lomedico, P. T. & Saunders, G. F. (1976) *Nucleic Acids Res.* **3**, 381–391.

11. Aviv, H. & Leder, P. (1972) *Proc. Natl. Acad. Sci. USA* **69,** 1408–1412.
12. Palmiter, R. D. (1974) *Biochemistry* **13,** 3606–3615.
13. Berger, S. L. & Birkenmeier, C. S. (1979) *Biochemistry* **18,** 5143–5149.
14. Reichlin, M. (1980) *Methods Enzymol.* **70,** 159–165.
15. Shapiro, S. Z. & Young, J. R. (1981) *J. Biol. Chem.* **256,** 1495–1498.
16. Wickens, M. P., Buell, G. N. & Schimke, R. T. (1978) *J. Biol. Chem.* **253,** 2483–2495.
17. Roychoudhury, R. & Wu, R. (1980) *Methods Enzymol.* **65,** 43–62.
18. Hanahan, D. (1983) *J. Mol. Biol.* **166,** 557–580.
19. Pelham, H. R. B. & Jackson, R. J. (1976) *Eur. J. Biochem.* **67,** 247–256.
20. Laemmli, U. K. & Favre, M. (1973) *J. Mol. Biol.* **80,** 575–599.
21. Kessler, S. W. (1975) *J. Immunol.* **115,** 1617–1624.
22. Schwartz, D. E., Tizard, R. & Gilbert, W. (1983) *Cell* **32,** 853–869.
23. Taylor, J. M., Illmensee, R. & Summers, J. (1976) *Biochim. Biophys. Acta* **442,** 324–330.
24. Grunstein, M. & Hogness, D. S. (1975) *Proc. Natl. Acad. Sci. USA* **72,** 3961–3965.
25. Hanahan, D. & Meselson, M. (1980) *Gene* **10,** 63–67.
26. Rigby, P. W. J., Dieckmann, M., Rhodes, C. & Berg, P. (1977) *J. Mol. Biol.* **113,** 237–251.
27. Southern, E. M. (1975) *J. Mol. Biol.* **98,** 503–517.
28. Wu, C. (1980) *Nature (London)* **286,** 854–860.
29. Hoopes, B. C. & McClure, W. R. (1981) *Nucleic Acids Res.* **9,** 5493–5504.
30. Kafatos, F. C., Jones, C. W. & Efstratiadis, A. (1979) *Nucleic Acids Res.* **7,** 1541–1552.
31. Ricciardi, R. P., Miller, J. S. & Roberts, B. E. (1979) *Proc. Natl. Acad. Sci. USA* **76,** 4927–4931.
32. Benton, W. D. & Davis, R. W. (1977) *Science* **196,** 180–182.
33. Dodgson, J. B., Stromer, J. & Engel, J. D. (1979) *Cell* **17,** 879–887.
34. Maxam, A. M. & Gilbert, W. (1980) *Methods Enzymol.* **65,** 499–560.
35. Putman, S. J., Coulson, A. F. W., Farley, I. R. T., Riddleston, B. & Knowles, J. R. (1972) *Biochem. J.* **129,** 301–310.
36. Pette, D., Luh, W. & Bucher, T. (1962) *Biochem. Biophys. Res. Commun.* **7,** 419–424.
37. Esnouf, M. P., Harris, R. P. & McVittie, J. D. (1982) *Methods Enzymol.* **89,** 579–583.
38. Ordahl, C. P., Tilghman, S. M., Ovitt, C., Fornwald, J. & Largen, M. T. (1980) *Nucleic Acids Res.* **8,** 4989–5005.
39. Frischauf, A., Garoff, H. & Lehrach, H. (1980) *Nucleic Acids Res.* **8,** 5541–5549.
40. Breathnach, R. & Chambon, P. (1981) *Annu. Rev. Biochem.* **50,** 349–383.
41. Furth, A. J., Milman, J. D., Priddle, J. D. & Offord, R. E. (1974) *Biochem. J.* **139,** 11–25.
42. Corran, P. H. & Waley, S. G. (1975) *Biochem. J.* **145,** 335–344.
43. Kolb, E., Harris, J. I. & Bridgen, J. (1974) *Biochem. J.* **137,** 185–197.
44. Amann, E., Brosius, J. & Ptashne, M. (1983) *Gene* **25,** 167–178.
45. Alber, T. & Kowasaki, G. (1982) *J. Mol. Appl. Genet.* **1,** 419–434.
46. Pahel, G., Bloom, F. R. & Tyler, B. (1979) *J. Bacteriol.* **138,** 653–656.
47. Babul, J. (1978) *J. Biol. Chem.* **253,** 4350–4355.
48. Straus, D., Raines, R., Kawashima, E., Knowles, J. R. & Gilbert, W. (1985) *Proc. Natl. Acad. Sci. USA* **82,** 2272–2276.
49. Thomson, J., Gerstenberger, P. D., Goldberg, D. E., Gociar, E., de Silva, A. O. & Fraenkel, D. G. (1979) *J. Bacteriol.* **137,** 502–506.
50. Sharp, P., Berk, A. J. & Berget, S. M. (1980) *Methods Enzymol.* **65,** 750–768.

The Journal of Neuroscience, July 1989, 9(7): 2468–2485

The Monoclonal Antibody B30 Recognizes a Specific Neuronal Cell Surface Antigen in the Developing Mesencephalic Trigeminal Nucleus of the Mouse

Didier Y. Stainier[2] **and Walter Gilbert**[1]

[1] Department of Cellular and Developmental Biology and [2] Department of Biochemistry and Molecular Biology, Harvard University, Cambridge, Massachusetts 02138

A monoclonal antibody, B30, obtained with whole cells from embryonic brain as an immunogen, recognizes a neuronal cell surface antigen that appears only in 2 distinct systems in the developing mouse brain: the trigeminal system and the cerebellum. In the trigeminal system, B30 labels the surface of neurons, including their axons and their transient dendrites, in 2 groups of cells: the centrally located mesencephalic trigeminal nucleus and the peripheral trigeminal ganglion. Immunoreactivity is detectable during axon outgrowth, peaks around the seventh postnatal day, and disappears around 2 weeks after birth. In the cerebellum, B30 labels 2 layers of cells during development. Perinatally, and for about a week after birth, the layer of premigratory granule cells stains. After their maturation, Purkinje cells start to stain and by 12 d postnatally all the Purkinje cell bodies, their axons, and their dendritic trees show strong immunoreactivity. Subsequently, and in the adult, this staining is lost from some cells to reveal bands of antigen positive and negative Purkinje cells. Initial biochemical characterization of the epitope shows that it is carried on 2 minor gangliosides.

Cell surface molecules have been postulated to be involved in axonal growth and guidance, target recognition, and synapse formation (Weiss, 1947; Sperry, 1963; Roseman, 1974; Hood et al., 1977). The identification of such molecules specific for subsets of neurons would not only provide us with a way to isolate these cells but also with a biochemical tool to test for specific recognition processes. Since the monoclonal antibody technique was introduced into neuroscience (Barnstable, 1980; Zipser and McKay, 1981), several groups have successfully raised monoclonal antibodies to regionally restricted cell surface antigens in both invertebrate (Goodman et al., 1984; McKay et al., 1984) and vertebrate species (Cohen and Selvendran, 1981; McKay and Hockfield, 1982; Fujita and Obata, 1984; Stallcup et al., 1985; Constantine-Paton et al., 1986; Moskal and Schaffner, 1986; Yamamoto et al., 1986; Mori et al., 1987). Those studies reveal a number of general and some more specific molecules associated with the extracellular matrix and the cell membrane and point to a combinational and hierarchical system determining growth cone guidance and target recognition (see Bastiani et al., 1987; Steller et al., 1987; and for a review, see Jessel, 1988).

We have generated monoclonal antibodies to single-cell suspensions from embryonic mouse and rat brains and isolated hybridomas that define subpopulations of neurons by cell surface immunoreactivity. Parts of this effort have been reported previously (Rayburn et al., 1987; Stainier et al., 1987). This paper describes a monoclonal antibody, mAb B30, which identifies small populations of cells in the developing mouse CNS. B30 defines neurons and their processes in the mesencephalic trigeminal nucleus (MesV) and also outlines neurons in the trigeminal ganglion. Furthermore, it labels specific layers in the cerebellum at various times in development.

Materials and Methods

Monoclonal antibody production and screening. Timed pregnant female rats and mice were obtained from Charles River and housed until the appropriate gestational age (embryonic day 15 or E15). Animals were killed either by cervical dislocation or euthanized with ether. Uteri were dissected from pregnant animals into ice-cold PBS (pH 7.4), individual embryos were removed, and the mid- and hindbrains were dissected free of other tissue and stripped of their membranes. Using fire-polished pasteur pipettes of decreasing bore size, a cell suspension was prepared and washed several times in PBS. This was then used as an immunogen at 10^6 cells per injection following the standard immunization schedule: one intraperitoneal (i.p.) injection every 3 weeks for 9 weeks and the spleen fused 4 d after the last injection. Dimethyl dioctadecyl ammonium bromide (DDA, Eastman Kodak Co.; Baechtel and Prager, 1982) was used as an adjuvant and administered i.p. 4 hr before the injection of the cells. Spleen cells were fused by conventional techniques using FOX-NY as the myeloma cell line (Galfre et al., 1977). (E15 rat mid- and hindbrain cells provided the immunogen for the fusion that led to the isolation of the B30 hybridoma.) The resulting hybrid cell lines were screeened first on chunks of embryonic mouse brain and subsequently on 120-μm-thick vibratome sections of newborn mouse brain using the following staining conditions: after incubating for 1 hr in 10% normal goat serum (NGS) diluted in PBS, the tissue was exposed to supernatant of hybridoma culture overnight at 4°C, then, after washing several times with 10% NGS/PBS (2 × 5 min), antibody binding was visualized by incubation with a fluoresceinated goat anti-mouse IgG + IgM (Boehringer-Mannheim) diluted 1:40 in PBS for 2 hr. After a final wash in PBS (2 × 10 min), the tissue chunks were examined.

The initial screen consisted of staining chunks in 96-well plates (see Results). Controls for nonspecific fluorescence are included in each plate and reveal faint background fluorescence on some morphologically distinct cells which can be eliminated with incubation in increasing concentrations of NGS (the background binding is attributed to Fc receptors on microglial cells). Statistically, in the original fusion, 95% of the wells

Received Sept. 17, 1988; revised Nov. 28, 1988; accepted Dec. 2, 1988.

We thank Dr. Helen Rayburn and Dr. Hen-Ming Wu, who took part in the early stages of this project. We also thank Beth Bennett for expert technical help and advice in tissue culture, Albert Ko for advice on TLC techniques, Dr. K. Lloyd for mAb R24, Lloyd Schoenbach for help with figures, and Dr. Carl Fulwiler for critical reading of the manuscript.

Correspondence should be addressed to Dr. Walter Gilbert, Department of Cellular and Developmental Biology, Harvard University, 16 Divinity Ave., Cambridge, MA 02138.

The Journal of Neuroscience, July 1989, 9(7) **2469**

showed hybridoma colonies (1–5/well); 10% of the wells screened showed cell surface immunoreactivity on a portion of cells in the CNS and another 15% reacted with various fibrillary structures.

Lines producing antibodies of interest were clones by limiting dilution and antibody subclass was determined by the Ouchterlony immunodiffusion technique (Miles Scientific).

Immunohistochemistry. Fresh brains of fetal and postnatal mice were embedded in 2.5% LGT agarose in PBS. We used a vibroslice from W.P.I. (New Haven, CT) to cut 100- to 120-μm-thick vibratome sections. The slicing was usually done at room temperature, though we found it easier to cut thinner sections in precooled buffer. All the immunohistochemical reactions were performed at room temperature (with gentle shaking) on free-floating sections in polystyrene culture dishes by sequential incubations in 10% NGS in PBS for 1 hr; B30 hybridoma culture supernatant overnight (12–18 hr); after several washes with 10% NGS/PBS (2 × 10 min) fluoresceinated goat anti-mouse IgM (Southern Biotech), diluted 1:100 for 2 hr. After several more washes in 10% NGS/PBS (2 × 10 min), the sections were slightly fixed if longer preservation was needed (1% paraformaldehyde for 30 min either before or after incubation with the secondary antibody) and mounted in fresh mounting medium (90% glycerol in PBS containing 1 mg/ml *p*-phenylenediamine and 0.1% sodium azide). They were examined in the culture dishes or between 2 coverslips separated by silicone grease. A Nikon Diaphot-TMD inverted microscope equipped with epifluorescence was used as well as a Lasersharp MRC-500 confocal microscope. Appropriate controls were done, including other identified mouse IgMs and extended incubations with the secondary fluorosceinated antibody. The resulting background fluorescence was negligible (see Results).

Cell culture and staining. E13 mouse trigeminal ganglia were dissected and triturated, and the cells were plated on collagen coated plates in L-15 medium with 10% CBS or in HB101 medium (Hana Biologics). NGF (50 ng/ml) was added to the cultures. Live cultures were stained as described above (first Ab for 40 min and second Ab for 20 min at room temperature), 2 and 6 d after plating. Cultures were returned to the incubator after observation.

Ganglioside preparation and TLC analysis. Freshly dissected postnatal day 7 (10 P7 mice) brain stem (+MesV containing areas of the midbrain), P7 occipital cortex (10 mice), and P12 (10 mice) cerebellum were homogenized in chloroform/methanol (2:1, vol/vol) and extracted with 0.12 M KCl (Suzuki, 1965). The extracts were dried by rotary evaporation, taken up in water, desalted through Bond-Elut columns (Analytichem International) and finally lyophilized for storage at −20°C.

Thin-layer chromatography of gangliosides was done on aluminium-backed TLC plates (silica gel 60, 0.2 mm thick, E. Merck). The plates were preactivated by heating them to 110°C for 30 min and developed in chloroform/methanol/0.25% aqueous $CaCl_2$ (60:40:10). Ganglioside standards were obtained from Calbiochem and were detected by resorcinol stain. R24 (anti-GD3) was a generous gift from Dr. K. Lloyd. After drying, the plates were dipped in 0.05% poly(isobutylmethacrylate) (Polysciences) in hexane for 90 sec. The dried chromatograms were sprayed with PBS before soaking in a blocking solution (PBS, 3% BSA, and 5% NGS) for 30 min (all the incubations were done at room temperature with gentle shaking). The chromatograms were overlaid with the first antibody (diluted hybridoma supernatant) for 3 hr. After repeated washes in PBS, the chromatograms were overlaid with alkaline phosphatase-conjugated second antibody (diluted 1:1000) for 1 hr. After repeated washes in PBS, BCIP, and NBT (BRL) were used at a 1:1 molar ratio in 0.1 M Tris-HCl (pH 9.5), 0.1 M NaCl, 50 mM $MgCl_2$ for the color reaction. Upon completion, the reaction was stopped in water.

Enzyme treatments and chromatography on DEAE-Sephadex A-25 (Pharmacia) were done as described by Schwarting et al. (1987).

Results

Techniques

We needed to develop quick and efficient methods to screen for monoclonal antibodies (mAbs) binding to cell surface antigens that would define rare subsets of neurons in the developing CNS. For the initial screenings, the embryonic brains were gently triturated with a polished pasteur pipette, and the resulting pieces were aliquoted into 96-well plates for immunochemical treatment with supernatant from the fusion plates. We then used an inverted microscope to examine individual wells. If greater resolution was needed, the sample could be taken out of the well and pressed gently between 2 coverslips for further observation. Various staining patterns appeared in the 96-well plates: staining of most cells, staining of most cells in some chunks, staining of some cells in some chunks, staining of fibrillary structures. Some 20% of the wells screened were selected for further immunohistochemical characterizations.

In order to display the surface staining of unfixed cells, we stained thick sections (120–150 μm) of fresh tissue. These sections enable us to screen easily a sizeable fraction of the developing CNS to localize rare antigens. Staining fresh tissue permits a clear visualization of the surface staining of cells and processes. In embryonic and perinatal tissue, the antibody penetrates through the sample, and one can follow axons through and across the section for long distances. In the adult, only the surface layers of the section stain. The vibratome slices were processed as floating sections in 35 mm culture dishes. This approach involving fresh sections was analyzed for various immunohistochemical artifacts. The cells' death in these vibratome sections occurs quite readily. Cells in the superficial layer are broken from the cutting action of the blade, and other cells will die over time from a lack of nutrients and the conditions of incubation. Controls that showed that the staining patterns are not due to some artifactual staining of dead or damaged cells included using other mouse mAbs as well as a variety of incubations in the secondary fluoresceinated antibody. Furthermore, a number of antibodies of the IgM isotype: B5, B14, B25, and C6 were characterized in parallel with B30 (also an IgM). B5 and B25 stain neuronal cells on their surface, while B14 and C6 stain intracellular antigens of fibrous appearance. None of the above antibodies shared any of the specific staining properties of B30 either in the trigeminal system or in the cerebellum (data not shown). Overnight incubation in a commercial mixture of mouse IgMs followed by overnight incubation in fluoresceinated goat anti-mouse IgMs antibodies typically resulted in a low overall background. Thus, the patterns we describe depend on staining with a specific first antibody. Furthermore, the thickness of the section makes it possible to observe cells staining at a depth away from the broken surface layer.

The trigeminal system

Primary sensory neurons of the trigeminal system divide themselves between the trigeminal ganglion and the MesV. The latter forms a narrow layer of scattered monopolar neurons at the lateral margins of the periaqueductal gray matter and extends from the level of the trigeminal motor nucleus caudally to the level of the superior colliculus rostrally. MesV neurons are the cell bodies of primary afferents and are the only such cells to be found within the adult vertebrate CNS. They are also the largest sensory neurons in the CNS. In the adult mouse brain, they number about 1500 (Hinrichsen and Larramendi, 1969). Although many more are generated during embryonic development, there is a period of extensive cell death around birth (Alley, 1974; Hiscock and Straznicky, 1986).

The trigeminal ganglion

B30 labels neurons of the trigeminal ganglion (TG) and their processes (Fig. 1*A*). This micrograph shows a parasagittal section of an E13 mouse head stained with B30. The central branches of the afferent neurons in the TG enter the pontine region of the brain stem (b), turn towards the spinal cord, and ramify in their various termination areas in the brain stem and the spinal

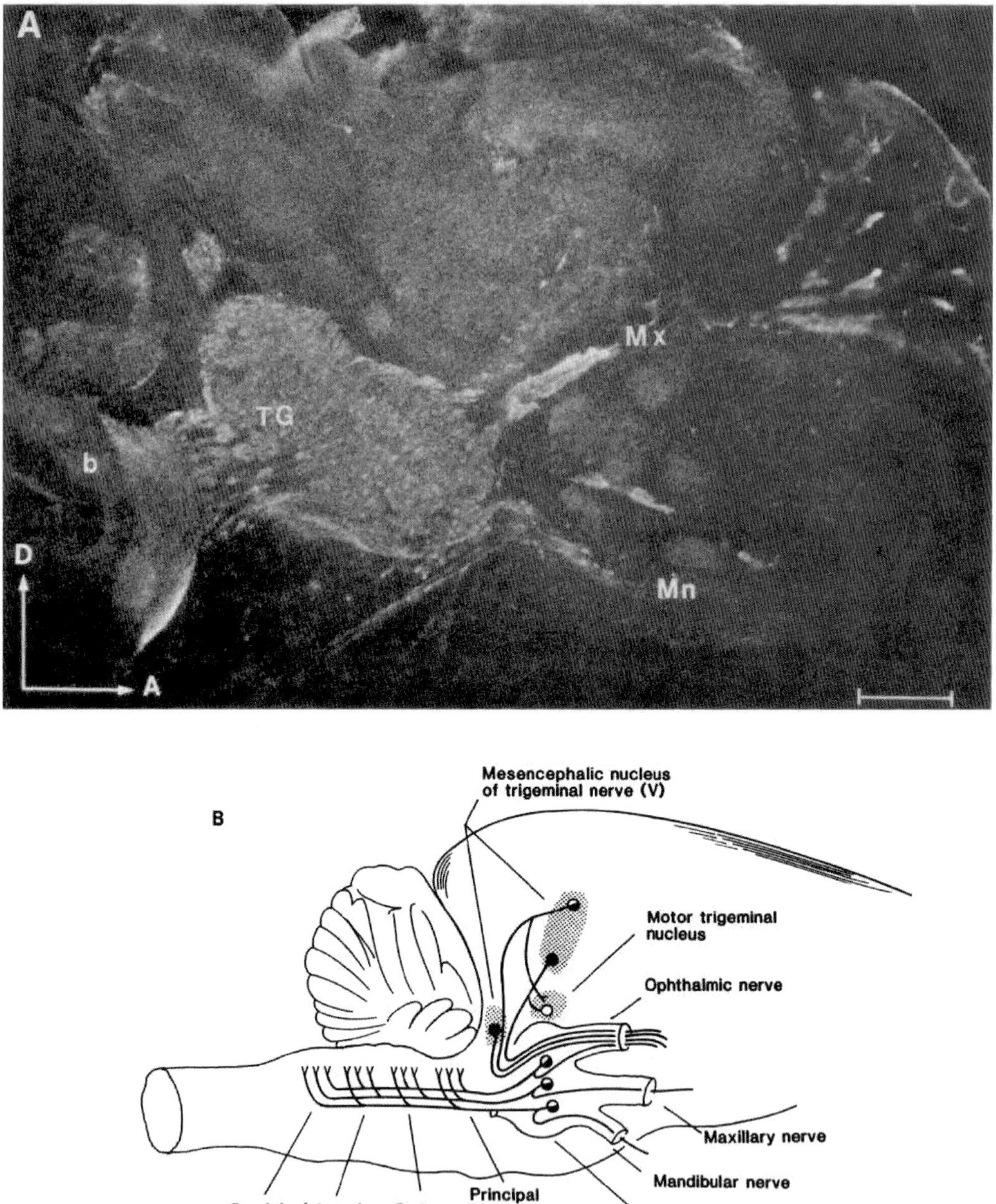

Figure 1. A, B30 labels neurons of the TG. Parasagittal section of embryonic day 13 (E13) mouse head stained with B30. Immunoreactivity outlines the ganglion, its connection to the brain stem (*b*) dorsally as well as parts of its maxillary (*Mx*) and mandibular (*Mn*) peripheral projections. *D,* dorsal; *A,* anterior. Scale bar, 300 μm. *B,* Central connections of the trigeminal nerve. Schematic representation in the sagittal plane. The sensory nuclei of the trigeminal nerve receive afferent input from the skin of the face and oral mucosa, from the ciliary muscle of the eye as well as from receptors in the facial musculature and the temporomandibular joint. The central branches of the afferent neurons in the TG enter the pontine region of the brain stem. They terminate in the principal sensory nucleus and the nuclei of the spinal trigeminal tract—the oral, the interpolar, and the caudal. Collaterals of MesV neurons go to the motor nucleus of the trigeminal nerve. B30 immunoreactivity defines neurons in both the TG and MesV as well as their axons (central and peripheral projections).

The Journal of Neuroscience, July 1989, 9(7) 2471

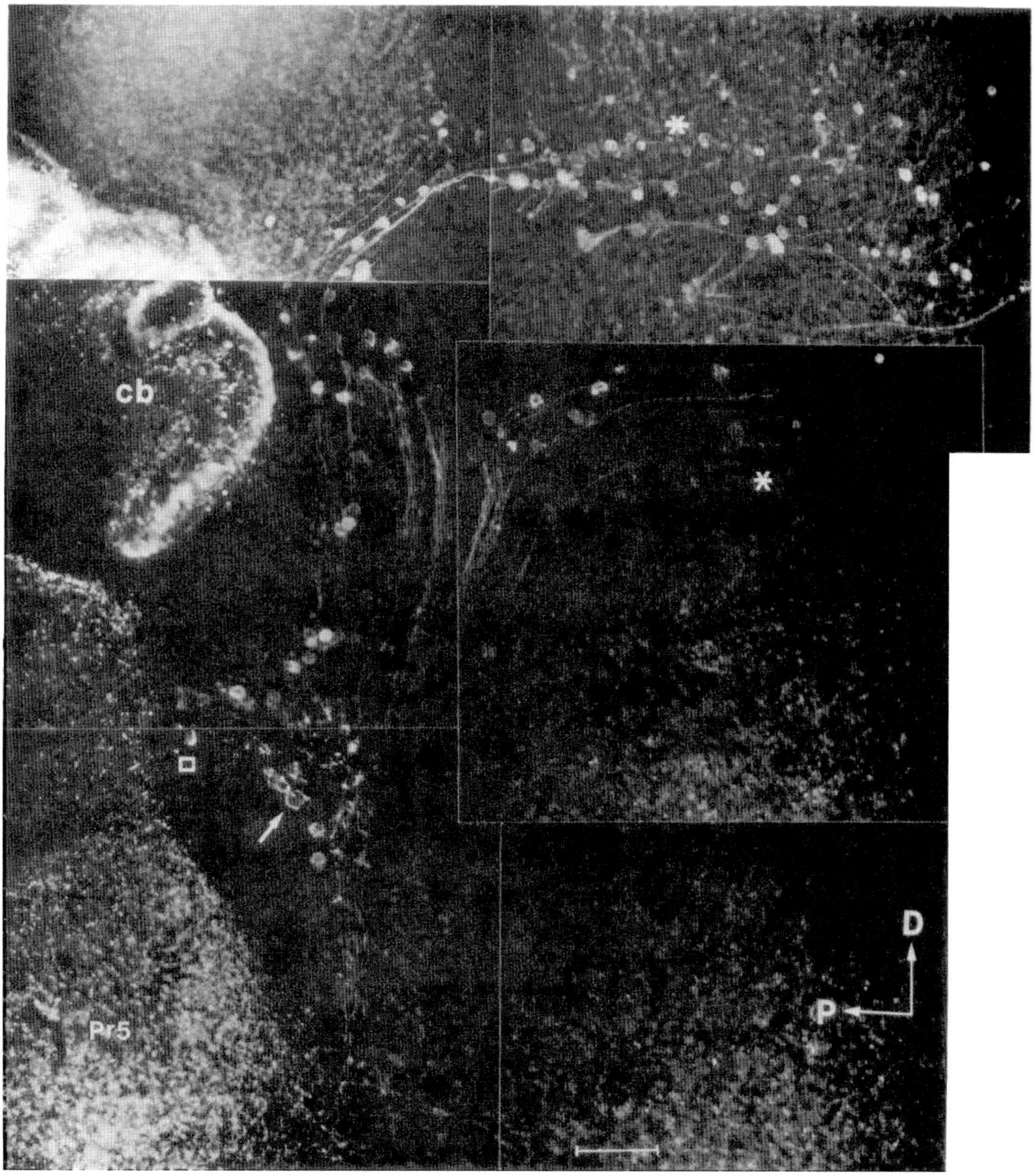

Figure 2. B30 labels neurons of the mesencephalic trigeminal nucleus. Parasagittal section of P4 mouse brain stained with B30. Immunoreactive axons can be followed for several millimeters. The nucleus can be divided into superior (*stars*) and inferior (*arrow*) parts; the cells of the superior part, at the level of the superior colliculus rostrally, send axons brushing past cells of the inferior part at the level of the trigeminal motor nucleus caudally. Some cells of the caudal group extend up into the cerebellar peduncles (*square*). Caudally immunoreactive areas apparent here include premigratory granule cells in the cerebellum (*cb*), spinocerebellar fibers, and the innervated principal sensory nucleus (*Pr5*). *D*, dorsal; *P*, posterior. Scale bar, 200 μm. Smaller cells of the dorsal group are about 20 μm; the larger cells of the ventral group, 35 μm.

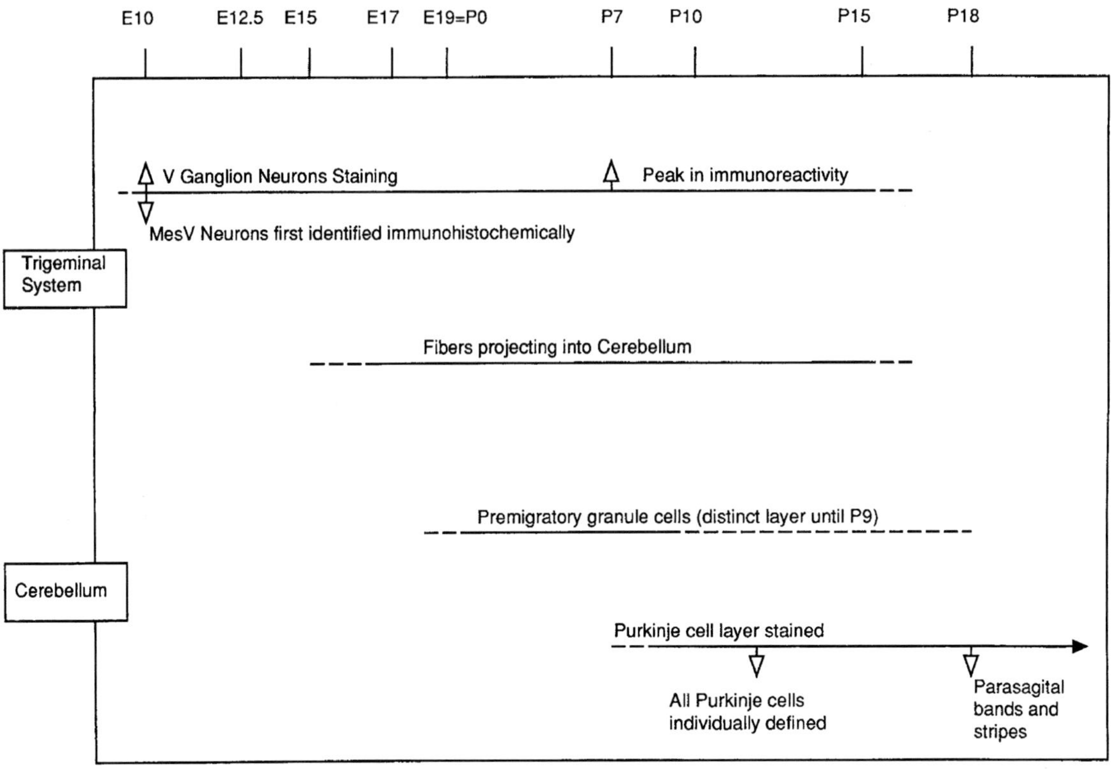

Figure 3. Chart of time course of B30 expression in the developing mouse CNS: it is limited to the trigeminal system (including the TG in the PNS) and the cerebellum. *Horizontal dotted lines* at ends of bars reflect uncertainty as to exact time point of onset or termination of B30 expression between sample days.

cord Peripherally, the trigeminal nerve branches out into 3 major nerves: the ophthalmic (not shown), the maxillary (Mx), and the mandibular (Mn). The specific B30 immunoreactivity allows us to look at the central regions that the TG innervates, thereby confirming results obtained by anatomical and electrophysiological methods. The main termination areas in the CNS are located in the principal sensory nucleus and the 3 nuclei of the spinal trigeminal tract—the oral, the interpolar and the caudal (see Fig. 1*B* for a schematized representation of the trigeminal system and its major connections).

The mesencephalic trigeminal nucleus

The montage in Figure 2 shows a parasagittal section of a P4 mouse brain stained with B30; neurons in the MesV as well as their axons and collaterals are stained, and the immunoreactive axons can be followed for several millimeters. The MesV nucleus is flat and generally lies in only 1 or 2 of the thick parasagittal sections. The nucleus exhibits 2 main groups of neurons in the superior part (stars) and one group in the inferior part (arrow). Also, axons from dorsorostrally located cells brush past cells in the ventrocaudal region. In this latter region, MesV neurons can be seen advancing right into the cerebellar peduncles (square). At this time, B30-immunoreactive premigratory granule cells also delineate the cerebellar folds (cb). Temporally, we have identified B30-positive MesV neurons in embryos as early as E10 (embryos stained as whole amounts; data not shown). B30 immunoreactivity in the MesV peaks around P7 and fades out after 2 weeks postnatally. In the cerebellum, premigratory granule cells stain as a distinct layer from E18 to P9. As we shall show later, Purkinje cells become immunoreactive starting at P7 and continue up to the adult stage. Figure 3 outlines the spatiotemporal distribution of B30 immunoreactivity.

Because of their central location and their supposed origin from the neural crest (Narayanan and Narayanan, 1978), MesV neurons have been an interesting and convenient group of neurons for anatomical, histological, and physiological investigations from the time of Meynert (1872). These cells establish their peripheral connections postnatally in line with the relatively late establishment of tooth development and innervation and the late appearance of secondary endings in the muscle spindles of the jaw. Synaptogenesis within the mesencephalic nucleus takes place at the same early postnatal period (Alley, 1973). At this time of hypothesized synapse formation within the nucleus, the neurons are very sharply outlined by B30 immunoreactivity. Staining of horizontal sections shows the semilunar arrangement of the nucleus, thereby confirming its identity. Figure 4 shows 2 crescent-shaped nuclei laterally surrounding the aqueduct (Aq), while most MesV axons run together rostrocaudally. In this figure, immunoreactive premigratory granule cells delineate the rostral part of the cerebellum (Cb). Within the nucleus, B30 immunoreactivity does not define geographical subpopulations of neurons, although there is a certain quanti-

The Journal of Neuroscience, July 1989, 9(7) 2473

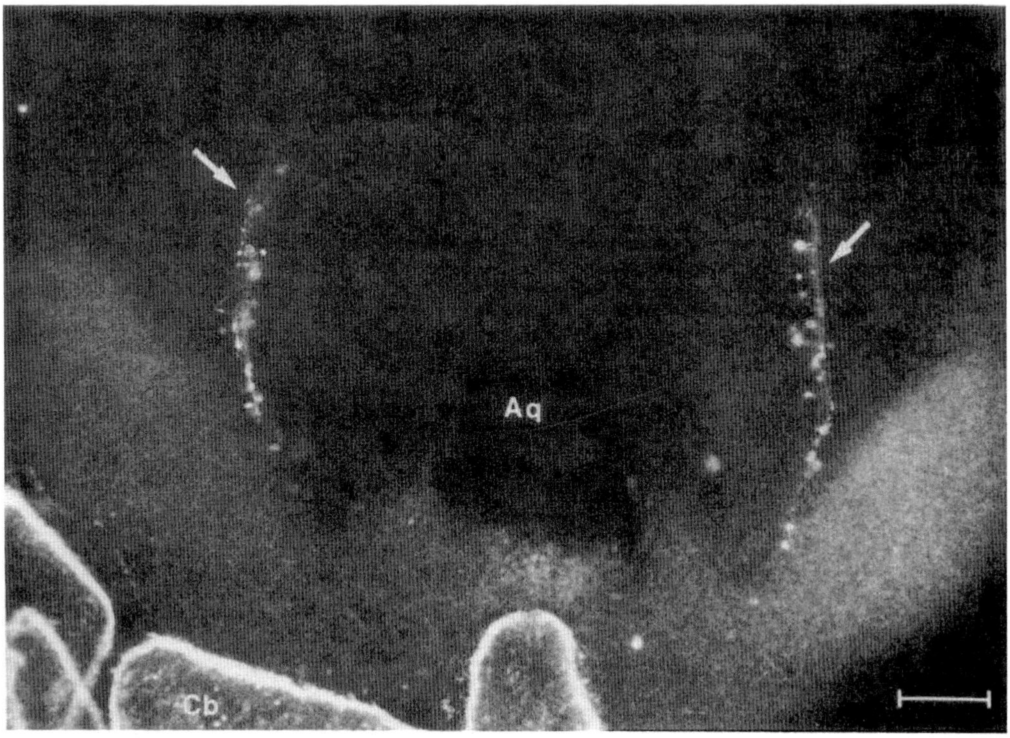

Figure 4. Semilunar arrangement of the trigeminal mesencephalic nucleus (*arrows*). Horizontal sections of P5 mouse brain stained with B30. Micrograph showing the arrangement of the MesV nucleus around the Aqueduct of Sylvius (*Aq*). The immunoreactivity observed caudally is associated with the cerebellum (*Cb*). Scale bar, 300 μm.

tative range of staining intensity. We estimate that 10–15% of the MesV neurons do not stain at P7, but this heterogeneity may be due to the presence of naturally dying or dead neurons.

We have followed MesV axons and collaterals that terminate in brain-stem regions such as the trigeminal motor nucleus and the reticular areas. Cerebellar projections of MesV have also been suggested from axonal transport studies (Saigal et al., 1980; Somana et al., 1980). Figure 5 shows B30-positive fibers (arrow), thought to originate from the spinal cord, projecting specifically to the 2 caudalmost cerebellar lobules. The inset is a low-power micrograph of the same field: the staining of premigratory granule cells delineates the entire cerebellum, while the B30-reactive fibers terminate specifically in the caudalmost area. We also have observed stained MesV neurons advancing right into the cerebellar peduncles. The B30-immunoreactive neuron that sits at the center of Figure 6 sends an axon towards the cerebellum (Cb). This axon was traced into the caudalmost cerebellar lobule (out of the field of this picture). So although we have not yet determined the origin of the B30-immunoreactive spinocerebellar projections, this immunohistochemical technique allowed us to confirm the hypothesized cerebellar projections of a few MesV neurons.

Perinatally, an area around the amygdala contains a few unresolved immunoreactive neurons and axons (data not shown).

Morphology of the MesV neurons

Our immunohistochemical data reveal interesting steps in the morphogenesis and organization of MesV neurons, confirming light microscopic studies done by Hinrichsen and Larramendi (1969), Alley (1974), and others.

Within the MesV nucleus, cell size variation is apparent; large cells (30–35 μm), pear-shaped and usually unipolar, occupy in general the ventral portion of the nucleus (Fig. 7*A*, arrow). Medium-sized (Fig. 7*A*, star) and small cells, more irregular in shape, are mostly unipolar though some are bipolar; they predominate in the rostral and ventromedial part of the nucleus. Perinatally, axons of dorsal cells can be seen to form *en passant* transient axodendritic contacts with more medial cells exhibiting a few dendrites (Fig. 7*B*, arrows). Ventrocaudally those same axons come by the cells of the inferior part of the nucleus at such close proximity that the soma and bypassing axons cannot be separated optically. These latter cells exhibit bouton-like structures, thus suggesting axosomatic connections (see Hinrichsen and Larramendi, 1968). Postnatally, the cell surface is also studded with small spines, which may be vestiges of resorbed dendrites or might reflect increases in its surface area to help the cell in its metabolism (Hinrichsen and Larramendi, 1969) (Fig. 7*C*). Alternatively, the close timing of spine formation and synaptogenesis in the MesV (starting around postnatal day 5 for a period of 10 d in the hamster; see Alley, 1973) suggests that these surface structures may participate in synapse formation. Neurons are sometimes seen closely juxtaposed (see Figs. 2, 7*A*). Extensive somasomatic contacts within these groups give rise to numerous plasmalemmal membrane specializations resembling gap junctions and suggesting electrotonic coupling (Hinrichsen, 1970; Baker and Llinás, 1971).

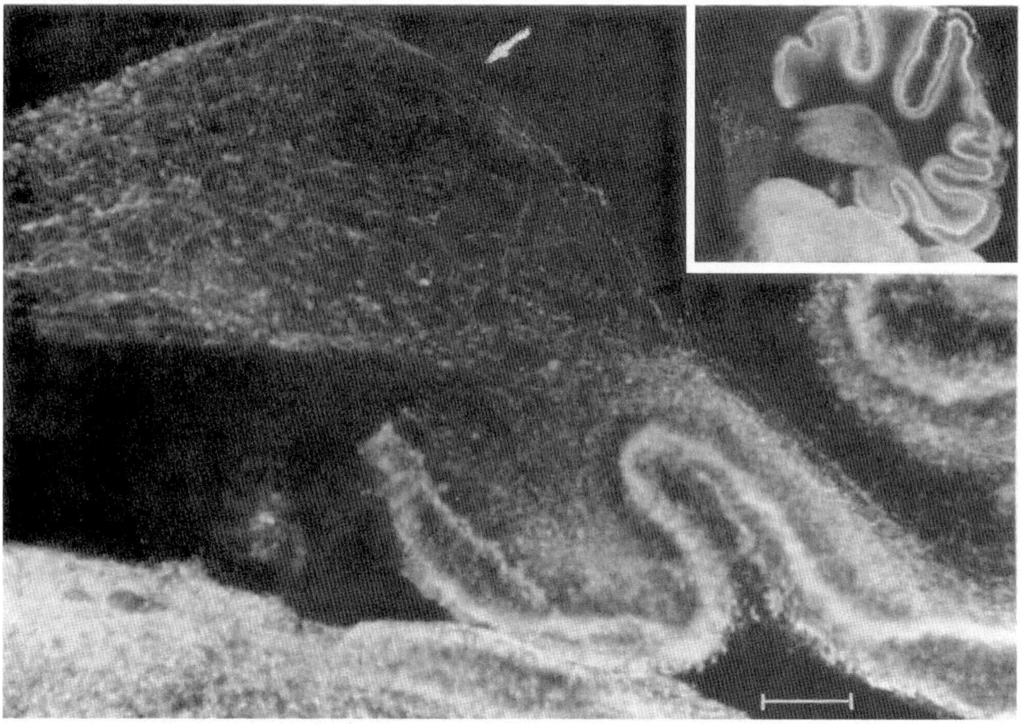

Figure 5. Immunoreactive fibers projecting into the cerebellum terminate specifically in the 2 caudalmost lobules. Parasagittal section of a P4 mouse cerebellum stained with B30. The cellular identity of their targets can not be determined at this stage as these axons do not establish synaptic connections until after 1 week postnatally. A confocal microscope is now being used to resolve the projection of these fibers on sections of older animals. Scale bar, 120 μm. *Inset,* Low power micrograph of the same field; MesV neurons can be observed advancing into the cerebellar peduncles while the B30 immunoreactive spinocerebellar fibers project specifically into the caudalmost lobules. The immunoreactivity in the region of the brain stem underlying the cerebellum outlines the termination areas of the central trigeminal projections in the spinal trigeminal nuclei.

Cerebellum

The B30-immunoreactive fibers reach well into the cerebellum by E15, hereby confirming findings by Mason (1985) and others. They observed that axons are present within cerebellar anlage for days before birth waiting for the differentiation of target cell dendrites. These axons establish synaptic connections during the second postnatal week.

The cerebellar folds are outlined by a thin layer of B30-reactive cells at E19. Figure 8 shows that at P5 only a specific subset of the external granule cell layer (EGL) is stained, i.e., the premigratory granule cells. They are seen just entering the molecular layer (ML). At postnatal day 7 (P7), B30 stains the premigratory granule cells as well as a layer just inside the ML: the monolayer where the Purkinje cells (PCs) are localized. Subsequently, individual PCs will become distinct. So at P12, as shown in Figure 9, *A*, *B*, all PCs, their axons and their dendritic trees are labeled. Purkinje axons (Fig. 9*A*, arrow) go through the internal granule cell layer (IGL) to the white matter where they gather to project mainly to the deep cerebellar nuclei. The dense staining of the ML in Figure 9*B* is due to the heavily branched dendritic trees of the PCs. The fibers projecting into the cerebellum are staining very lightly at this stage. By P18, granule cell migration is completed (Larramendi, 1969).

By P18, a mosaic or banded pattern of staining of PCs appears. At this time, B30 selectively stains only certain PCs, their axons and their dendritic trees to form a mosaic of immunoreactive bands, which persists into the adult stage. This pattern is best viewed in horizontal sections of a 3-week-old mouse cerebellum as shown in Figure 10, *A*, *B*. The immunoreactive PCs are clustered together in groups of varying sizes (usually from 5 to 10 PCs). These groups are most clear in the vermis, and they form parasagittal bands running rostrocaudally throughout the cerebellar cortex. The bands of B30-positive cells alternate with similar bands of B30-negative cells.

Figure 10*C* shows a pattern of thin immunoreactive stripes that extend across the ML. This pattern is seen mostly in the hemispheres and sometimes on the borders of the parasagittal bands described above. Each stripe is associated with an individual B30-positive Purkinje cell and lies 45–60 μm away from the next stripe. These narrow structures may be Purkinje trees in their last stage of maturation: the dendritic trees mature morphologically and become isoplanar, running in a plane perpendicular to the axis of the folia, thus giving this striped staining pattern. A similar phenomenon has been observed by Hawkes and Leclerc (1987) using an anti-microtubule associated protein antibody (anti-zebrin; see Discussion), and they referred to is as satellite bands.

The Journal of Neuroscience, July 1989, 9(7) 2475

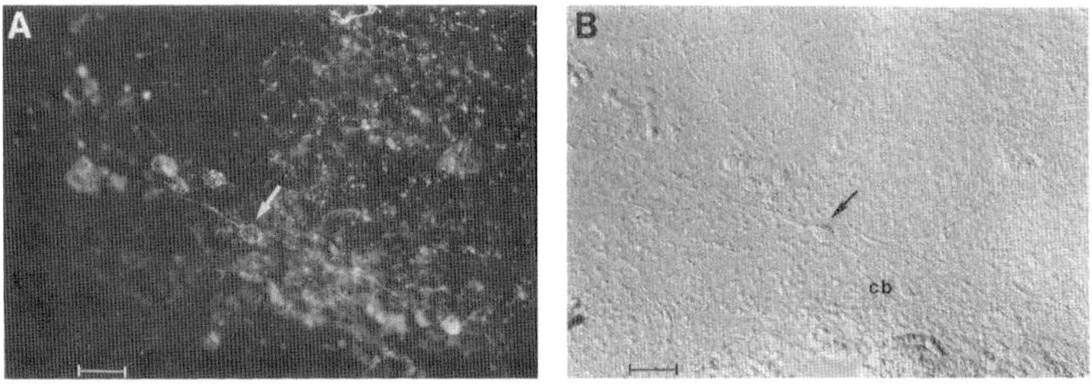

Figure 6. Projection of a MesV neuron into the cerebellum. A MesV neuron (*arrow*) advances into the cerebellar peduncles and sends an axon into the cerebellum (*cb*). *A*, Fluorescent exposure of a parasagittal section (P5 mouse) stained with B30. *B*, Hoffman phase micrograph of the same field. Scale bars, 55 μm.

Staining: antibody penetration in thick sections

To address the problem of antibody penetration in thick sections, we examined a normally processed 120-μm-thick parallel section of a P5 brain with a Lasersharp MRC-500 confocal microscope and optically dissected it every 17 μm. Immunoreactive MesV neurons and axons appeared in every optical section. Figure 7*D* shows a stereo pair, constructed by projecting the 8 sections with an offset of ± 1. This 3-dimensional view reveals a previously unobserved phenomenon in the MesV nucleus: a unipolar neuron sends an axon that apparently terminates on a passing axon (arrow).

Cell surface staining

We have been able to stain with B30 *in vivo*. Newborn mice were injected intracranially with B30 ascites (10 μl); 16 hr later, they were either killed and processed normally (sections incubated for 2 hr with fluoresceinated secondary antibody after a 6 hr incubation in 10% NGS) or injected intracranially with 10 μl of undiluted fluoresceinated secondary antibody. These latter animals were processed 12 hr after the second injection. Appropriate controls were done in parallel. In all cases, immunoreactivity defined a small area of trauma at the site of injection (around the fourth ventricle). However, B30 clearly defined a number of MesV neurons and their axons, fibers in the cerebellum reaching into the caudalmost lobules, and premigratory granule cells in the external granule cell layer. Figure 11*A* shows a parallel section of a P3 mouse after sequential injection of both primary and secondary antibodies. A group of stained MesV neurons appears caudally (star), while more rostrally a well-defined MesV neuron sends an axon laterally (arrow). Also, the section shown in Figure 7, *A*, *B* were both processed from brains injected with B30 ascites and stained, after sectioning, with the secondary antibody alone. These experiments directly confirm *in vivo* the cell surface immunoreactivity observed in the vibratome sections and also illustrate the relative permeability of the early postnatal brain.

Furthermore, live cultures of dissociated E13 mouse embryo TG were stained with B30. Figure 11*B* shows the cell surface labeling of a B30-immunoreactive neuron 6 d after plating.

B30 antigen

Binding of B30 to cells in culture was eliminated by pretreating the cultures with organic solvents (chloroform/methanol, 2:1), suggesting that the antigen was a lipid. Mild proteolytic treatments of similar cultures did not affect their immunoreactivity (in fact, staining was slightly enhanced), and Western blots of membrane proteins prepared from B30-immunoreactive areas of P5 mouse brains did not reveal any specific antigen (data not shown). Figure 12 shows an immunostaining of a thin-layer chromatography (TLC) plate of acidic glycolipid extracts of dissected P7 midbrain and brain stem (B30 trigeminal immunoreactive areas), P12 cerebellum, and P7 cortex. Two discrete bands were found in both B30-immunoreactive areas but not in the cortex. One of the bands migrated slightly above GM1, the other migrated between GD1a and GD3 in a chloroform/methanol/aqueous $CaCl_2$ solvent system. No other immunoreactivity suggestive of lipoproteins or phospholipids was observed on the TLC plate. Neuraminidase digestion of P12 cerebellar acidic glycolipids and chromatography on DEAE-Sephadex A25 showed that the antigens contain sialic acid (data not shown). These results imply that the B30 epitope is carried on by 2 distinct gangliosides. The migration pattern of the B30 antigen is similar to that of the Jones antigen (Schlosshauer et al., 1988), but they differ in the base lability of their epitopes (data not shown) and the relative distribution of immunoreactivity on TLC plate analysis of gangliosides from 2-week-old cerebellum (cf. Fig. 12, lane a, and Schlosshauer et al., 1988, fig. 4a, P14). Moreover, B30 immunocytochemistry in the developing rat CNS [where the Jones antigen was shown to be associated with neural cell and process migration (Mendez-Otero et al., 1988)] revealed no neuronal cell surface immunoreactivity except for that associated with the Purkinje cells (which parallels that observed in the mouse cerebellum).

Discussion

The monoclonal antibody B30 recognizes a cell surface antigen that is expressed selectively in the developing CNS of the mouse. Immunoreactivity is confined to 2 separate systems, the trigeminal system and the cerebellum. Spatiotemporally regulated, the

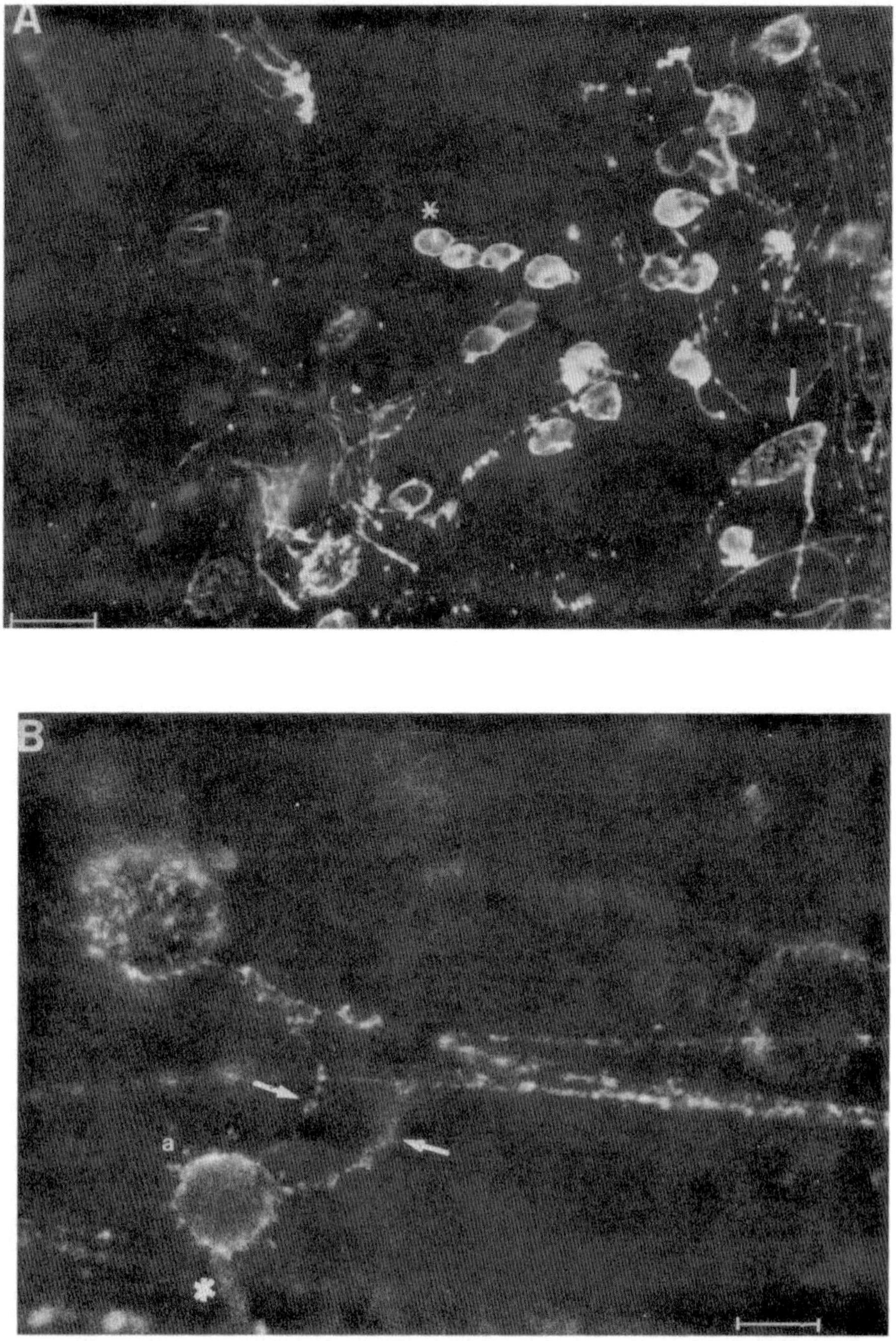

Figure 7. Morphological characteristics of MesV neurons. *A*, Cells, mostly unipolar, differ greatly in size. Large cells, 35–40 μm (*arrow*), predominate in the ventral part of the nucleus; smaller cells, 20–25 μm (*star*), predominate in the rostral part. Parasagittal section of a P7 mouse brain stained with B30; ventral region of the nucleus. Scale bar, 45 μm. *B*, Perinatally, short dendrites can be observed making axodendritic contacts. Parasagittal section of P3 mouse brain stained with B30. Neuron (*a*) sends 2 short dendrites (*arrow*) which contact passing axons. Its axon (*star*) projects out of the plane of the picture. Scale bar, 25 μm. *C*, Cell surface is also studded with small spines whose role is unclear at this time. Fluorescent

The Journal of Neuroscience, July 1989, 9(7) 2477

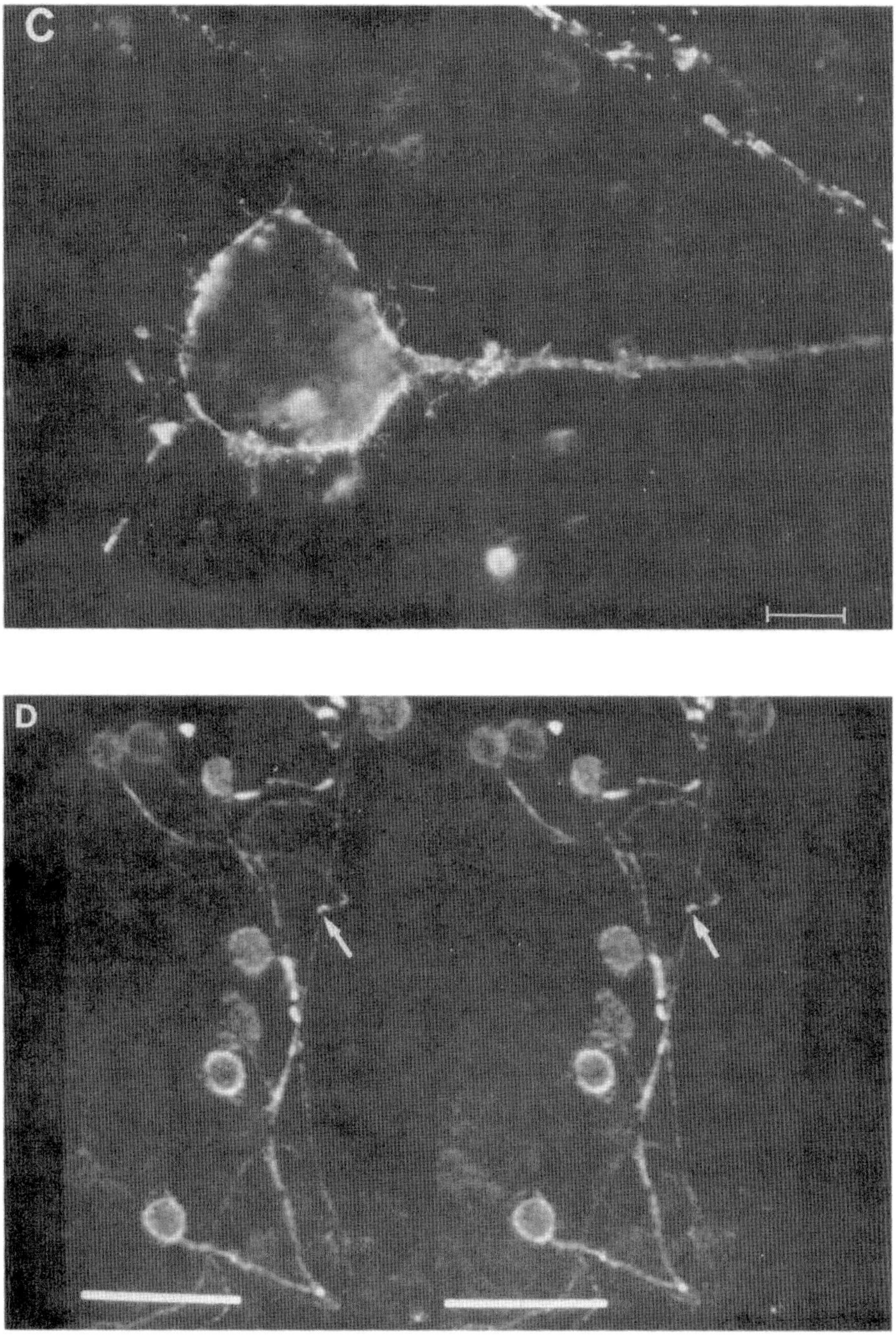

micrograph of a MesV neuron in a P7 mouse brain. Scale bar, 15 μm. *D*, Three-dimensional arrangement of MesV neurons. Horizontal section of a P6 mouse brain stained with B30 and observed with a Lasersharp MRC-500 confocal microscope. Stereo pair obtained by projecting 8 frames with opposing offsets. This method reveals several interesting features including an axoaxonal contact by which a specific axon apparently terminates on a passing vertical axon (*arrow*). Scale bar, 50 μm.

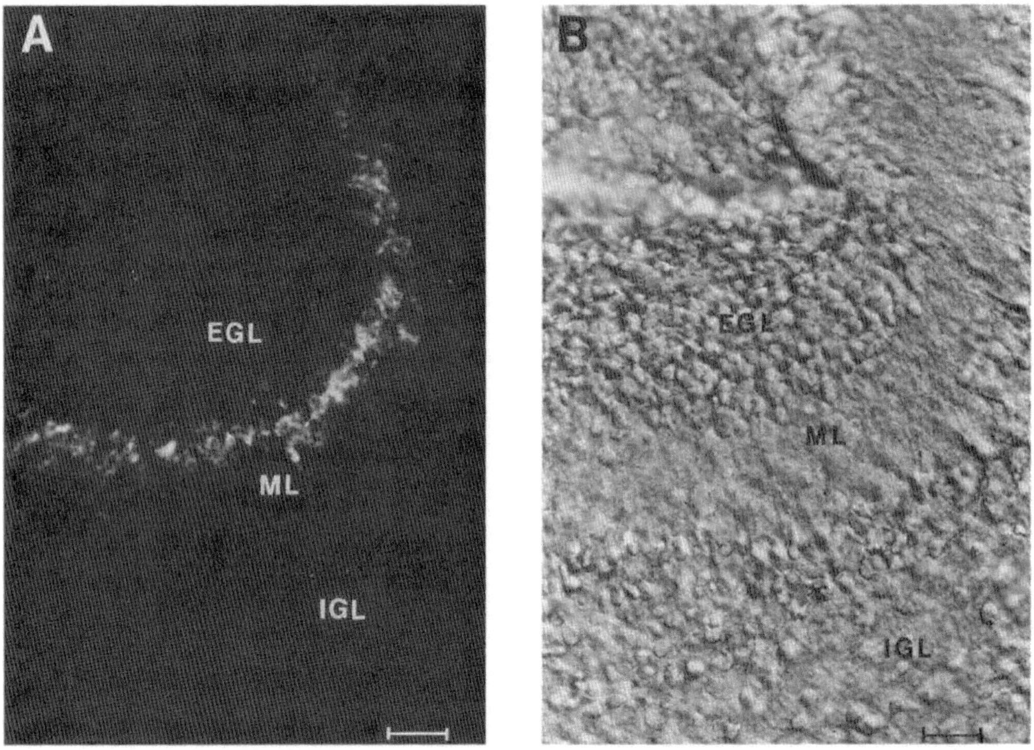

Figure 8. B30 labels the premigratory granule cell layer in early postnatal cerebellum. *A*, Parasagittal section of mouse P5 cerebellum stained with B30. At this stage, only a subset of the cells in the external granule cell layer (*EGL*) are staining. They form the layer of postmitotic cells just outside the molecular layer (*ML*). *B*, Hoffman phase image of the same field; notice the molecular layer (*ML*) and the internal granule cell layer (*IGL*) showing no immunoreactivity. Scale bars, 55 μm.

B30 antigenic determinant may be correlated with a synaptogenetic period both in the mesencephalic trigeminal nucleus and on Purkinje cells.

The trigeminal system

B30 stains neurons in the mesencephalic trigeminal nucleus, thus providing a molecular marker for a unique group of cells: MesV neurons constitute the only primary sensory neurons in the CNS. B30 staining allowed us to trace their axons, some of them extending into the cerebellum. The expression of the B30 antigen is transient and seems to be heterogenous from cell to cell, correlating it with the differentiation of the nucleus and of individual neurons. By *in vivo* staining experiments with postnatal animals and by staining cells in primary cultures, we were able to confirm the association of the antigen with the cell surface and to demonstrate its existence on live cells. Other primary sensory neurons of the PNS, specifically neurons in other cranial nerves and in dorsal root ganglia, also exhibit B30 immunoreactivity on both their peripheral and central projections (data not included). The characteristics and time course of their immunoreactivity seem to follow those of the trigeminal neurons (see Fig. 3).

The cerebellum

B30 defines different cerebellar layers during development, independently of the trigeminal system. Perinatally it labels a set of postmitotic cells ready to migrate, the premigratory granule cells. Purkinje cells, in a monolayer by P5, start staining around P7, and by P12 all the Purkinje cells, their axons, and their dendritic trees are outlined by B30. This corresponds in time to the maturation of their dendritic trees and the formation of synapses by the climbing fibers and by the parallel fibers onto those dendrites (Altman, 1972).

As some Purkinje cells lose B30 immunoreactivity, a banded pattern of staining of these cells appears around P18 and continues into the adult mouse. The existence of such bands of staining of Purkinje cells in an alternating pattern has led to much speculation that it might be related to the plasticity of certain synapses or to a specific afferent source (see Leclerc et al., 1987). There is extensive evidence, anatomical, biochemical and physiological, for a parasagittal zonation of the mammalian cerebellar cortex and also of cortical afferent and efferent projections (for a physiological review, see Scott, 1963; Bloedel and Courville, 1981). Bands similar to those obtained immuno-

The Journal of Neuroscience, July 1989, 9(7) 2479

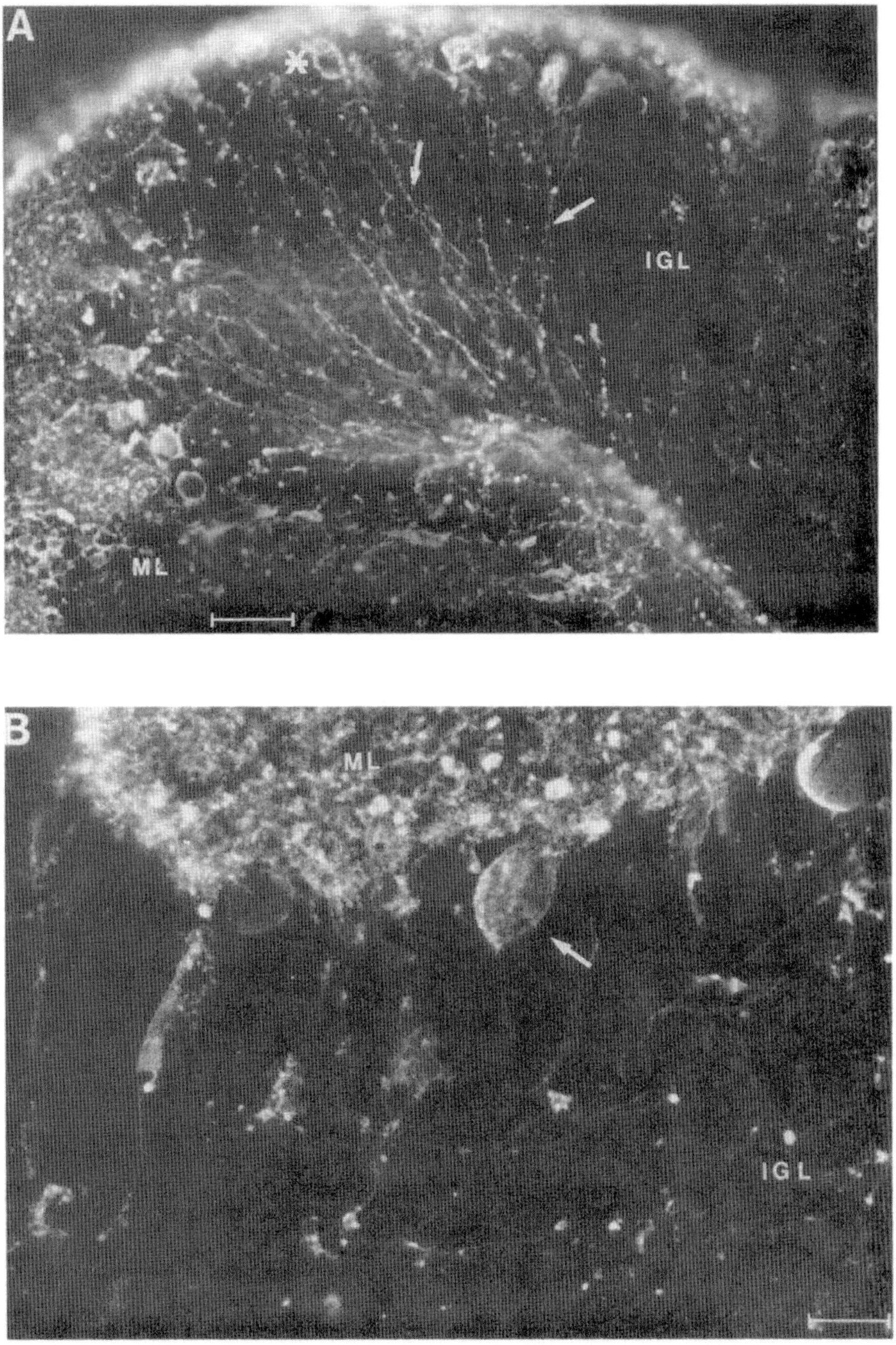

Figure 9. B30 labels all the Purkinje cells, their axons and their dendritic trees by P12. *A*, Parasagittal section of mouse P12 cerebellum stained with B30. Postmitotic Purkinje cells form a monolayer between the molecular layer (*ML*) and the internal granule cell layer (*IGL*) by P5. This monolayer starts staining around P7 and individual Purkinje cells (*star*) become distinct by P9. Purkinje axons (*arrow*) go through the IGL to the white matter where they gather to project mainly to the deep cerebellar nuclei. Scale bar, 60 μm. *B*, Fluorescent micrograph of B30 stained Purkinje cells and axons. Notice the dense staining of the molecular layer (*ML*) above the Purkinje cells due to their heavily branched dendritic trees. The IGL is devoid of any immunoreactivity except for the Purkinje axons. Scale bar, 30 μm.

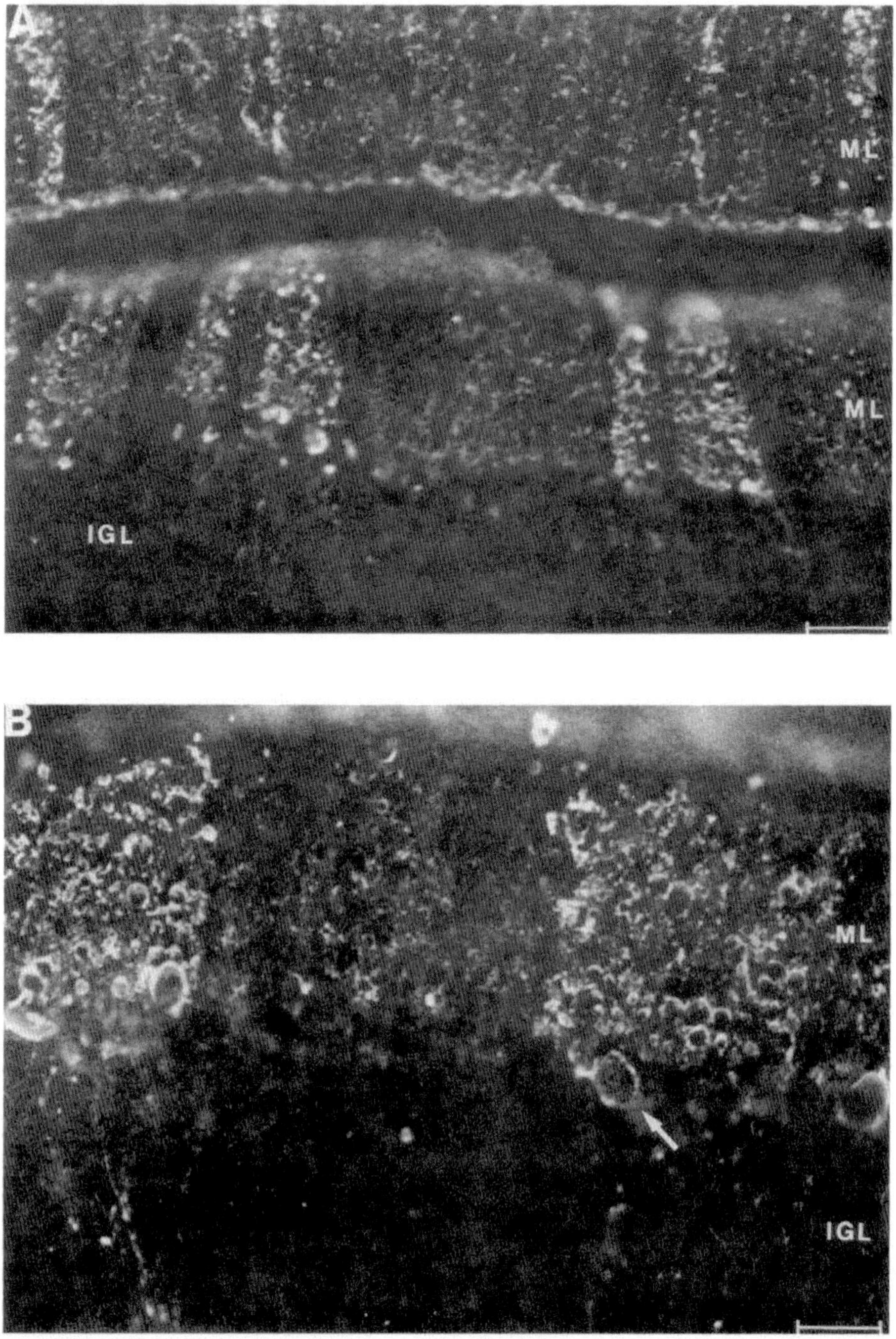

Figure 10. Heterogeneity in B30 immunoreactivity of Purkinje cells at P20. *A*, Horizontal section of P20 mouse cerebellum stained with B30. Parasagittal bands of B30 immunoreactivity appear alternating with similar bands of cells lacking B30 immunoreactivity. These bands are mostly seen in the vermis. Scale bar, 120 μm. *B*, Micrograph of a similar field as in (*A*). The bands are usually from 5 to 10 cells wide and include the Purkinje cell bodies (*arrow*), their axons and dendritic trees. Scale bar, 60 μm. *C*, Horizontal section of a P18 mouse cerebellum folium from the left hemisphere. Narrow, positively immunoreactive bands of cells, most often only one cell wide are found mostly in the hemispheres of late postnatal mouse cerebellum. Scale bar, 120 μm. *IGL*, internal granule cell layer; *ML*, molecular layer.

The Journal of Neuroscience, July 1989, 9(7) 2481

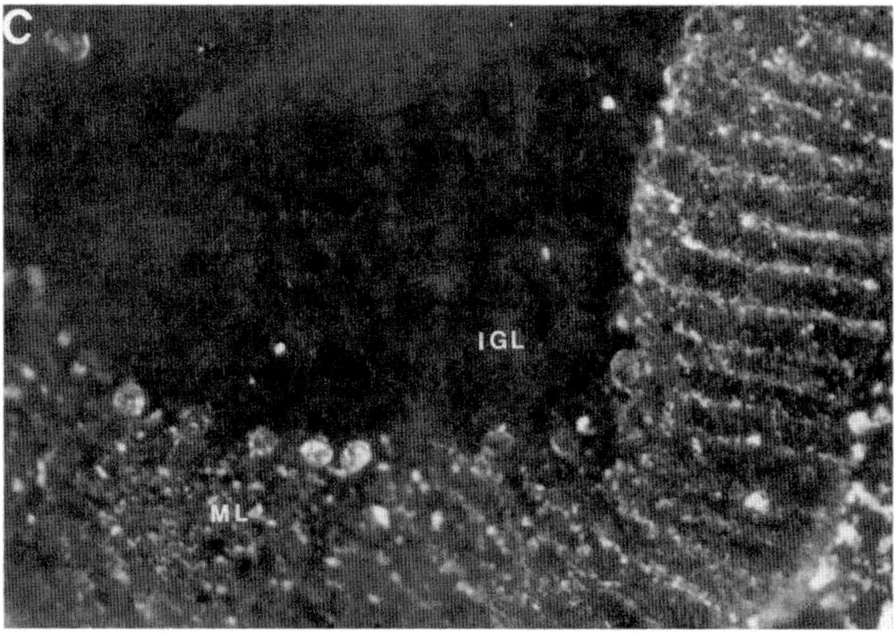

Figure 10. Continued.

histochemically with B30 have been seen in the cerebellar cortex of immature cats stained for acetylcholinesterase (Marani and Voogd, 1977) and have also been observed in the rat cerebellum by Hawkes and Leclerc (1987). The latter used an anti-microtubule associated protein antibody (anti-zebrin) and showed antigen expression spreading throughout the cerebellar cortex during the second postnatal week. Subsequently, immunoreactivity is suppressed in a subset of the Purkinje cells, thus developing the adult pattern of parasagittal antigenic bands. Preliminary experiments have shown that the B30-positive regions correspond to the anti-zebrin-positive areas (R. Hawkes, personal communication).

Technical considerations

We sought cell surface antigens that would appear on unfixed cells and that would define rare populations of neurons. Our initial approach aimed neither to bias the immune recognition process nor to alter the conformation of the antigens in any way. Using whole-cell suspensions from embryonic brains as immunogen was successful in enriching for cell surface antigens. In the original fusion, 10% of the wells screened contained a hybridoma line secreting antibodies against antigens located at the cell surface.

Our screening procedures tried to combine both efficiency and accuracy while extending our range of detection. Triturating the brain into very small pieces allowed us to look for rare activities while surveying a representative sample of the developing CNS. Staining fresh chunks of tissue enabled us to distinguish surface staining and to preserve some of the cell–cell relationships during the early screenings. Subsequently, brains of different embryonic and postnatal stages were sliced in order to verify and to localize the specific mAb binding. Thick fresh sections were used for immunohistochemistry: they preserve the morphological and chemical integrity of the cells and their environment, while still allowing enough penetration to give a good perspective of the neuronal interactions.

Functional considerations of the B30 antigen

B30 defines 3 homogeneous groups of CNS neurons during varying periods of their maturation: (1) MesV neurons from a time of early axon outgrowth and guidance (E10) through a time of synapse formation, postnatally; (2) premigratory, postmitotic granule cells in the early postnatal cerebellum; and (3) the Purkinje cells during a period of synapse formation and dendritic tree maturation.

In light of recent reports, it has become evident that not only glycoproteins on the cell surface, but also cell surface gangliosides may be involved in cellular interactions and adhesion (Cheresh et al., 1986; Sariola et al., 1988). Specifically, gangliosides are thought to redistribute actively into discrete areas of the cell, and this may cause synergism with cell surface receptors by creating an appropriate electrostatic environment (Cheresh and Klier, 1986; Cheresh et al., 1986; Burns et al., 1988). Moreover, biochemical data show that GD2 augments vitronectin receptor function in the presence of Ca^{2+} (Cheresh et al., 1987). The B30 epitope is carried on a minor ganglioside. Much like cell surface molecules involved in differential adhesion and axon guidance, surface molecules involved during final target recognition and formation of specific synapses (synaptogenetic molecules) may act in a combinatorial and synergistic fashion. Rare gangliosides could be used as part of a combinatorial recognition system, possibly reflecting different modifications of

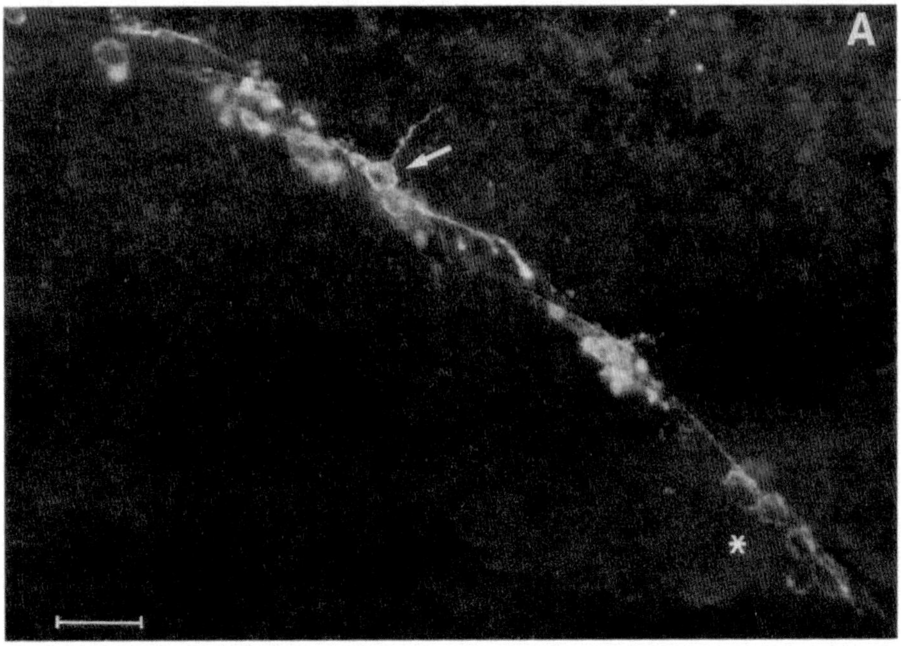

Figure 11. Cell surface association of B30 antigen. *A,* Parallel section of a P3 mouse following *in vivo* staining with B30. B. fluoresceinated secondary antibody were sequentially injected into the brain of a newborn mouse (see text for detail). Subsequently the brain was sliced into 120 μm parallel sections for examination. Notice a caudally situated group of MesV cells (*star*), and, more rostrally, a MesV neuron sending an axon laterally (*arrow*). Scale bar, 60 μm. *B,* B30 staining of live cultures. Embryonic day 13 (E13) mouse trigeminal ganglia were dissociated and single cells were put in culture. Fresh cultures were stained 6 days after plating. Scale bar, 60 μm. *C,* Hoffman phase micrograph of the same field. The *arrow* points to the B30 immunoreactive neuron. Scale bar, 60 μm.

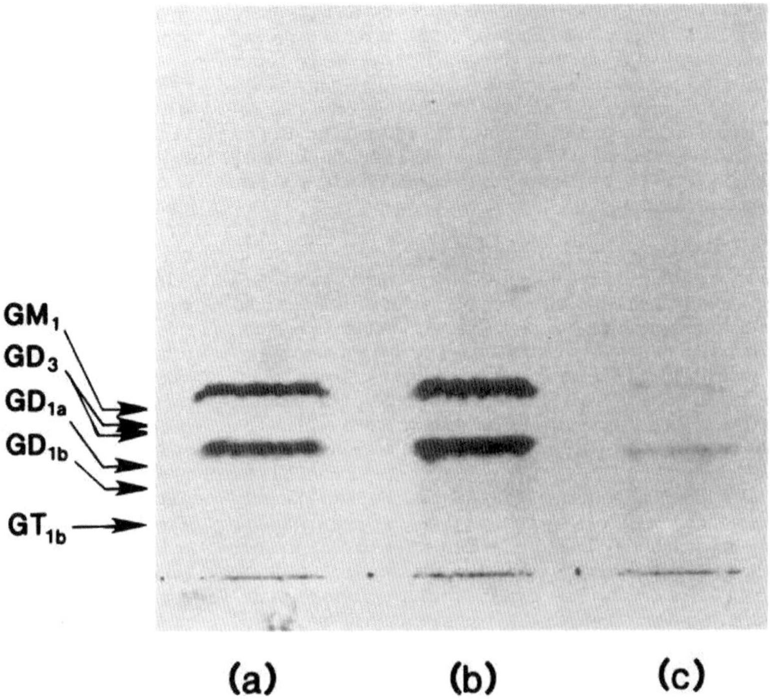

Figure 12. B30 immunostaining of a TLC plate of acidic glycolipids. Acidic glycolipids containing 0.5-1 μg sialic acid were applied to the plate and developed with chloroform/methanol/0.25% aqueous $CaCl_2$ (60:40:10). (*a*) Acidic glycolipids from P12 mouse cerebellum; (*b*) acidic glycolipids from P7 mouse midbrain and brain stems (dissected to include the B30 immunoreactive areas); (*c*) acidic glycolipids from P7 mouse cortex. Ganglioside standards from Calbiochem (GM_1, GD_{1a}, GD_{1b}, GT_{1b}) were run in parallel and stained with resorcinol; GD_3 doublet migration was determined by R24 binding (data not shown).

The Journal of Neuroscience, July 1989, 9(7) **2483**

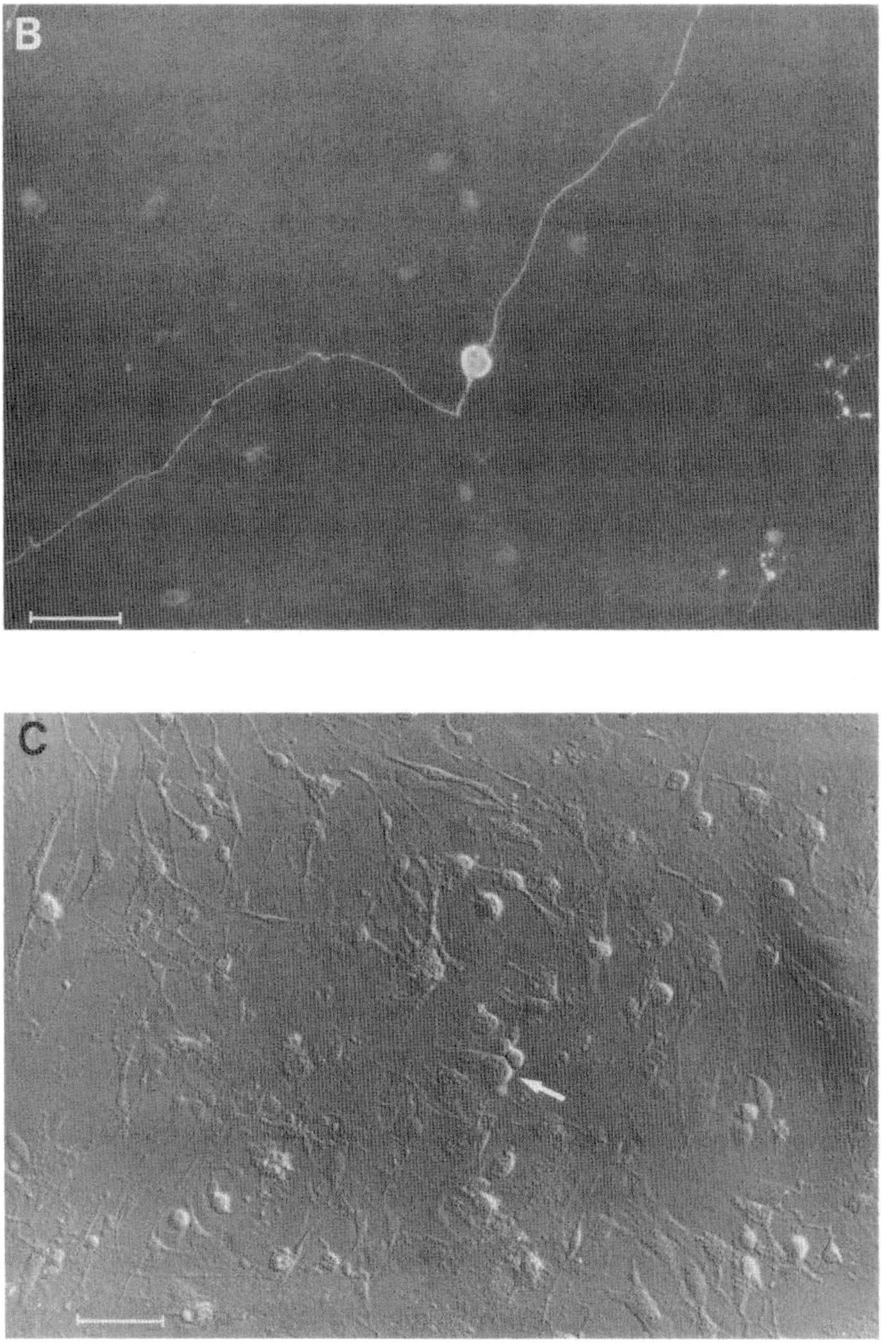

Figure 11. Continued.

an underlying molecule. Thus, the B30 antigen might be part of such guiding and synaptogenetic arrays, used differently in different systems.

References

Alley, K. E. (1973) Quantitative analysis of the synaptogenetic period in the trigeminal mesencephalic nucleus. Anat. Rec. *177:* 49–60.

Alley, K. E. (1974) Morphogenesis of the trigeminal mesencephalic nucleus in the hamster: Cytogenesis and neurone death. J. Embryol. Exp. Morphol. *31:* 99–121.

Altman, J. (1972) Postnatal development of the cerebellar cortex in the rat. II. Phases in the maturation of Purkinje cells and of the molecular layer. J. Comp. Neurol. *145:* 399–464.

Baechtel, S., and M. D. Prager (1982) Interaction of antigens with dimethyldioctadecylammonium bromide, a chemically defined biological response modifier. Cancer Res. *42:* 4959–4963.

Baker, R., and R. Llinás (1971) Electrotonic coupling between neurons in the rat mesencephalic nucleus. J. Physiol. (Lond.) *212:* 45–63.

Barnstable, C. J. (1980) Monoclonal antibodies which recognize different cell types in the rat retina. Nature *286:* 231–235.

Bastiani, M. J., A. L. Harrelson, P. M. Snow, and C. S. Goodman (1987) Expression of fasciclin I and II glycoproteins on subsets of axon pathways during neuronal development in the grasshopper. Cell *48:* 745–755.

Bloedel, J. M., and J. Courville (1981) Cerebellar afferent systems. In *Handbook of Physiology,* V. Brookes, ed., pp. 735–829, Williams & Wilkins, Baltimore.

Burns, G. F., C. M. Lucas, G. W. Krissansen, J. A. Werkmeister, D. B. Scanlon, R. J. Simpson, and M. A. Vadas (1988) Synergism between membrane gangliosides and Arg-Gly-Asp-directed glycoprotein receptors in attachment to matrix proteins by melanoma cells. J. Cell Biol. *107:* 1225–1230.

Cheresh, D. A., and F. G. Klier (1986) Disialoganglioside GD2 distributes into substrate-associated microprocesses on human melanoma cells during their attachment to fibronectin. J. Cell Biol. *102:* 1887–1897.

Cheresh, D. A., M. D. Pierschbacher, M. A. Herzig, and K. Mujoo (1986) Disialogangliosides GD2 and GD3 are involved in the attachment of human melanoma and neuroblastoma cells to extracellular matrix proteins. J. Cell Biol. *102:* 688–696.

Cheresh, D. A., R. Pytela, M. D. Pierschbacher, F. G. Klier, E. Ruoslahti, and R. A. Reisfield (1987) An Arg-Gly-Asp-directed receptor on the surface of human melanoma cells exists in a divalent cation-dependent functional complex with the disialoganglioside GD2. J. Cell Biol. *105:* 1163–1173.

Cohen, J., and S. Y. Selvendran (1981) A neuronal cell surface marker is found in the CNS but not in peripheral neurons. Nature *291:* 421–423.

Constantine-Paton, M., A. Blum, R. Mendez-Otero, and C. Barnstable (1986) A cell surface molecule distributed in a dorsoventral gradient in the perinatal rat retina. Nature *324:* 459–462.

Fujita, S. C., and K. Obata (1984) Monoclonal antibodies demonstrate regional specificity in the spinal funiculi of the chick embryo. Neurosci. Res. *1:* 131–148.

Galfre, G., S. C. Howe, C. Milstein, G. W. Butcher, and J. C. Howard (1977) Antibodies to major histocompatibility antigens produced by hybrid cell lines. Nature *266:* 550–552.

Goodman, C. S., M. J. Bastiani, C. Q. Doe, S. duLac, S. L. Helfand, J. Y. Kuwada, and J. B. Thomas (1984) Cell recognition during neuronal development. Science *225:* 1271–1279.

Hawkes, R., and N. Leclerc (1987) Antigenic map of the rat cerebellar cortex: The distribution of parasagittal bands as revealed by monoclonal anti-Purkinje cell antibody mabQ113. J. Comp. Neurol. *256:* 29–41.

Hinrichsen, C. F. L. (1970) Coupling between cells of the mesencephalic trigeminal nucleus. J. Dent. Res. *49:* 1369–1373.

Hinrichsen, C. F. L., and L. M. H. Larramendi (1968) Synapses and cluster formation of the mouse mesencephalic fifth nucleus. Brain Res. *7:* 296–299.

Hinrichsen, C. F. L., and L. M. H. Larramendi (1969) Features of trigeminal mesencephalic nucleus structure and organization. Am. J. Anat. *126:* 497–506.

Hiscock, J., and C. Straznicky (1986) The formation of axonal projections of the mesencephalic trigeminal neurones in chick embryos. J. Embryol. Exp. Morphol. *93:* 281–290.

Hood, L., H. V. Huang, and W. J. Dryer (1977) The area-code hypothesis: The immune system provides clues to understanding the genetic and molecular basis of cell recognition during development. J. Supramol. Struct. *7:* 531–559.

Jessel, T. M. (1988) Adhesion molecules and the hierarchy of neural development. Rev. Neuron *1:* 3–13.

Larramendi, L. M. H. (1969) Analysis of synaptogenesis in the cerebellum of the mouse. In *Neurobiology of Cerebellar Evolution and Development,* R. Llinás, ed., pp. 803–843, Am. Med. Assoc. Ed. & Res. Foundation, Chicago.

Leclerc, N., C. Gravel, and R. Hawkes (1987) Parasagittal zonation in the cerebellar cortex develops independent of afferent input. Soc. Neurosci. Abstr. *308:* 17.

Marani, E., and J. Voogd (1977) An acetylcholinesterase band pattern in the molecular layer of the cat cerebellum. J. Anat. *124:* 335–345.

Mason, C. A. (1985) Growing tips of embryonic cerebellar axons *in vivo.* J. Neurosci. Res. *13:* 55–73.

McKay, R., and S. V. Hockfield (1982) Monoclonal antibodies distinguish antigenically discrete neuronal types in the vertebrate central nervous system. Proc. Natl. Acad. Sci. USA *79:* 6747–6751.

McKay, R. D. G., S. V. Hockfield, J. Johansen, I. Thompson, and K. Frederickson (1984) Surface molecules identify groups of growing axons. Science *222:* 678–684.

Mendez-Otero, R., B. Schlosshauer, C. Barnstable, and M. Constantine-Paton (1988) A developmentally regulated antigen associated with neural cell and process migration. J. Neurosci. *8:* 564–579.

Meynert, T. H. (1872) The brain of mammals. In *Stricker's Manual of Histology* (English trans.), pp. 650–766, William Wood, New York.

Mori, K., S. C. Fujita, Y. Watanabe, K. Obata, and O. Hayaishi (1987) Telencephalon-specific antigen identified by monoclonal antibody. Proc. Natl. Acad. Sci. USA *84:* 3921–3925.

Moskal, J. R., and A. E. Schaffner (1986) Monoclonal antibodies to the dentate gyrus: Immunocytochemical characterization and flow cytometric analysis of hippocampal neurons bearing a unique cell-surface antigen. J. Neurosci. *6:* 2045–2053.

Narayanan, C. H., and Y. Narayanan (1978) Determination of the embryonic origin of the mesencephalic nucleus of the trigeminal nerve in birds. J. Embryol. Exp. Morphol. *43:* 85–105.

Rayburn, H., H. M. Wu, D. Stainier, E. Bennett, and W. Gilbert (1987) Monoclonal antibodies directed at the surfaces of embryonic brain cells. Soc. Neurosci. Abstr. *391:* 3.

Roseman, S. (1974) The biosynthesis of cell-surface components and their potential role in intercellular communication. In *The Neurosciences, Third Study Program,* F. O. Schmitt and F. G. Worden, eds., pp. 795–804, MIT Press, Cambridge, MA.

Saigal, R. P., A. N. Karamanlidis, J. Voogd, O. Mangana, and H. Michaloudi (1980) Secondary trigeminocerebellar projections in sheep studied with the horseradish peroxidase tracing method. J. Comp. Neurol. *189:* 537–553.

Sariola, H., E. Aufderheide, H. Bernhard, S. Henke-Fahle, W. Dippold, and P. Ekblom (1988) Antibodies to cell surface ganglioside GD3 perturb inductive epithelial-mesenchymal interactions. Cell *54:* 235–245.

Schlosshauer, B., A. S. Blum, R. Mendez-Otero, C. J. Barnstable, and M. Constantine-Paton (1988) Developmental regulation of ganglioside antigens recognized by the JONES antibody. J. Neurosci. *8:* 580–592.

Schwarting, G. A., F. B. Jungalwala, D. K. Chou, A. M. Boyer, and M. Yamamoto (1987) Sulfated glucuronic acid-containing glycoconjugates are temporally and spatially regulated antigens in the developing mammalian nervous system. Dev. Biol. *120:* 65–76.

Scott, T. G. (1963) A unique pattern of localization in the cerebellum. Nature *200:* 793.

Somana, R., N. Kotchabhakdi, and F. Walberg (1980) Cerebellar afferents from the trigeminal sensory nuclei in the cat. Exp. Brain Res. *38:* 57–64.

Sperry, R. W. (1963) Chemoaffinity in the orderly growth of nerve fiber patterns and connections. Proc. Natl. Acad. Sci. USA *50:* 703–710.

Stainier, D., H. M. Wu, H. Rayburn, E. Bennett, and W. Gilbert (1987) Monoclonal cell surface antibody defining mesencephalic trigeminal nucleus. Soc. Neurosci. Abstr. *391:* 3.

The Journal of Neuroscience, July 1989, 9(7) 2485

Stallcup, W. B., L. L. Beasley, and J. M. Levine (1985) Antibody against nerve growth factor-inducible large external (NILE) glycoprotein labels nerve fiber tracts in the developing rat nervous system. J. Neurosci. *5:* 1090–1101.
Steller, H., K. F. Fishbach, and G. M. Rubin (1987) Disconnected: A locus required for neuronal pathway formation in the visual system of *Drosophila.* Cell *50:* 1139–1153.
Suzuki, K. (1965) The pattern of mammalian brain ganglionides. III. Regional and developmental differences. J. Neurochem. *12:* 969–979.
Weiss, P. (1947) The problem of specificity in growth and development. Yale J. Biol. Med. *19:* 235–278.
Yamamoto, M., A. M. Boyer, J. E. Crandall, M. Edwards, and H. Tanaka (1986) Distribution of stage specific neurite-associated proteins in the developing murine nervous system recognized by a monoclonal antibody. J. Neurosci. *6:* 3576–3594.
Zipser, B. and R. McKay (1981) Monoclonal antibodies distinguish identifiable neurones in the leech. Nature *289:* 549–554.

Proc. Natl. Acad. Sci. USA
Vol. 86, pp. 5673–5677, August 1989
Biochemistry

One-sided polymerase chain reaction: The amplification of cDNA

[direct sequencing/tropomyosin/zebrafish (*Brachydanio rerio*)/European common frog (*Rana temporaria*)/sequence evolution]

OSAMU OHARA*, ROBERT L. DORIT†, AND WALTER GILBERT

Department of Cellular and Developmental Biology, Harvard University, 16 Divinity Avenue, Cambridge, MA 02138

Contributed by Walter Gilbert, March 29, 1989

ABSTRACT **We report a rapid technique, based on the polymerase chain reaction (PCR), for the direct targeting, enhancement, and sequencing of previously uncharacterized cDNAs. This method is not limited to previously sequenced transcripts, since it requires only two adjacent or partially overlapping specific primers from only one side of the region to be amplified. These primers can be located anywhere within the message. The specific primers are used in conjunction with nonspecific primers targeted either to the poly(A)$^+$ region of the message or to an enzymatically synthesized d(A) tail. Pairwise combinations of specific and general primers allow for the amplification of regions both 3′ and 5′ to the point of entry into the message. The amplified PCR products can be cloned, sequenced directly by genomic sequencing, or labeled for sequencing by amplifying with a radioactive primer. We illustrate the power of this approach by deriving the cDNA sequences for the skeletal muscle α-tropomyosins of European common frog (*Rana temporaria*) and zebrafish (*Brachydanio rerio*) using only 300 ng of a total poly(A)$^+$ preparation. In these examples, we gained initial entry into the tropomyosin messages by using heterologous primers (to conserved regions) derived from the rat skeletal muscle α-tropomyosin sequence. The frog and zebrafish sequences are used in an analysis of tropomyosin evolution across the vertebrate phylogenetic spectrum. The results underscore the conservative nature of the tropomyosin molecule and support the notion of a constrained heptapeptide unit as the fundamental structural motif of tropomyosin.**

The polymerase chain reaction (PCR) is a powerful method for the enrichment of specific target sequences. By carrying out repeated cycles of annealing, synthesis, and denaturation in the presence of a thermostable DNA polymerase [*Thermus aquaticus* DNA polymerase (Taq polymerase)], sequences bounded by a pair of short, sequence-specific primers can be amplified more than a millionfold (1). However, the dependence of the PCR technique on a pair of specific primers flanking the region of interest has generally limited its use to targets whose sequence is known *a priori*.

Frequently, one is interested in sequences that have not been as yet fully characterized. A method requiring only a single small region of known sequence (from which specific primers can be designed) while still taking advantage of the power of PCR would substantially facilitate the extraction and sequencing of DNA molecules.

We report here a general method for the amplification of previously unsequenced cDNA target molecules. Fig. 1*a* illustrates schematically the three-pronged approach we have developed. First, we gain access into a transcript by isolating and amplifying a core region, using two primers derived from related sequences. Then, based on this core sequence, we design primers that, in combination with a second nonspecific primer [complementary either to the 3′ poly(A) tail or to an enzymatically synthesized tail at the 5′ end], permit the amplification of the regions both upstream and downstream of the core sequence. The combination of a single specific primer and a nonspecific primer generally is not sufficient to yield a unique product. Consequently, we reamplify this mix using a second specific primer, adjacent to the first, and the same nonspecific primer. This results in the amplification of a unique sequence.

These amplified fragments can be cloned into vectors from which the insert can be subsequently sequenced (2); alternatively, they can be directly sequenced by using either the "genomic sequencing" (3) or the dideoxy chain-termination method (4). We have also used a rapid, direct, chemical sequencing strategy that capitalizes on the specificity of the PCR method. A single primer is labeled by using any of the currently available tagging methods (see *Materials and Methods*) and then used to reamplify a sample previously enriched for a particular sequence. This produces double-stranded target molecules, labeled exclusively at one of the 5′ ends, that then can be chemically sequenced (5, 6) and visualized directly on a sequencing gel.

To illustrate the power of these techniques, we extracted (in parallel) and sequenced the complete transcripts of skeletal muscle α-tropomyosin from both the European common frog (*Rana temporaria*) and the zebrafish (*Brachydanio rerio*).

Other methods that seek to extend the use of PCR to previously unknown sequences have recently been reported. The restriction digestion and subsequent circularization of genomic DNA (inverse polymerase chain reaction, IPCR) permits the amplification of unique sequences (7). Methods involving the tailing of cDNA [the "RACE" (8) and "A-PCR" (9) approaches] permit the subsequent cloning of uncharacterized or variable messages.

MATERIALS AND METHODS

Preparation of cDNA. Poly(A)$^+$ RNA was prepared from frog leg muscle and from complete zebrafish according to standard methods (10). Poly(A)$^+$ RNA was primed by using an oligo(dT) 20-mer. cDNA synthesis on 100 ng of poly(A)$^+$ RNA was accomplished by using Moloney murine leukemia virus (Mo-MLV) reverse transcriptase in a volume of 5 μl (11).

Amplification of Core Region. Following cDNA synthesis as described above, 100 pmol of each of a pair of specific (imperfect) oligomers was added, in water, bringing the volume to 75 μl. The mixture was boiled for 3 min and quenched on ice. Remaining PCR reagents were then added. Amplifications were done in 100 μl of 50 mM KCl, 10 mM Tris·HCl (pH 8.8), 1.5 mM $MgCl_2$, 3 mM dithiothreitol, 0.1 mg of bovine serum albumin per ml, 200 μM (each) dNTPs, and

The publication costs of this article were defrayed in part by page charge payment. This article must therefore be hereby marked "*advertisement*" in accordance with 18 U.S.C. §1734 solely to indicate this fact.

Abbreviations: PCR, polymerase chain reaction; Mo-MLV, Moloney murine leukemia virus.

*Present address: Shionogi Research Laboratories, Fukushima-ku, Osaka 553, Japan.
†To whom reprint requests should be addressed.

2.5 units of Taq polymerase. Thirty cycles of PCR were carried out: annealing at 45°C, 2 min; extension at 72°C, 3 min; denaturation at 94°C, 1 min; final extension, 5 min.

Amplification of 3′ Region. The amplifications for the 3′ region were carried out in a final 100-μl volume, using a 10-μl sample of the above "core amplification" mixture, to which 100 pmol of both a nested specific primer and the nonspecific $(dT)_{20}$ primer were added. Remaining reagents and PCR conditions are as above.

Synthesis of Specific cDNA Molecules for Amplification of 5′ Region. A specific oligomer was used to prime the Mo-MLV reverse transcriptase on the poly(A)$^+$ RNA. Initial annealing of the oligomer (1 fmol) was done in 5 μl containing 0.2 M NaCl, 40 mM Tris·HCl (pH 7.5), 5 mM EDTA, and 100 ng of poly(A)$^+$ RNA, incubated at 65°C for 5 min, placed at 40°C for 4 hr, ethanol precipitated, resuspended in 25 μl of reverse transcriptase mix containing 50 mM Tris·HCl (pH 8.3), 75 mM KCl, 10 mM dithiothreitol, 3 mM $MgCl_2$, 50 μg of Actinomycin D per ml, 0.5 mM (each) dNTPs, and 200 units of Mo-MLV reverse transcriptase (BRL), and incubated at 37°C for 1 hr.

Poly[d(A)] Tailing of Specific cDNA Molecules and Amplification of 5′ Region. Following the synthesis described above, the mixture was phenol extracted, ethanol precipitated twice, dried under vacuum, resuspended in 6.5 μl of water, boiled for 2 min, quenched on ice, and tailed in a final 10-μl volume using terminal deoxynucleotidyltransferase, according to standard protocols (12). The products were ethanol precipitated and resuspended in 100 μl of the PCR mixture as described in "Amplification of Core Region." PCR conditions were as follows: 40 cycles, annealing at 50°C, 30 sec; extension at 72°C, 1 min; denaturation at 94°C for 40 sec, in the presence of 100 pmol of the specific oligomer and $(dT)_{20}$. A 10-μl sample of the above was reamplified in 100-μl final volume with a nested, specific oligomer, internal to the first. PCR conditions were as follows: 55°C, 30 sec; 72°C, 1 min; 94°C, 40 sec; 30 cycles.

Direct Sequencing of PCR Products. Ten picomoles of a specific PCR primer was labeled with T4 polynucleotide kinase (NEB) and 200 μCi of [γ-^{32}P]ATP (6000 Ci/mmol; 1 Ci = 37 GBq) under conditions specified by the supplier (37°C for 30 min). The 10-μl kinase mixture was placed directly into 90 μl of standard PCR mixture containing 100 pmol of unlabeled second primer and 1 ng of template DNA. Five to 10 PCR cycles generate enough labeled DNA for direct chemical sequencing. Labeled fragment was gel purified in a 5% native acrylamide gel and recovered by absorbing onto a DEAE membrane (13).

RESULTS

The results of our three-pronged approach (Fig. 1*a*) are described below.

Sequencing the Core Region: Use of Imperfect Primers Derived from Homologous Sequences. To enter the sequence of the unknown tropomyosin cDNAs, we first identified regions of high amino acid sequence conservation in previously sequenced tropomyosin genes (14–18). We chose primers complementary to those conserved stretches that contained amino acids with low codon degeneracy and synthesized two sets of "21-mers," R1/R2 and R3/R4, derived from the rat tropomyosin nucleotide sequence (15, 16). Fig. 1*b* depicts the positions of all primers used in this study, and Table 1 collects the sequences of all the oligonucleotides described in this paper.

We prepared poly(A)$^+$ RNA from frog leg muscle and from a complete zebrafish homogenate. Tropomyosin transcripts appear to comprise >0.1% of the total muscle poly(A)$^+$ RNA pool (19). A complete array of cDNA molecules was then generated from the poly(A)$^+$ RNA mixture using Mo-MLV

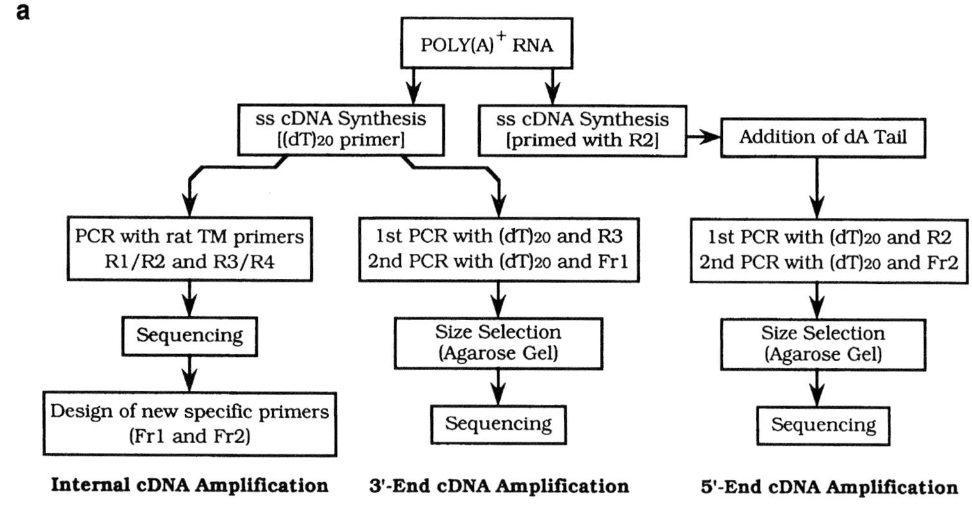

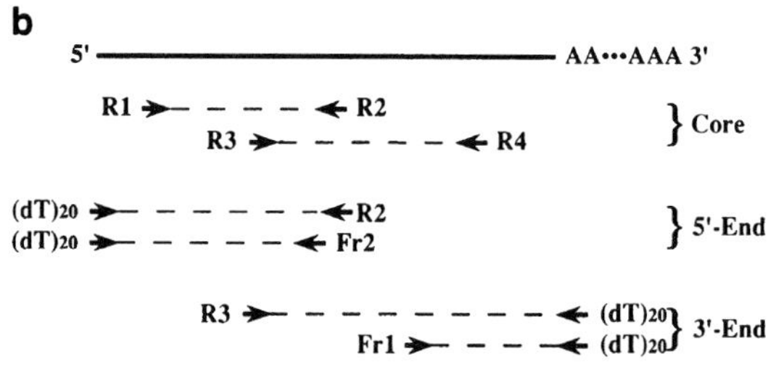

FIG. 1. Three-part strategy for cDNA amplification. (*a*) Steps involved in the amplification and eventual sequencing of complete cDNA transcripts. ss, Single-stranded; TM, tropomyosin. (*b*) Position on an idealized tropomyosin cDNA sequence of the various oligomers discussed in the text.

Table 1. Oligonucleotide primer sequences

	PCR primer
R1:	5′-ATGGACGCCATCAAGAAGAAG-3′ (1)
R2:	5′-CTCCATCTTCTCCTCATCCTT-3′ (425)
R3:	5′-ATCCAGCTGGTTGAGGAGGAG-3′ (273)
R4:	5′-GGTCTCAGCCTCCTTCAGCTT-3′ (710)
FR1:	5′-GAGCTTGAGGAAGAGTTGAAAA-3′ (573)
FR2:	5′-GTTCTCAATGACTTTCATGCCTC-3′ (395)
	Sequencing/probing primer
FR3:	5′-CTGTTCTGCTCTGTCCAAGG-3′ (71)
FR4:	5′-TATGCCCAGAAACTGAAGTA-3′ (780)

The prefixes indicate the source of the sequence information (R = rat; FR = frog). Numbers in parentheses at the right indicate nucleotide position on the target sequence where the 5′ end of the oligomer will anneal. Position 1 is the first nucleotide in the initiation methionine.

reverse transcriptase (11) primed with an oligo(dT) 20-mer. The single-stranded cDNA resulting from this synthesis became the template for PCR amplifications using the two sets of imperfect primers. As shown in Fig. 1*b*, these imperfect primer pairs each spanned ≈400 nucleotides, with a 100-nucleotide region of overlap. Thus we sought to amplify ≈700 nucleotides of the tropomyosin transcript core region.

The results of the amplifications on frog and zebrafish cDNA are shown in the left panel (lanes 1–4) of Fig. 2*a*. Although a major discrete band of the expected size appears, there might have been two or more isotypes of tropomyosin present in the original cDNA preparations, and the amplified product might have contained a mixture of different tropomyosin transcripts. To ensure that a homogeneous transcript was sequenced, the amplified products were cut out of the 2% agarose gel, eluted, and cloned into the "Bluescript" plasmid vector (Stratagene) using the G-C tailing method (12). We sequenced six colonies (20, 21) containing frog muscle cDNA amplified with the R3/R4 primer pair: all were identical and contained only the skeletal muscle α transcript. In contrast, five colonies derived from the amplification of zebrafish cDNA with the R3/R4 primer pair proved heterogeneous upon sequencing: three colonies carried skeletal muscle α transcripts, one carried smooth muscle β transcript, and one contained the skeletal muscle β transcript. Since our focus was exclusively on the major isotype of tropomyosin, and we favor methods that do not involve the cloning and sequencing of a single amplified molecule, given concerns about the fidelity of Taq polymerase (22), we used direct sequencing of the PCR product as frequently as possible. We obtained the complete sequence of the core region directly from the amplified fragments by using a combination of genomic sequencing and the direct PCR-mediated sequencing method. Even in those cases in which a mixture of templates was present, the direct method allowed us to read the sequence of the major cDNA isotype. Fig. 3 shows a representative fragment of a sequencing gel obtained by this method. Approximately 700 nucleotides of sequence were thus generated from each of the two tropomyosin cDNA sequences. Nonetheless, substantial portions of the transcript, both upstream and downstream of the amplified fragments, remained to be explored.

Characterizing the 3′ Region: Enrichment Amplification Using One-Sided Specific Primers. To extend our analysis into the 3′ region of the transcript, we amplified the cDNA mixture prepared for the previous step, using equimolar amounts of the single specific primer R3 (originally thought to be imperfectly matched) located in the previously derived core region, and the $(dT)_{20}$ primer. After 30 cycles of amplification, a large number of different sequences appeared, as shown by the smears in lanes 5 and 8 of Fig. 2*a*. This amplified mixture nevertheless did include the sequence of interest—the 3′ region of the tropomyosin transcript—since we could detect it on a Southern blot by probing with an internal oligomer (Fr4) (lanes 5 and 8, Fig. 2*b*).

To rescue the appropriate sequence, a 10% aliquot of the above amplification was reamplified directly, using 100 pmol of a new, nested specific oligomer (Fr1) and an equimolar amount of the $(dT)_{20}$ oligomer. Lanes 6 and 9 of Fig. 2*a* (and lanes 6 and 9, Fig. 2*b*) show that there is still a heterogeneous mixture of amplified fragments, most likely reflecting the distribution of poly$(A)^+$ tail lengths. The uppermost band on the 2% agarose gel (≈400–500 nucleotides long), which appeared to contain our target molecule, was cut out, eluted, and reamplified using the same pair of primers [$(dT)_{20}$ and Fr1]. This size-selection reduced the potential targets for amplification and increased the specificity of subsequent PCR rounds. Lanes 7 and 10, Fig. 2*a*, show that this resulted in a single band, which contained the desired sequence (as seen in lanes 7 and 10, Fig. 2*b*, probed with oligo Fr4). We sequenced the amplified product using both the genomic

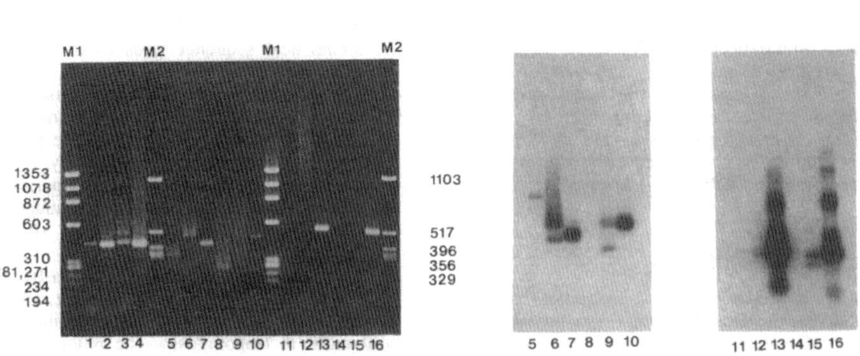

FIG. 2. (*a*) Agarose gels stained with ethidium bromide showing the results of amplifications of core, 3′, and 5′ regions. Lanes M1 and M2, DNA size markers (given in nucleotides) (φX *Hae* III digest and Bluescript *Hin*fI digest). Lanes 1–4, results of core region amplification: frog cDNA, primers R1/R2 (lane 1); primers R3/R4 (lane 2); fish cDNA, primers R1/R2 (lane 3); primers R3/R4 (lane 4). Lanes 5–10, 3′ end amplifications: frog cDNA, primers $(dT)_{20}$/R3 (lane 5); primers $(dT)_{20}$/Fr1 (lane 6); size-selected frog amplification, primers $(dT)_{20}$/Fr1 (lane 7); fish cDNA, primers $(dT)_{20}$/R3 (lane 8); primers $(dT)_{20}$/Fr1 (lane 9); size-selected fish amplification, primers $(dT)_{20}$/Fr1 (lane 10). Lanes 11–16, 5′ end amplifications: frog cDNA, primers $(dT)_{20}$/R2 (lane 11); primers $(dT)_{20}$/Fr2 (lane 12); size-selected frog amplification, primers $(dT)_{20}$/Fr2 (lane 13); fish cDNA, primers $(dT)_{20}$/R2 (lane 14); primers $(dT)_{20}$/Fr2 (lane 15); size-selected fish amplification, primers $(dT)_{20}$/Fr2 (lane 16). (*b*) Southern blot of lanes 5–10, probed with oligomer Fr4. (*c*) Southern blot of lanes 11–16, probed with oligomer Fr3.

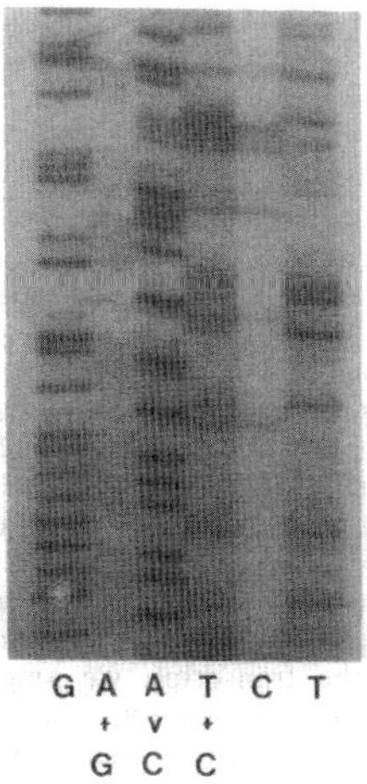

FIG. 3. Fragment of sequencing gel obtained by direct PCR-mediated method. The fragment shown corresponds to the internal core region of zebrafish tropomyosin cDNA (nucleotide positions 340–400). Sequencing reactions are as described in refs. 5 and 6.

sequencing and the direct sequencing methods previously described. In this way, we generated an additional 300 nucleotides [including about 100 residues of the poly(A) tail] in the case of the European common frog cDNA and ≈350 nucleotides of zebrafish tropomyosin sequence.

Characterizing the 5′ Region: Single Primer Extension, Tailing, and Amplification. By using a specific oligomer (R2) complementary to the rat tropomyosin sequence, we primed the Mo-MLV reverse transcriptase on the poly$(A)^+$ RNA to generate cDNA molecules containing the 5′ region of the transcript. We added poly(dA) tails using terminal deoxynucleotidyltransferase (12) and amplified these sequences by PCR, using the same specific primer (R2) and a $(dT)_{20}$ nonspecific primer complementary to the newly added tails. After 40 cycles of amplification, the PCR product could generally not be visualized with ethidium bromide staining (lanes 11 and 14, Fig. 2*a*) but did appear as a faint band on subsequent Southern blots probed with an internal probe (Fr3) (lanes 11 and 14, Fig. 2*c*; not visible in the reproduction). An aliquot of the above mixture, when reamplified with a new, nested internal primer (Fr2) and the nonspecific $(dT)_{20}$ primer, resulted in a discrete band. This band, however, was accompanied by a high molecular weight background (lanes 12 and 15, Fig. 2*a*; lanes 12 and 15, Fig. 2*c*). We cut out the discrete band and reamplified using the same primer pair [Fr2/$(dT)_{20}$]. In each case, this procedure yielded a major band, which Southern blotting (lanes 13 and 16 in Fig. 2*c*) and subsequent sequencing confirmed corresponds to the 5′ regions of the frog and zebrafish tropomyosin cDNAs, respectively.

DISCUSSION

Our two cDNA sequences permit a comparison of tropomyosin sequences across the deepest of vertebrate phylogenetic forks and substantially increase the time scale over which the evolution of tropomyosin can be surveyed. The comparison reveals an extremely stable and conservative structural protein. [The tropomyosin cDNA sequences have been deposited in GenBank (accession nos. M24634 and M24635). Table 2 summarizes some properties of the sequences.] Despite the broad phylogenetic distance (≈400 million years since the last common ancestor) separating *R. temporaria* from *B. rerio*, the two tropomyosin proteins differ at only 19 of 284 amino acids. This degree of conservation is probably due to the structural role this protein plays, interacting with other proteins (actin, troponin) in the muscle contraction apparatus (24, 25).

Table 2. Summary of zebrafish (*B. rerio*) and European common frog (*R. temporaria*) tropomyosin cDNA sequences and comparisons

	5′ untranslated region	Coding region	3′ untranslated region
B. rerio	87 nt	855 nt (284 aa)	75 nt
R. temporaria	109 nt	855 nt (284 aa)	72 nt
No. of nucleotide differences	*	130 nt Silent: 101 nt Replace: 29 nt (18 aa)	54 nt†
% difference		Silent: 35 Replace: 3.2‡	75.0

nt, Nucleotides; aa, amino acids.
*The 5′ untranslated region shows two regions of sequence conservation [positions −109 to −74 (frog), −87 to −52 (fish), and −6 to 1 (frog and fish)].
†The 3′ regions have been aligned without gaps.
‡Percentage differences have been calculated according to ref. 23. Shown are the weighted uncorrected averages. Corrected weighted averages for silent positions = 110.5%; for replacement positions = 3.8%.

A large number of different tropomyosins have been identified, from both muscle and cytoplasmic sources. Muscle tissue (smooth and striated) may contain both α and β subunits of tropomyosin; up to nine isoforms of a given tropomyosin may be present as a result of alternative splicing (14, 26). This heterogeneity of possible tropomyosin forms will be reflected in a mixture of cDNAs within a single organism. A procedure such as ours, which used whole organisms (in the case of zebrafish) or striated muscle (in the case of the frog) to prepare mRNA, makes it difficult *a priori* to determine which cDNA is being isolated. To determine the identity of the cDNA sequences, we compared them to all available vertebrate tropomyosin sequences: quail (*Coturnix coturnix*) skeletal α-tropomyosin (27), chicken (*Gallus gallus*) smooth muscle α-tropomyosin (28), rabbit (*Oryctolaqus cuniculus*) muscle α-tropomyosin (29), rat (*Rattus norvegicus*) skeletal α- and β-tropomyosins (15, 30), and human (*Homo sapiens*) skeletal muscle α-tropomyosin and nonmuscle tropomyosin (31). Despite the extensive conservation in the primary structure of this protein, interspecific variation in the amino acid sequence does exist. A number of these variable sites are diagnostic for particular tropomyosin types, occurring (for example) in all striated muscle α-tropomyosins but not in smooth muscle or β-tropomyosins. Of the 86 variable sites in our comparison, 32 are possibly "informative"—a residue at that site is shared by at least two but no more than $n - 1$ sequences in the comparison. The 12 sites that are informative for this case all suggest that the two cDNA sequences presented here are striated muscle α-tropomyosin messages.

Table 3 compares the striated-muscle α-tropomyosins from five vertebrates (human, rat, quail, frog, and zebrafish). All of these sequences are remarkably similar. Does the structural motif in tropomyosin—the putative NxxNAxB heptaresidue repeat that runs throughout the sequence associated with the double coiled-coil structure of the molecule (33)—constrain this variation? In the five proteins compared, there are 52 variable sites; of these, 29 are at "x" positions within the hepatapeptide motif, whereas 11 occur at "N" (hydrophobic) positions, 2 at "A" (acidic) positions, and 5 at "B" (basic) positions. This distribution is significantly different from the uniform random expectation ($P < 0.05$), with

Table 3. Amino acid sequence divergence of available skeletal muscle α-tropomyosins

	Zebrafish	Frog	Quail	Rat	Human
Zebrafish		20 (7.0)	19 (6.7)	22 (7.7)	20 (7.0)
Frog	10 (6.6)		13 (4.6)	16 (5.6)	18 (6.3)
Quail	12 (7.9)	6 (3.9)		12 (4.2)	19 (6.7)
Rat	14 (9.2)	10 (6.6)	7 (4.6)		25 (8.8)
Human	10 (6.6)	7 (4.6)	8 (5.3)	13 (8.6)	

Figures above the diagonal indicate amino acid differences over the entire molecule; figures in parentheses show the percentage divergence between sequences. Data below the diagonal reflect amino acid changes and percentage divergence after excluding the most variable and alternatively spliced exons [1, 2, 6, and 9 (32)]. Percentages below the diagonal are calculated by using exons 3, 4, 5, 7, and 8 (total amino acids: 152).

an excess of substitutions at "x" positions that accords with the prediction based on the heptapeptide motif: the second, third, and fifth positions are less stringently constrained. Nonetheless, the small amount of variation in these sequences reflects a strong purifying selection acting on practically every residue—in the overwhelming majority of positions even equivalent substitutions have been weeded out. The 52 variant amino acids in our comparison occur at 38 positions; 47 of these changes are equivalent substitutions.

Curiously, the pattern of values in Table 3 is not what one expects for a slowly evolving gene. The sequence divergences are not congruent with the phylogenetic tree based on the species involved. (For example, the divergence between rat and human tropomyosins is greater than the divergence between frog and human tropomyosins.) This suggests strongly that there are several genes for the skeletal muscle α-tropomyosins, which diverged from a common ancestral gene prior to the vertebrate colonization of the land. This possibility is further supported by the distribution of parallel substitutions at identical positions. At 10 variable positions the same amino acid change occurs in two sequences, and in no case does the similarity reflect common ancestry.

We believe this method for obtaining and characterizing RNA transcripts represents a significant improvement over previous approaches and holds great promise for the comparative sequencing of homologous transcripts from different species. Initial entry into the cDNA sequence can be achieved by using imperfect oligomers derived from related organisms. The need for specific primers from only one side of the region to be amplified expands the power of PCR, permitting the amplification of previously uncharacterized messages. This method does not require library construction and screening, laborious and often unpredictable steps in the isolation of complete transcripts, and, in contrast with previous approaches, requires a far smaller amount of starting poly(A)$^+$ RNA—only 300 ng of poly(A)$^+$ RNA was utilized in determining the full sequences of both frog and zebrafish tropomyosin cDNAs. Finally, the method lends itself to the processing in parallel of cDNAs derived from a variety of different sources.

We thank Milligen (Bedford, MA) for the use of their 6500 synthesizer, Perkin–Elmer/Cetus for the use of their thermal cycler, and members of the Gilbert laboratory for helpful discussions. This work was supported by the National Institutes of Health Grant GM 37997-02. O.O. was supported by Shionogi Research Laboratories.

1. Saiki, R. K., Gelfand, D. H., Stoffel, S., Scharf, S. J., Higuchi, R., Horn, G. T., Mullis, K. B. & Erlich, H. A. (1988) *Science* **239,** 487–491.
2. Scharf, S. J., Holm, G. T. & Erlich, H. A. (1986) *Science* **233,** 1076–1078.
3. Church, G. M. & Gilbert, W. (1984) *Proc. Natl. Acad. Sci. USA* **81,** 1991–1995.
4. Gyllensten, U. B. & Erlich, H. A. (1988) *Proc. Natl. Acad. Sci. USA* **85,** 7652–7656.
5. Maxam, A. M. & Gilbert, W. (1977) *Proc. Natl. Acad. Sci. USA* **74,** 560–564.
6. Rubin, C. M. & Schmid, C. W. (1980) *Nucleic Acids Res.* **8,** 4613–4619.
7. Ochman, H., Gerber, A. S. & Hartl, D. L. (1988) *Genetics* **120,** 621–623.
8. Frohman, M. A., Dush, M. K. & Martin, G. R. (1988) *Proc. Natl. Acad. Sci. USA* **85,** 8998–9002.
9. Loh, E. Y., Elliott, J. F., Cwirla, S., Lanier, L. L. & Davis, M. M. (1989) *Science* **243,** 217–220.
10. Ausebel, F. M., Brent, R., Kingston, R. E., Moore, D. D., Seidman, J. G., Smith, J. A. & Struhl, K., eds. (1987) *Current Protocols in Molecular Biology* (Wiley, New York), Units 4.3 and 4.5.
11. D'Alessio, J. M. & Gerard, G. F. (1988) *Nucleic Acids Res.* **16,** 1999–2014.
12. Eschenfeldt, W. H., Puskas, R. S. & Berger, S. L. (1987) *Methods Enzymol.* **152,** 337–342.
13. Dretzen, G., Bellard, M., Sassone-Corsi, P. & Chambon, P. (1981) *Anal. Biochem.* **112,** 295–298.
14. Ruiz-Opazo, N. & Nadal-Ginard, N. (1987) *J. Biol. Chem.* **262,** 4755–4765.
15. Helfman, D. M., Cheley, S., Kuismanen, E., Finn, L. A. & Yamawaki-Kataoka, Y. (1986) *Mol. Cell. Biol.* **6,** 3582–3595.
16. Mak, A. S., Smillie, L. B. & Stewart, G. R. (1980) *J. Biol. Chem.* **255,** 3647–3655.
17. MacLeod, A. R., Houlker, C., Reinach, F. C., Smillie, L. B., Talbot, K., Modi, G. & Walsh, F. S. (1985) *Proc. Natl. Acad. Sci. USA* **82,** 7835–7839.
18. MacLeod, A. R. (1982) *Eur. J. Biochem.* **126,** 293–297.
19. Helfman, D. M., Feramsico, J. R., Fiddes, J. C., Thomas, G. P. & Hughes, S. H. (1983) *Proc. Natl. Acad. Sci. USA* **80,** 31–35.
20. Church, G. M. & Kieffer-Higgins, S. (1988) *Science* **240,** 185–188.
21. Birnboim, H. C. & Doly, J. (1979) *Nucleic Acids Res.* **7,** 1513–1523.
22. Wilson, A. C. & Paabo, S. (1988) *Nature (London)* **334,** 387–388.
23. Perler, F., Efstratiadis, A., Lomedico, P., Gilbert, W., Kolodner, R. & Dodgson, J. (1980) *Cell* **20,** 555–566.
24. Smillie, L. B. (1975) *J. Mol. Biol.* **98,** (2), 281–291.
25. Pearlstone, J. R. & Smillie, L. B. (1982) *J. Biol. Chem.* **257,** 10587–10592.
26. Giometti, C. S. & Anderson, N. L. (1984) *J. Biol. Chem.* **259,** 14113–14120.
27. Hastings, K. E. M. & Emerson, C. P., Jr. (1982) *Proc. Natl. Acad. Sci. USA* **79,** 1553–1557.
28. Helfman, D. M., Feramisco, J. R., Ricci, W. M., Hughes, S. H. (1984) *J. Biol. Chem.* **259,** 14136–14143.
29. Putney, S. D., Herlihy, W. C. & Schimmel, P. (1983) *Nature (London)* **302,** 718–723.
30. Yamawaki-Katakoa, Y. & Helfman, D. M. (1985) *J. Biol. Chem.* **260,** 14440–14445.
31. Clayton, L., Reinach, F. C., Chumbley, G. M. & MacLeod, A. R. (1988) *J. Mol. Biol.* **201,** 507–515.
32. Colote, S., Widada, J. S., Ferraz, C., Bonhomme, F., Marti, J. & Liautard, J. P. (1988) *J. Mol. Evol.* **27,** 228–235.
33. McLachlan, A. D. & Stewart, M. (1975) *J. Mol. Biol.* **98,** 293–304.

Proc. Natl. Acad. Sci. USA
Vol. 87, pp. 923–927, February 1990
Neurobiology

Pioneer neurons in the mouse trigeminal sensory system

(mesencephalic trigeminal nucleus/monoclonal antibody/specific neuronal antigen/axon outgrowth)

DIDIER Y. R. STAINIER* AND WALTER GILBERT†

Departments of †Cellular and Developmental Biology and *Biochemistry and Molecular Biology, Harvard University, Cambridge, MA 02138

Contributed by Walter Gilbert, November 14, 1989

ABSTRACT **Pioneer neurons establish preliminary nerve pathways that are followed by later-growing axons. The existence of pioneers and their importance is well documented in invertebrate systems. In mammals, early neuronal development has generally been difficult to study because of the size and complexity of the embryos, and the lack of adequate markers. Here we look at the time of earliest axonal outgrowth in the mouse embryo by using specific monoclonal antibodies to stain wholemount preparations. During the period of formation and closure of the neuropore beginning at embryonic day 8.5, we can follow the earliest trigeminal sensory neurons extending axons along stereotyped pathways. In the trigeminal ganglion, an early wave of neurogenesis gives rise to a small number of neurons whose axons pioneer the different trigeminal tracts in the periphery. After a brief pause (12 hr), these primary axons branch out to innervate individual targets. Emerging a day later, secondary fibers extend along the pioneers. By contrast, in the central nervous system, neurons of the mesencephalic trigeminal nucleus extend toward the rhombencephalon independently, ignoring preexisting fibers. These results show the existence of an early set of axonal tracts in the mouse peripheral nervous system that may be used for the guidance of later-differentiating neurons.**

Pioneer neurons establish nerve pathways when distances are short and tissues can be traversed (1–3). In insect embryos, they are essential for guiding later-differentiating neurons toward their targets (2, 3). Very recently, McConnell *et al.* (4) showed that the subplate neurons in the developing mammalian cortex invade the thalamus early in fetal life, thereby providing a possible pathway for later-born cortical neurons. A pioneer-like behavior has also been attributed to the earliest neurons of the chicken peripheral trigeminal system by the immunohistochemical studies of Moody *et al.* (5) and by the retrograde tracing studies of Covell and Noden (6). However, these latter observations (5, 6) were limited by the use of cryostat sections to analyze projection patterns. Non-serial thin-section microscopy does not reveal the three-dimensional relationship of a putative pioneer with other axons.

In this study, we describe the pioneering property of some of the earliest differentiating neurons in the mouse. We use specific monoclonal antibodies (mAbs) to stain embryonic wholemounts and visualize the full extent of axonal outgrowth from a defined set of neurons. The wholemount preparations are optically dissected with the confocal microscope and the resulting images are then combined to generate a three-dimensional representation of the specimen. In the trigeminal ganglion, an early wave of neuronal differentiation gives rise to a small number of neurons, presumably derived from the epidermal placodes as has been shown in the chicken embryo (5, 6). These neurons pioneer the different trigeminal tracts in the periphery. (The term "pioneer" is used merely to suggest that these are the first detected axons in these pathways and that secondary or trailing fibers appear to grow along them. No claim is made at this point about the requirement of these axons for the guidance of subsequent fibers.) Later-differentiating neurons, derived mostly from the neural crest (5, 6), seem to project along the preexisting tracts. By contrast, in the central nervous system (CNS), neurons of the mesencephalic trigeminal nucleus (MesV) differentiate at a steady rate and project caudally in an independent manner; fasciculation does not take place.

We show the developmental sequence of early axonal outgrowth in the mouse trigeminal system, focusing on the ophthalmic projection, and report that the early-born ganglionic neurons pioneer the peripheral tracts.

MATERIALS AND METHODS

mAb Production and Characterization. mAb E1.9 was isolated from a fusion in which the immunogen consisted of whole cells from embryonic day 12 (E12) rat mid- and hindbrain. It recognizes a cytoplasmic epitope in the primary sensory and motor neurons during axonal outgrowth, appearing at E8.5 and disappearing by E12 in the mouse. More specifically, it stains neural crest-derived sensory neurons (e.g., neurons of dorsal root ganglia) as well as placode-derived sensory neurons (e.g., neurons of cranial nerves VII and VIII). Its full description will be published elsewhere.

Immunohistochemistry. Pregnant CD1 mice were obtained from Charles River Breeding Laboratories under a specific breeding schedule: mice were bred for 3 hr from noon to 3 p.m. on E0; our embryonic day ran from noon to noon. Animals were sacrificed by cervical dislocation or euthanized with halothane. Embryos staged according to Theiler (7) were slit longitudinally along the forming neuropore and processed as half-embryo wholemounts. Immunostaining was done as previously described (8). All the reactions were done at room temperature (with gentle shaking) in polystyrene culture dishes by sequential incubations in the following: 10% normal goat serum (NGS) in phosphate-buffered saline (PBS) for 1 hr; hybridoma culture supernatant overnight (12–18 hr); after three 10-min washes with 10% NGS/PBS, fluoresceinated goat anti-mouse IgM (Southern Biotechnology Associates, Birmingham, AL), diluted 1:100, for 2 hr. After three more 10-min washes in 10% NGS/PBS, the wholemounts were mounted between two coverslips separated by silicone grease, in fresh mounting medium. For E1.9 staining, the embryos were fixed with 1% paraformaldehyde for 10 min, rinsed several times with PBS, and incubated for 2 hr in 10% NGS/PBS. Saponin (0.04%) was included throughout the E1.9 immunostaining. We used a Lasersharp MRC-500 confocal microscope mounted with fluorescence optics to observe specific axonal pathways. Serial sections were collected and combined by projection. The resulting images

The publication costs of this article were defrayed in part by page charge payment. This article must therefore be hereby marked "*advertisement*" in accordance with 18 U.S.C. §1734 solely to indicate this fact.

Abbreviations: mAb, monoclonal antibody; CNS, central nervous system; MesV, mesencephalic trigeminal nucleus; E*n*, embryonic day *n*.

were stored in a WORM optical disk (Maxtor, San Jose, CA) and printed using a video printer. The data in this paper come from analyzing about 120 embryos, some as early as the 5-somite stage (E8) and others as late as E12.

RESULTS

We used two specific mAbs to describe the existence and behavior of pioneer sensory neurons in the mouse trigeminal system. mAb E1.9 recognizes a cytoplasmic epitope expressed in growing primary sensory and motor axons from E8.5 to E12 (unpublished data). mAb E1.9 immunoreactivity first appears in the central nervous system at E8.5 and in the trigeminal ganglion at E9; it disappears at E12. mAb B30 recognizes a rare ganglioside expressed on trigeminal sensory neurons shortly after differentiation (8). Carefully staged embryos (7) were dissected, treated immunohistochemically as wholemounts, and examined with a Lasersharp MRC-500 confocal microscope equipped with fluorescence optics.

During E9 (day of vaginal plug is E0), at the 14-somite stage, a few cells in the region of the primitive trigeminal ganglion are E1.9-positive (Fig. 1*a*). The filamentous appearance of 1.9 immunoreactivity does not yet define neuronal cell bodies, but short axonal projections span the width of the ganglionic space and fibers are stretched out, orienting toward the periphery. The initially low neuronal density stays constant throughout E9, as suggested by birthdating studies in the rat (9) and as revealed by E1.9 immunoreactivity. A second wave of neurogenesis starts at E10, lasts for about 48 hr, and populates the trigeminal ganglion. These secondary neurons are E1.9$^+$ as soon as they differentiate.

Major axonal growth is visible by E9.5, at the 21-somite stage; Fig. 1*b* shows that the ganglion's polarity is established as its dorsal tip has sent out a short (100 μm) projection. A few axons form this budding ophthalmic projection. Although at this point the leading axons and their growth cones stand close together, observation at a slightly later stage shows that they are growing independently of each other. Fig. 2*a* shows that by E10, at the 30-somite stage, the pioneer growth cones of the ophthalmic projection have reached past the eyecup. At the base of these first axons, a second wave of axogenesis is appearing. The leading axons are few in number (three or four); the path of their outgrowth is specific, restricted to a narrow strip between the retina and the telencephalon, yet they seem to be growing individually. These axons grew approximately 450 μm in 12 hr; this growth rate of just under 40 μm/hr is about twice as fast as that calculated by Davies (10) for secondary fibers in the maxillary nerve. The trailing

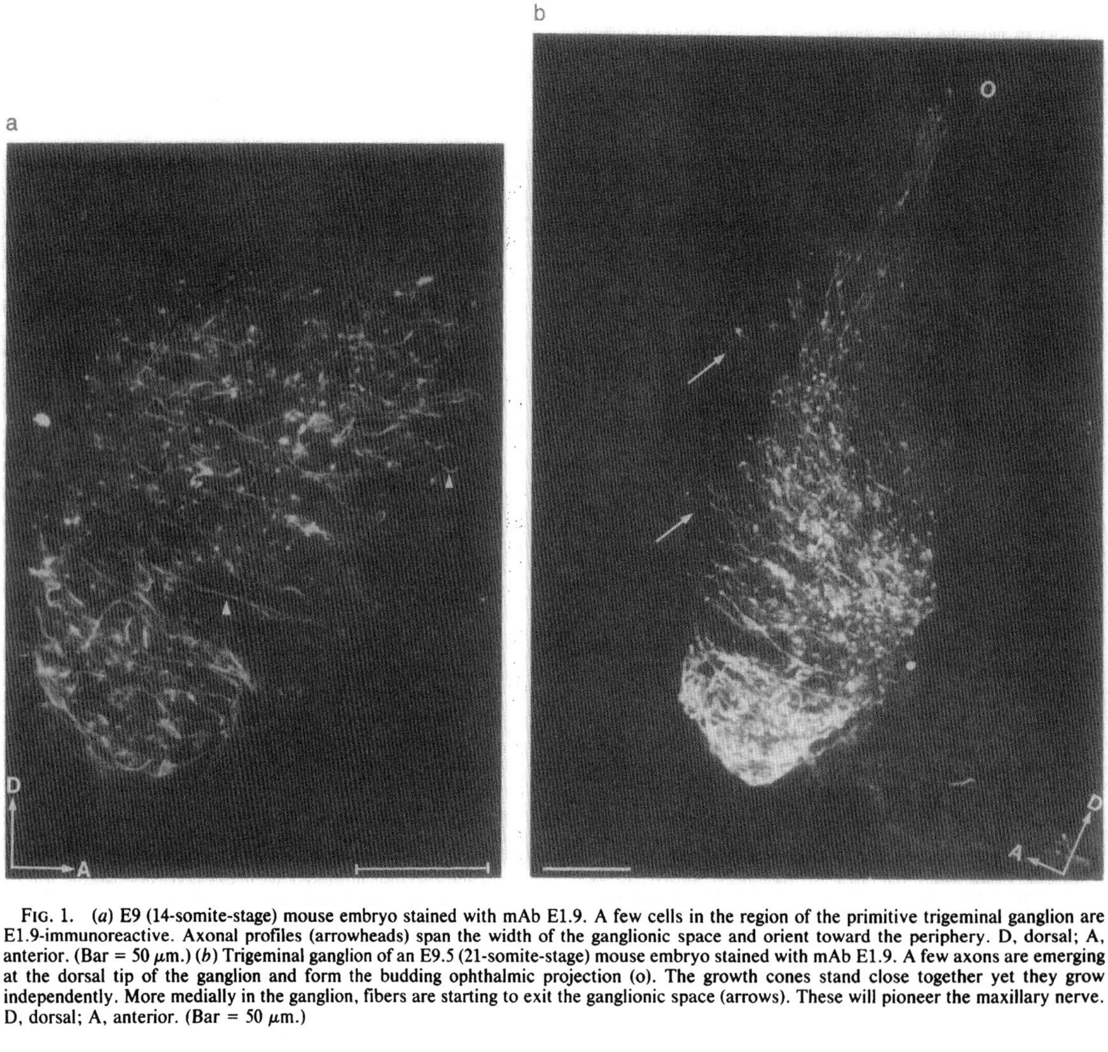

FIG. 1. (*a*) E9 (14-somite-stage) mouse embryo stained with mAb E1.9. A few cells in the region of the primitive trigeminal ganglion are E1.9-immunoreactive. Axonal profiles (arrowheads) span the width of the ganglionic space and orient toward the periphery. D, dorsal; A, anterior. (Bar = 50 μm.) (*b*) Trigeminal ganglion of an E9.5 (21-somite-stage) mouse embryo stained with mAb E1.9. A few axons are emerging at the dorsal tip of the ganglion and form the budding ophthalmic projection (o). The growth cones stand close together yet they grow independently. More medially in the ganglion, fibers are starting to exit the ganglionic space (arrows). These will pioneer the maxillary nerve. D, dorsal; A, anterior. (Bar = 50 μm.)

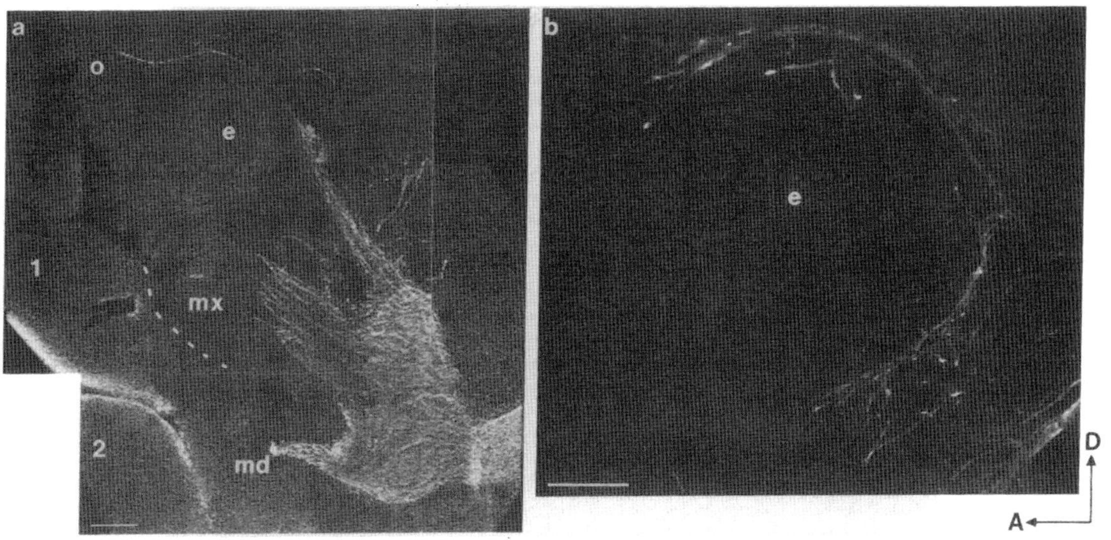

FIG. 2. (*a*) Trigeminal ganglion of an E10 (30-somite-stage) mouse embryo stained with mAb E1.9. The pioneer axons of the ophthalmic projection (o) have reached past the eyecup (e). A few axons (three or four) pioneer this narrow pathway. At the base of this projection, more fibers are emerging and are starting to grow along the initial axons. Some of the pioneering maxillary (mx) fibers are seen extending. At this early E10 stage, the leading maxillary fibers are about 100 μm away from their target area. The distal tip of the mandibular (md) projection has been severed accidentally. The first and second branchial bars are indicated by 1 and 2; dots delineate the first branchial bar pushed up against the maxillary process during mounting. (Bar = 50 μm.) (*b*) Trigeminal ophthalmic projection of an E10.5 (35-somite-stage) mouse embryo stained with mAb E1.9. The pioneer axons of the ophthalmic (o) branch have not grown past the point reached at the 30-somite stage (*a*). The leading axons have now started branching and more fibers have grown along them. Branching ophthalmic fibers now surround the eyecup (e) dorsally as well as ventrally. D, dorsal; A, anterior. (Bar = 50 μm.)

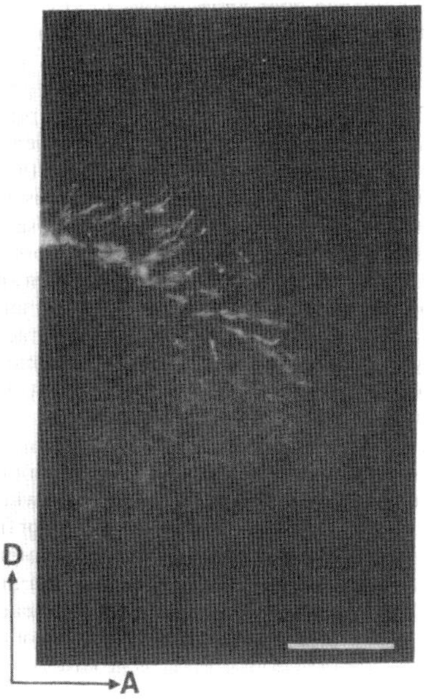

FIG. 3. Branching of pioneer fibers in the vicinity of the epithelium of the maxillary process, as seen in a 16-mm cryostat section of an E10.5 (37-somite-stage) mouse embryo stained with mAb E1.9. Pioneer fibers first reach the proximity of the maxillary epithelium during E10 and start branching at E10.5. D, dorsal; A, anterior. (Bar = 50 μm.) Our embryonic day ran from noon to noon and we took CD1 embryos in the early afternoon and around midnight. Davies and Lumsden (11) obtained embryos from 8-hr overnight mating of CD1 (Charles River Breeding Laboratories) mice; their embryonic day ran from midnight to midnight. We believe that most of their work was done with embryos slightly older than ours.

axons near the ganglion seem to be associating with the early fibers, and one can subsequently observe their growth along the pioneering axons. Fig. 2*b* shows that by E10.5, at the 35-somite stage, the ophthalmic projection has not grown much past the point reached at the 30-somite stage, in Fig. 2*a*, but that its leading axons have started branching, an early event in the innervation process.

Neurogenesis and axonogenesis in the maxillary and mandibular branches proceed in a similar fashion, although with a slightly later onset than in the ophthalmic branch. A limited number of pioneer axons lead the way, directly growing to the general area of their respective target (see for example the maxillary fibers in Fig. 2*a*). After a period of little distal growth, these pioneer axons start branching in the vicinity of the target. Meanwhile, a second wave of axons projects and grows, apparently fasciculating on the leading fibers. Later, these secondary axons themselves branch to innervate the full extent of the target area. Fig. 3 shows a few pioneer fibers branching in the vicinity of the epithelium of the maxillary process at the 37-somite stage (E10.5).

In the CNS, mesencephalic neural crest cells emerge and start migrating at the 4- to 7-somite stage (12). Some of them give rise to the primary sensory neurons that form the MesV (13). At E8.5, by the 10-somite stage, E1.9 immunoreactivity first outlines a few neurons that lie in the rostral part of the mesencephalon. This is the earliest neuronal outgrowth that we have detected in the CNS; previously, neurons were thought to first differentiate in the mouse CNS by E9–E10 (14). Fig. 4 shows two adjacent neurons that send short projections caudally towards the rhombencephalon. The location, time of birth, and direction of projection of the neurons in Fig. 4 are characteristic of MesV neurons. Indeed, mAb B30, which we showed to be specific for MesV neurons in the mouse CNS (8), outlines such caudally projecting neurons at a slightly later stage in their maturation. As more mesencephalic neurons differentiate (MesV neurons differ-

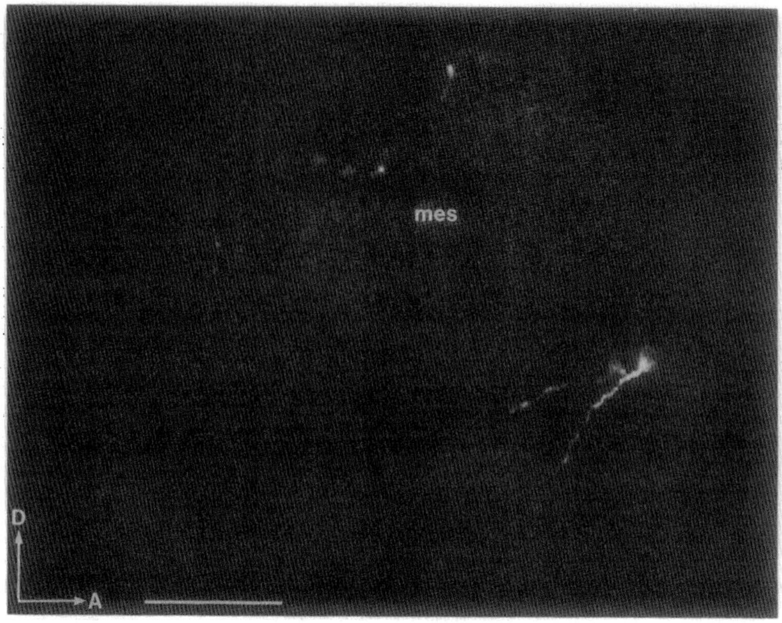

FIG. 4. Earliest detectable neurons in the mouse CNS. An E8.5 (10-somite-stage) mouse embryo stained with mAb E1.9 is shown at the level of the midbrain. Two neurons lie clustered in the mesencephalon (mes). Their location and time of birth are characteristic of MesV neurons. They also send short projections caudally towards the rhombencephalon. D, dorsal; A, anterior. (Bar = 50 μm.)

entiate at a slow but steady rate from E8.5 to E12), their axons also project caudally in an independent manner; fasciculation does not take place. This axonal separation remains constant and can be seen in postnatal animals (8). MesV neurons are clustered, but their axons remain distinct when projecting to the brainstem. When they exit the CNS at the level of the pons, these MesV axons pass through the trigeminal ganglion and grow along the pioneered mandibular tract to their peripheral targets.

DISCUSSION

In the 10-somite-stage mouse embryo, cells from the epidermal placodes start migrating into the region of the primitive trigeminal ganglion, differentiate, and give rise to the first trigeminal neurons shortly after E9 (15). Later-differentiating, neural crest-derived neurons populate most of the ganglion (5, 6, 9, 15). [^{3}H]Thymidine birthdating studies in the rat (9) as well as electron microscope observations (11) confirm that trigeminal neurons differentiate shortly after E9 and that the first fibers leave the ganglion at its dorsal tip at E9.5. These developmental studies and the observation that mAb E1.9 stains both neural crest-derived and placode-derived sensory neurons (e.g., neurons of the dorsal root ganglia and neurons of the VIIth and VIIIth ganglia, respectively) indicate that the E1.9-immunoreactive axons that leave the ganglion at its dorsal tip by E9.5 (Fig. 1*b*) are the first to grow out. Furthermore, E1.9 immunostaining of cryostat sections of E10.5–E12 trigeminal ganglion shows no E1.9-negative areas, indicating that mAb E1.9 stains all trigeminal neurons. In the E10 embryo, electron microscopy shows small fascicles of nerve fibers leaving the ganglion (11). Our study complements this observation by showing that within these fascicles, a few fibers have reached the target area while the others have just started growing, apparently by fasciculating on the early fibers. This is best exemplified by the ophthalmic fibers in Fig. 2*a*: the pioneer axons have extended 450–500 μm and reached their target area while the secondary fibers have grown 75–100 μm out of the ganglion.

Selective guidance cues allow axons to grow along very precise pathways to their targets (reviewed in ref. 16). In general, axons may follow pathways labeled by specific molecular cues (17), or they may respond to gradients of diffusible factors secreted by cells within the target (18–20). Two lines of *in vitro* work have provided evidence for the chemotropic guidance of axons: the observation that the floor plate orients the growth of rat spinal cord commissural axons (18) and a set of experiments done with the mouse trigeminal ganglion (19, 20). In this latter work, the observation of explants cocultured in collagen gels led Lumsden and Davies (19, 20) to conclude that trigeminal cutaneous target fields explanted at E10 and E11 release a diffusible factor which specifically directs neurite outgrowth from trigeminal ganglia of the same age (and that this factor is distinct from nerve growth factor).

Lacking specific markers for trigeminal neurons, Lumsden and Davies designed their experiments on temporal assumptions drawn from observations of silver-stained wholemounts and thin sections (11). By E9.5, a small number of fibers have emerged from the dorsal side of the ganglion to pioneer the ophthalmic projection. But, whereas Lumsden and Davies reported that fibers first reach the maxillary target epithelium early in E11 (they also described a few fibers contacting the epithelium of the mandibular process at E10.5), we observe that pioneer axons first reach the proximity of the maxillary target epithelium during E10 and start branching out at E10.5 (Fig. 3). During E11, the secondary fibers growing along these initial axons first approach their target. Thus, in culturing E10 and E11 explants, Lumsden and Davies were not looking at the earliest sensory outgrowth but at secondary outgrowth approaching tissue previously explored by the pioneer axons. Tropism may be involved in conjunction with fasciculation in guiding the secondary fibers. Alternatively, such a target-derived tropic effect *in vitro* may only be a by-product of the differentiation of the target cells after an initial axonal approach (one expects such factors to be secreted from the differentiating target cell to adjust the extent of its innervation, primarily through controlling local sprouting). In fact, there was a small but consistent increase in target-directed outgrowth in their collagen gel experiments on moving from E10 to E11 explants, which may reflect the ongoing differentiation of the target tissue in the presence of the pioneer axons [as has been suggested for embryonic chicken lumbosacral motoneurons (21)]. Tropic guidance of the earliest trigeminal sensory neurons remains an attractive hypothesis which should be tested in cocultures of E9–E9.5 explants.

Our results show that in the periphery, the initial trigeminal axons may constitute a "labeled" pathway available for the guidance of later fibers; thus, different mechanisms may be involved in guiding the pioneers versus the later axons. By contrast, in the CNS, MesV axons do not fasciculate; the initial axons of the MesV cannot directly serve as guides for subsequent MesV fibers.

We thank Dr. A. Ghysen for interesting discussions and Dr. C. Fulwiler and M. Grether for critical reading of the manuscript.

1. Bate, C. M. (1976) *Nature (London)* **260,** 54–56.
2. Raper, J. A., Bastiani, M. J. & Goodman, C. S. (1984) *J. Neurosci.* **4,** 2329–2345.
3. Klose, M. & Bentley, D. (1989) *Science* **245,** 982–984.
4. McConnell, S. K., Ghosh, A. & Shatz, C. J. (1989) *Science* **245,** 978–982.
5. Moody, S. A., Quigg, M. S. & Frankfurter, A. (1989) *J. Comp. Neurol.* **279,** 567–580.
6. Covell, D. A. & Noden, D. M. (1989) *J. Comp. Neurol.* **286,** 488–503.
7. Theiler, K. (1989) *The House Mouse* (Springer, Berlin), 2nd Ed.
8. Stainier, D. Y. & Gilbert, W. (1989) *J. Neurosci.* **9,** 2468–2485.
9. Altman, J. & Bayer, S. A. (1982) *Adv. Anat. Embryol. Cell Biol.* **74,** 1–89.
10. Davies, A. M. (1987) *Development* **100,** 307–311.
11. Davies, A. M. & Lumsden, A. G. S. (1984) *J. Comp. Neurol.* **223,** 124–137.
12. Chan, W. Y. & Tam, P. P. L. (1988) *Development* **102,** 427–442.
13. Narayanan, C. H. & Narayanan, Y. (1978) *J. Embryol. Exp. Morphol.* **43,** 85–105.
14. Jacobson, M. (1978) *Developmental Neurobiology* (Plenum, New York), pp. 57–114.
15. Verwoerd, C. D. A. & van Oostrom (1979) *Adv. Anat. Embryol. Cell Biol.* **58,** 1–75.
16. Dodd, J. & Jessell, T. M. (1988) *Science* **242,** 692–699.
17. Ghysen, A. (1978) *Nature (London)* **274,** 869–872.
18. Tessier-Lavigne, M., Placzek, M., Lumsden, A. G. S., Dodd, J. & Jessell, T. (1988) *Nature (London)* **336,** 775–778.
19. Lumsden, A. G. S. & Davies, A. M. (1983) *Nature (London)* **306,** 786–788.
20. Lumsden, A. G. S. & Davies, A. M. (1986) *Nature (London)* **323,** 538–539.
21. Lance-Jones, C. & Landmesser, L. (1981) *Proc. R. Soc. London Ser. B.* **214,** 1–18.

RNA editing as a source of genetic variation

Laura F. Landweber & Walter Gilbert

Department of Cellular and Developmental Biology,
Harvard University Biological Laboratories, 16 Divinity Avenue,
Cambridge, Massachusetts 02138, USA

KINETOPLASTID RNA editing alters mitochondrial RNA transcripts by addition and deletion of uridine residues[1], producing open reading frames that may be twice as long as the original RNA[2]. Although the *COIII* gene encoding cytochrome *c* oxidase subunit III in *Trypanosoma brucei* is edited along its entire length[2], the presumably homologous genes in two related trypanosomes, *Leishmania tarentolae* and *Crithidia fasciculata*, are only modestly edited at their 5′ ends[1]. We used a comparative approach to investigate the evolution of an edited gene and to determine how well editing creates conserved protein sequences. As RNA editing probably involves the pairing of several guide RNA molecules with the messenger RNA[3,4], we expected the edited proteins to be resistant to evolutionary change. Here we report that RNA editing is extensive in the mitochondria of four species of the insect parasite *Herpetomonas*, which is possibly an evolutionary precursor of *T. brucei* and *L. tarentolae*[5], and the discovery that RNA editing is a novel source of frameshift mutations over evolutionary time. The edited proteins accumulate mutations nearly twice as rapidly as the unedited versions.

We used a polymerase chain reaction (PCR) strategy which exploits the tandem arrangement of genes on the maxicircle to amplify the *COIII* gene from four species of *Herpetomonas*: *H. mariadeanei*, *H. samuelpessoai*, *H. muscarum muscarum* and *H. megaseliae*. *COIII* is flanked at its 3′ end by the well conserved cytochrome *b* sequence, providing a downstream PCR primer, and at its 5′ end by *MURF3*, which is less conserved and extensively edited *T. brucei*[6]. We therefore chose an upstream PCR primer from within the *COIII* DNA sequence of *T. brucei*[2]. Direct sequencing of these PCR products[7] gave the unedited *COIII* sequences corresponding to the last 230 bases of these genes (Fig. 3).

We determined the completely edited mRNA sequence for each of the four species of *Herpetomonas* by direct sequencing of RNA PCR products using primers based on the *T. brucei* complementary DNA sequence[2,8] and also by one-sided anchor PCR[9,10], extending either to the 5′ end of the mRNA or to the 3′ untranslated region including the poly(A) tail, followed by cloning and sequencing. As some of the 5′ anchor-PCR products from *H. mariadeanei*, *H. m. muscarum* and *H. megaseliae* were partially edited at the 5′ end (data not shown), we could identify PCR primers to amplify the rest of the mitochondrial *COIII* gene in these three species.

Figure 1 shows that there is extensive editing in the *COIII* transcript from each of the four species of *Herpetomonas*. Uridine additions (and rare deletions) create 90-93% of the 288 amino-acid codons. Uridine insertions also appear in the poly(A) tails, as in other pan-edited genes[2]. PCR amplification

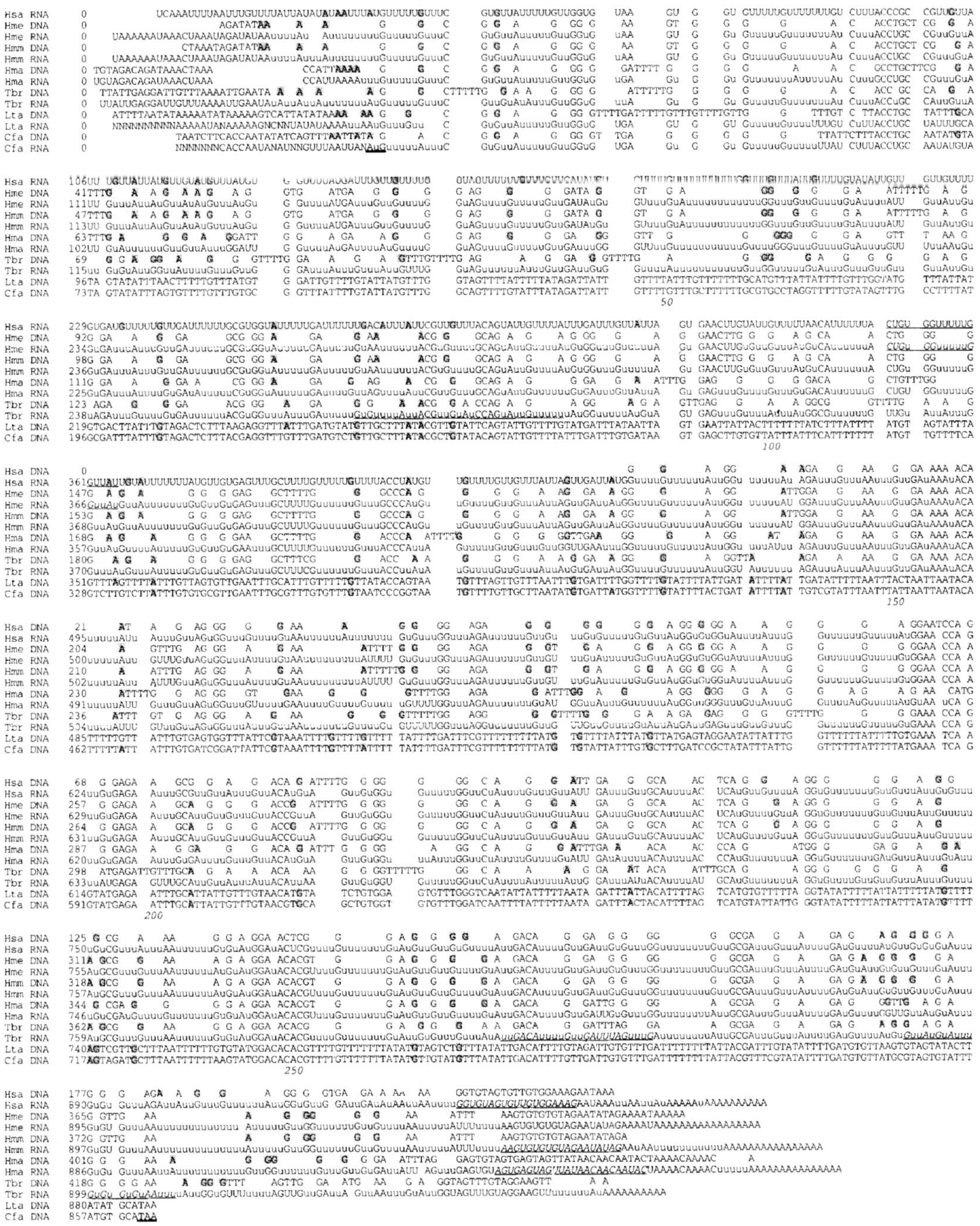

FIG. 1 Alignment of *COIII* DNA and edited RNA sequences for *H. samuelpessoai* (Hsa), *H. megaseliae* (Hme), *H. m. muscarum* (Hmm), *H. mariadeanei* (Hma), *T. brucei* (Tbr), *L. tarentolae* (Lta)[1,12] and *C. fasciculata* (Cfa)[1,13]. DNA sequences in upper case; uridines in mRNA that are added by RNA editing in lower case; encoded thymidines deleted from the mRNA indicated by gaps. Frameshifted nucleotides are highlighted. The start and stop codons are underlined and the corresponding positions on the amino-acid sequences (Fig. 2) are numbered below the *C. fasciculata* sequence. Editing adds 548 uridines at 198 sites and deletes 27 uridines at seven sites in *H. mariadeanei*, creating 59% of the 867-nucleotide open reading frame. Similarly, editing adds 569 uridines in *H. megaseliae* and 572 uridines in *H. m. muscarum* at 208 identical sites and deletes 12 uridines at seven sites in *H. megaseliae* and 13 uridines at eight sites in *H. m. muscarum*, creating 60% of the coding sequence. These last two closely related sequences differ from each other only at one uridine deletion, one silent A→G transition at position 510, one frameshift at position 109, which was a polymorphism in *H. m. muscarum*, and two insertions of extra uridines in the 5′ and 3′ untranslated regions. In the region in which we can compare both the genomic DNA and edited RNA for *H. samuelpessoai*, 310 new uridines appear at 122 sites and six disappear at three sites: uridine insertions create 61% of the coding sequence. Between Hme, Hmm and Hsa, 35, 37 frameshifts or suppressor frameshifts occur in the coding region; 50, 50, 52 occur between Hsa, Hmm,

with primers corresponding to the edited mRNA failed to detect any genomic copy that matched the edited sequence for each case (data not shown).

Table 1 compares the degrees of sequence similarity between species at the DNA, RNA and predicted protein levels. *C. fasciculata* and *L. tarentolae*, which are edited in only 6% of the amino-acid codons, show a higher degree of sequence conservation at the amino-acid level (90%) than at the mRNA level (86%), as would normally be expected with silent changes outweighing replacements. But the highest degree of sequence conservation for the pan-edited genes is at the RNA and DNA level, in spite of massive sequence deletion. The level of predicted protein similarity decreases significantly in pairwise comparisons between *Herpetomonas* and *T. brucei* or within *Herpetomonas*, owing to frameshift mutations that are introduced by changes in editing which are compensated by suppressor frameshift mutations at a nearby editing site, producing about half of the amino-acid replacements, which cluster in the frameshifted regions (Fig. 2). The unusual amino-acid substitutions leucine (UUG) to cysteine (UGU) to valine (GUU) occur frequently as a result of a local frameshift. In different species, editing often produces different mRNA sequences from the same unedited DNA sequences (Fig. 1), providing an additional source of genetic novelty. Figure 3 shows an alignment of the compressed cryptic DNA sequences which reveals sites that are conserved in the DNA but shifted to another location in the mRNA. We propose that each of these frameshifts in the edited mRNA is the result of two compensatory mutations in the guide RNAs, small maxicircle or minicircle transcripts that appear to mediate editing by base-pairing with specific regions of the edited transcript, allowing some G·U base-pairs. Complete editing proceeds 3′ to 5′ and requires a set of overlapping guide RNAs. Editing by each guide RNA creates an anchor sequence for binding the next guide RNA[3,4]. Thus RNA editing is a surprisingly inefficient mechanism for conserving amino-acid sequence.

The requirement for base-pairing imposes additional restrictions on the mRNA sequence such that synonymous substitutions may not be neutral. We do observe a nearly threefold lower rate of synonymous changes in the edited genes, either for the entire reading frame or just in the regions where there are no frameshifts (Fig. 1). Although the rate of A-to-G transitions is not significantly different for edited or unedited genes, presumably reflecting the ability of a U in the guide RNA to pair with either A or G, the U-to-C transitions are completely suppressed. Furthermore, only a fraction of the gene is actually encoded in the DNA. The number and location of DNA-encoded uridines that are deleted by editing also vary greatly among all of the species. Between species, the 5′ and 3′ untranslated regions are even more divergent in both the unedited and edited sequence, suggesting that constraints are relaxed outside the coding region.

Do the fully edited transcripts correspond to actual proteins? One difficulty with our interpretation is the lack of an in-frame AUG in the fully edited mRNAs from *H. mariadeanei*, *H. m. muscarum* and *H. megaseliae*, although a conserved AG exists at the same DNA position in every case (Fig. 3). Editing creates an in-frame AUG in the other kinetoplastid species, but inserts more than one uridine in each of these three examples, perhaps to make the non-canonical initiation codon UUG[11]. The predicted N-terminal amino-acid sequences are very highly conserved in all of the species, and editing creates a conserved stop codon and predicted amino-acid sequences of the same length (Fig. 2), strengthening the view that the genes are functional. We are confident that the sequences in Fig. 1 are the fully edited transcripts, because we sequenced several clones from each 5′ anchor PCR, and we used direct PCR sequencing to obtain most of the sequences, including the 5′ ends. There are also two examples in *L. tarentolae* of transcripts that do not contain AUG initiation codons[1], including one that is edited.

Do the edited proteins have a fast clock? The ratio of protein sequence distances to the actual pairwise mitochondrial ribosomal DNA distances (ref. 5; and L.F.L. and W.G., manuscript in preparation) uses the changes in the mitochondrial ribosomal RNA genes as a measure of the evolutionary distances between species. Pairwise comparisons of these normalized rates of protein sequence divergence give values from 1.10 to 1.47 between species with extensive editing and divergence (1.17 between *H. samuelpessoai* and *H. m. muscarum*; 1.35, 1.47 between *H. samuelpessoai*, *H. m. muscarum* and *H. mariadeanei*; 1.23, 1.10, Hme and Hma; 62 between Tbr and Hsa; 46, 48 between Hme, Hmm and Tbr; 64 between Tbr and Hma; and possibly 8 between Lta and Cfa. The GenBank accession numbers for these sequences are L10845–L10852.

METHODS. PCR amplification from either total DNA or first strand cDNA prepared with random hexamers, a *Not*I-d(T)$_{18}$ anchor primer (Pharmacia) or 3′-specific primer and murine reverse transcriptase (Pharmacia) from total RNA[14] was as described[15]. The sequences of RNA PCR primers are underlined; negative-strand primers are italicized. Single-stranded templates for direct sequencing[15] were produced either by λ exonuclease digestion[16] or by capture of a biotinylated strand onto streptavidin coated magnetic beads[7]. 5′ anchor-PCR products, using G-tailed cDNA and 5′ $[CUA]_4CTCGAGAATT(C)_{12}(GAT)$ primer, and 3′ anchor-PCR products[9,10] were visible on an ethidium-bromide-stained agarose gel after nested PCR with an internal primer[9] and the anchor[10]. Anchor-PCR products were gel-purified, cloned (T-vector, Novagen, and UDG cloning, BRL) and at least three clones were sequenced on both strands with *Taq* DNA polymerase (Promega fmol sequencing system).

```
          **+ * *********+******+ * +      *     ****+*+*  *
Hsa      0MFLFRVIFVGVSGVFVFLSLPAVVICYYVVCLCGFMICCFGSFLFVDMCF
Hme      0MFLFRVIFVGVSGVFVFLSLPAVVIVYYVICLCGFMICCFGSFVFVDMCF
Hmm      0MFLFRVIFVGVSGVFVFLSLPAVCIVYYVICLCGFMICCFGSFVFVDMCF
Hma      0MFLFRVIFVGVSGVFLFLSLPAVCIVFFVVIWIGFMILCFGSFVFVDLGF
Tbr      0MFLFRCIFVGVSGVFVFLSLPAIVIVYWLFCLLGFICLLFGSFLFVDCGF
Lta      0MFV-RVIFVGVSGVFVFLSLPAICIVYLTFCLCGLFCIMFGSFIFIDYCF
Cfa      0MFLFRVIFVGVSGVFVFLSLPAICIVYLVFCLCGLFCIMFGSFVFIDYCF

          + *     *++ *  *+* *  **+**     +* +****+**+ ** +*
Hsa     50VFFLVCLLFCILLLFCDVFVDFLRGIFDFLTFIRCLQYCFIWFVISELVL
Hme     50VFFFFGLLFCILLLLCDLFVDFLRGIFDFCNFLRVLQYCFMWFVFSELVL
Hmm     50VFFFFGLLFCILLLLCDLFVDFLRGIFDFCNFLRVLQYCFMWFVFSELVL
Hma     50VFFFVLGLFCILFLMCDLFCDIFRGIFDFVSFIRCLQYCFLWFVISEFVL
Tbr     50IFFFVGFCICLLLLLLDLFCDFLRGLFDFCVLLRCIQYCFLWFLCSEFVL
Lta     49ICFFACLLFCLVCLLCDLFVDSLRGLFDVCCFIRCIQYCFVWFIISELLL
Cfa     50VCFFACLGFCIVCLLCDLFVDSLRGLFDVCCFIRCIQYCFIWFVISELVL

          *++ * *+*+  +*+  ****** +*  * *** ++** ***+++ +***
Hsa    100FLTFFTVVFGYCIFLCCEFAFVFVLPMLFCCLLVDYGFVFYWFFIDLFNL
Hme    100FMSFFTVVFGYVIFLCCEFAFVFCLPMLFCCLLVDYGFVFYWFFMDLFNL
Hmm    100FMSFFTVVFGYVIFLCCEFAFVFCLPMLFCCLLVDYGFVFYWFFMDLFNL
Hma    100FVTFFAVVFGYVLFLCCEFAFVFCLPISFCCLLVEFGFCFYWFYLDLFNL
Tbr    100FMAFFVVLFGLCLFLCCEFAFVFCLPYMFCCLLCDYGFVFYWYFLDLFNL
Lta     99FLSLFYVVFSLVLFVSVEFAFVFVIPVMFSCLICDFGFVFYWYFIDIFNL
Cfa    100FISFFYVVFSLVLFVCVEFAFVFVIPVMFCCLICDYGFVFYWYFIVVFNL

          ***** *** *** **     *****      ** ++  **+** **+**
Hsa    150LINTFLLFVSGLFCNFLFFCVWFRFFVVCVFVLWCGILFGFLFLWNQLWE
Hme    150LINTFLLFVSGLFCNFFLFCVWFRFFVVCIFVLWCGILFGFLFLWNQLWE
Hmm    150LINTFLLFVSGLFCNFFYFCVWFRFFVVCIFVLWCGILFGFLFLWNQLWE
Hma    150LINTFLLFVSGLFLNFVLFLFWFRFFCMVICFLWLGLLFGFMFLCNQLWE
Tbr    150LINTFYLFVSGLFVNFFVLCFWFRFFCCCCFVLWLSLLFGFLFLWNQLWE
Lta    149LINTFLLFVSGLFVNFVLFLFWFRFFLCVLFMLWVGILFGFLFLWNQVWE
Cfa    150LINTFLLFVIGLFVNFVLFLFWFRFFLCVLFVLWSAIFIGFLFLWNQVWE

          * +**+*   **+******* +*+*** ** **+ *+*+   * * ** ***
Hsa    200FALLFVTCSCGVFGSILFVIDLLHFTHVVLGVFLLFIVFCRLFNFLCMDT
Hme    200FALLFVTVSCGVFGSILFVIDLLHFTHVLLGVFLLFIVFMRLFNFLCMDT
Hmm    200FALLFVTVSCGVFGSILFVIDLLHFTHVLLGVFLLFIVFMRLFNFLCMDT
Hma    200FVILFVTCSCGLFGSILFCIDILHFTHVFLGVFLMFICICRCFVFLCMDT
Tbr    200FALLFITLSCGVFGSILFLLDLLHFMHVFLGVLLLFICFMRLFNFLCMDT
Lta    199FALLFVTCICGVFGSILFLIDLLHFSHVFLGIFLLFLCFSRCFNFLCMDT
Cfa    200FALLFVTCSCGVFGSILFLIDLLHFSHVLLGIFLLFICFGRCFNFLSMDT

          ****** *  ****** **************   +**    *
Hsa    250RFVFLYVVVFYWHFVDCVWFFLLRFVYFDVLCCVYLCV  term
Hme    250RFVFLYVVVLYWHFVDCVWFFLLRFVYFDVLCCLYLCV  term
Hmm    250RFVFLYVVVLYWHFVDCVWFFLLRFVYFDVLCCLYLCV  term
Hma    250RFVFLYVVVLYWHFVDCVWFFLLRFVYFDVLVVMYLCV  term
Tbr    250RFVFLYCVCLYWHFVDLVWFFLLRFVYFDVLCVMYLCV  term
Lta    249RFVFLYVVCLYWHFVDCVWFFLLRFVYFDVLSVVYLYA  term
Cfa    250RFVFLYVVCLYWHFVDCVWFFLLRFVYFDVLCVVYLCA  term
```

FIG. 2 Alignment of the predicted amino-acid sequences for seven kinetoplastid species. Positions conserved in all seven species are marked with an asterisk; conservative amino-acid replacements are marked with a cross.

```
Hme    0                           AGATATAAAAGGC-----GGA-GGGGAA
Hmm    0                      CTAAATAGATATAAAAGGC-----GGA-GGGGAA
Hma    0  TGTAGACAGATAAACTAAACC-AT-TAAAAGGC-----GGA-GGGGGA
Tbr    0 TTATTGAGGATTGTTTAAAATTGAATAAAAAGGCTTTTTGGAAGGGG-A

Hme   22-----GTGGGGACACCTGCT-CGGATTTGAAGAAGAG----GTGATGAGG
Hmm   28-----GTGGGGACACCTGCT-CGGATTTGAAGAAGAG----GTGATGAGG
Hma   40TTTT-G-GGGAACGCCTGCTTCGGATTTG-AGGAGGATT--G-GA-GAAG
Tbr   48TTTTTG-GGGGACACC-GC--CAGA---GGAGGAGGGTTTTG-GA-AGAG

Hme   62---G----GAGGGGATAG----GT-GAGGGGGAATTTTTGAGGGAAGGAG
Hmm   68---G----GAGGGGATAG----GT-GAGGGGGAATTTTTGAGGGAAGGAG
Hma   83---G----GAGGGGA-GG----GTTGGGGGGGAGTTT--AAGGGAAGGAA
Tbr   89TTTGTTTTGAGAGGA-GGTTTTG--AGGGGAGGG-----GAGAGAGGGAA

Hme  100CGGGAGAGAAACGG-GCAGAGAGGGGA---G--GAACTTGGGAGCAACT-
Hmm  106CGGGAGAGAAACGG-GCAGAGAGGGGA---G--GAACTTGGGAGCAACT-
Hma  119CGGGAGAGAGACGG-GCAGAGGGAGAATTTG--GAG---GGGGACAGCT-
Tbr  131CGGGAGAGGAACGGACCAGAGAGGAGA---GTTGAG---GAAGGCGGTTT

Hsa    0                                               GG
Hme  143-G---GGGGAGAGGGGAGGCTTTTGGGCCCAG----GGGAAG--GAAGGG
Hmm  148-G---GGGGAGAGGGGAGGCTTTTGGGCCCAG----GGGAAG--GAAGGG
Hma  162-GTTTGG-GAGAGGGGAAGCTTTTGGACCCAATTTTGGGGGGTTGAAGGG
Tbr  175TG---AAGGAGAGGGGAGGCTTTCGGACC-AA----GGGAAG--GAAGGG

Hsa    2AGG--A--AGAGAAGGAAAAACAAT---A---GAGGGG-GAAA-----GG
Hme  183AGG--ATTGGAGAAGGAAAAACAA----GTTTGAGGGA-GAAATTTT-GG
Hmm  188AGG--ATTGGAGAAGGAAAAACAA----ATTTGAGGGA-GAAATTTTTGG
Hma  210AGG--AT-AGAGAAGGAAAAACAATTTTG---GAGGGGTGAAG-----GG
Tbr  215AGGTTA--AGAAAAGGAAAAACAATTT-GT--GAGGGA-GAAG-----GG

Hsa   32-----GGAGAGG----GGGGAGGGGGAAG----GGAGGAATCCAGG-GAG
Hme  221-----GGAGAGGT---GAGGAGGGGGAAG----GGGGGAA-CCAAG-GAG
Hmm  227-----GGAGAGGT---GAGGAGGGGGAAG----GGGGGAA-CCAAG-GAG
Hma  249TTTT-GGAGAGATTT-GGAGAGGGGGGAG----GAGAGAA-CATGG-GAG
Tbr  254TTTTTGGAGGGGTTTTGGGAAGAGAGGGGTTTTGGGGAAA-CCAGATGAG

Hsa   72A--A---GCGGAGACAGATTTTGGGG-----GGGCAGGATT-GAG-GCAA
Hme  261A--A---GCAGGGACCGATTTTGGGG-----GGGCAGGA---GAG-GCAA
Hmm  267A--A---GCAGGGACCGATTTTGGGG-----GGGCAGGA---GAG-GCAA
Hma  291A--A---G-GAGGACAGATTT-GGGG-----AGGCAGGATTTGAA-ACAA
Tbr  303ATTGTTTGCAGAAACAAA----GGGGTTTTTGGGCAAAG---GAATACAA

Hsa  110C--TCAGGA-GGG-GGAGGGCGAAAGGAGGAACTCG-GGAGGGGAGACAG
Hme  297C--TCAGGA-GGG-GGAGAGCGGAAAGAGGAACACGTGGAGGGGAGACAG
Hmm  303C--TCAGGA-GGG-GGAGAGCGGAAAGAGGAACACGTGGAGGGGAGACAG
Hma  328C--CCAG-ATGGGGAGAGAGCGAGGGGAGGAACACGTGGAGGGGAGACAG
Tbr  346TTTGCAG-A-GGGGGGAGAGCGGAAGGAGGAACACG-GGAGGGAAGACAG

Hsa  155GA---GGGGGGCGAGAGAGAG--GGGAGG--G-AGAAGGAGGGGTGAGAA
Hme  343GA---GGGGGGCGAGAGAGAG--GGGAGGTTG-A-AA---GGGGGAAATT
Hmm  349GA---GGGGGGCGAGAGAGAG--GGGAGGTTG-A-AA---GGGGGAAATT
Hma  375GATT-GGGGAGCGAGAGAGGGTTGAGAGG--G-A-AA---GGGGGGGAAT
Tbr  394GATTTAGGAAGCGAGAGAGAG--GAGAGG--GGA-AA---GGG------T

Hsa  197AAAAG-GTGTAGTGTTG-TGGAAAGAATAAA
Hme  383TAA---GTGT-GTGT--------AGAATATAGAAAATAAAAA
Hmm  389TAA---GTGT-GTGT--------AGAATATAGA
Hma  417TTAGGAGTGTAGTG-AG-TAGTTATAACAACAATACTAAAACAAAACA
Tbr  429TTAGTTGGAATGAAGAGGTAGTTTGTAGGAAGTTAA
```

FIG. 3 Conservation of cryptic DNA sequences. The unedited *COIII* genes for *Herpetomonas* and *T. brucei* are aligned. The sequences are underlined of the primers used to amplify the 5′ end of the gene and the DNA fragment described in ref. 2. We did not determine the 5′ end of the *H. samuelpessoai* DNA sequence because none of the 5′ anchor PCR products were partially edited at the 5′ end. The Hsa sequence begins 17 bp downstream from the PCR and sequencing primer. The sequence of the negative-strand cytochrome *b* primer is 5′-biotin-CCTAAACTAAA(AT)CC(AT)AC(CA)CCATA-3′.

1.30 between *H. samuelpessoai*, *H. m. muscarum*, *H. mariadeanei* and *T. brucei*), compared to a value of 0.69 between *L. tarentolae* and *C. fasciculata*. Thus the edited proteins evolve on average 84% faster (between 60% and 113% faster) than the 5′-edited versions because of frameshifting. We expected the requirement for multiple changes in the guide RNAs to inhibit the rate of amino-acid change for edited genes, as it does for synonymous changes. Instead, the extensively edited trypanosomatid mitochondrial genes appear to be driven in rapid genetic variation.

Note added in proof: The authors of ref. 12 suggest that the *L. tarentolae* COIII protein itself is unusually rapidly evolving, even among the other maxicircle encoded proteins. Thus the extensively edited COIII proteins probably evolve even faster than our calculation shows. □

Received 26 October 1992; accepted 23 February 1993.

1. Shaw, J. M., Feagin, J. E., Stuart, K. & Simpson, L. *Cell* **53,** 401-411 (1988).
2. Feagin, J. E., Abraham, J. M. & Stuart, K. *Cell* **53,** 413-422 (1988).
3. Blum, B., Bakalara, N. & Simpson, L. *Cell* **60,** 189-198 (1990).
4. Maslov, D. A. & Simpson, L. *Cell* **70,** 459-467 (1992).
5. Lake, J. A. *et al. Proc. natn. Acad. Sci. U.S.A.* **85,** 4779-83 (1988).
6. Koslowsky, D. J. *et al. Cell* **62,** 901-911 (1990).
7. Hultman, T., Stahl, S., Hornes, E. & Uhlen, M. *Nucleic Acids Res.* **17,** 4937-4946 (1989).
8. Volloch, V., Schweitzer, B. & Rits, S. *Proc. natn. Acad. Sci. U.S.A.* **88,** 10671-10675 (1991).
9. Ohara, O., Dorit, R. L. & Gilbert, W. *Proc. natn. Acad. Sci. U.S.A.* **86,** 5673-5677 (1989).
10. Loh, E. Y. *et al. Science* **243,** 217-220 (1989).
11. Young, I. G. *et al. Eur. J. Biochem.* **116,** 165-170 (1981).
12. de la Cruz, V. F., Neckelmann, N. & Simpson, L. *J. biol. Chem.* **259,** 15136-15147 (1984).
13. Sloof, P., van den Burg, J., Voogt, A. & Benne, R. *Nucleic Acids Res.* **15,** 51-65 (1987).
14. Chomczynski, P. & Sacchi, N. *Analyt. Biochem.* **162,** 156-159 (1987).
15. Kreitman, M. & Landweber, L. F. *Gene Anal. Tech.* **6,** 84-88 (1989).
16. Higuchi, R. G. & Ochman, H. *Nucleic Acids Res.* **17,** 5865 (1989).

ACKNOWLEDGEMENTS. We thank R. C. Lewontin for discussion, A. Fiks for technical assistance and V. Voloch for oligonucleotides. This work was supported by the NIH. L.L. is a Howard Hughes Medical Institute Predoctoral Fellow.

Section VII:
G4 DNA

364 LETTERS TO NATURE NATURE VOL. 334 28 JULY 1988

Formation of parallel four-stranded complexes by guanine-rich motifs in DNA and its implications for meiosis

Dipankar Sen & Walter Gilbert

Department of Cellular and Developmental Biology, Harvard University, Cambridge, Massachusetts 02138, USA

We have discovered that single-stranded DNA containing short guanine-rich motifs will self-associate at physiological salt concentrations to make four-stranded structures in which the strands run in parallel fashion. We believe these complexes are held together by guanines bonded to each other by Hoogsteen pairing. Such guanine-rich sequences occur in immunoglobulin switch regions[1], in gene promoters[2,3], and in chromosomal telomeres[4]. We speculate that this self-recognition of guanine-rich motifs of DNA serves to bring together, and to zipper up in register, the four homologous chromatids during meiosis.

We found this self-recognition of certain guanine-rich DNA sequences in the course of experiments examining the immunoglobulin switch regions. Using synthesized DNA fragments, we had begun mobility shift experiments seeking specific proteins that might bind to the switch regions. But the oligonucleotides aggregated and moved to novel positions on the gels; we investigated these complexes to discover that they were parallel sets of four strands.

The genes for the immunoglobulin heavy chains undergo switch recombinations to bring different constant regions next to the rearranged variable region exons during the differentiation of B lymphocytes to plasma cells[1]. The switch-regions, the sites of these recombinations, lie 5′ to the constant-region genes and are stretches of repetitive DNA 1-10 kilobases (kb) in size. They consist largely of tandem repeats of GC-rich DNA which are 20-50 base pairs (bp) long and have various degrees of similarity. The coding-strand consensus sequences[5] contain typically 65-70% purines, with guanines making up 50% of the total. Several repetitive motifs occur, such as GGGGT and GAGCT, and the conserved sequence (G)GGGGAGCTGGGG, found in Sγ1, Sγ2b and Sγ3.

We synthesized oligomers, 20-49 bases long, corresponding to the IgG switch regions. The guanine-rich strands are shown in the legend to Fig. 1. When these oligomers were stored in buffer containing monovalent salt, each spontaneously generated a complex of markedly lower electrophoretic mobility. Figure 1, lane 1, shows this complex (M4) for one of the oligomers, M. The pyrimidine-rich complementary strands, in contrast, show no tendency to form analogous complexes.

These complexes formed in a concentration-dependent way, and this fact, together with their low electrophoretic mobility, suggested that they were inter-strand aggregates rather than internally folded, base-paired single strands. Kinased mixtures of the oligomer and complex were readily separated from one another by preparative gel-electrophoresis (Fig. 1, lane 3).

To determine the structure of such a complex, we first examined its ability to bind its complementary strand. Figure 1 shows the results of mixing labelled samples of the complex of the oligomer M and its complementary strand R in various ratios and analysing portions on a non-denaturing 6% polyacrylamide gel. Four new bands, of even slower mobility than the complex, appear in lanes 4-8 (a portion of the complex breaks down to form the Watson-Crick dimer '2'). As increasing amounts of the complementary strand are added, the label in all four upper bands coalesces into the fourth and uppermost band, which we thus infer to be a M4R4 product. Lanes 10-14 show a mirror experiment with the label in the complementary strand. The

NATURE VOL. 334 28 JULY 1988 LETTERS TO NATURE 365

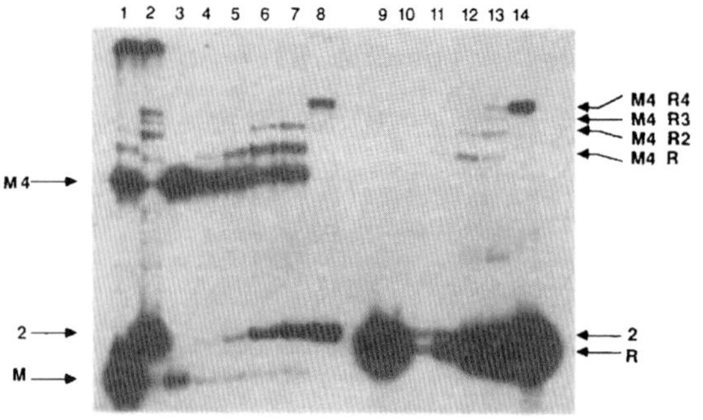

Sμ:	GAGCTGAGCTGGGGTGAGCT
Sγ3:	PGGACCAGGCTGGGACAGCTCTGGGGAGCTGGGGTGGGTGGGAGTGTGG
Sγ1:	GPPTCCAGGCTGAGCAGCTACAGGGGAGCTGGGGYAPPTGGGAPTPTPG
Sγ2b:	GGGACCAGACCTAGCAGCTPTGGGGGAGCTGGGGAAGGTPGGAPTPTGA
	(P = Purine, Y = Pyrimidine).
oligo P:	GGGACCAGACCTAGCAGCTATGGGGGAGCTGGGGAAGGTGGGAATGTGA
oligo Q:	TACAGGGGAGCTGGGGTAGA
oligo M:	TAGTCCAGGCTGAGCAGGTACGGGGGAGCTGGGGTAGATGGGAATGT
oligo N:	GTCCAGGCTGAGCAGGTACGGGGGGAGCTGGGGTAGATGGGAATGA

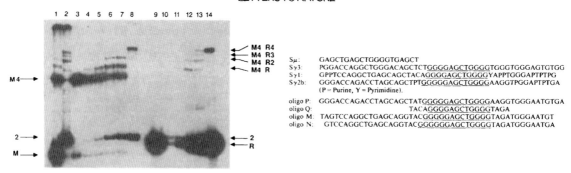

Fig. 1 Left, non-denaturing polyacrylamide-gel mobilities of the oligomer M and its complexes. M: oligomer M monostrand; M4: tetra-strand of M; R: complementary strand to M; 2: Watson-Crick double strand of M and R; M4R-M4R4: M4 complexed with one to four strands of R. Consensus sequences[5] of switch-region repeat units are shown on the right, where P is the 49b repeat of Sγ2*b*; Q contains the conserved motif GGGGAGCTGGGG; M and N are derived from Sγ1, except that N contains an extra guanine residue, to give a six-guanine stretch. **Methods.** Oligomers were made using a Milligen Autogen 6500 DNA synthesizer. Complexes were formed by storing the oligonucleotides at 100–500 μg ml^{-1} in 10 mM Tris *p*H 8, 90 mM NaCl, 1 mM EDTA buffer (TNE) for 24–48 h at 4 °C and then isolated on preparative gels. Association reactions were carried out in 10 μl of TNE buffer at 20 °C, for 30 min. Lane 1, 1 ng total of oligomer M and its complex M4; lane 2, 1 ng labelled M and M4 added to 5 ng cold complementary strand R; lane 3, 0.4 ng gel-purified M4; lanes 4–8, 0.4 ng labelled M4, mixed with 0.1, 0.3, 1.0, 3.0 and 10.0 ng respectively of unlabelled R; lane 9, 3 ng labelled complementary strand R; lanes 10–14, 0.1 ng unlabelled M4, mixed with 0.1, 0.3, 1.0, 3.0 and 10.0 ng respectively of labelled R. Samples were electrophoresed through a 6% polyacrylamide gel in 50 mM TBE buffer[6] at 20 °C.

formation of the four upper bands, each containing the complementary strand R, shows that the complex of M is a tetramer (termed M4), the upper bands thus representing the binding of 1-4 complementary strands to M4.

To elucidate the structure of the complex, the guanines of gel-purified M and M4 were methylated with dimethyl sulphate[6]. Piperidine-treated samples, run on a sequencing gel, revealed a complete methylation-protection of the sequence GGGGGAGCTGGGG (Fig. 2*a*, lanes M: 1 and 2). The unavailability of the N-7s of these guanines for methylation suggests their involvement in Hoogsteen bonding[7]. Figure 2*a* also shows a similar methylation-protection pattern for all of the complexes M4, N4, P4 and Q4. In P4 the protected region spans 21 bases with a total of 15 guanines protected, extending over the sequence GGGGGAGCTGGGGAAGGTGGG. But the GGGs at the 3′ end of M and N are not protected; the sequence of the other bases must also be relevant to the ability of the complex to extend.

The protection patterns of M4, N4 and P4 clearly demonstrate that the orientation of the four strands is *parallel*. No antiparallel orientation could completely protect the N-7s of every guanine within the motifs.

X-ray fibre diffraction studies and model-building on polyG (refs 8, 9) have suggested that the ribo-homopolymer could form a four-stranded structure, with the guanines Hoogsteen bonded, as in Fig. 2*b*. Our results demonstrate that short guanine-rich motifs, in a natural DNA sequence, can form stable four-stranded structures and are able to maintain their four-strandedness even when complexed to complementary strands. Figure 3 shows the structures we predict: M4R4 represents the uppermost electrophoretic band of Fig. 1, with the complex M4 bound to four complementary strands R.

Might this self-recognition and four-strand formation in parallel orientation by guanine-rich strands of DNA have a function *in vivo*?

We propose that these four-stranded regions, which we will call G4-DNA, have a role in the pairing of homologous chromosomes during meiosis. In the early stages of meiosis, after DNA synthesis, four copies of the DNA double helix, two pairs of sister chromatids, are brought together. In the stages of prophase 1 partially condensed homologous chromosomes appear precisely aligned, side by side, as a prelude to eventual exchanges of genetic information between the homologues (reviewed in refs 10, 11).

Synapsis is sometimes observed to begin at telomeres (reviewed in refs 4, 12) to generate 'bouquet' structures, although in other instances, it is initiated at intra-chromosomal loci[13,14]. The association of telomeres is consistent with our hypothesis, because all non-mitochondrial telomeres sequenced to date contain repeated guanine-rich motifs that are added to the chromosome subsequent to replication[4]. Furthermore, the single-strands containing the guanine-rich motifs are usually longer than their complementary strands and thus overhang them by 12–16 bases (reviewed in ref. 15).

Henderson *et al.*[16] have recently shown that guanine-rich synthetic oligomers from various telomeres can form guanine-mediated fold-back structures at low temperatures. But judging from the similarity of the telomere sequences to the switch sequences, we predict that telomere sequences will also form G4-DNA at physiological salt concentrations. Although we would not rule out a role for proteins in telomere-telomere interactions *in vivo*, we propose that the primary driving force for their association may come from this intrinsic property of parallel aggregation by self-recognizing guanine-rich sequences. Oka and Thomas[17] examined the aggregation of macronuclear DNA from *Oxytricha* and showed that it was dependent upon a telomeric overhang of --TTTTGGGGTTTTGGGG being single stranded. We interpret this observation as involving the formation of G4-DNA structures: four Watson strands interacting in contrast to Oka and Thomas' interpretation of two Watson and two Crick strands being involved.

Figure 4 sketches our proposal that G4-DNA elements tie together the four chromatids at meiosis. Once joined at the telomeres, or elsewhere, the tight association of the homologues (zippering up) proceeds via a recognition of 'self' and subsequent four-strand formation in parallel fashion of specific sequences along the chromatid. A number of loci along the chromosome, each containing a guanine-rich motif of unique size and sequence, together constitutes a chromosomal fingerprint to facilitate recognition and association. The guanine-rich strands of the motifs might be made accessible to each other by local unwinding or by the process described by Mirkin *et al.*[18] in which homopurine-homopyridine tracts of a duplex DNA undergo a superhelix-induced structural transition to a novel DNA conformation, the H form, leaving a polypurine strand free from base pairing. Most simply, however, the cell could control G4-DNA formation by synthesizing both melting proteins that would bind to C-rich DNA as well as G4-binding

366 LETTERS TO NATURE NATURE VOL. 334 28 JULY 1988

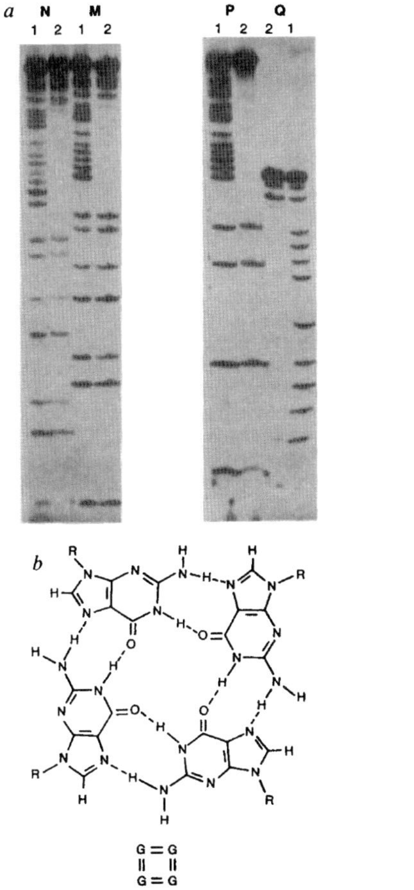

Fig. 2 *a*, N-7 methylation patterns of oligomers M, N, P and Q (lanes labelled 1), and their respective complexes (lanes labelled 2). Methylation reactions were adapted from Maxam and Gilbert[6]. Kinased DNA was treated with dimethyl sulphate and electrophoresed on a 6% non-denaturing acrylamide gel. Bands of complex and monomer were detected by autoradiography, eluted into TNE, precipitated, treated with piperidine, and analysed on a 20% sequencing gel. *b*, Model structure of a Hoogsteen hydrogen-bonded guanine tetrad[8,9].

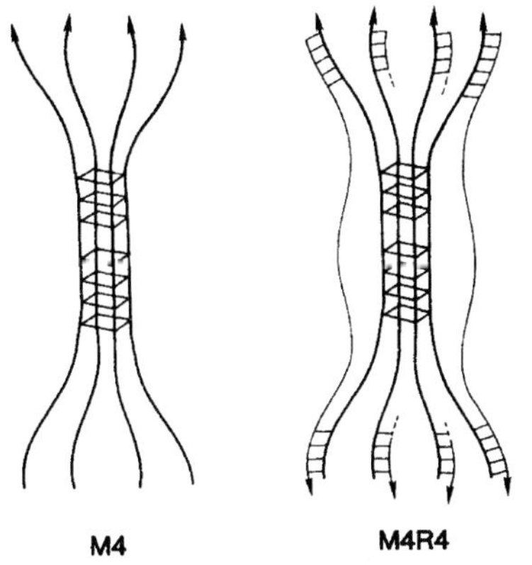

Fig. 3 Schematic model for the complexes M4 and M4R4.

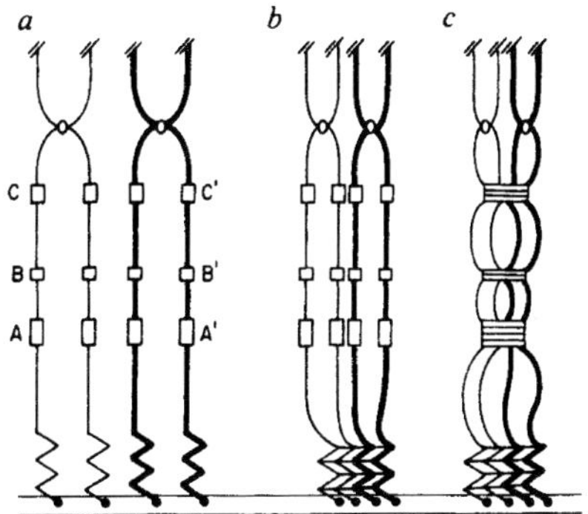

Fig. 4 A scheme for the participation of guanine-rich telomeres and internal chromosomal motifs in homologue recognition during meiotic prophase. Telomeres are shown by zigzagged lines, and guanine-rich 'recognition' motifs as A, B and C (with counterparts A', B' and C'). *a*, Homologous chromosomes are shown with their telomeres attached to the nuclear envelope. *b*, The four telomeres associate, by G4-DNA formation, to generate 'bouquets'. *c*, Recognition and synapsis occur mediated by stretches of G4-DNA along the homologues.

proteins that would stabilize G4-DNA, irrespective of sequence. Removing both types of proteins, or even synthesizing G4-melting proteins, would disrupt the pairing.

G4-DNA forming regions may replicate late as the last step before pairing. Hotta and Stern[19] observed that there is a late synthesis, during zygotene, of ~0.3% of the DNA, GC-rich, that is required in order that meiosis can proceed.

The pairing of non-homologues too might be initiated by G4-DNA formation from near-identical guanine motifs, but would not be successful over the entire length of dissimilar chromsomes. There is some cytological evidence for such loose pairing[20].

In summary, we attribute the interaction of *four* double helices at meiosis to an underlying *four*-fold recognition event mediated by short stretches of guanines, in a like-with-like interaction.

We thank Dr Matthew Meselson for discussions. This work was supported in part by the NIH. We thank Milligen for the DNA synthesizer.

Received 3 May; accepted 14 June 1988.

1. Shimizu, A. & Honjo, T. *Cell* **36**, 801-803 (1984).
2. Evans, T., Schon, E., Gora-Maslak, G., Patterson, J. & Efstratiadis, A. *Nucleic Acids Res.* **12**, 8043-8058 (1984).
3. Kilpatrick, M. W., Torri, A., Kang, D. S., Engler, J. A. & Wells, R. D. *J. biol. Chem.* **261**, 11350-11354 (1986).
4. Blackburn, E. H. & Szostak, J. W. *A. Rev. Biochem.* **53**, 163-194 (1984).
5. Nikaido, T., Yamawaki-Kataoka, Y. & Honjo, T. *J. biol. Chem.* **257**, 7322-7329 (1982).
6. Maxam, A. & Gilbert, W. *Meth. Enzym.* **65**, 499-560 (1977).
7. Hoogsteen, K. *Acta Crystallogr.* **12**, 822-823 (1959).
8. Arnott, S., Chandrasekaran, R. & Marttila, C. M. *Biochem. J.* **141**, 537-543 (1974).
9. Zimmerman, S. B., Cohen, G. H. & Davis, D. R. *J. molec. Biol.* **92**, 181-192 (1975).
10. John, B. *Chromosoma* **54**, 295-325 (1976).
11. Maguire, M. P. *J. theor. Biol.* **106**, 605-615 (1984).
12. Ashley, T. & Wagenaar, E. B. *Can. J. Genet. Cytol.* **16**, 61-76 (1974).
13. Gillies, C. B. *Trav. Lab. Carlsberg* **40**, 135 (1975).
14. Holm, P. B. *Trav. Lab. Carlsberg* **42**, 103 (1977).
15. Weiner, A. M. *Cell* **52**, 155-157 (1988).
16. Henderson, E., Hardin, C. C., Walk, S. K., Tinoco, I. & Blackburn, E. H. *Cell* **51**, 899-908 (1987).
17. Oka, Y. & Thomas, C. A. *Nucleic Acids Res.* **15**, 8877-8898 (1987).
18. Mirkin, S. M. *et al. Nature* **330**, 495-497 (1987).
19. Hotta, Y. & Stern, H. *J. molec. Biol.* **55**, 337-355 (1971).
20. Moens, P. B. *Cold Spring Harb. quant. Symp.* **49**, 699-705 (1984).

Proc. Natl. Acad. Sci. USA
Vol. 90, pp. 3157–3161, April 1993
Biochemistry

Identification and characterization of a nuclease activity specific for G4 tetrastranded DNA

(G quartet/DNA binding/telomere/meiosis/switch recombination)

ZHIPING LIU*†‡, J. DANIEL FRANTZ*, WALTER GILBERT*, AND BIK-KWOON TYE†

*Department of Cellular and Developmental Biology, Harvard University, Biological Laboratories, 16 Divinity Avenue, Cambridge, MA 02138; and †Section of Biochemistry, Molecular and Cell Biology, Cornell University, Ithaca, NY 14853

Contributed by Walter Gilbert, January 7, 1993

ABSTRACT **We have identified a nuclease activity that is specific for G4 tetrastranded DNA. This activity, found in a partially purified fraction for a yeast telomere-binding protein, binds to DNA molecules with G4 tetrastranded structure, regardless of their nucleotide sequences, and cleaves the DNA in a neighboring single-stranded region 5′ to the G4 structure. Competition with various G4-DNA molecules inhibits the cleavage reaction, suggesting that this nuclease activity is specific for G4 tetrastranded DNA. The existence of this enzymatic activity that reacts with G4 DNAs but not with single-stranded or Watson–Crick duplex DNAs suggests that tetrastranded DNA may have a distinct biological function *in vivo*.**

G4-DNA is a tetrastranded DNA structure in which the four strands are held together by guanine–guanine Hoogsteen base pairing (1). Such structures have a characteristic stack of guanine quartet planes and have the sugar strands in either a parallel or an antiparallel orientation (2, 3). They are formed from G-rich DNA sequences with strings of at least three contiguous guanines. These sequences can be found in evolutionarily and functionally conserved chromosomal regions, such as telomeres and the switch regions of immunoglobulin heavy-chain genes (4, 5). Although the existence of G4-DNAs *in vivo* remains to be established, we describe here an enzymatic activity specific for such G4-DNA. This finding suggests that these structures play a role *in vivo*.

In vitro studies with G-rich oligonucleotides from telomeres and the switch regions have identified four-stranded structures by their different mobilities in nondenaturing gel electrophoresis. G4-DNA generated from intramolecular folding, referred to as G-quartet DNA (3), in which the adjacent strands are in antiparallel orientation, migrates faster than its unfolded precursor. G4-DNA molecules generated from intermolecular interactions, with the adjacent strands in either an antiparallel or parallel configuration, migrate much more slowly than the unfolded oligonucleotides (1, 2). The structures of some of these tetrastranded DNAs were recently elucidated by x-ray crystal diffraction and two-dimensional NMR studies (6, 7). The formation of the four-stranded structure requires Na^+ ions, and the structure, once formed, can be further stabilized by K^+ ions (8).

In a fraction partially purified from a yeast extract, we have identified an enzyme that requires the presence of a G4 structure and cleaves the DNA in a neighboring single-stranded region. Although we do not fully understand the mechanistic details of the cleavage reaction and cofactor effects, we describe here a preliminary characterization of this G4-DNA-specific binding and cleavage reaction.

MATERIALS AND METHODS

Oligonucleotides. All oligonucleotides used in this study were made on an Applied Biosystems DNA synthesizer using phosphoramidite chemistry. The DNA sequences are listed in Fig. 1. Full-length products were isolated from 8 M urea/20% polyacrylamide gels. G4-DNAs were made according to published procedures and purified from a 6% polyacrylamide gel containing 50 mM KCl (8). G4-DNA probes were labeled at the 5′ end with T4 polynucleotide kinase and [γ-^{32}P]ATP, precipitated with 0.3 M sodium acetate in 70% ethanol, and stored in 10 mM Tris, pH 8/1 mM EDTA/50 mM KCl.

Preparation of the Protein Fraction. The yeast extract was prepared and fractionated as described (9). In brief, cells of *Saccharomyces cerevisiae* strain BJ2168 (*MATa pep4-3 prc1-407 prb-1122 ura3-52 trp1 leu2*) were harvested at midlogarithmic phase. The cell paste was frozen in liquid nitrogen, lysed in a Waring blender, and suspended in 100 mM KCl buffer A [50 mM Hepes, pH 7.5/1 mM EDTA, 1 mM dithiothreitol/10% (vol/vol) glycerol]. The crude lysate was cleared by ultracentrifugation and subjected to a 25–50% (percent saturation) ammonium sulfate fractionation. After dialysis, the ammonium sulfate fraction was purified further on a phosphocellulose column and a DNA affinity column, enriching for the DNA-binding activity of telomere-binding factor α (TBFα). The matrix of the DNA column was made from cellulose and a linearized plasmid containing a cloned human telomeric fragment carrying 27 repeats of GGGATT. Yeast proteins bound to the DNA column were eluted with a linear gradient of 0.10–0.60 M KCl in buffer A. After the DNA column, the protein concentration in the fractions was around 10 μg/ml, based on India ink staining on nitrocellulose filter with bovine serum albumin as standard. All the work presented here was done with the material from the DNA column fractions 27–30, as shown in Fig. 2, which we will simply refer to as the protein fraction.

Bandshift Assay. The assay for TBFα binding activity on an agarose gel was performed as described (9). To assay for the G4-DNA-specific binding and cleavage activity, aliquots of protein fractions were mixed with labeled G4-DNA probes (1–5 fmol or as specified) in a final volume of 20 μl containing binding reaction buffer (20 mM Hepes, pH 7.5/50 mM KCl/5 mM $MgCl_2$/5–10% glycerol containing *Escherichia coli* single-stranded DNA at 0.1 μg/ml), incubated at room temperature for 10 min or as specified, and loaded on a 5 or 6% polyacrylamide gel precooled to 4°C. The electrophoresis was carried out in Tris/borate/EDTA buffer containing 50 mM KCl in the cold room at about 7 V/cm, until the bromophenol blue dye had run two-thirds of the full-gel length. The gel was then dried and an autoradiogram was made. For quantitation, the dried gels were directly scanned and analyzed on a blot analyzer.

Analysis of the Cleavage Product. Cleavage reactions were carried out as described above. At the end of the incubation, 20 μl of 10 mM EDTA was added to stop the reaction. Then,

The publication costs of this article were defrayed in part by page charge payment. This article must therefore be hereby marked "*advertisement*" in accordance with 18 U.S.C. §1734 solely to indicate this fact.

Abbreviation: TBF, telomere-binding factor.
‡To whom reprint requests should be sent at * address.

P	(49mer):	5'-GGGACCAGACCTAGCAGCTATGGGGGAGCTGGGGAAGGTGGGAATGTGA-3'
TP	(49mer):	5'-TGGACCAGACCTAGCAGCTATGGGGGAGCTGGGGAAGGTGGGAATGTGA-3'
TP-S	(16mer):	5'-TGGACCAGACCTAGCA-3'
CTP-S	(18mer):	5'-GCTGCTAGGTCTGGTCCA-3'
LBG	(35mer):	5'-TAGGGTAGTGTTAGGGTAGTGTTAGGGTGTGGGTG-3'
Oxy-16	(16mer):	5'-TTGGGGTTTTGGGGTT-3'
P55	(55mer):	5'-ATATGATGGACCAGACCTAGCAGCTATGGGGGAGCTGGGGAAGGTGGGAATGTGA-3'
M2	(35mer):	5'-TAGTCCAGGCTGAGCAGGTACGGGGGAGCTGGGGT-3'
M2-S	(21mer):	5'-TAGTCCAGGCTGAGCAGGTAC-3'
HT-G	(29mer):	5'-TCGAGTTAGGGTTAGGGTTAGGGTTAGGG-3'
HT-C	(29mer):	5'-TCGACCGTAACCCTAACCCTAACCCTAAC-3'

FIG. 1. Oligonucleotides used in this work.

1 ml of 1-butanol was added to each reaction mixture in an Eppendorf centrifuge tubes. After brief vortex mixing, DNA was recovered as pellets by centrifugation in an Eppendorf centrifuge for 2 min at full speed. The organic liquid was discarded and the DNA pellets were air-dried for 2–5 min. To open a G4 structure while avoiding nonspecific strand scission, DNA samples were treated in one of the following ways. The DNA pellets were resuspended in 50 μl of 0.1 M NaOH, incubated at 60°C for 10 min, butanol-extracted, suspended in 5 μl of formamide loading dye containing 0.1 M NaOH, and, after being heated in a 90°C water bath for 1 min, loaded on a 20% sequencing gel containing either 8 M urea or 50% formamide. Alternatively, DNA pellets from the cleavage reaction were directly suspended in 5 μl formamide loading dye with 0.1 M NaOH, heated in a 90°C water bath for 5 min, and loaded on a 20% sequencing gel. After electrophoresis, autoradiograms were made from the gels with Kodak XAR-5 film. Sequencing ladders were generated according to the Maxam–Gilbert method (10).

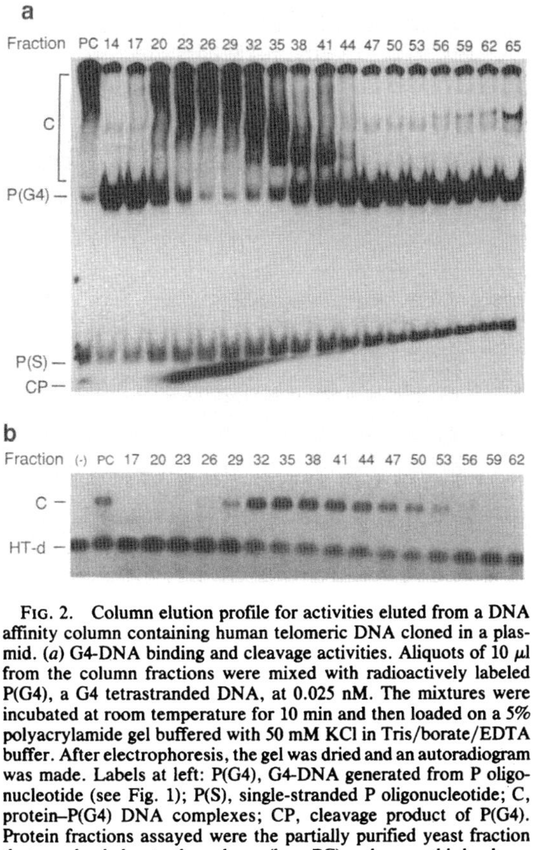

FIG. 2. Column elution profile for activities eluted from a DNA affinity column containing human telomeric DNA cloned in a plasmid. (*a*) G4-DNA binding and cleavage activities. Aliquots of 10 μl from the column fractions were mixed with radioactively labeled P(G4), a G4 tetrastranded DNA, at 0.025 nM. The mixtures were incubated at room temperature for 10 min and then loaded on a 5% polyacrylamide gel buffered with 50 mM KCl in Tris/borate/EDTA buffer. After electrophoresis, the gel was dried and an autoradiogram was made. Labels at left: P(G4), G4-DNA generated from P oligonucleotide (see Fig. 1); P(S), single-stranded P oligonucleotide; C, protein–P(G4) DNA complexes; CP, cleavage product of P(G4). Protein fractions assayed were the partially purified yeast fraction that was loaded onto the column (lane PC) and every third column fractions from 14 to 65. (*b*) Elution of TBFα activity. Aliquots of 10 μl from the same fractions as those assayed in *a* were used in an agarose gel bandshift assay. The probe (2 fmol per assay) was TCGA(CCCTAA)$_4$·TCGA(TTAGGG)$_4$ duplex DNA (HT-d) labeled by filling in with Klenow DNA polymerase.

RESULTS

Identification of a G4-DNA Binding and Cleavage Activity. Liu and Tye (9) previously identified a telomere-binding protein in yeast, TBFα, by means of its specific binding to human telomeric DNA. In exploring further the properties of this protein, we found that the partially purified TBFα fraction also contained a binding activity that is specific for G4-DNA. Fig. 2*b* shows the elution profile of TBFα from a DNA affinity column consisting of a cloned human telomeric DNA, assayed by its binding to a (CCCTAA)$_4$·(TTAGGG)$_4$ duplex. The elution peak of this binding activity centered at fraction 35. While assaying the same column fractions with a G4-DNA probe [P(G4), see Fig. 1 for a list of the oligonucleotides used in this work], we detected a binding activity eluted from the column (Fig. 2*a*, fractions 23–39) at a lower salt concentration. In these same fractions, another migrating species, CP, appears, moving faster than the single-stranded form of the P oligonucleotide. The simplest explanation is that CP is a cleavage product of the P(G4) molecule, since single-stranded probe, P(S), was unaffected. We conjectured that these column fractions contained a nuclease specific for G4-DNA.

Characterization of the Binding and Cleavage Activities. Fig. 3*a* shows that both the binding and cleavage activities require a protein component. Preincubation of the protein fraction at 65°C for 15 min completely inactivated the cleavage activities and partially inactivated the G4-DNA binding activity (lane 3). Treatment with a detergent (lane 4) or digestion with proteinase K (lane 5) inactivated both the binding and the cleavage activities. Furthermore, the inactivation caused by proteinase K was blocked by a serine-protease inhibitor (lanes 6). The single-stranded TP oligonucleotide exhibited no binding and cleavage (compare lane 8 with lane 2). These results suggest that the binding to and cleavage of TP(G4) are probably linked events and that the G4 structure present in the TP(G4) DNA plays a fundamental role in these reactions.

Fig. 3*b* shows a time course of the cleavage reaction analyzed in a 6% nondenaturing gel. Binding to G4-DNA probe seems to precede the cleavage event, since at an early time point (lane 3) most of the TP(G4) material is in the bound

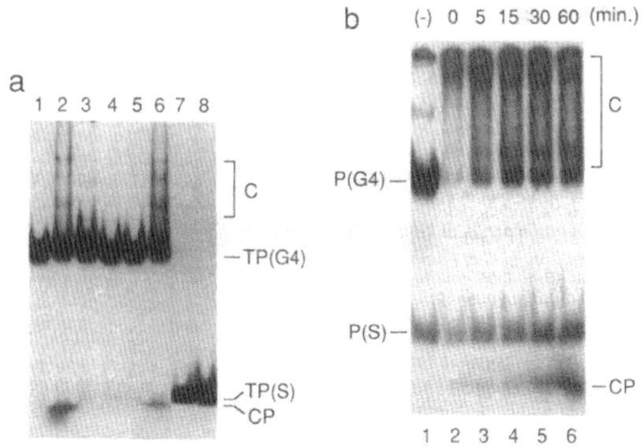

FIG. 3. Characterization of the G4-DNA binding and cleavage activities. (*a*) Inactivation of the cleavage activity. The G4-DNA probe used in the reactions shown in lanes 1–6 was generated from the TP oligonucleotide. The radioactively labeled DNA probe used in reactions shown in lanes 7 and 8 was a single-stranded TP oligonucleotide. TP(S), single-stranded TP oligonucleotide; TP(G4), G4-DNA generated from TP; C, protein–TP(G4) DNA complexes; CP, cleavage product. The assay was the same as described for Fig. 2*a* except that 5-μl aliquots from the DNA column fraction 29 were pretreated for 15 min as follows: lane 2, no treatment; lane 3, heating in a 65°C water bath; lane 4, incubation in the presence of 1% SDS at room temperature; lane 5, digestion with 0.5 μg of proteinase K at room temperature; lane 6, the same as lane 5 except that phenylmethanesulfonyl fluoride was added to a final concentration of 0.1 mM in the reaction mixture. No protein was added in reactions shown in lanes 1 and 7, while the same amount of protein was used in lanes 2 and 8. (*b*) Time course of the cleavage reaction. P(G4) probe, radioactively labeled at the 5′ end, was incubated with 5 μl of the protein fraction at room temperature for the time specified at the top, and then loaded on a 5% polyacrylamide gel and treated as described for Fig. 2*a*. No protein was added in the reaction shown in lane 1, and bands migrating more slowly than P(G4) were probably due to higher-order structure formed between P(G4) molecules (11).

complex and yet very little cleavage product has been produced, which suggests that the binding to the substrate is rapid whereas the ensuing cleavage is slow. The amount of cleavage product increases linearly with the reaction time over the first 30 min (data not shown).

The Cleavage Products. We analyzed the products of the cleavage reaction on sequencing gels. Fig. 4*a* shows that when the probe is TP(G4) labeled at the 5′ end, there is only one major product, which migrates like a dinucleotide with a 3′-OH end. The time course shown in lanes 4–9 suggests that the dinucleotide is either the initial cleavage product or the only product of the reaction, since we detect no other larger fragments that would correspond to cleavages more distal from the 5′ end. Fig. 4*b* shows the cleavage sequences for this and two other G4-DNA substrates, P55(G4) and M2(G4). The cleavage site shifts on P55(G4), compared with TP(G4), even though this molecule is essentially the same as TP(G4) except for a six-nucleotide extension at the 5′ end. There is no sequence consensus or invariant distance that serves as the signal for the cleavage reaction.

Specificity of the Cleavage Reaction. Since the cleavage reaction occurs in a single-stranded region, is the G4 structure required for the reaction? Fig. 5 shows that TP oligonucleotide in single-stranded form cannot be cleaved, in contrast to its G4 form (lane 4 and lane 2, respectively). Furthermore, a single-stranded and truncated TP oligonucleotide, TP-S, corresponding to the 17 nucleotides on the 5′ end of TP, is not cleaved (lanes 3 and 10), consistent with the notion that G4 structure is required for the cleavage reaction. We also converted this truncated TP into a duplex form by annealing it to its complementary strand, CTP-S. Fig. 5, lanes

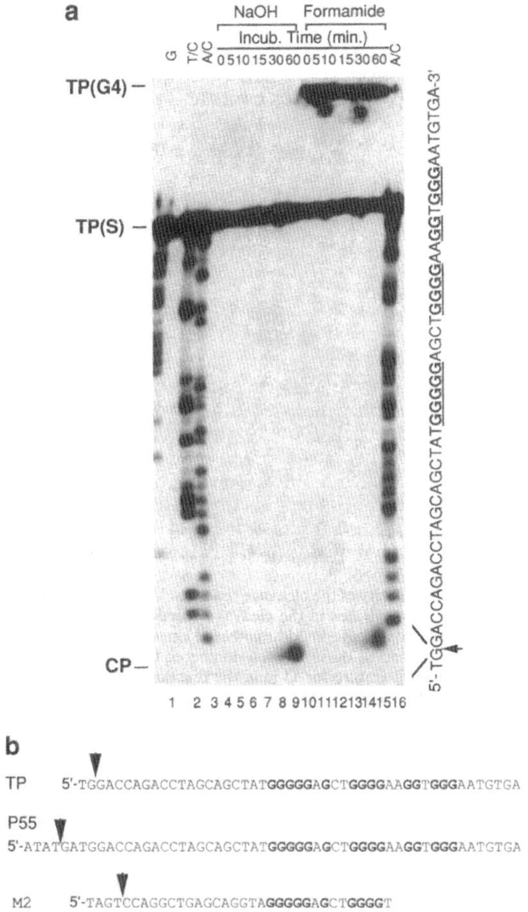

FIG. 4. Analysis of the 5′-end cleavage products on G4-DNAs generated from three different oligonucleotides. (*a*) Cleavage product on TP(G4). Aliquots of 2.5 μl of the protein fraction were incubated with 0.125 nM 5′-end-labeled TP(G4) for various lengths of time, as indicated at the top of the autoradiograph, and then phenol-extracted once. For each sample, the aqueous phase was split in half and the labeled DNA was recovered by butanol extraction. Half of the sample was directly suspended in 3 μl of formamide loading dye. The other half was treated with 0.1 M NaOH as described for Fig. 5, extracted with butanol to dried pellets, and suspended in 3 μl of formamide loading dye. All samples were heated in a 90°C water bath for 2 min, quickly cooled on ice, and then loaded on a 20% polyacrylamide gel containing 50% formamide as follows: lanes 4–9, alkaline-treated samples; lanes 10–14, samples not subjected to alkaline treatment. The incubation time for a particular reaction is indicated at the top. No protein was added in the reaction shown in lanes 4 and 10. Lanes 1, 2, 3, and 16 show Maxam–Gilbert sequencing ladder of 5′-end-labeled TP oligonucleotide. The actual sequence of TP is shown at right, in which the underlined G residues were involved in guanine-quartet formation. Arrow marks the cleavage site. (*b*) The 5′-end cleavage site observed on three different G4-DNAs. The oligonucleotide sequences are aligned based on their G-rich motifs that are engaged in four-stranded DNA formation, as indicated by the G residues in bold type. Arrowheads mark the deduced cleavage sites that give rise to the observed 5′-end cleavage products.

5 and 6, shows that this duplex DNA cannot be cleaved either. Finally, we annealed CTP-S to TP(G4), which converts the 5′ single-stranded region to a Watson–Crick duplex form but leaves the G4 structure intact. This hybrid shows no

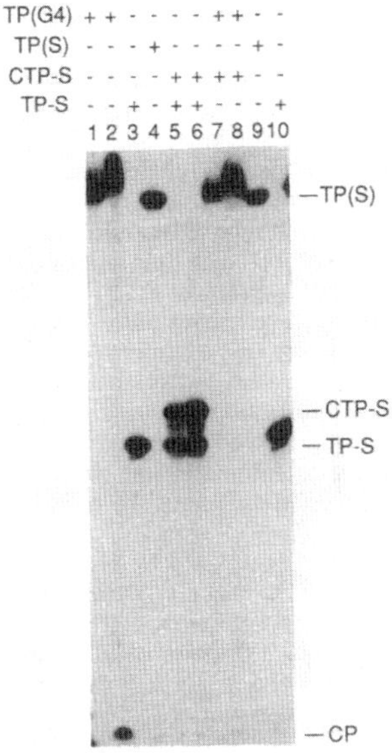

FIG. 5. Specificity of the cleavage reaction. Various DNA probes were tested as substrates in the cleavage reaction. No protein was added to reactions shown in odd-numbered lanes and 5 μl of protein fraction was added to those in even-numbered lanes. After incubation at room temperature for 15 min, the reactions were stopped by phenol extraction and the DNAs were recovered by butanol extraction. The G4-DNA was opened up by suspending the DNA in 50 μl of 0.1 M NaOH, incubating at 60°C for 10 min, and recovering by butanol extraction. Then the DNA samples were resuspended in 3 μl of formamide loading dye and loaded on a 15% polyacrylamide gel containing 8 M urea. After electrophoresis, radioactive DNA bands were visualized by autoradiography. All DNA probes were labeled at the 5′ end by using polynucleotide kinase with [γ-^{32}P]ATP. The DNA added (+) in each reaction is indicated at the top of the autoradiograph and is also listed below: lanes 1 and 2, TP(G4) at 0.063 nM; lanes 3 and 10, single-stranded TP-S at 0.25 nM; lanes 5 and 6, duplex DNA formed between CTP-S and TP-S at 0.25 nM (both oligonucleotides were labeled at the 5′ end); lanes 7 and 8, unlabeled CTP-S annealed to 5′-end-labeled TP(G4) at a ratio of 0.25 nM/0.063 nM; lanes 4 and 9, single-stranded TP oligonucleotide at 0.25 nM.

cleavage (lanes 7 and 8). These experiments demonstrate that both the G4 structure and the single-stranded region are required for the cleavage reaction; furthermore, a Watson–Crick duplex or single-stranded DNA alone is not an effective substrate.

Saturation Competition of Both the Binding and the Cleavage Activity by Various DNAs. What is the role of the G4 structure in the cleavage reaction? If binding to G4 were required for the cleavage reaction, one would expect that competition for the binding should correlate with an inhibition in the cleavage reaction. We reasoned that a competition in this enzymatic recognition could be achieved only at a high substrate concentration where the cleavage reaction proceeds at V_{max} and the addition of more substrate to the reaction, such as a binding competitor, will no longer simply increase the reaction rate. The cleavage reaction can be saturated by a G4-DNA concentration of 10 nM (data not shown). Then the further addition of a 50-fold molar excess of unlabeled competitor DNA produces both a competition for the binding and an inhibition of the cleavage only when G4 structure is present in the competitor molecules (Fig. 6*a*, lanes 3–6). The four-stranded DNA molecules generated from TP and M2 oligonucleotides share similar G4 structure but differ in their single-stranded regions, suggesting that the sequences in the single-stranded region do not affect the

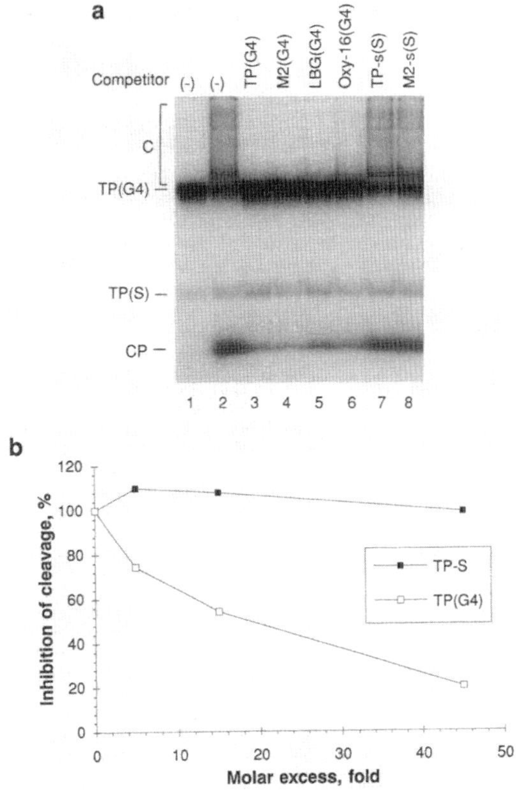

FIG. 6. Saturation competition of the binding and cleavage reactions on 5′-end-labeled TP(G4). (*a*) Each reaction mixture contained 10 nM labeled TP(G4) probe and 500 nM nonradioactive G4-DNAs or 2.0 μM nonradioactive single-stranded DNAs as indicated at the top. No specific competitor was added in reactions shown in lanes 1 and 2 except single-stranded *E. coli* DNA present in the reaction buffer at 5 μg/ml, and no protein was added to the reaction shown in lane 1. For lanes 2–8, aliquots (5 μl) of the protein fraction were mixed with the DNA probes in the reaction buffer, incubated at room temperature for 20 min, and loaded on a 6% polyacrylamide gel. TP(S), single-stranded TP oligonucleotide; TP(G4), G4-DNA generated from TP; CP, cleavage product. In this set of experiments, the final concentration of glycerol was 2.5%, and that of KCl was about 60 mM. (*b*) Inhibition of the cleavage activity by competition with TP-S and TP(G4) oligonucleotides. Cleavage reactions were carried out as described in *a*. The molar excess of TP-S was calculated on the basis of its ratio to the 5′ single-stranded region of TP(G4) probe in the reaction mixture. The percentage of cleavage activity was derived by dividing the amount of CP generated in reaction mixtures containing competitor with that from the no-competitor reaction mixture.

binding and that a 3′ single-stranded region on the G4 structure is not required for the binding and the cleavage reaction. G4-DNAs generated from LBG and Oxy-16, oligonucleotides derived from telomeric G-rich strands from yeast (*S. cerevisiae*) and ciliated protozoans (*Oxytricha*) respectively, have G4 structures different from that of TP(G4). They effectively competed with the TP(G4) probe for binding and inhibited the cleavage activity.

Two single-stranded oligonucleotides, corresponding to the truncated 5′ single-stranded regions of TP and M2 respectively, did not affect the binding or cleavage at a 50-fold molar excess (Fig. 6*a*, lanes 7 and 8). Fig. 6*b* shows that, over a range up to a 45-fold molar excess relative to the TP(G4) probe, TP-S did not affect the cleavage reaction, whereas

TP(G4) consistently inhibited the reaction. Thus, we conclude that protein binding to the G4 structure is required for the cleavage reaction and that the observed cleavage activity is specific for G4 tetrastranded DNA.

DISCUSSION

We describe here a nuclease activity that is specific for G4-DNAs and makes a cut in a single-stranded region 5′ to the G4 domain. The requirement for G4-DNA was established by examining the action of this enzyme on single-stranded and G4-DNA substrates and by saturation competition. In each case, we observed cleavage only when there was binding to a G4-DNA probe; with a single-stranded sequence corresponding to the cleavage substrate, we saw no binding and no cleavage. This line of interpretation is further supported by the results of saturation competition experiments, where we observed competition for binding and cleavage by several tetrastranded DNAs with different G4 motifs but no competition by the single-stranded region of the substrate. So far, we do not know the specificity of the cleavage reaction except that it occurs in a single-stranded region within a certain distance from the G4 structure, with no apparent sequence specificity. The nuclease activity described here could well be part of a general enzymatic machinery that interacts with G4 tetrastranded DNA.

This nuclease activity was found in the course of an investigation of proteins that bound to duplex telomeric DNA. In this search, Liu and Tye (9) identified a yeast protein, TBFα, that can bind to duplex telomeric DNA from yeast, human, and *Tetrahymena*. We conjecture that TBFα is involved in regenerating the telomere *in vivo*, because its general telomere-binding specificity correlates with the observation that heterologous telomeres can serve as substrates for terminal addition in yeast (12, 13). The coincidence that the G4-DNA binding and cleavage activity described here was found in a partially purified TBFα fraction suggests that this activity and G4-DNA might be involved in telomere regeneration. It is not clear how the G4-DNA binding and cleavage activity was retained on the human telomeric DNA column, although a tempting explanation is that TBFα is actually a part of this G4-DNA binding complex, which was disrupted during the elution before TBFα was dissociated from the telomeric DNA.

One possible function for this enzyme could be to remove the fold-back structures, or G-quartets, that might appear at the ends of a linear chromosome, in order to allow the extension of the G-rich strand by telomere terminal transferase. The physical ends of linear chromosomes are G-rich 3′ overhangs, synthesized by a telomerase. However, a telomeric G-rich oligonucleotide from *Oxytricha* readily forms, under physiological conditions, an intramolecularly folded structure that cannot be utilized by *Oxytricha* telomerase (14). This would inhibit telomere replication *in vivo*. The cleavage activity described here could, in principle, provide a release from such inhibition by G-quartet DNA, by removing this stable structure to make chromosomal ends accessible to telomere terminal transferase.

Another role for a nucleolytic enzyme specific for G4-DNA could be envisioned in the context of telomere–telomere recombination or switch recombination. Telomere–telomere recombination occurs between the heterologous telomeric ends of linear plasmids in yeast (15, 16). The recombined telomeric DNA segments share little sequence similarity except their G-richness. This is reminiscent of switch recombination of immunoglobulin heavy-chain genes: the switch regions where recombination occurs are invariably G-rich. Given the highly conserved G-richness in telomeres and switch regions, we hypothesize that these nonhomologous recombination events are mediated by a tetrastranded DNA structure.

G4-DNA has been postulated to play a role in meiotic chromosome pairing (1). This G4-DNA dependent nuclease activity may be part of a complex that permits the DNA to untangle. Overall, the existence of a nucleolytic activity specific for a G4-DNA region suggests that this unusual DNA structure may play a role *in vivo*.

We thank Dr. Tom Cech for discussion. This work was supported by National Institutes of Health Grants GM49842 (to B.-K.T.) and GM41895 (to W.G.). Z.L. was supported in part by a postdoctoral fellowship from the Irvington Institute for Medical Research.

1. Sen, D. & Gilbert, W. (1988) *Nature (London)* **334**, 364–366.
2. Sundquist, W. I. & Klug, A. (1989) *Nature (London)* **342**, 825–829.
3. Williamson, J. R., Raghuraman, M. K. & Cech, T. R. (1989) *Cell* **59**, 871–880.
4. Blackburn, E. H. (1991) *Nature (London)* **350**, 569–573.
5. Nikaido, T., Yamawaki-Kataoka, Y. & Honjo, T. (1982) *J. Biol. Chem.* **257**, 7322–7329.
6. Kang, C., Zhang, X., Ratliff, R., Moyzis, R. & Rich, A. (1992) *Nature (London)* **356**, 126–131.
7. Smith, F. W. & Feigon, J. (1992) *Nature (London)* **356**, 164–168.
8. Sen, D. & Gilbert, W. (1990) *Nature (London)* **344**, 410–414.
9. Liu, Z. & Tye, B.-K. (1991) *Genes Dev.* **5**, 49–59.
10. Maxam, A. & Gilbert, W. (1980) *Methods Enzymol.* **65**, 499–560.
11. Sen, D. & Gilbert, W. (1992) *Biochemistry* **31**, 65–70.
12. Shampay, J., Szostak, J. W. & Blackburn, E. H. (1984) *Nature (London)* **310**, 154–157.
13. Brown, W. R. (1989) *Nature (London)* **338**, 774–776.
14. Zahler, A. M., Williamson, J. R., Cech, T. R. & Prescott, D. M. (1991) *Nature (London)* **350**, 718–720.
15. Pluta, A. F. & Zakian, V. A. (1989) *Nature (London)* **337**, 429–433.
16. Wang, S. S. & Zakian, V. A. (1990) *Nature (London)* **345**, 456–458.

The Yeast *KEM1* Gene Encodes a Nuclease Specific for G4 Tetraplex DNA: Implication of In Vivo Functions for This Novel DNA Structure

Zhiping Liu and Walter Gilbert
Department of Molecular and Cellular Biology
The Biological Laboratories
Harvard University
Cambridge, Massachusetts 02138

Summary

We have previously reported the identification of a G4-DNA-dependent nuclease from S. cerevisiae that recognizes a tetrastranded G4-DNA structure and cuts in a single-stranded region 5′ to the G4 structure. We purify this activity to homogeneity and show it to be the product of the S. cerevisiae *KEM1* gene, which is also known as *SEP1*, *DST2*, *XRN1*, and *RAR5*. Since a homozygous deletion of the *KEM1* gene blocks meiotic cells at the 4N stage, the finding of these G4-dependent DNA binding and cleavage activities for the *KEM1* gene product supports the hypothesis that G4-DNA may play a role in meiosis.

Introduction

G4-DNA is a tetrastranded DNA structure in which the four strands are held together by guanine–guanine Hoogsteen base pairing (Sen and Gilbert, 1988), producing characteristic stacks of guanine quartet (G4) planes that have the sugar strands running in either a parallel or an antiparallel orientation (Sundquist and Klug, 1989; Williamson et al., 1989). G4-DNA arises from G-rich DNA sequences that contain substrings of contiguous guanines. Such guanine-rich sequences are found in a number of functionally and evolutionarily conserved genomic regions such as telomeres, the switch regions of the immunoglobulin heavy chain locus, dimerization domains of retrovirus, and certain gene promoters (Blackburn, 1991; Shimizu and Honjo, 1984; Hammond-Kosack et al., 1992). The formation of the four-stranded structure requires Na^+, and the structure, once formed, can be further stabilized by K^+ (Sen and Gilbert, 1990). The main feature of this novel DNA structure, a guanine quartet held together by Hoogsteen base pairing, has been confirmed by X-ray crystal diffraction and two-dimensional nuclear magnetic resonance studies, although the arrangement of the sugar backbones shows many variations (Kang et al., 1992; Smith and Feigon, 1992; Wang and Patel, 1993). In this paper, we provide evidence for the biological significance of this novel DNA structure through a reverse genetic approach.

There have been several reports describing proteins that interact with G4-DNA in vitro. The C-terminal basic region of retrovirus capsid protein has been shown to facilitate the folding of the 3′ end dimerization domains of the RNA genome into G4-RNA, although this protein is not a necessary requirement for the dimerization in vitro (Darlix et al., 1990; Sundquist and Heaphy, 1993; Awang and Sen, 1993). Similarly, the β subunit of Oxytricha telomere-binding protein helps to fold telomeric G-rich sequence into various tetraplex forms, suggesting that G4-DNA or G-quartet is likely to play a role in telomere functioning (Fang and Cech, 1993a, 1993b). However, the biological relevance of this chaperon-mediated folding process still remains to be seen. Other proteins, such as scavenger receptor and a protein isolated from hepatocyte chromatin, have been reported to selectively bind to G4-DNA (Pearson et al., 1993; Weisman-Shomer and Fry, 1993). Again, the biological significance of these binding events is not clear.

We have previously reported the identification of a nuclease activity specific for G4 tetrastranded DNA in a partially purified yeast extract (Liu et al., 1993). This activity recognized a stack of guanine quartets and cut in a single-stranded region on the 5′ side of the quartet structure. The specificity was demonstrated by showing that a variety of G4-DNAs would compete with the cleavage target, while single-stranded DNAs did not. The cleavage reaction requires the presence of a G4 structure and a Mg^{2+}. For each G4-DNA substrate, there is only one major cleavage site in a single-stranded region 5′ to the G4 structure; among different substrates, the cleavage positions ranged from 17–23 nucleotides away from the G4 structure and 2–4 nucleotides away from the 5′ end.

Here, we describe the purification of this G4-DNA-dependent nuclease activity. Several peptide sequences, derived from the purified material, gave near-identical matches with the protein sequence encoded by a yeast gene, known variously as *KEM1* (Kim et al., 1990), *SEP1* (Tishkoff et al., 1991), *DST2* (Dykstra et al., 1991), *XRN1* (Larimer and Stevens, 1990), and *RAR5* (Kipling et al., 1991). We further demonstrate that purified *SEP1* gene product has the G4-DNA-dependent nuclease activity and that this activity is absent in *KEM1*-deleted cells.

Results

Our strategy was to purify the G4-DNA-dependent nuclease activity using the nuclease activity as an assay and to study later the G4-DNA binding properties of the purified material (to avoid prejudicing the experiments). Thus, we did not use G4-DNA or double-stranded telomeric DNA columns but pursued a conventional purification, as summarized in Table 2 and as described in more detail in Experimental Procedures.

Assay and Unit Definition

We prepared G4-DNA by incubating an appropriate oligonucleotide (Table 1 lists the oligonucleotides used and shows the cleavage positions) in the presence of alkali metal cation at 60°C for three days and then by purifying the G4-DNA product by gel electrophoresis. The nuclease assay consisted of incubating the preformed 5′ end-labeled G4-DNA with a protein fraction in the presence of excess amounts of single- and double-stranded DNA competitors and then of separating the cleavage products by either poly-

Cell
1084

Table 1. Oligonucleotides Used in This Study

Oligonucleotide	Length	DNA Sequence
TP	(49-mer)	**5′**-TG↓GACCAGACCTAGCAGCTAT**GGGGG**AGCT**GGGGAAGG**T**GGG**AATGTGA-**3′**
M2	(35-mer)	**5′**-TAGT↓CCAGGCTGAGCAGGTAC**GGGGG**AGCT**GGGG**T-**3′**
TP-S	(16-mer)	**5′**-TGGACCAGACCTAGCA-**3′**
M2-S	(21-mer)	**5′**-TAGTCCAGGCTGAGCAGGTAC-**3′**
BQ3	(24-mer)	**5′**-CACGTAT**GGGGG**AGCT**GGGG**TAT (**biotin**) A-**3′**
OXY-4	(32-mer)	**5′**-TTTT**GGGG**TTTT**GGGG**TTTT**GGGG**TTTT**GGGG**-3′
TPOXY	(52-mer)	5′-TGGACCAGACCTAGCAGCTTTTT**GGGG**TTTT**GGGG**TTTT**GGGG**TTTT**GGGG**-**3′**

Arrows mark the previously identified cleavage sites by this G4-DNA-dependent nuclease (Liu et al., 1993).

acrylamide gel electrophoresis (PAGE) or trichloroacetic acid precipitation. One unit of the G4-DNA-dependent nuclease activity is defined as the conversion of 10 fmol of the 5′ end of TP oligonucleotide in G4-DNA form (TP[G4]) (Table 1) into an acid-soluble form in a reaction incubated at 37°C for 10 min at a substrate concentration of 0.63 nM.

Phosphocellulose Column Chromatography

The lysis of yeast cells and preparation of material for phosphocellulose column chromatography are described in Experimental Procedures. Column fractions, eluted with a linear KCl gradient, were assayed with several DNA probes to identify a G4-DNA-dependent cleavage activity. Figure 1 shows that several nuclease activities eluted from the column that have different substrate and cleavage specificities. Figure 1a shows three cleavage activities on a G4-DNA probe, TP(G4), that elute around fractions 30, 60, and 100. Figure 1b shows that two of these activities also appear on a fold-back probe, TP oligonucleotide and OXY-4 oligonuclетide hybrid (TPOXY) (Table 1), which has the same 5′ single-stranded region as TP(G4) but an antiparallel, instead of a parallel, G4 structure at its 3′ end. The cleavage activity in fractions 30–40 appears to be nonspecific in that it also cleaves OXY-4 (an antiparallel G4 probe with only four Ts on the 5′ end) and a single-stranded probe, TP-S (Figures 1d and 1c). Figure 1e shows a denaturing gel that compares the cleavage products generated by these nuclease activities, using a different G4-DNA probe, M2 oligonucleotide in G4-DNA form [M2(G4)] (Table 1). The three major cleavage activities generate different products. On a sequencing gel, fraction 100 yields 5′-pTAGT-OH-3′, the expected product (Liu et al., 1993), while the cleavage product generated from fraction 30 appears to be one nucleotide longer (data not shown). The active fractions around fraction 100 were pooled for further purification.

Q-Sepharose Column Chromatography of the G4-DNA-Dependent Nuclease

Figure 1f shows the elution profile from a Q-Sepharose column, using our standard gel assay. The preferential cleavage on G4-DNA is very clear, since in the same reaction only the G4-DNA but not the single-stranded form was cleaved (Figure 1f, lanes 14–17; also comparing lanes 11 and 19 of Figure 1c). Throughout the purification, we followed the major cleavage activity and ignored the binding activity since the binding pattern was complex and confusing even in the presence of large excesses of single- and double-stranded competitors.

Further High Pressure Liquid Chromatography Purification

After the phosphocellulose and Q-Sepharose columns, we further purified this activity through three high pressure liquid chromatography (HPLC) columns (Mono S, diethylaminoethyl, and Mono Q). Figure 2 shows the elution of the cleavage activity and polypeptides from the last column (Mono Q). Several peptides (Figure 2b, lanes 27–29) with apparent molecular weights around 140 kDa coelute with the cleavage activity, which also coincides with a G4-DNA binding activity. Table 2 summarizes the overall purification.

Is the G4-DNA-binding activity related to the 140 kDa proteins? We estimated the amount of G4-DNA-binding activity present in fraction 28 in a band shift assay by omitting Mg^{2+} from the reaction. About 60% of the protein in this fraction, in terms of mass, became bound to G4-DNA (data not shown). Thus, the G4-DNA-binding activity was not a trace contaminant. Furthermore, Figures 2c and 2d show that these 140 kDa peptides also copurified with the cleavage activity on a G4-DNA affinity column. These results further suggested that the 140 kDa protein is the G4-DNA-dependent nuclease. On a small-scale purification with a protease-deficient yeast strain, we observed only a single protein band in SDS–PAGE around the size range of 140 kDa (data not shown). Since we did the large scale purification with a commercial yeast product, a non-protease-deficient wild-type yeast harvested at stationary phase, we believe these peptides were the products of either proteolytic degradation or posttranslational modification(s) of a single protein. Based on this reasoning, we purified the major band through SDS–PAGE and submit-

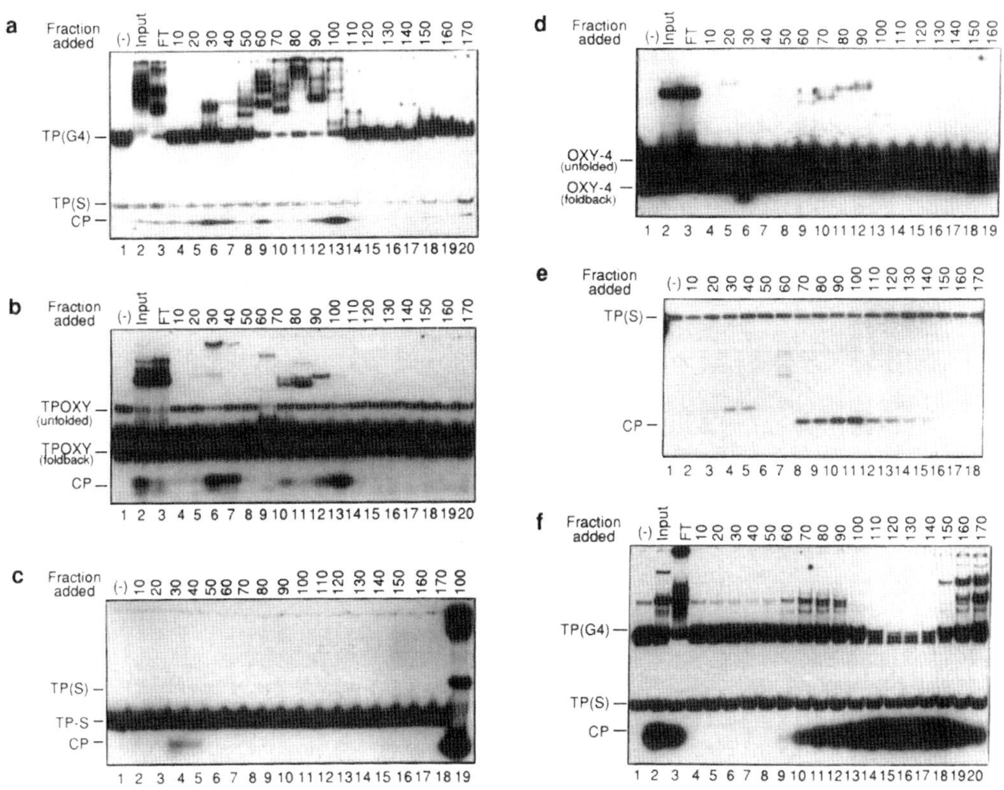

Figure 1. Fractionation of the G4-DNA-Dependent Nuclease on a Phosphocellulose Column

Fraction II material was loaded on a phosphocellulose column (bed volume 500 ml) equilibrated in buffer A with 150 mM KCl. After washing the loaded column with 500 ml of 150 mM KCl buffer A, the bound proteins were eluted at a flow rate of 100 ml/hr with a 2-liter linear salt gradient from 150–800 mM KCl, collected in 13 ml fractions. Column fractions were then assayed using different oligonucleotides, labeled with ^{32}P at the 5′ end, as substrates. Aliquots of 5 µl of the column fractions were added to each reaction unless otherwise noted.

(a) TP(G4) reactions. The substrate was 2 fmol of preformed TP(G4)/reaction. The sample added in lane 2 was 5 µl of a 1:10 diluted solution of fraction II.

(b) TPOXY reactions. The substrate was folded TPOXY (5 fmol/reaction), which was heated at 90°C for 2 min and cooled on ice for 10 min in the reaction buffer.

(c) TP-S reactions. Substrate (5 fmol/reaction) was a single-stranded oligonucleotide corresponding to the 5′ single-stranded region of TP. The reaction mixtures were resolved on a 10% polyacrylamide gel. The reaction shown in lane 20 was the same as that shown in lane 11, except that the substrate added was 5 fmol of TP(G4) instead of TP-S.

(d) OXY-4 reactions. Substrate OXY-4 (5 fmol/reaction) was folded into antiparallel form the same way as described in (b). Samples were resolved on a 7% polyacrylamide gel.

(e) M2(G4) reactions. The reactions were carried out as described in (a). At the end of incubation, the DNA was recovered by butanol extraction, heated in 50 µl of 1 M piperidine at 65°C for 10 min, recovered again by butanol extraction, dissolved in 3 µl of formamide loading dye, and resolved on a 8 M urea–20% sequencing gel.

(f) Elution of the G4-DNA-dependent cleavage activity from a Q-Sepharose column. The pooled fraction from the phosphocellulose column (which peaked at about 0.45 M KCl) was dialyzed in 100 mM KCl buffer A, loaded on a fast flow Q-Sepharose column (bed volume of 20 ml) equilibrated in 100 mM KCl, and eluted with a linear salt gradient from 100–600 mM KCl (total volume of 80 ml). The eluate was collected in 1 ml fractions. For the gel assay, aliquots of 3 µl of 1:50 diluted solution of the column fractions were mixed with 20 fmol of TP(G4) and then were treated as in (a). The fractions added to each reaction are indicated at the top.

Labels indicate the following: TP(G4), TP oligonucleotide in G4-DNA form; TP(S), single-stranded TP oligonucleotide; TP-S, truncated version of TP oligonucleotide; and CP, cleavage products.

ted it to the Harvard Microchemistry Facility for peptide sequencing.

Peptide Sequences and Their Alignments with KEM1 Protein

The sequencing effort yielded the peptide sequences of four tryptic fragments. Using these peptide sequences as queries, we searched the SWISSPROT database and found either perfect or near-identical matches with a known yeast protein, KEM1 (Kim et al., 1990). Figure 3 shows the sequence alignments to these peptides. Peptide 1 matches perfectly. For peptide 2, the mismatch of

Cell
1086

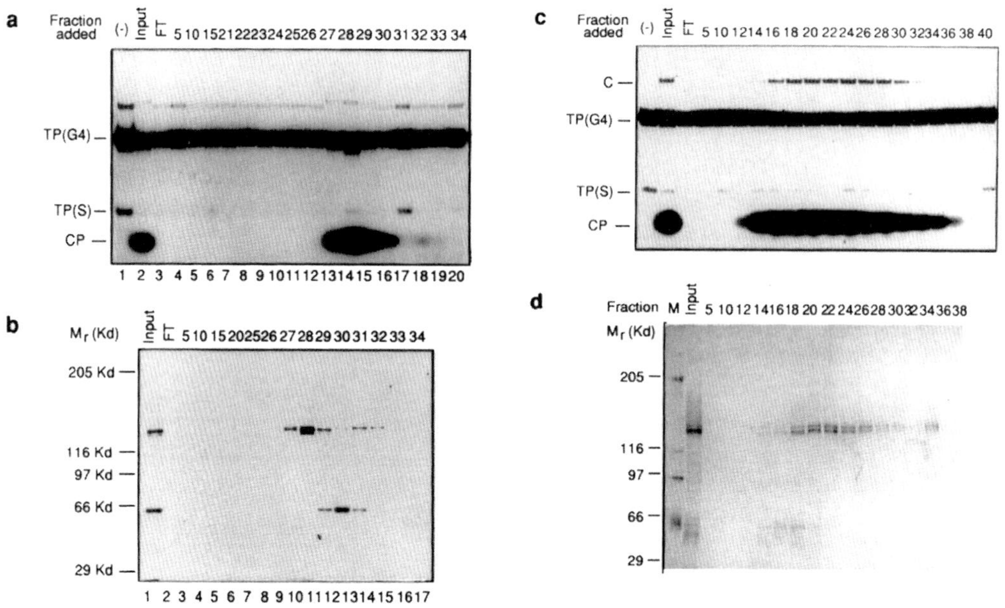

Figure 2. Purification of the G4-DNA-Dependent Nuclease on Mono Q HPLC and G4-DNA Columns

(a) Gel assay of the cleavage activity for the column fractions eluted from Mono Q with a linear salt gradient. Aliquots of the column fractions (3 μl of 1:20 diluted solution) were incubated with 20 fmol of TP(G4) under the reaction conditions as described in Experimental Procedures. The reaction mixtures were then resolved on a 5% polyacrylamide gel. The fractions added to each reaction are shown at the top. No protein was added in lane 1. Abbreviation: Input, fraction VI material loaded on the column.

(b) SDS–PAGE of the column fractions. Aliquots of 3 μl from the fractions were diluted into SDS loading dye and heated at 100°C for 10 min under reducing conditions. Samples were then resolved on a 6% SDS gel and visualized by silver staining (Merril et al., 1979). Numbers at the top indicate the fractions loaded in each lane. The migration positions of molecular weight markers were indicated on the left. Abbreviation: Kd, kDa.

(c) Cleavage assay for the G4-DNA column fractions. The preparation of the column and chromatography are described in Experimental Procedures. The column was eluted with a linear salt gradient. Aliquots of 3 μl from 1:5 diluted solution of the column fractions were assayed as described in (a). Abbreviation: Input, fraction VII material loaded on the column.

(d) SDS–PAGE of the G4-DNA column fractions. Sample preparation, electrophoresis, and silver staining were as described in (b), except that 10 μl aliquots of the column fractions were first precipitated with one-third of the volume of 50% trichlororacetic acid. The pellets were rinsed once with ice-cold acetone, dissolved in SDS loading buffer, neutralized with 1 M Tris base, and loaded on a 6% gel. The numbers at the top are the column fractions.

Abbreviations: FT, column flowthrough fraction; CP, cleavage products; C, protein–DNA bound complex; and M_r, relative molecular weight.

the C-terminal arginine is not significant since this arginine should have been removed after a complete tryptic digestion. Peptide 3 is a low-confidence sequence; yet, among all the proteins in the database, KEM1 is still the best-scoring sequence, and the mismatches here are probably due to peptide sequencing errors. Peptide 4 is a high-confidence sequence that matches a sequence in KEM1 protein perfectly, although no assignment could be made for the two amino acids at the N-terminal of this tryptic fragment. Based on these alignments and a previous report

Table 2. Purification Table of the G4-DNA-Dependent Nuclease

Fraction	Volume (ml)	Activity (U/μl)	Protein Concentration (mg/ml)	Total Protein (mg)	Specific Activity (U/μg)	Purification (fold)
Fraction I crude lysate	200	9.06	15.1	3020	0.60	1
Fraction II (0%–50% ammonium sulfate)	25	64.2	37.5	938	1.71	2.9
Fraction III (phosphocellulose)	152	8.62	0.27	41.0	31.9	53
Fraction IV (Q-Sepharose)	25	22.0	0.31	7.63	72.2	120
Fraction V (Mono S HPLC)	5.5	60.0	0.013	0.072	4620	7700
Fraction VI (diethylaminoethyl HPLC)	2.0	57.1	0.023	0.046	2500	4170
Fraction VII (Mono Q HPLC)	1.5	60.7	0.021	0.032	2890	4820

```
Peptide 1                SSFALLR
                         |||||||
KEM1            VPREAVLNPRSSFALLRSQKFDLGDRVVYIQDSGK
                  1150      1160      1170

Peptide 2            TSIAAVEDHIMKR
                     |||||||||||||
KEM1            SDSLTKTSIAAVEDHIMKYAANIEGHERKQLAKVP
                     1120      1130      1140

Peptide 3                KIFYM?IGKVAPSY
                         |||||  |:  |||
KEM1            DHLFQTIKPKKIFYMAIDGVAPRAKMNQQRARRFR
                70        80        90        100

Peptide 4              ??EYSNLTLNLLK
                         |||||||||||
KEM1            GYARRNPRGWEYSNLTLNLLKEYRQTFPDFFFRLS
                     1040      1050      1060
```

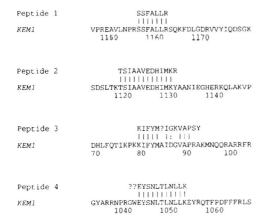

Figure 3. Alignment of Four Sequenced Peptides with the KEM1 Protein Sequence

Proteins in fraction VII were separated on a 6% SDS gel and blotted onto a nitrocellulose membrane. Protein bands were visualized by Ponceau S staining. One of the more abundant 140 kDa bands was excised and submitted for peptide sequencing. The sequences of peptides 1–3 were derived from a single run of an HPLC fraction containing three overlapping tryptic fragments. The signal ratio was 7:2:1 (peptide 1 to peptide 2 to peptide 3), with peptide 1 being a high-confidence sequence. The sequence of peptide 4 was obtained from a separate run. A question mark indicates that no unambiguous assignment could be made for the amino acid at that position. The peptide sequence similarity matches with the KEM1 protein sequence were derived by searching SWISSPROT database with the FASTA program (Pearson and Lipman, 1988). The numbers indicate the amino acid position in the KEM1 protein sequence.

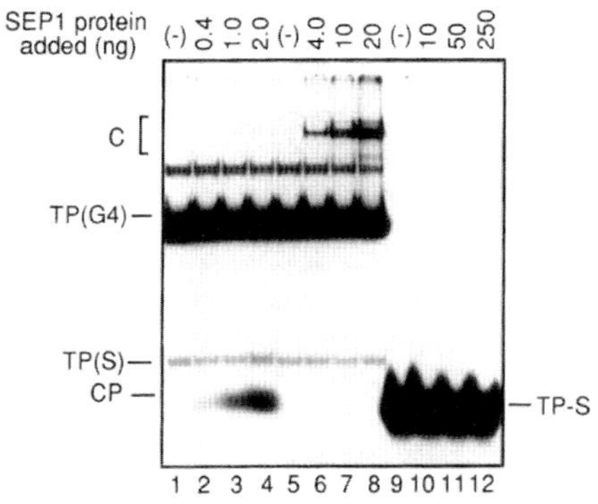

Figure 4. Recombinant SEP1 Protein Binds To and Cleaves TP(G4)

Purified SEP1 protein was incubated with 20 fmol of TP(G4) that was labeled at the 5′ end with ^{32}P. The amount of SEP1 added to each reaction is indicated at the top. Lanes 1–4, the reactions were incubated at 37°C for 15 min in 3 mM Mg^{2+}. Lanes 5–12, the reactions were incubated on ice for 15 min in the absence of Mg^{2+}. The reaction mixtures were resolved on a 5% polyacrylamide gel. The gel was then dried and autoradiographed. Abbreviations: TP(S), single-stranded TP oligonucleotide; TP(G4), G4-DNA form of TP; CP, the cleavage product; C, protein–DNA bound complex.

that *KEM1* is a single-copy gene in the yeast genome (Kim et al., 1990), the sequenced 140 kDa protein is the yeast KEM1 protein.

The *KEM1* gene is also known as *SEP1* (Tishkoff et al., 1991), *DST2* (Dykstra et al., 1991), *XRN1* (Larimer and Stevens, 1990), and *RAR5* (Kipling et al., 1991). It was identified by five groups through four different approaches, but its in vivo function is still not clear. The gene product of *KEM1/SEP1* is a 175 kDa protein of 1528 amino acids that migrates as a 160 kDa protein in SDS–PAGE. The difference in molecular weight of our purified material may be due to the loss of about 170 amino acids from its C-terminal caused by proteolytic degradation. However, is the gene product of *KEM1/SEP1* actually the G4-DNA-dependent nuclease?

Purified SEP1 Protein Binds to G4-DNA and Cleaves at the Same Site as the G4-DNA-Dependent Nuclease

SEP1 protein has been overexpressed and purified from yeast by Johnson and Kolodner (1991) (a sample of SEP1 was provided by A. Johnson and R. Kolodner). Figure 4 shows that the recombinant SEP1 protein binds to G4-DNA in the absence of Mg^{2+} (lanes 4–8) and cleaves the G4-DNA in the presence of magnesium (lanes 1–4). Under these conditions, the protein does not bind to or cleave a single-stranded DNA (Figure 4, lanes 9–12) that corresponds to the 5′ single-stranded region in TP(G4). The binding behavior of SEP1 to G4-DNA appears to be complex (Figure 4, lane 8), and an overnight incubation of the binding reaction at 4°C resulted in two bound complexes (see Figure 6b, lane 2), possibly the result of a 2-fold symmetry element on the G4-DNA binding two molecules of SEP1 protein. It appears that the relative binding affinity of this protein for G4-DNA is at least 100-fold higher than that for single-stranded DNA (Figure 4, comparing patterns shown in lanes 8 and 12). The same set of experiments with a different set of oligonucleotides, i.e., M2(G4) and M2-S, produced similar results (data not shown).

SEP1 cleavage product is the same as that produced by G4-DNA-dependent nuclease. Figure 5 shows the analysis of the products of cleavage reactions on a sequencing gel, using 5′ end-labeled TP(G4) as the substrate. Both the G4-DNA-dependent nuclease and SEP1 protein produce a dinucleotide with a 3′-OH group. This further establishes the identity of the SEP1 protein and the G4-DNA-dependent nuclease.

Comparison of Purified G4-DNA-Dependent Nuclease and the SEP1 Protein

We compared our purified fraction with the recombinant SEP1 protein by SDS–PAGE and Western blot analysis. Figure 6a shows that the purified recombinant SEP1 protein migrates as a single polypeptide with an apparent molecular weight of 160 kDa (lane 1). In our fraction, there are two or three polypeptides that migrate in the size range

Cell
1088

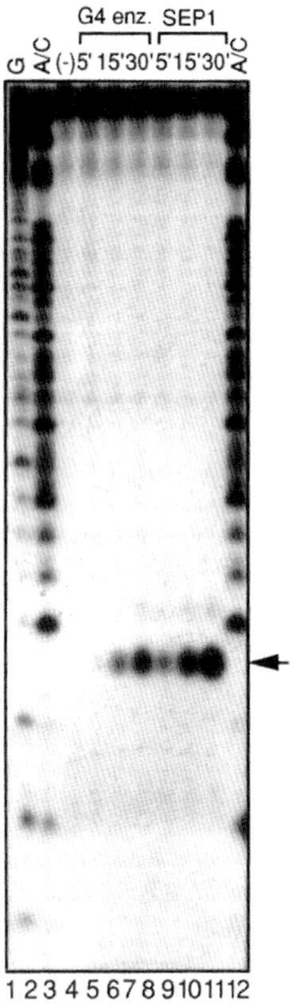

Figure 5. The Cleavage Products Generated by SEP1 Protein and the G4-DNA-Dependent Nuclease on TP(G4)

Proteins were incubated with 20 fmol of TP(G4) at 37°C for the times shown. The DNA and cleavage products were pelleted by butanol precipitation. The pellet was then dissolved in 50 μl of 1 M piperidine and heated at 65°C for 10 min to open the G4 structure. The samples were recovered by butanol precipitation and dissolved in 5 μl of formamide loading dye. They were heated at 100°C for 1 min and loaded on an 8 M urea–20% polyacrylamide gel (ratio of 19:1), electrophoresed, and autoradiographed. Lanes 1, 2, and 12 were Maxam–Gilbert sequencing ladders (Maxam-Gilbert, [1980]). Protein added to each reaction was as follows: lanes 4–6, 0.53 ng of the G4-DNA-dependent nuclease purified through a G4-DNA column; and lanes 7–9, 6.4 ng of recombinant SEP1 protein. No protein was added in lane 3. The arrow marks the cleavage products.

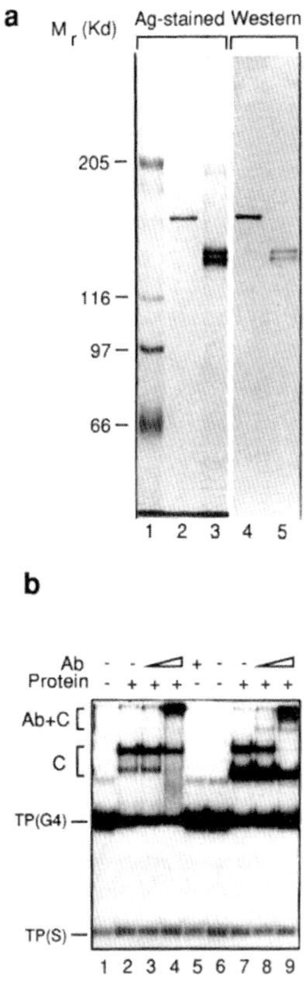

Figure 6. G4-DNA-Dependent Nuclease Cross-Reacts with an Anti-SEP1 Antibody

(a) Comparison of recombinant SEP1 protein and the G4-DNA-dependent nuclease by SDS–PAGE and Western blot analysis. Protein samples were heated at 100°C for 10 min in loading buffer with a reducing agent and then were loaded on a 6% SDS polyacrylamide gel. After the electrophoresis, the gel was cut in half. One-half was stained with silver (Merril et al., 1979). The other was blotted onto a piece of nitrocellulose membrane, probed with an anti-SEP1 antiserum (Towbin et al., 1979), and visualized with an alkaline phosphatase–conjugated antibody using NBT/BCIP (nitroblue tetrazolium/5-bromo-4-chloro-3-indolylphosphate toluidinium) as the substrate. Protein loaded was as follows: lanes 2 and 4, 16 ng of SEP1; and lanes 3 and 5, 32 ng of the G4-DNA-dependent nuclease purified from the G4-DNA affinity column. High molecular weight standards from Sigma were loaded in lane 1. Abbreviation: M_r, relative molecular weight.

(b) Supershift affected by anti-SEP1 antibody. Either recombinant SEP1 protein or G4-DNA nuclease was incubated with 20 fmol of TP(G4) probe without Mg^{2+} at 4°C for 1 hr; then, aliquots of diluted anti-SEP1 antiserum were added to the reactions and incubated overnight at 4°C. Samples were then resolved on a 5% nondenaturing gel. Protein added to the reactions: lanes 2–4, 8 ng of SEP1 protein; and lanes 7–9, 2 ng of G4-DNA-dependent nuclease purified through the G4-DNA column. Antibody added was as follows: lanes 3 and 8, 5 μl of 1:500 diluted anti-SEP1 antiserum; and lanes 4, 5, and 9, 5 μl of 1:50 diluted anti-SEP1 antiserum. Labels indicate the following: TP(G4), TP oligonucleotide in G4-DNA form; TP(S), single-stranded TP; C, complex formed between TP(G4) and SEP1 or the G4-DNA-dependent nuclease; Ab+C, tripartite complex resulted from the binding of anti-SEP1 antibody to C.

of 140 kDa (Figure 6a, lane 2). Both SEP1 and the 140 kDa proteins react with an anti-SEP1 antibody (Figure 6a, lanes 3 and 4).

To further confirm the identity of our enzyme with SEP1 protein, we did an antibody supershift experiment. After mixing the protein samples with G4-DNA and incubating for 1 hr, we added anti-SEP1 antiserum to the reaction mixture and incubated it overnight. Figure 6b shows that anti-SEP1 antibody recognizes the complex between SEP1 protein and G4-DNA and further retards its electrophoretic mobility (lane 1–4), demonstrating that SEP1 protein is in the protein–G4-DNA bound complexes. Further-

Table 3. Comparison of Cleavage Activity by G4-DNA-Dependent Nuclease and SEP1 on Different Substrates

Protein	TP(G4)	Oligo(A)$_{12-18}$	TP-S
G4 nuclease[a]	5.40[b]	1.1	0.18
SEP1[c]	0.35	0.11	0.018

[a] G4-DNA nuclease added to each reaction was 0.4 ng of protein purified through a G4-DNA column, as shown in Figure 2d (fraction 24).
[b] The number is measured as picomoles of 5′ end released from the cleavage reaction, as acid-soluble form, per nanogram of protein added. The reactions (trichloroacetic acid assays) were carried out as described in Experimental Procedures. The substrate in each reaction was 2 pmol of oligonucleotide labeled at the 5′ end with ^{32}P, and each assay contained nonspecific competitors of 20 pmol of pd(A)$_{12-18}$ and 1 μg of E. coli DNA. Each reaction was done in duplicate, and the number shown here is the average. The protein concentrations were normalized based on the silver binding assay.
[c] SEP1 protein was a recombinant protein ovexpressed in and purified from yeast (Johnson and Kolodner, 1991). To each reaction, 3.2 ng of protein was added.

more, this antibody recognizes the bound complex formed between the G4-DNA-dependent nuclease and TP(G4), demonstrating that SEP1 protein is at least part of this enzyme (Figure 6b, lanes 5–8). The slightly different migration position of the bound complexes and their different response to anti-SEP1 antibody may be attributed to the loss of a peptide fragment of about 170 amino acids from the C-terminal in the protein due to proteolysis.

We also compared these two proteins as nucleases on other substrates since the SEP1 protein has also been reported to be an RNAase. Table 3 shows that recombinant SEP1 protein does cleave single-stranded DNA and RNA. However, it is much more active on G4-DNA than on RNA or single-stranded DNA. Our G4-DNA nuclease preparation behaves similarly in that it is much more active on G4-DNA. The ratios of specific activities of the purified SEP1 and the G4-DNA-dependent nuclease track together, although the SEP1 preparation is only one-tenth as active. We attribute this difference to an inactivation of the SEP1 protein during storage (which is frequently observed; Arlen Johnson, personal communication). In addition, an inhibitor for the exonuclease activity of SEP1 protein, N-ethylmaleimide (Johnson and Kolodner, 1991), also blocks the G4-DNA-dependent cleavage activity of the recombinant SEP1 protein and the G4-DNA nuclease (data not shown), further demonstrating that these two proteins are the same.

KEM1-Deleted Cells Lack the G4-DNA-Dependent Nuclease

If the G4-DNA-dependent nuclease is the product of the *KEM1/SEP1* gene, one would expect that its activity would be missing from *kem1–null* (or *KEM1*-deleted) cells. Extracts were prepared from a *kem1–null* strain and a wild-type yeast strain (the latter was used as a benchmark for a mock purification from *kem1–null* cells to locate fractions that should contain the G4-DNA-dependent nuclease activity). We put both preparations through exactly the same treatments, i.e., ammonium sulfate fractionation, phosphocellulose column chromatography, and Mono Q HPLC. From the Mono Q fractions, we first identified, based on the cleavage assay, the fraction from wild-type cells that contained the G4-DNA-dependent nuclease activity, and then we compared it with the corresponding fraction derived from the *kem1–null* cells. Table 4 shows that the G4-DNA-dependent nuclease activity is indeed missing in *KEM1*-deleted cells, which is evident as a drastic decrease in the specific cleavage activity on TP(G4) even though the protein concentration of both fractions is about the same (approximately 10 μg/ml).

Table 4. Cleavage Activities in *kem1-null* and Wild-Type Yeast Cells

	Specific Activity (U/mg)	
Cell-Free Extract	TP(G4)	TP-S
Wild type	64.1 ± 9.0	23.7 ± 3.7
kem1-null	0.15 ± 0.07	7.17 ± 0.75

Discussion

We purified a G4-DNA-dependent nuclease to a simple set of components on silver-stained gels. By peptide sequence analysis, we showed that this enzyme is the product of a previously identified gene, called variously *KEM1* (Kim et al., 1990), *SEP1* (Tishkoff et al., 1991), *DST2* (Dykstra et al., 1991), *XRN1* (Larimer and Stevens, 1990), and *RAR5* (Kipling et al., 1991). Several lines of evidence demonstrate that this G4-DNA-dependent enzyme is indeed the KEM1/SEP1 protein. First, this cleavage activity is missing from the extract prepared from a *KEM1* deletion strain. Secondly, purified SEP1 protein binds to and cleaves G4-DNA and has a similar requirement for Mg^{2+}. Both purified SEP1 protein and our G4-DNA-dependent enzyme yield the same cleavage product on 5′ end-labeled TP(G4) molecules. Furthermore, both proteins cleave single-stranded DNA and RNA, as has been reported previously for SEP1 protein (Johnson and Kolodner, 1991), but both have a higher activity on G4-DNA. Given the additional fact that, in the absence of Mg^{2+}, SEP1 protein binds to TP(G4) but not the corresponding 5′ single-stranded region, TP-S, it seems clear that the gene product of *KEM1/SEP1* is a G4-DNA-dependent enzyme. Furthermore, an anti-SEP1 antibody recognizes the protein–DNA complex generated either from SEP1 protein or from our G4 enzyme binding to TP(G4) in a supershift gel assay, further establishing the identity of these two proteins.

Our purified protein is smaller than the full-length KEM1/SEP1 protein. This is likely due to proteolysis since for large-scale purification we used a commercial yeast source rather than a protease-deficient lab strain, and the cells were harvested at stationary phase during which extensive proteolysis could have occurred even before the lysis. Based on the matching positions of sequenced peptides with the KEM1 protein, it seems possible that a peptide fragment of about 170 amino acids was cleaved off from the carboxyl end of the G4-DNA-dependent nuclease protein since peptide 3 matches amino acids 79–90 of KEM1 protein. In fact, SEP1 was originally purified as a 132 kDa protein owing to the degradation of a peptide fragment from the C-terminal that is believed to be more susceptible to proteolysis (Tishkoff et al., 1991; Dykstra et al., 1991).

The *KEM1* gene was identified by mutations that block nuclear fusion during yeast mating (Kim et al., 1990). The SEP1 and the DST2 proteins were originally identified as DNA strand exchange proteins that promoted the pairing and strand transfer of a single-stranded DNA into a homologous DNA duplex (Dykstra et al., 1991; Tishkoff et al., 1991). *XRN1* was identified as a gene that encodes a 5′ to 3′ exoribonuclease through a reverse genetic approach (Larimer and Stevens, 1990). The *RAR5* gene was identified by mutations that enhance the mitotic stability of a minichromosome bearing a defective (or cryptic) autonomously replicating sequence element (Kipling et al., 1991). The DNA sequences of these five genes are identical.

What is the function of *KEM1/SEP1* in vivo? A large number of phenotypes have been documented for *KEM1* deletion yeast strains, resulting in various interpretations for its function. In light of our finding that the gene product of *KEM1* is a G4-DNA-dependent enzyme, we will provide an alternative interpretation of these results.

Diploid cells with a homozygous deletion of *KEM1* fail to sporulate, and for the few cells that do sporulate, the recombination frequency decreases drastically (Kim et al., 1990). Flow cytometry data show that the unsporulated cells are blocked at the 4N stage, which places the execution point of *KEM1* after premeiotic DNA synthesis and before the end of Meiosis I (Tishkoff et al., 1991). Epistasis analysis has demonstrated that the terminal phenotype of *dst2 spo13* double and *rad50 dst2 spo13* triple mutants is still *dst2*, i.e., unable to sporulate (Dykstra et al., 1991). In yeast, the *spo13* mutation allows meiotic cells to bypass division I and directly go through a single equational division, generating two diploid cells (Klapholz and Esposito, 1980). The fact that *spo13* can bypass the *rad50 rad52* double mutant but not the *rad50 dst2* suggests that the function of *KEM1* is outside the pathway of the *RAD50–RAD57* epistatic group, which is believed to be involved in homologous recombination and DNA repair in yeast (Friedberg et al., 1991).

Since *KEM1* encodes a G4-DNA-dependent nuclease, we interpret these results as the manifestation that G4-DNA is involved in meiosis. There are three possible scenarios. First, *KEM1* may act in a recombination pathway parallel to and independent of the *RAD50–RAD57* pathway. In this pathway, G4-DNA serves as a structure intermediate, as depicted in the model shown in Figure 7. This model rationalizes the G4-DNA-dependent activities and the strand exchange activity into a new recombination pathway that is stimulated by G4-DNA formation. The G4-DNA-dependent binding and cleavage reactions precede the strand transfer reaction that brings the cleaved strand into a homologous duplex region. Such a G4-DNA-mediated reaction could well be the underlying mechanism for

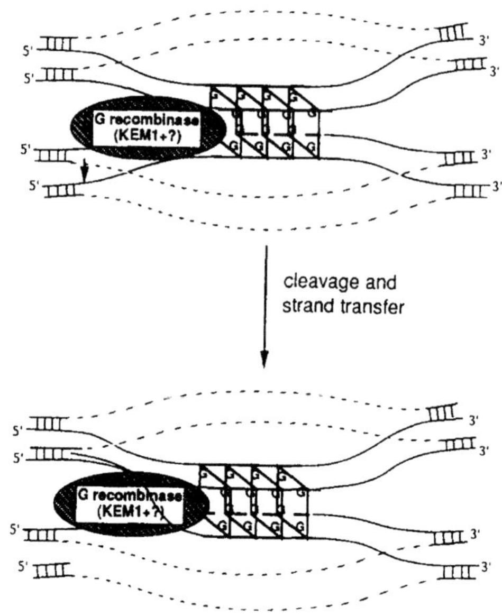

Figure 7. A Model To Account for the G4-DNA-Dependent Cleavage and Strand Transfer Activities in Meiotic Recombination

Homologous chromosomes are tied together through a G4 structure (bold lines connecting G to G in the quartets). The complementary strands are represented by dashed lines. The strands engaged in G4-DNA formation are oriented in parallel with the 5′ and 3′ ends indicated. We name the enzyme involved in this G4-DNA-mediated reaction as G recombinase, which is also likely to operate in other G-rich sequence-mediated processes, such as telomere–telomere and immunoglobulin isotype switching recombinations. The KEM1 protein is at least a part of this enzyme. The initial G4-DNA-dependent cleavage and strand transfer reactions are carried out by the KEM1 protein. The activity responsible for the subsequent religation has yet to be identified. The structure intermediate involved in telomere–telomere and switch recombination could well be an antiparallel G4 form, but the principle outlined here for recombination remains the same.

G-rich sequence-mediated recombination, such as telomere–telomere recombination or class switch recombination for immunoglobulins (Pluta and Zakian, 1989; Shimizu and Honjo, 1984). In support of this hypothesis, insertions of telomeric sequences into internal locations on yeast chromosomes create hotspots for meiotic recombination (White et al., 1993), and an interstitial telomeric sequence has been shown to be a hotspot for recombination in the Armenian hamster (Ashly and Ward, 1993). Thus, we predict that other G-rich sequences, such as the switch regions of immunoglobulin heavy chain locus and the 5′-linked polymorphic region of the human insulin gene (Hammond-Kosack et al., 1992), should be able to function as hotspot sequences for yeast meiotic recombination. A second alternative is that the enzyme serves to untangle the paired chromosomes at the pachytene stage. The fact that the *spo13* mutation does not bypass the meiotic block caused by a *KEM1* homozygous deletion can be explained as *KEM1* activity being epistatic to that of *SPO13*. A third

G4 Tetraplex DNA in Meiosis
1091

possibility is that the *KEM1* gene product is involved in the synapsis of two pairs of sister chromatids at Meiosis I, upstream of the *RAD50* function.

Overall, the meiotic block at the 4N stage for *KEM1–null* cells supports the hypothesis that G4-DNA is involved in meiosis (Sen and Gilbert, 1988), especially since we had anticipated this phenotype for cells that had this G4-DNA-dependent nuclease deleted (Liu et al., 1993).

Deletion of the *KEM1* gene in haploid cells also leads to several striking mitotic defects. The cells have a higher chromosome loss rate, are hypersensitive to an anti-microtubule drug, benomyl, and lose viability upon nitrogen starvation. Although these phenotypes have been attributed to a lesion in microtubule function (Kim et al., 1990), we think a more plausible interpretation is that the KEM1 protein is involved in processing chromosomes, especially at telomeres. In the absence of KEM1 protein, the cells enter mitosis prematurely, giving rise to these mitotic phenotypes. The telomere is a likely target for the action of the *KEM1* gene product because telomeric G-rich strands can readily fold into guanine-quartet forms under physiological conditions. A mitotic role for KEM1 protein then would be to process such a folded structure so that the telomere can function properly. Alternatively, this protein may participate in several different cellular processes, one of them being RNA metabolism.

In conclusion, we have demonstrated that the *KEM1* gene encodes a G4-DNA-dependent nuclease. The phenotypes associated with the *kem1–null* allele support the notion that G4-DNA is a biologically relevant structure likely to play a role in meiosis, G-rich sequence-mediated recombination, and telomere function.

Experimental Procedures

Oligonucleotides

All oligonucleotides used in this study were made on an Applied Biosystems DNA synthesizer based on phosphoramidite chemistry. The DNA sequences are listed in Table 1. Full-length products were isolated from 8M urea–20% polyacrylamide gels. G4-DNAs were made according to previously published procedures and were purified from a 6% polyacrylamide gel containing 5 mM KCl (Sen and Gilbert, 1992). G4-DNA probes were labeled at the 5′ end with T4 polynucleotide kinase and [γ-^{32}P]ATP, precipitated with 0.3 M NaOAc and 70% ethanol, and stored in TE buffer (10 mM Tris–1 mM EDTA [pH 7.5]) containing 50 mM KCl. To further remove the unincorporated radioactive nucleotide, the probe was purified through a Push column (Stratagene). Oligo(A)$_{12-18}$ and pd(A)$_{12-18}$ were purchased from Pharmacia.

Protein Purification

Large-scale purification was done with a commercial yeast source, and small-scale purification was done with protease-deficient strain BJ2168 (MAT**a**, *pep4-3 prc1-407 prb-1122 ura3-52 trp1 leu2*). Since both procedures are quite similar, we only give a description for the large-scale preparation. Fresh yeast cake (one-half pound) was broken into small pieces by hand and frozen in liquid nitrogen immediately. The frozen yeast pellets were lysed in a Waring blender, in liquid nitrogen, at top speed for 5 min. The lysed yeast powder was thawed in a cold room for about 2 hr; then, 220 ml of buffer A (50 mM HEPES [pH 7.5], 1 mM EDTA, 1 mM DTT, 10% glycerol) was added and was supplemented with 100 mM KCl, 5 mM $MgCl_2$, and protease inhibitors. The final concentration of protease inhibitors added was 1 μg/ml pepstatin A, 2 μg/ml leupeptin, 1 μg/ml anti-pain, and 0.2 mM PMSF. All the purification steps were carried out at 4°C, except for those involving HPLC. The lysate was cleared first by centrifugation at 8000 rpm for 20 min in a SA-600 rotor and then by an ultracentrifugation at 45,000 rpm for 60 min in a Ti 50.2 rotor. The supernatant from the last spin, named fraction I, was subjected to an ammonium sulfate fractionation (0%–50% saturation). Solid ammonium sulfate was added over a period of 20 min and was stirred for another 30 min. The precipitate was recovered by a centrifugation at 8000 rpm for 20 min in a SA-600 rotor and dissolved in buffer A supplemented with protease inhibitors and 50 mM KCl. This was fraction II. After dialysis against 800 ml of buffer A containing 50 mM KCl for 2 hr and after being adjusted to 0.3 M KCl, this fraction was loaded onto a phosphocellulose column (5 × 15 cm) preequilibrated in 0.3 M KCl at a flow rate of 100 ml/hr. The loaded column was washed with 500 ml of 0.3 M buffer A supplemented with protease inhibitors and was eluted with 0.5 M KCl buffer A. The pool of the active fractions, fraction III, was diluted with buffer A to bring the KCl concentration down to 100 mM and was loaded onto a fast flow Q-Sepharose column (1.5 × 9 cm). After washing the column with 50 ml of 0.1 M KCl buffer A, the bound proteins were eluted with a linear KCl gradient from 0.1–0.6 M (80 ml) at a flow rate of 75 ml/hr, and 1 ml fractions were collected. The active fractions (peaked around 270 mM KCl) were pooled, named fraction IV, and dialyzed against 500 ml of buffer H (20 mM HEPES [pH 7.5], 1 mM EDTA, 1 mM DTT, and 10% glycerol) containing 50 mM KCl. The dialyzed sample was injected into a Mono S column (Pharmacia, HR5/5) connected to a HPLC system (Waters 650) and eluted, at a flow rate of 1 ml/min, with a linear gradient from 0.15–0.50 M KCl. The active fractions, which peaked at 320 mM KCl, were pooled (fraction V). Fraction V was diluted with buffer H to bring the KCl concentration down to 100 mM, was injected into a DEAE column (Waters, Protein Pak Glass DEAE-5PW), and was eluted with a 0.1–0.4 M KCl gradient at a flow rate of 1 ml/min. The eluate was collected at 1 μl/fraction. The active fractions peaked around 230 mM KCl and were pooled (fraction VI). Fraction VI was purified further through a Mono Q column (Pharmacia, HR5/5) in the same way as described for the DEAE column.

The G4-DNA column was made from biotinylated BQ oligonucleotide BQ3(G4) and streptavidin–agarose. BQ3 oligonucleotide was put into parallel G4 form as described (Sen and Gilbert, 1992), and the G4 products were used directly without going through gel purification. The resin, 0.5 ml of streptavidin–agarose (GIBCO BRL), was equilibrated in 100 mM KCl buffer A and then was incubated with 25 nmol of BQ3(G4) in the same buffer at 4°C for 30 min. In a cold room, the G4-DNA resin was packed into a column and washed with about 5 ml of 1 M KCl and then 10 ml of 100 mM KCl buffer A. The G4-DNA-dependent nuclease fraction was loaded onto the column in 100 mM KCl buffer A at a flow rate of 2 ml/hr. The column was then washed with 5 ml of 100 mM KCl buffer A and was eluted with a linear salt gradient from 0.1–0.6 M KCl (10 ml). The eluate was collected in 0.23 ml fractions. The column can be stored in a cold room in 1 M KCl buffer A and reused several times without much loss in capacity.

Cleavage Activity Assay

The standard reaction buffer contains 20 mM HEPES (pH 7.5), 3 mM $MgCl_2$, 100 mM KCl, 10% glycerol, 1 μg sonicated Escherichia coli DNA, and 1 pmol unlabeled TP-S/5 to 20 fmol of 5′ ^{32}P-labeled oligonucleotide probe in a final volume of 20 μl. For the gel assay, the protein fraction was mixed with labeled probe as specified in the reaction buffer, incubated at 37°C for 15 min, and loaded on a 5% (or as specified) polyacrylamide gel (37.5:1). The gel was buffered in TBE (50 mM Tris, 50 mM boric acid, 1 mM EDTA) containing 5 mM KCl, cooled in a cold room to 4°C, and prerun at 120 V for 1 hr. The electrophoresis was carried out at 4°C and 120 V until bromophenol blue dye ran half way into the gel. The gel was dried and autoradiographed on Kodak XAR-5 films. Alternatively, the dried gel was scanned with a PhosphorImager.

For the trichloroacetic acid assay, the reactions were carried out the same way as described for the gel assay, except that the amount of the labeled substrate was brought to 50 fmol with the corresponding unlabeled oligonucleotide. After a 10 min incubation at 37°C, an equal volume of 20% trichloroacetic acid was added, and the mixture was held on ice for 10 min. The acid-insoluble material was spun down in an Eppendorf centrifuge at the top speed for 10 min; the cleavage products remained in the supernatant. The supernatant was mixed with 1 ml of a scintillation fluid (Aquasol, DuPont) and counted in a scintillation counter. One unit of the G4-DNA-dependent nuclease activity is defined as the conversion of 10 fmol of the 5′ end of TP(G4) into an acid-soluble form in a reaction incubated at 37°C for 10 min at a substrate concentration of 0.63 nM. To assay for an optimal cleavage activity, the Mg^{2+} concentration was increased to 10 mM, while the amount of substrate and single-stranded competitors, such as TP-S, was increased to 1 pmol and 10 pmol, respectively.

Miscellaneous
Protein concentrations were measured according to the method of Bradford (1976). For samples with low protein concentrations, a colorimetric assay based on sliver binding was used (Krystal et al., 1985). Western blot analysis was based on a published procedure (Towbin et al., 1979).

Acknowledgments

We thank Richard Kolodner and Arlen Johnson for purified recombinant SEP1 protein and anti-SEP1 antiserum, Gerald Fink for *KEM1* deletion strains, Lawrence Bogorad for unlimited access to his HPLC instrument, Bruce Stillman for yeast *pep4* cells, and James Wang for a suggestion on the purification and comments on the manuscript. We also thank the Harvard Microchemistry Facility for peptide sequencing and members of the Gilbert Lab for discussions. This work was supported by grants from the National Institutes of Health (to W. G.) and a postdoctoral fellowship from the Irvington Institute for Medical Research (to Z. L.).

Received March 31, 1994; revised May 6, 1994.

References

Ashly, T., and Ward, D. C. (1993). A "hot-spot" of recombination coincides with an interstitial telomeric sequence in the Armenian hamster. Cytogenet. Cell. Genet. *62*, 169–171.

Awang, G., and Sen, D. (1993). Model of dimerization of HIV-1 genomic RNA. Biochemistry *32*, 11453–11457.

Blackburn, E. H. (1991). Structure and function of telomeres. Nature *350*, 569–573.

Bradford, M. M. (1976). A rapid and sensitive method for the quantitation of microgram quantities of protein utilizing the principle of protein-dye binding. Anal. Biochem. *72*, 248–254.

Darlix, J.-L., Gabus, C., Nugeyre, M.-T., Clavel, F., and Barre-Sinoussi, F. (1990). *Cis* elements and *trans*-acting factors involved in the RNA dimerization of the human immunodeficiency virus HIV-1. J. Mol. Biol. *216*, 689–699.

Dykstra, C. C., Kitada, K., Clark, A. B., Hamatake, R. K., and Sugino, A. (1991). Cloning and characterization of *DST2*, the gene for DNA strand transfer protein β from *Saccharomyces cerevisiae* Mol. Cell. Biol. *11*, 2583–2592.

Fang, G., and Cech, T. R. (1993a). The β subunit of Oxytricha telomere-binding protein promotes G-quartet formation by telomeric DNA. Cell *74*, 875–885.

Fang, G., and Cech, T. R. (1993b). Characterization of a G-quartet formation reaction promoted by the β-subunit of the *Oxytricha* telomere-binding protein. Biochemistry *32*, 11646–11657.

Friedberg E. C., Siede, W., and Cooper, A. J. (1991). Cellular responses to DNA damage in yeast. In The Molecular and Cellular Biology of the Yeast *Saccharomyces*, Volume One, J. R. Broach, J. R. Pringle, and E. W. Jones, eds. (Cold Spring Harbor, New York: Cold Spring Harbor Laboratory Press), pp. 147–192.

Hammond-Kosack, M. C. U., Dobrinski, B., Lurz, R., and Docherty, K. (1992). The insulin gene–linked polymorphic region exhibits an altered DNA structure. Nucl. Acids Res. *20*, 231–236.

Johnson, A. W., and Kolodner, R. D. (1991). Strand exchange protein 1 from *Saccharomyces cerevisiae*. J. Biol. Chem. *21*, 14046–14054.

Kang, C., Zhang, X., Ratliff, R., Moyzis, R., and Rich, A. (1992). Crystal structure of four-stranded *Oxytricha* telomeric DNA. Nature *356*, 126–131.

Kim, J., Ljungdahl, P. O., and Fink, G. R. (1990). *kem* mutations affect nuclear fusion in *Saccharomyces cerevisiae*. Genetics *126*, 799–812.

Kipling, D., Tambini, C., and Kearsey, S. (1991). *rar* mutations which increase artificial chromosome stability in *Saccharomyces cerevisiae* identify transcription and recombination proteins. Nucl. Acids Res. *19*, 1385–1391.

Klapholz, S., and Esposito, R. E. (1980) Isolation of *spo12-1* and *spo13-1* from a natural variant of yeast that undergo a single meiotic division. Genetics *96*, 567–588.

Krystal, G., MacDonnald, C., Munt, B., and Ashwell, S. (1985). A method for quantitating nanogram amounts of soluble protein using the principle of silver binding. Anal. Biochem. *148*, 451–460.

Larimer, F. W., and Stevens A. (1990). Disruption of the gene *XRN1*, coding for a 5′–3′ exoribonuclease, restricts yeast cell growth. Gene *95*, 85–90.

Liu, Z., Frantz, J. D., Gilbert, W., and Tye, B.-K. (1993). Identification and characterization of a nuclease activity specific for G4 tetra-stranded DNA. Proc. Natl. Acad. Sci. USA *90*, 3157–3161.

Maxam, A., and Gilbert, W. (1980). Sequencing end-labeled DNA with base-specific chemical cleavages. Meth. Enzymol. *65*, 499–560.

Merril, C. R., Switzer, R. C., and Keuren, M. L. (1979). Trace polypeptides in cellular extracts and human body fluids detected by two-dimensional electrophoresis and a highly sensitive silver strain. Proc. Natl. Acad. Sci. USA *76*, 4335–4339.

Pearson, W. R., and Lipman, D. J. (1988). Improved tools for biological sequence comparison. Proc. Natl. Acad. Sci. USA *85*, 2444–2448.

Pearson, A. M., Rich, A., and Krieger, M. (1993). Polynucleotide binding to macrophage scavenger receptors depends on the formation of base-quartet-stabilized four-stranded helices. J. Biol. Chem. *268*, 3546–3554.

Pluta, A. F., and Zakian, V. A. (1989). Recombination occurs during telomere formation in yeast. Nature *337*, 429–433.

Sen, D., and Gilbert, W. (1988). Formation of parallel four-stranded complexes by guanine-rich motifs in DNA and its implication for meiosis. Nature *334*, 364–366.

Sen, D., and Gilbert, W. (1990). A sodium–potassium switch in the formation of four-stranded G4-DNA. Nature *344*, 410–414.

Sen, D., and Gilbert, W. (1992). Guanine quartet structure. Meth. Enzymol. *211*, 191–199.

Shimizu, A., and Honjo, T. (1984). Immunoglobulin class switching. Cell *36*, 801–803.

Smith, F. W., and Feigon, J. (1992). Quadruplex structure of *Oxytricha* telomeric DNA oligonucleotides. Nature *356*, 164–168.

Sundquist, W. I., and Heaphy, S. (1993). Evidence for interstranded quadruplex formation in the dimerization of human immunodeficiency virus genomic RNA. Proc. Natl. Acad. Sci. USA *90*, 3393–3397.

Sundquist, W. I., and Klug, A. (1989). Telomeric DNA dimerizes by formation of guanine tetrads between hairpin loop. Nature *342*, 825–829.

Tishkoff, D. X., Johnson, A. W., and Kolodner, R. D. (1991). Molecular and genetic analysis of the gene encoding the *Saccharomyces cerevisiae* strand exchange protein SEP1. Mol. Cell. Biol. *11*, 2593–2608.

Towbin, H., Staehelin, T., and Gordon, J. (1979). Electrophoretic transfer of proteins from polyacrylamide gels to nitrocellulose sheets: procedure and some applications. Proc. Natl. Acad. Sci. USA *76*, 4350–4354.

Wang, Y., and Patel, D. J. (1993). Solution structure of the human telomeric repeat $d[AG_3(T_2AG_3)_3]$ G-tetraplex. Structure *1*, 263–282.

Weisman-Shomer, P., and Fry, M. (1993). QUAD, a protein from hepatocyte chromatin that binds selectively to quanine-rich quadruplex DNA. J. Biol. Chem. *268*, 3306–3312.

White, M. A., Dominska, M., and Petes, T. D. (1993). Transcription factors are required for the meiotic recombination hotspot at the *HIS4* locus in *Saccharomyces cerevisiae* Proc. Natl. Acad. Sci. USA *90*, 6621–6625.

Williamson, J. R., Raghuraman, M. K., and Cech, T. R. (1989). Monovalent cation–induced structure of telomeric DNA: the G-quartet model. Cell *59*, 871–880.

Section VIII:
The RNA World

Origin of life

The RNA world

from Walter Gilbert

UNTIL recently, when one thought of the varied molecular processes at the origin of life, one imagined that the first self-replicating systems consisted of both RNA and protein. RNA served to hold information, whereas protein molecules provided all the enzymic activities needed to make copies of RNA and to reproduce themselves. The cycle that developed a self-replicating system out of the primitive soup of amino acids and nucleotides had two radically different components[1].

Now it seems possible that the informational and catalytic properties of these two components may be combined in a single molecular species. Last week in these columns Frank Westheimer[2] described the discovery of enzymic activities in the RNA molecules of *Escherichia coli*, in which ribonuclease-P cuts phosphodiester bonds during the maturation of the transfer RNA molecule[3,4], and of *Tetrahymena*, whose ribosomal RNA contains a self-splicing exon[5-7]. If there are two enzymic activities associated with RNA, there may be more. And if there are activities among these RNA enzymes, or ribozymes, that can catalyse the synthesis of a new RNA molecule from precursors and an RNA template, then there is no need for protein enzymes at the beginning of evolution. One can contemplate an RNA world, containing only RNA molecules that serve to catalyse the synthesis of themselves[8].

The self-splicing intron is an RNA element that can splice itself out of an RNA molecule. This reaction should be reversible, and the intron could splice itself back into an appropriate nucleotide sequence. Thus, in the RNA world, such introns could both remove and insert themselves into the background of replicating RNA molecules. The significance of this is not the simple insertion and removal of introns, but the fact that two introns, separated from each other by another RNA element, an exon, can combine with each other so as to remove as a unit both themselves and the intervening exon from one RNA molecule and to insert into another. Thus, self-inserting introns can create transposons to move exons around. This property provides RNA with a major evolutionary facility that it otherwise lacks — recombination, the ability to produce new combinations of genes. Of course the self-replicating molecules would in any case have evolved slowly by miscopying, that is, by mutation. But transposons provide the equivalent of sex — the infectious transmission of genetic elements from one organism to another. Recombination and sex are powerful devices to permit a more useful exon to pass from one replicating structure to an unrelated one.

This picture of the RNA world is one of replicating molecules that reassort exons by transposable elements created by introns. This process builds and remakes RNA molecules by chunks and also permits the useful distinction between information and function. Information storage needs to be one-dimensional, for ease of copying, but molecules with enzymic functions tend to be tight three-dimensional structures, whose forms are unrelated to the demands of any copying mechanism. (This dichotomy is most obvious today between the linear order along DNA and the structure of proteins.) In the RNA world, the structure that would be replicated has the full complement of introns. Some of the daughters, by splicing out all their introns, would convert to functional molecules, the ribozymes. A remnant of this process may be the structure of transfer RNA, where a compact secondary structure is broken up by the insertion of an intron.

The first stage of evolution proceeds, then, by RNA molecules performing the catalytic activities necessary to assemble themselves from a nucleotide soup. The RNA molecules evolve in self-replicating patterns, using recombination and mutation to explore new functions and to adapt to new niches. By using RNA cofactors, such as nicotinamide adenine dinucleotide and flavin mononucleotide they then develop an entire range of enzymic activities[9]. At the next stage, RNA molecules began to synthesize proteins, first by developing RNA adapter molecules that can bind activated amino acids and then by arranging them according to an RNA template using other RNA molecules such as the RNA core of the ribosome. This process would make the first proteins, which would simply be better enzymes than their RNA counterparts. I suggest that protein molecules do not carry out enzymic reactions of a different nature from RNA molecules but are able to perform the same reactions more effectively and rapidly, and hence will eventually dominate. These protein enzymes are encoded by RNA exons, thus they, in turn, are built up of mini-elements of structure.

Finally, DNA appeared on the scene, the ultimate holder of information copied from the genetic RNA molecules by reverse transcription. After double-stranded DNA evolved there exists a stable linear information store, error-correcting because of its double-stranded structure but still capable of mutation and recombination. RNA is then relegated to the intermediate role that it has today — no longer the centre of the stage, displaced by DNA and the more effective protein enzymes. But a few RNA enzymic activities still exist, the two described recently[3-7], and possibly others in the role of ribosomal RNA or in the splicing of eukaryotic messenger RNA. The relic of this process is the intron/exon structure of genes, left imprinted on DNA from the RNA molecules that earlier encoded proteins, a residue of the basic mechanism of RNA recombination. □

1. Eigen, M., Gardiner, W., Schuster, P. & Winkler-Oswatitsch, R. *Sci. Am.* **244**, 88 (1981).
2. Westheimer, F.H. *Nature* **319**, 534 (1986).
3. Guerrier-Takada, C., Gardiner, K., Marsh, T., Pace, N. & Altman, S. *Cell* **35**, 849 (1983).
4. Guerrier-Takada, C., & Altman, S. *Science* **223**, 285 (1984).
5. Kruger, K. *et al. Cell* **31**, 147 (1982).
6. Cech, T.R. *Int. Rev. Cytol.* **93**, 3 (1985).
7. Zaug, A.J. & Cech, T.R. *Science* **231**, 470 (1986).
8. Sharp, P.A. *Cell* **42**, 397 (1985).
9. White, H.B. III *J. molec. Evol.* **7**, 101 (1976).

Walter Gilbert is at the Biological Laboratories, Harvard University, 16 Divinity Avenue, Cambridge, Massachusetts 02138, USA.

Infrared detectors

Superlattices point ahead

from Gordon C. Osbourn

TECHNICAL developments in the past few years have made clear what the next generation of infrared detectors will be like. The need is for semiconductor materials appropriate for long-wavelength ($\geqslant$ 12 μm) imaging, at the focal planes of new telescopes and for various defence applications. Most effort has focused on the II–VI alloy system (Hg,Cd)Te, but work has also begun on a recently proposed III–V system consisting of many alternating layers of thin mismatched In(As,Sb) crystal layers, called strained-layer superlattices (SLSs). A crucial first step in their development has now been achieved with the successful growth of crystals of SLSs in the In(As,Sb) alloy system[1-3].

The materials from which long-wavelength detectors might be made are limited by the stringency of the requirements. These imaging devices will consist of two-dimensional arrays of many individual photovoltaic detector elements on a single wafer of a material which must have an energy bandgap less than or equal to 0.1 eV at about 77 K, the operating temperature of the infrared detector. This bandgap is smaller than that of all the III–V alloy

Proc. Natl. Acad. Sci. USA
Vol. 84, pp. 1978–1980, April 1987
Evolution

Basic protein enhances the incorporation of DNA into lipid vesicles: Model for the formation of primordial cells

(membrane/lysozyme/prebiotic/DNA encapsulation)

DANIEL G. JAY* AND WALTER GILBERT†

*Department of Neurobiology, Harvard Medical School, Boston, MA 02115; and †Biological Laboratories, Harvard University, 16 Divinity Avenue, Cambridge, MA 02138

Contributed by Walter Gilbert, December 15, 1986

ABSTRACT **DNA can be encapsulated into lipid vesicles formed by sonication. The presence of a basic protein, lysozyme, enhances the incorporation 100-fold above the level expected by random trapping. This is demonstrated by the ability of the lipid vesicles to protect DNA from digestion with DNase. Such an enhancement of nucleic acid incorporation into vesicles by basic polypeptides and the sharply increased concentration of these macromolecules in the internal volume may have been advantageous in prebiotic evolution.**

A critical process in the formation of the first cells was the isolation of macromolecules away from the external environment. Recent speculations about the RNA origin of life suggest that the first enzymes were RNA and that the first polypeptides played only a structural role (1). This led us to conjecture that one early role of protein might have been to help package nucleic acids within lipid membranes. The encapsulation of DNA by large lipid vesicles has been reported (2, 3), but the amount of DNA encapsulated was in simple proportion to the internal volume: the DNA must have been encapsulated by random trapping. The interaction of nucleic acid with phospholipid is an unfavored event because of the polyanionic nature of DNA or RNA, the hydrophobicity of lipid, and the negative charges associated with the phospholipid head groups. We propose that basic polypeptides might have mediated vesicle formation around nucleic acids by neutralizing charge, by condensing the linear strands, and by acting as an amphipathic template for lipid-bilayer formation. The experiments reported here show that basic proteins are efficient mediators of DNA encapsulation within lipid membranes: the presence of lysozyme causes DNA to be segregated into lipid vesicles 100-fold that expected by random trapping.

MATERIALS AND METHODS

Lysozyme, histone, bovine serum albumin, trypsin, and Ficoll were obtained from Sigma. Micrococcal nuclease was obtained from Boehringer Mannheim. Asolectin (a chloroform/methanol extract of soybean lipids) was obtained from Avanti Polar Lipids (Birmingham, AL). Linear [^{32}P]5′-DNA was a gift from S. Swanberg and D. Paul.

Vesicles were formed by a variation of the method of Dixon and Hokin (4). A solution of 50 μg of asolectin in 10 μl of chloroform was dried in the bottom of a Microfuge tube under a nitrogen stream, washed two times with 25 μl of ether, and dried under nitrogen for 1 hr at 40°C. Ten nanograms of linear [^{32}P]5′-DNA, 1000 base pairs long, and 10 μg of protein were added in a final volume of 10 μl of 10 mM Tris·HCl, pH 7.4/1 mM EDTA, and sonicated in a bath sonicator (Laboratory Supplies, Hicksville, NY) for 6 min during which time the mixture became opalescent, indicative of the formation of vesicles.

The vesicles were challenged with micrococcal nuclease by making the preparations 10 mM $CaCl_2$ and adding 0.3 unit of micrococcal nuclease. The samples were incubated for 60 min at 37°C. The digestions were terminated by the addition of 1 μl of 250 mM EDTA followed by the addition of 2 μl of 20% $NaDodSO_4$ and boiling for 3 min. Samples were analyzed by gel electrophoresis and autoradiography.

Flotation of vesicles by Ficoll-gradient centrifugation was done by an adaption of the method of Fraley *et al.* (3). A vesicle sample with a volume of 100 μl was diluted with 30% Ficoll in 10 mM Tris·HCl, pH 7.4/1 mM EDTA to a final volume of 500 μl. The sample was overlaid with a Ficoll step gradient in 10 mM Tris·HCl, pH 7.4/1 mM EDTA: 1 ml of 20% Ficoll, 3 ml of 10% Ficoll, and 1.5 ml of buffer alone in 13 × 51-mm ultraclear centrifuge tubes. Samples were spun in a 50.1 swinging bucket rotor at 40,000 rpm for 30 min and decelerated without braking. The gradients were fractionated into 130-μl samples, and radiolabeled DNA was detected by Cherenkov radiation.

Trapped internal volume was measured by preparing 10-μl samples of vesicles with 250,000 cpm of ^{14}C-labeled sucrose. The samples were diluted with 90 μl of 10 mM Tris·HCl, pH 7.4/1 mM EDTA and spun on an Airfuge centrifuge at 95,000 rpm for 1 hr. The supernatant was removed, and the pellet was washed with 100 μl of 10 mM Tris·HCl, pH 7.4/1 mM EDTA. The pellets were solubilized with 10 μl of 1% $NaDodSO_4$ and counted. Samples were done in triplicate, and samples of radiolabeled sucrose without vesicles were used as controls.

RESULTS AND DISCUSSION

DNase Analysis of Encapsulation. Our assay follows the ability of lipid, formed into vesicles by sonication, to protect labeled DNA against subsequent digestion with nuclease. We used DNA rather than RNA because of a concern about endogenous RNases in the lipid preparations. In preliminary experiments we tried histones; however, such basic proteins can protect DNA against DNase directly in the absence of lipid, so we turned to lysozyme, a basic protein that does not have specific DNA-binding properties.

[^{32}P]5′-DNA was added to a mixture of phospholipids, either alone or with protein, and the sample was sonicated to form vesicles. The protein was present in 1000-fold weight excess. After these vesicles were challenged with micrococcal nuclease, the DNA was assayed for protection from digestion by gel electrophoresis and autoradiography.

Fig. 1 shows the results of such experiments. The labeled DNA, alone, without nuclease treatment, runs as a single band (lane A). Thus the sonication treatment did not damage the DNA (observed for both 1000- and 4300-base pair-long fragments). Samples of DNA sonicated with lysozyme alone (lane B) or lipid alone (lane C) and subsequently treated with

The publication costs of this article were defrayed in part by page charge payment. This article must therefore be hereby marked "*advertisement*" in accordance with 18 U.S.C. §1734 solely to indicate this fact.

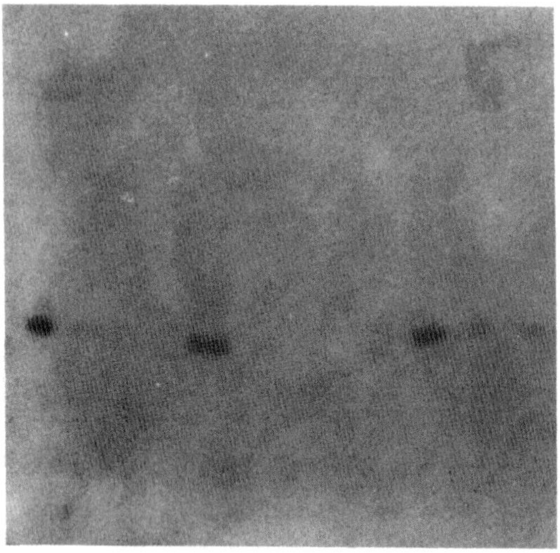

FIG. 1. Lysozyme-mediated vesicle encapsulation protects DNA from nuclease digestion. Linear [^{32}P]5′-DNA, 1000 base pairs in length, was sonicated with lipid and protein to form vesicles and challenged with micrococcal nuclease. The samples were run on a 0.7% agarose gel, dried, and autoradiographed with intensifier screen. Lane A, DNA; lane B, DNA and lysozyme with nuclease digestion; lane C, DNA and lipid with nuclease digestion; lane D, DNA, lipid, and lysozyme with nuclease digestion; lane E, DNA, lipid, and bovine serum albumin with nuclease digestion; lane F, DNA, lipid, and lysozyme treated with 1% Triton X-100 before nuclease digestion; lane G, DNA added after the sonication of lipid and lysozyme followed by nuclease digestion; lane H, one μg of trypsin added to DNA, lipid, and lysozyme after sonication followed by nuclease digestion; lane I, one μg of trypsin added before sonication to DNA, lipid, and lysozyme followed by nuclease digestion; lane J, 200 mM $MgCl_2$, DNA, lipid, and lysozyme with nuclease digestion.

nuclease were completely digested. However, when DNA was sonicated with lysozyme and lipid together (lane D), at least 50% of the DNA was protected (verified by densitometry scans of the autoradiograph). The sonication of DNA together with lipid and an acidic protein, bovine serum albumin, did not result in protection (lane E). These data show that under these conditions DNA is not efficiently encapsulated by lipid vesicle formation unless a basic protein (lysozyme) is present. Acidic protein (bovine serum albumin) cannot perform this function.

If the DNA was added to previously sonicated lysozyme/lipid mixtures, it was not protected from nuclease digestion (lane F), and the addition of the non-ionic detergent Triton X-100 (which disrupts vesicles) to lysozyme-lipid vesicles led to the digestion of the DNA (lane G). Incubation of the lipid, lysozyme, and DNA mixture with trypsin prevented protection only if the trypsin was added before sonication (lane I) but not afterward (lane H). These data demonstrate that the protection is dependent upon the integrity of the vesicles. Furthermore, the DNA and protein must both be present during vesicle formation.

The addition of 200 mM $MgCl_2$ to the lysozyme, lipid, and DNA mixture prevented the protection from nuclease (lane J). This suggests that the interaction of DNA with lysozyme is charge dependent, but more than a simple charge neutralization of the nucleic acid (which would be provided by the Mg^{2+} counter-ions) is required for lipid encapsulation of the DNA.

Flotation Analysis of Encapsulation. As an alternative demonstration that the DNA is associated with vesicles, we analyzed vesicles prepared with radiolabeled DNA and either lysozyme or bovine serum albumin by centrifugation on Ficoll step gradients. After centrifugation, we observed two opalescent bands—one at the interface between the sample and 20% Ficoll and the other between the 10% Ficoll and the buffer layers. The radiolabeled DNA was detected by Cherenkov counting each fraction from the gradient. Fig. 2 shows that in the lysozyme-prepared vesicles, 50% of the radiolabeled DNA comigrated with the first opalescent band, floating with the vesicle population on the Ficoll step gradient. However, the DNA remained at the bottom of the gradient when the vesicles were made with bovine serum albumin.

Both samples were examined by electron microscopy of thin sections from Airfuge-pelleted vesicles. They contained a heterogeneous mixture of vesicles including single lamellar and multilamellar structures. Vesicles prepared with lysozyme had diameters between 1000 and 2000 Å, whereas those made with bovine serum albumin ranged from 2500 to 4000 Å.

These data together show that lysozyme mediates an enhanced segregation of DNA into vesicles in a charge-dependent fashion and that once vesicles are formed, macromolecules are excluded. The observations that the protection is eliminated by detergent and that the DNA must be present before sonication argue against a nonspecific protection of the DNA by lipid.

Lysozyme Enhances DNA Encapsulation 100-Fold Above That Expected by Random Trapping. Fraley *et al.* (3) and Deamer and Barchfield (2) have reported encapsulation of DNA by the formation of large vesicles in which the fraction of DNA incorporated was equal to the fraction of total volume trapped by the vesicles. Half of the DNA added was thus incorporated, because the large average diameter of these vesicles (0.4 micrometers) resulted in 50% of the total volume being trapped.

The internal volume of the vesicles in our experiments was 0.5% of the total volume (determined by ^{14}C-labeled sucrose trapped volume studies and corroborated by calculations based on the total amount of lipid and the average vesicle diameter). Fifty percent of the DNA was protected, well above the expectation from random trapping during sonication. This implies that lysozyme provides some specific enhancement for vesicle encapsulation beyond charge neu-

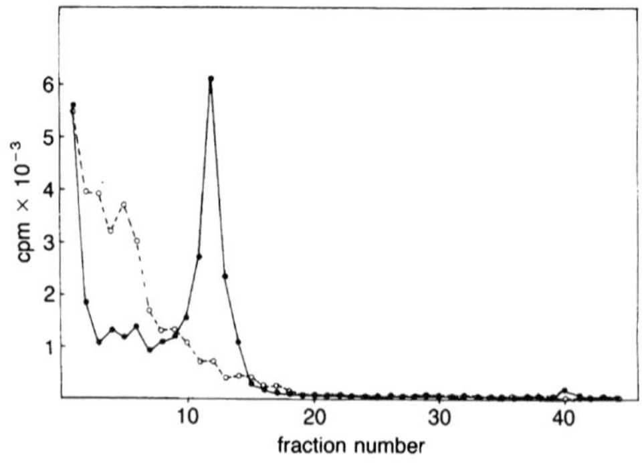

FIG. 2. Flotation of vesicles by Ficoll-gradient centrifugation. Ficoll gradient samples were prepared as described. Two opalescent bands were observed at the interfaces between the sample and 20% Ficoll and between 10% and 0% Ficoll; the location of the peak of ^{32}P radioactivity corresponds to the former opalescent band. The gradients were fractionated into 130-μl aliquots, and radiolabeled DNA was detected by Cherenkov radiation. Vesicles were prepared with [^{32}P]5′-DNA, lipid, and lysozyme (●) or with [^{32}P]5′-DNA, lipid, and bovine serum albumin (○).

tralization, which would only account for incorporation by random trapping. In fact, the concentration of DNA within the internal environment of the vesicles was increased 100-fold.

The preferential enclosure of DNA by lipid mediated by lysozyme may have a practical application. Vesicle fusion has been used to introduce DNA into animal cells (3, 5). Enhanced incorporation of DNA may increase the efficiency of this method of DNA transfection and permit one to use vesicles as targeting vehicles.

Significance for Prebiotic Evolution. If protein-mediated encapsulation occurred during early cell formation, why would such an event favor prebiotic evolution? Encapsulation segregates macromolecules away from the external environment. Protein-mediated encapsulation creates high local concentrations of protein and nucleic acids within the vesicular volume compared with random trapping. This would enhance the interaction of molecules with low affinities, potentiating the formation of aggregates with biological function. Permeable membranes or polypeptide pores would permit the entry of small molecules. The fusion and budding of vesicles might provide a means by which functional aggregates could be shuffled to obtain new prebiotic species. Most importantly, the sequestering of self-replicating nucleic acids into membrane-bound structures would provide the separate entities on which natural selection could operate.

The authors gratefully acknowledge Hal Mekeel for assistance with electron microscopy, David Paul and Stephen Swanberg for generously providing radiolabeled DNA, Jim Huettner and Gilbert Chin for helpful discussions, and Guido Guidotti and Jeremy R. Knowles for critical reading of the manuscript. D.G.J. thanks David Hubel and the Harvard Society of Fellows for support.

1. Gilbert, W. (1986) *Nature (London)* **319,** 618.
2. Deamer, D. W. & Barchfield, G. L. (1982) *J. Mol. Evol.* **18,** 203–266.
3. Fraley, R., Subramani, S., Berg, P. & Papahadjopoulos, D. (1980) *J. Biol. Chem.* **255,** 10431–10435.
4. Dixon, J. F. & Hokin, L. E. (1980) *J. Biol. Chem.* **255,** 10681–10686.
5. Loyter, A., Zakai, N. & Kulka, R. G. (1975) *J. Cell Biol.* **66,** 292–304.

Section IX:
Introns, Exons, and Gene Evolution

INTRONS AND EXONS: PLAYGROUNDS OF EVOLUTION

Walter Gilbert[1]

Department of Biochemistry and Molecular Biology
Harvard University, Cambridge, Massachusetts 02138

ABSTRACT Eukaryotic genes are mosaic structures: regions of DNA bearing the coding information separated by stretches that have no apparent function. I speculate that this interspersion plays an important evolutionary role.

The structure of genes in higher cells is not simple but complex, characterized by the coding information lying in noncontiguous segments along the DNA. In order for such a gene to function, the RNA polymerase first makes a long transcription product, then a splicing enzyme removes portions of that transcript and splices together the remaining segments to form a mature message, which will be translated in the cytoplasm. This genetic structure and process was first detected in adenovirus (1,2), and since then it has been observed for a large number of eukaryotic genes: the various genes for globins (3,4,5), for the immunoglobulin light chains (6,7), for ovalbumin (8,9), for lysozyme (10), for the immunoglobulin heavy chains (11,12), for the cytochrome b gene of yeast mitochondria (13), for some yeast transfer RNAs (14,15), and, as we shall hear later in this meeting, for the genes for preproinsulin (16). However, such a structure was not seen in the analysis of histone genes (17). How can we understand this profound difference in structure between higher cell genes and those to which we are accustomed in bacteria?

The new picture conceives of a genetic region of DNA expressed as a long transcription unit, which through splicing can rearrange and eliminate some of its sequences, in one or several ways, to produce one or several final gene products. That a single protein chain may not lie on a contiguous region of DNA will show up in a genetic analysis. The map positions of mutations that lie in the protein will cluster, the clusters will lie at a distance from each other, and in the

[1]American Cancer Society Professor of Molecular Biology, supported by NIH Grant GM09541.

ISBN 0-12-068350-4

spaces between may fall mutations that block RNA transcription or the necessary processing. Benzer's notion of the cistron, while still having an operational validity, has now become far more complex. Those regions of DNA, that lie within the final gene as defined genetically, but that lie between segments that encode parts of the mature RNA product, we call introns (for intra-cistronic regions) (18,6). The DNA regions that correspond to sequences expressed in the mature message, we call exons, these being both the coding regions translated into protein as well as non-translated regions which appear in the mature RNA gene product. The gene is thus an alternating series of exons and introns, the introns being eliminated from the transcript by splicing. This language is neutral and does not presume that the regions called introns, also known as intervening sequences, have been inserted into the coding region of some preexisting gene. The basic question, as one looks at these genes lying in pieces, is: Were the genes once whole but became separated by the insertion of elements whose purpose is to keep the exons apart? Or were the parts of the genes always separated and then put together by the introns to create the evolving genetic element?

Some properties of introns have emerged over the past year. The transcription unit very often has multiple introns, whose lengths range from as little as ten bases to ten thousand bases, both within the coding region and within untranslated parts of the ultimate message. No convincing structural feature has been seen at the intron-exon boundaries which would in itself completely characterize the splice point. The most general homology identified so far is to a short repeated sequence at the two ends of the intron, a CAGG tetranucleotide, and as observed by Chambon (19), the greatest consistency is that there is a GT at the left boundary of the intron and an AG at the right boundary, although even this has an exception. The intron sequence is generally pyrimidine rich. Although these elements of sequence are suggestive, they are not sufficient for recognition. (See the review by Crick (20) for a discussion of splicing.)

In general the average length of the intron is much greater than that of the exon; the genes are spread across the chromosomal DNA to a much greater extent than the final coding capacity would require. In large part this explains the excess DNA in higher organisms: their genes are not trimmed down efficient structures but are instead long transcription units, five to ten times larger than the aggregate of the exons. In addition, these genes lie apart at distances comparable to their lengths. In principle, a single transcription unit can be read out in different ways, a different subset of sequences, a different set of exons, spliced together to make a gene product. Introns for one reading may be exons for another.

What is the role of the introns? The first possibility that comes to mind is that they are regulatory elements, sequences which must be removed in order for specific gene products to be made. One such notion is that a specific, specialized, splicing enzyme could be used as a control element for a unique gene. Nuclear transcription would run all the time, but, when needed, the control gene product, a splicing enzyme, would call for and create a specific functional sequence. In this picture, of course, the sequences of an intron used as a recognition element should be different from all of the others; one should not expect to find homologies in the intron-exon boundaries. Viruses are the best candidates for such roles: eliminating host splicing and providing a new virus splicing pattern would be an effective way of taking over the cell. An alternative regulatory use might be an obligatory removal property: the introns could be elements which must be removed in order for any message to move from the nucleus to cytoplasm, a role certainly better played by these elements when they occur in the untranslated regions. This last model would envisage both a control and a structural role for the intron RNA seqences. As far as control is concerned, a major lesson of molecular biology has been that if any biochemical process can be used for control, it will be. Furthermore, the relationship between biochemistry and molecular biology is not a one-to-one mapping of biological processes on unique biochemical solutions. There is not one way of replicating DNA but many, differing in the details of the enzymology. There is not one way of controlling genes in bacteria, such as that imagined when only repressors were thought to play a role, but many, all having in common only the property that the recognition of structure at a point of possible biochemical control can permit a genetic element to exercize specific control. Thus I would not be surprised to find that some introns are used as control functions.

A different hypothesis is that the introns are simply adventitious structural features. The fact that they often lie between repeated sequences is reminiscent of the transposing elements of bacteria, either the insertion sequences or the transposons, because when these sequences enter new DNA they create a short repeating seqence, nine bases long in the case of IS1 (21,22), five long for IS2 (23), flanking the intrusion. Although we recognize these insertion elements in bacteria through their inactivation of genes, if the cell had a mechanism to splice between the terminal redundancies of these seqences as they appeared in a message, the DNA element would be invisible in terms of gene expression. Certainly the role of the omega sequence in the mitochondrial RNA (24,25,26) is reminiscent of such an element, and there are suggestions of transposable elements in yeast, if not yet in mammals.

However, introns that are added elements might well appear in different places in duplicate genes, if they add after the original events that caused the gene duplication. Secondly, if the introns are transposable elements, one would expect to find homology between them, yet in families of genes the intronic sequences drift faster than the exonic sequences, with the exception of the regions near the intron-exon borders. In the few cases that have been examined the intronic sequences represent unique DNA (27), consistent with their playing a role that does not depend on the specific nature of their sequences.

Let us pursue the view that the introns exist and are preserved because they play some critical role in evolution. One hypothesizes that the splicing patterns are induced by general enzymes. Given the existence of such preexisting enzymes in the cell, evolutionary arguments propose that the extra DNA carried in the introns is of sufficient use that it would be preserved over long periods. The lack of introns in histone genes, which occur as multiple repeats, would be a consequence of the need for a very high rate of synthesis: under extreme evolutionary pressure extra baggage should be dropped. In the general situation one would imagine a balance between the energy lost in maintaining excess DNA and the freedom to do certain things over evolutionary times. But what are those things?

One possibility, suggested by Philip Leder, is that the introns represent sequences added to structural genes to freeze duplications in place. The role of these sequences in such a model is that after a gene is duplicated, the diversification of DNA sequence permitted by the rapid drift of both intronic and flanking seqences would cut down recombination between the copies, which recombination could lead, by unequal crossing over, to the loss of one of those genes. Such a role for introns would suggest that they should only occur in genes which have been duplicated, members of multi-gene families. This is true for the hemoglobins and the immunoglobulins but not so clearly for ovalbumin and other proteins. Furthermore, to block recombination best and to stabilize a duplicated gene, one might expect introns to arise in different places in such genes rather than in the same places throughout the family, since this last fact argues that they antedate the duplications. This model suggests a negative role, that the introns <u>block</u> recombination in order to slow down gene loss and exchange. The alternative view is that introns play a positive role, that they speed up the search for new gene products either by opening new pathways for gene evolution or by enhancing recombination, which shuffles the parts of genes.

A general, non-specific mechanism involving splicing out long segments of RNA can be of strong evolutionary benefit. Alterations of the splicing pattern, mutations leading to incomplete splicing at a high frequency, provide ways to read out the same DNA region in many different patterns simultaneously. In this manner gene functions can drift without abolishing entirely the function of the preexisting gene. New gene products can be sought while the old gene product is still made at a reasonable rate. There is no necessity for the genes to duplicate before they drift to new functions; the extra DNA required for the new uses, on the intron-exon picture, is imbedded within the DNA of the genes rather than being carried as extra copies. Evolution's constant exploration can take place through subtle modifications of the splicing pattern while maintaining the underlying, still useful structure of the previous gene.

The other evolutionary thought is the realization that expanding the gene in size moves the exons apart along the DNA and thus will permit a higher rate of both legitimate and illegitimate recombination between them. The illegitimate recombination will shuffle the exons from different genes and bring them together in new combinations. It will permit gene duplications, or exon duplications, to fuse, not by an exact match of expressed sequence at the point of fusion but through a more random crossover that causes a duplication of the previous exon anywhere within several thousand bases and puts it on the same transcription unit. Such an event is likely to lead to a low production of the protein containing the duplicated exon sequence and this provides a handle, if this duplication of exon sequence is a benefit, for natural selection to polish the system.

Legitimate recombination between homologous sequences within introns will also play a role. Consider the evolution of a gene made up of introns and exons. A mutation in one exon might produce a better gene product, and that gene would begin to spread by selection through the population. At the same time, a mutation to better performance in another exon might also be spreading through the population. Can these two better exons get together in the same gene? They can by recombination -- and the separation of the two exons by a long intron will increase the amount of recombination between these mutations and thus the ease by which both changes can be brought within a single gene. This process will work best, and thus will select for, those cases in which the exons represent separate functional elements in the final gene product — since then the mutations to better function will always complement in cis.

Thus, if the introns are used to assort the exons, then the exons should represent useful solutions to the structure/function problem. They should not only represent domains, in the sense of complete structures, but also functional elements of the protein, components that can be used again and again in different proteins, to solve parts of the problem of overall function. This picture arose naturally from the first structures worked out at the DNA level for the light chain of the immunoglobulins (6,7). Here the coding sequence for the final gene product is interrupted by two introns, a short one separating the hydrophobic signal sequence at the amino terminus of the protein between amino acids -4 and -3 and a longer one dividing the gene into two halves which correspond to the structural domains of the immunoglobulin light chain. The immunoglobulin molecule consists of a series of repeated domains derived from a common ancestor. The light chain has two of these; the heavy chain of IgG or IgA has four, while that of IgM contains five. Figure 1 shows the position of the intron on an outline of the X-ray structure of the immunoglobulin light chain. Clearly it separates the two domains.

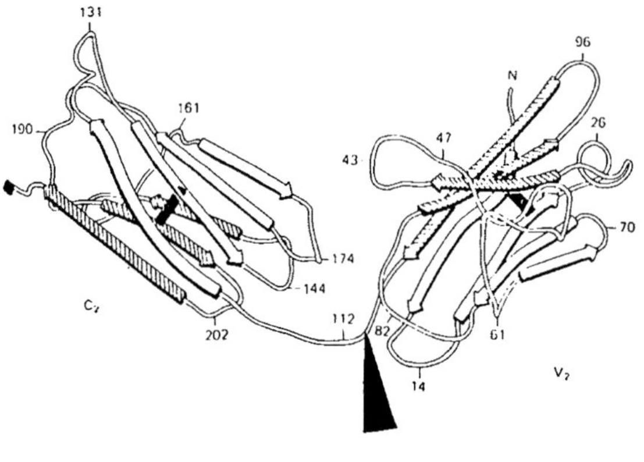

FIGURE 1. Schematic course of the backbone of an immunoglobulin light chain (28).

The initial nineteen amino acids in the coding seqence for the immunoglobulin light chain are a hydrophobic leader sequence. The first sixteen lie on a separate exon. One might wonder why only sixteen amino acids lie separated. Is this a violation of the idea that a functional unit must be carried on the exon? Israel Schechter and his coworkers (29) have shown that these first sixteen amino acids carry a complete hydrophobic signal and the cutting signal, the cutting occurring three amino acids over from the last glycine, irrespective of the intervening amino acids. Thus the region coded on this first exon will serve as a signal sequence if moved to other proteins.

Very recent work in the laboratories of Tonegawa (11) and Lee Hood (12) has demonstrated that the separation of domains by introns is a general property of immunoglobulin heavy chains. Introns separate the three constant region domains CH1, CH2, and CH3 in the cases of the gamma 1 and the alpha heavy chains. Tonegawa's laboratory has gone further and shown by DNA seqencing that a separate exon encodes the hinge region, a sequence of fifteen amino acids that lies between CH1 and CH2 to provide a flexible point in the molecule and to carry those cysteines that form the disulfide connections between the two heavy chains. Thus the immunoglobulins provide a complete example in which the exons code domains and functional elements. In these cases, when we examine the DNA, we find written upon it, in an exon-intron code, an insight into the functional and physical structure of the protein, which we could otherwise only obtain from biochemical and crystallographic analysis.

There are a number of proteins now being studied which have clearly repeating structures, such as serum albumin or ovomucoid; one awaits the correlation of the exons with those structures.

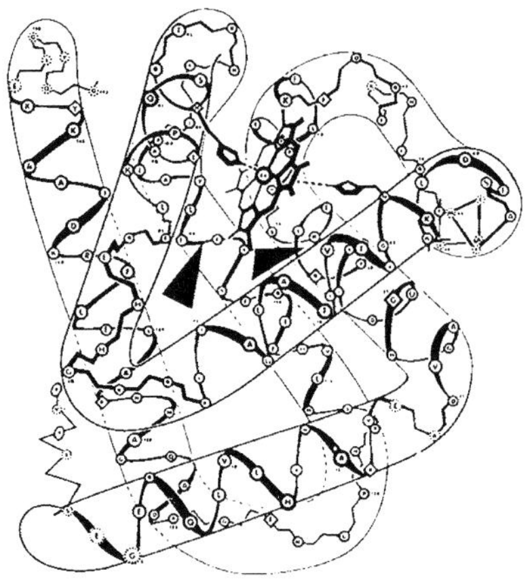

FIGURE 2. Schematic outline (30) of the structure of myoglobin, showing the positions of the introns.

What about the globins? These proteins have a simpler structure; there are no obvious domains to be represented by the three exons into which the genes for the alpha, the beta and the other globins are separated (3,4,5). However, figure 2 shows that the introns lie in alpha-helical runs in the globin molecule. This suggests that the "functional elements," those elements that might evolve separately, are the corners, the turns in the protein structure, exposed to the medium. The two flanking exons code for elements that wrap the product of the central exon, which carries the contacts that determine the heme binding site: the two histidines that coordinate to the iron and a total of 18 out of the 21 amino acids that contact the heme in the beta chain, 15 out of 19 in the alpha chain of globin (31). Figure 3 dissects the molecule to show a view of the central section and the two wings. One might conjecture that the mini-globin encoded in the central exon might provide a sufficient structure to bind the heme in a functional way, it may be a remnant of an early heme carrier.

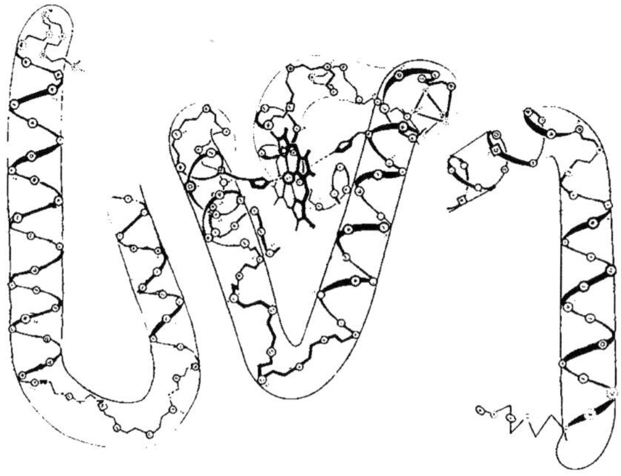

FIGURE 3. The products of the central and flanking exons of globin.

As we will report elsewhere in this meeting, we recently examined the two genes for insulin in the rat (16). Insulin is a small two-chain peptide hormone, whose two chains are connected together by an intermediate peptide in the precursor molecule, proinsulin. The purpose of the intermediate peptide (the C peptide) is simply to hold the chains together so that the disulfide bonds can form easily; at the last stage of maturation, a tryptic-like activity cleaves off the C peptide from proinsulin. Proinsulin itself is also the result of a cleavage; the direct gene product is preproinsulin, a molecule that carries an amino-terminal hydrophobic signal sequence, cleaved as the peptide chain passes through the membrane. The rat has two non-allelic genes for preproinsulin. One of these genes has two introns, one in the C peptide and the other in the untranslated region immediately before the gene. Although we do not see any element that distinguishes the hydrophobic presequence, the intron in the C peptide may divide the gene into portions that indicate the way in which two ancestral units came together. Figure 4 shows the position of the intron on a schematic structure for insulin (32). The inslin molecule consists of two chains lying on top of each other (cross-linked by the disulfides). The C peptide probably lies as a flexible loop down one side of the molecule, exposed to enzymatic attack at its ends. Thus, the intron in the C peptide divides the molecule into its two essential halves.

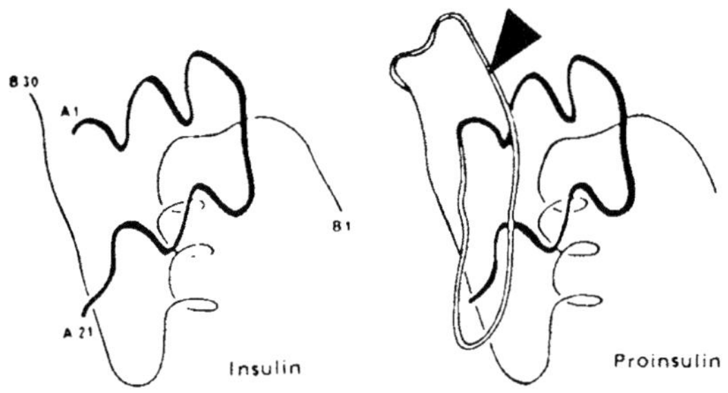

FIGURE 4. The position of the intron in proinsulin.

The second gene for insulin in the rat does not contain the intron in the coding sequence. It is entirely missing; the DNA sequence there running colinearily with the protein sequence. Both genes, however, share the first intron in the 5' untranslated region. What does this tell us? First, since the two genes in the rat function at similar rates, the intron that divides the coding region can not be essential, in cis, for the functioning of one of the genes. Second, this pair of genes demonstrates that in some sense introns come and go. Either the intron in gene II has been added after the gene was duplicated in the rodent family some 40 million years ago, or a precise deletion removed the intron from gene I in the same time span. By examining the insulin genes in other species, we should soon be able to decide which of these alternatives is correct, since these genes will have evolved from, and thus will reflect the structure of, a common ancestor. One hypothesizes that introns will be found to be lost occasionally, as they may have been between the hydrophobic region and the coding sequence of the insulins, and the gene with many introns will be the older form.

How likely is the loss of an intron? If there is no selection pressure, i.e. if the extra DNA in an intron is not an observable hindrance to growth, then an intron should be lost only through neutral drift. For neutral point mutations the rate of fixations is about 1% in 1 to 2 million years (33). Thus a specific base will mutate in 100 to 200 million years. If we examine the introns in globin (34). small additions and deletions occur at about one sixth the rate of base changes: an addition or a deletion at a specific nucleotide every 10^9 years. The frequency of a deletion covering a specific base should be the sum of all possible deletions covering (or ending at) that base, and so the frequency of one deletion, of specified endpoints, would be perhaps 100 to 1000 fold less than that of all deletions. Thus the neutral loss of an intron, by exact deletion, would occur over times of the order of 10^{11} to 10^{12} years, a rare event. Introns should breath in size, as a result of additions and deletions, but not easily vanish.

One might wonder how the splicing enzyme arose. Certainly there would be an evolutionary force for the co-evolution of the splicing enzyme and a spliced gene that provides an essential function. If at the moment that the gene that will make the splicing enzyme first mutates to that function there is in the cell a second gene, spliced by the first, that makes a product selected by evolution, then the two will be selected together. As soon as other genes mutate to produce products formed using the same splicing mechanism, the splicing enzyme would become so essential for the cell that it could not be lost. However, there is an even more general model that endows a strong evolutionary advantage on the splicing mechanism. Suppose a higher cell, in striving to make a number of proteins, has difficulty creating promotors, ribosome sites and all of the paraphernalia of exact protein synthesis. It might transcribe its DNA into long messages, but only manage to read out some limited regions of these messages into proteins. Suddenly a new ability emerges, a weak splicing activity that makes a number of connections and splices within the long transcripts. This provides a combinatorial mixture of expressed regions. The many newly spliced messengers enable the cell in a single mutational step to explore a large number of possibilities.

The ideas discussed here present the gene not as a coding structure interrupted by introns but as an intron background into which the exons have been either inserted by recombination or called forth by mutation. Like bubbles, the exons float on the sea of introns.

REFERENCES

1. Berget, S. M., Moore, C., and Sharp, P.A. (1977). Proc. Nat. Acad. Sci., U.S.A., 74, 3171.
2. Chow, L. I., Gelinas, R. E., Broker, T. R., and Roberts, R. J. (1977). Cell 12, 1.
3. Tilghman, S. M., Tiemeier, D. C., Seidman, J. G., Peterlin, B. M., Sullivan, M., Maizel, J. V., and Leder, P. (1978). Proc. Nat. Acad. Sci., U.S.A. 75, 725.
4. Jeffreys, A. J., and Flavell, R. A. (1977). Cell 12, 1097.
5. Leder, A., Miller, H. I., Hamer, D. H., Seidman, J. G., Norman, B., Sullivan, M., and Leder, P. (1978). Proc. Nat. Acad. Sci., U.S.A. 75, 6187.
6. Tonegawa, S., Maxam, A. M., Tizard, R., Bernard, O., and Gilbert, W. (1978). Proc. Nat. Acad. Sci., U.S.A. 74, 3518.
7. Brack, C., and Tonegawa, S. (1977). Proc. Nat. Acad. Sci., U.S.A. 74, 5652.
8. Breathnach, R., Mandel, J. L., and Chambon, P. (1977). Nature 270, 314.
9. Dugaiczyk, A., Woo, S. L. C., Lai, C. C., Mase, M. L., McReynolds, L., and O'Malley, B. W. (1978). Nature 274, 328.
10. Nguyen-Huu, M. C., Strathmann, M., Groner, B., Wurtz, T., Land, H., Giesecke, K., Sippel, A. E., and Schuetz, G. (1979). Proc. Nat. Acad. Sci., U.S.A. 76, 76.
11. Sakano, H., Rogers, J. H., Hueppi, K., Brack, C., Traunecker, A., Maki, R., Wall, R. and Tonegawa, S. (1979). Nature 277, 627.
12. Early, P. W., Davis, M. M., Kaback, D. B., Davidson, N., and Hood, L. (1979). Proc. Nat. Acad. Sci., U.S.A. 76, 857.
13. Slonimski, P. P., Claisse, M. L., Foucher, M., Jacq, C., Kochko, A., Lamouroux, A., Pajot, P., Perrodin, G., Spyridakis, A., and Wambier-Kluppel, M. L. (1978). In "Biochemistry and Genetics of Yeast" (M. Bacila, B. L. Horecker, and O. M. Steffani, eds.), pp. 339-401. Academic Press, New York.
14. Goodman, H. M., Olson, M. V., and Hall, B. D. (1977). Proc. Nat. Acad. Sci., U.S.A. 74, 5433.
15. Valenzuela, P., Vanegas, A., Weinberg, F., Bishop, R., and Rutter, W. J. (1978). Proc. Nat. Acad. Sci., U.S.A. 75, 190
16. Efstratiadis, A., Lomedico, P., Rosenthal, N., Kolodner, R., Tizard, R., Perler, F., Villa-Komaroff, L., Naber, S., Chick, W., Broome, S., and Gilbert, W. (1979). This

volume.
17. Schaffner, W., Kunz, G., Daetwyler, H., Telford, J., Smith, H. O., and Birnsteil, M. L. (1978). Cell 14, 655.
18. Gilbert, W. (1978). Nature 271, 501.
19. Breathnach, R., Benoist, C., O'Hare, K., Gannon, F., and Chambon, P. (1978). Proc. Nat. Acad. Sci., U.S.A. 75, 4853-4857.
20. Crick, F. (1979). Science 204, 264.
21. Calos, M. P., Johnsrud, L., and Miller, J. H. (1978). Cell 13, 411.
22. Grindley, N. D. F. (1978). Cell 13, 419.
23. Rosenberg, M., Court, D., Shimatake, H., Brady, C., and Wulff, D. L. (1978). Nature 272, 414.
24. Bos, J. L., Heyting, C., Borst, P., Arnberg, A. C., and von Bruggen, E. F. (178). Nature 275, 336-338.
25. Faye, G., Dennebouy, N., Kujawa, C., and Jacq, C. (1979). Molec. Gen. Genetics 168, 101-109.
26. Dujon, B., and Morimoto, R., private communication.
27. Miller, H. I., Konkel, D. A., and Leder, P. (1978). Nature 275, 772.
28. Schiffer, M., Girling, R. L., Ely, K. R., and Edmundson, A. B. (1973). Biochemistry 12, 4620.
29. Schechter, I., Wolf, O., Zemell, R., and Burstein, V. (1979). Fed. Proc. 38, 1839.
30. Dickerson, R. E. (1964). In "The Proteins, Vol. II" (H. Neurath, ed.), p. 634. Academic Press, New York.
31. Perutz, M. F., Muirhead, H., Cox, J. M., and Goaman, L. C. G. (1968). Nature 219, 131.
32. Blundell, T. L., Bedarkar, E., Rinderknecht, E., and Humbel, R. E. (1978). Proc. Nat. Acad. Sci., U.S.A. 75, 180-184.
33. Brown, W. M., George, M., and Wilson, A. C. (1979). Proc. Nat. Acad. Sci., U.S.A. 76, 1967.
34. van den Berg, J., van Ooyen, A., Mantei, N., Schamboeck, A., Grosveld, G., Flavell, K. A., and Weissmann, C. (1978). Nature 275, 37.

Cell, Vol. 18, 545-558, October 1979, Copyright © 1979 by MIT

The Structure and Evolution of the Two Nonallelic Rat Preproinsulin Genes

Peter Lomedico,* Nadia Rosenthal,† Argiris Efstratiadis,† Walter Gilbert,* Richard Kolodner‡ and Richard Tizard*
*The Biological Laboratories
Harvard University
Cambridge, Massachusetts 02138
†Department of Biological Chemistry
Harvard Medical School
Boston, Massachusetts 02115
‡Sidney Farber Cancer Institute
Boston, Massachusetts 02115

Summary

In the rat, there are two nonallelic genes for preproinsulin. The insulin end products are very similar and are equally expressed. We have isolated clones carrying these genes and their flanking sequences, and characterized them by DNA sequencing and electron microscopic analysis. We have established the primary structure of the preproinsulin mRNAs and the signal peptides of these two proteins. One of the genes contains two introns: a 499 bp intron interrupting the region encoding the connecting peptide and a 119 bp intron interrupting the segment encoding the 5′ noncoding region of the mRNA. The introns are transcribed and present in a preproinsulin mRNA precursor. The other gene possesses the smaller, but not the larger, of the two introns. Calculations based on the divergence of the two preproinsulin nucleotide and amino acid sequences indicate that these genes are the products of a recent duplication. Thus one of the genes gained or lost an intron since that time.

Introduction

Mature insulin, consisting of two polypeptide chains, B and A, connected by two disulfide bridges, is the end product of the maturation of preproinsulin, a single precursor polypeptide chain of 110 amino acids (Figure 1). Preproinsulin (Lomedico and Saunders, 1976; Chan, Keim and Steiner, 1976; Chan and Steiner, 1977) is synthesized, and sequentially processed to proinsulin and insulin, in the β cells of pancreatic islets. The processing events define the following regions of the molecule: NH_2-preregion-B chain-2 basic amino acids-C peptide-2 basic amino acids-A chain. The first step in the processing is the elimination of the first 24 amino acids (preregion), which serve as a hydrophobic signal sequence for the transfer of the nascent chain through the microsomal membranes of the rough endoplasmic reticulum (Milstein et al., 1972; Blobel and Dobberstein, 1975). The product, proinsulin, folds up, the disulfide bridges form, and proteolytic cleavages release the C peptide and the basic amino acids. Although insulin-like peptides have been detected in several invertebrate species (for a review, see Falkmer and Ostberg, 1977), only vertebrate insulins (and in some cases proinsulins) have been purified to homogeneity and sequenced (Humbel, Bosshard and Zahn, 1972; Steiner et al., 1974; Dayhoff, 1978). In the rat (and also in the mouse and in three fish species) there are two insulins, in contrast to other organisms. These insulins, I and II, are the end products of nonallelic genes, because they both occur in all individual pancreata (Smith, 1966) of randomly bred rats and are synthesized in almost equal proportions (52%:48%, I:II) (Clark and Steiner, 1969). The two rat proinsulins differ from each other by two amino acids in the B chain and by two amino acids in the C peptide (Figure 1). A partial amino acid sequence of the preregion of preproinsulin I has been determined from cloned double-stranded cDNA (ds-cDNA) (Ullrich et al., 1977; Villa-Komaroff et al., 1978) and from the immunoprecipitable products of a cell-free translation directed by a crude RNA preparation from pancreatic islets (Chan et al., 1976). This paper describes the structure of both rat preproinsulin genes, isolated from a rat chromosomal DNA library by using as probe cloned preproinsulin ds-cDNA.

Results

Characterization of Preproinsulin ds-cDNA Clones

To initiate these studies, we made a collection of preproinsulin ds-cDNA clones which provided the pure probes necessary for the screening of a rat chromosomal DNA library. Moreover, a knowledge of the mRNA sequences was essential to characterize the chromosomal gene structure, especially since the coding regions of eucaryotic structural genes are often interrupted by introns (for a review, see Dawid and Wahli, 1979). As we described elsewhere (Villa-Komaroff et al., 1978), we derived these cDNA clones from the poly(A)-containing cytoplasmic RNA of a transplantable rat insulinoma (Chick et al., 1977). We have characterized five of these clones, two of which, pI19 and pI47, have been previously described (Villa-Komaroff et al., 1978).

DNA sequencing of the insert of clone pI19 revealed that the cloned DNA corresponds to preproinsulin I and encodes the entire preproinsulin chain with the exception of the first three amino acid residues of the preregion (Figure 2). It is therefore missing a copy of the entire 5′ noncoding region of the mRNA. It is also missing the 3′ noncoding region, with the exception of 10 nucleotides immediately following the TGA terminator.

The restriction maps of three other clones (pI41, pI7 and pI20) showed differences from that of pI19 (Figure 2), and DNA sequencing showed that they correspond to preproinsulin II. A characteristic feature

Cell
546

of the preproinsulin II ds-cDNA map is the presence of a Sma site, which was useful for the discrimination of the two genes in blotting experiments involving chromosomal DNA.

The inserts of clones pI41 and pI7 were completely sequenced. The cloned DNA of pI41 extends in the 3′ direction up to the second position of the 57th codon of preproinsulin and contains twenty 5′ noncoding nucleotides upstream from the ATG initiator. The insert of pI7 does not overlap with that of pI41, but begins with the 60th codon of preproinsulin and ends at the nucleotide immediately following the TAG terminator. Partial restriction mapping and DNA sequencing showed that the insert of clone pI20 covers the sequence of both clones pI47 and pI7, although we have not defined its exact 3′ end point.

When we compared the sequences of the ds-cDNA clones, we observed that single base errors had been introduced somewhere in the cloning process. There are four G to A substitutions, two in pI41 as compared with pI20, at the first position of codon 38 and the second position of codon 47 of preproinsulin, one substitution in pI7 at the first position of codon 69, and one substitution in pI19 as compared with pI47 at the first position of codon 37. All these changes would have corresponded to amino acid replacements not found in the sequenced proteins. These substitutions were most probably copying errors of the AMV reverse transcriptase (incorporating T for C while transcribing an oligopurine tract), although they are much more frequent than expected. This enzyme, utilizing a natural DNA template, incorporates one incorrect nucleotide per 900 (Gopinathan et al., 1979). Transversions are 30 fold lower than transitions (L. Loeb, personal communication). Error frequencies with mRNA templates have not, however, been reported.

Mapping of the Rat Chromosomal Preproinsulin Genes

The sizes of restriction fragments of chromosomal DNA that bear the preproinsulin genes were displayed by the DNA blotting technique (Southern, 1975). Figure 3 shows an example of the pattern of fragments obtained by digestion of NEDH rat liver DNA with several different enzymes, hybridizing to Pst-excised insert of clone pI19, ^{32}P-labeled by nick translation. In all cases, the hybridization patterns using DNA from rat kidney, spleen or the rat insulinoma were identical to those obtained from rat liver DNA. The restriction enzymes Bam, Pst, Bgl II and Hind II+III each produced two hybridizing bands, as expected. Eco RI, however, produced three bands of 7.0, 3.7 and 0.8 kilobases (kb). Further digests with the enzymes in combination with Sma (which cleaves specifically gene II) identified fragments coming from the preproinsulin II gene, among which was the 3.7 kb Eco RI

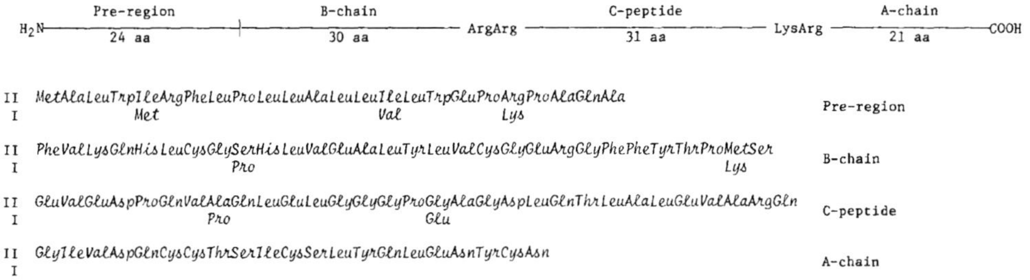

Figure 1. Primary Structure of Rat Preproinsulins I and II

The data for the proinsulin sequences are from Humbel et al. (1972) (A and B chains) and Steiner et al. (1974) (C peptide). The amino acid sequences of the preregions were deduced from our DNA sequencing data.

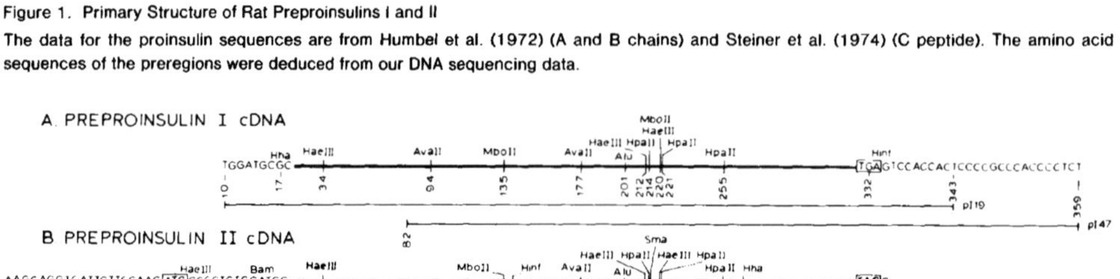

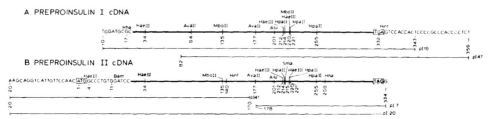

Figure 2. Restriction Maps of Preproinsulin I and II cDNA Clones

Each restriction endonuclease cleavage site is identified by a number indicating the last nucleotide of the message strand before the cut. Nucleotides are numbered beginning with the first nucleotide of the ATG initiator. Nucleotides in the 5′ direction from position one are identified by negative numbers, beginning with −1.

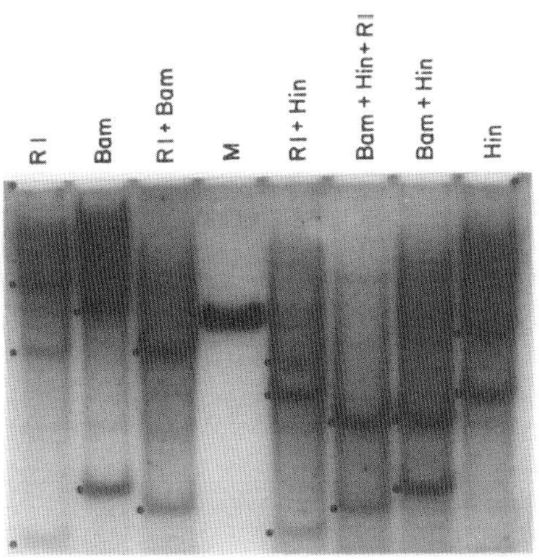

Figure 3. Detection of Preproinsulin Sequences in Rat Chromosomal DNA

High molecular weight rat liver DNA (30 μg per slot) was digested with restriction endonucleases, as indicated, and electrophoresed on a 1% agarose gel. It was then transferred to nitrocellulose and hybridized with ^{32}P-labeled DNA probe (nick-translated, Pst-excised insert of pI19). The autoradiogram of the filter is shown. The sizes of the hybridizing bands, indicated by dots, are listed in the legend to Figure 4. Lane M contained 0.1 ng of linearized pI19 DNA (hybridizing band) and molecular weight markers that were visualized by ethidium bromide staining.

fragment. Other double and triple digests allowed us to construct a map of the two genes and their flanking sequences (Figure 4), and thus to conclude that the 7.0 kb Eco RI fragment corresponds to gene I, while the 3.7 and 0.8 kb Eco RI fragments are derived from gene II. Since there is no Eco RI site in the coding region of the preproinsulin II cDNA clone pI20, the intragenic Eco RI site must lie within an intron interrupting this gene.

Isolation of Cloned Rat Chromosomal Preproinsulin Genes

We screened a rat chromosomal DNA library constructed by Sargent et al. (1979), who ligated partial Eco RI fragments of DNA isolated from the liver of a single Sprague-Dawley rat (Simonsen Labs, Gilroy, California) to the purified Eco RI arms of phage λ Charon 4A (Blattner et al., 1977). Using the ^{32}P-labeled pI19 probe, we screened (Benton and Davis, 1977) 800,000 recombinant plaques and identified three hybridizing clones.

One of these clones (λ Charon 4A-rI1 or simply rI1) contains a 21 kb insert composed of five Eco RI fragments: 7.0, 6.2, 3.5, 2.6 and 1.4 kb (Figure 5A). Hybridization with pI19 probe (Figure 5B) demonstrated that the 7.0 kb fragment carries preproinsulin sequences, and hence this phage carries the preproinsulin I gene.

The insert of the second clone (rI2) is 16.5 kb long and composed of four Eco RI fragments: 9.3, 3.7, 2.7 and 0.8 kb (Figure 6A). Hybridization (Figure 6B) demonstrated that both the 3.7 and 0.8 kb Eco RI fragments carry preproinsulin sequences, and therefore this phage carries the preproinsulin II gene.

The insert of the last phage (rI22) contains the same four Eco RI fragments seen in rI2, but in the opposite orientation with respect to the phage arms as shown by analysis with other restriction enzymes. Thus rI2 and rI22 must be independent transformants, and the 16.5 kb DNA fragment most probably corresponds to a single segment of the rat genome and is not an artifact of ligation. We note that there are no overlapping Eco RI fragments in the digests of rI1 and rI2, and hence we have no information about the linkage of these genes.

Analysis of the Cloned Preproinsulin II Gene

We constructed a detailed restriction map of clone rI2 (Figure 7) and sequenced the region containing the gene (Figure 8). The DNA sequence confirms the conclusion that this is the preproinsulin II gene and demonstrates the existence of an intron of 499 bp interrupting the coding region between the codons of amino acids 62 and 63 of preproinsulin (that is, between the codons of the sixth and seventh amino acids of the C peptide). Furthermore, there is a second intron in the region corresponding to the 5′ noncoding region of the mRNA. This was first indicated by the fact that the last three 5′ terminal nucleotides (AAG) known from the sequence of clones pI41 and pI20 (Figure 2) were not contiguous with the rest in chromosomal DNA. To establish more of the sequence of the 5′ noncoding region of the preproinsulin mRNA II and to position it on chromosomal DNA, we extended the sequence from clone pI41 toward the 5′ direction as follows. DNA of clone pI41 was digested with Bam and end-labeled. It was then digested with Hae III to yield a small DNA fragment (with 11 nucleotides in its 5′-labeled antimessage strand, roughly corresponding to amino acids 1–4 of the preregion.) The antimessage strand of this fragment, after denaturation, served as primer in a reverse transcription reaction using as template oligo(dT)-cellulose bound RNA from the rat insulinoma. The DNA product was sequenced directly, and 56 nucleotides upstream from the initiation codon ending with a 5′ G were determined. This result was reproduced in two independent experiments. We positioned the 5′ end of the mRNA with an uncertainty of 1 nucleotide by reasoning as follows. The uncleaved transcript, electrophoresed in parallel with the cleavage products of the sequencing procedure, migrated as a single band. It is therefore improbable that it

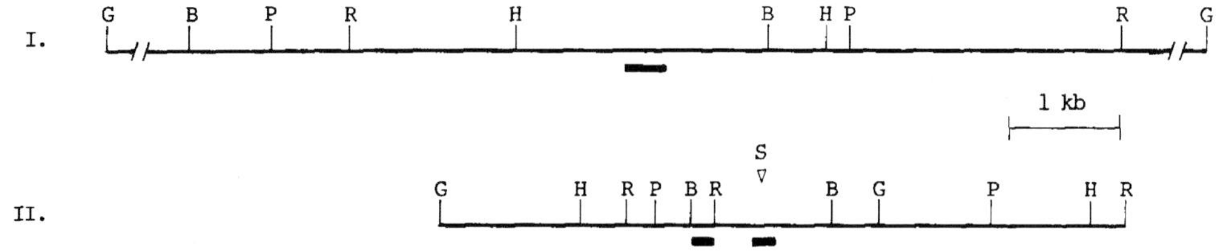

Figure 4. Restriction Maps of Rat Chromosomal DNA Fragments Containing Preproinsulin I and II Gene Sequences

The maps were constructed from the fragment lengths (in kb), indicated below, that were deduced from all our data: the autoradiograms of DNA blotting experiments, restriction analysis of cloned DNA and DNA sequencing.

	Gene I	Gene II
(R) Eco RI	7.0	3.7[a] + 0.8
(B) Bam	5.3	1.25[a]
(P) Pst	5.2	3.0[a]
(G) Bgl II	10.0	4.0[a]
(H) Hind II + III (Hin)	2.8	4.6
Eco RI + Bam	3.8	1.06 + 0.19[b]
Eco RI + Pst	4.5	2.5 + (0.6)[c]
Eco RI + Bgl II	7.0	1.5 + 0.8
Eco RI + Hin	2.8	3.4 + 0.8
Bam + Hin	2.3	1.25
Pst + Bgl II	5.2	2.0
Eco RI + Bam + Hin	2.3	1.06 + 0.19
Eco RI + Pst + Bgl II	4.5	1.5 + (0.6)

[a] These fragments are cleaved by Sma S and therefore belong to the gene II sequence. The rest of the fragments in this column were assigned to gene II because of a unique fit when the data were considered all together.
[b] This fragment was detected in an autoradiogram following transfer of DNA from a 5% polyacrylamide gel.
[c] This fragment was not detected in experiments involving chromosomal DNA, possibly because of poor binding to the filter.

The hybridization data locate gene I to the 2.3 kb Hin-Bam fragment, and gene II to the 1.06 and 0.19 kb Bam-Eco RI fragments; however, we show their positions (filled boxes) as determined by our final data.

represents a partial cDNA copy of this mRNA region. Even if a strong stopping point were present in this area because of extensive secondary structure, a minority of longer transcripts should have been observed. However, inspection of the chromosomal DNA sequence flanking the last readable nucleotide of the cDNA reveals that the region lacks secondary structure at the mRNA level.

The last nucleotide determined (G from its complementary C in the cDNA) is not necessarily the capping site (that is, the nucleotide corresponding to the one following the cap structure in the mature mRNA) because it is difficult, in general, to obtain information about the last nucleotide of the cDNA by this method (Baralle, 1977; Baralle and Brownlee, 1978). Unfortunately, we do not have preproinsulin mRNA in sequencible amounts and could not complement this information with RNA sequencing data. Thus we tentatively assign the nucleotide to the 5′ side of the last G (an A in chromosomal DNA) as the capping site. As is discussed below, the position of this site is at the predicted distance from a "Hogness box" identified in the 5′ flanking sequence.

The 3′ noncoding region is positioned by comparison to the sequence of that region reported by Ullrich et al. (1977) from preproinsulin I cDNA clones. We assume that preproinsulin mRNA II has the poly(A) added at the same point.

Analysis of the Cloned Preproinsulin I Gene

The DNA sequence of rI1 is not interrupted by an intron in the coding region. The sequence of the region upstream from the ATG initiator is very similar to that of rI2 (Figure 8). The similarity of the 5′ noncoding regions of the mRNAs as judged by the primer extension experiment demonstrates that the insulin I gene also possesses a 119 base pair (bp) intron in this region. The homology of the two 119 bp introns is greatest around the intron-exon junctions, while the body of the intron is less well conserved, as has been observed in other systems (for example, see van den Berg at al., 1978).

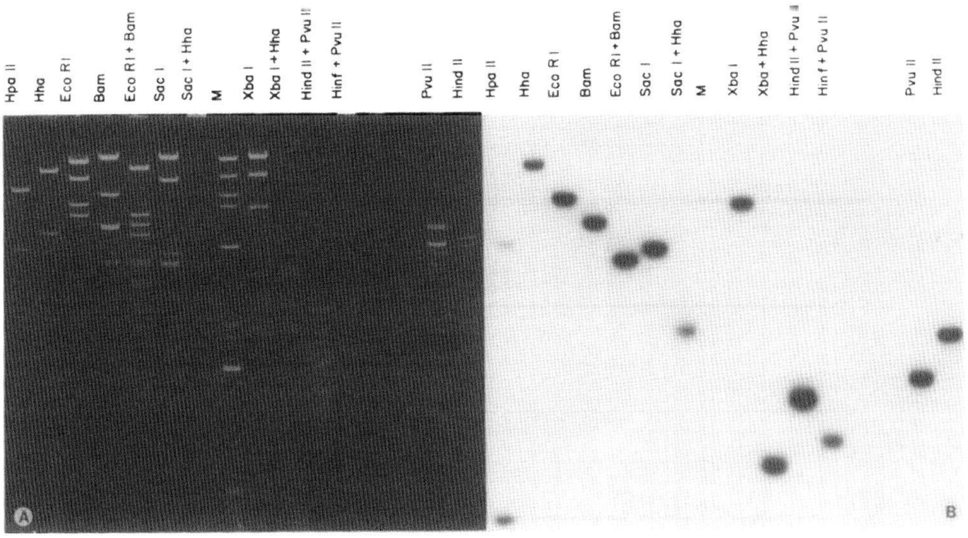

Figure 5. Restriction Endonuclease Analysis of Clone rI1 DNA

(A) DNA (0.5 μg per slot) was digested with various restriction endonucleases, and the products were electrophoresed on a 1% agarose gel and visualized by ethidium bromide staining. The lengths of molecular weight markers (in kb), lane M, are from top to bottom: λ Charon 4A × Eco RI, 19.8, 10.9, 8.0 and 7.0; pBR322 × Eco RI, 4.3; pBR322 × Eco RI × Pvu II, 2.3 and 2.1; pBR322 × Hinf, 1.6, 0.5 and smaller bands.

(B) Autoradiogram of a nitrocellulose filter carrying the transferred DNA bands from the gel in (A) after hybridization with the probe described in Figure 3. The sizes of the hybridizing bands are indicated in Figure 7.

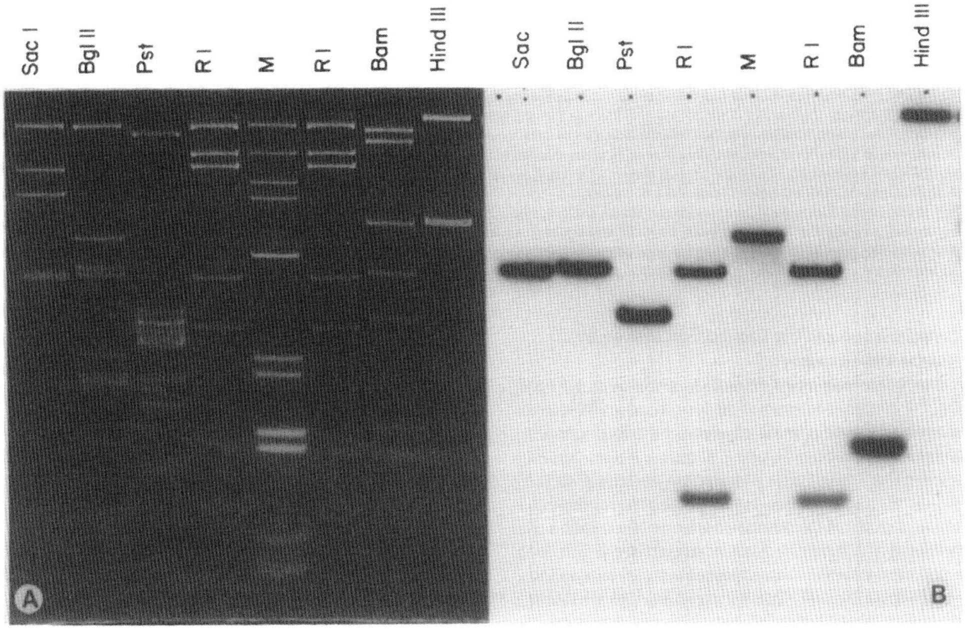

Figure 6. Restriction Endonuclease Analysis of Clone rI2 DNA

(A) and (B) are as in Figure 5. The bands of molecular weight markers (in kb), lane M, are from top to bottom: λ Charon 4A × Eco RI, 19.8, 10.9, 8.0 and 7.0; pI19 × EcoRI, 4.7; pBR322 × Eco RI, 4.3; pBR322 × EcoRI × Pvu II, 2.3 and 2.1; pBR322 × Taq, 1.4, 1.3, 0.6 and smaller bands.

Cell
550

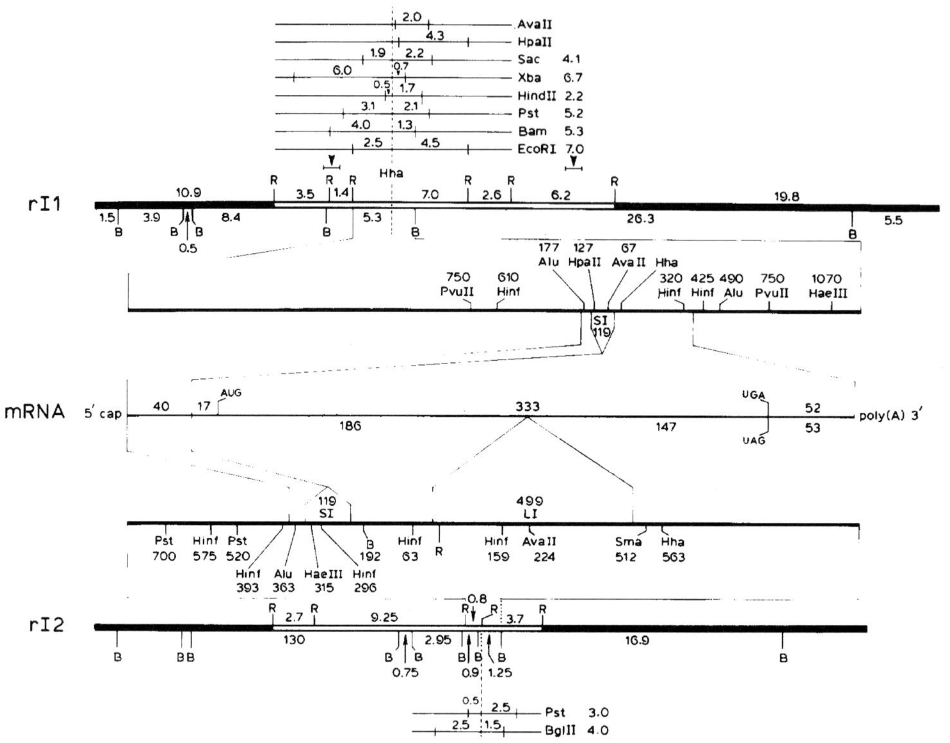

Figure 7. Detailed Restriction Maps of Clones rI1 and rI2

(rI1) The Eco RI (R) and Bam (B) sites and the distances (in kb) between these sites in the recombinant phage DNA are indicated. The short (right) and long (left) arms of the λ vector (filled boxes) are drawn in the opposite orientation so that the 5′ to 3′ direction of the message strand of the insert can be positioned from left to right. The lines indicated by thick arrows correspond to the positions of an inverted repeat in rI1 (Figure 10). The distances (in kb) of other restriction sites from an Hha site that lies in the 7.0 kb Eco RI fragment are also indicated (top). A blowup of the 3.8 kb Eco RI-Bam fragment containing the gene is shown. The distances (in bp) of various restriction sites from the same Hha site, as above, are indicated. Sites that can be found in Figure 2 are not shown.
(rI2) The Eco RI and Bam sites in the recombinant phage are indicated. Other sites in the insert or in a blowup of the region containing the gene are indicated using the Eco RI site that lies in the 1.25 kb Bam fragment as a reference point.
(mRNA) Correlation between the chromosomal gene sequences and the sequences present in mature mRNA. The positions of the small intron (SI) in rI1 and rI2 and the large intron (LI) in rI2 are indicated.

Characterization of the Cloned Sequences by Electron Microscopy

To examine sequence homologies extending further than the region corresponding to mature mRNA, we examined heteroduplexes between the two genes by electron microscopy. Figure 9 shows a heteroduplex between the 7 kb rI1 Eco RI fragment and the 4 kb rI2 Bgl II fragment. Measurements on ten molecules showed that this heteroduplex contains two regions of homology, 0.9 and 0.2 kb, separated by a 0.5 kb "insertion-deletion" loop presumably corresponding to the large intron of rI2. By comparing this structure with the restriction map (Figure 7), we were able to orient the 5′ and 3′ ends of the gene in the heteroduplex. Figure 7 shows that the large intron of rI2 dividing the 4 kb Bgl II fragments produces a short segment to the 3′ side which in the heteroduplex contains the 0.2 kb region of homology. The length of this homology seen in the electron micrograph is in excellent agreement with the sequencing data (Figure 8), since the two sequences diverge immediately after the end of the 3′ noncoding regions. From this analysis, we conclude that the 0.9 kb region of homology lies upstream from the intronic Eco RI site. The distance between the capping site and the left-hand boundary of the large intron is 365 bp. The homology of the 5′ flanking sequences of the two genes therefore extends for about 500 bp upstream from the capping site.

When we examined intact rI1 DNA, we observed

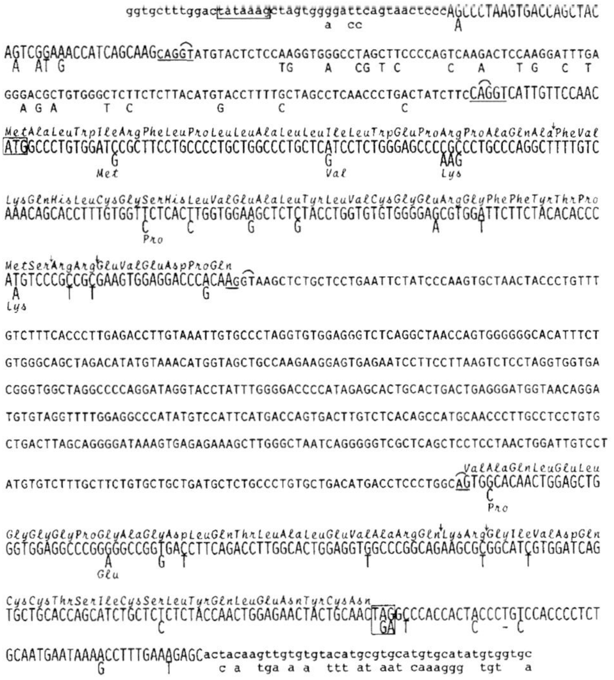

Figure 8. Nucleotide Sequences of the Rat Preproinsulin Genes

The nucleotide sequence of the mRNA strand of the preproinsulin II gene is displayed from the 5′ to the 3′ direction. Nucleotide substitutions in gene I sequence are indicated in a second line. Large and small capital letters represent sequences corresponding to mature mRNA and the introns, respectively. The large intron is absent from the gene I sequence. Gene flanking sequences are in lower case letters. Nucleotide redundancies at the intron-exon junctions are underlined, while the GT and AG dinucleotides at the beginning and end of the introns are overlined. The "Hogness box" in the 5′ gene flanking sequence and the initiation and termination codons are boxed. The amino acid sequence is displayed on a line above the coding sequence. The arrows indicate, in order, the boundaries of the preregion and the peptides B, C and A.

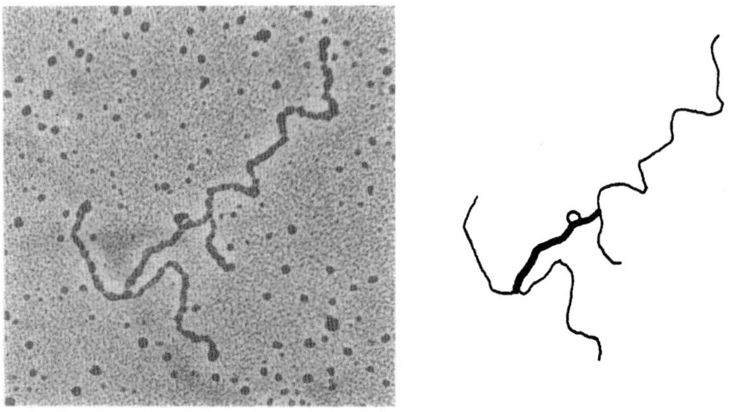

Figure 9. Electron Micrograph of a Heteroduplex Formed between the 7.0 kb rI1 Eco RI Fragment and the 4.0 kb rI2 Bgl II Fragment

The orientation is 5′ to 3′ from left to right (for details, see text). The short, upper-left and lower-right, single-stranded tails belong to the Bgl II fragment.

that it contained an inverted repeat (Figure 10A). The inverted sequences (each 1.1 kb) were separated from each other by about 12 kb of single-stranded DNA. To locate the inverted sequences with respect to the arms of the cloning vector, we analyzed heteroduplexes between rI1 and rI2 (Figure 10B) in which the homologous gene sequences were not annealed, while the inverted repeat formed a duplex stem. The inserts of the clones are in the same orientation; however, molecules in which the gene sequences were also annealed were uninterpretable due to tangling. From this analysis, we conclude that the distance from the outer boundary of each repeat to the insert-vector arm junction is 3 kb on the side of the

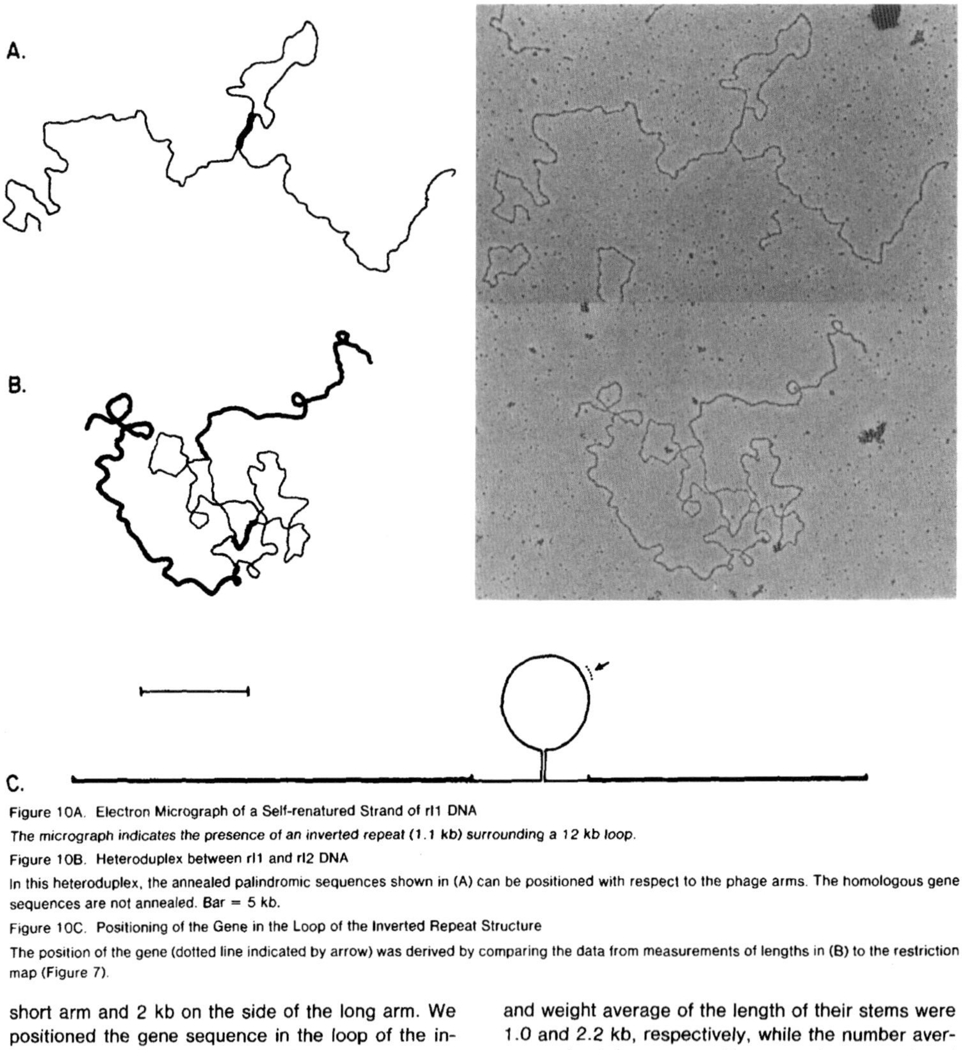

Figure 10A. Electron Micrograph of a Self-renatured Strand of rI1 DNA

The micrograph indicates the presence of an inverted repeat (1.1 kb) surrounding a 12 kb loop.

Figure 10B. Heteroduplex between rI1 and rI2 DNA

In this heteroduplex, the annealed palindromic sequences shown in (A) can be positioned with respect to the phage arms. The homologous gene sequences are not annealed. Bar = 5 kb.

Figure 10C. Positioning of the Gene in the Loop of the Inverted Repeat Structure

The position of the gene (dotted line indicated by arrow) was derived by comparing the data from measurements of lengths in (B) to the restriction map (Figure 7).

short arm and 2 kb on the side of the long arm. We positioned the gene sequence in the loop of the inverted repeat structure (Figure 10C) by comparing the electron microscopic data to the restriction map (Figure 7), and confirmed this location by measuring the distance between the region of homology of the two genes and the short arm of the vector in a heteroduplex in which the rI2 strand was intact, while the rI1 strand was broken in the nonhomologous region close to the 3′ end. Thus the palindromic sequence to the long arm side was eliminated and the gene regions were annealed.

In an electron microscopic study of the inverted repeats of rat chromosomal DNA (Szala et al., 1977), two kinds of structures were observed: hairpins and stem-loop structures. Only 10% of the examined molecules were of the latter type. The number average and weight average of the length of their stems were 1.0 and 2.2 kb, respectively, while the number average and weight average of loop lengths were 1.2 and 5 kb, respectively. Only one in 26 molecules had a 12 kb loop. Thus although the length of the palindromic sequence in rI1 is typical of rat foldback DNA, the stem and loop structure is rare.

R loop formation between rI2 DNA and nuclear RNA from the rat insulinoma, containing mature-sized mRNA molecules (as judged by RNA blotting), showed two types of structures (Figure 11). Most of the R loops observed (we examined 30 molecules) were typical of mature mRNA: one DNA strand was displaced, while the middle of the hybridized strand (about 0.5 kb) was looped out, presumably because the intronic sequence was spliced from the mRNA. Looping out at the region of the 119 bp intron was not

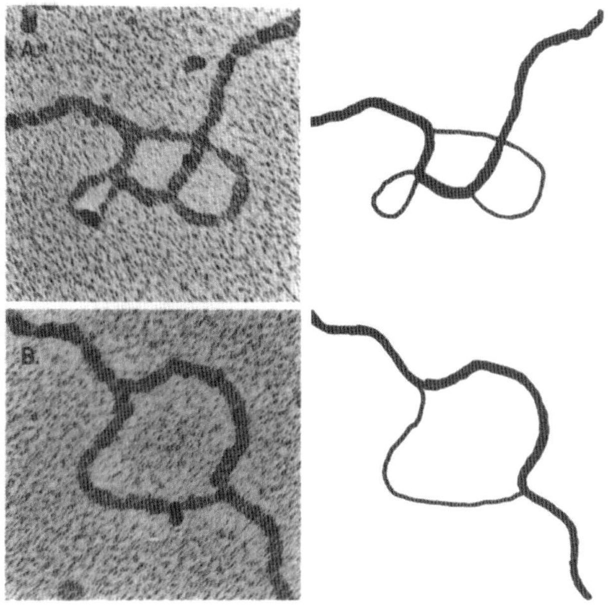

Figure 11. R Loops Formed between rI2 DNA and RNA Molecules Present in a Nuclear RNA Preparation

(A) shows an R loop with mature spliced mRNA. (B) shows an R loop with a putative mRNA precursor.

observed, probably because of the small size of this structure (see also Tilghman et al., 1978a). A few larger R loops (about 1 kb) were also present (we examined six molecules). They were smooth, without looping out of the hybridized strand. We conclude that as in the case of other eucaryotic mRNAs, globin mRNA (Tilghman et al., 1978b), ovalbumin mRNA (Roop et al., 1978) and immunoglobin mRNA (Schibler, Marcu and Perry, 1978; Gilmore-Hebert et al., 1978), preproinsulin mRNA is initially transcribed in the form of a precursor and goes through a maturation process. We cannot exclude the possibility that an even longer primary transcript might exist.

Discussion

The Structure of Preproinsulin mRNAs

The results described in this paper establish the primary structure of both mature preproinsulin mRNAs, with the exception of a tentative assignment for the 5′ terminal nucleotide. The 5′ noncoding region of both mRNAs, excluding the initiator, consists of 57 nucleotides. The 3′ noncoding region, excluding the terminator, consists of 52 nucleotides in preproinsulin I mRNA and 53 nucleotides in preproinsulin II mRNA.

The first 40 nucleotides of the 5′ noncoding region are not contiguous with the rest in chromosomal DNA. Such spliced "leader" sequences were first seen in viral mRNAs (for example, see Berget, Moore and Sharp, 1977; Chow et al., 1977). Similar "leaders" of 45 nucleotides and about 70 nucleotides have been described in ovalbumin and silk fibroin, respectively (Gannon et al., 1979; Tsujimoto and Suzuki, 1979). There is a possible stable secondary structure (a stem and loop) ($\Delta G = -9.4$ kcal) in the 5′ noncoding region of the mature preproinsulin mRNA, 9 bases before the AUG (Figure 12). The open loop can base-pair with the 3′ end of 18S rRNA (Hagenbüchle et al., 1978) forming another stable association ($\Delta G = -10.4$ kcal). The sequences involved in this base pairing are conserved in the two rat mRNAs. This interaction might play a role in ribosome binding and the efficiency of translation of preproinsulin mRNA. The splice point lies in the stem of this loop. If this structure were important, the splice being necessary to bring a ribosome binding site near the AUG or to form a recognition element for the initiation of protein synthesis, one would expect to find similar structures in other messages. Figure 12 shows that we can identify such stems and loops at the same distance before the AUG initiator in other spliced messengers (ovalbumin and fibroin mRNAs). Traces of a similar structure appear in the unspliced globin messengers.

With the exception of the large intron in rI2, the two gene sequences are very homologous. This homology terminates at the end corresponding to the 3′ end of the mRNA but extends for about 500 bp upstream from the capping site in the 5′ gene flanking region. We note that this homologous DNA segment is longer than the length of the encoded structural information.

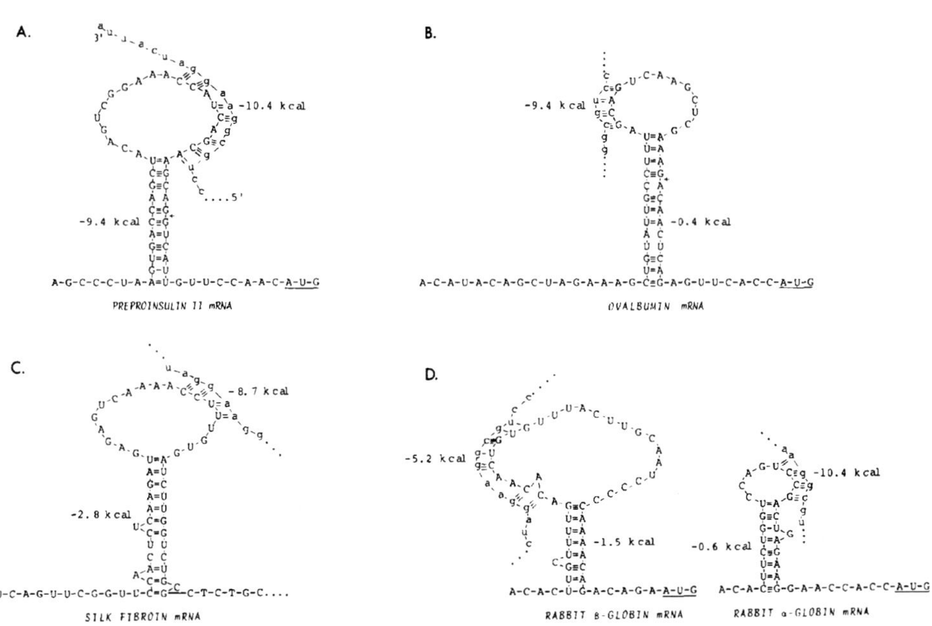

Figure 12. Possible Secondary Structure Features of (A) Preproinsulin mRNA, (B) Ovalbumin mRNA, (C) Fibroin mRNA and (D) Rabbit α- and β-Globin mRNAs

Sequences (B) and (D) are listed by Baralle and Brownlee (1978). Sequence (C) is from Tsujimoto and Suzuki (1979). The stabilities of the stems of the structures and the association with 18S rRNA were calculated by the method of Tinoco et al. (1973). The positions of the splice points in the preproinsulin and ovalbumin sequences are indicated by arrows.

A "Hogness box," TATAAAG, is present in both rI1 and rI2 sequences 23 bp upstream from the assigned capping site. Such a sequence is present in the 5′ flanking DNA of all eucaryotic structural genes sequenced thus far (Gannon et al., 1979); in general, the distance between the capping site and the seventh nucleotide of the box is 24 ± 1 nucleotides. It remains to be seen whether the long 5′ region of homology is transcribed in a longer mRNA precursor than the one identified or is conserved for some other reason. If this area is not transcribed, the "Hogness box" identified in this area might be related to initiation of transcription.

The DNA sequencing data establish the complete amino acid sequences of the preregion of both preproinsulins. Comparison of these two amino acid sequences reveals that there are three amino acid replacements. These replacements are very conservative, in the sense that in one case a basic amino acid replaces another basic residue, and the other replacements are between hydrophobic amino acids. Comparison with partial sequences of the preregions of bovine (Lomedico et al., 1977) and fish (Shields and Blobel, 1977) preproinsulins (Figure 13) shows that certain positions are invariable, especially two pairs of leucine residues in the middle of the hydrophobic core.

The Introns of Preproinsulin Genes

We have shown that the rat preproinsulin II gene contains two introns, while the preproinsulin I gene contains only one. Since both of these genes appear to function equally (Clark and Steiner, 1969), we must infer that the large intron of the gene II sequence does not serve a critical function (in cis) in insulin expression. It is possible, however, that the small intron, which is present in both genes, could play a regulatory role. To assign a regulatory function to the large intron, one has to postulate as yet unobserved subtle regulatory mechanisms operating in two different β cell populations, expressing each gene separately.

The sequences of the small intron proximal to the boundaries with the exons are conserved and have a 5 nucleotide terminal redundancy, while there is a redundancy of 2 nucleotides in the large intron of rI2. As in other cases (see Dawid and Wahli, 1979), the terminal redundancy prevents the definition of a unique splicing point. By appropriate selection of this point (Figure 8), however, both introns follow the general rule introduced by Chambon (Breathnach et

Two Nonallelic Rat Preproinsulin Genes
555

```
                                                                                   ↓
Rat I                     MetAlaLeuTrpIleArgPheLeuProLeuLeuAlaLeuLeuIleLeuTrpGluProArgProAlaGlnAlaPhe
Rat II                                Met                         Val             Lys
                                                                                   ↓
Bovine                       - Leu -  -  - Leu -  - LeuLeu - LeuLeu - Leu -  -  -  -  -  -  -  - Phe
                                                                               ↓
Anglerfish                - AlaLeu - Leu -  - Phe - LeuLeuValLeuLeuValVal -  -  -  -  -  - AlaVal
                                                                               ↓
Sea Raven         MetMet  -  - Leu -  -  -  -  -  - LeuLeu - LeuLeu - Leu -  -  -  -  -  -  -  -
```

Figure 13. Comparison of the Amino Acid Sequences of the Hydropobic Preregions of Rat I and II, Bovine (Lomedico et al., 1977), Anglerfish and Sea Raven (Shields and Blobel, 1977) Preproinsulins

al., 1978)—that is, they begin with a GT and end with an AG dinucleotide.

Did the large intron in rI2 serve at some early time to bring together the two flanking exons during evolution, as suggested elsewhere (Gilbert, 1978, 1979), or is it an adventitious element added to a previously existing gene? The answer to this question will follow if we can trace the evolution of these two genes.

Molecular Evolution of a Duplicated Gene

The similarity of the two rat preproinsulin genes suggests that they are the result of a recent gene duplication. Can we estimate the time since that moment assuming that the genes have drifted neutrally apart? Insulin itself changes very slowly, and thus its sequence variations are not a good measure of evolutionary time. Nonetheless, the two rat insulins are very closely related, with only 4% differences, while insulins that must have diverged at much earlier times show considerable changes. Rabbit and chicken insulins differ by 14%; human and fish insulins differ by 33% [this last number should be corrected for multiple changes (Dayhoff, 1978) in order to compare it to smaller percent changes for which correction is insignificant; it becomes 43%]. The rat genes have diverged only one tenth as much as human and fish, or one fourth as much as rabbit and chicken. A numerical estimate of divergence time can be obtained from these figures by assuming that neutral changes accumulate in a given protein at a constant rate (the evolutionary clock hypothesis; for example, see Wilson, Carlson and White, 1977). This rate can be expressed in UEP (unit evolutionary period), the time in millions of years (MY) required for the fixation of 1% changes in amino acid sequence between two lines. The UEP for insulin is 14 (Wilson et al., 1977). Thus the rat insulin differences, 4 ± 3% (Poisson standard deviation), correspond to 56 ± 42 MY, with corresponding estimates of 200 and 600 MY for the divergences of the other two pairs of genes, respectively.

In contrast to the insulin molecule, the C peptide of proinsulin drifts at a much faster rate (UEP of 1.9) and provides a better measure of recent evolutionary changes. Hence the differences in the C peptides of human/monkey (3%), human/pig (32%; 41% corrected), human/duck (55%; 94% corrected) and rat I/II (6% difference) correspond to divergence times of 6, 80, 180 and 11 ± 9 MY, respectively. These calculations argue that the rat preproinsulin gene duplication occurred sometime after the beginning of the mammalian radiation—that is, later than 85 MY ago (McKenna, 1975). Mice also have two insulins which are identical in sequence to the ones of the rat (Humbel et al., 1972), while other rodents (spiny mouse and syrian hamster) express a single insulin sequence (Dayhoff, 1978). The phylogenetic classification of these species (Simpson, 1959) is consistent with the idea that the insulin gene duplication occurred during rodent evolution after the divergence of the spiny mouse and syrian hamster and before the mouse-rat divergence.

The nucleic acid sequences enable us to examine the drift of silent, and presumably neutral, sites in the two genes. The small intron and the silent positions within the coding region have drifted to the same degree. Table 1 shows that the small introns, eliminating the 10 bp proximal to the exon junctions, which may be conserved, have 21 ± 5% substitutions. The silent sites have changed 19 ± 5% (corrected to 22 ± 6%), while replacement sites have remained quite conserved (3% changes). The intron clearly is as evolutionarily silent as those positions in the mRNA whose changes do not lead to amino acid replacements. The total drift of the intron or the silent positions is 23 ± 6% (corrected). How can we interpret this number? On the one hand, we can compare this figure to the amount of drift seen between the β-globin genes of rabbits and mice (van den Berg et al., 1978). Table 2 shows that again the introns and the silent positions drift the same amount, in this case 68 ± 6%. This is considerably greater than that of the rat I/II divergence. We conclude that the rabbit/mouse divergence occurred three times as long ago as the duplication of the rat genes. Unfortunately, none of the actual divergence times are known accurately. If the rabbit/mouse divergence were 85 MY ago, we would assign a time of 28 ± 8 MY for the rat gene duplication.

Alternatively, we can attempt to assign an absolute rate to nucleotide drift and fixation. Kafatos et al. (1977) calculated that the hypervariable region of the fibrinopeptides, a candidate for an unselected protein region, shows 55% nucleotide substitution averaged over the mammalian radiation. We turn this number into a nucleotide UEP first by correcting it to 98%,

Table 1. Nucleotide Differences between Rat Preproinsulin I and II

	Coding Region			
	Silent Substitution Sites		Replacement Sites	
	#	%	#	%
Signal peptide	1/19.5	5 ± 5	4/52.5	8 ± 4
B chain	5/21	24 ± 11	2/69	3 ± 2
C peptide	7/28.3	25 ± 9	2/76.7	3 ± 2
A chain	2/12	17 ± 12	0/51	0
Total	15/80.8	19 ± 5	8/249.2	3 ± 1
Total-corrected		22 ± 6		
	Noncoding Region			
	#	%	% (Corrected)	
Small intron				
119 bp	21/119	18 ± 4		
Intron body	21/99	21 ± 5	24 ± 6	
5′ noncoding region	5/57	9 ± 4		
3′ noncoding region	6/53	11 ± 5		
3′ sequence proximal to end of mRNA	26/43	60 ± 12		
5′ sequence proximal to beginning of mRNA	3/39	8 ± 4		

To calculate the rate of change at "silent sites," we totaled the possible single-step mutations in the protein, three possible changes at each base (3 × 330), and classified each of these changes as a silent change, if it did not change the amino acid, or a replacement. We averaged the two genes and then divided the total number of changes by three and recorded that as the number of "silent sites." The fractions (#) shown are the actual numbers of silent substitutions divided by the available "silent sites." A similar calculation was made for the replacement sites. The C peptide includes the associated four basic amino acids. The body of the intron includes the entire sequence except for the 10 bp proximal to each exon. The errors shown are the Poisson standard deviations.

Table 2. Comparison of Rabbit and Mouse β-Globin Genes

			Corrected
Silent changes	49/110	44 ± 6%	66 ± 9%
Replacement changes	35/330	11 ± 2%	12 ± 2%
Intron partial sequences eliminating 10 bases from the ends:			
35 + (5 gaps)/100	40 ± 6%		
68 + (12 gaps)/167	48 ± 5%		
Average	45 ± 4%		
Average-corrected	68 ± 6%		

These calculations are based on the data of van den Berg et al. (1978). The silent changes have been calculated as the observed number of silent changes divided by the number of silent sites (25% of the total number of sites). The intron sequences were compared as by van den Berg et al. (1978), except that each gap was scored as one change.

and then by assuming that the radiation occurred 85 MY ago. This then is 1% change in 0.9 MY of divergence, a UEP of 0.9. By this standard, the rat genes diverged 21 ± 5 MY ago. We conclude that the separation between these genes occurred 20–35 million years ago, after the different mammalian genera separated but around, or before, the time rats and mice diverged [35 million years ago (A. Wilson, personal communication) or less, as our data suggest]. Hence rats and mice have the same set of duplicated preproinsulin genes, but we expect that other mammals will have a single gene whose structure should correspond to the common stock. Similarly, the two preproinsulin genes in some fish probably represent a late independent divergence.

An evolutionary test is clear: the comparison of different preproinsulin genes will show whether new introns arise or old ones are lost. If the introns have served in the past to bring the exons that correspond to the different elements of insulin together, the ancestral gene will have more introns.

Experimental Procedures

DNA and RNA Preparation

The transplantable rat insulinoma (Chick et al., 1977) is derived from and carried in the NEDH rat strain (the history of which is described by Udupa, Warren and Chute 1974). DNA was prepared from the tumor and NEDH rat liver, spleen and kidney by the Blin and Stafford (1976) procedure.

The rat insulin cDNA plasmids were propagated in χ1776 and the DNA was isolated as described (Villa-Komaroff et al., 1978).

Recombinant λ Charon 4A phage were grown (using $DP50_{sup}F$) and purified as described by Blattner et al. (1977), and the DNA was prepared as detailed by Maniatis et al. (1978).

Cytoplasmic poly(A)-containing RNA from the rat insulinoma was purified as previously described (Villa-Komaroff et al., 1978). Nuclear RNA from the same source was purified essentially as described by Ross (1976).

Filter Hybridizations

Phage DNA was transferred to nitrocellulose filters using the plaque hybridization technique (Benton and Davis, 1977).

Restriction enzymes were purchased from New England Biolabs and used according to the recommended assay conditions. Restriction fragments were electrophoresed on agarose gels (Helling, Goodman and Boyer, 1974), and the DNA was transferred to nitrocellulose filters as described by Southern (1975) with the following modification: we transferred in 3 M NaCl/0.3 M Na citrate/1 M NH_4 acetate and then washed the filter in 1 M NH_4 acetate before air drying (W. Herr, personal communication). Filters were pretreated, hybridized and washed as described by Maniatis et al. (1978). Radioactive probes were generated by nick translation (Rigby et al., 1977) of the Pst-excised insert from pI19 using ^{32}P-α-dNTPs (>300 Ci/mmole; New England Nuclear) and E. coli DNA polymerase I (Boehringer-Mannheim) to specific activities of about 2×10^8 cpm/μg.

DNA Sequencing

5′ end-labeled fragments were prepared and sequenced by the Maxam and Gilbert (1977) procedure. 3′ end-labeled Sma fragments were generated using the Klenow fragment of E. coli DNA polymerase I (New England Biolabs) and ^{32}P-α-dCTP.

Electron Microscopy

DNA-DNA heteroduplexes were formed as described by Davis, Simon and Davidson (1971). R loops were prepared as described by Kaback, Angerer and Davidson (1979). Inverted repeats were examined as described (Kolodner and Tewari, 1979). All samples were mounted for electron microscopy by the formamide technique as described by Davis et al. (1971). Single-stranded $\phi\chi$ DNA and double-stranded $\phi\chi$ and pM2 DNAs were included as internal standards in all experiments.

Recombinant DNA Safety Procedures
The rat library was constructed by Sargent et al. (1979) under P3 containment using the approved EK2 host/vector λ Charon 4A/ DP50$_{sup}$F. P3 containment was used to screen this collection and to propagate the positive clones initially. After the NIH Guidelines were changed (January 1979), these phage were handled under P2/EK2 containment.

Acknowledgments

We thank T. Sargent, B. Wallace and J. Bonner for graciously making available their rat chromosomal DNA library; L. Villa-Komaroff and F. Perler for help with certain experiments; W. Chick for generously providing rat insulinoma; A. Wilson and L. Loeb for information; M. Koehler for help with the figures; and T. Maniatis and the colleagues in his laboratory for their hospitality and advice during the early stages of this project. This work was supported by a grant from the American Cancer Society, Massachusetts Division, to A.E. and by an NIH grant to W.G. P.L. is a fellow of the Juvenile Diabetes Foundation. A.E. is the Harvard Medical School Hsien Wu Investigator. W.G. is an American Cancer Society professor of molecular biology.

The costs of publication of this article were defrayed in part by the payment of page charges. This article must therefore be hereby marked "*advertisement*" in accordance with 18 U.S.C. Section 1734 solely to indicate this fact.

Received July 6, 1979

References

Baralle, F. E. (1977). Cell *12*, 1085.

Baralle, F. E. and Brownlee, G. G. (1978). Nature *274*, 84.

Benton, W. D. and Davis, R. W. (1977). Science *196*, 180.

Berget, S. M., Moore, C. and Sharp, P. A. (1977). Proc. Nat. Acad. Sci. USA *74*, 3171.

Blattner, F. R., Williams, B. G., Blechl, A. E., Thompson, K. D., Faber, H. E., Furlong, L. A., Grunwald, D. J., Kiefer, D. O., Moore, D. D., Schumm, J. W., Sheldon, E. L. and Smithies, O. (1977). Science *196*, 161.

Blin, N. and Stafford, D. W. (1976). Nucl. Acids Res. *3*, 2303.

Blobel, G. and Dobberstein, B. (1975). J. Cell Biol. *67*, 835.

Breathnach, R., Benoist, C., O'Hare, K., Gannon, F. and Chambon, P. (1978). Proc. Nat. Acad. Sci. USA *75*, 4853.

Chan, S. J. and Steiner, D. F. (1977). Trends Biochem. Sci. *2*, 254.

Chan, S. J., Keim, P. and Steiner, D. F. (1976). Proc. Nat. Acad. Sci. USA *73*, 1964.

Chick, W. L., Warren, S., Chute, R. N., Like, A. A., Lauris, V. and Kitchen, K. C. (1977). Proc. Nat. Acad. Sci. USA *74*, 628.

Chow, L. T., Gelinas, R. E., Broker, T. R. and Roberts, R. J. (1977). Cell *12*, 1.

Clark, J. L. and Steiner, D. F. (1969). Proc. Nat. Acad. Sci. USA *62*, 278.

Davis, R. W., Simon, M. and Davidson, N. (1971). In Methods in Enzymology, *21*, L. Grossman and K. Moldave, eds. (New York: Academic Press), p. 413.

Dawid, I. B. and Wahli, W. (1979). Dev. Biol. *69*, 305.

Dayhoff, M. O. (1978). Atlas of Protein Sequence, *5*, supplement 3 (Washington, D. C.: National Biomedical Research Foundation).

Falkmer, S. and Ostberg, Y. (1977). In The Diabetic Pancreas, B. W. Volk and K. F. Wellman, eds. (New York: Plenum Press), p. 15.

Gannon, F., O'Hare, K., Perrin, F., LePennec, J. P., Benoist, C., Cochet, M., Breathnach, R., Royal, A., Gaparin, A., Cami, B. and Chambon, P. (1979). Nature *278*, 428.

Gilbert, W. (1978). Nature *271*, 501.

Gilbert, W. (1979). In Eukaryotic Gene Regulation, ICN-UCLA Symposia on Molecular and Cellular Biology, *14*, R. Axel, T. Maniatis and C. F. Fox, eds. (New York: Academic Press), in press.

Gilmore-Hebert, M., Hercules, K., Komaromy, M. and Wall, R. (1978). Proc. Nat. Acad. Sci. USA *75*, 6044.

Gopinathan, K. P., Weymouth, L. A., Kunkel, T. A. and Loeb, L. A. (1979). Nature *278*, 857.

Hagenbüchle, O., Santer, M., Steitz, J. A. and Mans, R. J. (1978). Cell *13*, 551.

Helling, R. B., Goodman, H. M. and Boyer, H. W. (1974). J. Virol. *14*, 1235.

Humbel, R. E., Bosshard, H. R. and Zahn, H. (1972). In Handbook of Physiology, Endocrinology (Washington, D.C.: American Physiological Society), p. 111.

Kaback, D. B., Angerer, L. and Davidson, N. (1979). Nucl. Acids Res. *6*, 2499.

Kafatos, F. C., Efstratiadis, A., Forget, B. G. and Weissman, S. M. (1977). Proc. Nat. Acad. Sci. USA *74*, 5618.

Kolodner, R. and Tewari, K. K. (1979). Proc. Nat. Acad. Sci. USA *76*, 41.

Lomedico, P. T. and Saunders, G. F. (1976). Nucl. Acids Res. *3*, 381.

Lomedico, P. T., Chan, S. J., Steiner, D. F. and Saunders, G. F. (1977). J. Biol. Chem. *252*, 7971.

McKenna, M. C. (1975). In Phylogeny of the Primates, W. P. Luckett and F. S. Szalay, eds. (New York: Plenum Press), p. 21.

Maniatis, T., Hardison, R. C., Lacy, E., Lauer, J., O'Connell, C., Quon, D., Sim, G. K. and Efstratiadis, A. (1978). Cell *15*, 687.

Maxam, A. M. and Gilbert, W. (1977). Proc. Nat. Acad. Sci. USA *74*, 560.

Milstein, C., Brownlee, G. G., Harrison, T. M. and Mathews, M. B. (1972). Nature New Biol. *239*, 117.

Rigby, P. W. J., Dieckmann, M., Rhodes, C. and Berg, P. (1977). J. Mol. Biol. *113*, 237.

Roop, D. R., Nordstrom, J. L., Tsai, S. Y., Tsai, M.-J. and O'Malley, B. W. (1978). Cell *15*, 671.

Ross, J. (1976). J. Mol. Biol. *106*, 403.

Sargent, T. D., Wu, J., Sala-Trepat, J. M., Wallace, R. B., Reyes, A. A. and Bonner, J. (1979). Proc. Nat. Acad. Sci. USA *76*, 3256.

Schibler, U., Marcu, K. B. and Perry, R. P. (1978). Cell *15*, 1495.

Shields, D. and Blobel, G. (1977). Proc. Nat. Acad. Sci. USA *74*, 2059.

Simpson, G. G. (1959). Cold Spring Harbor Symp. Quant. Biol. *24*, 255.

Smith, L. F. (1966). Am. J. Med. *40*, 662.

Southern, E. M. (1975). J. Mol. Biol. *98*, 503.

Steiner, D. F., Kemmler, W., Tager, H. S. and Peterson, J. D. (1974). Fed. Proc. *33*, 2105.

Szala, S., Michalska, J., Paterak, H., Bieniek, B. and Chorazy, M. (1977). FEBS Letters *77*, 94.

Tilghman, S. M., Tiemeier, D. C., Seidman, J. G., Peterlin, M., Sullivan, M., Maizel, J. V. and Leder, P. (1978a). Proc. Nat. Acad. Sci. USA *75*, 725.

Tilghman, S. M., Curtis, P. J., Tiemeier, D. C., Leder, P. and Weissmann, C. (1978b). Proc. Nat. Acad. Sci. USA *75*, 1309.

Tinoco, I., Borer, P. N., Dengler, B., Levine, M. D., Uhlenbeck, O. C., Crothers, D. M. and Gralla, J. (1973). Nature New Biol. *246*, 40.

Tsujimoto, Y. and Suzuki, Y. (1979). Cell *16*, 425.

Udupa, K. B., Warren, S. and Chute, R. N. (1974). Health Physics *26*, 319.

Ullrich, A., Shine, J., Chirgwin, J., Pictet, R., Tischer, E., Rutter, W. J. and Goodman, H. M. (1977). Science *196*, 1313.

van den Berg, J., van Ooyen, A., Mantei, N., Schambock, A., Grosveld, G., Flavell, R. A. and Weissman, C. (1978). Nature *276*, 37.

Villa-Komaroff, L., Efstratiadis, A., Broome, S., Lomedico, P., Tizard, R., Naber, S., Chick, W. and Gilbert, W. (1978). Proc. Nat. Acad. Sci. USA *75*, 3727.

Wilson, A. C., Carlson, S. S. and White, T. J. (1977). Ann. Rev. Biochem. *46*, 573.

Cell, Vol. 20, 555-566, June 1980, Copyright © 1980 by MIT

The Evolution of Genes: the Chicken Preproinsulin Gene

Francine Perler and Argiris Efstratiadis
Department of Biological Chemistry
Harvard Medical School
Boston, Massachusetts 02115
Peter Lomedico and Walter Gilbert
Biological Laboratories
Harvard University
Cambridge, Massachusetts 02138
Richard Kolodner
Sidney Farber Cancer Institute
Boston, Massachusetts 02115
Jerry Dodgson
Department of Microbiology and Public Health
Michigan State University
East Lansing, Michigan 48824

Summary

We have characterized a clone carrying a chicken preproinsulin gene, which is present in only one copy in the chicken genome. The gene contains two introns: a 3.5 kb intron interrupting the region encoding the connecting peptide and a 119 bp intron interrupting the DNA corresponding to the 5′ noncoding region of the mRNA. This is similar to the structure of rat insulin gene II; therefore it represents the common ancestor. Since the rat insulin gene I lacks a 499 bp intron in the coding region, the rat genes have evolved by a recent gene duplication followed by loss of this intron in one copy. The divergences between insulin gene sequences, and also between globin genes, show that changes at introns and silent positions in coding regions appear very rapidly (7×10^{-9} substitutions per nucleotide site per year), but that the accumulation of changes in these sites saturates, although not completely, after about 100 million years. From this we conclude that not all of these sites are neutral and that they do not behave as accurate evolutionary clocks over long periods of time. However, nucleotide substitutions leading to amino acid replacements are an excellent clock. Our analysis indicates that this clock is driven by selection.

Introduction

The gene for preproinsulin provides an interesting example of evolution. Although most vertebrates have a single insulin, rats, mice, and three fish species (tuna, bonito and toadfish) have two insulins (Humbel, Bosshard and Zahn, 1972; Dayhoff, 1978). We recently determined the structure of the two nonallelic preproinsulin genes in the rat (Lomedico et al., 1979). They are both expressed equally in the endocrine pancreas (Clark and Steiner, 1969). Their sequence similarity suggests that they are the result of a moderately recent gene duplication. However, their structures differ. One of the genes contains a long intron interrupting the coding sequences; the other does not. This difference poses a clearly formulated problem: one of the genes has gained or the other has lost this intron during or since the duplication. To resolve this question we isolated and determined the structure of a preproinsulin gene from a distantly related species, the chicken.

The gene for preproinsulin is transcribed in the islet cells of the endocrine pancreas into an mRNA precursor which is then spliced to produce mature mRNA (Lomedico et al., 1979). This mRNA is translated into preproinsulin (Chan and Steiner, 1977), which is sequentially processed into proinsulin and finally to insulin. Proinsulin is generated by cleavage of the first 24 amino acids (preregion) from the amino terminus of preproinsulin. Proinsulin molecules hexamerize, stabilized by zinc atoms (Frank and Veros, 1970). In proinsulin, the two regions of the polypeptide chain that will become the two chains of mature insulin, the B and A chains, are connected by a middle peptide, the C peptide, flanked by two pairs of basic amino acids. The role of the C peptide is to bring together the two portions of the polypeptide chain so that disulfide bridges can form appropriately between the B and A segments. Moreover, it seems that its presence increases the solubility of the hexamers (Grant, Coombs and Franks, 1972). In the last stage of maturation, proteolytic enzymes cleave at the basic amino acids to release the C peptide and form mature insulin. When the connecting peptide is removed, insulin crystallizes as granules.

The rat gene for preproinsulin I has a 119 bp intron interrupting the segment corresponding to the 5′ noncoding region of the mRNA. In addition to this small intron, the gene for preproinsulin II contains a 499 bp intron interrupting the coding segment for the C peptide. Thus in gene II the second long intron divides the region corresponding to the polypeptide chain into two portions corresponding to the two regions brought together in mature insulin. Since both genes function equally, we conclude that the large intron in gene II cannot serve a critical function, in cis, in gene expression. However, if splicing is a necessary prerequisite for the appearance of translatable mRNA in the cytoplasm (Gruss et al., 1979; Hamer and Leder, 1979; Hamer et al., 1979), the small intron present in both rat preproinsulin genes could fulfill that requirement.

Since the rat genes are products of a duplication that occurred after the mammalian radiation (approximately 85 million years ago; Romero-Herrera et al., 1973; McKenna, 1975), we expect that other mammals, and certainly birds, will have a single preproinsulin gene whose structure should reflect the common ancestor. Here we describe the structure of the unique preproinsulin gene of the chicken. Its structure resem-

Cell
556

bles that of rat gene II: a long intron divides the coding region. Thus the older gene structure has more introns, as predicted by the hypothesis that the introns have a role in the assembly of genes (Gilbert, 1979).

Results

Isolation of a Cloned Chicken Chromosomal Preproinsulin Gene

In preliminary blot-hybridization experiments (Southern, 1975), we determined that a rat preproinsulin cDNA probe [Pst-excised insert of clone pI19 (Villa-Komaroff et al., 1978), ^{32}P-labeled by nick translation] has sufficient cross homology to the corresponding sequences in chicken chromosomal DNA to detect restriction fragments bearing the chicken preproinsulin gene under nonstringent hybridization conditions (50°C, 1 M NaCl). Thus we used these conditions to screen a chicken chromosomal DNA library (Dodgson, Strommer and Engel, 1979). For the generation of this library, DNA restriction fragments produced by partial Hae III plus Alu I digestion were ligated to Eco RI synthetic DNA linkers and joined to the purified Eco RI arms of phage λ Charon 4A (Blattner et al., 1977).

We screened (Benton and Davis, 1977) 200,000 recombinant plaques and identified five hybridizing clones, corresponding to three different sequences, as revealed by restriction endonuclease analysis. Two of the three different clones hybridize fortuitously to rat preproinsulin sequences. Following digestion with a variety of restriction enzymes, blot-hybridization analysis indicated that only one of the resulting bands in each case hybridizes with the rat cDNA probe. Moreover, if the rat cDNA plasmid pI47, which lacks the region corresponding to the preregion of the protein (Villa-Komaroff et al., 1978), is used as probe instead of pI19, there is no detectable hybridization. DNA sequencing of a fragment from one of these two clones, hybridizing to pI19, revealed a sequence of 17 nucleotides identical to nucleotides 26–42 of the preregion of rat preproinsulin. If one nuleotide gap is allowed in the rat sequence, the homology extends between nucleotides 26–48. The rest of the sequence flanking this area (a total of 80 nucleotides) is completely different (data not shown).

The third clone (λ Charon 4A-cI15, or simply cI15) contains a 13.9 kb insert composed of two Eco RI fragments of 9.5 and 4.4 kb, both hybridizing to the rat probe (Figure 1). Digestion with Bam produces several fragments, one of which (5.9 kb) hybridizes to the probe. Eco RI cleaves this fragment once and produces two fragments of 0.8 and 5.1 kb, both hybridizing to the probe. DNA sequencing of the 0.8 kb Bam-Eco RI fragment, end-labeled at the Bam site, showed that part of the sequence (Bam → Eco RI, 5′ → 3′) corresponds to the known amino acid sequence (Humbel et al., 1972) of the B chain of chicken insulin. We concluded that clone cI15 carries a chicken preproinsulin gene. To facilitate restriction mapping and DNA sequencing we subcloned the 5.9 kb Bam fragment into the Bam site of plasmid pBR322.

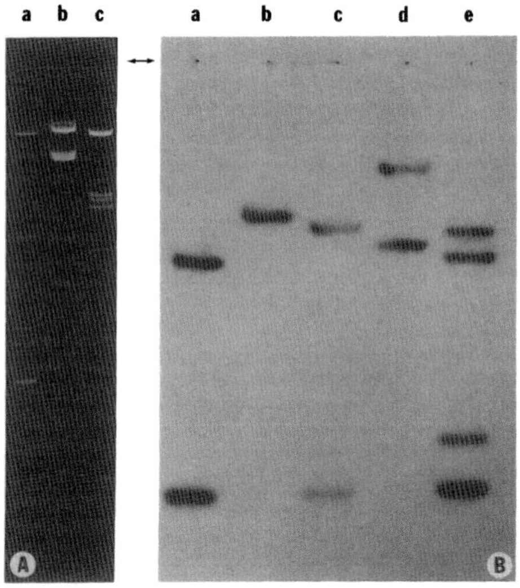

Figure 1. Restriction Endonuclease Analysis of Clone cI15

(A) cI15 DNA (0.5 μg per slot) was digested with Eco RI (lane b) or Bam (lane c), and the products were electrophoresed on a 1% agarose gel and visualized by ethidium bromide staining. The lengths of molecular weight markers in kb (lane a) are from top to bottom: (λ Charon 4A × Eco RI) 19.8, 10.9, 8.0 and 7.0; (pBR322 × Eco RI) 4.3; (pBR322 × Eco RI × Pvu II) 2.3 and 2.1; (pBR322 × Hinf) 1.6, 0.5 and smaller bands.

(B) Autoradiogram of a nitrocellulose filter carrying transferred cI15 DNA fragments (from a different gel than in A) after hybridization with ^{32}P-labeled DNA probe (nick-translated, Pst-excised insert of pI19). The hybridizing bands were generated by digestion with the following restriction enzymes: (lane b) Bam, 5.9 kb; (lane c) Bam and Eco RI, 5.1 and 0.8 kb; (lane d) Eco RI, 9.5 and 4.4 kb. The bands in lane a (control) are the two Eco RI hybridizing bands (3.7 and 0.8 kb) of the rat clone rI2 (Lomedico et al., 1979). Lane e shows, in addition to the bands in lane a, two more molecular weight markers, pI19 × Hind III (first band, always running at 4.3 kb) and pI19 × Hind III × Hinf (1.15 kb, third band).

Analysis of the Cloned Chicken Preproinsulin Gene

We constructed a restriction map of clone cI15 (Figure 2) and sequenced across the DNA regions encoding preproinsulin (Figure 3). The sequence establishes the primary structure of the coding region of chicken preproinsulin mRNA. However, it is difficult to establish the sites to which the cap and the poly(A) are attached because we lack independent information concerning the mRNA structure. We tentatively assigned these sites as follows. By comparing this sequence to those of the rat we identified the heptanucleotide TATAATT in the 5′ gene flanking region. This is a variant of the sequence TATAAAT, first described by Goldberg and Hogness (Goldberg, 1979), which is

The Chicken Preproinsulin Gene
557

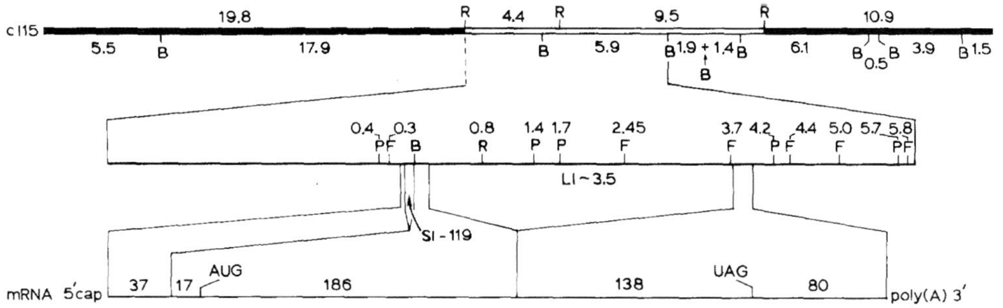

Figure 2. Restriction Map of Clone cl15

(cl15) The Eco RI (R) and Bam (B) sites and the distances (in kb) between these sites in the recombinant phage DNA are indicated. The long and short arms of the λ vector (filled boxes) are in the same direction as the transcriptional orientation (5' → 3') of the insert. The relative position of the 1.9 and 1.4 kb Bam fragments is not known. A blow-up of the region containing the gene sequence is shown. For simplicity, only the distances (in kb) of Eco RI, Pst (P) and Hinf (F) sites from the Bam site present in this segment (reference point) are indicated. Several Pvu II and Ava II sites exist in this area, but are not completely mapped. A Pvu II site exists 60 bp to the right of the reference point, and a second Pvu II site is present 2.2 kb from the first. Two Ava II sites are present 900 and 460 bp, respectively, to the left and right of the Hinf site that lies 3.7 kb from the reference point.

(mRNA) Correlation between the chromosomal gene sequence and the sequence present in mature mRNA. The positions of the small intron (SI) and the large intron (LI) are indicated; their sizes are in bp and kb, respectively.

```
                        cttctggttataattggtcatttattatgacttttaaAGCCTGATGAATAAAAT

ATTCCTTTCCTCTTCAGAAGGTCCATTTGCTTCTGTAGTCTTGTTTTCACGTCAAAGGAGCTGAGGGACATAAGATGC

CTGATGATAGCTTATTCCTCCCTTGCAACCCCCCCGTGTCTCCTTTGCTTCCTACCTCTAGGCCTCCCCCAGCTCATC

MetAlaLeuTrpIleArgSerLeuProLeuLeuAlaLeuLeuValPheSerGlyProGlyThrSerTyrAlaAlaAla
ATGGCTCTCTGGATCCGATCACTGCCTCTTCTGGCTCTCCTTGTCTTTTCTGGCCCTGGAACCAGCTATGCAGCTGCC
          Bam

AsnGlnHisLeuCysGlySerHisLeuValGluAlaLeuTyrLeuValCysGlyGluArgGlyPhePheTyrSerPro
AACCAGCACCTCTGTGGCTCCCACTTGGTGGAGGCTCTCTACCTGGTGTGTGGAGAGCGTGGCTTCTTCTACTCCCCC

LysAlaArgArgAspValGluGlnProLeu
AAAGCCCGACGGGATGTCGAGCAGCCCCTAGGTAAGTCAGTTTGACCATGACTACATTCATATGCTATATGATGCAAA
                TaqI

AAGCAACTGTCTATCTTTGATGGTGACACAAGGAATGTCCTTGGTGGGGAATG.... ~0.4kb ....GGTCTATCATT

CCTCCTTCATGGGTGATTTTCAAACAGTTTAAAAATTGCTTCCATGTCTTGTTTTTATCTACTGTGAGCTAAAAGCCC

TCATCAGCCCGAATTCTTTAGGTCACATATTCTAGCTCTCTGTCTACATAAACTGTTCTGCATTTGGCCCATACCATT
          Eco RI

ACGGAATGGTGATGGGTGGAGAAGCCTTGCCAGCTACAAGCAGAGCAAGCAAACTGGAAGAACAGCATTGTACGGTTT

TCACACCATGTTCCTATGCAGGAGGGTTGTCTTTTTCAAAGTAGCATCAAACCTGCTTTCATTGTCTCCCTTGG....

~2.5 kb ....TCTGAGAGCATTAGCTTTGGGAAATCTTTGCAAGTGCTCCTCATTTCCCAGGCAGCTCTTCACTTAC

ACACCTGGTATCTGAAAAGCGGGTCTCCCTGGGACTCAAAAGCTAGGGATGGTGCGGATGAACTATGACTTACCTTCT
                                 Hinf

             ValSerSerProLeuArgGlyGluAlaGlyValLeuProPheGlnGlnGluGluTyrGluLysVal
TTTCCCTTGGCAGTGAGCAGTCCCTTGCGTGGCGAGGCAGGAGTGCTGCCTTTCCAGCAGGAGGAATACGAGAAAGTC

LysArgGlyIleValGluGlnCysCysHisAsnThrCysSerLeuTyrGlnLeuGluAsnTyrCysAsn
AAGCGAGGGATTGTTGAGCAATGCTGCCATAACACGTGTTCCCTCTACCAACTGGAGAACTACTGCAACTAGCCAAGA

AGCCAGAAGCGGGCACAGACATACACTTACTCTATCGCACCTTCAAAGCATTTGAATAAACCTTGTTGGTCTACtgga

agacttgtgccattttatgtgtccatggtt
```

Figure 3. Nucleotide Sequence of the Chicken Preproinsulin Gene

The nucleotide sequence of the mRNA strand of the gene is displayed from the 5' to the 3' direction. Large and small capital letters represent sequences corresponding to the mature mRNA and the introns, respectively. Gene flanking sequences are in lowercase letters. Certain important restriction sites and the nucleotide redundancies at the intron-exon junctions are underlined, while the GT and AG dinucleotides at the beginning and end of the introns are overlined. The "Goldberg-Hogness box" in the 5' gene flanking sequence, the initiation and termination codons, and the "AATAAA box" in the region corresponding to the 3' noncoding region of the mRNA are boxed. The amino acid sequence is displayed on a line above the coding sequence. The arrows indicate, in order, the boundaries of the preregion and the peptides B, C and A.

always present upstream from the capping site of eucaryotic genes (for example see Gannon et al., 1979; Hardison et al., 1979; Lomedico et al., 1979). The tetranucleotide AGCC, which is identical to the first tetranucleotide of the rat preproinsulin II mRNA, occurs 22 nucleotides to the 3' side of the "Goldberg-Hogness box." In the rat, the distance between these two nucleotide blocks is 23 nucleotides. We assigned the A of the AGCC tetranucleotide as the capping site and searched for an intron in the area between this site and the initiator. By comparing the sequence of this segment with the rat preproinsulin sequences, and applying the generalization that an intron begins with GT and ends with AG (Breathnach et al., 1978), we identified a putative small intron with a three-nucleotide terminal redundancy at its boundaries and with a size identical to that of the rat intron (119 bp). A similar intron (178 bp), also identified by comparison to the rat genes, exists in cloned human insulin DNA (Goodman, 1980; A. Ullrich, personal communication), where its presence was verified by sequencing the cDNA product of a reverse transcription reaction, using as primer a fragment of cloned DNA (A. Ullrich, personal communication). Unfortunately, we were unable to perform such an experiment successfully because of the extremely low preproinsulin mRNA yield from whole chicken pancreas. However, we verified indirectly the presence of the small intron by S1 mapping (Berk and Sharp, 1977) as follows: cl15 DNA was cleaved with Pst and then subjected to limited digestion with Exo III. The anti-message strand was then synthesized using reverse transcription and α-^{32}P-deoxynucleotide triphosphates. Following Bam digestion, a 0.4 kb Pst-Bam fragment containing the region of interest (Figure 2) was purified by gel electrophoresis. This fragment was hybridized with poly(A)$^+$ chiken pancreatic RNA under conditions that favor DNA:RNA hybridization (Casey and Davidson, 1977) and the mixture was subjected to nuclease S1 digestion. Gel electrophoresis of the digestion products under denaturing conditions revealed the presence of two resistant fragments of 32 and 37 nucleotides predicted from the sequence (data not shown).

The DNA sequence reveals that the structure of the chicken gene is similar to that of the rat preproinsulin II gene. A second large intron interrupts the coding region between the codons of amino acids 62 and 63 of preproinsulin (that is, between the codons of the sixth and seventh amino acids of the C peptide) as in the rat gene II. Restriction mapping and heteroduplex analysis (see below) show that the chicken intron is 3.5 kb long, much larger than that of the rat. The intron has a two-nucleotide terminal redundancy at its boundaries with the exons and, by selection of the splicing point, begins with a GT and ends with an AG dinucleotide. A 786 bp intron at the same position has also been identified in human chromosomal preproinsulin DNA (Goodman, 1980).

The distance between the terminator and the AATAAA box, which is always present in the region of eucaryotic genes corresponding to the 3′ noncoding region of the mRNA, varies between the three known sequences; it is 32–33 nucleotides in the rat (Lomedico et al., 1979), 53 nucleotides in the human (Bell et al., 1979) and 60 nucleotides in the chicken. However, in the rat and human sequences, the distance between the end of this box and the poly(A) attachment site, inclusive, is the same (14 nucleotides). We tentatively assign the poly(A) attachment site of the chicken preproinsulin mRNA to the corresponding point.

Heteroduplex Analysis

To examine whether the sequence homology between the chicken and rat genes extends farther than the coding regions, we examined by electron microscopy heteroduplexes between cl15 and the rat clone rl22 (Lomedico et al., 1979). This clone carries the rat gene II sequence in the same transcriptional orientation with respect to the λ vector arms as cl15 (5′ → 3′, long Charon 4A Eco RI arm → short arm). About 35% of all heteroduplexes with hybridized λ arms had one small region of homology, and 80% of them had a second even smaller homologous region. Therefore, homology between the chicken and rat sequences is limited to the coding regions. Figure 4 shows one of the heteroduplexes. Measurements on seven molecules indicated that the first region of homology is approximately 150 bp long and lies at the predicted distance from the long λ arm, as concluded by comparison with the restriction map (Figure 2). It therefore corresponds to the distance between the initiator and the left-hand boundary of the large intron (186 bp). Given the limited accuracy of electron microscopic measurements for such small lengths, we conclude that the two sets of data are in agreement. The second region of homology is too small to be measured. Its length is estimated to be $\leq$100 bp. We believe that this region corresponds to the sequence encoding the connecting amino acids and the A chain (a total of 69 bp). It is unlikely that a segment corresponding to a portion of the C peptide is part of this heteroduplexed region, because the C peptides of the two sequences are of unequal lengths and highly divergent. The length of single-stranded DNA between the two regions of homology is 3.4 and 0.6 kb. The 3.4 kb unhybridized loop corresponds to the large chicken intron.

The Chicken Genome Has Only One Preproinsulin Gene

Although only one insulin has been deteted in chicken pancreas (Humbel et al., 1972), this does not exclude the possibility that one or more silent preproinsulin genes are present in the chicken genome or that chicken insulin is the product of expression of multiple genes. We answered this question by blot-hybridization. Figure 5 shows that the fragments produced by Bam or Bam + Eco RI digestion are the same in chromosomal DNA and the cloned sequences. For this experiment the probe was nick-translated plasmid pcl21 carrying the 5.9 kb Bam fragment of the recombinant phage (Figure 5A). If there is a single copy, then 60 pg of the 5.9 kb Bam fragment were present in the chromosomal DNA (30 μg) (Figure 5A, lanes c and d); about five times this amount was present in

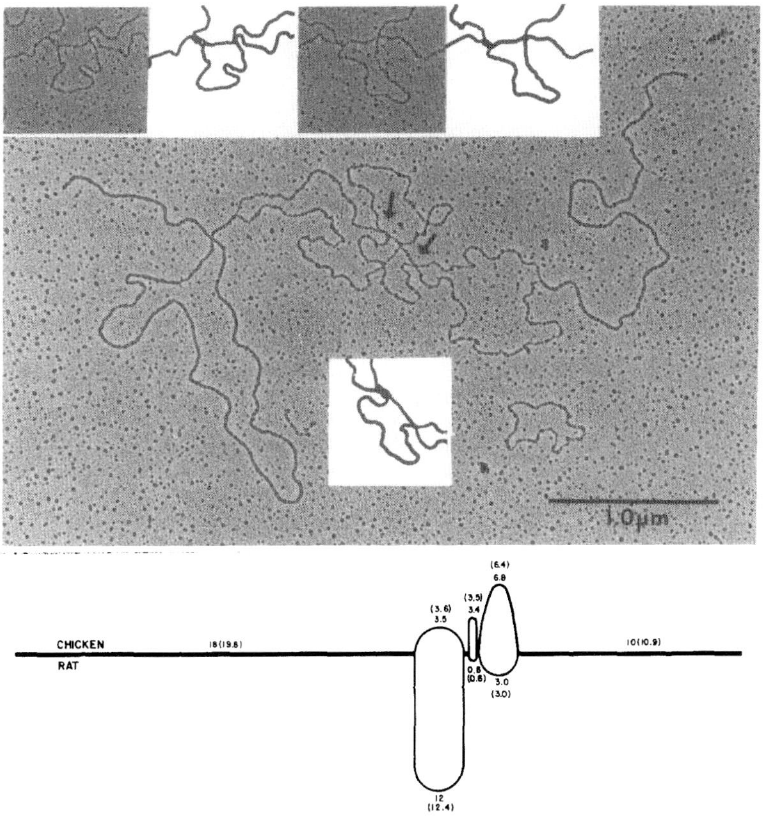

Figure 4. Electron Micrograph of Heteroduplexes Formed between the Chicken Clone cl15 and the Rat Clone rl22

The orientation from left to right corresponds to that of the restriction map (Figure 2). The arrows indicate the regions of homology between the rat and chicken sequences. The inserts in the upper left show portions of other heteroduplexes in which the loops between the arrows are spread more clearly. The diagram under the micrograph indicates the length (in kb) of each section of the heteroduplex as calculated from electron microscopic measurements using as standards single- and double-stranded ϕX174 DNA (the small circle in the lower right corner is single-stranded ϕX174). The lengths in parentheses indicate the sizes calculated from the restriction map.

the cloned DNA (Figure 5A, lanes a and b). Although quantitation by this method is difficult, we conclude from the relative intensity of the bands that only one preproinsulin gene is present in the chicken genome. Moreover, since both of the introns are present in the probe and only the predicted bands appear in the autoradiogram, the intron sequences of the chicken preproinsulin gene must be single-copy DNA. It could be argued, however, that the detection of hybridizing bands belonging to silent genes is below the sensitivity of this experiment, since only about 6% (333/5900) of the probe represents coding region. For this reason we repeated the experiment with a Bam-Taq I fragment probe (162 bp) corresponding to the preregion and the B chain (Figure 5B, lanes a and d, respectively). As predicted, only the 0.8 kb Bam-Eco RI band was hybridized.

Discussion

The Structure of Chicken Preproinsulin

The coding region of the mature chicken preproinsulin mRNA consists of 324 nucleotides, including the AUG initiator and the UAG terminator. This is shorter than the corresponding regions of the rat (Lomedico et al., 1979) and human (Bell et al., 1979) preproinsulin mRNAs (333 nucleotides), since the length of the chicken C peptide (28 versus 31 amino acids) is shorter. This is not a unique case, because with the exception of human, monkey, rat and horse C peptides (31 residues), other mammalian C peptides (pig, ox and dog) exhibit deletions (Steiner et al., 1974). Markussen and Sundby (1973) sequenced the C peptide of duck proinsulin, which is also shorter (26 amino acids), and proposed an alignment with the known

Cell
560

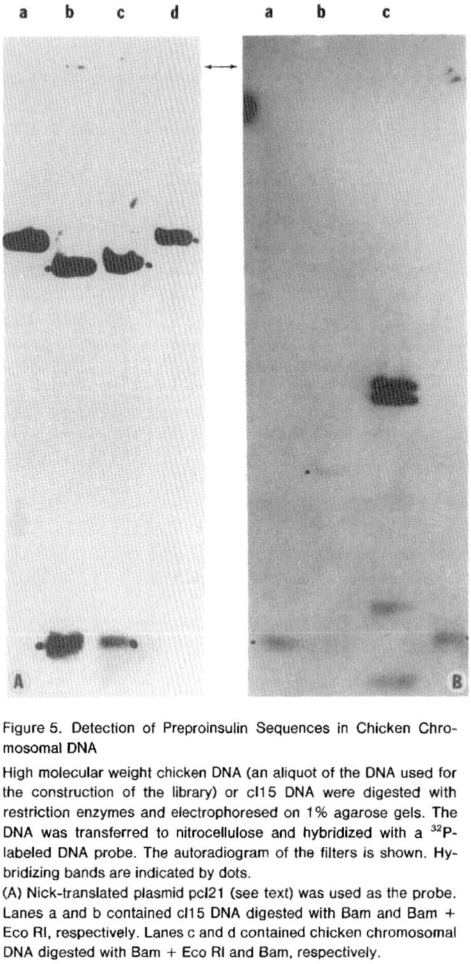

Figure 5. Detection of Preproinsulin Sequences in Chicken Chromosomal DNA

High molecular weight chicken DNA (an aliquot of the DNA used for the construction of the library) or cl15 DNA were digested with restriction enzymes and electrophoresed on 1% agarose gels. The DNA was transferred to nitrocellulose and hybridized with a ^{32}P-labeled DNA probe. The autoradiogram of the filters is shown. Hybridizing bands are indicated by dots.

(A) Nick-translated plasmid pcl21 (see text) was used as the probe. Lanes a and b contained cl15 DNA digested with Bam and Bam + Eco RI, respectively. Lanes c and d contained chicken chromosomal DNA digested with Bam + Eco RI and Bam, respectively.

(B) A nick-translated 162 bp Bam-Taq I fragment (Figure 3) was used as the probe. Lane a contained cl15 DNA digested with Bam plus Eco RI; a 0.8 kb fragment hybridizes. Lanes b and d contained chicken chromosomal DNA digested with Hinf (2.45 kb band) and Bam + Eco RI, respectively. Lane c contained, as markers, fragments of plasmid pl19 produced by various restriction enzymes.

mammalian C peptides by introducing gaps at certain amino acid sites. The duck and chicken C peptides are highly homologous, except that the chicken chain has two more amino acids at its carboxy terminus. Based on the available nucleotide sequence information we revised their alignment to that shown in Figure 6. Although this alignment seems convincing enough so that we can use it for our calculations (see below), it can be firmly established only when more sequences from other animal classes (for example, reptiles) become available.

It has been suggested (Tager and Steiner, 1972) that the ancestral mammalian C peptide had at least 31 amino acids. This was based on the observation that three different mammalian orders (primates, rodents and perissodactyls) have C peptides of this length that are homologous at their terminal segments. Since the chicken C peptide has a two-amino acid carboxy terminal extension, despite its internal deletions, we suggest that the more primitive C peptides (for example, those of reptiles) are even longer.

The DNA sequencing data establish the complete amino acid sequence of the preregion of chicken preproinsulin. Comparison with the complete or partial sequences of the preregions of other preproinsulins (Figure 7) indicates that the first part of this segment (14 residues in human, rat and chicken preproinsulin) is conserved, while the second part (10 residues) is divergent.

The Introns of Preproinsulin Genes

The human, rat II and chicken preproinsulin genes contain two introns of the following lengths: 178 and 786 bp (human), 119 and 499 bp (rat II) and 119 bp and 3.5 kb (chicken). The variation in length is much greater for the large introns, which is also the case for the globin genes. This, along with the observation that the functional rat I gene lacks the large intron but retains the small one, suggests that small and large introns in genes like those for preproinsulin and globin might be functionally different entities.

The most straightforward interpretation of the preproinsulin gene structures is that the rat gene II structure (two introns) reflects that of the single ancestral gene. During or after the duplication event that occurred since the mammalian radiation, one of the products of duplication lost the large intron in such a precise way that the coding region of the gene was left intact. Since the chance occurrence of such a precise deletion should be extremely rare (discussed by Gilbert, 1979), we have to conclude that some enzymatic mechanism operating in the germ line was responsible. Moreover, if the role of this intron was to bring together its two flanking exons at some early time during evolution (Gilbert, 1979), its loss should be neutral.

One key element of this argument is that the gene duplication is recent. We reject the null hypothesis of an ancient duplication event as follows. Let us assume that an ancestral preproinsulin gene (at the invertebrate stage of evolution) was duplicated to produce a functional gene and a pseudogene. Since the assumption that during or after this duplication the pseudogene lost the large intron changes the timing, but not the main conclusion, we will assume that the large intron was inserted into the gene while the pseudogene retained the structure of the ancestral gene. If this had been correct, all vertebrates (including chicken) would have a silent pseudogene, somehow activated in rats and mice after the mammalian radia-

```
                        Pro                  Glu
RAT I          G        CCA                  GAG     G  T                T
         GluValGluAspProGlnValAlaGlnLeuGluLeuGlyGlyGlyProGlyAlaGlyAspLeuGlnThrLeuAlaLeuGluValAlaArgGln
RAT II   GAAGTGGAGGACCCACAAGTGGCACAACTGGAGCTGGGTGGAGGCCCGGGGGCCGGTGACCTTCAGACCTTGGCACTGGAGGTGGCCCGGCAG
         GluAlaGluAspLeuGlnValGlyGlnValGluLeuGlyGlyGlyProGlyAlaGlySerLeuGlnProLeuAlaLeuGluGlySerLeuGln
HUMAN    GAGGCAGAGGACCTGCAGGTGGGGCAGGTGGAGCTGGGCGGGGGCCCTGGTGCAGGCAGCCTGCAGCCCTTGGCCCTGGAGGGGTCCCTGCAG
         AspValGluGlnProLeuValSerSerPro   LeuArgGly      GluAlaGlyValLeu   ProPheGlnGlnGluGluTyr   GluLysVal
CHICKEN  GATGTCGAGCAGCCCCTAGTGAGCAGTCCC---TTGCGTGGC------GAGGCAGGAGTGCTG---CCTTTCCAGCAGGAGGAATAC---GAGAAAGTC
                        AsnGly       His         Val  Glu          His                     Gln
DUCK                    AAYGGX       CAY         GTX  GAR          CAY                     CAR------
```

Figure 6. Alignment of Mammalian and Avian C Peptides

The sequences are from Lomedico et al., 1979 (rat I and II); Bell et al., 1979 (human); and Markussen and Sundby, 1973 (duck). Y is a pyrimidine, R is a purine and X is any base.

```
HAGFISH             - Leu - - - LeuAlaAla - - - Leu - LeuLeuLeu - - Ala - - - Ala - - ↓Ala
SEA RAVEN     MetMet - - Leu - - - - - - LeuLeu - LeuLeu - Leu - - - - - - - ↓-
ANGLERFISH           - AlaLeu - Leu - - Phe - LeuLeuValLeuLeuValVal - - - - - - AlaVal↓
BOVINE                 - Leu - - - Leu - - LeuLeu - LeuLeu - Leu - - - - - - - - ↓Phe
RAT I               MetAlaLeuTrpIleArgPheLeuProLeuLeuAlaLeuLeuIleLeuTrpGluProArgProAlaGlnAla↓Phe
RAT II                       Met                               Val       Lys
HUMAN                        Met  Leu                          Ala   Gly Asp    Ala  ↓
CHICKEN                      Ile  Ser                          ValPheSerGly GlyThrSerTyr ↓Ala
```

Figure 7. Comparison of the Amino Acid Sequences of the Preregions of Preproinsulins

Hagfish (Chan et al., 1979); sea raven and anglerfish (Shields and Blobel, 1977); bovine (Lomedico et al., 1977); rat I and II (Lomedico et al., 1979); and chicken. Dashes correspond to unknown amino acids, while empty spaces show identity with the rat I sequence.

tion (it was also independently activated in three fish species). Since the presence of the pseudogene was not documented in the chicken by blot-hybridization, we have to conclude that if it exists its structure has diverged to a great extent from that of the functional gene. This contradicts the fact that the two rat genes are very similar in sequence. Even if the two genes are linked (we have no information on this point), correction of the gene and activated pseudogene by coincidental evolution (Hood, Campbell and Elgin, 1975; Lauer, Shen and Maniatis, 1980; Zimmer et al., 1980) through a minimum of two rounds of unequal crossing over events (Smith, 1973) cannot be invoked. Not only would the recombination event be unlikely because of the presence of a long nonhomologous region in one of the sequences, but also, if it had occurred, the genes would not have retained their different intron structure.

Molecular Evolution of Preproinsulin Genes

What can we learn from a direct comparison of the DNA sequences of these preproinsulin genes?

Over evolutionary time protein sequences and gene sequences diverge. By comparing the products of related genes in different species or of diverging genes within the same species, one can observe that different proteins accumulate mutations at different rates. This accumulation is proportional to the divergence time (the evolutionary clock hypothesis; see Wilson, Carlson and White, 1977). There are two extreme views that attempt to explain this property. One is the neutralist position which argues that some 99% of the changes observed in amino acid sequences are neutral changes, which have no effect on the function of the protein, fixed in the population by random drift (for reviews see Kimura and Ohta, 1974; Wilson et al., 1977; Kimura, 1979). This hypothesis asserts that the rate of fixation is directly proportional to the underlying mutation rate, the different rate in different proteins then being due to the elimination of a greater or lesser fraction of the total possible changes because they are deleterious. Proteins with highly constrained structures would be expected to fix mutations at a slower rate because only a small fraction of the sites could be neutral (Dickerson, 1971). The other hypothesis is that the observed changes in protein structure have spread through the population by selection (Clarke, 1970; Richmond, 1970; Blundell and Wood, 1975). The constancy in the rate of fixation is then attributed to an approximate constancy in the rate of selection; proteins such as globin would mutate continually, although slowly, around a mean structure toward slightly better function within a continually varying cell milieu.

The DNA sequences of corresponding genes provide a wider set of data for these comparisons. Not only can we compare two genes in terms of how many base changes result in replacement of one amino acid by another, but we can also compare the changes at silent sites (those sites at which nucleotide changes do not replace amino acids). Furthermore, we can determine the rates of nucleotide change in introns or in the regions surrounding the gene. While changes at replacement sites can easily be acted upon by selection, the changes in the silent sites or in the introns are putatively neutral. On the neutralist basis we would expect the rate of appearance of such changes to be very high and the same in all genes, reflecting the true mutation rate. We previously made such a comparison for the two rat preproinsulin genes (Lomedico et al., 1979). In that case we concluded that the silent positions and the small intron drifted comparably (20–25% changes) and showed a rate of fixation some six times greater than all the replacement sites. Now we can look at the same phenomenon over a much longer time scale by comparing the rat genes with the chicken gene, the bird/mammal divergence occurring some 250–300 million years (MY) ago (Dickerson, 1971; Moore et al., 1976; Wilson et al., 1977). We can also compare another mammalian gene, the one encoding human preproinsulin (Bell et al., 1979).

Table 1 shows the results of these calculations. As described in Experimental Procedures, we have tried to correct the mutation rate more accurately for multiple events. These numbers, however, do not differ greatly from those derived from the simpler calculation we used before (Lomedico et al., 1979). The top section of Table 1 shows in general that the silent changes continue to increase as we go back to greater times of divergence. The C peptide area of the molecule provides us with a region in which there is a rapid accumulation of both replacement and silent substitutions. As is well known, the A and B chains are quite constrained (Kimura and Ohta, 1974; Wilson et al., 1977). The C peptide provides an internal standard to test the neutral hypothesis because its role does not depend greatly on its structure. Actually its replacement by a short synthetic bridge allows proper folding of the "proinsulin" molecule (Busse and Carpenter, 1976), although we do not know how such a change would affect other properties such as solubility.

Figure 8A shows that the accumulation of replacement changes in the C peptide is linear with time, occurring at about five times the corresponding rate in the A and B chains. The accumulation of silent changes in the C peptide is far more rapid but not as linear with time, possibly breaking at around 85 MY.

Table 1. Percentage Corrected Divergence of Gene Sequences

	Replacement Sites				Silent Sites				
Preproinsulin Genes	A and B Chains	C Peptide	C* Peptide	All-C	A and B Chains	C Peptide	C* Peptide	All-C	Small Intron
Rat I/II	1.8	3.2		3.5	32.3	18		23.5	21
Human/rat	5.2	19.2	21	7.5	76	91.4	110.2	64.8	
Human/chicken	8		62.5	13.5	122		139.5	133.3	
Rat/chicken	10.7		49.4	15.2	64		150	141	163
Globin Genes	α	β	α/β		α	β	α/β		β Small Intron
Human/rabbit	10.8	5.6			31.8	45.5			48.5
Human/mouse	8.4	12.9			83	49.4			52.6
Rabbit/mouse	10.9	12.1			81.7	63.8			64.6
Human/chicken	20.9	22.9			74.6	70.1			
Rabbit/chicken	22.8	23.3			63.9	80.8			
Mouse/chicken	20	26.7			87.2	78.8			
Human			46.3				89.5		
Rabbit			48.5				91.7		
Mouse			51.1				120		α/β 88.6
Chicken			51				87		
	Replacement Sites		Silent Sites	Small Intron	Large Intron				
Mouse $\beta^{major}/\beta^{minor}$	3.1		10.6	2.7	14				
Rabbit β alleles	1.5		0	0	0.35				
Growth Hormone Genes									
Human/rat	19.4		71						

The percentage corrected divergence of each pair of sequences was calculated as described in Experimental Procedures. The C* peptide for preproinsulin refers to the new alignment (Figure 6). Codons corresponding to gaps were excluded from the calculation. All-C includes the preregion (with the exception of the ATG initiator), the A and B chains and the four basic amino acids, but not the C peptide. In the α/β comparison we excluded the codons of one sequence corresponding to gaps in the other. The third exons of both sequences were aligned exactly. The data for globin genes are from R. M. Lawn et al., manuscript in preparation (human-β); Efstratiadis et al., 1977; Hardison et al., 1979; van Ooyen et al., 1979 (rabbit-β); Konkel et al., 1979 (mouse-β, minor and major); Richards et al., 1979 (chicken-β); Forget et al., 1979 (human-α); Heindell et al., 1978 (rabbit-α); Nishioka and Leder, 1979 (mouse-α). The chicken α-globin sequence (Salser et al., 1979) is unusual and does not correspond in its entirety to the two known amino acid sequences for chicken α-globins. However, the sequence represents a transcribable and therefore functional α-globin gene.

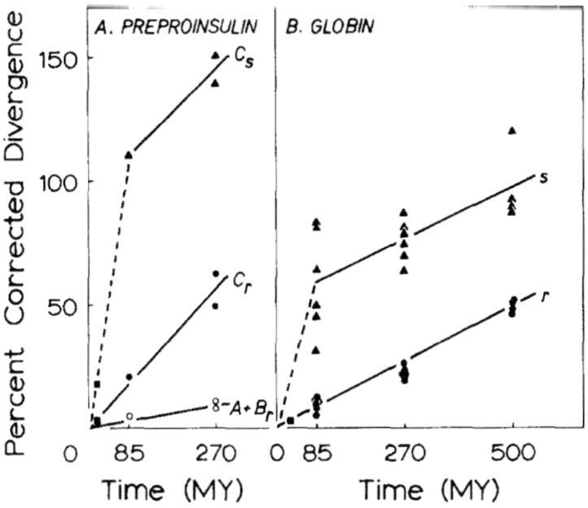

Figure 8. Plot of the Sequence Divergence of Silent and Replacement Substitutions against the Divergence Time

(A) The silent and replacement changes of the C peptide (Cs and Cr, respectively) correspond to the C* column of Table 1. A + B_r are the replacement changes in the A and B chains. The filled squares correspond to the changes in the C peptide of the rat genes I and II. (B) Silent (s) and replacement (r) changes in globin genes (Table 1). The filled square corresponds to the replacement changes in the mouse $\beta^{major}/\beta^{minor}$ genes.

The initial rate corresponds to a UEP (unit evolutionary period) of 0.7 or a mutation rate of 7×10^{-9} substitutions per nucleotide site per year. (UEP is the time in MY required for the fixation of 1% changes between two lines.) This high figure may still underestimate the true rate.

To explore this further we examined data from various globin genes. Table 1 shows pairwise comparisons between the α- and β-globin genes of human, rabbit, mouse and chicken, and comparisons of the α and β sequences within each species. This latter comparison is of two genes which have diverged some 500 MY ago (Dickerson, 1971). The silent sites exhibit wide spread in the rates. Where the comparison can be made, small introns and the silent sites show the same drift. Again the silent sites accumulate changes much more rapidly than the replacement sites. Although the data for the silent sites have been corrected for multiple events, the silent substitutions do not provide a good evolutionary clock extending over a long time scale. Figure 8B compiles the data for the globins and shows that the replacement sites provide an excellent clock. There is only a small amount of scatter, and the points lie on a line that extrapolates through the origin. The figure also indicates that the silent changes show a great deal of scatter. Their averages, however, define a line having the same slope as that of the replacements, which does not extrapolate through the origin. Initially the silent sites accumulate changes some seven times more rapidly than the replacement sites, but that accumulation saturates at 85–100 MY. Beyond that time the increase is not so rapid. The immediate implication is that there is selective pressure on some fraction of the sites. Thus we observe a rapid accumulation only in that fraction that is neutral. (Our preliminary analysis indicates that the other fraction consists of those silent changes in the same codon with one or two replacement substitutions.) The greater rate of appearance of silent changes in the C peptide region of the preproinsulin genes is obviously due to the existence of fewer constraints. In general, these constraints could be due to the imposition of required secondary structure at the mRNA level or could reflect specific anticodon requirements (see also Kafatos et al., 1977).

In a similar comparison between human and rat growth hormone (Seeburg et al., 1977; Roskam and Rougeon, 1979; Goeddel et al., 1979; Martial et al., 1979) the silent changes are like those for globin or insulin A and B chains, while the replacement sites behave like those for the C peptide (Table 1).

Thus not all of the silent sites are neutral. In the fraction that is neutral, the mutation rate is very great, so that the changes saturate in a divergence time comparable to that of the mammalian radiation.

We can use both the silent changes and the replacement changes as clocks over more recent time intervals. In Figure 8A we have put the divergence of the C peptide sequence of rat I and II genes, calculated from the replacement sites, on the line. This suggests a divergence time of approximately 15 MY. The corresponding silent changes are in agreement.

Consider the pair of mouse β^{major} and β^{minor} globin genes (Konkel, Maizel and Leder, 1979). The replacement changes (3.1%) lead us to an estimate of 30 MY (Figure 8B) for their divergence (the time since their last correction by coincidental evolution), although the silent changes (11%) are somewhat lower than expected. The changes in the large intron (14%) are close to the prediction. The figure from the small intron, however, is as low as the replacement changes (Table 1).

Is the clock provided by the replacement sites neutralist in origin? The high intrinsic rate of silent changes renders untenable the simple neutralist assumption that the low fixation rate of changes in globin genes is due to the existence of a small fraction of neutral sites where mutations can be fixed by drift. This interpretation would assert that the observed UEP of 10 (from Figure 8B) for replacements in globin is due to neutral changes occurring with a mutation UEP of 0.7 (that of the preproinsulin C peptide region), but only 7% of the possible changes at replacement sites could be fixed. Since the total number of possible replacement changes is about three times the number of replacement sites, this maximum number of neutral changes at replacement sites of globin corresponds to the possibility of three changes at 7% of the sites, or one change at 21% of the sites, or a combination. At most, the changes at the greater number of sites

(21%) should saturate at 11% of the sites (50% of 21%). Furthermore, the changes should saturate at these low percentages (less than 11%) in 85–100 MY, the same period required to saturate the silent changes. It is clear from Figure 8B or Table 1 that the clock extending back to the α/β divergence involves a higher percentage of replacement sites and does not saturate. We conclude that the evolutionary clock does not involve neutral changes occurring at a small number of replacement sites. Either we make more complex ad hoc assumptions to save the neutralist position (such as that each neutral change after fixation opens up new sites for further neutral change, which would have been deleterious before, so that the neutral fraction can drift over the surface of the molecule), or we conclude that the fixation of these replacement changes is due to a constant rate of selection.

The analysis offered enables us to point to a clear-cut example of selection in the globins. Two common rabbit β–globin alleles have been sequenced (Efstratiadis, Kafatos and Maniatis, 1977; Hardison et al., 1979; van Ooyen et al., 1979). These chains differ by four amino acids because of four single replacement substitutions. Silent changes are completely absent in the coding regions and the small intron, while there are two base changes in the 573 bp large intron. For the globin genes, the UEP of 10 means that there is one actual replacement substitution in the coding region (333 sites) every 3 MY of divergence. The neutral UEP of 1.4 for globin corresponds to one silent change in the 105 available sites every 1.3 MY, or to 2.3 silent changes for each replacement. Similarly, we expect a further 12 changes in the large intron for every replacement. Thus our expectation, based on drift, is at least 60 changes between the two rabbit alleles rather than the two observed. The four amino acid changes were fixed so rapidly by selection that there was no opportunity for silent changes to appear. It is likely that the amino acid changes are such that only their combination was selected. If there is any element of neutral drift we would conclude that this selection occurred approximately 0.5 MY ago (0.35% with a UEP of 1.4). However, the alternative interpretation of neutral mutations piggy-backing on the rarer, positively selected changes would permit us only to estimate that the event took place in the last 15 MY (1.5% with a UEP of 10). In any case, these two alleles of β–globin show that selection has occurred in the recent past.

In summary, then, we conclude that the driving force for fixation is selection operating on some fraction of the replacement changes. Each fixation carries along with the selected mutation other alterations, neutral in their effect, that have accumulated in that region of the DNA. These other neutral alterations include changes at silent sites, in introns, and unselectable changes at replacement sites. The less constrained the function of an amino acid sequence, such as the C peptide of insulin, the more easily it accumulates changes fixed by rare selectable events.

Experimental Procedures

The procedures used in this work have been described previously (Villa-Komaroff et al., 1978; Lomedico et al., 1979). In certain cases we sequenced (Maxam and Gilbert, 1977) 3′ end-labeled DNA fragments in addition to 5′ end-labeled fragments. For 3′ end-labeling, we end-filled by reverse transcriptase the ends of framents generated by restriction enzymes that leave 5′ extensions as follows. Following restriction endonuclease digestion (1–20 μg of DNA in a 10 μl reaction) the sample was heated at 65°C for 15 min. The volume was then adjusted to 20 μl and the end-filling reaction was carried out at 42°C for 2 hr in the presence of 100 mM Tris-HCl (pH 8.3), 100 mM KCl, 30 mM 2-mercaptoethanol, 5 μM (minimum) of all four α–^{32}P–dNTPs (300 Ci/mmole; New England Nuclear) and 15 units reverse transcriptase. Digestion with a second restriction enzyme (to generate uniquely labeled fragments) was usually carried out in sequence by first heat-inactivating the reverse trascriptase and then increasing the sample volume. Although 3′ end-labeled fragments have more than one labeled site at their end, the sequencing ladders are usually (but not always) unambiguous.

Heteroduplexes were formed as described by Davis, Simon and Davidson (1971). The spreading solution was 40% formamide, 50 μg/ml cytochrome c, 0.1 M Tris-HCl, 0.01 M EDTA (pH 8.5) and DNA. It was spread onto 10% formamide, 0.01 M Tris-HCl, 0.001 M EDTA (pH 8.5) at 23°C.

A Method for Calculating Sequence Divergence

We assume that the nucleotide substitutions (changes) are Poisson-distributed. Changes in the coding regions of each pair of sequences that we compare either result in amino acid replacements (replacement substitutions) or lead to the appearance of synonymous codons (silent substitutions).

First we count the number of potential silent or replacement sites. We consider the possible changes at each position of a codon and assign scores as shown in Table 2. We count separately the number of sites that can afford one, two or three changes and average each of the three categories between the two sequences that we compare.

Then we compare the two sequences codon by codon, score with one point each silent or replacement substitution, and categorize it according to the type of site in which it has occurred. For example, if the codons we compare are Pro CCC–Pro CCT, there is one change of category 3 of silent substitutions; if the codons are Val GTC–Ala GCC, there is one change of category 3 of replacement substitutions; the codons Leu CTC–Phe TTT have one change of category 3 of replacement substitutions (first codon position), a half change of category 3 of silent substitutions and a half change of category 1 of silent substitutions (third codon position); the change in the third position of codons Ile ATC–Thr ACG is half silent (category 3) and half replacement (category 1).

Finally, we divide the sum of silent or replacement substitutions in each category by the corresponding potential sites. Each percentage change (λ) is then corrected for multiple events by one of the following formulas:

Category 1: $3 \times \quad [-\tfrac{1}{2} \ln(1 - 2\lambda)]$ (1)

Category 2: $3/2 \times [-\tfrac{2}{3} \ln(1 - \tfrac{3}{2}\lambda)]$ (2)

Category 3: $[-\tfrac{3}{4} \ln(1 - \tfrac{4}{3}\lambda)]$ (3)

The brackets are correction formulas relating the observed frequency to the Poisson-distributed frequency of mutational hits for mutations restricted to one, two or three substitutions, respectively. The factors outside the brackets correct these restricted values up to a total mutation rate, assuming that transitions and transversions are equally probable (Kimura and Ohta, 1972; Salser, 1977).

For each pair of sequences the overall divergence is the average of two weight averages of the corrected percentages, using separately as weighing factors the number of sites or the number of changes in each category. Two categories were exceptionally small

Table 2. Scores of Silent and Replacement Sites of Codons

Met, Trp	0 0 0 **A B C** 3 3 3
Phe, Tyr, His, Gln, Glu Asn, Asp, Cys, Lys, Ser AGY	0 0 1 **A B C** 3 3 2
Ile	0 0 2 **A B C** 3 3 1
Val, Pro, Thr, Ala, Gly, Ser UCX Leu CUY, Arg CGY	0 0 3 **A B C** 3 3 0
Leu CUR, Arg CGR	1 0 3 **A B C** 2 3 0
Leu UUR, Arg AGR	1 0 1 **A B C** 2 3 2

The number of potential silent or replacement changes is shown above or below each codon (**ABC**), respectively. Y is a pyrimidine, R is a purine and X is any base.

and were dropped from the calculation: the 2-substitution category for the silent sites and the 1-substitution category for the replacement sites.

To calculate divergence in small introns, we used the following method. When two introns are of the same length (as in rat and chicken preproinsulin genes), they are aligned and compared directly. This comparison is simply indicative and not necessarily valid, since deletions and additions might have occurred but left by chance the same number of nucleotides in the intron. We note, for example, that although the small introns do not differ in size between rat and chicken, they are both smaller than the corresponding human intron. The sequence of the small globin introns that are of different lengths is written pairwise and then the entire length is screened for homologous blocks of nucleotides (≥4 nucleotides per block). Usually it becomes immediately obvious where a deletion has occurred in one of the two sequences, because by postulating it, good alignment is achieved for relatively long segments. In this primary alignment an effort is made not to maximize the homology but to minimize as much as possible single nucleotide deletions. In a second stage, the regions that lie between blocks and have the same length are compared and each homologous or nonhomologous nucleotide pair is scored as +1 or −1, respectively. Regions that lie between blocks and have unequal lengths are aligned by moving the shorter sequence, without ever interrupting it, relative to the longer one until maximum homology is achieved. Each nucleotide gap is scored as −1. The GT and AG dinucleotides bracketing the intron sequence are excluded from the comparison. The $\beta^{major}/\beta^{minor}$ large intron comparison was made simply by excluding all additions and deletions and calculating a point mutation divergence in the residual common sequence.

Acknowledgments

This work was supported by an NIH grant and a Basil O'Connor Starter Research grant from the National Foundation, March of Dimes, to A. E., and by an NIH grant to W. G. F. P. was suported by an NIH postdoctoral fellowship. P. L. is a fellow of the Juvenile Diabetes Foundation. A. E. is the Harvard Medical School Hsien Wu Investigator. W. G. is an American Cancer Society Professor of Molecular Biology.

The costs of publication of this article were defrayed in part by the payment of page charges. This article must therefore be hereby marked "advertisement" in accordance with 18 U.S.C. Section 1734 solely to indicate this fact.

Received February 28, 1980; revised April 8, 1980

References

Bell, C. I., Swain, W. P., Pictet, R., Cordell, B., Goodman, H. M. and Rutter, W. (1979). Nature *282*, 525.

Benton, W. D. and Davis, R. W. (1977). Science *196*, 180.

Berk, A. J. and Sharp, P. A. (1977). Cell *12*, 721.

Blattner, F. R., Williams, B. G., Blechl, A. E., Thompson, K. D., Faber, H. E., Furlong, L. A., Grunwald, D. J., Sheldon, E. L. and Smithies, O. (1977). Science *196*, 161.

Blundell, T. L. and Wood, S. P. (1975). Nature *257*, 197.

Breathnach, R., Benoist, C., O'Hare, K., Gannon, R. and Chambon, P. (1978). Proc. Nat. Acad. Sci. USA *75*, 4853.

Busse, W. and Carpenter, F. H. (1976). Biochemistry *15*, 1649.

Casey, J. and Davidson, N. (1977). Nucl. Acids Res. *4*, 1539.

Chan, S. J. and Steiner, D. F. (1977). Trends Biochem. Sci. *2*, 254.

Chan, S. J., Patzelt, C., Duguid R., Quinn, P. S., Labrecque, A., Noyes, B. E., Keim, P. S., Henrikson, R. L. and Steiner, D. F. (1979). In Miami Symposia, *16*, T. Russel, K. Breu, H. Faber and J. Schultz, eds. (New York: Academic Press), p. 361.

Clark, J. L. and Steiner, D. F. (1969). Proc. Nat. Acad. Sci. USA *62*, 278.

Clarke, B. (1970). Science *168*, 1009.

Davis, R. W., Simon, M. and Davidson, N. (1971). In Methods in Enzymology, *21*, L. Grossman and K. Moldave, eds. (New York: Academic Press), p. 413.

Dayhoff, M. O. (1978). Atlas of Protein Sequence and Structure, 5 (Washington, D.C.: National Biomedical Research Foundation), Supp. 3.

Dickerson, R. E. (1971). J. Mol. Evol. *1*, 26.

Dodgson, J. B., Strommer, J. and Engel, J. D. (1979). Cell *17*, 879.

Efstratiadis, A., Kafatos, F. C. and Maniatis, T. (1977). Cell *10*, 571.

Forget, B. G., Cavallesco, C., deRiel, J. K., Spritz, R. A., Choudary, R. V., Wilson, J. T., Wilson, L. B., Reddy, V. B. and Weissman, S. M. (1979). In Eukaryotic Gene Regulation, ICN-UCLA Symposia on Molecular and Cellular Biology, *14*, R. Axel, T. Maniatis and C. F. Fox, eds. (New York: Academic Press), p. 367.

Frank, B. H. and Veros, A. J. (1970). Biochem. Biophys. Res. Commun. *38*, 284.

Gannon, F., O'Hare, K., Perrin, F., LePennec, J. P., Benoist, C., Cochet, M., Breathnach, R., Royal, A., Garapin, A., Cami, B. and Chambon, P. (1979). Nature *278*, 428.

Gilbert, W. (1979). In Eukaryotic Gene Regulation, ICN-UCLA Symposia on Molecular and Cellular Biology, *14*, R. Axel, T. Maniatis and C. F. Fox, eds. (New York: Academic Press) p. 1.

Goeddel, D. V., Heyneker, H. L., Hozumi, T., Arentzen, R., Itakura, K., Miozari, G., Crea, R. and Seeburg, P. H. (1979). Nature *281*, 544.

Goldberg, M. (1979). Ph.D. thesis, Stanford University, Stanford, California.

Goodman, H. M. (1980). Abstracts, The Twelfth Miami Winter Symposium (New York: Academic Press), p. 14.

Grant, P. T., Coombs, T. L. and Franks, B. H. (1972). Biochem. J. *126*, 433.

Gruss, P., Lai, C., Dhar, R. and Khoury, G. (1979). Proc. Nat. Acad. Sci. USA *76*, 4317.

Hamer, D. H. and Leder, P. (1979). Cell *18*, 1299.

Hamer, D. H., Smith, K. D., Boyer, S. H. and Leder, P. (1979). Cell *17*, 725.

Hardison, R. C., Butler, E. T., III, Lacy, E., Maniatis, T., Rosenthal, N. and Efstratriadis, A. (1979). Cell *18*, 1285.

Heindell, H. C., Liu, A., Paddock, G. V., Studnicka, G. M. and Salser,

W. A. (1978). Cell *15*, 43.

Hood, L., Campbell, J. H. and Elgin, S. C. R. (1975). Ann. Rev. Genet. *9*, 305.

Humbel, R. E., Bosshard, H. R. and Zahn, H. (1972). In Handbook of Physiology: Endocrinology (Washington, D.C.: American Physiological Society), p. 111.

Kafatos, F. C., Efstratiadis, A., Forget, B. G. and Weissman, S. M. (1977). Proc. Nat. Acad. Sci. USA *74*, 5618.

Kimura, M. (1979). Sci. Am. *241*, 98.

Kimura, M. and Ohta, T. (1972). J. Mol. Evol. *2*, 87.

Kimura, M. and Ohta, T. (1974). Proc. Nat. Acad. Sci. USA *71*, 2848.

Konkel, D. A., Maizel, J. V., Jr. and Leder, P. (1979). Cell *18*, 865.

Lauer, J., Shen, C.-K. J. and Maniatis, T. (1980). Cell *20*, 119–130.

Lomedico, P. T., Chan, S. J., Steiner, D. F. and Saunders, G. F. (1977). J. Biol. Chem. *252*, 7971.

Lomedico, P., Rosenthal, N., Efstratiadis, A., Gilbert, W., Kolodner, R. and Tizard, R. (1979). Cell *18*, 545.

McKenna, M. C. (1975). In The Phylogeny of the Primates, W. P. Luckett and F. S. Szalay, eds. (New York: Plenum Press), p. 21.

Markussen, J. and Sundby, F. (1973). Eur. J. Biochem. *34*, 401.

Martial, J. A., Hallewell, R. A., Baxter, J. D. and Goodman, H. M. (1979). Science *205*, 602.

Maxam, A. M. and Gilbert, W. (1977). Proc. Nat. Acad. Sci. USA *74*, 560.

Moore, G. W., Goodman, M., Callahan, C., Holmquist, R. and Moise, H. (1976). J. Mol. Biol. *105*, 15.

Nishioka, Y. and Leder, P. (1979). Cell *18*, 875.

Richards, R. I., Shine, J., Ullrich, A., Wells, J. R. E. and Goodman, H. M. (1979). Nucl. Acids Res. *7*, 1137.

Richmond, R. C. (1970). Nature *225*, 1025.

Romero-Herrera, A. E., Lehmann, H., Joysey, K. A. and Friday, A. E. (1973). Nature *246*, 389.

Roskam, W. G. and Rougeon, F. (1979). Nucl. Acids Res. *7*, 305.

Salser, W. (1977). Cold Springer Harbor Symp. Quant. Biol. *42*, 985.

Salser, W. A., Cummings, I., Liu, A., Strommer, J., Padayatty, J. and Clarke, P. (1979). In Cellular and Molecular Regulation of Hemoglobin Switching, G. Stamatoyannopoulos and A. W. Nienhuis, eds. (New York: Grune and Stratton), p. 621.

Seeburg, P. H., Shine, J., Martial, J. A., Baxter, J. D. and Goodman, H. M. (1977). Nature *270*, 486.

Shields, D. and Blobel, G. (1977). Proc. Nat. Acad. Sci. USA *74*, 2059.

Smith, G. P. (1973). Cold Spring Harbor Symp. Quant. Biol. *38*, 507.

Southern, E. M. (1975). J. Mol. Biol. *98*, 503.

Steiner, D. F., Kemmler, W., Tager, H. S. and Peterson, J. D. (1974). Fed. Proc. *33*, 2105.

Tager, H. S. and Steiner, D. F. (1972). J. Biol. Chem. *247*, 7936.

van Ooyen, A., van den Berg, J., Mantei, N. and Weissmann, C. (1979). Science *206*, 337.

Villa-Komaroff, L., Efstratiadis, A., Broome, S., Lomedico, P., Tizard, R., Naber, S., Chick, W. and Gilbert, W. (1978). Proc. Nat. Acad. Sci. USA *75*, 3727.

Wilson, A. C., Carlson, S. S. and White, T. J. (1977). Ann. Rev. Biochem. *46*, 573.

Zimmer, E. A., Martin, S. L., Beveley, S. M., Kan, Y. W. and Wilson, A. C. (1980). Proc. Nat. Acad. Sci. USA, in press.

Cell, Vol. 40, 81-90, January 1985, Copyright © 1985 by MIT 0092-8674/85/010081-10 $02.00/0

Intron/Exon Structure of the Chicken Pyruvate Kinase Gene

Nils Lonberg and Walter Gilbert
Department of Biochemistry and Molecular Biology
Harvard University
Cambridge, Massachusetts 02138
and Biogen Research Corporation
Cambridge, Massachusetts 02142

Summary

The chicken pyruvate kinase gene is interrupted by at least ten introns, including nine introns within the coding region. We compare the structure of this gene with the three-dimensional protein structure of the homologous cat muscle enzyme. The introns are not randomly placed—they divide the coding sequence into fairly uniformly sized pieces encoding discrete elements of secondary structure. The introns tend to fall at interruptions between stretches of α-helix or β-sheet residues, and each of the six exons that contribute to the barrel-shaped central domain include one or two repeats of a simple unit, an α-helix plus a β strand. This structure suggests that introns were not inserted into a previously uninterrupted coding sequence, but instead are products of the evolution of the first pyruvate kinase gene. We have found some sequence homology between a segment of pyruvate kinase and the structurally homologous mononucleotide binding fold of alcohol dehydrogenase. The superposition of these two regions aligns an intron from the maize alcohol dehydrogenase gene four nucleotides from an intron in the chicken pyruvate kinase gene.

Introduction

Most genes found in higher eukaryotic organisms are broken up into short (usually 100 to 200 bp) coding segments (exons) interrupted by long stretches of noncoding DNA (introns). In contrast, the coding sequences of most prokaryotic and lower eukaryotic genes are contiguous. Gilbert (1978) suggested that the intron/exon structure of eukaryotic genes might be a record of their evolutionary history: these genes evolved by exploiting RNA splicing to recruit and combine small segments of coding sequence. Evidence in support of this hypothesis is provided by the structure of such genes as those coding for collagen (Tate et al., 1982) and serum albumin (Alexander et al., 1981), which reflect the modular repeated structures of the corresponding proteins, and those coding for immunoglobulins (Tonegawa, 1983), which are broken up into pieces encoding distinct functional domains. Doolittle (1978) extended this hypothesis to suggest that introns were present in the genomes of primitive organisms, and that prokaryotes and lower eukaryotes have simply eliminated introns in the process of streamlining their genomes. To test this hypothesis we have examined the structure of a gene that evolved prior to the prokaryote/eukaryote split. Such a gene might be expected to contain introns if they existed in the common ancestor of prokaryotes and eukaryotes. If the gene was originally assembled from smaller exonic modules, the location of these introns might reflect the structure of the protein.

The glycolytic enzyme pyruvate kinase (PK;ATP:pyruvate 2-0-phosphotransferase, EC 2.7.1.40) is an ancient protein that has been found to have an overall structure that is similar in bacteria, fungi, and animals—a tetramer composed of identical 50 to 60 kd subunits (Kayne, 1974). The yeast gene has been sequenced and contains no introns (Burke et al., 1983); presumably, the same is true for bacterial PK genes. We have investigated the structure of the chicken PK gene; yeast and chicken PK are homologous at 45% of their amino acids (Lonberg and Gilbert, 1983). However, unlike the yeast gene, the chicken PK gene is interrupted by at least ten introns, nine of which are within the coding sequence. We have looked at the position of these introns for clues to their origin. They do not appear to have been randomly inserted into the gene. Instead, introns are found in locations that suggest the PK gene evolved by combining smaller protein coding units and exploiting RNA splicing to express them as a single polypeptide.

Results

Isolation of Clones

We screened approximately 240,000 recombinant phage plaques from a chicken genomic library using a cloned chicken muscle pyruvate kinase cDNA probe. We picked and rescreened 25 positive clones; 13 hybridized weakly to the probe sequences, while the other 12 hybridized strongly. We chose 6 strongly hybridizing and 6 weakly hybridizing clones for restriction analysis.

Structural Gene

All of the strongly hybridizing clones included sequences overlapping what appears to be the single structural gene coding for pyruvate kinase in chickens. Southern blots of genomic DNA probed with cloned chicken muscle PK cDNA did not reveal the presence of other genes. Two clones, λ PK2 and λ PK3, span the protein coding region of the PK gene. A partial restriction map of the gene and the relative positions of these two clones are shown in Figure 1. The gene is divided into at least 11 exons. We have determined the sequence of the ten protein coding exons (Figure 2). These ten exons are distributed over approximately 13.6 kb of chromosomal sequence; they include all the sequence of the adult muscle mRNA except for approximately 64–65 nucleotides of the 5′ untranslated region. Because we do not know if the missing sequence is located on a single exon, we will refer to the protein coding exons as exons 1–10 (we will also refer to the nine introns that interrupt the protein coding sequence as introns 1–9).

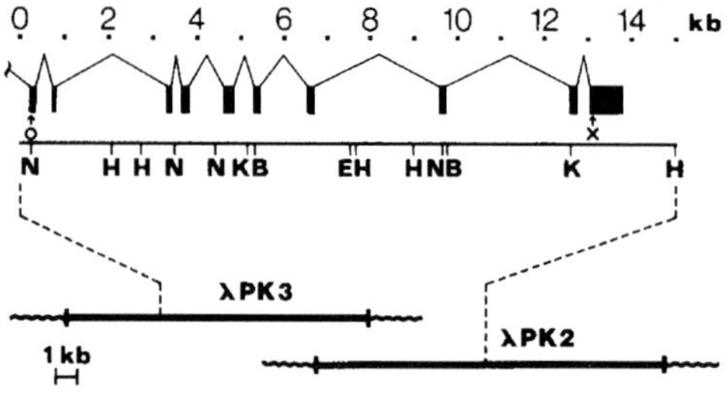

Figure 1. Restriction Map of PK Structural Gene

The solid blocks above the restriction map represent exons. The translation start (O) and stop (X) sites are indicated. The relative positions of the genomic clones λ PK2 and λ PK3 are shown below the restriction map (the scale is half that of the map). Thick lines represent inserts and wavy lines vector sequences. Bst EII, B; Eco RI, E; Hind III, H; Kpn I, K; Nco I, N.

Gene Structure vs. Protein Structure

To compare the gene structure of PK with the corresponding protein structure, we have used α-carbon coordinates for residues 13–529 of the cat muscle enzyme (Stuart et al., 1979; Hilary Muirhead, Linda Fothergill, and Emil Schiltz, personal communication). This enzyme is closely related to the chicken muscle enzyme; the two proteins are identical at 88% of their amino acid positions (Linda Fothergill and Emil Schiltz, personal communication).

PK is a tetramer of identical subunits. Each subunit consists of a short amino-terminal segment plus three domains: the central A domain, comprising approximately half the mass of the protein, flanked by the smaller B and C domains. The amino-terminal segment interacts primarily with the A domain of the neighboring subunit. A schematic drawing of a single PK subunit is shown in Figure 3a; this drawing is exploded into pieces coded for by individual exons in Figure 3b. The relationship of the intron positions to particular primary, secondary, and tertiary features of the protein structure is depicted in Figure 4. Figure 4 is intended as a guide to Figure 5, which shows stereo diagrams of the PK α-carbon skeleton, highlighting segments encoded by individual exons. We describe the structure of each of these segments below.

The central A domain of PK is a roughly symmetric barrel-shaped structure of eight β strands alternating with eight α helices. The α helices form a cylinder around a core of parallel β sheet, and introns divide the barrel into multiples of a simple unit of secondary structure—an α helix plus a β strand. Exons 1, 2, 3, 5, 6, and 7 contribute residues to the A domain; each of these six exons includes one or two α-helix/β-strand units. The first exon (exon 1) consists of a short α helix from the amino-terminal segment and the first β strand of the barrel. The next two exons (exons 2 and 3) each consist of one α helix and one β strand. Exon 3 leads into domain B, which interrupts the barrel between β strand 3 and α helix 3. Exon 4 forms the major part of domain B. Since the electron density in this region is much lower than that of the rest of the map (Stuart et al., 1979), the secondary structure assignments for domain B are uncertain; however, it appears that the intron following exon 4 is located between two contiguous β strands. This intron is followed by exon 5, which completes domain B and then forms two α-helix/β-sheet units in domain A. The exon ends with a short, one and a half turn helix in the loop between β strand 5 and α helix 5. The next exon (exon 6) again consists of two αβ units. The last exon of the A domain (exon 7) is composed of a single αβ unit, together with the first turn of the final α helix of the barrel. This helix is slightly beneath the barrel and is preceded by a narrow, relatively disordered helical segment. As a consequence of the repeated α-helix/β-strand intron motif, four of the five A domain introns fall in a single plane at the top of the barrel. The fifth intron is located at the bottom of the barrel, on the final α helix of the central A domain, at the point at which an imaginary plane dividing the A and C domains would intersect this helix.

Domain C consists of five α helices surrounding a central ribbon of five β strands. This domain includes a structure similar to the so-called "mononucleotide binding fold" (mnbf). The mnbf comprises one-half of the roughly symmetric nicotinamide adenine dinucleotide (NAD) binding domain, common to lactate dehydrogenase, malate dehydrogenase, alcohol dehydrogenase (ADH), and glyceraldehyde-3-phosphate dehydrogenase (Rossmann et al., 1974). The mnbf consists of three parallel β strands (β A, β B, and β C) connected via two α helices (α B and α C); two mnbf units are connected by either an α helix or a less ordered structure to form the NAD binding unit.

PK domain C is divided into three exons. The first exon (exon 8) consists primarily of two long α helices that form intersubunit contacts in the native tetramer (5). This exon ends with a β strand that corresponds to β A of the mnbf, and part of an α helix corresponding to mnbf α B. PK exon 9 includes the region corresponding to mnbf β B, α C, and β C; it concludes at the COOH terminus of a helix analogous to the helix that forms a bridge between two mnbf structures in some of the dehydrogenases. The last exon of PK (exon 10) consists of two β strands that extend the mnbf β sheet. In the dehydrogenase NAD binding domain, the β sheet is extended by a second mnbf unit. Although exon ten of PK lies in the same relative position, it is in no sense topologically equivalent to a second mnbf; the β strands of PK exon ten are antiparallel and extend toward the center of the sheet, whereas the dehydrogenase β strands are parallel and extend away from the center.

Comparison of PK and ADH

We have compared the intron/exon and protein structures of PK domain C to the intron/exon and protein structures of the NAD binding region of ADH. The intron/exon structure of maize ADH has been determined (Dennis et al., 1984), and the structure and location of the NAD binding region can be inferred from the known three-dimensional structure of the homologous horse liver enzyme (Branden et al., 1973, 1984). We superposed the α carbon coordinates of all possible contiguous 51 amino acid segments of cat muscle PK and all possible contiguous 51 amino acid segments of horse liver ADH. For each pair of segments, we calculated the transformation that minimizes the sum of the squared distances between respective α carbons. The unique best superposition aligns the PK mnbf (residues

Pyruvate Kinase Gene Structure
83

```
TTTGGGTGGCATCCGTGGGGAATGACATTTAAATAGAAGTGTGTCTTGCCTGTATCTAAGACTCTCAAGTAGGAACATAACGATAGTCTGACCCAGTAGCAGCTTTCTTGTTACCTCCAT

                                                                                                                SerLysHisHi
ATTCATCACGTGGGTGTGTTGGTGTCTCTGCATGGTGAGCGAGCAGGGGCAATCCCTAAACCCGGCACCGTTTTCTTCTCTTGCAG/ACACCGAACTCCAGTAACCATGTCGAAGCACCA

sAspAlaGlyThrAlaPheIleGlnThrGlnGlnLeuHisAlaAlaMetAlaAspThrPheLeuGluHisMetCysArgLeuAspIleAspSerGluProThrIleAlaArgAsnThrGl
CGATGCAGGGACCGCTTTCATCCAGACCCAGCAGCTGCACGCTGCCATGGCAGACACCTTTCTGGAGCACATGTGCCGCCTGGACATCGACTCCGAGCCAACCATTGCCAGAAACACCGG

yIleIleCysThrIleG
CATCATCTGCACCATCG/GTACGTGGCTCCCAGATCCCAGGAACTGTGGCTCACGTTAGGCTGTGAACGCCTGCGTTGGTGAAACTCATTGTGGGAATGTGGGATGGCTTCAGCCTTCAT
CCGTTGCCAGGGTTGTGTCTTGTCTGTTGAAACAACACAACTTTCCCATGCCTGCCTGCCCAGTTTTATTGCTAGGGGTAAATCGAGCGATGTCTGAACTGGTTGGGTTCAGATGTGCTC
CCTAAGACAAGGCAAAGGAGCTGCTTTCGTGTTGAGCAAGTCTGTATCCACTGTGCTTCACTTATCCCCTCCATGTGAAACAGCCTGACGTTTGTGGGTTATCTCATTATTTATATGAAG

                              lyProAlaSerArgSerValAspLysLeuLysGluMetIleLysSerGlyMetAsnValAlaArgLeuAsnPheSerHisGlyThrHi
TTAATATTCAAACTGTTTCCTCCTTTTTTAG/GCCCAGCCTCCCGCTCTGTGGACAAGCTGAAGGAAATGATTAAATCTGGAATGAATGTTGCCCGCCTCAACTTCTCGCACGGCACCCA

sGlu
CGAG/GTAAGGGTGGGGATGGGGCTGACACTGCCCTGAACCTGTAACTGCTCATCCCTTCCTTCCCTTCTGTAAGAGGAGACCG..... 2.3 kb .....TTTTGGGCTATGTAGCAC

                                           TyrHisGluGlyThrIleLysAsnValArgGluAlaThrGluSerPheAlaSerAspProIleThrTyrArg
AAGCAGTGGAGTAGAAGTTGGGCTGAAAGTTTCCCTTTCTTTTCCAG/TATCATGAGGGCACAATTAAGAACGTGCGAGAGGCCACAGAGAGCTTTGCCTCTGACCCGATCACCTACAGA

ProValAlaIleAlaLeuAspThrLysGlyProGluIleArgThrGlyLeuIleLysGly
CCTGTGGCTATTGCACTGGACACCAAGGGACCTGAAATCCGAACTGGACTCATCAAGGGA/GTAAGTTGCTCTTTTTATTTACCCCCACCCCCGGTGCCTGAGACTAGATAGTGTTGTTC
CCATGGCCTGCTTTTCCCAAACAAGGATTATCTTATGAGAACAAAATAACTGTTTGCTGTGTCTGGCATTTGTAGCAAGAGTCTGCACAGACTGCTGCAAGCTGGGGGACAGCTGGGTGC

                                   SerGlyThrAlaGluValGluLeuLysLysGlyAlaAlaLeuLysValThrLeuAspAsnAlaPheMetGluAsnCysAspG
TTGGTTTTTTTTCTGAAATCCTTGTTTTGACATGCAG/AGTGGCACAGCAGAGGTGGAGCTCAAGAAGGGCGCAGCTCTCAAAGTGACGCTGGACAATGCCTTCATGGAGAACTGCGATG

luAsnValLeuTrpValAspTyrLysAsnLeuIleLysValIleAspValGlySerLysIleTyrValAspAspGlyLeuIleSerLeuLeuValLysGluLysG
AGAATGTGCTGTGGGTGGACTACAAGAACCTCATCAAAGTTATAGATGTGGGCAGCAAAATCTATGTGGATGACGGTCTCATTTCCTTGCTGGTTAAGGAGAAAG/GTGTGTAGCAACCC
TCACAGGATTTGCTAGCTCTGCCTGTTCATTCTTCCCTAAGAATCTTTGTGGGTTTTTTTTTCACCTCCAGAATGGTTC..... 0.2 kb .....TTACAACATTAAAACCCAGTGTT
TTATGGTTAGAAGCTGAACTAGCTTTACTGGTAATGCAAGGGAAAGTTAATTCCCCAGTGTTTCCTTCATGGTGGTTCAAAAATCTGCTCTAGCCTGGAGGTATCTCCTGTAACCCCAGA
GCGTGCAGGTGAGGTGCTGGGATATTATGGGATGCTCATGTGCAGAAATGCAACCTGCTAGTTCTTGGAGGGAGGCAGCCCCCGTCTCCTGCATGGGGGAAAGGAGGCATCTTTCCCAA
AAGCAGCACAGTGGGTTTGTGCTGTGCTGCCATGGGTACATTACATGCTCACATGTTATGGGAGTCCAAGGAGAGCAGTGTTGTGCTTCTGTTGTGATTGCTGTAAACACTGTGGTTCTC

                                                                                        lyLysAspPheValMetThrGluValGluAsn
AGGCAGCTTTGGCAGGAGGGGGTGTGGGAGCTGGTGGTGGGGAGCTCGCCCTGTTATGCTCTGACTGTGGGAATACATCTGTTTCAG/GCAAGGACTTTGTCATGACTGAGGTTGAGAAC

GlyGlyMetLeuGlySerLysLysGlyValAsnLeuProGlyAlaAlaValAspLeuProAlaValSerGluLysAspIleGlnAspLeuLysPheGlyValGluGlnAsnValAspMet
GGTGGCATGCTTGGTAGTAAGAAGGGAGTGAACCTCCCAGGTGCTGCGGTCGACCTGCCTGCAGTCTCAGAGAAGGACATTCAGGACCTGAAATTTGGCGTGGAGCAGAATGTGGACATG

ValPheAlaSerPheIleArgLysAlaAlaAspValHisAlaValArgLysValLeuGlyGluLysGlyLysHisIleLysIleIleSerLysIleGluAsnHisGluGlyValArgAr
GTGTTCGCTTCCTTCATCCGCAAAGCTGCTGATGTCCATGCTGTCAGGAAGGTGCTAGGGGAAAAGGGAAAGCACATCAAGATTATCAGCAAGATTGAGAATCACGAGGGTGTGCGCAG/
GTGAGCTCTGCACGAGTGGCACTGCAAGTCCTTCACTCTGTAGCAGCAGAAGGACACAGCCCTTTCATGCTGTAGATTTCTGAGGAGCTACGTCACTCTTTGTAATGAGTGTGAGTTCAG
ACTTCAGCGTT..... 0.05 kb .....TAGACTGAAATGTCTCAGATGAACTCTTGCTCCTCTTAATCTGTTTATATATAGCTTAAGAAACTGTCCTATGGCTTGACCTTGTTTTCC
TACCTCCTACTGTGGTATTACAGGGTACCAAAATGGCACCTGGAAGGTCAGTGTGCCGATAGGGTGTCTGCGCATCAGGCTCTAGAGTGATGCTTCCTTTCAGAAAAGGGGGACAGAGC

                                                         gPheAspGluIleMetGluAlaSerAspGlyIleMetValAlaArgGlyAspLeuGlyIleGluI
AGAGGGTCCACCCTAAGAACAAAGGGAGTAATGAATCTCTCCTCTTGTTGTCAG/GTTTGATGAGATCATGGAGGCCAGCGATGGCATTATGGTGGCCCGTGGTGACCTGGGTATTGAGA

leProAlaGluLysValPheLeuAlaGlnLysMetMetIleGlyArgCysAsnArgAlaGlyLysProIleIleCysAlaThrGln
TCCCTGCTGAAAAAGTCTTCCTCGCACAGAAGATGATGATTGGGCGCTGCAACAGGGCTGGCAAACCCATCATTTGTGCCACTCAG/GTGCTCAGAACATGAAGCTAGCTTGTATAACGG
CTGGGTGTTTGTGGGTCTGGCAGGGTGTGAGTGTCAGCAGCT..... 0.8 kb .....TTAATTCAGAGGTTGTTAGGGAGCAGGCCCCTCTTGGCTCTGTAACTACAGGAGAATGCT

                                                                                                                       MetL
ACCAGAGGTCAGGTTATATCCTGACTTCTACCTTATGTGTGTATTGAGGGGCCTTAGTGGAAAATCAGCCACTTCTTCCTGCTGTATCCTGACTCTTGGACTTGTTTATCTGCAG/ATGT

euGluSerMetIleLysLysProArgProThrArgAlaGluGlySerAspValAlaAsnAlaValLeuAspGlyAlaAspCysIleMetLeuSerGlyGluThrAlaLysGlyAspTyrP
TGGAAAGCATGATCAAGAAACCTCGCCCGACCCGCGCTGAGGGCAGTGATGTTGCCAATGCAGTTCTGGATGGAGCAGACTGCATCATGCTGTCTGGGGAGACCGCCAAGGGAGACTACC

roLeuGluAlaValArgMetGlnHisAla
CACTGGAGGCTGTGCGCATGCAGCACGCT/GTGAGTAGCCATCCTGGTGCAAGTGGCAGTGCTCTGCCTTCCATCTGAGTCCACACGTAAAGATGTAAATACCTGGTGCTAGGTGGCAGC
ACCTAGTAGCATAAAGCAGCAGTTCTTTAGTTCAAGCTCTCAGCATGAGGTTTACTTCACCTAGTGTGGAACCTTGGACTGAATGCTGTGTGCTACTGTCAGGTGGAGCTGAGAGCTGCT
TTGCAGTGCTCAGAGACCTAAAACCTCAATTATACCACATTCATTTTGATTTAAGGCCACGGTAAAGGCTTTTTGGCTGCTACCTTGGTATGAATTTCTTTGAGTGCCTTGTGCATTTTA
GTGTTTTAAATGCTTTAAACTTGCCCAAGGTCAGGACGCTGCTTGTTAAGCAGTGTATCTCTAGATAAAAAGTGTTAGCTGACTCCTCTGGGTTAGGGGGGGGAGGGGGTGTTGGGGGTT
GGGAGAGGGTGGGGAGAAGGCTTGAGGTGTTGCATCCAGCAGCTGATGCATTTGGTGGTGGCAATGTTTGTTGCACGAATGCTGACAGCACCATGTTTTTACATGCCTGAACTGCCAGAA
AGCTCTGCTCTGCTTCATGTTAACCCAGCTTGTGTTTCAGTTCATCAAAAATCCTGATGCAAGGCATCTAGAAGTACTGGATTTCCCCACTTGCTAGGGCTGTGAATCACGTTAGTTTCA
CCTGCAGCTCAGGTGATACATCTGAAGGCACTGAGCTAGCATTTCTGTTTGCAGTGCATTATCTAAACATCTTTTACCTGGTCAGTATGGAGTGCTGTTTTCTCATGTGGCAAATAAAGC
ATGACCTTAACTACTGGGCAGTGACTGCTGGCTTACAAGCTGAATTCTAACTGTGGACTCAAAATGTTCTTCCTTAGTCTTTCAGTCCTACGTTGAGGGCTGAATCTTCCTACCAAGGTT
TGGAAACTTAGTGCCATGAATATTGCAATTTTGGCCTGAAAGCTTAGTTAAGTACAGCACAACCTGTCTGAGTTTGTGGCTCTGCACCCCCCCTGCATCTGCAAGGAACTTTTCCTGCAG
GGCTGGGGATAATGGTGCT..... 1.7 kb .....TTGCATGCATTACTATGACAATTAAGCAAATTAACGTGATCTCATGTTACTCTGGTCTACCTTTCCTTCTTTTCTTTCCTCTT

                                 IleAlaArgGluAlaGluAlaAlaMetPheHisArgGlnGlnPheGluGluIleLeuArgHisSerValHisHisArgGluProAla
TTCTTCTCTGCCCTTCTTCCTCGGACCTGCAG/ATTGCTCGTGAGGCTGAGGCCGCAATGTTCCATCGTCAGCAGTTTGAAGAAATCTTACGCCACAGTGTACACCACAGGGAGCCTGCT

AspAlaMetAlaAlaGlyAlaValGluAlaSerPheLysCysLeuAlaAlaAlaLeuIleValMetThrGluSerGlyAr
GATGCCATGGCAGCAGGCGCGGTGGAGGCCTCCTTTAAGTGCTTAGCAGCAGCTCTGATAGTTATGACCGAGTCTGGCAG/GTAGGGCTCAGAGAAGGGAGAGGCTTAGGCTGCGTAAGG
TAACCGATGCTCGCTCCCAGCATGCGGGTTAGATATGTGCTGCTTGCCAGTCACTGCGTGATGAGACCACGCTTAACATTGCAGGCATTCAGCAAGCAGCGTGCAGCTCTGGTATCTGCA
GGGCAGGGCCGCAGGTGGGTAGGAGCAGAGCAAGAGTGCCTGGCAGTTGGGCATGAATGGCAATTGAGTGTGGGCTCAGCAGGGTGAGGTGCCGCCAACTGGGGGAAGGAAAACTTGGAG
GATCAAATTGCTTGTTCATTAAGGTGGTGCTGAAGCAATTT..... 2.4 kb .....GCTGAGCCTCGTGCTGCAGGCAGGGGTGCCCCTCTGGTTGTGCCTGCGCTGCAGTGCCGTG

                              gSerAlaHisLeuValSerArgTyrArgProArgAlaProIleIleAlaValThrArgAsnAspGlnThrAlaArgGlnAlaHisLeu
CCTCACCCTGTGCCTGTTGTCCTTCTTGCAG/GTCTGCACACCTGGTGTCCCGGTACCGCCCGCGGGCTCCCATCATCGCCGTCACCCGCAATGACCAAACAGCACGCCAGGCACACCTG

TyrArgGlyValPheProValLeuCysLysGlnProAlaHisAspAlaTrpAlaGluAspValAspLeuArgValAsnLeuGlyMetAsnValG
TACCGCGGCGTCTTCCCCGTGCTGTGCAAGCAGCCGGCCCACGATGCCTGGGCAGAGGATGTGGATCTCCGTGTGAACCTGGGCATGAATGTCG/GTGAGTGCTGCGCTCCTGCCTTGCT
CTGGGGGCTGCGGTGGGGCCGCAGCAGGACCTGCAGCACGGTGTTGGCGCTGCTGTTGGGATGCGGGGACCGAGGTTGCATAAGCGCACCGCATCTAAGGGGGAAATGGGACGGTGGTT

                                                                                                                       lyLy
TTGTCACAGCAGCCTTGAAATAGCTGCCTGCAGAAAACCAGGTTTGCTTTAATAAGCGAGAGAAGGGGGGTGTGAGGAGACCTGAGTGCTGACGCTTTTTCCTCCTTTTCCCCAG/GCAA

sAlaArgGlyPhePheLysThrGlyAspLeuValIleValLeuThrGlyTrpArgProGlySerGlyTyrThrAsnThrMetArgValValProValPro***
AGCCCGTGGATTCTTCAAGACCGGGGACCTGGTGATCGTGCTGACGGGCTGGCGCCCCGGCTCCGGCTACACCAACACCATGCGGGTGGTGCCCGTGCCATGAGCTGCCGGCGCCCCCGC
CTCTTCCACCCCCGCGCCCCTTCCCCATGCATTAGGCCAGCAGTCGCTTGCGATGTTCTCCTCGTCCCTAGCGTGGATTTAGGTTGCTCAACACCACCGAGTGCAGCAGCGGGGGAAGAG
ACCTTAAATCACTCAGAATACTTGAAGTGTGTAGTTGTTTCCAGACCTCTCTCCCTGTAGTTAGCCATGTCAGAGCGGGGCCCTGTGCTCCCCACGTCCCCCTAGCTTAGCACTGCCCTC
AGGGTGTGGCTGTGCCTCTCCTTGCCCCGGCTGTGGGGTGATGGGGAAGGAGTGGGCGCTTATGGTAACGGTTCTGCTGCTGCACACCGCAGCTCCCCCAGGTTGGTGCCCTCTGTCCAT
GAGGGGTTATAGGTTGTCCCCTCTGGGCTTACACTCCAAAGGTGTGCTCCACGGTTCCTACAAGAATCCCAAACCAGGGTGCTGCTCGTTGTCCCCCTGGGTGCAGGTGGAGTCCTGCGG
GTTGTTGGGGTGCAGGATGGGTGGGGGACGCACGCTTCATCTCCATGCAGCCACCTGCCCTTCTATGGTGTCTGTCTGTGTTGTCGCCTTTCCTGAGAGCGTTCCTTTTGTCTGGGCGCC
CTGTTTGCTCCGGTGGGTGTATTGGGTTGTGGAGCACCACTGTGCACGTCAATAAAGAGCTGAGCACCCCTGCCCGCGGCTTCGTGTGCTGGTGCTGATGCTGCGCTACGGCGGGATCAC
GCGTGGGGTGGGGCGGACGGGAGGCCACCGCGGGAGGGCACCCGGACAGGGGACGCGGGCACTTTAAGAGGCGGGGCAACTACGCGAGGGGGCGGGGCTGCGGCGGGGTCTCCACCTAT
CCTACCGGAGGGCGGTCCGCTGGTGTGAAGGGATCGAGCTCGCGGCGCCGCTCAGTGACCAATGGGAGCGAGGGGCGGGGGCGGGGCTAGGCGGCGGACGGCCAATAGG
```

Figure 2. PK Gene Sequence

The sequence of the ten protein coding exons of the PK structural gene. Splice junction sequences are underlined. The predicted amino acid sequence of the mature muscle isozyme is displayed above the DNA sequence. A probable polyadenylation signal, AATAAA, is underlined.

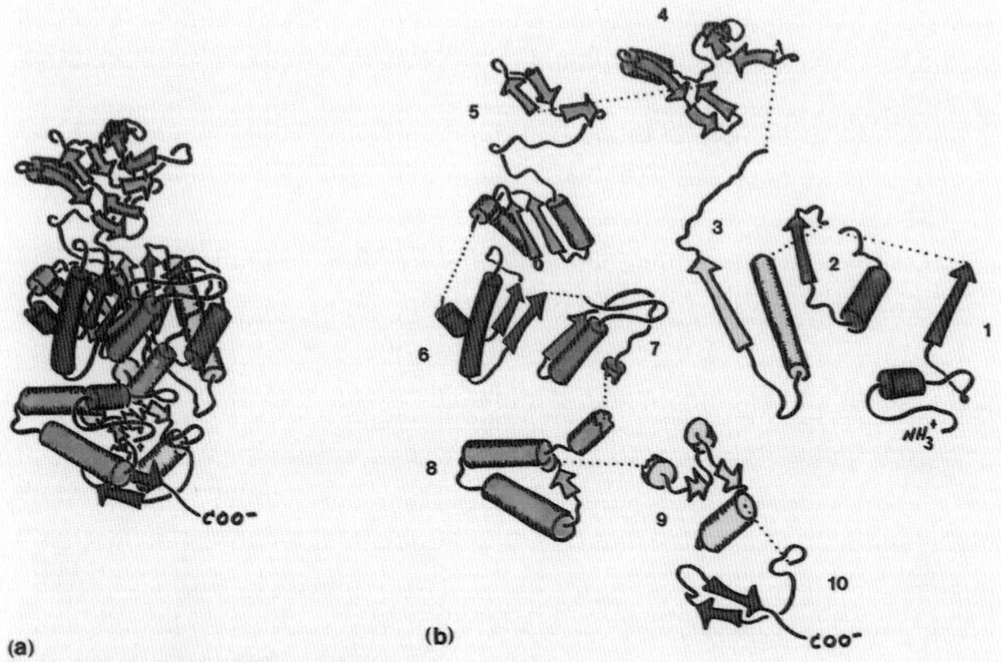

Figure 3. Schematic Diagram of PK Enzyme Structure

(a) Single PK subunit; (b) exploded view of PK subunit; pieces coded for by individual exons are separated at introns.

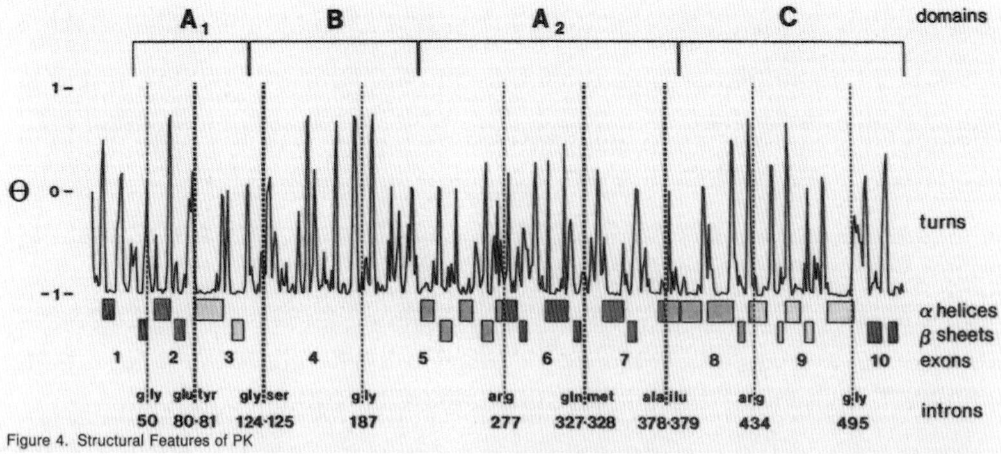

Figure 4. Structural Features of PK

The position of turns made by the polypeptide chain was determined by drawing least square lines through the α-carbon positions on either side of a residue, and calculating the cosine of the angle made by these two lines. At each residue the calculation was repeated including the central α carbon and two α carbons on each side, then three, etc., out to ten α carbons on each side. θ defined as the minimum value of the cosine at each residue, is plotted above; peaks indicate the positions of turns. This function corresponds fairly well to subjective assignments of turn residues. However, certain structural features of the protein are missed; for example, the elbow at residue 277 (the junction between exons 5 and 6; see Figure 5) is not identified by the θ function. The locations of the three structural domains are indicated above the plot (A_1 and A_2 refer to the two components of the A domain); α helices and β sheets below. In the absence of coordinates for backbone carbonyl oxygens, we have assigned α-helix and β-sheet residues on the basis of visual inspection of the α-carbon model. Assignments based on more objective criteria may shift the positions of secondary structural elements. No secondary structure assignments have been made for domain B because of the ill-defined nature of the electron-density map for this region (Hilary Muirhead, personal communication). The intron positions are marked by dashed lines. Amino acid residues that correspond to or abut introns are indicated below. Dashed lines interrupt amino acids at positions that reflect the intron positions within codons.

423–473) with the first mnbf of the ADH NAD binding domain (residues 194–244) so that the root mean square (r.m.s.) distance between α carbons is 2.6 Å. Figure 6 shows a cross-section of the full comparison of PK and ADH. The PK mnbf region is aligned with successive segments of ADH; the first but not the second mnbf of the ADH NAD binding domain matches the PK mnbf. The superposition of these two structures is shown in Figure 7a.

The amino acid sequence alignment produced by the structural superposition reveals some homology between PK and ADH. Figure 8 shows the mnbf and flanking sequences of yeast and chicken PK aligned with the corresponding sequences of horse and maize ADH; no gaps have been introduced to align the PK sequences with the ADH sequences. The homology between each of the four pairs of PK and ADH sequences is approximately 20% over the region spanned by exons 5 and 6 of maize ADH.

Structural superposition of the ADH and PK mnbf regions aligns the chicken PK sequence with the maize ADH sequence so that PK intron 8 is four nucleotides from ADH intron 5 (Figure 7b and 8). These introns fall roughly in the middle of the region of amino acid sequence homology between PK and ADH.

Possible Pseudogene

Half of the analyzed genomic clones include a region, spanning less than 2 kb, that hybridizes only weakly to the cloned muscle PK cDNA sequences. Using the three Pst I insert fragments of pPK102 and pPK201 (6) as individual probes, we found that this region is homologous to sequences within the 5′ third of the muscle PK message (the cross-hybridizing probe included 27 nucleotides of 5′ untranslated sequence and the first 657 nucleotides of coding sequence). We determined the first 282 nucleotides of sequence from the cross-hybridizing region of one of these genomic clones, $\lambda\psi$ PK1 (Figure 9), and found a segment of 92 nucleotides that can be aligned with the cDNA sequence of the chicken muscle enzyme. The two sequences match at 86 of the 92 positions when a single three-nucleotide break is introduced into $\lambda\psi$ PK1. The region of homology may extend in the 3′ direction; we did not determine any further sequence from this clone.

The PK-homologous sequence in $\lambda\psi$ PK1 spans the position of an intron in the PK structural gene. Thus it is most likely derived from a fragment of the PK message that has been reverse-transcribed and reincorporated into the genome. Furthermore, this region has suffered mutations without the constraints placed on a gene coding for a functional protein; the deviations from perfect homology do not conserve amino acid sequence or even cause conservative replacements.

Discussion

The strong sequence homology between yeast and chicken PK (Lonberg and Gilbert, 1983) suggests that the genes coding for these two enzymes are descended from a common ancestor. However, the two genes are dramatically dissimilar; the yeast gene includes a single, uninterrupted 1.5 kb open reading frame while the coding portion of the chicken gene is divided into ten exons dispersed over 13 kb of DNA. Which of these genes has diverged from the structure of the common ancestor? Has the chicken gene acquired introns over the course of its evolution, or did the original PK gene contain introns?

The position of the introns within the chicken PK gene may provide a clue to their origin. The nine introns that disrupt the coding sequence do not appear to be distributed randomly. Random placement of introns would lead to a geometric distribution of exon sizes, with a standard deviation approaching the mean. The ten protein coding exons of PK include an average of 159 bp of coding sequence. However, the standard deviation is only 47 bp, reflecting a uniform distribution of exon sizes. This is consistent with other known gene structures; as a general rule, while introns vary considerably in size, exons can be grouped into a few discrete size classes. Those approximately 140 bp in length are the most abundant (Naora and Deacon, 1982).

The uniform distribution of exon sizes is not the only indication that intron positions in the chicken PK gene are nonrandom—we also observe a correlation between intron/exon structure and protein structure. The introns tend to fall between discrete secondary structural elements. This correlation is most pronounced in the central A domain, where segments coded for by individual exons appear to be multiples of a simple "super secondary structural" unit, an α helix plus a β strand.

The apparent nonrandom placement of introns within the PK gene, at sites that correlate with features of the protein structure, argues against an insertional model for the origin of introns. Because biological information flows from DNA to protein, selection is the only process by which protein structure can dictate intron positions. For instance, introns might act as mutation hotspots, or as sites of misincorporation of amino acids due to RNA splicing errors. Because organisms would only be able to tolerate introns in noncritical regions of the gene, randomly inserted introns would be restricted to particular sites by selection. However, the introns in PK are not located primarily in nonconserved areas. For example, 13 codons are interrupted by, or abut, splice junctions in the chicken gene, and five (38%) of these code for residues that are identical in yeast and chicken. This is close to the overall homology (45%) between the two proteins. Therefore, selection can not explain the difference between the yeast and chicken PK genes.

Alternatively, the correlation between protein and gene structure could be a function of the early evolutionary history of the gene. Individual exons, coding for small peptides, could have been recruited by gene rearrangement, and/or gene duplication, and combined by RNA splicing to form the PK gene. The α-helix/β-sheet units of PK might be examples of primitive genes. A small gene encoding an α helix and a single strand of β sheet may not appear to confer much selective advantage. However, its peptide product may have functioned only as an aggregate; there is no need for the monomer to form a stable folded structure in solution. Entropic considerations would have made it unfavorable to rely on residues near the termini of these peptides to form secondary or tertiary structures that sta-

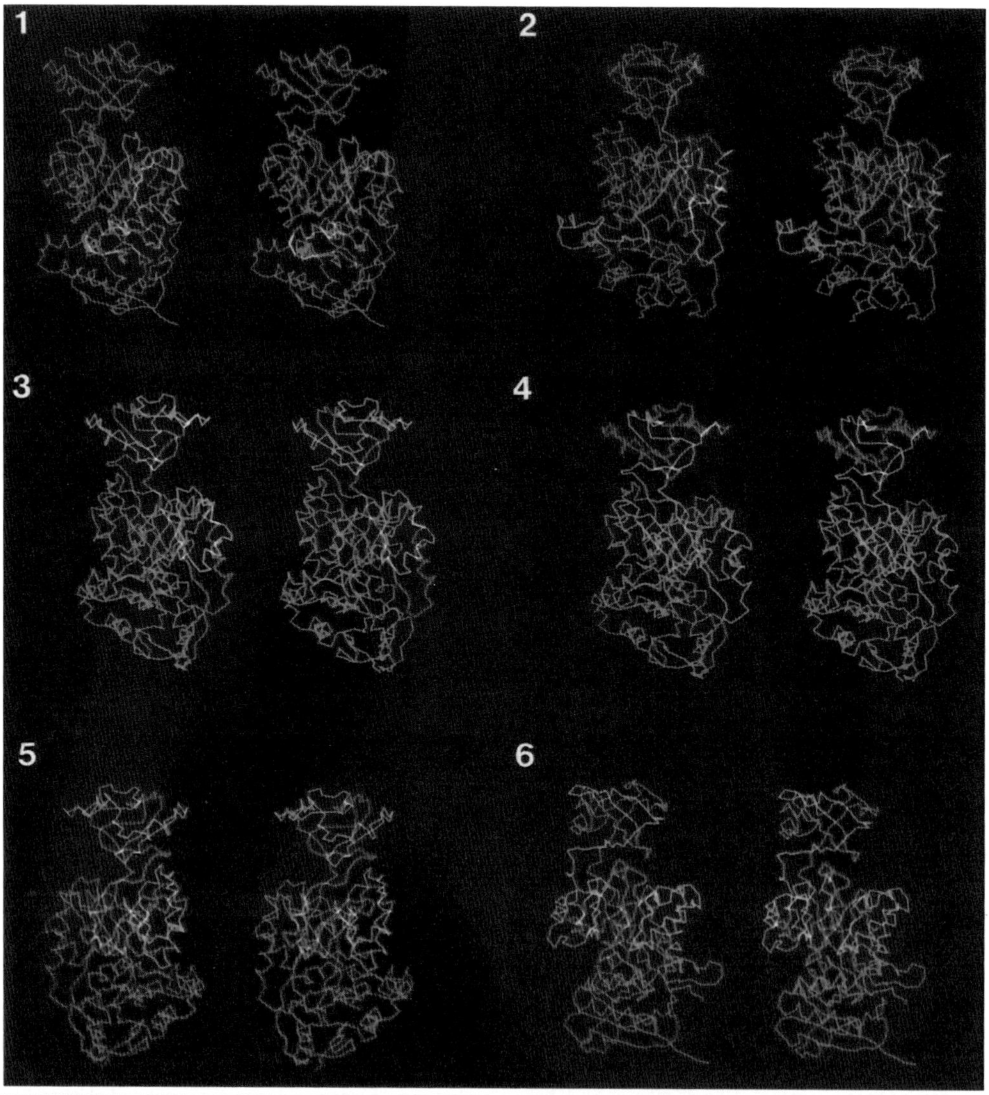

Figure 5. Stereo Images of PK α-Carbon Skeleton, Highlighting Segments Encoded by Individual Exons
Each exon is highlighted in orange, flanking exons are green, the rest of the protein is blue. Coordinates are for the cat muscle enzyme.

bilized the aggregate structure. These termini would therefore have been incorporated into loops between secondary structural elements when the peptides were tied together by RNA splicing. This explanation for the nonrandom placement of introns suggests that the intron/exon structure of the chicken PK gene is more ancient than the intronless structure of the yeast PK gene. The yeast PK gene could have lost its introns as the yeast genome became streamlined to facilitate rapid cell division. It has already been shown, for the rat insulin I gene (Perler et al., 1980), that mechanisms exist by which introns can be lost without altering coding sequences.

Gilbert (1978) proposed that exons might correspond to functional domains, and that illegitimate recombination within introns could assort these functions independently. Individual exons could be thought of as building blocks; a single block might be duplicated and used several times in one or more genes. If the PK gene evolved by the combination of exons, some of these exons might have been used more than once. To see if we could find an example

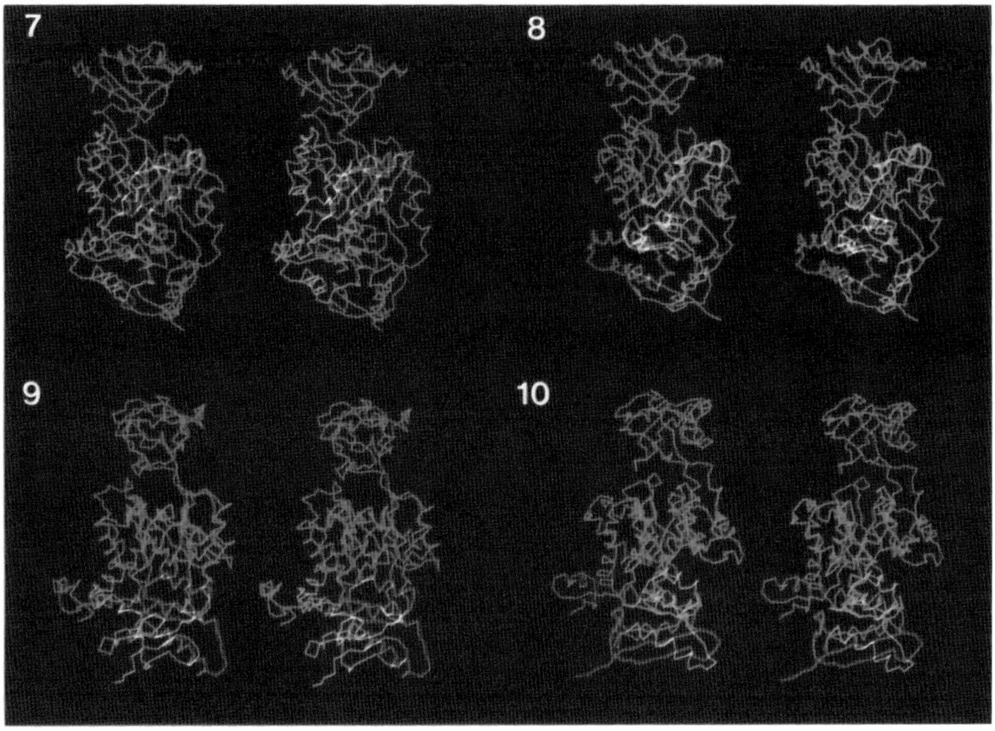

of such a building block, we compared individual PK exons with each other and with individual exons from ADH, which shares a common structural feature with PK. There is no evidence, other than secondary structure homology, to suggest that the repeated $\alpha\beta$ unit of the PK exons is due to the duplication of a single primordial exon. No obvious repeats are seen in the primary structure, and the intron/exon junctions do not fall at identical positions within codons. However, given the age of the PK gene, it is possible that primary sequence homology has been eroded by mutation and that the position of the introns within codons has shifted. The latter may have occurred in two ways: the primitive splicing mechanism may have been much less accurate than the present mechanism, with precise splice junctions having evolved subsequent to exon duplication, or single base changes near one splice junction could have activated cryptic splice sites near corresponding splice junctions. Such cryptic splice sites have been observed in thalassemias (Treisman et al., 1983). Because large shifts would disrupt large segments of coding sequence, it is unlikely that introns could have been moved by more than a few nucleotides at a time.

A better candidate for an example of exon duplication and recruitment comes from the mnbf region of PK. The mnbf is a common structural feature of a number of proteins, including the NAD binding dehydrogenases and the

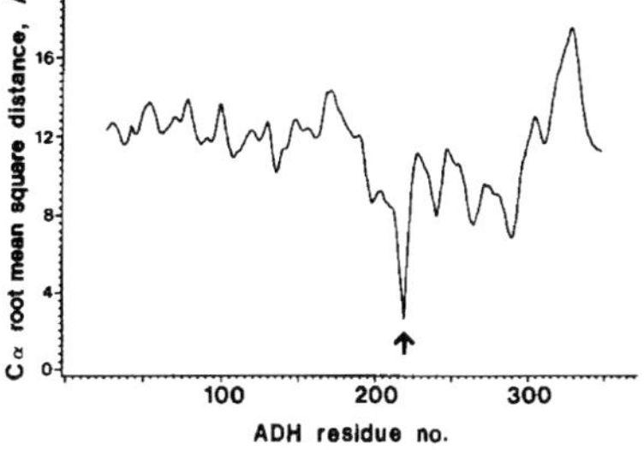

Figure 6. Root mean square distances between α carbons for 51 residue segment of PK superposed with successive 51 residue segments of ADH. Residues 423–473 of cat muscle PK are superposed with a sliding 51 residue window of horse liver ADH, to minimize the sum of the distances between aligned α carbons. The center of the 51 residue window of ADH is shown on the x-axis; for each superposition, the r.m.s. distance between respective α carbons is given on the y axis. The best superposition aligns PK residues 423–473 with ADH residues 194–244 so that the r.m.s. distance between α carbons is 2.6 Å (arrow). The mean r.m.s. distance for all possible superpositions between 51 residue segments of PK and ADH is 12.2 Å; the standard deviation is 2.3 Å.

Cell
88

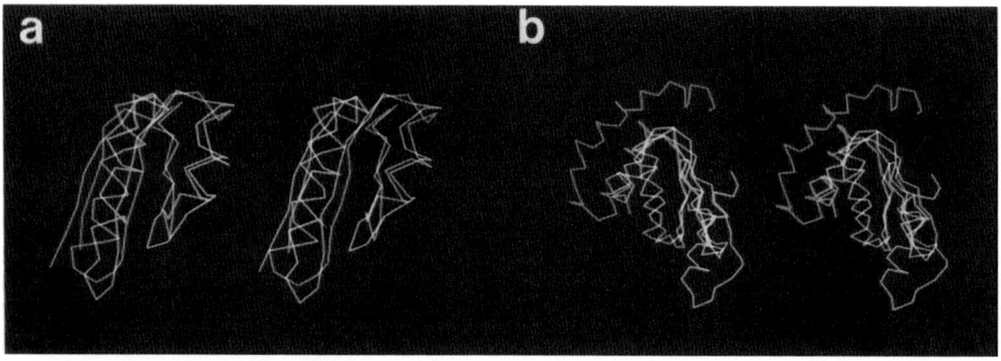

Figure 7. Comparison of PK mnbf with ADH mnbf

(a) Residues 423–473 of PK (blue) superposed with residues 194–244 of horse liver ADH (pink). The following rotation matrix and translation vector were used to transform the PK coordinates:

0.155	0.139	−0.978	56.414
0.181	0.969	0.166	−29.133
0.971	−0.203	0.125	−3.624

(b) Exons 8 and 9 of PK superposed with exons 5 and 6 of ADH using the above transformation. Dark blue, PK exon 8; light blue, PK exon 9; orange, ADH exon 5; pink, ADH exon 6. Computer models of PK are based on coordinates for the cat muscle enzyme; ADH coordinates are for horse liver ADH.

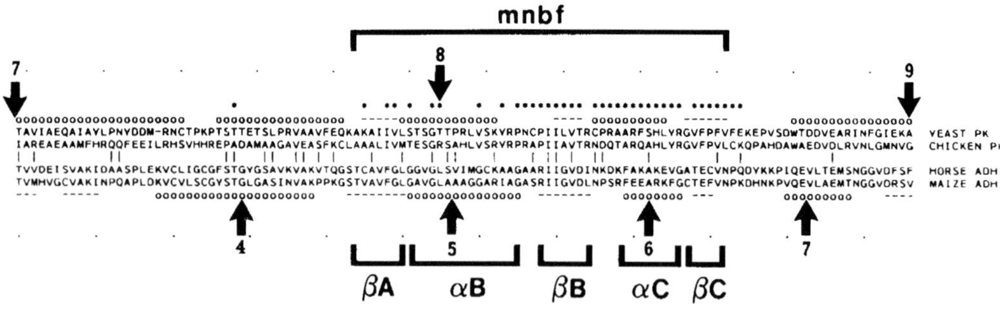

Figure 8. Sequence Alignment of PK and ADH mnbf Regions Based on Superposition of α-Carbon Skeletons

Residues 379–495 of chicken PK aligned with residues 351–466 of yeast PK, 150–266 of horse liver ADH, and 151–267 of maize ADH. Asterisks denote PK α carbons that are less than 3 Å from aligned ADH α carbons upon superposition. Vertical bars represent identities between one of the PK and one of the ADH sequences. Circles denote α helices and dashes, β strands. Intron positions are marked by arrows; numbers indicate the order of an intron from the NH_2 terminus. The positions of the mnbf, and its component secondary structural elements, are shown above and below the sequences. For each of the four pairs of aligned PK and ADH sequences, the number of sequence identities within the 54 amino acid segment corresponding to exons 5 and 6 of maize ADH (residues 180–233) is: yeast PK and maize ADH, 9; yeast PK and horse ADH, 11; chicken PK and maize ADH, 12; chicken PK and horse ADH, 11.

flavin mononucleotide binding flavodoxins (Rossmann and Argos, 1977). Rossmann et al. (1974) have proposed a common evolutionary origin for this structure and suggested it was present during precellular evolution. We have superposed the mnbf region of PK with a corresponding mnbf region of ADH so that the r.m.s. distance between α carbons is 2.6 Å. This superposition aligns the two proteins to reveal a weak sequence homology. The region of homology extends roughly over exons 5 and 6 of the maize ADH gene. The intron that divides these two exons is aligned four nucleotides from intron 8 of PK. Perhaps exons 5 and 6 of ADH represent the descendants of building blocks that were recruited by both the ADH and the PK genes. The introns flanking these two exons may have been lost or shifted in the PK gene, but the central intron remains in roughly the same position.

Pyruvate Kinase Gene Structure
89

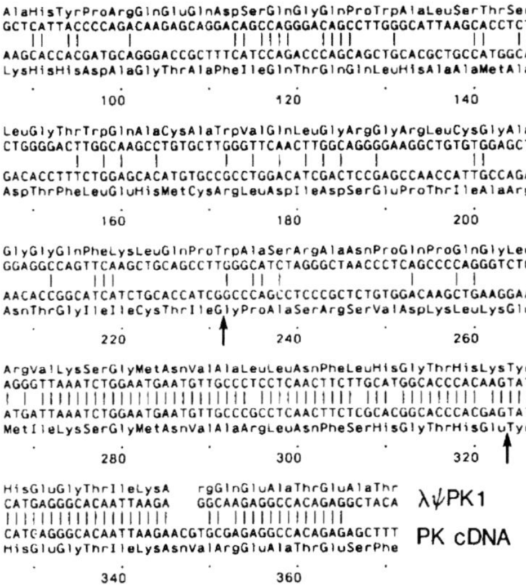

Figure 9. Comparison of Possible Pseudogene Sequence and PK cDNA Sequence

The sequence of part of λψ PK1 is aligned with the sequence of chicken muscle PK cDNA, a single break is introduced into the λψ PK1 sequence to maximize the homology. Arrows indicate the positions of introns in the PK structural gene.

Experimental Procedures

Isolation of Genomic Sequences

We screened a λ Charon 4A chicken library (Dodgson et al., 1979), provided by J. B. Dodgson, using the plaque hybridization procedure of Benton and Davis (1977). We used a nick-translated (Rigby et al., 1977) probe consisting of a mixture of the two Pst I insert fragments of pPK201 and the single Pst I insert fragment of pPK102 (Lonberg and Gilbert, 1983). These three fragments span 2.3 kb of the chicken muscle pyruvate kinase mRNA and include all of the coding sequence, 684 nucleotides of 3′ noncoding sequence, and 27 nucleotides of 5′ noncoding sequence.

Restriction Mapping

We constructed restriction maps of the isolated genomic clones using single and double restriction enzyme digests. We used Southern blot hybridization (Southern, 1975) to identify fragments containing particular exons. The final map accounts for all of the bands observed when Southern blots of total chicken liver DNA (gift of Jay Thomas) are probed with nick-translated cloned cDNA probes.

DNA Sequence Determination

We used the method of Maxam and Gilbert (1980) to determine the sequence of individual restriction fragments isolated directly from λ Charon 4A genomic clones and of fragments isolated from sequences that had been subcloned from the original λ clones into PUC8 (Vieira and Messing, 1982). We assembled the sequence data using the computer programs of Staden (1980).

Computing

Coordinates for the ADH protein structure were obtained from the Protein Data Bank (Bernstein et al., 1977). Coordinates for PK were obtained from Hilary Muirhead (personal communication).

Molecular models of PK and ADH were generated with a Digital Equipment Corp. VAX 11/780 computer and an Evans and Sutherland PS300 color graphics system, using the molecular graphics program FRODO (Jones, 1978), converted for the VAX by B. Bush and modified for the PS300 by J. W. Pflugrath and M. A. Saper.

We superposed segments of the ADH and PK protein structures, to minimize the sum of the squared distances between respective α carbons, using a program written by S. J. Remington (Remington and Matthews, 1980) and based on the matrix methods of Kabsch (1978) and McLachlan (1979).

Acknowledgements

We thank Stephen Harrison, Donald Straus, and Don Wiley for discussions throughout the course of this project. We also thank Hilary Muirhead for discussions and for communicating results prior to publication, Linda Fothergill and Emil Schiltz for communicating results prior to publication, and J. B. Dodgson for providing the chicken genomic library. This work was supported by a National Institutes of Health grant (GM09541-22) to W. G. and by Biogen Research Corporation.

The costs of publication of this article were defrayed in part by the payment of page charges. This article must therefore be hereby marked "*advertisement*" in accordance with 18 U.S.C. Section 1734 solely to indicate this fact.

Received October 18, 1984.

References

Alexander, F., Young, P. R., and Tilghman, S. M. (1984). Evolution of the albumin: α-fetoprotein ancestral gene from the amplification of a 27 nucleotide sequence. J. Mol. Biol. *173*, 159–176.

Benton, W. D., and Davis, R. W. (1977). Screening λ gt recombinant clones by hybridization to single plaques in situ. Science *196*, 180–182.

Bernstein, F. C., Koetzle, T. F., Williams, G. J. B., Meyer, E. F., Jr., Brice, M. D., Rodgers, J. R., Kennard, O., Shimanouchi, T., and Tasumi, M. (1977). The protein data bank: a computer-based archival file for macromolecular structures. J. Mol. Biol. *112*, 535–542.

Branden, C.-I., Eklund, H., Nordstrom, B., Boiwe, T., Soderlund, G., Zeppezauer, E., Ohlsson, I., and Akeson, I. (1973). Structure of liver alcohol dehydrogenase at 2.9-A resolution. Proc. Natl. Acad. Sci. USA *70*, 2439–2442.

Brandon, C.-I., Eklund, H., Cambillau, C., and Pryor, A. J. (1984). Correlation of exons with structural domains in alcohol dehydrogenase. EMBO J. *3*, 1307–1310.

Burke, R. L., Tekamp-Olson, P., and Najarian, R. (1983). The isolation, characterization, and sequence of the pyruvate kinase gene of saccharomyces cerevisiae. J. Biol. Chem. *258*, 2193–2201.

Dennis, E. S., Gerlach, W. L., Pryor, A. J., Bennetzen, J. L., Inglis, A., Llewellyn, D., Sachs, M. M., Ferl, R. J., and Peacock, W. J. (1984). Molecular analysis of the alcohol dehydrogenase (Adhl) gene of maize. Nucl. Acids Res. *12*, 3983–4000.

Dodgson, J. B., Strommer, J., and Engel, J. D. (1979). Isolation of the chicken β-globin gene and a linked embryonic β-like globin gene from a chicken DNA recombinant library. Cell *17*, 879–887.

Doolittle, W. F. (1978). Genes in pieces: were they ever together? Nature *272*, 581–582.

Gilbert, W. (1978). Why genes in pieces? Nature *271*, 501.

Jones, T. A. (1978). A graphics model building and refinement system for macromolecules. J. Appl. Crystallogr. *11*, 268–272.

Kabsch, W. (1978). A discussion of the solution for the best rotation to relate two sets of vectors. Acta Crystallogr. (sect. A) *34*, 827–828.

Kayne, F. J. (1974). Pyruvate kinase. In The Enzymes, P. D. Boyer, ed. 3rd ed., Vol. 8, (New York: Academic Press), pp. 353–382.

Lonberg, N., and Gilbert, W. (1983). Primary structure of chicken muscle pyruvate kinase mRNA. Proc. Natl. Acad. Sci. USA *80*, 3661–3665.

Maxam, A. M., and Gilbert, W. (1980). Sequencing end-labeled DNA with base-specific chemical cleavages. Meth. Enzymol. *65*, 499–560.

McLachlan, A. D. (1979). Gene duplication in the structural evolution of chymotrypsin. J. Mol. Biol. *128*, 49–79.

Naora, H., and Deacon, N. J. (1982). Relationship between the total size of exons and introns in protein coding genes of higher eukaryotes. Proc. Natl. Acad. Sci. USA *79*, 6196–6200.

Perler, F., Efstradiadis, A., Lomedico, P., Gilbert, W., Kolodner, R., and Dodgson, J. (1980). The evolution of genes: the chicken preproinsulin gene. Cell *20*, 555–566.

Remington, S. J., and Matthews, B. W. (1980). A systematic approach to the comparison of protein structures. J. Mol. Biol. *140*, 77–99.

Rigby, W. J., Diekmann, M., Rhodes, C., and Berg, P. (1977). Labeling deoxyribonucleic acid to high specific activity in vitro by nick translation with DNA polymerase. I. J. Mol. Biol. *113*, 237–251.

Rossmann, M. G., and Argos, P. (1977). The taxonomy of protein structure. J. Mol. Biol. *109*, 99–129.

Rossmann, M. G., Moras, D., and Olsen, K. W. (1974). Chemical and biological evolution of a nucleotide-binding protein. Nature *250*, 194–199.

Southern, E. (1975). Detection of specific sequences among DNA fragments separated by gel electrophoresis. J. Mol. Biol. *98*, 503–517.

Staden, R. (1980). A new computer method for the storage and manipulation of DNA gel reading data. Nucl. Acids Res. *8*, 3673–3694.

Stuart, D. L., Levine, M., Muirhead, H., and Stammers, D. K. (1979). Crystal structure of cat muscle pyruvate kinase at a resolution of 2.6 Å. J. Mol. Biol. *134*, 109–142.

Tate, V., Finer, M., Boedtker, H., and Doty, P. (1982). Procollagen genes: further sequence studies and interspecies comparisons. Cold Spring Harb. Symp. Quant. Biol. *47*, 1039–1049.

Tonegawa, S. (1983). Somatic generation of antibody diversity. Nature *302*, 575–581.

Treisman, R., Orkin, S. H., and Maniatis, T. (1983). Specific transcription and RNA splicing defects in five cloned β-thalassaemia genes. Nature *302*, 591–596.

Vieira, J., and Messing, J. (1982). The pUC plasmids, an M13mp7-derived system for insertion mutagenesis and sequencing with synthetic universal primers. Gene *19*, 259–268.

PERSPECTIVES

Genes-in-Pieces Revisited

Walter Gilbert

Mammalian genes are discontinuous, broken up along the DNA into alternating regions: coding sequences or exons, which are interspaced with other sequences, and introns that will be spliced out of the RNA transcript. What is the meaning of this arrangement?

In 1977, I conjectured that genes in eukaryotic cells arose as collections of exons brought together by recombination within intron sequences, and that the introns were the remnants of a process that speeded up evolution (*1, 2*). This hypothesis predicts that the exons code for useful portions of protein structure: functional regions, folding elements, domains, or subdomains—any segment that can be sorted independently during evolution (*3*). As relics of the recombination process that brought the exons together, the introns would be long, random sequences that would drift rapidly in sequence and size since the last act that assembled the gene.

Ford Doolittle (*4*) realized that there is no reason for this speeding of the evolutionary process to be restricted to eukaryotes, and he argued that the earliest organisms should have had split genes; the present day intron-less genomes of prokaryotes and of lower eukaryotes would then be the result of streamlining, a consequence of the evolutionary pressure for rapid replication. How well have these ideas fared?

Today we can draw some general conclusions about intron patterns. Introns are common in vertebrate genes, essentially absent in prokaryotes, and rare in lower eukaryotes such as yeast, with some dramatic exceptions such as the bithorax complex genes of drosophila. The distribution of exon sizes is rather narrow and peaks at 40 to 50 amino acid residues, but the sizes of introns scatter randomly and range from 50 to 10,000 to 20,000 bases in length. Since the introns are, on the average, so much larger than the exons, vertebrate genes have turned out to be much larger than we expected a decade ago, about 10 to 30 times larger than the coding sequence. The largest gene yet found is 200 kilobases long and 50-kilobase genes are not uncommon.

Where a protein has a structure containing repeated domains, the repeat is reflected in the intron distribution. For example, the basic domain of the immunoglobulins, the immunoglobulin fold, is carried on a single exon and repeated from one to five times as we compare β_2-microglobulin (*5*), the immunoglobulins (*6*), the histocompatibility antigens (*7*), and the T-cell receptors (*8*). The triple repeat of albumin (*9*), α-fetoprotein (*10*), or ovomucoid (*11*) clearly arose from a tripling of the underlying exon-intron structure. The helix of collagen was built up as a 40-fold repeat of an exon that bears a half turn (*12*). Thus the idea that introns serve to assemble the genes for proteins having repeating structures is well borne out.

Exons can often be correlated with functional elements of the encoded proteins. The hydrophobic signal sequences that tag a molecule for export are often carried on separate exons. Transmembrane, hinge, and cytoplasmic portions of immunoglobulins are similarly isolated. However, is there any pattern of the use of the same exon in different genes, when the same element of structure or function is required by different proteins? This would be the acid test of the theory. At last a dramatic example has been found.

In two important papers in this issue Südhof, Goldstein, Brown, and Russell (page 815), and Südhof, Russell, Goldstein, Brown, Sanchez-Pescador, and Bell (page 893), demonstrate the existence of exon shuffling. By analyzing the intron-exon structure of the gene for the low-density lipoprotein (LDL) receptor, the membrane receptor that binds the LDL particle and leads to its internalization within the cell, these workers found that the functional subdivisions of this protein are reflected in detail in the way this gene is broken up into 18 exons. More remarkably, however, Südhof *et al.* have shown that much of the LDL receptor gene is made up of exons recruited from other genes.

First, there is a stretch of 400 amino acids of the LDL receptor that shows 33 percent homology with the precursor for epidermal growth factor (EGF). This region of homology is encoded by eight contiguous exons in each gene. Of the nine introns involved, five are at identical positions, one has migrated a few codons, and three do not have mates. The most reasonable interpretation is that this region came as a whole from some common ancestral element and that the missing introns have subsequently been lost.

Second, a 40-amino-acid cysteine-rich stretch is repeated three times within this 400-amino-acid homologous segment. This sequence is encoded by a single exon, repeated thrice in both the LDL receptor and the EGF precursor genes, and arising once, as a separate exon, in the blood-clotting protein factor IX (*13*). Protein sequence homology suggests that this same exon will also be found in two other blood-clotting proteins, factor X and protein C (*14*).

Third, the LDL binding domain of the LDL receptor contains seven repeats of a 40-amino-acid sequence. Four of these occur on separate exons, while the other three occur on a single exon, as though two introns had been lost. This repeat element also occurs once in the complement factor C9; these authors predict that this will turn out to be a recurrence of that exon. The LDL receptor gene is thus a mosaic of exons derived from many diverse sources.

This work shows that introns have been used to assemble those genes that are the late products of evolution. But where did the introns come from?

There are two extreme alternatives. Either the introns are the vestigial linkers between useful coding sequences, left over from tying together simple reading frames at the beginning of evolution, or they arose by the insertion of DNA sequences into genes that were originally continuous. In the recent evolutionary period, we have only evidence for loss, as in the case of preproinsulin (*15*). We could interpret other anomalies, such as the ovalbumin—α-antitrypsin comparison (*16*) or the differences between the actin genes of plants (*17*), sea urchins (*18*), and vertebrates (*19*) as the loss of multiple introns. Can we distinguish between the possibility that preexisting introns have been lost by the lower eukaryotes and prokaryotes and the alternative, that introns have been created in

The author's address is 107 Upland Road, Cambridge, Massachusetts 02140.

the evolution that led to the vertebrates, either by insertion into a continuous gene or by the tying together of simpler gene elements?

A test is possible by examining old genes—genes whose products were in existence before the separation of the prokaryotes and eukaryotes. Such products are the enzymes for basic biochemical processes, which have the same three-dimensional structure in all cells. These ancient genes have a nonsplit structure in prokaryotes and in yeast. Do they have introns in the vertebrates? If so, are the introns randomly placed or do they correlate with structural elements?

The genes for three glycolytic enzymes whose three-dimensional structures are known have now been analyzed: glyceraldehyde phosphate dehydrogenase (GAPDH) (*20*), pyruvate kinase (PK) (*21*), and triose phosphate isomerase (TIM) (*22*), all from the chicken. All three genes have many introns; GAPDH has 11, PK has 9, and TIM has 6. The exons are quite regular in size, clustered about 30 to 40 amino acid residues in length, so the number of introns simply reflects the size of the protein. Three introns in GAPDH mark major domain borders as do two in PK, a third domain border in PK is not split. As we look more deeply into the tertiary structures of the domains encoded by the exons, we see that in PK and TIM most of the exons are compact pieces of the protein, modules in Go's sense (*23, 24*), each carrying one or two α-helical and β-sheet elements. The introns often mark turns or edges of secondary structure. The structures seem to be assembled out of the exon peptides; the intron positions are not random. But if these proteins were assembled to include the introns, then yeasts and prokaryotes, where the corresponding genes do not have introns, must have lost these dividers.

To attack this problem in another way we (*25*) have examined the TIM gene from a higher plant, maize, to see if its structure resembled that in the vertebrates. Maybe higher plants and animals would have similar gene structures that resemble the original gene in their single-cell common forebear more closely than do those genes of the lower eukaryotes that have evolved through so many more generations. Two-thirds of the maize TIM gene has been sequenced, revealing five introns. Three are at identical positions in corn and chicken, one has moved three codons over, and there is one extra intron in corn, at a bend in the last α-helix, an intron presumably lost from chicken TIM. The ancestral gene must have been already broken up in the eukaryotic progenitor cell before the time that the first algae and animal cells separated, probably more than a billion years ago, and the "lower" eukaryotes, such as yeasts and insects, must have lost introns as they evolved. The same argument applies to the prokaryotes; the organisms that went into symbiosis to form the eukaryotic cell probably all had genes made up of exons tied together by introns.

These ideas imply that the structure of the exon polypeptides must be telling us something profound about the "rules for creating proteins." Not only are proteins put together as mosaics of simpler structures, combinatorial assemblies of a smaller number of minigenes, but the folding principles may become apparent if we can understand the structure of the exon products and find the rules by which they were fitted together.

References

1. W. Gilbert, *Nature (London)* **271**, 501 (1978).
2. ———, *ibid., in Eucaryotic Gene Regulation: ICN-UCLA Symposia on Molecular and Cellular Biology*, R. Axel, T. Maniatis, C. F. Fox Eds. (Academic Press, New York, 1979), vol. 14, pp. 1–12.
3. C. C. F. Blake, *Nature (London)* **277**, 598 (1979); and see also *ibid.* **273**, 267 (1978) and *ibid.* **306**, 535 (1983).
4. W. F. Doolittle, *ibid.* **272**, 581 (1978).
5. J. R. Parnes and J. G. Seidman, *Cell* **29**, 662 (1982).
6. T. Honjo, *Annu. Rev. Immunol.* **1**, 499 (1983).
7. L. Hood, M. Steinmetz, B. Malissen, *ibid.*, p. 529.
8. M. Malissen *et al.*, *Cell* **37**, 1101 (1984).
9. T. D. Sargent, L. L. Jagodzinski, M. Yang, J. Bonner, *Mol. Cell. Biol.* **1**, 871 (1981).
10. F. A. Eiferman, P. R. Young, R. W. Scott, S. M. Tilghman, *Nature (London)* **294**, 713 (1981).
11. J. P. Stein, J. F. Catterall, P. Kristo, A. R. Means, B. W. O'Malley, *Cell* **21**, 681 (1980).
12. H. Boedtker and S. Aho, *Biochem. Soc. Symp.* **49**, 67 (1984).
13. D. S. Anson, K. H. Choo, D. J. G. Rees, F. Giannelli, K. Gould, J. A. Huddleston, G. G. Brownlee, *EMBO J.* **3**, 1053 (1984).
14. R. F. Doolittle, D.-F. Feng, M. S. Johnson, *Nature (London)* **307**, 558 (1984).
15. F. Perler, A. Efstratiadis, P. Lomedico, W. Gilbert, R. Kolodner, J. Dodgson, *Cell* **20**, 555 (1980).
16. M. Leicht, G. L. Long, T. Chandra, K. Kurachi, V. J. Kidd, M. Mace, E. W. Davie, S. L. C. Woo, *Nature (London)* **297**, 655 (1982).
17. D. M. Shah, R. C. Hightower, R. B. Meagher, *J. Mol. Appl. Genet.* **2**, 111 (1983).
18. A. D. Cooper and W. R. Crain, *Nucleic Acids Res.* **10**, 4081 (1982).
19. H. U. Ama *et al.*, *Mol. Cell Biol.* **4**, 1073 (1984).
20. E. M. Stone, K. N. Rothblum, R. J. Schwartz, *Nature (London)* **313**, 498 (1985).
21. N. Lonberg and W. Gilbert, *Cell* **40**, 81 (1985).
22. D. Straus and W. Gilbert, in preparation.
23. M. Go, *Nature (London)* **291**, 90 (1981).
24. ———, *Proc. Natl. Acad. Sci. U.S.A.* **80**, 1964 (1983).
25. M. Marchionni and W. Gilbert, unpublished.

15 April 1985

Cell, Vol. 46, 133–141, July 4, 1986, Copyright © 1986 by Cell Press

The Triosephosphate Isomerase Gene from Maize: Introns Antedate the Plant–Animal Divergence

Mark Marchionni and Walter Gilbert
Department of Cellular and Developmental Biology
Harvard University
Biological Laboratories
16 Divinity Avenue
Cambridge, Massachusetts 02138

Summary

We have cloned and characterized a cDNA and genomic DNA for the triosephosphate isomerase expressed in maize roots. The gene is interrupted by eight introns. If we compare this gene with that for the protein in chicken, which has six introns, we see that five of the introns are at identical places, one has shifted by three codons, and two are totally new. This great matching leads us to conclude that the introns were in place before the plant–animal divergence, and that the parental gene had at least eight introns, two of which were lost in the line that leads to animals.

Introduction

Genes of higher eukaryotes are discontinuous: regions of noncoding DNA (introns) separate the coding sequence into discrete segments (exons). Gilbert (1978) proposed that introns are vestigial DNA sequence, remnants of a recombination process that accelerated molecular evolution by assembling new linkage groups from the exons, creating novel gene products. That hypothesis of exon shuffling predicts that positions of introns should portray the evolutionary history of a gene; the exons themselves would encode distinct functional elements (Gilbert, 1978, 1979), stably folding peptides (Blake, 1978) or compact modules (Go, 1981, 1983). Evidence in support of these ideas derives from molecular studies on genes and proteins as diverse as immunoglobulins (reviewed by Tonegawa, 1983, and Honjo, 1983), collagen (Yamada et al., 1980), serum albumin–α-fetoprotein (Sargent et al., 1981; Eiferman et al., 1981), ovomucoid (Stein et al., 1980), globins (Go, 1981), intermediate filaments (Marchuk et al., 1984; Balcarek and Cowan, 1985), lysozyme (Jung et al., 1980), and crystallins (Moormann et al., 1983). Moreover, recent studies (Südhof et al., 1985a, 1985b) comparing the gene structure of low density lipoprotein receptor with those of epidermal growth factor and blood coagulation factors 9 and 10 provide a dramatic example of exon shuffling in recent evolutionary history.

Although there are a few exceptions (Kaine et al., 1983; Chu et al., 1984; Nellen et al., 1981; Miller, 1984), the conspicuous absence of introns in bacteria and yeast poses questions as to their age and origins. Were introns present as the first genes formed? Or have they more recently taken up residence in genomic DNA by invading continuous gene sequences? Doolittle (1978) suggested that the genomes of primitive cells contained introns but that rapidly dividing organisms lost them by streamlining in response to the selection pressure of rapid reproduction. The highly conserved, ubiquitous glycolytic enzymes most likely evolved completely before the archaebacteria–prokaryotic–eukaryotic division and thus represent extremely ancient genes. In vertebrates, introns punctuate nonrandomly the sequences encoding chicken pyruvate kinase (Lonberg and Gilbert, 1985), chicken glyceraldehyde phosphate dehydrogenase (Stone et al., 1985), chicken triosephosphate isomerase (Straus and Gilbert, 1985b), as well as human phosphoglycerate kinase (Michelson et al., 1985), suggesting that the introns were not inserted into preexisting genes. To push our view of one of these glycolytic enzyme genes back in time, we have characterized a gene encoding triosephosphate isomerase (EC 5.3.1.1, TIM) in maize in order to compare its structure to the chicken gene. A remarkable conservation of intron positions between these distant relatives indicates that the introns were in place before the plant–animal divergence, more than one billion years ago.

Results

Cloning and Structure of Maize TIM cDNA

During glycolysis, triosephosphate isomerase (TIM) catalyzes the interconversion of dihydroxyacetone phosphate and glyceraldehyde 3-phosphate. In addition to a cytosolic TIM, plant cells active in photosynthesis express a plastid enzyme encoded in the nucleus (Pichersky and Gottlieb, 1982). That isozyme functions in the dark reactions that fix carbon dioxide into sugars. As is true for other photosynthetic proteins, the chloroplast-specific TIM is light inducible (e.g., Nelson et al., 1984; Berry-Lowe et al., 1982). Hence, mRNA from roots grown in the dark should be enriched in the cytosolic form, which ought to be cognate to the chicken enzyme. Because TIM is less than 60% diverged across the most distant species (Pichersky et al., 1984; Straus and Gilbert, 1985b), cross-hybridization from chicken to maize root cDNA was a plausible approach to cloning the maize gene.

Using a chicken probe (Straus et al., 1985a), we screened 80,000 phage cDNA recombinants that represented mRNA extracted from maize root seedlings grown in the dark for 7 days. We chose hybridization and washing conditions of medium stringency (see Experimental Procedures) and detected 18 potential TIM clones. After partially mapping all of the inserts, we focused on two large (>800 bp) overlapping clones, subcloned each of these into the EcoRI site of pUC8 (Vieira and Messing, 1982), and sequenced (Maxam and Gilbert, 1980) both strands of the DNA.

Figure 1 displays a partial restriction map and the complete nucleotide sequence of a full-length maize root TIM cDNA clone, designated pMRT1. Within the 123 bases of 5′ untranslated sequence is a 37 base pyrimidine tract, which is flanked by two overlapping, imperfect, direct repeats of 20 bases each. A similar alternating pyrimidine

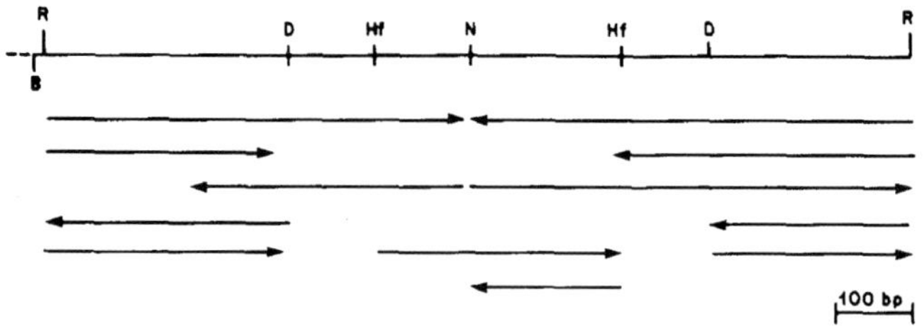


```
TGTCTTTCTTCCTCCACAAGCAGCGAAATCCAATCTAGAAGCTCCCCTCTCCCTCCCTCCCTCTCTCTCTCTCTCTTCGCCGTCCGAAGCTCCGCACCCC   100
AATCTAATCGACACCTCACCGAGATGGGCCGCAAGTTCTTCGTCGGTGGCAACTGGAAATGCAATGGAACCACAGATCAGGTCGAGAAGATTGTCAAAAC   200
                       M  G  R  K  F  F  V  G  G  N  W  K  C  N  G  T  T  D  Q  V  E  K  I  V  K  T
CCTGAATGAAGGACAGGTTCCCCCTTCAGATGTTGTTGAGGTCGTTGTCAGCCCTCCTTATGTCTTCCTTCCTGTGGTCAAGAGCCAGCTGCGCCAAGAG   300
 L  N  E  G  Q  V  P  P  S  D  V  V  E  V  V  V  S  P  P  Y  V  F  L  P  V  V  K  S  Q  L  R  Q  E
TTCCATGTTGCTGCTCAGAACTGCTGGGTGAAGAAGGGAGGTGCTTTCACTGGTGAAGTCAGTGCTGAGATGCTCGTCAACCTTGGTGTTCCCTGGGTCA   400
F  H  V  A  A  Q  N  C  W  V  K  K  G  G  A  F  T  G  E  V  S  A  E  M  L  V  N  L  G  V  P  W  V  I
TTCTTGGACACTCTGAAAGGAGAGCTCTGCTGGGAGAATCAAATGAATTTGTTGGAGACAAGGTTGCGTATGCCCTGTCTCAGGGACTAAAGGTCATTGC   500
  L  G  H  S  E  R  R  A  L  L  G  E  S  N  E  F  V  G  D  K  V  A  Y  A  L  S  Q  G  L  K  V  I  A
ATGTGTTGGTGAGACCCTTGAGCAGAGGGAGGCTGGGTCTACCATGGATGTTGTTGCTGCACAAACAAAAGCAATTGCTGAGAAGATCAAGGACTGGAGC   600
 C  V  G  E  T  L  E  Q  R  E  A  G  S  T  M  D  V  V  A  A  Q  T  K  A  I  A  E  K  I  K  D  W  S
AACGTAGTTGTTGCCTATGAACCAGTTTGGGCTATTGGAACTGGTAAAGTTGCCACCCCAGCTCAGGCTCAGGAAGTGCACGCCTCCCTGAGGGATTGGC   700
N  V  V  V  A  Y  E  P  V  W  A  I  G  T  G  K  V  A  T  P  A  Q  A  Q  E  V  H  A  S  L  R  D  W  L
TAAAGACCAATGCCAGCCCTGAGGTTGCTGAATCTACTAGGATCATCTACGGAGGCTCTGTAACTGCTGCGAACTGCAAAGAGCTAGCAGCACAGCCTGA   800
  K  T  N  A  S  P  E  V  A  E  S  T  R  I  I  Y  G  G  S  V  T  A  A  N  C  K  E  L  A  A  Q  P  D
TGTCGATGGTTTTCTTGTCGGTGGAGCTTCGTTGAAGCCTGAGTTCATCGACATCATCAACGCGGCCACCGTGAAGTCCGCTTAAGATGTTACGCTGAAG   900
 V  D  G  F  L  V  G  G  A  S  L  K  P  E  F  I  D  I  I  N  A  A  T  V  K  S  A
ACGAACATACTTTTTTTGCTCAACTGTGCTATGTAAGCTAGTAGCTTTTGCGCAGGAGCAGAGACTGTTTTGCCTGCCCCCAACTTTTAGCTTGAGCTTG  1000
CTAATAATGTTTACCTCTGGACGTATCAATAATGGTGCTTATGTAAAAAAAAAAAAAAAAAAAAAAAAAAAAAAAAAAAAAAAAAAAAAAGGAATTC
```

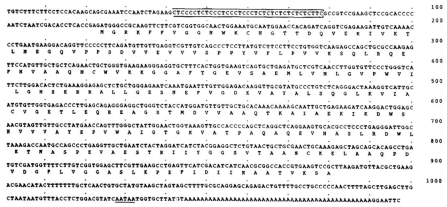

Figure 1. Structure of a Maize Triosephosphate Isomerase cDNA Expressed in Root

We subcloned into plasmid pUC8 the insert of a λgt10 phage, which we had isolated from a root cDNA library using a chicken TIM probe. We determined the nucleotide sequence of the full-length clone by the chemical cleavage method (Maxam and Gilbert, 1980). Boxed in the 5′ untranslated region is a 37 base pyrimidine stretch, and the polyadenylation signal is underlined 12 bases before the site of poly(A) addition. Indicated in the upper panel are the restriction sites we used for sequencing: R=EcoRI, B=BamHI, D=DdeI, Hf=HinfI, and N=NcoI.

tract found upstream of the alcohol dehydrogenase 1 gene has been implicated in maize roots as important for the induction of transcription in response to anaerobic stress (Hake et al., 1985; R. Ferl, personal communication). A polypeptide chain of 253 residues begins at nucleotide 124 and is followed by a 3′ untranslated stretch of 159 bases. Located 15 bases upstream of the site of polyadenylation is AATAA, which resembles the consensus sequence (Fitzgerald and Shenk, 1981) but lacks the final A.

TIM Is Encoded by a Small Multigene Family in Maize

In chicken, TIM is encoded by a unique gene (Straus and Gilbert, 1985b). Human DNA, however, has two processed pseudogenes (Brown et al., 1985) in addition to a single-copy functional TIM locus. Some wild flowers of the genus Clarkia have two unlinked TIM loci (Pichersky and Gottlieb, 1983). To enumerate the TIM genes in maize, we analyzed Southern blots of maize DNA with a specific nick-translated fragment (bases 584–661 in pMRT1; see Figure 1), which we chose with the following properties in mind. First, such a short probe, 78 bp, is unlikely to harbor sites for restriction enzymes that have 6 bp recognition sequences. Second, this segment of TIM cDNA sequence surrounds the active site residue Glu-165 and is most (83%) conserved in nucleotide sequence between maize and chicken. Finally, this sequence corresponds to a specific exon (number 6, see Figure 5). Thus each fragment detected should represent a distinct maize TIM gene. Since maize DNA is highly methylated, we used enzymes that lacked CXG in their recognition sequences. Figure 2 shows the hybridization of NcoI-digested maize DNA; nine fragments appear. The band at 1.35 kb corresponds to the 1374 bp NcoI fragment in the cloned gene described below. The pattern suggests that maize contains a total of at least nine genes and pseudogenes encoding TIM.

Cloning and Structure of an Active Maize TIM Gene

We have characterized a functional structural gene corresponding to our cDNA clone using overlapping clones, derived from two different genomic libraries. Initially, we labeled the insert from pMRT1 (Figure 1) and then probed an incomplete maize genomic library (10^6 EcoRI partial maize DNA fragments in Charon 4A), which was the kind gift of J. Sorenson (Upjohn Co., Kalamazoo, MI). We detected and recovered 28 hybridizing phage, then mapped and sorted them into groups according to their hybridiza-

Maize and Chicken TIM Genes Share Intron Positions
135

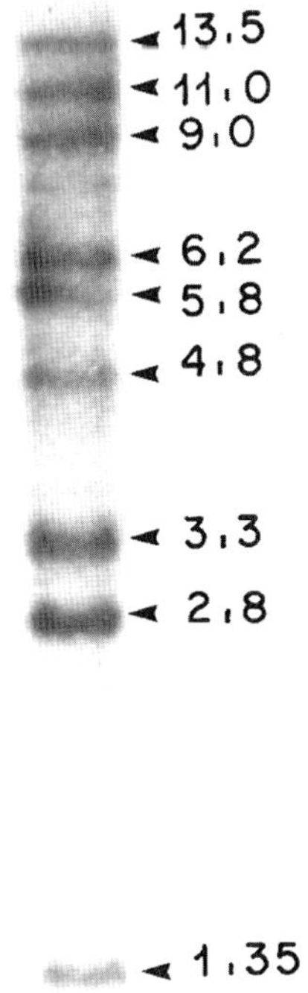

Figure 2. Gene Counting of Triosephosphate Isomerase in Maize
We digested to completion 10 μg of maize nuclear DNA using NcoI, electrophoresed the fragments in a 1% agarose gel, and blotted (Southern, 1975) and immobilized them onto an uncharged nylon membrane. We probed the filter with a nick-translated fragment comprising sequences of maize TIM exon 6, as described in the text. Autoradiography was for 3 days at −80°C (with an intensifying screen).

tion to discrete cDNA subprobes. Though no individual phage contained sequences complementary to the entire cDNA, we chose one, MT11, that reacted with a probe from the 3′ untranslated region and several probes containing coding sequences near the C terminus. As summarized in Figure 3, we subcloned overlapping NcoI and BamHI fragments of MT11 into plasmid vectors pKK233-2 (J. Brosius, unpublished) and pUC8 (Vieira and Messing, 1982) and determined their nucleotide sequences. Maize TIM sequences in MT11 adjoin the left arm of Charon 4A and encode the C-terminal 100 amino acid residues. Introns divide these coding sequences into four exons.

After mapping the remaining TIM clones of this collection, we concluded that none overlapped with MT11. From that group, however, we sequenced another phage that bore a portion of a second maize TIM gene differing by a few amino acid substitutions (17 replacements out of 102 residues compared) and by at least 50% in the intron sequences. However, the five intron positions in this partially characterized gene are identical to those of MT11.

To characterize the remaining portion of the expressed maize TIM gene, we screened a second genomic library (gift of J. Shen, Harvard University) constructed of MboII partials. To find our way through the forest of multiple genes, we identified specific restriction fragments and synthetic oligonucleotides (Table 1) that detected unique sequences on Southern blots of maize genomic DNA. In particular, both an N-terminal HaeIII–PvuII fragment (bases 128–287 in Figure 1) and a synthetic 20-mer (number 2) hybridized under stringent conditions solely to a 1.4 kb HinfI fragment. Therefore, we screened 4 × 10^6 recombinants with that specific nick-translated fragment (bases 128–287 in Figure 1) and detected 20 positive phage. Subsequently, we diagnosed them with a set of synthetic oligomers (Table 1) derived from sequences of 5′ untranslated DNA, the coding segments of the N-terminal 150 residues, and an intron of MT11. One phage, MT46, hybridized with each probe tested. We concluded that MT46 overlapped MT11.

We analyzed MT46 using a combination of genomic sequencing (Church and Gilbert, 1984) and direct chemical sequencing of the recombinant phage to determine all of the remaining exons and 90% of the remaining intron sequence (see Figure 3). Most often we exploited genomic sequencing as adapted for use with synthetic oligonucleotide probes (Tizard and Nick, unpublished data) to map seven of the remaining nine intron/exon boundaries. In Figure 4, we display an example of our use of this technique. When we analyzed a complete HaeIII digest of recombinant phage MT 46, we could discern more than 50 fragments in the ethidium bromide stained gel. We chemically sequenced the mixture of fragments, then electrophoresed, transferred, and cross-linked the DNA to a solid support, as described in Experimental Procedures. We visualized the sequences of individual HaeIII fragments by probing with oligonucleotides that border the enzyme cleavage site. After we determined the sequence using the first probe (Figure 4, left panel), the membrane was stripped and reprobed with another oligonucleotide (center panel). The complete removal of the first probe is documented well in this second autoradiograph by the absence of the unreacted band seen in the initial sequence ladder. Furthermore, we used that technique to confirm the overlap of those two genomic clones by sequencing from their shared NdeI site. In three cases (restriction sites for StuI, NcoI, and XbaI), however, we resorted to conventional chemical sequencing.

We designate this active gene as maize triosephosphate isomerase 1; Figure 5 summarizes its nucleotide sequence. The 3.8 kb gene is divided into nine exons by eight introns. All of the splice junctions conform to the GT/AG rule (Mount, 1982). Prominent among the features of this gene is the nonrandom interruption of coding sequence by introns. Whereas these introns range broadly in length from 92 to 630 bp, the exons fall into three discrete size classes. The first and last exons are relatively short, encoding only 13 and 15 residues, respectively. Exons 3 and 5 are the longest at 42 and 44 amino acids. Tightly clustered around a length of 26–32 residues is the preponderant class composed of exons 2, 4, 6, 7, and 8. Models that postulate the insertion of introns into previ-

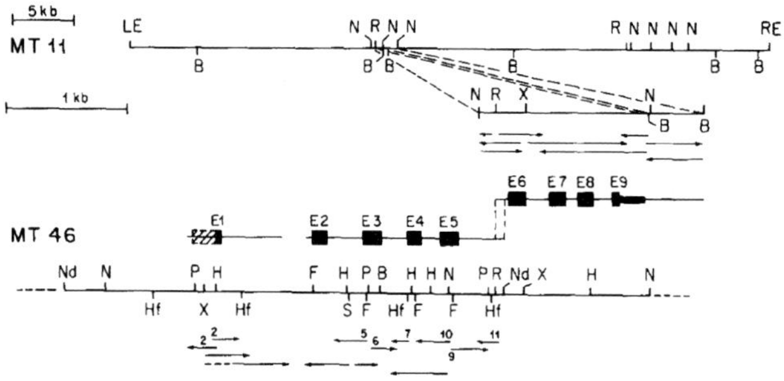

Figure 3. Cloning and Sequencing of the Structural Gene for Triosephosphate Isomerase of Maize

Using the insert of cDNA clone pMRT1 as a probe, we isolated the genomic clone MT11. The 5 kb scale applies solely to that phage clone, which contained sequences composing exons 6–9 and the 3′ untranslated region on two overlapping restriction fragments: a 1.2 kb NcoI piece and a 300 bp BamHI fragment. We subcloned those fragments into plasmids, then sequenced both strands by conventional chemical cleavage methods, as indicated by the arrows. We found an overlapping genomic clone, MT46, and as described in the text, we determined sequences of the 5′ untranslated region, exons 1–5 and most of the introns between them. Numbers above the arrows designate the probes (Table 1) that we used to visualize genomic sequences; elsewhere we sequenced labeled DNA fragments. Restriction sites are abbreviated as follows: Nd=NdeI, N=NcoI, Hf=HinfI, P=PvuII, X=XbaI, H=HaeIII, F=Fnu4H, S=StuI, Bs=BstNI, R=EcoRI, and B=BamHI.

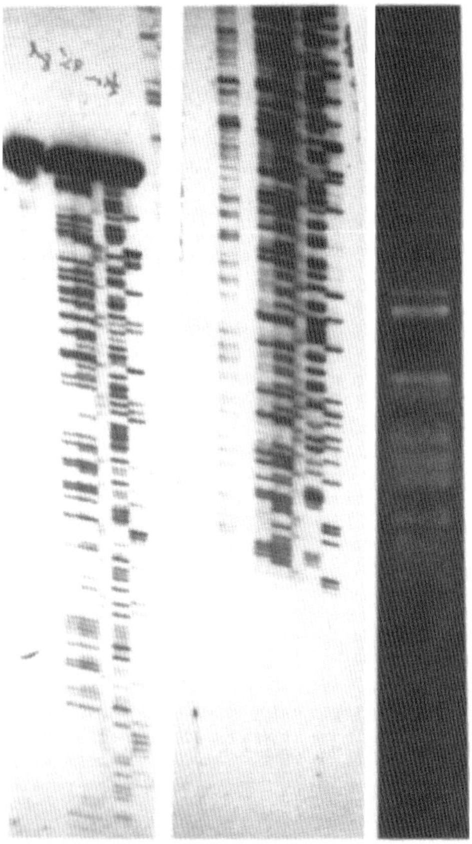

Figure 4. Genomic Sequencing of Maize TIM Clone MT46

We digested to completion 10 μg of the maize genomic clone MT46, then analyzed 1 μg of the DNA fragments on a 1% agarose gel (ethidium bromide stain). We treated the remaining 9 μg by chemical cleavage reagents, resolved the fragments by electrophoresis, and transferred them to uncharged nylon membranes, as described in Experimental Procedures. Initially, using probe 2 (see Table 1), we visualized (left panel) sequences of the lower strand of exon 1, extending for 115 bases up the autoradiogram to a HaeIII site in intron 1. We stripped probe 2 off the filter and restained with probe 7 (center panel), this time determining 155 bases of upper strand sequence within exon 4 and crossing into intron 3. Samples are loaded in the following order: G, A + G (missing in this gel), A > C, C + T, C, and T.

ously uninterrupted genes (Orgel and Crick, 1980; Cavelier-Smith, 1980) are not supported by these data. Rather, our findings on the structure of maize TIM 1 suggest that exon shuffling played an important role in the assembly of ancient genes.

Conservation of Intron/Exon Patterns in TIM

How old are the introns in TIM? Figure 6 shows that all six chicken introns are shared with maize. Five are located at exactly the same positions, while one has shifted over three codons. Moreover, there are two extra introns in maize, located near each of the termini. The conservation of intron positions between plant and animal TIM genes spreads across much of the molecule. This pattern cannot be explained easily by the separate insertion of introns into a continuous, preexisting gene. Rather, the identity of the positions of the five introns leads us to the conclusion that the ancestral gene was broken up at these positions before the time of plant–animal divergence. Thus we interpret intron six as a case of sliding and introns one and eight as lost in the chicken. Measurements of the rate of divergence of 5S ribosomal RNA genes (Ohama et al., 1984; Huysmans et al., 1983) suggest that plants and animals have evolved separately for approximately one billion years. Hence, the common forebear of plants and animals, a unicellular organism living at least one billion years ago, contained a TIM gene with a structure resembling the one present in corn.

Maize and Chicken TIM Genes Share Intron Positions
137

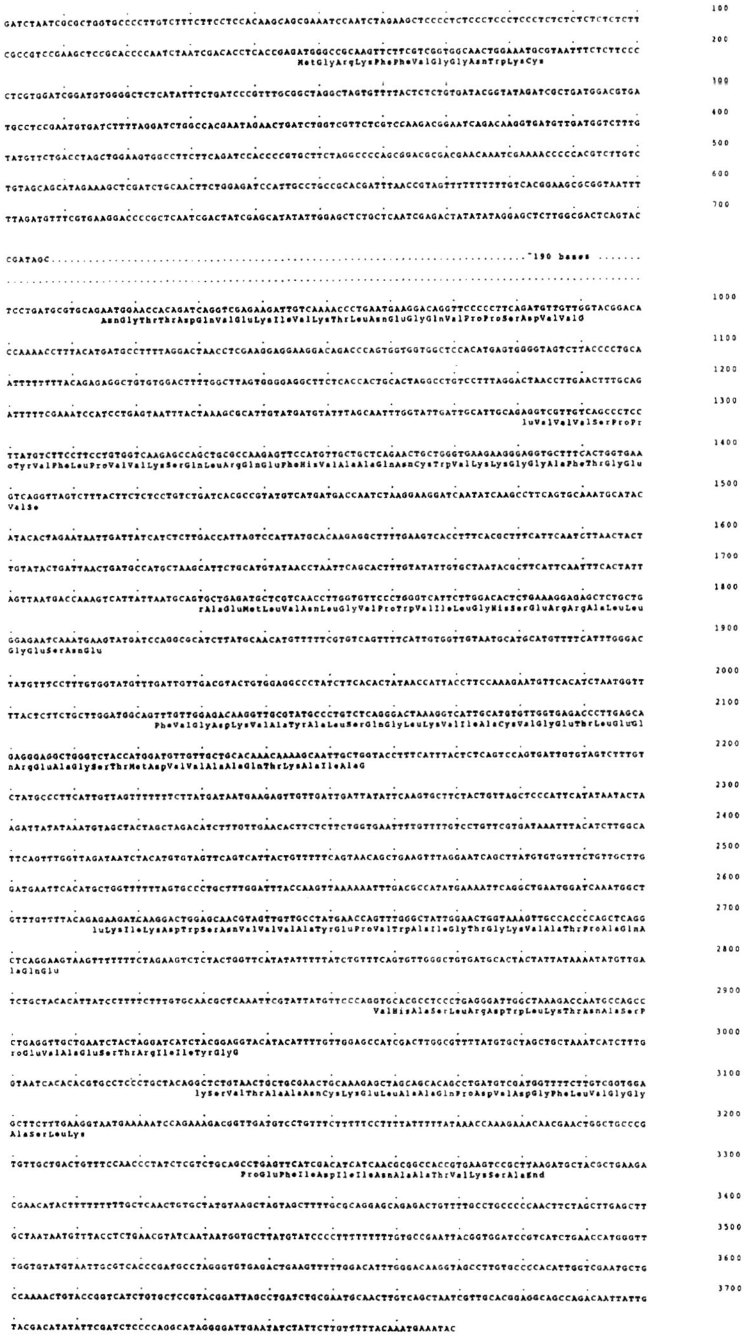

Figure 5. Nucleotide Sequence of an Active Gene for Triosephosphate Isomerase of Maize

Shown is a composite of the nucleotide sequences determined in two overlapping genomic clones, MT11 and MT46 (Figure 3). These sequences differ from maize root cDNA at a single nucleotide (compare pMRT1 at 830 in Figure 1 and position 3106 in this figure), probably reflecting a G/T transversion between the strains studied or an artifact of cloning the gene.

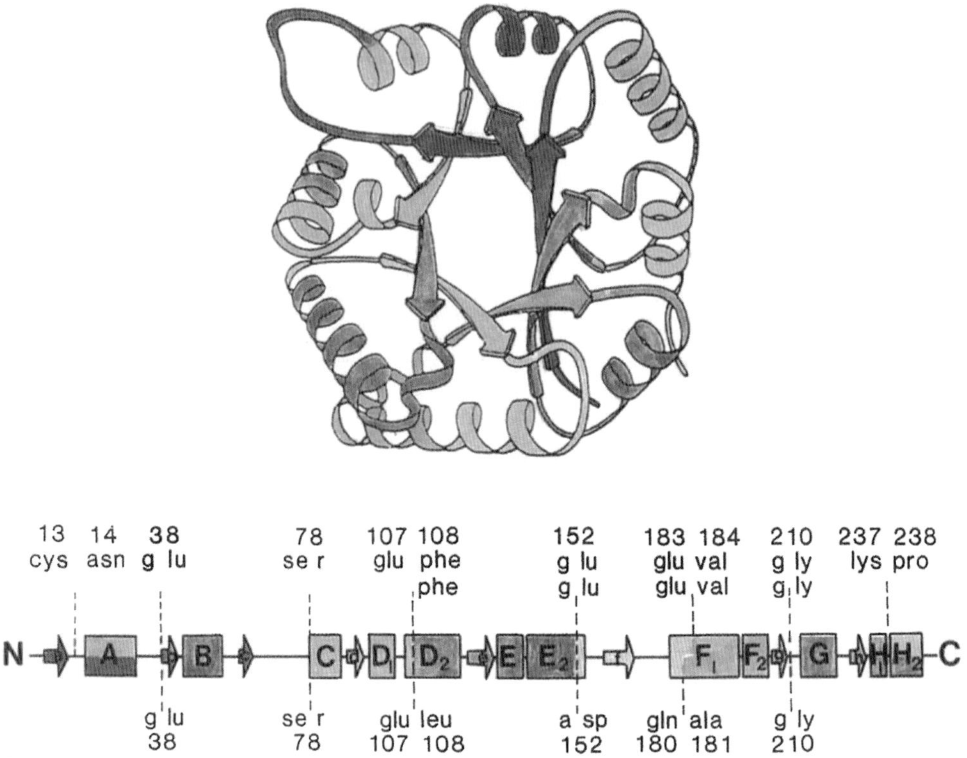

Figure 6. Phylogenetic Comparison of Intron Positions in Triosephosphate Isomerase

We have aligned the positions of the introns in maize (Figure 5) and chicken (Straus and Gilbert, 1985b) TIM genes onto the amino acid sequence of the chicken enzyme. We illustrate in this figure examples from two different genes in maize and from the single gene in chicken, above and below the linear intron/exon map, respectively. β-Strands (arrows a–h) and α-helices (cylinders A–H) are represented in the colors appearing in the schematic diagram above. There we have divided the TIM barrel according to the maize gene structure. The figure is redrawn from Jane Richardson.

Correlations of Gene and Protein Structure in TIM

TIM is a homodimer, with subunits ranging, in different species, from 248–253 amino acids in length. Three-dimensional structures of the yeast and chicken enzymes exhibit an indistinguishable conformation in their active sites and display virtual identity elsewhere in the protein (Alber et al., 1981a, 1981b). Each subunit folds into a pseudosymmetrical barrel (Banner et al., 1975), which is composed of an alternating α-helix/β-strand motif that is repeated 8 times. A core of parallel β-strands is surrounded by a concentric sheath of α-helices. Figure 6 shows that seven of nine introns fall between, and not within, the helices and strands constituting the TIM barrel. The terminal exons of chicken are divided by introns in maize. Intron 1 (Figure 6) divides β-strand a from helix A (three turns), coded by exon 2. The 8th intron in maize (between Lys-237 and Pro-238) occurs at a bend in the last helix (H).

Does our knowledge of the evolutionary history and anatomy of TIM permit an understanding of how that ancient gene might have been assembled? Brändén (unpublished data) has proposed that proteins with an α/β structure were constructed by fusing exons bearing α/β motifs. His model accounts for a high correlation of intron positions in loops located on the catalytic face (at the carboxyl end of β structures) of those enzymes. Six introns (1, 3, 4, 6, 7, and 8) in TIM obey those rules, but two do not: the intron located at the beginning of β-strand b (Glu-38) and the one near the end of helix E (Glu/Asp-152) are positioned on the amino side of the barrel.

Discussion

A fundamental puzzle in molecular biology today is that the genes of prokaryotes and eukaryotes are patently so different in structure. To explain the presence of introns in eukaryotic DNA (and consequently their absence in prokaryotes and some simpler eukaryotes) requires one of three alternative hypotheses: (1) the original genes were discontinuous, and prokaryotes and yeast lost their introns by streamlining their genomes (Doolittle, 1978); (2) eukaryotes have inserted introns continuously throughout evolution as selfish DNA established residence in their genomes (Orgel and Crick, 1980; Cavalier-Smith, 1980);

Maize and Chicken TIM Genes Share Intron Positions
139

Table 1. Synthetic Oligonucleotide Probes Used for Genomic Sequencing TIM Phage

Probe	Sequence	-Mer	Site	Position	Reads
1	CTAGAAGTTCCCCTCTCCCT	20	XbaI	36	3′ Lower
2	CCGCAAGTTCTTCGTCGGTG	20	HaeIII	127	3′ Lower
3	CTGTGGTTCCATTGCATTTCC	21	Sau3A	173	5′ Upper
4	GGGTTTTGACAATCTTCTCG	20	TaqI	181	3′ Upper
5	CTGCGCCAAGAGTTCCATGT	20	PvuII	288	3′ Lower
6	CCTGTGGTCAAGAGCCAGCTG	21	HhaI	292	5′ Lower
7	AGAGTGTCCAAGAATGACCC	20	BstNI	392	3′ Upper
8	GGAGAGCTCTGCTGGGAGAA	20	HinfI	435	5′ Lower
9	AGCAACAACATCCATGGTAG	20	AccI	538	3′ Upper
10	GGGAGGCTGGGTCTACCATG	20	NcoI	541	5′ Lower
11	GGTAAATCCAAAGCAGGGCAC	21	NdeI	179	5′ Upper

We designed probes 1–10 from the nucleotide sequence of pMRT1 (Figure 1), and probe 11 from intron sequence in genomic clone MT11. For analyses of immobilized phage plaques, Southern blots, and genomic sequence electrotransfers, we labeled probes at their 5′ ends and hybridized them to the appropriate filters. Commencing from the positions and restriction sites shown, we used these probes to visualize sequences of the complementary strand in the directions indicated.

or (3) introns were added at some time shortly after nucleated cells split from the bacteria, and thus precluded yeast from their invasion.

To test those models we have focused on a gene that was created early in evolution, before the divergence of prokaryotes and eukaryotes. Triosephosphate isomerase is a ubiquitous and highly conserved glycolytic enzyme, which has radiated to all organisms as a perfect catalyst. Because nearly 25% of the residues are invariant across vastly different lineages (see Straus and Gilbert, 1985b), the activity present in all surviving species descended from a common ancestral gene. The TIM gene of maize has eight introns; chicken has six. Five of these are in identical positions. Parsimony prescribes that it is preferable to assume a small number of losses rather than a large number of independent insertions in identical locations. In the one other published comparison of plant and animal genes (Shah et al., 1983), in actin the position of one intron is preserved, while all others are different. The pattern in that case could have been explained equally well by either intron gain, loss, or sliding (Fornwold et al., 1982; Davidson et al., 1982; Craik et al., 1983). Our results do not support the model that introns have been added continuously during the last billion years of evolution, because we would not have expected such strong identity between the two species. We conclude that the ancestral gene had eight introns, two were lost in the line that leads to chicken, while one slid over three codons. Precedent exists for the excision of an intron from functional genes in preproinsulin (Perler et al., 1980) and myosin heavy chain (Strehler et al., 1985). That the exons in TIM tend to be of regular size and correlate with recognized elements of protein structure is not in accord with either of the second two hypotheses. Thus, all these arguments give strong support to the idea that introns are as old as the genes themselves.

Experimental Procedures

Isolation of Clones

We obtained a λgt10 root cDNA library (gift of T. Theugh, Stanford University) constructed from the mRNA of maize seedlings grown in the dark for 7 days. Using a cloned chicken cDNA (Straus and Gilbert, 1985a) as a probe, we screened the library according to the plaque hybridization procedure of Benton and Davis (1977). Following a 1 hr prehybridization, we denatured the nick-translated (Rigby et al., 1977) insert, and hybridized (10^6 cpm/ml; 1 × 10^8 to 2 × 10^8 cpm/μg) to duplicate nitrocellulose filters in 5× SSC, 10× Denhardt's, 50 mM phosphate buffer (pH 7.2), 1 mM EDTA, 0.1% SDS, 10% dextran sulfate, and 250 μg/ml of denatured salmon testes DNA. After 12 hr at 65°C, we washed (65°C) filters once for 1 hr in 3× SSC, 0.2% SDS and thrice for 30 min in 2× SSC, 0.2% SDS. We exposed XAR-5 film (with intensifying screens) for three days at −80°C and subsequently recovered and plaque-purified the positive phage.

A W64A maize genomic library containing 15–20 kb EcoRI fragments cloned in Charon 4A was kindly provided by J. Sorenson (Upjohn Co., Kalamazoo, MI.), and a maize genomic library of MboII partial fragments cloned in EMBL IIIB was the generous gift of J. Shen (Harvard University). We grew phage of both genomic libraries in E. coli host strain LE392 and probed them with homologous nick-translated fragments derived from maize cDNA clones. For those homologous screenings, we did our final washes at 65°C in 0.5× SSC, 0.1% SDS.

To obtain the subclones used in DNA sequence determination, we ligated (T4 ligase; New England Biolabs) overlapping fragments containing maize TIM exons into a complementary site of linearized, dephosphorylated (Calf intestinal phosphotase, Boehringer Mannheim) vectors and transformed the ligation products into competent E. coli MC1061 (Casadaban and Cohen, 1980) or JM83 (Vieira and Messing, 1982). Subsequently, we selected clones on the basis of antibiotic resistance and colony hybridization (Grunstein and Hogness, 1975).

Nucleic Acids

We extracted maize nuclear DNA from the shoots of 7 day old seedlings as described (Rivin et al., 1982). We isolated DNA from CsCl-purified recombinant phage according to the method of Thomas and Davis (1974). Plasmid DNA was prepared by alkaline lysis and CsCl-EtBr centrifugation. Restriction enzymes, purchased from New England Biolabs and Boehringer Mannheim, were used as recommended by the suppliers. We mapped the restriction sites used for subcloning and sequence determination by electrophoresis in 6% acrylamide gels or 1% agarose gels, which we blotted (Southern, 1975) onto nitrocellulose or nylon membranes. We probed those filters with nick-translated restriction fragments or with synthetic oligonucleotide probes, which we labeled with crude [γ-^{32}P]ATP (7000 Ci/mmol; New England Nuclear and ICN), using T4 polynucleotide kinase (Boehringer Mannheim).

DNA Sequence Determination

All sequencing was done by the Maxam–Gilbert chemical cleavage method (1980) and included modifications suggested by Rubin and Schmid (1980) and by Bencini et al. (1984). To determine the DNA sequences of maize genomic clones directly, we took advantage of genomic sequencing (Church and Gilbert, 1984), but used synthetic oligonucleotide probes (Tizard and Nick, unpublished data) to analyze the electroblotted DNA. For each restriction site used in genomic sequencing, we analyzed 10 μg of phage DNA. After confirming that the digests were complete, we precipitated the fragments once with spermidine, then twice with ethanol. We dissolved the fragments in 25 μl of H_2O and dispensed 3 μl for each of the six reactions. We electrophoresed 2 ng of the unlabeled sequence fragments in each lane of a 6% polyacrylamide–7 M urea sequencing gel. We cast such gels in molds of 60 × 50 × 0.04 cm as discontinuous gradients (Biggin et al., 1983) of TBE buffer (1 M TBE = 1 M Tris-HCl, 1 M boric acid, and 30 mM EDTA, pH 8.3), according to a recipe developed by R. Tizard (Biogen Research Corp., Cambridge, MA). We poured these gels in four stages—12.5% 300 mM TBE with 5% sucrose, 12.5% 217 mM TBE, 12.5% 133 mM TBE, and 63% 50 mM TBE—and ran them for 4 hr at 90 W. Then, we followed the procedures developed by Church and Gilbert (1984) for electrotransfer, ultraviolet illumination, and hybridization of DNA immobilized on uncharged nylon membranes (Biodyne A, Pall and Zetabind, AMF-Cuno). However, as prescribed by Tizard and Nick (unpublished data), we supplemented the solutions used for hybridization and washing with 0.25 M and 0.125 M NaCl, respectively. We included 20 pmol of 5′-end-labeled 20-mer in a 10 ml hybridization for 4 hr at 45°C. In most cases, following 2 days of film exposure (without an intensifying screen) we could read the sequence beginning 25 bases past the restriction cut to beyond 250 bases. To visualize sequences from other restriction fragments, we eluted the hybridized probe in 0.05 N NaOH and restained with a different probe.

Computer Analysis

Sequence data was assembled using the programs of Staden (1980) and analyzed with programs from the University of Wisconsin Genetics Computer Group on VAX 780 and Microvax 2 computers.

Acknowledgments

Foremost we thank Donald Straus for numerous contributions throughout this investigation. For maize libraries we acknowledge Tanya Theugh, John Sorenson, and Jen Shen; for expertise in genomic DNA sequencing we thank George Church, Richard Tizard, and Harry Nick; for plasmid preparations we acknowledge Neil Malone; for synthesis of oligonucleotides we thank K. L. Ramachandran; and, for assistance in preparation of the manuscript, we thank Nancie Thurston. This work has been supported in part by Biogen Research Corp., Cambridge, Massachusetts. M. M. was also supported by PHS grant 5 F32 CA07048-03 from the National Cancer Institute.

The costs of publication of this article were defrayed in part by the payment of page charges. This article must therefore be hereby marked "*advertisement*" in accordance with 18 U.S.C. Section 1734 solely to indicate this fact.

Received March 31, 1986.

References

Alber, T., Banner, D. W., Bloomer, A. C., Petsko, G. A., Phillips, D., Rivers, P. S., and Wilson, I. A. (1981a). On the three-dimensional structure and catalytic mechanism of triosephosphate isomerase. Phil. Trans. Roy. Soc. (Lond.) B *293*, 159–171.

Alber, T., Hartman, F. C., Johnson, R. M., Petsko, G. A., and Tsernoglou, D. (1981b). Crystallization of yeast triose phosphate isomerase from polyethylene glycol. Protein crystal formation following phase separation. J. Biol. Chem. *256*, 1356–1361.

Balcarek, J. M., and Cowan, N. J. (1985). Structure of the mouse glial fibrillary acidic protein gene: implications for the evolution of the intermediate filament multigene family. Nucl. Acids Res. *13*, 5527–5543.

Banner, D. W., Bloomer, A. C., Petsko, G. A., Phillips, D. C., Pogson, C. I., Wilson, I. A., Corran, P. H., Furth, A. J., Milman, J. D., Offord, R. E., Priddle, J. D., and Waley, S. G. (1975). Structure of chicken muscle triose phosphate isomerase determined crystallographically at 2.5 Å resolution using amino acid sequence data. Nature *255*, 609–614.

Bencini, D. A., O'Donovan, G. A., and Wild, J. R. (1984). Rapid chemical degradation sequencing. Biotechniques *2*, 4–5.

Benton, W. D., and Davis, R. W. (1977). Screening lambda gt recombinant clones by hybridization to single plaques in situ. Science *196*, 159–176.

Berry-Lowe, S. L., McKnight, T. D., Shah, D. M., and Meagher, R. B. (1982). The nucleotide sequence, expression, and evolution of one member of a multigene family encoding the small subunit of ribulose-1, 5-bisphosphate carboxylase in soybean. J. Mol. Appl. Genet. *1*, 483–498.

Biggin, M. D., Gibson, T. J., and Hong, G. F. (1983). Buffer gradient gels and ^{35}S label as an aid to rapid DNA sequence determination. Proc. Natl. Acad. Sci. USA *80*, 3963–3965.

Blake, C. C. F. (1978). Do genes-in-pieces imply proteins-in-pieces? Nature *273*, 267.

Brown, J. R., Daar, I. O., Krug, J. R., and Maquat, L. E. (1985). Characterization of the functional gene and several processed pseudogenes in the human triosephosphate isomerase gene family. Mol. Cell. Biol. *5*, 1694–1706.

Casadaban, M. J., and Cohen, S. N. (1980). Analysis of gene control signals by DNA fusion and cloning in Escherichia coli. J. Mol. Biol. *138*, 179–207.

Cavalier-Smith, T. (1980). How selfish is DNA? Nature *285*, 617–618.

Chu, F. K., Maley, G. F., Maley, F., and Melfort, M. (1984). Intervening sequence in the thymidylate synthase gene of bacteriophage T4. Proc. Natl. Acad. Sci. USA *81*, 3049–3053.

Church, G. M., and Gilbert, W. (1984). Genomic sequencing. Proc. Natl. Acad. Sci. USA *81*, 1991–1995.

Crabtree, G. R., Comeau, C. M., Fowlkes, D. M., Fornace, A. J., Jr., Malley, J. D., and Kant, J. A. (1985). Evolution and structure of the fibrinogen genes. Random insertion of introns or selective loss? J. Mol. Biol. *185*, 1–19.

Craik, C. S., Rutter, W. J., and Fletterick, R. (1983). Splice junctions: association with variation in protein structure. Science *220*, 1125–1129.

Davidson, E. H., Thomas, T. L., Scheller, R. H., and Britten, R. J. (1982). The sea urchin actin genes, and a speculation on the evolutionary significance of small gene families. In Genome Evolution, G. A. Dover and R. B. Flavell, eds. (London: Academic Press), pp. 117–192.

Doolittle, W. F. (1978). Genes in pieces: were they ever together? Nature *272*, 581–582.

Eiferman, F. A., Young, P. R., Scott, R. W., and Tilghman, S. M. (1981). Intragenic amplification and divergence in the mouse alpha-fetoprotein gene. Nature *294*, 713–718.

Fitzgerald, M., and Shenk, T. (1981). The sequence 5′-AAUAAA-3′ forms parts of the recognition site for polyadenylation of late SV40 mRNAs. Cell *24*, 251–260.

Fornwald, J. A., Kuncio, G., Peng, I., and Ordahl, C. P. (1982). The complete nucleotide sequence of the chick alpha actin gene and its evolutionary relationship to the actin gene family. Nucl. Acids Res. *19*, 3861–3876.

Gilbert, W. (1978). Why genes in pieces? Nature *271*, 501.

Gilbert, W. (1979). Introns and exons: playgrounds of evolution. In Eucaryotic Gene Regulation: ICN-UCLA Symposia on Molecular and Cellular Biology, R. Axel, T. Maniatis, and C. F. Fox, eds. (New York: Academic Press), pp. 1–10.

Go, M. (1981). Correlation of DNA exonic regions with protein structural units in haemoglobin. Nature *291*, 90–92.

Go, M. (1983). Modular structural units, exons, and function in chicken lysozyme. Proc. Natl. Acad. Sci. USA *80*, 1964–1968.

Grunstein, M., and Hogness, D. (1975). Colony hybridization: a method for the isolation of cloned DNAs that contain a specific gene. Proc. Natl. Acad. Sci. USA *72*, 3961–3965.

Hake, S., Kelley, P. M., Taylor, W. C., and Freeling, M. (1985). Coordinate induction of alcohol dehydrogenase 1, aldolase, and other anaerobic RNAs in maize. J. Biol. Chem. *260*, 5050–5054.

Honjo, T. (1983). Immunoglobulin genes. Ann. Rev. Immunol. *1*, 499–528.

Huysmans, E., Dams, E., Vandenberghe, A., and DeWachter, R. (1983). The nucleotide sequences of the 5S rRNAs of four mushrooms and their use in studying the phylogenetic position of basidiomycetes among the eukaryotes. Nucl. Acids Res. *11*, 2871–2880.

Jung, A., Sippel, A., Grez, M., and Schutz, G. (1980). Exons encode functional and structural units of chicken lysozyme. Proc. Natl. Acad. Sci. USA *77*, 5359–5763.

Kaine, B. P., Gupta, R., and Woese, C. R. (1983). Putative introns in tRNA genes of procaryotes. Proc. Natl. Acad. Sci. USA *80*, 3309–3312.

Lonberg, N., and Gilbert, W. (1985). Intron/exon structure of the chicken pyruvate kinase gene. Cell *40*, 81–90.

Marchuk, D., McCrohon, S., and Fuchs, E. (1984). Remarkable conservation of structure among intermediate filament genes. Cell *39*, 491–498.

Maxam, A. M., and Gilbert, W. (1980). Sequencing end-labeled DNA with base-specific chemical cleavages. Meth. Enzymol. *65*, 499–560.

Michelson, A. M., Blake, C. C., Evans, S. T., and Orkin, S. H. (1985). Structure of the human phosphoglycerate kinase gene and the intron-mediated evolution and dispersal of the nucleotide-binding domain. Proc. Natl. Acad. Sci. USA *82*, 6965–6969.

Miller, A. M. (1984). The yeast MAT alpha 1 gene contains two introns. EMBO J. *3*, 1061–1065.

Moorman, R. J., den Dunnen, J. T., Mulleners, L., Andreoli, P., Bloemendal, H., and Schoenmakers, J. G. (1983). Strict co-linearity of genetic and protein folding domains in an intragenically duplicated rat lens gamma-crystallin gene. J. Mol. Biol. *171*, 353–368.

Mount, S. M. (1982). A catalogue of splice junction sequences. Nucl. Acids Res. *10*, 461–472.

Nellen, W., Donath, C., Moos, M., and Gallwitz, D. (1981). The nucleotide sequences of the actin genes from Saccharomyces carlbergensis and Saccharomyces cerevisiae are identical except for their introns. J. Mol. Appl. Genet. *1*, 239–244.

Nelson, T., Harpster, M. H., Mayfield, S. P., and Taylor, W. C. (1984). Light-regulated gene expression during maize leaf development. J. Cell Biol. *98*, 558–564.

Ohama, T., Kumazaki, T., Hori, H., and Osawa, S. (1984). Evolution of multicellular animals as deduced from 5S rRNA sequences: a possible early emergence of the Mesozoa. Nucl. Acids Res. *12*, 5101–5108.

Orgel, L. E., and Crick, F. H. C. (1980). Selfish DNA: the ultimate parasite. Nature *284*, 604–607.

Perler, F., Efstratiadis, A., Lomedico, P., Gilbert, W., Kolodner, R., and Dodgson, J. (1980). The evolution of genes: the chicken preproinsulin gene. Cell *20*, 555–566.

Pichersky, E., and Gottlieb, L. D. (1983). Evidence for the duplication of the structural genes coding plastid and cytosolic isozymes of triose phosphate isomerase in diploid species of Clarkia. Genetics *105*, 421–436.

Pichersky, E., Gottlieb, L. D., and Hess, J. F. (1984). Nucleotide sequence of the triose phosphate isomerase gene of Escherichia coli. Mol. Gen. Genet. *195*, 314–320.

Rigby, W. F. Diekmann, M., Rhodes, C., and Berg, P. (1977). Labelling DNA to high specific activity in vitro by nick translation with DNA polymerase I. J. Mol. Biol. *113*, 237–251.

Rivin, C. J., Zimmer, E. A., and Walbot, V. (1982). Isolation of DNA and DNA recombinants from maize. In Maize for Biological Research, W. F. Sheridan, ed., (Grand Forks, North Dakota: University Press), pp. 161–164.

Rubin, C. M., and Schmid, C. W. (1980). Pyrimidine-specific chemical reactions useful for DNA sequencing. Nucl. Acids Res. *8*, 4616–4619.

Sargent, T. D., Jagodzinski, L. L., Yang, M., and Bonner, J. (1981). Fine structure and evolution of the rat serum albumin gene. Mol. Cell. Biol. *1*, 871–883.

Shah, D. M., Hightower, R. C., and Meagher, R. B. (1983). Genes encoding actin in higher plants: intron positions are highly conserved but the coding sequences are not. J. Mol. Appl. Genet. *2*, 111–126.

Southern, E. M. (1975). Detection of specific sequences among DNA fragments separated by gel electrophoresis. J. Mol. Biol. *8*, 503–517.

Staden, R. (1980). A new computer method for the storage and manipulation of DNA gel reading data. Nucl. Acids Res. *8*, 3673–3694.

Stein, J. P., Catterall, J. F., Kristo, P., Means, A. R., and O'Malley, B. W. (1980). Ovomucoid intervening sequences specify functional domains and generate protein polymorphism. Cell *21*, 681–687.

Stone, E. M., Rothblum, K. N., and Schwartz, R. J. (1985). Intron-dependent evolution of chicken glyceraldehyde phosphate dehydrogenase gene. Nature *313*, 498–500.

Straus, D., and Gilbert, W. (1985a). Chicken triosephosphate isomerase complements an Escherichia coli deficiency. Proc. Natl. Acad. Sci. USA *82*, 2014–2018.

Straus, D., and Gilbert, W. (1985b). Genetic engineering in the Precambrian: Structure of the chicken triosephosphate isomerase gene. Mol. Cell. Biol. *5*, 3497–3506.

Strehler, E. E., Mahdavi, V., Periasamy, M., and Nadal-Ginard, B. (1985). Intron positions are conserved in the 5′ end region of myosin heavy-chain genes. J. Biol. Chem. *260*, 468–471.

Südhof, T. C., Goldstein, J. L., Brown, M. S., and Russell, D. W. (1985a). The LDL receptor gene: a mosaic of exons shared with different proteins. Science *228*, 815–822.

Südhof, T. C., Russell, D. W., Goldstein, J. L., Brown, M. S., Sanchez-Pescador, R., and Bell, G. I. (1985b). Cassette of eight exons shared by genes for LDL receptor and EGF precursor. Science *228*, 893–895.

Thomas, M., and Davis, R. W. (1975). Studies on the cleavage of bacterio-phage lambda DNA with EcoRI restriction endonuclease. J. Mol. Biol. *91*, 315–328.

Tonogawa, S. (1983). Somatic generation of antibody diversity. Nature *302*, 591–596.

Vieira, J., and Messing, J. (1982). The pUC plasmids, an M13mp7-derived system for insertion mutagenesis and sequencing with synthetic universal primers. Gene *19*, 259–268.

Yamada, Y., Avvedimento, V. E., Mudryj, M., Ohkubo, H., Vogeli, G., Irani, M., Pastan, I., and de Crombrugghe, B. (1980). The collagen gene: evidence for its evolutionary assembly by amplification of a DNA segment containing an exon of 54 bp. Cell *22*, 887–892.

Cell, Vol. 46, 151–154, July 18, 1986, Copyright © 1986 by Cell Press

On the Antiquity of Introns

Minireview

Walter Gilbert,* Mark Marchionni,*
and Gary McKnight†
* Biological Laboratories
Harvard University
Cambridge, Massachusetts 02138
† ZymoGenetics, Inc.
2121 North 35th St.
Seattle, Washington 98103

The organization of vertebrate genes into introns and exons poses two problems: what is the use of this alternation of coding and noncoding regions? and how did this arrangement arise? The exon shuffling hypothesis states that introns increase the ease, and hence the rate, of recombination events that move exons around and insert them into other genes. In this picture, exons are "useful" elements of protein structure or function, and introns—since there is no great evolutionary pressure to shorten and remove them—are the relics of the acts that formed the genes. Today, a growing body of evidence suggests that exon shuffling has played a major role in the structuring of genes that have arisen during the last 500 million years of evolution down the vertebrate lines. The most dramatic example of this is the use of a separate exon to disperse an epidermal-growth-factor-like domain to a series of unrelated proteins: the low density lipoprotein receptor, the epidermal growth factor precursor, and the blood clotting factors IX and X (Südhof et al., Science *228*, 893–895, 1985). However, the question of the origin of introns remains unsolved.

The two extreme views are that either introns were always with us, being used to assemble the first genes, or that they were added at some point to break up preexisting genes. The first hypothesis (Doolittle, Nature *272*, 581–582, 1978) presupposes the existence of a splicing apparatus in the first cells, but introns were lost from prokaryotes as their genomes became streamlined for rapid DNA replication. The intron structure was retained in eukaryotes, except for those organisms seeking special niches that involved pressure on their DNA content. The second hypothesis assumes that prokaryotic genes resemble the ancestral ones and that the addition of introns and the splicing mechanism was either concomitant with the emergence of the eukaryotic cell or occurred gradually over its ensuing evolution. Such models (Orgel and Crick, Nature *284*, 604–607, 1980) depict introns as arising from transposable elements, with ends that bear splicing sequences that could insert invisibly into the DNA.

The possible early origin of splicing as an RNA-catalyzed process (Cech, Cell *44*, 207–210, 1986), and the few examples of splicing in prokaryotes (Kaine et al., PNAS *80*, 3309–3312, 1983; Chu et al., PNAS *81*, 3049–3053, 1984) are straws that point toward the first hypothesis. However, a direct test is to study genes that came into existence before the prokaryotic/eukaryotic divergence.

The ubiquitous enzyme triosephosphate isomerase (TIM) carries out a basic step in glycolysis and gluconeogenesis and its sequence is highly conserved across all species. This protein evolved to its final three-dimensional structure (and its "perfect" enzymatic activity) before the divergence of the archaebacteria, the eubacteria, and the eukaryotes. The gene coding for TIM has no introns in E. coli or in yeast; however, it has five introns in Aspergillus nidulans (McKnight et al., Cell *46*, 143–147, 1986), six introns in chickens (Straus and Gilbert, Mol. Cell. Biol. *5*, 3497–3506, 1985) and humans (Brown et al., Mol. Cell. Biol. *5*, 1694–1706, 1986), and eight introns in maize (Marchionni and Gilbert, Cell *46*, 133–141, 1986). The dotted lines in Figure 1 indicate the intron positions in maize (top), chicken (middle), and Aspergillus (bottom). Can we explain this pattern?

We can trace the evolutionary pattern of these organisms by using a mutational clock based on either the very slowly changing 5S ribosomal RNA genes (Ohama et al., Nucl. Acids Res. *12*, 5101–5108, 1984) or a very slowly drifting protein. Figure 2 shows the tree based on the slow appearance of amino acid substitutions in TIM itself. Such an analysis shows that the prokaryotic and the eukaryotic cell diverged approximately 1.5 billion years ago. At a later point, about 1.2 billion years ago, the fungi, including yeast and Aspergillus, split off from the line that led eventually to plants and animals. About one billion years ago, the plants and animals split. Eventually some animals developed backbones, and about 400 million years ago the fish split from the warm-blooded animals. About 300 million years ago the birds and mammals took separate paths, while the basic mammalian radiation happened about 100 million years ago.

The striking agreement of five of the intron positions in TIM between maize and vertebrates suggests that all of these introns were in place before the division of plants and animals. The easiest interpretations of the maize in-

Figure 1

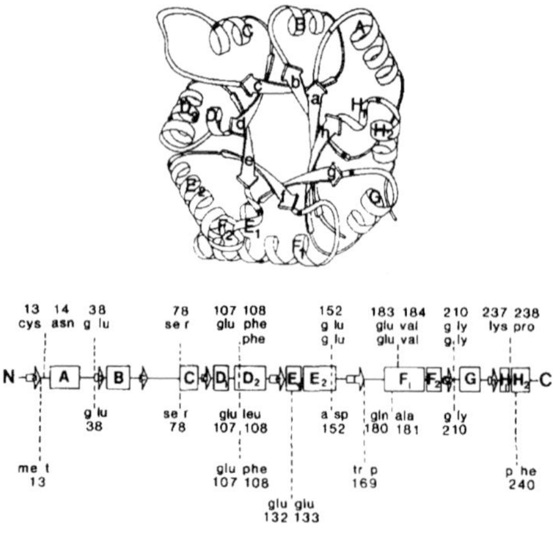

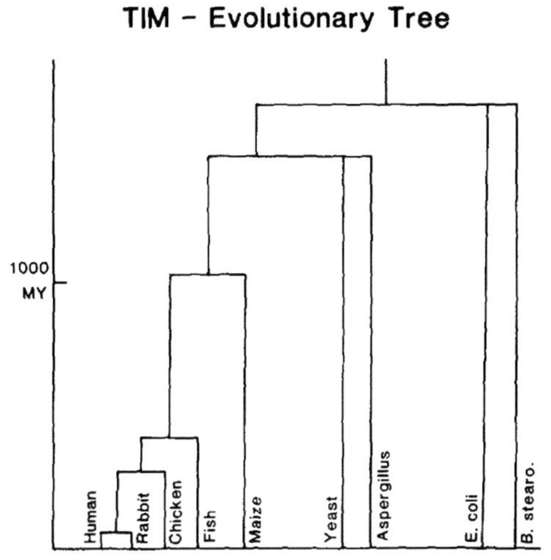

Figure 2

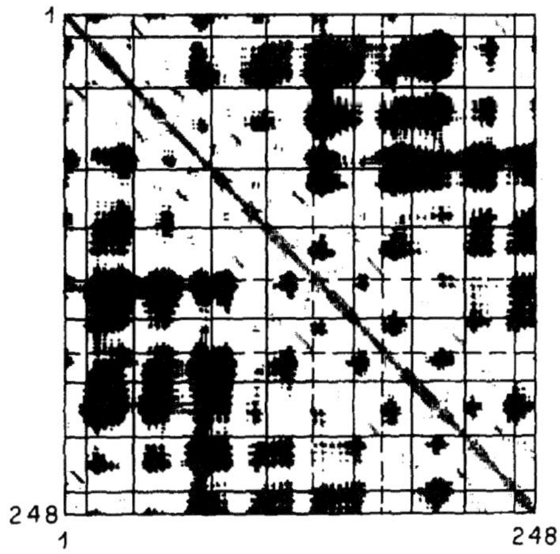

Figure 3

trons that do not match the vertebrate ones are that one intron has moved over three amino acids while the outermost two introns were lost.

The sliding of an intron's position might be the result of a mutation which blocks a splice junction, unmasking a cryptic splice site that adds or deletes small amounts of protein. The precise loss of an intron could occur through an exact deletion or through recombination with a DNA copy of a spliced gene transcript, as in pseudogene formation. There is one clear example of intron loss (Perler et al., Cell *20*, 555–566, 1980), but as yet there is no definitive proof of intron addition. Another likely example of intron loss is the alcohol dehydrogenase (ADH) gene of Arabidopsis, a small plant with a restricted DNA content; this gene has only six of the nine introns found in the ADH gene of maize (Chang and Meyerowitz, PNAS *83*, 1408–1412, 1986).

TIM is the first clear case of extensive intron conservation between plants and animals. Shah et al. (J. Mol. Appl. Genet. *2*, 111–126, 1983) compared the actin gene over this distance, but found that only one intron position out of three in plants and out of seven in rat is identical. That alone is not sufficient agreement to prove that the general mechanism is one of loss. Given the TIM case, however, we predict that the differences in the actin genes in plants and animals are all attributable to loss. Furthermore, we predict that *all* the intronless genes in the simpler eukaryotes that evolved after the plant/animal divergence will prove to be examples of loss.

This experiment has pushed the intron structure back at least a billion years. But might not introns have been added during the early evolution of the eukaryotic cell? A simple interpretation might argue that yeast split off before any introns had been introduced, that Aspergillus represents a later organism in which introns were being added, and that intron addition was complete by the time of the division of maize and animals. There is nothing in intron positioning alone that would rule this out. One intron in Aspergillus is identical in position with introns in maize and chicken, two are near the positions of the supernumerary introns in maize, but are one base and seven bases removed, and two are at totally new positions. The movement of an intron by only one or two bases involves the chance appearance of two cryptic splice sites simultaneously, rather than sequentially. However, the alternative argument would be that some process that added introns placed them inaccurately after the separation of Aspergillus from the line that leads to maize and chicken.

The model that genes were assembled from exons predicts that there should be a connection between exon structure and the three-dimensional structure of the protein. If we assume that all ten introns were present in the preexisting gene, then we note that the exons created by these introns are rather regular in size. Clearly, that result is contrary to the expectations arising from adding introns randomly to a preexisting structure. Furthermore, a number of introns are in regions of very high amino acid sequence conservation (e.g., intron 7 breaks a tryptophan residue at position 169, in the most conserved region). This is not what one would expect if the introns were added as mutagenic agents to a preexisting gene.

Mitiko Gō suggested that exons frequently represent modules—polypeptides that fold in a compact fashion (Nature *291*, 90–92, 1981). This measure of compactness can be visualized on a contour map of the distances between each pair of amino acids in the protein. On such a map, the modules are clusters representing residues that all lie within 28 Å of each other. Figure 3 shows such a plot for TIM, where the heavy lines correspond to the introns in maize, and the Aspergillus positions are superimposed as dotted lines. If we consider the nine exons enclosed by the boxes formed solely of solid lines along the diagonal, exons 1, 2, 4, 7, 8, and 9 avoid the dark regions (>28 Å), and hence are modules; but exons 3, 5, and 6 do not have this property. However, if we assume that the extra introns in Aspergillus were also there in the original gene, what had been exons 5 and 6 are now broken up by the Aspergillus introns (dotted lines) in just such a way as to eliminate the most distant regions and create four modules. Of the eleven exons left in this hypothetical gene, only one, exon 3, does not represent a module. We predict that in some other TIM gene there will be an intron that breaks up this exon.

This argument that introns break up the gene into pieces that define modular elements in the final protein strongly supports the view that the original protein was segmented at the beginning of evolution when the first TIM gene was assembled, and that in all lines lacking introns in this gene, the introns were lost in the course of evolution. The evolutionary force behind that loss would be the pressure on DNA content arising in organisms that specialized toward rapid DNA synthesis and quick cellular division. Only the genes in the slowly replicating cells of complex organisms still retain the full stigmata of their birth.

The Exon Theory of Genes

W. GILBERT
The Biological Laboratories, Harvard University, Cambridge, Massachusetts 02138

Since the intron/exon structure of genes was discovered 10 years ago (Berget et al. 1977; Broker et al. 1977), only a few generalities about the properties of introns have emerged. Most vertebrate genes, but not all, have an intron/exon structure. The length distribution of exons is rather narrow, peaking at about 40 or 50 amino acids. However, introns are an order of magnitude longer than the exons; their length distribution is very broad, the shortest introns being only 50 bases long, the longest extending out to some 50,000 bp. No essential function has been found that requires the presence of all the introns in a gene. If one compares genes from different species, separated by a sufficient evolutionary distance, the exon sequences of homologous genes drift slowly, the positions coding for amino acids being conserved, whereas the intron sequences drift as rapidly as third-base positions, indicating that they are evolutionarily silent. Nonetheless, there is a general role that introns might play, solely because of their length and position, by participating in genetic recombination and hence increasing the rate at which the exons reassort as independent elements. This is the concept of exon shuffling (Blake 1978, 1979, 1983; Gilbert 1978).

The argument that introns increase the rate of recombination is straightforward. We know that there exist recombinational processes that can create interchanges between contiguous parts of a gene. Such recombinational processes, often called illegitimate recombination, would involve the recombination between DNA sequences at a few matched bases. A single such recombination could be used to make a double-length gene out of a simple structure; a double recombination might be used to insert a fragment of one gene into another. Such recombinations are observed in microorganisms. An example in humans is hemoglobin Lepore. However, if the two regions to be recombined were to be separated by a 10,000-base intron in the finished gene, the illegitimate recombination that combines them need not take place exactly at the end of one exon or exactly at the beginning of a second, but anywhere within 10,000 bases after the end of one exon or within 10,000 bases before the beginning of the second. On a combinatorial basis alone, this recombination process is 10^8 times more rapid than that involving exact recombination. Thus, the introns represent hot spots for recombination; by their mere presence and length they increase the rate of recombination, and hence shuffling of the exons, by factors of the order of 10^6 or 10^8. Under this model, the presence of an intron/exon structure in recent genes reflects the most probable way for new genes to arise: through the coupling of exons by intron-mediated recombination.

This picture provides a way of creating new genes from the combination of exons, in a way more in keeping with the requirements of natural selection than that of the classic model for domain doubling. For example, the classic model suggests that the putting together of domains to make a protein with a repeated domain structure will involve an extremely rare recombination event that would produce a large amount of a correct gene product. The intron/exon model suggests that a frequent recombination event creates a long intron separating the exons of the two domains. This new gene need not produce, at least initially, a large amount of product, but is instead a trial gene that produces only a small amount of product. Natural selection will then work on variations of this gene selecting for splicing mutants that make splicing more effective and accurate, and thus increase the yield of the gene product. The traditional hypothesis involves a rare event leading to an immediately useful product. In contrast, the shuffling hypothesis involves a common event that occurs as the first of a series of small steps leading to a new, complex gene product.

There are now many examples of the shuffling of exons in the genes that have arisen throughout Metazoan evolution. The most striking of these is the LDL receptor, EGF precursor, and blood factors IX and X story (Südhof et al. 1985a,b). But where did this intron/exon structure originate?

The Exon Theory of Genes

We suggest that the first genes were assembled by recombination within introns linking exons serving as minigenes (Doolittle 1978; Gilbert 1979). The complete form of this hypothesis is that the first exons encoded very short polypeptide fragments, essentially statistically occurring open reading frames, ranging in length from 15 to 20 amino acids. The very first genes might simply have made such polypeptide products, which then assembled into multimeric structures with enzymatic activity. The splicing mechanism would also have permitted the assemblage, using *trans*-splicing, of multiexon protein chains, through the intermediate existence of multimeric RNA complexes. However, it is only recombination between introns that creates at the DNA level, at the genetic level, a single heritable gene, made up of exons, that will dictate a single, complex protein subunit. This picture asserts that the small peptides used in that original set of exons probably had structure in solution, or structure in an appropriate hydrophobic environment, that gave them the ability to serve as elements of form and function. The structures need not be as three-dimensionally stable as those of today's proteins, because we are looking at the slow biochemical processes occurring at the beginning of evolution.

The polypeptides encoded by the first exons would be small and compact and represent elements that fold up in space (modules in Mitiko Gō's sense, circumscribed by a sphere 28 Å in diameter [Gō 1981]).

Over the sweep of evolutionary time, introns are lost and more complicated exons are formed. The exon theory of genes hypothesizes that the only processes with appreciable rates are those of intron loss. Although the probability of intron loss by exact deletion is very rare, and the probability of intron loss by an approximate deletion that removes the intron and some of the surrounding material is rather rare, there is one process known that over evolutionary time removes introns—retroposition: A mature message is copied back into DNA and then part of all of that DNA is recombined into the chromosome. This process will remove introns singly and, as in pseudogene formation, will take an entire complex structure, remove all of its introns, and reinsert it into DNA. In general, this does not lead to a functional gene because the reinsertion is in some region of DNA that is not transcribed. However, if the reinsertion is into an intron within a previously existing gene, the reinserted element will serve as a complex exon, often carboxy-terminal, in the new gene. This process, slow over evolutionary time, combines the elementary exons into more complicated structures, which will be shuffled in their turn. The immunoglobulin fold, a preeminent example of a single exon used throughout an entire gene family, is a complex exon of this kind, since its length, at 120 amino acids, would be that of some 5–6 minimal exons. This offers an explanation for the additional introns found in CD4 and in N-CAM: They are remnants of the original immunoglobulin structure.

An extensive piece of evidence for this general exon theory is the gene structure of triosephosphate isomerase (TIM) (Straus and Gilbert 1985; Marchionni and Gilbert 1986). TIM, an enzyme of glycolytic metabolism, is an extremely ancient protein, which evolved completely before the divergence of the Eukaryotes from the Eubacteria and the Archaebacteria. This gene has six introns in animals and eight introns in plants. Five of these introns are in identical positions, and one is a closely matched position. This high identity argues that the gene was split before the divergence of plants and animals, approximately a billion years ago. This gene has also been sequenced in *Aspergillus* (McKnight et al. 1986), where it has five introns, and *Saccharomyces* and *Escherichia coli*, where it has none. Three of the five introns in *Aspergillus* agree in their positions with those of the plant gene, but two are at novel positions. Superficially, these numbers might suggest that the gene evolved from one of no introns in bacteria to added introns as one moved up through the fungi, alternative ones in alternative lines, to a set of eight introns in the lineage that led to the plants and animals. However, a more subtle interpretation of the positions of the introns (Gilbert et al. 1986) shows that all of the introns, including the ones in *Aspergillus* that are not represented in the plant/animal line, fall upon divisions of the protein that break the protein into modules, as defined by Mitiko Gō (1979): The exons represent compact elements of polypeptide structure, each lying within a sphere about 28 Å in diameter. If we assume that all ten of the TIM introns that have thus far been found in corn, chicken, and *Aspergillus* were already present in some ancestral TIM gene, then 10 of the 11 exons of that gene are modules in this sense, and one is not. That peculiar exon, we hypothesize, will turn out to be divided by an intron in the ancestral gene.

The exon theory suggests that the first genes had an intron/exon structure. The intron's role was to assemble and reassort the exons as individual elements during the formation of those genes. Over evolutionary time, simple exons are fused to make complicated ones. Can we push this model back before the origin of DNA?

The force of the intron/exon idea is that it is efficient to make genes out of building blocks that can be rearranged by the evolutionary process. Of course, this is true on a larger scale when one makes an entire organism out of smaller elements, or one makes complex proteins out of dissimilar subunits encoded by genes on different chromosomes. The assortment of elements onto different chromosomes, or even into separate genes, provides the ability to reassort those elements in each generation to provide for the variation that drives evolution. The intron structure permits the assortment of elements *within* genes over a much longer time span; still by a process rapid compared to the full sweep of geological history. This ability to reassort the genetic elements of the biological molecules might have been the role of the intron/exon structure at the very beginning of evolution.

The RNA World

Our view of the origin of life has been drastically changed in the last few years by the discovery of several enzymatic activities in RNA molecules, notably those involved in the making and breaking of phosphodiester bonds (the work of Cech, Altman, and their colleagues described in papers in this volume [Been et al.; Lawrence et al.]). This raises the possibility that RNA might show a wide variety of other enzymatic activities. Ulenbeck (Sampson et al., this volume) has given a simple demonstration that there can be metallo-RNA enzymes. White (1976) suggested several years ago that many of the enzymatic cofactors used to carry out chemical reactions might be the remnants of RNA molecules.

Thus, one might imagine a purely RNA-catalyzed

world in which RNA enzymes, ribozymes, using their own functionalities as well as cofactors (metals, NAD, FMN, etc.), carry out all of the enzymatic reactions needed for primitive life (Gilbert 1986). Here RNA would serve as the holder of genetic information, copied by ribozymes. The intron/exon hypothesis distinguishes between the genetic material and the ribozyme gene products. The genetic material would have an intron/exon structure, initially involving self-splicing introns. The base-pairing and secondary structure features of this molecule can be specialized as a substrate for replication. Then, when the genetic material with its complete intron/exon structure is copied by the ribozymes, many copies are made: Some of the daughter copies splice out their introns and combine their exons, thus producing functional ribozymes. Those ribozyme RNAs are specialized for enzymatic activity. They are three-dimensional structures that are not constrained by the need to be copied by an RNA-replicating mechanism.

How might such structures arise? The self-splicing intron, which can splice itself out of an RNA molecule, will presumably, at some rate, be able to splice itself back into an RNA molecule and serve as an insertion sequence. (This has not yet been demonstrated experimentally.) Thus, insertion sequences are implicit in the RNA enzymology that we already know. However, two insertion sequences around an intervening region of RNA serve to construct a transposon and will be able to carry that intervening region around from one RNA molecule to another. In this way, a structure with self-splicing introns can also be viewed, over a much longer time scale, as a pattern of transposons that permit the shuffling of exons. In addition, RNA recombination, once an intron/exon structure arises, will permit the shuffling of exons for the same reasons that it does at the DNA level.

This picture of the RNA world is one in which the intron/exon structure is used to make ribozymes, the ribozymes provide the full enzymology of an RNA cell, and short, presumably 1,000–10,000 long RNA molecules serve as genetic material. We hypothesize that this RNA state had a long history and existed all the way through the period of the first isolation of RNA molecules within membranes. If among the molecules made by the RNA are basic molecules for charge neutralization, a requirement for many of the RNA-RNA interactions, then the RNA molecules may be wrapped up within membranes, which would serve to isolate the genetic structures from the outside world and permit them to compete with each other so that natural selection could be effective.

The first proteins would be simple homopolymer chains or simple dipeptide/tripeptide structures involving very few amino acids. The first oligopeptides involved basic amino acids whose properties would enhance the likelihood of RNA being wrapped in membranes (Jay and Gilbert 1987) and thus improve the ability of cells to function. A second role of such early polypeptides would be to serve as pores through membranes. A third role would be to support the three-dimensional structure of ribozymes.

The transition to an RNA-protein world could be quite gradual. The ability to activate amino acids and insert them into oligopeptide chains can emerge slowly, a few amino acids at a time, because the first role of the oligopeptides is supportive rather than enzymatic. In this picture of the sweep of evolution, the role of protein enzymes is not required by some unusual chemical feature of enzymology. Rather, the hypothesis is that RNA molecules and cofactors are a sufficient set of enzymes to carry out all the chemical reactions necessary for the first cellular structures. Protein enzymes offer improvements in the rate of catalysis rather than in the types of reactions catalyzed. One can imagine a gradual supplanting of many of the ribozyme functions, first by protein-RNA combinations and later by complete protein structures. Because the oligopeptides will be coded by RNA messages that have undergone splicing, like the ribozymes, and that are derived from an underlying genetic material with an intron/exon structure, the proteins, too, are assembled from exons from the very beginning. Of course, the protein exons and the enzymes that they finally create are unrelated in an evolutionary sense to the ribozyme functions they replace. Many enzymatic activities will represent novel combinations of polypeptide exons that solve an underlying enzymological problem differently than does a ribozyme.

After the emergence of proteins, one would have a world of cells containing RNA as the genetic material, ribozymes involved in protein synthesis (rRNA and tRNAs, and possibly activating enzymes), ribozymes involved in RNA splicing, or a group II self-splicing pattern, as well as having protein enzymes and structural elements. In addition, lipid membranes would permit the separation of inside from outside and provide the boundaries between independent replicating entities. The genetic material would be rather small RNA molecules, with an intron/exon structure, with each cell having many copies of each gene. Such genetic molecules are small, because their effective size is determined roughly by the inverse of mutation rate.

DNA would arise as a genetic material after the introduction of the enzymatic process of reverse transcription and the other enzymatic processes needed to convert the ribonucleotide precursors into the deoxyribonucleotide precursors. DNA as a genetic material offers a double-strand structure whose error correction mechanisms permit the creation of very long genomes and hence a simpler transmission of information to the progeny. On this view, the DNA has a very late-developed, protein-based enzymology. This is consistent with the fast conversions of nucleotides into deoxynucleotides and the ultrafast synthesis rate of DNA. Because the genetic RNA molecules that were copied into DNA had an intron/exon structure, the DNA itself preserves that intron/exon structure.

The progenote cell's DNA more closely resembled that of the modern eukaryotic cell than that of a current prokaryote. Over evolutionary time the descendants of this original cell have diversified, branched, and specialized. One line gave rise to the Eubacteria, simplifying ribosomal structures and specializing its DNA and its cell structure for rapid multiplication, consequently losing the intron structure. Another line gave rise to the Archaebacteria, also losing the intron/exon structure. A third line maintained the intron/exon structure and introduced effective splicing mechanisms, probably making a transition from group II splicing, where the guide sequences are internal in the intron, to a development of a *trans*-splicing mechanism involving a fully developed spliceosome. The cell line that makes this transition is the ancestor of the eukaryotes. It eventually develops a nuclear membrane, maintains a complex ribosomal structure, and picks up prokaryotic symbiotes to create mitochondria and chloroplasts. This line leads to the multicellular organisms. But the genomes of modern metazoa still have retained an intron/exon structure that resembles and has arisen from that in the very first DNA-containing organism.

The Numerology of Evolution

The intron/exon hypothesis, the notion that genes are assembled from small modules, provides a clear resolution of the numerical paradox that is often seen as circumscribing the possibilities of evolution. How is it possible to find a protein structure consisting of 200 amino acids by some random walk through evolution? This question reflects the fact that a 200-amino-acid-long structure is one out of some 20^{200} possible structures, and there is neither time nor carbon in the universe to explore all those possible structures and find the relevant one. How then is it possible for natural selection to create the protein enzymes that we see around us? The exon theory provides a simple explanation. It suggests that the first genes are selected out of a group of 20^{20} possibilities, a far smaller universe of possibilities (15 kg of random proteinaceous material contains all possible 20-long sequences). In fact, it may be the case that only a small fraction of the 20^{20} possible 20-long sequences actually have relevant three-dimensional shapes. For example, suppose that only 10^6 such different shapes exist in the space of exons. Then the evolutionary path leading to a 200-amino-acid-long protein would be the selection of a first exon out of the space of 10^6 possibilities, followed by the subsequent choice of a second exon again as one of 10^6 shapes, then the choice of a third, and so forth. If done in order, the ten choices required to produce a 200-amino-acid-long final protein explore only 10^7 possibilities (rather than the 10^{60} possibilities if a simultaneous choice were being made), far less than the 10^{260} possibilities we contemplated originally. This argument shows that the numerology of evolution is not an insurmountable problem, and it implies further that evolution could have examined only an extremely tiny fraction of the possible structures of proteins. There may be very, very many different three-dimensional configurations that can solve a given enzymatic problem with the same efficiency.

What is the total number of shapes available in the world of exons? (Above, we hypothesized there to be only 10^6.) One can try to estimate that from our current knowledge by assuming that one can assemble from the gene products that have been studied so far, a group that represents random draws from an underlying world of exon shapes. If this were the case, one could compare the total number of exons studied with the number of repeats of exon structures to get some idea of the full universe of exon shapes represented in all proteins. If the sampling were random, then we would expect the number of repeats to be approximately $n^2/2N$, where N is the total number of exons, and n is the number of exons that have been observed. (This is essentially the birthday problem. $1/N$ is the probability one exon drawn from the pool will match a specified earlier one, and $n^2/2$ is the number of pairs of possible matches of n objects.) These numbers are difficult to determine, in part because we do not know how to compare shapes rather than amino acid sequence. But one can make very rough guesses. If one has observed about 3000 exons and has seen about 100 repeats, then the total universe of exons would be 10^5. These estimates are very crude but do suggest that there is a very small world of shapes from which all gene products were formed.

REFERENCES

Berget, S.M., A.J. Berk, T. Harrison, and P.A. Sharp. 1978. Spliced segments at the 5′ termini of adenovirus-2 late mRNA: A role for heterogeneous nuclear RNA in mammalian cells. *Cold Spring Harbor Symp. Quant. Biol.* **42:** 523.

Blake, C.C.F. 1978. Do genes-in-pieces imply proteins-in-pieces? *Nature* **273:** 267.

———. 1979. Exons encode protein functional units. *Nature* **277:** 598.

———. 1983. Exons—Present from the beginning? *Nature* **306:** 535.

Broker, T.R., L.T. Chow, A.R. Dunn, R.E. Gelinas, J.A. Hassell, D.F. Klessig, J.B. Lewis, R.J. Roberts, and B.S. Zain. 1978. Adenovirus-2 messengers—An example of baroque molecular architecture. *Cold Spring Harbor Symp. Quant. Biol.* **42:** 531.

Doolittle, W.F. 1978. Genes in pieces: Were they ever together? *Nature* **272:** 581.

Gilbert, W. 1978. Why genes in pieces? *Nature* **271:** 501.

———. 1979. Introns and exons: Playgrounds of evolution. *ICN-UCLA Symp. Mol. Cell. Biol.* **14:** 1.

———. 1986. Origin of life, the RNA world. *Nature* **319:** 618.

Gilbert, W., M. Marchionni, and G. McKnight. 1986. On the antiquity of introns. *Cell* **46:** 151.

Gō, M. 1979. Eucaryotic gene regulation. *ICN-UCLA Symp. Mol. Cell. Biol.* **14:** 1.

———. 1981. Correlation of DNA exonic regions with protein structural units in haemoglobin. *Nature* **291:** 90.

Jay, D. and W. Gilbert. 1987. Basic protein enhances the incorporation of DNA into lipid vesicles: Model for the formation of primordial cells. *Proc. Natl. Acad. Sci.* **84:** 1978.

Marchionni, M. and W. Gilbert. 1986. The triosephosphate isomerase gene from maize: Introns antedate the plant-animal divergence. *Cell* **46:** 133.

McKnight, G.L., P.J. O'Hara, and M.L. Parker. 1986. Nucleotide sequence of the triosephosphate isomerase gene from *Aspergillus nidulans:* Implication for a differential loss of introns. *Cell* **46:** 143.

Straus, D. and W. Gilbert. 1985. Genetic engineering in the precambrian: Structure of the chicken triosephosphate isomerase gene. *Mol. Cell. Biol.* **5:** 3497.

Südhof, T.C., J.L. Goldstein, M.S. Brown, and D.W. Russell. 1985a. The LDL receptor gene: A mosaic of exons shared with different proteins. *Science* **228:** 815.

Südhof, T.C., D.W. Russell, J.L. Goldstein, M.S. Brown, R. Sanchez-Pescador, and G.I. Bell. 1985b. Cassette of eight exons shared by genes for LDL receptor and EGF precursor. *Science* **228:** 893.

White, H.B., III. 1976. Coenzymes as fossils of an earlier metabolic state. *J. Mol. Evol.* **7:** 101.

How Big Is the Universe of Exons?

ROBERT L. DORIT, LLOYD SCHOENBACH, WALTER GILBERT

If genes have been assembled from exon subunits, the frequency with which exons are reused leads to an estimate of the size of the underlying exon universe. An exon database was constructed from available protein sequences, and homologous exons were identified on the basis of amino acid identity; statistically significant matches were determined by Monte Carlo methods. It is estimated that only 1000 to 7000 exons were needed to construct all proteins.

MOST GENES IN COMPLEX EUKARYOTES CONSIST OF short exons separated by long introns. In one view, genes are assembled, via intron-mediated recombination, from exon modules that code for functional domains, folding regions, or structural elements (*1*, *2*). Such models portray introns as a retained primitive feature. Alternatively, the phylogenetic distribution of introns has led to arguments that introns are a derived feature of eukaryotic genomes, the result of bursts of parasitic elements invading early (and continuous) eukaryotic coding regions (*3*, *4*).

The hypothesis of exon shuffling proposes that complex genetic information is built up by joining previously independent exons, thus giving rise to more complex proteins and to novel enzymatic functions. This view of the modular assembly of extant genes is supported by the common structural features of certain large gene superfamilies, such as the immunoglobulin-like superfamily (*5*), and by the examples of exon reuse observed in the mosaic structure of the LDL (low density lipoprotein) receptor and the EGF (epithelial growth factor) precursor (*6*). In other gene superfamilies, the older intron-exon gene structure is still apparent in certain representatives, while other members of the same family have lost introns (possibly through retroposition of a mature message) to produce genes with longer and more complicated exons, but with few or no remaining introns. An example of this pattern is the opsin superfamily, which includes genes with four introns as well as genes for beta-adrenergic receptors, which have no introns at all (*7*).

The ancient character of introns is also supported by data suggesting that introns antedate the divergence of plants and animals a billion years ago (*8*). Intron-exon structures may also predate the endosymbiotic incorporation of chloroplasts and mitochondria, which occurred about 2 billion years ago (*9*). Introns may, in fact, antedate the first branchings of life on Earth: the first protogenes may have already displayed intron-exon structure. The original exons may have been 15 to 20 amino acids long; processes of intron sliding and intron loss leading to more complex exons have produced the present day spectrum (*2*).

In this article, the frequency of exon shuffling events is surveyed in order to address the following question: How many different exons were required to generate the current protein diversity? We have identified homologous exons (those of common evolutionary origin) on the basis of amino acid sequence similarity. To the extent that every exon in an underlying universe of exons has an equal probability of being incorporated into a gene, we can then estimate the size of that underlying universe by determining how frequently homologous exons appear in nonhomologous genes.

We first constructed a database of all known exons. The available databases contain large numbers of homologous gene sequences; we eliminated such duplication in order to obtain a collection of exons derived solely from independent genes, unrelated by direct descent. We then made pairwise comparisons of all these independent exons to identify statistically significant sequence similarities, which, we argue, indicate exon homology.

Finally, using a simple sampling model, we took this number of exon repeats to estimate the size of the exon universe. If we survey n exons that have been drawn with replacement from an underlying set of size N, we expect the number of repeats to be given by the product of $n(n-1)/2$, the number of pairs of objects in the collection, and the probability that any pair will match, $1/N$. The number of single repeats is thus $n(n-1)/2N$. Accordingly the expectation for triple repeats is $n(n-1)(n-2)/6N^2$ and so forth (*10*).

The authors are with the Department of Cellular and Developmental Biology, Harvard University, 16 Divinity Avenue, Cambridge, MA 02138.

The exon database. The database of exons was drawn from the eukaryotic genes of known structure recorded in the GenBank and EMBL computer databases (*11*). Using the DNA sequence and the features table, we wrote computer programs that translated each exon. We then inspected the resulting collection of exons and corrected by hand those cases in which the exon boundary had been incorrectly specified (by typographical error). To obtain a distilled database containing exons drawn only from putatively nonhomologous proteins, we first purged the database of homologous genes from different species, retaining a representative human sequence wherever possible. We then removed closely related or duplicated genes, such as the multiple globin sequences (alpha, beta, embryonic, and myoglobin), again retaining but a single example, and culled repeating structures within single genes (such as multiple repeats of a single exon that make up the collagen gene, and the triply repeated domains in serum albumin and ovomucoid). Recurrent elements in a gene superfamily, such as the multiple repeats of the immunoglobulin fold in the immunoglobulin superfamily, or the multiple occurrences of the serine-protease domains in the family that includes the blood-clotting factors, were also pared down to single representative examples. Our initial exon sequence comparisons still included occasional exon pairs displaying more than 80 percent amino acid sequence similarity—one member of each such pair was discarded. Finally, we arbitrarily excluded all exons shorter than 20 amino acids, both because of the high sequence similarity that would be required to establish statistical significance and because this size class contains many signal sequences, which display unusually biased amino acid compositions. These extensive refinings eventually reduced the original database to less than half its size, leaving us with a purged collection containing 1255 exons. The final distribution of exon lengths peaks around 40 to 50 amino acids (Fig. 1).

Criteria for exon similarity and statistical significance. We compared the amino acid sequences of individual exons by making pairwise comparisons, scoring only exact amino-acid matches, and allowing no gaps. A given exon of length N (number of amino acid residues) was compared to all exons of length N to $N+10$, thus allowing for small variations in exon length resulting from insertions, deletions, or splice-site shifts. To optimize alignments, we allowed exons to slide up to five amino acids out of end register in either direction during each pairwise comparison and recorded the best percentage match (the number of matching amino acids times 100, divided by the length of the shorter exon). For computational

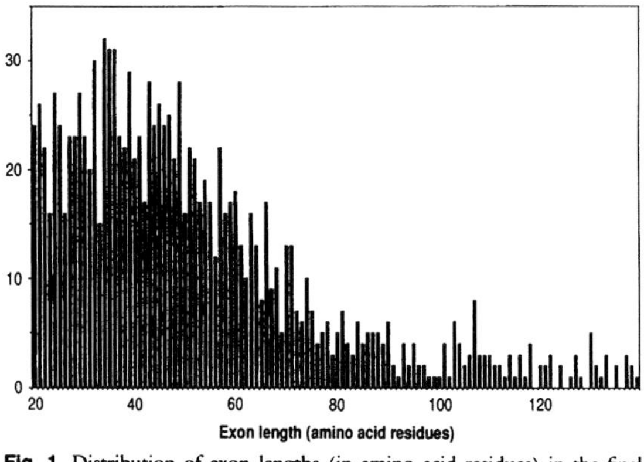

Fig. 1. Distribution of exon lengths (in amino acid residues) in the final reduced database. Exons were identified in GenBank (version 56) and EMBL (version 15). That collection was purged of repeats, homologous genes, and superfamily relationships by repeated rounds of analysis. Exons shorter than 20 amino acids were excluded from the analysis.

Fig. 2. Significance cutoffs as a function of exon length. The cutoff is defined as the highest value of the similarity statistic that occurs only once in 20 simulation runs. Asterisks indicate sequence similarity for each identified case of exon shuffling.

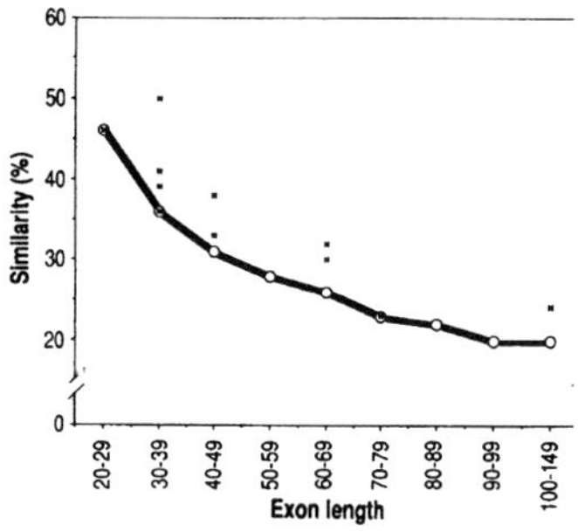

convenience, the exon database was arbitrarily divided into nine length classes (20 to 29, 30 to 39, and so on). Inset a of Fig. 4 shows the output of a representative similarity run for exons of length 40 to 49. The number of events recorded in the histogram corresponds to the number of exon pairs compared in the search. The histogram of similarity values (for exons of length 40 to 49 a.a.) is normally distributed about a mean similarity of 12 percent (with a variance of 8 percent). Roughly speaking, if all amino acids appeared at the same frequency, one expects an average match of 5 percent, improved about three standard deviations by the sliding algorithm. Our study involved a total of 215,166 pairwise exon comparisons, and about 3 million actual comparisons.

We developed criteria for the statistical significance of exon sequence matches by carrying out repeated Monte Carlo simulations, each time randomizing the sequence of every exon and comparing that randomized exon with the original data set. Each random sequence was constructed by sampling (with replacement) from an amino-acid pool derived from all the exons in the real data set. Thus, the amino acid compositions of the randomized and real data sets are identical, but a biased composition of an actual real exon is not likely to reappear in the randomized data set. The comparison program run on these randomized exons establishes the level of exon similarities expected by chance alone. By carrying out 20 different randomized runs, we determine for each range of exon sizes the highest similarity value that occurred only once in 20 runs, and we take this as a cutoff value (Fig. 2). Any match between two real exons that is greater or equal to the cutoff value will be statistically significant, since it is likely to occur by chance no more than once in 20 trials. This similarity cutoff is quite stringent; in order to be considered homologous, exons must display sequence similarities ranging from 46 percent identity for exons of length 20 to 29 down to about 20 percent identity for exons of length 100.

Fourteen exon pairs exhibited amino acid similarity greater or equal to the required cutoff values (Table 1). The similarity values of these matches, relative to the simulation cutoffs, appear in Fig. 2, and the matching pairs themselves are listed in Table 2. Some of these matches have been recognized before (shown by an asterisk [*]), while others are new. The known examples include the collagen-like domain of mannose-binding protein (*12*); the EGF-like domains documented in both Factor IX (*13*) and Factor XII (*14*); the collagen motif characteristic of the β-chain of complement C1q (*15*), and the thyroglobulin-like alternatively spliced exon (*6*) of the Ia antigen–associated chain (*16*). The examples of exon shuffling (Table 1) illustrate a number of themes. A motif, encoded by a single exon, may be performing a similar function in two otherwise unrelated proteins. For example, the first exons of collagenase and major urinary protein serve (at least in part) as signal peptides in both proteins (*17*, *18*). Exon 4 of the chloroplast *psbA* gene and exon 17 of band 3 protein function as membrane-spanning domains

(19, 20). In contrast, exon 2 of β-lymphotoxin (21) and exon 3 of the asialoglycoprotein receptor (12, 22) represent a single hydrophobic domain playing different roles—as a signal sequence in the first protein and as a transmembrane segment in the latter. Table 1 contains several exons derived from collagens, intermediate filaments, or other structural proteins. This pattern may reflect the limited number of basic motifs that can serve as connective or matrix proteins, as well as the evolutionarily conservative character of such protein sequences.

To verify that these matches represented genuine cases of exon shuffling, we compared the proteins from which the exons were derived in their entirety, seeking to maximize the alignment across the whole protein by allowing gaps (23). In all the cases that we describe, the amino acid similarity across the entire protein, or across any region (excluding the exon pair we identify) is significantly lower than that of our matched exon pair. We present two examples of exon matches in the context of their proteins (Fig. 3). In Fig. 3A the intron positions are in similar phase, clearly the surrounding sequences and exons of these genes are not related to the degree exhibited by the relevant exon sequences. In Fig. 3B the intron junctions have drifted in both position and phase.

Table 1. Identified cases of exon homology. The identity, length, and sequence similarity of exon pairs are shown, arranged by decreasing similarity. Asterisks indicate previously identified exon homologies.

Protein	Exon	Exon sizes (a.a. residues)	Similarity (%)
Human α-1 (II) collagen (32)	[X]	36	50*
Rat mannose-binding protein A (12)	[2]	38	
Human apolipoprotein B-100 (33)	[1]	24	46
Human EGF receptor (34)	[1]	29	
Human blood coagulation factor XII (14)	[7]	34	41*
Human factor IX gene (13)	[4]	37	
Human pro-α-1 type I collagen (35)	[47]	34	38
Human elastin (36)	[8]	41	
Mouse major urinary protein (18)	[1]	32	38
Rabbit collagenase (17)	[1]	34	
Chicken steroid inducible hsp (37)	[7]	40	38
Human neurofilament subunit NF-L (38)	[4]	47	
Human lymphotoxin (TNF-β) (21)	[2]	33	36
Rat asialoglycoprotein receptor (22)	[3]	38	
Schizophyllum 1G2 gene (fruiting) (39)	[1]	40	33
Human fibronectin (40)	[1]	49	
Chicken fps proto-oncogene (41)	[8]	40	33
Human neurofilament subunit NF-L (38)	[4]	47	
Mouse α-2 type IV collagen (42)	[5]	60	32*
Human complement C1q B-chain (15)	[1]	64	
Murine Ii gene, Ia antigen–associated (16)	[6b]	63	30*
Bovine thyroglobulin (43)	[18]	64	
Silkmoth chorion (44)	[2]	108	24
Mouse keratin, intermediate filament (45)	[7]	112	
C. reinhardtii chloroplast *psbA* gene (19)	[4]	77	23
Mouse band 3 (20)	[17]	84	
Human serum albumin (46)	[4]	70	23
Human K6b epidermal keratin (47)	[7]	73	

Table 2. Comparison of pairwise similarity values for the real and scrambled Monte Carlo simulations in the uppermost 5 percent of the distributions. The table displays exon length intervals, numbers of actual comparisons, the similarity value that specified the top 5 percent cutoff, the number of matches for both the Monte Carlo and the actual runs above the 5 percent cutoff, the excess of the real matches relative to the simulations, and Fisher's exact *P* value for that excess (one degree of freedom). The Monte Carlo value is the mean of 20 simulations.

Exon lengths	Comparisons (no.)	Cutoff (percent)	Monte Carlo	Real	Excess matches	*P*
20–20	53251	21	3801	4072	271	0.000
30–39	61144	19	3359	3625	266	0.001
40–49	54190	17	3557	3775	218	0.004
50–59	27191	16	1614	1633	19	0.372
60–69	11678	15	839	914	75	0.033
70–79	3693	14	336	372	36	0.084
80–89	1778	14	91	103	12	0.23
90–99	558	13	56	54	−2	0.46
100–149	1683	13	103	102	−1	0.50

These 14 exon matches predict an underlying exon universe of 56,000 sequences. Because we rely on amino acid sequence identity in our analysis, allow no gaps in the alignment process, and demand such a high degree of similarity for significance, we are likely to underestimate the number of homologous exon pairs, and hence overestimate the universe. Certain standard examples of exon shuffling, such as the LDL receptor, were missing from the database and are so are not in our table. We have also omitted certain well-known cases of exon shuffling, such as the serine protease domain, a conspicuous feature of a large family of proteins (24–26), and the various immunoglobulin motifs shared by the members of the immunoglobulin (Ig) superfamily (5). Both the serine protease and immunoglobulin domains span more than a single exon and thus do not meet the specific criteria of this study.

Cases where intron loss leads to the incorporation of a shuffled exon into a larger protein domain are also likely to be missed given the limited size of our search window (±10 amino acids). Finally we demand 30 to 40 percent sequence identity for most comparisons. Proteins (or protein domains) that have drifted very far in amino acid sequence may nonetheless retain their three-dimensional similarity: one can identify structural homologies in circumstances where only 10 percent of the amino acid sequence is conserved (27). Thus, many exons with common evolutionary origins will not be recognizable by amino acid sequence similarity alone. We believe that this calculation of 56,000 members could easily be a five- to tenfold overestimate of the size of the exon universe.

The wedge calculation. In examining the distribution of similarity values, we noticed an excess of real matches (relative to our simulations) at the high end of the similarity distribution. Is this excess statistically significant? In the earlier Monte Carlo calculations, the randomized sequences were chosen to match the overall amino acid composition of the exon database. Two real exons that share a highly skewed amino acid composition then would likely match above the significance criterion. In that first calculation, we considered such a match to be evidence of evolutionary homology. To test more stringently for any excess of real matches, we carried out new Monte Carlo simulations, this time scrambling the amino acid sequence of each exon (creating anagrams of the real exons) and hence preserving the compositional bias of each particular exon. We averaged 20 simulations to produce a baseline distribution against which to compare the real exon similarities. The data for exon comparisons in the 40 to 49 window are shown in Fig. 4. The full curve in the inset shows the distribution of the matches in the real data, and the enlargements show, for the right-hand tail of the similarity value distribution, the differences between the matches found with real data and those from scrambled exons. To estimate the significance of this excess, we took the top 5 percent of the distribution of the similarity statistic, compared the real and simulation distributions, and determined the significance of the excess (Fisher's exact test). In this top 5 percent includes all matches above 17 percent sequence similarity; the excess of real over scrambled is shown as the black tips of the bars. Table 2 shows the comparisons of the top 5 percent of the distribution for each exon size class, which we refer to as "the wedge." Significant excesses of real matches do exist for most of the exon size classes (*28*). The total excess sums to 830 matches over all significant intervals. This number of matches, arising in the sample of 1255 exons, predicts an underlying exon universe of just 950 exons.

This low number for a universe of fundamental shapes suggests that our database of 1255 exons includes examples of most of the original exon universe. The calculation demonstrates that the number of matches is significantly above the expectation based on stochastic sequence similarity, even after discounting shared compositional biases. (For example, two exons that consist of 50 percent leucines will tend to match for that reason alone; this calculation excludes such matches.) While this approach provides substantial statistical power, it does not identify specific pairs of homologous exons. One cannot deduce which matches in the wedge are biologically meaningful and which constitute the random background.

A further test also shows that the excess of matches in the wedge is likely due to exon shuffling. One might have argued that the excess of amino acid identities in the real sequences reflects some convergent or recurrent theme of protein structure, apparent in real sequences but absent in scrambled sequences. For example, some hidden sequence regularity in α helices or some correlation in dipeptide or tripeptide frequencies could cause protein sequences to match against each other at a frequency above random expectation. To test for a strong effect of such features, we carried out further Monte Carlo simulations, this time creating pseudo-exons by transposing a block of sequence from the front (NH_2-terminus) to the rear (COOH-terminus) of an exon and then comparing this pseudo-exon against the other members of the database. This rearrangement of sequence blocks within an exon would preserve sequence similarity due to local features but destroy any similarity that depended on the actual boundaries of the exon. Simulations transposing blocks of 15 to 25 amino acids for the 40 to 49 window gave results that agree with those of the scrambled exon simulation; there is the same significant excess of matches of real exons over both the "scrambled exon" simulations and the "block-transposed exon" simulations. The similarity between real exons thus does not derive from small stretches of local identity but depends instead on the position of the outer boundaries. We expect this outcome if the excess of matches displaying high scores comes about because of true exon homology, where the protein sequence within the exons has been preserved with respect to the positions of the introns.

Reliability of the estimates. We present two different methods to estimate the size of the exon universe. The first identifies 14 examples of exon homology and estimates an underlying exon universe of about 56,000 members. The second identifies a significant total excess of high pairwise similarities (in the top 5 percent of the distribution) corresponding to 830 cases of exon shuffling, thus reflecting an underlying universe of 950 exons. We believe that the first calculation overestimates the size of the exon universe, while the second calculation, although reflecting some significant aspect of exon structure, may be an underestimate. Our best expectation lies somewhere in between. The geometric mean of these numbers is about 7000. Our final expectation, on balance, is between 1000 and 7000 for the size of the exon universe.

Our conclusion may be exaggerated. We may not have succeeded in eliminating all of the biases inherent in the computer databases. The sequence similarity we observe might be the result of convergent evolution, although there is no a priori reason to suspect that any convergence would respect exon boundaries. Furthermore, this search could examine only eukaryotic sequences. If the prokaryotic proteins turn out not to be related to the exon peptide patterns apparent in the eukaryotic sequences, the number of total patterns in the universe would necessarily increase. Finally, the traditional

A

```
                         ↓2
MDDQRDLISNHEQLPILGNRPREPERCSRGALYTGVSVLVAL..LLAGQA
(Ii gene)    |  |   || |      |                  |
......FDFYQRRLVTLAESPRAPSPVWSSAYLPQCDAFGGWEPVQCHAA
      ↑2  (thyroglobulin)
                                  ↓1
TTAYFLYQQQGRLDKLTITSQNLQLESLRMKLPKSAKPVSQMRMATPLLM
|         |       |    |        |         |
TGHCWCVDGKGEYVPTSLTARSRQ.......IPQCPTSCERLRASG....
 ↑1                               ↑1
           ↓0                   ↓0
RPMSMDNMLLGPVKNVTKYGNMTQDHVMHLLTRSGPLEYPQLKGTFPENL
        ||   |                 |     |   |
........LLSSWKQAGVQAEPSPKDLF.IPTCLETGEFARLQASEAGTW
                                   ↑0
             ↓0                             ↓1
KHLKNSMDGVNWKIFESWMKQWLLFEMSKNSLEEKKPTEAPPKVLTKCQE
     |  ||                           |         |
CVDPASGEGV........................PPGTNSSAQCPSLCEV
                                           ↑1
                    EXON  6b
EVSHIPA.VYPGAFRPKC.DENGNYLPLQCHGRHCYCWCVFPNGTEVPHT
  |  |         | |  | |   | ||      ||||   | ||| |
LQSGVPSRRTSPGYSPACRAEDGGFSPVQCDPAQGSCWCVLGSGEEVPGT
                    EXON  18
         ↓1
KSRGRHNCSEPLDMEDLSSGLGVTRQELGQVTL
   ||
RVAGRQPA.........................
       ↑1
```

B

```
                      ↓1             EXON  3
MTKDYQDFQHLDNENDHHQLQRGPPPAPRLLQRLCSGFRLFLLSLGLSIL
(asialoglycoprotein receptor)
                         |  ||      |  | || |||
.......................MTPPERLFLPRVCGTTLHLLLLGLLLV
                       (lymphotoxin)  EXON  2
            ↓1                              ↓1
LLVVVCVITSQNSQLREDLRVLRQNFSNFTVSTEDQVKALTTQGERVGRK
||                     |                  |  |   |
LLPGA..................QGLPGVGLTPS......AAQTARQHPK
                        ↑0
                  ↓1
MKLVESQLEKHQEDLREDHSR...LLLHV...KQLVSDVRSLSCQMAALR
| |  | | |    |  | |    ||           |  |||
MHLAHSNL.KPAAHLIGDPSKQNSLLWRANTDRAFLQDGFSLSNNSLLVP
                ↑1
  ↓1
GNGSERICCPINWVEYEGSCYWFSSSVKPWTEADKYCQLENAHLVVVTSW
  |   |      |   |  |       |   |             |
TSG...IYFVYSQVVFSGKAYSPKATSSPLYLAHEVQLFSSQYPFHVPLL

  ↓0                                      ↓2
EEQRFVQQHMGPLNTWIGLTDQNGPWKWVDGTDYET...GFKNWRPGQPD
  |  |    |    |              |    |   |
SSQKMV..YPGLQEPWLHSMYHGAAFQLTQGDQLSTHTDGIPHLVLSPST

DWYG.HGLGGGEDCAHFTTDGHWNDDVCRRPYRWVCETELGKAN*
   |   |
VFFGAFAL*....................................
```

Fig. 3. Representative exon shuffling events. The exons shown in boldface are displayed as originally aligned by our search; the surrounding sequences and exons are simply displayed for contrast. Vertical arrows indicate the phase of the intron/exon boundaries. (**A**) Comparison between the alternatively spliced exon (6b) of the murine Ii gene (*16*) and exon 18 of bovine thyroglobulin (*43*). (**B**) Comparison of exon 3 of the rat asialogylcoprotein receptor (*22*) and exon 2 of human lymphotoxin (*21*). The surrounding protein sequences are aligned to maximize overall sequence similarity. Abbreviations for the amino acid residues are: A, Ala; C, Cys; D, Asp; E, Glu; F, Phe; G, Gly; H, His; I, Ile; K, Lys; L, Leu; M, Met; N, Asn; P, Pro; Q, Gln; R, Arg; S, Ser; T, Thr; V, Val; W, Trp; and Y, Tyr.

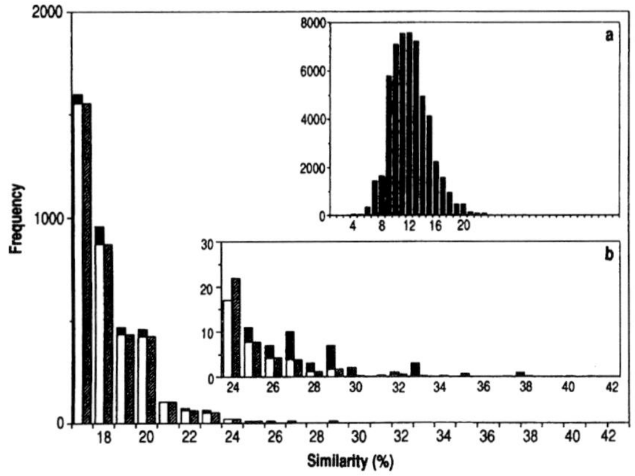

Fig. 4. Representative distribution of the pairwise similarity values for real and randomized exons of length 40 to 49. The figure shows the highest 5 percent of the distribution for the real and simulation comparisons: cross-hatched bars, simulation results; white bars, real exon comparisons. Excess of real matches (wedge) is displayed as black boxes. Inset **a** shows the total distribution of similarity scores for real exon pairwise comparisons. Inset **b** enlarges the rightmost tail of the distribution.

strategies of molecular biology may constrain the kinds of sequences found; systematic whole genome sequences may reveal novel classes of proteins and exons.

The surprisingly small size of our estimate emphasizes the finite character of the underlying exon universe. The number of possible 40-amino acid–long structures, 20^{40} or 10^{52}, is a much larger domain of shapes than the 10^3 to 10^4 that we here predict. Although rules restricting the folding of amino acid chains may have eliminated a large number of amino acid sequences, chance alone may account for which specific elements were in the initial set of exons that gave rise to modern proteins. With a sufficient set of three-dimensional shapes, stabilities, and rudimentary functions, the evolution of proteins could be set in motion.

Several recent studies on protein structure also suggest that the number of possible three-dimensional shapes is quite small. Jones and his co-workers observed that if one connects two points with a loop of 6 or 7 amino acids, only a limited number of Cα patterns will fit (*29*). Unger and co-workers (*30*) have recently shown, by examining the three-dimensional Cα structures of each set of six amino acids in the crystallographic database, that the structures of all hexamers can be clustered into only 80 types rather than 10^8. These observations suggest that the range of shapes in proteins is not as extensive as one might have feared. Recently, Sander and Schneider (*31*) have sought to establish the extent of amino acid sequence similarity that predicts structural (three-dimensional) "homology" between two protein sequences. Their analysis determines a curve of threshold similarity parallel, but slightly more stringent, than our cutoff curve (Fig. 2) and strongly supports the argument that our homologous exon pairs may indeed adopt similar three-dimensional configurations within the different proteins.

Our argument also helps to elucidate the processes of protein evolution. The complexity of modern proteins would have been generated by simply combinatorial arrangements of a limited number of units of structure and function. Particular functional units—DNA-binding motifs, for example, or metal-binding domains—reappear in different contexts to confer new functions on novel exon combinations.

The consequences of a combinatorial search through sequence space are profound. In contrast to a random amino-acid search, the modular building of proteins entails a faster but far more restricted exploration of possible solutions. A 200 amino acid protein may result from a linear search through only 25,000 possible combinations (five modules of 40 amino acids each; 5000 possible exon shapes), rather than the 20^{200} solutions that comprise a full amino-acid-by-amino-acid search.

History constrains all evolutionary phenomena. We have argued that modern protein diversity represents only a very limited exploration of sequence space, an exploration constrained by the success of earlier motifs. While we could argue that the corner of sequence space occupied by modern proteins represents the best of all possible worlds, a selective optimum reached after a careful evolutionary walk through all of sequence space, this seems extremely unlikely. The processes that result in protein diversification—exon reassortment initially, followed by gene duplication and divergence—sharply limit protein sequence diversity. Extant proteins may well lie at local, not global, optima.

REFERENCES AND NOTES

1. W. Gilbert, *Nature* **271**, 501 (1978); W. F. Doolittle, *ibid.* **272**, 581 (1978); C. C. F. Blake, *ibid.* **306**, 535 (1983); C. C. F. Blake, *ibid.* **277**, 598 (1979).
2. W. Gilbert, *Cold Spring Harbor Symp. Quant. Biol.*, **52**, 901 (1987).
3. T. Cavalier-Smith, *Nature* **315**, 283 (1985); J. Rogers, *ibid.*, p. 458.
4. D. A. Hickey, B. F. Benkez, S. M. Abukashawa, *J. Theor. Biol.* **137**, 41 (1989); D. A. Hickey and B. F. Benkez, *ibid.* **121**, 283 (1986).
5. T. Hunkapiller and L. Hood, *Adv. Immunol.* **44**, 1 (1989).
6. T. C. Südhof, J. L. Goldstein, M. S. Brown, D. W. Russell, *Science* **228**, 815 (1985); T. C. Südhof *et al.*, *ibid.*, p. 893.
7. R. A. F. Dixon *et al.*, *Nature* **321**, 75 (1986); T. Kubo *et al.*, *ibid.* **323**, 411 (1986).
8. M. Marchionni and W. Gilbert, *Cell* **46**, 133 (1986); W. Gilbert, M. Marchionni, G. McKnight, *ibid.*, p. 151.
9. K. Obaru, T. Tsuzuki, C. Setoyama, K. Shimada, *J. Mol. Biol.* **200**, 13 (1988); C. Setoyama, T. Joh, T. Tsuzuki, K. Shimada, *ibid.* **202**, 355 (1988); M. C. Shih, P. Heinrich, H. M. Goodman, *Science* **242**, 1164 (1988); F. Quigley, W. F. Martin, R. Cerff, *Proc. Natl. Acad. Sci. U.S.A.* **85**, 2672 (1988).
10. If the exons have a general probability distribution P_i (for the *i*th exon), then the expectation of doubles is $n(n-1)/2$ times ΣP_i^2 since ΣP_i^2 is the total probability that a pair matches. Similarly triples are $[n(n-1)(n-2)/6]\ \Sigma P_i^3$. If the distribution were exponential $P(x=(1/\sigma)\exp(-x/\sigma)$, then the estimate for the universe is 2σ, the number of exons that would account for 86 percent of the occurrences.
11. C. Burks *et al.* "GenBank: Current Status and Future Direction," in *Molecular Evolution: Computer Analysis of Protein and Nucleic Acid Sequences*, R. F. Doolittle, Ed. (Academic Press, New York, in press); in our work, exons were drawn from the GenBank (version 56) and EMBL (version 15) databases.
12. (RATMABPA) K. Drickamer and V. McCreary, *J. Biol. Chem.* **262**, 2582 (1987); M. E. Taylor, P. M. Brickell, R. K. Craig, J. A. Summerfield, *Biochem. J.* **262**, 763 (1989).
13. (HUMFIXG) D. M. Anson *et al.*, *EMBO J.* **3**(5), 1053 (1984); S. Yoshitake, B. G. Schach, D. C. Foster, E. W. Davie, K. Kurachi, *Biochemistry* **24**, 3736 (1985).
14. (HUMCFXII) D. E. Cool and R. T. A. MacGillivray, *J. Biol. Chem.* **262**, 13662 (1987).
15. (HUMC1QB1) K. B. M. Reid, *Biochem J.* **231**, 729 (1985).
16. (MMIIGC) N. Koch, W. Lauer, J. Habicht, B. Dobberstein, *EMBO J.* **6**, 1677 (1987).
17. (RABCN) M. E. Fini, I. M. Plucinska, A. S. Mayer, R. H. Gross, C. E. Brinckerhoff, *Biochemistry* **26**, 6156 (1987).
18. (MUSMUPBS) A. J. Clark, P. M. Clissold, R. Al Shawi, P. Beattie, J. Bishop, *EMBO J.* **3**, 1045 (1984); A. J. Clark, P. Ghazal, R. W. Bingham, D. Barrett, J. O. Bishop, *ibid.* **4**, 3159 (1985).
19. (CRECPSBA) J. M. Erickson, M. Rahire, J.-D. Rochaix, *EMBO J.* **3**, 2753 (1984); J. K. Mohana Rao, P. A. Hargrave, P. Argos, *FEBS Lett.* **156**(1), 165 (1983).
20. (MUSBAND3I) R. R. Kopito, M. A. Andersson, H. F. Lodish, *Proc. Natl. Acad. Sci. U.S.A.* **84**, 7149 (1987).
21. (HUMTNFB) G. E. Nedwin, *Nucleic Acids Res.* **13**, 6361 (1985).
22. (RATRHL) J. O. Leung, E. C. Holland, K. Drickamer, *J. Biol. Chem.* **260**, 12523 (1985).
23. The full-protein alignments were made with the algorithm of Needleman and Wunsch, as implemented in the GAP programs provided by the University of Wisconsin [J. Devereux, P. Haeberli, O. Smithies, *Nucleic Acids. Res.* **12**(1), 387 (1984)]. Where necessary, alignments were anchored on the exon pair we identified.
24. G. H. Swift *et al.*, *J. Biol. Chem.* **259**, 14271 (1984).
25. P. J. O'Hara *et al.*, *Proc. Natl. Acad. Sci. U.S.A.* **84**, 5158 (1987).
26. S. K. Hanks, A. M. Quinn, T. Hunter, *Science* **241**, 42 (1988)
27. C. C. Hyde, S. A. Ahmed, E. A. Padlan, E. W. Miles, D. R. Davies, *J. Biol. Chem.* **263**, 17857 (1988).
28. Another way of doing this calculation is to consider the excess in each percentage match and calculate a chi-squared ((observed-expected)^2/expected) for each of the entries (pooling values first for the small entries) then add all of these chi-square

values and demand significance for the number of degrees of freedom corresponding to the number of entries pooled. Both of these calculations suggest that the excess is significant at well above the 95 percent level.
29. T. A. Jones and S. Thirup, *EMBO J.* **5**, 819 (1986).
30. R. Unger, D. Harel, S. Wherland, J. L. Sussman, *Proteins: Struct. Funct. Genet.* **5**, 355 (1989).
31. C. Sander and R. Schneider, *Proteins*, in press.
32. (HUMCG1A1) K. S. E. Cheah, N. G. Stoker, J. R. Griffin, F. G. Grosveld, E. Solomon, *Proc. Natl. Acad. Sci. U.S.A.* **82**, 2555 (1985).
33. (HUMAPOB1) T. J. Knott *et al.*, *Science* **230**, 37 (1985).
34. (HUMEGFRG) S. Ishii *et al.*, *Proc. Natl. Acad. Sci. U.S.A.* **82**, 4920 (1985).
35. (HUMC1PA) M.-L. Chu *et al.*, *Nature* **310**, 337 (1984).
36. (HUMEL) Z. Indik *et al.*, *Proc. Natl. Acad. Sci. U.S.A.* **84**, 5680 (1987).
37. (GGHSP108) M. Forsgren, B. Raden, M. Israelsson, K. Larsson, L.-O. Heden, *FEBS Lett.* **213**, 254 (1987).
38. (HUMNFLG) J.-P, Julien *et al.*, *Biochim. Biophys. Acta* **909**, 10 (1987).
39. (SCO1G2) J. J. M. Dons *et al.*, *EMBO J.* **3**, 2102 (1984).
40. (HUMFN) A. R. Kornblihtt, K. Vibe-Pedersen, F. E. Baralle, *Nucleic Acids Res.* **12**, 5853 (1984).
41. (GGCFPSE) C.-C. Huang, C. Hammond, J. M. Bishop, *J. Mol. Biol.* **181**, 175 (1985).
42. (MUSCOLA2) M. Kurkinen, M. P. Bernard, D. P. Barlow, L. T. Chow, *Nature* **317**, 177 (1985).
43. (BTTHYR) J. Parma, D. Christophe, V. Pohl, G. Vassart, *J. Mol. Biol.* **196**, 769 (1987).
44. (BMOCH11A) K. Iatrou, S. G. Tsitilou, F. C. Kafatos, *Proc. Natl. Acad. Sci. U.S.A.* **81**, 4452 (1984).
45. (MUSKETEPI) P. M. Steinert, R. H. Rice, D. R. Roop, B. L. Trus, A. C. Steven, *Nature* **302**, 794 (1983); T. M. Krieg *et al.*, *J. Biol Chem.* **260**, 5867 (1985).
46. (HUMALBGC) P. P. Minghetti *et al.*, *J. Biol. Chem.* **261**, 6747 (1986).
47. (HUMKEREP) D. Marchuk, S. McCrohon, E. Fuchs, *Proc. Natl. Acad. Sci. U.S.A.* **82**, 1609 (1985).
48. We thank J. Willett, H. Spencer, and R. C. Lewontin for helpful statistical advice, J. Knowles for careful reading of the manuscript, and members of the Gilbert Lab for useful assistance and criticism. This work is supported by NIH grant GM37997-03.

24 August 1990; accepted 22 October 1990

"Calm down, Helen. We've been the focus of watch-dog groups before."

Gene, 135 (1993) 137–144
© 1993 Elsevier Science Publishers B.V. All rights reserved. 0378-1119/93/$06.00

GENE 07475

On the ancient nature of introns*

(Exons; modules; Go plot; triosephosphate isomerase; pyruvate kinase; retrotransposition)

Walter Gilbert[a] and Manuel Glynias[b]

[a]The Biological Laboratories, Harvard University, 16 Divinity Avenue, Cambridge, MA 02138, USA; and [b]Research Institute, The Cleveland Clinic Foundation, Department of Cardiovascular Biology, 9500 Euclid Avenue, Cleveland, OH 44195, USA. Tel. (1-216) 445-9745, Fax (1-216) 444-9263

Received by G. Bernardi: 2 July 1993; Accepted: 5 July 1993; Received at publishers: 2 August 1993

SUMMARY

We discuss some of the arguments for introns arising early or late in evolution. We outline the exon theory of genes and discuss the series of discoveries of introns in the gene (*TPI*) encoding triosephosphate isomerase (TPI) that have filled out a series of better fits to the Go plot, culminating in the 1986 prediction of an intron position that was finally discovered in 1992. We present a statistical argument that the 11-intron structure of *TPI* (based on attributing all of the introns to an ancestral gene and interpreting three cases of very close intron positions as examples of sliding) has a clear relationship to the protein structure. The exons of this 11-intron *TPI* are a better approximation to Mitiko Go's modules (Go, 1981) than are 99.9% of all alternative exon patterns corresponding to 11 introns placed randomly in the gene, and better than 96% of all alternative patterns in which the lengths of the exons are preserved while the introns are moved. We combine four tests relating exons to protein structure: (*i*) whether the exons are compact modules, (*ii*) whether the exons contain most of the close contacts in the protein, (*iii*) whether the exon configuration maximized buried surface area along the backbone, and (*iv*) whether the exons maximize their content of hydrogen bonds. On a joint measure for these tests, the native exon structure with 11 introns fits these tests better than 99.4% of all alternative structures obtained by permuting the exon lengths and intron positions. This statistical test supports the idea that intron positions are related to protein structure, and hence that this protein was most likely put together out of exons as units of structure, arising as its gene was assembled from exons.

INTRODUCTION

Since the discovery of the intron/exon structure of genes in 1977 (Berget et al., 1977; Brack and Tonegawa, 1977; Breathnach et al., 1977; Chow et al., 1977), 16 years ago, there has been a debate about the role and origin of this genetic structure. One view conceives of the original genes as continuous regions, like the genes in bacteria, and thinks of the introns as added elements, possibly transposons, that enter the DNA to break up the underlying gene structure (Rogers, 1985; 1989; Cavalier-Smith, 1991). The other view argues that the introns are extremely ancient, that the first genes were not continuous structures but were represented on the DNA as exons separated from each other by introns, and that the mechanism of evolution was recombination within those introns that provided new combinations of the exons and, hence, new gene and protein structures (Blake, 1978; Doolittle, 1978; Gilbert, 1978; 1979). The extreme positions are that the introns were new and added in the course of evolution or that the introns were very old and preceded evolution. Of course it is difficult to find any

Correspondence to: Dr. W. Gilbert, The Biological Laboratories, 16 Divinity Avenue, Cambridge, MA 02138, USA. Tel. (1-617) 495-0760; Fax (1-617) 496-4313; e-mail. gilbert@chromo.harvard.edu

*Presented at the COGENE Symposium, 'From the Double Helix to the Human Genome: 40 Years of Molecular Genetics', UNESCO, Paris, 21–23 April 1993.

Abbreviations: aa, amino acid(s); nt, nucleotide(s); *PyK*, pyruvate kinase-encoding gene; TPI, triosephosphate isomerase; *TPI*, gene encoding TPI.

serious arguments that could lead to a conclusion one way or the other. For the most part there has been more heat than light in this argument, but here we will explore some of the consequences of looking at genes as the product of an exon/intron structure.

EVOLUTION

That exons can be shuffled into new combinations appears clearly in those genes that arose late in evolution. Those genes that arose uniquely in the vertebrates often are put together from pre-existing exons by some process of exon shuffling. The interesting question arises if we look over all of evolution. The idea that the introns arose very early is the view that the original cell had a genomic structure that had an intron/exon makeup and that the line of evolution leading to the nucleus of the eukaryotic cell maintained that original intron/exon structure for a large number of genes, while other lines of evolution lost the introns. Fig. 1 is a cursory view of this sweep of evolution. On this hypothesis, the original splicing mechanism was gradually frozen into that used by the type-III introns in the eukaryotic nucleus. This extreme picture is that the lines that lead to the prokaryotes, the archaebacter, and a number of the protists, lost their introns by specializing down lines of descent that involved evolutionary pressure on their DNA content, and those lines ended up today with no introns. Along the way, the mitochondria and chloroplasts picked up by the eukaryotic

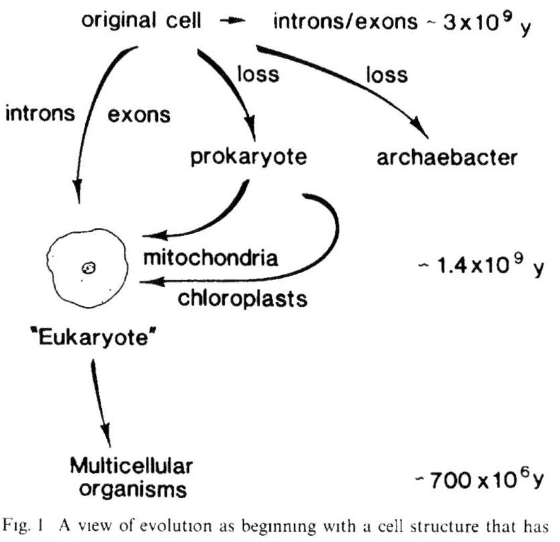

Fig. 1 A view of evolution as beginning with a cell structure that has an intron/exon genome. Down one line that leads to the eukaryotic nucleus, the intron structure is preserved, and the splicing mechanism is ultimately modified to the type-III nuclear splicing. Down the lines that lead to the prokaryotes and Archaebacter, there is a continuing process of loss of introns, some early, some possibly late y = year(s)

cell, represent intermediate stages in that process at which all the introns had not yet been lost as of the time of that absorption. This way of looking at evolution suggests that the current structure of the prokaryotes is not an image of the primitive world but is a highly evolved structure that has undergone a great deal of evolutionary modeling. As we look at the organisms around us, we never see a primitive organism. We see modern organisms, which may or may not resemble a particular primitive ancestor.

(a) Exon theory of genes

One hypothesis, The Exon Theory of Genes (Gilbert, 1987), is that the original genes, the ur-genes, corresponded to short polypeptides, 15–20 aa long. Those polypeptides were probably folding elements or functional domains, and the original enzymes were probably clusters of them. Later genes were then assembled out of the 'ur-genes' by the fusion of introns, and on this picture the introns are simply the residual glue. A gene today has introns because at some time in its history it was assembled by recombination within those introns. On this picture of evolution the major processes are shuffling and reassortment of the exons and the creation of more complicated exons from simple ones by retroposition (summarized in Fig. 2). By retroposition we mean a process by which an intron-containing gene is transcribed into RNA, the introns spliced out, and then that RNA is copied back into the DNA by a reverse transcriptase, and that DNA copy reinserted back into the genome. We know this process constructs pseudogenes; however, if the reinsertion were to be into an intron, into a transcription unit, the result will be a new, complex, C-terminal exon. Thus complex exons are formed no longer looking

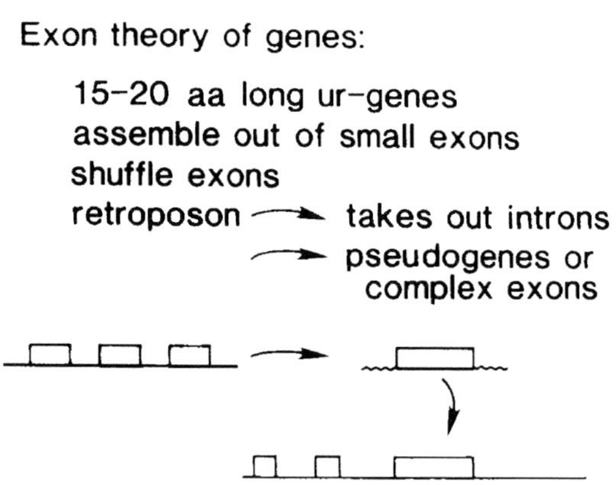

Fig. 2. The exon theory of genes This figure outlines the critical elements of the exon theory that the original primitive genes, 'ur-genes', encoding 15–20 aa long polypeptides became used as exons in more complicated genes encoding full-length proteins.

like the 15–20 aa long simple structures, and those complex exons again get shuffled over evolutionary time. The immunoglobulin fold is such a complex exon, 112 aa long, that was constructed out of some six pieces and then used as a unit.

This general hypothesis is that gene evolution is dominated by recombination within the introns, by sliding and drift of the introns, and by the loss of introns. Recombination within introns can lead to wholly new combinations of exons. Movements of intron positions can either put in a loop of new protein and create novel sequence or the intron sliding may simply be a passive series of mutations that must leave an intron present but not necessarily in the same place. The loss of introns occurs either by exact deletion, a very rare process, by inexact deletion, which takes out the intron and some surrounding coding sequence, a process that occurs with moderate frequency, or retroposition effects, which generate an intron-free DNA copy which can remove introns by gene conversion or by full insertion. This last is a rather rapid process on an evolutionary time scale.

(b) Introns of the triosephosphate isomerase-encoding gene (*TPI*)

How might one get some insight into whether this exon theory is true? Several years ago we attacked this problem by sequencing the gene for an extremely old protein, TPI, a fundamental component of glycolytic metabolism. We chose an ancient protein because on any model that suggests that evolution added introns to previously existing proteins and then used those introns to shuffle exons, to make vertebrate genes for instance, there is no reason to find an intron/exon structure for any protein that has not undergone evolutionary remodeling and construction in that recent past. TPI came into existence in the progenote, since its sequence is 40% identical across all organisms and its three-dimensional structure is conserved. This enzyme does not have a separate evolutionary history in the bacteria, separate from the plants and the vertebrates.

TPI has a lovely subunit structure: an eightfold barrel of eight β-sheet strands and eight α-helical returns. Fig. 3 shows this structure. Straus and Gilbert (1985) examined the intron/exon structure of this gene in the chicken and Marchionni and Gilbert (1986), in maize, to discover first that there were a large number of introns and second that of the six introns which were in the vertebrates five were at identical places in the plants (Fig. 3). One more intron was at a similar place, three aa over in maize, and the maize gene had two new introns at novel places. This coincidence of intron positions is an excellent argument that the *TPI* gene structure was broken up *before* the separation of plants and animals, that it had an intron/exon structure in some single-cell forbearer.

If we think of the overall pattern of evolution, we then know that the *TPI* gene structure is broken up with six introns in the vertebrates, with eight in maize, and, back a 10^9 years ago, with eight before the division of plants and animals. However, the gene is a single unit in bacteria and also in yeast. The question is then was the gene in pieces at the beginning of evolution, and introns lost down the unicellular lines, or were the introns added in the eukaryotic forbears of the plants and vertebrates?

McKnight et al. (1986) worked out the structure of the *TPI* gene in *Aspergillus* and found five introns. Fig. 3 shows the patterns of all these introns. One intron is in an identical position in *Aspergillus*, maize, and chicken. Two are at positions similar to those in plants, but differ in detail beginning one nt over and four nt over, respectively, and two are at totally novel positions. Fig. 3 shows the relation of the intron positions to the secondary structure. Just from this information alone, one cannot decide the question of whether these introns were added or primeval. Those who believe the introns are added look at the three introns that are one, four, and nine nt over and say these are *different* introns, and since there would not have been introns one nt apart in an original gene, the introns must be added in the separate lines. Those who believe the introns are primeval argue that each pair is the *same* intron, moved in the course of evolution. This is a typical 'lumpers' versus 'splitters' debate. People like Palmer (Palmer and Logsdon, 1991) and Rogers (1985; 1989) are 'splitters' and say these close introns are different because they do not occur in the identical places. We are 'lumpers', and say that these are examples of the same intron having moved over evolutionary time. The positioning of introns on the DNA alone does not give guidance to decide between the two views. In 1986, however, we made an argument (Gilbert et al., 1986) that the positions of these two novel introns (one of which is in the most conserved region of the protein and close to the active site) are not random but can be best understood in terms of a particular way of interpreting the protein structure due to Mitiko Go (Go, 1981).

(c) Modules and the Go plot

Mitiko Go (Go, 1981) argued that exons may correspond to *modules*, polypeptide sequences that are compact in space, circumscribable by a 28 Å sphere. She came to this argument by constructing a distance plot, a Go plot, in which each pair of aa residues in the protein is represented in a triangular array and the distances between the α-carbons of each pair of residues is plotted. (In the example of Fig. 4, the plot is colored black for distances $\geqslant 28$ Å and gray for distances $\leqslant 12$ Å.) Go

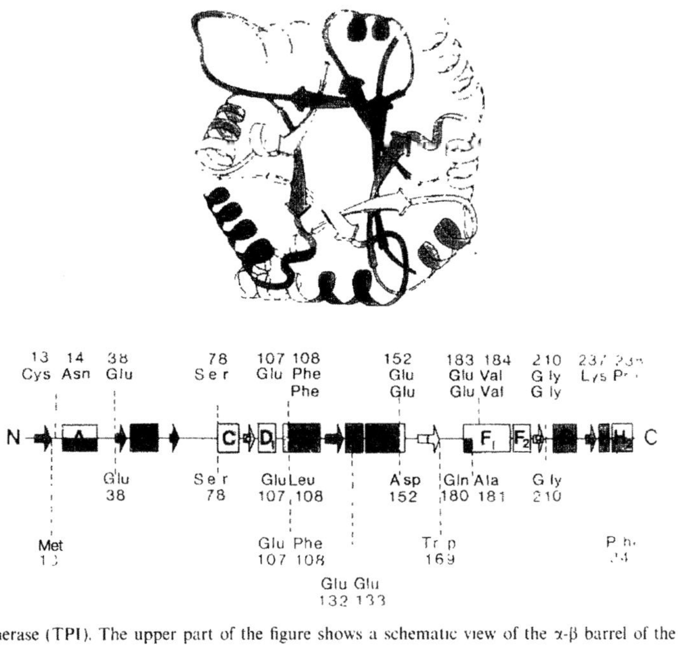

Fig 3. Triosephosphate isomerase (TPI). The upper part of the figure shows a schematic view of the α-β barrel of the subunit of TPI The lower part of the figure shows the secondary structure of TPI with the position of the introns indicated. Above, the secondary structure line are the eight introns of maize (at aa residues 13–14, 38, 78, 107–108, 152, 183–184, 210, and 237–238). The six introns marked and named immediately below the secondary structure line are those of chicken (at aa residues 38, 78, 107–108, 152, 180–181, and 210) The five introns whose names are at the very bottom of the figure are those of *Aspergillus* (at aa residues 13, 107–108, 132–133, 163, and 240).

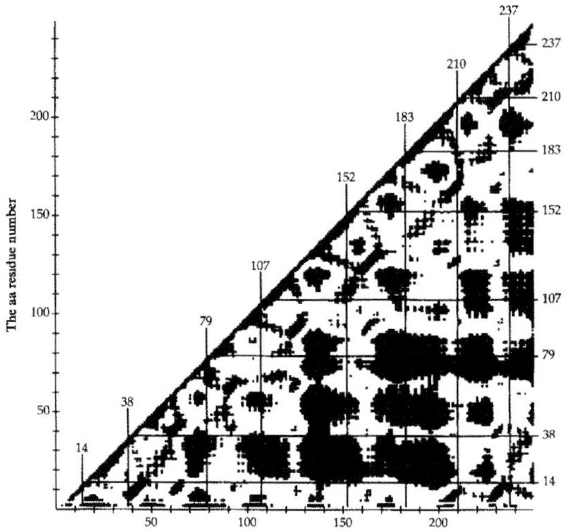

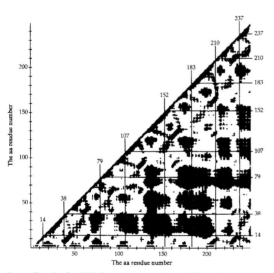

Fig. 4. Go plot for TPI showing the positions of the eight introns in maize. This is a plot of all distances in the three-dimensional structure between the α-carbons of each pair of aa residues. Distances less than or equal to 12 Å are shown in gray, distances greater than or equal to 28 Å are shown in black. The aa residue number runs along the *x* and *y* axes. Most of the triangles along the diagonal, which correspond to the exons, are free of black regions, but there are notable exceptions in the third, the fifth, and the sixth exons

(1981) first developed such a plot for globin and observed that two of the three exons were compact (triangular areas corresponding to the exon avoiding black regions) but that the third central exon was not compact, not circumscribable by a sphere 28 Å in diameter. She predicted the existence of an intron in an ancestral globin, that would result in breaking up this bilobate exon into two separate portions. Soon after, Jensen et al. (1981) discovered that predicted intron in leghemoglobin. (Several other globins also have an intron in this position; Moens et al. (1992).) This was the first prediction of an intron position, and the prediction was based on the argument that the exons should represent compact regions in space.

In 1986 we examined the Go plot for maize TPI. Fig. 4 shows that six exons are compact, but three are not. However, as Fig. 5 shows, the two novel positions in *Aspergillus* (shown in bold) break up two of the non-compact exons. The two novel *Aspergillus* introns are in very special places with respect to the three-dimensional structure of the protein; this led us first to argue that all these introns were old and then to observe that in this Go plot for TPI, ten of the exons were compact, and one was not. We predicted that there should be an 11th intron in *TPI*, at a position that would break this offending exon (Gilbert et al., 1986). Last year, Tittiger et al. (1993)

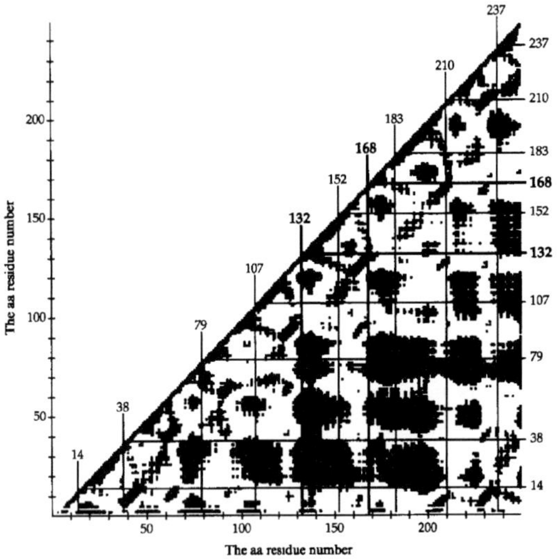

Fig. 5. A Go plot for TPI showing the eight introns from maize as well as the two novel introns, at aa positions 132 and 168, of the *Aspergillus* gene. The plot is described in the legend to Fig. 4. The novel intron at aa position 132, shown in bold, breaks the 'exon' previously defined between aa positions 107 and 152 in such a way as to make two compact modules out of it. The *Aspergillus* intron at aa residue 168 breaks the 'exon' previously defined between aa positions 152 and 183 in such a way as to make two compact modules.

discovered an intron in mosquito *TPI* which breaks that exon in just the way we predicted. Fig. 6 shows the Go plot for TPI with the mosquito intron marked in bold. This is the second prediction that this theory has made; the ability of a theory to make predictions is a very important aspect and the closest that one can come to a 'confirmation' of a theory. The fact that the intron at aa position 64 breaks the exon in such a way as to generate the modular structure is a strong argument that that intron is ancient, and that the primeval gene was broken into introns and exons.

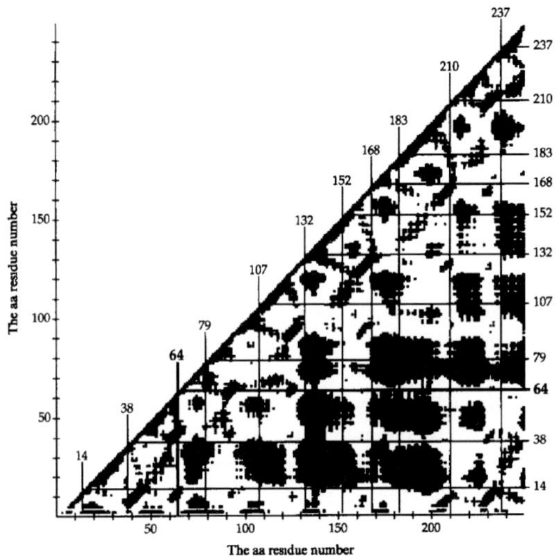

Fig. 6. A Go plot for TPI showing in addition the position of the intron in mosquito, marked in bold at aa position 64. This intron breaks the 'exon' previuosly defined by introns at aa positions 38 and 79 into two modules.

If the mosquito intron is ancient, then we learn further that the loss of introns is extremely easy, because in the mosquito gene only this one intron remains, while in another insect where the *TPI* gene is known, *Drosophila*, the intron at aa 64 is not present, only one intron remains at the position corresponding to aa 183. The total loss of all but one intron has occurred differently in different insect lines. Furthermore, there have been many independent losses of intron 64: *Drosophila*, in the vertebrates, in the plants, and in *Aspergillus*. However, that intron had to be maintained in the original animal line long enough to turn up in the insects. Intron loss then is an extremely probable event, so likely that one cannot use the pattern of intron loss to draw evolutionary trees.

(d) A statistical test of intron position

The force of the argument about modules is that if we assign all these introns to the original structure of the gene, we have divided that gene into enough small regions to avoid all the black regions on the Go plot. One might argue that all that has been done is to break up the gene into small regions, that if one breaks up a gene into small regions, those ultimate regions will be compact in space, thus the exons so defined will avoid all such dark regions on a Go plot, and that, therefore, there really is no content in the modular argument. However, we have devised a mathematical test of the module conjecture (M.G., L. Schoenbach and W.G.: On the ancient nature of introns, to be published (1993)). The alternative picture, in which the introns have been added in the course of evolution by a process like transposition, predicts essentially that the intron positions should occur at random with respect to the three-dimensional structure. If the introns had been transposons entering the DNA at specific DNA sequences, the separation between those DNA sequences, like the separation between restriction sites, would follow a random distribution, equi-probable along the DNA. In order to see how non-random the observed set of introns is, we choose a set of intron positions at random along the DNA, construct a pattern of exons, and ask how those exons overlap the black regions of the Go plot. Then we assign a score to each random set of introns that is equal to the number of pairs of residues that both lie within an exon and are separated by more than 28 Å. (This score is the area of black region contained within the exon triangles in the Go plot.) We then choose another random set and compute a new

142

score. Then we can ask how likely it is that a random set of introns leads to a better Go plot score than does the native set of exons. Table I shows the results of this calculation running through 250 000 random alternatives. The five-intron pattern of *Aspergillus TPI* is not particularly unusual by this criterion. However, the six-intron structure of the vertebrates is already moderately unusual. A composite gene with ten introns (the sum of maize and *Aspergillus* positions), is better than 99.3% of all possible random alternatives, and the last intron from the mosquito makes the set better than 99.9% of all random alternatives. Clearly, the intron positions in *TPI* are not at all what one would expect if they had been put in at random.

Now one might counter that even though our statistical argument is correct, the intron positions in *TPI* do not look random, even to the naked eye. In general, exon lengths in a gene are correlated in size. Fig. 7 shows a plot of the length of each of the exons in *TPI*, arranged in order from the smallest to the largest, the smallest being about ten aa long, the largest being about 28 aa long. This is quite different from a random distribution: they are all roughly similar in length: an average length of 20 aa with a standard deviation of 7. (This is generally true. Exon lengths overall are peaked around 30–40 aa.) If introns were random in position, one would expect an exponential distribution of exon lengths with a lot of very short exons and a few long ones. Fig. 7 also shows a plot of a typical random distribution of 'exons' for *TPI* with

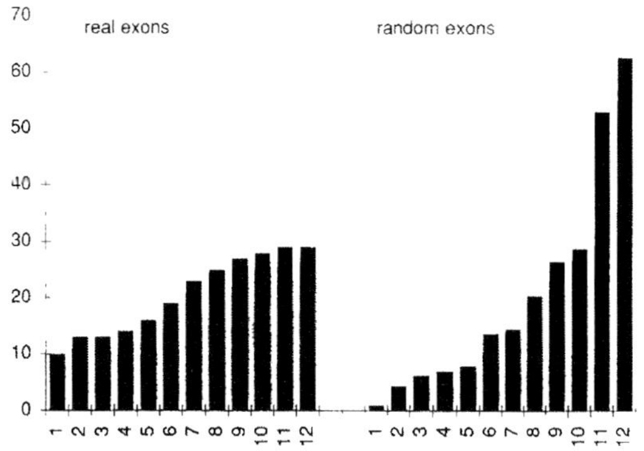

Fig 7. The distribution of TPI exon lengths. The left-hand panel shows the aa lengths of the 12 exons of TPI, in order of increasing length. The right-hand panel shows one particular set of random exons for TPI created during our calculation. The *x* axis lists exons (1–12) and the *y* axis specifies their length (in aa)

an average of 20 and a standard deviation (equal the average) around 19. The few larger exons then produce very bad scores in our test.

How can we get around this objection? We can do this calculation in a more subtle way, by simply randomly permuting all of the exon lengths. That process generates sets of random intron positions, whose exons have the same length distribution as the real ones, and we can ask again how many of the 28 Å distances are now included in each permuted set. Table II shows the results. The native set is still unusual. *TPI* with ten introns is better than 87% of the permutations, and the set with the 11th intron from mosquito is better than 96% of the random alternatives. Thus the pattern of introns is not what one would expect if it had been generated by a random pro-

TABLE I

The goodness-of-fit to the Go plot of TPI of different configurations of introns in the *TPI* gene compared to the same number of introns-arranged randomly in the gene

Native exon maps[a]	Native number of dots[b]	Percent of random alternatives with equal or lower scores[c]
TPI with 5 introns (*Aspergillus*)	870	53.6
TPI with 6 introns (chicken)	287	4.9
TPI with 8 introns (maize)	235	13.6
TPI with 10 introns (maize and *Aspergillus*)	65	0.72
TPI with 11 introns (including Culex)	22	0.12

[a]The sets of introns in different *TPI* genes, either actual (*Aspergillus*, chicken, and maize) or attributed (10-intron patterns or 11-intron patterns attributed to ancestral genes as described in section **d**.

[b]The number of distances ⩾28 Å contained within the set of exons defined by the actual set of introns listed in the first column (footnote a).

[c]The score (number of distances ⩾28 Å) was calculated for each set of random intron positions. The percent of 250 000 random alternatives that have scores better (i.e., equal to or lower) than the comparison score of the second column.

TABLE II

The goodness-of-fit to the Go plot of TPI of the *TPI* introns compared to permuted sets of exon maps[a]

Native exon maps	Native number of dots	Percent of permuted alternatives with equal or lower scores[b]
TPI with 5 introns (*Aspergillus*)	870	84.9
TPI with 6 introns (chicken)	287	70.2
TPI with 8 introns (maize)	242	84.2
TPI with 10 introns (maize and *Aspergillus*)	65	12.6
TPI with 11 introns (including Culex)	22	3.9

[a]See Table I.

[b]The exons were randomly permuted (shuffled) to produce a new test set of intron positions. The percent of 250 000 such random alternatives with scores less than or equal to the scores in the second column is given.

cess that ignored the three-dimensional structure of the protein.

We have generalized this calculation in a number of ways. One of the ways shows that the correct exons contain more close contacts than a permuted set. The Go plot in Fig. 6 shows in 'gray' all pairs of aa residues less than 12 Å apart. How do the exons break up the 'gray' pattern? In fact, the native pattern of exons has more close contacts (includes more of the 'gray') than any alternative permuted pattern. (The 10-intron set is better than 77% of the permutations, the 11-intron set is better than 96%.) We have looked at two other criteria. Do the original exons have more buried surface along the backbone than do alternative permutations? The answer is yes (96% and 98%). Does the original exon pattern have more inter-exon hydrogen bonds along the backbone than alternative permutations? Again, the answer is yes (82% and 94%).

We can put all of these criteria together into a single calculation, by assigning to the native set a single score which is the average of its percentile ranks on each of the four tests and comparing that score with the similar averages for each permuted set. This gives us a joint way of asking how significant it is to satisfy all four of these tests together: the hydrogen bond test, the buried surface test, a closeness test, and a distance test. (Since the tests are not correlated, this last score for the permuted sets is distributed roughly normally with a mean of 50% and a standard deviation of 20%.) For *TPI* the 10-intron set is better than 95% of the alternatives, and the 11-intron set is better than 99.4%. Thus the full set of intron positions, including mosquito, are highly unusual in respect to any picture of the introns being added randomly to the structure of the protein and reflects the three-dimensional structure of the protein. This way of looking at the world suggests (but as yet we have not been able to carry this through) that one may ultimately be able to use these ways of formulating the uniqueness of the positions of introns to predict the existence of novel introns and to analyze the folding structure of proteins.

The difficulty with applying this test to other proteins is obvious from some of the numbers we have given. If we consider a protein for which we have lost some subset of the introns, then the tests applied to the residual exons will not obey our criteria. For example, if we look at the gene (*PyK*) for pyruvate kinase, a gene which normally has nine introns, and use the 28 Å criterion, those nine introns are unusual in terms of random positions since they are better than 97% of the alternatives. However, if we control for exon size by permuting the exons there seems to be nothing unusual about the original set of intron positions, which has a score of 50%. However, in the Go plot, nine of the ten exons of *PyK* are modules in Mitiko Go's sense, and one of them is not. That one has a very pronounced bilobate structure. If we break that exon and predict an intron, then the intron positions in that 10-intron *PyK* are better than 96% of the permuted alternatives (Table III). This is a prediction, for the *PyK* gene, that there should be such an intron at a position near aa residue 222.

TABLE III

The goodness-of-fit to the Go plot of PyK of the introns in the *PyK* gene[a]

	Percent of random alternatives with equal or lower scores	Percent of permuted alternatives with equal or lower scores
PyK with 9 introns (actual)	2.4	0.02
PyK with 10 introns (predicted)	53	4

[a]See Tables I and II.

CONCLUSIONS

We have been able to do a calculation that suggests that the positions of the introns in *TPI* are not random in respect to the three-dimensional structure of the protein. Thus the exons are related to units of the structure of the protein. The likely reason is that this protein was built up from those exons at the origin of evolutionary time.

REFERENCES

Berget, S.M., Moore, C. and Sharp, P.A.: Spliced segments at the 5′ terminus of adenovirus-2 late mRNA. Proc. Natl. Acad. Sci. USA 74 (1977) 3171–3175.

Blake, C.C.F.: Do genes-in-pieces imply proteins-in-pieces? Nature 273 (1978) 267–268.

Brack, C. and Tonegawa, S.: Variable and constant parts of the immunoglobulin light chain of a mouse myeloma cell are 1250 nontranslated bases apart. Proc. Natl. Acad. Sci. USA 74 (1977) 5652–5656.

Breathnach, R., Mandel, J.L. and Chambon, P.: Ovalbumin gene is split in chicken DNA. Nature 170 (1977) 314–319.

Cavalier-Smith, T.: Intron phylogeny: a new hypothesis. Trends Genet. 7 (1991) 145–148.

Chow, L.T., Gelinas, R.E., Broker, T.R. and Roberts, R.J.: An amazing sequence arrangement at the 5′ ends of adenovirus-2 messenger RNA. Cell 12 (1977) 1–8.

Doolittle, W.F.: Genes in pieces: were they ever together? Nature 272 (1978) 581–582.

Gilbert, W.: Why genes in pieces? Nature 271 (1978) 501.

Gilbert, W.: Introns and exons: playgrounds of evolution. ICN- UCLA Symp. Mol. Cell. Biol. 14 (1979) 1–12.

Gilbert, W.: The exon theory of genes. Cold Spring Harbor Symp. Quant. Biol. 52 (1987) 901–905.

144

Gilbert, W., Marchionni, M. and McKnight, G. On the antiquity of introns. Cell 46 (1986) 151–154.

Go, M. Correlation of DNA exonic regions with protein structural units in haemoglobin. Nature 291 (1981) 90–93.

Jensen, E.O., Paludan, K., Hyldig-Nielsen, J.J., Jorgensen, P and Marcker, K.A.: The structure of a chromosomal leghaemoglobin gene from soybean. Nature 291 (1981) 677–679.

Marchionni, M and Gilbert, W.: The triosephosphate isomerase gene from maize: introns antedate the plant-animal divergence. Cell 46 (1986) 133–141.

McKnight, G.L., O'Hara, P.J. and Parker, M L: Nucleotide sequence of the triosephosphate isomerase gene from *Aspergillus nidulans*: implication for a differential loss of introns. Cell 46 (1986) 143–147.

Moens, L., Vanfleteren, J., De Baere, I., Jellie, A M., Tate, W. and Trotman, C N A. Unexpected intron location in non-vertebrate globin genes. FEBS Lett 312 (1992) 105–109.

Palmer, J.D. and Logsdon Jr., J M. The recent origins of introns. Curr Opin Genet. Dev 1 (1991) 470–477.

Rogers, J. Exon shuffling and intron insertion in serine protease genes. Nature 315 (1985) 458–459.

Rogers, J H. How were introns inserted into nuclear genes? Trends Genet. 5 (1989) 213–216.

Straus, D. and Gilbert, W. Genetic engineering in the precambrian: structure of the chicken triosephosphate isomerase gene. Mol. Cell. Biol 5 (1985) 3497–3506.

Tittiger, C., Whyard, S and Walker, V.K: A novel intron site in the triosephosphate isomerase gene from the mosquito *Culex tarsalis*. Nature 361 (1993) 470–472.

Tracing Biological Evolution in Protein and Gene Structures.
M. Gō and P. Schimmel, editors.

Tests of the exon theory of genes

Walter Gilbert[1], Manyuan Long[1], Carl Rosenberg[2] and Manuel Glynias[3]
[1]The Biological Laboratories, Harvard University, Cambridge, Massachusetts, USA; [2]Whitehead Institute/MIT Center for Genome Research, Cambridge, USA; and [3]Research Institute, The Cleveland Clinic Foundation, Department of Cardiovascular Biology, Cleveland, Ohio, USA

Abstract. The exon theory of genes proposes that the first genes were assembled by recombination within introns that connected exons encoding short polypeptides that served as elements of structure or function. The primary paths of evolution were then exon shuffling and intron loss.

We discuss the relationship between exons and "modules," compact regions of polypeptide chain. Analysis of the triosephosphate isomerase genes argues that the set of "ancestral" intron positions is significantly related to the three-dimensional structure of the protein and suggests that the original gene was assembled from exons.

An analysis of a purged database of intron-containing genes shows that the distribution of intron phase within codons is nonrandom, 48% of the introns being in phase zero, and that there is a highly significant excess of symmetric exons (beginning and ending in the same phase), symmetric pairs, triples, etc., over expectation. This is evidence for exon shuffling in eukaryotic cells. By examining that part of the exon database that is homologous to prokaryotic proteins, we find again an excess of symmetric exons and hence evidence for exon shuffling in the original construction of genes.

Key words: Gō plot, modules, retroposition, triosephosphate isomerase.

The existence of the intron/exon structure of genes poses a problem: what is the origin of the introns? Can one deduce whether the introns were present from the beginning of evolution or were added during its course? One view is that the introns are old and were used to piece together the first complex genes. This view is summarized as "the exon theory of genes" [1]. The alternative view is that the introns are recent and were added to the structure of pre-existing genes by some process such as the transposition of a DNA or an RNA element to break up a continuous coding region.

The exon theory of genes expresses the idea that evolution began with short polypeptide chains, of the order of a single open reading frame, 15–20 amino acids long. That size estimate is based on arguments attributing introns to ancestral genes. That size also corresponds to the length of polypeptide chain that one would expect to have some structure in solution. This theory hypothesizes that these small exons are assembled into genes, that the mechanism of making novel genes is to shuffle exons, and that, over evolutionary time, there are processes that make complex exons

Address for correspondence: Dr W. Gilbert, The Biological Laboratories, Harvard University, 16 Divinity Ave., Cambridge, MA 02138, USA.

238

from simple ones. The dominant process for this last theory is retroposition, which takes out introns by the same process that makes pseudogenes. A reverse transcriptase copies a spliced message back into DNA that then reinserts into the chromosome. If the reinsertion of the intronless copy is into some nonexpressed area of the genome, that copy becomes a pseudogene. If that reinsertion is into an intron in some previously existing gene, the copy will become a complex 3' exon which will then proceed to evolve on its own.

A clear example of such retroposition to create a complex exon was worked out by Manyuan Long and Charles Langley [2] for the ADH gene in *Drosophila*, which normally has a complex structure involving four exons. In two *Drosophila* species that originated about 2.5 million years ago, *D. yakuba* and *D. teissieri*, there is a novel copy of an intronless ADH coding sequence attached to a 5' set of exons of some other gene, to make a new functional gene, the *jingwei* gene. Over evolutionary time, retroposition is a very important process leading both to the loss of single introns by recombination with cDNA copies and to the total loss of introns and the formation of complex exons.

The general hypothesis of this theory is that evolution is driven by recombination within introns, driven by the sliding and movement of introns to add new regions of polypeptide chain to the ends of exons and thus to produce new protein sequence at those boundaries, and driven by the loss of introns by rather common, but generally unselected, processes. On this picture the exons correspond to portions of protein sequence that can separately contribute to structure and function in an evolutionary sense. The simplest model would be that the exon products would be stable folding units, capable of independent existence as well as shuffling. More subtle models would permit exons to donate functions to their aggregates, such as a leader sequence, a hinge, a transmembrane element, or even a necessary amino acid.

The general prediction of the attitude contained in the exon theory of genes is that proteins will turn out to be made up of elements that ultimately correspond to exons and also correspond to "modules". By "module" we mean a notion introduced by Mitiko Gō [3]: a region of polypeptide chain compactly folded in space and circumscribable by some maximum dimension, generally 28 Å.

There are three different models for intron evolution. On one extreme, gene evolution would correspond to a process in which the introns came into existence at the very beginning and have only been lost ever since. However, there are processes, as described by Tokio Tani (this symposium), of the addition of introns into previously existing genes, and a variant model in which introns play a dominant role in gene origination, but also are lost and gained afterwards which is also consistent with the exon theory of genes. A second type of model is that introns arose by addition to continuous genes at the origin of the eukaryotic cell or at the origin of the metazoan organisms. Simple variations on this model are that introns are only lost after that time or that they are both lost and gained. This second type of model is consistent with the use of exon shuffling to make vertebrate genes, but these models have no role for introns in prokaryotic proteins. The third type of model is that introns appeared only very late in evolution, even being added today, and there is no exon shuffling. These are three extreme views of the world. If the introns have been only lost, then, when we attribute all of the introns observed today in different homologous forms of a gene to an ancestral gene, all of those introns should match up to some aspect of the protein structure. However, if some introns have also been gained over evolutionary time, then the attributed "ancestral" gene will have both introns that are related to protein structure and some that are not. One of the deep problems in trying to construct tests of the exon theory is whether one can see signs of ancient introns against a background of loss and gain, or whether the loss and gain simply washes out any statistical test. Can we see past the present forms to the real biological phenomena?

Exons as compact structures

The argument that exons might represent compact structures in space has been used over the years to make several predictions as to the positions of novel introns. Mitiko Gō [3] first observed that the exons in globin could be described as compact structures, in the sense that the exon product folds back and forth in such a way that the polypeptide chain lies circumscribed by a sphere 28 Å in diameter, if there were three introns in globin rather than the two that had been observed in hemoglobin and myoglobin. This prediction was soon fulfilled by the discovery of a novel intron in leghaemoglobin [4], which was found to have a novel intron in the approximate position predicted. Her analysis used a two-dimensional plot (which we call a Gō plot), each point of which corresponds to a pair of amino acids in the protein structure. On this plot the distances between each pair of amino acids are indicated, usually by coloring the plot black for all amino acid distances greater than or equal to 28 Å. On such a plot, all pairs of amino acid residues in an exon are represented by a triangle along the diagonal of the plot. If that triangle includes no black regions or is just bordered by black regions, one concludes that all pairs of distances within the exon are less than 28 Å and that the exon is compact. If that triangle contains some black areas, one concludes that the exon is extended and could, by the addition of an intron, be broken into two portions, each of which would now correspond to smaller triangles which do not overlap the black regions. When we first examined such plots for the triosephosphate isomerase gene, we observed that three of the seven chicken exons contained black areas on the Gō plot, and we commented that two of these should probably be broken up by further introns [5]. A year later these introns were discovered, and we predicted in 1986 [6] that the last remaining exon was a further candidate for an intervening intron. In 1992 Claus Tittiger [7] discovered this intron in mosquito triosephosphate isomerase. Figure 1 shows the Gō plot for triosephosphate isomerase and the position of the novel intron in mosquito. This series of successful predictions makes us feel that there is some property in the world that is being encompassed in this way of looking at the exons as compact structures. Can we extend these ideas?

One can look at these predictions and say that they are examples of the matching

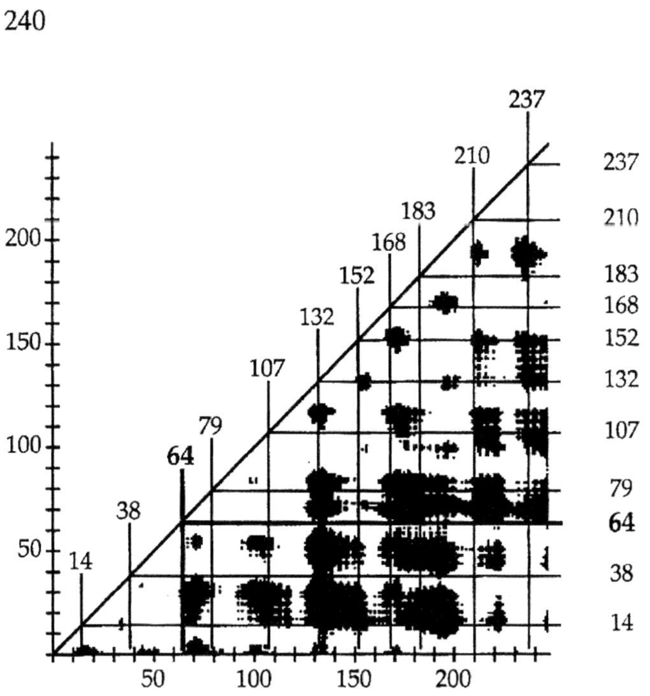

Fig. 1. A distance plot for triosephosphate isomerase. The black areas represent pairs of amino acid residues whose C^{α} carbons are more than 28 Å apart. The axes are the residues in the protein. The positions of the introns are shown, and the novel intron in mosquito is shown in bold.

of intron positions to a compactness definition of exons that reflects the three-dimensional structure of the protein. One could also look at the argument that has just been given and say: this fails to do anything significant because all that has been done is to break the protein up into small domains. If one breaks the plot up along the diagonal into small regions, those regions will eventually be compact, the corresponding "exons" will eventually be small in space. Can we find a method to show that the pattern of introns is very different from what one would find had one just broken the protein up into small pieces? The simplest approach is to break the protein up randomly and compare the distribution of hypothetical exons constructed by such a random pattern of "introns" with the real distribution. A measure of how good any breakup is in terms of compactness is found by counting the number of distances greater than 28 Å that are included in each exon (or in the case of a Gō plot, the dots contained in the triangle that corresponds to each exon). That procedure yields a numerical criterion that can be used to calculate what fraction of the alternative ways of breaking up the gene by random "introns" is better than the original breakup [8]. Table 1 shows that if we start with the observed gene, a triosephosphate isomerase (TPI) from chicken with five introns, or one from plants with eight introns, then many of the random alternatives have better scores, but as we add more and more introns to a hypothetical original gene, to get 10 introns (by combining corn and *Aspergillus*) or the 11 introns including one from mosquito, very few of the random alternatives are as good a fit to that distance plot as is the native set. The set of real positions are better than 99.9% of the random alternatives. This

Table 1.

Genes	Percentage of random alternatives with worse scores
TPI with 5 introns (*Aspergillus*)	46.4%
TPI with 6 introns (Chicken)	95.1%
TPI with 8 introns (Maize)	86.2%
TPI with 10 introns (Maize and *Aspergillus*)	99.3%
TPI with 11 introns (including *Culex*)	99.88%

is adequate to reject the null hypothesis that the introns are randomly placed in the gene, as we would expect if they had been added to DNA sequence by a process that did not look at protein structure. However, one might object to this null hypothesis and argue that random intron positions do not resemble intron positions in proteins. Random intron positions would mean that there are a lot of small exons and some very large ones, while in real genes most exons are about the same size. To take this into account, we can do a far more conservative calculation. We can maintain exon size in a variety of ways, the simplest is simply to permute the positions of the exons, to shuffle the exons, and ask how well does that alternative fit the distance plot. Table 2 shows that triosephosphate isomerase with 10 introns is better than 87% of the shuffled alternatives, while that of 11 introns is better than 96% of the shuffled alternatives. Once one adds the mosquito intron to the structure, one has a distribution of introns that is unusual, before that we cannot show significance by this very conservative test.

We generalized this idea by devising three more tests of exon structure [8]. A second test is compactness: are most of the close contacts between amino acid residues, those whose C^{α} positions are less than 12 Å apart, within the exons as opposed to between them. The third test counts and maximizes how many hydrogen bonds along the backbone are included within the exons, and the last test measures how much buried surface along the backbone is included within the exons. All four of these tests are related roughly to a notion of compactness along the protein chain, but they are not the same criterion. Each of these tests can be turned into a numerical score and tested against the original set of intron positions versus the random sets. These four tests do not correlate with each other on the randomized sets; they are not

Table 2.

Genes	Percentage of shuffled alternatives with worse scores
TPI with 5 introns (*Aspergillus*)	15.1%
TPI with 6 introns (Chicken)	29.8%
TPI with 8 introns (Maize)	15.8%
TPI with 10 introns (Maize and *Aspergillus*)	87.4%
TPI with 11 introns (including *Culex*)	96.1%

242

looking at the same parameter. We combine the four tests by ranking 10,000 alternative shuffled sets for each of the tests, averaging the ranks, and asking what rank is that average in that overall set of shuffles. This is a way of combining four tests in such a way as to show their independence and to get their distribution to approximate a Gaussian. The four tests show that the 10-intron structure for TPI is very significantly related to the protein structure, at the 95% level, and that relationship goes up above 99.4% as we add the last intron from mosquito.

These tests are reasonably robust. Logsdon and Palmer (unpublished) have found other introns in triosephosphate isomerase and they have used the positions of those other introns to argue against our first compactness test score, because introns at these extra positions deteriorate the score. However, the four tests are resistant to those added introns, and the score is still above 99%. That means that this set of introns including the added ones is still unusually related to the protein structure. Nonetheless, a true correlation between intron placement and protein structure can be obscured either by the loss of introns, so that one has too few introns to work with, or by the gain of a few adventitious introns, in which case one has too many.

We have been trying, but haven't yet succeeded to turn this way of looking at the gene and protein into a way to predict where introns might have been in the protein structure, but connections between looking at the protein structure statistically and predicting where the best intron positions might hypothetically be still do not work too well.

Last year Stoltzfus et al. [9] discussed a variety of tests of whether the exons corresponded to three-dimensional structures in proteins. In general, their efforts to find tests of the correlation between exons and three-dimensional structure were negative, and they took their inability to find such correlations as a disproof of the exon theory of genes. For the most part, they did not test the more usual interpretations of the exon theory of genes but examined only their own interpretations. The failure of their calculations reflects a failure of their interpretations, rather than the exon theory. They first discussed and disproved the concept that the exons would correspond to a secondary structure element such as an α-helix or a β-sheet, since they failed to find any correlation between introns and the boundaries of such secondary elements. They ignored the fact that, as early as 1979, it was pointed out that the introns in globin broke up the α-helices and that the exons were likely to represent more complex elements such as turns in the protein structure [10]. Stoltzfus et al. attacked at great length the idea of "centrality", the notion that introns lie close to the center of the proteins. For this notion they referenced one of our papers [5]. However, they failed to notice that this reference's only comment about "centrality" is that "centrality" does not work for triosephosphate isomerase. Lastly, they argued that calculations like the first test we discussed above fail when applied to a few other proteins. Their argument fails to take into account the problem posed by the biological reality. If introns existed only in the original protein gene and were only lost since, then one could establish an approximation of the original pattern by examining all of the currently existing forms of the gene. If the original pattern were one in which the introns were related to three-dimensional structure of the original gene, one would have some hope of finding some remnants of that relationship as one adds more and more introns into the structure. However, any ability to see a pattern would be defeated either if there was too great a loss or if there is any random process of intron addition, so that one has adventitious introns in addition to ancestral ones, or if there had been any alteration of the protein structure since its first assembly by the processes of evolution. Intron loss, and the possibility of intron gain make it extremely difficult to predict in advance which analyses of which gene structures will be revealing. Our own attitude is that the best approach is to identify a well-preserved fossil gene, not too ravaged by evolution, and to analyze the relationships of its structure to that of the protein. Only time will tell if enough clear examples will appear.

Intron positions in a database of exons

We have constructed a database of all genes that have an intron/exon structure and peptide sequence that can be identified automatically from their feature tables using GenBank. We capture about 95% of the genes that are there, about 5% have feature tables that cannot be read automatically. This process yielded, from GenBank 84, 9,300 sequences and 45,000 exons. These, of course, include very many repeats of things such as the globin sequences, the immunoglobulin sequences, and many fragmentary genes. We then purge that database automatically by various similarity measurements. In general, we purge the database by using FastA to construct a match between gene products, dividing the number of matching amino acids by the length of the shorter protein to provide a percentage match score, and for each match, above some cutoff, keeping only the entry with more exons. To identify automatically a set of reasonably independent proteins, we purged down to a similarity score of 20%. This yielded a set of 1,925 proteins and 13,042 exons. After dropping the first and last exons, we looked at the order of 1,600 genes with about 9,200 exons. Figure 2 shows the final exon distribution. As is always the case, this is a rather narrow distribution, peaking at an exon length of 35. This is not the 15 to 20 exon length we mentioned before, and one interpretation of this in the "exon theory of genes" is that these modern exons are already fusions of two to three primitive structures. (There are even in the database a few exons of length one and length two.)

Given all these exons, what can we say? The first thing we examined was the phase distribution of the introns. (The phase of an intron is its position in a codon: phase zero, one, or two corresponding to the intron lying before the codon, after the first base, or after the second base respectively.) If introns are added by any mechanism that looks at DNA sequence, one expects them to be added randomly with respect to the coding ability of the gene. One would expect them to be equally represented in all three phases. However, there is a great excess of introns in phase zero: 48% are in phase zero, 30% are in phase one, and 22% are in phase two. Rather than the 3,706 introns one expects in phase zero, there are 5,263, an excess of 1,557. These numbers are so large that this is wildly statistically significant. It is a surprise

244

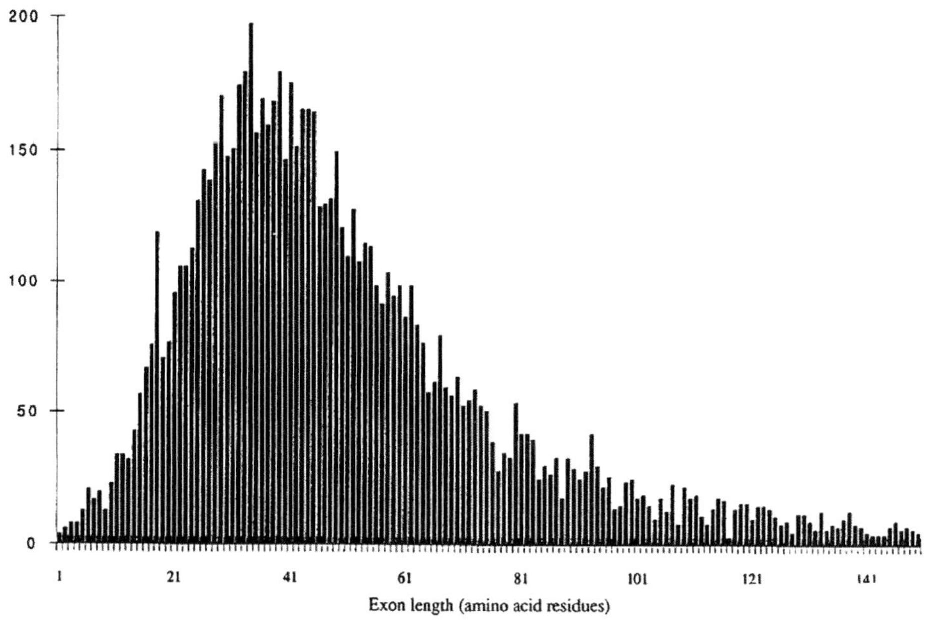

Fig. 2. The distribution of exon lengths in the final database.

to find an excess of intron positions relating to an aspect of protein coding, and the straightforward interpretation is that exon shuffling has produced an excess of phase zero introns, because exons ending in zero phase are more easily reused.

However, one might argue that perhaps some rule constraining the addition of introns accounts for this excess. If the introns entered into some sequence or sequences that were long enough to be themselves related to protein structure, this phase relationship would be only a consequence of the addition of introns. This argument is that the hypothetical target for introns (as transposons or as reverse splicing) is not short sequences, which one would expect to be random with respect to coding phase, but longer sequences which might be specific in respect to protein structure [11]. So here we simply note this unequal distribution in phase, and go on to examine another property that is not at all what we expect if introns were added. We examine symmetric exons, exons that begin and end in the same phase. Consider one process of exon shuffling in which exons add into the intron of some previously existing gene. That pre-existing intron is in some phase, let's say phase one, with respect to the surrounding exons. In order to add an exon to fit into this intron, that exon must have phase one at both ends; it must be symmetric. This is the observation that exon shuffling will work easier if the exons are symmetric, either having phase zero at both ends, phase one at both ends, or phase two at both ends. Thus, we asked is there an excess of symmetric exons or nonsymmetric exons? Furthermore, the criterion for inserting a pair of exons into an intron would be the same phase on the outside of the pair, but any phase in between. Thus we ask is there an excess of symmetric exons, symmetric pairs, symmetric triples, and so forth. In the reduced database there are 3,929 symmetric exons: 2,325 (0,0) exons, 1,085 (1,1) exons and 519 (2,2) exons. What do we expect? Since there is an excess of phase zero introns, we compare the observed number of symmetric exons to the expectation given the frequency of introns in each phase. (For example, we compare $P_1(j,k)$ with P_jP_k, where $P_1(j,k)$ is the probability of an exon being bounded by phase j, k, and P_j is the probability of an intron being in phase j). Then, even above the bias of intron phases, we observe a 12% excess of symmetric (0,0) exons, a 30% excess of symmetric (1,1) exons, and a 12% excess of symmetric (2,2) exons. Again, these numbers are large enough, the 12% excess difference is 250, the 30% excess is about 253, so that these are very, very significant differences. Table 3 shows that this excess is true both for symmetric exons, for pairs of exons asking if the outside is symmetric, for triples of exons, for quadruples, for quintuples. The excesses are consistent, but the absolute numbers fall as one looks at longer sets, so that eventually the excesses stop being significant. (We compare the pairs of exons with the prior probability based on the actual frequencies of all single exons, etc.). This is a very strong argument for exon shuffling, just seen by looking at the structure of the database. One might hypothesize that if there is a special rule that says there are phase zero introns in excess, maybe there should also be a rule that says there are phase (0,0) exons in excess. The striking thing, however, is that we observe a much greater, 30%, excess of (1,1) exons. This excess is direct evidence for repeated symmetric exons in the database, and that of course is a restatement of the exon shuffling hypothesis.

To restate the argument most bluntly: the great excess in (0,0) exons over the number that would be expected if the introns were random with respect to phase (2,325 versus 1,021), the null hypothesis, is directly an argument for exon shuffling. On this basis the excess of phase zero introns is a consequence of the excess of (0,0) exons. However, even if we accept the weaker argument, that perhaps a large part of this excess of (0,0) exons is due to the excess of phase zero introns due in turn to a (hypothetical) law of addition of introns so that they are more in phase zero, the fact then that the excess over this prior expectation is greatest for the (1,1) exons serves as independent evidence for exon shuffling (or, equivalently, for exons being related

Table 3. Symmetric exon patterns.

Exons	(0,0)	(1,1)	(2,2)	χ^2
1	2325/2075 (12%)	1085/832 (30%)	519/461 (13%)	200.37
2	1935/1829 (6%)	930/662 (40%)	421/373 (13%)	177.16
3	1593/1492 (7%)	715/518 (38%)	330/305 (8%)	132.76
4	1357/1230 (10%)	602/424 (42%)	311/252 (23%)	150.45
5	1060/1015 (4%)	461/350 (32%)	209/208 (0%)	53.64
6	899/842 (7%)	390/290 (34%)	183/172 (6%)	60.35
7	736/696 (6%)	318/240 (32%)	133/142 (–6%)	42.55

Note: Symmetric exons and exon sets in the purged database. Observed/expected (percentage difference) is shown. The expected values are calculated based on the prior expectation of frequencies for the composite exon sets. The χ square values are for the full set of nine intron combinations. The critical χ square value is 15.51 (df = 8) for the 0.05 level.

246

to the details of protein structure, since the phases, which are related to amino acid coding, are correlated at a distance.)

These numbers are independent of the purging. At every level of purging, from 60% down to 20% similarity, there is the same percentage excess of (1,1), (0,0), and (2,2) exons, and the same phase bias. If the distribution of the symmetric property had been what one expects for a random distribution of the ends of the exons, then duplications in the database, having an extra copy of a gene with a random distribution of exons, will not perturb the underlying randomness of the distribution. The fact that the distribution is nonrandom, an excess of symmetric exons, is not a property that could be created by having extra copies of genes in the database.

A considerable fraction of all exons are involved. Out of about 9,200 internal exons, if one asks how many exons in exon sets are involved, about 1,733 are involved in the excess. So at least 19% of the exons in the database were involved in exon shuffling.

Overall, the statistics of the distribution of symmetric exons provides evidence for exon shuffling. Our database is one of eukaryotic proteins, because those are the proteins that have an intron/exon structure. Thus the argument so far serves to prove exon shuffling for eukaryotic genes. However, we can ask how many of those proteins are homologous to prokaryotic proteins? We defined a set of those "ancient" proteins that are homologous to prokaryotic proteins by comparing our intron-containing database to the prokaryotic database (GenBank 85) using BLAST. We identify genes in our database that match prokaryotic proteins with BLAST scores above 75 (probabilities below 0.005 for the size of the prokaryotic database). These are genes like triosephosphate isomerase, or the "mitochondrial" genes that have introns in nuclear-encoded proteins, or other genes of ancient ancestry. We aligned each of the prokaryotic matches to the eukaryotic gene using the Smith-Waterman alignment, and defined a new database that contains only those introns that lie in the region of the eukaryotic protein that matches the prokaryotic one. This represents a class of genes (regions) that have introns but which arose as prokaryotic genes. These genes are not examples of exon shuffling in the eukaryotic world because these protein sequences came into existence in the progenote, and were present in the last common ancestor. We found 431 genes (and 1,587 internal exons) in the database to be homologous to prokaryotic genes, 22% of our database. The story for these genes is essentially the same. The intron phases are not random: 54% are in phase zero, 25%, in phase one, and 21%, in phase two. Again we can look at the symmetric exons in these bacterial-related genes. Again there is a dramatic excess of (0,0) exons over random expectation, and if we look at the excess over the biased expectation, Table 4 shows that there is a 31% excess of (1,N,1) exons (pairs of exons). However, there is lower significance, because the numbers begin to be small.

Now this is direct evidence that there has been exon shuffling in the ancestry of the genes with introns that are homologs of the prokaryotic genes. The implication of this is that ancient introns were used to assemble the prokaryotic genes. The advantage of this calculation is that it looks at a signal in the database that is derived from the act of exon shuffling, the *excess* of symmetric exons. If there is a

Table 4. Symmetric exon patterns for ancient genes.

	Nonpooled (nine exon types)				Pooled (two exon types)	
Exons	(0,0)	(1,1)	(2,2)	χ^2	(0,0)+(1,1)+(2,2)	χ^2
1	495/461 (12%)	109/96 (14%)	78/73 (7%)	10.01	682/630	6.96
2	404/359 (12%)	81/62 (31%)	50/52 (–4%)	20.96	535/473	13.45
3	315/270 (17%)	52/46 (13%)	34/39 (–13%)	16.96	401/355	10.07
4	243/206 (18%)	36/35 (3%)	26/23 (13%)	12.22	305/264	10.57

Note: Symmetric exons and exon sets for the regions corresponding to ancient genes. Observed/expected (percentage difference) is shown. The expected values are calculated based on the prior expectation of frequencies for the composite exon sets. The critical χ square value for the full set of nine intron combinations is 15.51 (df = 8) for the 0.05 level. For the two exon type calculation, the critical χ square value is 3.84 (df = 1) for the 0.05 level.

background of introns added to the database, it will not cancel the effect.

The excess of symmetric exons in this "ancient" part of the database has a simple interpretation in terms of original introns connecting exons that were shuffled at the origin of evolution and have simply survived. The excess has no interpretation in terms of an intron-added model.

Acknowledgements

This work was supported in part by the National Institutes of Health.

References

1. Gilbert W. The exon theory of genes. Cold Spring Harbor Symp Quant Biol 1987;52:901–905.
2. Long M, Langley C. Natural selection and the origin of *jingwei*, a chimeric processed functional gene in *Drosophila*. Science 1993;260:91–95.
3. Gō M. Correlation of DNA exonic regions with protein structural units in haemoglobin. Nature 1981;291:90–93.
4. Jensen EO, Paludan K, Hyldig-Nielsen JJ, Jorgensen P, Marcker KA. The structure of a chromosomal leghaemoglobin gene from soybean. Nature 1981;291:677–679.
5. Straus D, Gilbert W. Genetic engineering in the Precambrian: Structure of the chicken triosephosphate isomerase gene. Molec Cell Biol 1985;5(12):3497–3506.
6. Marchionni M, McKnight G, Gilbert W. On the antiquity of introns. Cell 1986;46:151–154.
7. Tittiger C, Whyard S, Walker VK. A novel intron site in the triosephosphate isomerase gene from the mosquito *Culex tarsalis*. Nature 1993;361:470–472.
8. Glynias M, Gilbert W. On the ancient nature of introns. Gene 1993;135:137–144.
9. Stoltzfus A, Spencer DF, Zuker M, Logsdon JM, Doolittle WF. Testing the exon theory of genes: the evidence from protein structure. Science 1994;265:202–207.
10. Gilbert W. Introns and exons: playgrounds of evolution. In: Axel R, Maniatis T, Fox CF (eds) Eucaryotic Gene Regulation. New York: Academic Press, 1979;1–12.
11. Fedorov A, Suboch G, Bujakov M, Fedorova L. Analysis of nonuniformity in intron phase distribution. Nucl Acid Res 1992;20:2553–2557.

Proc. Natl. Acad. Sci. USA
Vol. 92, pp. 12495–12499, December 1995
Evolution

Intron phase correlations and the evolution of the intron/exon structure of genes

(exon shuffling/ancient conserved regions)

MANYUAN LONG*, CARL ROSENBERG†, AND WALTER GILBERT*

*Department of Molecular and Cellular Biology, The Biological Laboratories, Harvard University, Cambridge, MA 02138; and †Whitehead Institute/Massachusetts Institute of Technology Center for Genome Research, Cambridge, MA 02139

Contributed by Walter Gilbert, September 13, 1995

ABSTRACT Two issues in the evolution of the intron/exon structure of genes are the role of exon shuffling and the origin of introns. Using a large data base of eukaryotic intron-containing genes, we have found that there are correlations between intron phases leading to an excess of symmetric exons and symmetric exon sets. We interpret these excesses as manifestations of exon shuffling and make a conservative estimate that at least 19% of the exons in the data base were involved in exon shuffling, suggesting an important role for exon shuffling in evolution. Furthermore, these excesses of symmetric exons appear also in those regions of eukaryotic genes that are homologous to prokaryotic genes: the ancient conserved regions. This last fact cannot be explained in terms of the insertional theory of introns but rather supports the concept that some of the introns were ancient, the exon theory of genes.

How did the intron/exon structure of genes evolve? Broadly speaking, there are two alternative theories for the origin of introns. The insertional theory of introns, also called the "introns-late" theory, invokes recent events of intron insertion into eukaryotic genomes (1–3). On the other hand, the "introns-early" hypothesis (4–7) takes the position that introns were used in the progenote to assemble genes and that the evolutionary pattern since was one of loss down many lines, some retention, and, possibly, some gain. On this model one expects some correlation of exons with functional elements of the protein gene products.

The introns-late hypothesis assumes that introns were added to preexisting genes at some time during evolution. One form would suggest that introns were added early in eukaryotic evolution and were used to shuffle parts of genes to account for the Cambrian explosion of metazoan species. Such models would expect exon shuffling to appear only in evolutionarily late gene products. A still more extreme model is that introns were added only very late in evolution and that no exon shuffling has occurred. In all introns-late models one expects no correlation between exons and elements of protein function for ancient genes, and observations of correlations between three-dimensional structure and exon structures (8–10), of correlations between intron positions in plants and animal genes (11–15), or of correlations between intron positions in genes that diverged in the progenote (16–19) are all viewed as statistical coincidences (20–22).

This study began with a search for different testable predictions derived from these two competing theories. Intron phase [the position of the intron within a codon, phase 0, 1, or 2 lying before the first base, after the first base, or after the second base, respectively (23)], is an important evolutionary character for whose distribution the two theories generate different predictions. Since the insertion of introns into previously uninterrupted genes, as the introns-late view advocates, does not affect the structure of the gene product, introns in each phase should have similar chances to survive. Thus, the simple form of the introns-late theory predicts a random phase distribution for introns. On the contrary, the introns-early theory predicts a nonrandom phase distribution for introns as a consequence of exon shuffling events, since exon shuffling works better if introns are in the same phase. Furthermore, exon shuffling should produce correlations in intron phases, since symmetric exons shuffle more easily, while insertional models predict that intron phases are uncorrelated. The prediction of a random distribution of intron phases and random correlations allows us to detect exon shuffling and to test the insertional theory of intron origins. In this report, we show that this population analysis indicates both an important role for exon shuffling throughout evolution and also support for the hypothesis of the ancient origin of introns.

METHODS

Exon Data Bases. We used GenBank (24) entries released in 1994 [release 84, which also includes European Molecular Biology Laboratory (EMBL) and DNA Database of Japan (DDBJ) entries] to construct a raw data base that contains information about the intron/exon structures, definitions of genes, locus names, and, finally, protein sequences. Then we calculated the intron positions and phases and the numbers and sizes of the exons for these 9276 sequences (45,095 exons) to create a processed data base. We discarded entries with obvious problems, such as wrong feature tables. In testing our extraction program, we identified all entries in GenBank that had the word *exon* in the definition line and sampled randomly about 10% of these entries to check how many include the "cds. . .join" and "cds. . .complement join" command that we used to identify exon-containing genes. We found that about 3% of exons lacked this reference, and we excluded them. A further 5% lacked peptide sequence and were discarded. Therefore, we recovered about 97% of all annotated genes in the data base. To discard duplicates and partial sequences, we compared all the protein sequences with BLASTP (25) and purged all related entries but the longest, using a similarity score of 99% (99% match over the length of the shorter sequence), yielding 6611 genes and 33,150 exons. We then further distilled the data base, using FASTA (26) (similarity scores are calculated from the FASTA alignment as the match percentage times the length of the overlapping region divided by the length of the shorter sequence) and purging down to a criterion of a 20% match to the shorter sequence, keeping the sequence with more exons each time (the purging yielded 4681 genes at the 80% level, 3704 at 60%, 2942 at 40%, and 1925 at 20%). The plant genes are from FASTA purging of a complete data base of plant intron-containing genes to the 20% simi-

The publication costs of this article were defrayed in part by page charge payment. This article must therefore be hereby marked "*advertisement*" in accordance with 18 U.S.C. §1734 solely to indicate this fact.

Abbreviation: ACRs, ancient conserved regions.

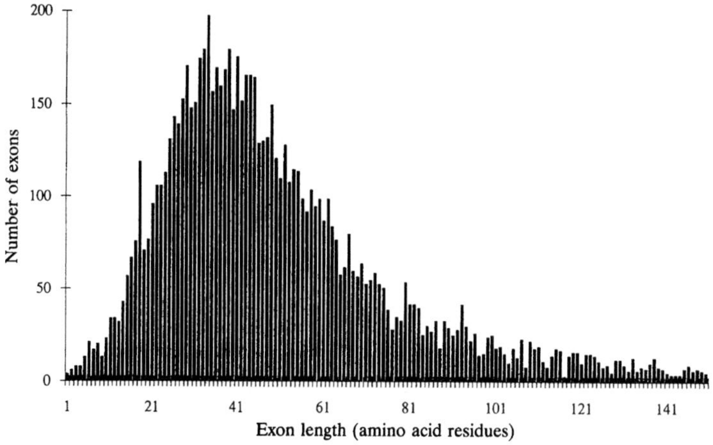

FIG. 1. Distribution of exon lengths. For the internal exons in the data base purged to 20% similarity, the histogram shows the numbers of exons at each length: 94% of all the exons are of length less than 150 amino acids, and there is a scattering of longer exons, the longest having 3205 amino acid residues.

larity level. The animal genes are drawn from the final data base.

Expected Frequencies of Intron Phase Combinations. The expected frequencies of sets of intron phase combinations, Bayesian expectations, are calculated from the observed frequencies of their component members. For example, given $f_1(i, j)$, the observed frequency of single exons whose bounding phases are i and j (5′ and 3′ respectively), the probabilities for the sets of length 2, $P_2\ (i,\ j) = \Sigma_m f_1(i,\ m){\cdot}f_1(m,\ j)/\Sigma_i\Sigma_j\Sigma_k f_1(i,\ j){\cdot}f_1(j,\ k)$. In general, the probability for a set of length n, $P_n(i, j)$, is related to the product of n f_1s, $P_n(i, j) = (\Sigma\cdots\Sigma)_{n-1}f_1\cdots f_1/(\Sigma\cdots\Sigma)_{n+1}f_1\cdots f_1$.

Exon Number Involved in Shuffling. For sets of singles, doubles, and triples, we counted only the symmetric exons, the doubles that have phase combinations: 010, 020, 101, 121, 202, and 212, and triples: 0120, 0210, 1021, 1201, 2012, and 2102. Their expectations were calculated as Bayesian expectations.

Ancient Conserved Regions (ACRs). Following the method of Green *et al.* (27), we identified genes in the final data base that match prokaryotic proteins (GenBank release 85) with BLAST scores above 75. We purged false alignments due to highly biased amino acid composition. Then we aligned each of the prokaryotic matches to the eukaryotic gene by using a Smith–Waterman alignment (the program used was sent generously by P. Green) and defined a new data base that contained only those introns that lie in the region of the eukaryotic protein that matches the prokaryotic one. This represents a class of genes (regions) that have introns but which arose as prokaryotic genes. There are 296 genes with 1496 introns and 1200 internal exons in this data base of homologs of prokaryotic genes.

RESULTS

Construction of Exon Data Base. To survey the intron phase distribution, we first built a data base of eukaryotic intron-containing genes extracted from GenBank release 84. To discard duplicates and partial sequences in this data base, we purged all related sequences by using BLASTP and FASTA. The final data base, purged to 20%, contains 1925 independent or quasi-independent protein sequences containing 13,042 exons and 11,117 introns (we use the prefix *quasi* here because one cannot infer with certainty that the sequences are truly independent; these low criteria for similarity still miss a few related genes). We found 1600 genes that have more than one intron, and there are 9192 internal exons. Fig. 1 shows a histogram of internal exon lengths: a distribution peaked at 35 residues.

Distribution of Intron Phases. The three classes of intron phases are far from evenly distributed. Table 1 shows that there is an excess of phase zero introns, as was found in a previous survey of a small sample (28). While this result from a large sample is consistent with the prediction of the exon theory of genes and contradicts the simple form of the insertional theory, there are alternative hypotheses for the insertional theory which cannot be rejected. To disfavor the appearance of phase one or phase two introns, one might assume that the insertion was into long nucleotide sequences which might be correlated with the local amino acid sequence of the gene product, such as has been suggested for an AG|GT sequence at the exon boundaries (28). However, we can distinguish between the two theories by studying intron phase correlations.

Intron Phase Correlations. The association of two adjacent introns in eukaryotic genes can be in any of nine different intron phase combinations, which can be classified into two groups—symmetric exons (0, 0), (1, 1), and (2, 2) and asymmetric exons (0, 1), (0, 2), (1, 2), (1, 0), (2, 0), and (2, 1) (the first number is the phase of the 5′ intron and the second number is the phase of the 3′ intron). Symmetric exons will spread more easily by exon shuffling (29) because the addition of a symmetric exon into an intron of the same phase does not disturb the reading frame. On the contrary, the intron insertional theory predicts a random correlation of intron phases across exons.

The simplest null hypothesis again is the random expectation that intron phases should have equal frequencies, 1/3, and that each pair of introns, each exon combination, should be equiprobable at 1/9. Compared to this, the intron phase correlation is very far from random, and there is a 125% excess of symmetric (0, 0) exons over the null expectation. This dramatic excess of symmetric exons could be taken as support for exon shuffling, with the corollary that the excess of phase zero introns would then be a result of the shuffling of symmetric (0, 0) exons. However, let us consider the more

Table 1. Proportions of the three intron phases

No. (%) of introns			
Phase zero	Phase one	Phase two	Total
5263 (48%)	3372 (30%)	2482 (22%)	11,117

Table 2. Observed and expected symmetric intron phase associations

No. of exons	(0, 0)	(1, 1)	(2, 2)	No. of exons	χ^2
1	2325/2075 (12%)	1085/832 (30%)	519/461 (13%)	9192	200.4
2	1935/1829 (6%)	930/662 (40%)	421/373 (13%)	7556	177.2
3	1593/1492 (7%)	715/518 (38%)	330/305 (8%)	6202	132.8
4	1357/1230 (10%)	602/424 (41%)	311/252 (23%)	5116	150.5
5	1060/1015 (4%)	461/350 (32%)	209/208 (0%)	4224	53.6
6	899/842 (7 %)	390/290 (34%)	183/172 (6%)	3503	60.4
7	736/696 (6 %)	318/240 (33%)	133/142 (−6%)	2893	42.6

The observed and expected intron associations are expressed as the observed number/the expectation and excess of observed is in parentheses. For one exon, the expectation is P_jP_iN, where P_i is the proportion of introns of phase i and N is the total number of exons. For sets of exons, the expectation is the Bayesian probability based on the observed proportions of internal exons (15). The χ^2 values are calculated on the total distribution (nine numbers). The asymmetric exon sets are all less than expectation.

conservative hypothesis, that the phase zero bias might have some cause other than exon shuffling, and compare the observed frequencies of exons with the expected frequencies based on a random association of introns [the expected frequency of exon (i, j) being then the product of the frequencies of introns of phase i and phase j].

The χ^2 test shows that the deviation of the 9192 exons from a random association of phase is very significant ($\chi^2 = 200$, a P value less than 10^{-40}). Thus intron phases are not combined in a random fashion. Furthermore, symmetric exons are significantly more frequent than expected and asymmetric exons are fewer. Table 2 shows that there is a 12% excess of (0, 0) exons over expectation and a 30% excess of (1, 1) exons. This is strong evidence for the exon shuffling hypothesis, because, even if there were some hitherto undiscovered rule about phase zero introns and correlations of phase zero introns, one would not expect a significant excess of (1, 1) exons; symmetry *per se* seems to be favored. One source of nonrandom bias is human error in the data base. We estimate that the feature table is wrong in a few percent of the entries. The most common error is to assign introns to phase zero rather than using ···|GT···AG|··· rules. Such a bias would not lead to an excess of (1, 1) correlations.

One further property of exon shuffling is that groups of exons are likely to shuffle together. Table 2 also surveys the distributions of the outside intron phases for groups of two to seven exons. The observed frequencies of each group are compared with Bayesian conditional probabilities based on the frequencies of the exons making up the group. Since the numbers of exons in the sets of exons falls as the sets get longer (next-to-last column in Table 2), the statistical significance of the deviation from expectation (last column in Table 2) falls as the length of the set of exons increases. However, the striking fact is that the fractional excess over expectation is maintained for each type of symmetrical set; (0, 0) sets are about 7% over expectation, while (1, 1) sets are all about 30% over expectation. This behavior of sets of exons is a further argument for exon shuffling.

Could this excess of symmetric exons and exon sets be due to alternative splicing as distinct from exon shuffling? Alternative splicing is dependent solely on the biological properties of individual genes (30, 31) rather than the distances between the introns. Alternative splicing frequently adds alternative sets of exons to the beginning or ends of genes. Such events do not restrict the phases or the phase combinations of exons. Alternative splicing events that add an additional internal exon to a previously functioning structure do require that the added exon be symmetric, but, of course, also require that the exon/intron structure be related to the structure and function of the gene product.

Role of Exon Shuffling in Evolution. How much of the data base would have to be involved in exon shuffling events to produce these observed deviations from randomness? We take the difference between the observed and the expected number of exons in symmetric sets as an estimate of the number of exons which had to be involved in exon shuffling. Since there may be an overlap in the use of exons between the exon sets of different length, we calculate only the excess symmetric exons over expectation, and the specific excesses of pairs and triples that do not contain any symmetric exons within each set. The differences between observations and expectations (d = O − E) are singles, $d = 561$; doubles, $d = 400$; and triples, $d = 124$. Then the number of exons involved in shuffling is $1 \times 561 + 2 \times 400 + 3 \times 124 = 1733$, which is 19% of the purged data base. This is a lower bound for the true fraction involved in shuffling, since we considered the excesses only for singles, certain pairs, and certain triples of exons to avoid the possibility of counting twice, and other factors like intron drift would weaken the evidence for symmetric exon patterns. (If all of the deviation from the 1/3 expectation of intron phases were to be due to exon shuffling, then at least 28% of the data base had to be involved.) Thus, quantitatively, exon shuffling is very important in the evolution of the intron/exon structure of genes.

This analysis has shown that the intron correlations in eukaryotic genes are not random, and it suggests a dominant role for exon shuffling. However, several authors (32–34) have argued that exon shuffling is a recent phenomenon, limited to vertebrate genes. Further analysis indicates that this may not be the case.

We analyzed the intron phase correlations for both plant and animal genes separately. Table 3 shows that the nonrandom

Table 3. Intron correlations in data base subsets

Subset	(0, 0)	(1, 1)	(2, 2)	No. of exons	χ^2
Plant	579/515 (12%)	118/88 (34%)	76/71 (7%)	1642	33.5
All animal	1869/1678 (11%)	992/774 (28%)	464/409 (13%)	7923	160.0
Animal without *Caenorhabditis elegans*	1189/1011 (11%)	778/604 (29%)	235/208 (13%)	5010	163.1

The proportions of phase zero, one, and two introns for the internal exons of plant genes are 56%, 23%, and 21% (total 2790 introns); the proportions for animal genes are 46%, 31%, and 23% (total 10,999 introns). The expectations for the symmetric exons are calculated on the basis of these proportions.

Table 4. Intron correlations in ACRs

No. of exons	Six-way test					Two-way test		
	(0, 0)	(1, 1)	(2, 2)	χ^2	P	(0, 0) + (1, 1) + (2, 2)	χ^2	P
1	407/389 (7%)	88/76 (16%)	67/59 (14%)	9.1	0.110	562/515 (9%)	7.1	0.0080
2	340/305 (11%)	62/51 (22%)	37/44 (−15%)	13.1	0.026	439/400 (10%)	6.5	0.0110
3	275/236 (17%)	43/39 (12%)	30/34 (−10%)	13.2	0.027	348/309 (13%)	8.4	0.0035
4	217/182 (19%)	30/30 (0%)	20/26 (23%)	11.5	0.042	267/238 (12%)	6.0	0.0140

Results are expressed as observed/expected (percent difference). The expected values are calculated on the basis of the prior expectation of frequencies for the composite exon sets. The critical χ^2 value in a test of three symmetric and three asymmetric summed combinations, or six-way test, is 11.07 (df = 5) for the 0.05 level; the critical χ^2 value in a two-way test at the 0.05 level is 3.84 (df = 1).

correlations of introns and the excess of symmetric exons holds not only for animal genes but also for plant genes. Thus, exon shuffling is not limited to vertebrate or animal genomes. [This statistical argument is also supported by a recent report of exon shuffling in plant genes (35).] The animal genes in the data base include a large number of genes from the *C. elegans* project whose intron positions were determined by a computer program, GENEFINDER (36). One might worry that these "hypothetical" genes might perturb the calculation. Table 3 also shows that all the *C. elegans* genes can be dropped from the data base without affecting the phase correlation excesses.

Intron Phase Correlations in ACRs. More importantly, we have investigated intron phase correlations in a subset of 296 genes from the final data base that are homologous to prokaryotic genes (27), using only those introns that lie in the region of match between the eukaryotic proteins and the prokaryotic ones. These ACRs represent gene structures that were in existence in the last common ancestor of the prokaryotes and the eukaryotes. Since these genes have no introns in the prokaryotes, the introns-late theory holds that all the introns in this portion of the data base are insertional. No shuffling could have occurred after the prokaryotic–eukaryotic divergence, since these gene regions are colinear. However, just as for the overall data base, the distribution of intron phases and phase combinations in these ACRs are not random. Among the 1496 introns, phase zero introns constitute 55%; phase one introns, 24%; and phase two introns, 21%. Furthermore, there are statistically significant excesses of symmetric exon sets. Table 4 shows the distribution of individual symmetric exon types for sets of one to four exons. The distributions for individual patterns of pairs, triples, and quadruples of exons reach statistical significance. However, in all cases, a two-way test shows clearly that symmetrical exons are in excess at the 1% probability level. Thus the intron/exon structure of these ACRs suggests that they were constructed by exon shuffling, which would have to have occurred in the progenote.

DISCUSSION

How robust is this analysis? One might argue that those intron correlations are due to some pattern of repeated genes in the data base: that the purging had not produced a set of independent entries. To test this possibility we repeated the calculation for data bases created at different purging levels. Fig. 2 shows that the intron phase frequencies and the excess of symmetric exons (only pairs are shown) are stable across the purging levels. Thus, the deviations from randomness are not a property of redundancy in the data base. The few quasi-independent sequences in the data base which may come from same-gene families do not bias the statistical analysis. Even if we purge the data base to the 10% level, yielding only 689 genes and 7115 exons (657 genes with 5737 internal exons), the phase correlation excesses still stand.

To give an insertional theory a best shot, we have assumed that the nonrandom intron phase distribution might be a result of introns inserted into specific sequences of bases which could be correlated with local amino acid sequence. One example of this is the suggestion that an AG|GT sequence in the exons at the junctions is a remnant of a preinsertion target. However, Stephens and Schneider (37) find no significant AG|GT sequence within the exons in a study of 1800 human introns, and only a slight excess of AG|G. [An alternative interpretation of this AG|G bias would be that such sequences were selected for, after the fact, by evolutionary pressure from the splicing mechanism, such as better matching to the small RNAs (38).] On such insertional models, one might argue that the base composition of the DNA would bias the third positions of codons and thus that variation in the GC bias would drive such "entry" sequences into different phase relations in different genes. To test if the intron phase correlations are the result of some genes being rich in phase zero introns while other genes are rich in phase one introns, a feature that would create correlations between the phases, we examined the distribution of introns within genes in the data base. Histograms of the distribution of introns within genes show a pattern broadly peaked around the average positions, except for spikes at 100% for the three pure phases. We have dropped those sets of exons (68 genes) and repeated the calculation of symmetric exon excesses to find again a 9% excess of (0, 0) exons and a 24% excess of (1, 1) exons.

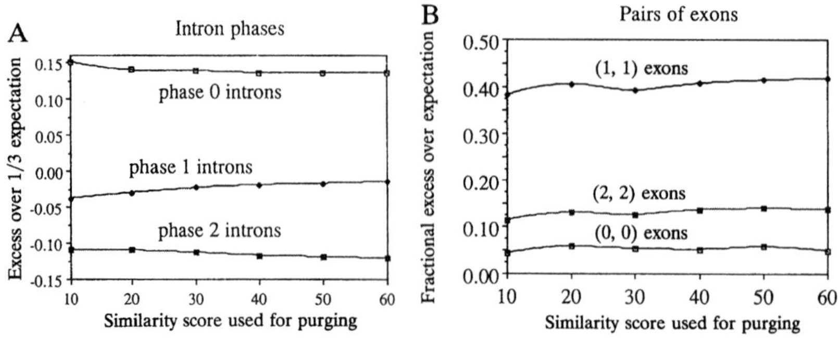

FIG. 2. Stability of the deviations from expectation at various purging levels. (*A*) Deviations of the intron phases from the 1/3 expectation. (*B*) Difference between the frequencies of symmetric pairs of exons and the expectation based on the observed exon frequencies.

This study of intron phase correlations has revealed a signal which we interpret as a mark of exon shuffling. This statistical approach has the advantage that such a signal is only weakened, not abolished, by the drift or loss of introns over evolutionary time or by the addition of a few novel introns. The argument is that the intron correlations in the eukaryotic versions of "ancient" genes, the excess of symmetric sets of exons, shows that these exon sets were once used in exon shuffling which had to occur in the progenote. Hence these exons were elements that were used to compose the original genes. These intron phase correlations are an independent support for the exon theory of genes. The argument from phase correlations thus adds to the previous arguments based on the correlation of intron positions with three-dimensional structure for triosephosphate isomerase (8–10), the coincidence of plant and animal introns (11–15), or the identification of ancient introns on the basis of the patterns of descent (16–19) to support the idea that some introns are extremely ancient. Unlike the previous arguments, the intron phase correlation calculation detects a clear, statistically significant signal for the existence of early introns against a background of loss and gain.

We thank Richard C. Lewontin, Nathan Goodman, Charles H. Langley, Stephen M. Mount, Daniel Weinreich, Andrew Berry, Leslie Gotlieb, and all the members of the Gilbert Lab, especially Keith Robison and Zhiping Liu, for useful discussions and helpful computing advice. We thank Phillip Green for the computer program for the Smith–Waterman algorithm and Eric Lander for allowing us access to his computing facility. We thank the National Institutes of Health for their support, Grant GM37997 to W.G. and Grant HG0098 to Eric Lander.

1. Cavalier-Smith, T. (1991) *Trends Genet.* **7,** 145–148.
2. Rogers, J. (1985) *Nature (London)* **315,** 458–459.
3. Rogers, J. (1989) *Trends Genet.* **5,** 213–216.
4. Blake, C. C. F. (1978) *Nature (London)* **273,** 267.
5. Doolittle, W. F. (1978) *Nature (London)* **272,** 581–582.
6. Gilbert, W. (1987) *Cold Spring Harbor Symp. Quant. Biol.* **52,** 901–905.
7. Gilbert, W. (1978) *Nature (London)* **271,** 501.
8. Gilbert, W. & Glynias, M. (1993) *Gene* **135,** 137–144.
9. Gilbert, W., Marchionni, M. & McKnight, G. (1986) *Cell* **46,** 151–153.
10. Go, M. (1981) *Nature (London)* **291,** 90–93.
11. Marchionni, M. & Gilbert, W. (1986) *Cell* **46,** 133–141.
12. Nawrath, C., Schell, J. & Koncz, C. (1990) *Mol. Gen. Genet.* **223,** 65–75.
13. Pardo, J. M. & Serrano, R. (1989) *J. Biol. Chem.* **264,** 8557–8562.
14. Shah, D. M., Hightower, R. C. & Meagher, R. B. J. (1983) *Mol. Appl. Genet.* **2,** 111–126.
15. Imajuku, Y., Hirayama, T., Endoh, H. & Oka, A. (1992) *FEBS Lett.* **304,** 73–77.
16. Iwabe, N., Kuma, K., Kishino, H., Hasegawa, M. & Miyata, T. (1990) *J. Mol. Evol.* **31,** 205–210.
17. Kersanach, R., Brinkmann, H., Liaud, M.-F., Zhang, D.-X., Martin, W. & Cerff, R. (1994) *Nature (London)* **367,** 387–389.
18. Obaru, K., Tsuzuki, T., Setoyama, C. & Shimada, K. (1988) *J. Mol. Biol.* **200,** 13–22.
19. Setoyama, C., Joh, T., Tsuzuk, T. & Shimada, K. (1988) *J. Mol. Biol.* **202,** 355–364.
20. Logsdon, J. M., Jr., & Palmer, J. D. (1994) *Nature (London)* **369,** 526.
21. Stoltzfus, A. (1994) *Nature (London)* **369,** 526–527.
22. Stoltzfus, A., Spencer, D. F., Zuker, M., Logsdon, J. M., Jr., & Doolittle, W. F. (1994) *Science* **265,** 202–207.
23. Sharp, P. A. (1981) *Cell* **23,** 643–646.
24. Burks, C., Cinkosky, M. J., Gilna, P., Hayden, J. E., Abe, Y., Atencio, E. J., Barnhouse, S., Benton, D., Buenafe, C. A., Cumella, K. & Burks, E. (1990) *Methods Enzymol.* **183,** 3–22.
25. Altschul, S. F., Gish, N. M., Miller, W., Myers, E. W. & Lipman, D. J. (1990) *J. Mol. Biol.* **215,** 403–410.
26. Pearson, W. R. & Lipman, D. J. (1988) *Proc. Natl. Acad. Sci. USA* **85,** 2444–2448.
27. Green, P., Lipman, D., Hillier, L., Waterson, R., States, D. & Claverie, J.-M. (1993) *Science* **259,** 1711–1716.
28. Fedorov, A., Fedorov, A., Suboch, G., Bujakov, M. & Fedorova, L. (1992) *Nucleic Acids Res.* **20,** 2553–2557.
29. Patthy, L. (1987) *FEBS Lett.* **214,** 1–7.
30. Hodges, D. & Bernstein, S. I. (1994) *Adv. Genet.* **31,** 207–228.
31. McKeown, M. (1992) *Annu. Rev. Cell Biol.* **8,** 133–155.
32. Palmer, J. D. & Logsdon, J. M., Jr. (1991) *Curr. Opin. Genet. Dev.* **1,** 470–477.
33. Patthy, L. (1991) *BioEssays* **13,** 187–192.
34. Stoltzfus, A., Spencer, D. F., Zuker, M., Logsdon, J. M., Jr., & Doolittle, W. F. (1994) *Science* **265,** 202–207.
35. Domon, C. & Steinmetz, A. (1994) *Mol. Gen. Genet.* **244,** 312–317.
36. Wilson, R. (1994) *Nature (London)* **368,** 32–38.
37. Stephens, R. M. & Schneider, T. D. (1992) *J. Mol. Biol.* **228,** 1124–1136.
38. Horowitz, D. S. & Krainer, A. R. (1994) *Trends Genet.* **10,** 100–106.

Proc. Natl. Acad. Sci. USA
Vol. 93, pp. 7727–7731, July 1996
Evolution

Exon shuffling and the origin of the mitochondrial targeting function in plant cytochrome c1 precursor

(glyceraldehyde-3-phosphate dehydrogenase/presequences/transit peptides/nucleus-encoded organelle genes)

MANYUAN LONG*†, SANDRO J. DE SOUZA*, CARL ROSENBERG‡, AND WALTER GILBERT*

*Department of Molecular and Cellular Biology, The Biological Laboratories, Harvard University, Cambridge, MA 02138; and ‡Whitehead Institute/Massachusetts Institute of Technology Center for Genome Research, Cambridge, MA 02139

Contributed by Walter Gilbert, April 16, 1996

ABSTRACT **Since most of the examples of "exon shuffling" are between vertebrate genes, the view is often expressed that exon shuffling is limited to the evolutionarily recent lineage of vertebrates. Although exon shuffling in plants has been inferred from the analysis of intron phases of plant genes [Long, M., Rosenberg, C. & Gilbert, W. (1995) *Proc. Natl. Acad. Sci. USA* 92, 12495–12499] and from the comparison of two functionally unknown sunflower genes [Domon, C. & Steinmetz, A. (1994) *Mol. Gen. Genet.* 244, 312–317], clear cases of exon shuffling in plant genes remain to be uncovered. Here, we report an example of exon shuffling in two important nucleus-encoded plant genes: cytosolic glyceraldehyde-3-phosphate dehydrogenase (cytosolic GAPDH or GapC) and cytochrome c1 precursor. The intron–exon structures of the shuffled region indicate that the shuffling event took place at the DNA sequence level. In this case, we can establish a donor–recipient relationship for the exon shuffling. Three amino terminal exons of GapC have been donated to cytochrome c1, where, in a new protein environment, they serve as a source of the mitochondrial targeting function. This finding throws light upon an old important but unsolved question in gene evolution: the origin of presequences or transit peptides that generally exist in nucleus-encoded organelle genes.**

Exon shuffling is a molecular evolutionary mechanism to create new genes with novel functions by combining exons from unrelated genes (1, 2). Early examples, reported 10 years ago, include the low density lipoprotein receptor (3) and proteins involved in blood coagulation and fibrinolysis (4). Since then, many events of exon shuffling have been identified, most of which took place in modern genes (5). Exon shuffling in ancient genes or ancient conserved regions—i.e., the genes or the regions of genes homologous between eukaryotes and prokaryotes (6)—was inferred from a statistical analysis of intron phases (7). Both actual examples and statistical analysis show that the exon shuffling played a very important role in gene evolution.

However, the available reports of exon shuffling are almost all from the genes from one evolutionarily recent lineage—the vertebrates. One exception is the case of two genes in sunflowers (8), but the function and evolutionary history of these plant genes are unclear. This limited distribution of exon shuffling examples has often led to the conclusion that exon shuffling is restricted to vertebrate genomes and hence occurred very recently (9–11). However, the analysis of intron phases in plant genes detects a strong signal of exon shuffling (7), although the statistical approach is unable to reveal specific examples.

Many nuclearly encoded genes for organellar proteins are thought to have been transferred from the organellar genomes (12). In two nuclear genes, the *coxII* gene encoding the subunit 2 of cytochrome *c* oxidase in mitochondria and the *rpl22* gene encoding the ribosomal protein CL22 in chloroplasts, additional protein coding regions were acquired in amino terminal end after the gene transfer events (13, 14). These regions have an organelle targeting activity. What is the origin of these additional protein domains, which created a new independent function? This is an old but unsolved question.

FIG. 1. Alignment of pea GapC1 (GapC) and potato cytochrome c1 (cyc1) protein sequences. The lowercase letters in the sequences and the numbers above or beneath these letters in the matched regions mark the intron positions and intron phases. A BLAST comparison of the first three exons of these two genes showed that the match in this region is very significant ($P = 1.1 \times 10^{-6}$).

The publication costs of this article were defrayed in part by page charge payment. This article must therefore be hereby marked "*advertisement*" in accordance with 18 U.S.C. §1734 solely to indicate this fact.

Abbreviation: GAPDH, glyceraldehyde-3-phosphate dehydrogenase.

†To whom reprint requests should be addressed. e-mail: long@nucleus.harvard.edu.

In this paper, we report an example of exon shuffling in two important nuclearly encoded plant genes, the GapC gene and the cytochrome c1 gene. In this case, we can determine the donor–recipient relationship in this shuffling process to show that glyceraldehyde-3-phosphate dehydrogenase (GAPDH) donated its three consecutive N-terminal exons to cytochrome c1 where these exons evolved to a mitochondrial targeting function in a new protein environment.

METHODS

Data Bases. The exon data base (7) based on GenBank release 84 contains 1925 independent or near independent proteins and 13045 exons. Matched exons were sought among these sequences. The nonredundant BLAST (15) search for the GAPDH and cytochrome c1 protein sequences was done over the data base Protein Data Bank (1995), Protein Information Resource (release 47.0), SwissProt (release 32.0), and GenBank (release 93) (16). The relevant sequence data were extracted from these data bases. The Protein Data Bank was also used to extract the three-dimensional structure files of GAPDH.

Alignment of the GAPDH and Cytochrome c1 Protein Sequences. To check the mosaic pattern caused by exon shuffling, the entire proteins of GapC in pea and cytochrome c1 in potato were aligned by using the program GAP in Gentetics Computer Group package (17). The gap weight and length weight were 10.0 and 3.0, respectively.

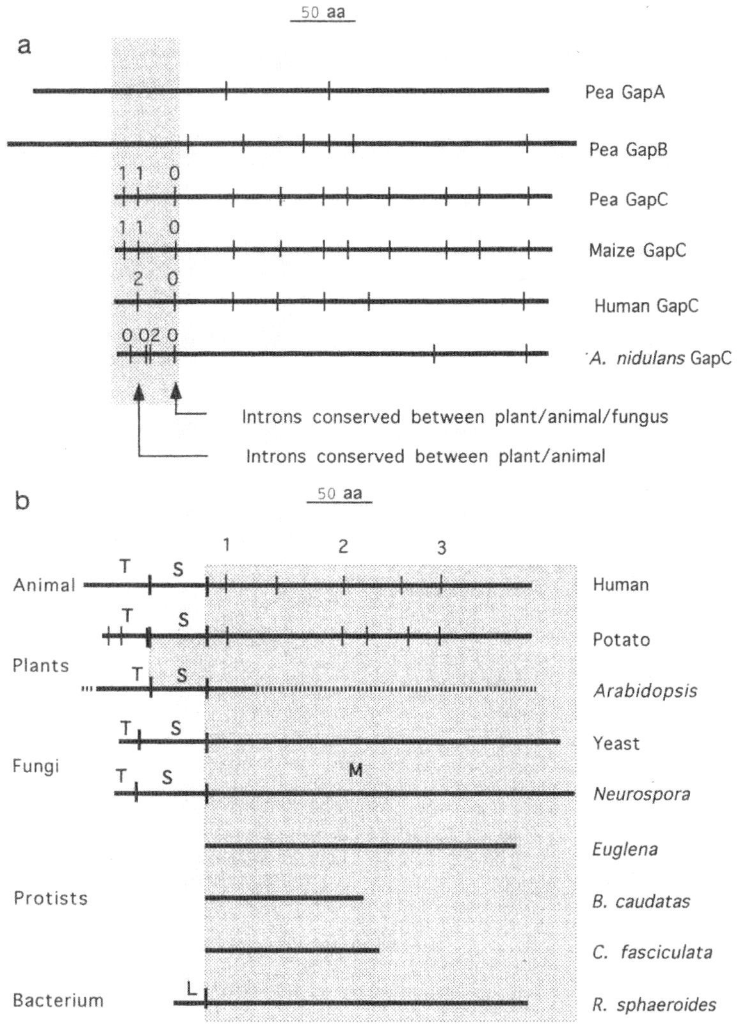

FIG. 2. (*a*) Schematic intron–exon structures of GapA, GapB, and GapC in pea, maize, human, and fungus. The vertical lines represent introns. The gray area includes the region involved in the exon shuffling event. The numbers over introns in the gray area indicate the phases of introns. The two conserved introns indicated by the two arrows were noticed in ref. 22. (*b*) Schematic structures of cytochrome c1 precursor in various organisms. The thin bars show intron positions. T, S, M, and L, divided by thick bars, show the mitochondrial targeting domain, intramitochondrial sorting domain, mature cytochrome c1 protein, and the leader region in bacterial proteins, respectively. The numbers show the position of three ancestral introns. The shaded area between potato and *Arabidopsis* in the presequences shows that this region is able to be aligned. The partial sequence of the *Arabidopsis* gene in the GenBank data base contains the presequence and small portion of mature protein, as shown in the solid line (the dashed lines represent the uncharacterized remainder of the gene).

Phylogenetic Analysis. The phylogenetic tree of the cytochrome c1 presequence (mitochondrial targeting domain) of potato and the corresponding regions of GAPDH proteins in various organisms was inferred by the neighbor-joining method (18) using a computer program in the phylogenetic inference package (PHYLIP version 3.5) (19). The replicate number in bootstrap analysis is 100. All the sequences were drawn from GenBank version 90.

Gō Plot Analysis. To assess the compactness of the shuffled region in GapC as an independent functional unit, we analyzed the three-dimensional structure of human GAPDH (20) by using Gō plots (21). The Gō plot is a distance plot, in which the distances between the alpha-carbon elements of each pair of amino acids in the protein is plotted as a black dot for distances equal to or longer than a specified length. A compact module is a triangle along the diagonal that does not contain the black region in the plot.

RESULTS

In a computer survey of an exon data base (7), we observed a high similarity (44% identity and 64% similarity over 41 amino acids) between the 5′ three consecutive exons of the pea Gapc1 and the potato cytochrome c1 precursor (22, 23). Fig. 1 shows an alignment of these two gene products by using the GAP program. Only the N-terminal three exons align significantly and the positions and phases of the first three introns in this cytosolic GAPDH gene, 3, 11, and 44 (1, 1, and 0), resemble those of the cytochrome c1 gene: 2, 13, and 41 (1, 1, and 0). The phase of an intron, evolutionarily very conserved, is defined as the position of intron within a codon (the intron phase 0, 1, and 2 indicate the introns lying between two codons, after the first base, and after the second base, respectively) (24).

GAPDH is a highly conserved gene. An ancient triplication produced the GapA, GapB, and GapC lineages (25). In eukaryotes, GapA and GapB proteins are involved in the photosynthetic Calvin cycle in chloroplasts, and GapC proteins function in glycolysis in the cytosol. The first 44 amino acids are conserved across the eubacterial, animal, and plant genes. In this region, the phase 0 intron at position 44 is conserved across plants, animals, and fungi, as noted previously (22), while the other two introns have more specific distributions. Fig. 2*a* shows the distribution of these introns across various organisms.

Cytochrome c1 is a protein component involved at the redox center of the respiratory systems of eukaryotic and prokaryotic organisms, where its role is to pass electrons to cytochrome *c*. The cytochrome c1 gene in eukaryotes is located in the nucleus (26). Because the protein must move to the electron transfer chain in mitochondria, almost all eukaryotic genes except protist ones (26) have complex presequences, as in the case of plant *coxII* (13). These presequences include a mitochondrial targeting domain and an intramitochondrial sorting domain that directs the protein into the mitochondrial intermembrane space (27). The presequences differ among fungi, plants, and animals in size and sequence but maintain similar functions (Fig. 2*b*).

A BLAST search for matches to the presequence (amino acids 1–41) of potato cytochrome c1, the mitochondrial targeting domain, found 79% of the 280 GAPDH sequences pulled out by the full pea GapC sequence (scores range from 58 to 1924) from the data bases. A phylogenetic analysis (Fig. 3) shows that the cytochrome c1 presequence branches with several plant cytoplasmic GAPDHs. Furthermore, only the plant GapC sequences have a corresponding intron pattern (Fig. 2*a*). These features suggest that the exon shuffling event occurred in the plant lineage.

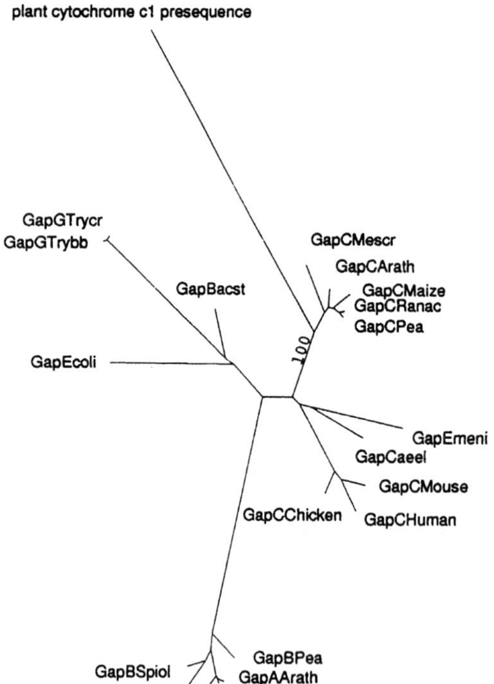

FIG. 3. Phylogenetic tree of the cytochrome c1 presequence (mitochondrial targeting domain) of potato and the corresponding regions of GAPDH proteins in various organisms inferred by the neighbor-joining method. The number in the plant branch is a bootstrap value in 100 replicates. [Abbreviations used are as follows: GapG, glycosome GAPDH; Arath, *Arabidopsis*; Bacst, *Bacillus stearothermophilus*; Caeel, *Caenorhabditis elegans*; Emeni, *Emerricella nidulans*; Mescr, *Mesembryanthemum crystallinum* (common ice plant); Ranac, *Ranunculus acer* (common buttercup); Spiol, *Spinacia oleracea* (spinach); Trycr, *Trypanosoma cruzi*; Trybb, *Trypanosoma brucei*.]

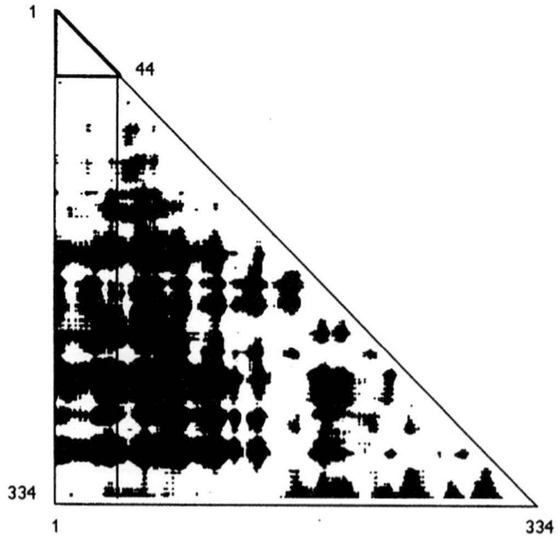

FIG. 4. Gō plot of human GAPDH protein. The exons from amino acid 1 to 44 form a compact region (the highlighted triangle) in which the distances between all pairs of amino acids are smaller than 28 Å (the distances longer than 28 Å are marked by black dots). The numbers indicate the positions of the first (residue 44) and last residues.

```
                           + +   -   +  +   +    + - +   + -
Potato cytochrome c1:    6 KIRIGFDGFGRINRFITRGAAQRNDSKLPSRND 38
                           |:::| |||||| | :|| |       : : ||
Human GAPDH:             3 KVKVGVDGFGRIGRLVTRAAFNSGKVDIVAIND 35
                           + +   -   +  +   +      + -     -

                           ----------^***********^^---------
```

FIG. 5. The charge distribution in the mitochondrial targeting domain of potato cytochrome c1 and the corresponding region in human GAPDH sequence. The pattern of secondary structure shown is derived from the human protein (. . . —. . . , extended conformation; ∧, turn; . . . ***. . . , α-helix). The potato cytochrome c1 sequence also contains a hypothetical amphiphilic α-helix (amino acid residues 7–24).

Which gene is donor in the shuffling? The amino terminal sequence of the plant GapC is ancient, shared with prokaryotic counterparts. However, for cytochrome c1, the shuffled region is present only in some plants (Fig. 2*b*). Thus cytochrome c1 was the recipient and GapC the donor in the shuffling event. Since the presequence of *Arabidopsis* cytochrome c1 differs from that of potato in the mitochondria targeting region (data from H. Hofte, 1993, GenBank accession no. Z25972), the shuffling event from GapC probably took place after the divergence between *Solanaceae* and *Brassicaceae* (Fig. 2*b*).

The donated region is an independent compact module in GapC in the sense of Mitiko Gō (circumscribable by a 28 Å sphere) (Fig. 4) and has an (extended conformation)-turn-(alpha helix)-turn-(extended conformation) structure (Fig. 5). The α-helix is amphiphilic: positive charged amino acids face the surface while the other side is buried against the hydrophobic core. Such a positively charged amphiphilic helix is crucial for the targeting activity in the organelle-specific proteins (28, 29). Finally, most of the charged amino acids are evolutionarily conserved (Fig. 5), which implies some functional similarity of this sequence between GapC and plant cytochrome c1.

DISCUSSION

We have shown that the plant GapC contributed its three consecutive amino-terminal exons to the cytochrome c1 precursor to generate a mitochondrial targeting function. This is a clear example of exon shuffling in plants, showing that exon shuffling is not limited to the vertebrate lineages. This observation provides support for the conclusion, based on a statistical analysis of intron phase correlations, that exon shuffling played an important role throughout the evolution of all genomes (5, 7). One reason that examples of exon shuffling were rare in non-vertebrate genes may be the biased data set containing more vertebrate sequences. As the data bases for nonvertebrate genes increase, we predict that more cases of exon shuffling in these organisms will appear.

The very similar intron–exon structure of the shuffled region of the two genes, the same number of introns with close or identical positions and phases, suggests that this region was a result of shuffling at the DNA sequence level rather than a consequence of retroposition.

This finding also shows a clear role for exon shuffling in the origin of presequences or transit peptides in the nuclearly encoded organellar proteins, as speculated previously (13, 14). Cytochrome c1 is a nuclearly encoded enzyme in eukaryotes. Although the presequences of human and plant cytochrome c1 are unrelated, the intron patterns suggest a common ancestral nuclear gene for the mature protein. Of five introns in this region of the two genes, one is identical in position (position 1), a second pair lies within two amino acids (position 2), and a third pair lies within five amino acids (position 3) (Fig. 2*b*). The intron patterns suggest the possibilities of a common ancestral nuclear gene or a common ancestral transfer from the mitochondrion to the nucleus. The ancestral cytochrome c1 gene in plants must have been targeted to the mitochondrion; thus this targeting sequence was replaced in the line leading to the potato by the GapC gene. This replacement may have been selected by some advantage in using the GapC promoter.

Does the donor shuffled sequence from GapC have an organelle targeting function? Although the shuffled sequence exists in all GAPDH genes, including GapA and GapB that are chloroplast-specific, both GapA and GapB have additional N-terminal elements, commonly believed to be responsible for the transit of the proteins to the organelle. Furthermore, GapC does not require organelle targeting because it functions in the cytosol. If some GapC were to occur in the mitochondrion, the "presequence" in the donor gene would have served as an organelle targeter. However, the existence of GapC proteins in an organelle has not been reported to date although GapC genes are probably of mitochondrial origin (25, 30), and the general belief is that all GapC is cytoplasmic. Thus we suggest that the donor shuffled sequence is important but not sufficient for organelle targeting activity. The fast evolution of the presequence after shuffling, with a substitution rate more than 10 times that in the GAPDH genes (Fig. 3), suggests that rapid adaptive evolution to specialize the sequence for efficient targeting took place after the shuffling event, as occurs after the origin of new genes (31, 32).

We thank R. C. Lewontin, W. Martin, R. Cerff, J. D. Palmer, W. F. Doolittle, D. Weinreich, T. Sang, and D.-M. Zhang for valuable discussions. We thank the National Institutes of Health for their support (Grant GM37997 to W.G.). S.J.d.S. is supported by Conselho Nacional de Desenvolvimento Cientifico e Tecnologico and the PEW–Latin American Fellows Program.

1. Gilbert, W. (1978) *Nature (London)* **271,** 501.
2. Gilbert, W. (1987) *Cold Spring Harbor Symp. Quant. Biol.* **52,** 901–905.
3. Sudhof, T. C., Russel, D. W., Goldstein, J. L., Brown, M. S. Sanchez-Pescador, R. & Bell, G. I. (1985) *Science* **228,** 893–895
4. Patthy, L. (1985) *Cell* **41,** 657–663.
5. Long, M., de Souza S. J. & Gilbert, W. (1995) *Curr. Opin. Genet. Dev.* **6,** 774–778.
6. Green, P., Lipman, D., Hillier, L., Waterson, R., States, D. & Claverie, J.-M. (1993) *Science* **2259,** 1711–1716.
7. Long, M., Rosenberg, C. & Gilbert W. (1995) *Proc. Natl. Acad. Sci. USA* **92,** 12495–12499.
8. Domon, C. & Steinmetz, A. (1994) *Mol. Gen. Genet.* **244,** 312–317.
9. Palmer, J. D. & Logsdon, J. M., Jr. (1991) *Curr. Opin. Genet. Dev.* **1,** 470–477.
10. Patthy, L. (1991) *BioEssays* **13,** 187–192.
11. Stoltzfus, A., Spencer, D. F., Zuker, M., Logsdon, J. M. & Doolittle, W. F. (1994) *Science* **265,** 202–207.
12. Palmer, J. (1985) *Annu. Rev. Genet.* **19,** 325–354.
13. Nugent, J. M. & Palmer, J. D. (1991) *Cell* **66,** 374–381.
14. Gantt, J. S., Baldauf, S. L.,Calie, P. J., Weeden, N. F. & Palmer, J. D. (1991) *EMBO J.* **10,** 3073–3078.
15. Altschul, S. F., Warren, G., Miller, W., Myers, E. W. & Lipman, D. J. (1990) *J. Mol. Biol.* **215,** 403–410.
16. Doolittle, R. F. (1990) *Molecular Evolution: Computer Analysis of Protein and Nucleic Acid Sequences* (Academic, New York).
17. Devereus, J., Haeberli, P. & Smithies, O. (1984) *Nucleic Acids Res.* **12,** 387–391.
18. Saitou, N. & Nei, M. (1987) *Mol. Biol. Evol.* **4,** 406–425.
19. Felsenstein, J. (1992) PHYLIP Manual (Univ. of California, Berkeley), Version 3.5.
20. Watson, H. C. & Campbell, J. C. (1988) *Brookhaven Protein Database* (Brookhaven National Laboratory, Upton, NY), No. 3GPD.
21. Gō, M. (1981) *Nature (London)* **291,** 90–93.

22. Kersanach, R., Brinkmann, H., Liaud, M. F., Zhang, D.-X., Martin, W. & Cerff, R. (1994) *Nature (London)* **367,** 387–389.
23. Wegener, S. & Schmitz, U. K. (1993) *Curr. Genet.* **24,** 256–259.
24. Sharp, P. A. (1981) *Cell* **23,** 643–646.
25. Cerff, R. (1995) in *Tracing Biological Evolution in Protein and Gene Structures*, eds. Go, P. & Schimmel, P. (Elsevier, Amsterdam), pp. 205–227.
26. Priest, J. W., Zachary, A. W. & Hajduk, S. L. (1993) *Biochim. Biophys. Acta* **1144,** 229–231.
27. van Loon, A. P. G. M., Brandli, A. W., Pesoid-Hurt, B., Blank, D. & Schatz, G. (1987) *EMBO J.* **6,** 2433–2439.
28. Roise, D., Horvath, S. J., Tomich, J. M., Richards, J. H. & Schatz, G. (1986) *EMBO J.* **5,** 1327–1334.
29. Schatz, G. & Dobberstein, B. (1996) *Science* **271,** 1519–1526.
30. Martin, W., Brinkmann, H., Savona, C. & Cerff, R. (1993) *Proc. Natl. Acad. Sci. USA* **90,** 8692–8696.
31. Long, M. & Langley, C. H. (1993) *Science* **260,** 91–93.
32. Ohta, T. (1994) *Genetics* **138,** 1331–1337.

Proc. Natl. Acad. Sci. USA
Vol. 93, pp. 14632–14636, December 1996
Evolution

Intron positions correlate with module boundaries in ancient proteins

(intron evolution/introns-early)

SANDRO JOSE DE SOUZA*, MANYUAN LONG, LLOYD SCHOENBACH, SCOTT WILLIAM ROY, AND WALTER GILBERT

Department of Molecular and Cellular Biology, Biological Laboratories, Harvard University, 16 Divinity Avenue, Cambridge, MA 02138

Contributed by Walter Gilbert, October 11, 1996

ABSTRACT **We analyze the three-dimensional structure of proteins by a computer program that finds regions of sequence that contain module boundaries, defining a module as a segment of polypeptide chain bounded in space by a specific given distance. The program defines a set of "linker regions" that have the property that if an intron were to be placed into each linker region, the protein would be dissected into a set of modules all less than the specified diameter. We test a set of 32 proteins, all of ancient origin, and a corresponding set of 570 intron positions, to ask if there is a statistically significant excess of intron positions within the linker regions. For 28-Å modules, a standard size used historically, we find such an excess, with $P < 0.003$. This correlation is neither due to a compositional or sequence bias in the linker regions nor to a surface bias in intron positions. Furthermore, a subset of 20 introns, which can be putatively identified as old, lies even more explicitly within the linker regions, with $P < 0.0003$. Thus, there is a strong correlation between intron positions and three-dimensional structural elements of ancient proteins as expected by the introns-early approach. We then study a range of module diameters and show that, as the diameter varies, significant peaks of correlation appear for module diameters centered at 21.7, 27.6, and 32.9 Å. These preferred module diameters roughly correspond to predicted exon sizes of 15, 22, and 30 residues. Thus, there are significant correlations between introns, modules, and a quantized pattern of the lengths of polypeptide chains, which is the prediction of the "Exon Theory of Genes."**

Do introns delineate elements of protein tertiary structure? This issue is crucial to the debate about the role and origin of introns (1–8): did introns appear at the beginning of evolution, creating the first genes by exon shuffling, or did they arise during evolution by the insertion of adventitious elements into genes? The "introns-early" view predicts that the exons should represent functional or folding elements of protein structure (1–4), whereas the "introns-late" view (5–8) expects that the insertion of introns might respect DNA sequence but should be uncorrelated with protein structure.

The Exon Theory of Genes (1), an expansion of the introns-early approach, hypothesizes that the first protein coding genomes had an intron–exon structure in which the introns served as hotspots of recombination to shuffle exons to create the first genes. The products of the original coding elements, the first exons, were short polypeptides 15–20 amino acids long that served as elements of folding or function. This theory holds that over time small exons were fused together by reverse transcriptase-mediated retroposition to make more complicated exons to be shuffled in turn. [A complete example of the creation of complex exons by retrotransposition has been worked out for the gene *Jingwei* in *Drosophila* (9)]. Two or three fusions on average would be needed to lead to today's exon distribution peaked at 35–40 residues (10).

The Exon Theory of Genes holds that the basic processes of gene evolution were exon shuffling, the sliding and drift of introns at exon boundaries, and the creation of complicated exons by the loss of introns. Intron loss is hypothesized to be very easy and to occur down all lines that specialize for rapid replication, such as the bacteria or *Saccharomyces cerevisiae* (11).

The critical prediction of the Exon Theory of Genes is that proteins will turn out to be assembled from small folding elements, modules in the sense of Mitiko Gō (regions of the polypeptide chain that are compact in space), which will be related to the products of exons. Although a variety of arguments for early introns have been advanced, some of which include the use of the module hypothesis to predict the existence of certain introns (12–14) while others involve the coincidence of intron positions in genes separated by great evolutionary distances (13), arguments for the introns-late view have recently appeared.

Stoltzfus and collaborators (6) attacked the general notion that exons were related to elements of protein structure by showing that introns were not correlated with the ends of protein secondary structure elements (α-helices and β-sheets) and challenged all efforts to show a connection between exons and modules. Palmer and coworkers (5) argued that introns arose late in evolution, based on the broad phylogenetic distribution of introns (lacking in bacteria and many protists present in higher eukaryotes), as well as on the specific distribution of novel introns in triosephosphate isomerase (8). These defenders of the introns-late view challenge all notions of intron sliding or drift (15) and assert that introns very close in position in homologous genes represent separate acts of addition. That novel introns can arise in general is supported by the finding of introns in the U6 RNA in fungal species (16–18), which clearly have arisen recently by reverse splicing followed by reverse transcription and gene conversion. However, some introns arising late does not prove that all introns were late. The problem is to detect whether or not some introns arose early. This paper introduces a statistical test to demonstrate that introns correlate with module boundaries in ancient proteins and shows that this correlation is neither due to a composition or sequence bias in the module boundaries nor to a surface bias in intron position. We argue that this correlation strongly suggests that there was exon shuffling in the progenote.

MATERIALS AND METHODS

Sample. The data (Table 1) consists of 32 ancient conserved proteins, which have homologs without introns in prokaryotes and with introns in eukaryotes, and their intron positions. Intron positions were defined by aligning the homologous sequences to the Protein Data Bank (PDB) reference sequence with CLUSTAL V and counting each position and phase

The publication costs of this article were defrayed in part by page charge payment. This article must therefore be hereby marked "*advertisement*" in accordance with 18 U.S.C. §1734 solely to indicate this fact.

Abbreviation: PDB, Protein Data Bank.
*To whom reprint requests should be addressed.

Table 1. The sample of 32 proteins

Protein	Abbreviation	PDB
Acid amylase	ACIDAMY	2AAA
Acyl-CoA dehydrogenase	ACYL	3MDD
Aldolase	ALDOL	1ALD
Aldose reductase	ALREDUC	1DLA
Alcohol dehydrogenase	ADH	1ADB
Alkaline phosphatase	ALK	1AJA
Amylase	AMY	1PPI
Aspartate aminotransferase	AAT	1AMA
Catalase	CAT	8CAT
Citrate synthase	CSYN	1CTS
Cu-superoxide dismutase	CUSOD	1SDY
Cytochrome *c*	CYT	1CCR
Dihydrofolate reductase	DHFR	1DHF
Elongation factor TU	EFTU	1EFT
Enolase	ENOL	1EBG
Glucose-6-phosphate dehydrogenase	G6PD	1DPG
Gluthatione *S*-transferase	GST	1GSS
Glyceraldehyde 3-phosphate dehydrogenase	GAPDH	3GPD
Glycogen phosphorylase	GLYPHOS	1GPA
Heat-shock protein 70	HSP70	1ATR
Hemoglobin	HEMO	2DHB
High pI amylase	HIAMY	1AMY
Lactate dehydrogenase	LDH	2LDX
Lysozyme	LYS	1LAA
Malate dehydrogenase	MDH	4MDH
Mn-superoxide dismutase	MNSOD	1MSD
Phosphoglycerate kinase	PGK	3PGK
Phosphofructokinase	PFK	3PFK
Phosphoglycerate mutase	PGM	3PGM
Pyruvate kinase	PK	From author
Triosephosphate isomerase	TPI	1TIM
Xylanase	XYLA	1CLX

The last column lists the PDB entry that yielded the coordinates. Where the PDB files were missing a few coordinates the α-carbon positions were filled-in by linear interpolation. Pyruvate kinase coordinates were supplied by H. Muirhead (University of Bristol, United Kingdom).

separately to yield 570 instances. These proteins represent all the full-length ancient proteins with known coordinates that appear in an intron data base of ancient proteins. The PDB files occasionally have missing residues, for which coordinates were not determined. For such missing residues, we supplied dummy coordinates for the α-carbon positions by linear interpolation. The intron data base, based on GenBank 90, is an updated version of a similar data base based on GenBank 84 and previously described (10).

Algorithm. INTER-MODULE is written in ANSI C and compiled with a Sun C compiler in SunOS 4.1 (on a Sun SPARCstation 10). The source code will be available in our web site (http://golgi.harvard.edu/gilbert.html).

RESULTS

Modules and Linker Regions. We define a module as a continuous region of polypeptide chain all of whose α-carbons lie less than a defined distance apart, the module "diameter." Such a region lies inside a geometric volume of constant diameter called a "Reuleaux Form." For a given diameter d, the Reuleaux Form of largest volume is a sphere of diameter d. In general, a Reuleaux Form in three dimensions can be circumscribed by a sphere of diameter $\leq \sqrt{(3/2)}d$. If the polypeptide chain fills the Reuleaux Form, the module would represent a compact element along the chain. On a triangular Gō plot (12) of the distances between each pair of α-carbons in a three-dimensional structure, if all distances greater than the defining size are shaded black, then any right triangle drawn along the diagonal that does not contain black regions will define a module. The longest-chain modules correspond to those triangles whose size is limited by touching black regions on both sides. Fig. 1 shows such a Gō plot: the five large triangles define the set of longest chain modules at 28 Å.

Although this definition enabled one to hypothesize the relationship between exons and modules and to predict the existence of certain introns (12–14), it does not provide an obvious way to predict specific boundaries for each module.

The problem is essentially that the longest chains at 28 Å (Fig. 1) overlap, and so, if one is to draw smaller triangles for non-overlapping modules, the Gō plot offers no guidance as to where to mark the boundaries. We turn this problem around by defining the overlaps between the longest modules as "linker regions" (Fig. 1). If an intron were to be placed in each linker region, the protein would be dissected into a set of modules each less than 28 Å in diameter.

This notion of linker regions immediately defines a simple statistical test, a χ^2 test, for the correlation of intron positions and module boundaries. If the introns were correlated with modules, one expects an excess of intron positions to fall within linker regions. If the introns have been added to previously existing DNA sequences, one expects the intron positions to be arranged randomly, and there should be no significant excess in the linker regions.

To define the linker regions objectively, we have written a computer program, INTER-MODULE, which first takes a Brookhaven Protein Databank file of coordinates, constructs a Gō plot of distances between α-carbons, and then, for a specified distance criterion, defines the set of linker regions. Where the triangles overlap nicely, as in Fig. 1, the definition of linker regions is straightforward. In general, the program begins with the longest chain N-terminal module and then, at that module's C-terminal residue, constructs the largest module (right triangle) possible by first extending its C-terminal boundary until that line touches a black area and then increasing its size until its N-terminal boundary touches a black area. The program then repeats for the next module. The overlaps of these modules define the linker regions.

Tests at a 28-Å Module Size. Is there a statistically significant excess of intron positions within the linker regions? We first examine 28-Å modules, since modules of this size were defined by Mitiko Gō for her analysis of hemoglobin and her prediction of a novel intron (12) and were used again for the prediction of introns in triosephosphate isomerase (13, 14). Table 2 shows

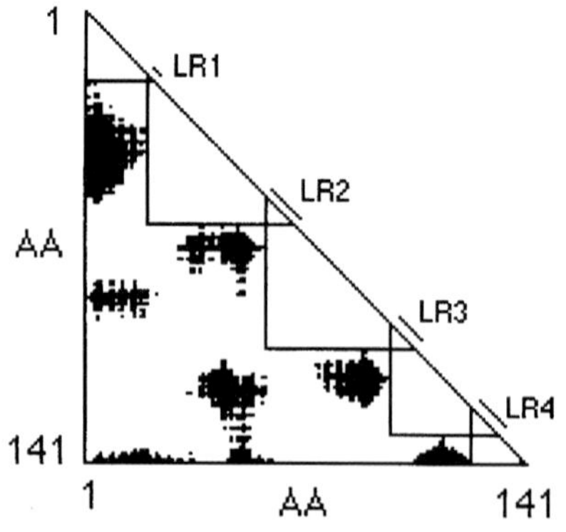

FIG. 1. Gō plot for hemoglobin. The black regions represent pairs of α-carbons that are separated by 28 Å or more in horse hemoglobin (2DHB). Five modules are identified by large triangles along the diagonal, whose size is limited by touching the black regions. The linker regions (LR) are defined as the region of overlap of those triangles.

that 216 of the 570 introns lie in the 28-Å linker regions, almost 34 more than expected on a random basis. This distribution has $\chi^2 = 9.0$; $P < 0.003$. We can reject the null hypothesis that introns are located randomly in genes; rather they show a preference for the boundaries of 28-Å modules in the protein products.

If the excess of introns in the linker regions is due to a signal from ancient introns, seen above a background of added or moved introns, then in a subset of clearly "old" introns that excess should be greater. Table 3 lists 20 "old" introns: introns conserved in position between at least three of the groups of vertebrates, invertebrates, plants, and fungi. (This identification of conserved introns accepts a small amount of sliding, up to four codons). Fourteen of the 20 introns are located within the linker regions, rather than the 6.4 expected; $P < 0.0003$. This sharply higher significance argues that the intron/module correlation is a consequence of the age of the intron.

Tests of Insertional Models. Could this excess of intron positions within linker regions be explained by some biased addition model? Such a model might hypothesize that there is a bias in DNA composition or sequence in the linker regions, possibly caused by an amino acid bias, that would serve to target an excess of intron additions to these regions. One such model would follow from the hypothesis of Craik *et al.* (19) that

Table 2. Intron positions in linker regions

Protein	Fraction	No. introns	E	O	O-E	*P*
AAT	0.24	30	7.3	10	+2.7	0.27
ACIDAMY	0.31	9	2.8	3	+0.2	0.99
ACYL	0.43	15	6.5	5	−1.5	0.43
ADH	0.22	38	8.5	9	+0.5	0.99
ALDOL	0.31	17	5.2	8	+2.8	0.14
ALK	0.30	10	2.9	1	−1.9	0.18
ALREDUC	0.30	16	4.8	7	+2.2	0.22
AMY	0.34	17	5.7	6	+0.3	0.99
CAT	0.29	20	5.9	7	+1.1	0.65
CSYN	0.26	4	1.0	1	0.0	1.0
CUSOD	0.35	23	8.1	4	−4.1	0.07
CYT	0.31	7	2.2	4	+1.8	0.14
DHFR	0.42	13	5.5	10	+4.5	0.01
EFTU	0.42	10	4.2	5	+0.8	0.68
ENOL	0.35	28	9.8	10	+0.2	0.99
G6PD	0.28	19	5.4	6	+0.6	0.75
GAPDH	0.40	46	18.6	24	+5.4	0.1
GLYPHOS	0.28	20	5.6	8	+2.4	0.22
GST	0.25	28	7.0	6	−1.0	0.70
HEMO	0.23	15	3.4	8	+4.6	0.005
HIAMY	0.36	4	1.4	1	−0.4	0.68
HSP70	0.36	31	11.1	16	+4.9	0.07
LDH	0.27	11	3.0	3	0.0	1.0
LYS	0.32	4	1.3	1	−0.3	0.99
MDH	0.22	23	5.1	3	−2.1	0.28
MNSOD	0.26	12	3.1	3	−0.1	0.99
PFK	0.26	26	7.5	14	+6.5	0.006
PGK	0.41	20	8.2	9	+0.8	0.75
PGM	0.35	5	1.8	2	+0.2	0.99
PK	0.44	16	7.0	6	−1.0	0.65
TPI	0.36	21	7.6	9	+1.4	0.52
XYLA	0.35	12	4.2	7	+2.8	0.09
Total		570	182.5	216		

A listing for each protein of the fraction of the sequence that lies in the linker regions, the total number of intron positions, the expected (E) and observed (O) number of intron positions within the linker regions, and the excess of observed over expected (O–E). The χ^2 value for the overall sum of E and O values, using a two-way test for excess inside and depletion outside, appears at the bottom: $\chi^2 = 9.0$, $P < 0.003$.

Table 3. Old introns

Protein	Intron position	Status
TPI	38	In
TPI	79	Out
TPI	108	Out
TPI	152	In
TPI	181/4	In
TPI	210	In
GST	148/51	In
HEMO	30	Out
HEMO	100	In
ALDOL	266/9	In
PFK	264/6	In
GAPDH	9	In
GAPDH	43	In
GAPDH	75/8/9	In
GAPDH	146/7	Out
PGK	22	In
PGK	91/2/3	In
CUSOD	23/4	Out
HSP70	69	In
ENOL	60/4	Out

A listing of 20 "ancient" introns identified as having matching positions in three out of four groups of vertebrates, invertebrates, plants, or fungi. Introns were considered homologous if they had slid up to four codons. The status column defines their character with respect to the linker regions for 28-Å modules (using the average position for slid introns). Fourteen of the 20 intron positions are within the linker regions rather than the 6.4 expected: $\chi^2 = 13.3$, $P < 0.0003$.

intron positions lie on the surface of proteins. If it were true that introns entered more frequently into codons for surface residues and if module boundaries were to lie on the surface of proteins, then intron positions and module boundaries would be correlated, but not in a causal fashion.

We find no support for such insertional models. The linker regions for the set of 32 proteins show no significant variation from the global average in amino acid or DNA composition. The frequency of hypothesized "proto-splice sites," such as AGGT or AGG (20), show no preference for linker regions (0.42% in linker regions vs. 0.44% in general for AGGT or 1.60% in linker regions vs. 1.89% for AGG). Furthermore, neither the linker regions nor the intron positions are unusually located on the protein surface. Using the program NACCESS (21) to calculate the relative accessibility [the percent accessibility of each residue in the protein compared with its solvent accessibility in an Ala-X-Ala tripeptide (22)], we find that the relative accessibility of the average residue is 26 ± 26%, of the linker region 20 ± 19%, and of the introns, 25 ± 24% (±SD). Fig. 2 shows the detailed distribution of relative accessibility values for residues of these three classes. The hypothesis of Craik *et al.* (19) that intron insertions are restricted to the surface of proteins is not supported by these data.

General Test at All Module Sizes. Since INTER-MODULE will predict linker regions for any module diameter, one is not restricted to a 28-Å criterion. Fig. 3*A* shows the excess of intron positions inside the linker regions for the range of module diameters from 6 to 50 Å. Fig. 3*B* plots the corresponding χ^2 values. There are three major peaks in both excess introns and statistical significance at module diameters centered at 21.7, 27.6, and 32.9 Å, all with *P* values around 0.001. The peak near 28 Å corresponds to the traditional module size.

How are these significant module diameters to be understood? In predicting linker regions, INTER-MODULE is effectively predicting a set of exons for each three-dimensional structure. We calculated the average internal exon length (and standard deviation), assuming "exons" to be defined as lying

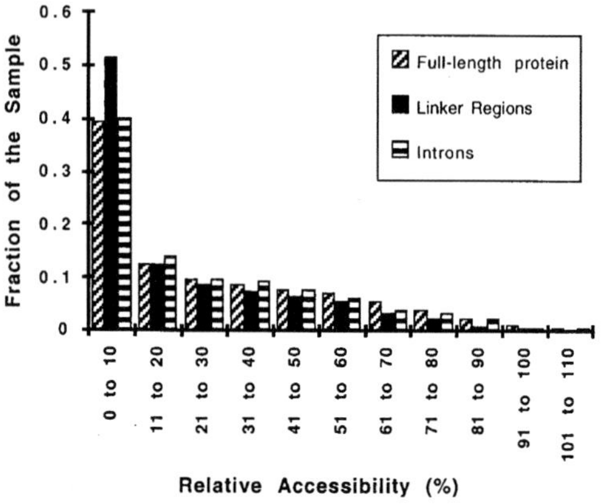

FIG. 2. Distribution of relative accessibility values for introns. A histogram of the relative accessibility as defined by NACCESS (ratio of solvent accessibility in the structure to that in the tripeptide Ala-X-Ala) for residues in general in the 32 proteins, residues in the linker regions, and residues that contain introns or that flank phase 0 introns.

between the midpoints of the linker regions, for each of these peaks. Fig. 4 shows that the three peaks correspond to exon sets that are roughly 15, 22, and 30 amino acid residues in length (15 ± 5, 22 ± 9, and 30 ± 14).

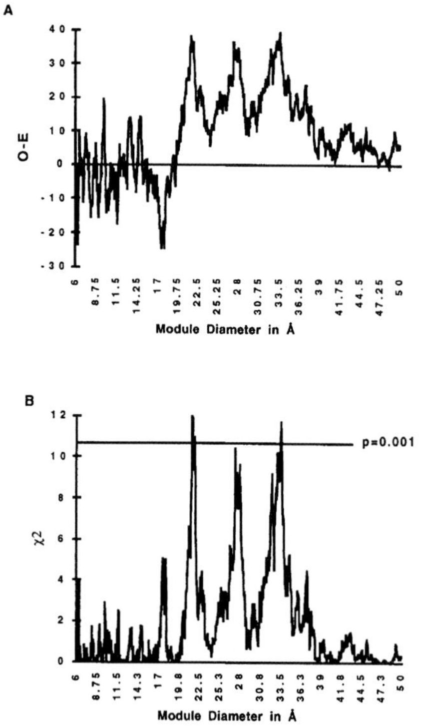

FIG. 3. Excess intron positions in linker regions as a function of module diameter. (*A*) The excess (Observed-Expected, O-E) values for each module diameter ranging from 6 to 50 Å in intervals of 0.05 Å. (*B*) The χ^2 values for the excess. There are peaks of significance. The peaks centered at 21.7, 27.6, and 32.9 Å have *P* values around 0.001. The peak values are at 21.4, 27.4, and 33.5 Å, with χ^2 = 12.0, 10.4, and 11.7, respectively.

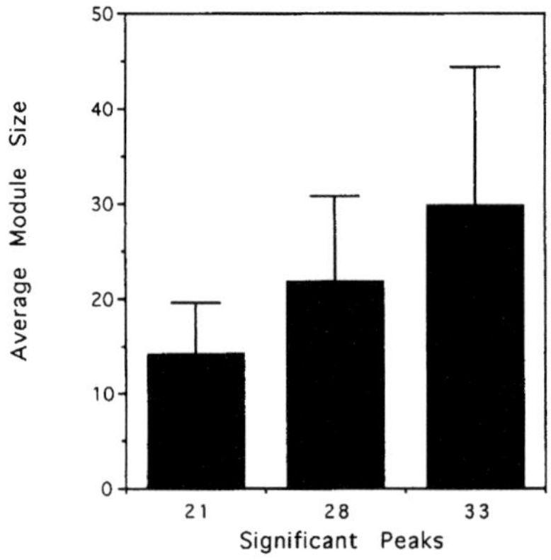

FIG. 4. Lengths of predicted exons for each of the peaks of significance. INTER-MODULE predicts the length of internal exons (defined between the mid-points of the linker regions) for the module diameters corresponding to the peaks of significance centered at 21.7, 27.6, and 32.9 Å. The plot shows the average and standard deviation of the internal exons. The three peaks correspond to distributions centered around 15, 22, and 30 amino acids.

DISCUSSION

This finding of a correlation of intron positions with the boundaries of modules, corresponding to exon sizes of 15, 22, and 30 amino acid residues, fulfills the prediction of the Exon Theory of Genes and suggests that we are seeing a residual signal of the ur-exons that combined to make the present exons.

Is this statistical signal what one would expect if these ancient protein genes had been assembled by exon shuffling in the progenote? Any signal of correlation would have been weakened over time by very easy intron loss (2, 3), by intron sliding or slipping (15), or by intron addition (16–18). Furthermore, any agreement between the three-dimensional structure of the original exons and that of their descendants would be weakened through changes in the protein shape arising by mutation and selection after assembly. Thus, one expects only a small fraction of the current introns and module boundaries to match. Nonetheless, the statistical signal itself is very strong.

Are there alternative explanations for this correlation of introns with three-dimensional structure other than the one of original introns? The introns-late school might argue that introns insert into specific nucleic acid sequences and these sequences, pre-intron targets (20), might be tied to specific amino acid patterns, and those patterns again tied, in some to-be-defined way, to the three-dimensional structure. One variant of this idea is the argument that introns "add" to sequences of biased composition, which might be associated with amino acids that lie on the surface of proteins.

We have shown that the linker regions are not biased in amino acid composition, in DNA composition, or in proto-splice sequences. Furthermore, the linker regions do not lie on the surface of the proteins, nor do the intron positions in our set show a surface bias.

Still another type of intron-insertion argument suggests that a bias might have arisen through natural selection. One assumes that introns entered genes randomly, but that only those organisms were selected within which the introns had inserted in such a way as to permit a useful module to be shuffled out of an ancient protein and used elsewhere as a target of natural selection. This model, however, does not work. The issue is one of the fixation of mutations. The selection that fixes the exon-based module in some novel protein does not fix the appropriate donor form of the gene in the population. If one argues that it is selection that has fixed the exon in the ancient gene itself, that statement corresponds to exon shuffling in the progenote, which is the introns-early conclusion.

The strongest argument against a biased insertion model is that the subset of putatively old introns shows enhanced localization, since on an insertion model for introns any subset should not behave differently. However, in defining ancient introns, we have accepted some use of sliding. We emphasize that our full intron correlation studies do not assume intron sliding and treat each intron position as a separate object.

How strong is the statistical argument? The argument for a correlation with 28-Å modules is straightforward. The module size was chosen in advance, by previous work, and the χ^2 value has a straightforward interpretation. However, when we vary the module size in the graph of Fig. 3*B*, we are doing almost a thousand calculations, and if these were only random fluctuations, one might still expect one of the points to vary out to $P = 0.001$. However, Fig. 3*A* shows that the excess of intron positions in the linker regions is robust, showing general peaks in that excess centered at 21.7, 27.6, and 32.9 Å in module diameter. The χ^2 plot shows that these peaks are broadly significant.

Why have other groups failed to find a correlation between exons and modules? Beyond the specific definition of modules, a further problem is that of sample size. Only 6% of the intron positions in our sample were involved in the excess at 28 Å. One needs a sufficiently large number of introns to see such an excess with high significance. Previously, Stoltzfus and coworkers (6) tested just four proteins and found indeed that exons in general coded for 28-Å modules, but that this correlation lacked significance. Logsdon and coworkers (8) tested only triosephosphate isomerase to find no correlation. However, Gō and Noguti (23), using the same four protein sample used by Stoltzfus *et al.* (6), claimed to reach statistical significance when they tested the position of introns in relation to the type of module boundaries defined by their analysis.

What is the significance of the "exon sizes"? We speculate that the sizes around 15 residues correspond to α-helices and α-helices with turns. Specific small peptides of these size ranges have such structures in solution (24–28). The longer sets would then be more complicated structures, involving turns to make the modules compact, and may represent the fusion of simpler elements. Preliminary analysis of the three-dimensional structures of 21-Å modules shows that many correspond to two secondary structure elements (such as helix/helix, helix/strand, strand/strand, and strand/helix) interconnected by a turn.

CONCLUSION

This paper demonstrates that intron positions are strongly correlated with the boundaries of modules around 22, 28, and 33 Å in diameter in the three-dimensional structure of current proteins. These sizes would correspond to a hypothetical exon pattern with exons about 15, 22, and 30 residues long, which supports the idea that short exons were used to assemble the ancient conserved proteins. A second argument that some introns are very old is the intron-phase correlation in ancient genes (10). The excess of phase symmetric exons, exon-pairs, and exon-triples in genes that came into existence in the progenote is also an argument for exon shuffling in the common ancestor. The Exon Theory of Genes, which holds that some introns are very old and were used to assemble genes in the common ancestor of all life, is now supported by two strong, independent statistical arguments that detect a signal of ancient introns over any possible background of loss and addition.

We are indebted to Helen Muirhead for providing coordinates of pyruvate kinase and Nancy Maizels and Bill Martin for valuable discussions. This work was supported by National Institutes of Health Grant GM 37997. S.J.d.S. was supported by Conselho Nacional de Desenvolvimento Cientifico e Tecnologico (CNPq–Brazil) and the PEW–Latin American Program.

1. Gilbert, W. (1987) *Cold Spring Harbor Symp. Quant. Biol.* **52,** 901–905.
2. Gilbert, W. & Glynias, M. (1993) *Gene* **135,** 137–143.
3. Long, M., de Souza, S. J. & Gilbert, W. (1995) *Curr. Opin. Genet. Dev.* **5,** 774–778.
4. de Souza, S. J., Long, M. & Gilbert, W. (1996) *Genes Cells* **1,** 493–505.
5. Palmer, J. D. & Logsdon, J. M., Jr. (1991) *Curr. Opin. Genet. Dev.* **1,** 470–477.
6. Stoltzfus, A., Spencer, D. F., Zuker, M., Logsdon, J. M., Jr., & Doolittle, W. F. (1994) *Science* **265,** 202–207.
7. Kwiatowski, J., Krawczyk, M., Kornacki, M., Bailey, K. & Ayala, F. J. (1995) *Proc. Natl. Acad. Sci. USA* **92,** 8503–8506.
8. Logsdon, J. M., Jr., Tyshenko, M. G., Dixon, C., Jafari, J. D., Walker, V. K. & Palmer, J. D. (1995) *Proc. Natl. Acad. Sci. USA* **92,** 8507–8511.
9. Long, M. & Langley, C. H. (1993) *Science* **260,** 91–95.
10. Long, M., Rosenberg, C. & Gilbert, W. (1995) *Proc. Natl. Acad. Sci. USA* **92,** 12495–12499.
11. Fink, G. R. (1987) *Cell* **49,** 5–6.
12. Gō, M. (1981) *Nature (London)* **291,** 90–93.
13. Gilbert, W., Marchionni, M. & McKnight, G. (1986) *Cell* **46,** 151–153.
14. Straus, D. & Gilbert, W. (1985) *Mol. Cell. Biol.* **5,** 3497–3506.
15. Cerff, R. (1995) in *Tracing Biological Evolution in Protein and Gene Structures*, eds. Gō, M. & Schimmel, P. (Elsevier, Amsterdam), pp. 205–228.
16. Tani, T. & Oshima, Y. (1991) *Genes Dev.* **5,** 1022–1031.
17. Tani, T. & Oshima, Y. (1989) *Nature (London)* **337,** 87–90.
18. Tani, T., Takahashi, Y., Urushiyama, S. & Oshima, Y. (1995) in *Tracing Biological Evolution in Protein and Gene Structures*, eds. Gō, M. & Schimmel, P. (Elsevier, Amsterdam), pp. 97–114.
19. Craik, C. S., Sprang, S., Fletterick, R. & Rutter, W. J. (1982) *Nature (London)* **299,** 180–182.
20. Dibb, N. J. & Newman, A. J. (1989) *EMBO J.* **8,** 2015–2021.
21. Hubbard, S. J. & Thornton, J. M. (1993) NACCESS Computer Program (Dept. of Biochem. and Mol. Biol., University College London).
22. Hubbard, S. J. & Thornton, J. M. (1991) *J. Mol. Biol.* **220,** 507–515.
23. Gō, M. & Noguti, T. (1995) in *Tracing Biological Evolution in Protein and Gene Structures*, eds. Gō, M. & Schimmel, P. (Elsevier, Amsterdam), pp. 229–236.
24. Bairaktari, E., Mierke, D. F., Mammi, S. & Peggion, E. (1990) *Biochemistry* **29,** 10097–10102.
25. Scanlon, M. J., Fairlie, D. P., Craik, D. J., Englebretsen, D. R. & West, M. L. (1995) *Biochemistry* **34,** 8242–8249.
26. Moroder, L., D'Ursi, A., Picone, D., Amodeo, P. & Temussi, P. A. (1993) *Biochem. Biophys. Res. Commun.* **190,** 741–746.
27. Mendz, G. L., Barden, J. A. & Martenson, R. E. (1995) *Eur. J. Biochem.* **231,** 659–666.
28. Maciejewski, M. W. & Zehfus, M. H. (1995) *Biochemistry* **34,** 5795–5800

Proc. Natl. Acad. Sci. USA
Vol. 94, pp. 7698–7703, July 1997
Colloquium Paper

This paper was presented at a colloquium entitled "Genetics and the Origin of Species," organized by Francisco J. Ayala (Co-chair) and Walter M. Fitch (Co-chair), held January 30–February 1, 1997, at the National Academy of Sciences Beckman Center in Irvine, CA.

Origin of Genes

(intron/exon/module/evolution)

WALTER GILBERT*, SANDRO J. DE SOUZA, AND MANYUAN LONG

Department of Molecular and Cellular Biology, Biological Laboratories, Harvard University, 16 Divinity Avenue, Cambridge, MA 02138

ABSTRACT **We discuss two tests of the hypothesis that the first genes were assembled from exons. The hypothesis of exon shuffling in the progenote predicts that intron phases will be correlated so that exons will be an integer number of codons and predicts that the exons will be correlated with compact regions of polypeptide chain. These predictions have been tested on ancient conserved proteins (proteins without introns in prokaryotes but with introns in eukaryotes) and hold with high statistical significance. We conclude that introns are correlated with compact features of proteins 15-, 22-, or 30-amino acid residues long, as was predicted by "The Exon Theory of Genes."**

The role of introns and exons in the history of genes has been the subject of debate between two extreme positions. One side holds that introns were used to assemble the first genes, an "introns-early" view (1, 2), and the other side maintains that introns were added during evolution to break up previously continuous genes, an "introns-late" view (3, 4). This discussion has a significant impact on our conceptions about the way genes were constructed in the first cells. Unfortunately, the two sides make opposing judgments about each piece of evidence, and no decisive evidence has yet been agreed upon.

For example, in the context of phylogenies, that bacteria have no introns whereas vertebrates have many introns is interpreted differently by the two sides. One view is that introns were there originally and were simply lost; the alternative view is that they were gained. In homologous genes, one often finds introns in similar but not identical positions between genes separated by great evolutionary distances. The early-intronists say that these positions represent the same original intron, possibly moved slightly in position (intron drift or sliding). The late-intronists say that it is obvious that introns could not have existed in such closely neighboring positions in a single original gene and that, because introns could not have moved, these near coincidences must be evidence of insertion.

There have been efforts to correlate introns with the three-dimensional structure of proteins. The introns-late view denies that there are any such correlations and asserts that introns behave as though they were inserted randomly into the structure of genes (5). Alternatively, the early-intron position generally affirms such a connection but, up to now, has not been able to muster any strong statistical evidence. Recently, however, we have defined such a correlation in a way that yields strong statistical support (6).

There are three possible scenarios for the evolutionary history of introns. One is that there were introns at the very beginning of evolution and that during evolution they were lost or, possibly, mostly lost and some added. This complex of ideas is "The Exon Theory of Genes" (2). The extreme alternative view is that introns were added very late in evolution, even in the last few million years, and thus have nothing to do with the rearrangement of pieces of genes. There is no exon shuffling on this picture. A third, intermediate view, popular in its own right, is that the introns arose at the initiation of multicellularity. In this picture, the Cambrian explosion used introns to create exon shuffling and a profusion of new genes. The idea of exon shuffling is that introns are as hot spots for genetic recombination, which is a property that introns would have solely because of their length. Introns affect the rate of homologous recombination between exons in a way that scales with length, but, more importantly, they affect nonhomologous recombination as the square of their length. Consider a new gene made by a new combination of regions of earlier genes by an unequal crossing-over event, a rare event at the DNA level, that matches small, similar sequences between two DNAs. To make a new protein that contains the first part of one protein with the second part of another requires such a rare, and in frame, event. However, if the regions that encode parts of the protein are separated by 1,000–10,000-base-long introns along the DNA, a process of unequal crossing-over occurring anywhere within that intron between the exons will create a new combination of exons. There is a combinatorial number of ways to find the matching of short sequences to initiate the unequal crossing over, and thus the recombination process will go a million to a hundred million times faster in the presence of an intron. This is a great enhancement of the rate of creation of new genes.

The Exon Theory of Genes (2) is a specific statement of the idea that the first genes were made of small pieces. The crucial elements of that theory are that the very first genes and exons represented small polypeptide chains ≈15–20 amino acids long, that the basic method used by evolution to make new genes was to shuffle the exons, and that a major trend of evolution was then to lose introns and to fuse small exons together to make complicated exons. (The first enzymes probably were aggregates of such short gene products, but these ur-exons were soon tied together by an intron/exon system so that the proteins would have a covalently connected backbone.)

The dominant evolutionary processes are thus to be recombination within introns, the sliding and drift of introns to change amino acid sequence around their borders, and the loss of introns, which can change the gene structure but does not affect protein structure. The strength of this concept is its argument that one searches sequence space not by amino acids and point mutations but by larger elements. We might compare a protein to a sentence. It is easier to understand the sentence as made up of words rather than simply as a string of letters.

*To whom reprint requests should be addressed.

How are introns lost? The most direct way is retroposition. A spliced RNA transcript of a gene with an intron/exon structure is copied back into cDNA by a reverse transcriptase, and that DNA is inserted into the chromosome within an intron of a previously existing gene. Splicing can now make that element serve as a complex exon in a new gene. (This process makes pseudogenes, if the reinsertion does not fall under a promoter.) A clear example of this process was worked out in the *jingwei* gene system in *Drosophila* (7).

The argument that the first exons were 15–20 amino acids long does not have direct support in today's exon distribution, which peaks at lengths of 35–40-amino acid residues (8). In terms of exon fusion, we expect that there has been, on average, two or three acts of fusion in going from the original 15–20-amino acid long exons to the pieces that are being shuffled today.

That protein evolution begins with 15–20-residue polypeptides, essentially small ORFs, whose products are just long enough to have some shape in solution (or as an aggregate) provides an answer to the classic problem of how long proteins evolved. Although it is impossible to find one of $(20)^{200}$ sequences by a random process (there is not enough carbon in the universe), all short fragments 15–20-residues long can be found in a few mols of material.

Although we have described the Exon Theory of Genes as involving DNA-based introns and exons, the theory flows naturally out of an RNA world view (9) that (*i*) pictures RNA genetic material creating (by splicing) RNA enzymes to do all of the biochemistry, (*ii*) introduces then activated amino acids, one by one, to build up oligopeptides to support ribozyme function, and, finally, (*iii*) uses 20 amino acids, short exons, and mRNA splicing to create protein enzymes. This RNA world picture is supported by the ribosome's RNA-based peptide-bond catalysis, by the spliceosome's RNA enzyme-based splicing mechanism, and by the essential RNA involvement in DNA synthesis and the biosynthesis of the DNA precursors (10).

How can one devise any proofs or disproofs of these attitudes about the origin of genes? The polar views make different predictions, which can be tested (8, 11). The theory that introns are present today because there was exon shuffling in the original genes makes certain predictions about intron position and phase whereas theories that the introns were added to DNA sequence by a random process make different predictions. One example of such an introns-added theory is the hypothesis of a transposable element that bears splicing signals on its ends. If such an element were to insert into a gene, its RNA transcript would be spliced out, and the gene product would be unaffected. An element of this kind could spread through the genome as selfish DNA and put introns everywhere.

Intron Phase Predictions

The first set of predictions involves intron phase, the position of the intron within a codon. Even though there is no signature on the message after an intron has been spliced out, the intron position along the DNA can be referenced to the ultimate protein sequence. An intron can lie either between the codons, phase 0, after the first base, phase 1, or after the second base, phase 2. This is an evolutionarily conserved property if the intron remains present in the gene. If the introns had been inserted into the DNA, there would be no "phase" preference at the point of insertion. That insertion, as a DNA process, could take note of DNA sequence but not protein sequence. If, on the other hand, the exons had been shuffled and exchanged, the simplest model would have all introns in the same phase so that every combination between exons would work. Thus, introns-early predicts phase bias, and introns-late predicts (in its simplest form) equal numbers in each phase.

A second property is phase correlation. Consider an exon bounded by introns. If the two introns had been inserted into a continuous gene, there could be no necessary relation between the phases of the intron that lies before and the intron that lies after the exon. The two events of insertion, and hence the phases, should be uncorrelated. On the other hand, if the exon had been inserted into a previously existing intron, then the phases of the intron on either side should be the same so that the reading frame will continue across it. That is, exon shuffling suggests that exons should be multiples of three bases. Intron addition makes no such commitment.

How might one test these predictions? We constructed a database of exons by going to GenBank, identifying all genes with introns, purging that set to remove related genes, and getting a set of quasi-independent genes: 1636 genes with 9192 internal exons from GenBank 84 (8). We then looked at a special subset of those eukaryotic genes: those that have homologous sequences in the prokaryotes. In our database, there were 296 such genes with 1496 introns (8). These genes have the following essential property: They are prokaryotic genes that are colinear with a region of an eukaryotic gene. The prokaryotic gene has no introns; the region of eukaryotic gene has introns. Any introns-late model requires that all of these introns be inserted because there cannot have been exon shuffling for these sequences. Although one might argue in general, for eukaryotic sequences, that they could have been made by exon shuffling, these particular parts of eukaryotic sequence cannot be so made because they are orthologous and thus derive from the cenancestor. In an introns-late picture, all the introns in these homologous regions must be derived. They must have been inserted, and so they should show no phase bias and no phase correlations. According to the introns-early model, there should be such correlations because some or all of these introns originated in the progenote where these genes were assembled by exon shuffling.

In fact, this subset of introns in ancient, conserved regions does show a phase bias: 55% are in phase 0, 24% are in phase 1, and 21% are in phase 2. (The alternative model predicts 33%, 33%, and 33%.) Still more interesting, from a biological viewpoint, there is an excess of symmetric exons, symmetric pairs of exons, triples, and quadruples of exons. Table 1 shows these data (8). All of these sets show an excess of multiples of three, significant at about the 1% level. This is the first strong argument for the existence of ancient introns. The excess of symmetric exons in these ancient conserved regions is predicted in a simple way by the idea that the introns were used to assemble the first gene, but it is not predicted by an insertional model without special biochemical pleadings that forces this result to happen. (One might, in principle, argue that introns inserted into special sequences on the DNA, like AG|GT, and that these sequences might show a bias relative to amino acid sequence to generate a phase bias. To this one might then add the *ad hoc* assumption that the splicing mechanism sees both ends of the exon and likes it to be a

Table 1. Intron correlations in ancient conserved regions

	Observed/expected			
Sets of exons	Symmetric	Asymmetric	χ^2	*P*
1	562/515 (9%)	725/772	7.1	0.008
2	439/400 (10%)	530/569	6.5	0.011
3	348/309 (13%)	400/439	8.4	0.004
4	267/238 (12%)	312/341	6.0	0.014

The symmetric exons or exon sets begin and end in the same phase. The asymmetric sets begin and end in different phases. The expectations for the single exons were calculated from the observed intron phases. The expectations for the sets of exons were calculated from the prior observed frequencies of the subset exons. The percentage difference with the symmetric exon or exon sets is calculated as (observed − expected)/expected.

multiple of three.) A simpler interpretation is that there was exon shuffling in the formation of the first genes.

Introns Correlate with Protein Structure

If genes had been put together from small shuffled pieces, proteins should be made up of repeated small elements, possibly elements of folding, although the evolutionary argument only requires elements that can be subject to natural selection. Evolution requires only function; it does not require biochemical structure. The prediction of the Exon Theory of Genes is that there should be such elements that evolution has selected, which we call modules, and that they should be coextensive to exons.

By module, we mean a region of polypeptide chain that can be circumscribed in space by a maximum diameter. If one traces the Cα positions along the backbone of the polypeptide chain in space and requires that all of the pairwise distances be less than some maximum value, then this region of the chain must fold back and forth in space. Putting a maximum length over the Cα distances, roughly speaking, means that, as that region of chain travels through space, it can be circumscribed by a sphere of that diameter.

How might one define the boundaries of such modules? The module notion was first introduced by Mitiko Go in the early 1980s and was used to predict the existence of novel introns. She used it (12) to suggest that there should be a novel intron in globin and that the intron was later discovered in the leghemoglobin of plants. We used that same idea to predict the existence of positions in which one might find introns in triosephosphate isomerase (13, 14). One difficulty with this notion, and a challenge that the introns-late supporters have made, is that this concept of compactness does not provide a sharp view of where the boundaries are to be. The "spheres" overlap, and one does not have a clean definition of where one module stops and the next begins. We have converted this difficulty into a virtue (6) by suggesting that one should take the overlap regions between the spheres that surround the folded portions and, rather than asking for a single intron to be added at a precise position, define these overlaps as "boundary regions" within which introns might lie. Such boundary regions are designed such that, if one put an intron into each of those regions along the gene, the gene product would be dissected into modules less than the specified diameter. This notion is well defined, and one can now write a computer program to define those regions.

Fig. 1 shows this definition of the boundary regions for globin. By constructing a distance plot, a Go plot, of all pairwise distances between Cα positions along the protein and marking all distances >28 Å in black, one can easily see that the five large triangles along the diagonal identify the longest segments of polypeptide chain that lie in 28-Å modules and that the four small overlap triangles define the boundary regions. The gene thus is divided into two portions, one that corresponds to modules and another that corresponds to boundary regions.

This yields a very simple statistical test. The boundary regions are approximately one-third of a gene: Do introns lie preferentially in these regions, or are their positions random? Again, we considered ancient conserved regions, choosing ones that correspond to three-dimensional structures homologous to bacterial genes without introns and to eukaryotic sequences with introns, to ask: Do those intron positions in the eukaryotic homologs tend to lie in the boundary regions or do they not? The two theories predict quite opposite effects. These are all derived introns on the introns-late model, they were added to the preexisting gene, and their positions should be random. The early-intron model predicts that these positions should fall in the boundary regions.

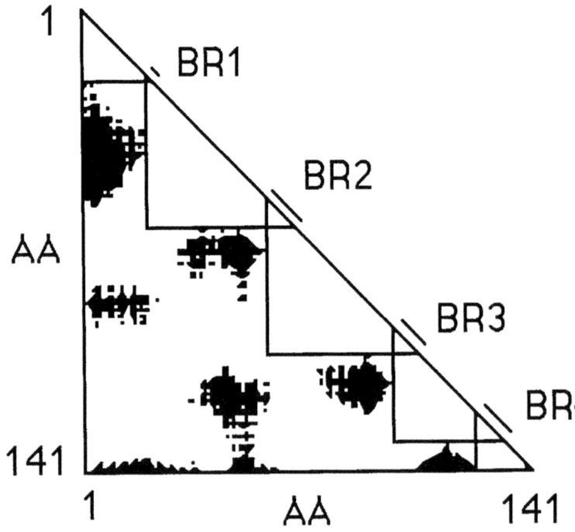

FIG. 1. Go plot for horse hemoglobin. The black spots represent pairs of amino acids whose α-carbons are separated by 28 Å or more. The five large triangles correspond to modules. Boundary regions (BR) are defined by the overlap of these triangles.

Using 28 Å to define the modules, a size that was used before to define modules for triosephosphate isomerase or globin, we examined a set of 32 ancient proteins and a corresponding set of 570 intron positions. The random expectation was to find 182.5 introns in the boundary regions, but we found 214. That is a 17% excess, not a big number, but there are so many positions that the χ^2 is 8 and the P value is less than 0.005.

One might wonder if there could be some other reason, rather than ancient introns, that introns might lie within the boundary regions. One possibility might be the existence of some special sequences in the boundary regions, or some sequence biases, that could serve as targets for insertion. We have examined the sequences in the boundary regions and do not find any particular sequence or compositional bias at the amino acid level or the DNA level. Occasionally, people conjecture that introns might have targeted sequences like AGG or AGGT, "proto-splicing" sequences, but there is no excess of those sequences in the boundary regions. Craik and his coworkers once suggested that introns might lie on the surface of proteins (15), thus one might think that the boundary regions perhaps are on the surface of the proteins and that is why introns are in those regions. However, in this set of proteins, neither the boundary regions nor the introns are biased toward the surface (6).

So far, we have not been able to identify any bias-dependent model that would put introns into the boundary regions. The hypothesis we are testing, the Exon Theory of Genes, says that intron positions should lie within these boundary regions. Even though some introns may have been added in the course of evolution, even though some introns may have been lost, even though some introns may have moved, and even though the protein structure may have altered since it was put together, one can still see an excess there.

A further argument that the excess of intron positions in the boundary regions is due to intron antiquity is found in the examination of an "ancient" subset of the intron positions. We examined those introns that have the same, or similar, positions in three of the four groups: plants, vertebrates, invertebrates, and fungi. Of the 20 introns in this subset, 13 lie in the boundary regions whereas only 6.5 are expected. That is a 100% excess, as opposed to the 17% excess overall. Thus, in a group that is selected to be ancient, we found a higher bias. (That bias was significantly different from the 17%; the χ^2 for the difference between 100% and 17% was 6.5, a P value $\approx$ 0.01.) This finding is further support for the idea that the underlying signal is due to ancient introns. If the pattern is simply one of biased insertion, then any subset should simply have a value ranging around the 17% excess.

The 28 Å size was purely arbitrary, a particular one that we had used historically. It worked, and we had chosen that size before we knew that this analysis would work, but there was no profound reason for that size. Because we have a computer program that can take any diameter and decompose the protein into modules corresponding to that diameter, we can ask: Is there some optimal decomposition? Fig. 2 shows the results of varying the module diameters from 6 Å, which is one amino acid apart along the chain, out to 50 Å and plotting the χ^2 values for the significance of the excess of introns within the boundary regions. Fig. 2 shows three peaks of significance: one peak corresponding to an ≈21-Å diameter, one of an ≈28-Å diameter, and one of an ≈33-Å diameter. The peaks rise to probabilities ≈ 0.001. This is a strong statistical argument that there are three differently sized structural elements in these proteins that are correlated with intron positions. (One might worry that the curve shows a statistical calculation repeated a thousand times; if the phenomenon had been purely random, at least one of the points should have yielded a *P* value of 0.001. If one examines the underlying distribution of the excess of intron positions, one sees that it is robust: Smooth peaks appear in the excess of the observed intron positions over the expectations.)

Thus, we conclude that intron positions are correlated with modules of three different diameters: 21 Å, 28 Å, and 33 Å. Can we understand these modules in a more informative way? We can ask about the average length of the polypeptide chain contained within each of these modules, which is equivalent to asking for the average length of the hypothetical exons predicted by the computer program. Fig. 3 shows that the 21-Å modules have an average length of 15 amino acid residues; the 28-Å modules have an average length of 22 residues; and the

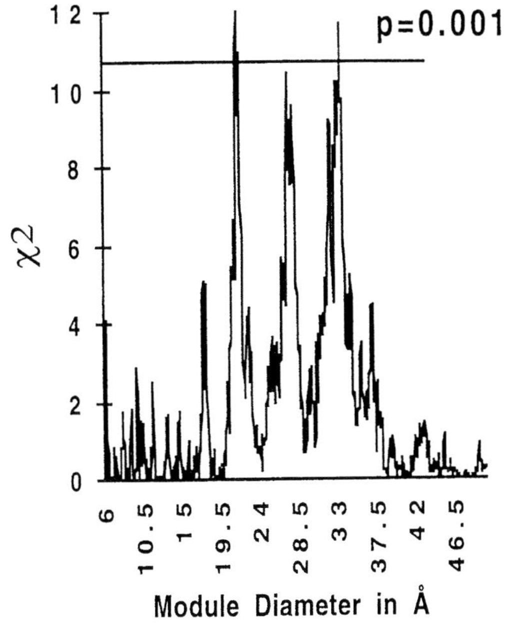

FIG. 2. χ^2 distribution for the matching of intron positions to the boundary regions of 32 ancient proteins as a function of module diameter. The 570 intron positions were drawn from version 90 of GenBank. There are three major peaks of significance around module diameters of 21, 28, and 33 Å.

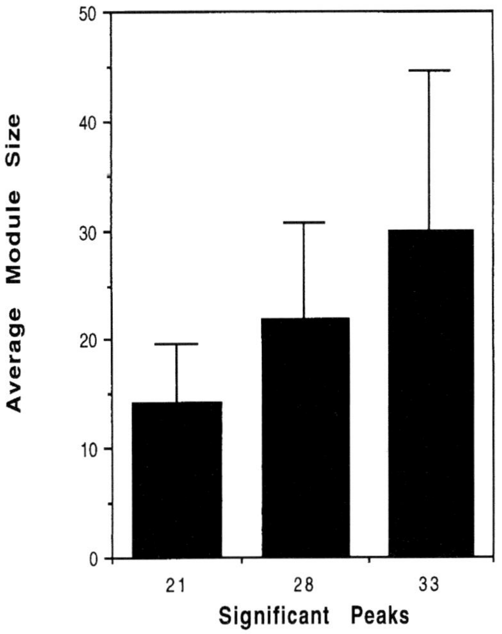

FIG. 3. Lengths of predicted modules for the peaks of significance around 21, 28, and 33 Å. The three peaks correspond to distributions centered around 15, 22, and 30 amino acid residues in length.

33-Å modules have an average length of 30 residues (with a considerable spread). We have given a very strong statistical argument, with *P* values ≈ 0.001, that introns define elements of protein structure with sizes of 15, 22, and 30 residues. This feature is exactly what the Exon Theory of Genes suggested back in 1987.

Recently, we went back to the database. Since the calculation was first done, there are 90 more introns, 662 in total, so we can redo the calculation to see if it is better or worse. Most of the novel intron positions have come in through the *Caenorhabditis elegans* project, so they represent great evolutionary distances from many that were in the database before. With the new data, the peaks improve in statistical significance. Fig. 4 shows that the peak at 21 Å rises to a $\chi^2 \approx 19$, and both it and the peak at 28 Å rise to a *P* value less than 0.0001.

Currently, we are analyzing the shapes that make up these peaks. The peak at 21 Å, for the set of 32 proteins, arises from a set of 822 modules. Because we know the structures of the proteins, we know the three-dimensional structures of each of these modules, and we can search for signs of exon shuffling. The hypothesis that we are testing not only says that there should be correlations of ancient introns with these modules but also that there should be a pattern of reuse of these elements. What we expect to find is that some 21-Å regions, some 28-Å regions, and some 33-Å regions will have been used over and over again. Once we have a classification of shapes that are reused, we will ask for further evidence that those shapes correspond to shuffled exons. Such evidence would be that those modules that have been reused are ones correlated with introns or that those modules that have been reused show sequence similarities that would suggest a divergent evolution. At this time, we know these patterns only very crudely. The most common module is an α-helix followed by a turn and a strand; ≈8–10% of all shapes at 21 Å are of that form. Then there are strand–turn–helix shapes, helix–turn–helix, and strand–turn–strand shapes repeating in the 21 Å peak. The other peaks contain more complicated shapes.

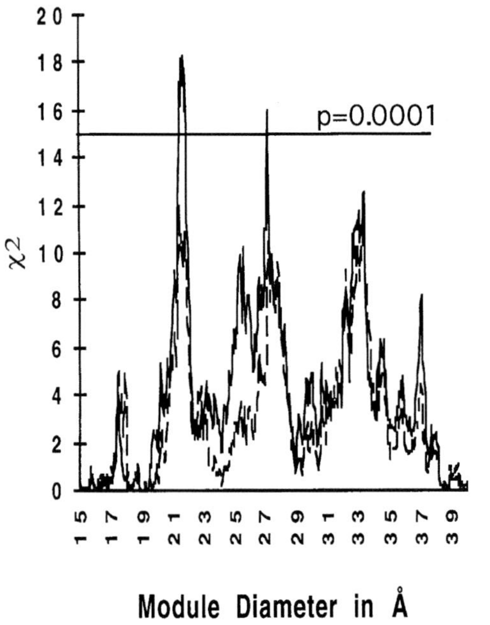

FIG. 4. The same analysis shown in Fig. 2 (dashed line) was repeated using a database of intron positions based on GenBank version 96 (662 intron positions, continuous line). The peaks around 21, 28, and 33 Å now reach χ^2 values around 19, 15, and 13, respectively.

DISCUSSION

We have reviewed here two strong, statistical arguments that there were ancient introns used to shuffle exons in the first genes. Both arguments detect a signal of the presence of ancient introns in today's intron spectrum, over a background that could be due to new introns, to moved introns, or to mutation and change of the protein structures. Both arguments, intron phase correlations and intron correlation to modules, were applied to ancient conserved regions of gene sequence. These are regions of sequence conserved between prokaryotes and eukaryotes; thus, these genes, on any theory, came into existence early in evolution, possibly in the progenote, certainly in the cenancestor, the last common ancestor. These regions are colinear between the prokaryotic forms and their eukaryotic homologs.

It is for these ancient conserved regions especially that the two theories make the most divergent predictions. All forms of introns-late theories assert that these genes came into existence before there were spliceosomal introns. Hence, all of the introns in their eukaryotic counterparts had to be inserted during the course of evolution; they must be derived characters because the prokaryotic form, on those theories, was created as a continuous whole. No exon shuffling can have intervened for these eukaryotic counterparts because they are colinear to the prokaryotic forms. Conversely, all introns-early theories predict that these proteins actually were assembled from exons in the progenote or later by exon shuffling. During the evolution of the prokaryotes, these theories predict that all of the introns were lost. Only in the eukaryotic forms did (some of) these introns survive.

Thus, for these introns, one theory says all were added, and thus should obey random statistics, and the other theory predicts that the current introns will show correlations due to their ancient origin. The databases of gene sequences have so increased in size that one can show that these traces of ancient introns have sharp statistical significance. As the databases continue to increase in the future, these tests will become even more convincing.

Issues of Selection and Adaptation

Are the introns under selection? In general, we argue that they are not. The hypothesis that the role of introns was to speed up evolution by increasing the recombination between exons is not based on the idea that they therefore were selected for that use. Such an idea would be a wrong teleological view, i.e., that they are present because they aid future selection. Rather, our view is that they are present because the easy path in the past that lead to the creation of a gene used them and that they have not yet been removed by selective pressure. Although the introns are not under any selective pressure in general, where there has been pressure on DNA size there would have been loss of introns, such as in prokaryotes, *Arabidopsis*, or other small genome organisms. *Drosophila*, for example, recently has been shown to have a high deletion frequency for unneeded DNA (16) associated with genome slimming, which suggests that many current introns in *Drosophila* may be adaptive and be maintained by such features as enhancers or gene expression timing. Many introns in *Drosophila* are very small, which may reflect the deletion pressure for loss of sequence that still does not go to completion because of the difficulty of removing the intron exactly. Gerald Fink (17) suggested that, in *S. cerevisae*, a special mechanism (in that case a runaway reverse transcriptase) led to the loss of introns as a result of bombarding the genome with cDNA copies of spliced messengers.

Could natural selection on added introns create the observed correlation between introns and protein features? Such models fail because, for these ancient conserved genes, they involve selection for a future purpose. One such model, for example, hypothesizes that, as introns are being added to these ancient (continuous) genes, a well formed exon is shuffled off for use in some other gene and hence selected for. In reality, selection could fix that novel exon in the new gene in the population, but that selection would fail to fix the correct ancestral (donor) form of the gene in the population. (If the organisms had sex, then the donor form is unlinked and hence not fixed. If the organism had only one linkage group, so that the donor form would be fixed by piggybacking, so too would all of the wrongly inserted introns everywhere in the genome.)

CONCLUSION

We have examined a large set of introns in ancient conserved regions. All of these introns should have been derived, late features if the first genes had been continuous. We found, however, that these introns show patterns of correlation to the gene sequence and to the protein structure of the gene products that are consistent with the predictions of The Exon Theory of Genes.

We thank the National Institutes of Health for support (Grant GM 37997). S.J.d.S. was supported by Fundacao de Amparo a Pesquisa do Estado de Sao Paulo and the PEW–Latin American Fellows Program.

1. Doolittle, W. F. (1978) *Nature (London)* **272,** 581–582.
2. Gilbert, W. (1987) *Cold Spring Harbor Symp. Quant. Biol.* **52,** 901–905.
3. Palmer, J. D. & Logsdon, J. M. J. (1991) *Curr. Opin. Genet. Dev.* **1,** 470–477.
4. Cavalier-Smith, C. C. F. (1978) *J. Cell Sci.*
5. Stoltzfus, A., Spencer, D. F., Zuker, M., Logsdon, J. M. J. & Doolittle, W. F. (1994) *Science* **265,** 202–207.
6. de Souza, S. J., Long, M., Schoenbach, L., Roy, S. W. & Gilbert, W. (1996) *Proc. Natl. Acad. Sci. USA* **93,** 14632–14636.
7. Long, M. & Langley, C. (1993) *Science* **260,** 91–95.

8. Long, M., Rosenberg, C. & Gilbert, W. (1995) *Proc. Natl. Acad. Sci. USA* **92,** 12495–12499.
9. Gilbert, W. (1986) *Nature (London)* **319,** 618.
10. Gesteland, R. F. & Atkins, J. F., eds. (1993) *The RNA World,* (Cold Spring Harbor Lab. Press, Plainview, NY).
11. Long, M., de Souza, S. J. & Gilbert, W. (1995) *Curr. Opin. Genet. Dev.* **5,** 774–778.
12. Go, M. (1981) *Nature (London)* **291,** 90–93.
13. Straus, D. & Gilbert, W. (1985) *Mol. Cell. Biol.* **5,** 3497–3506.
14. Gilbert, W., Marchionni, M. & McKnight, G. (1986) *Cell* **46,** 151–154.
15. Craik, C. S., Sprang, S., Fletterick, R. & Rutter, W. J. (1982) *Nature (London)* **299,** 180–182.
16. Petrov, D. A., Lozovskaya, E. R. & Hartl, D. L. (1996) *Nature (London)* **384,** 346–349.
17. Fink, G. R. (1987) *Cell* **49,** 5–6.

Proc. Natl. Acad. Sci. USA
Vol. 95, pp. 5094–5099, April 1998
Evolution

Toward a resolution of the introns early/late debate: Only phase zero introns are correlated with the structure of ancient proteins

(introns-early/introns-late/modules)

SANDRO J. DE SOUZA, MANYUAN LONG, ROBERT J. KLEIN, SCOTT ROY, SHIN LIN, AND WALTER GILBERT*

Department of Molecular and Cellular Biology, The Biological Laboratories, Harvard University, Cambridge, MA 02138

Contributed by Walter Gilbert, February 25, 1998

ABSTRACT **We present evidence that a well defined subset of intron positions shows a non-random distribution in ancient genes. We analyze a database of ancient conserved regions drawn from GenBank 101 to retest two predictions of the theory that the first genes were constructed by exon shuffling. These predictions are that there should be an excess of symmetric exons (and sets of exons) flanked by introns of the same phase (positions within the codon) and that intron positions in ancient proteins should correlate with the boundaries of compact protein modules. Both these predictions are supported by the data, with considerable statistical force (P values < 0.0001). Intron positions correlate to modules of diameters around 21, 27, and 33 Å, and this correlation is due to phase zero introns. We suggest that 30–40% of present day intron positions in ancient genes correspond to phase zero introns originally present in the progenote, while almost all of the remaining intron positions correspond to introns added, or moved, appearing equally in all three intron phases. This proposal provides a resolution for many of the arguments of the introns-early/introns-late debate.**

The rapid expansion of knowledge of DNA sequences, rising a factor of 10 every 5 years, has now reached the point where one can survey with great statistical power the intron spectrum of genes. This has enabled us to create critical tests of speculations about the role of introns and their history by studying ancient conserved genes, whose protein products are conserved between prokaryotes and eukaryotes. Such genes have no introns in their prokaryotic forms but introns in their eukaryotic homologs. Introns-late models must predict that all introns in these genes were inserted into previously continuous genes that correspond to ancestral forms that were similar to the current prokaryotic genes (1, 2). Thus, such models predict that these introns should not respect intron phase (the position within a codon), should not show phase correlations, and should not be related to the three-dimensional structure of the protein products of these genes. Alternatively, introns-early models, which hypothesize that introns were used in the progenote to assemble the first genes (3–6), look upon some or all of these introns as residues of that process and expect these introns to have been associated with the process of exon shuffling and, hence, to show restrictions on intron phase, to show phase correlations, and to be related to the three-dimensional structure of the proteins.

Over the last several years, we have published two statistical arguments that suggest that introns share properties of the type predicted by an introns-early theory. One of these arguments is that introns in genes for ancient conserved proteins are correlated in phase (the position within the codon) so that exons, or sets of exons, tend to begin and to end in the same phase, to be multiples of three bases. This argument was shown to hold at about the $P = 0.01$ level (7). A second argument showed that intron positions were correlated with an aspect of the three-dimensional structure of ancient proteins, specifically that intron positions were associated with compact modules of diameters 21, 27, and 33 Å, with P values less than 0.01 (8). Both of these regularities are predictions of any theory that holds that some or all of the introns were used in the progenote to assemble the genes for these proteins by exon shuffling; neither of these regularities is predicted by theories which hold that the introns were inserted into DNA by processes that are unrelated to the ultimate structure of the gene product.

However, in the last year two papers have appeared that continue the argument that introns are late. One by Cho and Doolittle (9) tries to study a possible coincidence of intron positions in gene pairs that represent duplications that occurred in the progenote, ancient paralogous genes to ask whether the pattern of intron positions in those genes is more suggestive of intron addition or intron loss. A second paper studying the intron distribution in a large gene family argues that the pattern observed is more one of addition or movement than loss (10).

The continuing increase of DNA sequences in the public databases, increasing by a factor of two every 18 months, has led us to reinvestigate this problem using much more data. In this paper, we shall show that the statistical regularities mentioned above can now be analyzed in greater detail with much higher statistical confidence. We reaffirm the basic regularities that we saw before, but now, since there is more data, we can go further in the analysis of the correlation of introns with three-dimensional structural elements. This further analysis shows that the strong correlation is carried by introns that lie between codons (in phase zero), while the introns that lie within the codons (phase one and phase two) do not show strong correlations with three-dimensional structure. This analysis suggests an explicit description of intron positions in terms of both ancient introns and later additions in a way that resolves the conflict between the two viewpoints. We conclude that about 35% of the introns present in ancient genes are ancient, lie primarily in phase zero between codons, and are related to compact elements of protein structure, modules, ranging in diameter between 21 and 33 Å. About 65% of the introns have been added to pre-existing genes, equal fractions in each of the other phases uncorrelated to structure. This division explains why certain analyses see a large fraction of introns as being added to previously existing genes, while the theory that the original genes were constructed through introns remains the simplest and strongest way of predicting the observed regularities.

The publication costs of this article were defrayed in part by page charge payment. This article must therefore be hereby marked "*advertisement*" in accordance with 18 U.S.C. §1734 solely to indicate this fact.

Abbreviation: ACR, ancient conserved region.

*To whom reprint requests should be addressed at: Department of Molecular and Cellular Biology, The Biological Laboratories, 16 Divinity Ave., Harvard University, Cambridge, MA 02138. e-mail: gilbert@chromo.harvard.edu.

PROCEDURES

Intron Database. We used GenBank release 101 to construct a database containing all entries with an intron/exon organization. We then purged it down to a criterion of a 20% match to the shorter sequence, keeping the sequence with more introns each time, using a program, GBPURGE, written for a DEC Alpha based on a FASTA comparison. The ancient conserved region (ACR) database was constructed, and the expected frequencies of intron phase combinations were calculated as described before (7).

Dataset for Structural Analysis. Forty-four ancient proteins (list below) with 988 intron positions were used in this study. Intron positions were defined by searching the full intron database with the Protein Data Bank reference sequence using FASTA. One difference between the approach used here and that in de Souza *et al.* (8) and Gilbert *et al.* (11) is that a specific program creates two files, one containing the structure coordinates and one containing the sequence of the reference protein in FASTA format (used to search the intron database), using the original Protein Data Bank file as a template. This additional step made sure that the coordinate and FASTA sequence files exactly correspond to each other. The source code of this program will be available on our web site (http://golgi.harvard.edu/gilbert.html).

List of 44 Ancient Proteins. The names inside parentheses correspond to the Protein Data Bank accession codes; the 12 additional proteins are starred: aspartate aminotransferase (1ama); acid amylase (2aaa); acyl-CoA dehydrogenase (3mdd); adenosine deaminase* (1add); alcohol dehydrogenase (1adb); adenylate kinase* (3adk); aldehyde dehydrogenase* (1ad3); aldolase (1ald); alkaline phosphatase (1aja); aldose reductase (1dla); amylase (1ppi); aspartate transcarbamoylase* (1raa); aspartyl-trna synthetase* (1asy); catalase (8cat); citrate synthase (1cts); cu++ superoxide dismutase (1sdy); cytochrome *c* (1ccr); dihydrofolate reductase (1dhf); dihydrolipoamide dehydrogenase* (3lad); elongation factor tu (1eft); enolase (1ebg); glucose 6-phosphate dehydrogenase (1dpg); glyceraldehyde 3-phosphate dehydrogenase (3 gpd); glutamate dehydrogenase* (1hrd); glycogen phosphorylase (1 gpa); glutathione reductase* (1 gra); glutathione *S*-transferase (1gss); hemoglobin (2dhb); high pi amylase (1amy); heat shock protein 70 (1atr); lactate dehydrogenase (2ldx); lysozyme (1laa); malate dehydrogenase (4mdh); mn++ superoxide dismutase (1msd); nucleoside diphosphate kinase* (1ndl); ornithine transcarbamoylase* (1ort); porphobilinogen deaminase* (1pda); phosphofructokinase (3pfk); phosphoglycerate kinase (3pgk); phosphoglycerate mutase (3pgm); pyruvate kinase (from author); thioredoxin* (2trx); triosephosphate isomerase (1tim); xylanase (1clx).

RESULTS

Intron Phase Correlations Revisited. We extracted a subdatabase of genes with introns from GenBank 101 to obtain a database with 25,666 entries. We purged this to eliminate sequences that matched by more than 20% of the shorter sequence, using a FASTA matching program, and saved the versions that have more introns to produce a reduced database of 5,772 members. A large fraction of the genomic genes now come from the *Caenorhabditis elegans* sequencing project. The intron/exon structure of these genes is somewhat problematic because it is predicted by computer. These genes account for 42% of our original database. We constructed purged databases with or without the *C. elegans* material: 5,772 genes with and 1,997 genes without. We identified ACRs as those regions of eukaryotic sequence homologous and colinear to prokaryotic genes using as a criteria a BLAST score greater than 75. We thus obtained a database of introns that lie within the ACRs which we could analyze for intron-phase correlations. All of these introns must have been added to the pre-existing gene on an introns-late model, since there can be no exon shuffling in the history of these particular regions of the eukaryotic genes. However, introns-early models predict that some or all of these introns could be the result of exon shuffling that created the ancestral form of these genes in the progenote.

As we previously observed (7), there is a bias in the intron phase distribution for ACR introns: 54% lie in phase zero, 25% lie in phase one, and 21% lie in phase two. Above this bias, there are correlations in phase between introns on either side of exons producing an excess of symmetric exons. Table 1 shows these excesses, as well as listing the behavior of symmetric pairs, triples, quadruples, and quintuples of exons. For each of these cases, the expected values are calculated based on the observed frequencies for the components. The expectation for symmetric exons is based on the frequency of introns in each phase; the expectation for the symmetric pairs of exons is calculated using the observed frequency for the component exons, beginning and ending in all phases; and similarly for the other sets. Table 1 shows that the excesses of symmetric exons and exon sets above expectation are extremely significant. Table 2 shows the same calculation for the ACR dataset derived from the database that includes *C. elegans*; this larger dataset shows an even greater statistical significance.

The approximately 10% excess of symmetric zero–zero exons and 15% excess of one–one exons (and exon sets) is not the expectation of any insertional model but is consistent with an exon shuffling history. These excesses are above the biased expectations based on the observed intron phase frequencies, in which the greatest number of introns are in phase zero, and the largest excess over that biased expectation is for the one–one symmetric exons. We stress that the actual deviations from randomness are very large. If the original expectation for intron phases had been random, as it would be on the simplest addition model, one-third in each phase, then there are almost three times more zero–zero exons than expected, a 200% excess.

Correlations between Intron Position and Module Boundaries. We have reanalyzed the correlation between intron positions and the three-dimensional structure of the protein products of ancient conserved genes, using a larger set of genes and a more extensive set of intron positions. We expanded the group of 32 proteins (8) to a set of 44 ancient conserved proteins and 988 intron positions from GenBank 101. Where the three-dimensional structure was available, if we had not already used those genes, we added new genes that had been

Table 1. Intron correlations within ancient conserved regions (*C. elegans* sequences excluded)

Length	(0,0)	(1,1)	(2,2)	Number	χ^2	*P*
1	1046/934 (12%)	237/210 (13%)	165/140 (18%)	3241	41.5	1×10^{-6}
2	879/779 (12%)	172/154 (19%)	115/113 (11%)	2599	32.4	1×10^{-4}
3	739/656 (13%)	143/116 (23%)	84/90 (−6%)	2105	39.8	5×10^{-6}
4	616/556 (11%)	111/88 (26%)	64/69 (−7%)	1702	26.4	8×10^{-4}
5	486/456 (7%)	82/71 (15%)	57/53 (8%)	1371	8.4	0.4

The data are given for each exon type as observed number/expected number, with the percent excess of observed over expectation in parentheses. The column labeled "Number" lists the total number of exons or of exon sets of the given length. There are 910 ACR regions in this database, and the overall phase bias is 54, 25, and 21% for phases zero, one, and two.

Table 2. Intron correlations within ancient conserved regions using the entire purged database

Internal Exons	(0,0)	(1,1)	(2,2)	No. of exons	χ^2	P
1	1506/1388 (8%)	392/321 (22%)	300/272 (10%)	5133	51.7	3×10^{-8}
2	1181/1059 (12%)	279/234 (19%)	226/203 (11%)	3857	46.7	5×10^{-8}
3	931/828 (13%)	202/171 (18%)	141/149 (−5%)	2921	40.2	1×10^{-6}
4	731/640 (14%)	153/129 (19%)	108/111 (−3%)	2231	29.9	2×10^{-4}
5	556/507 (10%)	107/100 (7%)	85/78 (9%)	1711	16.7	3×10^{-2}

The data are given for each exon type as observed number/expected number, with the percent excess of observed over expectation in parentheses. The column labeled "Number" lists the total number of exons or of exon sets of the given length. There are 1,916 ACR regions in this database, and the overall phase bias is 52, 25, and 23% for phases zero, one, and two.

used in two recent papers that argued for a pattern of later-moved introns.

We analyzed the three-dimensional structures with a program, INTERMODULE (8). Briefly, we define a module as a segment of the polypeptide chain such that all the distances between the C-alpha carbons are bounded by some maximum diameter. The program dissects each three-dimensional structure into a minimally overlapping set of modules of the specified diameter. The overlaps between the modules, which we call boundary regions, provide a series of regions in which we expect to find an excess of intron positions. These "boundary regions" are such that if an intron were to be placed into each of the boundary regions, the gene product would be dissected into a set of modules all less than the specified diameter. We accumulate a list of all intron positions in genes homologous to the sequence of the known structure, counting each different intron position in the nucleic acid sequence once. We then calculate whether there is an excess of intron positions in the boundary regions over the random expectation, which is that the introns were added to the DNA in a manner that did not respect protein structure. We test the significance of the excess with a simple χ^2 calculation. Fig. 1 shows the output of this calculation for this set of 44 protein and 988 intron positions and displays the χ^2 values for each possible module diameter from 15 Å to 40 Å. Fig. 1 shows that there is a statistically significant excess of intron positions in boundary regions for a range of module sizes, with striking peaks around diameters of 21, 28, and 33 Å [as we observed in de Souza *et al.* (8)]. Minor peaks appear near 25 and 37 Å. The major peaks reach χ^2 values of 11, 13, and 9 and probability values around $P = 0.001$. We interpret this curve as showing that introns tend to mark the boundaries of modules of different sizes in this set of proteins. The linear amino acid sequences that correspond to these module diameters are about 15 amino acids long for the smallest modules and range up to an average length of 30 amino acids for the 33-Å modules. Fig. 1 shows that intron positions are correlated with short elements of polypeptide structure in these 44 ancient conserved proteins. The phenomenon is robust; if one examines the excess of intron positions in these regions, one sees a moderately smooth curve that tracks with the χ^2 result.

Now that we have so much more data, we can examine the separate components of the statistical signal. Slightly more than half the intron positions correspond to phase-zero introns, and so we break the data into a phase-zero portion and a phase-one plus phase-two portion. Fig. 2 shows the χ^2

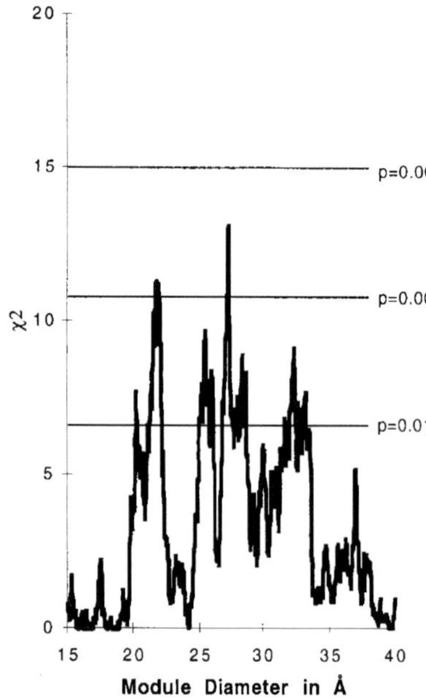

FIG. 1. χ^2 values for the excess of intron positions inside the boundary regions as a function of module diameter. The major peaks are around 21, 27, and 33 Å. The 988 intron positions were drawn from release 101 of GenBank.

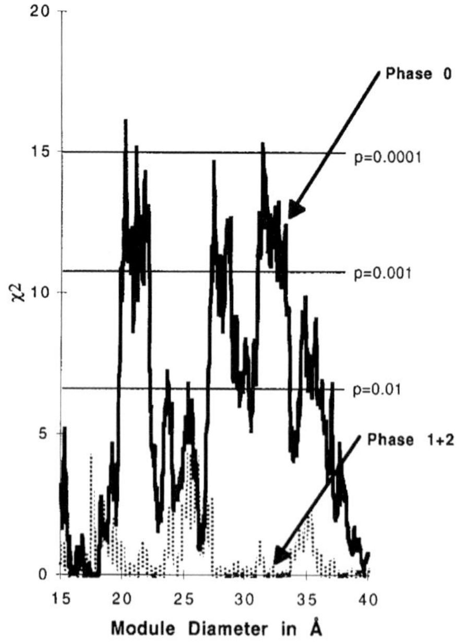

FIG. 2. The χ^2 values for the excess of intron positions inside the boundary regions as a function of module diameter for phase zero and for phases one and two separately. The INTERMODULE calculation was done for phase zero intron positions only (554 positions) (black) and for a set of both phase one and phase two intron positions (434 positions) (gray).

distribution for the excess of phase-zero intron positions in boundary regions as compared with the phase-one and phase-two intron positions. The statistical effect is carried entirely by the phase-zero introns. Phase one and phase two introns do not show enough preference for the boundary regions to be statistically notable. Furthermore, the statistical signal is stronger for the phase-zero data alone, which means that we have taken a random background out of the calculation. The statistical signal now reaches a χ^2 value of 16.5, a P value smaller than 0.0001 for the 21-Å diameter modules. In general, the P values are 10-fold better for the phase zero introns.

Of the 988 introns, 56% are phase zero, 23% are phase one, and 21% are phase two. We suggest that, since the simplest model for intron addition is that equal numbers of introns are added in all three phases, one should interpret the phase two introns as a measure of the background rate of addition and estimate that an equal number of introns were added in phase zero and phase one (added or randomly moved). The excess intron positions over this background, the 35 percentage points of the intron positions in phase zero, are candidates for being ancient introns. About two percentage points of the intron positions that lie in phase one are candidates for being ancient. To put this another way, we suggest that about 65% of all introns are new, added equally in all three phases, and that about 35% of all introns are candidates for being old, are correlated with three-dimensional structure of ancient proteins, and show the excess phase correlations. Of these, almost all are phase zero; 60% of all phase zero intron positions represent old positions, and about 10% of the phase one positions represent old positions.

This analysis can be taken further by asking whether one can see any variation in the different kingdoms in terms of this distribution of phase zero introns. For the 44 proteins, 405 intron positions (224 in phase zero) arise in genes sequenced from the vertebrates. There are 238 positions (150 in phase zero) in genes from the plants, 287 positions (163 in phase zero) in genes from the invertebrates, and 149 positions (72 in phase zero) in the genes from the fungi. When we look at these groups, we observe that vertebrate introns lack a correlation with protein modules. Fig. 3 shows as a percentage the excess of phase zero introns over the expectation for vertebrate and nonvertebrate introns. Vertebrate introns do not show much of an excess of phase zero introns, although there is a small excess around 33 Å. Fig. 4 shows that the overall χ^2 values improve for the set of nonvertebrate phase zero introns. The curve identifies patterns of modules at diameters of 21, 28, and 33 Å with P values < 0.0001.

DISCUSSION

This finding that the correlation of intron positions with the modular structure of ancient proteins is carried primarily by the phase zero introns, along with the interpretation that about 65% of the present introns are added or moved, while about 35% of the introns are correlated with the three-dimensional structure and are candidates for introns left over from exon shuffling in the progenote, provides a resolution of many of the arguments about introns early versus introns late. We identify one fraction of the introns as candidates for introns-late and another fraction as specific candidates for introns-early. This compromise does not, however, satisfy an introns-late view because it argues that a fraction of the introns are early, which has the implication that exon shuffling was involved in the construction of the first genes (5).

The lack of a general signal (there is a small excess of intron positions around 33 Å) for correlation in the vertebrates is interesting and puzzling. There is roughly the same excess of phase zero introns in the subset of vertebrate intron positions as there is in the other subsets. This would suggest that intron positions are subject to different dynamics in accordance with

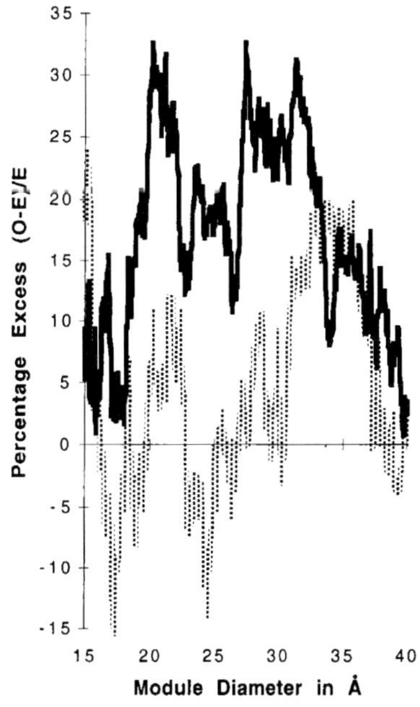

FIG. 3. The percentage excess of intron positions above expectation ((0-E)/E) for the datasets of nonvertebrate (black) and vertebrate (gray) phase zero intron positions.

their phylogenetic distribution. The genes that we added to our original set of 32 proteins tended to weaken the statistical signal rather than strengthen it. The reason, we now understand, lies in that the bulk of the intron positions in those genes came from vertebrate sequences. We expect future work with other organisms will only strengthen the statistical signal.

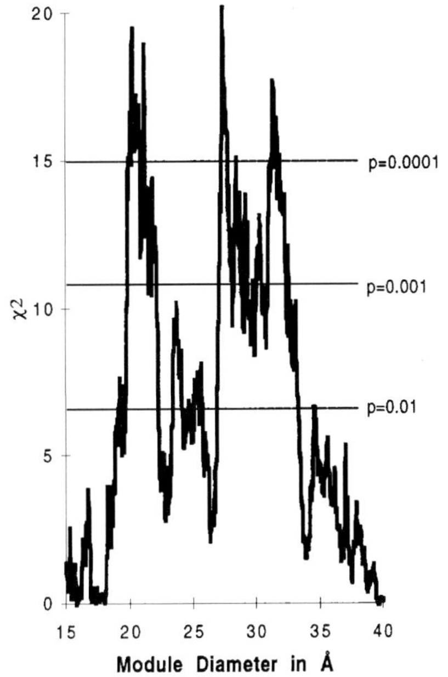

FIG. 4. χ^2 values for the excess of nonvertebrate phase zero intron positions (389 positions) inside the boundary regions as a function of module diameter.

How can this emphasis on phase zero intron positions be reconciled with the apparently greater excess of (1, 1) symmetric exons than (0, 0) symmetric exons? As we pointed out earlier, and as Tables 1 and 2 show, there is a much greater absolute number of symmetric (0, 0) exons than (1, 1) exons. In Table 1, there are 1,046 (0, 0) exons, 237 (1, 1) exons, and 165 (2, 2) exons. There are four times as many (0, 0) exons as (1, 1) exons and six times as many as (2, 2) exons, which represents, on our model, the "background" assumption. The greater background number of (1, 1) exons is part of the reason that we think there were a few phase one introns originally.

We have argued that about one-third of all introns are candidates for being ancient and fall mostly in phase zero (about 10% may be in phase one). The other two-thirds occur roughly equal in all three phases and are candidates for introns added in all three phases or moved randomly into all three phases. We can further use our data to estimate how many introns would have been lost during evolution. If the conjecture that the first genes were constructed of exons, in the length patterns that we infer, is correct (lengths ranging from roughly 15 to 30 amino acids), we might expect the 44 proteins, whose aggregate length is 15,400 amino acids, to have about 670 introns originally in phase zero. Since we now see about 330 introns in phase zero correlated with three-dimensional structure, we suggest that half of the original introns were lost. This figure is also consistent with the argument that half the introns were lost to generate the current exon spectrum from the original shorter spectrum.

Considerations on Theory. How does one test a theory? The general goodness of a theory lies in how well satisfied and how varied are its predictions. Does a theory encompass many aspects of the observed world or is it simply an *ad hoc* restatement of an observation? The theory that the intron/exon structure of genes is a consequence of the first genes being assembled by exon shuffling at the beginning of evolution makes a variety of different predictions, some of which have now been tested and shown to hold with high statistical support. This theory predicts that introns should tend to be in the same phase, that intron phases should be correlated across groups of exons with a preference for symmetric patterns, and that introns should be related to three-dimensional structure of proteins. We showed here that these three predictions are met for phase zero intron positions in ACRs; the actual fraction of intron positions involved we estimate to be about 30 to 40% of all intron positions in the ACRs.

The theory of early exon shuffling also makes further predictions which have not yet been tested. One of these predictions is that the modules should show a pattern of reuse across protein structures that would follow from their having been used by exon shuffling as elements to assemble these proteins. This prediction is that a small set of modules were used over and over again. Still a further prediction is that a pattern of introns will be found to repeat at the boundaries of modules reused by shuffling. These two further predictions have not yet been tested and are independent of the three tests that have already been done.

Further Alternative Theories. Are there alternative theories that could explain the observations? We discuss some theories of intron addition below. These theories have an *ad hoc* character and can often be shown not to hold by a consideration of other properties.

The excess of introns in phase zero in the intron phase distribution might be explained by the introns adding to shadow sequences in previously continuous genes, sequences like AGGT or AGG that have been hypothesized to serve as targeting sites for the insertion of introns (12). Such a theory can be tested by examining the conservation of exon sequences at intron boundaries, to see how strong such "shadow sequences" actually are, and by examining the distribution of putative target sequences to see if they match the phase distribution. We have given a discussion (13) that shows that the current distribution of such sequences does not mimic the intron phase distribution.

A separate theory to explain intron phase correlations, that exons are often multiples of three bases, is to postulate that the splicing mechanism measures the size of an exon and tends to measure that size in multiples of three. That such a mechanism might possibly exist could be supported biochemically by such arguments as those by Robberson *et al.* (14), that splicing mechanisms can recognize both ends of the exon, and by the observation in some RNA viruses (15) of a preferred packaging of RNA by a physical process in multiples of six. In such a model, one might hypothesize that the nucleosome restricts exons to multiples of three (by an unknown mechanism). Although such a model would predict an excess of symmetric exons, it would not predict any further excesses of symmetric exon pairs, triples, etc.

Consider the alternative theory that there was a set of target sequences, present a billion years ago but which have mutated since, that were correlated with amino acid sequences in such a way that the inserted introns tend to lie between amino acids and the corresponding amino acid sequences lie in the regions between the modules, the boundary regions. Such a theory has a superficial plausibility. However, it would be untestable by the phase position, the symmetric exon, or the module data because it is hypothesized to agree with these findings. However, this example of an *ad hoc* theory does not predict the excess of symmetric sets of exons nor does it predict any reuse of modules in different proteins.

There are a variety of theories that attempt to correlate positions of added introns to the boundaries of modules by invoking some form of evolutionary selection pressure. One class of such theories suggests that introns are effectively mutagenic upon their addition to a previously existing gene and, hence, would tend to survive in loops in the proteins or in other regions of low conservation. However, the boundaries of modules, as we have described them, lie frequently within alpha helices or beta strands. Introns are often found in regions of extremely high conservation, and, in general, introns lie in regions of high conservation in proportion to the extent of such regions.

Another hypothesis is that, as introns add to pre-existing genes, if a pair of introns falls around a module, that module might be shuffled out and used in another gene and, hence, selected by evolution. Such models preserve the modules created by intron addition because they are used by shuffling. However, these models invoke a wrong view of evolution. The selection procedure that fixes in the population the shuffled module in the largest gene does not fix the original version of the donor gene, since the donor gene, in general, is genetically unlinked. The ancient conserved genes that we have studied here have to be donor genes on these models.

Another evolutionary argument suggests that when introns add to pre-existing genes, the increased homologous recombination that the intron creates, between the parts of the protein that lie outside its termini, is deleterious if the intron lies inside a module because recombination breaks up co-adapted sites inside the module (M. Meselson, unpublished manuscript). This model, however, does not explain why the exon/module correlation exists only for phase zero introns as well as the excess of symmetric exons.

Although these particular *ad hoc* theories fail, there may still be some alternative theory that accounts for all the data. The nature of the alternative is yet unknown, the best tests of the Exon Theory of Genes lie in its further predictions.

Overall, our final picture is that about 35% of today's intron positions in ancient proteins represent ancient introns, that over the course of evolution about half of the original introns were lost, and that, again over evolutionary time, a number of introns have been added in all three phases corresponding roughly to 65% of the intron positions in today's databases.

S.J.d.S. was supported by Fundacao de Amparo a Pesquisa do Estado de Sao Paulo (Sao Paulo, Brasil) and the PEW-Latin American Fellows Program.

1. Cavalier-Smith, T. (1991) *Trends Genet.* **7,** 145–148.
2. Palmer, J. D. & Logsdon, J. M. J. (1991) *Curr. Opin. Genet. Dev.* **1,** 470–477.
3. Doolittle, W. F. (1978) *Nature (London)* **272,** 581–582.
4. Gilbert, W. (1979) *Introns and Exons: Playgrounds of Evolution.* (Academic Press, New York).
5. Gilbert, W. (1987) *Cold Spring Harbor Symp. Quant. Biol.* **52,** 901–905.
6. de Souza, S. J., Long, M. & Gilbert, W. (1996) *Genes Cells* **1,** 493–505.
7. Long, M., Rosenberg, C. & Gilbert, W. (1995) *Proc. Natl. Acad. Sci. USA* **92,** 12495–12499.
8. de Souza, S. J., Long, M., Schoenbach, L., Roy, S. W. & Gilbert, W. (1996) *Proc. Natl. Acad. Sci. USA* **93,** 14632–14636.
9. Cho, G. & Doolittle, R. F. (1997) *J. Mol. Evol.* **44,** 573–584.
10. Rzhetsky, A., Ayala, F. J., Hsu, L. C., Chang, C. & Yoshida, A. (1997) *Proc. Natl. Acad. Sci. USA* **94,** 6820–6825.
11. Gilbert, W., de Souza, S. J. & Long, M. (1997) *Proc. Natl. Acad. Sci. USA* **94,** 7698–7703.
12. Dibb, N. J. & Newman, A. J. (1989) *EMBO J.* **8,** 2015–2021.
13. Long, M., de Souza, S. J., Rosenberg, C. & Gilbert, W. (1998) *Proc. Natl. Acad. Sci. USA* **95,** 219–223.
14. Robberson, B. C., Cote, G. J. & Berget, S. M. (1990) *Mol. Cell. Biol.* **10,** 84–94.
15. Calain, P. & Roux, L. (1993) *J. Virol.* **67,** 4822–4830.

ELSEVIER

Gene 238 (1999) 85–91

GENE
AN INTERNATIONAL JOURNAL ON GENES AND GENOMES

www.elsevier.com/locate/gene

Centripetal modules and ancient introns

Scott William Roy [a], Michiko Nosaka [b], Sandro J. de Souza [c], Walter Gilbert [a,] *

[a] *Harvard University, The Biological Laboratories, 16 Divinity Avenue, Cambridge, MA 02138, USA*

[b] *Sasebo National College of Technology, 1 Okishin-Tyou, Sasebo City, Nagasaki 857-1117, Japan*

[c] *Ludwig Institute for Cancer Research, Rua Prof Antonio Prudente 109, 4 andar 01509-010 Sao Paulo, SP, Brazil*

Received 12 February 1999; received in revised form 22 June 1999; accepted 6 July 1999; Received by G. Bernardi

Abstract

We have created an algorithm which instantiates the centripetal definition of modules, compact regions of protein structure, as introduced by Go and Nosaka (M. Go and M. Nosaka, 1987. Protein architecture and the origin of introns. Cold Spring Harbor Symp. Quant. Bio. 52, 915–924). That definition seeks the minima of a function that sums the squares of C-alpha–carbon distances over a window around each amino acid residue in a three-dimensional protein structure and identifies such minima with module boundaries. We analyze a set of 44 ancient conserved proteins, with known three-dimensional structures, which have intronless homologues in bacteria and intron-containing homologues in the eukaryotes, with a corresponding set of 988 intron positions. We show that the phase zero intron positions are significantly correlated with the module boundaries ($p=0.0002$), while the intron positions that lie within codons, in phase one and phase two, are not correlated with these 'centripetal' module boundaries.

Furthermore, we analyze the phylogenetic distribution of intron positions and identify a subset of putatively 'ancient' intron positions: phase zero positions in one phylogenetic kingdom which have an associated intron either in an identical position or within three codons in another phylogenetic kingdom (a notion of intron sliding). This subset of 120 'ancient' introns lies closer to the module boundaries than does the full set of phase zero introns with high significance, a p-value of 0.008.

We conclude that the behavior of this set of introns supports the prediction of a mixed theory: that some introns are very old and were used for exon shuffling in the progenote, while many introns have been lost and added since.

Keywords: Evolution; Exon; Progenote

1. Introduction

The origin of introns has often been the subject of a debate concerning the idea that introns arose very early and were used to assemble the first genes (for review, see Gilbert et al., 1997), and the idea that they arose later and were added as insertional elements interrupting previously-continuous genes (for review, see Logsdon, 1998). The former view is that the earliest genes were formed through the process of shuffling genetic elements with short polypeptide products. This 'exon shuffling' process would make the ancestral genes through the formation of novel recombinations within the progenote. The alternative view is that an unknown process produced, in the progenote, the long proteins that we see today and then introns were added during the course of evolution to break up the pre-existing continuous genes.

One force behind the idea that introns are ancient is that introns would have sped up illegitimate recombination in the creation of new genes. Exon shuffling can rapidly make a new combination of two pieces of old genes by a rare illegitimate recombination within an intron. If the introns involved are 1000 to 10 000 bases long, that recombination event is of the order of a million to a hundred million times more probable than it would be in the absence of introns.

This debate does not have a clear resolution today. The bacteria have no spliceosomal introns, the protists have few introns while the higher metazoa and the plants have very many introns. The two sides of this debate look upon these facts either as evidence of the gain of introns in the more complex organisms, or as evidence of the loss of introns in simple organisms which have short generation times. A second point of

* Corresponding author. Tel.: +1-617-495-0760; fax; +1-617-496-4313.

E-mail address: gilbert@chromo.harvard.edu (W. Gilbert)

0378-1119/99/$ – see front matter
PII: S0378-1119(99)00292-9

contention involves introns that occur at similar, but not necessarily identical, positions in homologous genes separated by great evolutionary distances. This fact is interpreted by the two sides of the argument in two opposite ways: the 'introns-late' group claims that such examples are evidence of insertion because one would not expect to find introns separated by a few bases in an ancestral gene; the 'introns-early' group look on these cases as evidence of movement of the intron position in one branch of the evolutionary tree (sliding). Thirdly, the two sides have argued about whether or not introns correlate with the three-dimensional structure of proteins. The idea that introns arose very early and were used to shuffle the pieces of ancestral proteins is generally associated with the notion that those pieces of a protein which can be shuffled must have some modular structure which enables them to combine in different patterns to make more complex proteins. Hence, the idea of early exon shuffling is associated with the idea that the original exons should have represented functional or folding units of the protein (Blake, 1978). So those who believe in early introns generally affirm a correlation of introns with the three-dimensional structure of proteins. Alternatively, if the introns have been added to the DNA late in evolution, then they have been added by some function acting at the DNA level and, in general, one would not expect to see a correlation between intron positions and the three-dimensional structure of proteins. Therefore, those who believe that introns were added late usually deny any correlation with the structure of proteins (Stoltzfus et al., 1994).

'The Exon Theory of Genes' (Gilbert, 1987) expresses the idea that the original exons corresponded to small pieces of polypeptide chain, about 15 to 20 amino acids long, that were essentially open reading frames with useful shapes. Exon shuffling then assembled whole genes out of these smaller pieces. Over evolutionary time, simple exons fused together to make complicated exons. (That process of fusion by retroposition is known to exist (Long and Langley, 1993).) In summary, the basic evolutionary processes were recombination within introns in the progenote, the sliding or movement of intron boundaries (which could change the amino acid sequence near the boundary) and the loss of introns over the course of evolution.

The argument that the original exons might have been 15 to 20 amino acids long and that introns are lost over evolutionary time would mean that there has been a very large loss of introns to produce the present spectrum of exons in today's complicated genes. The current spectrum of exons is peaked sharply at 30 to 40 amino acids in length. Therefore, there would have to have been two to three acts of fusion on average in the history of modern exons.

In summary, there are three different ways of looking at the present intron distribution. One extreme is that the introns were added recently, there was no exon shuffling and all introns have been added in the last few hundred million years. Another popular model is that the introns arose about the time of the Cambrian explosion, in the late pre-Cambrian. In this view introns arose with metazoan organisms and exon shuffling was used to create the genes involved in the metazoan diversification. (The intron spectrum in plants would have had to arise separately.) Such introns might be lost after that time, or lost and some added. This is a view of exon shuffling restricted to the higher eukaryotes. The third notion is that the introns might have arisen at the very beginning of evolution and then some were lost or some added. This third view corresponds to 'The Exon Theory of Genes.' — its critical statements are that introns arose in making the progenote and that some survive to this day.

2. Background

How can one test any of these ideas? Our approach has been to look at genes which are homologous between prokaryotes and eukaryotes. We call the regions of protein similarity in such genes 'ancient conserved regions' (or ancient conserved proteins). These regions are colinear between eukaryotic genes and prokaryotic genes. They have no introns in the prokaryotes but do have introns in the eukaryotic homologues. These ancient conserved regions originated in the progenote and were present in the last common ancestor, the cenancestor, of the prokaryotes and the eukaryotes. In an introns-late model, all the introns found in these regions must have been inserted. They can have no properties related to exon shuffling (since there was no shuffling in the progenote in this model) because the ancient conserved regions came into existence as a continuous element which is still colinear between the intron-containing version and the ancestral gene. However, in an introns-early model, some or all of these introns could be left over from exon shuffling in the history of the cenancestor, all being lost in the prokaryotes. So in one model, these introns could have properties related to the assembly of the three-dimensional structure of the gene product, while in the other model all of these introns must be inserted and cannot have been involved in any past recombination.

Over recent years, several arguments have emerged supporting the idea that introns have an unusual correlation with the structure of the polypeptide gene product for these ancient conserved proteins. One argument by Long et al., (1998) shows that there is an excess of exons which are multiples of three bases — exons tend to begin and end with the same phase. (Phase is the position of the intron within the ultimate reading frame resulting from the splicing out of the intron. Phases zero, one and two refer to introns between the codons, or after the first and second base respectively.) All models of inserted introns would predict that the insertion at one position would be unrelated to the insertion which made the other boundary of the exon. The observed correlation of intron phase, significantly above chance, suggests that there was exon shuffling in the history of these ancient conserved proteins.

The second argument (de Souza et al., 1996) shows that intron positions are significantly correlated with modules in the three-dimensional structure of these ancient proteins. The definition of 'module' concerns the idea that it is a compact region of the polypeptide chain which could be surrounded in space by some maximum diameter. A computer program analyzes the three-dimensional structure of gene products into compact regions surrounded by spheres and defines the overlaps between these spheres as possible 'boundary regions' between these compact pieces. The 'boundary regions' are such that if an intron were to be placed in each 'boundary region,' the protein would be dissected into modules of the defined diameter. The question then is, is there an excess of introns in these boundary regions? The 'boundary regions' correspond to about one third of the protein sequence and there is a simple statistical test to ask if there is an excess.

The latest version of this calculation (de Souza et al., 1998) used a set of 44 ancient conserved proteins and a set of 988 intron positions drawn from GenBank 101. They found three peaks in the correlation of intron positions with the boundary regions, peaks reaching χ^2-values corresponding to p-values of about 0.001. These peaks of correlation occur for module diameters of 21 Å, 27 Å and 33 Å. These diameters for modules correspond to enclosed lengths of polypeptide chains about 15, 22 and 30 residues long, respectively. To go further, de Souza et al. (1998) analyzed the phase distribution of the set of introns. Only the phase zero intron positions show good correlation with the module boundary regions, in fact a ten-fold better correlation than the entire intron set, while the introns in phases one and two do not show any correlation. About 55% of the intron positions are in phase zero and about 20% each are in phases one and two. We conjectured that there is a background of added introns in all three phases, perhaps equally, because one would expect that an intron would add to a DNA sequence and not be sensitive to a reading frame. That would suggest that the excess of introns in phase zero is correlated with three-dimensional structure and possibly old, while about 60% of the positions were added, 20% into each phase.

3. Centripetal modules

Here we analyze a different definition of the modular elements of three-dimensional structure. This definition, due originally to Go and Nosaka (1987), is a centripetal definition of 'modules.' One takes the sum of the squares of the C-alpha distances between each pair of amino acids over a window around each residue:

$$F_i=[1/(2k+1)] \sum_{j=i-k}^{j=i+k} |r_i-r_j|^2,$$

(The windows are modified appropriately at the ends of the protein.) This function F_i has a series of minima. Go and Nosaka used values of k from 40 to 90 (windows 81 to 181 residues wide). Later, in analyzing bacterial barnase, Go and Nosaka (1987) used k-values of 20–45 (Noguti et al., 1993). We have chosen the k-values of the original paper as it addressed intron correlation in eukaryotes, while the others are concerned solely with structure. Go and Nosaka (1987) defined module boundaries as lying at the minima of the centripetal profile. Fig. 1 shows such a pattern for enolase. The different traces define slightly different minima and maxima. In order to define a set of 'minima' or 'module boundaries' in an objective way, we have developed an algorithmic approach. We wrote a computer program, CENMODULE, to define a set of minima of the centripetal function, given a PDB file. The program considers a set of six windows for $k=40, 50,\ldots 90$ and defines an alternating series of minima and maxima for each value of k, by requiring that the function varies by at least five percent of its average value in order to define a relative minimum or maximum. If a set of six minima all lie between the innermost two neighboring maxima, the program averages their values and defines a minimum at this average value. If not all the curves have minima between these two maxima, the point is not used.

Fig. 1 shows the positions of the boundaries of modules defined by such a program. The arrows show the positions of phase zero and phase one or two introns which lie in homologous genes. One sees that many introns lie close to the boundaries, but others do not. Is there any preference for the intron positions to lie near these boundaries? To test whether or not intron positions are related to the boundaries, we measured the distance in amino acid residues between each intron position and the closest boundary, and then averaged these distances for all introns in the real set. We then selected randomly-generated hypothetical intron positions along the 44 genes so that the chance of selecting a given position in a given gene is the same for every residue in every gene. We then compared the average distance for the real set with the average distance for a set (of the same size) of these randomly-generated hypothetical intron positions. We then computed the number of such sets out of 10 000 sets with an average distance less than or equal to that of the real set of introns. This Monte Carlo method directly gives us a p-value. We analyzed the 44 ancient conserved proteins and 988 intron positions used by de Souza et al. (1998).

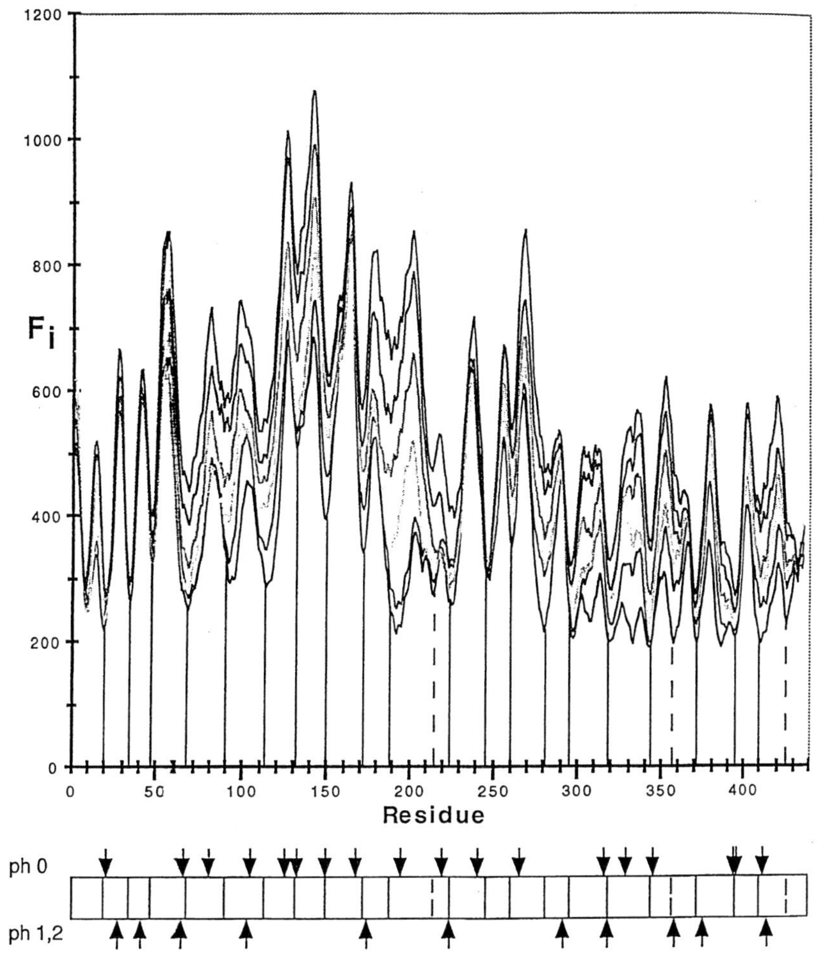

Fig. 1. The pattern of centripetal modules for enolase. The solid lines represent module boundaries defined at both $k=40$–90 and $k=50$. The dashed lines represent the additional module boundaries defined only at $k=50$. The intron positions are shown as arrows.

Table 1 shows that the intron positions lie a little closer to the module boundaries than do those for a random set, but even that small difference is significant since in 10 000 random sets of intron positions only 123 lie as close or closer to the boundaries. For the subset of phase zero introns, only two out of 10 000 random sets lie as close to the boundaries, while for the phase one and phase two subsets essentially half of the random sets are better and half are worse. This is an excellent p-value, $p=0.0002$, for the phase zero subset, while essentially no preference is shown by phase one or phase two introns. Thus this alternative way of defining modules, as compact structures defined by the centripetal averaging function, does define three-dimensional structures which are correlated with phase zero intron positions.

Table 1
Average distance from module boundaries for $k=40$–90[a]

Phase	Real	Random	Number of sets ≤ real out of 10 000
All	7.35	7.85	123
0	6.96	7.86	2
1	8.05	7.86	6632
2	7.62	7.86	3324

[a] Comparison of the average distance between intron positions and module boundaries for 988 intron positions (553 in phase 0; 227 in phase 1; and 208 in phase 2). The number of random sets out of 10 000 lying as close or closer to the module boundaries gives p-values of 0.012, 0.0002, 0.66 and 0.33, respectively.

After this exploration, averaging over six different windows, we found that one window size was sufficient. Table 2 shows that module boundaries defined just for $k=50$ work very well. In this case, we changed CENMODULE so that it required a valley of at least

Table 2
Average distance from module boundaries defined by a $k=50$ window

Phase	Real	Random	p-value based on 100000 random sets
All	6.21	6.79	0.0005
0	5.91	6.79	0.00003
1	6.81	6.79	0.54
2	6.34	6.79	0.13

10% of the average value to define a minimum. On Fig. 1, there are three new boundaries, shown in dashed lines, that arise with this $k=50$ definition. The total set of modules has an average size of about 27 amino acid residues. Table 2 shows the p-values. The phase zero subset has a p-value of 3×10^{-5} while the phase one and phase two subsets show no significant correlation. (We also examined the set of $k=20$–45 windows used by Noguti et al. (1993) to analyze barnase. That set of modules shows only a weak correlation with intron positions ($p=0.05$) and the phase zero subset has a p-value of only 0.08.)

4. Phylogenetic arguments

The phase zero intron positions are correlated with the three-dimensional structure of these ancient proteins. The hypothesis that some of these phase zero introns were used in the progenote to construct the genes for ancient conserved proteins predicts such a correlation. However, one might argue that intron positions happen to be associated with three-dimensional structure for some other reason, for instance insertion into 'pre-splicing' sequences. To counter such arguments we asked if there is any other property of those introns which might suggest that they are ancient. One such property is their phylogenetic distribution — one would expect that older introns would be in positions broadly shared between ancient phylogenetic groups.

Table 3 shows the distribution of these intron positions into the broad phylogenetic divisions: vertebrates, invertebrates, fungi, plants and protists. 29 of the phase zero positions are found in more than one eukaryotic kingdom. On average, these 29 introns lie closer to module boundaries (5.17 residues) than do the random ones (6.79 residues) or even the general phase zero introns (5.91 residues), but the number of introns is too small to yield any statistical significance. We generalize this identification of putative ancient positions by introducing a notion of intron sliding, an old notion that if introns are in close positions in different phylogenetic groups this might correspond to an ancestral position. If there is a phase zero intron in one kingdom and an intron in any phase within three amino acids of that position in one other kingdom, we identify the phase zero position as an 'ancient' intron — this definition of 'sliding' was used by de Souza et al. (1996). That adds 91 phase zero positions to our test, with average distance 4.93 residues, still closer to the boundaries. Table 4 shows all this data.

Table 3
The phylogenetic profile of the sample of intron positions

Subset	Number of positions[a]	Phases		
		0	1	2
Total positions	988	553	227	208
Found in:				
Vertebrates	406	225	97	84
Invertebrates	288	163	67	58
Fungi	149	72	44	33
Plants	236	149	41	46
Protists	40	20	12	8

[a] Introns in the same position in different kingdoms are counted separately in the respective kingdoms but are included only once in the total of positions.

To test the significance of these 120 candidates, we altered our previous programs slightly. Since we are testing the significance of a group of introns (those that appear ancient) as a subset of the larger set (phase zero introns), we selected random subsets, not from all possible positions but just from the original set of phase zero introns. Thus, each phase zero intron has the same chance of being selected into a given random subset. Only 83 out of 10000 subsets of 120 introns have equal or better averages, and thus the subset of 'ancient slid' introns is more highly correlated with module boundaries with a p-value of 0.008. This is a striking result. We have selected from the phase zero intron positions, which already show some correlation with modules, a subset whose members are putatively descended from ancient positions by an independent criterion, phylogenetic matching. This subset is significantly better correlated with the module boundaries than are random subsets of equal size. This is a strong argument that the correlation of phase zero intron positions with modules is due to the existence of ancient introns.

Of course, a hypothesis of sliding might also be interpreted as the phase zero position being the 'slid' position while a phase one or phase two position is the 'ancient' position. One can test such a model by asking

Table 4
Intron positions shared between distinct phylogenetic groups

	Average distance to module boundaries
Random	6.79
Phase zero	5.91
29 shared across kingdoms	5.17
91 phase zero positions are near a position in another kingdom:	
91 slid	4.93
Total 120 'ancient'	4.99

Table 5
Candidates for 'ancient' positions by kingdom

Kingdom	Positions	Average distance
Animals		
All	343	6.24
'Ancient'	61	5.39
Vertebrates		
All	224	6.28
'Ancient'	43	6.02
Invertebrates		
All	163	5.90
Ancient'	34	3.68
Fungi		
All	72	4.86
Ancient'	25	4.48
Plant		
All	149	5.46
'Ancient'	59	4.91
Protists		
All	20	6.00
'Ancient'	6	4.50

Table 6
Candidates for 'ancient' positions by kingdom — without vertebrate positions

Kingdom	Positions	Average distance	p of subset
Animals (except vertebrates)			
All	163	5.90	
'Ancient'	34	3.14	0.00032
Fungi			
All	72	4.86	
'Ancient'	17	2.53	0.0052
Plant			
All	149	5.46	
'Ancient'	38	3.45	0.00032
Protists			
All	20	6.00	
'Ancient'	2	4.50	0.411

if it has any support — does the subset of phase one or two introns which match other phase one or two introns in other kingdoms within three residues show a correlation with the module boundaries? However, that special subset is not at all close to the module boundaries (average distance 9.86). The only intron positions which show a closeness to the boundaries are those associated with a phase zero neighbor.

Table 5 shows the candidate 'ancient' intron positions by kingdom. There is a much better correlation of intron positions with the boundaries for all phylogenetic subsets except for the vertebrates which do not correlate very well. For the sliding argument, the vertebrate introns raise a problem because they represent so many positions in all phases that if they were arranged without respect to structure (as was suggested by de Souza et al., 1998) one sixth of all the other phase zero introns would be called 'ancient' because there would be a vertebrate intron near them. This would obscure the statistical signal. To test this, we dropped all the vertebrate introns from the problem and considered only the remaining positions. Table 6 shows that for each phylogenetic group separately, except for the protists where there are only two 'ancient' introns, the 'slid' subset of phase zero positions lies even closer to the boundaries and the p-values are all very significant.

Furthermore, once we have dropped the vertebrate positions, we obtain an aggregate set of 76 'ancient' positions with an average distance of 3.64 residues out of 388 phase zero positions with an average distance of 5.66 residues. The p-value for this set as a subset of all phase zero introns, excluding vertebrates, is 3×10^{-6}. Only three out of 1 000 000 sets of 76 drawn from the 388 are as close or closer to the boundaries.

We have described this approach as 'dropping the vertebrate introns'. More correctly, we are not using the vertebrate introns to define a putative subset of 'ancient' introns. That subset is now defined as having matching or slid members in the four kingdoms: protists, fungi, plants and animals (invertebrates). To test whether or not this subset (of all phase zero introns) is statistically significant we tested this not against all phase zero intron positions but against the (more correlated) set of introns without the vertebrate positions, of which it is also a subset.

How many 'ancient' intron positions have we, in fact, identified? In this set of 76 positions, there are pairs of phase zero positions which lie close to each other. In this case, we hypothesize that one intron position represents the ancient position and the other a 'slid' position, though they are both identified as 'ancient' by our criteria. Selecting one member only of each such pair, we can identify 57 intron positions as being putatively descended from separate ancient introns.

Given the result that 'ancient' introns lie close to module boundaries, we can go further and estimate what fraction of each of the different phylogenetic intron sets are old by our definition. We can do this because our measure of matching to the boundary, the average distance from the boundaries, is a linear combination. If we assume that the phase zero introns that are added in the course of evolution lie randomly with respect to the module boundaries, then we can interpret any measured average position for a set of introns as a mixture of a set of ancient positions, which have an average distance of about 3.5 residues from the boundaries, and a random set which has an average distance of 6.79 residues from the boundaries.

Table 7 shows such a set of estimates for 'old' introns based on this calculation. For the non-vertebrate positions, we estimate that there might be a total of about 134 ancient positions and 254 new positions in the total set of intron positions. We have identified explicitly 57 of those 134 ancient positions. For the exclusively verte-

Table 7
Candidates for 'ancient' positions by kingdom — without vertebrate positions

Kingdom	Positions	Average distance	'Old' estimate
Animals (vertebrates)			
All	165	6.48	20
Animals (except vertebrates)			
All	163	5.90	
'Ancient'	34	3.68	44
Fungi			
All	72	4.86	
'Ancient'	17	2.53	42
Plant			
All	149	5.46	
'Ancient'	38	3.45	60
Protists			
All	20	6.00	
'Ancient'	2	4.50	5

brate positions, there are about 20 ancient positions and about 145 new — a much smaller fraction of old positions. Thus, overall, we estimate that there are about 150 old positions in the set, of which we have identified one third. This last is a smaller estimate for 'ancient' introns than that given by de Souza et al. (1998), who suggested that of the 553 phase zero intron positions about 300 might be ancient (the excess of the phase zero positions over the phase one and phase two positions). That estimate was based on the argument that intron positions might have been added randomly, and equally, in all three phases. We now have a sharper way of making an estimate for 'ancient' introns, which suggests that a smaller fraction of phase zero intron positions are actually ancient. This would be the case if intron addition is biased and more introns were added in phase zero than in phase one and phase two.

5. Conclusion

Overall, the arguments given here show that there is a correlation of phase zero introns with modules, compact three-dimensional structure elements, in a set of ancient conserved proteins. These are protein sequences which are colinear between prokaryotes and eukaryotes, which have no introns in the prokaryotes but do have introns in their eukaryotic homologues. We have shown this correlation with two different definitions of modules: one involving the description of a module as a region of the polypeptide chain bounded by a maximum distance between pairs of C-alpha carbons; and the other defining the region of the polypeptide chain as delineated by minima of a function that sums the squares of distances between pairs of C-alpha carbons. Both of these ways of looking at modules show that there is a correlation of phase zero intron positions with module boundaries, but no correlation for phase one or phase two positions. There is an excess of phase zero positions over either phase one or phase two positions. This different intron behavior, by phase, is difficult to accommodate in a pure introns-late model in which introns have been added to previously existing continuous genes. It is very easy to accommodate the behavior in a mixed model in which some of the introns are ancient, used in exon shuffling in the creation of the first genes from exons whose products are modules and lie primarily between the codons of the original exons. Superimposed on that original intron spectrum, we then expect to see the loss of many of these ancient introns and the addition of introns in all three phases.

We have developed here a further separate line of argument which identifies a putative set of ancient introns by phylogenetic arguments and shows that that putative set is more correlated with module boundaries than is the larger set of phase zero introns, with a *p*-value of 0.008. The phase zero intron positions in this special subset are candidates for being truly ancient, in that they lie in identical or similar positions across very broad phylogenetic distances, in fact in different kingdoms. Since these positions show a sharper and highly significant correlation with the module boundaries, this further supports the idea that some introns are very ancient and were used to assemble ancient proteins.

References

Blake, C.C.F., 1978. Do genes-in-pieces imply proteins-in-pieces? Nature 273, 267.

De Souza, S.J., Long, M., Schoenbach, L., Roy, S.W., Gilbert, W., 1996. Intron positions correlate with module boundaries in proteins. Proc. Natl. Acad. Sci. USA 93, 14632–14636.

de Souza, S.J., Long, M., Klein, R.J., Roy, S., Gilbert, W., 1998. Toward a resolution of the introns early/late debate: only phase zero introns are correlated with the structure of ancient proteins. Proc. Natl. Acad. Sci. USA 95, 5094–5099.

Gilbert, W., 1987. The exon theory of genes. Cold Spring Harbor Symp. Quant. Biol. 52, 901–905.

Gilbert, W., de Souza, S.J., Long, M., 1997. Origin of Genes. Proc. Natl. Acad. Sci. USA 94, 7698–7703.

Go, M., Nosaka, M., 1987. Protein architecture and the origin of introns. Cold Spring Harbor Symp. Quant. Bio. 52, 915–924.

Logsdon Jr., J.M., 1998. The recent origins of spliceosomal introns revisited. Curr. Opin. Genet. Dev. 8 (6), 637–648.

Long, M., Langley, C., 1993. Natural selection and the origin of *jingwei*, a chimeric processed functional gene in *Drosophila*. Science 260, 91–95.

Long, M., de Souza, S.J., Rosenberg, C., Gilbert, W., 1998. Relationship between 'proto-splice sites' and intron phases: evidence from dicodon analysis. Proc. Natl. Acad. Sci. USA 95, 219–223.

Stoltzfus, A., Spencer, D.F., Zuker, M., Logsdon, J.M., Doolittle, W.F., 1994. Testing the exon theory of genes: the evidence from protein structure. Science 265, 202–207.

Noguti, T., Sakakibara, H., Go, M., 1993. Localization of hydrogen bonds within modules in barnase. Proteins 16 (4), 357–363.

Large-scale comparison of intron positions in mammalian genes shows intron loss but no gain

Scott W. Roy*[†], Alexei Fedorov[†‡], and Walter Gilbert*[§]

*Biological Laboratories, 16 Divinity Avenue, Cambridge, MA 02138; and [‡]Department of Medicine, Medical College of Ohio, 3120 Glendale Avenue, Toledo, OH 43614-5809

Contributed by Walter Gilbert, April 17, 2003

We compared intron–exon structures in 1,560 human–mouse orthologs and 360 mouse–rat orthologs. The origin of differences in intron positions between species was inferred by comparison with an outgroup, *Fugu* for human–mouse and human for mouse–rat. Among 10,020 intron positions in the human–mouse comparison, we found unequivocal evidence for five independent intron losses in the mouse lineage but no evidence for intron loss in humans or for intron gain in either lineage. Among 1,459 positions in rat–mouse comparisons, we found evidence for one loss in rat but neither loss in mouse nor gain in either lineage. In each case, the intron losses were exact, without change in the surrounding coding sequence, and involved introns that are extremely short, with an average of 200 bp, an order of magnitude shorter than the mammalian average. These results favor a model whereby introns are lost through gene conversion with intronless copies of the gene. In addition, the finding of widespread conservation of intron–exon structure, even over large evolutionary distances, suggests that comparative methods employing information about gene structures should be very successful in correctly predicting exon boundaries in genomic sequences.

When it was discovered 25 years ago that eukaryotes, unlike prokaryotes, had split gene structures, it came as quite a shock. Where did these so-called introns come from? What uses did they have? How did they propagate? How labile were their positions and sequences? These are questions that have proved very difficult to solve despite occupying the minds and laboratories of a generation of biologists.

As the exon–intron structure of genes reaches its silver anniversary, much about this structure is still a mystery and the subject of intense research. In this postgenomic world, introns have proven themselves a singular pest in our attempts to predict gene structures from raw genome sequence. They have almost no consensus sequence over their lengths; they can be absurdly long; and they are the substrates of a bewildering array of different alternative splicing patterns. One promising avenue of improvement of the algorithms would further exploit comparisons with other genomic sequences. If intron–exon structures are highly conserved between two species, the viability of an exon prediction should be able to be evaluated by comparison with the orthologous gene copy in another organism.

However, before such comparisons can be used, it is important to know more about the degree of conservation of gene structure between related species. Two ill understood processes that may change the intron–exon structure of genes are intron loss, in which the intervening noncoding sequence between two exons is jettisoned, and intron gain, in which an intron appears *de novo*. Apparent instances of both have been described. The first such event was characterized by Perler *et al.* (1) in 1980. Rats have two insulin genes, one with a two-exon-one-intron structure and the other with a three-exon-two-intron structure (in which one intron matches the single intron of the other gene copy). The genes are paralogs as a result of a recent duplication. To determine which structure was ancestral, the authors screened a genomic chicken library to find insulin genes in chicken. They found a sole copy, with a three-exon-two-intron structure and thus inferred that the other copy had lost one of its introns.

Compelling evidence for intron gain was slower in coming. In 1995, Logsdon *et al.* (2) sequenced the TPI genes for a host of metazoans and found that the gene took on many different structures in different species. In some cases, an intron position in one species was not shared with any other species, such that its phylogenetic distribution could be explained by either a single insertion or by up to 15 losses. Thus it was generally accepted that intron gain also occurs.

Since that time, many instances of intron gain and loss have been described (refs. 3–5; see refs. 6 and 7 for thorough reviews). However, such studies have been for the most part case studies, uncovering one or two instances in a single gene. Thus, it is hard to infer from the literature the relative importance of the two processes and thus to develop a qualitative sense of the shaping processes of intron evolution and of the general conservation of intron–exon structures between orthologs.

To begin to fill this void, we undertook two global studies of the differences in intron–exon structure, one between mouse and rat and the other between mouse and human. Comparing discordant intron structures to those of an outgroup (human for the mouse–rat comparison; *Fugu* for the human–mouse comparison), we were able to infer whether a given difference was caused by a gain in the intron-containing lineage or by a loss in the other.

Comparing 10,020 introns in human–mouse orthologs and 1,459 in mouse–rat, we found convincing evidence for five intron losses in the mouse lineage since its divergence from humans and one intron loss in the rat lineage since its divergence from mouse. In each of the characterized losses, the intron was exacted precisely without change to the flanking coding region. In each case, the corresponding intron in the intron-containing gene copy is very short, suggesting a bias for loss of short introns. We found no instances that resembled intron gain, suggesting that the mechanisms of intron gain are nonfunctional in mammals. The fraction of introns that have been lost since each evolutionary divergence is thus tiny, on the order of 0.06%.

Methods

Exon–Intron Databases. All 4,310 known mouse and 1,800 known rat genes with characterized intron–exon structures were obtained from the latest release (132) of GenBank. All human genes, both confirmed and predicted, were obtained from the human genome annotation available on the National Center for Biotechnology Information web site. We used the EID programs of Saxonov *et al.* (9) to generate databases of the intron–exon structures and sequences of all genes for each organism.

Gene Pairs. We did reciprocal BLASTP searches between all rat and all mouse genes and between all mouse and all human genes. For the rat–mouse comparison, this yielded 360 unique gene pairs

[†]S.W.R. and A.F. contributed equally to this work.

[§]To whom correspondence should be addressed. E-mail: gilbert@nucleus.harvard.edu.

 www.pnas.org/cgi/doi/10.1073/pnas.1232297100

Table 1. Apparent (but not actual) cases of intron discordance

	No. of analyzed introns							
Compared species	GenBank mRNA	Sliding	Alignment boundary	Alternative splicing	Boundary sliding	Low homology	GenBank errors	No *Fugu* homolog
Mouse–rat	108	79	17	4	6	11	14	—
Human–mouse	277	89	58	43	41	16	14	9

GenBank mRNA, apparent species-specific intron position due to GenBank record that is a mosaic of genomic and mRNA sequences; sliding, introns in one species at a position near to an intron position in the other species; alignment boundary, intron position is found at the boundary of the BLAST alignment; alternative splicing, an extra exon in one species leads to an extra, and thus unmatched, intron position; boundary sliding, coding sequence appears to have expanded or contracted at the boundary of the intron position; low homology, introns found in areas of low sequence homology; GenBank errors, unlikely exon–intron structures in GenBank records (see text); no *Fugu* homolog, human–mouse discordances whose ancestral states cannot be inferred because of the lack of a convincing *Fugu* ortholog.

with 1,459 intron positions. For the mouse-human comparison, there were 10,020 intron positions in 1,576 gene pairs.

Automated Intron Comparisons. For each pair of genes, we used our CIP program (10) to map the intron positions of each gene onto the corresponding BLAST protein alignment. All alignments in which there was at least one unique intron position in one of the species were inspected by eye. Those instances in which there was an intron present at a position in one species with no corresponding intron in the other in a region of good alignment were marked as instances of discordant intron–exon structure.

For each instance of discordant intron–exon structure, an ortholog from an outgroup was sought. For the rat–mouse comparison, we used human; for mouse–human, we used *Fugu* (individual gene structures were obtained from www.jgi.doe.gov). The three sequences were aligned by using the default options of CLUSTALW. There were several cases in which CLUSTALW placed one of the two genes from the more closely related species (human or mouse for human–mouse–*Fugu*; mouse or rat for rat–mouse–human) as the outgroup. We eliminated these as probable cases of paralogs. We then again used our CIP program to align the mouse gene to that of the outgroup for intron position comparison. If the outgroup had an intron at the discordant position, the difference was attributed to intron loss. If not, the difference was attributed to gain.

Manual Confirmation. Initially, our program identified 543 cases of mouse–human and 259 cases of rat–mouse intron discordance. However, on detailed inspection of the alignments, we found that all but a few cases were explained by several recurrent patterns (Table 1).

The most common case involved long stretches of alignment in which one organism had relatively large numbers of introns and the other had none. We checked many of these cases and found that all were due to GenBank records that joined genomic data to mRNA or cDNA data, with intron positions marked only in the genomic regions. Such cases accounted for more than half of the discordant intron positions (277 human–mouse; 108 mouse–rat).

The next most common pattern appeared as an intron position that had moved a few or several bases with respect to the coding sequences. Fifty-four human–mouse (71 mouse–rat) of these were identified by CIP as "sliding" based on a difference in positions of two codons or less. A further 35 human–mouse (8 mouse–rat) were identified by eye at distances of up to six codons. Although there is still a lively debate about the frequency of such sliding events in nature, in this instance we take most of these cases to be caused by database errors and/or misprediction of intron–exon boundaries by the various prediction programs.

A third pattern showed an intron at the boundary of an alignment, just as would be expected in the case of gene truncation or elongation by an intron-mediated process. This also does not fit the pattern of simple intron loss or gain because the sequence on one side of the intron is not homologous to the sequence from the orthologous gene. There were 58 such human–mouse (17 mouse–rat) cases.

Many other cases were apparent instances of alternative splicing, where a matched intron position flanked a region of alignment gap that extended exactly to a second unmatched intron position, after which the ungapped alignment between the two protein sequences was restored. This is exactly the pattern expected in the case of a comparison between different isoforms of the same transcript in different species. Forty-three human–mouse (4 mouse–rat) cases showed this pattern.

Equally important were apparent instances of "boundary sliding," where one (but not both) boundaries of an intron appeared to have moved by an integral number of codons, converting some intron sequence to coding sequence (in the case of an intron contraction) or vice versa (in the case of an intron expansion). Without further investigation, it is impossible to determine which of these cases represent actual instances of intron boundary sliding and which are attributable to various prediction or annotation errors. Thus, we simply note that such a computational error would be easy to make, and thus we expect at least some of these to be mispredictions. There were 41 human–mouse (6 mouse–rat) cases.

A further 16 unmatched human–mouse (11 mouse–rat) introns fell in regions of very low amino acid sequence conservation, a pattern that is not expected from a simple gain or loss. Therefore, we discarded these cases as probable gene fusions or results of other genome dynamics. Finally, nine mismatched human–mouse intron positions had no corresponding orthologous region in *Fugu*, rendering their further investigation impossible.

This manual inspection of the database eliminated all but 18 human–mouse and 14 mouse–rat cases of discordant intron positions, which we then analyzed further.

Confirmation of GenBank Records. For all instances of discordant introns, we checked the GenBank records for the genes involved to ensure that the record seemed to contain a feasible gene structure. We thus identified several GenBank entries that contained "introns" of lengths 3, 2, 1, and even 0 bp, which we discarded. In addition, we were able to eliminate as unlikely a few instances in the human genome where the entire intron is a series of N's, instances that we attribute to misapprehension of the relationship between adjacent contigs in the assembly. This analysis yielded 12 human–mouse and 5 mouse–rat cases of apparent intron loss and no cases of gain.

We then did an online BLAST search of the genome sequence for each of the 17 genes that appeared to have lost an intron. In seven mouse–human and four mouse–rat cases, we found that

Table 2. Results of intron comparison

Compared species	No. of gene pairs	Total introns	Results	
			Gain	Loss
Mouse–rat	360	1,459	0	1 (Rn)
Human–mouse	1,560	10,020	0	5 (Mm)

the genomic copy of the gene harbored an intron at exactly the same position as the orthologous gene. Thus, these initial identifications seem to be caused by GenBank errors in which intron positions were left out of the annotation.

This final analysis yielded six cases of well supported intron loss, five in mouse since divergence with human and one in rat since divergence with mouse, in which the GenBank record and the genomic copy agree as to the absence of an intron position (Table 2).

Results and Discussion

The first striking result of this analysis was the number of genes for which all intron positions matched exactly. Of 1,590 orthologous human–mouse pairs, 1,410 showed no deviations at all in intron alignment. Of 360 mouse–rat pairs, 307 were identical. Thus, even taking into account the multitude of problems with such a large-scale analysis (see below), 85% of mouse–rat orthologs have unambiguously maintained the entirety of their genomic structures for at least 30 million years (My) and 90% of mouse–human orthologs have done the same for 75 My.

Furthermore, of the remaining 543 discordant positions in 180 human–mouse orthologs (239 in 53 for mouse–rat), half were easily explained by one basic problem with many GenBank records. In each of these cases, there appeared to have been massive intron loss in one of the two organisms. However, upon closer inspection, we found that the apparent absence of introns was caused by an erroneous gene structure that was a mosaic of mRNA and genomic DNA. Thus, in this case, although some intron positions were included in the GenBank record, most were not, due to the sequence in the record having been partially derived from an mRNA.

However, this does not exclude the possibility that some of these cases are indeed examples of massive intron loss. Such instances have been well documented (11, 12). However, because with our protocol it is hard to distinguish such events of massive loss from GenBank errors or from retrotransposed pseudogenes, we have discarded these examples from the analysis. Our analysis is thus only sensitive to examples of simple intron loss or gain, events in which up- and downstream intron positions are unchanged.

This leaves only 264 human–mouse (131 mouse–rat) discordant positions of any possible biological interest (Table 1). Among these, 89 human–mouse (79 mouse–rat) occur at sites where the intron appears to have shifted over a few bases in one organism relative to the other, putative cases of so-called intron sliding. Although a few of these examples may actually represent true changes in the gene structure, it is hard to reconcile such an apparently high rate of intron movement, the most plausible models of which invoke intron excision and subsequent reinsertion followed by reverse transcription, with such low rates of the two separate processes loss and gain. As such, we think it much more likely that most of these cases are attributable to various computer and human errors in which intron positions have been mismarked. This notion is supported by the marked reduction in sequence similarity between orthologs in the region between the intron positions. It appears that parts of intron have been called coding and vice versa in one of the orthologs.

There are 41 human–mouse (6 mouse–rat) instances of apparent contraction or expansion of an intron, in which there is extra coding sequence directly to one side of the intron. This phenomenon is referred to as "boundary sliding." Again, without further investigation, it is impossible to detect whether such instances reflect true differences or failures of the gene prediction programs.

A further 43 human–mouse (4 mouse–rat) positions appear to be alignments between different spliceosomal isoforms of the same transcript. In these cases, the inclusion of an extra exon in one organism has led to an apparent intron discordance. Such instances could either reflect a true species-specific exon or, more likely, a failure of BLAST to find the proper orthologous transcript or a lack of the proper transcript in the database. Lacking further confirmation, it is hard to say whether such discordances actually reflect biological differences, but they are clearly not intron losses or gains.

In 58 human–mouse (17 mouse–rat) cases, the discordant intron lies within three codons of the end of the alignment. In these cases, the intron marks the boundary of the region of homology, exactly as one would expect in an intron-mediated gene fusion or truncation. However, such a pattern would also be expected if true coding sequence has been marked as an intron or vice versa, thus disrupting the sequence homology. Again, without further investigation, such instances are hard to interpret, though we are more optimistic that a substantial fraction of these instances may in fact represent interesting biological changes.

Finally, 16 human–mouse (11 mouse–rat) discordant introns were found in regions of bad mouse–human (mouse–rat) alignment, and a further 9 human–mouse cases were found in regions of human–mouse alignment that had no ortholog in *Fugu*. The former are worthy of further investigation, though they certainly do not resemble the simple intron gain or loss that we sought. Some fraction of the latter may represent real cases of intron movement, though at present it is impossible to tell because of the lack of an outgroup.

This left 18 cases of human–mouse intron discordance and 14 cases of mouse–rat intron discordance. In each case, we examined the GenBank files for the genes. In every instance where comparison with an outgroup appeared to show an intron insertion, the given GenBank file was very suspect. We found multiple instances of introns with lengths from 0 to 3 bp, all of which we discarded as extremely unlikely. We also found cases in the human genome assembly where the entire intron was populated by N's, suggesting that this was caused by a failure of the assembler to recognize that the two flanking regions were in fact adjacent in the genome. In no case did we find a viable example of intron insertion. This is interesting in light of the finding of the newly inserted intron found in the SRY gene of marsupials and suggests that either this insertion is an extremely rare case or that some silencing of the intron insertion machinery occurred since the divergence of placental and marsupial mammals (13).

However, we did find several cases of loss (Fig. 1). In each of these cases, there is an intron at a given position in exactly one of the aligned species, and that intron is found in the genomic copy of the outgroup as well. Interestingly, as Table 3 shows, each of the corresponding introns is very short, with an overall mean of 205 bp, compared with an overall human mean of >2,500 bp and median of 1,820 bp (calculated from our intron–exon database). As Fig. 1 shows, all six protein sequence alignments show extremely strong sequence conservation. In addition, the regions directly flanking the discordant intron positions show equally strong conservation. There are no gaps in the alignments, suggesting that the loss was exact.

It is important to note that these instances are not cases of massive intron loss, for instance, through the genomic incorporation of the product of a retroposition. This is evident in Fig. 1, where it can be seen that neighboring intron positions are conserved and that the intron loss event is indeed specific to a single internal intron in each case.

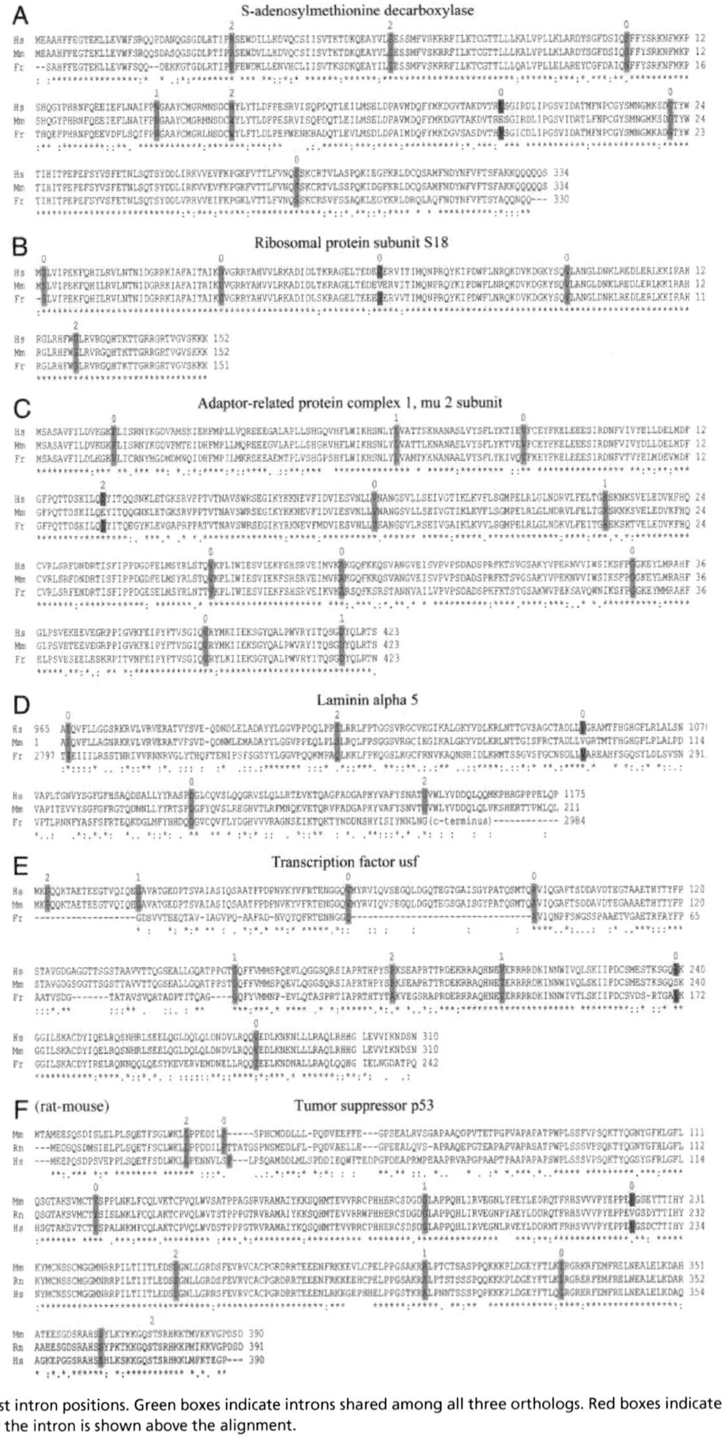

Fig. 1. Alignments with lost intron positions. Green boxes indicate introns shared among all three orthologs. Red boxes indicate introns that have been lost in one species. The phase of the intron is shown above the alignment.

It is striking that, in each case, the intron loss has not been accompanied by the creation of a gap in the alignment and that sequences flanking the lost intron have retained extremely high sequence similarly. The most straightforward model explaining this pattern invokes recombination between the genomic copy of a gene and a product of reverse transcription of a processed

Table 3. Characterization of deleted introns

Species	Gene identifier (EID release 132)	Gene name	Fig. 1 panel	Length of corresponding intron, nt
Mouse	128985_AB025024	*S*-adenosylmethionine decarboxylase	*A*	291 (Human); no (Rat)
Mouse	129712_AF100956	Ribosomal protein subunit S18	*B*	81 (Human); no (Rat)
Mouse	131047_F139406S11	Adaptor-related protein complex 1, mu 2 subunit	*C*	113 (Human); ? (Rat)
Mouse	131297_MMAJ6993	Laminin alpha 5	*D*	107 (Human); no (Rat)
Mouse	132494_MMU41741	Transcription factor usf	*E*	245 (Human); 223 (Rat)
Rat	133770_RATP53TS08	Tumor suppressor p53	*F*	393 (Mouse)

mRNA copy of the gene. The predictions of this model fit our data in two important ways. First, such a process is expected to cause exact deletions, because the intron will have been precisely excised in the mRNA, leaving the coding region intact. Second, such a process would be expected to favor the deletion of shorter introns, because the matching of sequence required to permit such double crossover events would be much easier. Although such a short-intron bias might also be expected from a bias in the length of spontaneous genomic deletions, such deletions would not be expected to be exact, as are the instances here.

These results are an important step in the debate over the relative roles of various processes in the shaping of the modern intron–exon structures of genes. First, they show that introns can be lost exactly, without alteration to nearby coding sequences. Second, they show that intron–exon structures can change in mammals. Third, although not statistically significant, they suggest that the rate of intron loss in rodents may be higher than in humans ($P = 0.07$, binomial distribution), and the fact that no case of gain or loss was found in humans suggests that human introns may have remained static for up to 75 million years. One explanation for a higher rate of loss in rodents could be rodents' shorter generation time. An analogous effect on the rate of synonymous site distribution has been demonstrated by Gillespie (14). Also intriguing is the observation that the intron corresponding to each loss in rodents is shorter than would be expected, suggesting that the mechanism of intron loss may favor short introns.

These results also cast doubt on two previously touted models. The most popular model of intron gain postulates retrotransposon properties of introns as responsible for their propagation (15, 16). However, if retrotransposon activity were a central mechanism for intron spread, one would expect to see many new introns in a group such as mammals, where so many transposable elements are so active. That we see no gains whatsoever suggests that introns do not move by simple retrotransposition. A second model is that of Rogers (17) and of Brenner and colleagues (3). These authors suggest that tandem duplications of exons could lead to the use of cryptic splice signals within the exons, thus creating an extra intron in the middle of a previously intact exon. Again, this model is difficult to reconcile with the lack of new introns in mammals, whose genomes exhibit thousands of tandem duplications.

Lastly, it is important to note that the intron losses observed here are not completely akin to the polymorphic loss recently found in the jingwei gene by Llopart *et al.* (18). Whereas the losses characterized here are exact, with no change in coding sequence, the post-intron-loss jingwei allele has an extra four codons relative to the ancestral allele as a result of an incomplete intron deletion. Thus, the jingwei example is unlikely to have been generated by recombination with a cDNA; the two different types of losses may be completely disparate in their mechanisms. In addition, the polymorphic state of the jingwei intron loss allowed the authors to demonstrate that the new allele shows a signature of positive selection, whereas it is impossible in the examples found here to infer the possible forces leading to the fixation of the post-intron-loss alleles.

The finding of six separate exact intron deletions in rodent lineages in the absence of any additions suggests that the process of intron loss may have a higher rate than that of intron gain in mammals. The pattern observed favors a model in which introns are lost through gene conversion with products generated by reverse transcription of mRNA copies of a gene, creating, after a double recombination, genomic gene copies in which the intron is exactly deleted and a pattern by which the loss of shorter introns is favored.

These results are also important in informing comparative gene prediction. If orthologs between human and mouse have virtually identical intron–exon structures, then cases of ambiguous assignment of intron boundaries should be resolvable by comparison with the other species. Although the intron–exon structure is nearly identical, silent positions have saturated between the genomes. Thus, when choosing, for instance, between multiple GT motifs as the beginning of an intron, comparison with the orthologous gene sequence will often strongly favor one as conserved between species. The inclusion of more genomic sequences from other species will only further strengthen such methods.

A.F. was supported by startup funds from the Bioinformatics Laboratory at the Medical College of Ohio.

1. Perler, F., Efstratiadis, A., Lomedico, P., Gilbert, W., Kolodner, R. & Dodgson, J. (1980) *Cell* **20,** 555–566.
2. Logsdon, J. M., Jr., Tyshenko, M. G., Dixon, C., D.-Jafari, J., Walker, V. K. & Palmer, J. D. (1995) *Proc. Natl. Acad. Sci. USA* **92,** 8507–8511.
3. Venkatesh, B., Ning, Y. & Brenner, S. (1999) *Proc. Natl. Acad. Sci. USA* **96,** 10267–10271.
4. Gotoh, O. (1998) *Mol. Biol. Evol.* **15,** 1447–1459.
5. Wada, H., Kobayashi, M., Satoh, R., Miyasaka, H. & Shirayama, Y. (2002) *J. Mol. Evol.* **54,** 118–128.
6. Logsdon, J. M., Stoltzfus, A. & Doolittle, W. F. (1998) *Curr. Biol.* **8,** R560–R563.
7. Logsdon, J. M., Jr. (1998) *Curr. Opin. Genet. Dev.* **8,** 637–648.
8. Paquette, S. M., Bak, S. & Feyereisen, R. (2000) *DNA Cell Biol.* **19,** 307–317.
9. Saxonov, S., Daizadeh, I., Fedorov, A. & Gilbert, W. (2000) *Nucleic Acids Res.* **28,** 185–190.
10. Fedorov, A., Merican, A. F. & Gilbert, W. (2002) *Proc. Natl. Acad. Sci. USA* **99,** 16128–16133.
11. McCarrey, J. R. & Thomas, K. (1987) *Nature* **326,** 501–505.
12. Hendriksen, P. J., Hoogerbrugge, J. W., Baarends, W. M., de Boer, P., Vreeburg, J. T., Vos, E. A., van der Lende, T. & Grootegoed, J. A. (1997) *Genomics* **41,** 350–359.
13. O'Neill, R. J., Brennan, F. E., Delbridge, M. L., Crozier, R. H. & Graves, J. A. (1998) *Proc. Natl. Acad. Sci. USA* **95,** 1653–1657.
14. Gillespie, J. H. (1991) *The Cause of Molecular Evolution* (Oxford Univ. Press, Oxford).
15. Cavalier-Smith, T. (1985) *Nature* **315,** 283–284.
16. Palmer, J. D. & Logsdon, J. M., Jr. (1991) *Curr. Opin. Genet. Dev.* **1,** 470–477.
17. Rogers, J. H. (1990) *FEBS Lett.* **268,** 339–343.
18. Llopart, A., Comeron, J. M., Brunet, F. G., Lachaise, D. & Long, M. (2002) *Proc. Natl. Acad. Sci. USA* **99,** 8121–8126.

Genetica **118:** 267–278, 2003.

The universe of exons revisited

Serge Saxonov[1,2] & Walter Gilbert[1,*]
[1]*Department of Molecular and Cellular Biology, The Biological Laboratories, 16 Divinity Avenue, Harvard University, Cambridge, MA 02138, USA (Phone: +1-617/495-0760; Fax: +1-617/496-4313; E-mail: gilbert@nucleus.harvard.edu);* [2]*Present address: Stanford Medical Informatics, 251 Campus Drive, Medical School Office X215, Stanford, CA 94305, USA (E-mail: saxonov@stanford.edu);* **Author for correspondence*

Key words: ACR, BLAST, evolution, exon, gene-structure, intron

Abstract

We study the distribution of exons in eukaryotic genes to determine whether one can detect the reuse of exon sequences and to use the frequency of such reuse to estimate how many ancestral exon sequences there might have been. We use two databases of exons. One contained 56,276 internal exons from putatively unrelated genes (less than 20% sequence identity) and the second contained 8917 internal exons from regions of these genes that are homologous and colinear with prokaryotic genes; these are ancient conserved regions (ACRs). At the 95% significance level we find 3500 exon-sequence matches in the large database and 500 matches in the ACR database. These matches correspond to groups of similar sequences. The size–rank relationship for these groups follows a power law, the size falling off as the inverse square root of the rank. This form of the power law distribution leads us to make an estimate for the size of a possible universe of ancestral exons. Using the data corresponding to the ACR regions, that universe is estimated to be about 15,000–30,000 in size.

Introduction

If the exons in genes represented ancestral units, one might be able to estimate how many such elements existed by studying the frequency of reuse of these units (Dorit, Schoenbach & Gilbert, 1990). However, the issue of the origin of introns is still controversial (Cho & Doolittle, 1997; de Souza et al., 1998; Logsdon, Stoltzfus & Doolittle, 1998; Roy et al., 1999). The two sides of the argument, introns-early and introns-late, disagree mainly over whether spliceosomal introns existed prior to the split of the eukaryotic and prokaryotic lineages. The introns-late view holds that all of the introns currently seen in eukaryotes have been inserted at various times during the evolution of those lineages. The alternative position interprets some introns as ancestral structures, that have been retained in eukaryotes but lost in prokaryotes. 'Exon shuffling' is an efficient method of creating novel functional genes from pre-existing sub-units (Gilbert, 1978). If introns were ancient and existed from the beginning of life, the formation of the original genes could also have involved exon shuffling (Doolittle, 1978). Consequently, the debate about the origin of introns is also a debate over the origin of genes and the beginning of evolution; one such theory of evolution is the 'Exon theory of genes' (Gilbert, 1987).

The availability of massive amounts of new genomic sequence now permits statistical arguments about intron distributions in ancient genes. Recent work (de Souza et al., 1998; Roy et al., 1999) may offer a resolution to the controversy. This work suggests that, while a majority of introns have been recently inserted, the antiquity of *some* phase zero (lying between codons) introns seems highly probable. This new data supports the view that exon shuffling played an important role in the construction of ancient genes, suggesting that we can entertain the implications of the exon theory of genes with increasing confidence. The essential idea of that theory is that the enormous variety of protein structures is generated from a limited pool of exons, each of which codes for a small functional or structural domain. This suggests an explanation for the rapid evolutionary appearance of new forms, such as the 'explosion' of the Cambrian era. Those evolutionary explosions are considerably more difficult to explain if evolution is always forced to create new structures by slowly changing existing genes. In addition to providing a plausible framework for recent molecular evolution, the theory suggests a method for the creation of the original genes for the first complex proteins. The initial proteins were essentially aggregates of short polypeptides, about 20 amino acids long, which corresponded to the original exons. One conjectures that these original units represented self-folding modules: folding elements or functional domains that could be clustered together to form enzymes. The genes for these enzymes are then later assembled out of these short elements as exons linked by introns. Furthermore, over evolutionary time, many introns were lost leading to the creation of longer, complicated exons that could in turn be shuffled. Exon shuffling is then a major force creating new protein shapes and new functions.

As a consequence of this theory, one would expect to find exons reused within and between different genes. About a decade ago, Dorit et al. (Dorit, Schoenbach & Gilbert, 1990) attempted a systematic search for examples of exon shuffling by comparing exons from unrelated genes. Furthermore, they tried to estimate the size of the 'exon universe,' how many exons were needed to assemble the current diversity of genes, by counting the frequency with which exons were reused. Very limited data was available, only 1255 exons, and they further assumed that every exon had the same probability of being chosen for gene construction. (This assumption of a uniform probability distribution necessarily leads to an underestimate for the size of the exon universe.) In the current work, we return to this analysis with two very much larger sequence sets. We shall show that the probability of exon use is not constant, but has an interesting mathematical form. Our goal is to investigate the patterns of exon reuse and to make new estimates for the size of the 'exon universe.'

We have worked with two sets of exons: one derived from the entire known eukaryotic genome, the other, from those regions of eukaryotic sequence that are homologous to prokaryotic sequence. We have labeled the latter such regions as ACRs, because, barring horizontal transfer, they are likely to represent sequences that were present before the prokaryote/eukaryote split and have since been conserved. This set of exons is most interesting to analyze because it allows us to uncover features that would otherwise be obstructed by sequence divergence. Also the ACR exons, created by introns that lie in ACR sequences which are colinear between prokaryotic (without introns) and eukaryotic (with introns) gene products, provide a stringent test for introns-early, introns-late theories. These exons (introns) are all created by intron addition, on an introns-late theory, since the ancestral form of the gene is given by the prokaryotic form, and there can be no exon shuffling for these regions. Conversely, an introns-early theory holds that some, or all, of these exons were involved in exon shuffling in the last common ancestor of the prokaryotes and eukaryotes. We have searched for examples of exon shuffling within our two samples by searching for homologous exons. We show that there is a statistically significant reuse of exon sequences in these samples. Furthermore, certain exons are used considerably more than others, with a distribution following a power law with the exponent close to $-1/2$. By relying on this form of this distribution we are then able to make estimates for the size of the exon universe.

Materials and methods

Exon database

To generate sets of exons we employed the exon–intron database (EID) constructed from GenBank 112 (Benson et al., 1999; Saxonov et al., 2000). The initial set of genes in the EID was reduced by removing genes with clearly erroneous structures (marked with 'EEEE' in the database). The sample was further purged to a level of less than 20% amino acid identity level using BLAST 2.0 comparisons without low-complexity filters (Altschul et al., 1997). To aid in purging, we generated a list of experimentally confirmed exon boundaries, using GenBank's EST database. Consequently, in the comparisons we preferentially kept genes with confirmed intron positions, greater number of exons, greater lengths, and more GT. . .AG-rule conforming introns (in order of priority). The final set was composed of 11,552 putatively unrelated eukaryotic genes that contained introns.

ACR database

The database of ancient conserved regions (ACRs) was constructed by searching every gene from the

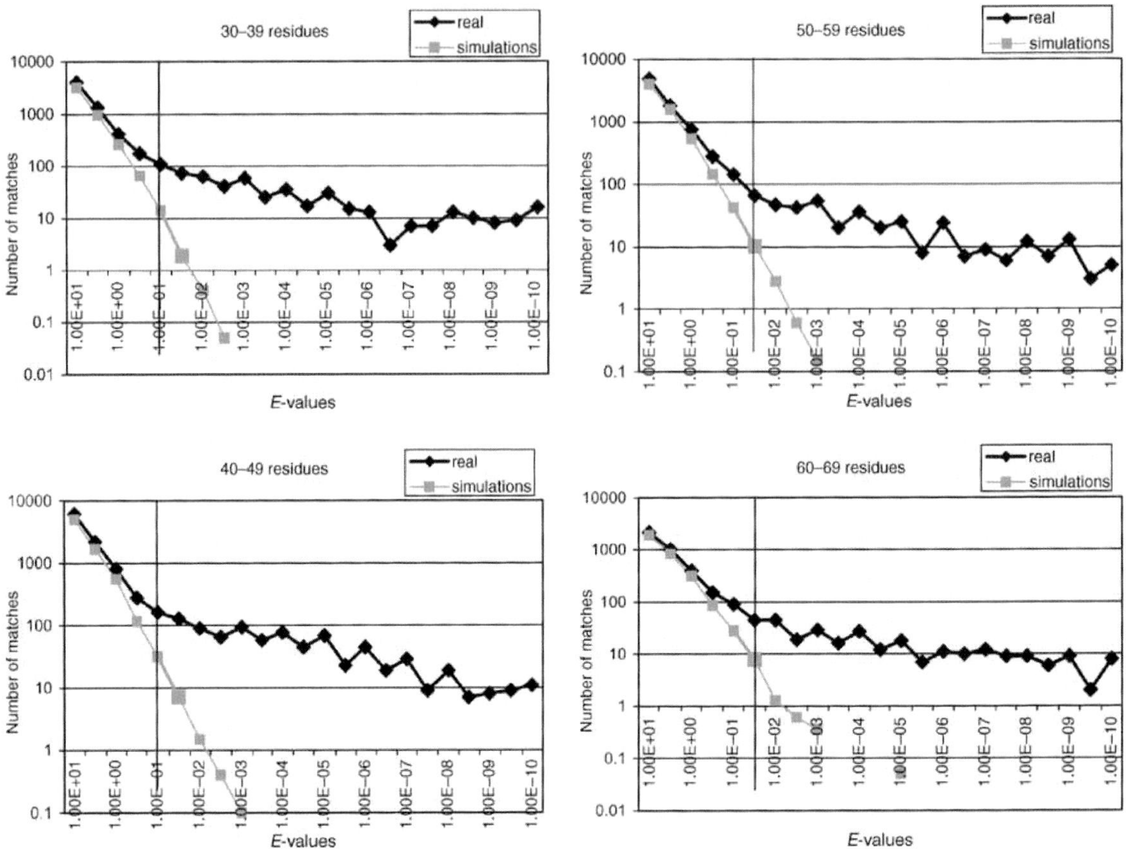

Figure 1. Distributions of exon matches for four length-types. Vertical lines represent the cutoffs used.

purged database against the bacterial subset of GenBank 112. The search was performed using BLAST 2.0. We extracted from the eukaryotic sequences the regions that were homologous and colinear to prokaryotic sequences using as a criterion a BLAST E-value of less than 10^{-5}. The final database contained 2680 regions.

Exon comparisons

We performed two analogous sets of comparisons: one for exons in the entire purged database, and one for exons in the ACR database. From both databases, we extracted all the internal exons of length 15–119 amino acid residues. There were 56,276 such exons in the entire purged database and 8917 in the ACR database. Then a BLAST 2.0 search was performed to compare each exon with a set of all the exons of the same length or slightly shorter. Thus, an exon of length m residues was searched against all the exons of length $m - 5$ through m. In addition, since every exon of length $m + 5$ would in turn be compared to shorter exons, exons of length m end up being compared to the whole range of $m - 5$ through $m + 5$. We allowed for a length variation of up to five residues in order to account for insertions, deletions, or splice site shifts. Due to the difficulty of establishing a significance criterion, only exons of length greater than or equal to 20 residues were used as queries. Consequently, exons of length 15 were only compared to those 20 residues long, those of length 16 – only to exons of length 20 and 21, and so on. In the interest of computational efficiency, it proved more practical to remove duplicate matches (of the same length) from the output of the blast searches rather than to construct individually-tailored databases for every exon before performing the searches.

When conducting comparisons, it must be kept in mind that BLAST may produce alignments between the beginning of one exon and the end of

270

Expectations:

	20-29	30-39	40-49	50-59	60-69	70-79	80-89	90-99	100-109	110-119
1.00E+01	1157.75	3203.10	5045.95	4005.90	1902.40	943.80	562.15	346.70	227.25	139.40
3.16E+00	330.25	966.50	1680.10	1571.20	832.10	408.25	238.25	147.60	97.35	64.55
1.00E+00	72.30	257.05	547.00	531.45	310.45	152.05	96.55	58.25	42.10	28.85
3.16E-01	11.45	63.90	117.45	144.60	84.30	48.20	31.40	20.10	12.60	9.80
1.00E-01	1.85	13.90	31.85	42.05	27.95	17.10	11.25	7.95	5.30	3.45
3.16E-02	0.40	1.90	7.05	10.15	7.85	4.60	3.30	2.05	1.65	2.00
1.00E-02	0.05	0.40	1.50	2.75	1.25	1.55	0.70	0.45	0.90	0.45
3.16E-03		0.05	0.40	0.60	0.60	0.25	0.30	0.15	0.40	0.05
1.00E-03			0.10	0.15	0.35	0.10	0.35	0.15	0.25	0.30
3.16E-04						0.05	0.05			0.05
1.00E-04										0.05
3.16E-05						0.05				
1.00E-05					0.05					

Real Exons:

	20-29	30-39	40-49	50-59	60-69	70-79	80-89	90-99	100-109	110-119
1.00E+01	1539	4055	6198	4859	2192	1033	640	361	233	162
3.16E+00	485	1331	2206	1819	999	424	248	165	109	66
1.00E+00	173	411	802	755	392	173	88	68	49	29
3.16E-01	73	172	279	276	150	60	41	32	17	10
1.00E-01	71	107	164	143	90	32	28	14	6	8
3.16E-02	35	73	129	66	45	15	14	12	5	5
1.00E-02	29	63	90	47	45	8	4	10	4	
3.16E-03	35	41	65	42	19	4	4	4	3	3
1.00E-03	28	58	94	54	29	8	11	5	1	5
3.16E-04	14	25	58	20	16	7	1	7	3	1
1.00E-04	25	35	77	36	27	9	5	5		1
3.16E-05	10	17	45	20	12	5	3	1	3	2
1.00E-05	16	30	69	25	18	6	4	2		3
3.16E-06	5	15	23	8	7	3	4	4	2	
1.00E-06	5	13	45	24	11	6	2	4	4	
3.16E-07	3	3	19	7	10	2	1	1	3	2
1.00E-07	4	7	29	9	12	7	7	4	2	
3.16E-08	6	7	9	6	9	1	1		2	1
1.00E-08	7	13	19	12	9	1	2			
3.16E-09	3	10	7	7	6	2	1	4		
1.00E-09	2	8	8	13	9	1	2		2	1
3.16E-10	2	9	9	3	2	1	2	3	1	2
1.00E-10	3	16	11	5	8	2	4	5	1	1
<=3.16E-11	11	70	54	66	86	57	28	85	99	24

Figure 2. Distribution of matches of exons from the entire purged database. The matches are grouped into 10 length-types. Expectations were calculated by dividing the number of simulated matches by 20. Horizontal lines denote significance cutoffs.

another. Since we are looking for examples of exon shuffling, we eliminated such matches by removing any alignment where either of the corresponding intron positions in the two sequences differed by more than five amino acid residues. The set of matches reported by BLAST was then divided into 10 classes based on the length of the longer exon in the match. Thus, we obtained sets with query lengths of 20–29, 30–39 and so on through 110–119.

Removal of exons with biased compositions

Despite low-complexity filters used by BLAST, a small fraction of the matches had spuriously high homology scores that arise from very biased amino acid compositions. To eliminate such matches from our analysis, we determined the amino acid composition for the entire exon database. Using the proportions of amino acids found in the database as expectations, we used a chi-squared-like formula to give 'bias' scores to all the exons. Specifically, the bias score was calculated using

$$S = \sum_{i=1}^{20} \frac{(o_i - e_i)^2}{e_i}, \qquad (1)$$

where i enumerates the 20 amino acid residues, o_i represents the frequency of the ith amino acid in the exon and e_i is the frequency of the ith amino acid in the entire database. Then, for all exons of the same length, we identified the 2% with the highest scores and discarded them. As a result, the exon database was reduced to 55,097 entries while the ACR database was pared down to 8882. By examining high-scoring matches in simulations, we concluded that this 2% cutoff is quite safe because whereas a 1% cutoff eliminates a relatively large fraction of simulated matches, the additional elimination when using 2, 3, 4, or 5% cutoffs is rather small.

Monte-Carlo calculations to establish criteria for significance

The E-values given by the BLAST programs are not sufficient to provide tests of significance for the short comparisons that were performed. To determine measures of significance, we performed Monte-Carlo simulations of our exon comparisons. We generated 20 simulated versions of each exon by randomly picking (with replacement) amino acids from that exon. Each simulated version was of the same length and, roughly, of the same composition as the original exon. Every simulated exon was then compared to the real dataset in the same way as the real comparisons. By averaging 20 simulation runs, we produced a baseline distribution against which we could compare the real exon matches. In general, the Monte-Carlo background falls off as $1/E$ as a function of E-value (Figure 1). We used as a cutoff an E-value such that less than 5% of all the hits (at that value and lower values) are spurious. For example, from Figure 2, for lengths 20–29 the cutoff would be 3.16E−01 with 387 matches of which 14 are spurious.

Results

We analyzed two sets of exons. We created a set of 'independent' genes by purging a database of genes with exons (the EID, see Methods, constructed from GenBank 112), to a level of less than 20% amino acid identity using BLAST 2. This yielded 11,552 putatively unrelated eukaryotic genes that contained 56,276 internal exons of length 15–119 residues. We constructed a second set, an ACR database, by comparing each gene in the purged database with the bacterial subset of GenBank 112. We extracted from the eukaryotic sequences the regions that were homologous and colinear to prokaryotic sequences with a BLAST 2.0 E-value of less that 10^{-5}. The ACR database contained 2680 regions and 8917 internal exons (between 15 and 119 amino acids).

To eliminate those exons that had extreme compositional biases, which would match fortuitously well because of those biases, we used χ^2 measure of deviation in amino acid frequencies and dropped the extreme 2% of the exons in each size class, leaving 55,097 entries in the purged database and 8882 in the ACR database.

We compared each of the exons to others of smaller size, up to −5 residues, demanding also that the intron positions lie within five residues of each other for an acceptable match. As mentioned above, since longer exons were analyzed in their turn, this approach yielded comparisons to the whole range of lengths from −5 to +5 residues. We pooled the results in size bins: 20–29, 30–39, etc., and in bins of E-values, each bin being half an order of magnitude (1.00E−10 to 3.16E−10, 3.16E−10 to 1.00E−9, etc.). Figure 2 shows the numbers of matches for the purged database and Figure 3 shows those for the ACR database.

Generally, the E-value given by BLAST is considered to be the best indicator for judging a homology (Brenner, Chothia & Hubbard, 1998). It is normalized for length and incorporates all the information in the alignments (such as gaps and similarities between non-identical amino acids.) However, we cannot use that E-value directly as an expectation here, because we use unusually short sequences, while the E-value estimate was optimized for long sequences, and we filter the search results to remove duplicate matches, biased sequences and matches where intron positions are far apart. Thus an E-value of 10^{-4} reported by BLAST will not mean that we should expect to get one match of that quality if we performed 10^4 searches. To determine the appropriate cutoffs for significance, we performed Monte-Carlo simulations. We generated 20 simulated versions of each exon by randomly picking (with replacement) amino acids from that exon. Each simulated version was of the same length and of, roughly, the same composition as the original exon.

272

Expectations:

	20-29	30-39	40-49	50-59	60-69	70-79	80-89	90-99	100-109	110-119
1.00E+01	223.15	457.50	530.35	330.05	138.00	66.85	32.70	18.75	12.05	6.90
3.16E+00	110.85	263.25	308.40	194.25	76.80	38.55	18.70	10.60	6.85	4.15
1.00E+00	25.15	62.85	100.60	72.00	31.80	14.80	9.20	5.30	3.45	2.00
3.16E-01	5.05	14.50	25.95	17.85	8.85	4.20	2.35	1.25	1.05	0.75
1.00E-01	1.15	3.20	6.75	5.90	2.45	1.65	1.10	0.70	0.25	0.20
3.16E-02	0.25	0.70	1.95	1.95	0.60	0.65	0.20	0.10	0.30	0.05
1.00E-02		0.05	0.35	0.35	0.30	0.30		0.05		0.05
3.16E-03			0.10	0.20		0.10		0.05		
1.00E-03					0.05	0.05				

Real Exons:

	20-29	30-39	40-49	50-59	60-69	70-79	80-89	90-99	100-109	110-119
1.00E+01	290	573	684	378	160	97	32	18	16	7
3.16E+00	134	376	426	254	79	42	25	21	11	4
1.00E+00	63	124	182	112	51	17	8	5	2	3
3.16E-01	31	45	54	22	16	10	6			1
1.00E-01	11	33	27	15	10	2	1	5	1	
3.16E-02	17	22	14	6	8	1				
1.00E-02	4	17	11	10	2					
3.16E-03	1	19	2	5	3	1				1
1.00E-03	6	23	9	13	5	1		1		
3.16E-04		13	8	3	3	1	2			
1.00E-04	3	21	4	3	4					1
3.16E-05	2	8	6	7	2					
1.00E-05		5	4	4	4			1		1
3.16E-06	1	7	4		2	1				
1.00E-06	2	5	3	1	4	1				1
3.16E-07	1		2	2	1			1		
1.00E-07	1		2		1				1	
3.16E-08	1		4							1
1.00E-08			2	1	2					
3.16E-09	3		1						1	
1.00E-09	2		1	1	1					
3.16E-10	1		2		1					
1.00E-10	1		1	2						
<=3.16E-11		5	7	2	9	1		1	1	1

Figure 3. Distribution of matches of exons from the ACRs. The matches are grouped into 10 length-types. Expectations are calculated by dividing the number of simulated matches by 20. Horizontal lines denote the cutoffs.

Every simulated exon of length n was then compared to all the real exons of length $n - 5$ through n, with consequent removal of duplicate matches and biased sequences. By averaging 20 simulations, we produced a baseline distribution against which we could compare the real exon matches. Figures 2 and 3 further show the distributions of simulated matches across the entire range of E-values reported by BLAST.

Because BLAST uses a heuristic algorithm, it misses a fraction of homologies – a fraction that is greater for larger E-values. For that reason, we cannot compare the overall shape of the distributions of real and simulated matches, but we can have a reliable view of the two distributions at small E-values. Figure 1 presents these comparisons for four length bins. For large E-values, the distribution of real matches follows that of simulated comparisons rather closely. However, for smaller E-values, the number of simulated matches drops off as $1/E$ while the number of real matches remains at a higher level. Clearly, for low E-values there is a significant excess of real matches over the random expectation. Since the behavior varies somewhat in different length classes, we used different cutoffs in different classes. Each cutoff was established by selecting the highest E-value (more exactly the bin with the highest E-value) for which the number of real matches below the cutoff is

Table 1. Significance cutoffs for all the length classes

Length class	All genes		ACRs	
	Cutoffs	Number of matches	Cutoffs	Number of matches
20–29	0.31600	387 (13.75)	0.10000	57 (1.40)
30–39	0.10000	620 (16.25)	0.10000	178 (3.95)
40–49	0.10000	1024 (40.90)	0.03160	87 (2.40)
50–59	0.03160	470 (13.65)	0.03160	60 (2.50)
60–69	0.03160	380 (10.10)	0.03160	52 (0.95)
70–79	0.03160	145 (6.60)	0.00316	6 (0.15)
80–89	0.03160	100 (4.70)	0.01000	2
90–99	0.03160	156 (2.80)	0.03160	4 (0.20)
100–109	0.03160	135 (3.20)	0.01000	3
110–119	0.01000	46 (0.90)	0.10000	6 (0.30)
Total		3463 (112.85)		455 (11.85)

Numbers in parentheses represent the random background as deduced from Monte-Carlo simulations.

at least twenty times greater than that expected from the simulations. The vertical lines in Figure 1, and the horizontal lines in Figures 2 and 3, show where the cutoffs were placed.

Table 1 shows the cutoffs and the number of significant matches in each length class. An examination of the alignments reveals that some exons are more prevalent than others. To make sense of the distribution of matches, we grouped exons based on the matching patterns. We used two algorithms – both variations on the simple greedy grouping, where if A matches B and B matches C, all three A, B, and C are necessarily in the same group. In both algorithms, each group would start from a pair of exons in a single match. The group would then grow through the addition of new exons that match members of the group. Now, the problem with a simple greedy algorithm is that it carries a tendency to join disparate large groups together because of occasional weak or spurious matches. To try to avoid this, our first algorithm employed a restriction on the length of exons in each group, to prevent the groups from extending unnaturally. The restriction was that new exons could not be more than five residues longer than the longest of the original two exons, nor more than five residues shorter than the shortest of the original pair. The second algorithm used a different constraint on group growth – each new exon had to match at least two (not one) of the exons already present in the group.

Since all the matches above our cutoffs are statistically significant, the vast majority of them represent real biological homologies. It is thus not unreasonable to suppose that each exon group is composed of the descendents of a particular ancestral exon. In this sense, the size of each group would be directly related to the frequency with which the corresponding ancestral exon was used in the construction of genes in our database. Therefore by examining the distribution of group sizes one can get a view of the frequency distribution of the ancestral exons. To analyze this distribution we plot their frequencies (i.e., group sizes) as a function of rank r, where rank is the position of the exon in a list ordered in terms of the exon frequency. Figure 4 shows such graphs for the two grouping algorithms using logarithmic scales for both axes.

The striking feature of the plots is that they all appear essentially linear. In addition, the slopes of the lines fall close to -0.5. The linearity of the graphs implies that the distributions follow a power-law scaling $f \propto r^{\beta}$, where exponent β is the slope of the line on the log–log plot. Such a pattern can be described as Zipf-like after a linguist who first noticed the power scaling behavior of frequency versus rank among words in natural languages (Zipf, 1949). (Specifically, Zipf found that frequencies of words in a large text follow power-scaling with the exponent of -1. A similar pattern was observed for other natural phenomena, such as city populations and river basin sizes.)

Simulations

Is it possible that the power-scaling phenomenon is an artifact of the way we chose to group exons, rather than a valid biological property? To test this we performed a series of simulations. The idea is to produce

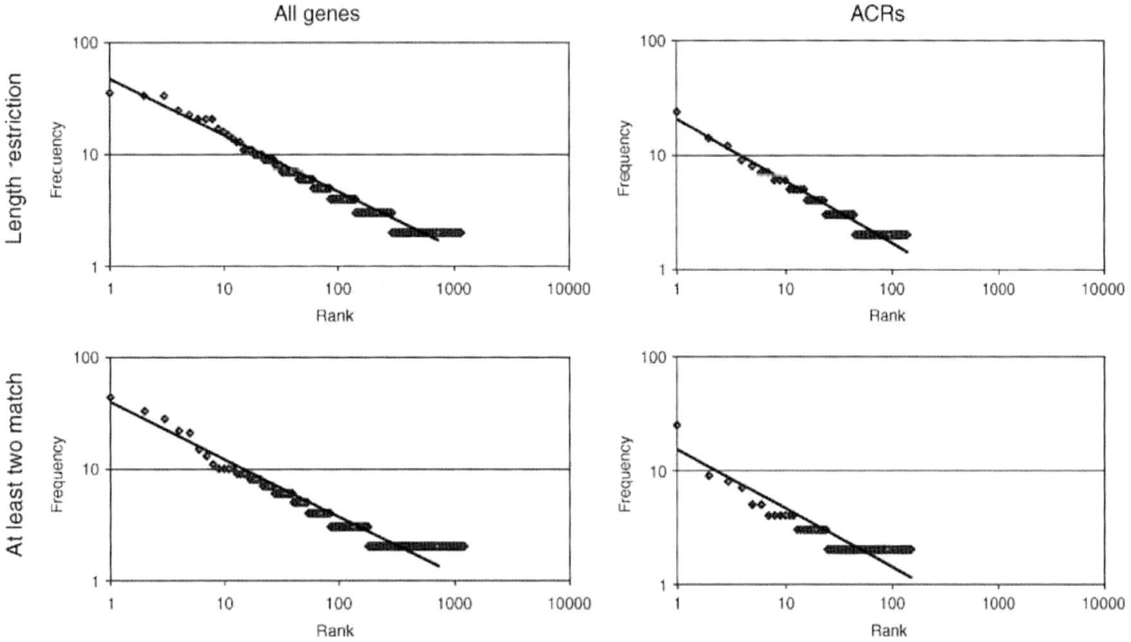

Figure 4. Exon distributions for the two grouping algorithms. The trendlines are drawn through the midrank of each frequency.

a random graph of size n by starting with a set of n points and randomly generating matches between points in the set. Each individual pair of points would have a match with some probability p. After that, we can apply a greedy algorithm to collect points into groups based on the generated matches. The sizes of the groups can then be ordered and graphed against their rank on log–log scales. It is obvious that for small p-values groups are unlikely to form; while for large p-values all the points will be absorbed into a single large group. The range of p-values at which there is some non-trivial behavior is quite small. Our simulations suggest that groups worthy of interest form when p is close to $1/n$. Furthermore, for values less than $1/n$ the behavior seems to be rather close to power-scaling, whereas when p is greater than $1/n$ a single large group dominates (however, the rest of the groups do fall into a power scaling pattern).

Employing $n = 55{,}100$ (the approximate number of exons in the sample of all the eukaryotic genes), Figure 5(a) and (b) shows some of the simulated distributions of groups for the range of p-values with non-trivial behavior. Clearly, simulations that exhibit power scaling sweep through a range of slopes. Figure 6(a) plots the slopes of the simulations versus the p-values. (The slopes were calculated after discarding the largest group to accommodate simulations with larger p-values.) Similarly Figure 6(b) plots the sizes of the largest groups in simulations against the p-values. Both plots suggest that the simulations behave reasonably away from a critical value at $1/n$ and that a simulation with $p = 0.000015$ would produce the same slope of -0.5 that we obtain from the distribution of exons. However, Figure 7(a) shows that at that p-value, the simulation curve falls well above the true exon curve. Figure 7(b) shows another comparison of the exon curve with a simulation curve, where the largest group in the simulation roughly corresponds to that in our exon set. Here, the slopes are clearly divergent. Finally, based on the number of matches in the set of exons we can estimate the probability of a match between any two exons and compare that probability to the p-value we encountered in the simulations. With our cutoffs, we obtained a total of 3463 matches which implies that the probability of a match is $3463/(55{,}100^2/2) = 0.00000228$. This estimate is about an order of magnitude smaller than the p-values at which simulations produce the patterns comparable to (but still very different from) that of the exon distribution. Similar results hold for the ACR comparisons. Overall, we conclude that it is impossible to recreate the power-scaling pattern of the exon matches through our type of simulations.

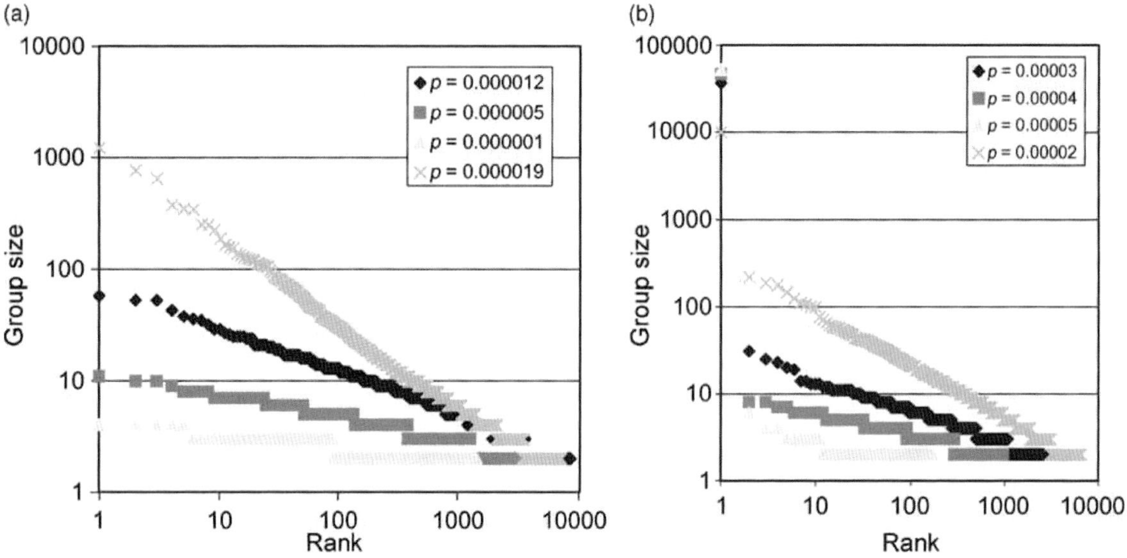

Figure 5. Distribution of groups in simulation experiments at different p-values. (a) Smaller p-values resulting in power-scaling behavior. (b) Larger p-values resulting in the domination by a single group.

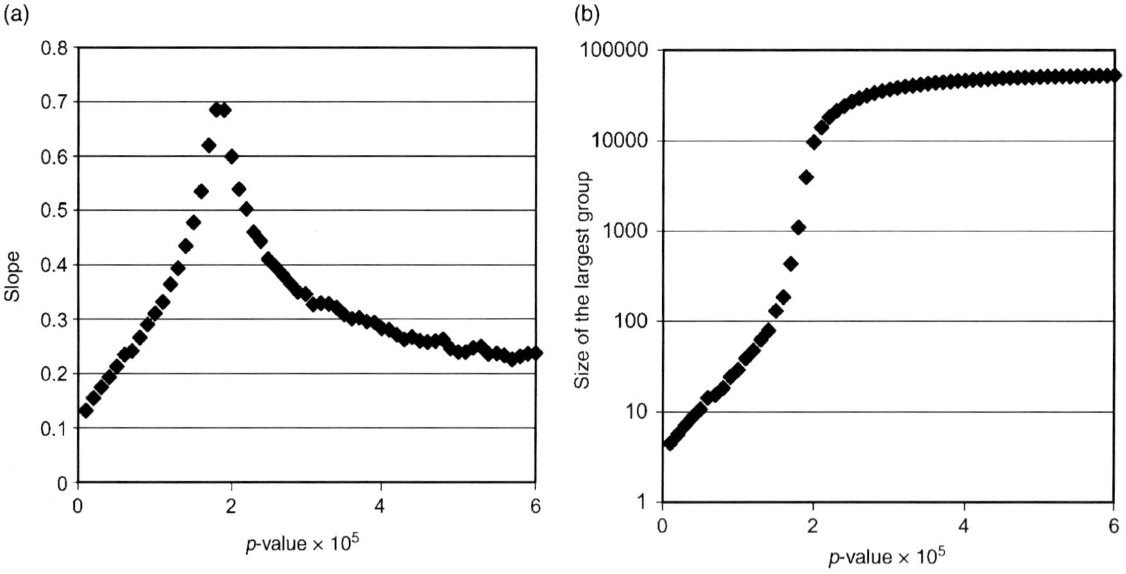

Figure 6. Characteristics of group size distributions at different p-values with 55,100 points. The results are averages of 20 simulations at each p-value. (a) Slopes of the distributions on log–log plots (excluding the largest group). (b) Sizes of the largest groups in simulations.

Now, while it appears unlikely that the Zipf-like distribution of exons arose as an artifact of the grouping algorithm, its origin remains unclear. One can speculate that the peculiarities of molecular evolution tend to produce families of sequences that fall into a Zipf-like pattern with the slope of $-1/2$. Presumably the connections within our exon groups are not entirely random and some phylogenetic constraints necessarily apply to them. It may, in fact, be possible to simulate the behavior of the exons by a random graph, if we could incorporate some tree-like properties in the generation of the subgroups.

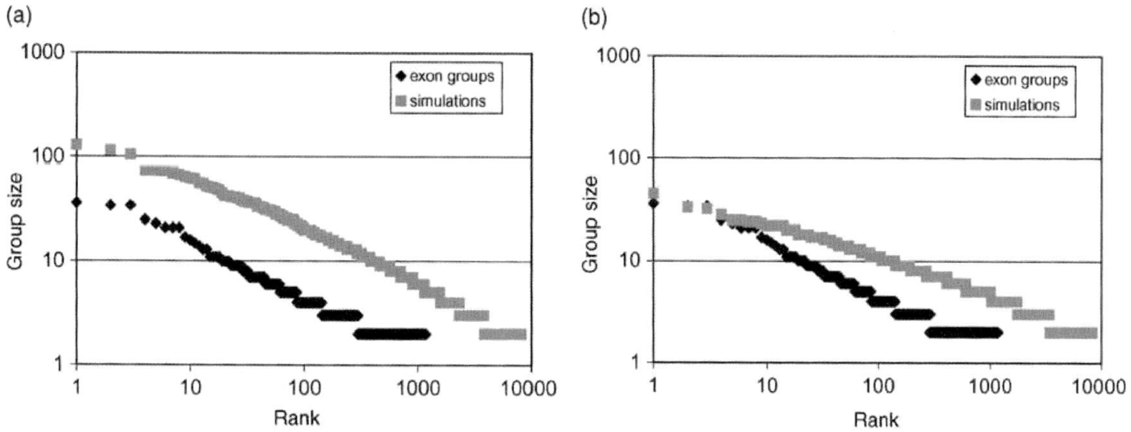

Figure 7. Comparison of two simulations with real exon groups. Exon groups were formed using the length-restricted algorithm. (a) Simulation p-value = 0.000015, so that the slope is close to 1/2. (b) Simulation p-value = 0.000011, so that the y-intercept is the same as in the exon groups.

Table 2. Estimates for the size of the exon universe (N), based on the power scaling distributions

	Grouping method	First group	β	A	N
All genes (55097 exons)	Length match ≥2	48	−0.50	0.000871	330000
		40	−0.52	0.000726	750000
ACRs (8882 exons)	Length match ≥2	21	−0.54	0.002364	95000
		15	−0.52	0.001689	130000

First group sizes are based on trendlines fitted to the midpoints of individual frequencies. The estimates are based on formula 4 (see text). If one were to assume the slope of −1/2, the estimates become 330,000 for all genes and 49,000 for ACRs. The two grouping methods are discussed in the text.

Universe of exons

The power-scaling pattern can be employed to make an estimate of how many ancient exons were involved in producing the current diversity of genes. If we assume that the Zipf-like behavior is a property of the probability distribution of ancient exons, then the individual probabilities can be written down as

$$p_r = A \times r^{\beta}, \tag{2}$$

where A is a normalization constant, r is the rank and β is the slope on a log–log plot. Now, $\sum_{r=1}^{N} p_r = 1$, where N is the last index, or equivalently the size of the exon universe. We can approximate the above sum by an integral to get

$$\int_1^N p(r)\,\mathrm{d}r \approx 1 \to A\int_1^N r^{\beta}\,\mathrm{d}r = A\left(\frac{1}{\beta+1}r^{\beta+1}\right)\Bigg]_1^N \approx 1, \tag{3}$$

which leads to

$$\frac{A}{1+\beta}N^{1+\beta} \approx 1 \to N \approx \left(\frac{1+\beta}{A}\right)^{\frac{1}{(1+\beta)}}, \tag{4}$$

where we have assumed that N is much greater than 1. We thus get an expression for N in terms of the normalization constant and the slope. We can then estimate A by making use of the fact that $p_1 = A$ and that probability associated with the highest ranked exon can be approximated with its frequency. That frequency corresponds to the size of the highest ranked group divided by the number of exons in the sample. If the true probability distribution corresponds to $\beta = -0.5$, then we might estimate N for the two top patterns of Figure 4 as 330,000 for all genes and as 49,000 for the ACRs. However, these numbers are quite sensitive to the value for β. The fitted trendlines of Figure 4 yield the estimates in Table 2.

Formula (4)'s strong dependence on the slope (β) leads to a variation in the estimates for N. However, while the range of these estimates is rather large, it stays within a single order of magnitude. The ACR estimates are smaller than the all-genes counterparts because of the comparatively large number of exon matches in the ACR sample. In general, if the probability of match were kept constant, one would expect that the number of matches between elements of a set would grow as the square of the set size. However, the size of the all-genes sample is 6.2 times larger than that of the ACR sample, while the number of *matches* in the all-genes sample is only 7.6 (see Table 1) times greater than that in the ACR sample – much less than the expected factor of 38. This discrepancy can be explained by the fact that ACRs involve conserved regions of sequence. Forces of molecular evolution that change homologous exons to make them appear unrelated will have acted less on the exons derived from conserved regions. Consequently, the estimate obtained from the ACR database should allow us to look further back in time and thus represent a closer approximation to the size of the exon universe.

Our first estimate then for N, the universe of exons, based on the ACR data choosing a cutoff so that the matches are significant at the 5% level, is a range 50,000–100,000. However, if we have missed matches that are biologically significant, that would underestimate N. We can take that into account by asking for an estimate of the largest number of matches consistent with Figure 3. That would be the total difference between the curves of Figure 1, for instance. For the ACRs that is 1606 matches (vs. 455 using the cutoff). Since the number of matches should rise as the square of the group sizes, this would correspond to groups about 1.9 times larger or the N about 3.5 times smaller, or an estimate of 15,000–30,000 for N.

Dorit et al. (Dorit, Schoenbach & Gilbert, 1990) tried to estimate the number of ancestral exons by examining a collection of 1255 exons and assuming equi-probable ancestral exons. Their first estimate was an N of 56,000. They then tried to estimate (the 'wedge' calculation) a maximum number of matches and a minimum size for N, which was 950. They ultimately settled on 7000 as an estimate. Our present estimates are higher, but in the same range of 10^4 to 10^5 exons.

Even though estimates of the order of 100,000 may appear quite large, they pale in comparison with 20^{40} – the number of all possible exons of length 40, the peak of the distribution of real exons. If evolution were to construct a protein composed of 300 amino acids by a simple amino acid search it would need to consider 20^{300} possibilities. If however, it were to construct the protein by a linear search through the universe of 10^5 exons, it would only need to try 10^6 possibilities. Thus, a relatively small size of the exon universe allows for a faster exploration of the set of potential proteins, while limiting the variety of the possible shapes.

Is our estimate for the number of ancestral exons a 'good' one? Unfortunately, we expect that many of the current intron positions in the ACR proteins do not reflect ancestral characteristics. The findings of de Souza et al. (de Souza et al., 1998) and Roy et al. (Roy et al., 1999) suggest that only 25% of current intron positions (half of all phase zero positions) could be ancestral while three-quarters of all current positions are new. Furthermore, the *Exon theory of genes* would suggest that half to three-quarters of the original intron positions should have disappeared. Thus there is only about a 1/16 chance that a given exon is correctly bounded by ancestral introns. Since our calculation uses an approximate matching of length, permitting the ends of the exons that match to vary by up to ±5 residues, if the average exon is 40 long then about 1/4 of the time one would have an intron in a 'pseudo-correct' position solely by chance to add the 1/4 correct. Thus there is about 1/4 probability that an exon will be recognized. Thus the groups should be about four times larger, and the estimate for the universe about 16 times smaller, or only a few thousand 40-long shapes.

Conclusion

We find a significant number of matches of exon sequences in ACR genes. These matches produce groups whose frequency distribution follows a power law, falling off roughly as the inverse square root of the rank. If we take this power law as proscriptive, then we would estimate a universe of 50,000–100,000. If we estimate the total number of matches above the random background, that estimate drops to 15,000–30,000. If we try to estimate how many true matches one might have missed, if there were ancestral introns that have been lost over time, these estimates for an 'ancestral universe' drop by an order of magnitude. We have focused here on the ACR genes, because these sequences may show the least remodeling by evolution and thus the sequence comparisons may carry us further back in time.

References

Altschul, S.F., T.L. Madden, A.A. Schaffer, J. Zhang, Z. Zhang et al., 1997. Gapped BLAST and PSI-BLAST: a new generation of protein database search programs. Nucl. Acids Res. 25: 3389–3402.

Benson, D.A., M.S. Boguski, D.J. Lipman, J. Ostell, B.F. Ouellette et al., 1999. GenBank. Nucl. Acids Res. 27: 12–17.

278

Brenner, S.E., C. Chothia & T.J. Hubbard, 1998. Assessing sequence comparison methods with reliable structurally identified distant evolutionary relationships. Proc. Natl. Acad. Sci. USA 95: 6073–6078.

Cho, G. & R.F. Doolittle, 1997. Intron distribution in ancient paralogs supports random insertion and not random loss. J. Mol. Evol. 44: 573–584.

de Souza, S.J., M. Long, R.J. Klein, S. Roy, S. Lin et al., 1998. Toward a resolution of the introns early/late debate: only phase zero introns are correlated with the structure of ancient proteins. Proc. Natl. Acad. Sci. USA 95: 5094–5099.

Doolittle, W.F., 1978. Genes in pieces: were they ever together? Nature 272: 581–582.

Dorit, R.L., L. Schoenbach & W. Gilbert, 1990. How big is the universe of exons? Science 250: 1377–1382.

Gilbert, W., 1978. Why genes in pieces? Nature 271: 501.

Gilbert, W., 1987. The exon theory of genes. Cold Spring Harb. Symp. Quant. Biol. 52: 901–905.

Logsdon Jr., J.M., A. Stoltzfus & W.F. Doolittle, 1998. Molecular evolution: recent cases of spliceosomal intron gain? Curr. Biol. 8: R560–563.

Roy, S.W., M. Nosaka, S.J. de Souza & W. Gilbert, 1999. Centripetal modules and ancient introns. Gene 238: 85–91.

Saxonov, S., I. Daizadeh, A. Fedorov & W. Gilbert, 2000. EID: the exon–intron database – an exhaustive database of protein-coding intron-containing genes. Nucl. Acids Res. 28: 185–190.

Zipf, G.K., 1949. Human Behavior and the Principle of Least Effort. Addison-Wesley, Redwood City, CA.

The pattern of intron loss

Scott W. Roy* and Walter Gilbert

Department of Molecular and Cellular Biology, Harvard University, 16 Divinity Avenue, Cambridge, MA 02138

Contributed by Walter Gilbert, November 11, 2004

We studied intron loss in 684 groups of orthologous genes from seven fully sequenced eukaryotic genomes. We found that introns closer to the 3′ ends of genes are preferentially lost, as predicted if introns are lost through gene conversion with a reverse transcriptase product of a spliced mRNA. Adjacent introns tend to be lost in concert, as expected if such events span multiple intron positions. Directly contrary to the expectations of some, introns that do not interrupt codons (phase zero) are more, not less, likely to be lost, an intriguing and previously unappreciated result. Adjacent introns with matching phases are not more likely to be retained, as would be expected if they enjoyed a relative selective advantage. The findings of 3′ and phase zero intron loss biases are in direct contradiction to an extremely recent study of fungi intron evolution. All patterns are less pronounced in the lineage leading to *Caenorhabditis elegans*, suggesting that the process of intron loss may be qualitatively different in nematodes. Our results support a reverse transcriptase-mediated model of intron loss.

evolution | genome evolution

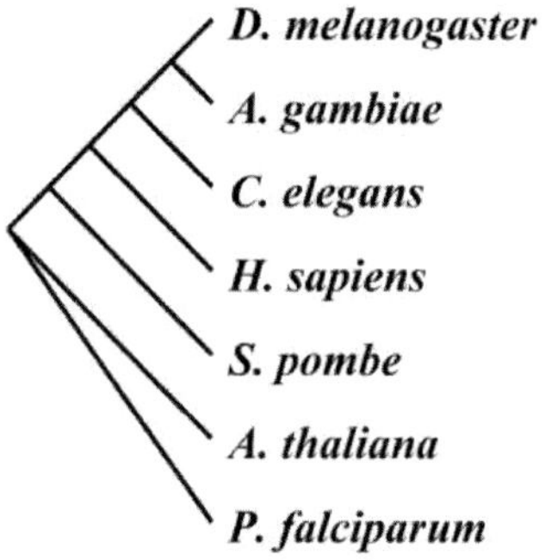

Fig. 1. The most likely relationship between the analyzed species. The deepest node is not resolved, with *Arabidopsis thaliana* clustering either with *Plasmodium falciparum* or with the other species.

Two generalities of the intron-exon structure of eukaryotic genes remain unexplained. First, introns in intron-sparse species or in intron-sparse genes cluster near the 5′ ends of genes (1, 2). Second, intron positions within codons are doubly biased: introns tend to lie between the third base of one codon and the first base of the subsequent codon (phase zero) rather than between the first and second (phase one) or second and third (phase two) bases of a codon; and adjacent introns tend to be of the same phase (3–9).

The 5′ skew could be due to mutation-biased intron loss, selection-biased intron loss, or biased intron gain. If intron loss proceeds by means of gene conversion of the genomic copy of a gene by the reverse transcriptase product of a spliced transcript (RT-mRNAs) (10–15), the 3′ bias of RT products (12) could cause a higher rate of loss for 3′ introns (1, 2). Alternatively, possible preferential retention of 5′ introns could reflect their greater selective importance, possibly due to a greater concentration of regulatory elements (reviewed in ref. 16). Finally, intron gain could favor 5′ ends of genes for some unappreciated reason.

We analyzed intron losses in 684 groups of orthologous genes from seven eukaryotic species, previously analyzed by Rogozin *et al.* (17). Fig. 1 shows the most likely phylogeny for the species (18). Results calculated assuming the alternative coelomata grouping (19, 20) are similar and provided in Tables 3 and 4 and Figs. 6–9, which are published as supporting information on the PNAS web site. For each lineage, we defined introns known to be ancestral to the lineage (KAL) based on presence in both the sister group of the lineage and an outgroup. For example, introns present in a dipteran (*Drosophila melanogaster* and *Anopheles gambiae*) or *Caenorhabditis elegans* as well as a non-animal are KAL for the lineage leading to *Homo sapiens*. We found that 3′ KAL introns are preferentially lost for every lineage analyzed except *C. elegans*, suggesting that the clustering of introns near the 5′ of genes in intron-sparse genomes is due to biased intron loss.

To determine whether this 3′ bias is due to RT-mRNA-mediated loss or to differential selection, we analyzed the pattern of loss among adjacent KAL introns. If introns are lost through gene conversion by RT-mRNAs, adjacent introns may sometimes be lost in concert when a gene conversion event spans multiple intron positions (14, 21, 22). For each lineage, we compared the pattern of intron loss with the expectation assuming independent intron loss, and found a signal of concerted loss of adjacent introns for diptera and *Saccharomyces pombe*, but not *C. elegans*, evidence that the 5′ bias is due to RT-mRNA-mediated intron loss.

The two phase skews (abundance of phase zero introns and correspondence of adjacent intron phases) could be due to the legacy of gene formation, to insertional bias, to selection, or to alternative splicing. First, the biases could be echoes of gene formation through combination of exons or groups of exons (3–7, 9, 23–26) mediated by ancient phase zero introns (23). Second, intron insertion might be phase-biased (5, 8, 16, 27–34), perhaps due to preferential insertion into sites which themselves happen to be phase-biased (although, see ref. 7). Third, selection could favor phase zero introns due to transcript fidelity, usefulness in exon shuffling, or resilience to intron boundary sliding (17, 35), and the avoidance of downstream frame shifts in cases of errant exon exclusion could favor adjacent introns of the same phase (35). Finally, conditionally spliced exons that must be flanked by same phase introns to avoid downstream frame shifts could cause the bias.

A relative selective advantage for phase zero introns predicts preferential retention of phase zero introns. Instead, we find the opposite-phase zero KAL introns are more likely to be lost in all analyzed lineages except *C. elegans*. Among introns known to be present at the animal/fungi split, introns in phase zero are retained in fewer animal/fungi taxa than those in phases one and two. Thus, the phase bias persists despite, not because of, intron loss biases. KAL introns with adjacent KAL introns of matching phase are no more likely to be retained, suggesting that adjacent intron phase correspondence is not due to selection.

The pattern of intron loss for 684 eukaryotic groups of orthologs provides evidence that (*i*) introns are lost through gene

Abbreviation: KAL, known to be ancestral to the lineage.

*To whom correspondence should be addressed. E-mail: scottroy@fas.harvard.edu.

Table 1. Summary of the data

Lineage	Sister taxa	Total introns	Shared with Sister	Shared with Non-sister	KAL introns (retained plus lost)	Genes with P ≥ 2; r ≥ 1
D. melanogaster	*A. gambiae*	725	382	489	451 (295 + 156)	11
A. gambiae	*D. melanogaster*	675	382	451	489 (295 + 194)	3
Diptera	*C. elegans*	1,016	234	609	634 (198 + 436)	32
C. elegans	Diptera	1,468	234	634	609 (198 + 411)	29
H. sapiens	Diptera, *C. elegans*	3,345	933	907	551 (339 + 112)	4
S. pombe	Animals	450	223	158	927 (131 + 796)	30
A. thaliana	Animals, *S. pombe*	2,933	908	97	119 (73 + 46)	0
P. falciparum	Animals, *Sp. At*	450	143	—	—	—

Sister taxa, the data set species diverging at the base of the given lineage; introns, total number of introns in the descendent(s) of the lineage; introns shared with sister, number of introns shared between species in the first two columns; introns shared with non-sister, number of introns shared between species in the first column and species not in the second column; KAL introns, number of KAL introns retained plus number lost in lineage, as defined in the text; genes with P > 2; r ≥ 1, in the last column, number of genes that have lost two or more KAL introns as well as retained one or more KAL introns along the lineage. P, KAL introns lost; r, KAL introns retained.

conversion by RT-mRNAs; (*ii*) adjacent introns are often lost in concert; (*iii*) phase zero introns are lost at higher rates than phase one and two introns, thus selection does not appear to drive phase zero intron abundance; (*iv*) adjacent introns of the same phase are not preferentially retained, thus selection does not appear to drive adjacent intron phase correspondence; and (*v*) the lineage leading to *C. elegans* is exceptional in its intron loss pattern, evidencing none of the biases observed elsewhere.

Methods

Data Set and Programs. We downloaded amino acid level sequence alignments and corresponding intron positions with presence-absence matrices for each intron position in the conserved regions of 684 groups of orthologous genes, compiled by Rogozin *et al.* (see ref. 17 for details), from the National Center for Biotechnology Information (which can be accessed at ftp://ftp.ncbi.nlm.nih.gov./pub/koonin/intron_evolution). Introns present at the exact same position in two orthologs were assumed to be homologous. *S. cerevisiae* was excluded because it has extremely few introns and interrupts the lineage running from the animal-fungus split to *S. pombe.* Analyses were performed by using PERL scripts.

KAL Introns. For each lineage, we defined KAL introns as those present in a sister group to the lineage, as well as an outgroup, and which are thus assumed to be present at the base of the lineage [e.g., for the *C. elegans* lineage, those introns present in a dipteran (sister group) as well as *H. sapiens* or a non-animal (outgroup)]. Introns present in the studied taxa (e.g., *C. elegans*) and the sister but no outgroup, or in the studied taxa and an outgroup, but not the sister, are also known to be ancestral to the lineage. However, such introns are only known to be ancestral by virtue of their presence in the studied taxa. If they had been lost in this taxa, they would not be known to be ancestral. Thus, such introns are themselves a retention-biased set, inappropriate for studying the pattern of loss, and were excluded.

Choice of Lineages. The data are summarized in Table 1. We selected lineages with large numbers of KAL introns for further analysis. For analyses of the pattern of adjacent intron loss, only genes with at least one retained KAL intron and at least two lost KAL introns are informative (right column of Table 1). We therefore chose the general diptera lineage, rather than the specific *A. gambiae* or *D. melanogaster* lineages, because it has more informative genes. When speaking of intron loss in diptera, "retained" denotes presence in one or both dipteran species, and "lost" denotes absence in both.

Probability of Sums over a Group of Genes. For many tests, we compare the sum of a function for each gene over a group of genes to the null expectation. For a function $F = \Sigma_{j=1}^{n} f_j$, where f_j is some function for the jth gene of n total, $\Pr\{F = X\} = \Pi_j \Pr\{f_j = x_j\}$ summed over all sets of x values for which $\Sigma_j x_j = X$. Expressions for $\Pr\{f_j = x_j\}$ for a range of functions are given below. P is then the probability that F is as divergent from the expectation as is the real value.

Adjacent Intron Loss. The probability that a gene which randomly loses l and retains r KAL introns through single-intron loss events loses exactly d pairs of lost adjacent introns (where n lost adjacent introns count as $n - 1$ lost pairs) is

$$\Pr\{d|l, r\} = \frac{\binom{l-1}{d}\binom{r+1}{l-d}}{\binom{r+l}{l}}. \quad [1]$$

The probability that it loses t triples of adjacent KAL introns is:

$$\Pr\{t|l, r\} = \left\{ \begin{array}{ll} \dfrac{\sum_{i=0}^{\frac{l}{2}} \binom{r+1}{i,\, l-2i}}{\binom{r+l}{l}} & \text{for } t = 0 \\ \dfrac{\sum_{i=1}^{\frac{l-t}{2}} \binom{i+t-1}{t}\binom{r+1}{i,\, l-t-2i}}{\binom{r+l}{l}} & \text{for } t > 0 \end{array} \right\}. \quad [2]$$

The probability that it loses q quartets of adjacent KAL introns is

$$\Pr\{q|l, r\} = \left\{ \begin{array}{ll} \dfrac{\sum_{i=0}^{\frac{l}{3}} \sum_{j=0}^{\frac{l-3i}{2}} \binom{r+1}{i,\, j,\, l-3i-2j}}{\binom{r+l}{l}} & \text{for } q = 0 \\ \dfrac{\sum_{i=1}^{\frac{l-q}{3}} \binom{i+q-1}{q} \sum_{j=0}^{\frac{l-q-3i}{2}} \binom{r+1}{i,\, j,\, l-q-3i-2j}}{\binom{r+l}{l}} & \text{for } q > 0 \end{array} \right\}, \quad [3]$$

where terms including negative factorials are defined to be zero throughout.

If introns are lost independently, the probability that a gene will lose e separate clusters of one or more adjacent introns is

$$\Pr\{e|l, r\} = \frac{\binom{l-1}{e-1}\binom{r+1}{e}}{\binom{r+l}{l}}. \qquad [4]$$

If instead adjacent introns tend to be lost in concert, the lost introns will fall in fewer, larger clusters. The chance that all lost KAL introns will be adjacent (forming one cluster) is

$$\Pr\{e = 1|l, r\} = \frac{\binom{l-1}{0}\binom{r+1}{1}}{\binom{r+l}{l}} = \frac{r+1}{\binom{r+l}{l}}. \qquad [5]$$

Because these calculations require multiple KAL introns to have been lost in a gene, numbers for the *H. sapiens* lineage were too small for analysis (four genes lost two KAL introns; none of the genes lost more).

Loss by Phase. To test whether introns of different phases in the same gene are equally likely to be lost, we use only genes with KAL introns both in and out of phase zero and both lost and retained. The chance that such a gene loses x out of z total phase zero KAL introns is

$$\Pr\{x|z, l, r\} = \frac{\binom{z}{x}\binom{r+l-z}{l-x}}{\binom{r+l}{l}}. \qquad [6]$$

3′ Loss Bias. For some groups of orthologs, the termini are unalignable, obscuring the true end of the ancestral protein. Therefore, we discarded genes in which the last conserved region lies 20% or more of the length of the alignment from either end of the alignment.

Results

5′-3′ Position. We calculated the correlation between intron position and retention or loss (1, 0 respectively) for the KAL introns for each lineage. We measured intron position either as number of codons from the 3′ end of the gene or as percentage of total coding sequence length from 3′ to 5′ (Fig. 2). There is no significant correlation for the former measure; for the latter, there are significant positive correlations (greater 3′ loss; $P < 0.05$) for *H. sapiens* and *S. pombe*. However, these metrics are problematic. Intergene differences in numbers of transcripts, rates of reverse transcription per transcript, distributions of reverse transcription product lengths, and rates of gene conversion may obscure the signal. To avoid such problems, we then compared introns within the same gene. For each lineage, we counted each gene as an independent test. If a 3′ intron loss bias exists, the number of genes with a positive correlation between retention and distance from the 3′ should be greater than the number with a negative correlation; if not, the numbers should be equal. As Fig. 2*C* shows, for each lineage, more genes have a positive than a negative correlation. This bias is statistically significant for three of the four lineages (diptera, $P = 0.032$; *H. sapiens*, $P = 0.032$; *S. pombe*, $P = 2.8 \times 10^{-6}$; *C. elegans*, $P = 0.31$ by χ^2).

Adjacent Intron Loss. For each lineage, we calculated probability distributions for the total numbers of pairs, trios, and quartets of adjacent KAL introns lost, assuming independent single-intron loss, and compared these distributions to the observed values. For instance, a gene losing two of its three KAL introns will either retain the 5′ intron (100), the middle intron (010), or the 3′ intron (001),

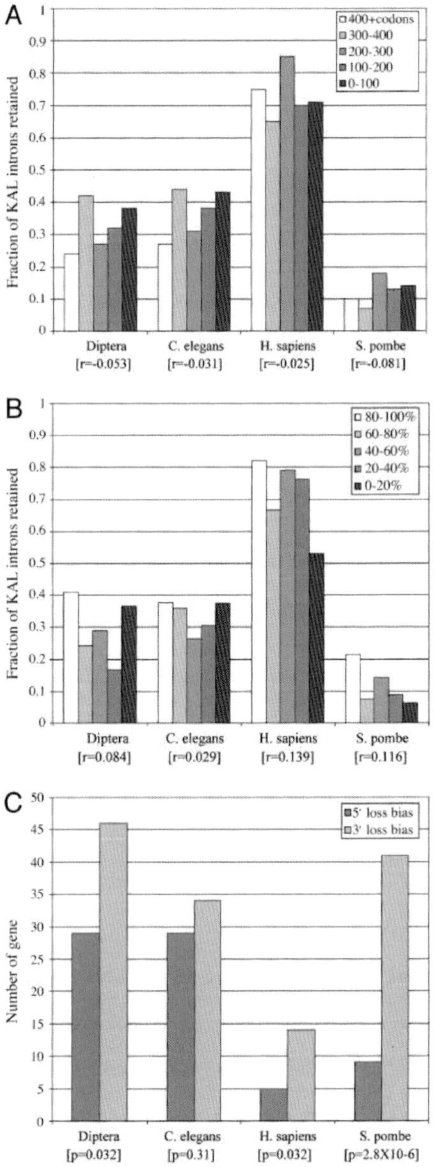

Fig. 2. 3′ intron loss bias. (*A*) The relationship between intron position measured as the length of coding sequence from the 3′ end of the gene and the fraction of introns retained. (*B*) The relationship between the intron position and the fraction of introns retained is shown. For each lineage, each bar represents a quintile of gene length: the rightmost bar gives the fraction retained for KAL introns in the C-terminal 20% of the gene, the second rightmost the fraction for introns from 20–40% of the length from 3′ to 5′, etc. (*C*) Detection of 3′ loss bias per gene. For each lineage, the numbers of genes showing a positive correlation between distance from 3′ and intron retention (3′ loss bias, light bars) and negative correlation (5′ bias, dark bars) are given.

thus the chance of losing adjacent introns is 2/3. A gene losing three of five KAL introns may lose no adjacent pairs of introns (01010), one adjacent pair (00101, 00110, 10010, 01001, 01100, and 10100),

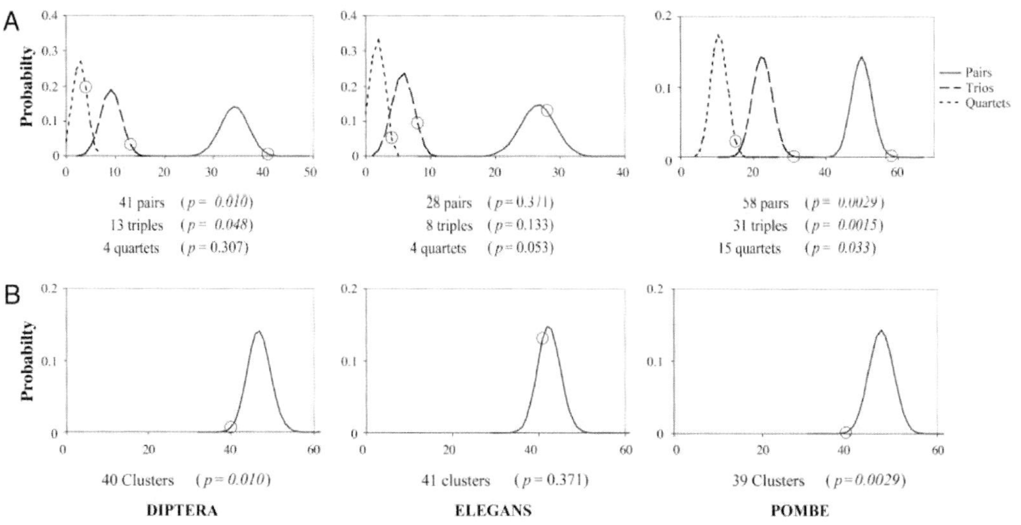

Fig. 3. Adjacent introns are lost in concert. (*A*) The probability distributions for total numbers of pairs, trios, and quartets of adjacent KAL introns lost for all analyzed genes for each lineage, with the observed values given below. (*B*) The probability distribution for the total numbers of clusters of one or more lost adjacent introns for all analyzed genes, with the observed values given below.

or two adjacent pairs (00011, 10001, and 11000), with respective probabilities of 1:6:3. This gene also has a 3/10 chance (00011, 10001, and 11000) of losing an adjacent trio. For each lineage, we calculated probability distributions for numbers of lost adjacent pairs, trios, and quartets of KAL introns for each gene and used these values to calculate overall probability distributions for the total numbers for all genes. Fig. 3*A* gives probability distributions and observations. The *S. pombe* and diptera lineages show significant excesses of adjacent pairs and trios of lost KAL introns, and *S. pombe* shows a significant excess of quartets (diptera pairs: P = 0.010, trios, P = 0.048, quartets, P = 0.307; *S. pombe*: P = 0.0029, P = 0.0015, P = 0.033). *C. elegans* shows slight but nonsignificant trends.

We then asked generally about clustering of KAL intron losses along the gene. For instance, if a gene independently loses three KAL introns of five, the relative probabilities of one, two, and three clusters of one or more adjacent lost KAL introns is 3:6:1 (11000, 10001, and 00011 vs. 01100, 00110, 10100, 10010, 01001, and 00101 vs. 01010). For each lineage, we calculated probability distributions for each gene, and from these distsributions the overall probability distributions for total numbers of clusters for all genes (Fig. 3*B*). All lineages show fewer clusters (greater clustering) than expected, reaching significance for diptera (P = 0.010) and *S. pombe* (P = 0.0029). There was also a significant excess of genes in which all lost KAL introns were adjacent (one cluster) in diptera (P = 0.013) and *S. pombe* (P = 0.0013), but not *C. elegans* (P = 0.425).

Phase. For each lineage, we calculated the fractions of phase zero and phase one and two KAL introns retained (Fig. 4*A*). In the diptera lineage, 87/327 (26.7%) phase zero and 111/307 (36.2%) phase one and two KAL introns were retained (P = 0.0061). For humans, the numbers are 152/219 (69.4%) and 187/232 (80.6%; P = 0.0041), for *S. pombe*, 52/493 (10.5%) and 79/434 (18.2%; P = 0.00059). Only *C. elegans* gives a nonsignificant result (P = 0.22).

To ensure this result was not due to phase zero introns happening to lie in loss-prone genes, we analyzed the loss pattern by phase for each gene separately. For each informative gene (those with phase zero and non-phase zero KAL introns and that have lost some but not all of their KAL introns) for each lineage, we calculated expectations and probability distributions for losing a given number of phase zero introns given the number of total introns lost. Although the data set is small because of a paucity of informative genes, the general trend holds. For the lineage leading to diptera, 54 phase zero introns were lost vs. 49.0 expected (P = 0.096); for *H. sapiens*, 11 observed vs. 6.7 expected (P = 0.012); for *S. pombe*, 55 observed vs. 50.5 expected (P = 0.086).

Another way to look at the data is to analyze introns known to be present at the fungi-animal split (by virtue of presence in fungi and/or animals as well as plants and/or *Plasmodium*). There are 503 such introns in phase zero and 451 in phase one or two. For each intron we asked how many animal-fungi taxa it was found in (1, 2, 3, or 4) among *S. pombe, H. sapiens, C. elegans* and diptera. Fig. 4*b* gives the results. Only 32.0% phase zero introns were found in more than one fungi-animal taxon, compared with 53.8% for phase one and two introns ($P = 2.79 \times 10^{-6}$ by a χ^2 test) and only 0.8% compared with 5.8% were found in all four taxa (P = 0.0046). The overall difference between the distributions is significant ($P = 1.0 \times 10^{-5}$ by χ^2, 3 df).

Adjacent Intron Phase. For each lineage, we divided KAL introns based on matching or nonmatching adjacent intron phase (for upstream and downstream adjacent KAL introns, separately) and asked whether introns whose adjacent KAL introns are in the same phase were preferentially retained. No pattern of preferential retention of introns which match the adjacent KAL intron in phase is observed (Table 2).

Introns Shared with *Plasmodium*. If *Plasmodium* is an outgroup to the plant-animal divergence, introns shared between *Plasmodium falciparum* and modern plants, animals, or fungi have been maintained for an extremely long time down multiple disparate lineages. Alternatively, if *Plasmodium* is more closely related to plants, it has lost an impressive 92% percent of its introns since that divergence (S.W.R. and W.G., unpublished data). In either case, introns shared between *Plasmodium* and animals/fungi/plants are expected to be particularly resilient to intron loss. Among introns present in animals and/or fungi as well as *Plasmodium* and/or *Arabidopsis* (and thus known present at the animal/fungi split), 104 are present

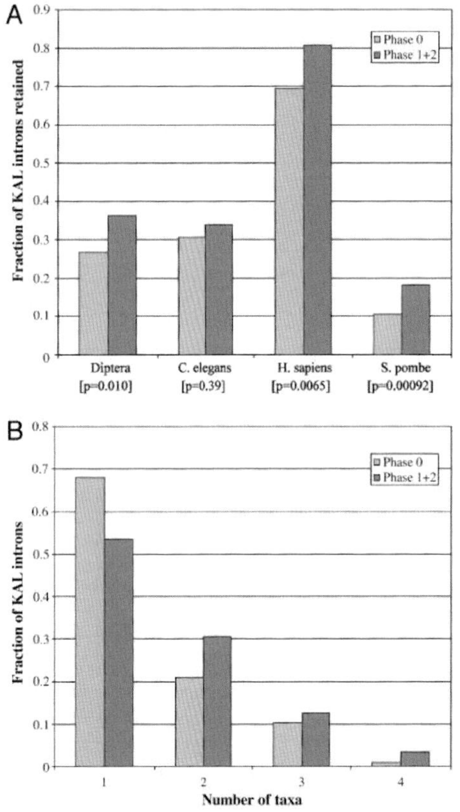

Fig. 4. Phase-biased intron loss. (*A*) The fraction of phase zero, phase one, and two KAL introns retained along each lineage. (*B*) The distribution of the number of fungi/animal taxa (diptera, *C. elegans*, *H. sapiens*, and *S. pombe*) in which an intron is retained for introns shared between fungi/animals and *Arabidopsis*/*Plasmodium* for phase zero and phase one and two introns.

Table 2. The effect of adjacent intron phase on intron loss

		Percent introns retained		
		Adjacent intron phase		
		Same	Different	*P*
Diptera				
Ph0	5′	14 (202)	23 (188)	0.99
	3′	23 (202)	16 (159)	0.08
Ph1	5′	31 (35)	13 (112)	*0.01*
	3′	43 (35)	27 (128)	0.06
Ph2	5′	25 (32)	31 (123)	0.81
	3′	23 (32)	32 (136)	0.46
All	5′	17 (269)	22 (423)	0.95
	3′	27 (269)	25 (423)	0.29
C. elegans				
Ph0	5′	17 (139)	22 (127)	0.87
	3′	22 (139)	24 (106)	0.65
Ph1	5′	19 (26)	22 (69)	0.70
	3′	35 (26)	32 (81)	0.50
Ph2	5′	18 (22)	17 (83)	0.55
	3′	36 (22)	16 (92)	*0.04*
All	5′	18 (187)	20 (279)	0.92
	3′	26 (187)	24 (279)	0.35
H. sapiens				
Ph0	5′	67 (24)	80 (35)	0.93
	3′	75 (24)	86 (21)	0.90
Ph1	5′	100 (8)	67 (9)	0.12
	3′	100 (8)	87 (23)	0.39
Ph2	5′	63 (8)	85 (20)	0.96
	3′	88 (8)	80 (20)	0.55
All	5′	73 (40)	80 (64)	0.86
	3′	83 (40)	84 (64)	0.72
S. pombe				
Ph0	5′	4 (137)	9 (122)	0.96
	3′	9 (137)	10 (104)	0.60
Ph1	5′	9 (23)	11 (65)	0.74
	3′	4 (23)	21 (81)	0.99
Ph2	5′	5 (21)	12 (84)	0.93
	3′	5 (21)	23 (86)	0.99
All	5′	5 (181)	10 (271)	0.99
	3′	8 (181)	17 (271)	1.00

5′/3′, whether the effect of the upstream (5′) or downstream (3′) adjacent intron is being tested; same, adjacent intron is of the same phase; different, adjacent intron is of different phase. Sample sizes are given in parentheses. *P* values were calculated with a one-tailed Fisher's exact test.

in *Plasmodium*, and an additional 835 are present in *Arabidopsis*. For each such intron, we asked how many animal/fungi taxa it is present in. The data are shown in Fig. 5. Only 36.8% of non-*Plasmodium* introns, but 53.8% of *Plasmodium* introns, are present in multiple fungi/animal taxa ($P = 0.00031$); only 1.4% of non-*Plasmodium* introns, but 5.9% of *Plasmodium* introns are present in all four taxa ($P = 0.0055$). The overall distributions are significantly different ($P = 0.00014$ by χ^2, 3 df).

Discussion

Our results provide two lines of evidence for a model of intron loss through gene conversion with RT-mRNAs. First, 3′ introns are more likely to be lost in the lineages leading to diptera and *S. pombe* as predicted either by a model of RT-mRNA-mediated intron loss or by even rates of loss, followed by stronger selection against loss of 5′ introns [possibly due to important intronic regulatory elements, which appear to be enriched in 5′ introns (16)]. Thus, the observation of a greater 3′ bias among introns found in diverse species (36) appears to be due to biases in the pattern of intron loss, not gain. These results are in tension with other recent studies. Patthy and Banyai (37) found instances of intron loss in even the 5′ regions of very long multidomain genes in *D. melanogaster* and *C. elegans*. Cho *et al.* (38) found no 3′ intron bias in the cytoplasmic polyadenylation element-binding genes of nematodes, leading them to suggest than introns might instead be lost by genomic deletion. If in fact introns are lost by genomic deletion, the only other plausible proposed mechanism for intron loss that we are aware of, we should see a fairly high ratio of inexact deletions in which a small number of codons are added or lost from the flanking coding sequence (39) to exact deletions, a prediction that has yet to be tested. However, our finding of a lack of 3′ loss bias in *C. elegans* suggests that perhaps intron loss in nematodes is subject to different forces, possibly resolving these tensions.

Second, we find that introns that have been lost tend to be adjacent along the gene, suggesting concerted loss of adjacent introns. Alternatively, clustering of intron losses along the gene could reflect differential intron loss rates along the length of the gene. However, in the absence of such evidence, this result also supports the RT-mRNA-mediated loss model.

A second set of results provides evidence against creation of the observed phase biases by selection or biased loss. The tendency of introns to fall between codons (phase zero) has been explained by some as more positive (or less negative) selection on these introns.

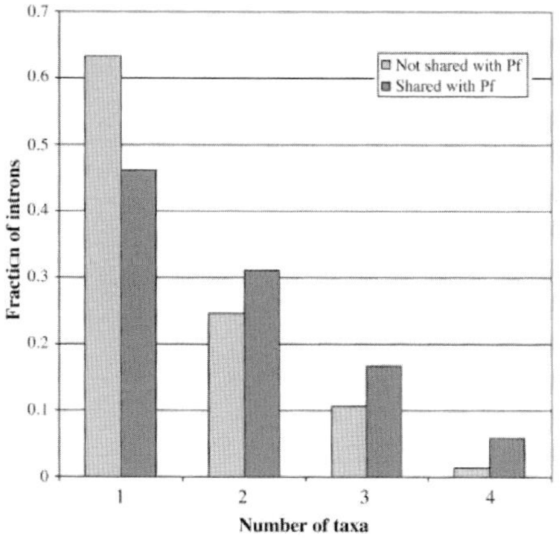

Fig. 5. Introns shared with *Plasmodium* are preferentially retained. The distribution of number of fungi/animal taxa in which an intron is retained for introns shared between fungi/animals and *Arabidopsis*/*Plasmodium* for introns shared and not shared with *Plasmodium*.

However, phase zero introns are shown here to be more, not less, likely to be lost. This result is quite general across the eukaryotes studied. These results eliminate selection, as well as a mutational loss bias, as possible explanations for the observed bias. Correspondence of adjacent intron phases also appears non-selection-driven, because we find that adjacent KAL introns whose phases match are no more likely to be retained.

Our findings of phase zero and 3′ biases in the pattern of intron loss contradict the findings of Nielsen *et al.* (40), who found neither bias in a similar very recent genome-wide study of orthologous groups in four fungal taxa. They analyzed the influence of the 5′-3′ position on intron loss by dividing introns in each gene into five quintiles based on position along the gene and comparing probabilities of loss between groups. However, differences between genes in gene lengths, expression levels, reverse transcription rates, distributions of reverse transcription product lengths, and rates of gene conversion cause problems for such intergene comparisons. These problems may be overcome by comparing introns within the same gene, as we do here. Supporting this interpretation, whereas our intragene analysis shows a strong 3′ loss bias, our intergene analysis, which is very similar to that of Nielsen *et al.* (40), does not. An intragene analysis of their data could determine whether choice of method can explain this discrepancy.

The discrepancy over phase bias in intron loss is harder to explain. Whereas we find that phase zero introns are preferentially lost, their table 1 in ref. 40 shows that phase zero introns are instead preferentially retained in the lineage leading to *Neurospora crassa* ($P = 0.048$ by a Fisher's Exact test on phase zero and non-phase zero, conserved and raw losses) and show no significant bias in the other two lineages. These patterns are surprising and deserve further attention.

A final intriguing result is the deviation in the intron loss pattern in the lineage leading to *C. elegans*. Of the three observed patterns observed here (3′ loss bias, a signal of adjacent intron loss, and phase zero intron loss bias) none is observed in *C. elegans*. This result suggests that intron loss may occur through qualitatively different mechanisms in nematodes. Kent and Zahler (41) offer the interesting observation that the 5′ and 3′ boundaries of unique introns between *C. elegans* and *Caenorhabditis briggsae* show greater similarity to each other than do the boundaries of control introns, and suggested that introns may be lost by nonhomologous recombination between intron boundaries [although their result could also be explained by preferential intron insertion into sites with a consensus sequence of AG|GT (17, 27–34)]. Comparative analyses of multiple nematode genomes should help to better understand these deviations.

The pattern of intron lost in 684 groups of orthologs diverse eukaryotic taxa suggests that (*i*) introns are lost through gene conversion by a retrotransposed copy of a spliced transcript of the gene often spanning multiple intron positions, (*ii*) the phase biases of introns are not due to differential selection or loss biases, and (*iii*) intron loss in the lineage leading to *C. elegans* does not exhibit the biases found in other lineages, suggesting that the dynamics of intron evolution may be qualitatively different in nematodes.

1. Sakurai, A., Fujimori, S., Kochiwa, H., Kitamura-Abe, S., Washio, T., Saito, R., Carninci, P., Hayashizaki, Y. & Tomita, M. (2002) *Gene* **300,** 89–95.
2. Mourier, T. & Jeffares, D. C. (2003) *Science* **300,** 1393.
3. Fedorov, A., Suboch, G., Bujakov, M. & Fedorova, L. (1992) *Nucleic Acids Res.* **20,** 2553–2557.
4. Long, M., Rosenberg, C. & Gilbert, W. (1995) *Proc. Natl. Acad. Sci. USA* **92,** 12495–12499.
5. Long, M., de Souza, S. J., Rosenberg, C. & Gilbert, W. (1998) *Proc. Natl. Acad. Sci. USA* **95,** 219–223.
6. Fedorov, A., Fedorova, L., Starshenko, V., Filatov, V. & Grigor'ev, E. (1998) *J. Mol. Evol.* **46,** 263–271.
7. Long, M. & Rosenberg, C. (2000) *Mol. Biol. Evol.* **17,** 1789–1796.
8. Wolf, Y. I., Kondrashov, F. A. & Koonin, E. V. (2001) *Trends Genet.* **16,** 333–334.
9. Roy, S. W., Lewis, B. P., Fedorov, A. & Gilbert, W. (2001) *Trends Genet.* **17,** 496–501.
10. Bernstein, L. B., Mount, S. M. & Weiner, A. M. (1983) *Cell* **32,** 461–472.
11. Lewin, R. (1983) *Science* **219,** 1052–1054.
12. Weiner, A. M. (1986) *Annu. Rev. Biochem.* **55,** 631–661.
13. Fink, G. R. (1987) *Cell* **49,** 5–6.
14. Long, M. & Langley, C. H. (1993) *Science* **260,** 91–95.
15. Derr, L. K. (1998) *Genetics* **148,** 937–945.
16. Fedorova, L. & Fedorov, A. (2003) *Genetica (The Hague)* **118,** 123–131.
17. Rogozin, I. B., Wolf, Y. I., Sorokin, A. V., Mirkin, B. G. & Koonin, E. V. (2003) *Curr. Biol.* **13,** 1512–1517.
18. Aguinaldo, A. M., Turbeville, J. M., Linford, L. S., Rivera, M. C., Garey, J. R., Raff, R. A. & Lake, J. A. (1997) *Nature* **387,** 489–493.
19. Knoll, A. H. & Carroll, S. B. (1999) *Science* **284,** 2129–2137.
20. Wolf, Y. I., Rogozin, I. B. & Koonin, E. V. (2004) *Genome Res.* **14,** 29–36.
21. Frugoli, J. A., McPeak, M. A., Thomas, T. L. & McClung, C. R. (1998) *Genetics* **149,** 355–365.
22. Wada, H., Kobayashi, M., Sato, R., Satoh, N., Miyasaka, H. & Shirayama, Y. (2002) *J. Mol. Evol.* **54,** 118–128.
23. Gilbert, W. (1987) *Cold Spring Harbor Symp. Quant. Biol.* **52,** 901–905.
24. Patthy, L. (1999) *Gene* **238,** 103–114.
25. de Souza, S. J., Long, M., Schoenbach, L., Roy, S. W. & Gilbert, W. (1996) *Proc. Natl. Acad. Sci. USA* **93,** 14632–14636.
26. Roy, S. W. (2003) *Genetica (The Hague)* **118,** 251–266.
27. Dibb, N. J. & Newman, A. J. (1989) *EMBO J.* **8,** 2015–2021.
28. Tarrio, R., Rodriguez-Trelles, F. & Ayala, F. J. (2003) *Proc. Natl. Acad. Sci. USA* **100,** 6580–6583.
29. Sverdlov, A. V., Rogozin, I. B., Babenko, V. N. & Koonin, E. V. (2003) *Curr. Biol.* **13,** 2170–2174.
30. Sadusky, T., Newman, A. J. & Dibb, N. J. (2004) *Curr. Biol.* **14,** 505–509.
31. Qiu, W. G., Schisler, N. & Stoltzfus, A. (2004) *Mol. Biol. Evol.* **21,** 1252–1263.
32. Coghlan, A. & Wolfe, K. H. (2004) *Proc. Natl. Acad. Sci. USA* **101,** 11362–11367.
33. Paquette, S. M., Bak, S. & Feyereisen, R. (2000) *DNA Cell Biol.* **19,** 307–317.
34. Logsdon, J. M., Jr. (1998) *Curr. Opin. Genet. Dev.* **8,** 637–648.
35. Lynch, M. (2002) *Proc. Natl. Acad. Sci. USA* **99,** 6118–6123.
36. Sverdlov, A. V., Babenko, V. N., Rogozin, I. B. & Koonin, E. V. (2004) *Gene* **338,** 85–91.
37. Banyai, L. & Patthy, L. (2004) *FEBS Lett.* **565,** 127–132.
38. Cho, S., Jin, S. W., Cohen, A. & Ellis, R. E. (2004) *Genome Res.* **14,** 1207–1220.
39. Llopart, A., Comeron, J. M., Brunet, F. G., Lachaise, D. & Long, M. (2002) *Proc. Natl. Acad. Sci. USA* **99,** 8121–8126.
40. Nielsen, C. B., Friedman, B., Birren, B., Burge, C. B. & Galagan, J. E. (2004) *PloS Biol.* **2,** e422.
41. Kent, W. J. & Zahler, A. M. (2000) *Genome Res.* **10,** 1115–1125.

Section X:
Paradigm Shift and Computing

Towards a paradigm shift in biology

The steady conversion of new techniques into purchasable kits and the accumulation of nucleotide sequence data in the electronic data banks leads one practitioner to cry, "Molecular biology is dead — Long live molecular biology!".

THERE is a malaise in biology. The growing excitement about the genome project is marred by a worry that something is wrong — a tension in the minds of many biologists reflected in the frequent declaration that sequencing is boring. And yet everyone is sequencing. What can be happening? Our paradigm is changing.

Molecular biology, from which has sprung the attitude that the best approach is to identify a relevant region of DNA, a gene, and then to clone and sequence it before proceeding, is now the underpinning of all biological science. Biology has been transformed by the ability to make genes and then the gene products to order. Developmental biology now looks first for a gene to specify a form in the embryo. Cellular biology looks to the gene to specify a structural element. And medicine looks to genes to yield the body's proteins or to trace causes for illnesses. Evolutionary questions — from the origin of life to the speciation of birds — are all traced by patterns on DNA molecules. Ecology characterizes natural populations by amplifying their DNA. The social habits of lions, the wanderings of turtles and the migrations of human populations leave patterns on their DNA. Legal issues of life or death can turn on DNA fingerprints.

And now the genome project contemplates working out the complete DNA pattern and listing every one of the genes that characterize all of the model species that biologists study — ourselves even included.

At the same time, all of these experimental processes — cloning, amplifying and sequencing DNA — have become cook-book techniques. One looks up a recipe in the Maniatis book, or sometimes simply buys a kit and follows the instructions in the inserted instructional leaflet. Scientists write letters bemoaning the fact that students no longer understand how their experiments really work. What has been the point of their education?

The questions of science always lie in what is not yet known. Although our techniques determine what questions we can study, they are not themselves the goal. The march of science devises ever newer and more powerful techniques. Widely used techniques begin as breakthroughs in a single laboratory, move to being used by many researchers, then by technicians, then to being taught in undergraduate courses and then to being supplied as purchased services — or, in their turn, superseded.

Fifteen years ago, nobody could work out DNA sequences, today every molecular scientists does so and, five years from now, it will all be purchased from an outside supplier. Just this happened with restriction enzymes. In 1970, each of my graduate students had to make restriction enzymes in order to work with DNA molecules; by 1976 the enzymes were all purchased and today no graduate student knows how to make them. Once one had to synthesize triphosphates to do experiments; still earlier, of course, one blew one's own glassware.

Yet in the current paradigm, the attack on the problems of biology is viewed as being solely experimental. The 'correct' approach is to identify a gene by some direct experimental procedure — determined by some property of its product or otherwise related to its phenotype — to clone it, to sequence it, to make its product and to continue to work experimentally so as to seek an understanding of its function.

The new paradigm, now emerging, is that all the 'genes' will be known (in the sense of being resident in databases available electronically), and that the starting point of a biological investigation will be theoretical. An individual scientist will begin with a theoretical conjecture, only then turning to experiment to follow or test that hypothesis. The actual biology will continue to be done as "small science" — depending on individual insight and inspiration to produce new knowledge — but the reagents that the scientist uses will include a knowledge of the primary sequence of the organism, together with a list of all previous deductions from that sequence.

How quickly will this happen? It is happening today: the databases now contain enough information to affect the interpretations of almost every sequence. If a new sequence has no match in the databases as they are, a week later a still newer sequence will match it. For 15 years, the DNA databases have grown by 60 per cent a year, a factor of ten every five years. The human genome project will continue and accelerate this rate of increase. Thus I expect that sequence data for all of the model organisms and half of the total knowledge of the human organism will be available in five to seven years, and all of it by the end of the decade.

To use this flood of knowledge, which will pour across the computer networks of the world, biologists not only must become computer-literate, but also change their approach to the problem of understanding life.

The next tenfold increase in the amount of information in the databases will divide the world into haves and have-nots, unless each of us connects to that information and learns how to sift through it for the parts we need. This is not more difficult than knowing how to access the scientific literature as it is at present, for even that skill involves more than a traditional reading of the printed page, but today involves a search by computer.

We must hook our individual computers into the worldwide network that gives us access to daily changes in the database and also makes immediate our communications with each other. The programs that display and analyse the material for us must be improved — and we must learn how to use them more effectively. Like the purchased kits, they will make our life easier, but also like the kits, we must understand enough of how they work to use them effectively.

The view that the genome project is breaking the rice bowl of the individual biologist confuses the pattern of experiments done today with the essential questions of the science. Many of those who complain about the genome project are really manifesting fears of technological unemployment. Their hard-won PhDs seem suddenly to be valueless because they think of themselves as being trained to a single marketable skill, for a particular way of doing experiments. But this is not the meaning of their education. Their doctorates should be testimonials that they had solved a novel problem, and in so doing had learned the general ability to find whatever new or old techniques were needed; a skill that transcends any particular problem. **Walter Gilbert**

Walter Gilbert is Carl M. Loeb university professor in the Department of Cellular and Developmental Biology at Harvard University, in the Biological Laboratories, 16 Divinity Avenue, Cambridge, Massachusetts 02138 USA.

Large scale bacterial gene discovery by similarity search

Keith Robison[1,2], Walter Gilbert[1] & George M. Church[2,3]

DNA sequencing efforts frequently uncover genes other than the targeted ones. We have used rapid database scanning methods to search for undescribed eubacterial and archean protein coding frames in regions flanking known genes. By searching all prokaryotic DNA sequences not marked as coding for proteins or stable RNAs against the protein databases, we have identified more than 450 new examples of bacterial proteins, as well as a smaller number of possible revisions to known proteins, at a surprisingly high rate of one new protein or revision for every 24 initial DNA sequences or 8,300 nucleotides examined. Seven proteins are members of families which have not been described in prokaryotic sequences. We also describe 49 re-interpretations of existing sequence data of particular biological significance.

[1]Department of Cellular and Molecular Biology, Harvard University, 16 Divinity Avenue Cambridge, Massachussetts 02138, USA
[2]Department of Genetics, Harvard Medical School
[3]Howard Hughes Medical Institute, 200 Longwood Avenue, Boston, Massachussetts 02115, USA

Correspondence should be addressed to K.R.

More than 100 million basepairs (bp) of DNA sequence are available from the public databases, a quantity approximately equivalent to the *Drosophila* genome. With such vast amounts of data, it would not be surprising if some interesting biological sequence features were escaping notice. This would be expected to be especially true for eubacteria and archeans, which possess tightly-packed genomes with little intergenic DNA. Starting with more than 18 million bp of prokaryotic DNA sequence, we have executed a systematic search for previously undetected protein-coding genes using the similarity search program BLASTX[1,2]. Our strategy was to remove all DNA regions known to encode proteins or structural RNAs, and to use BLASTX to translate the remaining DNA in all six reading frames and search the resultant translations against the public protein sequence databases. Whilst we expected to find some new genes, we were surprised to uncover more than 450 genes which had previously escaped detection. Furthermore, seven of these genes belonged to gene families which had not been previously identified outside of the eukaryotes. Others belonged to large gene families with critical roles in bacterial metabolism, such as transporters and transcriptional regulators. Our results suggest that periodic exhaustive reappraisal of the sequence databases can be a productive method of biological discovery.

Reduction of data

Table 1*a* shows the progression of the data through our search scheme. The initial rejection of RNA- and protein-coding regions eliminated 73% of the nucleotides from the original set (GenBank Release 74), and further reduction was accomplished by focusing on chromosomal DNA (by eliminating plasmid sequences) and combining segments of identical sequence. Because each non-coding region was handled individually and many entries had multiple non-coding regions, the final dataset had more segments (but fewer nucleotides) than the starting set.

In any exhaustive screen for genes, a significant challenge lies in eliminating uninteresting positives which escaped the screening procedures. Not surprisingly, most sequences either matched nothing or yielded matches containing only simple amino acid repeat motifs (Table 1*b*). Inspection of the apparently significant matches revealed that the majority represented cases of known but unannotated reading frames. Further analysis eliminated other matches as duplicates, matches to bacterial repetitive elements, matches to prokaryotic contaminants reported as eukaryotic sequences (K.R., manuscript submitted) or known frameshift or truncation mutants.

Novel bits and pieces

The BLAST programs are based on a Poisson model[1-3] which allows for objective threshold score selection. Because these scores are of small magnitude, it is convenient to refer to them by their negative log, which we will term pPoisson (in analogy to pH). Table 2 presents a summary for each new protein found using a cutoff pPoisson of 5.40, a value chosen based on the size of our query set (see Discussion). Figure 1 examines the relationships between each new protein and the best match in the database. We have estimated the similarity of each new protein by calculating the percent identity (summed over all BLAST[2] alignment blocks) with its best match in the database. As shown by this plot, most of non-frameshifted new sequences are only short fragments

Table 1 Sequence data reduction

a Selection of search set	Sequences	Nucleotides
Release 75 bacterial entries	10,807	18,859,529
Non-coding regions	15,890	5,153,250
After duplicate region elimination	14,201	4,728,257
Chromosomal regions	12,582	4,301,046
b Initial classification of sequences using BLASTX	Sequences	Associated nucleotides
No match	9,508	2,387,240
Exact match	1,921	1,049,672
Vector match	119	51,629
Possibly novel	1,034	524,505
c Summary of reading frames found		
Novel, unframeshifted	401	
Novel, frameshifted	70	
Extension of known frame	13	
Frameshift truncation of known frame	36	
	520 (1 new/24 queries or 8.3 Kb)	

from the ends of sequenced segments. However, a surprising number of long open reading frames (ORFs) were found. The plot also demonstrates that a number of sequences identical or nearly identical to known sequences were found. These are either highly conserved central biological components, or the query sequence and the database match are from closely related species. The plot shows some "whole" sequences to be abnormally small. We used an entirely objective criterion for being whole (bounded by stop codons with no BLASTX data suggesting a frameshift), and so it would be expected that some of our "whole" ORFs are not truely complete but only appear so due to sequencing errors.

A question of interest is whether we are repeatedly finding members of the same protein families. Because the fragments themselves are very short, simply cross-searching the new proteins would be likely to miss many family relationships. However, if two new proteins both match the same database entry, it is likely that they belong to the same family. As all of the sequences were searched against the same database, we can use the database entry accession numbers as unambiguous identifiers. For each new protein we generate a list of all matches above the 5.40 pPoisson threshold, and repeatedly combine any pair of lists sharing a common entry until no further combinations can be made. The identities of the database entries can then be carefully examined to identify what family each list represents. Families represented more than three times are shown in Table 3.

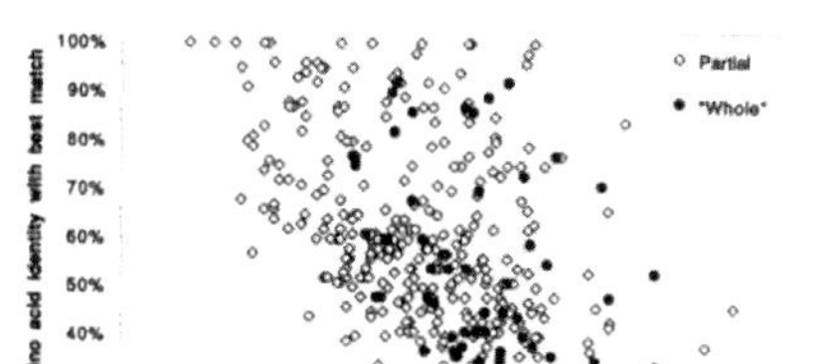

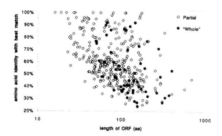

Fig. 1 Comparison of novel, non-frameshifted ORFs against its highest-scoring BLAST match in the database. The X-axis is the length of the ORF; Y-axis displays the percent identity between the novel ORF and the database entry (an estimate of protein relatedness). ORFs were classified as "Whole" if they were bounded by two stop codons and no evidence for a frameshift could be found in the BLASTX data.

Proteins novel to the prokaryotic domain

Many proteins and some protein families appear to be restricted to particular phylogenetic divisions. We have identified seven proteins which had been previously observed only in eukaryotes (numbers 31, 95, 138, 185, 334, 402 and 433) and one (320) which matches only mitochondrial enzymes (in a sense prokaryotic in origin) but had not been observed previously in bacteria.

Identification of non-AUG translational start sites

Translation initiation generally occurs at AUG codons, but initiation at UUG, CUG and GUG have been observed in a number of prokaryotic systems[4]. In the absence of direct biochemical or genetic data, protein similarity searches offer a hope of identifying non-AUG translation start sites. Figure 2 shows four cases in which two or more

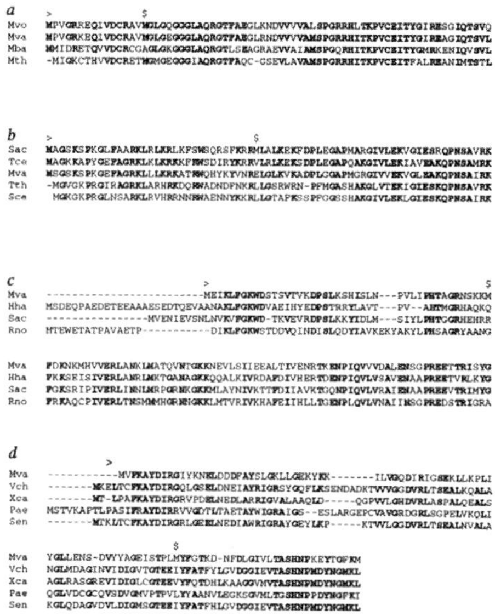

Fig. 2 Identification of probable UUG translational start sites in Archean genes by similarity. The probable UUG start is indicated with >, and the AUG proposed as the start in the original reference with $. Residues identical in at least three aligned sequences are in bold. *a*, N terminus of methyl-coenzyme M reductase operon C protein from *M. voltae*[26] aligned with those of *Methanococcus vannielii*[27], *Methanosarcina barkeri*[28] and *Methanobacterium thermoautotrophicum*[29]. *b*, N terminus of rs12 from *Sulfolobus acidocaldarius*[30] aligned with those from *M. vannelii*[31], *Thermococcus celer*[32], *Tetrahymena thermophila*[33] and the analogous protein, rs28, from *Saccharomyces cerevisae*[34]. *c*, N terminus of small-subunit ribosomal protein 7 (rs7) from *M. vannelii*[31] aligned with rs7 from *Halobacterium halobium*[35] and *Sulfolobus acidocaldarius*[36] and rs5 from *Rattus norvegicus*[37]. *d*, N terminus of an *M. vannelii* unidentified reading frame (URF) downstream from several translational apparatus genes[31] aligned with phosphomannomutases from *Vibrio cholerae*[38], *Xanthomonas campestris*[39], *Pseudomonas aeruginosa*[40] and *Salmonella enterica*[41]. No significant similarity between the *M. vannelii* URF and phosphomannomutases is observed downstream of the shown alignment, despite the aligned area representing only the N-terminal quarter of any of the phosphomannomutases.

Table 2 Novel open-reading frames passing the 5.40pPoisson cutoff

Query sequence					Matching sequence	
No.	Access.	Spe.	Len.	pPoiss	Spe.	Protein
Archeans						
1	X03774	**Hal**	59W	13.64	Eco	Plasmid R1 protein Hok
2	X63837	**Hsa**	116C	13.08	**Hha**	putative bactio-opsin activator
3	M34778	**Mfe**	85C*	16.70	Bla	threonine synthase
4	M26976	**Mfe**	46C	9.72	Bsu	hisC homolog
5	M84821	**Mfo**	244W*	19.82	Wsu	formate dehydrogenase D
6	X04021	**Mva**	57C	11.02	**Mst**	y36K (ORF)
7	X02516	**Mvo**	139N	30.96	**Hha**	RNA polymerase subunit B'
8	X14818	**Sac**	187N	27.07	Eco	translation initiation factor 2
9	J05184	**Sac**	273C*	26.44	Mel	acyl-CoA dehydrogenase
Cyanobacteria						
10	X17482	Acy	295W	51.85	Aci	indole-3-glycerol phosphate synthase
11	X17482	Acy	405C	30.02	Bsu	acetolactate synthase, large subunit
12	X54196	Ani	132C	5.82	Eco	hemolysin D
13	X58847	Ans	85W	6.74	Cso	ribosomal protein S1
14	X65511	Ans	53N	12.92	Eco	nfrC (cytoplasmic protein of unknown function)
15	M20806	Fdi	55N	28.01	Pan	ORF W
16	X60313	Pho	65N	18.54	Aqu	ORF in petBD operon
17	X58719	Scp	146W	11.11	Bsu	ysuL (ORF)
18	M99378	Scp	172W	36.48	Cac	ORF
19	X53695	Scp	158C	13.34	Eco	signal transduction protein barA
20	M95289	Scp	32C	14.08	Scp	orf 3' of cpeB
21	X59809	Scp	35C	12.47	Scp	orf257
22	X59809	Scp	45N	24.40	Ssp	ORF 198
23	X15646	Spl	110W	25.55	Avi	ORF
24	S36216	Spl	111W	16.80	Hpa	modification methylase
25	S43783	Spl	110W	69.89	Scp	nitrogen regulatory protein P-II
26	X53839	Svu	77N	9.60	Scp	transcriptional repressor smtB
27	S67470	Svu	293N	93.15	Sys	cytochrome oxidase subunit I
28	S69957	Sys	54C	7.66	Dno	ATP-dependent protease ClpB
29	M57518	Sys	189N	11.08	Eco	glycine cleavage system protein T
30	S39528	Sys	170N	64.28	Zmo	enolase
Gram positive Eubacteria, high G+C division						
31	X62276	Ais	174C	16.60	Aim	homoserine acetyltransferase
32	M11276	Ars	242C*	47.19	Atu	betaglucosidase
33	M25819	Cgl	148N	23.62	Bsu	membrane-associated protein
34	M89931	Cgl	76N*	14.60	Pae	branched-chain amino acid carrier protein
35	M58020	Mle	128C	34.24	Ecl	enoylpyruvate transferase
36	X56657	Mle	77C	21.92	Eco	6-O-methylguanine-DNA-methyltransferase
37	M95576	Mle	166W*	80.92	Mtu	grpT (grpE-like protein)
38	X06422	Mtu	127W	6.54	Bsu	glnR regulatory protein
39	M94109	Mtu	74C	24.96	Cgl	ORF X
40	M30046	Mtu	218N*	33.96	Eco	phosphate transport protein pstC
41	S37797	Mtu	89I	9.42	Mtu	probable antigen
42	S61892	Mtu	88C	11.57	Pae	non-hemolytic phospholipase C
43	J03838	Mtu	310N	12.54	Psp	alkyl sulfatase
44	Z12001	Rfa	108W	7.80	Eco	transcriptional regulatory uhpA
45	X52980	Sco	51C	12.40	Spo	ATP-dependent transporter
46	M18244	Slm	56N	35.23	Sal	ORF 2
47	M92297	Slv	114C	35.89	Sat	tyrosinase trans-activator
48	Y00142	Slv	111C	22.27	Sli	probable transcriptional activator
49	M80346	Smt	92C	11.59	Ser	cytochrome P450
50	X53401	Sri	88C	8.14	Eco	nitrate-responsive regulator narP
51	M57297	Ssc	115C	11.33	Eco	succinate-semialdehyde dehydrogenase
52	S43326	Stc	48C	13.49	Sco	yw12(ORF)
53	D90006	Stm	137N	17.01	Sau	antiseptic resistance protein
Gram positive Eubacteria, low G+C division						
54	X62835	Bac	311W	21.60	Eco	lipase-like enzyme
55	M24074	Ban	80N	10.74	Eco	adenine phosphoribosyltransferase
56	M93419	Bas	87C	39.68	Bsu	excinuclease ABC subunit C
57	D00312	Bas	48I	10.44	Eco	ycfA (ORF)
58	D10487	Bat	127W*	7.10	**Tpe**	yr33 (ORF)
59	X61286	Bbr	57C	14.40	Bsp	beta-mannanase
60	D00863	Bbr	165N	34.22	Eco	yihA (ORF)
61	X61658	Bbr	65C	12.41	Eco	inner membrane protein envZ
62	M13148	Bca	23C	11.59	Bst	tyrosyl-tRNA synthetase
63	X55703	Bca	25C	8.00	Bsu	chorismate mutase
64	X53507	Bce	256C	98.85	Bas	neopullananase
65	X07723	Bci	75N	13.66	Bsu	stage II sporulation protein J
66	X52880	Bci	77C*	34.85	Bsu	GMP synthase
67	M16657	Bci	199C*	19.57	Ctc	xylanase
68	M74818	Bco	165W	78.44	Eco	mannose-specific PTS enzyme II
69	X53930	Bfi	56C*	17.38	Bca	DNA Polymerase I
70	X12797	Bla	25N	6.35	Bsu	homoserine kinase
71	Y00140	Bla	28C	13.12	Cgl	homoserine dehydrogenase
72	M37169	Bli	262W*	134.08	Bme	glucose 1-dehydrogenase
73	X00413	Bli	50W	26.01	Bsu	ribosomal protein L33
74	M12777	Bma	106N	33.44	Sau	plasmid pUB110 ORF
75	Y00154	Bme	24C	9.16	Bsu	grpE-like protein
76	D90243	Boh	88N	37.41	Bsp	neopullulanase
77	D90243	Boh	160N	6.57	Spn	maltose transport protein malX
78	M60211	Bpo	122W	25.37	Cth	yamY (ORF)
79	X57094	Bpo	59C	18.01	Fsc	enniatin synthetase
80	K00544	Bpu	133N	28.59	Bsu	glucose-starvation inducible gene gsiAA
81	S76941	Bpu	46C	9.89	Bsu	xylose repressor
82	S76941	Bpu	128C	15.64	Eco	yihO (ORF)
83	X05793	Bpu	630C	5.72	Eco	yihP (ORF)
84	D00712	Bre	160C	9.85	Zra	acetoacetyl-CoA reductase
85	M29291	Bsh	183C	29.19	Eco	threonine dehydratase, catabolic
86	X64809	Bsh	48C	8.02	Lmo	p60
87	M34237	Bst	72C	5.92	Bsp	Bleomycin resistance protein
88	D00539	Bst	113N*	39.68	Bsu	histidinol-phosphate aminotransferase
89	M24493	Bst	113W*	26.10	Bsu	sigma-54 factor
90	M57457	Bst	36C	7.15	Pla	major tail protein V
91	M34826	Bsu	156W	14.06	Aox	6-hydroxy-D-nicotine oxidase
92	X15659	Bsu	68C	20.00	Bsu	stage II sporulation protein J
93	M16189	Bsu	60C	16.66	Bsu	transcriptional regulatory degU
94	M37388	Bsu	241C*	40.01	Bsu	glucose-starvation inducible protein gsiAA
95	M34053	Bsu	48C	7.49	Can	cysteine synthase
96	M74010	Bsu	97N	21.15	Ech	L-asparaginase
97	M59757	Bsu	138C	11.27	Eco	yceC (ORF)
98	X02988	Bsu	147W	14.51	Hin	very-short-patch-repair endonuclease
99	X53307	Bsu	76N	7.49	Lca	PTS factor II, lactose-specific
100	D00854	Bsu	193W	75.57	Lde	ORF 5' to glnA
101	M77238	Bsu	152C	24.92	Lla	probable ATP-dependent protease
102	L03181	Bsu	96N*	30.77	Lla	2-isopropylmalate synthase
103	M19299	Bsu	127C	10.82	PBS	xre 3' orf
104	M99611	Bsu	100C	7.02	Pce	2,2-dialkylglycine decarboxylase
105	J03006	Bsu	268W	31.70	Sty	nirC protein
106	L04470	Bsu	140W	27.96	Sxy	arsenic efflux pump
107	D90189	Bsu	137N*	42.74	Sys	peptidyl-prolyl cis-trans isomerase
108	X14178	Bth	220N*	51.28	Csa	beta-glucosidase
109	M30503	Bth	75C	5.62	Eco	possible integrase/ recombinase xprB
110	X55436	Bth	95N	11.01	Hom	aldehyde dehydrogenase type III
111	M35107	Cac	260N	14.12	Bsu	transcriptional regulator degU
112	X65276	Cac	58C	16.96	Eco	replication protein dnaB
113	M30196	Cbo	104C	35.29	Cby	non-toxic component of toxin complex
114	M92906	Cbo	58C	24.00	Cby	non-toxic component of toxin complex
115	M95180	Ccl	41C	24.74	Bsu	DNA polymerase III, alpha chain
116	L02868	Clo	124C	16.80	Ckl	membrane protein
117	M11214	Cpa	49N	8.24	Eco	PTS factor III, glucose-specific
118	X62915	Cpe	90C*	21.59	Cac	grpE-like protein
119	X62914	Cpe	141W	5.60	Pae	secretion protein
120	X12575	Csa	99C	25.64	Eco	6-O-methylguanine DNA methyltransferase
121	M34459	Csa	57C	17.92	Tsa	beta-xylosidase
122	D13268	Cst	375W	33.47	Eco	purine synthesis repressor
123	M57579	Cth	85C*	30.80	Ara	inner membrane lactose transport protein
124	S60770	Efa	140C	13.21	Bme	ORF near glucose dehydrogenase

Query sequence					Matching sequence	
No.	Access.	Spe.	Len.	pPoiss	Spe.	Protein
125	S60770	Efa	70C	12.04	Bsu	membrane bound protein
126	M98865	Erh	23C	6.62	Bsu	grpE-like protein
127	M26929	Lca	151C*	34.35	Bsu	O-6-methylguanine DNA alkyltransferase
128	X66723	Lhe	102W	14.03	Eco	ORF
129	M84770	Lla	37C	7.46	P54	integrase
130	X61210	Lmo	34C	17.6	Liv	ribose-phosphate pyrophosphokinase
131	M83946	Lpa	94C	33.15	Bst	ribosomal protein L12
132	M32342	Mga	48C	8.60	Bna	acyl carrier protein precursor
133	X61529	Mge	50C	28.64	Mpn	DNA gyrase subunit B
134	X61511	Mge	74C	18.13	Tth	aspartyl-tRNA synthetase
135	M21590	Mmy	57C	22.64	Bst	ribosomal protein S15
136	X60513	Msm	124N	38.96	Mmu	pyruvate carboxylase
137	X60513	Msm	105C	7.26	Shy	carboxyphosphonoenolpyruvate phosphonomutase
138	X06414	Myc	153C	11.80	Hpo	ORF 4
139	X06414	Myc	39N	6.01	Sty	methionyl aminopeptidase
140	M76403	Pas	242N*	16.24	Eco	nirC protein
141	X64389	Sau	164W	36.77	Bsh	bioX protein
142	M63176	Sau	82N	19.60	Eco	DNA ligase
143	X16457	Sau	84C	26.38	Rno	3-ketoacyl-CoA thiolase B
144	M69236	Sct	61C	21.19	Efa	ORF
145	M36849	Smu	119W	25.59	Eco	rbs repressor
146	X13136	Spn	29N	8.43	Bsu	recombination protein recM
147	M14339	Spn	155C	18.82	Eco	probable permease
148	X05577	Spn	141C	17.51	Eco	streptothricin acetyltransferase
149	M86905	Spy	61N	8.14	Spy	exotoxin type B precursor
150	M74770	Xmy	119N	50.47	Myc	ribosomal protein L16
Gamma purple bacteria, enterics						
151	X66978	Cfr	128N	34.42	Eco	tryptophan-specific transport protein
152	Z14002	Cfr	35N	14.14	Eco	nicotinate-nucleotide pyrophosphorylase
153	J03433	Eae	143N	26.64	Bsu	als operon regulatory protein
154	J03433	Eae	47N	9.85	Kte	alpha-acetolactate synthase
155	V00255	Eca	91C	42.26	Eco	mioC (ORF)
156	M59909	Eca	124C*	64.96	Eco	lipoprotein-28 precursor
157	M36651	Eca	103C*	31.54	Eco	exonuclease VII
158	M62739	Ech	129W	74.89	Eco	ATP synthase E chain
159	Y00549	Ech	47C	27.19	Eco	nitrogen regulatory protein P-II
160	M14509	Ech	76W	19.55	Eco	ferrisiderophore reductase
161	X63207	Ech	39C	15.54	Eco	plasmid F traM protein
162	M31456	Ech	30C	12.24	Eco	L-ribulose-phosphate-4-epimerase
163	M31308	Ech	90C*	46.12	Eco	rep helicase
164	M88012	Ecl	134N	74.39	Cvi	groEL chaperonin
165	X61919	Eco	64C	19.82	Ace	phosphoribulokinase
166	M58000	Eco	177C*	15.82	Aeu	cation efflux system protein
167	X03691	Eco	76C	28.43	Bst	glutamine transport protein, ATP-dependent
168	X61239	Eco	98W	10.60	Bst	ORF
169	M34333	Eco	61W	6.02	Bst	ORF
170	X57947	Eco	117N	14.47	Bsu	yrtP (ORF)
171	X07091	Eco	183C	17.48	CCp	ORF near psaC
172	M64541	Eco	29C	8.89	Cpa	glyceraldehyde-3-phosphate dehydrogenase
173	S37527	Eco	114N	7.01	Cpe	yngB (ORF)
174	J03732	Eco	39C	20.77	Ech	2-keto-3-deoxygluconate oxydoreductase
175	L03308	Eco	72C	24.25	Eco	hypE-homolog
176	M93362	Eco	73N	14.59	Eco	thermoregulatory protein envY
177	J01619	Eco	103W	12.89	Eco	pap fimbrial protein precursor
178	M30139	Eco	64C	11.74	Eco	plasmid killer protein pnd2
179	M61713	Eco	49N	7.74	Eco	pilus assembly protein
180	M23240	Eco	135C	6.85	Eco	preprotein translocase secA subunit
181	J04039	Eco	77C	6.07	Eco	tRNA uracil-5-methyltransferase
182	M14018	Eco	29N	5.74	Eco	plasmid R1 ORF yprC
183	M69116	Eco	194C*	15.47	Eco	enterobactin synthetase component E
184	M64977	Eco	72M	14.27	Eco	3-deoxy-manno-octulosonate cytidylyltransferase
185	X53983	Eco	105N	10.49	Gma	early nodulin
186	K01197	Eco	133C	77.04	Hin	glycinamide ribonucleotide transformylase
187	X02306	Eco	103N	23.44	Hin	aldose 1-epimerase
188	M93984	Eco	190N	17.82	Hin	ORF
189	M36649	Eco	100N	15.20	Hin	fimbriae structural chain
190	M76389	Eco	414W*	42.54	Hin	ORF near com101A gene
191	X57560	Eco	110N	37.15	Hse	nif-specific regulatory protein
192	X07520	Eco	91C	5.77	Kpn	nif-specific regulatory protein
193	M73320	Eco	62N	8.96	Oar	11-beta-hydroxysteroid dehydrogenase
194	X12982	Eco	100C	17.64	Pae	carbamate kinase
195	M33020	Eco	229C*	30.70	Pae	protein secretion factor pilC
196	M55249	Eco	76W	34.20	PB1	PB1 protein 2
197	M55249	Eco	67W	27.52	PB1	PB1 protein fil
198	M64472	Eco	77N	13.51	Pmi	major fimbrial subuint
199	X62121	Eco	167C	116.33	PP1	gene 16 protein
200	M55249	Eco	221C*	60.17	PP2	terminase
201	M88334	Eco	120N	16.09	Psp	ORF
202	M15273	Eco	96C	6.41	Scp	nitrate transporter
203	M96235	Eco	181C	6.74	Sli	chloramphenicol resistance protein
204	J03725	Eco	76N	35.25	Sma	yphL (ORF)
205	D90227	Eco	70C	9.77	Smu	galactose-6-phosphate isomerase
206	M24856	Eco	111C	65.36	Sty	yprO (ORF)
207	M21451	Eco	116W	57.62	Sty	ypts (ORF)
208	X04704	Eco	70N	40.37	Sty	branched chain amino acid carrier protein
209	M12858	Eco	18N	6.96	Sty	ymt2 (ORF)
210	X62530	Eco	297C	5.82	Sty	lipopolysaccharide heptosyltransferase-1
211	M27273	Eco	75N*	21.02	Sty	fructokinase
212	M31532	Eco	32I	11.96	Zmo	alcohol dehydrogenase I
213	M87280	Ehe	404W	90.82	Hau	ydlM (ORF)
214	D90087	Eur	117C*	51.80	Eco	enoylpyruvate transferase
215	M36068	Kae	78N	24.16	Brb	amiE 3' region ORF
216	S81065	Kox	27C	13.05	Kpn	pullulanase secretion envelope pulD
217	X53433	Koz	182C	9.37	Eco	yihW (ORF)
218	M28676	Kpn	47N	19.40	Eco	transcriptional regulstor nadR
219	X16817	Kpn	33N	14.42	Eco	demethylubiquinone methyltransferase
220	X13303	Kpn	84C	12.14	Sco	ORF J12
221	D11025	Sfl	85N	32.48	Eco	multicopy suppressor of htrB
222	X13131	Sfl	29C	16.82	Eco	UDP-glucose pyrophosphorylase
223	X66849	Sfl	281W	73.01	Hin	glycinimide ribonucleotide transformylase
224	J03391	Sin	107C	49.80	PP4	integrase
225	M94066	Sma	169C	115.28	Eco	sigma-70 factor
226	X60821	Sma	170C	102.43	Eco	yaaA (ORF)
227	M68877	Sma	78C	6.03	Sma	fimbrial protein smfE
228	M27219	Sma	32N	13.68	Sty	flagellar operon sigma factor
229	D10257	Spu	78N	49.85	Sty	oxaloacetate decarboxylase alpha chain
230	X54548	Sty	88C	57.52	Eco	deoxyGTP triphosphohydrolase
231	M29687	Sty	95N	55.59	Eco	tRNA modification enzyme miaA
232	S75596	Sty	102N	50.77	Eco	yjaB (ORF)
233	X07689	Sty	82N	49.33	Eco	catalase hydroperoxidase I
234	M64606	Sty	68C	46.52	Eco	nirD protein
235	M68936	Sty	62N	37.59	Eco	cell division protein sufI
236	M62725	Sty	62N	27.12	Eco	spermidine/putrescine transport protein
237	J01798	Sty	51W	25.07	Eco	ubiX/dedF (ORF)
238	M26046	Sty	45C	24.06	Eco	ribonuclease HII
239	S53120	Sty	68W	23.96	Eco	fumarate hydratase
240	M89481	Sty	46N	21.06	Eco	D-glutamate biosynthetic protein murI
241	M24021	Sty	51C	20.82	Eco	DNA topoisomerase III
242	M57431	Sty	42N	19.48	Eco	cold shock protein 7.4
243	S42598	Sty	39C	12.85	Eco	attaching and effacing protein
244	X63336	Sty	35N	11.70	Eco	yafA (ORF)
245	M10894	Sty	24C	11.02	Eco	gamma-glutamyl phosphate reductase
246	X05382	Sty	22C	9.80	Eco	osmotically inducible lipoprotein
247	M29701	Sty	16N	6.41	Eco	URF downstream of dnaE
248	X12374	Sty	217C*	125.70	Eco	phosphoenolpyruvate carboxykinase
249	X12569	Sty	130C*	37.30	Eco	ycdA
250	D10015	Sty	52N*	18.00	Eco	putative ATP-dependent RNA helicase rhlB/mmrA

Query sequence					Matching sequence	
No.	Access.	Spe.	Len.	pPoiss	Spe.	Protein
251	M84575	Sty	96N	6.57	Kae	ribitol dehydrogenase
252	M55342	Sty	108C	25.59	PLa	y194 (ORF)
253	D12589	Sty	68N	13.92	PP1	phage P1 phd gene product
254	X53368	Yen	63C	22.36	Sty	negative regulator of flagellin synthesis
255	X67771	Yen	40N	24.29	Yps	low calcium response protein lcrR
256	S65012	Ype	46C	32.77	Yen	ylc4 (ORF)
Gamma purple bacteria, other						
257	X66859	Aci	169N	53.21	Eco	yhbF (ORF)
258	M74510	Bap	113N	60.70	Eco	RNA Polymerase alpha subunit
259	M76589	Bap	19N	7.03	Eco	ribosomal protein L20
260	M76589	Bap	112C	75.89	Kpn	translation initiation factor IF3
261	M88613	Cbu	89C*	24.38	Eco	sigma-32 factor
262	M13765	Dno	57N	11.92	Bsu	enolpyruvylshikimate phosphate synthase
263	M33459	Hin	35N	8.30	Bja	hupD gene product
264	M33457	Hin	45I	10.31	Bsh	DAPA aminotransferase
265	M94855	Hin	158W	12.10	Bsu	poly(glycerol-phosphate) alpha-glucosyltransferase
266	M99049	Hin	25C	6.11	Cvi	ORF
267	M37487	Hin	112N	60.54	Eco	glycerol facilitator protein
268	X52124	Hin	147C	54.60	Eco	RNA polymerase beta-prime chain
269	M27280	Hin	203C	37.33	Eco	protease IV
270	M18878	Hin	72C	35.49	Eco	tolB protein
271	M33454	Hin	71I	24.47	Eco	phosphate transport protein pstC
272	X57315	Hin	53C	10.15	Eco	cytochrome C
273	M33441	Hin	27I	9.77	Eco	methionyl tRNA-synthetase
274	X56903	Hin	170C*	58.57	Eco	rRNA methylase
275	J03359	Hin	109C*	35.68	Eco	N-acetylglucosamine-6-phosphate deacetylase
276	M33458	Hin	90C	28.28	Efa	K/Cu transporting ATPase A
277	M87490	Hin	66C	26.52	Sty	3-isopropylmalate dehydratase
278	D10668	Hpa	106C	49.57	Eco	methionine synthase
279	S75161	Hso	52N	15.82	Eco	stringent starvation protein
280	S62141	Lmi	167N	30.08	Eco	L-serine dehydratase 1
281	D12922	Lpn	142N	27.92	Bst	acetylornithine transaminase
282	M31830	Lpn	29N	7.80	Eco	DNA Pol III gamma/delta subunit precursor
283	L06160	Pae	295C	15.54	Bfi	DNA photolyase
284	L06161	Pae	119N	15.38	Eco	integron In7 ORF341
285	M74132	Pae	35N	10.62	Eco	yigM (ORF)
286	L06157	Pae	199N*	72.85	Eco	D-fructose-6-phosphate amidotransferase
287	M63283	Pae	203C*	50.47	Eco	yaaA (ORF)
288	M27175	Pae	213N	52.18	Pae	killer protein of pyocin s1
289	L06160	Pae	66N	15.77	Pae	aminoglycoside 3'-N-acetyltransferase
290	L06157	Pae	140C	5.85	Pmi	urease accessory protein ureE
291	M59036	Pae	162C	10.66	Sty	ORF
292	M94078	Pae	87C	6.59	Sty	uroporhyrin-III C-methyltransferase
293	X14913	Pfl	147N	7.89	Nta	cinnamyl-alcohol dehydrogenase
294	X54523	Pfl	32C	7.74	Pfl	endoglucanase A
295	M59210	Pha	219C*	64.66	Eco	periplasmic arginine binding protein
296	M30186	Pmi	52N	7.14	Efa	K/Cu transporting ATPase A
297	M68901	Pmu	39N	11.41	Eco	nitrate/nitrite response regulator narP
298	X12975	Pph	63N	13.02	Sqa	dogfish CFTR homolog
299	M83673	Pps	136C	83.54	Psp	ORF near bph operon
300	M19460	Ppu	60N*	16.80	Aci	catechol 1,2-dioxygenase
301	M57613	Ppu	175W	30.41	Eco	leucine-responsive regulatory protein
302	M60276	Ppu	112N	74.09	Psp	dmpR transcriptional regulator
303	J05293	Ppu	143W	37.70	Psp	positive regulatory gene
304	M35140	Ppu	140N	37.62	Sty	proline permease component
305	J03681	Psy	409W*	24.03	Bsh	DAPA aminotransferase
306	X17150	Psy	59N	22.40	Kae	urea amidohydrolase
307	M55911	Psy	161C	91.36	Ppu	trpE 5' region ORF
308	M28524	Psy	42C	13.15	Psy	probable regulator of pathogenicity
309	M73971	Psz	62C	22.21	Yen	phosphoserine aminotransferase
310	M81087	Tfe	53N	9.80	Bme	tms protein
311	S64997	Tfe	46C	12.00	Tfe	lysR-type protein rbcR
312	X16050	Val	27N	7.66	Eco	yieA (ORF similar to tms proteins)
313	X62635	Val	160C*	17.64	Eco	putrescine transport protein
314	X55363	Vch	94C	14.24	Eco	erythronate-4-phosphate dehydrogenase
315	X15438	Vch	279N*	95.68	Eco	tRNA (uracil-5-)-methyltransferase
316	M74035	Vch	50C	15.55	Val	fructokinase
317	X06758	Vfi	41N	8.80	Vha	luciferase beta chain
318	X06758	Vfi	86C	36.35	Xlu	acyl transferase
319	D90101	Vpa	122C	62.82	Bsu	ytdK (ORF)
320	S75324	Xca	93N*	25.68	MSs	succinyl-CoA:alpha-ketoacid coenzyme A transferase
321	S63844	Xca	79C	7.13	Vpr	neutral protease precursor
Purple bacteria, other						
322	X14716	Aau	152N*	37.19	Rca	ynf1 (ORF in nifB 3' region)
323	X62690	Abr	121C*	30.37	Pae	modification methylase
324	X67216	Abr	255N	29.60	Psp	carboxymethylene-butenolidase
325	M94886	Abr	179C	6.02	Sco	ORF in whiB 3' region
326	M96433	Aeu	75N	22.85	Ach	hydrogenase accessory protein HupA
327	M69036	Aeu	151W*	68.00	Ain	bacterioferritin
328	M59353	Aeu	67C	22.89	Eco	yhfA (ORF)
329	J05003	Aeu	59C	21.47	Eco	ATP-dependent protease, binding subunit
330	M69036	Aeu	187C*	59.48	Eco	glutathione synthetase
331	X52414	Aeu	114N	31.44	Ppu	todJ; todF product hydratase
332	M63007	Ain	120C	20.06	Sty	chemoreceptor
333	M32871	Ama	69C	24.39	Pfl	lepA homolog
334	D00521	Apo	81C	12.60	Mmu	xanthine dehydrogenase
335	S40757	Ara	131C	21.82	Cth	beta-galactosidase
336	Z15003	Atu	115C	17.42	Smu	sucrose phosphorylase
337	M20404	Bab	139C	10.74	Eco	peptidoglycan synthesis protein murI
338	M31765	Bja	48N	11.03	Bja	ORF
339	M60874	Bja	47N	9.74	Bsu	aconitate hydratase
340	X54638	Bja	72N*	16.70	Eco	ybaD (ORF in nusB 5' region)
341	M16751	Bja	116N	39.37	Eni	catabolic 3-dehydroquinase
342	M16493	Bpa	126N	7.03	Atu	virulence factor B4.1
343	M14378	Bpe	205N	6.52	Atu	virulence factor B4.1
344	X14199	Bpe	56N	9.85	Psp	transcriptional activator sdsB
345	M16492	Bro	205N	6.08	Atu	virulence factor B4.1
346	X03720	Brs	87W	17.00	Bja	nodulation protein
347	M35190	Cco	51C	8.29	Bsu	pyrrolidone-carboxylate peptidase
348	M26955	Ccr	73N	8.82	Ccr	flagellar hook protein
349	M91448	Ccr	75N	10.51	Hin	ORF (homolog of Eco dskA)
350	M63448	Cje	67C	15.19	Ype	ferric uptake regulation protein
351	X58242	Cru	209W	18.11	Eco	ribonuclease III
352	Z12609	Eik	177C	23.64	Eco	yaeB (ORF)
353	Z12609	Eik	124C	36.31	Fno	ATP-dependent transporter valA
354	X56977	Mbo	145N*	49.52	Eco	threonine deaminase
355	X07856	Mes	81N	51.32	Mex	moxR protein
356	X55394	Mtr	55C	7.92	Cac	60 KD chaperonin
357	M60861	Mxa	127C	6.07	Eur	phytoene dehydrogenase
358	X04835	Ngo	41C	10.12	Lca	tryptophan synthase beta chain
359	S85411	Ngo	64C	16.92	Syp	zinc finger protein
360	X68062	Nme	120W	9.74	Bsu	phospho-2-dehydro-3-deoxyheptonate aldolase
361	X68062	Nme	121W	36.04	Eco	yhbF (ORF)
362	X05934	Pad	170C	23.32	Eco	beta lactamase
363	M17522	Pad	73N	13.62	Eco	ribokinase
364	J05282	Pce	262W*	15.89	Ans	keyoacyl reductase
365	M58494	Pce	45C	11.70	Psz	nosA protein
366	M58445	Pol	205W	119.17	Pae	ORF
367	M58445	Pol	95C	57.44	Pae	phaC1 5' region ORF
368	D90355	Ppa	84N*	14.66	Bme	glucose 1-dehydrogenase A
369	X16732	Psa	71N	22.30	Eco	maltose permease lamB
370	S75887	Psm	91C	29.92	Pae	methylmalonate-semialdehyde dehydrogenase
371	M99632	Pso	270C	43.07	Rfr	nodulation protein
372	M99631	Pso	569C	50.27	Yen	yscC (ORF)
373	M18278	Psp	81C	9.54	Bme	glucose 1-dehydrogenase
374	M61114	Psp	124N	37.62	Eco	probable transport protein witA
375	M80212	Psp	146C*	16.72	Ppu	positive regulatory protein
376	M15201	Psp	170C	60.34	Psp	ytcB (ORF)
377	S77431	Psp	46C	12.48	Psp	tnpA (ORF)
378	S37018	Psp	110W*	26.54	Psp	hydantoin racemase
379	D10069	Psp	49C	13.00	Psz	nosA protein

Query sequence					Matching sequence	
No.	Access.	Spe.	Len.	pPoiss	Spe.	Protein
380	M22077	Psp	124C	11.05	Sco	putative ketoacyl reductase
381	K02081	Pst	83W	11.60	PB1	late control protein B
382	X12585	Pvu	23C	9.52	Eco	transcription anti-terminator nusG
383	X51816	Pvu	129C	76.28	Pmi	urease operon protein ureD
384	X51816	Pvu	88N	46.80	Pmi	urease accessory protein ureE precursor
386	L02348	Rca	95C	15.80	Aeu	NAD-reducing hydrogenase beta subunit
386	X63462	Rca	134C	23.72	Eco	protein export membrane protein secF
387	M12776	Rca	55C*	23.34	Tve	ORF
388	X63291	Rhs	85N	34.55	Atu	regulatory protein chvI
389	M16710	Rhs	42N	15.59	Rhs	nitrogenase molybdenum-iron protein
390	S38912	Rle	136C*	26.04	Ain	nifQ protein
391	X65619	Rlo	96W	15.34	Rhs	nodulation protein A
392	M57565	Rme	112C	36.16	Bsu	flagellar protein FliP
393	M94191	Rme	84N	32.42	Ccr	usg protein
394	M96584	Rme	46C	12.57	Eco	yajF (ORF)
395	V01215	Rme	59N	19.96	Rle	nitrogenase alpha subunit
396	M11268	Rme	144N*	46.80	Rlo	nodulation protein I
397	K01467	Rme	82C	13.92	Rne	nfe1 protein
398	M83823	Rsp	60C	6.55	Ccr	aminopeptidase N
399	X56157	Rsp	89C	19.19	Eco	probable transport protein
400	M83823	Rsp	216N*	18.44	**Hvo**	histidinol-phosphate aminotransferase
401	M14732	Rsp	41C	6.62	Rca	ypuM (ORF)
402	Z11601	Sta	527W*	9.30	Cde	galactose oxidase
403	M95300	Sta	51C	23.55	Mxa	pyridoxamine 5'-phosphate oxidase
404	S37355	Wsu	167C	5.49	Pvu	nitrogen regulation protein NR(I)
405	X51509	Wsu	199W	37.96	Wsu	fumarate reductase cytochrome b subunit
406	M62957	Zmo	158W	37.07	Pae	glutathionine reductase
407	M18802	Zmo	112C	10.01	Rsp	transketolase
Spirochaetes						
408	M28681	Bbu	62C	20.59	CCp	ribosomal protein L5
409	M58431	Bbu	121C	8.28	Sce	sec63/npl1 protein
410	M57256	Bhe	336W	10.33	Bhe	outer membrane lipoprotein
411	M86838	Bhe	109N	13.05	Bst	permease protein glnQ
412	M96579	Lbo	258C*	18.35	Bsu	DNA primase
413	D00598	Tde	179N	30.89	Bca	DNA polymerase I
414	M32401	Tpa	119N	25.40	Ath	adenine phosphoribosyltransferase
415	M17716	Tpa	70C	20.05	Eco	alanine tRNA-synthetase
Other Eubacteria (Bacteriodes, Chlamydia, Deinococcus, Thermus, Unknown)						
416	M63029	Bfr	84C	17.62	Eco	bacterioferritin comigratory protein
417	M34831	Bfr	63W	13.96	Pvu	sugE protein 2 homolog
418	M69217	Cpn	34N	14.00	Cps	hypB chaperonin homolog
419	M69227	Cpn	50C	21.36	Ctr	grpE-like protein
420	S83995	Cpn	79N*	28.82	Eco	RNA polymerase, beta chain
421	S90059	Ctr	28N	8.68	Bst	elongation factor G
422	M31739	Ctr	27N	8.34	Cps	hypB chaperonin homolog
423	M31119	Ctr	116C*	36.92	Eco	succinyl-CoA synthetase; alpha chain
424	X60678	Ctr	92N	28.74	Sci	ribosomal protein S2
425	M59306	Dra	41C	16.03	Mmu	beta enolase
426	S77498	Har	279N*	15.54	Eco	glycine cleavage system T protein
427	X56033	Taq	215C*	34.04	Cgl	aspartate kinase
428	X16595	Taq	35N	14.64	Tth	phosphoglycerate kinase
429	X07804	Tba	82W	16.62	Bme	ORF
430	D00728	Tba	201C*	101.64	Bsu	putative cytochrome A assembly factor
431	S53348	Tth	55W	19.39	Bst	ribosomal protein L33
432	M26923	Tth	126W	23.02	Cac	ORF
433	M32108	Tth	91N	29.35	Ddi	thymidine kinase complementing protein
434	X06657	Tth	42N	14.52	Eco	ribosomal protein S10
435	K01444	Tth	180C*	5.59	Pbl	3-isopropylmalate dehydratase
436	Z12118	Tth	117N	8.13	Spe	aklavinone C-11 hydroxylase
437	X64557	Tth	143N*	34.04	Tfe	yntR (probable transport protein)
438	X16399	Xsp	148W	87.80	Eco	universal stress protein A

The GenBank accession number for the nucleotide sequence and species of origin are noted (see below for species codes). Species codes for archeans are in bold; organelle and eukaryote codes are underlined. Phylogenetic classifications were obtained from the Ribosomal Database Project[25]. The length of the ORF in amino acids is listed along with the type of ORF (N, N-terminal; C, C-terminal; W, whole; I, internal; *, frameshifted). ORFs were classified as internal if no stop codons were found, and whole if the ORF was bounded by stop codons. Also listed is the Poisson score, species for the match, and matching protein. Where possible, protein nomenclature was taken from SwissProt[22].

Species codes

Aau	Azorhizobium caulinodans
Abr	Azospirillum brasilense
Ach	Azotobacter chroococcum
Aci	Acinetobacter calcoaceticus
Acy	Anabaena cylindrica
Aeu	Alcaligenes eutrophus
Aim	Ascobolus immersus
Ain	Azotobacter vinelandii
Ais	Actinomyces viscosus
Ama	Anaplasma marginale
Ami	Actinoplanes missouriensis
Ani	Anacystis nidulans
Ans	Anabaena sp.
Aox	Arthrobacter oxidans
Apo	Acetobacter polyoxogenes
Aqu	Agmenellum quadruplicatum
Ara	Agrobacterium radiobacter
Ars	Arthrobacter sp.
Ath	Arabidopsis thaliana
Atu	Agrobacterium tumefaciens
Bab	Brucella abortus
Bac	Bacillus acidocaldarius
Ban	Bacillus anthracis
Bap	Buchnera aphidicola
Bas	Bacillus sp.
Bat	Bacillus thermoglucosidasius
Bbr	Bacillus brevis
Bbu	Borrelia burgdorferi
Bca	Bacillus caldotenax
Bce	Bacillus cereus
Bci	Bacillus circulans
Bco	Bacillus coagulans
Bfi	Bacillus firmus
Bfr	Bacteroides fragilis
Bhe	Borrelia hermsii
Bja	Bradyrhizobium japonicum
Bla	Brevibacterium lactofermentum
Bli	Bacillus licheniformis
Bma	Bacillus macerans
Bme	Bacillus megaterium
Bna	Brassica napus
Boh	Bacillus ohbensis
Bpa	Bordetella parapertussis
Bpe	Bordetella pertussis
Bpo	Bacillus polymyxa
Bpu	Bacillus pumilus
Brb	Brevibacterium sp.
Bre	Brevibacterium sterolicum
Bro	Bordetella bronchiseptica
Brs	Bradyrhizobium sp.
Bsh	Bacillus sphaericus
Bst	Bacillus stearothermophilus
Bsu	Bacillus subtilis
Bth	Bacillus thuringiensis
Bun	Bacteroides uniformis
Cac	Clostridium acetobutylicum
Can	Capsicum annuum
Cbo	Clostridium botulinum
Cbu	Coxiella burnetii
Cby	Clostridium butyricum
Ccl	Clostridium cellulovorans
Cco	Campylobacter coli
CCp	Cyanelle Cyanophora paradoxa
Ccr	Caulobacter crescentus
Cde	Cladobotryum dendroides
Cfr	Citrobacter freundii
Cgl	Corynebacterium glutamicum
Cje	Campylobacter jejuni
Ckl	Clostridium kluyveri
Clo	Clostridium longisporum
Cpa	Clostridium pasteurianum
Cpe	Clostridium perfringens
Cpn	Chlamydia pneumoniae
Cps	Chlamydia psittaci
Cru	Cowdria ruminantium
Csa	Caldocellum saccharolyticum
Cst	Clostridium stercorarium
Ctc	Clostridium thermocellum
Cth	Clostridium thermosulfurogenes
Ctr	Chlamydia trachomatis
Cvi	Chromatium vinosum
Ddi	Dictyostelium discoideum
Dmo	Desulfurococcus mobilis
Dno	Dichelobacter (Bacteriodes) nodosus
Dra	Deinococcus radiodurans
Dvu	Desulfovibrio vulgaris
Eae	Enterobacter aerogenes
Eca	Erwinia carotovora
Ech	Erwinia chrysanthemi
Ech	Erwinia chrysanthema
Ecl	Enterobacter cloacae
Eco	Escherichia coli
Efa	Enterococcus (Streptococus) faecalis
Ehe	Erwinia herbicola
Eik	Eikenella corrodens
Eni	Emericella (Aspergillus) nidulans
Erh	Erysipelothrix rhusiopathiae
Eur	Erwinia uredovora
Fdi	Fremyella diplosiphon
Fno	Francisella novicida
Fsc	Fusarium scirpi
Gma	Glycine max
Gox	Gluconobacter oxydans
Hal	Hafnia alvei

Har	Unidentified Halophile
Hau	Herpetosiphon aurantiacus
Hha	Halobacterium halobium
Hin	Haemophilus influenzae
Hom	Homo sapiens
Hpa	Haemophilus parainfluenza
Hpo	Hansenula polymora
Hsa	Halobacterium salinarium
Hse	Herbaspirillum seropedicae
Hvo	Haloferax volcanii
Hso	Haemophilus somnus
Kae	Klebsiella aerogenes
Kox	Klebsiella oxytoca
Koz	Klebsiella ozaenae
Kpn	Klebsiella pneumoniae
Kte	Klebsiella terrigena
Lbo	Leptospira borgpetersenii
Lca	Lactobacillus casei
Lde	Lactobacillus delbrueckii
Lhe	Lactobacillus helveticus
Lla	Lactococcus lactis
Lmi	Legionella micdadei
Liv	Listeria ivanovii
Lmo	Listeria monocytogenes
Lpa	Lactobacillus paracasei
Lpn	Legionella pneumophila
Mba	Methanosarcina barkeri
Mbo	Moraxella bovis
Mbr	Methanobacterium bryantii
Mcl	Methylococcus capsulatus
Mel	Megasphaera elsdenii
Mes	Methylobacterium specialis
Mex	Methylobacterium extorquens
Mfe	Methanothermus fervidus
Mfo	Methanobacterium formicicum
Mga	Mycoplasma gallisepticum
Mge	Mycoplasma genitalium
Mle	Mycobacterium leprae
Mmu	Mus musculus
Mmy	Mycoplasma mycoides
Mno	Moraxella nonliquefaciens
Msm	Mycobacterium smegmatis
Mso	Methanothrix soehngenii
MSs	Mitochondrion Sus scrofa
Mst	Methanobrevibacter smithii
Mth	Methanobacterium thermoautotrophicum
Mtr	Methylosinus trichosporium
Mtu	Mycobacterium tuberculosis
Mva	Methanococcus vannielii
Mvo	Methanococcus voltae
Mxa	Myxococcus xanthus
Myc	Mycoplasma capricolum
Ngo	Neisseria gonorrhoeae
Nme	Neisseria meningitidis
Nsp	Nocardia sp.
Nta	Nicotiana tabacum
Oar	Ovis aries
Pad	Paracoccus denitrificans
Pae	Pseudomonas aeruginosa
Pan	Pseudoanabaena sp.
Pas	Peptostreptococcus asaccharolyticus
PB1	Phage 186
PP1	Phage P1
PP2	Phage P2
PP4	Phage P4
P54	Staphylococcus phage L54a
PLa	Phage Lambda
PT4	Phage T4
Pbl	Phycomyces blakesleeanus
Pce	Pseudomonas cepacia
Pfl	Pseudomonas fluorescens
Pha	Pasteurella haemolytica
Pho	Prochlorothrix hollandica
Pmi	Proteus mirabilis
Pmu	Pasteurella multocida
Pol	Pseudomonas oleovorans
Ppa	Pseudomonas paucimobilis
Pph	Photobacterium phosphoreum
Pps	Pseudomonas pseudoalcaligenes
Ppu	Pseudomonas putida
Psa	Pseudomonas saccharophila
Psm	Pseudomonas marginalis
Pso	Pseudomonas solanacearum
Psp	Pseudomonas sp.
Pst	Providencia stuartii
Psy	Pseudomonas syringae
Psz	Pseudomonas stutzeri
Pte	Pseudomonas testosteroni
Pvu	Proteus vulgaris
Rca	Rhodobacter capsulatus
Rfa	Rhodococcus fascians
Rfr	Rhizobium fredii
Rhs	Rhizobium sp.
Rle	Rhizobium leguminosarum
Rlo	Rhizobium loti
Rme	Rhizobium meliloti
Rno	Rattus norvegicus
Rsp	Rhodobacter sphaeroides
Sac	Sulfolobus acidocaldarius
Sat	Streptomyces antibioticus
Sau	Staphylococcus aureus
Sce	Saccharomyces cerevisiae
Sci	Spiroplasma citri
Sco	Streptomyces coelicolor
Scp	Synechococcus sp.
Sct	Streptococcus thermophilus
Sen	Salmonella enterica
Ser	Saccharopolyspora erythraea
Sfl	Shigella flexneri
Shy	Streptomyces hygroscopicus
Sin	Salmonella infantis
Slm	Streptomyces limosus
Slv	Streptomyces lividans
Sma	Serratia marcescens
Smt	Streptomyces thermotolerans
Smu	Streptococcus mutans
Spe	Streptomyces peucetius
Spl	Spirulina platensis
Spn	Streptococcus pneumoniae
Spo	Schizosaccharomyces pombe
Spu	Salmonella pullorum
Spy	Streptococcus pyogenes
Sqa	Squalus acanthias
Sri	Streptomyces rimosus
Ssc	Streptomyces scabies
Sta	Stigmatella aurantiaca
Stc	Streptomyces cinnamonensis
Stm	Streptomyces macromomyceticus
Stu	Solanum tuberosum
Sty	Salmonella typhimurium
Svu	Synechococcus vulcanus
Sys	Synechocystis sp.
Sxy	Staphylococcus xylosus
Tac	Thermoplasma acidophilum
Taq	Thermus aquaticus
Tba	Thermophilic bacterium
Tce	Thermococcus celer
Tde	Treponema denticola
Tet	Tetrahymena thermophila
Tfe	Thiobacillus ferrooxidans
Tin	Thermoactinomyces intermedius
Tma	Thermotoga maritima
Tpa	Treponema pallidum
Tpe	Thermophilum pendens
Tsa	Thermoanaerobacter saccharolyticum
Tth	Thermus thermophilus
Tve	Thiobacillus versutus
Val	Vibrio alginolyticus
Vch	Vibrio cholerae
Vel	Vibrio eltor
Vfi	Vibrio fischeri
Vpa	Vibrio parahaemolyticus
Vpr	Vibrio proteolyticus
Wsu	Wolinella succinogenes
Xca	Xanthomonas campestris
Xlu	Xenorhabdus luminescens
Xmy	Mycoplasma-like organism
Xsp	Bacterial sp.
Yen	Yersinia enterocolitica
Ype	Yersinia pestis
Yps	Yersinia pseudotuberculosis
Zma	Zea mays
Zmo	Zymomonas mobilis
Zra	Zoogloea ramigera

database entries support a hypothetical UUG start site for a previously described Archean open reading frame. Particularly striking are two examples in which homologous sequences from eukaryotes participate in identifying non-AUG start sites for two Archean ribosomal proteins (Fig. 2*b*, *c*). In Fig. 2*d*, a non-AUG start for an Archean URF is supported by several eubacterial phosphomannomutases. Curiously, only the N terminus of this URF shows such similarity; the remainder of the sequence shows no similarity to any known protein. Searches of the other two forward reading frames downstream of the observed similarity found no evidence for a frameshift. While this could be an artefactual match, we think it unlikely as the similarity is so strong (estimated pPoisson of 8.00 for the original BLASTX[1] match) and over such a great evolutionary distance. We propose instead that this represents the shuffling of a protein domain during evolution[5], although the possibility that it represents a cloning artifact cannot be ruled out. Because our dataset is both small and from multiple organisms, it is not feasible to use it to identify auxillary signals specific to non-AUG translation starts. Also, evidence from *E. coli* suggests that the context of non-AUG start codons is not

Table 3 Protein families detected three or more times in the novel ORFs, as determined by database match cluster analysis

	Family	ORF Nos from Table 2							
11	Transcriptional regulators; luxR/uhpA-type	44	50	93	111	125	191	192	297
		302	308	388					
8	ABC-type ATP-dependent transporters	45	167	202	236	298	353	399	411
6	Insect-type alcohol dehydrogenases	84	170	174	193	251	380		
4	Transcriptional regulators; 2-component	19	61	65	92				
3	Transcriptional regulators; lysR-type	153	311	344					
3	Glutathionine reductase	276	296	406					
*3	*A.tumefaciens* virulence factor B4.1	342	343	355					

28 more clusters contained two new sequences. In the cluster marked with an asterisk, all three sequences are from different species but the same genus. Where possible, family nomenclature was taken from SwissProt[22].

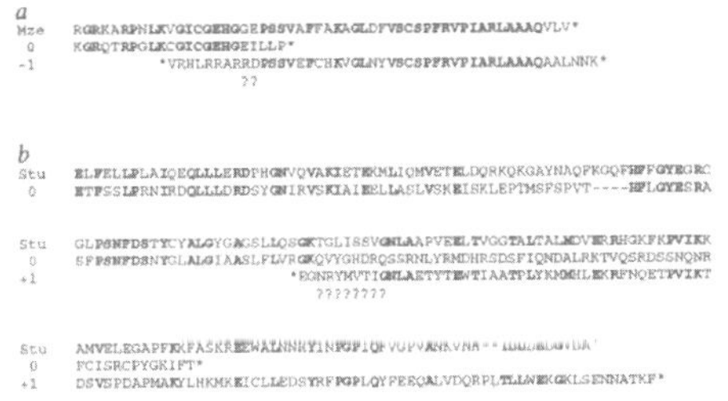

Fig. 3 Identification of truncated reading frames by similarity. Two possible reading frames from the bacterial DNA sequences are displayed in alignment with a database entry. Gaps are indicated by –. Frameshift sequencing errors are predicted to lie in the regions indicated by ??; and the aligned residues used in this prediction are in bold. *a*, C-terminus of Maize pyruvate, orthophosphate dikinase (PPDK)[42] aligned with the C-terminus of the reported reading frame (frame 0) for PPDK from *Bacteriodes symbiosis*[11] and frame –1 from the same region. The frame –1 alignment has a pPoisson of 9.77. *b*, C-terminus of the beta subunit of pyrophosphate-fructose 6-phosphate 1-phosphotransferase from *Solanum tuberosum* (potato)[12] aligned with a previously reported C-terminal URF (frame 0) from *Chlamydia trachomatis*[43] and reading frame +1 from the same region. The similarity between these two sequences had not been previously reported. The frame 0 alignments have pPoisson values of 7.59 (left of gap) and 5.36 (right of gap), and the main frame +1 alignment had a value of 5.92.

distinctively different from that of AUG start codons[6].

Identification of frameshifted reading frames

We found 70 proteins in which two or more database entries predict a frameshift in a novel coding sequence by detecting matches in two different reading frames. While we cannot rule out the possibility that they represent cases of translational "reprogramming"[7], post-transcriptional editing[8], or other interesting biological phenomena, most of these frameshifts are probably due to sequencing artefacts or frameshift mutations idiosyncratic to the bacterial strain sequenced. Sequencing error rates have been estimated to be in the range of 0.3 to 3%[9,10]. The 3% estimate was derived from analysis of error rates in vector contamination of the databases[10], and is probably the most relevant to this study (since it also concerned sequence features not detected by the original authors). At such an error rate, even relatively short reading frames would be expected to be broken up by frameshift errors. This, added to the fact that breaking up a reading frame reduces its visibility, make it unsurprising that we found so many (15% of total) frameshifted coding regions.

Sequencing errors may also lead to an under-appreciation of distant evolutionary relationships. Two such cases are shown in Fig. 3. Pyruvate phosphate dikinase (PPDK) from *Bacteriodes* shows greater than 50% sequence identity with the maize gene product[11], but the reported product is truncated at both termini with respect to the plant enzyme. As shown in Fig. 3*a*, a possible translation with high similarity to the maize PPDK's C-terminus is separated from the reported *Bacteriodes* coding frame by a frameshift. A second example (Fig. 3*b*) shows an ORF from *Chlamydia* for which no similarity had been reported[12]. As well as identifying this sequence as the only known homologue of potato pyrophosphate-fructose 6-phosphate 1-phosphotransferase (PPi PFK) α- and β-subunits (which are closely related), we have again discovered a frameshift which truncates the C-terminus of this ORF.

Discussion

Identification of coding regions within 3' or 5' ends of sequence data is frequently not trivial. Fragments may be too short to identify a single likely open reading frame to the exclusion of the five other possibilities, and methods based on measures such as codon bias often require relatively long sequences or suffer from edge effects. Also, sequencing insertion or deletion errors may break short reading frame fragments into pieces smaller than the detection limits for codon bias based methods[13]. Identification via protein database matches[1,14,15] is a useful but limited method, as reading frames can be found only if a homologue exists in the databases, and only if the region present matches that homologue.

Various shortcomings of the databases provide the greatest impediment to this type of search (Table 1). While efforts to reduce the number of redundant sequence entries and to ensure accurate and complete annotation have been announced[16], the size of the databases and their dynamic nature will continue to make this challenging. One way of accelerating re-scanning for missed features is automated pre-screening to eliminate vector sequences and previously reported genes which are not universally annotated (K.R., unpublished results).

Match probabilities calculated by BLASTX assume the case of a single query compared against an entire database[3]. In a comprehensive search such as this, a much higher threshold is required for significance than with a single query. Since the probability of a random match appearing in our final data is approximately equal to the chosen probability cutoff times the number of queries, we chose a 95% confidence pPoisson cutoff of 5.40 ($-\log(0.05 * 1 / 12{,}582)$). However, BLAST matches below this cutoff frequently indicate interesting biological relationships. In order to explore this "twilight zone", we used an initial pPoisson cutoff of 2.00. Such twilight zone matches may represent biologically significant alignments which have been underscored. There are two reasons to anticipate such cases. First, every scoring matrix functions best when used at a particular evolutionary distance, and performance of a matrix progressively degrades when used to score alignments at other distances[17]. We originally used the PAM120 matrix[18], because it is predicted to be the versatile PAM matrix for detecting relevant matches over all evolutionary distances[17]. Second, much of our dataset consists of short flanking regions, and we therefore expect to find many short N- or C-terminal regions of proteins. As a result, many of our matches could probably be extended to yield higher scores if the missing sequence information were available.

To search for underscored alignments, we first re-scored the 151 matches falling in the twilight zone using the PAM40 (Dayhoff)[18] and BLOSUM62 (ref. 19) matrices. The PAM40 matrix is better suited than the PAM120 for finding short alignments at small evolutionary distances[17]. BLOSUM62 is a recently developed matrix derived from local alignments with superior BLAST performance to PAM120 (ref. 19). As shown in Table 4, pPoisson values better than 5.40 were found for nine queries with PAM40 and 19 with BLOSUM62 (two passed using both matrices).

We used the data from all three matrices to look for alignments which were probably underscored due to missing sequence information. If we divide each database entry into *N* equal segments, the true matches would be expected to fall into either the initial segment or the final segment, while random matches would be expected to be evenly distributed. We can also use the orientation and position of each each ORF containing a BLAST match to classify it as C-terminal or N-terminal. Therefore, identifying a match to the appropriate terminal segment of a database entry improves our random match estimate by a factor of 1/N. This test implicitly assumes that protein segments are either C-terminal or N-terminal and do not shuffle, but this does not invalidate the test as shuffled segments are no more likely to pass the test than a random match. We tried this test using the results of the PAM40, PAM120, and BLOSUM62 searches of twilight matches and with N equal to 5, 10 or 20. As shown in Table 4, the scores for 13 sequences passed after application of this test (six of these sequences also passed by rescoring).

Why are so many genes escaping notice during the initial analysis of DNA sequences? Several reasons can be put forward. Efficient tools for searching nucleotide sequences against protein databases have only recently become widely available. However, this cannot explain the overlooking of long ORFs. Nor can these hidden genes be explained by the lack of available homologous sequences at the time of publication, as later workers could easily find the homologues with TBLASTN[2] or TFASTA[20] and deposit the newly found sequence in SwissProt or a similar database. Our primary conclusion is that many sequences in the databases are severely under-analysed. Rather than searching for hidden proteins with BLASTX or unreported homologues with TBLASTN and TFASTA, many biologists stick to their original expectations.

Methodology

Computation and analysis. Computation was performed on Sun SPARCstations running UNIX or using the GenInfo network BLAST[1,2] server provided by the National Center for Biotechnology Information. Programs for excerpting and translating GenBank[16] entries and parsing BLAST[2] output were written in C++ using the MOLBIO++ class library (K.R., manuscript in preparation). Except as noted, BLAST searches used Dayhoff's PAM120 substitution matrix[18].

Non-coding region extraction. DNA sequences with annotated coding regions were obtained from Release 74 of the GenBank[1] database (15 December 1992). For each input sequence, regions annotated as coding for protein, transfer RNA or ribosomal RNA were eliminated, and remaining segments longer than 42 nucleotides were extracted. Exactly identical sequences were combined using the program nrdb (W. Gish, NCBI). Entries marked as plasmid or transposon were segregated from apparent bacterial chromosome entries.

Novel protein identification. Possible coding regions were identified using BLASTX[1]. Each sequence was compared against NCBI's "nr"

Table 4 Analysis of "twilight" matches

	Query	Sequence			Rescoring			Matching sequence	
No.	Access	Spe	Len.	End	P40	P120	B62	Spe	Protein
439	X52070	Slv	65N	–	4.59	5.24	**7.00**	Ami	xylulose kinase
440	M59707	Mno	40C	C5	**8.33**	5.19	**5.60**	Mno	type 4 pilin
441	X61383	Vel	27C	–	**7.82**	5.10	3.34	Vch	14 KDa inner membrane protein
442	S38356	Bsu	67E	N10	–	4.96	**5.48**	Bth	putative recombinase
443	X03897	Bsu	232C	–	2.19	4.85	**5.96**	Eco	ydiA (ORF)
444	X06188	**Dmo**	177C	–	–	4.85	**9.05**	Eco	ORF in trpE 3' region
445	D90043	Bme	56C	C10	4.24	4.82	**5.52**	Bfr	rprX protein (ORF)
446	M13467	Bas	47N	–	**5.55**	4.82	**7.25**	Dvu	prismane protein
447	M67471	Lmo	30C	C5	4.89	4.70	4.92	Atu	chloramphenicol acetyltransferase
448	M88134	Hin	34C	C10	4.15	4.66	5.19	Eco	N-acetylglucosamine-6-phosphate deacetylase
449	J03568	Pte	61C	–	3.27	4.60	**6.43**	Sty	ORF
450	M64624	Rsp	684N	N5	2.35	4.49	4.96	Eco	cynR activatory protein
451	M38266	Eco	64E	–	3.80	4.48	**6.47**	Ype	hemP protein
452	X52504	**Hha**	100N	N10	**7.33**	4.39	5.28	Ppu	hypothetical protein
453	M63308	Eco	54C	C10	–	4.37	**5.89**	Bsu	membrane-associated protein
454	S75301	Bun	18C	–	**6.27**	4.35	4.21	Bfr	ORF in tetX 5' region
455	J04114	Bja	20C	C20	**6.82**	4.16	5.12	Rca	hupU protein
456	M84028	Hin	106C	–	–	4.13	**11.04**	Sco	tetracycline resistance protein homolog
457	D90244	Nsp	59C	–	–	4.09	**5.41**	Eco	yajE (ORF)
458	M28472	Mcl	111C	–	–	4.04	**9.08**	Bsu	regulatory protein tenI
459	M58699	Eco	26C	–	**6.02**	3.92	2.44	Ani	thioredoxin M
460	X04441	Pfl	76N	C5	4.82	3.89	4.66	Eco	probable regulatory protein pssR
461	M13256	Bli	34N	N10	**7.47**	3.80	4.80	Eco	purine nucleotide synthesis repressor
462	M33453	Hin	19C	C20	4.44	3.72	4.14	Eco	gamma-glutamyl phosphate reductase
463	M33443	Hin	66I	–	–	3.59	**7.11**	Sty	nitroreducatase
464	X14631	**Mbr**	125C	–	–	3.43	**5.66**	Cpe	yngB (ORF)
465	D00852	Pfl	154N	–	–	3.28	**5.92**	Bst	glutamine-binding protein
466	M12599	Psp	34C	C10	3.12	3.14	4.96	Mmu	t-complex protein
467	J01656	Eco	104C	C10	**6.44**	3.11	3.22	Eco	phosphoenolpyruvate carboxykinase
468	M69185	Eco	72N	–	3.85	3.09	**5.57**	Hom	CFTR Cl- transport protein
469	M36539	Bst	19C	C10	4.68	2.41	2.96	Eco	maltose transport protein
470	M11218	**Mvo**	301C	C5	–	2.36	**5.70**	Eco	ygjC (ORF)
471	Z11539	Ans	40N	N5	2.89	2.07	**5.43**	PT4	polynucleotide kinase

The 'End' column shows whether the protein match is N- or C-terminal (as defined in the text) at 1/20, 1/10 or 1/5. The BLASTX poisson scores using the PAM40 (P40), PAM120 (P120), and BLOSUM62 (B62) matrices are shown. '–' in the P40 column indicates a pPoisson score below the 2.00 cutoff. Scores in bold pass the 5.40 pPoisson cutoff; underlined pPoisson scores pass the end test (4.10 required at 1/20, 4.40 at 1/10 and 4.70 at 1/5). Other columns, see Table 2.

database consisting of Translated GenBank (release 75 plus updates to 31 December 1992)[16], PIR (release 35.0)[21], and SwissProt (release 24.0)[22], We initially saved the top 10 database matches which both scored better than 60 and a significance[3] of pPoisson >2.00. Matches against repetitive sequences were initially screened by eliminating matches to a list of sequences known to be composed primarily of repetitive amino acid motifs. Matches with better than 90% identity to known proteins of the same species as the query were initially classified as exact matches and were discarded after manual confirmation that they represented known proteins which had not been annotated in an original entry. Further manual culling eliminated spurious matches missed by the automatic filters. Cases of vector sequence contamination, unmarked plasmid sequences, and of the database annotation lacking coding regions mentioned in the original references were separated and referred to the GenBank management. Several cases of protein database contamination with vector sequence[23] or insertion elements (K.R., manuscript submitted) were also discovered and forwarded to the appropriate database(s).

Match classification. The remaining sequences were classified by whether they predicted a new coding region or proposed an extension of a known coding region, and these categories were then further classified as containing or not containing a frameshift. Sequences were classified as frameshifted if above-threshold BLAST matches to the same database sequence were detected in more than one coding frame[1]. For each non-frameshifted coding region, the reading frame containing the BLASTX match was extracted from the original entry and translated. For fragments appearing to contain the N-terminus of a protein, the probable translational start was identified manually. Probable translations for apparently frameshifted coding regions were assembled by hand. The sequences identified are being deposited in SwissProt[22], and are also available via anonymous ftp from golgi.harvard.edu (directory pub/robison/bctorfs). The sequences and summaries of the top BLAST matches may be accessed via Internet gopher at twod.med.harvard.edu, or via a HTTP hypertext browser (such as. Mosaic) at http://twod.med.harvard.edu/bctorfs. Further inquiries should be sent to krobison@mito.harvard.edu. Multiple alignments were generated using CLUSTALV[24] with the default parameter sets, followed by manual reduction of excessive end-gaps.

Match clustering. Multiply-represented protein families in the dataset were identified using a "greedy" clustering algorithm implemented with UNIX shell scripts, as described in the Results.

Acknowledgements

The authors wish to thank W. Gish (NCBI) for advice regarding the BLAST server, L. Schoenbach for assistance in preparing the figures, E. Bunce and D. Jacobson for assistance with the electronic servers, and W. Lam for constructive criticism. K.E.R is a National Defense Science and Engineering Fellow. This work was supported by the U.S. Department of Energy and National Institutes of Health.

Received 8 December 1993; accepted 11 February 1994.

1. Gish, W. & States, D. Identification of protein coding regions by database similarity search. *Nature Genet.* **3,** 266–272 (1993).
2. Altschul, S.F., Gish, W., Miller, W., Myers, E.W. & Lipman, D.J. Basic local alignment search tool. *J. molec. Biol.* **214,** 1–8 (1990).
3. Karlin, S. & Altschul, S.F. Methods for assessing the statistical significance of molecular sequence features by using general scoring schemes. *Proc. natn. Acad. Sci. U.S.A.* **87,** 2264–2268 (1990).
4. Osawa, S., Jukes, T.H., Watanabe, K. & Muto, A. Recent evidence for evolution of the genetic code. *Microbiol. Rev.* **56,** 229–264 (1992).
5. Roth, J.R., Lawrence, J.G., Rubenfield, M., Kieffer-Higgins, S. & Church, G.M. Characterization of the cobalamin (vitamin B12) biosynthetic genes of *Salmonella typhimurium*. *J. Bact.* **175**, 3303–3316 (1993).
6. Stormo, G.D., Schneider, T.D. , Gold, L. & Ehrenfeucht, A. Use of the 'Perceptron' algorithm to distinguish translational initiation sites in *E. coli*. *Nucl. Acids Res.* **10**, 2997–3011 (1982).
7. Gesteland, R.F., Weiss, R.B. & Atkins, J.F. Recoding, Reprogrammed genetic decoding. *Science* **257,** 1640–1641 (1992).
8. Cech, T.R. RNA editing: World's smallest introns? *Cell* **64,** 667–669 (1991).
9. Krawetz, S. A. Sequence errors described in GenBank: a means to determine the accuracy of DNA sequence interpretation. *Nucl. Acids. Res.* **17**, 3951–3957 (1989).
10. Kristensen, T., Lopez, R. & Prydz, H. An estimate of the sequencing error frequency in the DNA sequence databases. *DNA Seq.* **2**, 343–346 (1992).
11. Pocalyko, D.J., Carroll, L.J., Martin, B.M., Babbitt, P.C. & Dunaway–Mariano, D. Analysis of sequence homologueies in plant and bacterial pyruvate phosphate dikinase, Enzyme I of the bacterial phosphoenolpyruvate:sugar phosphotransferase system and other PEP–utilizing enzymes. *Biochem.* **29,** 10757–10765 (1990).
12. Carlisle, S.M. *et al.* Pyrophosphate–dependent phosphofructokinase: Conservation of protein sequence between the alpha– and beta–subunits and with the ATP–dependent phosphofructokinase. *J. biol. Chem.* **265,** 18366–18371 (1990).
13. Fickett, J.W. & Tung, C.S. Assessment of protein coding measures. *Nucl. Acids Res.* **20**, 6441–6450 (1992).
14. Posfai, J. & Roberts, R.J. Finding errors in DNA sequences. *Proc. natn. Acad. Sci. U.S.A.* **89,** 4698–4702 (1992).
15. States, D.J. & Botstein, D. Molecular sequence accuracy and the analysis of protein coding regions. *Proc. natn. Acad. Sci. U.S.A.* **88,** 5518–5522 (1991).
16. Benson, D., Lipman, D.J. & Ostell, J. GenBank. *Nucl. Acids Res.* **21**, 2963–2965 (1993).
17. Altschul, S.F. Amino acid substitution matrices from an information theoretic perspective. *J. molec. Biol.* **219**, 555–565 (1991).
18. Dayhoff, M.O., Schwartz, R.M. & Orcutt, B.C. in *Atlas of Protein Sequence and Structure* (ed. Dayhoff, M.O) **5,** 345–352 (National Biomedical Research Foundation, Washington D.C., 1978).
19. Henikoff, S. & Henikoff, J.G. Amino acid substitution matrices from protein blocks. *Proc. natn. Acad. Sci. U.S.A.* **89**, 10915–10919 (1992).
20. Pearson, W.R. & Lipman, D.J. Improved tools for biological sequence comparison. *Proc. natn. Acad. Sci. U.S.A.* **85,** 2444–2448 (1988).
21. Barker, W.C., George, D.G., Hunt, L.T. & Garavelli, J.S. The PIR protein sequence database. *Nucl. Acids Res.* **19,** 2231–2236 (1991).
22. Bairoch, A. & Boeckmann, B. The SWISS–PROT protein sequence data bank. *Nucl. Acids Res.* **19,** 2247–2249 (1991).
23. Claverie, J.-M. Identifying coding exons by similarity search: Alu–derived and other potentially misleading protein sequences. *Genomics* **12,** 838–841 (1992).
24. Higgins, D.G., Bleasby, A.J. & Fuchs, R. CLUSTAL V: improved software for multiple sequence alignment. *CABIOS* **8,** 181–191 (1992).
25. Larsen, N. *et al.* The ribosomal database project. *Nucl. Acids Res.* **21 (Suppl),** 3021–3023 (1993).
26. Klenin, A., *et al.* Comparative analysis of genes encoding methyl coenzyme M reductase in methanogenic bacteria. *Molec. gen. Genet.* **213,** 409–420 (1988).
27. Cram, D.S. *et al.* Structure and expression of the genes, mcrBDCGA, which encode the subunits of component C of methyl coenzyme M reductase in *Methanococcus vannielii*. *Proc. natn. Acad. Sci. U.S.A.* **84,** 3992–3996 (1987).
28. Bokranz, M. & Klein, A. Nucleotide sequence of the methyl coenzyme M reductase gene cluster from *Methanosarcina barkeri*. *Nucl. Acids Res.* **15,** 4350–4351 (1987).
29. Bokranz, M., Baeumner, G., Allmansberger, R., Ankel–Fuchs, D. & Klein, A. Cloning and characterization of the methyl coenzyme M reductase genes from *Methanobacterium thermoautotrophicum*. *J. Bacteriol.* **170,** 568–577 (1988).
30. Puehler, G., Lottspeich, F. & Zillig, W. Organization and nucleotide sequence of the genes encoding the large subunits A, B and C of the DNA–dependent RNA polymerase of the archaebacterium *Sulfolobus acidocaldarius*. *Nucl. Acids Res.* **17,** 4517–4534 (1987).
31. Lechner, K., Heller, K. & Boeck, A. Organization and nucleotide sequence of a transcription unit of *Methanococcus vannielii* comprising genes for protein synthesis elongation factors and ribosomal proteins. *J. molec. Evol.* **29,** 20–27 (1989).
32. Klenk, H.P., Schwass, V. & Zillig, W. Nucleotide sequence of the genes encoding the L30, S12 and S7 equivalent ribosomal proteins from the archaeum *Thermococcus celer*. *Nucl. Acids Res.* **19,** 6047–6047 (1991).
33. Nielsen, H., Andreasen, P.H., Dreisig, H., Kristiansen, K. & Engberg, J. An intron in a ribosomal protein gene from *Tetrahymena*. *EMBO J.* **5,** 2711–2717 (1986).
34. Alksne, L.E. & Warner, J.R. A novel cloning strategy reveals the gene for the yeast homologueue to *Escherichia coli* ribosomal protein S12. *J. biol. Chem.* **268,** 10813–10819 (1993).
35. Leffers, H., Gropp, F., Lottspeich, F., Zillig, W. & Garrett, R.A. Sequence, organisation, transcription and evolution of RNA polymerase subunit genes from the archaebacterial extreme halophiles *Halobacterium halobium* and *Halococcus morrhuae*. *J. molec. Biol.* **206,** 1–17 (1989).
36. Auer, J., Spicker, G., Mayerhofer, L., Puehler, G. & Boeck, A. Organisation and nucleotide sequence of a gene cluster comprising the translation elongation factor 1–alpha from the extreme thermophilic archaebacterium *Sulfolobus acidocaldarius*: Phylogenetic implications. *Syst. appl. Microbiol.* **14,** 14–22 (1990).
37. Kuwano, Y., Olvera, J. & Wool, I.G. The primary structure of rat ribosomal protein S5, a ribosomal protein present in the rat genome in a single copy. *J. biol. Chem.* **267**, 25304–25308 (1992).
38. Stroeher, U.H., Karageorgos, L.E., Morona, R., & Manning, P.A. Serotype conversion in *Vibrio cholerae* O1. *Proc. natn. Acad. Sci. U.S.A.* **89,** 2566–2570 (1992).
39. Koeplin, R. *et al.* Genetics of xanthan production in *Xanthomonas campestris*: the xanA and xanB genes are involved in UDP–glucose and GDP–mannose biosynthesis. *J. Bacteriol.* **174,** 191–199 (1992).
40. Zielinski, N.A., Chakrabarty, A.M. & Berry, A. Characterization and regulation of the *Pseudomonas aeruginosa* algc gene encoding phosphomannomutase. *J. biol. Chem.* **266,** 9754–9763 (1991).
41. Lee, S.J., Romana, L.K. & Reeves, P.R. Sequence and structural analysis of the rfb (O antigen) gene cluster from a group C1 *Salmonella enterica* strain. *J. gen. Microbiol.* **138,** 1843–1855 (1992).
42. Matsuoka, M. *et al.* Primary structure of maize pyruvate,orthophosphate dikinase as deduced from cDNA sequence. *J. biol. Chem.* **263,** 11080–11083 (1988).
43. Belunis, C.J., Mdluli, K.E., Raetz, C.R.H. & Nano, F.E. A novel 3-Deoxy-D-manno-octulosonic acid transferase from *Chlamydia trachomatis* required for expression of the genus–specific epitope. *J. biol. Chem.* **267,** 18702–18707 (1992).

Section XI:
Physics

IL NUOVO CIMENTO VOL. III, N. 3 1° Marzo 1956

On Generalised Dispersion Relations II.

ABDUS SALAM
St. John's College - Cambridge

W. GILBERT
Trinity College - Cambridge

(ricevuto il 16 Gennaio 1956)

Summary. — Considerations of an earlier paper are generalised to write down non-forward scattering, spin-flip dispersion relations. The contribution from the « bound state » is discussed.

---

1. – This paper continues the discussion of an earlier paper [1]. A meson of 4-momentum k ($k^2 = \mu^2$) is scattered by a nucleon of initial 4-momentum p, ($p^2 = \varkappa^2$) spin λ; no other particles are emitted in the finl state, the meson and the nucleon momenta being k' and p', and spin λ'. From energy conservation,

$$k' = p + k - p' , \qquad k\cdot(p - p') = p\cdot p' - \varkappa^2 . \tag{1}$$

In the earlier paper the scattering amplitude M_R defined covariantly, was written as

$$M_R(\lambda'\lambda) = \bar{u}^{\lambda'}(p')\{L' + i\gamma k M'\}u^{\lambda}(p) , \tag{2}$$

where L' and M' depend on $k\cdot(p+p')$ and $p\cdot p'$.

To obtain dispersion relations it was found necessary to work in a special

---

([1]) A. SALAM: *Nuovo Cimento*, **3**, 424 (1956).

([2]) Notation: $p = \boldsymbol{p}$, $p_4 = ip_0$, $\gamma_i = \begin{pmatrix} & \sigma_i \\ \sigma_i & \end{pmatrix}$, $\gamma_4 = \begin{pmatrix} 1 & \\ & -1 \end{pmatrix}$. Define $i\gamma p = i\boldsymbol{\gamma}\cdot\boldsymbol{p} + i\gamma_4 p_4$; $p\cdot q = p_0 q_0 - \boldsymbol{p}\cdot\boldsymbol{q}$. Thus $(i\gamma p)^2 = p^2$.

frame of reference defined by $\boldsymbol{p}+\boldsymbol{p}'=0$. If $\boldsymbol{p}=0, 0, P$, $\boldsymbol{k}=Q, 0, -P$ $p\cdot p'-\varkappa^2=2P^2$, and $k\cdot(p+p')=2k_0p_0$. Further (³) in this frame

$$(3)\qquad \begin{cases} u^1(p')\{1, i\gamma k\}u^1(p)=\dfrac{p_0}{\varkappa}, k_0 \\[2ex] u^2(p')\{1, i\gamma k\}u^1(p)=0, \dfrac{PQ}{\varkappa}. \end{cases}$$

In reference (¹), a dispersion relation was obtained for $M_R(11)=$ $=(p_0/\varkappa)L'+k_0M'$. It is in fact equally simple to obtain relations for L' and M' separately.

Consider $M_R(21)$, the spin-flip amplitude in this frame. $M_R(21)=-M_R(12)=$ $=(PQ/\varkappa)M'$. Now

$$(4)\qquad (p'2|[j(0), j(x_0, -x_1, x_2, x_3)]|p1) = (p'1|[j(0)j(x)]|p2)$$

$$(5)\qquad (p'2|[j(0), j(x_0, x_1, x_2, -x_3)]|p1) = -(p2|[j(0)j(x)]|p'2).$$

These results follow by using the parity operator; for example

$$P_{x_3}j(x)P_{x_3}^{-1}=-j(x_0, x_1, x_2, -x_3);\qquad P_{x_3}|p_0, \boldsymbol{p}, 1\rangle = i|p_0, -\boldsymbol{p}, 1\rangle.$$

From $M_R(12)=-M_R(21)$, and using (4) and (5),

$$M_R(21)=-\tfrac{1}{2}\int\theta(-x)\exp[-ik_0x_0]\sin Qx_1\{J_1(P, x)\sin Px_3+J_2(P, x)\cos Px_3\}\,d^4x.$$

Here

$$J_1(x)=\{(p'2|p1)-(p1|p'2)\}-\{(p'1|p2)-(p2|p'1)\}$$

$$J_2(x)=i\{(p'2|p1)+(p1|p'2)\}-i\{(p'1|p2)+(p2|p'1)\}$$

in an obvious notation. J_1, J_2 are real. From the contour (⁴) integral

$$\int\frac{dk_0'\exp[-ik_0'x_0]}{k_0'^2-k_0^2}\left(\frac{\sin Q'x_1}{Q'}\right),$$

(³) The (positive energy) free particle spinors are defined as

$$u_\alpha^\lambda(p)=\frac{1}{N(p)}(i\gamma p-k)_{\alpha\lambda},\qquad \lambda=1, 2;\qquad \alpha=1, 2, 3, 4.$$

We obtain $N(p)$, the «invariant» normalization from $u^{\lambda'}(p')u^\lambda(p)=\delta_{\lambda'\lambda}$. Thus $N^{-2}(p)=2\varkappa(\varkappa+p_0)$.

(⁴) M. L. GOLDBERGER: *Phys. Rev.*, **99**, 979 (1955).

we infer

$$\frac{\mathrm{Re}\, M'(k_0, P)}{k_0} = \frac{2\ \mathrm{p.v.}}{\pi}\int_0^\infty \frac{\mathrm{Im}\, M'(k_0', P)\,\mathrm{d}k_0'}{k_0'^2 - k_0^2}\,. \tag{6}$$

Clearly $k_0 M'(k_0, P)$ would satisfy a dispersion relation as in eq. (27) of reference (1). Thus also

$$\mathrm{Re}\, L'(k_0, P) = \frac{2\ \mathrm{p.v.}}{\pi}\int_0^\infty \frac{\mathrm{Im}\, L'(k_0', P) k_0'\,\mathrm{d}k_0'}{k_0'^2 - k_0^2}\,. \tag{7}$$

L_R' and M_R' are functions of invariant scalar products $k\cdot(p+p')$ and $p\cdot p'$. Relations (6) and (7) can thus be expressed entirely in terms of invariant quantities.

2. – It is perhaps useful to set down the transformation formulae from the frame used, to the conventional centre of mass frame, defined from the requirement $\boldsymbol{k}^c+\boldsymbol{p}^c=0$. The superscript « c » distinguishes c.m. quantities.

Let $p_\mu^c = a_{\mu\nu}p_\nu$. The matrix $(a_{\mu\nu})$ is given by

$$a_{11} = a_{44} = \cosh\chi\,, \qquad a_{14} = -a_{41} = i\sinh\chi\,, \qquad a_{22} = a_{33} = 1$$

where $\mathrm{tg}\, h\chi = Q/(p_0+k_0)$.

Define

$$S\gamma_\nu S^{-1} = a_{\mu\nu}\gamma_\mu\,.$$

For the $a_{\mu\nu}$ in (4),

$$S = \exp[(i/2)\,\gamma_1\gamma_4\chi] = c + i\gamma_1\gamma_4 s\,.$$

Here

$$c = \cosh\chi/2\,, \qquad s = \sinh\chi/2\,.$$

If $u^{(s\lambda)}(p^c) = Su^\lambda(p)$, clearly

$$\bar{u}^{\lambda'}(p')\{1, i\gamma k\}u^\lambda(p) = u^{(s\lambda')}(p'^c)\{1, i\gamma k^c\}u^{(s\lambda)}(p^c)\,. \tag{8}$$

Now

$$Su^\lambda(p)|_\alpha = \frac{1}{N(p)}\,S(i\gamma p - \varkappa)|_{\alpha\lambda} = \frac{1}{N(p)}\,(i\gamma p^c - \varkappa)S|_{\alpha\lambda}\,. \tag{9}$$

Considered as a 4×4 matrix, only two of the columns of $(i\gamma p^c - \varkappa)$ are linearly independent. Therefore

(10) $$Su^1(p) = [(p_0 + \varkappa)(p_0^c + \varkappa)]^{-\frac{1}{2}}\{c(p_0 + \varkappa)u^1(p^c) - Ps\,u^2(p^c)\}$$

(11) $$Su^2(p) = [(p_0 + \varkappa)(p_0^c + \varkappa)]^{-\frac{1}{2}}\{Ps\,u'(p^c) + c(p_0 + \varkappa)\,u^2(p^c)\}.$$

Define

(12) $$M_R^c(11) = u^1(p^c)\{L_R' + i\gamma k^c M_R'\}\,u'(p^c)\,, \quad \text{etc.}$$

From (10) and (11),

(13) $$M(11) = \alpha M^c(11) + \beta M^c(21) \text{ [5]}$$

(14) $$M(21) = -\beta M^c(11) + \alpha M^c(21)$$

where

$$\beta = \frac{P \sinh \chi}{p_0^c + \varkappa}\,, \qquad \alpha^2 + \beta^2 = 1\,.$$

$M^c(11)$ corresponds to no spin-flip, and $M^c(21)$ to the spin-flip amplitudes in the c.m. system (the latter with no azimuth change). In terms of the phase shifts, therefore

(15) $$M^c(11) = \left(\frac{4\pi}{\varkappa}\right)\frac{p_0^c + k_0^c}{k_c}\sum\,(la_{l-} + (l+1)a_{l+})P_l(\cos\theta_c)\,,$$

(16) $$M^c(21) = \left(\frac{4\pi}{\varkappa}\right)\frac{p_0^c + k_0^c}{k_c}\sum\,(a_{l-} - a_{l+})\sin\theta_c\,\frac{\mathrm{d}}{\mathrm{d}(\cos\theta_c)}P_l\,(\cos\theta_c)\,.$$

Here $a = e^{2i\delta} - 1/2i$, and δ_{l-} and δ_{l+} refer to phase shifts, corresponding to $j = l - \frac{1}{2}$, and $j = l + \frac{1}{2}$. k_c and θ_c are the 3-momentum and the angle in the c.m. system. To obtain k_c and θ_c, note $p\cdot p' - \varkappa^2 = k_c^2(1 - \cos\theta_c) = 2P^2$, and $k\cdot p = k_0 p_0 + P^2 = k^c p_0^c + k_c^2$.

3. – The case of charged mesons has been completely solved by GOLDBERGER [4]. Write

$$M_{R\alpha\beta} = \delta_{\alpha\beta}\bar{u}(L' + i\gamma k M')u + \tfrac{1}{2}[\tau_\alpha, \tau_\beta]\,\bar{u}(L'' + i\gamma k M'')u\,.$$

---

[5] On account of our choice of axes ($\boldsymbol{p} = 0,0,P$; $\boldsymbol{k} = Q, 0, -P$), $M(11) = M(22)$ and $M(12) = -M(21)$. The same holds for the trasformed quantities M^c.

Then L' and M'' satisfy a relation of the type (6); M', L'' a relation of the type (7). L' and M' refer to the phase shift combination $\frac{1}{3}a(\frac{1}{2}) + \frac{2}{3}a(\frac{3}{2})$, and L'' and M'' to $a(\frac{1}{2}) - a(\frac{3}{2})$.

In relations (6) and (7), $k_0 \geqslant (\mu^2+P^2)^{\frac{1}{2}}$ is the physical range; $(\varkappa\mu - P^2)/2p_0 \leqslant k_0 < (\mu^2+P^2)^{\frac{1}{2}}$ is the « unphysical range ». Below $(\varkappa\mu - P^2)/2p_0$, the so called « bound state » contributes to Im M_R. The exact energy dependence of this term can be computed if we adopt the convention,

$$\Gamma_5(p_1^2 = \varkappa^2,\ p_2^2 = \varkappa^2,\ (p_1 - p_2)^2 = \mu^2) = gZ^{-1}\gamma_5 = g_1\gamma_5\ , \quad (i\gamma p_1 + \varkappa = i\gamma p_2 + \varkappa = 0),$$

for the renormalization of the coupling constant in meson theory. The « bound state » contribution then involves the (unknown) renormalized coupling constant g_1, and has the form,

$$\text{Im } M_{R\alpha\beta}(\lambda'\lambda) = -\left(\frac{g_1^2}{4\pi}\right)\cdot 4\pi^2\tau_\alpha\tau_\beta\{\bar{u}^{\lambda'}(i\gamma k)u^{\lambda}\}\delta(\mu^2 + 2P^2 - 2k_0p_0)\ .$$

The renormalized coupling constant defined from $\Gamma_5(p_1 = p_2) = g_1'\gamma_5$ then differs from the above by $O(\mu^2/\varkappa^2)$.

After the work described here was completed, the authors, learnt from Prof. M. L. GOLDBERGER that he has obtained identical results in collaboration with Professors NAMBU and OEHME, and is applying the relations to calculate S phase shifts.

RIASSUNTO (*)

Si generalizzano le considerazioni di un precedente lavoro per scrivere le relazioni dello scattering non in avanti, e della dispersione per inversione di spin. Si discute il contributo dello « stato legato ».

(*) *Traduzione a cura della Redazione.*

PHYSICAL REVIEW VOLUME 108, NUMBER 4 NOVEMBER 15, 1957

New Dispersion Relations for Pion-Nucleon Scattering

Walter Gilbert*
Lyman Laboratory of Physics, Harvard University, Cambridge, Massachusetts

(Received August 5, 1957)

New relations expressing the imaginary part of the scattering amplitude as a coupling constant term and an integral over the physical spectrum of the real part of the amplitude are developed. These relations are applied to compute the meson-nucleon coupling constant from the experimental phase shifts, yielding the value $f^2=0.084$, and are used to compute the s-wave scattering lengths from the p-wave data.

INTRODUCTION

We shall develop a new, more convergent form for the dispersion relations for meson-nucleon scattering. These relationships will express the imaginary part of the amplitude in terms of a coupling constant term and an integral of the real part of the amplitude in the physical region. These new relations differ from the conjugate Hilbert transforms that connect the imaginary part of an amplitude to an integral of the real part in that the integrals are restricted to the experimentally accessible region above the threshold. These expressions have an additional convergence factor, the réciprocal of the laboratory momentum, and so provide relationships between the phase shifts less dependent upon the high-energy behavior of the theory or upon additional constant terms than the usual dispersion relations. We shall make two applications of these new relations: a coupling constant determination and a computation of the s-wave scattering lengths. In this paper we shall restrict ourselves to the dispersion relations for scattering into the forward direction, although the argument could be applied to the more general relations off the forward direction.

DISPERSION RELATIONS

Many authors have given the dispersion relations applicable to meson-nucleon scattering.[1–5] We shall collect in this section the relations and the formulas that we shall need.[6] The invariant T matrix for the pion-nucleon scattering process, considered as a two-by-two matrix in the nucleon isotopic spin space, we divide into two parts, even and odd, in the meson isotopic indexes,

$$T=\delta_{\alpha\beta}T^e+\tfrac{1}{2}[\tau_\alpha,\tau_\beta]T^o. \quad (1)$$

These even and odd amplitudes are related to the isotopic spin decomposition of the T matrix by

$$T^e=\tfrac{1}{3}(T^1+2T^3),\qquad T^o=\tfrac{1}{3}(T^1-T^3), \quad (2)$$

where the superscripts "1" and "3" refer to the $T=\frac{1}{2}$ and $T=\frac{3}{2}$ states, respectively. We shall separate out the nucleon spin dependence of the T matrix as

$$T(p,p',q)=\bar{u}(p)u(p')f(pp',pq)+\bar{u}(p)\gamma qu(p')g(pp',pq) \quad (3)$$

on the energy shell. The two invariant amplitudes, f and g, separately obey dispersion relations. The g amplitude is, to the order of μ^2/κ^2, the spin-flip amplitude, and since the π meson is pseudoscalar, the bound-state terms involving the coupling constant occur only in the dispersion relations for the g amplitude. We need only the relations in the forward direction, which we take to be, using z for $pq/\kappa\mu$, the laboratory energy of the meson divided by the meson's mass:

$$\mathrm{Re}\, f^o(\bar{z})=\bar{z}\frac{2}{\pi}P\int_1^\infty \frac{dz}{z^2-\bar{z}^2}\mathrm{Im}\, f^o(z), \quad (4)$$

$$\mathrm{Re}\, g^o(\bar{z})=\frac{bG^2/\kappa\mu}{\bar{z}^2-b^2}+\frac{2}{\pi}P\int_1^\infty \frac{zdz}{z^2-\bar{z}^2}\mathrm{Im}\, g^o(z), \quad (5)$$

$$\mathrm{Re}\, f^e(\bar{z})-\mathrm{Re}\, f^e(z')=\frac{2}{\pi}(\bar{z}^2-z'^2)P\int_1^\infty \frac{zdz\,\mathrm{Im}\, f^e(z)}{(z^2-\bar{z}^2)(z^2-z'^2)}, \quad (6)$$

$$\mathrm{Re}\, g^e(\bar{z})=-\frac{\bar{z}G^2/\kappa\mu}{\bar{z}^2-b^2}+\frac{2}{\pi}\bar{z}P\int_1^\infty \frac{dz}{z^2-\bar{z}^2}\mathrm{Im}\, g^e(z). \quad (7)$$

Here $b=\mu/2\kappa$, the position of the bound state singularity, and G is the renormalized, unrationalized pseudoscalar coupling constant. We shall also use the rationalized pseudovector coupling constant

$$f^2=b^2g^2=b^2G^2/4\pi. \quad (8)$$

We need to know the relation between this invariant decomposition of the scattering amplitude and the usual phase-shift expansion. We shall introduce direct and spin-flip amplitudes as, in the center-of-mass

---

* National Science Foundation Predoctoral Fellow.
[1] M. L. Goldberger, Phys. Rev. **99**, 979 (1955).
[2] Goldberger, Miyazawa, and Oehme, Phys. Rev. **99**, 986 (1955).
[3] A. Salam, Nuovo cimento **3**, 424 (1956), and A. Salam and W. Gilbert, Nuovo cimento **3**, 607 (1956).
[4] R. Oehme, Phys. Rev. **100**, 1503 (1955) and **102**, 1174 (1956).
[5] R. H. Capps and G. Takeda, Phys. Rev. **103**, 1877 (1956).
[6] We take $\hbar=c=1$ and use a timelike metric. pp' is the invariant product of two four-vectors. We are considering the scattering of a nucleon of mass κ, four-momentum p', by a meson of mass μ, four-momentum q', and isotopic spin β. The final state is characterized by p, q, and α. The nucleons are described by Dirac spinors $u(p)$ obeying the equation $(\gamma p-\kappa)u(p)=0$, $\bar{u}u=2\kappa$. In calculations we take $\kappa=6.7\mu$.

system,

$$T=D+i\boldsymbol{\sigma}\cdot\hat{n}\,\sin\theta\left(\frac{\mu^2\eta^2E}{\kappa}\right)S, \tag{9}$$

where θ is the angle of scattering in the center-of-mass system, $\hat{n}$ is a unit vector perpendicular to the plane of scattering, η is the center-of-mass momentum divided by the meson's mass, and E is the total center-of-mass energy. We shall also use the symbol $\beta=p_0+\kappa$, the sum of the nucleon energy and mass in the center-of-mass system, and $\alpha=E+\kappa$. Then we have the algebraic relations

$$\begin{aligned} E&=(\kappa^2+\mu^2+2\kappa\mu z)^{\frac{1}{2}},\\ \beta E&=\kappa(\alpha+\mu z), \qquad (10)\\ (z^2-1)^{\frac{1}{2}}\kappa&=\eta E.\end{aligned}$$

If the spinors in (3) are specialized to the center-of-mass system, the invariant functions $f(z)$ and $g(z)$ can be related to the direct and spin-flip amplitudes:

$$f(z)=\frac{\alpha}{2\beta E}D(z)-\mu zS(z), \tag{11}$$

$$g(z)=[D(z)/2\beta E]+S(z). \tag{12}$$

The phase-shift expansions for the direct and spin-flip amplitudes may be determined from the unitary condition on the invariant T matrix. In the forward direction the expansions are, with our normalization,

$$D(z)=\frac{8\pi E}{\mu\eta}\sum_{l=0}^{\infty}[la_{l-}+(l+1)a_{l+}], \tag{13}$$

$$S(z)=\frac{8\pi\kappa}{\mu^3\eta^3}\sum_{l=1}^{\infty}(a_{l-}-a_{l+})\frac{l(l+1)}{2}. \tag{14}$$

Here $a_{l\pm}$ is $e^{i\delta}\sin\delta$ for the state $j=l\pm\frac{1}{2}$ if the scattering is elastic. We shall normally cut off these expansions after the p-wave terms and assume that the scattering is elastic at all energies of interest. We use the usual notation for the s and p waves, a_1 and a_3 referring to $e^{i\delta}\sin\delta$ for the $T=\frac{1}{2}$ and $\frac{3}{2}$ s waves, $a_{2T,2J}$ for the p waves. We shall use a superscript "0" to denote the scattering lengths, taking the low-energy behavior of the phases to be $\delta\sim\delta^0\eta$ for the s-waves and $\delta\sim\delta^0\eta^3$ for the p waves.

The optical theorem is, with this normalization,

$$\operatorname{Im} D=2\mu\eta E\sigma, \tag{15}$$

or for the even and odd isotopic index amplitudes:

$$\operatorname{Im} D^e=\mu\eta E(\sigma^++\sigma^-), \tag{16}$$

$$\operatorname{Im} D^o=\mu\eta E(\sigma^--\sigma^+), \tag{17}$$

σ^+ and σ^- being the total cross sections for the scattering of positive and negative pions off protons. The relations (4) and (5), (6) and (7) may be combined using (10), (11), and (12) to yield Goldberger's[1,2] relations for the direct amplitude:

$$\operatorname{Re} D^o(z)=\frac{2\bar{z}bG^2}{\bar{z}^2-b^2}+\frac{2}{\pi}\bar{z}P\int_1^\infty\frac{dz\,\operatorname{Im} D^o(z)}{z^2-\bar{z}^2}, \tag{18}$$

$$\operatorname{Re} D^e(z)-\operatorname{Re} D^e(1)=(\bar{z}^2-1)\left\{\frac{2b^2G^2}{(\bar{z}^2-b^2)(1-b^2)}+\frac{2}{\pi}P\int_1^\infty\frac{zdz\,\operatorname{Im} D^e(z)}{(z^2-\bar{z}^2)(z^2-1)}\right\}. \tag{19}$$

The dispersion relations that we have written are consistent with the assumption that the cross sections σ^+ and σ^- become constant and equal at high energies and, further, that the spin-flip amplitude vanishes at high energies. That the cross sections become constant and the spin-flip amplitude vanish would follow from the existence of a finite range for the interaction and the interaction becoming completely inelastic at high energies.

NEW RELATIONS

Now we shall develop a more convergent form of the dispersion relations for each of the amplitudes introduced in the previous section, basing our arguments on the form of the usual dispersion relations for these amplitudes. We shall only discuss the forward direction, but a similar development could be made for the general case. Consider, as an example, the function $g^e(z)$. We have previously written relation (7) for this function which corresponds to the following properties:

(1) $g^e(z)$ is analytic in the upper half of the complex z plane,

(2) $g^e(z)$ is less than z at infinity in the upper half of the complex plane,

(3) the real part of $g^e(z)$ is an odd function on the real axis, the imaginary part of $q^e(z)$ is an even function on the real axis, and

(4) $\operatorname{Im} g^e(z)$ is zero on the real axis for $-1<z<1$ with the exception of a delta-function contribution

$$\tfrac{1}{2}\pi(G^2/\kappa\mu)[\delta(z-b)+\delta(z+b)].$$

We consider the function $g^e(z)/(z^2-1)^{\frac{1}{2}}$. By $(z^2-1)^{\frac{1}{2}}$ we mean that branch of the function that is analytic in the upper half plane, positive for z real, $z>1$, negative for z real, $z<-1$, and positive imaginary for z real, $-1<z<1$. The imaginary part of this function is positive in the upper half-plane, and the function has no zeros above the real axis.[7] This function is just the dimensionless laboratory momentum of the incoming

[7] We limit our discussion of the behavior of these functions to the upper half-plane as a matter of convenience; all of these functions are analytic in the entire plane with cut lines along the real axis from ± 1 to $\pm\infty$. The behavior of these functions on the real axis is always understood to be the behavior as a limit from the upper half-plane.

meson. Now the quotient function, $g^e(z)/(z^2-1)^{\frac{1}{2}}$, will have these properties:

(1) it is analytic in the upper half-plane,
(2) it goes to zero at infinity,
(3) the limit of the function onto the real axis from above has an even real part and an odd imaginary part, and
(4) the real part of this function is zero between $z=-1$ and $z=1$ on the real axis with the exception of two delta-function singularities, the real part of this function being $\mathrm{Im}\,[g^e(z)]/(1-z)^{\frac{1}{2}}$ in this region.

From the boundedness and symmetry properties we can write, using Cauchy's theorem,

$$\mathrm{Im}\left[\frac{g^e(\bar{z})}{(\bar{z}^2-1)^{\frac{1}{2}}}\right]=\frac{2}{\pi}\bar{z}P\int_0^\infty \frac{dz}{z^2-\bar{z}^2}\mathrm{Re}\left[\frac{g^e(z)}{(z^2-1)^{\frac{1}{2}}}\right], \tag{20}$$

in which Im [ ] and Re [ ] mean the imaginary and the real parts of the limit on the real axis from above of the analytic function inside the brackets. No trouble arises when the contour of integration is brought to the real axis. The integration is to be thought of, at first, as including semicircles above the points $z=\pm 1$. The contribution of these semicircles vanishes as the radius shrinks to zero, since the singularity in

$$\mathrm{Re}\,[g^e(z)/(z^2-1)^{\frac{1}{2}}]$$

at $z=\pm 1$ is integrable. Using the relation between the real part of $g^e(z)/(z^2-1)^{\frac{1}{2}}$ and the imaginary part of $g^e(z)$ in the nonphysical region, we have:

$$\mathrm{Im}\left[\frac{g^e(\bar{z})}{(\bar{z}^2-1)^{\frac{1}{2}}}\right]=\frac{2}{\pi}\bar{z}P\int_1^\infty \frac{dz}{z^2-\bar{z}^2}\frac{\mathrm{Re}\,g^e(z)}{(z^2-1)^{\frac{1}{2}}} +\frac{\bar{z}}{\bar{z}^2-b^2}\frac{G^2}{\kappa\mu(1-b^2)^{\frac{1}{2}}}. \tag{21}$$

The left-hand side of this relation is

$$\mathrm{Im}\,[g^e(\bar{z})]/(\bar{z}^2-1)^{\frac{1}{2}} \quad \text{for } \bar{z}>1,$$

and

$$-\mathrm{Re}\,[g^e(\bar{z})]/(1-\bar{z}^2)^{\frac{1}{2}} \quad \text{for } 0<\bar{z}<1.$$

There is an infinite discontinuity in this relation as the threshold is approached from below, since $\mathrm{Im}\,g^e(z)$ goes to zero as η at threshold while $\mathrm{Re}\,g^e(z)$ is a constant in the neighborhood of the threshold.

Similar relations can be written for the other amplitudes. We introduce the symbol $\xi=(z^2-1)^{\frac{1}{2}}$ and drop the factor $(1-b^2)^{\frac{1}{2}}$ that occurs in the bound state terms. Then some of the other possible relations are

$$\mathrm{Im}\left[\frac{g^o(\bar{z})}{\xi}\right]=\frac{2}{\pi}P\int_1^\infty \frac{zdz}{z^2-\bar{z}^2}\frac{\mathrm{Re}\,g^o(z)}{\xi}-\frac{bG^2/\kappa\mu}{\bar{z}^2-b^2}, \tag{22}$$

and

$$\mathrm{Im}\left[\frac{f^e(\bar{z})}{\xi}\right]=\frac{2}{\pi}P\int_1^\infty \frac{zdz}{z^2-\bar{z}^2}\frac{\mathrm{Re}f^e(z)}{\xi}, \tag{23}$$

where we do not have to perform a subtraction in order to obtain convergence provided that the $f^e(z)$ amplitude is less singular than z at infinity. The $f^o(z)$ amplitude obeys relation (21) without the bound-state term. For the direct amplitudes we have:

$$\mathrm{Im}\left[\frac{D^o(\bar{z})}{\xi}\right] =\frac{2}{\pi}\bar{z}P\int_1^\infty \frac{dz}{z^2-\bar{z}^2}\frac{\mathrm{Re}\,D^o(z)}{\xi}-\frac{\bar{z}}{\bar{z}^2-b^2}2bG^2, \tag{24}$$

$$\mathrm{Im}\left[\frac{D^e(\bar{z})}{\xi}\right]-\mathrm{Im}\left[\frac{D^e(z')}{\xi'}\right] =(\bar{z}^2-z'^2)\left\{-\frac{2b^2G^2}{(\bar{z}^2-b^2)(z'^2-b^2)} \frac{2}{\pi}P\int_1^\infty \frac{zdz}{(z^2-\bar{z}^2)(z^2-z'^2)}\frac{\mathrm{Re}\,D^e(z)}{\xi}\right\}, \tag{25}$$

or, in terms of cross sections, using (16) and (17), $\bar{z}>1$,

$$\sigma^+-\sigma^-=\frac{bG^2}{\kappa\mu\bar{z}}+\frac{2}{\pi}\frac{\bar{z}}{\kappa\mu}P\int_1^\infty \frac{dz}{z^2-\bar{z}^2}\frac{\mathrm{Re}\,D^o}{\xi}, \tag{26}$$

$$(\sigma^++\sigma^-)_{\bar{z}}-(\sigma^++\sigma^-)_{z'}=\frac{(\bar{z}^2-z'^2)}{\kappa\mu}\left\{-\frac{2b^2G^2}{\bar{z}^2z'^2} \frac{2}{\pi}P\int_1^\infty \frac{zdz}{z^2-\bar{z}^2}\frac{\mathrm{Re}\,D^e}{\xi}\frac{1}{z^2-z'^2}\right\}. \tag{27}$$

The integrals over the real parts of the amplitudes would be difficult to handle near threshold, since the integrand is so singular there. The integrands can be modified, however, to remove either the principal-part singularity or the singularity at the threshold. Since ξ has the analytic properties mentioned earlier, the following integral is an immediate consequence of Cauchy's theorem:

$$\frac{2}{\pi}P\int_1^\infty \frac{zdz}{z^2-\bar{z}^2}\frac{1}{\xi}=\begin{cases}0 & \text{for } \bar{z}>1\\ 1/(1-\bar{z}^2)^{\frac{1}{2}} & \text{for } \bar{z}<1.\end{cases} \tag{28}$$

We shall use this and similar integrals to isolate the dominant contributions to these singular integrals when we apply these new relations.

COUPLING CONSTANT DETERMINATION

The new relations are more rapidly convergent than the usual ones and therefore are better adapted for the analysis of the experimental data since the high-energy effects can be neglected. Of all the relations, the new relation (21) for the even-isotopic-index spin-flip amplitude is best suited for the determination of the coupling constant. The relation is rapidly convergent. The coupling constant is weighted with a large factor with respect to the phase shifts, being given essentially additively rather than as a difference of experimental quantities. We shall use relation (21) at threshold. In this limit, the dominant contribution from the integral arises from the neighborhood of the singularity. We use the fact that

$$\lim_{\bar{z}\to 1+} \frac{2}{\pi} P \int_1^\infty \frac{dz}{(z^2-\bar{z}^2)} \frac{1}{z\xi} = -1, \tag{29}$$

which follows from (28), to reduce the singularity. Thus (21) becomes

$$\frac{G^2}{\kappa\mu} + \mathrm{Re}\, g^e(1) = \lim_{\bar{z}\to 1} \frac{\mathrm{Im}\, g^e(\bar{z})}{\bar{\xi}} + \frac{2}{\pi}\int_1^\infty \frac{dz}{\xi^3}\left[\mathrm{Re}\, g^e(z) - \frac{1}{z}\mathrm{Re}\, g^e(1)\right]. \tag{30}$$

No difficulty arises from the limit of the principal-part function since the integral, after the subtraction, has an integrable singularity at threshold. The term arising from the imaginary part of the amplitude is related to the square of the s-wave scattering lengths. It is

$$\lim_{z\to 1} \frac{\mathrm{Im}\, g^e(z)}{\xi} = \frac{2\pi}{3\mu(\kappa+\mu)}[(\delta_1{}^0)^2 + 2(\delta_3{}^0)^2],$$

which is about 0.1% of the coupling constant term, if one uses Orear's values for the s-wave lengths.[8] Since the contribution to $g^e(z)$ from the direct amplitude is about one percent, we shall ignore it, although it could be taken into account using (12). The integral is rapidly convergent; dropping all higher waves in $g^e(z)$, we rewrite (30) as

$$3f^2 = \delta_{33}{}^0 - \delta_{31}{}^0 + \tfrac{1}{2}(\delta_{13}{}^0 - \delta_{11}{}^0) - [\lambda_{33} - \lambda_{31} + \tfrac{1}{2}(\lambda_{13} - \lambda_{11})], \tag{31}$$

in which

$$\lambda = \frac{1}{\pi}\int_1^\infty \frac{dz}{\xi^3}\left(\frac{\sin 2\delta}{\eta^3} - \frac{2\delta^0}{z}\right). \tag{32}$$

We shall evaluate the coupling constant using the Anderson fitted phase shifts.[9] See Table I. The contribution from the scattering lengths is

$$3f^2 \sim 0.301,$$

and the correction from the integral is -0.050, resulting in a value for the coupling constant,

$$f^2 = 0.084.$$

TABLE I. The scattering lengths, δ^0, and the integrals, λ, defined by (32) based upon the p-wave phases given by Anderson.[a]

State	δ^0	λ
33	0.248	0.0404
31	−0.042	−0.0060
13	0.0048	0.0029
11	−0.0175	−0.0033

[a] See reference 9.

In this determination, the 33 state contributes 83% and the 31 state, 14%. An alternative, but less convergent, way to calculate the coupling is to use the usual dispersion relation for $g^e(z)$, (7), which we write analogously to (30)

$$\frac{G^2}{\kappa\mu} + \mathrm{Re}\, g^e(1) = \frac{2}{\pi}\int_1^\infty \frac{dz}{\xi^2}\,\mathrm{Im}\, g^e(z).$$

Only the 33-state contribution to the integral is important. This yields

$$3f^2 = 0.301 - 0.052 \quad \text{or} \quad f^2 = 0.083.$$

Both of these values for the coupling constant are in good agreement with the value found by Haber-Schaim,[10] $f^2=0.082$. Our result is less than the value $f^2=0.10$ found by Davidon and Goldberger[11] using the same phases and a combination of relations (5) and (7). The relation we have used, however, is more convergent than theirs by a factor of z^2 and gives the coupling constant essentially additively in terms of the experimental data.

Another way to compute the coupling constant would be to modify Eq. (21) so that it could be used to represent the experimental data as a straight line whose intercept would be the coupling. This would also provide a check on the consistency of the data. Equation (21) can easily be brought to the form:

$$\frac{G^2}{\kappa\mu} + \bar{z}^2\left[\mathrm{Re}\, g^e(1) - \frac{2}{\pi}\int_1^\infty \frac{dz}{\xi^3}\chi(z)\right] = \frac{\bar{z}}{\bar{\xi}}\,\mathrm{Im}\, g^e(\bar{z}) + \frac{2}{\pi}\bar{z}^2\bar{\xi}^2 \int_1^\infty \frac{dz}{z^2-\bar{z}^2}\frac{1}{\xi^3}\left[\chi(z) - \frac{\bar{z}}{z}\chi(\bar{z})\right], \tag{33}$$

where

$$\chi(z) = \mathrm{Re}\, g^e(z) - \frac{1}{z}\mathrm{Re}\, g^e(1).$$

[8] J. Orear, Phys. Rev. **101**, 288 (1956).
[9] H. L. Anderson, *Sixth Annual Rochester Conference on High-Energy Physics*, 1956 (Interscience Publishers, Inc., New York, 1956).
[10] U. Haber-Schaim, Phys. Rev. **104**, 1113 (1956).
[11] W. C. Davidon and M. L. Goldberger, Phys. Rev. **104**, 1119 (1956).

If the right-hand side of (33) were plotted against z^2, the data should fit a straight line. Alternatively, we could use the fact that the threshold value of $\bar{z}/\xi \operatorname{Im} g^e(\bar{z})$ is insignificant compared to the coupling constant term and plot

$$\frac{G^2}{\kappa\mu}=-\frac{\bar{z}}{\xi^3}\operatorname{Im} g^e(\bar{z})-\frac{2}{\pi}\bar{z}^2\int_1^\infty \frac{dz}{z^2-\bar{z}^2}\frac{1}{\xi^3}\left[\chi(z)-\frac{\bar{z}}{z}\chi(\bar{z})\right], \quad (34)$$

which should be a constant.

S-WAVE SCATTERING LENGTHS

We shall now use these new relations to calculate the s-wave scattering lengths from the p-wave phase shifts. More exactly, given that the s waves are small, we shall show that they can be approximately calculated from the p waves alone, but that if the experimental data about the second maximum in the pion-nucleon scattering cross section, the $T=\frac{1}{2}$ maximum, is used, the scattering lengths can be calculated quite accurately. We shall not have to make any very extreme assumptions about the high-energy behavior of the theory; we need only the assertion that the cross sections for the scattering of positive and negative mesons by protons approach the same value at high energies. We use the two relations for the odd-isotopic-index direct scattering amplitude specialized to threshold. Relation (18) yields

$$\delta_1{}^0-\delta_3{}^0=6\frac{\kappa}{\kappa+\mu}f^2+\frac{3}{4\pi^2}\frac{\mu}{\kappa+\mu}\int_1^\infty\frac{dz}{\xi^2}\operatorname{Im} D^o, \quad (35)$$

and the new relation (24) yields

$$(\delta_1{}^0)^2-(\delta_3{}^0)^2=-6f^2-\frac{3}{4\pi^2}\frac{\mu}{\kappa}\lim_{\bar{z}\to1+}P\int_1^\infty\frac{dz}{z^2-\bar{z}^2}\frac{\operatorname{Re} D^o}{\xi}. \quad (36)$$

Expression (35) can be used to compute the difference of the s-wave scattering lengths, and this difference along with the experimental p-wave phase shifts, can be inserted into (36) to compute the sum of the scattering lengths.

Equation (35) was used by Goldberger, Miyazawa, and Oehme[2] to calculate this difference of scattering lengths from the experimental cross sections. Since the s waves are small, they and the small p waves may be neglected under the integral in (35). Exhibiting the important contributions to the difference, we have

$$\delta_1{}^0-\delta_3{}^0=6\frac{\kappa}{\kappa+\mu}f^2-\frac{4}{\pi}\frac{\kappa}{\kappa+\mu}\int_1^\infty\frac{dz}{z^2-1}\frac{E}{\kappa}\frac{\sin^2\delta_{33}}{\eta}$$
$$+\frac{3}{4\pi^2}\frac{\mu^2\kappa}{\kappa+\mu}\int\frac{dz}{\xi}(\sigma^- -\sigma^+), \quad (37)$$

where the integral is to be extended over the second maximum in the pion-nucleon cross section. The first term in the Born approximation, in pseudoscalar theory, for the difference of the scattering lengths. For a coupling constant $f^2=0.084$, it is 0.44. The p-wave integral is the dominant correction to the Born approximation. If this is integrated using Anderson's[9] values for the 33-phase shift along with the experimental values of Mukhin and Pontecorvo[12] above the resonance, the result is $\delta_1{}^0-\delta_3{}^0=0.20$ The contribution from the $T=\frac{1}{2}$ maximum brings this value up into agreement with the experimental value[8] $\delta_1{}^0-\delta_3{}^0=0.27$. Turning to the second relation, we see that a large part of the integral comes from the neighborhood of the singularity. We modify the integrand to eliminate this singular behavior. Since

$$\lim_{\bar{z}\to1+}P\int_1^\infty\frac{dz}{z^2-\bar{z}^2}\frac{1}{\xi}=-1,$$

we can subtract the threshold value of Re D^o from the integrand:

$$(\delta_1{}^0)^2-(\delta_3{}^0)^2=-6f^2+\frac{2}{\pi}\frac{\kappa+\mu}{\kappa}(\delta_1{}^0-\delta_3{}^0)$$
$$-\frac{3}{4\pi^2}\frac{\mu}{\kappa}\int_1^\infty\frac{dz}{\xi^3}[\operatorname{Re} D^o(z)-\operatorname{Re} D^o(1)]. \quad (38)$$

To the extent that the s waves are small and linear in the momentum, this subtraction takes into account, within 10%, the s-wave contribution to the integral. The rest of the integral may be approximated by the p-wave terms. The sum of the s-wave scattering lengths is then

$$\delta_1{}^0+\delta_3{}^0=\frac{2}{\pi}\frac{\kappa+\mu}{\kappa}+\frac{\Delta-6f^2}{\delta_1{}^0-\delta_3{}^0}, \quad (39)$$

where

$$\Delta=\frac{1}{\pi}\int_1^\infty\frac{dz}{\xi^2\eta^2}[2\sin2\delta_{33}+\sin2\delta_{31}$$
$$-2\sin2\delta_{13}-\sin2\delta_{11}]. \quad (40)$$

Using the Anderson phases,[9] Δ may be evaluated numerically, yielding $\Delta=0.315$. Various values for the scattering lengths are collected in Table II. The first

TABLE II. The s-wave scattering lengths computed from values of the coupling constant, an integral, Δ, over the p-wave phases, and the difference of scattering lengths using Eq. (39).

	f^2	Δ	$\delta_1{}^0-\delta_3{}^0$	$\delta_1{}^0$	$\delta_3{}^0$
(1)	0.084	0.315	0.27	0.15	−0.12
(2)	0.082	0.315	0.27	0.17	−0.10
(3)	0.084	0.315	0.20	−0.01	−0.21
(4)	0.084	0.350	0.20	0.10	−0.10

[12] A. I. Mukhin and B. M. Pontecorvo, J. Exptl. Theoret. Phys. (U.S.S.R.) **31**, 550 (1956) [translation: Soviet Phys. (JETP) **4**, 373 (1957).

and second lines are in good agreement with Orear's values,[8] but the scattering lengths are sensitive to the coupling constant. The third line of Table II uses the p-wave approximation for the difference of the lengths, producing rough agreement. The small p waves do not influence the qualitative picture. In line four, only the 33-phase shift is assumed to be different from zero, $\Delta=0.350$, and the scattering lengths are in qualitative agreement with experiment.

We should point out that for the first relation to hold, $\sigma^+-\sigma^-$ must vanish at high energies. For the new relation (35) to hold, all that is required is that the cross sections become constant at high energies or, more weakly, that neither the real nor the imaginary part of $D^o(z)$ increases as fast as z^2. The significance of this calculation is that a large part of the s-wave scattering may be treated as a relativistic effect induced by the p-wave interaction.

Reprinted from THE PHYSICAL REVIEW, Vol. 115, No. 3, 731–735, August 1, 1959
Printed in U. S. A.

Structure of the Vertex Function

STANLEY DESER,* WALTER GILBERT,† AND E. C. G. SUDARSHAN‡
Lyman Laboratory of Physics, Harvard University, Cambridge, Massachusetts
(Received October 15, 1958; revised manuscript received April 6, 1959)

An integral representation as a function of invariants is found for the Fourier transform of the matrix element between the vacuum and a one-particle state of the retarded commutator of two currents. A special case is a spectral representation for the vertex as a function of momentum transfer. The threshold in this representation is lower than that found in the usual perturbation theory.

INTRODUCTION

WE shall study the structure of the matrix element of the commutator, the retarded commutator, and the time-ordered product of two field operators taken between the vacuum and a single-particle state. Our technique will be to manipulate these functions in the physical region in such a way as to obtain an integral representation for the function in terms of the invariant momentum parameters that characterize its Fourier transform.

We shall discover that, for a special case of this representation, one can continue analytically certain of the invariant parameters out of the physical region and automatically obtain a spectral relation for the vertex function as a function of the momentum transfer.

The basic representation we shall find for the Fourier transform of the retarded commutator of two currents is

$$\int e^{ikx}\theta(x)dx\langle[j_1(x),j_2(0)]|p\rangle = \int d\mu d\beta \frac{H(\mu,\beta,p^2)}{k^2+2\beta pk-\mu+i\epsilon(pk+\beta p^2)}, \quad (1)$$

in which the variables are limited by

$$0\geqslant\beta\geqslant-1,\quad \mu\geqslant\max\{f(\beta),\,-\beta^2p^2\},$$

where $f(\beta)$ is a function, which we will specify later, determined by the mass spectra of the intermediate states, and $H(\mu,\beta,p^2)$ is uniquely determined by the Fourier transform in the physical region.

THE REPRESENTATION

We shall begin by deriving a representation for the matrix element of the commutator of two currents. We

* Now at Brandeis University, Waltham 54, Massachusetts.
† National Science Foundation Postdoctoral Fellow.
‡ On leave of absence from Tata Institute of Fundamental Research, Bombay, India.

shall discuss only boson fields, the extension to the general case being trivial. We consider a matrix element of the form

$$h(x^2,px)=\langle[j_1(x),j_2(0)]|p\rangle, \qquad (2)$$

in which $j_1(x)$ and $j_2(x)$ are the currents associated with two fields and the state $|p\rangle$ is a state of energy momentum p_μ; $p^2=\kappa^2$; $p_0>0$. This state may be either a single-particle state or the state produced by operating on the vacuum with a single Fourier component of a field. As a consequence of Lorentz invariance we know that this matrix element depends on the coordinate only through the variables x^2, px. It is also a function of p^2 and of other eigenvalues; the weight functions will depend implicitly on these parameters.

We introduce the Fourier transform of the matrix element with respect to the invariants:

$$h(x^2,px)=\int d\alpha d\beta \exp(i\alpha x^2)\exp(i\beta px)g(\alpha,\beta). \qquad (3)$$

The condition that the commutator vanish for space-like intervals, whenever $x^2<0$, is equivalent to the requirement that $g(\alpha,\beta)$ be analytic as a function of α in the lower half α-plane. The part of the Fourier transform, $g(\alpha,\beta)$, that is determined by values of the invariants in the region $p^2x^2\geqslant(px)^2$ is arbitrary but does not enter into the momentum-space Fourier transform that we shall take since this is extended only over real values of the coordinate-space vector. We express the exponential $\exp(i\alpha x^2)$ as the Fourier transform of a Gaussian, absorbing the numerical factors into the weight function:

$$h(x^2,px)=\int dkd\alpha d\beta \exp(-i\bar{k}x)\times\exp(-i\bar{k}^2/4\alpha)\exp(i\beta px)\frac{g(\alpha,\beta)}{\alpha|\alpha|}. \qquad (4)$$

Since $g(\alpha,\beta)$ is analytic in the lower half α-plane, we represent it as

$$g(\alpha,\beta)=\int_0^\infty d\lambda^2\exp(i\lambda^2/4\alpha)\tilde{H}(\lambda^2,\beta). \qquad (5)$$

Then the α-integration can be performed, yielding

$$h(x^2,px)=\mathcal{P}\int d\bar{k}\frac{\exp(-ikx)\exp(i\beta px)}{\bar{k}^2-\lambda^2}\tilde{H}(\lambda^2,\beta)d\lambda^2 d\beta. \qquad (6)$$

We use the identity

$$\mathcal{P}\int dk\frac{e^{-ikx}}{k^2-\lambda^2}=-i\pi\epsilon(x_0)\int dk\,\epsilon(k_0)e^{-ikx}\delta(k^2-\lambda^2), \qquad (7)$$

and introduce the Hilbert transform of $\tilde{H}(\lambda^2,\beta)$ in order to remove the $\epsilon(x_0)$ behavior produced by (7):

$$\tilde{H}(\lambda^2,\bar{\beta})=\mathcal{P}\int H(\lambda^2,\beta)\frac{d\beta}{\bar{\beta}-\beta}. \qquad (8)$$

The result of these substitutions is the expression

$$h(x^2,px)=\int dkd\beta\,\epsilon(pk)\exp(-ikx)\times\exp(i\beta px)\delta(\bar{k}^2-\lambda^2)H(\lambda^2,\beta), \qquad (9)$$

or, for the Fourier transform of the commutator,

$$f(k^2,pk)=\int e^{ikx}dx\,h(x^2,px)=\int_0^\infty d\lambda^2\int_{-\infty}^\infty d\beta\,\epsilon(pk+\beta p^2)\times\delta((k+\beta p)^2-\lambda^2)H(\lambda^2,\beta). \qquad (10)$$

For this representation for the commutator, the retarded Green's function may easily be derived; it is

$$\int e^{ikx}\theta(x)dx\,h(x^2,px)=\int_0^\infty d\lambda^2\int_{-\infty}^\infty d\beta\frac{H(\lambda^2,\beta)}{(k+\beta p+i\epsilon p)^2-\lambda^2}. \qquad (11)$$

LIMITS ON THE PARAMETERS

We turn now to the problem of restricting the ranges of the parameters in the representation. We have, as yet, imposed only the condition that the commutator vanish for space-like intervals. Further restrictions arise from the mass-spectra of the intermediate states. If we expand the commutator over a complete set of eigenvectors of the energy-momentum vector, the coordinate-space integration yields a δ-function and we have

$$f(k^2,pk)=(2\pi)^4\sum\delta(k-n_1)\langle j_1|n_1\rangle\langle n_1|j_2|p\rangle-(2\pi)^4\sum\delta(k+n_2-p)\langle j_2|n_2\rangle\langle n_2|j_1|p\rangle. \qquad (12)$$

Thus the Fourier transform of the commutator will vanish unless either

$$k=n_1 \quad\text{or}\quad p-k=n_2,$$

where n_1 and n_2 are possible intermediate momenta. In terms of invariants,

$$f(k^2,pk)\neq 0 \quad\text{if}\quad \begin{cases} k^2\geqslant M_1^2 \\ 2pk\geqslant 2\kappa M_1 \end{cases} \quad\text{or}\quad \begin{cases} 2pk\leqslant -M_2^2+p^2+k^2 \\ 2pk\leqslant 2\kappa(\kappa-M_2), \end{cases} \qquad (13)$$

where M_1 and M_2 are the lowest masses occurring in each ordering of the commutator. We know the Fourier transform only for real vectors k_μ; thus the function $f(k^2,pk)$ is given only for values of the invariants that lie in the physical region $(pk)^2 \geqslant p^2k^2$. In the nonphysical region, for those real values of the invariants that correspond to a momentum vector k_μ with imaginary components, the function will be determined by the representation (10). In the region $\mathcal{R}$, which consists of all of the (k^2,pk)-plane outside the nonphysical region and outside region (13), $f(k^2p,k)$ vanishes.

For convenience we introduce a new variable $\mu = \lambda^2 - \beta^2 p^2$ in the representation. Then

$$f(k^2,pk) = \int d\mu d\beta \, H(\mu,\beta)\epsilon(pk+\beta p^2) \times \delta(k^2+2pk\beta-\mu). \quad (14)$$

We allow μ and β to run from $-\infty$ to ∞ and incorporate the causality condition by having $H(\mu,\beta)$ vanish in the region $\mu < -\beta^2p^2$. Each point in the (k^2,pk) plane corresponds to a line integral of the function $\epsilon(pk+\beta p^2)H(\mu,\beta)$ over the straight line $\mu = k^2+2pk\beta$ in the (μ,β) plane.

The parabolic boundary of the nonphysical region in the (k^2,pk) plane, $(pk)^2 = p^2k^2$, generates a set of lines that have as their envelope the curve A: $\mu = -\beta^2p^2$, and touch the envelope at the point $\beta = -pk/p^2$ at which the ϵ-function changes sign. All other points in the physical region correspond to lines on which the ϵ-function changes sign in the interior of the region in which $H(\mu,\beta)$ is zero. The boundary $k^2 = M_1^2$, $2pk > \kappa M_1$ of $\mathcal{R}$ corresponds to the set of lines of positive slope passing through the point $\beta = 0$, $\mu = M_1^2$ and lying between the line $\beta = 0$ and a line tangent to the curve A. The other boundary $-2pk = M_2^2 - p^2 - k^2$ corresponds to a set of lines produced by the clockwise rotation of a line through the point $\beta = -1$, $\mu = M_2^2 - p^2$ from the line $\beta = -1$ to a line tangent to the curve A. The other points in the (k^2,pk) plane at which $f(k^2,pk)$ vanishes correspond to all other lines lying entirely in the region swept out by these two sets of lines. If $H(\mu,\beta)$ vanishes sufficiently rapidly for large β, then a necessary condition for all these line integrals to vanish is that the function $H(\mu,\beta)$ vanishes in the region covered by these lines. The support of $H(\mu,\beta)$ lies in the region $\mathcal{S}$, the complement of the region covered by lines generated by points in $\mathcal{R}$. Only in this region can $H(\mu,\beta)$ be arbitrary. An alternative derivation of a necessary condition involving a different characterization of $H(\mu,\beta)$ is given in the Appendix.

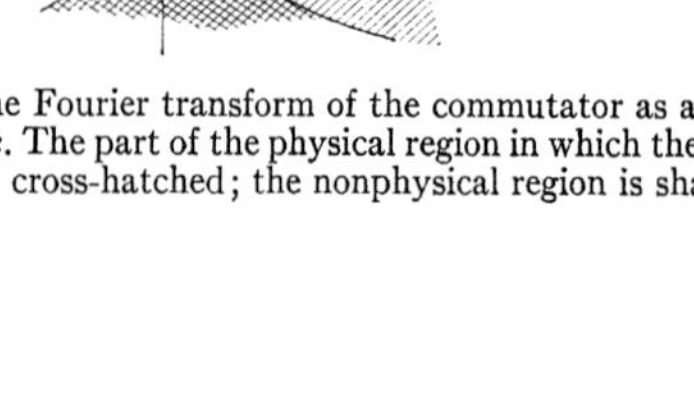

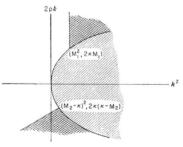

FIG. 1. The Fourier transform of the commutator as a function of k^2 and $2pk$. The part of the physical region in which the function is nonzero is cross-hatched; the nonphysical region is shaded.

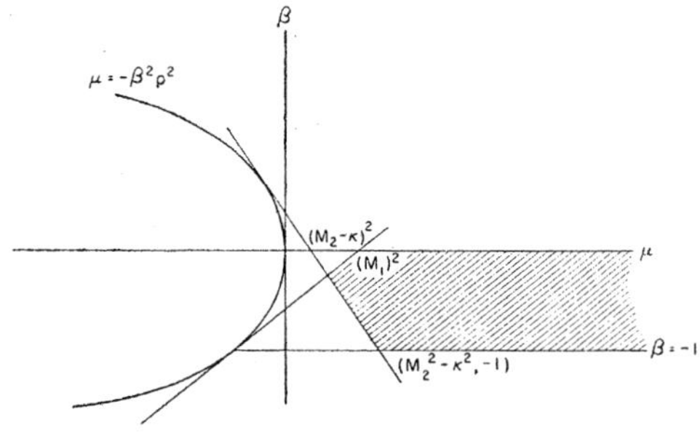

FIG. 2. The (μ,β)-plane. The weight function $H(\mu,\beta)$ is nonzero in the shaded region $\mathcal{S}$.

A typical situation is illustrated in Fig. 1, and the corresponding domain for $H(\mu,\beta)$ is shown in Fig. 2.

The two physical regions produced by the two orderings of the commutator may or may not be disjoint.

Case A: If $M_1 + M_2 \geqslant \kappa$, the two regions are disjoint. The region $\mathcal{R}$ will touch the boundary of the nonphysical region, and the region $\mathcal{S}$ will include only values of μ bounded below by the lines arising from the vertices of the region $\mathcal{R}$. Then $H(\mu,\beta)$ is nonvanishing only if

$$\mu \geqslant \max\{M_1^2 + 2\kappa M_1\beta,\ (M_2-\kappa)^2 + 2\kappa(\kappa - M_2)\beta\}, \quad (15)$$
$$0 \geqslant \beta \geqslant -1.$$

Case B: If $M_1 + M_2 < \kappa$, the two physical regions overlap. Then the region $\mathcal{R}$ is bounded by two straight lines. Their intersection generates a line which is a lower bound for μ, but which always cuts the parabola $\mu = -\beta^2p^2$. The region $\mathcal{S}$ is then

$$\mu \geqslant \max\{M_1^2 + \beta(M_1^2 - M_2^2 + \kappa^2),\ -\beta^2p^2\}, \quad (16)$$
$$0 \geqslant \beta \geqslant -1.$$

Case A is the normal case. For a vertex arising from the interaction of stable particles, the stability condition will make the sum of the intermediate masses greater than any of the external masses. In this case the sign of the ϵ-function associated with a point (k^2,pk) depends only on which physical region contains the point. The support of $f(k^2,pk)$ generated by the representation is bounded by straight lines, all of which lie wholly in the complement of region $\mathcal{R}$. Thus in case A we have shown that $f(k^2,pk)$ vanishes in the nose of the nonphysical region, up to the line defined by

$$\mu = M_1(M_2 - \kappa), \quad \beta = -(M_1 + \kappa - M_2)/2\kappa, \quad (17)$$

which corresponds to the tip of the region $\mathcal{S}$. Furthermore, for extreme values of the relative masses, the representation demands that the contributions from certain matrix elements vanish. If $M_2-\kappa>M_1$, then the term $\sum\langle j_1|n\rangle\langle n|j_2|p\rangle$ in which the sum is over all eigenvalues consonant with a fixed n_μ, vanishes for $M_1^2\leqslant n^2\leqslant(M_2-\kappa)^2$. Similarly, if $M_1>M_2+\kappa$, then $\sum\langle j_2|n\rangle\langle n|j_1|p\rangle$ vanishes for $M_2^2\leqslant n^2\leqslant(M_1-\kappa)^2$.

An analogous representation can easily be derived for the matrix element of a commutator between two single-particle states with the same momentum.

GREEN'S FUNCTIONS

We have derived the representation (1) with limits (15) or (16). Now we shall write more symmetric representations, in terms of the momenta k_1 and k_2 associated with each current. In case A, we shall write the time-ordered Green's function. Since in this case the time-ordered Green's function differs from the retarded function only by the sign of the $i\epsilon$ in the denominator of (1) associated with the physical region arising from the reversed order of operators in the commutator, the function we want is given by dropping the sign factor from the $i\epsilon$ term. If we define

$$\int e^{ik_1x}e^{ik_2y}dxdy\langle(j_1(x),j_2(y))_+|p\rangle =\delta(k_1+k_2-p)G_+(k_1^2,k_2^2,p^2), \quad (18)$$

then, changing the limits on μ to incorporate the term $-\beta p^2$ arising from the change in the momentum variables,

$$G_+(k_1^2,k_2^2,p^2)=\int\frac{d\mu d\xi_1d\xi_2\,\delta(\xi_1+\xi_2-1)H(\xi_1,\xi_2,\mu,p^2)}{\xi_1k_1^2+\xi_2k_2^2-\mu+i\epsilon},$$
$$\mu\geqslant\max\{\xi_1M_1^2+\xi_2(M_1-\kappa)^2,\ \xi_2M_2^2+\xi_1(M_2-\kappa)^2\},$$
$$\xi_1,\ \xi_2\geqslant0;\quad p^2\leqslant(M_1+M_2)^2. \quad (19)$$

For case B we use the similar retarded function: the only change in (19) is that $i\epsilon$ becomes $i\epsilon\{k_1^2-k_2^2+p^2(\xi_1-\xi_2)\}$ and the limits are

$$\mu\geqslant\max\{\xi_1M_1^2+\xi_2M_2^2,\ p^2\xi_1\xi_2\},$$
$$\xi_1,\ \xi_2\geqslant0;\quad p^2\geqslant(M_1+M_2)^2. \quad (20)$$

SPECTRAL REPRESENTATIONS

If we specialize the representation, putting a second particle on the energy shell, we obtain a spectral representation for the vertex as a function of momentum transfer. Consider two identical particles: let $k=p-p'$ where $(p')^2=p^2=\kappa^2$ (then $2pk=k^2$), and introduce a new variable $\sigma=\mu/(1+\beta)$ in (1); then, in the physical region where $k^2<0$, the vertex is

$$\Gamma(k^2)=\int_{\sigma_0}^{\infty}\frac{d\sigma\tilde{H}(\sigma)}{k^2-\sigma-i\epsilon}, \quad (21)$$

where

$$\tilde{H}(\sigma)=\int_{\mathcal{S}}d\beta(1+\beta)H(\sigma(1+\beta),\beta), \quad (22)$$

and the value of σ_0 is the minimum value of σ in the region $\mathcal{S}$ and is determined by (17):

$$\sigma_0=2\kappa M_1(M_2-\kappa)/(\kappa+M_2-M_1). \quad (23)$$

Here M_1 is the lowest mass that the odd-particle current can couple to the vacuum, and M_2 is the lowest mass that can be created by the current of the particle of mass κ. For stability $M_2>\kappa$. In the pion-nucleon vertex, $M_1=3m_\pi$, $M_2=\kappa+m_\pi$, and $\sigma_0=3m_\pi^2\kappa/(\kappa-m_\pi)$. In the first-order electromagnetic vertex, $M_1=2m_\pi$, $\sigma_0=2m_\pi 2\kappa/(2\kappa-m_\pi)$. The relation (21), which has been derived for negative momentum transfer, defines an analytic function for all values of the momentum transfer with a cut along the positive real axis. We can observe that the weight function is real and that, above $\sigma=(2\kappa)^2$, it is related to the amplitude for pair creation in an external field.

The thresholds for these spectral or dispersion relations are lower than those of the usual perturbation theory. However it has been shown by Karplus, Sommerfield, and Wichmann,[1] Nambu,[2] and Oehme[3] that the use of a wider set of intermediate masses in perturbation theory will lower the threshold. Our thresholds are lower bounds to the true threshold. It is quite possible that use of the fact that the intermediate states are made up of several particles will raise the thresholds.

ANALYTICITY IN THE MASSES

We can use this representation to study another aspect of the analytic behavior of the vertex. Let us treat p^2 as a time-like momentum transfer and consider the function (19) or (20) as a function of the mass $k_1^2=k_2^2=M^2$. Then the function is analytic for negative M^2 and can be continued as a function of M^2 up to the minimum value of μ. In the case of equal intermediate masses, this minimum value occurs in (19) when $p^2=M_1^2=M_2^2$ and is $\frac{1}{2}M_1^2$. Thus in the meson-nucleon vertex one can continue only up to $M^2\leqslant(\kappa+m_\pi)^2/2$. This would be on the mass shell only if $m_\pi\geqslant(\sqrt{2}-1)\kappa$, which is the condition obtained by Bremermann, Oehme, and Taylor.[4] We do not need a continuation of this type to obtain (21), since we have worked in the physical region and have obtained the spectral representation directly.

ACKNOWLEDGMENTS

We wish to thank Professor Julian Schwinger for suggesting such representations and for several discussions. One of us (W.G.) thanks the University of California Radiation Laboratory, Berkeley for its hospitality where this work was completed.

[1] Karplus, Sommerfield, and Wichmann, Phys. Rev. **111**, 1187 (1958).
[2] Y. Nambu, Nuovo cimento **9**, 610 (1958).
[3] R. Oehme, Phys. Rev. **111**, 1430 (1958).
[4] Bremermann, Oehme, and Taylor, Phys. Rev. **109**, 2178 (1958).

APPENDIX

We shall give an alternative derivation of the representation (1) and prove that the function $H(\mu,\beta)$ vanishes outside the region $\mathcal{S}$ if $f(k^2,pk)$ vanishes in $\mathcal{R}$. It is convenient to give the proof in two stages: First we show that if the function $f(k^2,pk)$ vanishes in the region $\mathcal{R}_0$: $(k-p)^2<0$, $k^2<0$, then $H(\mu,\beta)$ vanishes outside the region $\mathcal{S}_0$: $\mu\geqslant-\beta^2p^2$, $0\geqslant\beta\geqslant-1$. No assumption concerning the mass spectrum, beyond the axiom of the completeness of positive-energy states, is involved in requiring the vanishing of $f(k^2,pk)$ in $\mathcal{R}_0$. It is now a simple matter to incorporate the detailed mass-spectrum restrictions and to show that if $f(k^2,pk)$ vanishes in $\mathcal{R}$, then $H(\mu,\beta)$ vanishes outside $\mathcal{S}$.

Since the Fourier transform $\tilde{f}(x)$ of the function

$$f(k-\tfrac{1}{2}p)=f(k^2,pk)$$
$$=\int dx e^{i(k-\frac{1}{2}p)x}\langle[j_1(x/2),\, j_2(-x/2)]|p\rangle$$

vanishes for space-like x and the function itself vanishes in $\mathcal{R}_0$, according to Theorem 2 of Jost and Lehmann[5] it has a representation

$$f(k^2,pk)=\int_{u^2<p/^24} d^3u\int_0^\infty dt^2\left\{\varphi_1(\mathbf{u},t^2)+\varphi_2(\mathbf{u},t^2)\frac{\partial}{\partial k_0}\right\}$$
$$\times\Delta_t(k-\tfrac{1}{2}p-\mathbf{u}),$$

with suitable weight functions φ_1 and φ_2, where Δ_t is the invariant commutator function for a "mass" t. In this case, the weight functions φ_1 and φ_2 depend only on $\mathbf{u}^2$. We may now make use of the following identity given by Dyson[6]:

$$\int\Delta_t(q-\mathbf{u})\delta(\mathbf{u}^2-b^2)d^3u=\pi\int_{-b}^{b}I_0[t(b^2-v^2)^{\frac{1}{2}}]\Delta_t(q-\tilde{v})dv,$$

where $\tilde{v}$ is a four-vector with vanishing space components, and time component equal to v. Since in the Lorentz frame chosen p has only a time component, κ, we may put

$$\tilde{v}=\alpha p,$$

and obtain

$$f(k^2,pk)=\pi\int_0^\infty dt^2\int_0^{p^2/4}db^2\int_{-b/\kappa}^{b/\kappa}d\alpha\, I_0[t(b^2-\alpha^2p^2)^{\frac{1}{2}}]$$
$$\times\left\{\kappa\varphi_1(b^2,t^2)+\varphi_2(b^2,t^2)\frac{\partial}{\partial\alpha}\right\}\Delta_t(k-(\alpha+\tfrac{1}{2})p).$$

If we now use the identity

$$\Delta_t(q)=\epsilon(q_0)\int^\infty d\lambda^2\,\Delta_t(\sqrt{\lambda^2})\delta(q^2-\lambda^2),$$

and introduce the parameters

$$\beta=-(\alpha+\tfrac{1}{2}),\quad \mu=\lambda^2-\beta^2p^2,$$

it follows that

$$f(k^2,pk)=\int_1^0 d\beta\int_{-\beta^2p^2}^\infty d\mu\, H(\mu,\beta)\epsilon(pk+\beta p^2)$$
$$\times\delta(k^2+2pk\beta-\mu),$$

where

$$H(\mu,\beta)=\pi\int_0^\infty dt^2\int_0^{p^2/4}db^2\,\Delta_t[(\mu+\beta^2p^2)^{\frac{1}{2}}]$$
$$\times I_0\{t[b^2-(\beta+\tfrac{1}{2})^2p^2]^{\frac{1}{2}}\}\theta(b^2-(\beta+\tfrac{1}{2})^2p^2)$$
$$\times\{\kappa\varphi_1(b^2,t^2)-\varphi_2(b^2,t^2)\partial/\partial\beta\}.$$

This concludes the first stage of the proof and yields a completely general representation including causality and the positive-definite nature of the energy spectrum.

For arbitrary functions φ_1 and φ_2, $H(\mu,\beta)$ will, in general, not be a distribution in the Schwartz sense. It is here that we must confine ourselves to the narrower class of Schwartz distributions. (This corresponds to an implicit restriction on φ_1 and φ_2, which in the Jost-Lehmann representation could be taken as arbitrary functions.) For these distributions we can apply the arguments of Jost and Lehmann to incorporate explicit mass restrictions and show that the function $H(\mu,\beta)$ which vanishes outside $\mathcal{S}_0$ vanishes in the region outside $\mathcal{S}$, provided $f(k^2,pk)$ vanishes in $\mathcal{R}$. For any point in $\mathcal{R}$, the line integrals of $\epsilon(pk+\beta p^2)H(\mu,\beta)$ over all lines that do not intersect the *convex* region $\mathcal{S}$ vanishes if $f(k^2,pk)$ vanishes in $\mathcal{R}$; but for all the lines generated by points belonging to $\mathcal{R}$, $\epsilon(pk+\beta p^2)$ changes sign inside the parabola where $H(\mu,\beta)$ vanishes. Consequently we can replace $\epsilon(pk+\beta p^2)H(\mu,\beta)$ by an appropriately modified $\bar{H}(\mu,\beta)$ for suitable sets of these lines, which cover the entire region between $\mathcal{S}$ and $\mathcal{S}_0$. A lemma of Jost and Lehmann[5] which states that for a function which is nonvanishing only in a bounded domain, a necessary and sufficient condition that all line integrals over straight lines that do not intersect a convex region vanish is that the function vanishes outside the convex region, then permits us to conclude that $\bar{H}(\mu,\beta)$ itself vanishes in the region between $\mathcal{S}$ and $\mathcal{S}_0$. Consequently $H(\mu,\beta)$ vanishes everywhere outside $\mathcal{S}$.

---

[5] R. Jost and H. Lehmann, Nuovo cimento **5**, 1598 (1957).
[6] F. J. Dyson, Phys. Rev. **111**, 1717 (1958), Eq. (14).

Section XII:
The Manuscript for Selected Paper 1

Unstable Ribonucleic Acid Revealed by Pulse

Labelling of <u>Escherichia Coli</u>

by

Francois Gros*, W. Gilbert, H. Hiatt*, C. G. Kurland

R. W. Risebrough, and J. D. Watson

The Institut Pasteur, Paris*, and the Departments of Physics

and The Biological Laboratories of Harvard University

(a)

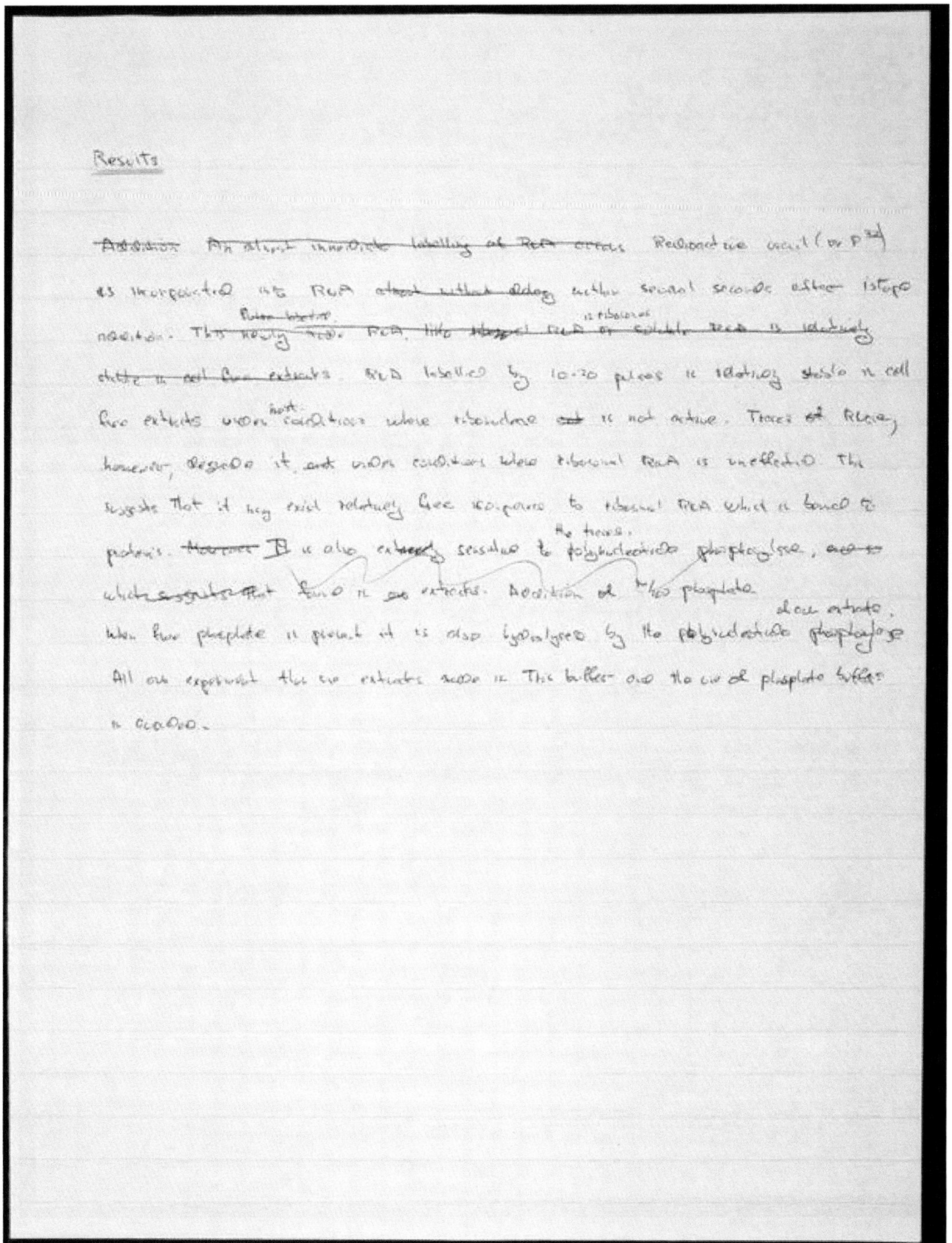

Results

~~Addition An almost immediate labelling of RNA occurs~~ Radioactive uracil (or P^{32}) is incorporated into RNA ~~almost without delay~~ within several seconds after isotope addition. This ~~newly made~~ pulse-labelled RNA, ~~like messenger RNA or soluble RNA is relatively stable in cell free extracts~~. RNA labelled by 10–20 pulses is relatively stable in cell free extracts under most conditions where ribonuclease is not active. Traces of RNase, however, degrade it, under conditions where ribosomal RNA is unaffected. This suggests that it may exist relatively free compared to ribosomal RNA which is bound to protein. ~~However~~ It is also extremely sensitive to the polynucleotide phosphorylase, ~~and so~~ which ~~suggests that~~ found in ~~our~~ extracts. Addition of the phosphate when free phosphate is present it is also hydrolysed by the polynucleotide phosphorylase of our extracts. All our experiments thus our extracts made in Tris buffer and the use of phosphate buffer is avoided.

(b)

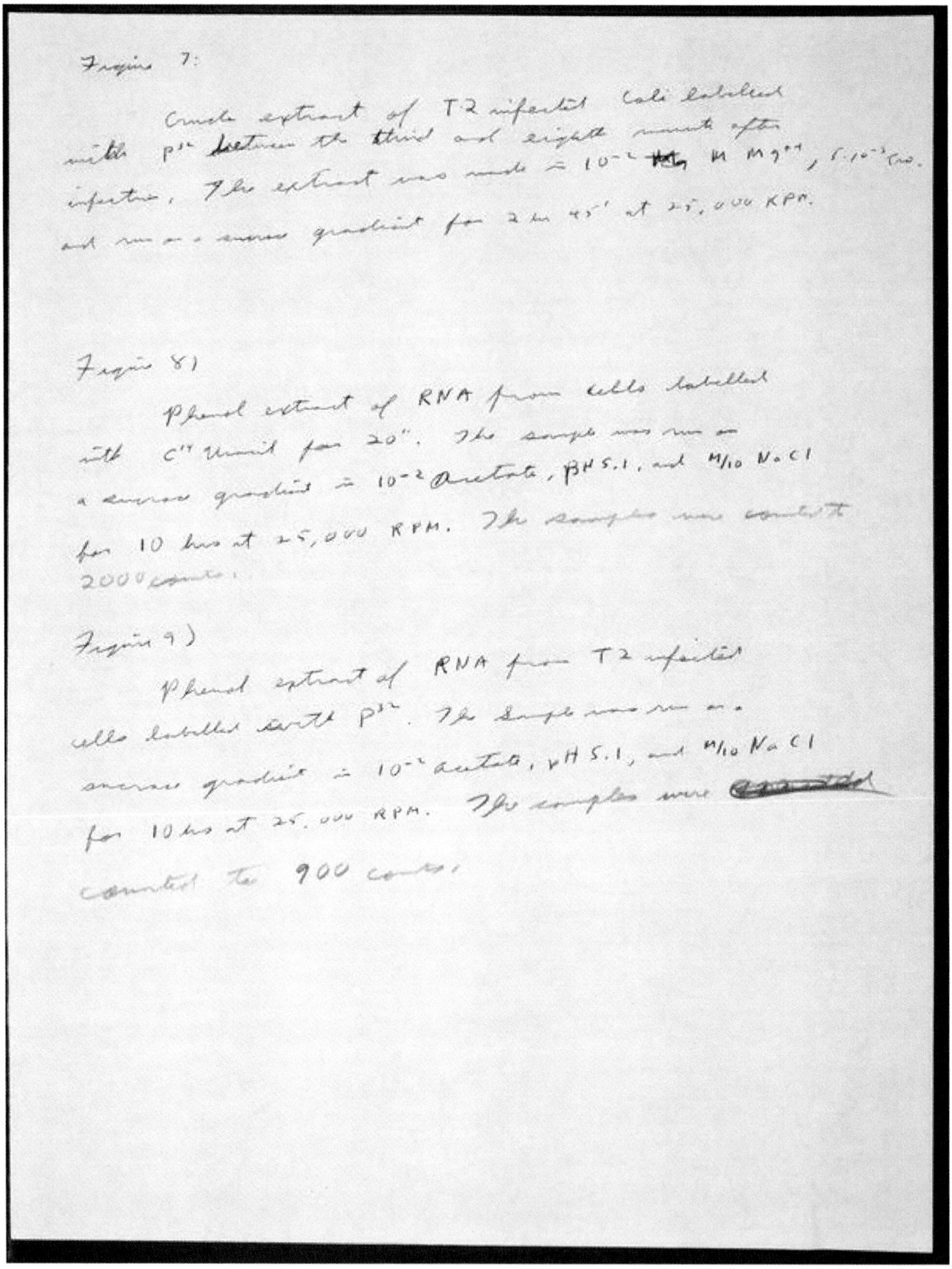

Figure 7:

Crude extract of T2 infected Coli labelled with P^{32} between the third and eighth minute after infection. The extract was made in 10^{-2} M Mg^{++}, $5 \cdot 10^{-3}$ Tris. and run on a sucrose gradient for 2 hr 45' at 25,000 RPM.

Figure 8)

Phenol extract of RNA from cells labelled with C^{14} Uracil for 20". The sample was run on a sucrose gradient in 10^{-2} acetate, pH 5.1, and M/10 NaCl for 10 hrs at 25,000 RPM. The samples were counted to 2000 counts.

Figure 9)

Phenol extract of RNA from T2 infected cells labelled with P^{32}. The sample was run on a sucrose gradient in 10^{-2} acetate, pH 5.1, and M/10 NaCl for 10 hrs at 25,000 RPM. The samples were counted to 900 counts.

(c)

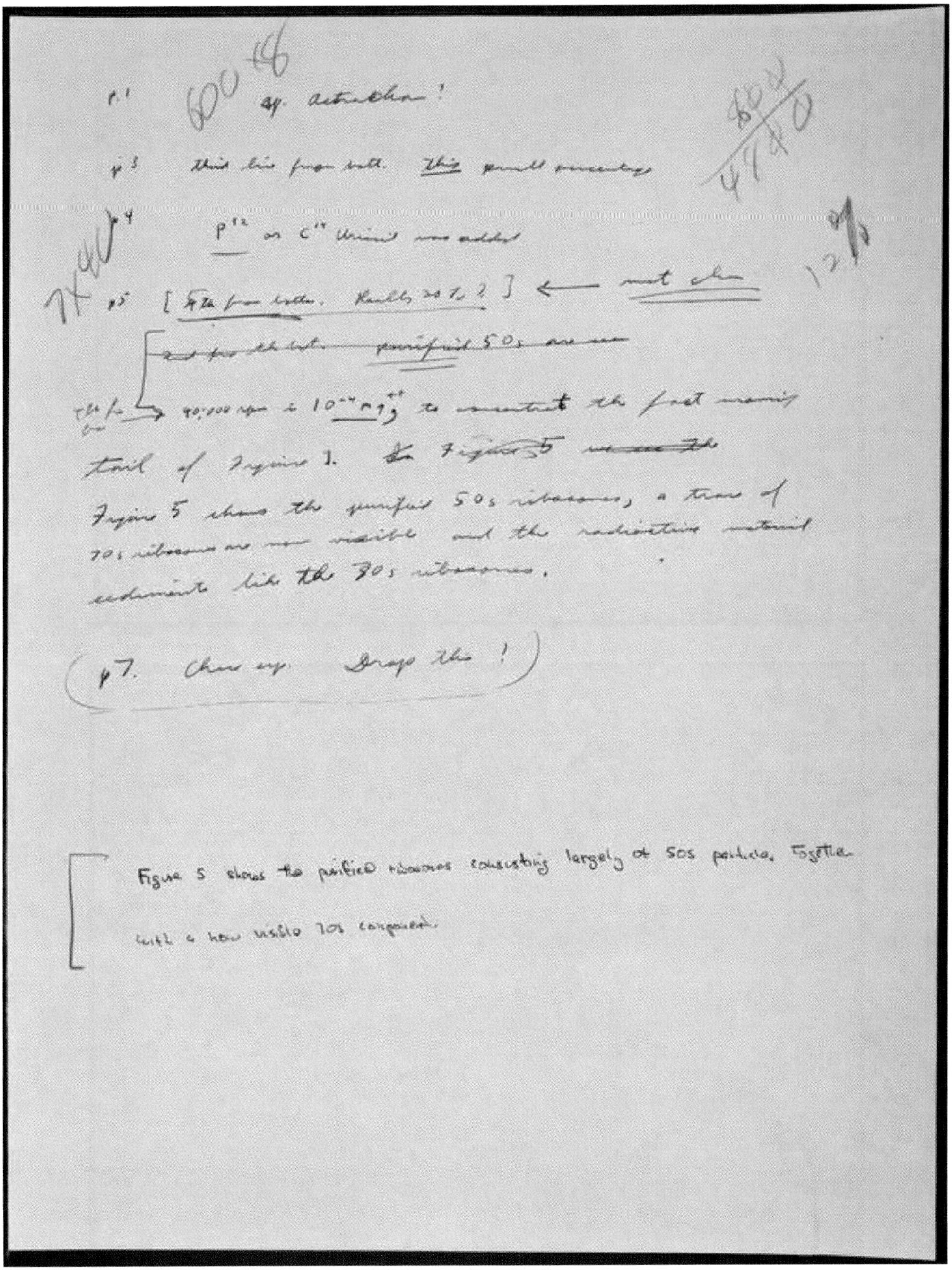

p.1 activation?

p.3 this line from ... This small ...

p.4 P^{32} or C^{14} uracil was added

p.5 [Data from ... Really 20%?] ← not clear

→ 40,000 rpm in 10^{-4} Mg^{++}, to concentrate the fast moving tail of Figure 1.

Figure 5 shows the purified 50s ribosomes, a trace of 70s ribosomes are now visible and the radioactive material sediments like the 70s ribosomes.

(p7. Clear up. Drop this!)

Figure 5 shows the purified ribosomes consisting largely of 50S particles, together with a now visible 70S component.

(d)

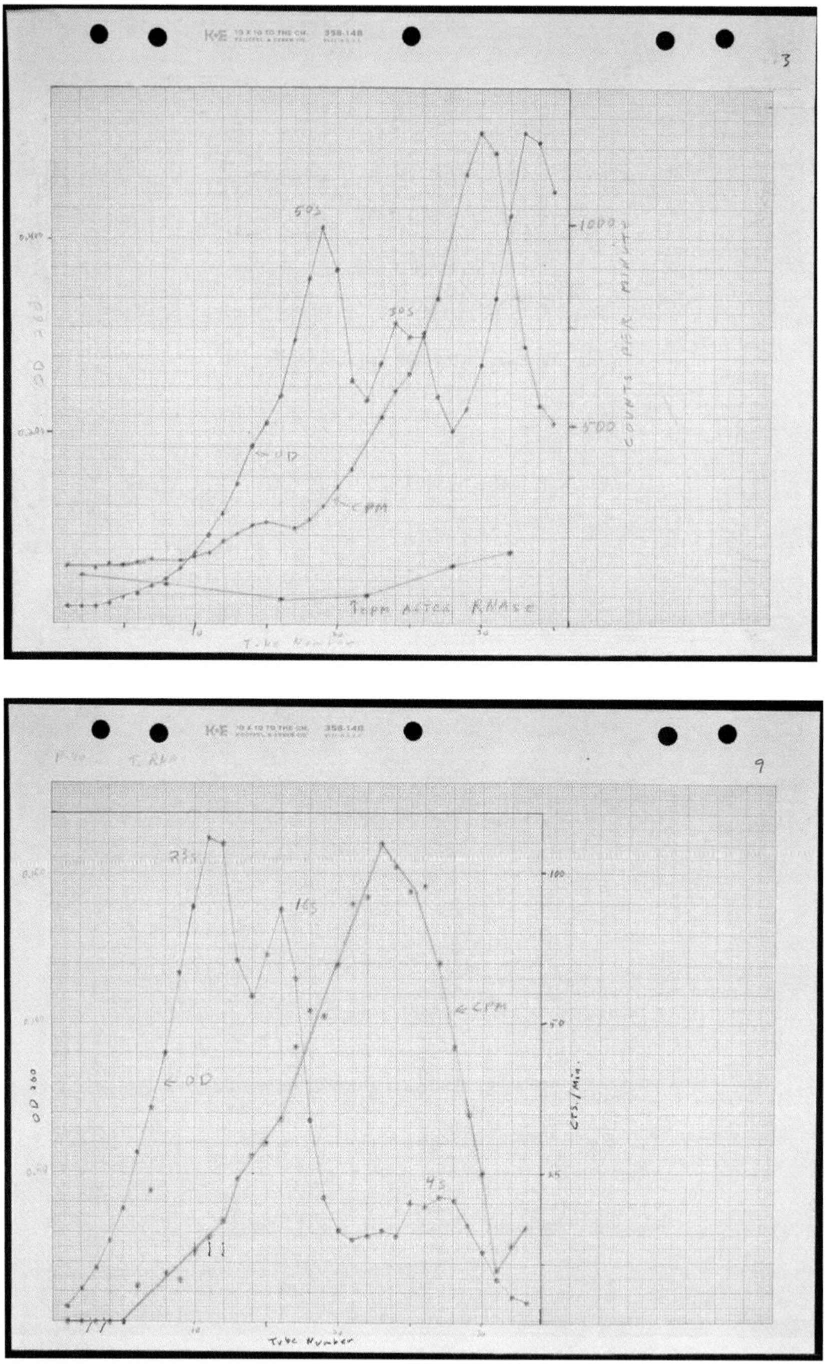
3
50S
30S
← OD
← CPM
1000
500
10
20
30
9
23S
16S
4S
← OD
← CPM
OD 260
100
50
Tube Number

(e)

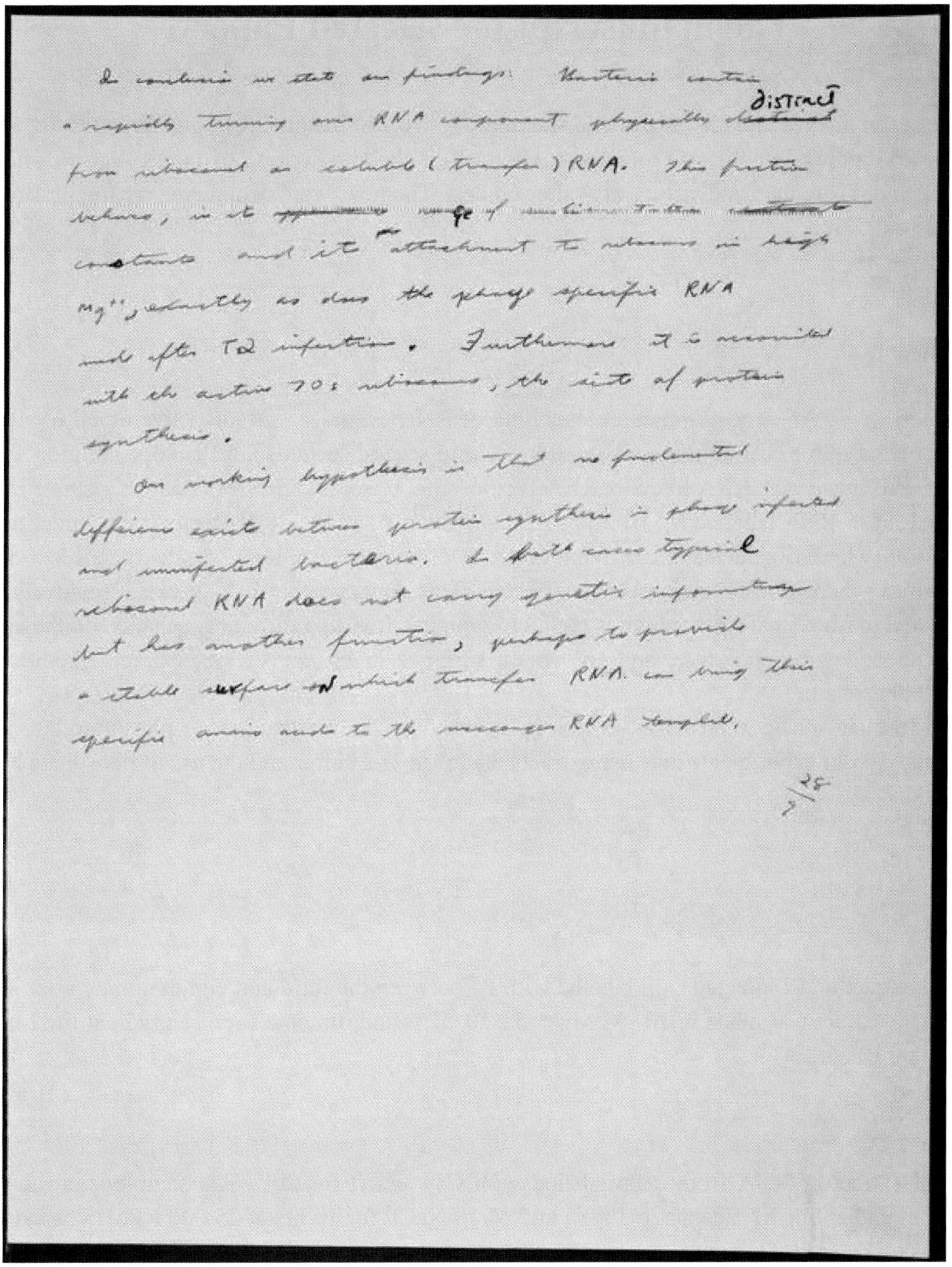
In conclusion we state our findings: Bacteria contain
a rapidly turning over RNA component physically DISTINCT
from ribosomal or soluble (transfer) RNA. This fraction
behaves, in its [illegible] range of sedimentation constants
and its attachment to ribosomes in high
Mg++, exactly as does the phage specific RNA
made after T2 infection. Furthermore it is associated
with the active 70s ribosomes, the site of protein
synthesis.
Our working hypothesis is that no fundamental
difference exists between protein synthesis in phage infected
and uninfected bacteria. In both cases typical
ribosomal RNA does not carry genetic information
but has another function, perhaps to provide
a stable surface on which transfer RNA can bring their
specific amino acids to the messenger RNA template.
7/25

(f)

The manuscript for selected Paper 1

These seven images include (a) the title and author page; (b) Results; (c) figure legends; (d) some notes; (e) Figures 3 and 9; (f) some discussion. (b) is the handwriting of James Watson whereas (c), (d) and (f) are the handwriting of Walter Gilbert. The texts were deciphered by Jürgen Brosius as follows:

Image (b)

Results

~~Addition~~ ~~An almost immediate labelling of RNA occurs~~ Radioactive uracil (or P^{32}) is incorporated into RNA ~~almost without delay~~ within several seconds after isotope addition. ~~Xxxx xxxxx This newly xxx RNA like xxx RNA is ribosomal or soluble. RNA is relatively stable in cell free extracts.~~ RNA labelled by 10-20 pulses is relatively stable in cell-free extracts under most? conditions where ribonuclease ~~xxx~~ is not active. Traces of RNase, however, degrade it ~~xxx~~ under conditions where ribosomal RNA is unaffected. This suggests that it may exists relatively free compared to ribosomal RNA which is bound to proteins. ~~It is also extremely sensitive to the xxxxx polynucleotide phosphorylase. and so which suggests that xxxx xx xxx extracts. Addition of M/100 phosphate~~

When free phosphate is present it is also hydrolyzed by the polynucleotide phosphorylase xxxx extracts. All our experiments thus use extracts made? in Tris buffer and the use of phosphate buffer is avoided.

Image (c)

Figure 7:

Crude extract of T2 infected Coli labeled with P^{32} between the third and eighth minute after infection. The extract was made in 10^{-2} M Mg^{++}, 5 • 10^{-3} Tris and run on a sucrose gradient for 2 hr 45' at 25,000 RPM.

Figure 8)

Phenol extract of RNA from cells labeled with C14 uracil for 20". The sample was run on a sucrose gradient in 10^{-2} acetate, pH 5.1 , and M/10 NaCl for 10 hrs at 25,000 RPM. The samples were counted to 2000 counts.

Figure 9)

Phenol extract of RNA from T2 infected cells labeled with P^{32}. The sample was run on a sucrose gradient in 10^{-2} acetate, pH 5.1 , and M/10 NaCl for 10 hrs at 25,000 RPM. The samples were ~~xxxxxx~~ counted to 900 counts.

Image (d)

p. 1 sp. Astrachan

p. 3 third line from bott. this xxxxxx percentage

p. 4 P^{32} or C^{14} Uracil was added

p.5 [x th from bottom Results xx to ?] ß must check

[~~and for the xxx.~~ ~~purified 50s are xxx~~

xxx xxx --> 40,000 rpm in 10^{-4} Mg^{++}, to concentrate the fast moving tail of

xxx Figure J?. ~~xx Figure 5 xx xx the~~

Figure 5 shows the purified 50s ribosomes, a trace of 70s ribosomes are now visible and the radioactive material sediments like the 70s ribosomes.

p. 7. Chase exp. Drop this ?

[Figure 5 shows the purified ribosomes consisting largely of 50s particles. Together with a now visible 70s component.]

Image (f)

In conclusion we state our findings: Bacteria contain a rapidly turning over RNA component physically ~~distinct~~ distinct from ribosomal or soluble (transfer) RNA. This fraction behaves in its ~~xxxxxxxxxxx~~ range of sedimentation ~~constants~~ constants and its attachment to ribosomes in high Mg^{++}, exactly as does the phage specific RNA made after T2 infection. Furthermore it is associated with the active 70s ribosomes, the site of protein synthesis.

Our working hypothesis is that no fundamental difference exists between protein synthesis in phage infected and uninfected bacteria. In both cases typical ribosomal RNA does not carry genetic information but has another function, perhaps to provide a stable surface on which transfer RNA can bring their specific amino acids to the messenger RNA template.

Part Two

Memoirs by Walter Gilbert

Section I: Messenger RNA and Protein Synthesis

My introduction to Molecular Biology was the discovery of messenger RNA. One spring day in 1960, Jim Watson said to me, "Something very exciting is going on in my lab." He and François Gros were trying to show that there was a special class of RNA molecules in bacteria: unstable copies of the DNA that served to take the genetic information of a gene from the DNA to the ribosomes, the "factories" in the cell that made proteins. The hypothesis was that such an RNA molecule carried the information like a computer tape to dictate the sequence of amino acids in a protein and then, after being used a few times, was destroyed so that a different protein could be made using a new messenger. I visited the laboratory to watch Jim and François's experiment. Jim gave me six papers to read that night, and the next day I joined in, working on the experimental search for messenger RNA.

At that time, I was an Assistant Professor of Theoretical Physics in the Physics Department at Harvard. I worked on the Quantum Theory of Fields, a mathematical theory that tried to explain and predict the properties of the nuclear particles, the forces that bound protons and neutrons to each other, and the behavior of the Pi mesons that created that force and could be made and studied in the cyclotrons of that day. (This was long before the discovery of the quarks.) I taught graduate courses in Electromagnetic Theory, Advanced Quantum Mechanics and Field Theory, and General Relativity. Although I had been an undergraduate at Harvard College, the only biology course I had ever had was in high school, where we dissected earthworms. Entering college, I thought I would be a chemist, but I chose a combined major in chemistry and physics and found myself more and more interested in physics. I went to graduate school at Harvard in Physics for one year and then went to Cambridge University in England as a research student (i.e. graduate student), thinking that it was a center of theoretical physics. I came to Cambridge as an eager American dreaming of being a student of Dirac (though I came much too late). I called the man who was to be my first supervisor as soon as I arrived and said eagerly, "I'm here! What should I do?" "Go away," he said. "Term has not begun."

Cambridge at that time was an exciting place to do theoretical physics. I joined the group of young research students (in mathematics, since that was the classification of theoretical physics at Cambridge) who all worked in a large office in the Arts School building. My thesis supervisor, Abdus Salam, a jovial Pakistani, had just come to Cambridge and had made major discoveries in Quantum Field Theory. He was six years older than I was. We wrote papers together in my two years at Cambridge before I went back to Harvard for a fourth year as a graduate student. I thought

I would do a Harvard Ph.D. as a graduate student of Julian Schwinger, but I was supposed to take more courses as a student and soon discovered that I couldn't bear to sit through lecture courses anymore. So, I wrote up my work and sent it off to Cambridge for a D. Phil. At that time, one was supposed to submit a bound thesis for review. If it did not pass, one could come back a year later and try again — no correction of typos or mistakes was possible until then. One night in that spring of 1957, I went over to MIT, and Vicky Weisskopf and Hans Bethe examined me on my thesis. I went home happy. The next year I became a National Science Foundation postdoc at Harvard, and the year after I was offered a special position, the Bayard Cutting Instructorship, commonly called Julian Schwinger's Assistant. In the summer of 1959, I became an Assistant Professor and began teaching graduate students in Theoretical Physics.

I had met Jim Watson while I was a graduate student in Cambridge. In the spring of 1956, my wife and I went to a party organized by Marietta Robertson in honor of her father, H. P. Robertson, a Caltech Astronomer who was visiting Cambridge at that time. Jim and I met, spent several hours talking, and became friends. I had heard Francis Crick lecture about DNA at a physics club, Del-squared-V, so I had learned about the structure, which had been worked out three years earlier. Like many physicists of that period, I had read Schrödinger's book *What is Life?*, which discussed the problem of how to store genetic information, hypothesized to be in the form of an aperiodic crystal. The solution to this problem was the Watson-Crick DNA structure. Jim and I saw a lot of each other that spring in Cambridge, since Celia, my wife, and I often had him over for dinner. Back at Harvard, Jim began his Assistant Professorship in the Biology Department, and Celia became his first technician as he started his lab. Our friendship continued over the next few years while I worked in Physics and Jim developed his laboratory with graduate students and collaborators.

Our experiments looking for messenger RNA were based on the background knowledge of the time. It was thought that the order of bases in DNA, corresponding to a gene, should dictate, in some way not yet understood, the order of amino acids in the protein product of that gene. We knew Francis Crick's statement about the 'central dogma' — DNA makes RNA makes protein — which had been first based upon observations of higher cells where the DNA was restricted to the nucleus, while RNA appeared in the cytoplasm in the form of particles called ribosomes that were made up of RNA and protein, which appeared to be the factories that synthesized new proteins (as shown by radioactive labeling with radioactive amino acids). Originally, one had thought that the RNA in the ribosome carried the information to make a protein, but the properties of the ribosomes finally argued against that. Their RNA was stable in bacteria, and the sequence of the RNA did not look like the sequence of the DNA in general, as measured by the ratios of the bases. Monica Riley, working with Arthur Pardee, François Jacob, and Jacques Monod, had shown that a gene on DNA inserted into a cell dictated a new, constant rate of protein synthesis, which stopped when the gene was removed (by radioactive decay, in her case). That suggested, if RNA was to be involved, that the RNA had to be unstable.

Furthermore, a line of experiments a few years before had shown that when the T-even phage infected a bacterial cell, net RNA synthesis stopped, but there was the synthesis of an unstable RNA that matched by base ratios the DNA of the phage. That RNA was a candidate for a messenger that carried the information for the phage proteins to the bacterial ribosomes where protein synthesis could occur. So, we set out to show that there was an unstable RNA in *E. coli* (our favorite bacterium) that could be such a messenger.

Bob Risebrough, a graduate student of Jim's, studied the RNA made by Phage T2, while Jim, François Gros, and I searched for an unstable bacterial RNA. Our hope was to show a set of similar properties between the phage RNA and the bacterial RNA, to show that they were both messenger RNAs.

Our experiments were direct but primitive. In the early experiments that spring and summer, François would be swirling two liters of growing *E. coli* in a 20-liter flask, Jim would be holding a stop watch, and I would hold a vial with 20 millicuries of radioactive phosphorus, P^{32}, in solution as the phosphate. Jim would shout, "Go," I would pour the radioactive phosphate into the 20-liter flask, François would swirl the flask strongly for 20 seconds until Jim shouted, "Stop," and François would pour the bacterial culture into a 50-liter container holding ice and sodium azide to stop all the incorporation and all processes. We would then proceed to the cold room, centrifuge down the bacteria, collect the pellet, and grind the cells with alumina by hand with a mortar and pestle to break the bacterial walls in order to make an extract of their contents. This extract would be treated with DNase to remove all the DNA (we could buy RNase and DNase easily), and, after the debris had been removed with a low speed centrifugal spin, samples of it were layered onto a 25-milliliter tube containing a gradient of sucrose (to stabilize the liquid so that the particles of a given size could be centrifuged down as a layer). We could examine either the raw extract, containing ribosomes and proteins, or purify just the RNAs, eliminating all proteins. After the high-speed centrifugation in the ultracentrifuge to separate molecules of RNA by size, we would carefully carry the plastic tubes across the laboratory, plug the top of each tube with a cork connected to a mouthpiece, puncture the bottom of the tube with a short hypodermic needle, and with breath control collect a series of drops into a rack of test tubes. One set of samples would be diluted and the optical adsorption measured in the ultraviolet, at 260 nm, to follow the RNA concentration along the tube. The second parallel set would be precipitated with acid (5% tricloracetic acid), the precipitate collected on a small filter, and the filter glued to a small aluminum planchette. Then one of us would take all the planchettes, about 33 from each centrifuge tube (there were three tubes in each experiment), upstairs to the "counting room." We measured the radioactivity by hand. We had a Geiger counter shielded by lead bricks and attached to a series of argon tubes that could accumulate the counts. One put a planchette up to the counter, closed it in with lead bricks, and started the counter and a stopwatch. At the end of a minute, one stopped the counter, wrote down the number of counts in a lab notebook, and went on to the next sample. After about two hours, one of us would come down with the results, and then we would plot the data to see what we could learn.

We could show that the pulse-labeled RNA in bacteria was similar to the T2 phage RNA in all respects. The two sedimented in the centrifuge at a molecular weight range in between those of the ribosomal RNAs and the small RNAs (then called soluble RNA, sRNA, but now called transfer RNA, tRNA). In ionic conditions more similar to those inside cells, the labeled RNAs associated with the ribosomes, and under conditions that identified just those ribosomes that were synthesizing proteins, the labeled RNAs stuck to just those ribosomes before being released by more severe treatment. These experiments demonstrated that the pulse-labelled RNA had all the characteristics that we expected for messenger RNA. Later experiments showed that the bacterial pulse-labelled RNA was similar to the bacterial DNA in the ratios of its bases and different from ribosomal RNA.

We continued the messenger experiments throughout the summer and fall of 1960. François Gros had been visiting Jim's lab and returned to Paris to continue work there. Jim took me to the Gordon Conference on Nucleic Acids early that summer — my first biological meeting, which taught me the current state of molecular biology — and the Phage Meeting in Cold Spring Harbor at the end of the summer. We did not talk about the RNA experiments to anyone that summer, since we were struggling to get the experiments right and make them convincing. Jim and I would occasionally describe the problem to ourselves as similar to looking for the neutrino (which had only recently been observed directly, 27 years after it was first hypothesized), a search for something recondite and super-hidden. As we continued the experiments, the technology available rapidly improved. The first automatic counters would take a load of 100 planchettes and count each one for a specified time and print out the results. Now we could leave the experiments counting and go out to dinner in the square, coming back to the lab at night to plot the results and to talk about them. Then we got automatic scintillation counters, which could measure the energy of the electron emitted by the radioactive tracers, and now we could efficiently use other isotopes for labeling, Carbon (C^{14}) and Hydrogen (H^3), and do double labeling experiments, following nucleic acids and proteins simultaneously. Something as simple as having an automatic machine count each sample for a specified number of counts, thus to a specified accuracy, and print out the results after an overnight unattended run made the experiments much easier. As time passed, we learned how to do better experiments. Our first experiments were much harder than they needed to be. We were labeling with phosphate and were growing the bacteria in medium containing 0.2 molar phosphate buffer, thus diluting the radioactive material. Changing to a tris buffer made it possible to lower the phosphate concentration, needed for growth, to 0.0001 molar, more than a thousand-fold increase in labeling efficiency, and the experiments could be done in a small test tube.

I found the life in the experimental laboratory very different from my life as a theoretical physicist. We theorists largely slept late, gathered in the afternoon, and then worked in a mostly solitary fashion, often late at night. Our struggle with ideas was generally a lonely one, thinking in isolation about mathematical problems. In the laboratory, there was constant interaction with other people both day and night: discussion, suggestions, argument, the play of new ideas from the literature. At that time, new work was generally received in preprint form or by word of mouth. I would be soaking up new techniques and ideas, asking the students for advice, and questioning how and why something was done. We hung out with each other, went back to the lab after dinner, and stayed there often until the early morning.

We continued the experiments through the autumn and, late in the year, wrote up the paper on messenger RNA that was published in *Nature* in May of 1961 [Paper #1, this volume]. Our paper appeared together with that of Matt Meselson, Sydney Brenner, and François Jacob, who showed that the T2 RNA, made after phage infection, associated with old bacterial ribosomes, which were the sites of new protein synthesis. Hence, it was messenger RNA that carried the information from the phage DNA to the protein factories. Our paper showed both this and, further, that there was similar RNA in the uninfected bacterial cell, which was an unstable messenger RNA.

Our work continued throughout the spring of 1961. We wrote a longer paper for the Cold Spring Harbor Symposium that summer. At that meeting, Sydney Brenner presented the Brenner, Meselson, and Jacob work, and François Gros presented our group's work. While the existence of messenger RNA was generally accepted, there were many skeptics, and the debate was often heated. Sydney, at one point in the discussion period, turned poetic and began talking of how beautiful this finding of an evanescent instruction tape was, how a hermetic secret of nature had been discovered. Suddenly, Erwin Chargaff's voice shouted out at the back of the hall, "Don't forget. Hermes was the god of thieves!"

The problem was that we would really have liked to prove that an RNA product of a specific gene could be used to direct the synthesis of the corresponding protein. The messenger experiments were indirect and hence lacked the certainty that the biochemists would have liked to see. The goal of all the work would be an understanding of how a sequence of bases in the nucleic acid coded for a sequence of amino acids in the protein, the genetic code. There were *in vitro* extracts of cells, both bacterial and animal, that seemed to make proteins, and people in many laboratories were trying to study how they worked.

Later that summer, there was a major Biochemistry meeting in Moscow. Jim and I went to that meeting. Russia had been closed to the West before then, and the situation between the USSR and America had been tense since the end of the war. That summer, the Russians opened up to several thousand biologists, and we seized the opportunity to see what the science was like, as well as to travel to Moscow and beyond. The most exciting story at the meeting emerged as a surprise. One morning, Jim and I were having breakfast at the hotel, and a young man came up to us to tell us of his work. He was Marshall Nirenberg, and while sitting at the breakfast table, he told us that he had used the *in vitro* system from *E. coli* to synthesize proteins. Moreover, he had been able to use a synthetic RNA, polyuridylic acid, which consists of only base, U, repeated many times, to direct

The first row is Alfred Tissieres, Francis Crick, and then I and Jim Watson in light-colored suits.

the synthesis of polyphenylalanine — a polymer of one repeated amino acid, phenylalanine. This was the first break in the genetic code. A series of Us, probably three but possibly four, must stand for phenylalanine. We were delighted and amazed to hear this. He was to give a 10-minute talk (all he could get permission to do) later that morning, and we went to hear it. The news got to Francis Crick, and he arranged for Marshall to speak at a plenary session, where he gave a dramatic talk. This discovery, that poly-U dictated the synthesis of polyphenylalanine, was both the first step in deciphering the genetic code and also clear proof of the messenger RNA hypothesis that an RNA molecule carried the information for protein synthesis to the ribosomal factories.

The world of scientific discovery is dominated by chance. The *in vitro* extracts of cells that showed evidence of protein synthesis in the test tube had been developed by Mahlon Hoagland and Paul Zamecnik, and Hoagland had used it to show that there was a small RNA molecule — then called sRNA, now called tRNA — that was used as an intermediate to carry the amino acid in protein synthesis. This gave life to an idea of Crick's, that one needed 20 adaptors to connect each of the 20 amino acids to a nucleic acid sequence. The extracts that Hoagland and Zamecnik developed were being used in many laboratories at that time to study the mechanism of protein synthesis, often by adding plant virus RNAs to the extract to see if they would code for proteins. Alfred Tissieres, then working as a guest in Jim's laboratory, was doing such experiments. He thought he was beginning to see some positive results, and he asked Paul Doty for a sample of synthetic RNA. Paul Doty's laboratory, in the Chemistry Department at Harvard, was making and studying the properties of polymers of the RNA bases. They had a lot of poly-A, poly-adenylic acid, and so they gave him a sample. Alfred put poly-A into the *in vitro* extract, but there was no incorporation. We now know that poly-A codes for polylysine. Why did he not find it? The extracts were processed by precipitating all the protein with boiling 5% trichloroacetic acid (TCA). Most proteins and especially polyphenylalanine are insoluble in 5% TCA. However, polylysine is soluble in 5% TCA and would only precipitate if one were to use 10% TCA. In fact, Alfred did not look too hard at this lack of incorporation of label, because that experiment was done as a negative control. At that time, we thought that the homo polymers — poly-U, poly-A, poly-C, and poly-G — should not code for anything. This was an attractive idea floating around at that time: the commaless code (Francis Crick, J. S. Griffith, and L. E. Orgel. "Codes Without Commas." *PNAS* 43 (1975): 416–421). There are 64 possible triplets of four bases. Is there a set of triplets that could be uniquely read into amino acids just by recognizing the triplet anywhere in the RNA? In order for a string of repeating triplets to have a unique way of being read, one first eliminates the pure repeats, since in two repeats, say UUUUUU, one could not tell where the correct codon lay. That leaves 60 triplets. Then, for each triplet that stands for an amino acid, the two cyclic permutations must not stand for anything, so only 20 triplets are possible. In detail, this commaless code has 20 sense triplets and 44 nonsense triplets, which would have made mutations to nonsense frequent. The striking numerical coincidence to the 20 amino acids found in proteins made this code seem plausible, so this attractive (but wrong) idea often influenced the way one planned or thought about experiments at that time. That is why the Nirenberg announcement was so unexpected.

The correct interpretation of the code would be worked out in 1961 by Crick and Brenner's study of the proflavine mutations in the rII gene of phage T4, which showed that the code was a series of triplets read from a starting point. (Francis Crick, Leslie Barnett, Sydney Brenner, and

R. J. Watts-Tobin. "General Nature of the Genetic Code for Proteins." *Nature* 192, 4809 (30 December 1961): 1227–1232.)

A detour: The rII gene of phage T4 was extremely useful for fine genetic analysis, since mutations in the gene could be recognized by plaque size on one strain of *E. coli*, fail to grow on another strain, and grow without any selection as well as the wild type on a third strain. Thus, Seymour Benzer used these properties to measure recombination frequencies between mutants, yielding the wild type, down to extremely rare events. Crick and Brenner used this system to study mutations induced by proflavine. They observed that such mutant phages could be made to revert to the wild type, often induced also by proflavine, by a secondary mutation at another position in the gene. Such mutations are called suppressors, since they reverse the effect of the first mutation. They observed that they could find many different suppressor mutations for a given proflavine mutation. They reasoned that, since proflavine is a flat molecule, rather the size of a base pair, perhaps it slid between the base pairs in DNA and caused a deletion or addition when the DNA was replicated. Based on that hypothesis, the first mutation might be a (–), a minus or a deletion, while each suppressor might be a (+), an addition. If the code were read from one starting point, then a (–) would cause the reading to fall out of phase, and a later (or earlier) (+) mutation would restore the phase, and the correct reading of the rest of the protein would ensue. Thus, one would expect, if the mutations were in a region of the protein that was not important, that one could find many (+) mutations for a given (–), and each of the (+) mutations would work with many (–)s, and vice-versa. They composed phages with double and triple mutations of the same type. The double mutations were still mutants, but the triple mutations (three minuses or three pluses) showed the wild type behavior. Thus, taking out three bases restored the reading phase, and they concluded that the code was read in a multiple of three, probably three bases for each codon, read from a starting point.

That experiment and interpretation was not obvious. Richard Feynman, a great physicist, had become interested in biology through the influence of Matt Meselson and was trying his hand playing with those proflavine-induced mutations in the rII region of phage T4. He came to Harvard during that time, probably the fall of 1960 or the spring of 1961, and gave a seminar on his work. He had found suppressor mutations induced by proflavine. Since he thought that the proflavine bound to the outside of the DNA, he interpreted all the mutations as amino acid changes in the protein. So he suggested that he was observing the correction of the mutant effect by a change of an amino acid elsewhere in the protein. He interpreted the genetic map of the mutations as showing, along the DNA, where the loops of the protein came back to touch each other, so that the two mutant amino acids could interact. Sitting in the audience, we all thought this was an elegant idea. But it was wrong. (He went back to physics.)

During the fall of 1961, I returned to physics research. However, by midyear I realized that I really enjoyed doing molecular biology, so I went back to experiments. Jim gave me a laboratory and an assistant, and I began to do experiments on the mechanism of protein synthesis. (During the next several years, while I worked in biology, I taught graduate physics courses and had graduate students in theoretical physics.)

I worked on a question: how does DNA do anything? The first part of the answer was the messenger — a DNA region is first copied to make a messenger RNA. But then the question becomes this: how does the messenger work? How are proteins made?

I realized I could attack that problem most simply and elegantly by using poly-U as a messenger and asking how far I could dissect the *in vitro* system that would synthesize polyphenylalanine. I could separate the ribosomes from the supernatant enzymes by centrifugation and analyze each separate molecular weight band by adding back the supernatant enzymes and poly-U and radioactive phenylalanine to see what size of particle supported the synthesis — or what sub-particle was needed [Papers #2 and #3, this volume]. These experiments were very clean and showed that both the 30S and 50S subparticle were needed, that their combination into 70S particles worked, and that the messenger bound to many ribosomes — a process that made a polyribosome, a series of ribosomes running along a messenger molecule. The most important experiment showed how the polypeptide chain grew. The answer was that the growing chain terminated at its carboxyl end in a tRNA, and as a new amino acid was brought up on a new molecule of tRNA, the amino group on that new amino acid displaced the previous ester bond and tRNA at the end of the polypeptide chain. Thus, the chain, now longer by one amino acid residue, was attached to that new tRNA.

This was a surprising discovery at the time. One was not really sure in which direction the polypeptide chain grew. Howard Dintzis had just argued that the chain grew from the amino-terminal toward the carboxyl end. But the first time he had done the experiment (which we had heard about at the Gordon Conference in 1960), he had gotten the opposite answer. By repeating and improving the experiment, he published the correct result. It is a curiosity of the history of science that all three of the synthetic directions in biology — the synthesis of DNA, 5′ to 3′ or the reverse; the synthesis of RNA, 5′ to 3′; and the synthesis of proteins, amino to carboxyl or the reverse — were all done the wrong way around in the first experiments. Since each of these determinations has a 50% likelihood of being right purely by chance, the scientists were unusually unlucky in their first experiments. However, by checking and repeating their work, the right answers emerged.

For the peptide chain, chemists thought that the amino acids might be transferred to the ribosomal RNA — transferred to the 3′-hydroxyls on the RNA as an intermediate step before being linked together. No one knew. When I first analyzed the radioactive product of the poly-U reaction, I realized that the polypeptide chain was attached to something. An amino acid's linkage to its transfer RNA is through an ester bond that is very labile and that hydrolyses at pH 9. I tried this, but the polypeptide did not come off the RNA. I tried pH 10, and the bond broke. I realized that the amino group on the amino acid was positively charged (binds a proton) and would weaken the bond, the longer polypeptide chain moves that positive charge away, and the ester bond is more stable. I could then do a double label experiment, putting radioactive phosphorus, P^{32}, into the tRNA that I added to the *in vitro* system and isolating the covalent complex of labeled polyphenylalanine and tRNA before taking it apart into the separate chains. This experiment clearly showed the mechanism of chain growth.

I published these results, my first independent papers in Biology, in the *Journal of Molecular Biology* and knew I had done well. I then tried a harder experiment, which was not successful. I realized that I had a way of isolating phenylalanine tRNA, labeled with P^{32}, from the gradient after separating it out bound to polyphenylalanine. The poly-U experiment had shown that the codon for phenylalanine was UUU. Thus, I thought that the anti-codon would be AAA. I tried to look for that anti-codon in the phenylalanine tRNA. I could break up the tRNA with Ribonuclease T1, which cleaved after G residues, and with Pancreatic Ribonuclease, which cleaved after C and U residues. Therefore, I expected this treatment to yield at least an AAAG or AAAC or AAAU, and perhaps a longer oligonucleotide with even more As. So I set up 2-D fingerprinting and looked for a long oligo after digesting my isolated phenylalanine tRNA. No long oligo was there. I gave up in disgust. Francis Crick had not yet published the wobble hypothesis — which was that since the codons for phenylalanine were UUU and UUC, the anti-codon would be GAA (not AAA, as the G paired by wobble with both U and C) — and the experiment I had tried would invariably have failed.

I continued working with the *in vitro* system. Luigi Gorini had discovered that the antibiotic streptomycin, which blocked protein synthesis and killed bacteria, could suppress certain mutations in streptomycin-resistant strains grown in the presence of streptomycin. Our conversations with Gorini and Julian Davies (then a postdoc with Gorini) led me to test the effect of streptomycin in the *in vitro* system. To our amazement and delight, the drug led to the misreading of the messenger RNA [Paper #4, this volume]. These experiments showed that the ribosome had an important role in holding the coding sequence in the message to the anticodon in the tRNA, and that that contact could be affected by other agents. We did not realize at the time how inaccurate the *in vitro* system was. The high level of magnesium that we used in the experiments led to a relaxation in the reading of the messenger — a necessary relaxation, in fact, because we did not start the peptide chains correctly. The correct mechanism was only discovered later by Jerry Adams. A correct, accurate system runs in 0.1 millimolar magnesium ion; we were using 10 millimolar.

One might wonder why I did not get a Ph.D. in biology if I wanted to do biology. My doctorate was in mathematics. Why was this relevant? I realized, during the period that I was working with Jim on messenger RNA, that the real function of the Ph.D. is not the specific problem addressed but the more general problem of learning how to decide for oneself what is true. The long part of the doctoral program lies in that question of training the mind. In an experimental science, the questions lie in the controls. How can you construct an experiment that will show something, and how can you be sure that the experiment is correct? What are alternative explanations that should be ruled out? How do you prevent yourself from misleading yourself? A high level of repetition is required. In control experiments, often changing variables that one does not expect to matter is needed in order to determine what is real and what is accidental. Fooling oneself through wishful thinking is the greatest danger.

Learning by doing, by apprenticeship to active scientists, is the real way one learns science. The only course I ever took in Biology was my high school course. My courses in Chemistry were taken as part of my undergraduate Chemistry and Physics major. Once in the lab, I learned from the other students and from papers (and their critique).

We all desired to make real proteins in the *in vitro* system. In principle, one should be able to isolate messenger RNA from *E. coli* and use it to make specific bacterial proteins. Alfred Tissieres and I tried that experiment — trying to isolate the RNA from a bacterium that was making beta-galactosidase and trying to synthesize the enzyme *in vitro*. I remember our excitement late one night when we did the experiment and added the color-forming reagent to each tube of the experiment. If the enzyme had been made (in these colorless solutions in each test tube), the enzyme should cleave the substrate to make a yellow color. One tube, the correct one, turned yellow! A wonderful, dramatic result. But it never could be repeated.

We are drinking champagne to celebrate Jim's Nobel. I am on the left, and then there is Jim Watson and Matt Meselson.

Jim Watson received a Nobel Award in 1962. Swedish television arrived in the lab the day before the announcement to interview him, not explaining why they were there. We all celebrated in the tea room.

The tea room was central to Jim's concept of a laboratory environment. The idea had been imprinted onto him in Cambridge as a central gathering place for the group, and we had tea and cookies at four o'clock every day with the students and postdocs. People also ate lunch here and used the large tables to work on manuscripts, being able to lay out cut-up versions to reorder the paragraphs. We held seminars in this room (equivalent in space to three laboratory rooms). When the group was at its largest, we had the students and postdocs give one-hour seminars. By meeting three times a week, each person gave a couple of seminars a year. The criticism was strong. We wanted the students to practice beforehand and to do a clear job of explaining the problem, their approaches, their successes, and their failures. (In one case, the seminar was so bad that the student was sent back to do it again at the next meeting — and even a third time to get it right.) It was brutal training, but the result was that the students were unafraid to speak at meetings and abroad.

Section II: How DNA Is Controlled

The second question we asked was: how is DNA controlled? How are genes turned on and off? At the time I entered biology, the work of the French group, François Jacob and Jacques Monod and their co-workers, had found the paradigm system: the control of the genes needed for the metabolism of lactose (milk sugar) in the bacterium *E. coli.* Their model discovery was that there was a specific gene whose role was to control other genes. In this case, the control was repression (turning off) of the other genes. (At that time, we thought that this would be the general case. Later, genes whose products turned on other genes were found. In higher cells, the major patterns of control are gene products that function at enhancers in combinatorial ways to turn on other genes, as well as many other processes by which specific gene products influence and control other DNA regions. These methods include binding to them, modifying them, modifying the histones that cover them, and even affecting the functioning of RNA by making RNA products that bind to and alter the functioning of messenger RNAs.) The critical element of their thought then (and now) was that there was a *specific* gene involved in the control. The presence of the enzymes was not called forth by some general balance or imbalance of chemical flows or by some external natural effect on the cell, but rather involved an actual gene product, the result of Darwinian evolution, to exercise the control.

As a bit of background, the genes for the enzymes of lactose metabolism, the *lac* genes, are a set of three genes: the *z* gene (*lacZ*) for the enzyme beta-galactosidase, which cleaves lactose (a disaccharide, i.e. two sugars linked together) into its component sugars glucose and galactose, which then can be metabolized separately by the cell; the *y* gene (*lacY*) for a permease function that concentrates lactose inside the cell; and the *a* gene (*lacA*), which acetylates lactose and which seemed (and still seems) to have no necessary function. In the absence of lactose, the wild type bacterium had only a trace amount of the enzyme, about 10 molecules per cell, while in the presence of the sugar as a carbon source, the bacterium made about 10,000 molecules of enzyme, a few percent of all its protein. This effect is due to the presence of the *i* gene (*lacI*), which makes a repressor that turns off all three genes. In the presence of a metabolite of the sugar, lactose, the repressor is inactivated, and the genes turn on. Jacob and Monod proposed that all three gene products, *z*, *y*, and *a*, were made from a common messenger RNA, synthesized from a start called a promoter (the *p* gene) and controlled by a site called an operator, the target of the repressor (the *o* gene). All these had genetic definitions through the presence of mutations in different strains of *E. coli*. (p^- mutations blocked the synthesis in *cis*. o^c mutations produced all three proteins in *cis* and were dominant. i^- mutations lead to full synthesis (constitutive expression) but were recessive to the i^+ form in *trans*.) All of this was worked out in the '50s and summarized in a masterful 1961 review (François Jacob and Jacques Monod. "Genetic regulatory mechanisms in the synthesis of proteins." *J. Mol. Biol.* 3, 3 (1961): 318–356), which I believe I read in preprint form at the end of 1960.

This knowledge of the ability to turn a gene in the bacterium on or off had been hardearned through the 1950s. Some of the first thoughts hypothesized that the presence of the sugar would form the enzyme around itself. After antibodies to beta-galactosidase had become available, the French group had looked for a precursor molecule in the cell and thought that they had found one, named *Pz* for "precursor to *z*" (beta-galactosidase). Later experiments showed that this was wrong, and when radioactive labeling techniques became available, they could show that the enzyme was synthesized *de novo* after the sugar became available.

More strikingly, the inducing molecule turned out not to be simply related to the sugar or the enzyme substrate, or even to have any affinity for the enzyme. Chemical modifications of galactosides, which were not substrates, were better inducers, the best being IPTG, or isopropyl thiogalactoside. In fact, as they realized that there were several genes arranged in an operon under this control, they could show that the other genes would not be induced by lactose if the beta-galactosidase was not present. The basal level of enzyme, about 10 molecules per cell, is essential to make the inducer from lactose.

When their genetic studies showed that there was a separate gene for induction, the *i*-gene, their thought was that this gene product turned on the enzyme synthesis in the presence of the inducer. The i^- mutations were thought to turn on the beta-galactosidase synthesis. Leo Szilard, a nuclear physicist who became interested in biology after the war, on a visit to Paris pointed out that they really did not know whether the control worked by turning on the synthesis of the enzyme or by turning it off. That thought led to experiments testing mutations in the system in diploid constructions. (This was very difficult originally, since only transitory diploids could be made. Over the years, these tests became easier, as plasmids were discovered and many ways of constructing stable diploid strains were found.) To their surprise, the i^- mutations were recessive to the wild type i^+. (And i^s mutations, where the repressor no longer recognizes the inducer, are dominant negatives in *trans*.) The control was negative; the *i*-gene made a repressor. However, this history still remains in the name of the gene for the *lac*-repressor — it is the *i*-gene, not the *r*-gene.

I had become an assistant professor of Physics in the fall of 1959. That five-year appointment meant that I came up for tenure by 1964. (Actually, I was reviewed for tenure in 1963.) The Harvard clock ran very quickly in those days. A Harvard letter to collect references at that time read, "We are thinking of making a senior appointment in Biology. Who would you suggest?" No names of possible candidates would be suggested. It was clear to me that I was not likely to be promoted in Physics, even though I had several graduate students in Physics and had been publishing in Physics, since most of my work and my clear future directions were in Biology. I began to make inquiries to friends on the West Coast as to whether there might be a job available. But, luckily for me, my friends at Harvard convinced the University to make a tenured appointment in Biophysics (which was a committee, not a department, at that time), and after review I was offered that appointment. I found myself on the senior staff of three groups — in the department of Physics, in the department of Biology, and in the committee of Biophysics as the sole senior staff member. (I viewed Biophysics as a way for Physics undergraduates to do doctorates in Biology, specifically Molecular Biology, without having to have an extensive background in Biology courses.) Now that I had tenure, I was free to work on a really difficult problem, on which I could have spent years and possibly not have had any success.

What sort of molecule was the repressor? When I visited Jacob and Monod in Paris in the summer of 1961 on my way back from the meeting in Moscow, they told me of experiments that suggested that the *lac* repressor was an RNA molecule, not a protein like most of the other gene products. (At that time, most known genes made enzymes or structural proteins. There were a few genes for tRNAs, perhaps 20, and some genes for ribosomal RNAs.) They had done experiments by transferring the genes for the *lac* repressor and beta-galactosidase into a bacterium lacking those genes by transitory bacterial mating, while blocking protein synthesis but letting RNA synthesis continue. They inferred from the cell's later inability to make the enzyme, even after the block was removed, that the repressor had been made during the period of protein inhibition. Those experiments soon turned out to be wrong because of a later-discovered further control mechanism, called catabolite repression. So the questions remained. What were repressors? And how did they function?

I thought at that time that the simplest model would be that the repressor gene product would be a protein, which would function by binding to DNA and interfering with the action of the RNA polymerase. I reasoned that, since my fingers could read the DNA sequence from the outside of a model of the double helix, so also could a clever protein use contacts and hydrogen bonds to read the sequence. (This turned out to be correct.) But many people thought that the DNA had to unwind for the sequence to be read to locate control sites, and for this an RNA molecule could be used. As late as 1971, long after our experiments, Francis Crick (Francis Crick. "General Model for the Chromosomes of Higher Organisms." *Nature* 234 (1971): 25–27) still thought that the control in higher cells involved the massive unwinding of DNA and the reading of sequence through clefts in proteins. (He thought that proteins would not have protuberances that could fit into the DNA.) As time passed in the early 60s and no one could identify control molecules, the problem of their nature became ever more interesting.

In one long, early experiment, I made agarose beads containing DNA and poured solutions of radioactive proteins, S^{35}-labeled, through long columns containing the DNA. I would try to elute DNA binding proteins with the inducer, IPTG. These experiments made me spend days in the cold room, working long hours with six-foot columns, hoping to find some pattern of specific binding and elution. My wife would come to the lab in the afternoon, pushing the children in a carriage, and they would shout up the building, "Daddy, have you found it?" No luck.

I was not totally alone while seeking this grail. Benno Mueller-Hill came to Harvard as my first postdoc and wanted to work on the repressor. Benno brought genetics to the lab and worked on finding new mutations in the *i*-gene. By finding a nonsense suppressible mutant, an *amber* mutant, he proved that the repressor was a protein. Mark Ptashne graduated from Matt Meselson's laboratory and became a Junior Fellow in Harvard's Society of Fellows. He planned to work on the other paradigm example of a repressor, the control gene of phage Lambda. Now that he had complete independence as a Junior Fellow, he devised his own line of attack, which was a double label approach to identify the phage repressor by comparison of two phage lines bearing different repressors. Mark, Benno, and I talked all the time about this problem as we did our separate experiments.

Our first major excitement occurred after the publication of a paper by the French Group that claimed to identify the true inducer. The assertion was that when the enzyme cleaved lactose to release glucose, it generally did so by transferring the galactose moiety of the sugar to a new acceptor, water. However, it made a new compound in the presence of glycerol, galactosylglycerol. They claimed that this molecule was the true intercellular inducer and was 1,000 times more potent than IPTG. While IPTG would induce the formation of the enzyme at half-maximum rates at 2×10^{-4} molar concentrations, the new compound was claimed to work at 10^{-7} molar concentrations. One could argue that the induction curve was quadratic and showed 1000 fold induction. The single-site affinity would then be about 33-fold less, so IPTG would bind to the repressor with an affinity around a six-micromolar concentration, while galactosylglycerol would bind at only six-nanomolar concentration. Since 10 molecules a cell (a possibility for the amount of repressor) would be 10 nanomolar, one should be able to see an excess binding of radioactive inducer to repressor by equilibrium dialysis at concentrations of the repressor that already existed in the pellet of concentrated bacterial cells (for example, at half-maximal binding, perhaps 16 nanomolar inside the dialysis bag and 6 nanomolar outside). We set out to do this experiment.

I made radioactive galactosylglycerol by buying radioactive glycerol and transferring the galactose to it from lactose with the enzyme. I purified it by paper chromatography and set out to use it. Of course, I checked whether or not it would work as claimed. I could show that it induced beta-galactosidase in a strain without a lactose permease at 100-nanomolar concentrations (in that strain, the compound would be destroyed by the beta-galactosidase that it would induce). So I took another strain that did not have the *lac* genes to test whether or not the compound had the same concentration inside the cell as outside. It was not concentrated, so I concluded that it was truly a super-inducer and began to do experiments with extracts of cells. I broke 50 grams of *E. coli* cells, fractionated and concentrated the cellular contents, put samples into dialysis sacs, dialyzed overnight against the radioactive galactosylglycerol, and finally counted matched volumes from within and outside the sacs to see if any excess of the radioactive material could be found. Total failure.

Why did this experiment fail? Because the super-inducer claim was wrong. I was tricked by a failure in my controls. I had not made truly iso-genic strains that differed only in the genes I needed to test. The standard laboratory strain of *E. coli* without the *lac* genes, which I had used to test whether the galactosyl-glycerol was concentrated, had a further, unknown defect. It was unusual in that it did not have a common, constitutive galactose permease (unknown at that time), which in almost all other strains of *E. coli* will concentrate galactose and galactosylglycerol 1,000-fold. Had I done the controls correctly, I would have found this out a year earlier (and discovered the galactose permease).

During this time, I was often asked to travel to give seminars. I would talk about the lactose control and the shape of the gene induction and our search for the repressor. I almost called my seminar, "Why haven't we found it."

Nonetheless, this first effort to find the repressor by equilibrium dialysis led us to try harder. Benno made a mutation in the repressor that was more sensitive to the inducer, so we expected that mutant to bind IPTG more tightly than the wild type. I began to fractionate and concentrate the proteins from the mutant strain and to dialyse them against radioactive IPTG. The first experiment that showed some sign of success had only a 4% excess of IPTG within the dialysis bag (not really believable, but I counted that sample exhaustively) to provide the stimulus to improve the concentration of protein. Soon, however, I could make protein fractions that bound an excess of 50 to 100 percent more IPTG within the bag: a clear signal. There followed several days of life-or-death controls: growing 50 liters of cells overnight, breaking 50-gram lots of mutant or control cells in the morning, and assaying each night. The mutant cell protein fraction would bind IPTG each time, while negative controls (such as a suppressible i^- or an i^s strain) would not, and furthermore, I could detect the difference in binding affinity between our novel mutant and the wild type repressor. We could even estimate the molecular weight of the repressor protein. Benno Mueller-Hill and I published our identification of the *lac* repressor as a protein in PNAS [Paper #5, this volume]. It was a biochemical tour de force to isolate a protein that only existed as a few copies per cell and that had no enzymatic activity. Indeed, I had a faculty member come up to me to ask, "Did you see that those two crazy people claimed to find the lactose repressor?" I could only say, "That's my paper; Benno and I did that."

At the same time, Mark Ptashne was working on the repressor of phage Lambda. During this period, we were close friends, talking frequently about our searches for these repressors. We both were fascinated by and drawn to the problem of identifying the products of control genes. He identified that phage repressor as a protein by isolating it using a double labeling experiment: comparing the proteins made after infection with phage Lambda labeled with one radioactive isotope with those made after infection with a phage Lambda carrying the repressor for the different but related phage 434. As he purified the phage proteins, he could finally see a peak in the pattern of radioactivity coming off a column used to separate proteins, which was labeled with only isotope. That was the repressor. He then could show that labeled protein bound to phage DNA, binding to the appropriate region at which the control would be exerted.

In order to study the binding of the *lac* repressor to DNA, we had to devise a partial purification of the protein using its IPTG binding, duplicate that purification blindly using radioactive sulfur to label all the cell's proteins, and then detect the binding of radioactive label to DNA containing the *lac* operator sequence. These experiments needed a 1000-fold purification of the repressor, which in the binding experiments still was only three parts in a thousand of the label put on the gradient. Nonetheless, these experiments [Paper #6, this volume] were successful, and we could show that mutant operators blocked the binding and that IPTG released the repressor protein from its double-stranded DNA target.

I think that out of all of the work I have done in Biology, I am most proud of the repressor discoveries. These experiments involved a multi-year dedication to a most difficult problem, with many false starts and much futile effort. While the work was supported by general grants to Jim Watson and my laboratory, none of the grants would mention the repressor work — it was too far out to be supported. It is interesting to look back on the production of the laboratory in those days. In 1967, the laboratory published seven papers. My name was on one of them, and Jim's name was on none. (The lab published 20 papers in 1966 and eight in 1968.) Jim did not put his name on any of the students' papers, and I followed him in this practice, only recording cases in which I seriously was involved in the experiments. I continued this custom for many years, only changing at the very latter part of my career, when I would finally put my name on papers to make sure that I really paid attention to them.

This period of the 60s, when Jim and I were both most active in the laboratory, involved the most extreme dedication to work. I would be in the laboratory essentially every day of the year, coming back in even on Christmas day to check on something. Jim and I would be in the laboratory after dinner every night. In the early 60s, we often would just have dinner in the square, talking about Biology, and then come back to the lab together to see what new results there were. Jim would take pride, as we walked up to the Biological Laboratories, that only our labs on the third floor were all lit inside the dark building, as the students and postdocs worked into the night. The rest of the faculty had gone home, but he and I and the students were still there, working and talking.

While I continued to work on the purification of the repressor, Benno made a mutation in the repressor gene that made 10-fold more repressor, the i^q mutation (*q* for quantity) [Paper #7, this volume], and this made it easier to make the protein. Additionally, by using a phage that carried the repressor gene but which was also defective in not being able to lyse the cell, we could induce the phage and have it multiply inside the bacterium, so that one could have hundreds of copies of the operator and the repressor gene in each cell. The i^q mutation had a striking effect when put into a cell that also carried an i^s gene (an uninducible repressor mutation, which would be *lac*$^-$). The diploid could be induced. We interpreted this as complementation between the subunits of the repressor. I had been able to show that the repressor was a tetramer and so had four binding sites for IPTG. We inferred that the diploid contained molecules of repressor, with three good subunits and one bad, and that the good subunits made the DNA binding feature of the molecule sensitive to IPTG, hence inducibility and *lac*$^+$. In this paper, when we assert that the repressor is a tetramer, we cite as a reference (reference #5) my own unpublished work. (I never got around to publishing those experiments on the purification. In those days, one could cite unpublished work. Today, the journals insist that references be only to published work. This has meant that the ability to reference conversations, and hence to trace the background of influences on the course of research, has been lost. Unfortunately.) Not everybody believed everything in the published research. Charley Yanofsky, a professor at Stanford University, whose work was on the tryptophan genes and their control in *E. coli* and who was a major figure in Molecular Biology, never believed this complementation experiment and thought we had made some mistake.

Benno left Harvard and set up his own laboratory in Cologne. When I visited him soon after that, he showed me a bottle with a heap of white powder in it. His student had purified 50 milligrams of *lac* repressor. I burst out laughing — to go from the trace of 10 molecules per cell to a large bottle full of fluffy powder! Unimaginable! His student, a protein chemist, went on to sequence the repressor protein for his thesis.

Over the next years, even though my students and I worked on other problems, I continued to be interested in the interaction of the *lac* repressor with its operator DNA. In the 1968–69 academic year, I spent a sabbatical year in Paris at the Institut Pasteur, trying to work on DNA replication [Paper #11, this volume], but nothing came of those experiments. My technician, Christina Weiss, with whom I had worked on the repressor problem, married Jeffrey Roberts, my graduate student, and left with him to go to Cornell. When I came back to Harvard, I hired a new technician, Allan Maxam, who had been undecided as to whether he would continue in science and had been working in a coffee house. Allan and I started to work to characterize the *lac* operator, first as a DNA fragment and then as an object whose chemical sequence should be determined.

I began to isolate the *lac* operator as a fragment of DNA protected by the *lac* repressor protein against DNase digestion. Suzanne Bourgeois and Arthur Riggs at the Salk Institute had discovered that the repressor bound to DNA would trap that DNA on a nitrocellulose filter and that the bound DNA would be released from the filter by IPTG. This gave one a way of analyzing the interaction of the repressor with DNA, as well as a way of isolating a DNA fragment that had the repressor binding site on it. I would go upstairs to a special room, where I would label a test-tube of cells with 100 millicuries of radioactive phosphorus, induce the nonlysing phage that carried the *lac* genes, and grow cells to phosphate exhaustion in a minimal medium. Working in thick lead-containing gloves from behind a Plexiglas shield, I would spin down the cells, sonicate them to open them and break up the DNA into 1,000-base pieces, and phenol extract all the proteins. Then, I would take the solution of DNA and RNA and, using repressor protein, bind a tiny fraction of the radioactivity to a filter before eluting the bound DNA with IPTG. Thus, I could prepare radioactive DNA fragments that carried the operator and bound to the repressor. By putting the repressor back on the DNA and digesting all the exposed DNA with DNase, filtering to catch the DNA bound to the repressor, and eluting again, I could collect a small, roughly 27-base-pair-long piece of labeled, double-stranded DNA. To characterize the fragment, we developed the gel technology, the trisborate-EDTA buffers (EDTA to bind any trace of magnesium or heavy metals and to inhibit any DNases) and the urea gels to analyse denatured DNA. This was all an extensive effort.

Now that I had the operator in my hands as a DNA fragment, I wanted to know its sequence. We had spent a long time, by the time Allan Maxam first came to the lab, trying to devise a DNA sequencing method. We worked at isolating pyrimidine and purine tracts from DNA and learning how to characterize them. (Pyrimidine tracts resist acid, while purine tracts resist hydrazine.) However, this work did not yield enough information to develop a sequencing method. (Years later, this chemical work paid off in supplying background knowledge used in developing a true sequencing method.) After I had isolated radioactive operator by DNase digestion, I turned to isolating larger amounts of unlabeled operator that we could sequence, in principle, by copying it into RNA with the RNA polymerase, which would yield radioactive RNA fragments that we could sequence by the methods developed by Fred Sanger — digesting the RNA with enzymes and identifying the resulting small fragments on two-dimensional chromatography/electrophoresis patterns. Gilbert, W. and Maxam, A. (1973) [Paper #8, this volume] describes our results. After two years of work, we could identify 24 bases of the sequence of the lac operator. We were thrilled to see symmetry in the sequence, which suggested that two subunits of the repressor interacted with the operator DNA.

Nancy Maizels, my graduate student, sequenced the first 63 bases of the *lac* messenger RNA for her Ph.D. thesis and showed that the RNA sequence included a copy of the operator sequence (N. Maizels. "The Nucleotide Sequence of the Lactose Messenger Ribonucleic Acid Transcribed from the UV5 Promoter Mutant of *Eschericia coli.*" *PNAS* 70, 12 (1973): 3585–3589). This showed that the promoter region, where the polymerase binds, lies just before the operator. Furthermore she could sequence and identify two operator constitutive mutations and show that they all fell within the operator sequence [Paper #9, this volume]. We later sequenced and discussed 11 different o^c mutations.

We had set up a small internal room near my office as an electrophoresis room. Five-foot-long paper electrophorus sheets wet with conductive buffer were folded inside three-foot-high tanks, with each end of the paper in a buffer bath but with the tank filled with aviation fuel to insulate and cool the sheet and prevent drying-out. A large, 10,000-volt power supply drove the electrophoresis. The small room was filled with tanks and had an emergency device to quench any fire that might break out (as well as quenching any unlucky person). The hall outside my office was lined with 50-gallon drums of aviation fuel to replenish the tanks.

Throughout this period in the early 70s, the restriction enzymes that cut DNA at defined sequences were discovered. Since we were doing DNA work, we needed as many of these different enzymes as we could get our hands on. Each student and postdoc in the laboratory was responsible for supplying one of these enzymes. They had to grow the appropriate bacterial strain and completely purify the corresponding enzyme to maintain the laboratory supply. Thus, each molecular biologist learned as a student enough protein chemistry to perform purification and characterization of an enzyme. (When I was involved in Biogen, a few years later in the early 80s, all the enzymes could be purchased, and no molecular biologist of that immediately following era knew how to purify an enzyme.)

Although we had determined the sequences at the beginning of the messenger for the *lac* operon and had ways of manipulating DNA with restriction enzymes, we did not pursue other *lac* sequences during this time, since the sequence of the *lac* promoter (and the entire sequence between the end of the *i* gene and the beginning of the *z* gene) was worked out by Reznikoff and his coworkers (R. Dickson, J. Ableson, W. Barnes, and W. Reznikoff. "Genetic Regulation: the *Lac* Control Region." *Science* 187, 4171 (1975): 27–35). Thus, all the sequences I cared so much about were known, and I was pursuing other interests. However, around early 1975, Andrei Mirzabekov got permission to travel in the United States, a rarity for a Russian scientist at that time. He came to see me, to urge me to explore the interaction between the *lac* repressor protein and DNA. He pointed out to me that he had been able to show that the reagent dimethyl sulfate, a methylating reagent, would put a methyl group onto the guanines in DNA in the major groove (at N7) and methylate the adenines in DNA where they are exposed in the minor groove (at N3). He had used this chemistry to show some aspects of histone binding to DNA, and he wanted me to study the binding of the repressor to the major and minor grooves of the operator DNA. I could not see any way of doing this, so I did not take the bait. Six months later, at the end of his time in the United States, he came back to urge me again to try this experiment. He, Allan Maxam, Jay Gralla, and I had lunch in a little cafeteria in the next building to the Bio Labs, and during that lunch, a way of doing the experiment was devised. I knew I could obtain a restriction fragment, 53/55 bases long, cut by two different enzymes and bearing the *lac* operator. Because the two ends were made by different enzymes, I knew I could put a P^{32} label at the 5′-end of each strand separately. I knew that the methylation of the DNA base, either methyl-G or methyl-A, would weaken the bond between the base and the sugar so that heat would release the base, leaving just the bare sugar. I also knew that the bare sugar could be destroyed by alkali treatment, cleaning off the phosphates along the backbone and leaving the rest of the DNA strand intact. Thus, on an end-labeled restriction fragment, if one G were methylated, the DNA strand could be broken there and release a shorter, end-labeled piece with a size that could be measured on a gel. Since I knew the complete sequence of this DNA fragment, I expected that I should be able to tell from the size which guanine had been methylated. Hence, if I treated the repressor-operator complex with dimethyl sulfate, I should have been able to observe whether the protein blocked the methylation and, hence, infer whether the protein had touched this region of DNA. I set out to do that experiment [Paper #10, this volume]. All of the DNA fragments, repressor, and restriction enzymes were available in the lab, and we made our own γ-labeled ATP of very high specific activity. After treatment with dimethyl sulfate and quenching the reaction, I heated the DNA to release any modified bases, sealed samples into capillary tubes with alkali, and heated them in a boiling water bath to destroy any bare sugar. Then I electrophoresed the samples to filter them by size and exposed the gel to a sheet of X-ray film to get an image of the radioactive bands. The first time the experiment was a failure (probably samples leaked), but the second time the experiment was a clear success:

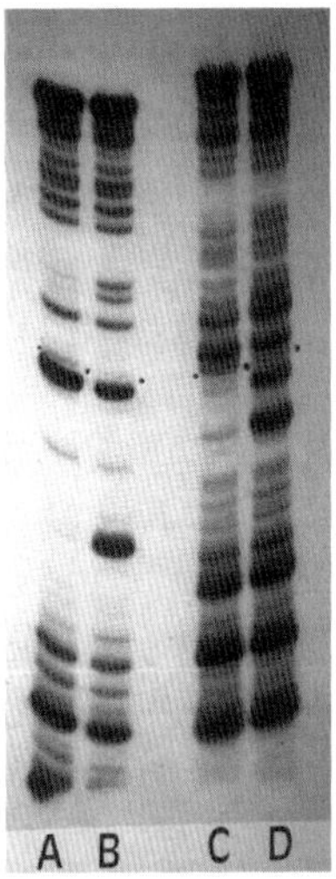

Lane B shows the pattern without the repressor on the shorter strand. Lane A shows the pattern with the repressor. The guanines react more rapidly; thus, the dark bands are guanines, and the light bands are adenines. The dots show the position of the marker dye, which runs with fragments 27 basis long.

The sequence of this 53-base-long strand of the restriction fragment was known to be

p^{32}ctGtttcctG$_{10}$tGtGAAAttG$_{20}$ttAtccGctc$_{30}$AcAAttccAc$_{40}$AcAAcAtAcG$_{50}$AGc

where all the purines that would interact with dimethyl sulfate are in the upper case.

The pattern in Lane B on the gel shows

p^{32}------------G-GAAA--G$_{20}$--A---G---$_{30}$A-AA----A-$_{40}$A-AA-A

The first protected band, going up from the bottom (smaller fragments run further down the gel from the top), is the guanine at position 20 in the sequence. Further up the gel, one sees that two adenines are protected at positions 33 and 34 in the strand [Paper #10, this volume]. Furthermore, one can see enhanced reactivity, where a pocket between the protein and the DNA provides an environment that will weakly bind the dimethyl sulfate molecule. This experiment demonstrates the binding of the protein to DNA.

We used this technique, which shows where a protein touches DNA, to explore other binding sites for the *lac* repressor along the DNA molecule and for the CAP (catabolic activation protein) factor binding site at the *lac* promoter (W. Gilbert, J. Majors, and A. Maxam. "How proteins recognize DNA sequences" In "Organization and Expression of Chromosomes" Dahlem Workshop Report (1976): 167–178). Ten years later, we used this technique *in vivo* to study the binding of the repressor in bacterial cells (H. Nick and W. Gilbert. "Detection *in vivo* of protein–DNA interactions within the *lac* operon of *Escherichia coli.*" *Nature* 313 (1985): 795–797) and to study an enhancer in mammalian cells (A. Ephrussi, G. Church, S. Tonagawa, and W. Gilbert. "B-lineage-specific interactions of an immunoglobulin enhancer with cellular factors *in vivo.*" *Science* 227 (1985): 134–140). It was quite enchanting that we could see the footprint of a protein on the DNA in a living cell as complicated as a mammalian one.

Section III: The Rolling Circle Model

In 1967, David Dressler, a postdoc in a neighboring laboratory down the hall, was working on the replication of phage PhiX174, a single-stranded circle of DNA. He and I often talked, and we came up with a model for how the phage might replicate. After synthesizing a second strand and making a closed double-stranded DNA circle, we hypothesized that the first strand was opened and peeled off the template in a continuous fashion as the DNA polymerase ran round and round the templating circle, producing a very long strand that could be cut up into single-stranded circles. We were very excited by this idea, since it solved one serious problem of DNA replication: how does one start the copying of a full-length DNA strand? By a form of continuous synthesis. This idea could be shown to work for PhiX and for phage Lambda, each of which could become a circle, as well as for phage T4, where it was known that the phage particle contained a linear molecule with a circular genetic map — each particle contained a molecule longer than one complete map, and each particle contained a different, circularly permuted piece of DNA. The solution to the T4 problem was to suggest that the phage DNA became a circle through recombination within the molecule, and then the circular molecule was copied by the rolling circle mechanism into a long DNA molecule that had repeating copies of the original circle along it. Oversized lengths were then measured out and packaged into phage particles by a "headfilling" mechanism, with each phage head containing more than the minimal circular amount of DNA and hence having within it a DNA molecule with redundant ends.

The rolling circle also suggested a mechanism for the transfer of DNA during bacterial mating and made a prediction about which strand should be passed from the donor to the recipient. This prediction turned out to be correct. We, of course, tried to extend this model to the circular chromosome of bacteria and tried to see rolling circles everywhere. Too much. The beautiful picture of replicating *E. coli* that John Cairns made showed that the circular bacterial chromosome, some 2 mm long, replicated in a bidirectional way from an origin at which the strands are separated to permit replication to commence on both strands. This turned out to be the general answer in higher cells, that long DNA molecules are replicated from many bidirectional origins. The ends of the molecules are handled by special telomere sequences and a special enzyme, telomerase, to solve the problem of copying DNA all the way out to the end for the lagging strand.

François Jacob also thought of this way of endlessly copying circular DNA molecules. Maxime Schwartz, who came from the Pasteur Institute, was spending a postdoctoral period as a Harvard Junior Fellow, working in Jim's and my laboratory at this time. He told us that Jacob had a novel

DNA model idea but said that it was terribly secret, and that he, Maxime, had been sworn to secrecy. This seemed very impressive, but when I met Jacob soon after this, he was happily telling everyone about his DNA model, which he called the Toilet Paper Roll Model. So I learned then, and saw many times when I was involved in the commercial world later, that a stress on secrecy is usually a cover for ignorance, sloth, or malfeasance.

Section IV: DNA Sequencing

When I did the experiment that showed the contacts between the *lac* repressor and the DNA operator, I expected to have to work out which fragment was which only with great difficulty, deciphering the bands on the gel using preexisting knowledge of the sequence. However, when I saw the pattern on the gel (shown in paper #10, this volume), I realized that the pattern was so clear, showing Gs and As in a ladder, that this technique was a sequencing method. It had taken Allan Maxam and me two years to work out 24 bases of the operator sequence, essentially one base a month. Here on the gel, one could clearly see the A at position 31, a space for a C, and then two As at 33 and 34 in an afternoon's work. (Even more of the sequence can be worked out from these two patterns. The autoradiograph shows the purines from G_{12} to A_{46}. The entire sequence could be worked out by knowing just the As and Gs from both strands.)

So I started out, with Allan Maxam, developing a full sequencing method based on the chemical cleavage of end-labeled DNA fragments. We quickly cleaned up the patterns of breakage at As and Gs. In fact, we sequenced a 200-base-long fragment by running from both ends, but we never published that. We then used hydrazine, which we had explored years before, to attack the pyrimidines, the Cs and Ts, and make a pyrimidine ladder. Allan then spent 9 months testing whether adding other chemicals to the reactions would affect the cleavage pattern and finally found that adding sodium chloride to the hydrazine reaction would suppress the reactivity of the Ts and give us a C-specific cleavage. Thus, we could determine all four bases and read the sequence as a ladder up the autoradiograph from a gel. The final sequencing pattern had an acid lane that showed both A and G, a dimethyl sulfate lane that showed strong Gs, a hydrazine-with-salt lane that showed the Cs, and a hydrazine lane that showed both Cs and Ts. Allan replaced the hydroxide cleavage of the revealed sugar with a hot piperidine cleavage, which was cleaner.

As we began to work on this chemical-based sequencing method, I went to see Jeremy Knowles, an excellent organic chemist who had just taken up an appointment at Harvard. I asked him for advice about how to distinguish the pyrimidine bases in DNA by chemical attack. He told me they were too similar chemically for one to be able to choose between them. Nonetheless, we became great friends and lunched together several times a week for 20 years.

There is an important conceptual trick behind the DNA sequencing. The problem of breaking the DNA chain at a specific base is divided into two parts. The first reaction, which distinguishes between the bases, is run only to a very slight extent. In the initial work, only one out of 100 bases was modified, while in the later work, it was only one in 1,000. Under these conditions, the chemical reaction can show its greatest specificity, which would be lost if we were to take these reactions to quantitative completion. Then, after the base has been driven off the sugar, the bare sugar is now chemically very different, and it can be cleaved quantitatively without harming the rest of the strand.

I had talked about the methylation protection experiment in August of 1975 at the Benzon symposium. We had a complete sequencing method by the spring of 1976, and I talked about it at the Nucleic Acids Gordon Conference that summer and gave away mimeographed instructions. Everybody started sequencing DNA. By the end of the decade, about half a million bases had been sequenced. Ten years later, by mid-1986, eight million bases were recorded in GenBank, the government database of DNA sequences. (We got around to publishing our method in 1977 [Paper #12, this volume]. Fred Sanger's dideoxy method came into use simultaneously in 1976, and the two methods were used roughly equally in different laboratories for a decade.)

As an example of this change in technology, in the early 70s, Nancy Maizels sequenced 63 bases for her Ph.D. thesis, as I mentioned earlier. The next Ph.D. for sequencing involved 1080 bases: Phillip Farabough worked out the gene for the *lac* repressor (P. Farabaugh. "Sequence of the *lacI* Gene." *Nature* 274 (1978): 765–769). The protein sequence had been worked out in Benno's laboratory by laborious conventional methods, and I was delighted that Phillip (comparatively) quickly found two errors in the published amino acid sequence. Greg Sutcliffe chose to sequence the entire small plasmid that we used to carry novel genes, pBR322, with 4362 bases, for his doctorate (G. Sutcliffe. "Complete Nucleotide Sequence of the *Escherichia coli* Plasmid pBR322." *Cold Spring Harbor Symposium* 43 (1979): 77–90). This was the last Ph.D. that I gave for sequencing. (20 years later, Sorel Fitz-Gibbon, a graduate student in Jeffrey Miller's laboratory, got her doctorate for sequencing the 2.2-megabase genome of a bacterium, but in addition, she had to interpret the sequence.)

The small room near my office no longer contained the electrophoresis tanks but now became a darkroom for the efficient development of the large X-ray films we used to display sequences. We made carrier-free P^{32} gamma-labeled ATP (with carrier-free P^{32} and a spinach enzyme) to use to end-label the DNA fragments for the sequencing, and we often used two-week exposures (films and gels wrapped and stored in freezers) so that half the radioactive atoms would have decayed, since we were fighting the small numbers of DNA molecules that we would get from biological sources (long before PCR).

At the Gordon Conference on DNA in 1976, where I talked about DNA sequencing, I shared a room with Susumu Tonagawa, and we talked extensively about immunology. He was trying to make clones of the immunoglobulin genes in order to understand the genetic basis of antibody diversity. We became friends, and when he found a clone containing the gene for the variable region of the Lambda light chain, he brought the clone to my laboratory in the spring of 1977 to work out the sequence. At that time, the protein sequence of many antibodies was known, as well as the 3-D structure. The antibodies contained a light chain of about 25,000 daltons, about half of which was a variable region (V-region) with high amino acid variability corresponding to different antigens (and different specificities for the same antigen). The second half was a constant region (a C-region, not showing variation across different antigens). They also contained a heavy chain, which also had a variable region and three constant domains. The light- and heavy-chain variable regions fitted together to create a surface to bind the antigen, while the COOH-terminal constant regions of the heavy chains were tied together to make dimers (IgG for example) or pentamers (IgM) in the functional molecules. The mystery remained. What is the gene structure? How are the variable amino-acid sequences created?

At Harvard, Susumu learned how to sequence DNA, and he, Allan Maxam, and I worked on the sequence of the light chain variable-region gene, which we finally reported about 9 months later [Paper #13, this volume]. We were delighted to see an amino acid coding sequence emerge as we sequenced the restriction fragments from the clone. The coding sequence for 98 amino acids appeared, matching the known V-region sequence (with amino acid changes in the hyper-variable regions of the specific antibodies whose protein sequence was known). However, Allan Maxam discovered that there was a 93-base-long sequence that did not code for any amino acids at the beginning of the gene and that separated 15 amino acid codons from four codons that preceded the known start of the V-chain. Those 15 codons looked like a hydrophobic leader sequence, which would direct the protein chain through the cell membrane. However, the protein chain was interrupted! How could this be? We did not understand the sequence of this gene fragment at all.

Soon, I went to the Cold Spring Harbor Symposium, as I did every summer. That year, the meeting in early June was on chromatin. I remember the extraordinary excitement at the evening session, when I was surprised and astounded by the talks demonstrating RNA splicing, seen in adenovirus, by Richard Roberts's and Phillip Sharp's groups, described for the first time at that meeting. Regions of uncoding RNA lay within virus transcripts and were spliced out to make shorter messenger RNAs. Fantastic! This provided an explanation of the 93-base-long gap in our V-gene sequence.

I traveled in Europe much of that summer, and everywhere the new conversations were about gene structure. Talking to Piotr Slonimski in Paris, I learned about yeast mutants in the mitochondria that were explainable by invoking a splicing of the messenger RNA. There were mutants in strange places in SV40 that also could be interpreted by a splicing.

These discussions with geneticists made me think about recombination and the general idea that moving parts of a gene far apart along the DNA would increase the rate of homologous recombination. Between like sequences, spreading two regions of a gene apart by 1,000 bases would increase the homologous recombination rate perhaps 100-fold compared to the recombination at a small region in the gene. This process permitted a faster evolution of better variants that arose by combining two different changes. However, even more importantly, such extra sequences would dramatically increase the rate of non-homologous recombination (between unlike sequences, two such 1,000-base regions of extra DNA between coding regions would increase the rate about a million-fold). This would permit the shuffling of useful parts of genes to make new proteins, and the hydrophobic leader sequence on the V-gene, separated from the rest of the sequence and perhaps originally donated by some other secreted protein, was a perfect example of this idea.

When I visited Susumu in Basel that summer, these ideas had become fleshed out, and I had given the names "exons" and "introns" to the coding and non-coding regions of the gene. The name "intron" was based on the intragenic region, a separator within the gene that increased recombination; the name "exon" was based on the expressed region, which appeared in the final spliced messenger. (Not everybody accepted these names. There were several years of arguments about what to call the extra non-coding DNA.) Susumu and I argued about what the final gene for the immunoglobulin would look like. We expected the V-region gene to be moved next to a C-region gene. However, I argued that the DNA regions only had to be brought close together into a transcription unit, and then splicing would connect the hydrophobic leader to the 98-amino-acid V-region, which could be spliced in turn into a C-region. Susumu argued the other side, that all this extra DNA would be removed as a mature gene was made. We bet a case of champagne on the outcome. Later that fall, when he had proved the structure of the mature gene (which had introns), he paid up.

When I came back to Harvard in the fall of 1977, I gave seminars on the intron-exon notion and gene-shuffling. We wrote up the Lambda light chain paper, and I put a short essay on this notion into it [Paper #13, this volume]. Then, I decided that the idea of gene shuffling should have broader exposure and pulled out this section to make a *Nature* News and Views piece [Paper #14, this volume]. My thought at that time was that the intron-exon structure of eukaryotic genes and the vastly higher rate of evolution of useful protein structures that would be created thereby, involving the manifold reuse of useful domains and sub-domains, would explain the Cambrian explosion. Ford Doolittle was having a sabbatical year in my laboratory. He realized that this high rate of evolution might be most useful for the first organisms, and he said so in response to my paper. I urged him to publish this, which he did (W. Ford Doolittle. "Genes in pieces: were they ever together?" *Nature* 272 (1978): 561–582). I thought then (and think now) that this was a most charming idea and adopted it in all my future work on evolution.

The level of DNA sequencing and the cloning of genes expanded greatly during the late 70s. Fred Sanger and I began to get awards together, such as the Lasker Award and the Gairdner Award. Some began to expect a Nobel Award. In 1980, I thought that if we were to get a Nobel, it would be in Physiology or Medicine. So when that award was announced in October to Snell, Benacerraf, and Dausset for the MHC discoveries, I assumed that nothing would happen that year. I remember talking to my friend, Jeremy Knowles, the night before the Chemistry Prize was to be announced. "Who do you think will win the Chemistry Award this year?" I asked him. He looked at me oddly, but we had a conversation about possible winners. The next morning I went to my lab early. As I got to my desk in my office, the phone rang. I picked it up, and it was a reporter. My first thought was "He wants me to comment on whoever has won the Nobel." Then I learned that I had won, with Fred Sanger and Paul Berg. The prize was given for the recombinant DNA technology. Paul stood in for all those who had moved DNA pieces from one organism to another and who had cloned DNA fragments in bacteria, while Fred and I had made it possible to understand the DNA fragments. Naturally, I was delighted. I went home to share the news with my family, including my father and mother, who were with us then. By the time I got there, they had learned the news on the radio.

I went back to the lab and celebrated with my group. We then organized a giant champagne party that night in Jeremy Knowles's house (mostly organized by Dianne Wirth, who was a postdoc in the lab), inviting people from all over Boston, including the doctor who had just done a heart bypass operation on my father.

I did not hear from the Nobel Committee until a telegram arrived three days later. In those days, the news was spread by the reporters; nowadays, the Nobel Foundation telephones the

Prizewinners. I took off with my wife to Nantucket, just to be alone for a few days. Then there was all the excitement in preparing for a trip to Stockholm.

The prize festival is totally delightful. It is the major event of the winter season. Arriving and being given a car and an escort for the week was thrilling. I took my two children, then college-age, and my father and mother, so it was a wonderful family trip. I was to give a special Nobel lecture [Paper #15, this volume], which was not just a look back on the reason for the prize — the discovery of a rapid sequencing method — but, in my case, since I was focused on the more recent ideas, a talk about gene structure and evolution.

The prize-awarding ceremony was most elegant. We were all splendid in rented tails and came out to the blare of trumpets. The prizes were given out by the King. Then, there was a dinner for a thousand people served in the town hall, followed by a ball. Later that night, after the first ball, there were balls thrown by the students (who had been ushering and helping, dressed in tails, with chests full of medals made of Coca-Cola bottle tops). The chemistry and physics laureates were inducted by the students into the Society of the Little Green Jumping Frog — and jumped accordingly.

Receiving a Nobel Prize can be quite devastating. Some people spend a year traveling and talking about the prize work. I did not take those invitations, but instead came back and tried to continue work in my lab at Harvard. However, I had been involved in starting a company, Biogen (which I shall describe in Section V), and in the summer of 1981, I became Chief Executive Officer of the company. I took leave from Harvard for two years and then resigned from there, but during this four-year period, I continued to have Harvard graduate students, working either in the lab at Harvard or, later, in a lab I maintained at Biogen. During this period, George Church developed genomic sequencing [Paper #16, this volume]. He discovered that one could immobilize the totality of a cell's DNA — cut with a restriction enzyme, broken by the chemical sequencing reactions, and separated by size on a gel — onto a nylon membrane and then reveal the sequence by probing the membrane with a radioactive oligonucleotide that hybridizes near the restriction cut. By probing with a different oligo, one sees more of the sequence. This not only shows the sequence from bacterial and mammalian cells without cloning, but also reveals methyl-Cs and can display protein–DNA contacts. We used this method to sequence a gene from *E. coli* [Paper #17, this volume] and, in the early 90s, tried to sequence an entire bacterium, *Mycoplasma capricolum*, which has a genome of 1,200 kb, as part of the genome program. I was very enthusiastic about trying first to sequence a bacterium and then a human by direct genomic sequencing (with possible automation), but, while we had made progress, the grant was not renewed, so we only, finally, after 2 years of actual sequencing, had 214 kb of bacterial sequence when we had to stop (P. Bork, C. Ouzounis, G. Casari, R. Schneider, M. Dolan, W. Gilbert, and P. Gillevet. "Exploring the *Mycoplasma capricolum* genome: a minimal cell reveals its physiology." *Molecular Microbiology* 16 (1995): 955–967).

When I left Biogen in 1985, it was about half a year before I was reappointed at Harvard, this time into the Cellular Biology Department. During this time, I went to a meeting organized by Robert Sinsheimer, whose purpose was to discuss sequencing the entire human genome. I thought the idea was absurd when I went to the meeting — three billion bases of sequence? Really? But as we talked about where the technology was in 1985, only a decade after our discoveries, I realized that it was only a question of industrial scale and that the problem was technologically doable. On the plane home, I wrote a short essay to Sinsheimer, outlining the structure of a 300-person institute, which, I suggested, could solve the problem in 30 years. I went to the Gordon Conference in 1986 and talked about the human genome, suggesting that the entire cost of getting a sequence would be about three billion dollars. (About this much was actually spent — about $300 million a year for 10 years. The Department of Energy got very interested early — they liked three billion dollar projects — but they thought it would be $3 billion a year. When they learned that it was only $3 billion total, they were less interested.) I became a strong supporter of the Human Genome Project [Paper #18, this volume]. I would give seminars, talking up the value of sequencing the human genome (and other comparison genomes) completely and accurately. I argued that the complete genome, created in an industrial-scale laboratory but placed into a reference database, would galvanize many aspects of the individual researchers' ability to study genes. I would argue that we could have the genome in 10 years, but it would take 100 years to fully understand it. I would say to the graduate students and postdocs, "We are going to break your rice bowl. Today, you can get a thesis by isolating and characterizing a new gene. Tomorrow, your students will just go to the Internet and look it up. They will have to do experiments on a scale you cannot conceive." And so it has come to pass.

Section V: Recombinant DNA

The idea of making new combinations of DNA molecules *in vitro* began to emerge in the early 70s. Restriction enzymes were discovered, and one could isolate DNA fragments from different organisms. Plasmids were found and transferred from one bacterium to another *in vitro*. They could be opened up with a restriction enzyme and a piece of DNA, isolated from another organism, and put into that gap — first by adding poly-A and poly-T tails to the different sides and letting those anneal together, and then, when Herbert Boyer discovered Eco R1, an enzyme that left a fourbase "sticky end" when it cut, by putting together any two Eco R1 ends to shuffle pieces of DNA.

At the Nucleic Acids Gordon Conference in 1973, we talked about those experiments and wondered if putting pieces of an animal tumor virus, SV40 (as Paul Berg was proposing), into *E. coli*, a bacterium that had been isolated from the human gut, would create organisms that might be harmful if they escaped the laboratory. A group of us, lead by David Baltimore, drafted a letter to the National Academy raising these issues about the recombinant DNA technology and calling for a review of this new field. Those discussions led to the Asilomar Meeting in early 1975, where a large review of this problem led to a call for national guidelines for such experiments, which were issued by the NIH in 1976. (These conversations led to the classification of experiments into low, medium, and high risk levels, with special laboratories prescribed for each. It was learned that the laboratory strains did not do well in nature and, furthermore, that special strains were available that could not live outside the laboratory at all. As time went on, it was learned that breaking the most virulent viruses into small pieces made work with them simpler and safer.)

I did not go to the Asilomar Conference, because I was at Harvard writing a grant against a deadline to ask for money to create a special laboratory in the Biological Laboratory building to work with animal viruses. That was the hot topic at that time, because animal viruses often caused not only disease but also, in some cases, tumors, and one hoped that that connection would cast light on the general problem of how cancer arises. A laboratory with special air handling and other features would be necessary for such work. By 1976, this special laboratory was also to be used to do recombinant DNA research, and our proposal led to wild discussions first at Harvard and then publicly in Cambridge.

At Harvard, Ruth Hubbard and George Wald opposed the construction of the laboratory, first in discussions within the departments and then before the administration. There, they lost the argument. I remember the dean of the faculty looking at Ruth and saying, "You say these experiments should not be done because they test the unknown. I thought the point of science was to explore the unknown."

The argument then spilled over into the city. There was a tempestuous public meeting in front of the city council in the summer of 1976, just as the NIH guidelines for recombinant DNA research were issued, in which scientists on both sides presented. The city council declared a moratorium on such research in Cambridge and set up a citizen's committee to look further into the matter. I remember, in that summer and fall, periods in which I, with the students and postdocs in the lab, would set up booths at city fairs to explain the science to the people and also do presentations in our laboratory to the citizen's committee. Ultimately, we were successful, and the city council resolved that recombinant DNA work could go on in Cambridge under the NIH guidelines.

All during this period in the early 70s, I was interested in this problem of finding animal genes and putting them into bacteria. A graduate student, Forest Fuller, as a first project, tried to make DNA from the messenger RNA of the hemoglobin genes of a mouse to put them into *E. coli*. He called this project, "Can I make bacteria bleed?" but he was not successful. Around 1975, I became interested in a more practical project: could one find the genes for insulin and, ultimately, figure out a way to make insulin in bacteria? I knew that the problem of making large amounts of insulin, then isolated from pig and cow pancreases, was a growing logistics problem and that the animal insulin was not precisely identical to the human insulin, which, in principle, could cause problems in treatment. Along with Argiris Efstradiatis and Lydia Villa-Komaroff, we started working on this problem, first trying to isolate the insulin messenger to use as a probe for the insulin gene DNA in clones. In late 1976, we learned that Bill Rutter's group at the University of California had scooped us. However, we continued the recombinant work.

One problem with all this work, trying to put DNA fragments into plasmids and inserting the plasmids into bacteria to grow pure clones containing those molecules, was that there were only limited ways of connecting the DNA pieces together. As I was thinking about this problem in 1975, I remembered that I had read a paper five years earlier that came from Khorana's laboratory with Vittorio Sgaramella as first author, in which it was mentioned that the phage T4 ligase would put together DNA fragments with blunt ends. (Both *E. coli* ligase and phage ligase would seal the chains of correctly matched sticky-end fragments, but only phage ligase was claimed to be able to fuse the flat ends of fully base-paired fragments together.) The claim had been thrown out in passing and not followed up on in the literature. I tried the experiment, putting DNA fragments together at room temperature with T4 ligase, and it worked dramatically. This was extraordinarily exciting. We could make arbitrary DNA combinations; we should have been able to put promoters in front of DNA copies of animal genes and make any protein in bacteria. I could not sleep for weeks. Keith Backman showed that we could make a portable promoter: a fragment with newly created Eco R1 ends, which could be easily attached to other genes [Paper #19, this volume]. During this period in the mid-70s, I traveled frequently to give seminars. In an hour, I would describe first the *lac* repressor and its contacts to DNA; then speak for 20 minutes on the chemical DNA sequencing; and then for another 20 minutes on the promise of recombinant DNA, the portable promoter, and the prospects for making useful proteins in bacteria. I spent about half my time traveling in those years. I had become an American Cancer Society Professor in Molecular Biology at Harvard (in an effort to bring in some soft money to allow for the expansion of the Biochemistry Department), and I was often referred to as the American Airlines Professor.

In the course of my time in Klaus Rajewsky's laboratory in Cologne, I learned that antibodies would bind to plastic sheeting. When I returned to Harvard with Stephanie Broome, we developed a way of detecting proteins made in colonies of bacteria, done by superimposing a sheet of plastic coated with antibodies against the protein of interest onto the colonies of bacteria growing on agar in a Petri dish. By treating the colonies to release their intracellular proteins, with a duplicate copy of the Petri dish colonies (made by replica plating) held in reserve, one could catch the molecules of interest and reveal them with a radioactive counter-antibody. Thus, one could develop a film that would show which colony was uniquely of interest among all the hundreds of colonies on the plate [Paper #20, this volume].

Our efforts to make insulin in bacteria were finally successful in the spring of 1978. I had met Bill Chick at the meeting on insulin that Lilly held in 1976, and through him, we had access to rat insulinoma tissue (tumors of the beta cells that produce insulin), from which we could get messenger RNA enriched in proinsulin messenger. (Proinsulin is the protein product of the gene for insulin, later cleaved by enzymes in the body to make the two-chain, mature, functional insulin.) We thought of making the proinsulin protein sequence as a fusion to a bacterial protein that had a hydrophobic leader sequence, which would serve to lead the protein out of the intracellular space, through the bacterial membrane, and into the space just inside the outer wall of the bacterium — the periplasmic space. We conjectured that this would protect the novel protein against destruction by proteases inside the cell, which seek and eliminate misfolded and broken proteins, and provide a partial purification. The antibody screening method worked, and we could isolate a clone that synthesized rat proinsulin [Paper #21, this volume]. This was the first eukaryotic protein made in bacteria (though this one was only folded correctly, as seen by antibodies — years later, we showed that we could demonstrate a biologically active protein). When we announced this result, it produced a firestorm of press coverage. I fled to a Gordon Conference and avoided a lot of the talking. We conceived that this method of secreting novel proteins into the periplasmic space would be generally useful and filed a patent on it.

Patenting was a new idea for molecular biologists in 1978. When we developed DNA sequencing in 1976, we did not patent the method. In those years, my attitude about patents would have been that scientific discoveries should be in the public domain. (This attitude was well expressed by the Curies, who did not patent the discovery of radium because of that belief. The companies that then began to manufacture radium for medical use did patent their methods, made fortunes, and forced the Curies to beg for samples of radium in order to continue their own research.) However, by 1978, the NIH had begun to appeal to the grantees to patent any useful discoveries and to have the inventors or the universities exploit the patents. These were IPAs, Institutional Patent Assignments — before this, all patents on grants were assigned to the government and generally not pursued. Harvard let chemists own their own patents but decided that the University would own all patents in biology and medicine. This new policy was finally adopted by congress in the Bayh-Dole Act of 1980. A technology-transfer office at Harvard came around to try to convince us that we should patent anything that looked interesting, so we complied and filed on the secretion of eukaryotic proteins in bacteria. I remember fruitless conversations, during which I and the tech-transfer office tried to interest companies in the patent in order to support research in my laboratory at Harvard. It never came to anything.

The secretion patent had an interesting life. Harvard licensed it to Biogen when we started the company. At a later time, the existence of the patent played a critical role in Schering's making a large investment in the nascent company: a scientist within Schering told me that he had told senior management that that patent offered an alternative way of making useful proteins in bacteria and would serve as a counterbalance to Genentech. They invested. Biogen never used the patent itself, but, long after my time there, they won a patent suit against another company, which had made a product by secretion, and collected $30 million in royalties, passing $3 million to Harvard.

My interest in making human proteins in bacteria led me to get involved in starting one of the first biotechnology companies. In 1978, two venture capitalists, Dan Adams and Ray Schaffer, came to see me. They were working for International Nickel (INCO), which, at that time, was a very large international mining company. They had made an investment in a new company, Genentech, and had hired as a consultant Phillip Sharp, who had advised on that investment. Through Phillip, they had come to me, after which they traveled around Europe to talk to other scientists in this new field of recombinant DNA. They then set up a meeting in Geneva of a group of nine scientists to talk about the possibilities of starting a company. The scientists were Ken Murray and Brian Hartley from England, Heinz Schaller and Peter Hofschneider from Germany, Phillipe Kourilsky from France, Charles Weissmann and Bernard Mach from Switzerland, and Phillip Sharp and myself from the United States.

As we talked, we found that we all thought the new technology had great promise in making human proteins of medical interest in bacteria, thus creating a source of new pharmaceutical products, but we were skeptical about a company and doubtful about the wisdom of commercial involvement. Should scientists get involved in companies? We were torn between our vision of ourselves as servants of only a notion of truth and the search for knowledge and, on the other side, the practical question of how that knowledge could be made relevant in the world and how drugs could be brought forward for human treatment. The venture capitalists did not really know anything practical about the formation and running of companies. Their conception was that Genentech looked interesting and that there might be room in the world for a second company in this field, so they looked to Europe to set up another recombinant DNA company. They thought of it as a patent-holding company (organized in the Netherlands Antilles, for tax reasons), and they had raised a total of $750,000, of which only $375,000 was in cash and the rest was in the promises of services in kind from INCO.

We had another meeting in Germany and then a third in Paris, at which we sent the venture-capitalists out of the room and had a serious discussion about a possible company. The VCs waited outside, hoping that at least one of us would go with the company so that their effort would not collapse. Inside the room, it turned out that eight of us would be a part of the company; Phillip Kourilsky was prevented from joining us by his home institution in Paris. The eight elected me chairman of our group, and we created the scientific board of Biogen. Since we distrusted the business people, we gave the scientific board strong powers: control of the distribution of the budget of the company and control of the stock offerings of the company. Furthermore, the scientific board would elect at least one-third of the board of directors. (Another third was from the venture capitalists, and the last was ultimately made up of representatives of strategic partners.)

We began by supporting research in the home laboratories of the scientific board members. Our first problem was to decide which commercial targets might be obtained. One possibility was asparaginase, which, isolated from bacteria, was then being tried as a treatment for cancer. I called up Merck to suggest modifying the bacteria to produce 10 to 100 times more protein, as in our efforts with the *lac* repressor, but the man in charge just laughed at me and said his real problem was how to get rid of tank cars of the ammonium sulfate solution used in the purification of asparaginase. (I did not realize at that time the difficulty in changing any purification of a drug during clinical trials.) Charles Weissmann wanted to work on alpha-interferon — the human molecule might have been uniquely useful in fighting viral infections, since only the human variant would work in people. Money from Biogen made it possible for Charles to fly to Finland, to the one laboratory, Kari Cantell's, that was making human interferon from human cells. He obtained RNA, flew to Germany to use a laboratory that would let him attempt to clone human material, and then finally got to work in his laboratory in Zurich. I had a telephone line put into my office at Harvard so that I could talk freely about the Biogen projects without incurring any charges to grants or the University. Ken Murray and Heinz Schaller began to work on the hepatitis B virus to make vaccines from virus protein, made both in bacteria and in yeast — the yeast patents were critical to Biogen in later years.

As chairman of the scientists, I found myself playing a more and more critical role in the company. Dan Adams was chair of the Board of Directors and CEO. Dan, Ray, and I went to raise money for the company. I still remember how hard it was to raise money from European sources. When we proposed an investment to the head of a large European company, he said, "What do you mean, take a 10% interest in your company? If we like it, we buy it!" However, we were more successful with American companies. We convinced Schering to make a strategic investment of $8 million in the company, at a valuation of $32 million, in return for a board seat and an agreement to develop three products through the cloning and expression. The leading product was to be alpha-interferon. At that time, INCO invested another $1.25 million, and we began to set up our own laboratories, initially in Geneva, Switzerland. Phil Sharp and I then convinced Monsanto to invest $20 million as a partner in chemistry and agriculture. A British hotel chain invested $10 million, and a private placement in 1981 brought in another $20 million.

The scientific board and the board of directors met frequently. At the first board of directors meeting, Dan Adams had brought Robert Swanson, the CEO of Genentech, and wanted him to be a director. I thought this most inappropriate, since they would be our main industrial competitor in the recombinant DNA field, and convinced the board that he should not join it. In January 1979, we had a joint meeting of both boards in Martinique to celebrate our first (partial) year. I remember Dan Adams saying to me, probably as we were drinking together, "You and I are the two indispensable men in this company!" A dangerous and always wrong thought. Soon after that, the CEO of INCO called me down to his office in New York City to show me the reasons that they were going to fire Dan Adams, and I, returning to Cambridge, then summoned Dan to my office at Harvard and fired him from Biogen. I found myself the last man standing as acting head of Biogen, but I was able to negotiate and complete the Schering investment. INCO wanted to take over Biogen completely, but I refused and kept the company independent. Now that we had some serious money, we hired Robert Cawthorne as CEO and charged him with creating company laboratories and staff in Geneva. I became the chair of the board of directors (along with a co-chair, Robin Nicholson, from INCO; the business people did not really trust me, as a scientist, to function alone).

In January 1980, at our joint meeting in Guadeloupe, I was delighted to learn that Charles Weissmann had been successful in cloning and expressing human alpha-interferon. I had been traveling in Japan over Christmas and the New Year (my wife and I had spent New Year's Eve on a mountain above Kyoto) and was not aware of Charles's progress over the Christmas period — finding the gene, discovering how to make the protein, and filing critical patents. Excited, wild discussions ensued about how to announce this discovery. Charles wanted to announce it in a scientific context, not just a news conference. Phillip Sharp set up a public seminar for Charles at MIT, and we chartered a small plane to fly Phillip, Charles, my wife, and me to the mainland to make connections that would get us to Boston, so that Charles could give his seminar and we could have a news conference (on a secret topic TO BE ANNOUNCED). After the seminar, which was very exciting, we went to a hotel for the news conference. Charles tried to explain to the reporters why the work was important. Seeing that he was having trouble, I took the microphone from him in order to summarize his work. Charles's discovery was on the front page of the *New York Times* but, unfortunately, illustrated with my photograph.

Why was this discovery so exciting? Only the human form of the protein will work in humans, meaning that the cloning gave the promise of making large amounts of the protein for medical use. At that moment, not only was there the promise of treating human viral infections, but also striking, anecdotal stories of dramatic effects of interferon on cancers. This protein held a "miracle drug" status. Interferon ultimately became a billion-dollar-a-year drug for Schering.

Now, Charles had to make industrial amounts of pure material. We set up a subsidiary Biogen laboratory for Charles in Zurich. He rented giant, 50,000-liter fermenters in Kundl in Austria to grow the *E. coli* expressing interferon, purified the protein, and, finally, took the pure material himself to Schering in New Jersey to be used in the clinical trials. Charles carried on a great medical tradition. He had made a drug for human use and tried it on himself before giving it to any patient! He injected himself with a massive dose and found himself lying on the floor, gasping and suffering from flu-like symptoms. But he recovered and gave the material to be used. He did not tell anyone about this experience at that time, only bringing it up years later.

Schering was slow to get the material into the hands of the doctors. Robin Nicholson and I went to see the head of all clinical work at Schering to convince him to start trials, telling him that the doctors in the field were eager to try the drug. (We knew that from Charles's contacts. Schering, however, had not asked them.) The head was shocked that we knew this sort of information. His view was that knowledge should only flow down the chain of command. However, after that, they slowly began to do trials. The questions of how to use interferon and what disease it might affect were very difficult. In fact, a lot of the early trials were wrong. A claimed dramatic effect on genital herpes turned out to be bias on the part of the investigators. A large trial on the treatment of Hepatitis B virus failed to show any efficacy (another wrong result — a later repeat showed that alpha-interferon worked on this disease, and this ultimately became the major use of interferon in Japan and the Far East). Finally, Schering showed that alpha-interferon cured hairy-cell leukemia and got its first registration. It was a very minor use, but once it was licensed, the uses of the drug expanded, and it became a major pharmaceutical until other, better treatments finally came in. All during this time, I was talking daily by phone to the head of the interferon clinical group at Schering, eager for results.

By the summer of 1981, Biogen was working on many projects in Geneva and moving to set up laboratories in Cambridge in the United States. (We picked Cambridge because the discussions of 1976 meant that that city had regulations in place to govern recombinant DNA work, and we knew we would not have to educate the city government from scratch.) I thought that the company was moving too slowly and proposed to the board of directors that I come into the company as CEO. Their first response was no. So I told them I would leave the company. They then said yes, and I became CEO, over the head of Rob Cawthorne and Bob Fildes, who had come to run the Cambridge wing of the company. (I took leave from Harvard to run the company, and then, when the two years of leave they granted were up, I resigned from Harvard. All during this period, I had graduate students and postdocs, working first at Harvard and then in a research laboratory I maintained at the company in Cambridge.)

As an academic, my first approach to becoming CEO was to try to read up on that job, thinking there might be guidance through books on business. I quickly learned that there was every conceivable approach and type of person playing that role in different companies. The essential qualities for a start-up CEO were the entrepreneurial drive to make things happen, a willingness to do anything needed, wild impatience, and the ability to make decisions and take responsibility. I found many similarities to running a large research laboratory at the university and running the company: both had to be financed, both had to achieve goals, and both involved getting people to perform. But there was one major difference. As a scientist, I could wait until all the evidence was in before deciding on publishing. As a business person, I had to make decisions based on incomplete evidence, since timely decisions were essential. The company burns money: every extra day is wildly expensive, so quick action is needed. Even a wrong decision, if quickly changed, is better than endless waiting for the complete story.

In any case, I enjoyed that role — the one truly free role in a company (although one serves at the behest of a board of directors). I would not have done it in a non-technical company, but I found I was interested in the science, the patent law, the business deals, and the growth of the company.

I kept my family in Cambridge and set up living quarters at the Hotel de la Paix in Geneva. They packed up my clothes when I was not there, and we had a telephone line put into my favorite room, overlooking the lake. I would get on the plane at Logan, carrying a paperback to read, and get off in Geneva to spend a week before returning. I spent about ten days each month in Geneva for four years, living on international time. European and American company customs are very different — we supplied cars to management in Europe, not in America. In Geneva, I had a company Jaguar, which spent most of the time being repaired while I was not there. In Cambridge, I had my own, old car. In Geneva, the chief financial officer would smash up his company car every year; in Cambridge, his equivalent was far more sober.

The company grew. Since my coming in blocked the two people who were running the two sides, they moved on. Robert Cawthorne left soon to be CEO of Rorer in Philadelphia and later to become CEO of Rhône-Poulenc Rorer. Robert Fildes left to become CEO of Cetus. I brought Julien Davies in to run the laboratory in Geneva and Richard Flavell to run the Cambridge one. Mark Skaletsky became COO. I grew the company from about 40 people to about 400.

In 1983, I took the company public. Genentech had gone public very dramatically in 1981. Biogen was too young then, but in 1983, at the next wave of public offerings, we convinced a group of banks to do an IPO. I did the road show alone, giving 70 talks over a three-week period across Europe and the United States. The structure of public offerings was different then. One presented to large groups of possible investors. I spoke to 600 people at a lunch in London, several hundred in Paris, and groups all across the United States, repeating the same pitch over and over again. We were successful and raised about 53 million dollars for the company, on a valuation of around $500 million. I was presented with an actual check.

Ten years later, when I was involved in the IPO for Myriad, the Biogen IPO was still one of the largest in Biotechnology. By that time, however, the structure of IPOs had changed radically. Now they were dominated by institutional investors who wanted to see the company alone, with no one else present to hear their questions. In that IPO, one went from group to group, facing three or four people. There would be 20 minutes of talk, 20 minutes for questions, and 20 minutes to get to the next institution.

Although our IPO was successful, it was at the top of the market. A computer company IPO just the day before had failed, and the market began to drop as we went out. Our stock price gradually fell over the next year, and the board of directors got more and more restive, finally deciding to fire me as CEO. I took a six-month leave from the board while they brought in Jim Vincent as the new CEO, and then I stayed on the Biogen board for the next seven years, until I became involved in the starting of Myriad. Jim Vincent took the attitude that you were either completely with him or completely against him, so I left Biogen completely.

I was reappointed at Harvard in 1985, now in the Cellular and Developmental Biology department, and brought people back with me to Harvard to set up a new laboratory, working on many topics, including neurobiology, the genome project, and the exon theory of genes.

Section VI: A Variety of Topics

My laboratory worked on a large number of topics over the years. The graduate students, for the most part, had independent topics, on which they worked alone, toward their theses. Sometimes a postdoc would appear, and a novel idea would emerge. Or I would travel.

I was also very interested in immunology. In 1977, I spent part of a sabbatical year in Klaus Rajewsky's laboratory in Cologne, performing an experiment in immunology of a very peculiar kind. We were trying to test Niels Jerne's idea that the pattern of expression of the various antibodies directed at an antigen was maintained by a network of antibody/anti-antibody interactions. The hypothesis stated that there was effectively an image of the shape of the antigen maintained within the animal in the form of a set of anti-antibodies — anti-ideotype antibodies — that as a whole might mimic the antigen and help maintain the set of antibodies that countered the antigen. Karen Fischer-Lindahl and I did experiments to test this "network theory" by sewing mice together — parabiosing — so that they would share their circulation and immune cells in an attempt to see if, when later they are separated, the circulating antibodies would dominate the new response to antigen. The experiments were not really successful [Paper #23, this volume], but working with mice was a great variation from the microorganisms I had been handling. This experience in immunology led to the immune detection methods we used in the insulin search.

When I came back to Harvard after Biogen, I was interested in neurobiology, and Didier Stanier was trying to find monoclonal antibodies that would recognize patterns of markers on mouse brain cells. We were seeking some pattern of cell-specific addressing, but we found a marker that lit up only a few neurons in the whole brain [Papers #25 and #27, this volume]. We were interested in the genome project, as well as evolution and genetics in general. Paper #22 studies the mechanisms by which mutations arise. Paper #24 shows that a vertebrate enzyme is so well conserved that it will function in bacteria. Paper #26 derives a useful technique in recombinant DNA research. Laura Landweber, in coming to the laboratory, wanted to do an entirely unique project using novel organisms totally foreign to the laboratory and worked on RNA editing. One result is in Paper #28.

Section VII: G4 DNA

We were totally surprised by Dipankar Sen's finding that DNA oligonucleotides with runs of guanines could bind together to produce four-stranded parallel structures, held together by Hoogsteen pairing of the guanines [Paper #29, this volume]. He first saw strangely moving bands on gel electrophoresis, but through careful reasoning and the addition of complementary strands, we deduced that these indeed were four-stranded complexes, and methylation protection experiments showed that the strands ran in parallel. (Many other ways in which guanine runs could quadruplex together — anti-parallel or parallel, intra- or inter-molecule — were later discovered.) This G4 (guanine quadruplex) DNA was unusually stabilized by a potassium ion, now not melting even if boiled. We conjectured that this way of aligning four DNA double helices must play a role in meiosis, where four copies of a chromosome bind together. However, this idea is still unproven.

Dipankar went on to characterize more fully this DNA form, showing the various ionic involvements with these structures as they formed differently in different ions. In later years, my laboratory found proteins that bound to such G4 regions and characterized one both as a binder and as an enzyme that could cut near such a region. That was identified as the product of a yeast gene needed for progression through meiosis [Papers #30 and #31, this volume].

Section VIII: The RNA World

As a genetically oriented molecular biologist, my thoughts about enzymology in the 80s would have been that proteins were the enzymes, since they have shapes that bind to the transition state of a reaction and hence strain the constituents to lower the free energy barrier between the initial and final states. I had not thought about the problem of how life began, although as a convinced evolutionist, I would have thought of a single common ancestor arising in a puddle somewhere. Our knowledge of prebiotic chemistry would have been the Miller experiment, suggesting that organic compounds could arise on an early earth or come in from meteor strikes. The thought was that comets might be 1 molar in formaldehyde and contain all sorts of other chemical products at high concentrations, and as such, the world's water came in from an early cometary bombardment.

In the early 80s, two types of RNA enzymes were discovered: the RNase P, which cleaved and matured tRNA, and a self-splicing intron in a ribosomal RNA, which could remove itself from the RNA without the aid of a protein. There was a general discussion about the implications of the existence of these enzymes. Philip Sharp, in 1985, wrote about all the types of introns and discussed the possibility that there might have been early RNA introns (before the separation of fungi, plants, and animals) but gave equal weight to the thought that the introns could have been added later. Frank Westheimer prepared a News and Views article for *Nature*, talking up the chemistry of the two RNA enzymes, and I recall seeing his paper in the manuscript and discussing it with him. That inspired me to conjecture, in a clear fashion, that all life might have begun with RNA, with no protein-based enzymes. I also gave a name to that theory: the RNA world [Paper #32, this volume].

The essential idea of the RNA world is that RNA molecules might be able to perform all of the types of enzymatic activities needed to support their replication, and a self-replicating system (RNA-based RNA replication of the RNA enzyme that can copy its own type) would be able to undergo mutation and Darwinian evolution. Variant organisms will multiply and compete, and so ever more effective variations will overtake the others and fix in the population. My excitement about this idea was also couched in my belief that the intron structure of genes led to exon shuffling and thus enhanced evolutionary rates, and the RNA world paper is centered about the idea that self-splicing introns would recombine RNA pieces and permit complex RNA molecules to be built up of small exonic parts, which in turn could be shuffled into new patterns by the RNA introns acting as transposons.

This is picture of a complete RNA chemistry at the beginning, which then leads later to RNA-coded and -catalyzed peptide formation to create protein fragments that serve as supports of ribozyme activity, which then leads to the synthesis of whole proteins with enhanced enzymatic activity. In this outline of evolution, DNA appears late to provide a more stable and longer-lasting storage molecule. The signs of this history remained, at that time, in the RNA cofactors still essential for some enzymes, in the fact of RNA precursors for the DNA nucleotides, and in the RNA initiation of the Okazaki fragments that are strung together in the second-strand synthesis displayed in all double-stranded DNA.

Dan Jay, then a Junior Fellow at Harvard, in the course of our many conversations over the dinners at the Society of Fellows that year, came up with the idea that one early role of peptides might be charge-neutralization by basic peptides of the acidic RNA, allowing nucleic acids to be more easily encapsulated within lipid membranes to make the early cells. That led him to do experiments to show that the addition of such basic proteins led to dramatic increases in the rate of packaging of the nucleic acids within the lipid membrane [Paper #33, this volume]. This was a most pleasing working-out of a theoretical conception.

How have these ideas fared over the last 30 years? The most dramatic finding was that the active center of the ribosome, which catalyzes the formation of the peptide bond, is a pure RNA structure (H. F. Noller, V. Hoffarth, and L. Zimniak. "Unusual resistance of peptidyl transferase to protein extraction procedures." *Science* 256 (1992): 1416–1419. There is a full review of the ribozyme structure studies: P. B. Moore and T. A. Steitz. "The roles of RNA in the synthesis of protein." *Cold Spring Harb Perspect Biol.* 3, 11 (2011))!

I thought originally that many more RNA enzymes would be found *in vivo*. While that has not come to pass, there have been many RNA enzymes synthesized *in vitro*, and a whole world of riboswitches (RNA folds in the beginnings of messenger RNAs that bind small molecules and affect the rate of translation), which may have evolved from RNA forms that could have bound small molecules in the RNA world, has been discovered (R. R. Breaker. "Riboswitches and the RNA world." *Cold Spring Harb Perspect Biol.* 4, 2 (2012)). Recently, a universe of non-coding RNAs, both small and large, has begun to be explored. The size of it rivals that of the protein universe in higher organisms. So far, these non-coding RNAs seem to have a role in modifying the expression pattern of genes in a pleiotropic way.

I had hoped that there would be an RNA-based RNA polymerase being studied in the laboratory by this time. Such a true replication function has not yet been found, but the longest RNA-based read has now gotten to 206 templated nucleotides (L. L. Martin, P. J. Unrau, and U. F. Müller. "RNA synthesis by *in vitro* selected ribozymes for recreating an RNA world." *Life* 5, 1 (2015): 247–268).

There has been a great deal of progress in the exploration of the prebiotic synthesis of organic molecules. At the time of my original paper, one's thinking was only of the Miller experiment, which suggested that amino acids might be made by lightning in a primitive earth atmosphere. Since then, other work, first that of Eschenmoser and then particularly the extensive studies of Southerland, has shown that chemical pathways can lead to biologically relevant chemical entities.

Section IX: Introns, Exons, and Gene Evolution

This section describes the long march of arguments over whether the first genes might have had an intron-exon structure, having been assembled from small exons, representing units of structure or function, by combination and shuffling at a distance represented by the introns. I realized early that the existence of introns, moving the exons hundreds or thousands of bases apart along the DNA (the longest is 50,000 bases) would increase the rate of homologous recombination between adventitious mutations in different exons a hundred- to a thousand-fold and would increase the rate of illegitimate recombination, connecting the exons from different genes, some hundred thousand- to multi-million-fold. However, beyond this ability to enhance evolution in the higher organisms, where the introns were first discovered, arose the possibility that this enhanced evolution went all the way back to the first genes and led to the common ancestor, a suggestion of Ford Doolittle's that I adopted and ran with.

The idea that introns were very early met a great deal of resistance over the years and often sparked very heated debate. People were used to the idea that bacteria looked like primitive organisms and hence that their genes must have been primitive. Furthermore, the laboratory strains of yeast were thought to be a standard for the eukaryotes, and their genes mostly had no introns except for a few near the 5′ ends. Therefore, many people believed the introns were added late.

The alternative picture is that bacteria are super highly evolved organisms. Their short generation time means that they have had very many more generations than the eukaryotes. According to this picture, they lost all their introns through specializing in fast DNA replication long before the eukaryotes appeared.

It took many years before everyone realized that the laboratory species of yeast was not typical and that it had undergone a genome duplication, followed by major intron loss compared to other yeast species, which were fully endowed with introns. In fact, the mechanism of intron loss, the recombination with DNA copies of spliced mRNA, most likely involves often partial copying of the mRNA into DNA, which are 3′ fragments. Since the DNA copying will begin at the poly-A tail, the introns are lost preferentially from the 3′ end of the gene.

After the "Genes in pieces" essay [Paper #14, this volume], the "Playgrounds of evolution" paper [Paper #34, this volume] sketched the idea that exons represented elements of protein structure. When we isolated the preproinsulin genes from the rat, we found two genes, one with two introns and one with only one [Paper #35, this volume]. Both genes had introns in the 5′ noncoding region, but only one had an intron in the C-peptide region. We estimated the time of this gene duplication from the rate of change of synonymous (silent) nucleotide sites as having taken place some 20–30 million years ago, long after the mammalian radiation. This discovery posed an evolutionary question: was the intron dividing the insulin-coding region lost or gained?

A year later, we answered this question by isolating the preproinsulin gene from chicken. The chicken would have diverged from the mammalian ancestor perhaps 350 million years ago, so this is an ancient comparison. That gene had two introns at the same position as those in the mammalian gene, and hence the rat duplication had lost an intron [Paper #36, this volume]. (This paper goes on to discuss the evolutionary clock by studying silent and replacement changes and argues for a selection basis for the observed rate of accumulation of changes.)

The idea of ancient exon shuffling as a way of making genes suggests that the exons should bear useful fragments of proteins — useful in the sense that they could appear in novel combinations. A possible form for this ancient exon product was suggested by Mitiko Go (Mitiko Go. "Correlation of DNA exonic regions with protein structural units in haemoglobin." *Nature* 291 (1981): 90–92), arguing that the ur-exon products were "modules," domains of polypeptide chain that could be circumscribed by a sphere 28 Å in diameter. Using this notion, she predicted that the gene for globin, which was known to have two introns, should also have a third intron to break the polypeptide chain into four modules. (I was sent her manuscript as a referee for *Nature*. I recommended prompt publication. I knew at that time that a group had found an intron in the place she predicted in soybean leghemoglobin, on which they later published (E. O. Jensen, K. Paludan, J. J. Hyldig-Nielsen, P. Jørgensen, and K. A. Marcker. "The structure of a chromosomal leghaemoglobin gene from soybean." *Nature* 291 (1981): 677–679).) Such findings showed that introns had to precede the plant–animal divergence, pushing them back more than a billion years.

Along these lines, we pursued the structure of the basic enzymes of biochemistry conserved across all organisms. The conserved structure in bacteria had no introns. The conserved gene in plants and animals had many introns. These fell on the boundaries of the Go modules [Papers #39–41, this volume], especially if one added in the introns in different copies of the genes in distant species.

This work continued as more and more data accumulated [Papers #42–53, this volume]. These papers continue to explore the idea that the introns were early, that there was really a striking correlation between introns and the Go modules, and that, while there was some gain of introns, the major pattern of eukaryotic evolution was one of loss of introns.

Section X: Paradigm Shift and Computing

Paper #54 is titled with the cry, “Molecular Biology is Dead! — Long Live Molecular Biology!” In 1991, it was a reflection on how the molecular/DNA approach had taken over all of biology, hence causing molecular biology to disappear as a separate field. Furthermore, it prefigures and predicts the constant and ever rapider change in technology that drives science forward. The first human genome sequence took several years of actual work. The third generation of machines and novel sequencing techniques can now do it in half an hour. Single-cell sequencing to examine the spectrum of RNA molecules begins to allow one to study today a thousand, tomorrow a million different cell types in the brain. The cry of science is “What is new?” “What have we found that we did not expect?”

Paper #55 demonstrates that there is much hidden information in the DNA databases. As they grow large, comparisons of different organisms reveal novel proteins. Computational biology begins to make novel biological discoveries.

Section XI: Physics and My Early Life

I was born on March 21, 1932, in Richardson House, which was the private side of the Boston Lying-In Hospital. My father was an instructor and later an assistant professor of Economics at Harvard. My mother had gone to Radcliffe. She married my father in front of a justice of the peace on her way up to college and so arrived at Radcliffe as that strange, dangerous creature — a married woman. This was 1923. My father was still in college then. She later shared stories of the dean taking her aside to pass on words of wisdom, such as "Never leave your husband alone with a serving woman" (not realizing that, as poor students, they did not have servants).

My parents had met when he was 13. She was living in an anarchist colony in Stelton, New Jersey, and he was bicycling around giving radical speeches. She was going to a Montessori school at the colony then, but she went on to high school in New Jersey and then to Radcliffe. She majored in Psychology and went on to graduate school at Harvard in Psychology, working toward a Ph.D. (In those years, the degree was rare. That department gave out a Ph.D. about every other year.) However, something went wrong with the subject she was researching, and she ceased being a student. She had me 20 months later, followed by my sister. We were brought up as the children of a psychologist, taking intelligence tests every year, as a game, as our mother followed our development. To some extent, she encouraged us into different directions, me into science and my sister into art and humanities. (Her thought was to prevent competition between her children, and this succeeded well — however, at the end of college, after majoring in History and Literature at Radcliffe, my sister decided to become a doctor, and after an extra year taking the pre-medical requisites, she went to medical school and has since had a long and fruitful career in medicine.)

We lived in Cambridge and Watertown while my father was an instructor at Harvard. Being one of the first Keynesian economists in the country, his lectures on Economics were standing-room-only as he explained the depression and what could be done about it. My mother thought the public schools were very bad, so she home-schooled us, as Massachusetts law at that time permitted anyone with a master's degree to teach. We learned to read very early and would go to the library every few days to come home with a great pile of books. When I went to Washington, D.C., at the age of seven, I was frustrated by the Public Library not allowing me to take out books from the adult section. I had read all the children's books by that time and had moved on.

We went to Washington, D.C., in September of 1939. On the train trip down, we spent the night in New York City with my father's mother. The radio news was filled with Hitler's march into Poland. My father had left Harvard to go into the Roosevelt Administration, going first as part of a group of economic experts called the "Harry Hopkins Brain Trust." Then, during the war, he became chief economist for the Office of Price Administration, involved in pushing government policy to force industry to amplify its wartime efforts and writing speeches for Roosevelt. The massive industrial expansion that produced the planes, ships, and tanks needed to fight the war was driven by the government — often despite the complaints from industry, which was more inclined toward business as usual rather than the 100-fold increase in production needed. We lived in D.C. for four years, then moved to Virginia, and then went back to D.C.

My memory is that I may have had a chemistry set before I was seven. By the time I was 12, I was reading my uncle's college text on chemistry (probably dating from the 20s) and using my allowance and money I made from a newspaper route to buy chemicals for a laboratory at home. In those days, I could buy, by mail, concentrated acids and other chemicals from the Fisher catalog. I had an accident one day when I was trying to make hydrogen gas from zinc and HCl. I was in my little laboratory, near our kitchen in a house in Virginia, and my mother was sitting across the room. I added the acid to the zinc metal in a small Erlenmeyer flask, stoppered with a glass jet, and after the bubbling began, I tried to light the gas coming from the jet. The flask exploded. My mother took me to the local hospital — I had a cut on my left wrist that would require several stitches. I was unnaturally silent as we rode to the hospital and during the procedure — since I was in deep thought. Finally, I announced, "I know what I did wrong." I still have that scar on my wrist as a permanent admonition to plan better. I should have let the gas bubble through a water trough, which would have allowed me to test that all the oxygen had been flushed from the flask by igniting the bubbles as they emerged from the water.

We lived in Virginia during the war, from 1943 to 1947, and I went to junior high school there. The high schools there were pretty bad, so I wanted to go to school in Washington, D.C. However, eighth grade in D.C. required Latin, and the Virginia schools had not taught Latin yet, so that was out. As a result, I went to Sidwell Friends School in Washington, D.C., which in 1945 was a private school that did not require Latin. I did French and German there as languages and took whatever science courses they offered. The atomic bomb was dropped in the summer of 1945, and I read the Smyth report and sought more information about nuclear physics and relativity. George Gamow's books *Mr. Tompkins in Wonderland* and *Mr. Tompkins Explores the Atom* came out around then. During my later years in high school, I would play hooky and go down to the Library of Congress to sit and read all day — a wonderful reading room and fantastic access to all kinds of literature. (The Library of Congress gets two copies of each copyrighted book or manuscript, which were available on call in the reading room.) I would read about nuclear energy, Van De Graaff generators, X-ray machines. I wrote a thesis on Eugene O'Neill and read his yet unpublished plays in manuscript form in the Library of Congress. My grades were fine, but the school finally had to send a letter to my mother — I was away too much. Would I please attend more classes?

I went to Harvard, entering in 1949, going up from D.C. to Boston by train, and finding my room in a dorm in the Yard. The first thing we had to do was to open a local bank account and hire a bedder (a woman who was employed to make the beds of the Harvard students). We had to find our courses. Although I had come to the university thinking I would be a chemist, I took the basic physics course as a freshman and decided to become a Chemistry and Physics major, since a combined major was offered. I found inorganic analysis terribly boring, and the organic chemistry was being taught in the last days of Louis Feiser (his last class) and so had not been modernized (the next year the class was entirely up to date). By that time, I was more interested in Physics and was taking the graduate courses before I graduated. I spent a year at Harvard in the Graduate School and then went to Cambridge in 1954.

Through Celia's father, I visited Einstein once in the summer of 1954 before leaving for England. We had a long and wonderful conversation — which I only vaguely remember. There just had been a paper published suggesting that a light–light interaction would shift the energy of star-light and produce a red shift (denying the velocity interpretation). I remember that Einstein gave me a simple, brilliant argument as to why the paper had to be wrong. Sadly, I do not remember the details.

In the summer of 1955, we went down during the vacation to go to meetings in Italy and Switzerland. We went to a meeting on general relativity in Berne, Switzerland. When we got there, we were refused entrance, but Celia sweet-talked the secretary into letting this graduate student in. Here is a picture of us at one of the dinners:

I am standing at the left. The woman next to me is Celia, my wife. The man next, looking out, is Otto Nathan, an economist and a long-time friend of Einstein, as well as the executor of his estate. The man next to him is Wolfgang Pauli, the great physicist.

At Cambridge, working with Abdus Salam, we tried to understand the interaction of the Pi mesons (thought of then as providing the forces between the nucleons). The methods of quantum electrodynamics, which depended upon the weakness of the electromagnetic coupling so that a perturbation expansion was useful, were not working for a field theory of the mesons and nucleons, since their interaction was very strong. We became interested in the idea of trying to predict aspects of the observations that could be deduced from very general physical principles, such as causality.

This approach was called a bootstrap. My first publication, with Abdus [Paper #56, this volume], was to explore how such a notion of causality (that physical interactions could not propagate faster than light, which in this case meant that the commutator of field variables had to vanish outside the light cone) could be used to show that the amplitude for the scattering of particles (the S-matrix) had a certain analytic form, and so that the general function could be calculated by knowing only a physically measurable part. I was naturally proud of this paper, but the publication process in those days was very slow. The European theoretical physicists published often in an Italian journal, *Nuovo Cimento*, which would hold the paper about two years before it appeared.

My first independent paper [Paper #57, this volume] used these ideas directly on the pion–nucleon problem. When one scattered pions on nucleons (pions on protons in the hydrogen nucleus), one saw a large bump as the energy increased, producing a peak in the scattering in the second angular momentum state, the p-wave state. (This was called a resonance, later recognized as corresponding to the presence of a new particle.) At the lowest energy of the scattering, all the effects are in the spherically symmetrical state, the s-wave state. I was able to show that the s-wave scattering could be thought of as a necessary relativistic correction to the dominant p-wave resonance by calculating the strength of the s-wave scattering, using only the presence of the p-wave resonance in the forward scattering cross-section. Furthermore, I could use the forward scattering cross-section measurements to calculate an accurate value for the strength of the pion–nucleon interaction, the coupling constant in such a theory. This paper showed how the experimental data hung together, one part of the data predicting another part.

I wrote the last paper included here [Paper #58, this volume] with Stanley Deser and George Sudarshan while I was a postdoc. It is an effort to extend these notions of restricting the structure of the amplitude describing the interaction of three particles.

As an Assistant Professor of Physics, I continued to publish in Physics and had graduate students do Ph.D.s with me, but my interests changed to Biology.

Part Three

Memories of the Gilbert Lab Alumni

In Praise of Wally Gilbert: If You Start Right Now

Benno Muller-Hill*

In the autumn of 1964, I met Walter Gilbert for the first time. I was a postdoc in Howard Rickenberg's lab in Bloomington, Indiana, and my time was running out. At the International Congress of Biochemistry in New York, I had asked Jim Watson if it would be possible for me to work with him. I told him that I had tried and failed to isolate the lac repressor. Jim told me that he had no such position but that Wally Gilbert, whose name I had never heard, might have an opportunity for me. Wally called me, and I came to Cambridge for an interview. I told him that I had worked on the specificity of the lac repressor and failed to isolate it. This, I learned, was a problem we both wanted to solve. Wally offered me the position.

When I came to the Bio Labs in the spring of 1965, I entered an amazing world. There were about ten graduate students and one postdoc in the Watson-Gilbert lab, all working on different and remarkable problems.

Jim was rarely present. He was writing his book at home. To speak with him was not easy. Wally, by contrast, worked twelve hours a day in the lab. He arrived between eleven and noon and left around midnight. He always had time for extensive discussions and loved to discuss the logic of experiments. "See an effect, push an effect" was one of his favorite adages. He is one of the very few people I have ever met in science with no fear of trying a new method or a machine he had never used before. He simply loved learning.

Experiments, once discussed, had to be done. "If you start right now," he would say, "you will get the bacterial culture around midnight. Sonify it right away and put it on the column, you will have the fractions the next morning," and so on. It took me years until I discovered that Wally's statement is a version of Hillel's dictum: "If not now, when?"

Wally followed Jim's example by choosing to be the co-author of his collaborators' work only when he had performed some of the experiments. For example, he suggested that I isolate a non-sense mutation of the lacI gene and thereby prove that the lac repressor is a protein. Although I took his advice, he refused to take credit in the form of co-authorship. Even more generously, he made me a co-author of his paper in which he showed that the lac repressor bound to the lac operator. I had spent so much time in vain trying to repress *in vitro* the transcription of the lac

* Institute of Genetics, University of Cologne, Cologne, Germany

promoter — an experiment which could not have worked, since it would have needed the addition of CAP/CRP protein.

Wally worked day and night, and yet he was devoted to his family of three children, one of whom died of cancer. Celia, his wife, was capable of the strength needed to go through such tragedy. It was through both Wally and Celia that I met Celia's father, I. F. Stone, the journalist who wrote and published *I. F. Stone's Weekly*. He was a most admirable person, an activist against the Vietnam War and a vocal democrat who later wrote a momentous book about the trial of Socrates. I encountered not only science but also American culture through Wally.

Our connection did not end when I left the Bio Labs and went to Cologne in the spring of 1968. Wally often came to Cologne. He stayed in our apartment for weeks when doing experiments in Klaus Rajewsky's lab. Year after year, he came to the Cologne Spring Meeting. My wife Rita and I were guests at Wally and Celia's house in Cambridge. I am grateful for all that. My continuing wish is that now, when all experiments are done, the conversation will go on.

Mentor, Colleague, Friend

Lydia Villa-Komaroff*

The Insulin Cloning Project

I had been doing experiments on Wally Gilbert's insulin cloning project for several weeks before I first met the man himself. I was a postdoctoral student in Fotis Kafatos' lab and had recently returned to the Harvard BioLabs, after a dismal and unproductive year-long stint at Cold Spring Harbor Laboratory (CSHL). Argiris "Arg" Efstratiadis, my lab-mate and colleague in Fotis Kafatos' lab, had already joined the Gilbert lab team attempting to clone insulin cDNA. Arg told Wally he thought I could contribute, and that was good enough for Wally.

The objectives of the project were to clone the rat proinsulin gene, to achieve expression of the protein in bacteria, and finally to do the same with human proinsulin. Rat insulin was first because cloning human DNA was forbidden in the United States. (Hereafter, although generally talking about proinsulin, I'll use "insulin" for short.) The project was obviously an important one, both scientifically and commercially. I thought it had a good chance of succeeding. The team had a rich source of rat insulin RNA in a rat pancreatic tumor (an insulinoma) supplied by Bill Chick and Steven Nabor at the Joslin Clinic in Boston. In addition, the members of the team had the right mix of skills. Arg was a virtuoso of RNA purification and cDNA synthesis, Peter LoMedico was the master of *in vitro* protein synthesis, and Stephanie Broome was quietly developing a radioimmunoassay that could be used to detect small amounts of any protein produced by bacterial colonies. And while everyone could sequence, Richard Tizard was available to sequence when everyone was busy and when it had to be done *this minute*.

There was, however, a need for someone to insert double-stranded DNA into a plasmid, then introduce the hybrid plasmid into the pathetically weakened strain of bacteria (*E. coli* — 1776) that had been approved for recombinant DNA use. While my time at CSHL had been unproductive in terms of crafting successful experiments or publications, I had learned how to handle the finicky bacterial strain. In addition, I had come back with a treasure trove of enzymes not generally available and a burning desire to end my time as a postdoc with a successful project. Here was a great project, along with a great team led by a legend. What I didn't appreciate, at the time I joined the

*Intersections SBD Consulting, Chestnut Hill, Massachusetts, USA

project, were the scientific difficulties the team had gone through. Later, I learned that the graduate student who had been the designated cloner had been asked to leave the lab.

I didn't meet Wally until we started to get results. He would be waiting, Buddha-like, outside the darkroom where we developed the films, which allowed us to visualize the radioactive molecules we were trying to follow. This was not neglect. This was time management and delegation, two of the many things at which Wally excels. Arg talked to Wally frequently, and I talked to Arg all the time. The goal was clear. We knew what had to be done, and Wally left us to it.

Arg and I hammered out the approach we would use. First, we would reverse-transcribe mRNA from the rat insulinoma and insert the cDNA into the plasmid pBR322. This plasmid contained genes for both tetracycline resistance and ampicillin resistance (penicillinase). We planned to insert the cDNA into a particular restriction site (a *PstI* site) present in the penicillinase gene and nowhere else in the plasmid. This site would make it easy both to insert and to excise the cDNA.

Arg and I thought that insertion into the *PstI* site would have another virtue as well: inserting the cDNA into the penicillinase gene would disrupt production of penicillinase. As a result, colonies of bacteria that contained pBR322 plasmids with the cDNA would retain their tetracycline resistance but would lose their ampicillin resistance. Colonies that grew on tetracycline plates but not on ampicillin plates would likely contain a cDNA insert. Some of those inserts would be cDNA for insulin.

Once we had identified colonies with insulin inserts, we would then turn to the challenge of getting the bacteria to produce and secrete insulin. Getting expression was going to be difficult. Since we did not have much control over the length of the tails or the DNA copies of the tumor cell RNA, Arg and I expected that we would have to use all the tricks available to us to reconfigure the DNA encoding insulin and get production of the protein.

Wally liked the plan. However, he thought that inserting cDNA for insulin into the penicillinase gene might not completely destroy penicillinase production. If so, we might see some colonies that grew slowly on an ampicillin plate. And if that happened, it would mean that the bacteria were producing a hybrid protein that contained penicillinase activity — and the insulin protein. Arg and I thought that was probably a pipe dream.

We began to clone cDNA into pBR322. Greg Sutcliffe had recently sequenced pBR322, which was how we knew that there was a convenient *PstI* restriction site in the penicillinase gene and that it was the only *PstI* site in the plasmid. In order to insert cDNA into the *PstI* site, we needed the enzyme terminal transferase. That was one of the enzymes in my treasure trove from CSHL.

The enzyme had been prepared by Fred Bollum. The story we heard was that he had injected most of his prep into a goat in an attempt to raise antibodies. No one else was able to purify the enzyme at the time; it wasn't until later that it became apparent that thymus from American cows was not a good starting material, since the enzyme was lost in the large amount of fat present in these fat animals. Bollum had used thin Mexican cows as his source of thymus.

Rick Firtel had been a postdoc with Harvey Lodish, one of my thesis advisors. Reconstituting the *PstI* site required adding short tails of dGTP to the plasmid and dCTP to the insert. This required a good preparation of terminal transferase, but also just the right reaction conditions. Arg knew that Rick, now in his own lab at UC San Diego, had been optimizing these "tailing" reactions, so I called Rick. He provided his recipes not only for tailing, but also for preparing the enzyme PstI.

Once we had optimized the tailing reaction for our conditions, we were ready to start inserting DNA into bacteria. We got about 2300 colonies that were tetracycline resistant, and we began to try to identify colonies that contained an insulin insert. This required a lot of tedious, repetitive, and exacting work by Arg and me, as well as the protein synthesis expertise of Peter LoMedico. As Arg and Peter identified clones that had inserts, I would plate the bacterial clones onto both tetracycline and ampicillin plates, as usual. The plates often remained in the incubator longer than usual before I would check them. I have a very clear memory of a day in the lab when I saw that bacterial colonies that we knew had an insulin insert were not only growing on tetracycline plates, but were also growing, albeit very slowly, on ampicillin plates. My college biology professor, Dr. Gairdner Moment, had a favorite piece of advice: "Cherish your exceptions!" I was staring at an exception.

I felt it could mean only one thing: the tailed DNA had not completely abolished penicillinase activity and secretion. If so, maybe that clone was expressing a hybrid protein with the functional properties of both penicillinase and insulin! Wally's "pipe dream" may have come true.

At the time, we didn't know just how lucky we were. It turned out that most, if not all, of our amp-resistant colonies were resistant not because of a hybrid protein that retained penicillinase activity. Instead, some of the plasmids in the cells had lost their insert and regenerated an intact penicillinase.

Of the over 1,000 bacterial colonies we obtained, experiments revealed that 48 were clones containing insulin-coding inserts. So we had successfully cloned the rat insulin gene, and the next step was to see if any of these insulin-DNA clones were producing insulin.

We moved into hyper-activity using Stephanie's radio-immunoassay to prove that insulin was being synthesized in the bacterial cells. When we developed the first films of the radio-immunoassay, there was a single clone that was positive for both insulin and penicillinase. We had hit the jackpot!

This time, however, Wally wasn't waiting outside the darkroom. He was at seminar in the next building. I ran over to get him, and we hurried back to the lab, saying little. We put the films up on the viewbox, and he stared at them for a minute. Then he said, "Interesting spots." And then, for the first of a few precious times, I saw that manifestation of his highest accolade — the big, Cheshire cat grin right out of Lewis Carroll, accompanied by "Very good".

Stephanie also found and optimized an old procedure (osmotic shock) that allowed us to determine if the hybrid protein was secreted. To our great delight, it was. Our bacteria were making about 100 molecules of hybrid protein per cell, a miniscule but momentous amount.

When we gathered to write the paper, Wally produced the first draft. Arg, Stephanie, and I tore it apart, leaving the first paragraph and the discussion alone mostly much intact. We also sat down to decide authorship. As the eldest of six children, a woman, and a Mexican-American, I had to resist my Mexican-American eldest daughter's tendency to put the interests of others before my own. But I'd come a long way in being able to advocate for myself, and I made my case for being first author, because I had filled in a crucial missing piece.

On the other hand, everyone on the paper had made an important contribution. In particular, Arg had put in almost two years of effort on the project and had been instrumental in designing the overall approach, as well as in the essential cDNA synthesis. Arg made the case that he should be first author. Arg and I were good friends and respected each other, but I pushed back. I said that both of us had an equally good case for being first author, but I needed first authorship more. Arg was on a roll, and I'd had a very unproductive past year. Arg thought about that then and agreed.

Throughout the discussion of authorship, Wally said nothing. He just listened, but his presence loomed large. The only indication he gave that he approved of our decision came months later when he asked me to join him as co-author on an article for *Scientific American*.

Wally left the writing session late in the evening, leaving Arg, Stephanie, and me to finish it. He told us to meet him at the law firm of Fish and Neve the next morning. We were to immediately apply for a patent. We worked all night, and I got home as my husband Tony and a guest from out of town were leaving for work. "Does Lydia often work all night?" the guest asked Tony.

I slept a couple of hours and then went to the law offices. Over the next weeks, we worked with the lawyer, Jim Haley, as the final patent was written. While there were 8 co-authors on the paper describing the cloning and expression, four inventors were named on the patent: Gilbert, Broome, Villa-Komaroff, and Efstratiadis. Jim Haley spent a great deal of time interviewing all the players to be sure that the inclusions and exclusions were appropriate. Wally determined the order of inventors, which reflects the reality that the idea was his. The order also recognized the key role that Stephanie's work played — without the expression assay, we could not have claimed expression. And if we had only achieved cloning of the insulin gene, we would have been runners-up, as the University of San Francisco group had cloned rat insulin the year before. Moreover, without Stephanie's osmotic shock procedure, we would not have been able to show that the insulin was secreted.

After a period of intense and entertaining media interest, we returned briefly to a somewhat more normal routine. I took a position as assistant professor at the newly established University of Massachusetts Medical School in Worcester in August of 1978. Almost immediately, I requested an unpaid leave of absence for the month of September: Wally wanted to move quickly to clone and express *human* insulin.

Experiments using recombinant DNA techniques to clone and express human molecules required the most stringent high-containment (P4) facility. Wally decided that our best option was the Porton Down military facility in England. Arg, Stephanie, and I made lists and prepared all the reagents we would need. I packed trunks full of the necessary equipment, supplies, and reagents — along with the mRNA for human insulin from our colleagues at the Joslin Clinic — and flew to England.

We went to Porton Down with high expectations and high spirits. In Cambridge, while we had worked closely together, we all had separate work schedules, lab spaces, and friends, family, and other activities to retreat to when tension or stress got to us. A quick walk to Harvard Square could ease the intensity. There was no such escape in England. Work in a P4 facility was hard. We were all dressed in hazmat suits. Experiments were conducted behind transparent plastic walls, performed by sticking one's hands into thick gloves that were sealed to the wall. We were in an outside space, and experimental materials were in an inside space. Working that way, our hands were "all thumbs": things went slowly.

During this stressful period, Wally was the team anchor. He made media in small bottles using a pressure cooker and poured plates when we learned that we could not use the facilities' autoclaves. He remained calm when we were frazzled and provided steady leadership. Through the intensity, we developed a kind of group camaraderie. We were soldiers together in a foxhole, engaged in combat.

After several weeks' work, we were pretty sure we had a human insulin clone. However, as we proceeded to perform the verification experiments, two events silenced our elation. First, we heard that a team from Genentech, including David Goeddel and Arthur Riggs, had beat us to the cloning and expression of human insulin. Worse, we learned that a rat insulin clone had contaminated one of our reagents: our insulin clone was rat, not human. After four exhausting weeks, we returned home empty-handed. The story of the race to clone insulin is told accurately and entertainingly in *Invisible Frontiers: The Race to Synthesize a Human Gene* by Stephen S. Hall.

Despite this disappointment, we were part of the birth of biotechnology. Biogen had been founded several months before we went to England, with Wally serving as Co-founder and Chairman of the Board. I declined a position at Biogen — both my graduate work and postdoctoral work had been done in some of the best labs in the world, and I wanted to know if I could stand on my own.

Wally quietly arranged for me to be a consultant to Biogen, a position I held for 8 years. That meant travel to Board meetings both in Boston and Geneva, where Biogen had a European lab at the time, as well as a much needed research grant for my lab. I had a front-row seat as Biogen grew, and I received a remarkable tutorial in business. Wally acted as my sponsor as well as a mentor. He wasn't much for offering advice unless asked a specific question, but he provided many more opportunities for his students and postdocs than many lab leaders who were more vocal about their mentorship.

The Board of Transkaryotic Therapies

I left bench research in 1996 and went to Northwestern University in Illinois to become Vice President for Research. I returned to Boston in 2003 to be Vice President for Research at the Whitehead Institute. Michael Astrue, CEO of Transkaryotic Therapies, asked me to join the TKT Board in the late fall of 2003. I'd met Mike through Biogen; he joined Biogen in 1992 as VP and General Council, but I suspected that Wally played a role in the invitation, as Wally was on the TKT Board.

TKT was founded in 1988 and went public in 1996. Its goal was to create enzymes as therapies for patients with inherited enzyme deficiency diseases. It was an exciting time to join the Board. Mike Astrue had taken over as CEO in early 2003, after the resignation of Richard Selden, the founding CEO. Mike had stabilized the company and restored morale.

TKT's main product was Repragal, a human alpha-galactosidase A produced by a genetically engineered human cell line — a treatment for Fabry's Disease. It was not yet approved in the U.S., but it was doing well in Europe. TKT also had a second promising product: iduronate-2-sulfatase (I2S), enzyme replacement therapy for the treatment of Hunter syndrome. In addition, Astrue was working to regain European rights to Dynepo, TKT's version of human erythropoietin. As a result, the stock price was on the rise. Everyone associated with TKT had the sense that TKT could become another Biogen.

TKT had a 7-member board. Wayne Yetter, Board Chairman, had been an executive at Pfizer and a CEO of Novartis, and he was a member of several boards. Dennis Langer, who joined the board when I did, was President of Dr. Reddy's Laboratories and had held high-level positions at several large pharmaceutical companies. Jonathan Leff and Rodman Moorhead were Managing Directors at Warburg, Pincus & Co. Rod was a founder of TKT and had been on the board since its inception. Wally, Mike, and I rounded out the board. Clearly, I had the least business experience.

In September 2005, Shire Pharmaceuticals made a bid to buy the company. The offer from Shire split the board. Jeff, Rod, and Wayne were in favor of the sale. Mike and Wally were opposed. Dennis and I were neutral, or at least we didn't express an opinion one way or the other. Wally coolly distilled the situation to its essential elements and clearly laid out what needed to be done and what information he thought was lacking. No one I've ever met does that as well as Wally. His calm and rational assessment was extremely important in our deliberations — Mike was very emotional in his opposition, and board meetings and calls became increasingly fractious.

Mike felt that Wayne and Matt had orchestrated the offer. That is, he thought the Board Chair had stimulated a buyout offer without consulting the CEO. That made the relationship between CEO and Board Chair untenable. Mike sought to have the Board elect a new Chair. He called several members of the Board, including me, and explained his position. During the call, he asked if I would be willing to be Chair. I replied that Wally was the most qualified and that Dennis had more experience than I did.

A board meeting was scheduled on Martin Luther King Jr. Day of 2005. I expected a relatively short meeting at which Wally would become Board chairman. However, the meeting — later memorialized in a decision from the Chancery Court of Delaware — lasted most of the day. After much discussion, the Board agreed that Mike's accusations were unwarranted. Wayne offered to step down as Chair, knowing that the Board could not function effectively given the rift between him and Mike. The Board agreed and proceeded to discuss who should be successor. When it came to choosing a new Chair, those members of the board that favored acquisition by Shire were opposed to Wally becoming chairman, since he had expressed his opposition to the merger. After much discussion, I was elected Chair of the Board.

Perhaps some board members felt that my inexperience would make it easier for them to drive the process. I knew I had a lot to learn. Because Mike soon took his concerns public, it was clear that I would need to put a process in place for considering Shire's offer that would withstand scrutiny. Even if I disagreed with him about the Shire offer, I knew that Wally would provide me with dispassionate advice.

I also turned to a second mentor. Don Jacobs, the long-time Dean of Northwestern's Kellogg School of Management, had seen to it that I took the Advanced Executive Course at Northwestern's Kellogg School of Management when I was at Northwestern. Don explained that lawyers and bankers in an acquisition deal generally favored doing a deal, because their compensation would typically include a cut of the action. They would be inclined to favor an acquisition even if the long-term prospects of the company were excellent. Their financial advice could not be dispassionate.

Don advised me to get legal and banking advice from people whose compensation would be the same, irrespective of the outcome of the transaction. The Board had already approved Cowen and Co as TKT's financial advisor for this transaction under the conditions that were standard at the time. So, with Board approval, Banc of America Securities was engaged to provide independent financial advice.

As for lawyers, others on the Board agreed with me that we needed alternative M&A lawyers to augment our outside counsel. After we did some homework, I called an outstanding M&A lawyer on a Saturday afternoon, Faiza Saeed from Cravath, Swaine and Moore. We agreed to a flat fee, and she was in Boston within 48 hours.

The next six months were a whirlwind. I was in constant communication with individual board members and the board as a whole. During the course of that 6 months, the company's market cap rose greatly. In July we voted 5-2 in favor of the acquisition at $37 a share, a 50% premium over the share price at the time of the offer.

In the end, Wally voted against the acquisition and I voted in favor of it. Despite that disagreement, Wally made it clear that there were no hard feelings: it was a decision that you could argue either way, and we simply disagreed.

Shareholders, including Carl Icahn, sued for appraisal rights, feeling the company had been undervalued. In the end, after initial arguments in court, the aggrieved shareholders settled with Shire at the original price.

Cytonome

In the summer of 2005 I left my position at the Whitehead Institute. When I told Wally, he told me that I'd handled the TKT board well and suggested I talk to his son John. He felt that I could help with board issues in the company John had founded, called Cytonome. When John and I talked, it seemed to me that I had more to offer inside the company than on the board. In the end, I took a position as Chief Scientific Officer (CSO) and a position on the Board. I told John I'd join the company for a year, but I ended up staying 8 years.

I joined Cytonome because of John's vision and his impressive and deep knowledge of business, engineering, and bone marrow transplantation. Cytonome spun out of Coventor, a company John had co-founded in 1996. Coventor had been a company that developed micro-electromechanical system (MEMS) software and also had a service component. In addition, Coventor had intellectual property in microfluidics.

In 2002, Coventor made the decision to focus on software. Coventor assigned microfluidic IP that John had been instrumental in developing to John, and he used it to found Cytonome. After a year of research and analysis, John decided that the best use of the microfluidic IP was in the field of bone marrow transplantation and cell therapy. He foresaw a need for high-speed, sterile sorting of cells that could be used therapeutically. He developed a clever microfluidic switch and envisioned a multi-channel chip that would enable the construction of a high-speed cell sorter where the fluidics was separated from the optics, thereby allowing sterile sorting. It was, and is, a beautiful concept.

When I joined the company in 2005 as CSO, John had raised initial capital from an angel investor, put together a very strong advisory board, and recruited a talented set of engineers. However, he was at odds with his board and some of his team over management issues, which were not John's strength. Two board members left, and John realized that a change was in order. He and I changed positions — he to CSO, me to CEO, a position I accepted on the condition that Wally join the board. After a few months, it became apparent that it would be best if John left the company.

In retrospect, I did not handle that transition well. Indeed, I feared that it might result in a breech with Wally. I needn't have worried. Wally was able to fully support John while recognizing what was best for the company and supporting my actions as CEO. At the same time, he made it clear that he was not happy with how the company was executing its engineering plan. He held several long meetings with the engineers, putting hard questions to them. It was another remarkable display of Wally's ability to focus on the essential elements of a situation — in this case, largely problems in the realm of engineering and manufacture — and point to potential solutions. Although the company has not yet achieved its dream of a cell sorter for human cell therapy, it does have one very successful revenue-producing product and a second about to be launched.

Wally

Of Wally's many remarkable attributes, his intellect is perhaps the most widely appreciated. There is no need to catalogue his contributions to biology/biochemistry and to physics. I would cite just one that strikes me as particularly noteworthy. In 1978, shortly after the discovery of splicing, Wally published a paper in *Nature,* "Why genes in pieces?", that sought to explain the potential evolutionary and biological advantages of splicing. That was pure Wally. When something is a true surprise, he isn't simply surprised. He immediately assumes there must be a reason, speculates what it may be, and thinks about how to prove it.

Another remarkable quality is Wally's curiosity — moreover, the range of things about which he is curious. In recent years, Wally has developed a long-standing passion for photography — an art altered by new hardware and software technologies. As an artist, Wally approaches photography with the same fierce intelligence, curiosity, and focus as he approaches science. His work is recognized all over the world. Not long ago, I was at an event, talking with some people I'd just met, and someone else mentioned Wally's name. "Oh," said one person, "do you mean Wally Gilbert the artist?" (I realize that this anecdote may also reveal how sequestered scientific fame can be.)

How can a person who is so curious find the time to pursue his curiosities? By a highly disciplined management of his time. I'm sure that at the end of each day Wally can recall some time that he has, by his lights, wasted — because life in a society sometimes requires that. But it seems to me that, more than anyone I know, Wally succeeds at spending his time on things he thinks are important (even when not particularly interesting), and the things that are interesting (even if they don't prove important).

Wally is the most intellectually confident person I know who is not also egotistical. He doesn't have to be the center of attention. Wally can take over a room without saying a word, but when the spotlight belongs to someone else, he never diverts attention to himself. If someone disagrees with him, he waits to hear the reason — and concedes the point, if he is convinced, which is not very often.

Wally puts the argument before the person. He doesn't prejudge people and instead judges a person's argument. As a Mexican-American woman, I've found that there are people who make judgments about the value of my opinions before they know what they are or what my background and credentials are. Not Wally. Wally wants the information first. He can very quickly determine a person's weaknesses, and he still appreciates their strengths. He makes a stringent assessment and has very high expectations, but they are the same standards and expectations for everyone — without any preconception about the person. It's hard to surprise him with an insight or an idea, but when you do, he gives you a side glance that makes you feel like you've just won Olympic gold.

Finally, you can't talk about Wally without discussing his relationship with Celia, which goes back to when they were in grade school. Those of us who have been privileged to know the pair are in awe of Celia — a poet, an artist, and an individual with wit to rival Dorothy Parker — and of their mutually supportive relationship. Wally attends most of Celia's readings and is a strong advocate for her art. When he is at a reading or one of her art exhibits, she is the focus and he the supportive spouse. Wally enjoys talking about the accomplishments of his family. His immense curiosity means that he learns about their interests and becomes an enthusiastic fan.

In sum: A person could not have a more stimulating and supportive mentor and friend than Wally Gilbert.

Acknowledgment

The author gratefully acknowledges Anthony Komaroff for editorial comments.

The *Bio Labs Midnight Hustler* Humorously Documents Life and Science in the Gilbert Lab, 1977–1980

Karen Talmadge*

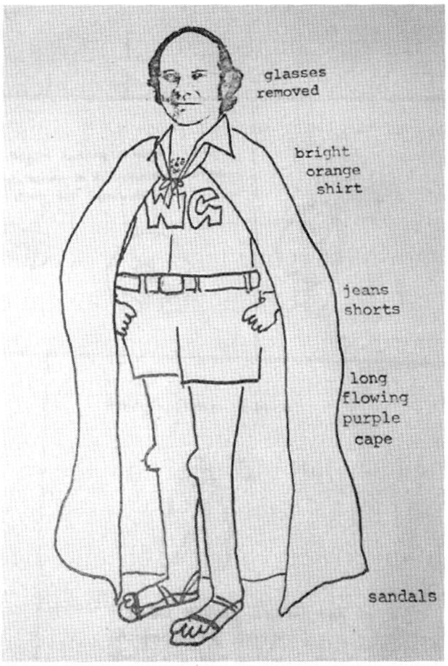

Figure 1: WG-Man, from "The Case of the Post-humous Post-Doc," an Agatha Crick-y mystery, issue 2, May 1977.

The man walked purposefully down the dimly lit hall. Dressed in a crumpled green suit, he had a proprietary tread. … Suddenly, he ducked into the darkroom and shut the door. Five minutes (with agitation), six minutes, seven minutes, and he emerged, dressed in sandals, jeans shorts, an orange t-shirt, a long, flowing, purple cape, and glasses removed. The Green Hornet! By day, mild-mannered biochemist and molecular biologist, selflessly serving humanity and seeking the cure for cancer (or the Nobel Prize, whichever comes first). By night, the GREATEST CRIME SOLVER OF THEM ALL!!![a] (Figure 1)

* Nabu Strategic Advisors, Los Altos Hills, California, USA

[a]The text is the first installment of "The Case of the Posthumous Post-Doc", an "Agatha Crick-y" mystery, in which her famous sleuth, WG-Man, discovers his post-doc is dead, from Issue 1, 1 April 1977. The figure is from the second installment of the series, issue 2, May 1977.

Can a pretend newspaper document, a productive research era, and the impact of the laboratory's leader provide a creative outlet for pressure? In the alternate, can wasting time and resources on childish pranks with a voluminous paper trail be compatible with BIG, ASTOUNDING science?

The *Bio Labs Midnight Hustler (BLMH)*, which will appear seven times between 1977 and 1980, presents satiric stories, songs, poems, cartoons, photographs, puzzles, and advertisements concerning life and science in the Gilbert Lab. It serves as a major creative outlet, as well as The Complaint Department. The humour is usually childish, often ridiculous, sometimes savage, generally in very poor taste, and occasionally very, very funny.

Obviously, "the Bio Labs" refers to Harvard's Biological Laboratories at 16 Divinity Avenue. An imposing building from the 1930s, the two life-size bronze rhinoceri flanking the large front doors, along with the frieze of animals carved in brick around the top of the building, boldly declare the research focus of that era. Inside, Wally's free-wheeling Biochemistry and Molecular Biology Department research empire sprawls along the northeast wing of the third floor, sharing (in that era) its equipment and facilities with David Dressler's lab next door. Less obvious, I hope, is that *Midnight Hustler* is the title of a 1971 porn film.[b]

From 1975–1981 in Wally's lab, first as his graduate student and then as his post-doc, I witness the transformation of our field: from basic science to creating the new biotechnology industry; from "unlocking the mysteries of life" to "the future of medical science". When I begin graduate school, *Nova*, the PBS science program, is paying close attention to Wally's work. Before I finish, the *Wall Street Journal* will be as well.

One day, as an early glimmer of the transformations to come, Wally runs into the Tea Room[c] sometime in late 1975 or early 1976. His film still dripping, he shows everyone lanes of cascading bands and excitedly explains how this will become Maxam-Gilbert DNA sequencing. At this time, Allan Maxam is Wally's Research Associate. He contributes enormously to the science and the culture of the lab. We call him our Main(e) Man, as he is from Maine, and he is the graduate students' most valuable lab resource after Wally.

In a wonderful book about this era, *Invisible Frontiers*,[d] Stephen Hall recounts how the proposal to build a P3 containment facility in the Bio Labs explodes into a fierce controversy over its potential use for cloning human genes in June 1976.[e] With anger and emotion on all sides, opponents warn of frightening consequences. Even I, as a beginning graduate student, being active in helping to present the science to the public and the Cambridge City Council, receive an angry anonymous call on my home phone. After a dramatic Cambridge City Council meeting in July, which I attend, the Council passes a moratorium on planning the containment facility.[f]

---

[b]http://www.imdb.com/title/tt1092604/, last accessed 27 November 2016.

[c]Our communal meeting and break room, where kitchen staff make tea and put out cookies weekdays at 4 pm. This is a continuing tradition from the Watson lab and Watson-Gilbert lab eras; both Jim Watson (Biochemistry and Molecular Biology's original Department Chair) and Wally conducted post-doctoral research at Cambridge University in England, although Wally's field was physics at that time.

[d]Hall SS, *Invisible Frontiers,* The Atlantic Monthly Press, New York, 1987.

[e]Hall SS, *Invisible Frontiers,* The Atlantic Monthly Press, New York, 1987, p. 42.

[f]Culliton BJ *Recombinant DNA: Cambridge City Council votes moratorium*, Science 193:4250, 1976.

This has a major impact on our lab, because Wally is gunning to clone the insulin gene. It is hard to overstate the enormity of that goal, a new and exciting possibility arising from the basic studies of bacterial gene function (with Wally one of its leading scientists of the last decade).

We feel the competition keenly, from multiple labs (many at UCSF), as well as from an unexpected source we learn about that June. The entrepreneur Bob Swanson, with Herb Boyer of UCSF, forms one of the first biotechnology companies, Genentech, to pursue the commercial potential of newly emerging recombinant DNA techniques.[g]

In the fateful early months of 1977, the Cambridge City Council lifts its moratorium on planning a P3 facility,[h] the Maxam-Gilbert chemical sequencing method paper appears,[i] and Fred Sanger's lab at Cambridge University publishes the "approximately 5,375-nucleotide sequence" of the bacteriophage phi X174 using their "plus-minus" enzymatic sequencing method.[j]

While Sanger's lab will ultimately develop the more reliable enzymatic technique after a decade of modifications, the Maxam-Gilbert chemical technique is a revolution in its time, because it immediately brings sequencing to anyone who can purify DNA, use chemical reagents in a fume hood, and run polyacrylamide gels — that is, all molecular biologists.

As news of the method spreads, Allan Maxam fields constant daily phone calls and helps/collaborates with many distinguished scientists who visit the lab, including Bill Haseltine and David Botstein, to name two who appear in the *BLMH*.

Like graduate students and post-docs in Wally's lab before and after, we feel the pressure to generate data, because Wally wants answers, and so do we. We all dread (because there are no data yet…) or hope (I have the data!) to hear those special words that mean Wally's attention is now laser-focused on you: "*What's up?*"

Even more stressful than Wally's attention, though, is his absence. As his reputation continues to grow, he travels even more — lecturing, moderating, reviewing, consulting, organizing, teaching. Everyone feels his absence from the lab, but especially Allan, whose work is even more disrupted by students and outside scientists seeking advice when Wally is out of town.

In the midst of this ever-growing competitive cloning and sequencing maelstrom, Allan begins to plot with his fellow prankster and room 388 lab-mate, Gilbert graduate student Phil Farabaugh, to create a pretend newspaper making fun of ourselves, of our situation, and of Wally. Phil goes from bench to bench, inviting the other lab members to join. The era of the *Bio Labs Midnight Hustler* begins.

Through its seven issues, Allan Maxam, Debra Peattie, and I have a hand in every one, as does Winship Herr once he joins the lab, after the first two issues. As graduate students and post-docs come and go, though, a large group of "correspondents" bring broad wit and creativity from around the building. I list below everyone who contributes in some manner at some point.[k]

---

[g] https://www.gene.com/media/company-information/chronology, last accessed 27 November 2016.

[h] Hall SS, *Invisible Frontiers,* The Atlantic Monthly Press, New York, 1987, pg. 54.

[i] Maxam AJ, Gilbert W. *A new method for sequencing DNA*. Proc Natl Acad Sci. USA 74:560–64, 1977.

[j] Sanger F *et al.* Nucleotide sequence of phiX 174 DNA. *Nature* 265, 687–95, 1977.

[k] Gilbert lab graduate students Forrest Fuller, Winship Herr, Lorraine Johnsrud, Nils Lonberg, Debra Peattie, Ulrich Siebenlist, Greg Sutcliffe, Karen Talmadge; Gilbert lab post-docs Jürgen Brosius, Peter Lomedico, Mark Pasek, Steve Stahl, Dyann Wirth; Kafatos Lab grad student Nadia Rosenthal; Kafatos Lab post-docs Argiris Efstradiatis, Lydia Villa-Komaroff; Lydia Villa-Komaroff's husband, Tony Komaroff; Dressler Lab grad student Kirston Koths; Ausubel Lab grad student, Gary Ruvken; Ausubel lab post-doc Sharon Long; Gilbert lab gurus Allan Maxam, Walter Gilbert. My apologies to anyone I have forgotten or overlooked.

The *BLMH's* editorial standards, elucidated with time, are bold:

- "All the news that fits, we print." (Issue 2)
- "For slander and libel, the Hustler's your bible." (Issue 4)
- "Code: GFY. Description: Editorial principle." (Issue 7)
- "It is against our policies to check boxes." (Issue 7)

The recurring themes are not surprising:

- Wally's future Nobel Prize
- Anyone else's Nobel Prize ambitions
- The constant scientific pressures
- The background commercial pressures
- The long hours
- Anyone or anything that irks us for any reason

We never think that Wally might not approve. He is the first to laugh with glee at our silly and/or rude gift exchanges for the holidays. He wears orange turtlenecks and drives a maroon Chevy convertible. His wife, Celia Stone Gilbert, is a brilliant poet, artist and writer. Wally happily dons a bright pink lab coat (which earlier jokesters so kindly dyed for him) when he works at the bench (see Figures 2a and 2b from 1979). Equally telling, when the host at a formal Nob Hill restaurant stiffly hands him a tie, Wally immediately puts it on, right over his turtleneck. "This only makes the restaurant look silly," he comments as we sit down.

Figures 2a and 2b: Wally working in the lab in his pink lab coat, July 1979. Photos poorly taken by me.

For all my life, I will also vividly remember Wally's response to data that do not support his hypothesis. (I do not remember the experiment itself.) His face breaks into its widest grin, and he exclaims, in essence, "I was wrong! This is wonderful! This means that [something else] could be true!! We have to look into this!!!" I am using multiple exclamation points to reflect the excitement in his voice and the delight on his face. Wally is never afraid — of unexpected findings, of challenging avenues, of asking questions, of wearing a pink lab coat, of being wrong — and this greatly influences us all.

The *BLMH* contributors write many of the features on the Gilbert Lab's latest model IBM Selectric typewriter, and its amazing innovation for that time, magnetic card memory, allows us to (gasp!) edit and print. Of course, the magnetic cards are physical, we have to load them into the typewriter, and each one only holds about 2/3 of a page of printed text.

Late at night, the *BLMH* "editors" physically cut out and glue the submitted articles/photos/poems/etc. onto pages of paper to create the layout. We copy each page, lay all the pages around the Tea Room, collate and staple each issue (copy machines do not perform these functions yet), and leave the stack in the room. We return the next day to see people from all over the building reading their copies, laughing. (Labs at other universities sometimes get copies as well, discussed later.)

Here are some of the highlights (?) from each issue. Where I know the author, I generally do not reveal him or her, to protect the guilty.

Issue 1, April 1, 1977, and Issue 2, May 1977

The first issue comes out on April Fool's Day, 1977 (of course), and the second one follows about six weeks later.

On its very first page, *BLMH* correspondents breathlessly congratulate Dr. William Haseltine's successful return to the Gilbert lab[l] to sequence regions of the Rous sarcoma virus:[m]

> *On the Road to Stockholm*
>
> *(UPI) Cambridge, MA, April 1, 1977 — The American Association of Pseudointellectuals today lauded Dr. William A. Haseltine of the Sidney Farber Cancer Center, Boston, Massachusetts. After a search of many years, Dr. Haseltine has finally found a way to get his name in the New York Times, and in so doing cure cancer in chickens. The scope of this achievement has stunned the scientific community here.*

Overworked graduate students already appear, and this is the just the first page of the first issue:

> *The chemical [to cure cancer] is obtained from the bodies of graduate students who died from the strain of working with Dr. Haseltine and administered to the chickens daily.*

---

[l]A former graduate student in the Watson-Gilbert lab, Bill is beginning his faculty position at the Harvard Medical School's (then-named) Sidney Farber Cancer Center.

[m]Haseltine WA, Maxam AM, Gilbert W Rous sarcoma virus genome is terminally redundant: the 5' sequence. Proc Natl Acad Sci USA 74:989–93, 1977.

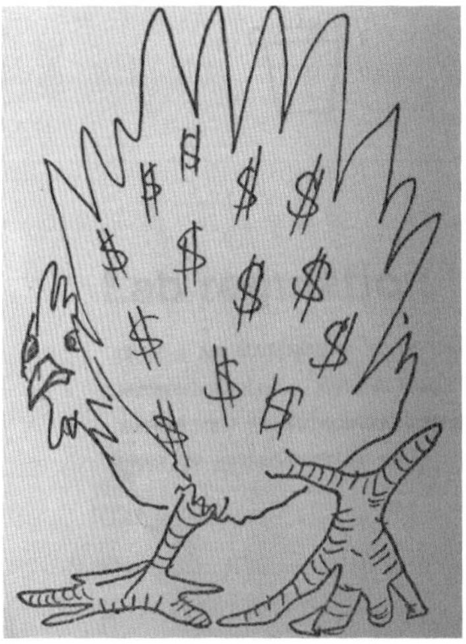

Figure 3: Future iced broiler, With lead stories, issue 1, April 1977.

In the accompanying story, "Fowl Play by Molecular Biologist and NCI Chicken Virus Supplier in Massive Get Rich Quick Scheme" from the "Chicago Mercantile Exchange", intrepid *BLHM* business reporters reveal a plunge in the price of iced broilers due to oversupply created by Dr. Haseltine's RSV experiments. A delightful chicken cartoon (Figure 3) accompanies the article, which also ominously notes:

In other trading, platinum futures closed slightly higher, due to demand by DNA sequencers.

Wally's attendance at an ICN-UCLA conference in Salt Lake City in March 1977 becomes *WG Says SALT Talks*[n] *Productive.*

"The first order of business", Gilbert reported, "was sodium chloride, and there was general agreement that 0.06666 molar concentrations are not required by restriction enzymes." This was the salt breakthrough that everyone has been waiting for, and had eluded previous SALT negotiators. When questioned by a Hustler columnist as to why it had taken so long to reach an understanding of the salinity of buffers, Gilbert's only reply was, "Your column needs more salt."

The competition to clone the insulin gene makes its first appearance in "*DNA Butt-Scuttle*" (yes, as in "butt-end ligation"), breezily "reporting" on the latest papers by "*Berb Hoyer and Rill Butter*".

Noting that the daughter of a prominent figure in the Women's Liberation movement is doing an undergraduate project in the lab, we dream up a "Classifried Ad" from "Brinkperson Instruments, Inc.", offering the:

...latest in non-sexist automatic pipetting, The Pipette Individual!

[n]A play on the SALT (Strategic Arms Limitation Treaty) talks underway 1969–1979 between the U.S. and the then-Soviet Union.

We pay homage to the lab's deep history with the *lac* operon with our "Quotation of the Week":

My i^{Q} *is –35.*

We mark the seminal publication of the Maxam-Gilbert sequencing technique, and its future commercial applications, by copying the title and authors from the journal, then make a slight alteration to the title: "*A new method for* ***promoting*** *DNA*".

Both issues include installments of the "Tale of the Posthumous Post-Doc" by "Agatha Crick-y" (see the beginning of this chapter and Figure 1, above). They both also contain sage advice on getting along with your lab mates by "Dear Debbiee" (also spelled "Debby" and "Debbie" throughout). Here is a short sample of the letters and replies, this one subtly referencing the race between the Gilbert and Ptashne labs for the *lac* and *lambda* repressors, while dismissing the fields of *Drosophila* and *lambda* phage genetics, all in a crisp 36 words:

Dear Debbie:
I really like to play squash, but every time I raise the subject, I get half an hour of Drosophila genetics, which I find disgusting. What should I do?
$\lambda dbio\ 30\text{-}7v_{1}v_{s}326\underline{nin}5\underline{S}_{7}$

Dear $\lambda dbio\ 30\text{-}7v_{1}v_{s}326\underline{nin}5\underline{S}_{7}$*:*
Find another racket.

In the lead article of second issue, our crack team of *BLMH* sportscasters hilariously narrates a pretend day of biochemical sports competition:

Gilbert Hustlers Outmuscle Boyer Cartel in Dual Meet

The long awaited battle of the biochemical titans took place this weekend in Cambridge. [...] The first contest Friday night featured DNA sequencing and, as expected, the Gilbert methylators swept the field. Sequencing captain, Allan Maxam, executed a perfect strand separation, the first in league history. [...] The third round Sunday was highlighted by brilliant performances. Who can forget Jeffrey Miller[o] *successfully streaking an eight-sectored plate for single colonies in 28 seconds, or Rich Tizard's*[p] *world record gel pouring feat...? How about Gilbert himself, burying Boyer in the climactic paper-pushing contest with the four feet square blueprints for the new biochemistry building?*[q]

Alfred Vellucci, the Mayor of Cambridge, and a major figure in the drama of the City Council's containment facility planning moratorium, receives our special treatment:

---

[o] A former Harvard Junior Fellow in the Watson-Gilbert lab.

[p] A Research Assistant with Wally.

[q] In a 1995 interview with science historian, Sally Hughes, Mary Betlach, a Research Assistant with Herb Boyer in 1977, says: "Oh, yes, Harvard. I remember we felt the competition with Wally [Walter] Gilbert's group at Harvard. They put out these little newsletters, 'The Midnight Hustler 3,' talking like we were sports teams or something. Those competitions were all very good-natured." [KT comment: Good-natured? Ha! She clearly never came to our lab.] https://archive.org/stream/earlycloningdna00betlrich/earlycloningdna00betlrich_djvu.txt, last accessed 15 January 2017.

Terrorists Hold Mayor, Halt Insemination

The latest in a series of Hanafi-style kidnappings[r] *took place Friday in Cambridge, Massachusetts. A group of youths, later identifying themselves as the Recombinant Youths, seized the Mayor's office, and are holding Mayor Vellucci and several workers hostage. A spokesperson for the group, talking to reporters by phone, said that the group is protesting the film "Demon Seed." "We are tired of the negative stereotyping of scientists and science by the movie industry." They feel that the movie industry should start focusing on relevant social problems, such as terrorism.*

We dedicate an entire page to the strategies lab members adopt to be able to ask Allan Maxam a question, known as "Main Man snaring":

As most of you well know, the Main Man (Homo mainus) is a very clever and elusive creature, whose evasive tactics have been refined over months of avoiding stupid questions from would-be sequencers. [...] Bob reminisced, "I remember long hours hanging around 388, on the prowl for the Main Man. My favourite and most successful method was to hide under his desk at 11 in the morning before he came in, and when I saw him, to reach out and grab his ankles and not let go!"

With geneticist and former lab member Jeffrey Miller, who is visiting the lab, our careful contributors dutifully delineate the impact of his supply purchases on Harvard's financial future:

For the first time in the memory of sources close to the University Comptroller, Harvard University is operating in the red. Apparently, this situation was precipitated when Visiting Professor Jeffrey Miller, former Junior Fellow and now at the Université de Génève, was told by American Cancer Society Professor, Walter Gilbert, "Oh, just have Pat authorize your req."

In the past weeks, Dr. Miller, in a one-man binge, bought out the complete stock of Bachem Biochemicals... In addition to these purchases, chiefly of 5-bromo-4-chloro-3-indolyl-D-galactoside and phorbol ester, Dr. Miller has obtained a controlling interest in three multi-national corporations.

A special correspondent to the *BLMH* conjures a story of a phantasmic Wally involving his convertible, his love of spicy Chinese food, and his even greater love of science, in "Fear and Loathing in the Bio Labs":

The maroon pig screeched to a stop at a red light on the Mass. Ave. strip. Almost out of ether. The Doctor of Gonzo Science had barely had enough label to make it through a meal at the Hunan. More garlic! More pepper! The waiters were clearly trying to water down his ginger intake. Snakes!

The convertible lurched as he wheeled one-handed into the parking lot. The acrylamide buzz of the early afternoon had subsided. He craved a new data experience.

The final page of this issue imagines the whimsical world of *E. coli*, abruptly transformed (ho ho!) by the horrors of recombinant DNA (Figure 4).

[r] In March 1977, Hanafi Movement members storm three U.S. government buildings, take hostages, and demand (among other things) the destruction of "Mohammad, Messenger of God," a movie they consider sacrilegious. https://en.wikipedia.org/wiki/1977_Hanafi_Siege, last accessed 27 November 2016.

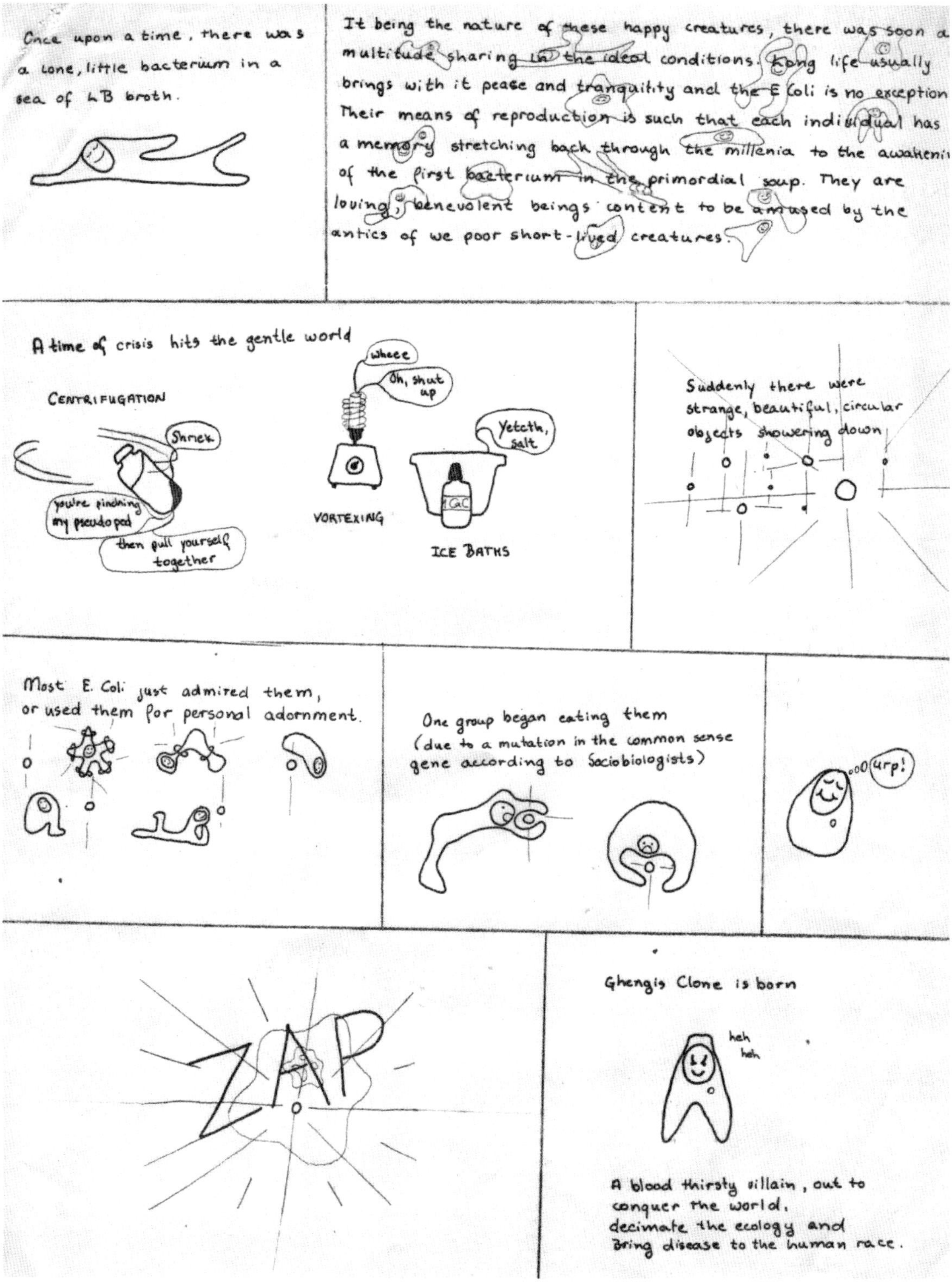

Figure 4: Cartoon of “Ghengis Klone” Last page of issue #2.

I will also note the tiny, plaintive, *cri de coeur* at the bottom of the first page of this second issue, as we pine for our peripatetic purple patron (aka our Waldo di Waldi):

It's 9:00 a.m., do you know where your lab boss is?

Issue 3, December 1977

The third issue of the *BLMH* leads with a lab-related version of "The Cat in the Hat Comes Back" on the cover.

The snow was falling, the temperature was low,
We were all working hard, as usual, you know.
The day was peaceful, even a bit quiet,
Hard to come by here, one can't deny it.
When, suddenly, the calm was shattered forthwith
As a familiar cry down the hallway did drift.
Our mouths hung open and our glazed eyes shone
As we heard the sounds of, "Waaalllyyyy, phone!"

Between the second and third *BLMH* issues, researchers at UCSF report that they have cloned rat insulin cDNAs.[s] The *BLMH* memorializes this with great economy through an official lab requisition form requesting "1 insulin cDNA clone" for the price of "100 grand" from "Genetech, Frisco, CA" [sic], charging it to the "Gilbert NIH Large" grant. Elsewhere, in a big box, we quote Vince Lombardi: "A good loser is a psychopath!"

The *BLMH* notes that David Dressler is this year's recipient of the Avogadro Award, which allows him to:

...rightly bear the titles of "The Count" and "Mr. Mole".

Our *BLMH* "Reporter about Town" breezily brings our readers the latest gossip on a certain "Scientist about Town" (Wally's photo drawn with a Santa hat and beard, as this is the holiday issue):

For those of us with the travelling urge, you've just missed the awarding of the Nobel Prizes in Sweden for 1977, but there will more next year, and you couldn't have gone this year anyway unless you had an outfit in blue chiffon and brocade.[t] This reporter has heard from a little bird that those in the know next year will be in turtlenecks and wrinkled green suits.[u]

[s]Ullrich A *et al.* Rat insulin genes: constructions of plasmids containing the coding sequences. *Science* 196:1313–1319, 1977.

[t]Rosalyn Yalow accepting her prize for her work on radioimmunoassays.

[u]Prototypical Wally attire at the time; the Nobel prize prediction is only early by two years.

We enjoy teasing David Botstein of MIT (who spends a few months in our lab learning DNA sequencing) for his joyously boisterous personality and booming voice in:

392 NOISE HAZARD INVESTIGATED
Workers complain of Bottered Ear Syndrome

Cambridge, MA. The U.S. Dept. of Health issued a strongly worded warning of impending health hazard for the Harvard community in general and the Gilbert Lab in particular due to the presence of dangerously high levels of noise emanating from Bio Labs room 392. Workers in and around the lab have complained of chronic ear trouble since the noise problem began last September. [...] Typical replies to this reporter's enquiries were, "Huh?", and "Can't hear you." "How many db's can we stand?" groaned one lab official. A spokesperson for the MIT Dept. of Biology dismissed the warnings as "irresponsible...and, unless they are withdrawn immediately, we are prepared to boycott all results and experimental techniques coming out of Harvard." The interview ended abruptly when the MIT official's hearing aid malfunctioned.

In an exciting *BLMH* exclusive, Harvard Security "intercepts" a letter from "Jane Goodall" to "Louis Leakey", containing her unpublished reflections on a "tribe known as Molecular Biologists" "here at Fair Harvard," as she observes them performing their "Collaboration Dance." Figure 5 presents the complete letter and more background on this amazing discovery.

Finally, in a special section on Molecular Religion, "Platerita di Alma" reports that Howard Goodman and William Rutter successfully cloned the human soul. Among the many fascinating technical details, they discover that soul cDNAs can be readily recognized because, instead of a cap, they have a halo. In the end, they find that the sequence of the soul cDNA is a simple repeat of GAGAGAGA. "This extraordinary breakthrough will certainly revolutionize religion," our author opines.

Issue 4, June 1978

In March 1978, Wally heads an international group of distinguished scientists who agree to help form what became the biotechnology company, Biogen.[v]

Two months later, he publishes a letter to *Nature* about the ground-breaking work from other labs demonstrating that coding regions ("expressing sequences") in mammalian genes are not always contiguous. Wally proposes that these non-coding "intervening sequences" could facilitate recombination during meiosis as well as evolution.[w] More controversially, he also proposes the names "introns" and "exons".

The following month, Villa-Komaroff, Efstradiatis, Broome, Tizard and Gilbert submit their paper on cloning and expression of a rat preproinsulin cDNA in *E. coli*, while also demonstrating that insulin antigen is secreted into the bacterial periplasmic space by translation of a hybrid gene encoding pre-β-lactamase fused in frame near its 3' end to most of the rat proinsulin gene.[x] Greg Sutcliffe's

[v] Hall SS *Invisible Frontiers*, Atlantic Monthly Press, New York, 1987, pp. 209–10.

[w] Gilbert W. *Why Genes in Pieces*? Nature. 1978 Feb 9;271(5645):501.

[x] Villa-Komaroff L, Efstradiatis A, Broome S, Tizard R, Naber SP, Chick WL, Gilbert W. *A bacterial clone synthesizing proinsulin.* Prod Natl Acad Sci USA 75:3727–31, 1978.

Eds. note: BLMH security forces intercepted this letter while opening mail as part of an ongoing investigation into the purchasing of coffee with Gilbert NIH grant funds.

Dr. L. B. Leakey
Olduvai Gorge
S. Africa 03197

My Dear Louis,

As the sun sets here in Cambridge on this cold winter's day, I can't help but recall that wonderful afternoon we spent together sipping tea and looking at your pretty skulls. I remember what you said about culture being a pattern and that to understand any aspect of the pattern it is necessary to have some understanding of what the people being studied do, what is actually expected of them, and what is encouraged. Oh, for those quiet reflective moments we shared. I must tell you I have discovered among the westerners right here at fair Harvard a tribe called Molecular Biologists who share many of the attributes of my Kwakiutl of Vancouver Island, i. e. competitive rivalry for prestige, they value frenzy, emotional excess and mortification of the flesh. Could this be megalaomaniac paranoia? A central ritual common to this group is called the "Collaboration Dance". It may be performed by as few as two members or as many as 15 participants. When large groups perform the dance a well defined hierarchy exists, sometimes distinguishable from division of labor during the dance, but most often revealed at the close of the dance when a paper is written with the order of authors usually reflecting the hierarchy. Sometimes the order is alphabetical; however, experienced dance members often change their names so as to assure first position on such manuscripts. Occasionally this ordering process is equal in time to the dance itself. It is generally taboo to discuss the ordering process prior to the initiation of the dance. The dance unfolds with a preliminary gathering of participants. There is a ritualistic exchange of refreshments such as tea or coffee. Some groups share cookies or smoke together. During this early process participants pay homage to one another through verbal exchange of compliments regarding past achievements while carefully circling each other gracefully, occasionally making eye contact. The major part of the dance consists of each member performing a task in full view of all participants or privately. Another exchange, this time of small test tubes containing colorless liquids occurs. To the keen observer the phenomenon we refer to as legitimacy and coercion is practiced with one member (the object of legitimacy) attempting to achieve compliance from another. One dance member attempts to make compliance with his/her wishes more attractive than the consequences of refusing to comply. The final resort I have noted among the Molecular Biologists is "Do what I want or I will kill you" followed by "O. K., kill me". I must close now, Louis.

As always,

Your friend,

Jane Goodall

Figure 5: Letter by "Jane Goodall" From issue #3.

Figure 6: Original Klone poster being modified on its way to becoming the *BLMH* cover for issue #4.

paper on the DNA sequence of the ampillicin resistance gene on the cloning plasmid, pBR322, also appears that month,[y] and his work on sequencing the entire plasmid is well underway.

Meanwhile, a *BLMH* contributor returns from Harvard Square with a hot pink poster advertising the local appearance of a metal rock band called Klone. A new series of pranks is born, including the fourth issue of the *BLMH*.

At the top of the flyer, we use our lettering set for artwork to write, "*The Bio Labs Midnight Hustler, in conjunction with N. Sulin Big Bucks, presents*" above the actual next line, the band's name. In the middle, we alter the club date to make it the *BLMH* publication date. At the bottom, we add, "*Also Appearing, Iggy and the Introns*". We paste a photo of Argiris Efstradiatis' face onto the band member wearing black (Arg always wears black), while the band member without a shirt gets a photo of Wally's face. Ta Da! We have our cover for Issue 4. Figure 6 shows the original

[y]Sutcliffe JG *Nucleotide sequence of the ampicillin resistance gene of the plasmid pBR322.* Proc Natl Acad Sci USA 75:3737–41, 1978.

poster on its way to becoming the cover (copied in black and white, probably due to a lack of colour copying at the time).

Still playing with Wally's naming proposal, *BLMH* wags alter a full-page magazine advertisement for condoms, substituting "Introns" for the condom's brand name, while proposed names by other scientists become the inferior brands:

> *Some guys do better than others. [...] They're in control. [...] They use INTRONS. [...] Accept no substitutes. Insertons, intervenons, spligons, and interruptons are unauthorized imitations."*

This issue also contains multiple expressions of the fierce pressure and competition we are feeling.

When Wally, in a lab meeting, articulates his mentoring approach and philosophy, mesmerized *BLMH* correspondents report:

> *Gilbert Gives Gabfest*
>
> *Guru Walter Gilbert last month revealed his philosophy on the leadership of a research group at the frontiers of science. Gathered around, eyes wide, ears open, his students and colleagues listened attentively as he wove a tapestry of individual guidance on a background of desperate medical need. [...]*
>
> *"The relationship I develop with each student has a three-fold nature. At the start, I carefully nurse and nurture the budding scientist. Then comes a period of benign neglect, when the student will send out his or her own roots to create the contacts needed to take best advantage of this fertile research soil. When the student is a stable, working part of the lab, I welcome him or her as a full-fledged member of the scientific club."*

However, the "article" ends with a sentiment Wally emphatically did not express: "I only ask one thing of my students, and that is there is NO F****** OFF... — which includes no eating meals or going home to sleep."

Intrepid *BLMH* reporters are also the first to reveal the creation of yet another commercial enterprise out of the Gilbert lab, "Maxamgen". The "company" "announces" it has cloned the genes for THC biosynthesis in bacteria. In a photo with the title "Dr. Maxam at work in his laboratory," we catch Allan by surprise as he performs actual lab work (a sonication), appropriately dressed in his lab coat and protective glasses. "Quality control is strict at Maxamgen," we write under the next photo, where Allan plays along, pretending to drink the sonicate.

The most pointed expression of the pressures we feel appears next to the Maxamgen article, taken from an advertisement for the same system our lab uses to secure large compressed gas cylinders against walls:

> ***GROUP LEADERS!!***
>
> *Not getting enough work out of your Grad Students?*
>
> *Not enough of that "belly-up-to-the-bench" action to get those exciting results that impress the Big Shots at meetings?*
>
> *You need **Bench Bondage!**"*

In the ad's photo, a Gilbert graduate student is pipetting while the advertised belt-and-clamp system secures her in place at her bench. Part of the joke is that this student might have won the "graduate student who is least likely to need Bench Bondage" award. Looking carefully at the photo (Figure 7), her eyes are focused a short distance beyond the test tube she is holding, while her lips are tightly closed, but perhaps ever so slightly up at the corners. She could be thinking, "I am trying to be serious, but I want to laugh." Or, perhaps more likely, she is thinking, "Why did I ever say yes to this?"

"Bench Bondage" is one of the most remembered pieces in any of the *BLMH* issues and a favourite of Wally's. The original page with pasted photos for "Maxamgen Accounces Breakthrough" and "Bench Bondage" appears in full in Figure 7.

Another classic, a special correspondent to the *BLMH* lauds Lydia Villa-Komaroff's first-author insulin paper with her own Dewar's profile, referencing an iconic ad series from the 1970s to the 1990s.[z] A portrait shows Lydia in the lab, and the text provides "her" answers to the standard Dewar's profile ad questions. Some of my favourite "Lydia" responses (with the full ad in Figure 8)

Hobbies: *Jogging, skiing, and drinking, not necessarily in that order*
Quote: *"Creating new life forms is damn hard work!" [...]*
Drink: *For the first drink, Dewar's, of course. "After that, what does it matter?"*

In our first nod to newly created Biogen, a *BLMH "Exposé"* uses six photographs to document a pretend break-in into Wally's assistant's office by a menacing figure wearing a white lab coat and white cloth over his or her face. Sporting an oversized name tag that reads, "*Genentech,*" the figure removes and photographs a file with an oversize label saying, "*Biogen*."

The lab's "poets" (clearly not Celia) create a page of silly limericks teasing all six of the post-docs in the lab at the time ("Post-docs on Parade"). We also include (less rude) limericks for Allan and Wally:

There once was a scientist named Allan
Whose popularity forever did tax him.
A sequencer divine,
Too modest, he pined,
"I just can't figure out what attracts them."

There once was a scientist named Walter
Whose experiments just never would falter.
Then the monster loomed,
And Chicago was doomed,
By a plasmid that insulin altered.

We memorialize Greg Sutcliffe's long evenings typing the 8,722-base pBR322 sequence (4,361 nucleotides in both directions)[aa] with a candid photo of him working on his thesis. Sitting

[z] http://www.dramming.com/2014/12/17/dewars-new-ad-campaign-lesson-learned/, last accessed 8 January 2017.

[aa] Sutcliffe JG. *Complete nucleotide sequence of the Escherichia coli plasmid, pBR322.* Cold Spring Harbor Symp Quant Biol 43, Part 1:77090, 1979.

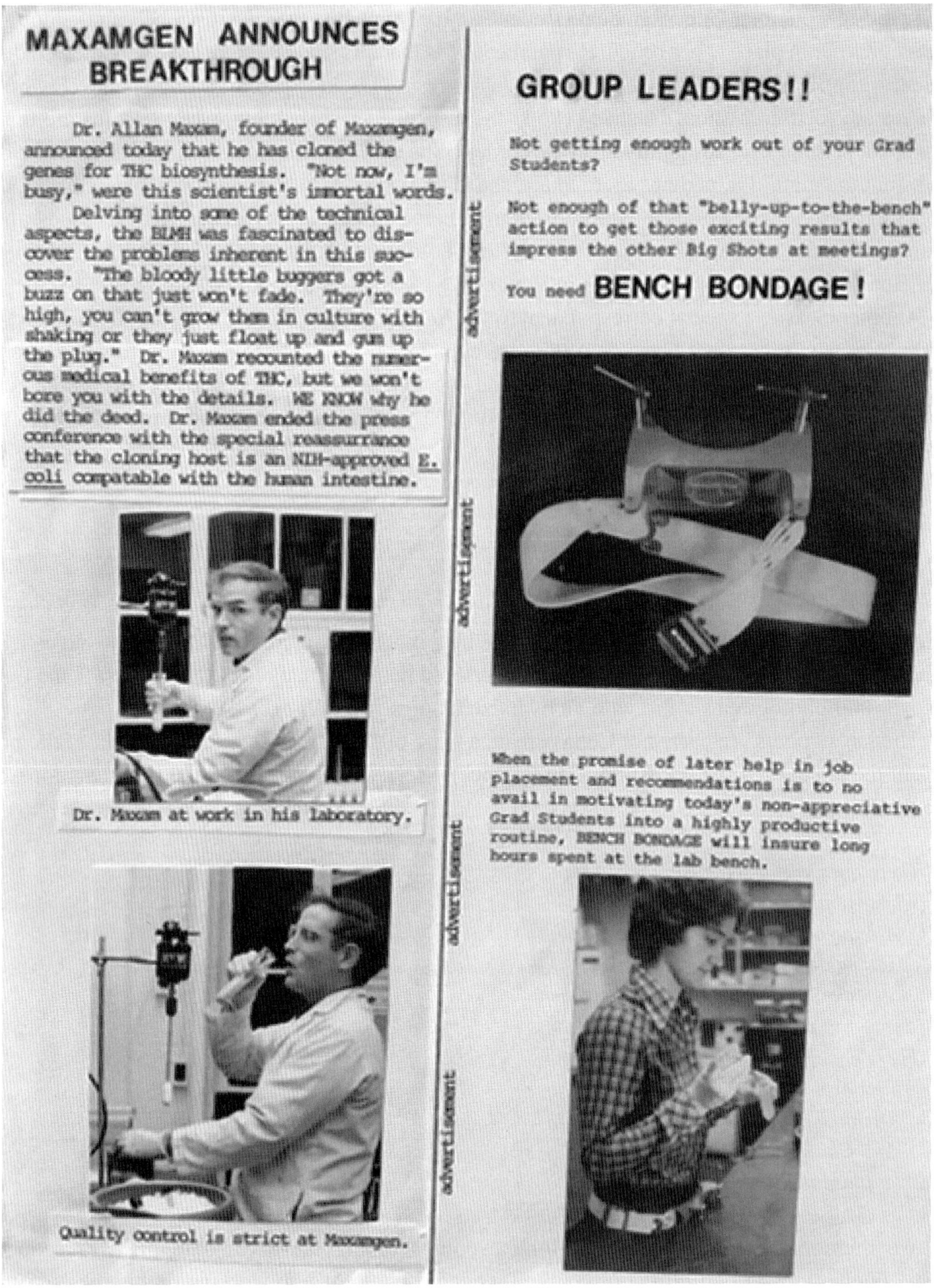

MAXAMGEN ANNOUNCES BREAKTHROUGH

Dr. Allan Maxam, founder of Maxamgen, announced today that he has cloned the genes for THC biosynthesis. "Not now, I'm busy," were this scientist's immortal words.

Delving into some of the technical aspects, the BLMH was fascinated to discover the problems inherent in this success. "The bloody little buggers got a buzz on that just won't fade. They're so high, you can't grow them in culture with shaking or they just float up and gum up the plug." Dr. Maxam recounted the numerous medical benefits of THC, but we won't bore you with the details. WE KNOW why he did the deed. Dr. Maxam ended the press conference with the special reassurrance that the cloning host is an NIH-approved _E. coli_ compatable with the human intestine.

Dr. Maxam at work in his laboratory.

Quality control is strict at Maxamgen.

advertisement advertisement advertisement advertisement

GROUP LEADERS!!

Not getting enough work out of your Grad Students?

Not enough of that "belly-up-to-the-bench" action to get those exciting results that impress the other Big Shots at meetings?

You need **BENCH BONDAGE!**

When the promise of later help in job placement and recommendations is to no avail in motivating today's non-appreciative Grad Students into a highly productive routine, BENCH BONDAGE will insure long hours spent at the lab bench.

Figure 7: Original page from issue #4, containing "Maxamgen Announces Breakthrough" and "Bench Bondage".

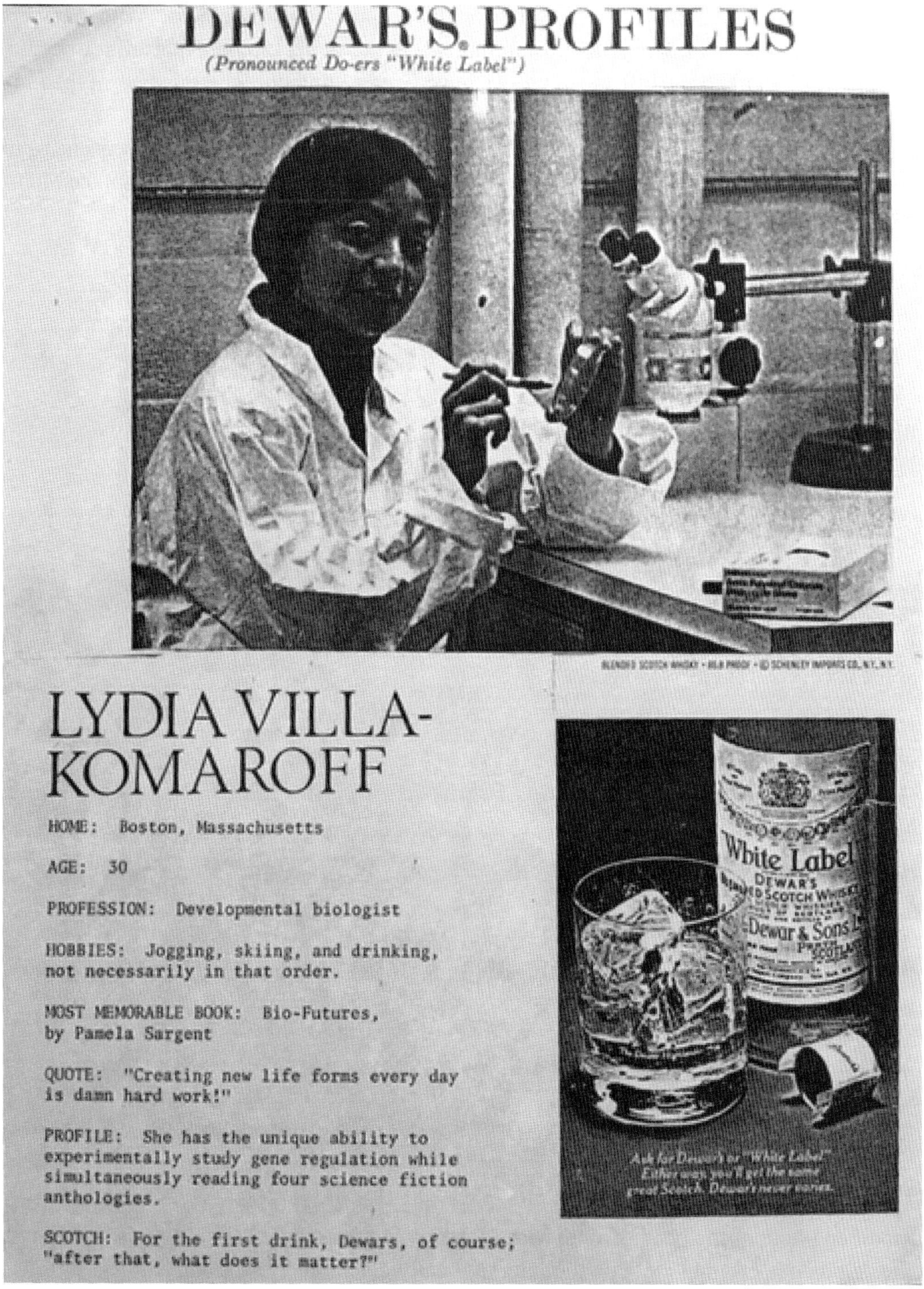

Figure 8: "Dewar's Profile" of Lydia Villa-Komaroff from *BLMH* issue #4.

at the desk in Wally's administrative office in front of the famous IBM Selectric, with papers piled high and files strewn about, Greg, not posing, intently reviews his typing. Appearing to come out of his mouth and go up to the office ceiling, we have Greg saying out loud this latest typed region of the "DNA sequence": *AGASOBCGFYCAGGFYACT*

The last page of the issue contains "The *BLMH* Songbook," with verses about DNA sequencing to the tunes of "I Get Around," by the Beach Boys, and "Hey, Hey, We're the Monkees," by (yup) the Monkees.

Some excerpts from the "Beach Boys":

I get around, round, round, round, I get around
Get around, round, round, I get around

I am getting bugged running up 'n down the same old gene
I gotta find a new clones where the frags come clean

We always take my 'zine cause it's never been beat
Yeah, we've never missed yet, with every C so neat

I get around, round, round, round, I get around
Get around, round, round, I get around

None of runs steady 'cause it wouldn't be right
To have to load your frags on a Saturday night

Some excerpts from the "Monkees":

Here we come...
Walking through the lab...
Get the funniest looks from
Everyone we grab...

Hey, hey, I'm a monkey,
People say I monkey around
But we're too busy sequencing
To put anybody down.

We're the young generation
And we've got something to clone,
We got to butt-end ligate
So Wally'll leave us alone

The Introns and Exons poster, July 1978

The modified photo of the band, Klone, with its joke line on the cover about Iggy and the Introns, becomes a raucous evening creating the internationally celebrated "rock bands," The Introns and The Exons.

We have to rush, because we want to create a poster for the Nucleic Acids Gordon Conference about to take place in a few weeks, being chaired by the famous Dr. Walter Gilbert, who, as I already mentioned, has proposed those very names.

Once again the brain child of Allan Maxam, most of the *BLMH* contributors at that time[ab] gather in the Tea Room around 9 p.m. We are wearing various outlandish outfits, loosely inspired by the photo of "Klone," and bring lots of beer, a bottle of Glenfiddich, and other "refreshments" of that era.

We turn up the music in the lab across the hall and set up our band's "equipment" — stacking power supplies and positioning a light box on a chair illuminating a sequencing gel film. We have two guitars, which various people pretend to play. Debra Peattie and I grab Geiger counters and pretend to sing into the probes. With the room's 6-foot high DNA model[ac] behind us, we strike multiple poses in multiple combinations of people. The stacks of equipment get higher, and the accompanying rock music gets louder.

Kirston Koths,[ad] our ace photographer, produces scores of photos recording our various poses. We glue our two favourites on poster-sized paper. At the top, we letter on, "The two greatest bands in Molecular Biology." We call the upper band The Introns, the lower band The Exons, and inscribe at the bottom:

For the most MIND-BLOWING meeting ever, bring the
Introns and Exons to your next Gordon Conference.
Contact our agent at the Harvard Biological Laboratories.

Ulli Siebenlist and Greg Sutcliffe drive from Cambridge to the Nucleic Acids Gordon Conference in New Hampshire with the DNA model (for Ulli's talk) and the "poster," which Ulli puts up. Whatever the conference attendees think about the poster (Wally loves it), they vote against using Introns and Exons as scientific terms, although Wally's names quickly become accepted in real life, because they are short yet descriptive. Figure 9 is a photo of the poster, framed and hanging in my home office.

Issue 5, Early 1979

In autumn 1978, Wally, Argiris Efstradiatis, Lydia Villa-Komaroff, and Stephanie Broome travel to the UK to use a P4 containment facility (required at that time by the NIH guidelines) to clone the human insulin gene cDNA, without success.

Avoiding the NIH containment requirement by using synthetic DNA, scientists at UCSF, City of Hope National Medical Center, and Genentech publish back-to-back papers on the creation of two synthetic human insulin genes, encoding the A and B chains,[ae] in December 1978 and their expression in *E. coli*[af] in January 1979.

[ab] Argiris Efstradiatis, Winship Herr, Peter Lomedico, Allan Maxam, Mark Pasek, Debra Peattie, Nadia Rosenthal, Greg Sutcliffe, Steve Stahl and Karen Talmadge.

[ac] The model, built by Gilbert grad student, Ulli Siebenlist, contains the T7 promoter.

[ad] Dressler lab grad student.

[ae] Crea R, Kraszewski A, Hirose T, Itakura K. *Chemical synthesis of genes for human insulin.* Proc Natl Acad Sci USA 75:5765–9, 1978.

[af] Goeddel DV, Kleid DG, Bolivar F, Heyenecker H, Yansura DG, Crea R. Hirose T, Kraszewski A, Itakura K, Riggs AD. Proc Natl Acad Sci USA 76:106–10, 1979.

Figure 9: Introns and Exons poster brought to the 1978 Nucleic Acids Gordon Conference chaired by Wally. From left to right, the "Introns": Winship Herr, Allan Maxam, Karen Talmadge, Mark Pasek, Debra Peattie, Steve Stahl, Greg Sutcliffe; the "Exons": Peter Lomedico, Debra Peattie, Karen Talmadge, Argiris Efstradiatis, Nadia Rosenthal (on the floor), Winship Herr.

The famous brilliant pretend newspaper, *Not the New York Times*, appears in Fall 1978 during a long *New York Times* strike.[ag] Its name banner, a perfect imitation of the actual *New York Times* font, becomes the title of our fifth *BLMH*, "published" in early 1979. We add, "But the *Bio Labs Midnight Hustler*" immediately underneath.

The front page also contains a photo of Wally playing with a yo-yo in the Tea Room ("Our Leader"), as well as a photo of Allan making a rude gesture ("Our Motto"). Both faces have boxes over their eyes, pretending to make the photos anonymous. Figure 10 shows a copy of this cover.

The lead story about the trip to the UK becomes, of course, "*Mission Impossible*":

> *An alert but unassuming, mild-mannered biochemist slipped quietly into the phone booth. He took a small tape recorder from his coat pocket, set it on a phone book, and let it play.*
>
> *"Hello, Wally. George Church is a seemingly innocuous graduate student in your laboratory. In reality, he is a member of a worldwide conspiracy [...] to create international chaos through the insidious induction of the disease syndrome known as diabetes.*[ah]
>
> *Your mission, if you accept it, will be to go to England with your Impossible Mission Team, to work in a top-secret military laboratory and to clone the gene for human insulin. If you fail, the Secretary will deny all knowledge of your actions. This tape will self-destruct in ten seconds. Good luck, Wally."*

After discussing this exciting challenge with his Mission Impossible Team, Wally renders his verdict: *"Aw, f*** it," he said, "it's too much trouble!"*

In "Maxamgen Does it Again", Dr. Maxam announces that his "company" has cloned the gene for human intelligence, which they detect by screening "each transformant for the ability to perform three simple tasks: running a maze, tying its shoes, and getting a PhD in biochemistry from Harvard University."

We create a fake journal, the "T1 Digest", dedicated to sequencing artifacts (and we have many examples to share).

One enterprising contributor approaches the Nobel Prize race more generally (and with unusual subtlety for the *BLMH*) through a pretend "United States Census Bureau Socioeconomic Survey," showing, by month, the number of bottles of vintage Dom Perignon champagne purchased during the year from a then-trendy Harvard Square wine shop.

As you can see in Figure 11, there is a >4 fold increase in vintage Dom Perignon champagne sales in (presumably very early) October, when Nobel Prizes are announced. The contributor also slyly promises a follow-up Graph B, where Nobel Prize dreams meet reality, depicting unopened Dom Perignon bottle returns over the same time frame.

With gusto and ridiculous bias, this *BLMH* issue presents two pages crammed with 15 jokes, altered to make fun of UCSF scientists. The title of the first page is "UCSF — America's brightest hope for cancer cure". The second page title sidebar says, "Why you should donate one-third of your lifetime earnings to UCSF cancer research."

[ag] http://www.nytimes.com/2008/11/15/nyregion/15about.html?_r=0, last accessed 19 September 2016.

[ah] Why George Church? Because he is the newest grad student in the lab at the time, and, as a soft-spoken vegan with a gentle voice and manner, he is the polar opposite of a diabolical villain.

Not The New York Times

BUT THE

BIO LABS MIDNIGHT HUSTLER®

OUR LEADER

OUR MOTTO

® This is a registered trademark of the Gilbert Group, Harvard University

Figure 10: Cover of Issue 5 from early 1979, *Not The New York Times, But The Bio Labs Midnight Hustler.*

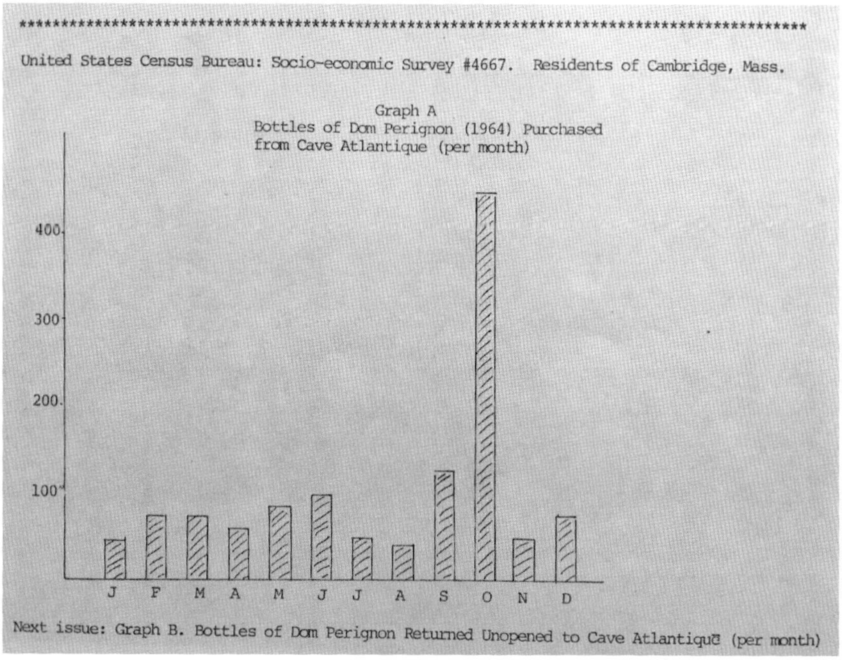

Figure 11: Monthly sales of 1964 Dom Perignon champagne at a Harvard Square wine shop, from Issue #4, June 1978.

An example:

What do you find written on the bottom of UCSF reagent bottles?
"OPEN OTHER END"
What do you find written on the top of UCSF reagent bottles?
"SEE OTHER END FOR INSTRUCTIONS"

Another side-splitter:

Did you know that, if a UCSF graduate student moves to Harvard, he or she diminishes the intelligence at both institutions?

All the jokes appear in Figures 12 and 13.[ai]

Recycling our humour, the last page of this issue is the photo of the "Introns", this time with the label "*Editorial Board of the Hustler*."

[ai]Meeting me for the first time, at the 1980 Nucleic Acids Gordon Conference, John Fiddes, a former graduate student with Fred Sanger (!) who at that time was finishing his post-doc with Howard Goodman (!!), asks me, "Did you have anything to do with all those UCSF jokes? What was that about?" John and I marry the following year.

UCSF –
America's brightest hope for cancer cure...

'Honestly, what are my chances of cloning this gene?' asked a greenhorn postdoc, soon after arriving in a well-known UCSF laboratory.
'Excellent,' asserted the UCSF professor whose lab that was, 'nine out of ten postdocs screw up all their experiments, and the nine who preceded you failed to get this gene.'

Why are UCSF grad students and postdocs constantly sought by the cloning conglomerates Biogen and Genentech?
That's obvious - because even under drugs, booze, and torture they can never remember just what it is they are working on.

Late one night the guard at the UCSF Medical Center main entrance stopped a student attempting to walk out with two huge plastic bags full of telephones.
'Could you explain why you are taking these phones out of UCSF,' he asked.
'Certainly,' she replied. 'I just passed an audition with the UCSF Symphony Orchestra and they told me to bring two sacks of phones with me for tonight's rehearsal.'

Did you hear about the UCSF tumor virologist who hasn't been around the lab since he joined the Mafia?
They made him an offer he couldn't understand.

Three UCSF graduate students, vacationing last summer in Uganda to get away from the turmoil of recombinant DNA research, got caught up in a movement to overthrow Idi Amin there. They were sentenced to be guillotined. As the first one waited for the blade to fall, it stuck, and he was released according to the old custom. The same thing happened to the second, and he too was released.
As the third UCSF grad student looked up waiting for the blade he shouted 'Hold on, I think I see what's making it stick.'

Have you heard about the energy-conscious UCSF researcher who has been working out the E. coli <u>lac</u> operator sequence for the last five years?
He turns his sequencing gels down to 100 volts at night to conserve electricity.

Did you know that if a UCSF graduate student moves to Harvard he or she diminishes the level of intelligence at both institutions?

A Martian landed on top of the UCSF Medical Center at ten o'clock one morning, right beside a UCSF recombinant DNA researcher having his beer break.
'Take me to your leader,' said the Martian, 'I'm from Mars.'
'I'm very pleased to meet you,' replied the UCSF molecular biologist, 'I've often meant to write and tell you how much I enjoy your candy bars.'

Figure 12: First page of UCSF jokes, Issue 5, Early 1979.

Why you should donate one third of your lifetime earnings to UCSF Cancer Research —

A woman from UCSF went to a psychiatrist and told him the guy she was living with, also at UCSF, believed he was urinating gamma-labeled ATP.
'Don't worry,' said the psychiatrist, 'I'll cure him of that.'
'Oh I don't want you to cure him,' replied the UCSF sequencer, 'I just want you to fix him so it comes out carrier-free!'

Exhausted after a twenty hour day spent searching in vain for a human needle in the rat haystack, two UCSF scientists decided to get relief in a corner of the P3 lab.
'This is the best stuff on earth,' said one to the other, 'it will get you high as a kite in seconds.'
Hoping for that, the other took a long pull, held it, and let it go into the safety cabinet. Behind the glass, laminar flow lifted the sheet of smoke ever higher, upward to the electronic nose at the top.
Suddenly all heaven and hell broke loose in a clanging of bells, blinking of lights, and whooshing of aromatic gases. Doors slammed shut, other doors flew open, voices boomed over the P.A., and people came running into the UCSF P3 from all directions.
'Right on!' exclaimed the rejuvenated gene-splicer, after what seemed an eternity, but was really only seconds, 'I haven't been buzz-bombed like this for weeks!'

Two new UCSF graduate students arrived in the Boyer lab at the same time, one from Harvard and the other from Princeton. For their rotation work, he set the Princeton graduate to Vortexing solutions and the Harvard grad to scraping old gels off their glass plates.
After a few weeks the student from Harvard complained to Boyer that he wanted to run the Vortexer, just like the Princeton guy. To which Herb Boyer replied, 'What in your background makes you think you can handle sophisticated scientific instruments?'

What do you find written on the bottom of UCSF reagent bottles?
'OPEN OTHER END'
What do you find written on the top of UCSF reagent bottles?
'SEE OTHER END FOR INSTRUCTIONS'

One of the better UCSF postdocs once set out to swim across the Bay. However, when he was about three-quarters of the way across he felt he couldn't make it, so he turned around and swam back.

One UCSF postdoc to another, over lunch:
'Joe, you synthesized the DNA for that HiIQ gene over a year ago. How come you haven't got it cloned and expressed yet?'
'Well, Sarah, that's a sad story. It's taking us much longer than expected to synthesize the pBR322 vector DNA. We can't clone until we finish making the plasmid.'

UCSF's Microbiology Department is reputed to have yet another facility for recombinant DNA research, a million dollar, space-age P4 laboratory devoted exclusively to the study of bacterial genetics.

Figure 13: Second page of UCSF jokes, Issue 5, early 1979.

Issue 6, August 1979

The cover of the sixth issue uses the banner from Larry Flynt's magazine *Hustler*, with separate photos of Wally and graduate student Nadia Rosenthal pasted so that they appear to be dancing together (Figure 14).

Former Cambridge Mayor, Alfred Vellucci, and the recombinant DNA controversy in Cambridge from 1976 reappear in a report about three baby bronze rhinos escaping from their cradles in the Bio Labs courtyard. Explaining how the baby rhinos got so big, Wally confesses to the accidental addition of an "epidermal growth factor-bearing plasmid" into their feed. The rhinos are eventually brought back down to size by Mark Ptashne giving them "juvenile hormone". (Who says the stories have to make sense?) The article also reveals the formation of Pope John Paul II's new biotech company, "Genuflech."

BLMH's pusillanimous prosifiers pretensiously parody one of the most elegant abstracts Wally pens in this era, "Exons and Introns: Playgrounds of Evolution" for a 1979 ICN-UCLA Symposium. The *BLMH* title becomes "Exxons and Biogens: Playgrounds of Economic Evolution," and I believe that the text closely follows the abstract:

> *It is less than three years ago that the scientific community first recognized a natural splicing mechanism between molecular biologists and large corporations. The natural (that is, pre-spliced) state can be viewed as molecular biologists floating like bubbles on a sea of dollar bills, although*

Figure 14: Issue 6 cover page.

the analogy is somewhat crude. The spliced product eliminates billions of dollars (which go to research, development and internal bribes) and leaves a molecular biologist fused to mere millions.

Since this discovery, Davos, Guadeloupe, Zurich, and South San Francisco have become the playgrounds for this economic evolution.

We create a pretend journal, the "T1 Digest", dedicated to sequencing artifacts. In the inaugural article, entitled "Sequence of the L1 deletion in the lac promoter region", the paper's Figure 1 supposedly shows a film containing bands representing the wild type sequence using the Maxam-Gilbert technique. Next to this (the paper's Figure 2) is a second, similar, film, with bands at the top and the bottom of each lane, but other absent. This is described as, "the sequence of the L1 deletion" — a blank region in the film. (Ho ho ho!) The first (and only) edition of the T1 Digest is shown in full in Figure 15.

The core of this issue is a "yearbook" featuring photos of the 28 current members of Wally's and David Dressler's labs,[aj] complete with votes for the typical high school yearbook categories, including *Most Likely to Succeed*. There are no prizes of any sort for guessing that Wally is named *Most Likely to Succeed*; this "award" is accompanied by my out-of-focus photo of Wally receiving a Secret Santa gift of "The Nobel Prize Game" at an earlier holiday party (Figure 16).

Issue 7, Mid-1980

In what becomes our last issue, *BLMH* editors use an NIH grant coversheet as our cover. Figure 17 contains a copy of the original coversheet; here are examples of the light-hearted answers in some of the boxes:

1. Title of grant application? *Bio Labs Midnight Hustler*
2. Response to specific program announcement? *Yes*. If yes, state RFA number and/or announcement title. *Support for dirty magazines*

3a. Principal Investigator? *Man, Maine M.*
3b. Social Security Number? *Same as Biogen account number*
3c. Mailing address? *Across the river*
3d. Position title? *Graduate Hustler*
3e. Department, Service, Laboratory or Equivalent? *Equivalent*
3f. Telephone? *Hasn't got one yet*
3g. Major subdivision? *Harvard Yard*

Did "Maine Man" really fill out the pretend grant coversheet? I won't tell (although the handwriting does).

[aj]As noted earlier, the two labs share common areas and equipment in this era. This includes the professors, staff, undergraduate and graduate students, and post-docs.

THE T1 DIGEST
Vol 1, No 1

****

Announcement

The editors of The T1 Digest wish to announce the formation of a new journal of electron microscopy, Acta Artifacta, to be edited by David Dressler of Harvard University.

****

Articles

Sequence of the L1 deletion in the lac promoter region

Waclaw Szybalski, Department of Oncology, University of Wisconsin

Summary
L1 is a large deletion which maps in the promoter region of the lac operon. I have sequenced a restiction fragment containing the L1 deletion and its sequence is .

Results
Figure 1 below displays the Maxam-Gilbert cleavage products of an end-labelled restiction fragment containing the lac promoter region. Figure 2 shows the same fragment with the L1 deletion.

Fig. 1

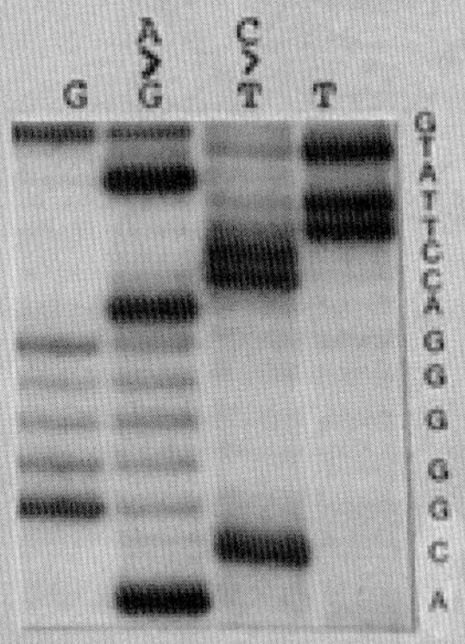

Fig. 2

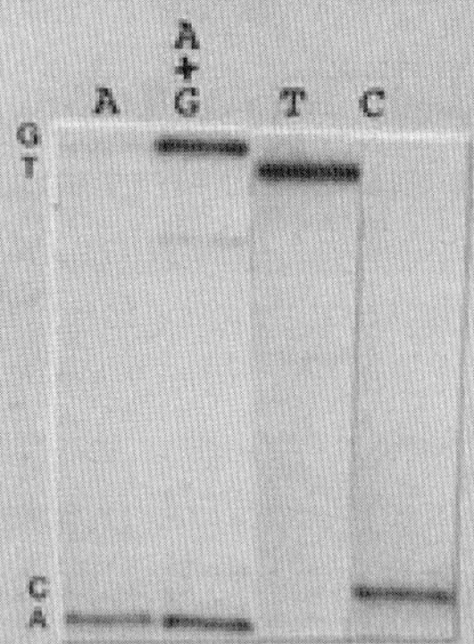

The deletion and flanking sequences are written and compared below to the wild type sequence.

L1 GT CA
WT GTATTCCAGGGGGCA

Discussion

Figure 15: T1 digest.

Figure 16: Wally with his Nobel Prize Game, his Secret Santa gift, December 1978. He played and won! Photo taken poorly by me.

Kicking off the issue, we publish a delightful Letter to the Editor from Gilda Radner's *Saturday Night Live* character Emily Litella, writing on a hot scientific topic of the time, with Ms. Litella's trademark malapropisms:

> *Dear Editor,*
>
> *What's all this I've been reading about shellfish DNA? Several authors in Nature (17 April 1980) try to explain why human cells should have shellfish DNA. Why are they so surprised? This is another case of molecular biologists who don't know any biology. Open any embryology textbook and you will see that the human embryo looks much more like fish than people; in fact, the 32-cell stage embryo looks just like a baby sea scallop. Why weren't these facts brought out in these papers? After all, how else could we breathe underwater for nine months if we didn't start out with fish DNA? […] I think it's an outrage!*
>
> *Yours sincerely,*
> *Gilda Radner (written signature)*
> *Emily Litella, BS (typed signature)*
> *Unemployed*

The *BLMH* replies:

> *Ms. Litella, that's selfish, not shellfish, DNA. And, for the sake of brevity, we will answer some of your other recent letters. Introns and exons have nothing to do with the major oil companies, and SV-40 is not a new lubricant they are trying to push. Now leave us alone!*
>
> *P.S. HLA is not a terrorist group.*

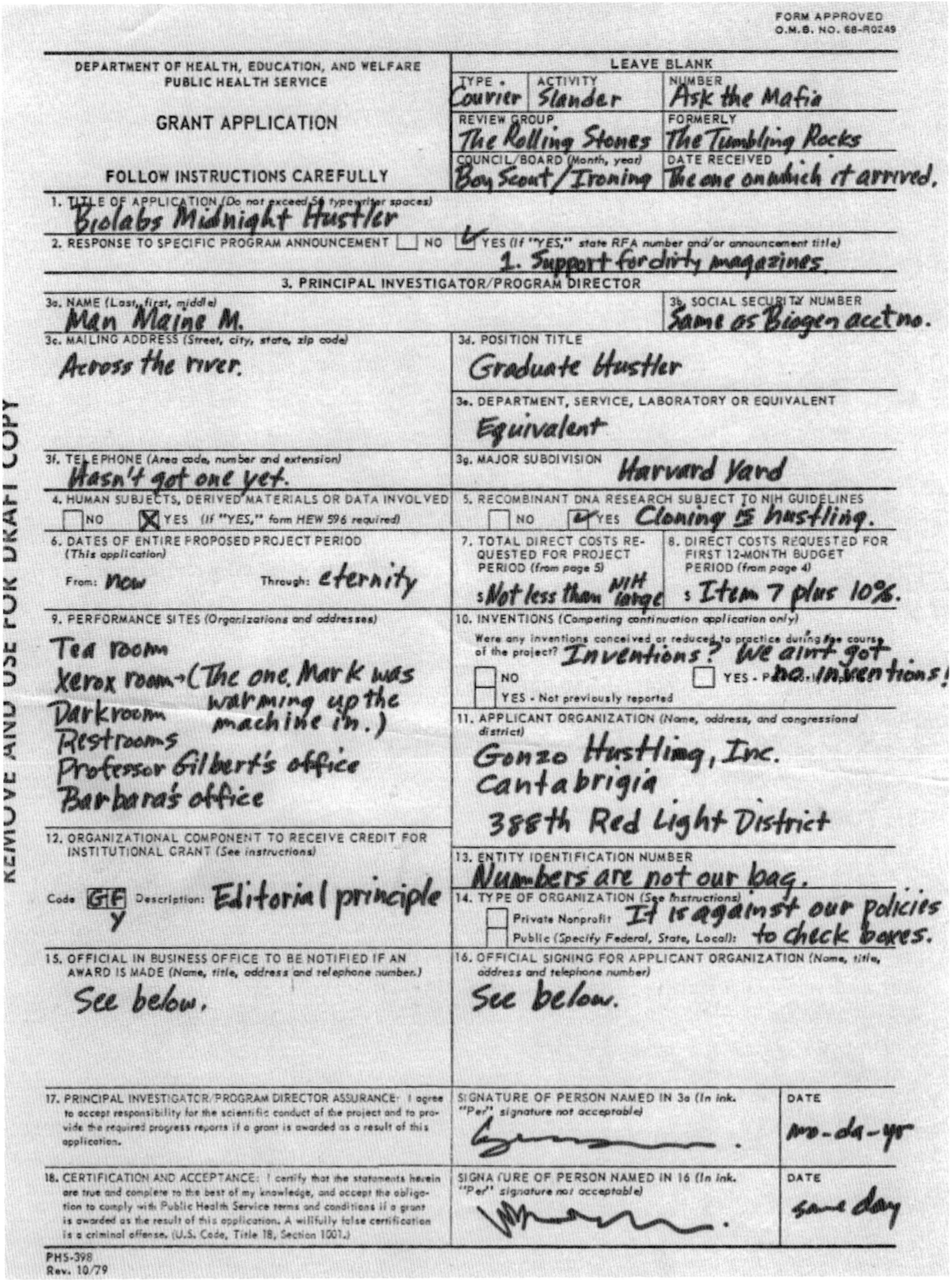

FORM APPROVED
O.M.B. NO. 68-R0249

DEPARTMENT OF HEALTH, EDUCATION, AND WELFARE
PUBLIC HEALTH SERVICE

GRANT APPLICATION

FOLLOW INSTRUCTIONS CAREFULLY

LEAVE BLANK

TYPE: Courier | ACTIVITY: Slander | NUMBER: Ask the Mafia

REVIEW GROUP: The Rolling Stones | FORMERLY: The Tumbling Rocks

COUNCIL/BOARD (Month, year): Boy Scout/Ironing | DATE RECEIVED: The one on which it arrived.

1. TITLE OF APPLICATION (Do not exceed 56 typewriter spaces)
Biolabs Midnight Hustler

2. RESPONSE TO SPECIFIC PROGRAM ANNOUNCEMENT ☐ NO ☑ YES (If "YES," state RFA number and/or announcement title)
1. Support for dirty magazines.

3. PRINCIPAL INVESTIGATOR/PROGRAM DIRECTOR

3a. NAME (Last, first, middle)
Man Maine M.

3b. SOCIAL SECURITY NUMBER
Same as Biogen acct no.

3c. MAILING ADDRESS (Street, city, state, zip code)
Across the river.

3d. POSITION TITLE
Graduate Hustler

3e. DEPARTMENT, SERVICE, LABORATORY OR EQUIVALENT
Equivalent

3f. TELEPHONE (Area code, number and extension)
Hasn't got one yet.

3g. MAJOR SUBDIVISION
Harvard Yard

4. HUMAN SUBJECTS, DERIVED MATERIALS OR DATA INVOLVED
☐ NO ☒ YES (If "YES," form HEW 596 required)

5. RECOMBINANT DNA RESEARCH SUBJECT TO NIH GUIDELINES
☐ NO ☑ YES Cloning is hustling.

6. DATES OF ENTIRE PROPOSED PROJECT PERIOD (This application)
From: now Through: eternity

7. TOTAL DIRECT COSTS REQUESTED FOR PROJECT PERIOD (from page 5)
$ Not less than NIH large

8. DIRECT COSTS REQUESTED FOR FIRST 12-MONTH BUDGET PERIOD (from page 4)
$ Item 7 plus 10%.

9. PERFORMANCE SITES (Organizations and addresses)
Tea room
Xerox room → (The one Mark was warming up the machine in.)
Darkroom
Restrooms
Professor Gilbert's office
Barbara's office

10. INVENTIONS (Competing continuation application only)
Were any inventions conceived or reduced to practice during the course of the project? Inventions? We aint got no inventions!
☐ NO
☐ YES - Not previously reported
☐ YES - Previously reported

11. APPLICANT ORGANIZATION (Name, address, and congressional district)
Gonzo Hustling, Inc.
Cantabrigia
388th Red Light District

12. ORGANIZATIONAL COMPONENT TO RECEIVE CREDIT FOR INSTITUTIONAL GRANT (See instructions)
Code: GFY Description: Editorial principle

13. ENTITY IDENTIFICATION NUMBER
Numbers are not our bag.

14. TYPE OF ORGANIZATION (See instructions)
☐ Private Nonprofit
☐ Public (Specify Federal, State, Local):
It is against our policies to check boxes.

15. OFFICIAL IN BUSINESS OFFICE TO BE NOTIFIED IF AN AWARD IS MADE (Name, title, address and telephone number.)
See below.

16. OFFICIAL SIGNING FOR APPLICANT ORGANIZATION (Name, title, address and telephone number)
See below.

17. PRINCIPAL INVESTIGATOR/PROGRAM DIRECTOR ASSURANCE: I agree to accept responsibility for the scientific conduct of the project and to provide the required progress reports if a grant is awarded as a result of this application.
SIGNATURE OF PERSON NAMED IN 3a (In ink. "Per" signature not acceptable)
DATE: mo-da-yr

18. CERTIFICATION AND ACCEPTANCE: I certify that the statements herein are true and complete to the best of my knowledge, and accept the obligation to comply with Public Health Service terms and conditions if a grant is awarded as the result of this application. A willfully false certification is a criminal offense. (U.S. Code, Title 18, Section 1001.)
SIGNATURE OF PERSON NAMED IN 16 (In ink. "Per" signature not acceptable)
DATE: same day

PHS-398
Rev. 10/79

REMOVE AND USE FOR DRAFT COPY

Figure 17: Cover page of Issue 7, Mid-1980.

Next to that letter is an actual advertisement for "Male Size Enlargement Techniques" with a substitute photo showing an (unnamed) male graduate student (and good sport) sitting at the console of the Dressler lab's electron microscope, with the instrument itself extending in a large column about two feet wide and about five feet above the console. The photo's legend: *"Satisfied customer shows off tool extender."*

We gently gibe a new post-doc with a penchant for pinching lab equipment in "Harvard Scientists Reveal Mysterious Magnet in Bio Labs 382":

> *Scientists in the research group headed by Dr. Walter Gilbert announced today the discovery of a mysterious magnetic field emanating from room 382 of the Biological Laboratories. Careful calibrations have defined the source more accurately to be the bench of [an unnamed] Gilbert post-doctoral fellow... The phenomenon was first observed when...small pieces of research equipment, formerly stored in various rooms of the Gilbert research complex, began to accumulate on [the post-doc's] bench. Interviewed yesterday [...] [the post-doc] confessed it was convenient "to have so much lab equipment at my fingertips" and joked, "Well, my [spouse] tells me I have a magnetic personality!"*

Callous correspondents cruelly carp over the potential commercial applications of the expression vectors that are part of my thesis work studying protein secretion.[ak] This is the first page of my supposed draft manuscript, complete with up-to-the minute "Star Wars" references:

CONSTRUCTION OF CLONING VECTORS (PAT. PENDING)
(production/profits/Schering-Plough/Inco)
Karen Talmadge and Walter Gilbert
Jr. Director and Director, Biogen, N.V.

> *Biogen, N.V., announces three new additions to its Clone-ease™ line of cloning products, the plasmids (pat. pending) Secretaframe 0®, "Secretaframe +1®", and "Secretaframe +2T®". We digested HinII-cut pKT41 with the double-stranded exonuclease, Chewbacka VectorI®, and created unique cloning sites by incubating the Chewbacka VectorI®-treated plasmids with Cut'N'Splice™ and a magic ingredient (pat. to be applied for). When total DNA from any organism is treated with new, expanded Clone-Ease™ kit, the gene product of interest is automatically cloned and secreted into 50 mg Pack-a-Protein pre-packaged vials from any Biogen patented host, including E. Cloni® and L-sells®. Call the Linkerline™, 617-495-3959,[al] for further information.*

Below the summary is the real Figure 1 from a paper I publish with Wally, bearing the legend, "Deletion map of pBR322 penicillinase gene and sequences of derivative plasmid signal sequence

[ak]Talmadge K, Stahl S, Gilbert W. *Eukaryotic signal sequence transports insulin antigen in Escherichia coli.* Proc Natl Acad Sci USA 77:3369–73, 1980.

[al]Direct phone line into lab 390, where I work.

(a)

(b)

Figure 18: (a) Arriving at my birthday party in San Francisco, July 2017: From left, childhood friend Adela Barcia, Roger Brent, Karen Talmadge, Celia Gilbert, Wally Gilbert, (b) A late night photo circa 1977, documenting the usual chaos in Wally's office, and particularly on his chalk board, which he had to erase to write anything new. From left, Karen Talmadge, Roger Brent, and post-doc from the then-U.S.S.R., Konstantin (Kostya) Scriabin, with an unidentifiable person at the bottom right of the photo.

regions…" in six different expression vectors. Wally paper clips a note (strangely, in my handwriting) covering the DNA sequences, with his instructions:

> *K.T.,*
> *Stonewall it — our patents were rejected!*
> *W.G.*

Just starting work in Wally's lab on muscle pyruvate kinase genes, and fresh from a centrifuge misadventure, a new graduate student (and very good sport) allows *BLMH* photographers to daringly document the notorious "Embden-Meyerhof Gang" in action. We see a photo of "Embden" at the bench, with a cute-as-a-button (and doomed) newborn chick standing next to a large lab blender (photo title "Embden mutilates chick"). Both Embden and the chick have black boxes over their eyes (to protect the innocent and the guilty, not necessarily in that order). The next photo documents the scarred centrifuge tub and mangled centrifuge rotor ("Meyerhof mutilates centrifuge").

With the title "*Are You a Cool Dude*?" and under three archaic dictionary definitions of "dude" (a "dandy or fop", a "city slicker or tourist", a "jazz aficionado"), we display a large photo of a graduate student from a different lab, visiting us to show off his Bar Mitzvah-worthy 3-piece suit (not his normal lab attire) and also swigging from a bottle, perhaps of whisky (not his normal lab beverage). Arrows point to various aspects of the photo with explanations of how this graduate student is demonstrating his quintessential dudeness (dudocity?). For example, *"Mind is a storehouse of useful Cool Expressions, such as 'KMA'"* (an arrow points to his head); *"Adoration by some local chicks"* (an arrow points to two female lab members nearby, laughing); *"Red sneakers, unmatched socks (data not shown)"* (an arrow points to his legs, as his feet are not in the photo).

"*Gilbert Gets Swede*" is the final Nobel Prize joke, only months before Wally actually receives the Nobel Prize in 1980:

> *After years of trying, Dr. Walter Gilbert announced today the acquisition of a genuine Swedish postdoc. Dr. Gilbert, who has been burned twice this year with Scandinavian-sounding names, appeared satisfied that Göran is, indeed, the genuine article. "He has already taught me to say 'Thank you' in Swedish," Dr. Gilbert exclaimed happily.*

So, can a pretend newspaper document a productive research era and provide a creative outlet for its pressures?

In a delightful 1995 *TIBS* article,[am] Jan Witkowski reveals the existence of TWO "humorous" "magazines" from biochemistry labs — the *BLMH* and *Brighter Biochemistry (BB)*, written by the Department of Biochemistry faculty of the Sir William Dunn Institute, in that other Cambridge, where they published (literally, each issue is printed and bound, possibly due to the lack of copy machines) eight issues, from 1923 to 1931.

[am] Witkowski J. *How I learned to stop worrying and love Cell.* Trends Biochem Sci 20:163–8,1995.

Both "magazines" contained similar types of content — poems, cartoons, pretend advertisements, pretend scientific reports. The key differences were a matter, shall I say, of tone. One publication was respectful and genteel, with a deft wit and a light touch; the other was the *Bio Labs Midnight Hustler*.

Witkowski suggests that the different status of the authors (faculty versus students) and the different eras in which they are writing (the 1920s versus the 1970s) may help explain this. To this list, I will add the differences between the British and American cultures, with the former valuing social position and hierarchy more than the latter, which continues today.

Witokowski quotes Joseph Needham, a member of the Sir William Dunn Institute Biochemistry Department faculty, rhapsodizing in 1949 about how *Brighter Biochemistry* bears:

> *...witness to the spirit of the Institute...full of brightness, not only of intellect and experiment, but of comradeship, alive awareness of the world outside biochemistry, and a warm inspiration owed and universally acknowledged to the leader and founder."*[an]

In contrast, Witkowski calls Wally's lab in the 1970s "...a hothouse — a forcing place for ambitious young molecular biologists to make their mark or perish"[aq] and quotes Stephen Hall's description in *Invisible Frontiers* of the atmosphere Wally sought to create in his lab:

> *...the open, egalitarian spirit of the 60s, and pointedly tried to eliminate all the unnecessary obstacles to making discoveries, which the students interpreted to include authority, formality, rules...*[aq]

Witkowski concludes overall that *BB* and *BLMH* provide data on how humor plays a role "in maintaining the cohesive unity necessary for the scientific enterprise."[ao]

I am not sure that progress in science requires a level of cohesiveness as much as, say, success in warfare or synchronized swimming. And, if humour is particularly important for science, shouldn't there be more *BBs* and *BLMHs*? In the Gilbert Lab, as in others, we shared equipment, methods and reagents. We attended weekly lab meetings to provide feedback and ideas, and we sometimes worked together. When we did, though, as in other labs, we developed a careful delineation of focus and responsibility, because, of course, documenting our individual contributions was critical for career advancement, as was first authorship on papers (see, *e.g.*, the letter from "Jane Goodall" to "Louis Leakey" in *BLMH* issue 5).

Did humour help? Of course, as did our lab meetings, our afternoon teas, our group meals at various Chinese restaurants, and our annual holiday parties with their secret Santa gifts (e.g. Wally's Nobel Prize game, Figure 16).

I suspect that the sustained effort required to create seven issues of the *BLMH* relates more to the extraordinary circumstances of those specific years (1977–1980) in the Gilbert Lab. These include the intersection of novel external challenges — the recombinant DNA controversies and the immediate impact of the nascent biotechnology industry — along with Wally's vertiginous and obviously well-deserved ascents into the scientific and biotechnology pantheons. With this as the backdrop to our research, the *BLMH* "correspondents" could select from a veritable buffet of opportunities for parody and fun.

[an]Witkowski J. *How I learned to stop worrying and love Cell.* Trends Biochem Sci 20:163–8, 1995, page 164.
[ao]Witkowski J. *How I learned to stop worrying and love Cell.* Trends Biochem Sci 20:163–8, 1995, page 168.

(a)

(b)

Figure 19: (a) During the birthday party: From left, Jeffrey Miller, Michele Calos, Karen Talmadge, Wally Gilbert, (b) From left, Jeffrey Miller and Michele Calos, circa 1978.

The lab membership at the time was also important. The *BLMH's* multiple authors (at least 15, see footnote 11) were not the only pranksters in the history of the Gilbert Lab (see, *e.g.*, Wally's pink lab coat from an earlier era, Figures 2a and 2b). I just suspect that we had an unusually high percentage of practical jokers, so that, when Phil Farabaugh went around to drum up enthusiasm for creating the *BLMH*, he shifted the equilibrium from inaction ("What a fun idea") to action

("Let's do this!"). We also had a relatively robust number of graduate students and post-docs habitually working late into the night, whether on their experiments or on "publishing" the *BLMH* (another mass action effect).

Why did the issues stop mid-1980? I left the lab in 1981, after a post-doctoral year with Wally, and bestowed my collection of random *BLMH* ideas and candid photos (poorly taken by me) to one of the "editors" who remained (I did not remember doing this until he brought them to me). I suspect that the remaining "editors" recognized that the forces fueling a lot of the humour — the recombinant DNA controversies, the race to clone and express the insulin gene, the birth of the biotechnology industry, and, most notably, the race for Wally's Nobel Prize — were now in the past (with all the battles ultimately won). And/or they intelligently focused on their research.

As I write this section, I have been noodling on Needham's quotation. His final words about *BB* are that it arises from "...a warm inspiration owed and universally acknowledged to the leader and founder."[ap] Reading and rereading the seven issues of the *BLMH*, I can see, with the clarifying lens of time, that they are love letters to Wally — for his brilliance and for all that he taught us as our lab generation navigated the rapid changes of that era. Similar to *Brighter Biochemistry*, we are expressing our deep respect for our leader, just not in such deeply respectful words and ways.[aq]

We are all acolytes of the Gospel of Gilbert: "It can be just as hard to study the small questions, so let's tackle the big ones." And we find ourselves applying Wally's teachings, not just in our science, but also in our approaches to many problems in life.[ar] As Wally's students even now, we strive for his intellectual rigour, we reflexively use and teach his myriad of methods for working efficiently, and we surprise ourselves as we achieve some of his fearlessness. He leaps from physics to molecular biology; wins a Nobel Prize in his new field; starts one of the most successful biotech companies (about to celebrate its 40th anniversary!), which he leads in its early years to some of its biggest selling drugs even now; becomes a venture investor; and, most recently, having reached the age where Harvard requires all its Professors to retire, he dives into a burgeoning career in computer art. Did I mention the Nobel Prize? With Wally as a constant role model and source of inspiration, I also have made some (much, much, smaller) leaps. Who says his students cannot go to law school; go back to science; move into business; co-found, fund, and lead an orthopedic spine medical device company; and then help lead a non-profit to a new structure and new funding models? Not Wally. (I am looking forward to my as-yet-unidentified future efforts in the arts.)

And, can we be productive while being frivolous? For your consideration, here are some of the research areas and papers from four of the main *BLMH* gluers and copiers ("editors") during the *BLMH* era: the sequence of the *lac* repressor gene[as] and the molecular nature of its spontaneous mutation hot spots[at]; novel chemical methods for sequencing RNA[au]; chemistries to probe mRNA

[ap] Witkowski J. *How I learned to stop worrying and love Cell.* Trends Biochem Sci 20:163–8, 1995, page 164.

[aq] Our respect for Allan Maxam is also very clear.

[ar] Wally also taught us to write titles that reveal the findings, pose the key experimental questions at the beginning of the paper, and write in the active voice. Thank you, Wally (I think)!

[as] Farabaugh PJ. *Sequence of the lacI gene*. Nature 274:765–9, 1978.

[at] Farabaugh PJ, Schmeissner U, Hofer M, Miller JH. *Genetic studies of the lac repressor. VII. On the molecular nature of spontaneous hotspots in the lacI gene of Escherichia coli.* J. Mol. Biol. 126:847–57, 1978.

[au] Peattie D. *Direct chemical method for sequencing RNA*. Proc Natl Acad Sci USA 76:1760–4, 1979.

secondary structure[av] and RNA-ribosomal[aw] interactions; studies on the structure and function of prokaryotic, eukaryotic and hybrid signal sequences to direct protein secretion and processing in *E. coli*[an,ax,ay]; murine leukemia virus AKV sequences[az]; studies on germ-line murine leukemia virus reintegrations[ba]; and the effect of thymic leukemias on murine immunoglobulin gene structures.[bb] Meanwhile, Allan Maxam is a co-author on 13 papers published 1976–1980,[bc] none of them otherwise cited in this paragraph.

We can never thank you enough, Wally, for the enormous privilege (and our amazing good fortune) of being your students.

Acknowledgments

Thank you to Nils Lonberg for sharing his memories and giving me all the *BLMH* materials I left with him for possible future *BLMHs*. Thank you as well (in alphabetical order) to Roger Brent, Argiris Efstradiatis, Phillip Farabaugh, Forrest Fuller, Walter Gilbert, Winship Herr, Kirston Koths, Debra Peattie, Ulrich Siebenlist, J. Gregor Sutcliffe, and Lydia Villa-Komaroff for sharing their memories and/or copies of BLMHs or other helpful documents and photos. A special thank you to Jürgen Brosius, Debra Peattie and Lydia Villa-Komaroff for carefully reviewing this chapter; to Jürgen Brosius for inviting me to give a talk about the *BLMH*'s at the 2005 Cold Spring Harbor symposium for Wally's 70th birthday, followed by cajoling Manyuan Long into inviting me to write this chapter; and to Wally Gilbert, Manyuan Long and Jurgen Brosius, for their support and incredible patience. An unanticipated benefit of writing this chapter, I reconnected with many former lab members, resulting in a fantastic addition to my blow-out 65th birthday party, organized for me by my daughter — a table filled with some of the contributors to, and gracious subjects of, the *BLMHs*, headlined by Wally Gilbert and Celia Gilbert, and also including (in alphabetical order) Roger Brent, Stephanie Broome, Michele Calos, Kirston Koths, Jeffrey Miller and Nils Lonberg. Figures 18a and 19a are photos from the July 2017 party in San Francisco, paired with Figures 18b and 19b, photos of some of the same people from the *BLMH* years, 1977–1980.

[av] Peattie D, Gilbert W. *Chemical probes for higher-order structure in RNA*. Proc Natl Acad Sci USA 77:4679–82, 1980.

[aw] Peattie DA, Douthwaite S, Garrett RA, Noller HF. *A "bulged" double helix in a RNA-protein contact site*. Proc Natl Acad Sci USA 78, 7331–5, 1981.

[ax] Talmadge K, Kaufman J, Gilbert W. *Bacteria mature preproinsulin to proinsulin*. Proc Natl Acad Sci USA 77:3988–92, 1980.

[ay] Talmadge K, Brosius J, Gilbert W. *An 'internal' signal sequence directs secretion and processing of proinsulin in bacteria*. Nature 294:176–8, 1981.

[az] Herr W, Corbin V, Gilbert W. *Nucleotide sequence of the 3" half of AKV.* Nucl Acids Res 10:6931–44, 1982.

[ba] Herr W, Gilbert W. *Germ-line MuLV reintegrations in AKR/J mice*. Nature 296:865–8, 1982.

[bb] Herr W, Perlmutter AP, Gilbert W. *Monoclonal AKR/J thymic leukemias contain multiple JH immungoglbin gene rearrangements*. Proc Natl Acad Sci USA 80:7433–6, 1983.

[bc] https://www.ncbi.nlm.nih.gov/pubmed?term=(Maxam%20A%5BAuthor%5D)%20AND%20(%221976%2F01%2F01%22%5BDate%20-%20Publication%5D%20%3A%20%221982%2F12%2F31%22%5BDate%20-%20Publication%5D), last accessed 15 January 2017.

Genes in Pizzas

Jürgen Brosius*

In August of 1977, when I first arrived in the U.S., I made a stop in Cape Cod, where I was kindly invited by Robert A. Zimmermann (University of Massachusetts, Amherst) to spend a few days at his vacation home before I continued west for my first postdoctoral position in Harry Noller's lab (University of California, Santa Cruz). It was precisely at this time that the scientific community was jolted by new discoveries; it had just been demonstrated that eukaryotic genes are split and mature messenger RNAs are patched together from larger heterogeneous nuclear RNA (hnRNA), often discarding the majority of the primary transcripts, while retaining the fused remains as template for translation [1, 2]. I must admit that I was dumbfounded to find my little molecular world — as I knew it from text books such as Jim Watson's *Molecular Biology of the Gene* — turned on its head. My naiveté concerning evolutionary concepts at the time did not help either. On the West Coast, I soon faced more practical challenges when faced with the task of sequencing an entire ribosomal RNA operon using the chemical method developed by Allan Maxam and Walter Gilbert [3], which, at the time, produced more reliable results compared to Fred Sanger's enzymatic sequencing approach.

Only a few months later, in February of 1978, *Nature* published a News and Views article by Wally Gilbert, entitled "Why genes in pieces? [4]" It offered not only reasonable explanations for this apparent quirk of nature, but some core assertions that were remarkable with respect to their foresight and accuracy. Ensuing research would substantiate these ideas for subsequent decades. This groundbreaking perspective had yet to raise the issue of whether introns are, in evolutionary terms, relatively late or early. Neither was the exon theory postulated, suggesting that exons correlate with defined protein domains or modules. Nevertheless, it was remarkable and seminal for a number of reasons:

(1) Gilbert introduced a catchy nomenclature: For the spliced-out intervening sequences he suggested introns, and for the parts that made up the mature mRNAs, the term exon was proposed. As we all know, the names stuck.
(2) He proposed the sharing of exons or groups of exons between genes via, e.g., illegitimate recombination and thus generation of diversity, a process termed exon shuffling, which was

* Institute of Experimental Pathology (ZMBE), University of Münster, Münster, Germany

substantiated by subsequent analysis of intron-containing genes [5] and whole genome sequence comparisons [6], at least for younger genes in multicellular organisms.

(3) He realized that the gene was not collinear with its product but instead a mosaic and more appropriately mirrored by the primary transcript: "The notion of the cistron, the genetic unit of function that one thought corresponded to a polypeptide chain, must be replaced by that of a transcription unit containing regions which will be lost from the mature messenger..."

(4) He suggested that the intron-exon structure of a gene would favour alternative splicing, a way to generate several protein variants out of a single gene: "...the dogma of one gene, one polypeptide chain disappeared. A gene, a contiguous region of DNA, rather corresponds to one transcription unit but that transcription unit can correspond to many polypeptide chains, of related or differing functions." Out of the disappointingly limited 20,000-human-gene repertoire, a copious number of protein isoforms and variants can be generated by alternative processing of the same transcription unit. As we know today, this not only applies to mRNAs encoding polypeptides but also to non-protein coding RNAs.

(5) Perhaps most importantly, Gilbert recognized that the intron/exon arrangement of genes gives evolutionary tinkering more room and leeway, especially since, initially and in most cases, the novel alternative splice product constitutes only a fraction of the original mRNA and consequently is chiefly neutral or slightly deleterious, even if the resulting protein product is (initially) functionally compromised. Over time, splice sites can be altered increasing the share of the new product or the novel exons can be lost again. The structural arrangement of genes with copious amounts of initially neutrally evolving sequences constitutes a large playground for the trial-and-error mode of evolution. "Single base changes ... not only can change protein sequences by the alteration of single amino acids but now, if they occur at the boundaries of the regions to be spliced out, can change the splicing pattern, resulting in the deletion or addition of whole sequences of amino acids. During the course of evolution relatively rare single mutations can generate novel proteins much more rapidly than would be possible if no splicing occurred." This has been borne out by numerous findings that neutrally evolving parts of retroposed elements, often including other intronic flanking regions, can be exonized [7–9]. In extreme cases "...one product's intron becomes another's exon." All these events are perfect examples of exaptation at the genomic level [10, 11].

After my first postdoctoral fellowship, I decided to make a stop on the East Coast on my (as it turned out rather protracted) way back to Europe. Fortunately, I was accepted in Wally's lab and began in September of 1980 at the Harvard Biolabs and stayed until late 1982. In this concentration of sharp-witted and creative individuals where everything and everyone was the target of wicked humor, I soon found out that my favorite paper was termed "genes in pizzas". (Figure 1) After my first month in the lab, Wally was awarded the 1980 Nobel Prize for Chemistry (Figure 2).

Those not on the photograph but in the Gilbert lab during my tenure: Göran Akusjärvi, Robert Bauchwitz, Barbara Bowen, Georges Brefort, Gary Brennan, Victoria Corbin, Ann Costa, Mary Erfle, Neal Farber, Toren Finckel, Nanda Gopinath, Douglas Hanahan, Adriana Holy, Martin Kreitman, Mark Marchionni, Harry Nick, Debra Peattie, Gary Ruvkun, Charles Slater, John R. Storella, Venkatesan Sundaresan, and Karen Talmadge.

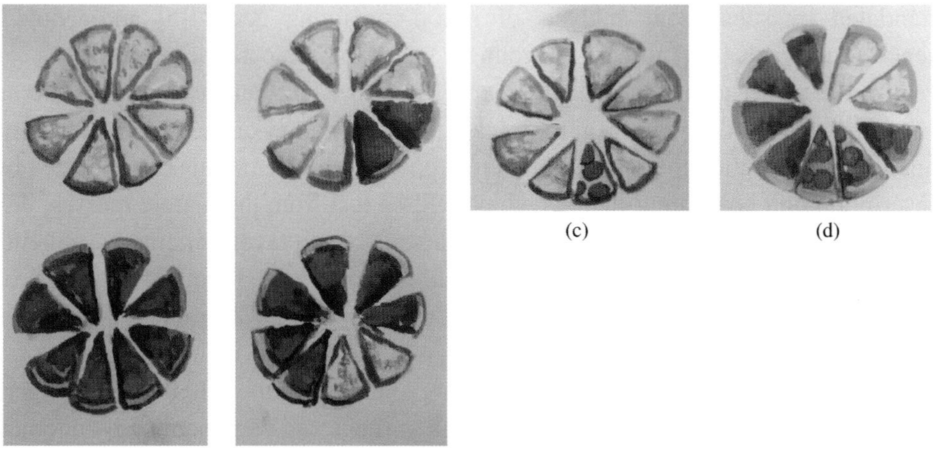

(a) (b) (c) (d)

Figure 1: Genes in Pizzas/Scrambled Pizzas: (a) two ancestral pizzas, (b) slice shuffling by recombination, (c) slice acquisition by retroposition, (d) composite pizza. Sketch by Cila R.B. Brosius.

The early Eighties were exciting times in the Gilbert lab, not only because of the close witnessing of the awarding of a Nobel Prize, but also due to the transition of genetic engineering from the laboratory to the fledging sector of biotechnology.

During that time, a tremendous diversity of projects was under investigation in Wally's lab. They included chromatin structure, chemical RNA sequencing and -probing, expression and subcellular localization of insulin, interferons and human serum albumin in *E. coli*, regulatory elements (promoters, terminators, enhancers), splicing in yeast mitochondria and insulin genes, virus and provirus structure/function, transcriptional activation using murine leukemia virus and Rous sarcoma virus as models, adenovirus, DNA methylation patterns, immunoglobulin gene rearrangements, genomic/multiplex sequencing, in vitro mutagenesis methods, protein/DNA interactions, erythropoetin gene cloning, *Leishmania* genes, evolution of alcohol dehydrogenase genes in *Drosophila* populations. The only concerted approaches had the goal of elucidating domain structure of enzymes in light of the exon theory of genes. Nils Lonberg, Donny Straus, and later Mark Marchionni worked on genes encoding enzymes from the Emden-Meyerhof pathway (triosephosphate isomerase and pyruvate kinase) and hence were addressed as the Emden-Meyerhof gang. Also, there was a student project on the lysozyme gene structure. Even Wally embarked on experiments from time to time. I remember our warm room full of flasks and contraptions, when Wally tried to tweak *E. coli* into hydrogen production. This variety of research projects had two effects: By interacting with your lab-mates, you could learn a lot outside your own line of research and there was hardly any internal competition, which lead to an amiable and relaxed yet highly productive atmosphere in the lab.

One remarkable feature of the Gilbert group was that Wally cared little about formal education and background as a prerequisite for accepting someone into the lab under his supervision, as long as that person was genuinely interested in science; it made no matter whether he or she came from a hippy community on the streets or a commercial fishing boat. This is testament that Wally, akin

Figure 2: Depicted are: (1) Jesse Hochstadt, (2) John Birmingham, (3) Wally Gilbert, (4) Jürgen Brosius, (5) William Poindexter Kennedy, (6) Carl E-Ching Wu, (7) Harris D. Bernstein, (8) Ingrid Akerblom, (9) Richard L. Cate, (10) Jairam Lingappa, (11) Nils Lonberg, (12) Dyann F. Wirth, (13) Aaron Perlmutter, (14) Donald Straus, (15) George M. Church, (16) Huntington Potter, (17) Bernard Dujon, (18) Winship Herr, (19) Stephanie Broome, (20) Helen Donis-Keller, (21) Dennis Schwarz, (22) Richard Tizard, (23) Sara Stein, (24) Barbara Wallner, (25) David Dressler. Credit: Harvard News.

to himself, preferred to have individuals in the lab who ignored artificial boundaries and limitations. Perhaps the best example was Barbara Bowen, who for many years had been Wally's competent and reliable secretary. When she showed interest in working in the laboratory, she simply traded typewriter etc. for a set of pipets. Could there be better proof for a highly successful transition than a publication in a journal such as *Science* [12]?

The cauldron for this active exchange of ideas and cross-fertilization was the seminar room, usually addressed by us as the tearoom. Every weekday around 4 p.m. a bell rang in the hallway.

Figure 3. Gilbert lab members during a small reunion on Cape Cod. Upper row: Debra Peattie, Neal Farber, Richard Tizard, Richard Cate. Bottom row: Jürgen Brosius, Barbara Wallner, **Ann Costa**, Emily Tizard, Mark Marchionni (around summer 2001).

Ann Costa (see Figure 3), a wonderful and warm-hearted human being who washed our dishes, prepared several pots of tea and provided some cookies. Unless you were in an experiment that could not be interrupted, everyone congregated into the tearoom, and we talked. We discussed our ideas, our failed and successful experiments, new developments in science, politics, anything. This tea time was not only popular with the Gilbert lab but also with students and postdocs from other groups in the BioLabs, who often visited the third floor for a cup and a chat. Regular lab seminars were held as well but were not nearly as popular as the informal exchange of ideas. During seminars, we usually were much more defensive and Winship Herr once drew a parallel with Francisco Goya's painting *The Third of May 1808*, depicting a Spanish freedom fighter mowed down by a firing squad.

Being a member of the Gilbert lab at that time felt like being a member of a large family. We often went out together for dinner — usually to Harvard Square's Yenching Szechuan-style restaurant (Figures 4, 5). A slight challenge was to find a time where everyone had their experiments at a stage where one could take a break. The famous "it'll take me two minutes" by Rich Cate meant that we easily could squeeze in another one-hour experiment. Occasionally, we went to see a movie together. Unforgettable was my induction of the Gilbert lab into the then vibrant German movie scene including Rainer Fassbinder's *Satan's Brew*, *Fear Eats the Soul* or Frank Ripploh's *Taxi zum Klo*. During the latter, our group of 1–2 dozen burst out in laughter when a drag queen that happened to be named "Wally" appeared on the screen. I am fortunate to have met such a large number of extraordinary individuals whom — until this day — I am proud to consider as my dear friends. Even "younger siblings," colleagues that joined Wally's lab long after I had left as part of this family, became part of my scientific and personal life. For example, with Manyuan Long and Sandro de Souza we organized

Figure 4: Members of the Gilbert lab at a Chinese Restaurant (possibly Yenching) in summer 1982. Top row: Ting Wu, Dawna Provost (Mark's girlfriend), Mark Marchionni, George M. Church, Aaron Perlmutter. Second row: Jürgen Brosius, Barbara Bowen, Marilyn Perlmutter (Aaron's wife), Ann Costa. Third row: Rich Cate, N.N. (probably a summer student), Mary Erfle, Carol Straus (Donny's wife), Nils Lonberg, Donald Straus, Adrian Krainer. Horizontal: Ingrid Akerblom.

Figure 5: Yenching Restaurant revisited, May 30 2014: Rich Cate, Jürgen Brosius, and Wally Gilbert.

Figure 6: The co-organizers of the 2005 Symposium "Tasting the Art of Science: Symposium to Celebrate the Scientific Explorations of Walter Gilbert" at Cold Spring Harbor Laboratories. Depicted here at a 2018 meeting in Chengdu, Szichuan, China. From left Sandro de Souza, Manyuan Long, Jürgen Brosius.

(Figure 6) a memorable Symposium with the title "Tasting the Art of Science: Symposium to Celebrate the Scientific Explorations of Walter Gilbert" in honour of Wally's achievements at the Cold Spring Harbor Laboratories in 2005 (Figures 7–27).

Another hotspot for the free and relaxed exchange of ideas — not only between Gilbert lab members, but also from other Harvard labs, the nearby MIT, and Biogen — was the Irish pub The Plough and Stars near Central Square (Figure 22). A number of collaborations were forged in this landmark over a Guinness or another brew. Even after I left for New York, the Plough was the site for the initiation of a fruitful collaboration involving Tomás Kirchhausen (Harvard), the Biogen protein chemistry group, and my lab at Columbia University resulting in the cloning of several genes encoding components of coated vesicles.

I am certain that I can speak for all of us during that exciting era at the BioLabs when I say that Wally's unorthodox way of thinking influenced us all. We are grateful to have interacted with and learned from him. Finally, I would like to thank WGBH radio station for their classical music programs (at the time), the NPR news, and especially Eric Jackson who widened my exposure to jazz with his program "Eric In The Evening". One could not have imagined a better pipetting companion during those long evenings and weekends in the lab.

Acknowledgment

Thanks to Cila R.B. Brosius for Figure 1 and Stephanie Klco-Brosius for editorial comments.

Figure 7: Standing back row from left to right: Nicholas Armstrong, Pamela Yelick, Didier Stainier, Carlos Alvarez, Alexei Fedorov, Carl Rosenberg; standing center row from left to right: Walter Gilbert, Nancy Munroe, Wan Lam, Li Jun Ma, Umesh Bhatia, Manyuan Long; sitting or squatting from left to right: Mitiko Go, Han Chang, Keith Robison, Toshinori Endo. Credit for Figures 7–27: Fran Perler, Perls of Wisdom Biotech Consulting.

Figure 8: Cocktail hour I (foreground, left: George M. Church and pink Polo shirt: Jan Witkowski).

Figure 9: Cocktail hour II (yellow Polo shirt with paper in left hand: Charles Weissman).

Figure 10: Carl Rosenberg, Alexei Fedorov, Umesh Bhatia, Manyuan Long.

Figure 11: Kate, John, and Celia Gilbert, Jim Watson.

Figure 12: Wally Gilbert, Peter Lomedico, Joan Steitz.

Figure 13: From left to right: George M. Church, Scott Roy, Laura Landweber.

Figure 14: From left to right top row: Jürgen Brosius, Karen Talmadge, Carl Wu, Ingrid Akerblom, Rich Cate, Jay Lingappa, Nils Lonberg, Donald Straus, George Church, Huntington Potter; bottom row: Bernard Dujon, Winship Herr, Stephanie Broome, Helen Donis-Keller, Wally Gilbert, Barbara Wallner, Neal Farber.

Figure 15: From left to right: Robert Bauchwitz and Neal Farber next to the introns/exons poster.

Figure 16: Wally Gilbert and Marty Kreitman.

Figure 17: Laura Landweber and Ford Doolittle.

Figure 18: From left to right: George M Church and Gary Ruvkun.

Figure 19: From left to right: Manyuan Long and Osamu Ohara.

Figure 20: From left to right: Jay Lingappa, Richard Tizard, Barbara Wallner.

Figure 21: At the table, left to right: Robert Kamen, Gary Ruvkun, Wally Gilbert, Nancy Maizels, Winship Herr, Gordon Carmichael.

Figure 22: From left to right: Michael Freeman and Didier Stainier.

Figure 22: Rich Cate, Donny Straus, and Barbara Wallner, May 30 2014 in front of the Plough and Stars, 912 Massachusetts Ave.

Figure 23: From left to right: Wan Lam, Keith Robison, Nicholas Marsh Armstrong, Wen Wang, Meena Sakharkar.

Figure 24: From left to right: Steven Benner and Bernard Dujon.

Figure 25: From left to right: Richard Cate, Barbara Wallner, Richard Tizard, Karen Talmadge, Stephen Stahl.

Figure 26: Lydia Villa-Komaroff and Wally Gilbert.

Figure 27: Francine Perler and Wally Gilbert.

1. Chow, L.T., *et al.*, An amazing sequence arrangement at the 5' ends of adenovirus 2 messenger RNA. Cell, 1977. **12**(1): p. 1–8.
2. Berget, S.M., C. Moore, and P.A. Sharp, Spliced segments at the 5' terminus of adenovirus 2 late mRNA. *Proc Natl Acad Sci U S A,* 1977. **74**(8): p. 3171–5.
3. Maxam, A.M. and W. Gilbert, A new method for sequencing DNA. *Proc Natl Acad Sci U S A*, 1977. **74**(2): p. 560–4.
4. Gilbert, W., Why genes in pieces? *Nature*, 1978. **271**(5645): p. 501.
5. Long, M., C. Rosenberg, and W. Gilbert, Intron phase correlations and the evolution of the intron/exon structure of genes. *Proc Natl Acad Sci U S A*, 1995. **92**(26): p. 12495–9.
6. Franca, G.S., D.V. Cancherini, and S.J. de Souza, Evolutionary history of exon shuffling. *Genetica*, 2012. **140**(4-6): p. 249–57.
7. Lev-Maor, G., *et al.*, The birth of an alternatively spliced exon: 3' splice-site selection in *Alu* exons. *Science*, 2003. **300**(5623): p. 1288–91.
8. Krull, M., J. Brosius, and J. Schmitz, Alu-SINE exonization: en route to protein-coding function. *Mol Biol Evol*, 2005. **22**(8): p. 1702–11.
9. Krull, M., *et al.*, Functional persistence of exonized mammalian-wide interspersed repeat elements (MIRs). *Genome Res*, 2007. **17**(8): p. 1139–45.
10. Brosius, J., Retroposons — seeds of evolution. *Science*, 1991. **251**(4995): p. 753.
11. Brosius, J. and S.J. Gould, On "genomenclature": a comprehensive (and respectful) taxonomy for pseudogenes and other "junk DNA". *Proc Natl Acad Sci U S A*, 1992. **89**(22): p. 10706–10.
12. Berman, S.A., *et al.*, Localization of an acetylcholine receptor intron to the nuclear membrane. *Science*, 1990. **247**(4939): p. 212–4.

Recollections, Indebtedness and Awe

Marty Kreitman*

I entered graduate school in Organismic and Evolutionary Biology at Harvard in the fall of 1978 to work on population genetics with the great Richard Lewontin. DNA sequencing had been invented only two years earlier, and the Gilbert Lab was just gearing up to sequence a plasmid, pBR322, their most ambitious undertaking to date. Introns had been discovered the previous year, though the name had only just been coined by Gilbert. At the time that I entered Lewontin's lab — where protein gel electrophoresis seemed nearly as much a subjective and fickle artform as a scientific method — the application of sequencing technology specifically, and the molecular biology and cloning revolution more generally, had not yet penetrated the lab members' collective consciousness. The first phylogenetic DNA sequence comparisons remained well off, though it would become essential to Kimura's neutral theory of molecular evolution. Only Allan Wilson's group at Berkeley was conducting studies of DNA sequence variation and evolution at the time, but this was with mtDNA, not nuclear DNA. MtDNA could be easily purified and analyzed with restriction enzymes and gel electrophoresis to quantify differences. Southern blotting could have brought evolutionary analysis of nuclear DNA into play at that time, but pioneering studies in population genetics using this approach didn't appear until the early '80s with the pioneering work of Chip Aquadro and Charles Langley. Cloned DNA to use as a probe simply was not yet available even for *Drosophila*, though that was about to change.

I can't say I had many meaningful interactions with Wally as a member of his lab until the year I completed my work, when he generously — and, I'm certain, for purely intellectual reasons — spent time over a period of weeks discussing my data. We sat in the Tea Room (created by Jim Watson when he had occupied the space), reading relevant chapters in Warren Ewen's book, *Mathematical Population Genetics*, the Bible of molecular population genetics. Afterwards, Wally afforded me his friendship, giving me a fuller opportunity over the ensuing years to discover the person beyond the scientist, which I will come around to shortly. But first, I want to set the stage for my appearance in the Gilbert Lab, since it would be impossible to separate the man from his work and from the scientific playground he created for his students.

*Department of Ecology & Evolution, The University of Chicago, Chicago, Illinois, USA

One thing I will say about Wally at the outset, in part because I discovered this trait in him early on, is the purity of his intellectual devotion to scientific reason and discovery. Let me set the backdrop. The rage in the Lewontin Lab as I entered it in 1978 was to better characterize the resolving power of protein gel electrophoresis to detect amino acid change. Such characterization would better allow the estimation of population genetics parameters, which come from estimating the standing pool of variants in a population sample. Gel electrophoresis detects only a fraction of protein variants, a fraction that was crudely estimated based on simple reasoning rather than experiment. I agreed at the beginning of 1979 to conduct a potentially definitive calibration experiment, which, though scientifically boring, would, I thought, be a quick route to a publication and potential thesis. The idea, which was not my own, was to calibrate gel electrophoresis with a vast array of protein variants of the *E. coli* lac repressor. To carry out the study, I would need to grow bacterial strains carrying alleles of interest, then purify the repressor protein sufficiently to be able to recognize its telltale band on a gel following electrophoresis and general protein staining. At the time, I had never worked with *E. coli* and had absolutely no knowledge, training, or skills to successfully carry out these procedures. To be sure, few of these skills existed in the Lewontin Lab at all.

But they did in Wally's lab, remnants of his pioneering work that won the race to purify the lac repressor in the preceding decade. I can't recall whether Lewontin cleared the way in advance for me to seek advice and carry out the work in Wally's lab. I believe I simply asked Wally whether I could do so; he thought the project was feasible, and he agreed to provide support. He pointed me to Michelle Calos, who had worked previously with the repressor. She and others generously helped me get started, though I ultimately failed spectacularly and gave up after six months, realizing that the effort to succeed would vastly exceed the scientific value of the work. I learned that anything Wally thought to be scientifically reasonable was fair game in his lab and that he was generous in sharing resources.

As I was failing with the lac repressor experiment, Mark Passos, a postdoc in the lab, impressed on me the possibility of measuring genetic variation directly in the genetic material itself now that DNA sequencing was possible. It was a tall task, though, and I knew nothing of the methodology. I therefore devoted much of the summer of 1979 to reading the literature in molecular biology, focusing on methods. And, as luck would have it, the very first gene was cloned that summer in *Drosophila melanogaster*, the alcohol dehydrogenase locus. Lewontin was excited by the possibility of my cloning and sequencing *Adh* alleles, but there were obstacles. Firstly, I could not possibly carry out the study in the Lewontin Lab. Lewontin was, in fact, ambivalent about letting this work go on in his lab for ethical reasons. He had previously opposed recombinant DNA research as it became possible, and he testified against it at the City of Cambridge hearings, noting that "experts" who were testifying to the safety of recombinant research, including of course Gilbert, also had the most to gain intellectually and financially from conducting recombinant research. Cambridge did in fact ban such research temporarily, causing, for example, the young T. Maniatis to flee to Cal Tech, while Wally set up his newly formed biotech company Biogen in Switzerland rather than in Cambridge. Under the circumstances, it would be difficult to imagine Wally being sympathetic enough to let me conduct this project in his lab — perhaps the only place in the world where I could succeed — much less finance the entire project!

Nevertheless, Dick Lewontin arranged a brief meeting with Wally. They were not friends, but already I was aware that the two had a keen respect for each other. Lewontin, when once asked whom he thought were his intellectual peers at Harvard, proffered Gilbert, Meselson … and then struggled. As it turned out, the respect was mutual. Despite Lewontin's devastating attack on recombinant research, to Wally this was simply a scientific argument he knew he would win. The politics of the matter were at best an intellectual distraction. After describing my project to Wally, he readily agreed to give me a place in his lab. I interpreted his willingness as a begrudging nod to Lewontin's unequalled intellect, a handshake between two similarly reared NYC Jews (as I also am). In fact, Wally later confessed to me that he had agreed simply because he thought the science would be worthwhile. He harbored absolutely no hard feelings towards Lewontin, he told me. Theirs was, if anything, a disagreement in the politics of science, a subject not so important to him as to interfere with the pursuit of knowledge. Such rationality is the essence of Wally's scientific genius.

I learned much from Wally about scientific practice. I have already mentioned his willingness to sit with me, simply thinking about my data and what could be said with confidence about it. In fact, he afforded this attention to every student who had results or a manuscript. Wally could spend days in discussion and argument, pushing the limits of what could be claimed about a result in order to establish where that boundary between knowledge and speculation lay. Wally possessed an unshakable confidence in his own ability to think through the implications of a scientific result. There was also a corollary that became a dictum in my own research: no reference could be made to an unpublished result or observation. If something was important enough mention in print, then its basis must be included for all to evaluate. To him, the ability to scrutinize (and of course potentially repeat or falsify) an experiment or result was and remains the foundation of the scientific method of inquiry.

The Gilbert Lab had perhaps 20 graduate students and postdocs conducting research on 20 different topics during this period, each being carried out with nearly complete independence. Indeed, I believe people joined his lab not only for its prominence (Jim Watson started the lab, which Wally then inherited when he became Director of the Cold Spring Harbor Lab), but also for its freelance spirit. But it was also a place with a fairly severe work ethic, with the lab operating 24 hours a day. My initial mentor, Winship Herr, to whom I will remain eternally grateful, in fact didn't even appear in the lab until around dinnertime or later. His day would often end in the morning with breakfast at an all-night diner. I remember this well, because I was incapable of doing anything without his instruction in the early months of my project. I would wait and wait and wait for his appearance all day long, only to be told something trivially obvious at 8pm, like: "Add 30 lambda of 3M NaOAc and then EtOH to the tubes and spin them in a centrifuge." Winship also impressed on me the importance of controls, which I often got wrong. Nighttime was when the fun in Wally's lab reached its peak. This was undoubtedly the tone set by Wally himself: he routinely returned to the lab after dinner when he was younger to work until midnight, as I am told. Even as a Nobel laureate, he could often be found working — and talking — well into the evening when he was in town.

The late-night tradition did, however, create a crisis for some. Wally was often travelling, spending most of his time in Geneva at Biogen. He would routinely call the lab first thing in the morning his time to get updates on research projects of immediate interest to him. Mine was never one of them, so I escaped these phone interrogations entirely. His call would be between 1am and 3am Boston time. Now, in point of fact, most of the lab members were present and available to talk to him at that time. However, the anxiety of not having a result to report prompted many to insist that they were not there when Wally summoned them to the phone. This was the dilemma for almost everyone but me!

Wally could be demanding, especially to students who were making slow progress. For a period, my lab-mate, Donny Strauss, was one of them. He ended up completing a beautiful comparative study of the intron–exon structure of the gene encoding trios-phosphate isomerase in mammals, but cloning the gene proved difficult. (Donny had a fascinating approach to conducting experiments. He liked to strip protocols down to their barest essence, so much so that they were pretty much guaranteed to fail. He would then add controls, building them back up until they finally worked. This method slowed him down considerably, though I believe it ultimately brought him much deeper insight into molecular methods than the approach I employed, which generally succeeded the first time through the careful execution of protocols. Things worked for me, but I was never confident that I knew why.) One winter, Donny went skiing and blew out his knee. Hobbled with crutches, Donny nevertheless persevered in the lab under both self-imposed and external pressure to get his gene cloned. Wally appeared one Sunday morning after a trip and immediately interrogated Donny about his most recent progress. When the interrogation was over and Donny was out of sight, I tried to defend him to Wally, pointing out that he was working under physical duress. Wally was taken aback — he hadn't even noticed and apologized forthwith. Wally could certainly focus, sometimes to a frightening degree.

I spoke of the freelance spirit of the lab, but there were some who tried to impose a pecking order nevertheless, which I was fortunate to quickly learn about. It had, in fact, little if anything to do with the person's ability, productivity, or knowledge. Once Winship departed after guiding me for the first nine months, I was forced to seek advice from others. It was a postdoc, Rich Cate, who saved me, discreetly telling me I needed to be more vigilant about the people from whom I took advice if I wanted to continue making progress. According to him, I had to be especially cautious about taking "free" advice I hadn't sought. He told me to ask Aaron for protocols for making lambda libraries and screening them; I never looked back, and I learned a valuable lesson in the process.

When I joined the lab, Rich Tizard, a super-tech, was completing the 4,361-bp pBR322 sequence. Soon thereafter, the Sanger lab completed lambda phage, a sequence longer by ten-fold; Rich moved to Biogen once the company moved back to Cambridge in 1982. For a couple of years, I suppose, I was the main practitioner of chemical sequencing in the lab. Aside from the inherent dangers of working with millicurie quantities of radioactive 32P to label DNA fragments and with mutagenic chemicals to modify or break DNA, sequencing was definitely an art. The most difficult challenge was almost entirely physical — manipulating meter-long glass plates to prepare sequencing gels. We even played around with 2-meter-long gels, but (fortunately) they proved impractical. Chemical sequencing was not for the faint-of-heart.

The only serious progress in sequencing technology that occurred in Wally's lab during my tenure came from the hands and mind of the incomparable George Church, whom I was fortunate to befriend. One day I went to George, who worked in the adjoining lab room, with a question, only to find him oddly excited about a very blurry sequencing autoradiograph. He had, it turned out, just invented multiplex sequencing, the first tangible progress in his life-long quest for improved DNA sequencing. It was a pleasure for me to observe Wally and George share this passion and mutual respect for one another. In addition, I benefited by quickly adapting George's multiplex method for four-cutter RFLP analysis of the Adh locus. At Princeton, my first graduate student, the inestimable Andrew Berry, together with a horde of undergraduates, ran a factory to analyze thousands of wild-type lines to determine the haplotype structure Adh, the process of natural selection for the Adh protein polymorphism, and the genetic architecture of populations of *D. melanogaster* along the East Coast of the U.S.

As for chemical sequencing, its practice perhaps reached its zenith in my project when I deposited some 28,000 bp of sequence into GenBank in 1983. Manyuan Long told me it represented 2% of GenBank at the time. That same year, I traveled to the MRC in Cambridge, England, to work in Bart Barrell's lab for a couple of months. There, Sanger sequencing development was continuing. Notably, the trip was paid for not by Wally but by Dick Lewontin! I brought back M13 shotgun cloning and Sanger sequencing technologies to the Biolabs, perhaps hastening the demise of Maxam-Gilbert sequencing. Also, Roger Staden had developed a large software package (written in Fortran) to assemble DNA sequences from shotgun clones, which he generously gave me to install at Harvard. They were a lightyear ahead in their sequencing technology, in part because Wally had bigger fish to fry, nurturing Biogen rather than pursuing sequencing technology; as science goes, no one looked back.

My friendship with Wally began once I left the lab. Returning for a visit to Harvard as a postdoc, I ran into Wally in Harvard Square. He seemed genuinely delighted to see me and invited me back to his home (his new manse) for lunch, a great privilege. I would routinely visit him whenever I returned thereafter, and he always reciprocated by making time to discuss my research and personal experiences as an Assistant Professor at Princeton. Soon after moving to the University of Chicago, I had the opportunity to discuss with Wally a design for a DNA sequence polymorphism database. Surprisingly, he took an active interest, even though the first human genome sequence was still far off. He arranged for me and a student to meet him in Cleveland at a startup company he had invested in, which was developing an electronic laboratory notebook. Wally, in fact, spearheaded quite a few venture capital investments in science technology startup companies. As always, one could sense the delight he took in applying his prodigious mind to a new subject, and we had a spirited discussion about how to structure the database and how to represent both single nucleotide polymorphism and haplotypes in it. I remember arguing about the merits of a reference sequence — what it actually represented when few if any individuals in a population might possess it — long before a reference sequence even existed. More recently, I had the pleasure of viewing his artwork with him at an opening at a gallery in Los Angeles (I was on sabbatical at UC Santa Barbara) and his two-day visit to the University of Chicago a couple of years ago.

My ties to Wally run deeper still. At Princeton, I employed a precocious freshman undergraduate named Laura Landweber to try PCR soon after it was invented. She spent nights injecting fresh Klenow fragments into the reaction each cycle and moving tubes between different temperature water baths, since thermostable Taq was not yet available and no commercial PCR machine yet existed (we invented two generations of our own). Laura was interested in RNA evolution and the origin of life. Following in my footsteps, she became a graduate student of both Dick and Wally at Harvard. In addition, he also took on Manyuan Long as a postdoc. Manyuan worked on Adh in *Drosophila* as I had, discovering that a young Adh pseudogene in two closely related species was actually a chimeric functional gene that employed exon shuffling. This and subsequent work with Wally led us to hire Manyuan as an Assistant Professor at the University of Chicago, where he has been an immense success. Manyuan has proudly carried forward Wally's astute prediction that new genes are formed in pieces by the process of exon shuffling, now a major tenet in molecular evolutionary biology.

I am most proud of Wally's place in my academic pedigree and our joint connection to Jim Watson. Wally, of course, met Jim in Cambridge, England, and Jim's infectious enthusiasm for molecular biology and evolution brought Wally into the field when they met up again at Harvard. Wally inherited Jim's lab, as I previously mentioned, so I guess you could say I did my graduate work in his borrowed space.

As it turns out, the same is true for my lab at the University of Chicago. In my second year here, a frumpy gentleman turned up wandering around my lab in the Zoology building one morning. John McDonald (of McDonald-Kreitman fame) inquired whether he was lost or needed help. The gentleman basically ignored John's offer and continued to wander around the lab. John then figured this person might be an alumnus. The conjecture was confirmed, and John was pleased to let him to continue his meandering through the lab. After a period of time, the gentleman returned to John at his bench and inquired what kind of research went on in this lab. I suppose he must have noticed our ABI sequencing machine at the very least. John, thinking this person unlikely to have scientific training, gently explained that we work on genetic variation. "What is genetic variation," he asked, "and how do you study it?" John patiently described that the genetic material, DNA, could be thought of as a long string of beads, with the beads coming in one of four colors. One complete string was a chromosome, and all individuals had the same complement of chromosomes with nearly the same order of colored beads, except in places in which the bead might have mutated to one of the other three colors. By DNA sequencing, we could identify these color-polymorphisms. The man gave no indication of knowing about the genetic material, sequencing, or polymorphism, but kept asking simple questions forcing John to respond.

Some ten or fifteen minutes into this elementary tutorial, a second postdoc, Bill Ballard, entered the lab and introduced himself to John's visitor. The visitor returned the favor. "Oh, I'm Jim Watson. This room used to be a student laboratory where I took my very first biology lab as an undergraduate." Needless to say, John, a pureblooded Irishman, turned crimson in embarrassment. Watson was obviously delighted by his devilishness. Later I met Jim, and, like Wally, he also befriended me. Upon returning to Cold Spring Harbor following this trip, Jim endowed an annual lecture in molecular evolution at the University of Chicago, named in honor of his mother, Jean. More recently, I helped Jim gift the University a statue of Charles Darwin that he had commissioned by the sculptor Pablo Eduardo. Jim had grown up only a couple of miles from the University of Chicago and was awarded a scholarship enabling him to attend the university. He recalls lectures by Sewall Wright, for instance, which impressed on him the central role evolution plays in biology. A highlight came a couple of years ago when we invited Jim to join Wally, who was selected to give the endowed lecture. The experience was perhaps the highlight of my own career, spending two days with the founders of modern biology, basking in their genuine respect and affection for each other.

Wally's generosity launched my career, and for that act of kindness (he would say it was a simple decision about supporting a worthwhile scientific project) I will be forever grateful. But that kind act aside, it would be difficult to describe the breadth of my awe of Wally. I first met him as he was about to receive a Nobel Prize in Chemistry, and once I began to understand the wide-ranging scientific projects going on in his lab and saw him in action as a scientist, his sheer brilliance was obvious. But he was also making the transition from preeminent scientist to biotech entrepreneur, in essence becoming a businessman rather than scientist. Indeed, he often showed up during this transition in a suit and tie, which of course we teased him about (behind his back). He even once described himself as a businessman first and scientist second. And indeed he was, raising unprecedented sums of money for Biogen. This was not his first change of career, of course, but his second, as he was first an accomplished theoretical physicist (good enough to be hired by Harvard!) before switching to molecular biology. And then, as if three remarkably successful careers were not enough, he then transitioned in his "retirement" to becoming a successful artist.

I believe Wally's special gift, endowed as he is with a fabulous brain, is his mindfulness. More than anyone I have ever known, he lives "in the moment" seemingly always. The impression every time I have seen or interacted with Wally is that at that moment in time there was nothing else he would rather be doing. Wally always seems to be living the dream just as he wishes it to play out. This special characteristic is especially obvious when observing him at his art gallery presentations, where he delights in talking to visitors about his creations and the thought processes he uses to create them, giving not the slightest hint in dress, demeanor or talk that he is also an accomplished theoretical physicist, Nobel Laureate, and founder of the biotech industry. This is not false modesty but rather pure mindfulness at work, the wellspring of his limitless self-confidence. For me, the simple act of kindness that gave life to my academic career blossomed into pure admiration for a remarkable human being. In that sense, I have been doubly blessed by his presence in my life. Thanks, Wally.

Wally Taught Me to Look Farther

Manyuan Long*

In the summer of 1992, after a few years study of population genetics and molecular evolution and thesis research on novel gene evolution at the University of California at Davis, I was thinking about where to go for my postdoctoral training. Among a number of postdoctoral offers, I noticed one strikingly different from those of other famous evolutionary biology departments with which I had been familiar. It was sent by a scholar that I had read extensively but had never considered the possibility of working with.

This letter came from Walter Gilbert, the Biological Laboratories, Harvard University, 16 Divinity Avenue, Cambridge, Massachusetts, signed by Nancie Monroe working for Wally during his absence while he traveled. I was a little bit reluctant, since many labs that had offered me post-doctoral positions were making excellent strides in evolutionary research. Joining them implicated a good future, including a secure position and the opportunity to participate in important scientific developments.

What drove me to investigate the program of gene evolution in Wally's lab, however, was my being cautioned against it. A professor at UC Davis warned me that Wally's lab was highly competitive, wondering if I could survive in such an environment. "He could fire all people in his lab in 10 minutes if he found that they were not smart," said the professor, a former Harvard postdoc, in an exaggerated tone. To his surprise, his comment only piqued my interest in the place and made me more eager to see what Wally was doing. I remembered an ancient Chinese saying I loved so much: one should feel no regret when dying at dusk if he hears a profound truth from a wise person at dawn! ("朝闻道, 夕死可矣", Confucius 551–479 BC).

I decided to read more of his recent publications to decide if I should take the chance or avoid the risk. I was immediately attracted by a short essay Wally published in 1991 (Towards a paradigm shift in biology. *Nature* 349: 99). Wally's essay did not teach me what to do day-to-day, but rather painted a grand picture for what biology would be as a consequence of the Human Genome Project. According to what Wally said, the stories I had been familiar with in graduate school and traditional research projects would soon become outdated. Moreover, the

* Department of Ecology and Evolution, The University of Chicago, Chicago, Illinois, USA

approaches to which we were then accustomed — with a few Eppendorf tubes and a thermal cycler to sequence a gene and find any evolutionary and functional information — would be professionally obsolete. The future would come in the form of genome sequencing with computational and theoretical analyses.

Back in the early 1990s, I knew nothing about the details of genomic sequencing, nor the newly arising bioinformatics found in a few labs in the USA. But the sense of encouragement I felt was vast. I was reminded of my early days as a young postman after high school with a heavy load in a large green bag, standing atop a hill in mountainous southwestern China: I felt that I saw farther, glimpsing the scenery beyond many distant hills before me. While the faraway landscapes were unclear, they were so beautiful in my imagination. I overcame my hesitation and flew to Boston.

After settling down in Somerville, on a humid and hot day in July of 1993, I rushed to Wally's lab on the fourth floor of the Biological Laboratories building at Harvard. The lab was crowded with more than two dozen people performing genome sequencing and other molecular experiments. Lloyd Schoenbach was simulating the distribution of introns with respect to their positions in protein structure. Wally asked me how good my computational skills were, and I answered I knew little but could learn. My first project was to construct an exon–intron database by exploiting the publically available GenBank database. After sitting in a Harvard extension school's evening course for programming a couple times, I felt that the teaching was slow and abandoned the classroom to find a quicker way.

Instructed by a friend in another Harvard lab, Daniel Weinreich (currently a professor of Brown University), I went to the Co-op at Harvard Square and picked up the thinnest, rather than the thickest, text book on the C language available to begin self-teaching. A couple months later, I successfully constructed my first exon–intron database from all the published genes in GenBank, the most popular database of gene sequences in the 1980s and 1990s. My exon–intron database described exon–intron structures in 9,276 gene sequences, exciting me for its biggest amount of information regarding the gene structure. During this learning period, I found that the programming was a beautiful experience, and that I could direct the flow of a gigantic amount of data. In today's genomic era, which proves the prediction Wally made almost three decades ago about the future of biology, I feel fortunate that I learned from Wally in the 1990s and was prepared for computational analysis in a new era.

The first thing I did using this database, under Wally's supervision, was to characterize intron phase position, a concept proposed by Phillip Sharp (1981, *Cell* 23: 643–646). It is defined as the positioning of the introns relative to codons. The first thing I observed was the dominance of phase 0 introns, which are located between two intact codons, indicating that introns in eukaryotes tend not to break codons apart. This was an amazing experience compared to my previous graduate research, which had been more based on non-mechanistic analysis of population-level genetic variation in *Drosophila*, from which I had hypothesized that the three types of intron phases should be equally probable.

Upon seeing this result, Wally immediately wrote down an equation to predict the intron phase association within a gene that has more than two introns, based on observed frequencies of the three types of introns. The equation said that the frequency of association of two adjacent introns should be equal to the product of the observed phase frequency of individual introns that flank an exon. I was inspired by this equation and added my hypothesis regarding intron evolution: if intron origination was a random insertion event, as was conventionally believed, we should predict that the relative abundance of nine-intron combinations would follow the equation. This theory can be extended to introns that flank two or more exons, constrained by the phase consistency in the internal exons.

When we returned to the database to compute the actual intron association frequency from tens of thousands of genes in GenBank, both Wally and I were shocked: the hypothesis of random intron insertion could not explain the data regarding intron distribution in eukaryotic genes. Moreover, we found that there were excess numbers of intron associations that were symmetric, i.e. the introns on two sides of an exon were of the same phase! Furthermore, of the three possible symmetric intron phase associations (0,0), (1,1) and (2,2), the association (1,1) showed up in excess. These findings coincide with the intron phase distributions in previously observed genes generated by exon shuffling, suggesting that exon insertion would be constrained to a length of a multiple of three, such that the downstream exons of donor genes would not have been impacted by a frameshift. Thus, it is clear that most eukaryotic genes have been shaped by exon shuffling, a notion that had been hypothesized around four decades ago by Wally in his influential short essay in 1978 (Why genes in pieces? *Nature* 271: 501).

On a cold day in the winter of 1994 at Harvard, while I was working on the intron database, the telephone rang. I picked up the phone and heard Wally's voice.

"Hi, Manny. I am over Colorado. Thirty thousand feet. On a flight to LA."

Wally told me that, on the plane, he had just thought of a way to detect the possibility of introns that might have originated at the time life first began in the RNA world. As evolutionary biologists, we are used to talking about things from long ago, millions of years in the past. However, very few evolutionary biologists at the time wanted to say anything explicitly about what might have occurred at the beginning of life, about four billion years ago, that would probably be in the RNA world. I had not thought that far yet — I just had nothing to say then about the origin of life.

Wally stated his rationale, the first I had ever heard, for his idea of detecting explicit signals from the earliest life. He proposed the use of ancient conserved regions (ACRs) in genes, identified by Phil Green *et al.* (1993, *Science* 259, 1711–1716) to detect ancient signals of exon shuffling. Because the ACRs are shared by eukaryotic and prokaryotic homologues, any detected signals of exon shuffling in the ACRs of eukaryotic genes must have appeared in the progenote, before the divergence of eukaryotes from prokaryotes. Otherwise, the colinearity in the ACRs between prokaryotes and eukaryotes would have been violated. I immediately implemented this idea in a new code and analyzed an ACR database I created by identifying genes in the exon–intron database that match prokaryotic proteins. I was overwhelmed when detecting so strong a signal from exon shuffling in the ACRs billion years ago: we for the first time were able to date exon shuffling events and involved introns to so distant a past, long before the emergence of eukaryotes. This was the first explicit evolutionary event that could be traced to the beginning stage of life, the one of the RNA world. I am proud to have had the opportunity to collaborate with Wally to explore what no evolutionary geneticists before could have opportunity to explore. We saw the light in the remote corner of early life.

Many in the field were surprised by the foresight of Wally, who could see forward twenty or thirty years or even longer. I was curious and wanted to find out how it was possible. I witnessed the unexpected broadness of academic interests of Wally, which went far beyond his interest and expertise in theoretical physics and the full spectrum of biology. In January of 1996, the winter in Boston was harsh. Wally, Sandro [de Souza], and I escaped to Irvine, California, to attend the first Gordon Conference of Molecular Evolution. In the meeting, Wally gave a talk on intron–exon evolution based on the data Sandro and I created from the database analysis. After the meeting, Wally invited Sandro and I to tour the Getty Museum with him. We soon arrived in front of the gorgeous museum, as I wondered what Wally wanted to do with us. He suggested that we look at all of the exhibitions there while he went to a training class on Middle Eastern archeology in the museum. This was the first time I became aware of Wally's vastly varied interests (Figure 1).

Three years later, when Wally visited my lab in Chicago, he showed me and my students during a dinner party a small piece of an antique ring from the Sumerian Kingdom, which he recognized as authentic. My son Rolland, trained by the Oriental Institute of Chicago, told me that distinguishing ancient Middle Eastern artifacts was not easy. In 2013, Marty and I (Figure 2) invited Wally to give a seminar to the general audiences of the Chicago community. Wally gave a talk entitled "From Science to Art". Wally summarized his major activities in both life science and the arts and discussed their connections. Wally told audiences that science and art shared a great similarity in that they both created something new. But the evaluation of science and art are different: in science, the truth has to be agreed upon all; in art, the work should sell well.

What occurred to the prediction of paradigm shift by Wally in 1991, which drove me to his lab?

Figure 1: Walter Gilbert and Manyuan Long in the Getty Museum, California. January 30, 1996.

Figure 2: Wally Gilbert visited two former students at the University of Chicago, October 17, 2013. From Left: Manyuan Long, Wally Gilbert and Marty Kreitman.

We all know today that the paradigm really shifted soon after the Human Genome Project was completed in the early 2000s. What I learned in Wally's lab has become a powerful approach in my study of novel gene evolution at Chicago. Many former graduate students and postdoctoral fellows in my laboratory inherited and further developed this approach in their laboratories and made important discoveries. I feel fortunate I made a good decision to work with Wally in the early 1990s.

A Letter to Wally Gilbert

Sandro J. de Souza*,†

Dear Wally,

It has been almost twenty years since I left your lab. Since then, I have returned to my home country, co-coordinated a cancer genomics project that was once considered one of the top projects in the field, formed dozens of young Brazilian scientists, written a book on evolution for laymen, and am now leading a bioinformatics initiative in a region not previously known for scientific excellence. Although I have been quite busy during all these years, I can't stop thinking about the period of time I spent in your lab. There is a word that linguists say reflects something unique about Brazilian Portuguese: saudade. It was made famous by the singer João Gilberto and his guitar in the song "Chega de Saudade" ("No More Saudade"), which is considered the birth of Brazilian jazz (Bossa-Nova). "Saudade" is hard to translate. It refers to a recollection of feelings that in the past were associated with positive meanings and experiences. In my case, I miss you, I miss the lab, and I miss the whole environment in and around BioLabs and Harvard Square.

I arrived at your lab in January of 1995 with the idea of looking for introns in the triose phosphate isomerase (TPI) gene in insect species. In 1993, Virginia Walker's group found an intron in *Culex tarsalis* TPI (Titiger *et al.*, 1993) at the same position that you had predicted (Gilbert *et al.*, 1986) years earlier, based on a putative correlation between gene and protein structures. The project was going fine when, by mid-August of 1995, I had an idea that changed my life (Figure 1). I was playing with Mitiko Go modules and intron positions when it came to me that we could have a computer program that would predict a region in the amino acid sequence where the boundary of a given module would fall. This strategy would not predict the exact position of the module boundary but rather a region very likely to contain the boundary. We called these regions linker regions (later in the papers, we decided to use the term "boundary regions"). The exact module boundary was hard to define. With the linker regions we could easily test a possible correlation between intron position and modules. If the boundary regions for a given protein correspond to 10% of the length of that protein, then we would expect 10% of introns falling in those regions if the intron distribution is random. A simple chi-square test would allow us to test the distribution of hundreds of intron positions in dozens of proteins considered ancient (having arisen before the prokaryote/eukaryote divergence), with their 3D structure determined.

* Bioinformatics Multidisciplinary Environment (BioME), Digital Metropolis Institute, UFRN, Natal, Brazil
† Brain Institute, UFRN, Natal, Brazil

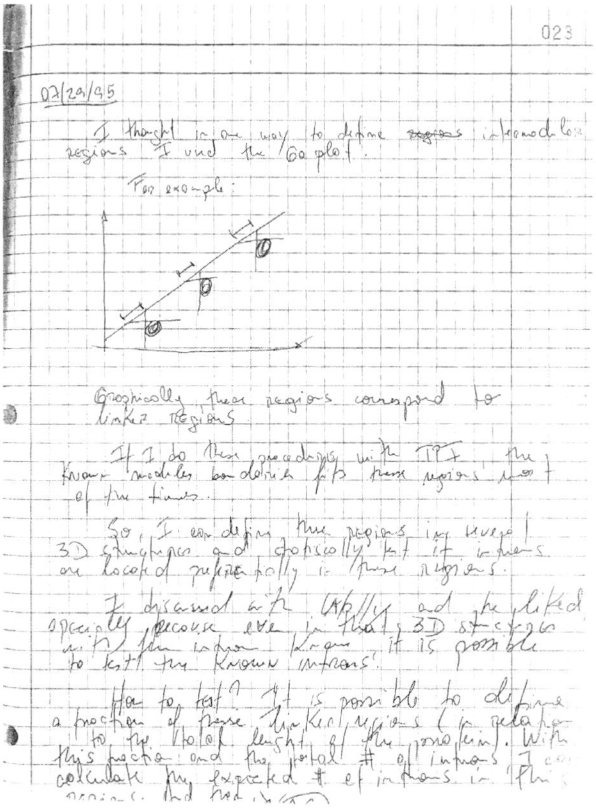

Figure 1: Scan of a page from my lab notebook (copy) registering the idea about the linker regions. Date was July 29, 1995.

I manually tested a few proteins using a database of intron positions developed by Manyuan Long. The preliminary findings were exciting, but we needed computer programs to automatically define the linker regions and test the distribution of intron positions regarding module boundaries. I had then the pleasure to work with the late Lloyd Schoenbach, who wrote a script in VB (Visual Basics) that had a very nice graphical display showing the Go plot, the linker regions, and the intron positions (Figure 2). It was clear then that I had to master programming. Bioinformatics was in its inception, and I realized that I should learn something new that promised to revolutionize biology (as you anticipated in your visionary one-page *Nature* paper in 1991 (Gilbert, 1991)). I first taught myself C/C++ but shifted to Perl following a suggestion from Keith Robinson (a Ph.D. student co-supervised by you with George Church). I had also the pleasure to interact with a young and brilliant Harvard undergraduate called Scott Roy. Excitement about our projects and results were the norm in the lab. The first fruit of that simple idea, a paper, described the method and showed a positive correlation between intron positions and module boundaries for a set of ancient proteins with their structures solved (de Souza *et al.*, 1996).

At the end of my postdoc, we published what I considered a seminal paper in the field (de Souza *et al.*, 1998). Using the same approach as in de Souza *et al.*, 1996, we showed that the

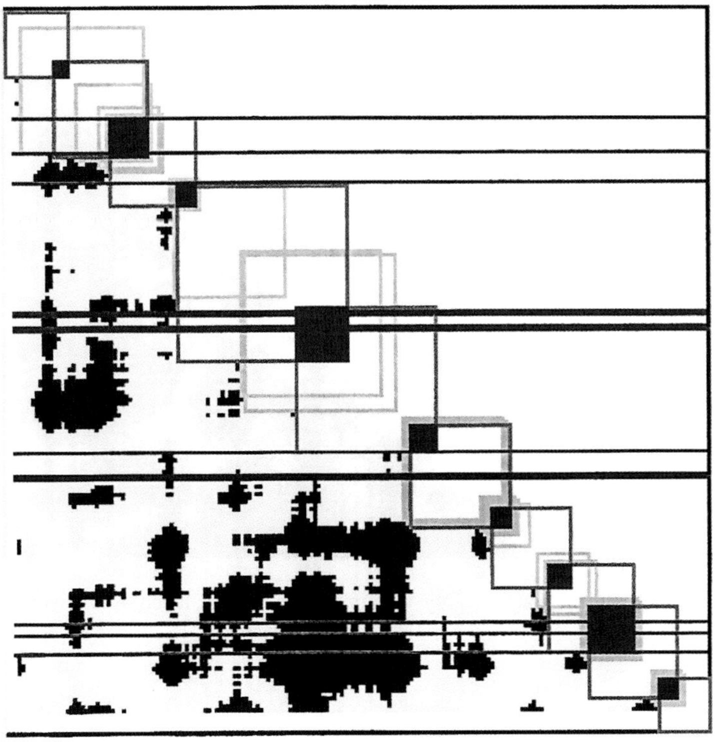

Figure 2: Output from Lloyd Schoenbach's VB program showing the linker regions (blue squares) and intron positions (horizontal lines) superimposed onto a Go plot.

association between intron positions and modules were restricted to phase 0 introns, those introns that are located between codons. In that paper we proposed a resolution for the introns-early/introns-late debate, arguing that primordial introns were basically phase 0 introns with subsequent insertions of all types of introns after the split between eukaryotes and prokaryotes. That was the basis for what I later called the Synthetic Theory of Intron Evolution (de Souza, 2003) — synthetic in the sense that it encompasses concepts from both introns-early and introns-late theories.

Life at the lab was not only about science. I remember a Gordon Research Conference in Ventura when you drove Manyuan and I to the J. Paul Getty Museum in Los Angeles (Figure 3). We had our ski days (you never went) at Wachusset, the beer and burger at O'Sullivan, the marguerites at José's. I am very grateful to you and people at the lab, including Nancy Munroe, Manyuan Long, Lloyd Schoenbach, Scott Roy, Anne Costa, Wan Lam and Carlos Alvarez. It was a terrific time. I am full of "saudade."

With warmest regards,

Sandro

Figure 3: Wally and I at the Getty Villa in Malibu. Picture taken by Manyuan Long.

Bibliography

de Souza SJ (2003). The emergence of a synthetic theory of intron evolution. *Genetica* 118:117–121.

de Souza SJ, Long M, Klein RJ, Roy S, Lin S, Gilbert W (1998). Toward a resolution of the introns early/late debate: only phase zero introns are correlated with the structure of ancient proteins. *Proc. Natl. Acad. Sci USA* 95:5094–5099.

de Souza SJ, Long M, Schoenbach L, Roy SW, Gilbert W (1996). Intron positions correlate with module boundaries in ancient proteins. *Proc. Natl. Acad. Sci. USA* 93:14632–14636.

Gilbert W (1991). Towards a paradigm shift in Biology. *Nature* 349:99.

Gilbert W, Marchionni M, McKnight G (1986). On the antiquity of introns. *Cell* 46:151–153.

Tittiger C, Whyard S, Walker VK (1993). A novel intron site in the triosephosphate isomerase gene from the mosquito Culex tarsalis. *Nature* 361:470–472.

Part Four
Artworks by Walter Gilbert

Walter Gilbert's Statement

I began making digital images as art when I discovered that I could make large prints from images taken with a small digital camera and that these prints carried an emotional and aesthetic impact. My earliest work was of fragments of the visual world, either portions of natural scenes or of man's architectural or industrial artifacts. My first one-person exhibition in 2004 included a 48″ × 72″ image made from a two-megapixel camera.

I was invited to Poland, by Jan Kubasiewicz and Josef Piwkowski, to do an installation at the Norblin Site in Warsaw. These photographs of decaying machinery were installed in Warsaw in the Summer of 2007 as twenty-six 12′ × 8′ hangings and thirty 36″ × 24″ prints, face-mounted on plexiglas. This show was exhibited again in Lodz and in Poznan.

After photographing dancers in the ballet, I went on to explore abstractions, first in a "Vanishing" series, which was based on a natural form, the outline of a human head. The many patterns produced in that series all shared some aspect of a biological or natural curve, which still was manifest even in the smallest cropping of those images.

In my later work the basic element was a straight, shaded line, which I used to create geometric patterns. The "Geometric Series" explored patterns in color or black-and-white created from overlapping squares or triangles or just from lines, taken either simply or in intersecting groups.

I make many images by hand on the computer. The computer simply holds the intermediate forms as I superpose the many layers I create to build up the image. The images begin in black and white, and then I color them in the computer. I generate these colors either by accessing the colors available or, in a more complicated fashion, by using the ability to change the global input–output functions for each color and intensity separately. When the layers containing the colored images interact with each other, still more color patterns appear. The computer is a digital workspace, driven by my hand and eye.

My most recent work involves photographs moved to extreme values in color space yielding strange color contrasts further superimposed on each other. These images exemplify my delight in light and form, and my search for a three-dimensional effect on a two-dimensional surface. I search for depth beyond the picture plane and for mystery.

Four Faces, 2009

36″ × 24″

C-Print Face-Mounted on Plexiglas

Three Heads, 2009
36″ × 24″

C-Prints Face-Mounted on Plexiglas

Vanishing Pattern Black and White #3, 2010

36″ × 24″

C-Print Face-Mounted on Plexiglas

Three Doors — Madrid, 2004

27″ × 18″

C-Print Mounted on UltraBoard with a Luster Laminate

Grease #1 — Warsaw, 2006

24″ × 36″

C-Print Face-Mounted on Plexiglas

Red Decay #1, 2006

27″ × 18″

C-Print Mounted on UltraBoard with a Luster Laminate

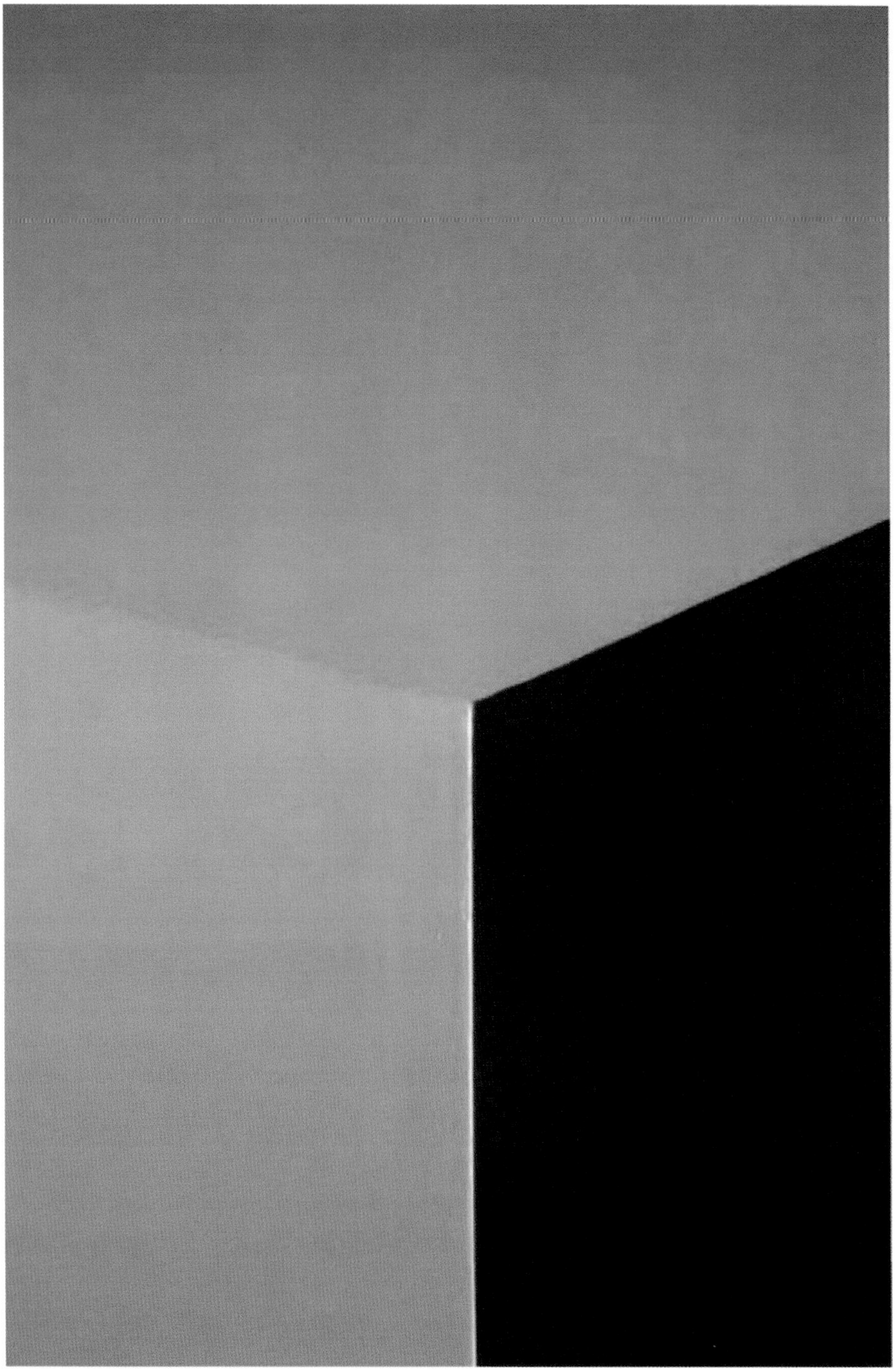

Corner — Los Angeles, 2006

36″ × 24″

C-Print Face-Mounted on Plexiglas

Dawn — Paris, 2011

36″ × 24″

C-Print Face-Mounted on Plexiglas

Watertowers — New York #1, 2011

36″ × 24″

C-Print Face-Mounted on Plexiglas

Transformation #32, 2015

30″ × 20″

Digital Print on Aluminum

Bricks and Windows — Red, 2018

36″ x 12″

Printed on Aluminum

Fanlights — Blue, 2018

36″ × 12″

Printed on Aluminum

Appendix

Walter Gilbert

Curriculum Vitae

Personal Data:

Born:	March 21, 1932, in Boston, Massachusetts, U.S.A.
Married:	Celia Stone Gilbert, two children
Website:	http://wallygilbert.com

Education:

1953	B.A. Harvard College, *summa cum laude*, Chemistry and Physics
1954	M.A. Harvard University, Physics
1957	Ph.D. Cambridge University, Mathematics (Thesis: On Generalized Dispersion Relations for Pion-Nucleon Scattering)

Appointments:

1953–57	NSF predoctoral fellowships held at Harvard University and Cambridge University
1957–58	NSF postdoctoral fellowship held at Harvard in Physics
1958–59	Assistant to Professor Schwinger at Harvard and Lecturer in Physics
1959–64	Assistant Professor of Physics, Harvard University
1964–68	Associate Professor of Biophysics, Harvard University
1968–72	Professor of Biochemistry, Harvard University
1972–81	American Cancer Society Professor of Molecular Biology, Harvard University
1978–83	Chair, Scientific Board of Directors, Biogen N.V.
1979–81	Co-Chair, Supervisory Board of Directors, Biogen N.V.
1981–84	Chair, Supervisory Board of Directors and Principal Executive Officer, Biogen N.V.
1985–86	Professor of Biology, Harvard University

1986–87	H.H. Timken Professor of Science, Dept. of Cellular and Developmental Biology, Harvard University
1987–93	Chair, Dept. of Cellular and Developmental Biology, Harvard University
1987–02	Carl M. Loeb University Professor, Dept. of Molecular and Cellular Biology, Harvard University
1992–	Vice Chair, Board of Directors, Myriad Genetics, Inc.
1994–14	Member, Board of Scientific Governors, The Scripps Research Institute
1996–02	Chair, Board of Directors, NetGenics, Inc.
1996–16	Chair, Board of Directors, Paratek Pharmaceuticals, Inc.
1998–10	Member, Board of Directors & Scientific Advisory Board, Memory Pharmaceuticals, Inc.
1999–04	Chair, Board of Directors and member of Scientific Advisory Board, Pintex Pharmaceuticals, Inc.
2000–04	Member, Board of Directors and Scientific Advisory Board, Transkaryotic Therapies, Inc.
2001–05	Member, Board of Directors, HospitalCareOnline.com, Inc.
2001–	Managing Director, BioVentures Investors
2003–04	Carl M. Loeb University Research Professor, Dept. of Molecular and Cellular Biology, Harvard University
2004–	Carl M. Loeb University Professor Emeritus, Harvard University
2015–	Director, Amylyx, Inc.

Awards:

United States Steel Foundation Award in Molecular Biology of the National Academy of Sciences, 1968

Elected American Academy of Arts and Sciences, 1968

Guggenheim Fellowship, 1968–69, Paris

Ledlie Prize, Harvard University, 1969 (with M. Ptashne)

Elected National Academy of Sciences, 1976

V.D. Mattia Lectureship, Roche Institute of Molecular Biology, 1976

Smith Kline & French Lecturer, U.C. Berkeley, 1977

Warren Triennial Prize, Massachusetts General Hospital, 1977 (with S. Benzer)

Louis and Bert Freedman Award, New York Academy of Sciences, 1977

Prix Charles-Leopold Mayer, Academie des Sciences, Institut de France, 1977 (with M. Ptashne and E. Witkin)

Doctor of Science (Honorary), University of Chicago, 1978

Doctor of Science (Honorary), Columbia University, 1978

Harrison Howe Award of the Rochester Branch of the American Chemical Society, 1978

Doctor of Science (Honorary), University of Rochester, 1979

Louisa Gross Horwitz Prize, Columbia University, 1979 (with F. Sanger)

Gairdner Foundation Annual Award, The Gairdner Foundation, 1979

Albert Lasker Basic Medical Research Award, The Albert and Mary Lasker Foundation, 1979 (with F. Sanger)
Prize Biochemical Analysis, The German Society for Clinical Chemistry, 1980 (with A. Maxam, F. Sanger, and A. Coulsen)
Sober Award, The American Society of Biological Chemists, 1980
Nobel Prize for Chemistry, 1980 (with F. Sanger and P. Berg)
Doctor of Science (Honorary), Yeshiva University, 1981
Senior U.S. Scientist Humboldt Award, 1987
Foreign Member of the Royal Society, 1987
Honorary Fellow, Trinity College, Cambridge, England, 1991
New England Entrepreneur of the Year Award, 1991
Ninth National Biotechnology Ventures Award, 1997
The Fourth Annual Biotechnology Heritage Award, 2002 (with P. Sharp)

Publications:

In theoretical physics:

1. On generalized dispersion relations, II, A. Salam and W. Gilbert. *Nuovo Cimento 3*, 607 (1956).
2. Yang-Fermi ambiguity, W. Gilbert and G. Screaton. *Phys. Rev. 104*, 1758 (1956).
3. New dispersion relations for pion-nucleon scattering, W. Gilbert. *Phys. Rev. 108*, 1078 (1957).
4. Structure of the vertex function, S. Deser, W. Gilbert and E. Sudarshan. *Phys. Rev. 115*, 731 (1959).
5. Structure of the forward scattering amplitude, S. Deser, W. Gilbert and E. Sudarshan. *Phys. Rev. 117*, 266 (1960).
6. Integral representation of two-point functions, S. Deser, W. Gilbert and E. Sudarshan. *Phys. Rev. 117*, 266 (1960).
7. Connection between gauge invariance and mass, D. Boulware and W. Gilbert. *Phys. Rev. 126*, 1563 (1962).
8. Broken symmetries and massless particles, W. Gilbert. *Phys. Rev. Lett. 12*, 713 (1964).

In molecular biology:

9. Unstable ribonucleic acid revealed by pulse labelling of *Escherichia coli*, F. Gros, H. Hiatt, W. Gilbert, C.G. Kurland, R.W. Reisbrough and J.D. Watson. *Nature 190*, 581–585 (1961).
10. Molecular and biological characterization of messenger RNA, F. Gros, W. Gilbert, H.H. Hiatt, G. Attardi, P.F. Spahr and J.D. Watson. *Cold Spring Harbor Symp. Quant. Biol. 26,* 111–132 (1961).
11. Polypeptide synthesis in *Escherichia coli*. I. Ribosomes and the active complex, W. Gilbert. *J. Mol. Biol. 6*, 374–388 (1963).

12. Polypeptide synthesis in *Escherichia coli*. II. The polypeptide chain and S-RNA, W. Gilbert. *J. Mol. Biol. 6*, 389–403 (1963).
13. The binding of S-RNA by *Escherichia coli* ribosomes, M. Cannon, R. Krug and W. Gilbert. *J. Mol. Biol. 7*, 360–378 (1963).
14. Protein synthesis in *Escherichia coli*, W. Gilbert. *Cold Spring Harbor Symp. Quant. Biol. 28*, 287–297 (1963).
15. Studies on methylated bases in transfer RNA, U.Z. Littauer, K. Muench, P. Berg, W. Gilbert and P.F. Spahr. *Cold Spring Harbor Symp. Quant. Biol. 28* 157–159 (1963).
16. Streptomycin, suppression and the code, J. Davies, W. Gilbert and L. Gorini. *Proc. Natl. Acad. Sci. 51*, 883–890 (1964).
17. Isolation of the *lac* repressor, W. Gilbert and B. Müller-Hill. *Proc. Natl. Acad. Sci. 56*, 1891–1898 (1966).
18. The *lac* operator is DNA, W. Gilbert and B. Müller-Hill. *Proc. Natl. Acad. Sci. 58*, 2415–2421 (1967).
19. Mutants that make more *lac* repressor, B. Müller-Hill, L. Crapo and W. Gilbert. *Proc. Natl. Acad. Sci. 59*, 1259–1264 (1968).
20. DNA replication: The rolling circle model, W. Gilbert and D. Dressler. *Cold Spring Harbor Symp. Quant. Biol. 33*, 473–484 (1968).
21. Genetic repressors, M. Ptashne and W. Gilbert. *Sci. Amer. 222(6)*, 36–44 (1970).
22. The lactose repressor, W. Gilbert and B. Müller-Hill. In *The Lactose Operon* (J. Beckwith and D. Zipser, eds.), Cold Spring Harbor Lab, New York, 93–109 (1970).
23. Repressors and genetic control, W. Gilbert. In *The Neurosciences: Second Study Program* (F.O Schmitt, ed.), The Rockefeller Univ. Press, New York, 946–954 (1970).
24. MSI and MSII made on ribosome in idling step of protein synthesis, W.A. Haseltine, R. Block, W. Gilbert and K. Weber. *Nature 238*, 381–384 (1972).
25. The *lac* repressor and the *lac* operator, W. Gilbert. In *CIBA Foundation Symp. 7: Polymerization in Biological Systems*, Elsevier Publishing Co., Amsterdam, 245–259 (1972).
26. The nucleotide sequence of the *lac* operator, W. Gilbert and A. Maxam. *Proc. Natl. Acad. Sci. USA 70*, 3581–3584 (1973).
27. Sequences of controlling regions of the *E. coli* lactose operon, W. Gilbert, N. Maizels and A. Maxam. *Genetics 79,* 227 (1975).
28. Lactose operator sequences and the action of *lac* repressor, W. Gilbert, J. Gralla, J. Majors and A. Maxam. In *Protein-Ligand Interactions* (H. Sund and G. Blauer, eds.), Walter de Gruyter, Berlin, 193–210 (1975).
29. Contacts between the *lac* repressor and DNA revealed by methylation, W. Gilbert, A. Maxam and A. Mirzabekov. In *Control of Ribosome Synthesis* (N. Kjeldgaard and O. Maaloe, eds.), Munksgaard, Copenhagen, 139–148 (1976).
30. Preferential protection of the minor groove of non-operator DNA by *lac* repressor against methylation by dimethyl sulphate, A.M. Kolchinsky, A.D. Mirzabekov, W. Gilbert and L. Li. *Nucl. Acids Res. 3(1)*, 11–18 (1976).
31. Starting and stopping sequences for the RNA polymerase, W. Gilbert. In *RNA Polymerase* (R. Losick and M. Chamberlin, eds.), Cold Spring Harbor Lab, New York, 193–205 (1976).

32. How proteins recognize DNA sequences, W. Gilbert, J. Majors and A. Maxam. In *Organization and Expression of Chromosomes*, Dahlem Workshop Report, Berlin, 167–178 (1976).
33. Construction of plasmids carrying the *c*I gene of bacteriophage lambda, K. Backman, M. Ptashne and W. Gilbert. *Proc. Natl. Acad. Sci. USA 73*, 4174–4178 (1976).
34. A new method for sequencing DNA, A.M. Maxam and W. Gilbert. *Proc. Natl. Acad. Sci. USA 74*, 560–564 (1977).
35. Rous sarcoma virus genome is terminally redundant: The 5′ sequence, W.A. Haseltine, A.M. Maxam and W. Gilbert. *Proc. Natl. Acad. Sci. USA 74*, 989–993 (1977).
36. Promoter region for yeast 5S ribosomal RNA, A. Maxam, R. Tizard, K.G. Skryabin and W. Gilbert. *Nature 267*, 643–645 (1977).
37. Mapping adenines, guanines and pyrimidines in RNA, H. Donis-Keller, A.M. Maxam and W. Gilbert. *Nucl. Acids Res. 4(8)*, 2527–2538 (1977).
38. A promoter region for yeast 5S RNA, W. Gilbert, A.M. Maxam, R. Tizard and K.G. Skryabin. In *Eucaryotic Genetic System* (J. Abelson and G. Wilcox, eds.), Academic Press, 15–23 (1977).
39. Contacts between the *lac* repressor and thymines in the *lac* operator, R. Ogata and W. Gilbert. *Proc. Natl. Acad. Sci. USA 74*, 4973–4976 (1977).
40. Why genes in pieces? W. Gilbert. *Nature 271*, 501 (1978).
41. Sequence of a mouse germ-line gene for a variable region of an immunoglobulin light chain, S. Tonegawa, A.M. Maxam, R. Tizard, O. Bernard and W. Gilbert. *Proc. Natl. Acad. Sci. USA 75*, 1485–1489 (1978).
42. Immunoglobulin screening method to detect specific translation products, S. Broome and W. Gilbert. *Proc. Natl. Acad. Sci. USA 75*, 2746–2749 (1978).
43. Molecular basis of base substitution hotspots in *Escherichia coli,* C. Coulondre, J.H. Miller, P.J. Farabaugh and W. Gilbert. *Nature 274*, 775–780 (1978).
44. A bacterial clone synthesizing proinsulin, L. Villa-Komaroff, A. Efstratiadis, S. Broome, P. Lomedico, R. Tizard, S.P. Naber, W.L. Chick and W. Gilbert. *Proc. Natl. Acad. Sci. USA 75*, 3727–3731 (1978).
45. An amino-terminal fragment of *lac* repressor binds specifically to *lac* operator, R.T. Ogata and W. Gilbert. *Proc. Natl. Acad. Sci. USA 75*, 5851–5854 (1978).
46. DNA-binding site of *lac* repressor probed by dimethylsulfate methylation of *lac* operator, R.T. Ogata and W. Gilbert. *J. Mol. Biol. 132*, 709–728 (1979).
47. Introns and exons: Playgrounds of evolution, W. Gilbert. In *Eucaryotic Gene Regulation* (R. Axel, T. Maniatis and C.F. Fox, eds.), Academic Press, 1–12 (1979).
48. The structure and transcription of rat preproinsulin genes, A. Efstratiadis, P. Lomedico, N. Rosenthal, R. Kolodner, R. Tizard, F. Perler, L. Villa-Komaroff, S. Naber, W. Chick, S. Broome and W. Gilbert. In *Eucaryotic Gene Regulation* (R. Axel, T. Maniatis and C.F. Fox, eds.), Academic Press, 301–315 (1979).
49. The structure and evolution of the two nonallelic rat preproinsulin genes, P. Lomedico, N. Rosenthal, A. Efstratiadis, W. Gilbert, R. Kolodner and R. Tizard. *Cell 18*, 545–558 (1979).

50. Pleiotropic mutations within two yeast mitochondrial cytochrome genes block mRNA processing, G.M. Church, P.P. Slonimski and W. Gilbert. *Cell 18*, 1209–1215 (1979).
51. Hepatitis B virus genes and their expression in *E. coli*, M. Pasek, T. Goto, W. Gilbert, B. Zink, H. Schaller, P. MacKay, G. Leadbetter and K. Murray. *Nature 282*, 575–579 (1979).
52. The structure of rat preproinsulin genes, P. Lomedico, N. Rosenthal, R. Kolodner, A. Efstratiadis and W. Gilbert. *Ann. N.Y.Acad. Sci.* 425 (1980).
53. Contacts between *Escherichia coli* RNA polymerase and an early promoter of phage T7, U. Siebenlist and W. Gilbert. *Proc. Natl. Acad. Sci. USA 77*, 122–126 (1980).
54. Useful proteins from recombinant bacteria, W. Gilbert and L. Villa-Komaroff. *Sci. Am. 242(4),* 74–84 (1980).
55. Sequencing end-labeled DNA with base-specific chemical cleavages, A.M. Maxam and W. Gilbert. In *Methods in Enzymology*, Vol. 65, Part I, 499–560 (K. Moldave and L. Grossman, ed.), Academic Press (1980).
56. *E. coli* RNA polymerase interacts homologously with two different promoters, U. Siebenlist, R.B. Simpson and W. Gilbert. *Cell 20*, 269–281 (1980).
57. The evolution of genes: The chicken preproinsulin gene, F. Perler, A. Efstratiadis, P. Lomedico, W. Gilbert, R. Kolodner and J. Dodgson. *Cell 20*, 555–566 (1980).
58. Yeast mitochondrial intron products required *in trans* for RNA splicing, G. Church and W. Gilbert. *Miami Symp.*: Mobilization and Reassembly of Genetic Information, 17, 379–394 (1980).
59. Eukaryotic signal sequence transports insulin antigen in *Escherichia coli*, K. Talmadge, S. Stahl and W. Gilbert. *Proc. Natl. Acad. Sci. USA 77*, 3369–3373 (1980).
60. The synthesis of insulin in bacteria: A model for the production of medically useful proteins in prokaryotic cells, L. Villa-Komaroff, S. Broome, S.P. Naber, A. Efstratiadis, P. Lomedico, R. Tizard, W.L. Chick and W. Gilbert. In *Birth Defects: Original Article Series*, Vol. XVI, No. 1, 53–68 (March of Dimes Birth Defects Found.) (1980)
61. Bacteria mature preproinsulin to proinsulin, K. Talmadge, J. Kaufman and W. Gilbert. *Proc. Natl. Acad. Sci. USA 77*, 3988–3992 (1980).
62. Chemical probes for higher-order structure in RNA, D.A. Peattie and W. Gilbert. *Proc. Natl. Acad. Sci. USA 77*, 4679–4682 (1980).
63. Construction of plasmid vectors with unique *Pst*I cloning sites in a signal sequence coding region, K. Talmadge and W. Gilbert. *Gene 12*, 235–241 (1980).
64. Tissue-specific exposure of chromatin structure at the 5′ terminus of the rat preproinsulin II gene, C. Wu and W. Gilbert. *Proc. Natl. Acad. Sci. USA 78*, 1577–1580 (1981).
65. Expression of active polypeptides in *E. Coli*, W. Gilbert. *Recomb. DNA Tech. Bull. 4(1),* 4–5 (1981).
66. DNA sequencing and gene structure, W. Gilbert. *Science 214,* 1305–1312 (1981).
67. The production of immunologically active surface antigens of hepatitis B virus by *Escherichia coli*, P. Mackay, M. Pasek, M. Magazin, R.T. Kovacic, B. Allet, S. Stahl, W. Gilbert, H. Schaller, S.A. Bruce and K. Murray. *Proc. Natl. Acad. Sci. USA 78*, 4510–4514 (1981).

68. Parabiosis as a model system for network interactions, K. Fischer Lindahl, W. Gilbert and K. Rajewsky. In *The Immune System,* Vol. 2, 24–32, Basel (1981).
69. Proinsulin from bacteria, K. Talmadge and W. Gilbert. In *Proceedings of the International Plasmid Meeting* (S. Levy, R. Clowes and E. Koenig, eds.), Plenum Publishing, 411–419 (1981).
70. An "internal" signal sequence directs secretion and processing of proinsulin in bacteria, K. Talmadge, J. Brosius and W. Gilbert. *Nature 294,* 176–177 (1981).
71. Germ-line MuLV reintegrations in AKR/J mice, W. Herr and W. Gilbert. *Nature 269,* 865–868 (1982).
72. Cellular location affects protein stability in *Escherichia coli,* K. Talmadge and W. Gilbert. *Proc. Natl. Acad. Sci. USA 79,* 1830–1833 (1982).
73. Nucleotide sequence of the 3′ half of AKV, W. Herr, V. Corbin and W. Gilbert. *Nucl. Acids Res. 10(21),* 6931–6944 (1982).
74. Somatically acquired recombinant murine leukemia proviruses in thymic leukemias of AKR/J mice, W. Herr and W. Gilbert. *J. Virology 46(1),* 70–82 (1983).
75. Isolation and mapping of cDNA hybridization probes specific for ecotropic and nonecotropic murine leukemia proviruses, W. Herr, D. Schwartz and W. Gilbert. *Virology 125,* 139–154 (1983).
76. Nucleotide sequence of rous sarcoma virus, D.E. Schwartz, R. Tizard and W. Gilbert. *Cell 32,* 853–869 (1983).
77. Comparison of the methylation patterns of the two rat insulin genes, R.L. Cate, W. Chick and W. Gilbert. *J. Biol. Chem. 258(10),* 6645–6652 (1983).
78. Primary structure of chicken muscle pyruvate kinase mRNA, N. Lonberg and W. Gilbert. *Proc. Natl. Acad. Sci. USA 80,* 3661–3665 (1983).
79. Monoclonal AKR/J thymic leukemias contain multiple *J*H immunoglobulin gene rearrangements, W. Herr, A.P. Perlmutter and W. Gilbert. *Proc. Natl. Acad. Sci. USA 80,* 7433–7436 (1983).
80. Genomic sequencing, G.M. Church and W. Gilbert. *Proc. Natl. Acad. Sci. USA 81,* 1991–1995 (1984).
81. Free and integrated recombinant murine leukemia virus DNAs appear in preleukemic thymuses of AKR/J mice, W. Herr and W. Gilbert. *J. Virol 50(1),* 155–162 (1984).
82. Antibodies of the secondary response can be expressed without switch recombination in normal mouse B cells, A.P. Perlmutter and W. Gilbert. *Proc. Natl. Acad. Sci. USA 81,* 7189–7193 (1984).
83. B lineage-specific interactions of an immunoglobulin enhancer with cellular factors *in vivo,* A. Ephrussi, G.M. Church, S. Tonegawa and W. Gilbert. *Science 227,* 134–140 (1985).
84. Intron/exon structure of the chicken pyruvate kinase gene, N. Lonberg and W. Gilbert. *Cell 40,* 81–90 (1985).
85. Detection *in vivo* of protein-DNA interactions within the *lac* operon of *Escherichia coli,* H. Nick and W. Gilbert. *Nature 313,* 795–798 (1985).
86. Cell-type-specific contacts to immunoglobulin enhancers in nuclei, G.M. Church, A. Ephrussi, W. Gilbert and S. Tonegawa. *Nature 313,* 798–801 (1985).

87. Chicken triosephosphate isomerase complements an *Escherichia coli* deficiency, D. Straus and W. Gilbert. *Proc. Natl. Acad. Sci. USA 82*, 2014–2018 (1985).
88. Rous sarcoma virus encodes a transcriptional activator, S. Broome and W. Gilbert. *Cell 40*, 537–546 (1985).
89. Active site of triosephosphate isomerase: *In vitro* mutagenesis and characterization of an altered enzyme, D. Straus, R. Raines, E. Kawashima, J.R. Knowles and W. Gilbert. *Proc. Natl. Acad. Sci. USA 82*, 2272–2276 (1985).
90. Genes-in-pieces revisited, W. Gilbert. *Science 228*, 823–824 (1985).
91. Genetic engineering in the Precambrian: Structure of the chicken triosephosphate isomerase gene, D. Straus and W. Gilbert. *Mol. Cell. Biol. 5(12)*, 3497–3506 (1985).
92. The genomic sequencing technique, G.M. Church and W. Gilbert. *Prog. Clin. Biol. Res. 177,* 17–21 (1985).
93. Detection of cytosine methylation in the maize alcohol dehydrogenase gene by genomic sequencing, H. Nick, B. Bowen, R.J. Ferl and W. Gilbert. *Nature 319*, 243–346 (1986).
94. Origin of life: The RNA world, W. Gilbert. *Nature 319*, 618 (1986).
95. Evolution of antibodies: The road not taken, W. Gilbert. *Nature 320*, 485–486 (1986).
96. Reaction energetics of a mutant triosephosphate isomerase in which the active-site glutamate has been changed to aspartate, R.T. Raines, E.L. Sutton, D.R. Straus, W. Gilbert and J.R. Knowles. *Biochem. 25*, 7142–7154 (1986).
97. The kinetic consequences of altering the catalytic residues of triosephosphate isomerase, R.T. Raines, D.R. Straus, W. Gilbert and J.R. Knowles. *Phil. Trans. R. Soc. Lond. A 317,* 371–380 (1986).
98. The triosephosphate isomerase gene from maize: Introns antedate the plant-animal divergence, M. Marchionni and W. Gilbert. *Cell 46*, 133–141 (1986).
99. On the antiquity of introns, W. Gilbert, M. Marchionni and G. McKnight. *Cell 46*, 151–153 (1986).
100. The exon theory of genes, W. Gilbert. *Cold Spring Harbor Symp. Quant. Biol. 52,* 901–905 (1987).
101. Genome sequencing: Creating a new biology for the twenty-first century, W. Gilbert. *Issues in Science & Technology III(3),* 26–35 (1987).
102. Basic protein enhances the incorporation of DNA into lipid vesicles: Model for the formation of primordial cells, D.G. Jay and W. Gilbert. *Proc. Natl. Acad. Sci. USA 84,* 1978–1980 (1987).
103. Formation of parallel four-stranded complexes by guanine-rich motifs in DNA and its implications for meiosis, D. Sen and W. Gilbert. *Nature 334*, 364–366 (1988).
104. Human genome sequencing, W. Gilbert. *Basic Life Sci. 46*, 29–36 (1988).
105. Differential expression of acetylcholine receptor mRNA in nuclei of cultured muscle cells, S. Bursztajn, S.A. Berman and W. Gilbert. *Proc. Natl. Acad. Sci. USA 86*, 2928–2932 (1989).
106. The monoclonal antibody B30 recognizes a specific neuronal cell surface antigen in the developing mesencephalic trigeminal nucleus of the mouse, D.Y. Stainier and W. Gilbert. *J. Neurosci. 9(7)*, 2468–2485 (1989).

107. One-sided polymerase chain reaction: The amplification of cDNA, O. Ohara, R.L. Dorit and W. Gilbert. *Proc. Natl. Acad. Sci. USA 86*, 5673–5677 (1989).
108. Direct genomic sequencing of bacterial DNA: The pyruvate kinase I gene of *Escherichia coli*, O. Ohara, R.L. Dorit and W. Gilbert. *Proc. Natl. Acad. Sci. USA 86*, 6883–6887 (1989).
109. Detection of mutations and DNA polymorphisms using whole genome Southern Cross hybridization, G. Ruvkun, W. Gilbert and H.R. Horvitz. *Nucl. Acids Res. 18(4)*, 809–815 (1990).
110. DNA trapping electrophoresis, L. Ulanovsky, G. Drouin and W. Gilbert. *Nature 343,* 190–192 (1990).
111. Localization of an acetylcholine receptor intron to the nuclear membrane, S.A. Berman, S. Bursztajn, B. Bowen and W. Gilbert. *Science 247*, 212–214 (1990).
112. Pioneer neurons in the mouse trigeminal sensory system, D.Y.R. Stainier and W. Gilbert. *Proc. Natl. Acad. Sci. USA 87*, 923–927 (1990).
113. A sodium-potassium switch in the formation of four-stranded G4-DNA, D. Sen and W. Gilbert. *Nature 344*, 410–414 (1990).
114. Factors released by ciliary neurons and spinal cord explants induce acetylcholine receptor mRNA expression in cultured muscle cells, S. Bursztajn, S.A. Berman and W. Gilbert. *J. Neurobiol. 21 (3)*, 387–399 (1990).
115. Simultaneous visualization of neuronal protein and receptor mRNA, S. Bursztajn, S.A. Berman and W. Gilbert. *BioTechniques 9(4)*, 440–449 (1990).
116. How big is the universe of exons? R.L. Dorit, L. Schoenbach and W. Gilbert. *Science 250,* 1377–1382 (1990).
117. Towards a paradigm shift in biology, W. Gilbert. *Nature 349,* 99 (1991).
118. Murine memory B cells are multi-isotype expressors, C.J. Wu, J.T. Karttunen, D.H.L. Chin, D. Sen and W. Gilbert. *Immunol. 72*, 48–55 (1991).
119. The B30 ganglioside is a cell surface marker for neural crest derived neurons in the developing mouse, D.Y.R. Stainier, D.H. Bilder and W. Gilbert. *Dev. Biol. 144*, 177–188 (1991).
120. Gene structure and evolutionary theory, W. Gilbert. In *New Perspectives on Evolution*, Wiley-Liss, Inc., 155–163 (1991).
121. Spatial domains in the developing forebrain: Developmental regulation of a restricted cell surface protein, D.Y.R. Stainier, D.H. Bilder and W. Gilbert. *Dev. Biol. 147*, 22–31 (1991).
122. The structure of telomeric DNA: DNA quadriplex formation [Review], D. Sen and W. Gilbert. *Curr. Opin. Struct. Biol. 1*, 35–438 (1991).
123. Neuronal differentiation and maturation in the mouse trigeminal sensory system, *in vivo* and *in vitro*, D.Y.R. Stainier and W. Gilbert. *J. Comp. Neurol. 311*, 300–312 (1991).
124. DNA sequencing, today and tomorrow, W. Gilbert. *Hospital Practice 26(10)*, 165–174 (1991).
125. The limited universe of exons [Review], R.L. Dorit and W. Gilbert. *Curr. Opin. Struct. Biol. 1*, 973–977 (1991).
126. Zebrafish embryology and neural development [Review], C. Fulwiler and W. Gilbert. *Curr. Opin. Cell Biol. 3*, 988–991 (1991).

127. Cationic switches in the formation of DNA structures containing guanine-quartets, D. Sen and W. Gilbert. In *Structure and Function, Vol. 1: Nucleic Acids*, (R.H. Sarma & M.H. Sarma, eds.) Adenine Press, 43–52 (1992).
128. Novel DNA superstructures formed by telomere-like oligomers, D. Sen and W. Gilbert. *Biochem. 31*, 65–70 (1992).
129. Guanine quartet structures, D. Sen and W. Gilbert. *Methods in Enzymol. 211*, 191–199 (1992).
130. A fate map for the first cleavages of the zebrafish, D. Strehlow and W. Gilbert. *Nature 361*, 451–453 (1993).
131. Identification and characterization of a nuclease activity specific for G4 tetrastranded DNA, Z. Liu, J.D. Frantz, W. Gilbert and B.K. Tye. *Proc. Natl. Acad. Sci. USA 90*, 3157–3161 (1993).
132. RNA editing as a source of genetic variation, L.F. Landweber and W. Gilbert. *Nature 363*, 179–182 (1993).
133. The boundaries of partially edited transcripts are not conserved in kinetoplastids: Implications for the guide RNA model of editing, L.F. Landweber, A.G. Fiks and W. Gilbert. *Proc. Natl. Acad. Sci. USA 90,* 9242–9246 (1993).
134. On the ancient nature of introns, W. Gilbert and M. Glynias. *Gene 135,* 137–144 (1993).
135. Phylogenetic analysis of RNA editing: A primitive genetic phenomenon, L.F. Landweber and W. Gilbert. *Proc. Natl. Acad. Sci. USA 91*, 918–921 (1994).
136. The fates of the blastomeres of the 16-cell zebrafish embryo, D. Strehlow, G. Heinrich and W. Gilbert. *Development 120*, 1791–1798 (1994).
137. Large scale bacterial gene discovery by similarity search, K. Robison, W. Gilbert and G.M. Church. *Nature Genetics 7*, 205–214 (1994).
138. The genetic data environment an expandable GUI for multiple sequence analysis, S.W. Smith, R. Overbeek, C.R. Woese, W. Gilbert and P.M. Gillevet. *CABIOS 10(6)* 671–675 (1994).
139. The yeast *KEM1* gene encodes a nuclease specific for G4 tetraplex DNA: Implication of *in vivo* functions for this novel DNA structure, Z. Liu and W. Gilbert. *Cell 77*, 1083–1092 (1994).
140. Retinoic acid is necessary for development of the ventral retina in zebrafish, N. Marsh-Armstrong, P. McCaffery, W. Gilbert, J.E. Dowling and U.C. Drager. *Proc. Natl. Acad. Sci. USA 91*, 7286–7290 (1994).
141. A yeast gene product, G4p2, with a specific affinity for quadruplex nucleic acids, J.D. Frantz and W. Gilbert. *J. Biol. Chem. 270(16)*, 9413–9419 (1995).
142. Absence of polymorphism at the ZFY locus on the human Y chromosome, R.L. Dorit, H. Akashi and W. Gilbert. *Science 268*, 1183–1185 (1995).
143. Gene disruption of a G4-DNA dependent nuclease in yeast leads to cellular senescence and telomere shortening, Z. Liu, A. Lee and W. Gilbert. *Proc. Natl. Acad. Sci. USA 92*, 6002–6006 (1995).
144. Large scale genomic sequencing: Optimization of genomic chemical sequencing reactions, M. Dolan, A. Ally, M.S. Purzycki, W. Gilbert and P.M. Gillevet. *BioTechniques 19(2)*, 264–273 (1995).

145. Exploring the *Mycoplasma capricolum* genome: A minimal cell reveals its physiology, P. Bork, C. Ouzounis, G. Casari, R. Schneider, C. Sander, M. Dolan, W. Gilbert and P.M. Gillevet. *Mol. Microbiol. 16(5)*, 955–967 (1995).
146. Tests of the exon theory of genes, W. Gilbert, M. Long, C. Rosenberg and M. Glynias. In *Tracing Biological Evolution in Protein and Gene Structures*, Proceedings of the 20th Taniguchi International Symp., Div. of Biophysics (M. Go and P. Schimmel, eds.), Elsevier Science B.V., 237–247 (1995).
147. A novel yeast gene product, G4p1, with a specific affinity for quadruplex nucleic acids, J.D. Frantz and W. Gilbert. *J. Biol. Chem. 270(35)*, 20692–20697 (1995).
148. Intron phase correlations and the evolution of the intron/exon structure of genes, M. Long, C. Rosenberg and W. Gilbert. *Proc. Natl. Acad. Sci. USA 92,* 12495–12499 (1995).
149. Evolution of the intron-exon structure of eukaryotic genes [Review], M. Long, S.J. de Souza and W. Gilbert. *Curr. Opin. Genet. Devel. 5,* 774–778 (1995).
150. More *haemophilus* and *mycoplasma* genes, K. Robison, W. Gilbert and G.M. Church. *Science 271*, 1302–1303 (1996).
151. Discovery of amphibian Tc1-like transposon families, W.L. Lam, P. Seo, K. Robison, S. Virk and W. Gilbert. *J. Mol. Biol. 257*, 359–366 (1996).
152. Exon shuffling and the origin of the mitochondrial targeting function in plant cytochrome c1 precursor, M. Long, S.J. de Souza, C. Rosenberg and W. Gilbert. *Proc. Natl. Acad. Sci. USA 93,* 7727–7731 (1996).
153. Introns and gene evolution, S.J. de Souza, M. Long and W. Gilbert. *Genes to Cells 1,* 493–505 (1996).
154. Active transposition in zebrafish, W.L. Lam, T.-S. Lee and W. Gilbert. *Proc. Natl. Acad. Sci. USA 93,* 10870–10875 (1996).
155. Novel $G^q\alpha$ isoform is a candidate transducer of rhodopsin signaling in a *Drosophila* testes-autonomous pacemaker, C.E. Alvarez, K. Robison and W. Gilbert. *Proc. Natl. Acad. Sci. USA 93,* 12278–12282 (1996).
156. Intron positions correlate with module boundaries in ancient proteins, S.J. de Souza, M. Long, L. Schoenbach, S.W. Roy and W. Gilbert. *Proc. Natl. Acad. Sci. USA* 93, 14632–14636 (1996).
157. Jak1 kinase is required for cell migrations and anterior specification in zebrafish embryos, G. Conway, A. Margoliath, S. Wong-Madden, R.J. Roberts and W. Gilbert. *Proc. Natl. Acad. Sci. USA 94,* 3082–3087 (1997).
158. Delta-interacting protein A and the origin of hepatitis delta antigen, M. Long, S.J. de Souza and W. Gilbert. *Science 276*, 824–825 (1997).
159. Dealing with database explosion: A cautionary note, U. Bhatia, K. Robison and W. Gilbert. *Science 276*, 1724–1725 (1997).
160. Ligands for ErbB-family receptors encoded by a neuregulin-like gene, H. Chang, D.J. Riese II, W. Gilbert, D.F. Sterns and U.J. McMahan. *Nature 387*, 509–512 (1997).
161. Origin of genes, W. Gilbert, S.J. de Souza and M. Long. *Proc. Natl. Acad. Sci. USA 94,* 7698–7703 (1997).

162. A novel zebrafish gene expressed specifically in the photoreceptor cells of the retina, H. Chang and W. Gilbert. *Biochem. & Biophys. Res. Comm.* 237, 84–89 (1997).
163. The yeast splice site revisited: New exon consensus from genomic analysis, M. Long, S.J. de Souza and W. Gilbert. *Cell 91*, 739–740 (1997).
164. The correlation between introns and the three-dimensional structure of proteins, S.J. de Souza, M. Long, L. Schoenbach, S.W. Roy and W. Gilbert. *Gene 205*, 141–144 (1997).
165. Relationship between "proto-splice sites" and intron phases: Evidence from dicodon analysis, M. Long, S.J. de Souza, C. Rosenberg and W. Gilbert. *Proc. Natl. Acad. Sci. USA 95,* 219–223 (1998).
166. Toward a resolution of the introns early/late debate: Only phase zero introns are correlated with the structure of ancient proteins, S.J. de Souza, M. Long, R.J. Klein, S. Roy, S. Lin, and W. Gilbert. *Proc. Natl. Acad. Sci. USA 95,* 5094–5099 (1998).
167. Introns and the RNA world, W. Gilbert and S.J. de Souza. In *The RNA World, Second Edition*, Cold Spring Harbor Laboratory Press, 221–231 (1999).
168. Centripetal modules and ancient introns, S.W. Roy, M. Nosaka, S.J. de Souza and W. Gilbert. *Gene 238*, 85–91 (1999).
169. EID: The exon-intron database — an exhaustive database of protein-coding intron-containing genes, S. Saxonov, I. Daizadeh, A. Fedorov and W. Gilbert. *Nucl. Acids Res. 28(1),* 185–190 (2000).
170. The genomic structure of C14orf1 is conserved across eukarya, C. Ottolenghi, I. Daizadeh, A. Ju, S. Kossida, G. Renault, M. Jacquet, A. Fellous, W. Gilbert and R. Veitia. *Mammalian Genome 11*, 786–788 (2000).
171. Footprints of primordial introns on the eukaryotic genome, S.W. Roy, B.P. Lewis, A. Fedorov and W. Gilbert. *TIG 17(9)*, 496–499 (2001).
172. Intron distribution difference for 276 ancient and 131 modern genes suggests the existence of ancient introns, A. Fedorov, X. Cao, S. Saxonov, S.J. de Souza, S.W. Roy and W. Gilbert. *Proc. Natl. Acad. Sci. USA 98,* 13177–13182 (2001).
173. Regularities of context-dependent codon bias in eukaryotic genes, A. Fedorov, S. Saxonov and W. Gilbert. *Nucl. Acids Res.* 30(5), 1192–1197 (2002).
174. Do introns favor or avoid regions of amino-acid conservation? T. Endo, A. Fedorov, S.J. de Souza and W. Gilbert. *Mol. Biol. Evol. 19(4),* 521–525 (2002).
175. The signal of ancient introns is obscured by intron density and homolog number, S.W. Roy, A. Fedorov and W. Gilbert. *Proc. Natl. Acad. Sci. USA 99,* 15513–15517 (2002).
176. Large-scale comparison of intron positions among animal, plant, and fungal genes, A. Fedorov, A.F. Merican and W. Gilbert. *Proc. Natl. Acad. Sci. USA 99,* 16128–16133 (2002).
177. Book review of *Watson and DNA: Making a Scientific Revolution* by V.K. McElheny, W. Gilbert. *Nature 421*, 315–316 (2003).
178. Phylogenetically older introns strongly correlate with module boundaries in ancient proteins, A. Fedorov, S. Roy, X. Cao and W. Gilbert. *Genome Res. 13.* 1155–1157 (2003).
179. Large-scale comparison of intron positions in mammalian genes shows intron loss but no gain, S.W. Roy, A. Fedorov and W. Gilbert. *Proc. Natl. Acad. Sci. USA 100*, 7158–7162 (2003).

180. The universe of exons revisited, S. Saxonov and W. Gilbert. *Genetica 118*, 267–278 (2003).
181. Mystery of intron gain, A. Fedorov, S. Roy, L. Fedorova and W. Gilbert. *Genome Res. 13*, 2236–2241, (2003).
182. The pattern of intron loss, S. W. Roy and W. Gilbert. *Proc. Natl. Acad. Sci. USA 102*, 713–718 (2005).

In his photographic art, he has had many one-person shows:
***Selected exhibitions*:**

"From Science to Art," City Gallery, San Diego, CA	2019
"Retrospective," LabCentral, Cambridge, MA	2018
"Towers," Viridian Gallery, Chelsea, NYC	2017
"Doors to Nowhere," Salon R, Cambridge, MA	2017
"Broken City," Khaki Gallery, Boston, MA	2016
"Journeying," Permanent exhibition, AGH University, Krakow, Poland	2016–
"Broken City" Viridian Gallery, Chelsea, NYC	2016
"Patterns & Recognition," Seoul National University Bundang Hospital, curated by Chang and Jae Kim	2015–2016
"Transformations," Viridian Artists, Chelsea, NYC	2014
"Patterns & Recognition," The Howard Hughes Medical Institute, Janelia Farm, VA	2014
"Wally Gilbert," CJ Gallery, Art San Diego 2013, San Diego, CA	2013
"Wally Gilbert: A Room of Light," Milton Art Museum, Canton, MA	2013
"Wally Gilbert: Black & White," Khaki Gallery, Boston, MA	2013
"Digital Constellations," Lindau City Museum, Lindau, Germany	2013
"Wally Gilbert: New Black and White Images," Viridian Artists, Chelsea, NYC	2013
"Wally Gilbert", CJ Gallery, Art San Diego 2012, San Diego, CA	2012
"En-Lighten," Khaki Gallery, Boston, MA	2012
"Journeying," The Artemis Gallery, Krakow, Poland, curated by Wieslawa Piotrowska-Sowadska	2012
"Pattern & Recognition," The Art Gallery, Antelope Valley College, Lancaster, CA	2012
"Squares, Triangles, and Lines," Galerie im Einstein, Berlin	2011
"Projekt Norblin," New Art Wet Music Foundation, Bydgoszcz, Poland	2011
"Squares and Triangles," Viridian Artists, Chelsea, NYC	2011
"Vanishing," CJ Gallery, San Diego, CA	2010
"Vanishing Profiles," Khaki Gallery, Boston, MA	2010
"The Norblin Project and Other Images," CJ Gallery and OCIO DESIGN GROUP, San Diego,CA	2010
"Wally@Wainwright," Wainwright Bank, Cambridge, MA	2010
"Vanishing," BAAK Gallery, Cambridge, MA	2009
Norblin Installation, Poznan, Poland, curated by Jan Kubasiewicz and Zuk Piwkowski	2009
"The Norblin Project and other Images," CJ Art Gallery, San Diego, CA	2009
"IN COLOR & BEYOND," Khaki Gallery, Boston, MA	2009

"Fresh Fruit," Mayyim Hayyim Gallery, Newton, MA	2009
"Stillness and Motion," Viridian Artists, Chelsea, NYC	2008
"LEEKS & CHAINS," Khaki Gallery, Wellesley, MA	2008
"The Norblin Project and Other Images," CJ Art Gallery, San Diego, CA	2007
BAAK Gallery, Cambridge, MA	2007
Norblin Installation, Galeria PATIO, Lodz, Poland, curated by Zuk Piwkowski, Jan Kubasiewicz, and Aurelia Mandziuk	2007
Norblin Site Installation, Warsaw, Poland, curated by Jan Kubasiewicz and Zuk Piwkowski	2007
"The Norblin Project: Images of Decay," American Center for Physics, College Park, MD	2007
"IN COLOR," Khaki Gallery, Wellesley, MA	2007
"The Norblin Project: Images of Decay," LACDA, Los Angeles, CA	2006
"The Norblin Project: Images of Decay," Viridian Artists, Chelsea, NYC	2006
Jock Colville Hall, Churchill College, University of Cambridge, Cambridge, UK	2006
Ann Janss Gallery, Los Angeles, CA	2005
Doran Gallery, Massachusetts College of Art, Boston, MA, curated by Jan Kubasiewicz	2004